The Impact of Technology on Sport II

BALKEMA – Proceedings and Monographs
in Engineering, Water and Earth Sciences

The Impact of Technology on Sport II

Edited by

F.K. Fuss

School of Biomedical and Chemical Engineering, Nanyang Technological University, Singapore

A. Subic

School of Aerospace, Mechanical and Manufacturing Engineering, RMIT University, Melbourne, Australia

S. Ujihashi

Graduate School of Information Science and Engineering, Tokyo Institute of Technology, Tokyo, Japan

LONDON / LEIDEN / NEW YORK / PHILADELPHIA / SINGAPORE

Taylor & Francis is an imprint of the Taylor & Francis Group, an informa business

© 2008 Taylor & Francis Group, London, UK

Typeset by Charon Tec Ltd (A Macmillan Company), Chennai, India
Printed and bound in Singapore by Markono Print Media Pte Ltd

All rights reserved. No part of this publication or the information contained herein may be reproduced, stored in a retrieval system, or transmitted in any form or by any means, electronic, mechanical, by photocopying, recording or otherwise, without written prior permission from the publishers.

Although all care is taken to ensure integrity and the quality of this publication and the information herein, no responsibility is assumed by the publishers nor the author for any damage to the property or persons as a result of operation or use of this publication and/or the information contained herein.

The book includes refereed contributions presented at the Asia-Pacific Congress on Sports Technology held at Nanyang Technological University in September 2007.

The book is edited by:
F.K. Fuss, A. Subic and S. Ujihashi

Published by: Taylor & Francis/Balkema
P.O. Box 447, 2300 AK Leiden, The Netherlands
e-mail: Pub.NL@tandf.co.uk
www.balkema.nl, www.taylorandfrancis.co.uk, www.crcpress.com

ISBN-13: 978-0-415-45695-1

Cover Design by F.K. Fuss

Contents

Preface　xix
Contributors List　xxi

1 Sports Technology and Engineering

Activation and Liability of Sports Engineering Activities Around the World　3
S. Ujihashi

2 Life Cycle, Environmental Engineering and Eco-Design

Ecodesign of Alpine Skis and Other Sport Equipment – Considering Environmental Issues in Product Design and Development　15
W. Wimmer & H. Ostad-Ahmad-Ghorabi

Integrating the Design for Environment Approach in Sports Products Development　25
A. Subic & N. Paterson

Going Green: The Application of Life Cycle Assessment Tools to the Indoor Sports Flooring Industry　37
A.W. Walker

3 Sports Medicine, Exercising and Clinical Biomechanics

Sports Technology in the Field of Sports Medicine　45
B. Tan

Biomechanical Factors in Mild Traumatic Brain Injuries Based on American Football and Soccer Players　51
M. Ziejewski, Z. Kou & C. Doetkott

Evaluation of Non-Pharmacologic Intervention for Parkinson Patient Using Gait Analysis: A Case Study　59
R. Ganason, S. Joseph, B.D. Wilson & V. Selvarajah

The Time-Course of Acute Changes in Achilles Tendon Morphology Following Exercise　65
S.C. Wearing, J.E. Smeathers, S.R. Urry & S.L. Hooper

Development of Portable Equipment for Cardiorespiratory Fitness
Measurement through a Sub-Max Exercise 69
K.R. Chung, S.Y. Kim, G.S. Hong, J.H. Hyeong & C.H. Choi

Genetic Programming for Knowledge Extraction from Star Excursion
Balance Test 75
S.N. Omkar, K.M. Manoj, M. Dipti & K.J. Vinay

An Analysis of Sun Salutation 81
S.N. Omkar

Development of a Force Feedback System for Exercising 87
J.Y. Sim, A.C. Ritchie & F.K. Fuss

Formalised Requirements Documentation of a Novel Exercise System
Using Enterprise Modelling Concepts 93
A.A. West, J.D. Smith, R.P. Monfared & R. Harrison

4 General Motion Analysis

A Low Cost Self Contained Platform for Human Motion Analysis 101
N. Davey, A. Wixted, Y. Ohgi & D.A. James

An Efficient and Accurate Method for Constructing 3D Human Models
from Multiple Cameras 113
C.K. Quah, A. Gagalowicz & H.S. Seah

Marker-Less 3D Video Motion Capture in Cluttered Environments 121
C.K. Quah, A. Gagalowicz & H.S. Seah

5 Apparel and Sport Surfaces

Improving the Understanding of Grip 129
S.E. Tomlinson, R. Lewis & M.J. Carré

Ionised Sports Undergarments: A Physiological Evaluation 135
A.R. Gray, J. Santry, T.M. Waller & M.P. Caine

The Influence of Hues on the Cortical Activity – A Recipe for Selecting
Sportswear Colours 141
J. Tripathy, F.K. Fuss, V.V. Kulish & S. Yang

The Development and Environmental Applications of the Goingstick® 149
M.J.D. Dufour & C. Mumford

Analysis of the Influence of Shockpad Properties on the Energy
Absorption of Artificial Turf Surfaces 155
T. Allgeuer, S. Bensason, A. Chang, J. Martin & E. Torres

6 Gait, Running and Shoes

Development of a Novel Nordic-Walking Equipment Due to
a New Sporting Technique 163
A. Sabo, M. Eckelt & M. Reichel

Pattern Recognition in Gait Analysis 169
A.A. Bakar, S.M.N.A. Senanayake, R. Ganason & B.D. Wilson

Analysis of Three-Dimensional Plantar Pressure Distribution Using
Standing Balance Measurement System 175
Y. Hayashi, N. Tsujiuchi, T. Koizumi & A. Nishi

The Use of Micro-Electro-Mechanical-Systems Technology to
Assess Gait Characteristics 181
J.B. Lee, B. Burkett, R.B. Mellifont & D.A. James

Smart Floor Design for Human Gait Analysis 187
S.M.N.A. Senanayake, D. Looi, M. Liew, K. Wong & J.W. Hee

Effect of Rocker Heel Angle of Walking Shoe on Gait Mechanics
and Muscle Activity 193
S.Y. An & K.K. Lee

Kinematics of the Foot Segments During the Acceleration Phase of
Sprinting: A Comparison of Barefoot and Sprint Spike Conditions 199
B.J. Williams, D.T. Toon, M.P. Caine & N. Hopkinson

7 Ball Sport – Golf

Non-Linear Viscoelastic Properties of Golf Balls 207
F.K. Fuss

The Influence of Wind Upon 3-Dimensional Trajectory of Golf Ball
Under Various Initial Conditions 223
T. Naruo & T. Mizota

The Influence of Groove Profile; Ball Type and Surface Condition
On Golf Ball Backspin Magnitude 229
J.E.M. Cornish, S.R. Otto & M. Strangwood

Experimental and Finite-Element Analyses of a Golf Ball Colliding
with Simplified Clubheads 235
K. Tanaka, H. Oodaira, Y. Teranishi, F. Sato & S. Ujihashi

The Dynamic Characteristic of a Golf Club Shaft 241
M. Shinozaki, Y. Takenoshita, A. Ogawa, A. Ming, N. Hirose & S. Saitou

Advanced Materials in Golf Driver Head Design 247
H.G. Widmann, A. Davis, S.R. Otto & M. Strangwood

The Influence of Different Golf Club Designs on Swing Performance
in Skilled Golfers 253
N. Betzler, G. Shan & K. Witte

Skill Analysis of the Wrist Turn in a Golf Swing to Utilize Shaft Elasticity 259
S. Suzuki, Y. Hoshino, Y. Kobayashi & M. Kazahaya

Motion Analysis Supported by Data Acquisition During Golf Swing 265
A. Sabo, M. Baltl, G. Schrammel & M. Reichel

Automatic Diagnosis System of Golf Swing 271
M. Ueda, Y. Shirai, N. Shimada & M. Oonuki

Kinematic Analysis of Golf Putting for Elite and Novice Golfers 277
J.S. Choi, H.S. Kim, J.H. Yi, Y.T. Lim & G.R. Tack

Development of Wireless Putting Grip Sensor System 283
H.S. Kim, J.S. Choi, J.H. Yi, Y.T. Lim & G.R. Tack

Leadership Behavior as Perceived by Collegiate Golf Coaches and
Players in Taiwan and the Relationship to Basic Personality Traits 289
B.F. Chang, I.C. Ma & Y.J. Jong

8 Ball Sport – Cricket

Non-Linear Viscoelastic Properties and Construction of Cricket Balls 297
B. Vikram & F.K. Fuss

Non-Linear Viscoelastic Impact Modelling of Cricket Balls 303
B. Vikram & F.K. Fuss

Aerodynamics of Cricket Ball – An Understanding of Swing 311
F. Alam, R. La Brooy & A. Subic

Contents

Analysis of Cricket Shots Using Inertial Sensors *A. Busch & D.A. James*	317
High Speed Video Evaluation of a Leg Spin Cricket Bowler *A.E.J. Cork, A.A. West & L.M. Justham*	323
An Analysis of the Differences in Bowling Technique for Elite Players During International Matches *L.M. Justham, A.A. West & A.E.J. Cork*	331
Engineering a Device which Imparts Spin onto a Cricket Ball *L.M. Justham, A.A. West & A.E.J. Cork*	337

9   Ball Sport – Baseball

The Effect of Wood Properties on the Performance of Baseball Bats *K.B. Blair, G. Williams & G. Vasquez*	345
Describing the Plastic Deformation of Aluminum Softball Bats *E. Biesen & L.V. Smith*	351
A Research About Performance of Metal Baseball Bats *Y. Goto, T. Watanabe, S. Hasuike, J. Hayasaka, M. Iwahara, A. Nagamatsu, K. Arai, A. Kondo, Y. Teranishi & H. Nagao*	357
Analysis of Baseball Bats Performance Using Field and Non-Destructive Tests *R.W. Smith, T.H. Liu & T.Y. Shiang*	363
Modelling Bounce of Sports Balls with Friction and Tangential Compliance *K.A. Ismail & W.J. Stronge*	371
Three-Dimensional Kinetic Analysis of Upper Limb Joints During the Forward Swing of Baseball Tee Batting Using An Instrumented Bat *S. Koike, T. Kawamura, H. Iida & M. Ae*	377

10   Ball Sport – Soccer

The Flight Trajectory of a Non-Spinning Soccer Ball *K. Seo, S. Barber, T. Asai, M. Carré & O. Kobayashi*	385
A Study on Wake Structure of Soccer Ball *T. Asai, K. Seo, O. Kobayashi & R. Sakashita*	391

Experimental Investigation of the Effects of Surface Geometry
on the Flight of a Non-Spinning Soccer Ball 397
S. Barber, K. Seo, T. Asai & M.J. Carré

Influence of Foot Angle and Impact Point on Ball Behavior in
Side-Foot Soccer Kicking 403
H. Ishii & T. Maruyama

Intelligent Musculosoccer Simulator 409
S.M.N.A. Senanayake & T.K. Khoo

Intelligent System for Soccer Gait Pattern Recognition 415
A.A. Bakar & S.M.N.A. Senanayake

11 Ball Sport – Tennis and Badminton

Parallel Measurements of Forearm EMG (Electromyography) and
Racket Vibration in Tennis: Development of the System for
Measurements 423
A. Shionoya, A. Inoue, K. Ogata & S. Horiuchi

Parallel Measurements of Forearm EMG (Electro-Myogram) and
Tennis Racket Vibration in Backhand Volley 429
A. Shionoya, A. Inoue, K. Ogata & S. Horiuchi

An Experimental and Computational Study of Tennis Ball
Aerodynamics 437
F. Alam, W. Tio, A. Subic & S. Watkins

Testing of Badminton Shuttles with a Prototype Launcher 443
J.C.C. Tan, S.K. Foong, S. Veluri & S. Sachdeva

Trajectories of Plastic and Feather Shuttlecocks 449
S.K. Foong & J.C.C. Tan

Sports Technique Impartation Method Using Exoskeleton 455
H.B. Lim, T. Ivan, K.H. Hoon & K.H. Low

Badminton Singles Simulation from the Data Obtained in
Physical Education Class 461
K. Suda, T. Nagayama, T. Arai & M. Nozawa

12 Ball Sport – Basketball, Bowling, and Hockey

A Comparison Between Bank and Direct Basketball Field Shots
Using a Dynamic Model   467
H. Okubo & M. Hubbard

Performance Analysis with an Instrumented Bowling Ball   473
L.S.A. Khang & F.K. Fuss

Accelerometry in Motion Analysis   479
S.M.N.A. Senanayake, A.W.W. Yuen & B.D. Wilson

Human Perception of Different Aspects of Field Hockey Stick
Performance   485
M.J. Carré & M.A. McHutchon

A Biomechanical Analysis of the Straight Hit of Elite Women
Hockey Players   491
S. Joseph, R. Ganason, B.D. Wilson, T.H. Teong & C.R. Kumar

13 Aquatics – Boating and Fishing

Difference in Force Application Between Rowing on the Water
and Rowing Ergometers   501
A.C. Ritchie

Effect of Oar Design on the Efficiency of the Rowing Stroke   509
A.C. Ritchie

The Effect of Rowing Technique on Boat Velocity: A Comparison of
HW and LW Pairs of Equivalent Velocity   513
M.M. Doyle, A.D. Lyttle & B.C. Elliott

Oar Forces from Unobtrusive Optical Fibre Sensors   519
M. Davis & R. Luescher

Materials Modelling for Improving Kayak Paddle-Shaft Simulation
Performance   525
P. Ewart & J. Verbeek

Proximity Safety Device for Sportscraft   531
S.G. O'Keefe, B.A. Muller, M.K. Maggs & D.A. James

Analysis and Animation of Bar Float Response to Fish Bite for
Sea Fishing 537
S. Yamabe, H. Kumamoto & O. Nishihara

14 Aquatics – Swimming

Biomechanics of Front-Crawl Swimming: Buoyancy as a Measure of
Anthropometric Quantity or a Motion-Dependent Quantity? 547
T. Yanai

Effect of Buoyant Material Attached to Swimsuit on Swimming 561
M. Nakashima, Y. Motegi, S. Ito & Y. Ohgi

Analysis of the Optimal Arm Stoke in the Backstroke 569
S. Ito

Swimming Stroke Analysis Using Multiple Accelerometer Devices
and Tethered Systems 577
N.P. Davey & D.A. James

A Semi-Automatic Competition Analysis Tool for Swimming 583
*X. Balius, V. Ferrer, A. Roig, R. Arellano, B. de la Fuente,
E. Morales, X. de Aymerich & J.A. Sánchez*

Computational Fluid Dynamics – A Tool for Future Swimming
Technique Prescription 587
M. Keys & A. Lyttle

Swim Power – An Approach Using Optical Motion Analysis 593
A. Ong & M. Koh

15 Athletics and Jumping

Kinematic Analysis of the Best Throws of the World Elite Discus
Throwers and of the Olympic Winner in Decathlon 601
S. Vodičková

Enhancing Measurement Acuity in the Horizontal Jumps:
The Rieti'99 Experience 607
X. Balius, A. Roig, C. Turró, J. Escoda & J.C. Álvarez

Effect of Joint Strengthening on Vertical Jumping Performance 613
H.-C. Chen & K.B. Cheng

Strength of Thigh Muscles and Ground Reaction Force on
Landing from Vertical Drop Jumps 619
C. Kim & J.C.C. Tan

Biomechanical Analysis of Landing from Different Heights with
Vision and without Vision 625
H.-S. Chung & W.-C. Chen

Optimal Viscoelastic Model to Estimate Vertical Ground Reaction
Force from Tibial Acceleration During Hopping 631
Y. Sakurai & T. Maruyama

A Portable Vertical Jump Analysis System 637
B.H. Khoo, S.M.N.A. Senanayake, D. Gouwanda & B.D. Wilson

The Effect of Shoe Bending Stiffness on Predictors of Sprint
Performance: A Pilot Study 643
D.T. Toon, N. Hopkinson & M.P. Caine

Three-Dimensional Analysis of Jump Motion Based on Multi-Body
Dynamics – The Contribution of Joint Torques of the Lower Limbs
to the Velocity of the Whole-Body Center of Gravity 649
S. Koike, H. Mori & M. Ae

The 3-D Kinematics of the Barbell During the Snatch Movement
for Elite Taiwan Weightlifters 655
H.T. Chiu, K.B. Cheng & C.H. Wang

16 Climbing and Mountaineering

Stress Distribution at the Finger Pulleys During Sport Climbing 663
M.A. Tan, F.K. Fuss & G. Niegl

Biomechanics of Free Climbing – A Mathematical Model for
Evaluation of Climbing Posture 671
N. Inou, Y. Otaki, K. Okunuki, M. Koseki & H. Kimura

The Fully Instrumented Climbing Wall: Performance Analysis,
Route Grading and Vector Diagrams – A Preliminary Study 677
F.K. Fuss & G. Niegl

Physiological Response to Different Parts of a Climbing Route 683
G. Balasekaran, F.K. Fuss & G. Niegl

Effect of Acetazolamide on Physiological Variables During
High Altitude in 15-Year Olds 689
G. Balasekaran, S. Thompson, J. Grantham & V. Govindaswamy

Examination of the Time-Dependent Behavior of Climbing Ropes 695
*I. Emri, M. Udovč, B. Zupančič, A. Nikonov, U. Florjančič,
S. Burnik & B.S. von Bernstorff*

Development of a Sharp Edge Resistance Test for Mountaineering Ropes 701
M. Blümel, V. Senner & H. Baier

17 Martial Arts and Archery

Effective Bodily Motion on Punching Technique of Shorinji-Kempo 709
T. Hashimoto, H. Hasegawa, H. Doki & M. Hokari

The Optimum Driving Model of Jump Back Kick in Taekwondo 715
C.-L. Lee

Non-Linear Viscoelasticity of Karate Punching Shields 721
J.K.L. Tan & F.K. Fuss

Biomechanical Analysis of Tai Chi Difficulty Movement "Teng Kong
Zheng Ti Tui" 727
Y.K. Yang, W. Xie, D. Lim & J.H. Zhou

Archery Bow Stabiliser Modelling 733
I. Zanevskyy

Mathematical Model of the Aiming Trajectory 741
C.-K. Hwang, K.-B. Lin & Y.-H. Lin

Archery Performance Analysis Based on the Coefficients of AR2
Model of Aiming Trajectory 747
K.-B. Lin, C.-K. Hwang & Y.-H. Lin

18 Motor Sport and Cycling

Formula SAE: Student Engagement Towards World Class Performance 755
S. Watkins & G. Pearson

Service Strength of High Tech Bicycles 767
M. Blümel, V. Senner & H. Baier

Effects of Venting Geometry on Thermal Comfort and Aerodynamic
Efficiency of Bicycle Helmets   773
F. Alam, A. Subic, A. Akbarzadeh & S. Watkins

Design of Leisure Sports Equipment and Methods of Sports Science –
An Example from Cycling   781
M. Müller, G. Mecke, H. Böhm, M. Niessen & V. Senner

19 Disability Sport

Disability Sports in Singapore – Paralympic Movement and
Greater International Responsibility   789
K.G. Wong

Performance Analysis in Wheelchair Racing   795
A.P. Susanto, F.K. Fuss, K.G. Wong & M.S. Jaffa

20 Winter Sports

Skiing Equipment: What is Done Towards More Safety, Performance
and Ergonomics?   803
V. Senner & S. Lehner

Airfolied Design for Alpine Ski Boots   813
L. Oggiano, L. Agnese & L.R. Sætran

Ground Reaction Forces Measurement Based on Strain Gauges in
Alpine Skiing   819
S. Vodičková & F. Vaverka

Fusion Motion Capture: Can Technology Be Used to Optimise Alpine
Ski Racing Technique?   825
M. Brodie, A. Walmsley & W. Page

Study on the Optimization of a Snowboard   833
Q. Wu & S. Ganguly

Comparative Experiments for Snowboard Vibration Characteristics   839
S. Kajiwara, D. Taniguchi, A. Nagamatsu, M. Iwahara & A. Kondou

Automated Inertial Feedback for Half-Pipe Snowboard Competition
and the Community Perception   845
J.W. Harding, K. Toohey, D.T. Martin, C. Mackintosh,
A.M. Lindh & D.A. James

Estimation of Dirt Attraction on Running Surfaces of Cross-Country Skis 851
L. Kuzmin & M. Tinnsten

Aerodynamic Forces Computation from High-Speed Video Image of
Ski Jumping Flight 857
M. Murakami, N. Hirai, K. Seo & Y. Ohgi

Approach for a Systematic Optimization of a Twoseater Bobsleigh 863
M. Müller & V. Senner

21 Youth Sports

The Design and Development of Electronic Playground Equipment to
Increase Fitness in Children 871
P.P. Hodgkins, S.J. Rothberg, P. Mallinson & M.P. Caine

The Testing of Electronic Playground Equipment to Increase
Fitness in Children 877
P.P. Hodgkins, S.J. Rothberg, P. Mallinson & M.P. Caine

Visualization of the Hazards Lurking in Playground Equipment
Based on Falling Simulations Using Children Multi-Body Models 883
*Y. Miyazaki, S. Watanabe, M. Mochimaru, M. Kouchi,
Y. Nishida & S. Ujihashi*

Personality Traits and their Relationship to Leisure Motivation and
Leisure Satisfaction in Southern Taiwan University Students 889
C.H. Chen

22 Coaching Technology and Sport Education

Video Technology and Coaching 897
B.D. Wilson, A. Pharmy & M. Nadzrin

How Can Voice-Enabled Technologies Help Athletes and Coaches to
Become More Efficient? 905
C. Stricker & P.H. Rey

The Singapore Sports School – Developing Elite Youth Athletes 911
G. Nair

Measuring Sports Class Learning Climates – The Development
of the Sports Class Environment Scale 917
T. Dowdell, L.M. Tomson & M. Davies

Developing Multimedia Courseware in Teaching Exercise
Physiology for Physical Education Major 921
S. Sethu, A.S. Nageswaran, D. Shunmuganathan & M. Elango

Author Index 925

Preface

Today, more people are participating in sports than ever before. With the increased interest and participation in sports, and the extensive media coverage of sporting events worldwide, sport has evolved into a global business worth around $600 billion in total. The world sporting goods market is estimated at $120 billion retail, with footwear accounting for $30 billion, apparel $50 billion and equipment $40 billion. In addition, the sporting goods industry has diversified over the years to accommodate the different interests and needs of the athletes and consumers in general. It has also promoted and helped develop new sports that have in turn served as a catalyst for new types of products.

Sport as we know it today can hardly be separated from technology. Transfer and integration of knowledge from a wide range of disciplines and industries has generated a rapid technological change in sports in modern times. New technologies have made sports faster, more powerful and enjoyable in many ways. Sports technologies have the capacity to enhance performance, prevent injuries or in some cases even changed the type and pattern of injury experienced by athletes. They can also make a sport more or less interesting to the spectators. Therefore, research has focused over the years on understanding the consequences of increasingly complex sports technologies from a systems perspective, and on developing new technologies and techniques that can improve not only the performance and enjoyment but also safety and overall wellbeing of athletes.

Technological change and its consequences have reinforced the need for a more proactive role of governing bodies and professional associations in sports. Rules governing the development and use of sports technology are multifaceted and take into consideration not only performance and safety but also broader social, economical and political issues. They may be driven by the desire to maintain a particular sporting tradition or heritage, to maintain the existing capital infrastructure and avoid over-performance, or to promote interest and appeal of the sport. In addition, as effects of global warming have become widely recognized in recent years, sustainability has become a much desired attribute in the development of new sports technologies for the future.

Clearly, sport technology has to be seen from the holistic, as well as inter- and transdisciplinary point of view. Product development requires close collaboration between engineers, athletes, sports scientists, and business managers. It requires an in-depth understanding of engineering disciplines, life and sport sciences, as well as economics. ***The Impact of Technology on Sport II*** has in its core precisely this philosophy and approach. It aims to provide a deeper insight into the current status of sports technology and to present recent developments in this area from the perspective of different disciplines, industrial practice, academia and athletes. This book brings together peer reviewed contributions from researchers around the world, including Japan, Australia, New Zealand, Singapore, Malaysia, Korea, Taiwan, India, United Kingdom, Germany, Austria, Czech Republic, Norway,

Sweden, Slovenia, Switzerland, Spain and USA. The book includes **135** refereed papers that have been presented at the 3rd Asia-Pacific Congress on Sports Technology in Singapore in September 2007. The Congress was jointly organised by the Nanyang Technological University, RMIT University and the Japan Sports Engineering Association (JSEA). The book has been edited and its contents divided into the following interrelated sections.

Sports Technology and Engineering; Life Cycle, Environmental Engineering and Eco-Design; Sports Medicine, Exercising and Clinical Biomechanics; General Motion Analysis; Apparel and Sport Surfaces; Gait, Running and Shoes; Ball Sport – Golf; Ball Sport – Cricket; Ball Sport – Baseball; Ball Sport – Soccer; Ball Sport – Tennis and Badminton; Ball sport – Basketball, Bowling, and Hockey; Aquatics – Boating and fishing; Aquatics – Swimming; Athletics and Jumping; Climbing and Mountaineering; Martial Arts and Archery; Motor Sport and Cycling; Disability Sport; Winter Sports; Youth Sports; Coaching Technology and Sport Education.

We gratefully acknowledge the authors and the referees who have made this publication possible with their research work and written contributions. We would also like to thank the Singapore, Melbourne and Tokyo Secretariats for their hard work and continuous support. Finally, we hope that a book on the transdisciplinary subject of sports technology, as diverse in topics and approaches as this one, will be of interest to sports technology researchers, practitioners and enthusiasts whatever their scientific background or persuasion.

Franz Konstantin Fuss, Aleksandar Subic and Sadayuki Ujihashi
Editors, September 2007

Contributors List

Ae M.
Institute of Health and Sport Science, University of Tsukuba University of Tsukuba, 1-1-1 Tennodai, Tsukuba, Ibaraki, 305-8574, Japan

Agnese L.
Politecnico di Torino, dipartimento di ingegneria aerospaziale.C.so Duca degli Abruzzi 24 10100 Torino. E-mail lukepra@libero.it

Akbarzadeh A.
School of Aerospace, Mechanical and Manufacturing Engineering, RMIT University; 264 Plenty Road, Bundoora, Melbourne, VIC 3083, Australia

Alam F.
School of Aerospace, Mechanical and Manufacturing Engineering, RMIT University; 264 Plenty Road, Bundoora, Melbourne, VIC 3083, Australia, E-mail: firoz.alam@rmit.edu.au

Allgeuer T.
Dow Europe GmbH, Bachtobelstr. 3, Horgen, 8810, Switzerland, E-mail:TAllgeuer@dow.com

Álvarez J.C.
Real Federación Española de Atletismo (RFEA), Avenida de Valladolid, 81 esc. dcha.1°, 28008 Madrid, Spain

An S.Y.
Department of Sports Science Kookmin University, 86-1 Jungneung- dong, Sungbuk-gu, Seoul, Korea

Arai K.
Hosei University, Department of Mechanical Engineering, Faculty of Engineering, Kajino-cho Koganei city Tokyo 3-7-2, Japan

Arai T.
Graduate School of Decision Science and Technology, Tokyo Institute of Technology, 2-12-1 Ookayama, Meguro, Tokyo, 152-8552 Japan

Arellano R.
Facultad de Ciencias de la Actividad Física y del Deporte. C/Carretera de Alfacar, s/n. 18011 Granada, Andalucía, Spain, E-mail: arellano@ugr.es

Asai T.
University of Tsukuba, 1-1-1 Tennodai, Tsukuba 305-8573, Japan

Baier H.
Technische Universität München, Institute for Lightweight Structures, Bolzmanstraße 15, 85748 Garching, Germany

Bakar A.A.
Monash University Sunway Campus, No. 2, Jalan Universiti, 46150 Petaling Jaya, Malaysia, E-mail: ababu1@student.monash.edu

Balasekaran G.
Nanyang Technological University, National Institute of Education, Physical Education and Sports Science, Singapore, E-mail: govindsamy.b@nie.edu.sg

Balius X.
Olympic Training Centre (CAR) of Catalunya, Avinguda Alcalde Barnils, 3-5, 08034 Sant Cugat del Vallès, Catalunya, Spain, E-mail: xbalius@car.edu

Baltl M.
Austrian Golf Association, Vienna, Austria

Barber S.
University of Sheffield, Mappin Street, Sheffield, S1 3J, UK

Bensason S.
Dow Europe GmbH, Bachtobelstr. 3, Horgen, 8810, Switzerland

Betzler N.
Napier University, School of Life Sciences, 10 Colinton Road, Edinburgh, EH10 5DT, United Kingdom, E-Mail: n.betzler@napier.ac.uk

Biesen E.
Washington State University, 201 Sloan, Spokane St, Pullman, WA 99164-2920 USA, E-mail: ebiesen@gonzaga.edu

Blair K.B.
Sports Innovation Group LLC, Arlington, MA USA, E-mail: kbb@sportsinnovationgroup.com

Blümel M.
Technische Universität München, Institute for Sports Equipment and Materials, Connollystraße 32 80809 München, E-mail: matthias.bluemel@gmx.de

Böhm H.
Department Sports Equipment and Materials, Technische Universität München, Munich, Germany

Brodie M.
Institute of Food Nutrition and Human Health, Massey University, Box 756, Wellington, New Zealand, E-mail: m.a.brodie@massey.ac.nz

Burkett B.
Centre for Healthy Activities Sport and Exercise (CHASE), University of the Sunshine Coast Queensland Australia, E-mail jbl001@student.usc.edu.au

Burnik S.
Faculty of Sport, University of Ljubljana, Ljubljana, Slovenija

Busch A.
Centre for Wireless Monitoring and Applications, Griffith University, Brisbane, Australia, E-mail: a.busch@griffith.edu.au

Caine M.P.
Sports Technology Research Group, Loughborough University, Loughborough, Leicestershire, LE11 3TU, England

Carré M.J.
Department of Mechanical Engineering, University of Sheffield, Mappin St, Sheffield, S1 3JD, UK, E-mail: m.j.carre@shef.ac.uk

Chang A.
The Dow Chemical Company, 2301 N Brazosport Blvd, Freeport, Texas, 77541, USA

Chang B.F.
Department of Physical Education, National Taichung University, Taiwan, E-mail: bifon0905@hotmail.com

Chen C.H.
Department of Sports & Recreation Management, Chang Jung Christian University, Taiwan. No.396 Chang Jung Rd., Sec.1 Kway Jen, Tainan, 711, Taiwan, E-mail: chenphd@mail.cjcu.edu.tw

Chen H.-C.
Institute of Physical Education, Health, and Leisure Studies, National Cheng Kung University, Tainan, Taiwan

Chen W.-C.
Graduate Institute of Sports Science, National College of Physical Education and Sports, Tau-Yuan, Taiwan, R.O.C. E-mail:ellmars@ms28.hinet.net

Cheng K.B.
Institute of Physical Education, Health, and Leisure Studies, National Cheng Kung University, Tainan, Taiwan, E-mail: kybcheng@mail.ncku.edu.tw

Chiu H.T.
Institute of Physical Education, Health and Leisure Studies, Cheng-Kung University, Tainan, Taiwan, Email: htchiu@mail.ncku.edu.tw

Choi C.H.
Fusion Technology Center, Korea Institute of Industrial Technology, 35-3, Hongcheon, Ipjang, Cheonan, Chungnam, South Korea

Choi J.S.
Biomedical Engineering, Konkuk University, Chungju, Korea, E-mail:jschoi98@gmail.com

Chung H.-S.
Graduate Institute of Physical Education, National College of Physical Education and Sports, Tau-Yuan, Taiwan, R.O.C. E-mail:j791203@ms41.hinet.net ; j791203@hotmail.com

Chung K.R.
Fusion Technology Center, Korea Institute of Industrial Technology, 35-3, Hongcheon, Ipjang, Cheonan, Chungnam, South Korea

Cork A.E.J.
Sports Technology Research Group, Wolfson School of Mechanical and Manufacturing Engineering, Loughborough University, Loughborough, Leicestershire, LE11 3TU, England, E-mail: A.E.J.Cork@lboro.ac.uk

Cornish J.E.M.
University of Birmingham, Edgbaston, UK, E-mail: jec191@bham.ac.uk

Davey N.P.
Centre for Wireless Monitoring and Applications, Griffith School of Engineering, Griffith University, Nathan Campus, Kessels Road, Nathan, QLD, Australia, 4111, E-mail: N.Davey@griffith.edu.au; Centre of Excellence for Applied Sport Science Research, Queensland Academy of Sport, Level 1 Queensland Sport and Athletics Centre (QSAC), Kessels Road, Nathan QLD 4111

Davies M.
School of Education and Professional Studies, Griffith University, Brisbane, Australia

Davis A.
University of Birmingham, Edgbaston, UK

Davis M.
Australian Institute of Sport. Biomechanics Department. Leverrier Crescent Bruce ACT Australia 2617, mark.davis@ausport.gov.au

de Aymerich X.
Instituto Vasco de Educación Física (SHEE/IVEF). Carretera de Lasarte, s/n. 01007 Vitoria-Gasteiz, Navarra, Spain

de la Fuente B.
Facultad de Ciencias de la Actividad Física y del Deporte. C/Carretera de Alfacar, s/n. 18011 Granada, Andalucía, Spain

Dipti M.
Indian Institute of Science, Bangalore, India

Doetkott C.
Research Analyst, Information Technology, North Dakota State University, Fargo, ND 58105-5285

Doki H.
Faculty of Engineering and Resource Science, Akita University 1-1 Tegata gakuen-machi, Akita-shi, Akita, 010-8502, Japan

Dowdell T.
School of Education and Professional Studies, Griffith University, Brisbane, Australia

Doyle M.M.
Western Australian Institute of Sport, Stephenson Avenue, Mt Claremont WA 6010, Australia, E-mail: mdoyle@wais.org.au

Dufour M.J.D.
Cranfield University, Cranfield, Bedford, MK43 0AL, UK, E-mail: m.dufour@cranfield.ac.uk

Eckelt M.
University of Applied Science, Technikum Wien, Sports-Equipment Technology, Vienna, Austria

Elango M.
Dept. of Physical Education, The MDT Hindu College, Tirunelveli, TamilNadu, South India, E-mail: msu_sports@rediffmail.com

Elliott B.C.
The University of Western Australia, School of Human Movement and Exercise Science, Stirling Hwy, Crawley WA 6009, Australia

Emri I.
Faculty of Mechanical Engineering, Center for Experimental Mechanics, University of Ljubljana, Ljubljana, Slovenija, E-Mail: igor.emri@fs.uni-lj.si

Escoda J.
Olympic Training Centre (CAR) of Catalunya, Avinguda Alcalde Barnils, 3-5, 08034 Sant Cugat del Vallès, Catalunya, Spain

Ewart P.
The Department of Engineering, Materials Division, The University of Waikato, Private Bag 3105, Hamilton, New Zealand, E-mail: p.ewart@waikato.ac.nz

Ferrer V.
Olympic Training Centre (CAR) of Catalunya, Avinguda Alcalde Barnils, 3-5, 08034 Sant Cugat del Vallès, Catalunya, Spain

Florjančič U.
Faculty of Mechanical Engineering, Center for Experimental Mechanics, University of Ljubljana, Ljubljana, Slovenija

Foong S.K.
Natural Sciences and Science Education, National Institute of Education, Nanyang Technological University, 1 Nanyang Walk, S637616, E-mail: seekit.foong@nie.edu.sg

Fuss F.K.
Sports Engineering Research Team, Division of Bioengineering, School of Biomedical and Chemical Engineering, 70 Nanyang Drive, Nanyang Technological University, Singapore 637457, E-mail: mfkfuss@ntu.edu.sg

Gagalowicz A.
INRIA, Domaine de Voluceau, BP105 78153 Le Chesnay, France, Email:andre.gagalowicz@inria.fr

Ganason R.
Human Performance Laboratory, Sports Biomechanics Centre, National Sports Institute of Malaysia, E-Mail: rohan_ganason@yahoo.com

Ganguly S.
Department of Mechanical Engineering & the Center for Nonlinear Dynamics and Control, Villanova University, 800 Lancaster Avenue, Villanova, PA 19085, USA

Goto Y.
Hosei University, Department of Mechanical Engineering, Faculty of Engineering, Kajino-cho Koganei city Tokyo 3-7-2, Japan, E-mail: yuta.goto.vb@gs-eng.hosei.ac.jp

Gouwanda D.
Monash University Sunway campus, No. 2, Jalan Universiti, 46150 Petaling Jaya, Malaysia

Govindaswamy V.
Computer Science and Engineering, University of Texas at Arlington, Texas, USA

Grantham J.
Nanyang Technological University, National Institute of Education, Physical Education and Sports Science, Singapore; Sports Medicine Centre, Qatar National Olympic Committee, Doha, Qatar

Gray A.R.
Progressive Sports Technologies Ltd., Innovation Centre, Loughborough University, Loughborough, Leicestershire, LE11 3EH, UK, E-mail: ash@progressivesports.co.uk

Harding J.W.
Applied Sport Research Centre, Australian Institute of Sport, Australia; Centre for Wireless Monitoring and Applications, Griffith University, Australia; Olympic Winter Institute of Australia, Australia

Harrison R.
Wolfson School of Mechanical and Manufacturing Engineering, Loughborough University, Loughborough, Leicestershire, LE11 3TU, England

Hasegawa H.
Faculty of Engineering and Resource Science, Akita University 1-1 Tegata gakuen-machi, Akita-shi, Akita, 010-8502, Japan, E-mail: hhasegaw@mech.akita-u.ac.jp

Hashimoto T.
Graduate School of Engineering and Resource Science, Akita University 1-1 Tegata gakuen-machi, Akita-shi, Akita, 010-8502, Japan

Hasuike S.
Hosei University, Department of Mechanical Engineering, Faculty of Engineering, Kajino-cho Koganei city Tokyo 3-7-2, Japan

Hayasaka J.
Hosei University, Department of Mechanical Engineering, Faculty of Engineering, Kajino-cho Koganei city Tokyo 3-7-2, Japan

Hayashi Y.
Department of Mechanical Engineering, Doshisha University, Kyotanabe, Kyoto, 610-0321, Japan

Hee J.W.
Monash University Sunway campus, No. 2, Jalan Universiti, 46150 Petaling Jaya, Malaysia, E-mail: jwhee1@student.monash.edu

Hirai N.
Graduate School of Systems and Information Engineering, University of Tsukuba, Japan.

Hirose N.
Mamiya-OP Co., Ltd., 3-5-1 Bijyogi, Toda-shi, Saitama, Japan

Hodgkins P.P.
Sports Technology Research Group, Loughborough University, Loughborough, Leicestershire, LE11 3TU, England, E-mail: p.p.hodgkins@lboro.ac.uk

Hokari M.
Faculty of Engineering and Resource Science, Akita University 1-1 Tegata gakuen-machi, Akita-shi, Akita, 010-8502, Japan

Hong G.S.
Fusion Technology Center, Korea Institute of Industrial Technology, 35-3, Hongcheon, Ipjang, Cheonan, Chungnam, South Korea

Hoon K.H.
School of Mechanical and Aerospace Engineering, Nanyang Technological University, 50 Nanyang Avenue, Singapore 639798, E-mail: mkhhoon@ntu.edu.sg

Hooper S.L.
Centre of Excellence for Applied Sports Science Research, Queensland Academy of Sport, PO Box 956, Brisbane, 4109, Australia, E-mail: sue.hooper@srq.qld.gov.au

Hopkinson N.
Wolfson School of Mechanical and Manufacturing Engineering, Loughborough University, Leicestershire, LE11 3TU, UK

Horiuchi S.
Asia University, Musashi-sakai, Musashino, Tokyo, Japan, E-mail: h-sho@ta2.so.net.ne.jp

Hoshino Y.
Graduate school of Engineering, Hokkaido University,kita13Nishi8, Sapporo, 060-8628, Japan, E-mail: hoshinoy@eng.hokudai.ac.jp

Hubbard M.
University of California, Davis, CA 95616, USA, E-mail: mhubbard@ucdavis.edu

Hwang C.-K.
Department of Electrical Engineering, Chung Hua University, Hsin-Chu, Taiwan 300, ROC, E-mail: simon@chu.edu.tw, bear@chu.edu.tw

Hyeong J.H.
Fusion Technology Center, Korea Institute of Industrial Technology, 35-3, Hongcheon, Ipjang, Cheonan, Chungnam, South Korea

Iida H.
Polytechnic University, 4-1-1 Hashimotodai, Sagamihara-City, Kanagawa, Japan

Inou N.
Tokyo Institute of Technology, Department of Mechanical and Control Engineering, I3-12, 2-12-1, O-okayama, Meguro-ku, Tokyo 152-8552, JAPAN, E-mail: inou@mech.titech.ac.jp

Inoue A.
Nagaoka University of Technology, 1603-1 Kamitomioka, Nagaoka, Niigata, Japan

Ishii H.
Graduate School of Decision Science and Technology, Tokyo Institute of Technology, W9-4, 2-12-1, O-okayama, Meguro-ku, Tokyo, 152-8552, Japan, E-mail: ishii.h.ac@m.titech.ac.jp

Ismail K.A.
Department of Engineering, University of Cambridge, Cambridge, CB2 1PZ, UK, E-mail: kai25@cam.ac.uk

Ito S.
Department of Mechanical Engineering, National Defense Academy, 1-10-20 Hashirimizu, Yokosuka, Kanagawa 239-8686, Japan, E-mail: ito@nda.ac.jp

Ivan T.
School of Mechanical and Aerospace Engineering, Nanyang Technological University, 50 Nanyang Avenue, Singapore 639798, E-mail: ivan0001@ntu.edu.sg

Iwahara M.
Hosei University, Department of Mechanical Engineering, Faculty of Engineering, Kajino-cho Koganei City Tokyo 3-7-2, Japan

Jaffa M.S.
Singapore Disability Sports Council, National Stadium (West Entrance), 15 Stadium Road, Singapore 397718, E-mail: jaffams@sdsc.org.sg

James D.A.
Centre for Wireless Technology Griffith University, Brisbane, Queensland, Australia, E-mail: d.james@griffith.edu.au; Centre of Excellence for Applied Sport Science Research, Queensland Academy of Sport, Brisbane, Queensland, Australia

Jong Y.J.
Department of Physical Education, National Chiayi University, Taiwan. E-mail: jong9275@mail.ncyu.edu.tw

Joseph S.
Human Performance Laboratory, Sports Biomechanics Centre, National Sports Institute of Malaysia, E-mail: sajujoseph@msn.com

Justham L.M.
Sports Technology Research Group, Wolfson School of Mechanical and Manufacturing Engineering, Loughborough University, Loughborough, Leicestershire, LE11 3TU, England, E-mail: A.E.J.Cork@lboro.ac.uk, A.A.West@lboro.ac.uk, L.Justham@lboro.ac.uk

Kajiwara S.
Hosei University, Graduate School in Mechanical Engineering

Kawamura T.
Institute of Health and Sport Science, University of Tsukuba, 1-1-1 Tennodai, Tsukuba, Ibaraki, 305-8574, Japan

Kazahaya M.
Faculty of Engineering, Kitami Institute of Technology, 165 Koen-cho, Kitami, 090-8507, Japan

Keys M.
Western Australian Institute of Sport, Challenge Stadium, Mt Claremont, WA 6010, Australia. E-mail: wais@wais.org.au; Department of Human Movement and Exercise Science/Department of Civil Engineering, The University of Western Australia., Crawley, WA 6907, Australia

Khang L.S.A.
Sports Engineering Research Team, Division of Bioengineering, School of Biomedical and Chemical Engineering, 70 Nanyang Drive, Nanyang Technological University, Singapore 637457

Khoo B.H.
Monash University Sunway Campus, No. 2, Jalan Universiti, 46150 Petaling Jaya, Malaysia

Khoo T.K.
Monash University Sunway Campus, No. 2, Jalan Universiti, 46150 Petaling Jaya, Malaysia

Kim C.
Physical Education and Sports Science, National Institute of Education, Nanyang Technological University, 1 Nanyang Walk, 637616, Singapore, E-mail: jkkim@nie.edu.sg

Kim H.S.
Biomedical Engineering, Konkuk University, Chungju, Korea

Kim S.Y.
Fusion Technology Center, Korea Institute of Industrial Technology, 35-3, Hongcheon, Ipjang, Cheonan, Chungnam, South Korea, E-mail: sayup.kim@gmail.com, sayub@kitech.re.kr

Kimura H.
Tokyo Institute of Technology, Department of Mechanical and Control Engineering, I3-12, 2-12-1, O-okayama, Meguro-ku, Tokyo 152-8552, Japan

Kobayashi O.
Tokai University, 1117 Kita-Kaname, Hiratsuka 259-1292, Japan

Kobayashi Y.
Graduate school of Engineering, Hokkaido University, kita13Nishi8, Sapporo, 060-8628, Japan, E-mail: kobay@eng.hokudai.ac.jp

Koh M.
Republic Polytechnic, Singapore, Republic Polytechnic, 9 Woodlands, Ave 9, Singapore 738964, Email: koh_teik_hin@rp.sg

Koike S.
Institute of Health and Sport Science, University of Tsukuba, 1-1-1 Tennodai, Tsukuba, Ibaraki, 305-8574, Japan, E-mail:koike@taiiku.tsukuba.ac.jp

Koizumi T.
Department of Mechanical Engineering, Doshisha University, Kyotanabe, Kyoto, 610-0321, Japan E-mail: tkoizumi@ mail.doshisha.ac.jp

Kondo A.
Hosei University, Department of Mechanical Engineering, Faculty of Engineering, Kajino-cho Koganei city Tokyo 3-7-2, Japan

Kondou A.
Hosei University, Graduate School in Mechanical Engineering

Koseki M.
Tokyo Institute of Technology, Department of Mechanical and Control Engineering, I3-12, 2-12-1, O-okayama, Meguro-ku, Tokyo 152-8552, Japan

Kou Z.
Department of Radiology, Wayne State University School of Medicine, Detroit, MI 48201

Kouchi M.
National Institute of Advanced Industrial Science and Technology, 2-41-6, Aomi, Koutou-ku, Tokyo, Japan

Kulish V.V.
School of Mechanical and Aerospace Engineering, Nanyang Technological University, 50 Nanyang Avenue, Singapore 639798

Kumamoto H.
Kyoto University, Yoshida Honmachi, Sakyo-ku, Kyoto, 606-8501, Japan, E-mail: yamabe@ sys.i.kyoto-u.ac.jp, kumamoto@i.kyoto-u.ac.jp

Kumar C.R.
National Women Hockey Coach, Malaysia

Kuzmin L.
Department of Engineering, Physics and Mathematics, Mid Sweden University, Teknikhuset (Q), Plan 3, Akademigatan 1, SE-831 25 Östersund, Sweden, E-mail: leonid.kuzmin@miun.se

La Brooy R.
School of Aerospace, Mechanical and Manufacturing Engineering, RMIT University; 264 Plenty Road, Bundoora, Melbourne, VIC 3083, Australia

Lee C.-L.
National Taiwan Normal University, 2F, No. 233, NongAn St., Taipei 140 Taiwan, E-mail: chen_lin325@yahoo.com.tw

Lee J.B.
Centre for Healthy Activities Sport and Exercise (CHASE), University of the Sunshine Coast Queensland Australia, E-mail jbl001@student.usc.edu.au

Lee K.K.
Department of Sports Science Kookmin University, 86-1 Jungneung- dong, Sungbuk-gu, Seoul, Korea, E-mail: kklee@kookmin.ac.kr

Lehner S.
Technical University Munich, Department of Sport Equipment and Materials, Connollystr. 32, 80809 Munich, Germany

Lewis R.
Department of Mechanical Engineering, University of Sheffield, Mappin Street, Sheffield, S1 3JD, UK

Liew M.
Monash University Sunway Campus, No. 2, Jalan Universiti, 46150 Petaling Jaya, Malaysia

Lim D.
Biomechanics Department, Singapore Sports Council, Singapore

Lim H.B.
School of Mechanical and Aerospace Engineering, Nanyang Technological University, 50 Nanyang Avenue, Singapore 639798, E-mail: limh0048@ntu.edu.sg

Lim Y.T.
Sports Science, Konkuk University, Chungju, Korea, E-mail: ytlim@kku.ac.kr

Lin K.-B.
Yuanpei University of Science and Technology, Hsin-Chu, Taiwan, ROC, E-mail: lingowbin@ypu.edu.tw

Lin Y.-H.
Department of Electrical Engineering, Chung Hua University, Hsin-Chu, Taiwan 300, ROC, E-mail: simon@chu.edu.tw, bear@chu.edu.tw

Lindh A.M.
Applied Sport Research Centre, Australian Institute of Sport, Australia

Liu T.H.
Institute of Sports Equipment Technology, Taipei Physical Education College, Taiwan

Looi D.
Monash University Sunway Campus, No. 2, Jalan Universiti, 46150 Petaling Jaya, Malaysia

Low K.H.
School of Mechanical and Aerospace Engineering, Nanyang Technological University, 50 Nanyang Avenue, Singapore 639798, E-mail: mkhlow@ntu.edu.sg

Luescher R.
Australian Institute of Sport. Physiology Department. Leverrier Crescent Bruce ACT Australia 2617, Raoul.luescher@ausport.gov.au

Lyttle A.D.
Western Australian Institute of Sport, Stephenson Avenue, Mt Claremont WA 6010, Australia

Ma I.C.
Office of Physical Education, Chung Shan Medical University, Taiwan, E-mail: maichieh@yahoo.com.tw

Mackintosh C.
Applied Sport Research Centre, Australian Institute of Sport, Australia

Maggs M.K.
Centre for Wireless Monitoring and Applications. Griffith University, Nathan, Brisbane QLD 4111, Australia

Mallinson P.
PlayDale Playgrounds Ltd, Haverthwaite, Ulverston, Cumbria, LA12 8AE, England, E-mail: paul@playdale.co.uk

Manoj K.M.
Indian Institute of Science, Bangalore, India

Martin D.T.
Applied Sport Research Centre, Australian Institute of Sport, Australia

Martin J.
The Dow Chemical Company, 2301 N Brazosport Blvd, Freeport, Texas, 77541, USA

Maruyama T.
Department of Human System Science, Tokyo Institute of Technology, W9-4, 2-12-1 O-okayama, Meguro-ku, Tokyo, 152-8552, Japan, E-mail: maruyama@hum.titech.ac.jp

McHutchon M.A.
Department of Mechancial Engineering, University of Sheffield, Mappin St, Sheffield, S1 3JD, UK

Mecke G.
Department Sports Equipment and Materials, Technische Universität München, Munich, Germany

Mellifont R.B.
Centre for Healthy Activities Sport and Exercise (CHASE), University of the Sunshine Coast Queensland Australia

Ming A.
The University of Electro-Communications, 1-5-1 Chofugaoka, Chofu-shi, Tokyo, Japan,
E-mail: ming@mce.uec.ac.jp

Miyazaki Y.
Kanazawa University, Kakuma-machi, Kanazawa, Ishikawa, Japan, E-mail: y-miyazaki@
t.kanazawa-u.ac.jp

Mizota T.
Fukuoka Institute of Technology, 3-30-1, Wajiro-Higashi, Higashi-Ku, Fukuoka, Japan,
E-mail: mizota@fit.ac.jp

Mochimaru M.
National Institute of Advanced Industrial Science and Technology, 2-41-6, Aomi, Koutou-ku,
Tokyo, Japan

Monfared R.P.
Wolfson School of Mechanical and Manufacturing Engineering, Loughborough University,
Loughborough, Leicestershire, LE11 3TU, England

Morales E.
Facultad de Ciencias de la Actividad Física y del Deporte. C/Carretera de Alfacar, s/n.
18011 Granada, Andalucía, Spain

Mori H.
Master's Program in Health and Physical Education, University of Tsukuba, 1-1-1
Tennodai, Tsukuba, Ibaraki, 305-8574, Japan, E-mail: mori@lasbim.taiiku.tsukuba.ac.jp

Motegi Y.
Tokyo Institute of Technology, 2-12-1-W8-2 Ookayama, Meguro, Tokyo 152-8552, Japan

Muller B.A.
Centre for Wireless Monitoring and Applications. Griffith University, Nathan, Brisbane
QLD 4111, Australia

Müller M.
Department Sports Equipment and Materials, Technische Universität München, Munich,
Germany

Mumford C.
New Zealand Sports Turf Institute, 163 Old West Road, PO Box 347, Palmerston North
4440, New Zealand

Murakami M.
Graduate School of Systems and Information Engineering, University of Tsukuba, Japan

Nadzrin M.
Centre for Biomechanics, National Sports Institute, Bukit Jalil, 57000, Kuala Lumpur, Malaysia

Nagamatsu A.
Hosei University, Department of Mechanical Engineering, Faculty of Engineering, Kajinocho Koganei city Tokyo 3-7-2, Japan

Nagao H.
Mizuno Co. Ltd.,1-12-35 South Kouhoku, Suminoe-ku Osaka, Japan

Nagayama T.
Graduate School of Decision Science and Technology, Tokyo Institute of Technology, 2-12-1 Ookayama, Meguro, Tokyo, 152-8552 Japan, E-mail: suda@hum.titech.ac.jp

Nageswaran A.S.
Department of Physical Education, H. H. The Rajah's College, Pudukottai, TamilNadu, South India, E-mail: asngeswa@rediffmail.com

Nair G.
Sports Science Academy, Singapore Sports School, Singapore; E-mail: gobinathan@sportsschool.edu.sg

Nakashima M.
Tokyo Institute of Technology, 2-12-1-W8-2 Ookayama, Meguro, Tokyo 152-8552, Japan

Naruo T.
Mizuno Corporation, 1-12-35, Nanko-Kita, Suminoe-Ku, Osaka, Japan, E-mail: tnaruo@mizuno.co.jp

Niegl G.
Department of Anthropology, University of Vienna, Austria, E-mail: guenther.niegl@univie.ac.at, and MedClimb, OEAV/ÖGV (Austrian Mountaineering Federation), Vienna, Austria

Niessen M.
Department Performance Diagnostics, Technische Universität München, Munich, Germany

Nikonov A.
Faculty of Mechanical Engineering, Center for Experimental Mechanics, University of Ljubljana, Ljubljana, Slovenija

Nishi A.
Department of Mechanical Engineering, Doshisha University, Kyotanabe, Kyoto, 610-0321, Japan

Nishida Y.
National Institute of Advanced Industrial Science and Technology, 2-41-6, Aomi, Koutou-ku, Tokyo, Japan

Nishihara O.
Kyoto University, Yoshida Honmachi, Sakyo-ku, Kyoto, 606-8501, Japan, E-mail: nishihara@i.kyoto-u.ac.jp

Nozawa M.
Graduate School of Decision Science and Technology, Tokyo Institute of Technology, 2-12-1 Ookayama, Meguro, Tokyo, 152-8552 Japan

O'Keefe S.G.
Centre for Wireless Monitoring and Applications. Griffith University, Nathan, Brisbane QLD 4111, Australia, E-mail: s.okeefe@griffith.edu.au

Ogata K.
Nagaoka University of Technology, 1603-1 Kamitomioka, Nagaoka, Niigata, Japan

Ogawa A.
The University of Electro-Communications, 1-5-1 Chofugaoka, Chofu-shi, Tokyo, Japan

Oggiano L.
Norwegian University of Science and Technology, Faculty of Engineering Science and Technology N-7491 Trondheim, Norway, E-mail: luca.oggiano@ntnu.no

Ohgi Y.
Graduate School of Media and Governance, Keio University, 5322 Endo, Fujisawa, Kanagawa 252-8520, Japan, E-mail: ohgi@sfc.keio.ac.jp

Okubo H.
Chiba Institute of Technology, 2-17-1 Tsudanuma Narashino, 2750016, Japan, E-mail: hiroki.okubo@it-chiba.ac.jp

Okunuki K.
Tokyo Institute of Technology, Department of Mechanical and Control Engineering, I3-12, 2-12-1, O-okayama, Meguro-ku, Tokyo 152-8552, Japan

Omkar S.N.
Department of Aerospace Engineering, Indian Institute of Science, Bangalore, India, E-mail: omkar@aero.iisc.ernet.in

Ong A.
Republic Polytechnic, Singapore, Republic Polytechnic, 9 Woodlands, Ave 9, Singapore 738964, Email: alex_ong@rp.sg

Contributors List

Oodaira H.
Tokyo Institute of Technology, 2-12-1-W8-14, Ookayama, Meguro-ku, Tokyo 152-8552, Japan

Oonuki M.
SRI R&D Ltd 1-1-2, Tsutsui-cho, Chou-ku Kobe Hyogo 651-0071, Japan

Ostad-Ahmad-Ghorabi H.
Vienna University of Technology, Institute for Engineering Design, Getreidemarkt 9, E307, 1060 Wien Austria, E-mail: ostad@ecodesign.at

Otaki Y.
Tokyo Institute of Technology, Department of Mechanical and Control Engineering, I3-12, 2-12-1, O-okayama, Meguro-ku, Tokyo 152-8552, Japan

Otto S.R.
R & A Rules Ltd, St Andrews, Fife, UK, E-mail: steveotto@randa.org

Page W.
Institute of Food Nutrition and Human Health, Massey University, Box 756, Wellington, New Zealand

Paterson N.
School of Aerospace, Mechanical and Manufacturing Engineering, RMIT University, Cnr Plenty Rd and McKimmies Lane, Bundoora, Melbourne, Victoria 3083, Australia

Pearson G.
School of Aerospace, Mechanical and Manufacturing Engineering, RMIT University, Cnr Plenty Rd and McKimmies Lane, Bundoora, Melbourne, Victoria 3083, Australia

Pharmy A.
Centre for Biomechanics, National Sports Institute, Bukit Jalil, 57000, Kuala Lumpur, Malaysia

Quah C.K.
Nanyang Technological University, School of Computer Engineering, Singapore, E-mail: quah_ck@pmail.ntu.edu.sg

Reichel M.
University of Applied Science, Technikum Wien, Sports-Equipment Technology, Vienna, Austria

Rey P.H.
AISTS – International Academy of Sports Science and Technology, PSE-C, CH-1015 Lausanne, Switzerland; HES-SO Valais – University of Applied Sciences Western Switzerland, Valais, CH-3960 Sierre, Switzerland, E-mail: paul-henri.rey@aists.org

Ritchie A.C.
School of Mechanical and Aerospace Engineering, Nanyang Technological University, 50 Nanyang Avenue, Singapore 639798, E-mail: macritchie@ntu.edu.sg

Roig A.
Olympic Training Centre (CAR) of Catalunya, Avinguda Alcalde Barnils, 3-5, 08034 Sant Cugat del Vallès, Catalunya, Spain

Rothberg S.J.
Sports Technology Research Group, Loughborough University, Loughborough, Leicestershire, LE11 3TU, England

Sabo A.
University of Applied Science, Technikum Wien, Sports-Equipment Technology, Vienna, Austria, E-mail: anton.sabo@technikum-wien.at

Sætran L.R.
Norwegian University of Science and Technology, Faculty of Engineering Science and Technology N-7491 Trondheim, Norway, E-mail: lars.satran@ntnu.no

Saitou S.
Mamiya-OP Co., Ltd., 3-5-1 Bijyogi, Toda-shi, Saitama, Japan

Sakashita R.
Kumamoto University, Kurokami, 2-40-1, Kumamoto, 860-8555, Japan

Sakurai Y.
Department of Human System Science, Tokyo Institute of Technology, W9-4, 2-12-1 O-okayama, Meguro-ku, Tokyo, 152-8552, Japan, E-mail: sakurai.y.ac@m.titech.ac.jp

Sánchez J.A.
Instituto Nacional de Educación Física de Galicia. Avda. Del Che Guevara 121, 15179 Bastiagueiro-Oleiros (A Coruña), Galicia, Spain. E-mail: jasanmol@udc.es

Santry J.
Progressive Sports Technologies Ltd., Innovation Centre, Loughborough University, Loughborough, Leicestershire, LE11 3EH, UK

Sato F.
Mizuno Corporation, 1-12-35, Nanko-Kita, Suminoe-ku, Osaka 559-8510, Japan

Schrammel G.
University of Applied Science, Technikum Wien, Sports-Equipment Technology, Vienna, Austria

Seah H.S.
Nanyang Technological University, School of Computer Engineering, Singapore, E-mail: ashsseah@ntu.edu.sg

Contributors List

Selvarajah V.
Baton Adventures, Malaysia, E-Mail: victo_80@hotmail.com

Senanayake S.M.N.A.
Monash University Sunway campus, No. 2, Jalan Universiti, 46150 Petaling Jaya, Malaysia, E-mail: senanayake.namal@eng.monash.edu.my

Senner V.
Technical University Munich, Department of Sport Equipment and Materials, Connollystr. 32, 80809 Munich, Germany, E-mail: senner@sp.tum.de

Seo K.
Faculty of Education, Art and Science, Yamagata University, 1-4-12 Kojirakawa, Yamagata990-8560, Japan, E-mail: seo@e.yamagata-u.ac.jp

Sethu S.
Dr. Sivanthi Aditanar College of Physical Education, Tiruchendur, TamilNadu, South India, E-mail: sksethu@rediffmail.com

Shan G.
University of Lethbridge, Department of Kinesiology, 4401 University Drive, Lethbridge, AB, T1K 3M4, E-mail: g.shan@uleth.ca

Shiang T.Y.
Institute of Sports Equipment Technology, Taipei Physical Education College, Taiwan

Shimada N.
Ritsumeikan University 1-1-1 Nojihigashi, Kusatsu, Shiga, 525-8577 Japan

Shinozaki M.
Setagayaizumi High School, 9-22-1 Kitakarasuyama, Setagaya-ku, Tokyo, Japan, E-mail: sinozaki@rm.mce.uec.ac.jp

Shionoya A.
Nagaoka University of Technology, 1603-1 Kamitomioka, Nagaoka, Niigata, Japan, E-mail: shionoya@nagaokaut.ac.jp

Shirai Y.
Ritsumeikan University 1-1-1 Nojihigashi, Kusatsu, Shiga, 525-8577 Japan

Shunmuganathan D.
Department of Physical Education, Manonmaniam Sundaranar University, Tirunelveli, TamilNadu, South India, E-mail: msu_sports@rediffmail.com

Sim J.Y.
Sports Engineering Research Team, Division of Bioengineering, School of Biomedical and Chemical Engineering, 70 Nanyang Drive, Nanyang Technological University, Singapore 637457

Smeathers J.E.
Institute of Health and Biomedical Innovation, Queensland University of Technology, Cnr Blamey Street & Musk Avenue, Brisbane, 4059, Australia

Smith J.D.
Wolfson School of Mechanical and Manufacturing Engineering, Loughborough University, Loughborough, Leicestershire, LE11 3TU, England

Smith L.V.
Washington State University, 201 Sloan, Spokane St, Pullman, WA 99164-2920 USA, E-mail: lvsmith@wsu.edu

Smith R.W.
Institute of Sports Equipment Technology, Taipei Physical Education College, Taiwan, E-mail: richard@tpec.edu.tw

Strangwood M.
University of Birmingham, Edgbaston, UK, E-mail: jec191@bham.ac.uk

Stricker C.
AISTS – International Academy of Sports Science and Technology, PSE-C, CH-1015 Lausanne, Switzerland; HES-SO Valais – University of Applied Sciences Western Switzerland, Valais, CH-3960 Sierre, Switzerland, E-mail: claude.stricker@aists.org

Stronge W.J.
Department of Engineering, University of Cambridge, Cambridge, CB2 1PZ, UK

Subic A.
School of Aerospace, Mechanical and Manufacturing Engineering, RMIT University, Cnr Plenty Rd and McKimmies Lane, Bundoora, Melbourne, Victoria 3083, Australia, E-mail: Aleksandar.Subic@rmit.edu.au

Suda K.
Graduate School of Decision Science and Technology, Tokyo Institute of Technology, 2-12-1 Ookayama, Meguro, Tokyo, 152-8552 Japan, E-mail: suda@hum.titech.ac.jp

Susanto A.P.
Sports Engineering Research Team, Division of Bioengineering, School of Biomedical and Chemical Engineering, 70 Nanyang Drive, Nanyang Technological University, Singapore 637457

Suzuki S.
Faculty of Engineering, Kitami Institute of Technology, 165 Koen-cho, Kitami, 090-8507, Japan, E-mail: zuki@mail.kitami-it.ac.jp

Tack G.R.
Biomedical Engineering, Konkuk University, Chungju, Korea, E-mail: jschoi98@gmail.com, bombs1@kku.ac.kr, jeong2yi@kku.ac.kr, grtack@kku.ac.kr

Takenoshita Y.
The University of Electro-Communications, 1-5-1 Chofugaoka, Chofu-shi, Tokyo, Japan

Tan B.
Changi Sports Medicine Centre, Changi General Hospital, 2 Simei Street 3, Singapore 529889, E-mail: benedict_tan@cgh.com.sg

Tan J.C.C.
Physical Education and Sports Science, National Institute of Education, Nanyang Technological University, 1 Nanyang Walk, 637616, Singapore, E-mail: ccjtan@nie.edu.sg

Tan J.K.L.
Sports Engineering Research Team, Division of Bioengineering, School of Biomedical and Chemical Engineering, 70 Nanyang Drive, Nanyang Technological University, Singapore 637457

Tan M.A.
Sports Engineering Research Team, Division of Bioengineering, School of Biomedical and Chemical Engineering, 70 Nanyang Drive, Nanyang Technological University, Singapore 637457

Tanaka K.
Tokyo Institute of Technology, 2-12-1-W8-14, Ookayama, Meguro-ku, Tokyo 152-8552, Japan, E-mail: ktanaka@mei.titech.ac.jp

Taniguchi D.
Hosei University, Graduate School in Mechanical Engineering, E-mail: daiki.taniguchi.7p@eng.hosei.ac.jp

Teong T.H.
Sports Centre, University of Malaya

Teranishi Y.
Mizuno Corporation, 1-12-35 South Kouhoku, Nanko-Kita, Suminoe-ku, Osaka 559-8510, Japan

Thompson S.
Nanyang Technological University, National Institute of Education, Physical Education and Sports Science, Singapore

Tinnsten M.
Department of Engineering, Physics and Mathematics, Mid Sweden University, Teknikhuset (Q), Plan 3, Akademigatan 1, SE-831 25 Östersund, Sweden, E-mail: mats.tinnsten@miun.se

Tio W.
School of Aerospace, Mechanical and Manufacturing Engineering, RMIT University; 264 Plenty Road, Bundoora, Melbourne, VIC 3083, Australia

Tomlinson S.E.
Department of Mechanical Engineering, University of Sheffield, Mappin Street, Sheffield, S1 3JD, UK, E-mail: s.tomlinson@sheffield.ac.uk

Tomson L.M.
School of Education and Professional Studies, Griffith University, Brisbane, Australia

Toohey K.
Department of Tourism, Leisure, Hotel & Sport Management, Griffith University, Australia

Toon D.T.
Wolfson School of Mechanical and Manufacturing Engineering, Loughborough University, Leicestershire LE11 3TU, England, E-mail: d.toon@lboro.ac.uk

Torres E.
Dow Europe GmbH, Bachtobelstr. 3, Horgen, 8810, Switzerland

Tripathy J.
Sports Engineering Research Team, Division of Bioengineering, School of Biomedical and Chemical Engineering, 70 Nanyang Drive, Nanyang Technological University, Singapore 637457

Tsujiuchi N.
Department of Mechanical Engineering, Doshisha University, Kyotanabe, Kyoto, 610-0321, Japan, E-mail: ntsujiuc@ mail.doshisha.ac.jp

Turró C.
Olympic Training Centre (CAR) of Catalunya, Avinguda Alcalde Barnils, 3-5, 08034 Sant Cugat del Vallès, Catalunya, Spain

Udovč M.
Faculty of Mechanical Engineering, Center for Experimental Mechanics, University of Ljubljana, Ljubljana, Slovenija

Ueda M.
SRI R&D Ltd 1-1-2, Tsutsui-cho, Chou-ku Kobe Hyogo 651-0071, Japan

Ujihashi S.
Tokyo Institute of Technology, 2-12-1, Oh-okayama, Meguro-ku, Tokyo, Japan, E-mail: ujihashi@mei.titech.ac.jp

Urry S.R.
Institute of Health and Biomedical Innovation, Queensland University of Technology, Cnr Blamey Street & Musk Avenue, Brisbane, 4059, Australia

Vasquez G.
Massachusetts Institute of Technology, Cambridge MA, USA, E-mail: vasquezm@mit.edu

Vaverka F.
Palacky University, Faculty of Physical Culture, Olomouc, Czech Republic

Verbeek J.
The Department of Engineering, Materials Division, The University of Waikato, Private Bag 3105, Hamilton, New Zealand

Vikram B.
Sports Engineering Research Team, Division of Bioengineering, School of Biomedical and Chemical Engineering, 70 Nanyang Drive, Nanyang Technological University, Singapore 637457

Vinay K.J.
Indian Institute of Technology (Madras), Chennai, India

Vodičková S.
Technical University of Liberec, Faculty of Education, Department of Physical Education, Hálkova 6, 46008 Liberec, Czech Republic, E-mail: vodickova.sona@volny.cz

von Bernstorff B.S.
BASF Aktiengesellschaft, Ludwigshafen, Germany

Walker A.W.
The Louis Laybourne Smith School of Architecture and Design, The University of South Australia, Adelaide, Australia, E-Mail: sandy.walker@unisa.edu.au

Waller T.M.
Progressive Sports Technologies Ltd., Innovation Centre, Loughborough University, Loughborough, Leicestershire, LE11 3EH, UK

Walmsley A.
Institute of Food Nutrition and Human Health, Massey University, Box 756, Wellington, New Zealand

Wang C.H.
Institute of Physical Education, Health and Leisure Studies, Cheng-Kung University, Tainan, Taiwan

Watanabe S.
Tokyo Institute of Technology, 2-12-1, O-Okayama, Meguro-ku, Tokyo, Japan

Watanabe T.
Hosei University, Department of Mechanical Engineering, Faculty of Engineering, Kajino-cho Koganei city Tokyo 3-7-2, Japan

Watkins S.
School of Aerospace, Mechanical and Manufacturing Engineering, RMIT University, Cnr Plenty Rd and McKimmies Lane, Bundoora, Melbourne, Victoria 3083, Australia, E-mail: simon@rmit.edu.au

Wearing S.C.
Centre of Excellence for Applied Sports Science Research, Queensland Academy of Sport, PO Box 956, Brisbane, 4109, Australia; Institute of Health and Biomedical Innovation, Queensland University of Technology, Cnr Blamey Street & Musk Avenue, Brisbane, 4059, Australia, E-mail: s.wearing@qut.edu.au

West A.A.
Sports Technology Research Group, Wolfson School of Mechanical and Manufacturing Engineering, Loughborough University, Loughborough, Leicestershire, LE11 3TU, England, E-mail: A.A.West@lboro.ac.uk

Widmann H.G.
University of Birmingham, Edgbaston, UK

Williams B.J.
Wolfson School of Mechanical and Manufacturing Engineering, Loughborough University, Leicestershire, LE11 3TU, UK, Email: b.j.williams@lboro.ac.uk

Williams G.
Rawlings Sporting Goods, St. Louis, MO, USA, E-mail: gwilliams@rawlings.com

Wilson B.D.
National Sports Institute, Center of Biomechanics, Bukit Jalil, Sri Petaling, 57000, Kuala Lumpur, Malaysia Email: barrywilson@nsc.gov.my

Wimmer W.
ECODESIGN company – engineering & management consultancy GmbH, Neubaugasse 25/2/3, 1070 Vienna, Austria, E-mail: wimmer@ecodesign-company.com

Witte K.
University of Magdeburg, Institute of Sport Science, Postfach 4120, 39106 Magdeburg, Germany, E-Mail: kerstin.witte@gse-w.uni-magdeburg.de

Wixted A.
Centre for Wireless Technology Griffith University, Brisbane, Queensland, Australia

Wong K.
Monash University Sunway Campus, No. 2, Jalan Universiti, 46150 Petaling Jaya, Malaysia

Wong K.G.
Executive Director, Singapore Disability Sports Council, National Stadium (West Entrance), 15 Stadium Road, Singapore 397718, E-mail: kevin@sdsc.org.sg

Contributors List

Wu Q.
Department of Mechanical Engineering & the Center for Nonlinear Dynamics and Control, Villanova University, 800 Lancaster Avenue, Villanova, PA 19085, USA, E-mail: qianhong.wu@villanova.edu

Xie W.
Biomechanics Department, Singapore Sports Council, Singapore

Yamabe S.
Kyoto University, Yoshida Honmachi, Sakyo-ku, Kyoto, 606-8501, Japan, E-mail: yamabe@sys.i.kyoto-u.ac.jp

Yanai T.
Chukyo University, School of Life System Science and Technology, 101 Tokodachi, Kaizu-cho, Toyota 470-0393 Japan, E-mail:tyanai@life.chukyo-u.ac.jp

Yang S.
Sports Engineering Research Team, Division of Bioengineering, School of Biomedical and Chemical Engineering, 70 Nanyang Drive, Nanyang Technological University, Singapore 637457

Yang Y.K.
School of Mechanical and Aerospace Engineering, Nanyang Technological University, Singapore, and Singapore National Wushu Federation, Singapore

Yi J.H.
Biomedical Engineering, Konkuk University, Chungju, Korea

Yuen A.W.W.
Monash University Sunway Campus, No 2, Jalan Universiti, 46150 Petaling Jaya, Malaysia

Zanevskyy I.
Casimir Pulaski Technical University, KWFiZ, Malczewskiego 22, Radom 26-600, Poland, E-mail: izanevsky@onet.eu

Zhou J.H.
Sports Biomechanics Division, Chengdu Sports University, China

Ziejewski M.
Mechanical Engineering and Applied Mechanics Department, North Dakota State University, Fargo, ND 58105-5285

Zupančič B.
Faculty of Mechanical Engineering, Center for Experimental Mechanics, University of Ljubljana, Ljubljana, Slovenija

1. Sports Technology and Engineering

ACTIVATION AND LIABILITY OF SPORTS ENGINEERING ACTIVITIES AROUND THE WORLD

S. UJIHASHI

Tokyo Institute of Technology, Tokyo, Japan

In 1989, "Sports Engineering" was invented in Japan as a technical discipline and the annual "Sports Engineering Symposium" was launched as a domestic meeting the following year. Subsequently the "Japan Sports Engineering Association (JSEA)" was established in 1991 as a domestic academic organization due to the unexpected level of success of the symposium. The primary purposes of the symposium and the association were to encourage research activities on the hardware of sports, to unify the people who were interested in the field, and also to collect the information on the research of the hardware of sports. In 1996, Steve Haake launched "The International Conference on the Engineering of Sports" in Sheffield, which was the first international sports engineering conference organized. The establishment of the ISEA (International Sports Engineering Association) and the publication of the journal "Sports Engineering" followed in 1998. After these happenings, the development of an organization of the people who were interested in Sports Engineering and other related fields and an increase in sports engineering research was fulfilled to that of the current condition. In this article, the brief history of sports engineering and other related fields is introduced and it is shown how much the related research has contributed to the improvement of sports performance and the sports industry. Furthermore it is insisted that sports engineering activities must not be for the satisfaction of researchers themselves, that is, for their hobby, but must contribute to the performance, safety, pleasure and happiness of the people who love sports.

1 Introduction

Since the idea of "Sports Engineering" was born, almost 20 years passed and during this period the activities of Sports Engineering have grown significantly. In this article the process is overviewed and it is discussed what Sports Engineering is for and how the academic societies and the researchers should be involved in Sports Engineering.

In particular Sports Engineering must contribute to, and encourage, the growth of the sports industry and not only focus on the performance of sports. Thus, the current condition of the sports industry is also introduced and it is shown what the industry is expecting from the activities of Sports Engineering.

2 Beginning of Sports Engineering

Before 1989, the research achievements on the hard ware of sports were scattered in the various fields of literature and were difficult to find. "Hobby researchers" whose main field was different often produced these achievements, as they were interested in a particular sport as a hobby. And then their work on the hardware of sports was not considered as a regular research job.

Table 1. JSEA Sports Engineering Symposia.

Year	Venue	Papers	Delegates
1990.10	Seminar House/Tokyo	30	124
1991.10	Sangyo-Shinko-Kaikan/Kawasaki	31	188
1992.10	Sangyo-Shinko-Kaikan/Kawasaki	31	154
1993.11	Kyurian/Tokyo	45	143
1994.11	Sangyo-Shinko-Kaikan/Kawasaki	46	157
1995.10	Tokyo Institute of Technology/Tokyo	46	170
1996.10	Coop-in Kyoto/Kyoto	65	178
1997.10	Chubu University/Ena, Gifu	52	134
1998.10	Ashiya/Hyogo	65	150
1999.10	Tukuba/Ibaragi	47	118
2000.11	Kochi-Kaikan/Kochi	57	134
2001.11	Japan Institute of Sports Sciences/Tokyo	41	117
2003.9	Yasuda Women's University/Hiroshima	47	98
2004.11	Awaji Yume Butai/Hyogo	61	125
2005.9	Tokyo Institute of Technology/Tokyo	37	100
2006.11	Ishikawa Industrial Promotion Center/Kanazawa	70	167
2007.11	Tsukuba University/Tsukuba, Ibaraki	–	–

Table 2. Statistics of the presented papers at the JSEA Sports Engineering Symposium (1990–2001).

Ranking	Topic	No. of papers	Composition ratio
1	Golf	101	0.183
2	Skiing	68	0.123
3	Tennis	60	0.180
4	Shoes	16	0.029
5	Bicycle cycling	14	0.025

The primary purpose of the JSEA (Japan Sports Engineering Association) was to unify the researchers who were involved in the research on the hardware of sports, and also to consolidate the information on the research results. For these purposes the "Sports Engineering Symposium" was started in 1990 and JSEA was founded in 1991 as an academic organization. Table 1 and Table 2 show the statistics of the JSEA symposia and Figure 1 shows the proceedings in the past.

On the other hand, Steve Haake, of the University of Sheffield, started "The International Conference on the Engineering of Sport" in 1996. Initially his attempt was not recognized

Figure 1. Proceedings of the JSEA Sports Engineering Symposium.

by the JSEA and the first encounter with Steve Haake was in 1997. The JSEA invited him to the "Sports Engineering Symposium" in Japan after recognizing his attempt. This was a good opportunity to discuss this common idea and the JSEA became confident in the future of sports engineering activities because there was another person who has the same idea in the world to work with.

3 Conferences and Societies

In 1998 the ISEA (International Sports Engineering Association) was launched officially at the 2nd International Conference on the Engineering of Sport, in Sheffield. Then the two obvious organizations of sports engineering, that is, JSEA and ISEA became available to the world and each association ran their own conference.

Later in 2003, Aleksandar Subic from the RMIT University started another stream of conferences. This conference became "The Asia-Pacific Congress on Sports Technology" at the 2nd Congress held in Tokyo in 2005, with the wider discipline of "Sports Technology" rather than sports engineering. This idea was that not only sports engineering but also sports science related disciplines, such as sports biomechanics, sports medicine, sports management and others, must be included in order to improve the performance in sports.

Many conferences and societies have been, and are, related to sports in the world, however, the above idea regarding the hardware of sports was unique and was not available in the past. These activities succeeded to certify the works on the hardware of sports in the same way as the other traditional engineering research works.

Table 3 and 4 show the statistics of the ISEA conferences and journals, and Figure 2 shows the organizers of the ISEA conferences 1996–2004.

After the formation of JSEA and ISEA, national sports engineering societies such as ASTA (Australasian Sports Technology Association), KSEA (Korean Sports Engineering Association), etc. have been founded. However generally speaking, the maintenance of a national society by itself is not easy, and therefore the possibility of a united organization of national societies has been discussed, but an effective unification model has not yet been established.

Table 3. The ISEA (International Sports Engineering Association) international conference on the engineering of sport in the past.

Year	Place	Papers	Countries	Delegates
1996	Sheffield (UK)	47	10	75
1998	Sheffield (UK)	65	11	110
2000	Sydney (AUS)	57	9	80
2002	Kyoto (JPN)	151	15	210
2004	Davis (USA)	194	16	259
2006	Munich (Germany)	181	21	370
2008	Biarritz (France)	–	–	–

Table 4. Topic and number of papers published by the journal "Sports Engineering" (vol. 5–9).

Topic	Total	Topic	Total
Biomechanics	24	Kayaking	2
Measurement	9	Rugby	2
Skiing	9	Swimming	2
Football	7	Badminton	1
Surface	5	Basketball	1
Tennis	5	Billiards	1
Bicycle cycling	5	Boat sailing	1
Golf	4	Bow archery	1
Baseball	3	Fishing	1
Bungee jump	2	Helmets	1
Cricket	2	Mountaineering	1
Hockey	2	Others	5

4 Sports Performance

The development of the hardware of sports has contributed significantly in the improvement of records, safety, comfort and other aspects.

In the case of athletics, FRP (Fibre Reinforced Plastic) materials used in the pole vault increased the world record to 6 m 14, as shown in Table 5, and the synthetic running tracks of polyurethane produced incredible results in all sprinting and jumping events, as shown in Table 6. In other events such as tennis, golf etc, the material revolution or development of new designs and manufacturing technologies has totally changed the equipment and also playing style. However, the excessive introduction of new technology has sometimes forced changes in the playing rules regarding the equipment.

Figure 2. Chairs of the ISEA Sports Engineering Conference.

Table 5. Impact of the material in the pole vault.

Material	Grip height	Best record (Athlete)	Year
Wood (hickory)	No data	3 m 40 (no data)	~1900
Bamboo	3 m 80 ~ 4 m 10	4 m 77 (no data)	1942
Metal	3 m 80 ~ 4 m 20	4 m 83 (Davis, USA)	1961
FRP	4 m 70 ~ 5 m 00	6 m 14 (Sergey Bubka, UKR)	1994

Table 6. Impact of synthetic surfaces in athletics (Men 100 metres).

Surface material	Best record	Athlete	Year
Clay	10.0 seconds	Bob Hayes (USA)	1964 (Tokyo)
Poly-Urethane	9.77 seconds	Asafa Powell (JAM)	2005 (Athens)

In tennis, for example, the material of the racket has changed from wood to CFRP, via a short period of metal. Because of the lightness and strength of CFRP, the maximum face area of rackets became almost twice the one of the wood era and as a result tennis became a much easier game for beginners to play.

On the other hand, in the professional game, tennis became less fun because the speed of ball was much higher than before and the receivers were in difficulties to return first services. As a result players with a fast service were more likely to control tennis games.

In order to solve this problem the ITF (International Tennis Federation) set a new rule regarding balls in 2001 as shown in Figure 3. Larger balls were introduced by the rule and their usage was recommended on fast surfaces to reduce speed after bouncing.

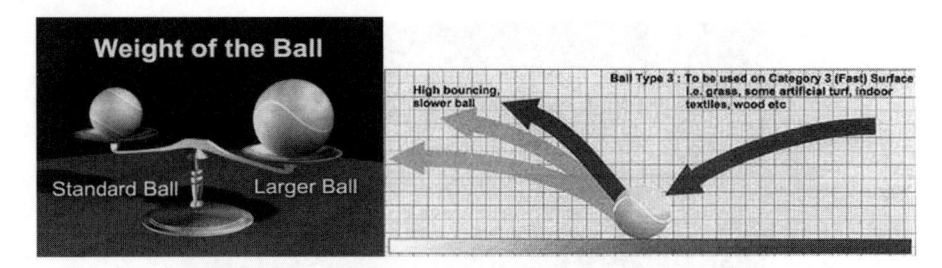

Figure 3. Introduction of larger ball in tennis. Reprinted with permission by the International Tennis Federation (ITF, 2007); left side: *Illustration to show the introduction of the Type 3 ball. It is 6% larger than the standard Type 2 ball but weighs the same*; right side: *Type 3 ball*.

USGA, R&A reach agreement on spring-like effect Uniform policy to be adopted in 2008
Golf's two governing bodies settled a major dispute on drivers Thursday, agreeing to adopt a uniform policy in 2008 and allowing different rules for tournament and recreational players until then. The compromise dealt specifically with the spring-like effect on thin-faced drivers such as Callaway Golf's ERC, an issue that sharply divided the U.S. Golf Association and the Royal & Ancient Golf Club. The USGA decided in 1998 that the coefficient of restitution for drivers -- how quickly the ball leaves the face of the club -- should be 0.83. The R&A, which makes the rules for golf everywhere in the world except for the United States and Mexico, declined to impose any limits on drivers.

Figure 4. New Rule on the COR of Golf Club Heads (GolfDigest, 2007).

Using golf as another example, in 1998 the USGA (United States Golf Association) banned clubs with a higher coefficient of restitution than 0.83 and the R&A (Royal & Ancient Golf Club of St Andrews) decided to introduce this new rule from 2008.

The material of drivers has changed from wood to titanium. This material change enabled the structure of drivers to be thin-walled and eventually the COR went up significantly by the aid of a spring-like effect. Also CFRP has been introduced as the shaft material of clubs. Due to the material innovation, recent golf clubs can produce much longer carries and as a result, the new technologies in golf equipment may spoil the design concept of golf courses. The USGA became anxious that the new technology may change golf as more powerful players are more likely to control games. To avoid this trend, the USGA invented the new rule (Figure 4) which is more challenging.

This change of the golf rule caused great controversy between governing bodies and manufacturers. Nevertheless, the manufacturers were forced to stop the development of high technology products.

The harmonisation between sports equipment and their playing rules became more important than before as indicate by the above examples.

5 Sports Industry

The sports market is considered a part of the leisure market as shown in Table 7. In Japan the expenditure for leisure is about 16% of the gross expenditure and the fraction of sports

Table 7. The market size of leisure in Japan; GNE: Gross National Expenditure, $\times 10^{12}$ JPY (Ujihashi, 2006).

Year	GNE	Leisure				
		Sport	Hobby	Entertainment	Travel	Total
1989	399.046	4.700	10.576	39.537	10.604	65.417
1990	435.361	5.179	10.558	44.862	12.171	72.770
1991	454.486	5.696	10.617	49.701	12.725	78.739
1992	476.064	6.005	10.646	52.955	12.480	82.086
1993	479.761	5.898	10.939	54.486	11.974	83.297
1994	483.201	5.746	11.157	56.097	11.725	84.725
1995	486.921	5.748	11.325	56.778	11.828	85.679
1996	500.310	5.693	11.937	55.366	12.146	85.142
1997	521.861	5.576	11.879	59.919	11.878	89.252
1998	515.834	5.330	11.820	58.503	11.362	87.015
1999	511.837	5.117	11.809	57.610	11.018	85.554
2000	513.534	4.886	11.822	57.226	11.124	85.058
2001	496.776	4.788	11.730	55.178	10.972	82.668
2002	489.618	4.599	11.697	56.139	10.813	83.248
2003	490.543	4.525	11.488	55.315	10.438	81.766
2004	496.058	4.380	11.632	54.775	10.554	81.341
2005	502.456	4.297	11.161	53.949	10.686	80.093

is only about 7% of the leisure, that is, about only 1% of gross expenditure. In order to develop the sports industry, more people must participate in sports, i.e. people must spend more time and money for sports. However Japanese people, particularly young generation are not keen to play sports but are very likely to enjoy watching professional sports. Japan faces big problems in the near future as the population of the young generation is decreasing significantly and also their body power is inferior to the former generation.

In order to encourage sports participation, the development of the infrastructure for sports must be enhanced and the life style must be changed.

The sports industry is divided into two major parts, the software and the hardware as shown in Figure 5. The software comprises various sports schools, travel and tickets for sports events etc. and the hardware encompasses equipment, facilities, and infrastructure etc.

In the last 15 years the trend is an obvious decrease in the expenditure for sports. Around 1990 it was predicted that the sports industry market size would grow up to JPY 10,000 billion. However this prediction did not materialize and the current market size is almost half of the expectation in the past. Many reasons are considered however the problem is whether sports engineering activities can improve the current situation.

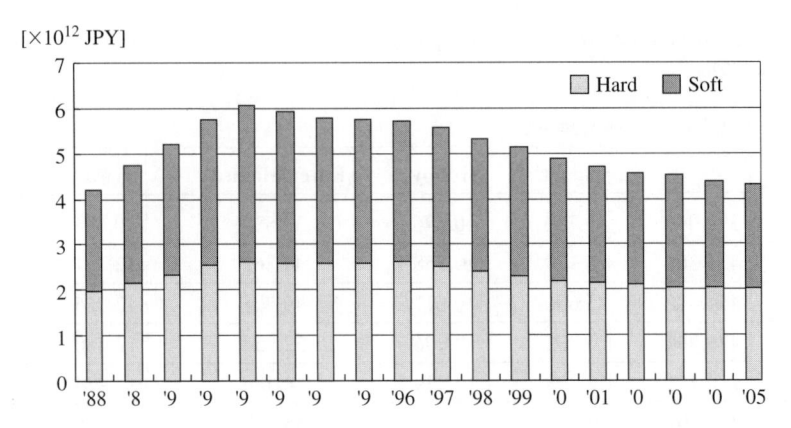

Figure 5. The market size of sports industry in Japan (Ujihashi, 2006).

6 Liability of Sports Engineering

The activities of Sports Engineering started in Japan about 20 years ago and expanded internationally very quickly. Regarding these 20 years as the first stage, the role was enlightenment of the importance of Sports Engineering and this was fulfilled successfully. In the next stage, Sports Engineering must perform as a real academic body that will support sports from the view point of manufacturing of the hardware of sports.

The role of Sports Engineering would be threefold:

1. Contributing to making sports more exciting, comfortable and safe
2. Supporting the manufacturers of the sports hardware
3. Environmental considerations that are the same as engineering in other fields

For the above roles researchers in academia must actively promote collaboration with the sports industry and try hard to produce a real contribution to sports. Thus, we should avoid "hobby research" which came from individual curiosity like research that was conceived with the thought "I wanted to know why this is". Unfortunately there are considerable numbers of papers produced by "Hobby researchers" in the past publications, even in Sports Engineering.

At the same time, as engineers we must of course regard environmental considerations: the avoidance of mass production and mass consumption is highly important in sports too. Price is a most effective tool to reduce the amount of consumption. Engineers must adopt the philosophy that a price reduction to beat a competitor is almost a criminal strategy.

7 Conclusion

The size of the sports industry is not so big in comparison to other industries like the automobile, electrical appliance, chemical industries etc. The size of academic societies or university departments is also proportional to the corresponding industry. Therefore the activities of sports engineering cannot expand without limit. An important point would be to act properly as a unique academia to support sports. At the same time, the creation of international cooperation between national societies would be a good strategy to maintain the societies themselves. "Sports Technology", as an expanded keyword of "Sports Engineering", might be another strategy to assist the survival of "Sports Engineering".

Anyway for activities related to sports, the most important aspect is to maintain people's health by promoting participation in sports and therefore the ultimate purpose of Sports Engineering is to also follow this philosophy.

References

GolfDigest (2007) USGA, R&A reach agreement on spring-like effect: uniform policy to be adopted in 2008 (online, May 9, 2007). Available: http://www.golfdigest.com/equipment/index.ssf?/equipment/20020509usga.html (accessed: 14 June 2007).

ITF (2007) History of Rule 3 – The Ball (online). Available: http://www.itftennis.com/technical/rules/history/ (accessed 14 June 2007).

Ujihashi S. (2006) *Sports Engineering*. Tokyo Institute of Technology, Department of Mechanical Engineering, Tokyo.

2. Life Cycle, Environmental Engineering and Eco-Design

ECODESIGN OF ALPINE SKIS AND OTHER SPORT EQUIPMENT – CONSIDERING ENVIRONMENTAL ISSUES IN PRODUCT DESIGN AND DEVELOPMENT

W. WIMMER[1] & H. OSTAD-AHMAD-GHORABI[2]

[1]*Ecodesign Company – Engineering & Management Consultancy GmbH,*
Vienna, Austria
[2]*Vienna University of Technology, Institute for Engineering Design,*
Vienna, Austria

This paper shows how to achieve environmental improvements of sport equipment. Two case studies are presented. For an alpine ski a classical product redesign is shown using available methods and tools. For a new Golf Swing Analyzer it is shown how to integrate Ecodesign in a new product development without having a reference product to evaluate.

1 Introduction

Product innovations are essential in a globalized market. Delivering a unique selling proposition can be realized by developing eco-products. A rising awareness about the environmental performance of products leads to competitive advantage for those having eco-products on the market. Ecodesign is a way to achieve competitive advantage. Ecodesign is a methodology for the design of products minimizing each product's environmental impact through all of its life cycle stages and life cycle costs respectively.

The basis for eco-product development is a good analysis through Life Cycle Thinking (LCT). By applying LCT all life cycle stages of a product are evaluated aiming at finding environmental improvement potentials. Table 1 lists the different life cycle stages. In this paper these five life cycle stages will be considered to obtain Ecodesign improvement strategies for sport equipments.

2 Product Improvement

In the following an alpine ski will be taken as a case study to demonstrate the product improvement approaches and strategies. In a first step the product structure has to be understood. Figure 1 shows a cross-section of such an alpine ski.

The bottom of the ski consists of a PE running surface (4 in Fig. 1), which is bonded to a laminate and framed by a wraparound steel edge (5). The interior of the ski is injected with foam (10), which constitutes the shape of the ski and ensures bonding of all components with each other. The graphic design of the ski is printed to the inner side of the transparent foil of the top surface (1). For reinforcement the inner side of this foil is bonded to a lattice

Table 1. Five life cycle stages of a product to be considered in LCT.

Raw materials	Manufacture	Distribution	Use	End of life
Polystyrene, Glass, Steel, Aluminum, …	Injection molding, Machining, Welding, …	Air, Rail, Road, … Packaging, …	Electricity, Batteries, …	Recycling, Incineration, …

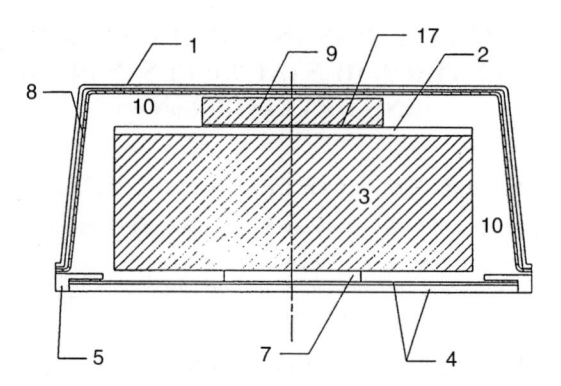

Figure 1. Cross section of an alpine ski (Wimmer et al., 2001).

(8). In the tip, center and tail areas the ski features a wide-meshed plastic inlay (9) between foil (1) and wood core (3), which is fixed to the laminate (2) by means of an adhesive lattice (17). In the tip area the wood core has been replaced by an ABS plastic layer (not shown). At the tail of the ski, running surface, steel edge, and shell meet in the tail protector.

2.1. Product Life Cycle Data

To obtain life cycle data for alpine skis a reference ski was taken into account, which weighs 1.8 kg. 35% of the weight comes from the wraparound steel edges, 20% from the wood core and 10% from the surface foil.

The rest are mostly glues, foams, rubbers and fleece used for assembling the ski. The manufacturing stage consists of the following steps:

1. Preparing parts and components: The steel edges are scoured and sandblasted. The foils which are delivered in reels are cut and the waist shape is milled.
2. Printing the surface: The foils are printed by screen printing technique.
3. Pressing process: The prepared parts and components are put and pressed together. The pressure for this process amounts 5bar and the required temperature is 120°C.
4. Grinding: Different parts of the ski are grinded.

Further, it is assumed that for distribution a lorry is used and that the distribution distance is 1000 km of average and shrinking foil is used for packaging.

In the use stage of the alpine ski additional materials are needed such as wax.

The end of life stage of the reference alpine ski is modeled to be a mixture of incineration to gain back energy and material recycling.

2.2. Environmental Evaluation and Environmental Profile

To obtain an environmental profile of the alpine ski and to be able to extract Ecodesign strategies for product improvement, the "Ecodesign Toolbox" developed at the Vienna University of Technology (VUT) was applied (Pamminger et al., 2006). The Ecodesign Toolbox is a six step approach which leads to green product concepts. The six steps are: product description, process analysis, product analysis, stakeholder analysis, process & product improvement and Green Product Concept according to Figure 2.

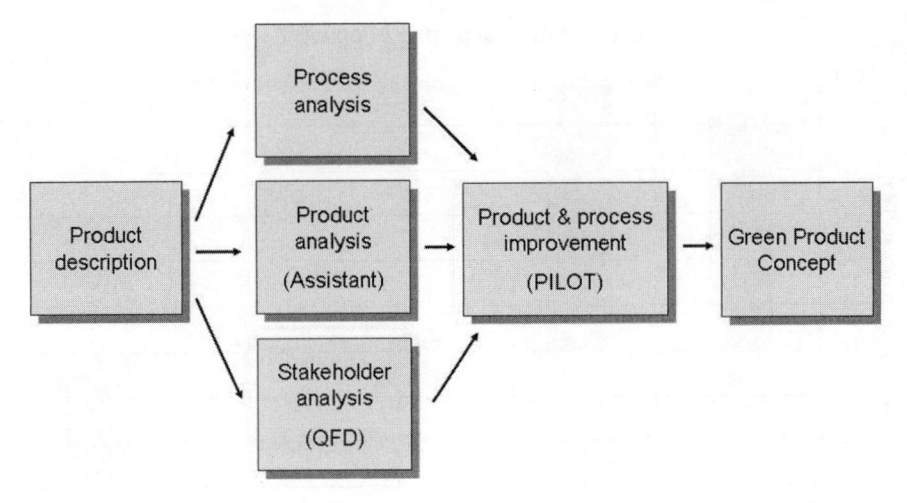

Figure 2. Six steps of the Ecodesign Toolbox.

The product description of the alpine ski and the process analysis was briefly introduced in 2.1. The stakeholder analysis prepares environmental laws and regulations for product development.

The Ecodesign Product Investigation, Learning and Optimization Tool for Sustainable Product Development (PILOT) is a tool which was developed at the VUT (PILOT, 2007), (Wimmer & Züst, 2002). The PILOT can be used online to achieve an environmental profile of a product. The environmental profile is derived by asking specific product data of the life cycle stages of a product and calculating via energy values as environmental impact indicators. By assigning energy values to the different materials and processes through the entire life cycle of the alpine ski, materials and processes contributing significantly to environmental impact can be tracked.

The environmental profile of the alpine ski is shown in Figure 3.

Figure 3 clearly shows that the first two life cycle stages, namely raw materials and manufacture, contribute most to the environmental impact of the alpine ski.

The modelled distribution scenario or the use stage constitute little to the environmental impact of the product.

2.3. Product Improvement Strategies

Following the next step of the Ecodesign Toolbox, product and process improvement strategies shall be obtained by using the Ecodesign checklists of the PILOT (Wimmer et al., 2004). For the ski manufacturer a special version of the PILOT has been developed providing ski-specific Ecodesign issues and checklists for each stage in the company's specific product development process. Figure 4 shows the PILOT which is adapted to skis.

Applying the Ecodesign SKI-PILOT results in the design strategies and design improvements listed in Table 2.

In a following step the achieved design improvements had been evaluated against cost saving potential and later those improvements resulting in an environmental improvement

Environmental profile of an alpine ski

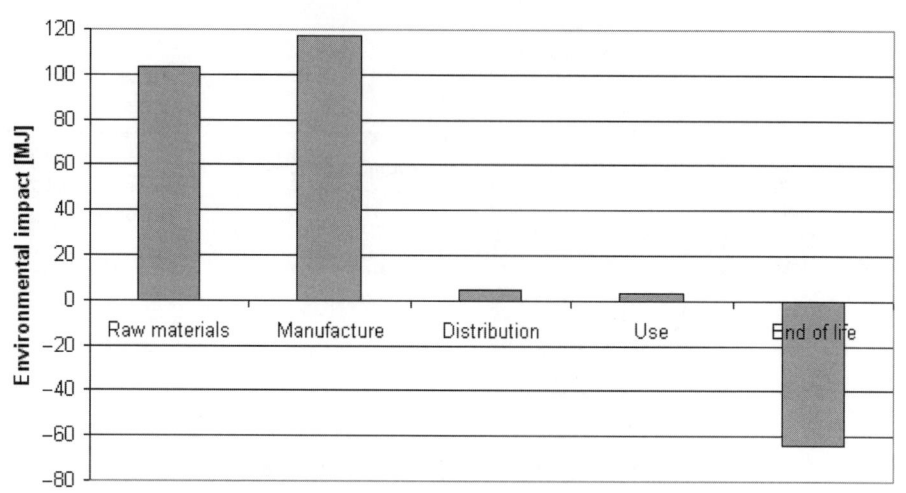

Figure 3. Environmental profile of the reference alpine ski.

Figure 4. Ecodesign SKI-PILOT adapted to the manufacturers product development process.

and in a cost reduction have been selected for implementation. All in all this procedure can be integrated in any continuous improvement process of a company.

3 Product Development Process

In some cases there are no reference products as outlined in chapter 2. The question is then how to deal with environmental issues in such product development tasks.

The product development process can be seen as an optimization in between sometimes conflicting targets. Finding a certain material for a certain price to fulfill a certain function is difficult enough. Now adding the environmental dimension (e.g. considering the energy to produce the material but also considering the recycling behavior of the material) additionally may become challenging.

Table 2. Ecodesign strategies and design improvements for the alpine ski.

General Ecodesign strategies selected with the PILOT checklists:	Product design improvements in detail:
Avoid or reduce the use of problematic materials and components	Reduction of printing width at the top surface foil
Prefer the use of recycled materials (secondary materials)	Increase of the portion of recycled material in the running surface
Reduce material input by integration of functions	New design of the bond laminate-running surface, the injected foam also provides for bonding (no additional adhesive agents)
Use low material input, low emission production technologies	Modification of design for optimized cutting to size of material from wider raw material strips
Avoid waste and emissions in the production process	Less paint coat and reduction of bonding surfaces, reduction of foam waste by redesigning foam injecting nozzle
Close material cycles in the production process	Modification of product design to ensure single material cutting chips, recycling of the chips
Recycle/reuse waste for new materials	Recycling of waste from single material PE running surface
Waste sorting/separation whenever possible	Modification of the bond between laminate and PE running surface
etc…	etc…

Common to most development processes is a certain sequence of developing product specifications first, then deriving the functional structure and using creativity techniques to develop several possible product concepts before one concept is selected for embodiment design through an evaluation and assessment procedure, see Figure 5.

Introducing environmental thinking in the early phase of product development is most important. When laying down the product specifications environment should be on board already. The environmental target values sorted out shall be understood by the design team. This is essential in order to come up with a good overall environmental performance of a product. But the question is how to derive the correct environmental targets for a certain product? This should be demonstrated with the new development of a Golf Swing Analyzer.

3.1. *Understanding the Environmental Impacts*

In case of a newly developed product, estimations of potential impacts along the product's life cycle have to be done. This goes together with defining the system boundaries for the product and its interactions with the environment. Relevant questions are how far those upstream processes, until the product leaves the manufacturer, should be considered and how the use phase and end of life phase are modeled.

Figure 5. Integrating Ecodesign in the product development process (Wimmer *et al.*, 2004).

In general a Material, Energy and Toxicity (MET) matrix is a good way to develop an environmental overview of a product – even with vague data. For the new development of the Golf Swing Analyzer the MET-matrix showed that either the materials used in the device or the energy needed to operate will have the main environmental influence. Manufacturing processes or distribution (transport and packaging) but also end of life will have minor impacts only.

Regarding the materials, the Golf Swing Analyzer (basic outline of components) consists of an Aluminum housing containing two lenses, an electronic control unit and some connecting cables. To work properly a tripod is needed. Out of these components the electronic control unit is environmentally most significant. One design task was to keep it as small as possible. The other design task is to have a robust but light weight housing and tripod. The energy supply side is a more challenging task since there are more options to choose. Assuming the device (as many other devices as well) could be operated with batteries or alternatively with rechargeable batteries.

Based on the assumptions that the Golf Swing Analyzer will be used from a golf club intensively on a driving range around 120 uses per year are realistic. Each use consists of a half hour training. Two scenarios are generally available to deliver the energy needed: Regular alkaline batteries, NiMH rechargeable batteries with external power supply unit. A third possibility applies in the case of the Golf Swing Analyzer. Since operating requires a Laptop for storing and calculating the measurements of the recorded golf swing the idea of using USB power supply came up. The environmental evaluation given in Figure 6 shows clearly which one is the best solution.

3.2. *Understanding Environmental Regulations*

Additionally to the environmental evaluation of a product several environmental directives have to be considered in the development of electronic sport equipment. Those are the Directive for the Restriction of the use of certain Hazardous Substances (Directive 2002/95/EC, 2003) and the Directive on Waste Electrical and Electronic Equipment (Directive 2002/96/EC, 2003) as well as the new and upcoming EuP-Directive setting Ecodesign requirements for energy-using products. (Directive 2005/32/EC, 2005).

Comparison of energy supply systems

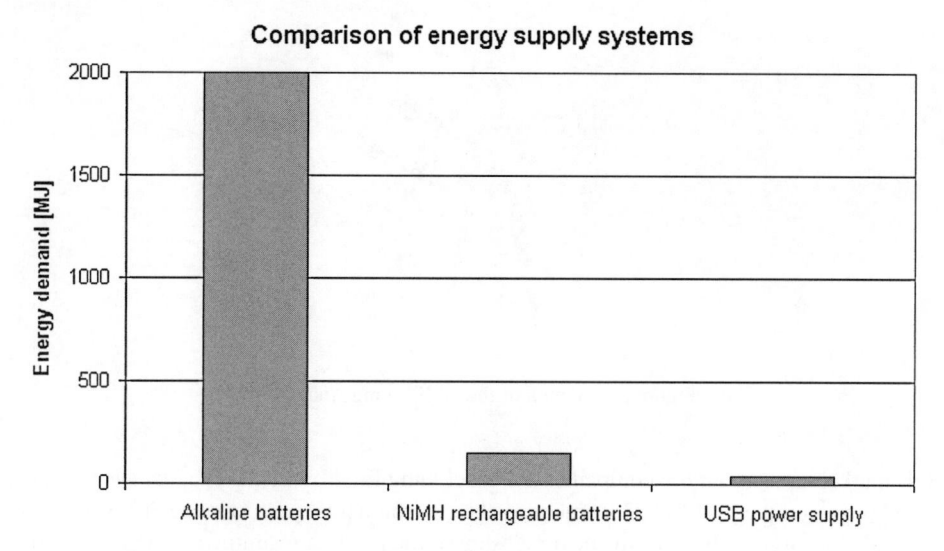

Figure 6. Environmental evaluation of the different energy supply systems for the Golf Swing Analyzer.

EuP is currently under development but will bring Ecodesign requirements to be fulfilled in order to achieve CE-marking of a product. Evidence for the compliance with the directive is the CE marking on the product through conformity assessment. The assessment can be made either through internal design control or integration into the Environmental Management and Audit Scheme (EMAS) with the design function included within the scope of that registration.

RoHS restricts the use of cadmium, lead, mercury, hexavalent chromium, and two bromated flame retardants. WEEE requires achieving a certain recycling percentage at the end of life of a product (EEE validity check, 2007).

3.3. Developing Green Flagship Products

Summarizing the above considerations in the early stage of product development the environmental situation of the product becomes clearer and design targets can be set in the beginning of the product development process:

- Low energy consumption through USB supply
- Light weight design
- Lead free soldering and RoHS compliant components
- Design for recycling and recycling instructions according to WEEE

3.4. Communicating the Environmental Performance

Once product improvements and a better environmental performance has been achieved the market needs to be informed about these achievements. A credible way of documenting the environmental performance is needed.

There are several possibilities available to do so. International standard (ISO 14020, 2000) foresees three types: Doing self declared environmental declarations, applying for an existing eco – labeling program or developing an Environmental Product Declaration. For the

Figure 7. Sketch of the Golf Swing Analyzer.

B2B market more and more companies are developing Environmental Product Declarations (Ecodesign-company, 2007). Documenting the environmental performance this way sets in principle the target values for the next redesign of the product – aiming at a better environmental performance.

4 Conclusion

What is needed in order to achieve a competitive advantage through Ecodesign is a full integration of environmental issues into the product development process. This requires staff training to the mostly new issue of understanding key environmental performance indicators in product development, this further requires company database for material and processes to do Life Cycle Thinking of products. Additionally appropriate Ecodesign methods combined with a new creativity are needed to enable eco-product development. At the end environmental communication is needed to market the eco-products and to ensure competitive advantage through Ecodesign.

Two examples have been provided in this paper – many more are available to learn from others and to find an own way of getting ready for the future challenges of achieving a good environmental performance of products.

References

Directive 2002/95/EC of the European Parliament and of the council of 27 January 2003 on the restriction of the use of certain hazardous sub-stances in electrical and electronic equipment (RoHS).

Directive 2002/96/EC of the European Parliament and of the council of 27 January 2003 on waste electrical and electronic equipment (WEEE).

Directive 2005/32/EC of the European Parliament and of the council of 6 July 2005 establishing a framework for the setting of Ecodesign requirements for energy-using products.

Ecodesign-company. Available: http://www.ecodesign-company.com/solutions (Accessed 15 April, 2007).

EEE validity check (2007), Available: http://www.ecodesign.at/pilot/eeg. (Accessed: 18th of April, 2007).

ISO 14020 (2000) Environmental labels and declarations – General principles.

Pamminger R., Wimmer W., Huber M. and Ostad Ahmad Ghorabi H. (2006) The ECODE-SIGN Toolbox for the development of green product concepts – First steps for product improvement. In: *Conference Proceedings of Going Green – CARE INNOVATION*, Vienna, Austria.

PILOT (2007) Available: http://www.ecodesign.at/pilot. (Accessed: 18th of April, 2007).

Wimmer W., Schneider H., Fischer P. and Pieber A. (2001) Identification of environmental improvement options with the new Ecodesign Product Investigation, Learning and Optimisation Tool (PILOT). In: Culley S., Duffy A., McMahon C. and Wallace K. (Eds.) *ICED 01: Design methods for performance and sustainability.* Glasgow, 21–23 August, UK.

Wimmer W. and Züst R. (2002) *ECODESIGN Pilot – Product-Investigation-, Learning- and Optimization- Tool for Sustainable Product Development.* Kluwer Academic Publishers, Dordrecht, The Netherlands.

Wimmer W, Züst R. and Lee K.M. (2004) *ECODESIGN Implementation – A Systematic Guidance on integrating Environmental Considerations into Product Development.* Springer, Dordrecht, The Netherlands.

INTEGRATING THE DESIGN FOR ENVIRONMENT APPROACH IN SPORTS PRODUCTS DEVELOPMENT

A. SUBIC & N. PATERSON

School of Aerospace, Mechanical and Manufacturing Engineering, RMIT University, Bundoora, Melbourne, Australia

As part of the global trend towards sustainable development many industries are adopting "green" design practices in order to improve their products from the perspective of the "triple bottom line". This paper discusses the design for environment approach to sports products design and presents a case study of composite tennis racquets with life cycle assessment (LCA) used to evaluate their environmental load and design features. The flexibility of the LCA approach is examined and LCA variations are suggested that may be applicable for specific situations. Furthermore, this work identifies the importance of uncertainty management within LCA and presents preliminary findings that suggest how uncertainty management can enhance the value of LCA to the design engineer.

1 Introduction

Sustainable development has been widely advocated in society and industry for some time and this trend continues to increase with the realization of the emerging impact of global warming. The sports industry as a whole has particular responsibilities in this regard, especially with respect to the developing countries as identified by the IOC (IOC, 2005). Materials and processes used for sports products carry with them potential environmental risks. For example ski boots use PVC (Poly Vinyl Chloride) based materials, athletic footwear uses petroleum-based solvents and other potentially damaging compounds such as sulphur hexa-fluoride in air bladders for cushioning and impact shock protection. Also, composites, such as carbon fibre reinforced polymers that are increasingly used in tennis racquets and other sports equipment provide greater strength or stiffness to weight performance but cannot be readily recycled at an acceptable cost or value.

It is estimated that around 80% of the environmental burden of a product is determined during the design stage. Hence, in modern design, environmental issues are given high priority, which has resulted in the development and application of new design tools and practices encompassed broadly by the design for environment approach (Subic, 2005). Central to this is the life cycle design approach, which is about developing more environmentally benign products and processes based on detailed understanding of the environmental hazards, risks and impacts over their entire life cycle including production, usage and disposal stages.

Governments, particularly in Europe, have been providing incentives to companies willing to adopt "greener" product design and manufacturing approaches. For example, Germany's product take-back laws have prompted European and US industries to reduce packaging and start designing products with disassembly and recycling in mind. France, Netherlands and Australia have special government agencies to foster clean technologies. In the US the Federal government has initiated a number of energy efficiency programs and has prescribed the use of recycled and/or recyclable materials in product design. Clearly there is a growing concern world wide for the environment whereby it is not acceptable any longer for products to be incinerated or dumped in landfill after their useful life.

In order to assess the environmental impact of a product it has become necessary for informed, science based weighting of the wide variety of environmental damages using standardised criteria and approaches. This allows the different products and phases of the lifecycle to be compared on the same scale. Lifecycle assessment (LCA) is the method typically used to calculate the environmental impact and/or damages of a product across the entire lifecycle, from production through to disposal, including the auxiliary processes and emissions that are inherently part of the products environmental footprint. There are 5 generic phases in the product life cycle including: *production, packaging, transport, use*, and *disposal*.

The following sections describe an LCA of composite tennis racquet designs following the form identified as best practice by the ISO (ISO, 1997). General and specific issues arising from this study and relating to LCA methodology are discussed briefly at the end of this paper. It is important to note that the ISO acknowledge the flexibility of the method and that different approaches are often adopted depending on the circumstances. Nevertheless, it is recommended that a standardised framework be followed (Subic & Paterson, 2006).

2 LCA of Composite Tennis Racquets

Composite tennis racquets consist of several different materials depending on the model and the manufacturer, some of which are highly protected trade secrets. This is one of the main reasons that information on detailed LCA of sports products is not readily available in the public domain. However, it is probable that most sporting goods companies may have involved environmental thinking at some stage, although most likely focused on the packaging, as noted in this work. Ideally environmental thinking, possibly in the form of LCA, should be conducted as early as possible during the design phase to achieve the benefits rather than as an after thought as for example part of the packaging requirements. In fact, novel ideas which aid environmental design as well as cost savings are more likely to result from consideration of the entire life cycle and particularly, how the different phases can influence one another.

2.1. *Goal, Scope and Boundaries*

Full LCA of a product make take considerable time (e.g. LCA of a car may take years) and requires a large commitment of resources from the companies involved. Alternatively, a so called screening LCA may be employed as in this case, for comparative assessment of products or design concepts to identify weaknesses and areas for possible improvement.

The main objective in this work was to conduct a screening LCA with the incorporation of uncertainty management using standard LCA software. A screening LCA is a streamlined LCA which can then be improved later with further iterations by incrementally adding new data as it is made available. It is the authors' opinion that this type of approach is most suited to developing the uncertainty management techniques that are desperately required in LCA, especially in early stages of design.

Four different tennis racquets (with two different designs and two different manufacturing techniques used) were sourced from one sporting goods manufacturer. The goal was to identify the materials and processes in the racquet construction in general and most importantly those materials and processes that give rise to the major differences in the racquets, allowing their different environmental impacts to be assessed. A current LCA inventory database IdeMat was used to define the inventory and environmental impacts of the racquets.

The boundaries for the analysis were set. The use phase of the life cycle was excluded from the analysis because there is no inherent energy (i.e. fuel) used by the product, apart from the energy generated by the player. The secondary transport phases in the case of tennis racquet production have also been excluded. Transportation involves the materials and waste products of transport itself. These are accounted for in the primary transport phase, where raw materials are taken to materials manufacturers. The data is included in addition to the by-products of the material extraction in the LCA software used. Determining the transport scenarios for the entire logistical process of an international company with multiple distribution routes and hubs is almost impossible. Furthermore, Abele *et al.* (2005) suggest that the transport phase is often excluded because it is too difficult to influence logistical procedures from a design perspective. The disposal phase is excluded because the complexity of disposal scenarios for composite products means there is no data available. Moreover, the data for municipal waste disposal of fibre reinforced composites is unavailable in current software. Other possible disposal scenarios such as recycling are important to consider, but similarly, require further work before any data on composite products can be incorporated successfully in the analysis.

In addition to the main boundaries of the analysis the functional unit of the product system must also be defined. The functional unit is a measure of the function of the studied system and it provides a reference to which the inputs and outputs can be related. Hence, two different products can be compared if they have the same functional unit. The functional unit of a tennis racquet requires assumptions to be made about how long it will be used for. The average casual player was considered. Consequently the average racquet's useful lifespan was assumed to be 5 years, with each racquet requiring 4 replacement grips and 2 sets of replacement strings. A racquet will probably last much longer than this, but it is usual for new models to replace existing racquets before their useful life has been exceeded.

2.2. Inventory

The manufacturing techniques used for tennis racquets are generally the same, with minor changes introduced from manufacturer to manufacturer, each having their own trade secrets. Although it is not possible to obtain or publish the details of these processes fully, the main steps that make up the majority of the construction were retrieved from literature (RacquetTech, 2003). In order to gather more accurate information on the materials it was necessary to adopt a reverse engineering approach. Although this is unusual in standard LCA, it is necessary in a situation where the materials are unknown.

Each tennis racquet had the dimensions and weights measured. Components that could be removed easily such as the grip, strings, end cap and bumpers were removed from the racquet, weighed and measured. Some replacement grips and strings were also measured for comparison. The remaining frames were then sectioned by sawing though the frames. The samples were sent to the Open University Forensic Materials Engineering group (UK) for analysis. Further sample preparation was carried out and a scanning electron microscope with X-ray diffraction was used to identify some materials, along with differential scanning calorimetry and Fourier transform infra red spectroscopy to identify others. A final inventory

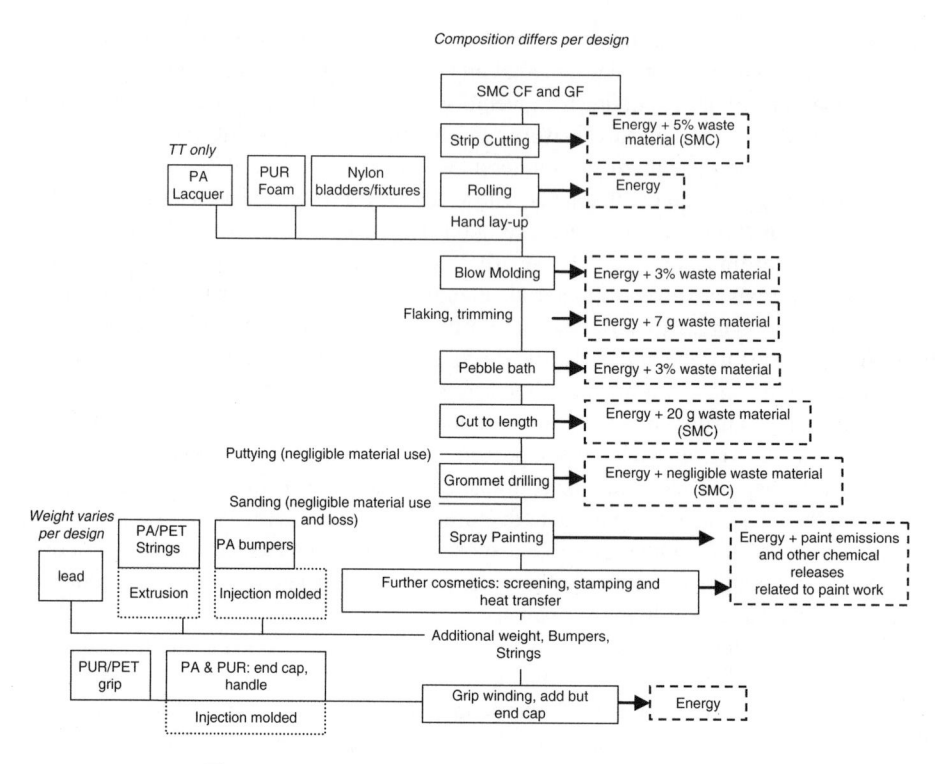

Figure 1. Inventory tree for composite tennis racquets.

was derived by choosing appropriate materials within the LCA database to match the results obtained from tests and literature. This meant that if the specific material or process identified was not available in the database the closest analogue was chosen. Figure 1 shows the final inventory tree including the processing steps used for construction of the racquets and estimates of the waste.

Table 1 shows the weights and types of materials found in the 4 different racquets, including estimations for the waste material and accurate data for packaging. There are many similarities between the materials used in the different racquets, as might be expected. For clarity, the materials that exhibit differences are indicated in bold.

There are 5 main areas where the racquets differ in material weights. The bumpers for example are a plastic component used to protect the frame and attach the strings. The lead alloy tape is used as additional weight in various places around the frame. It is acknowledged that this tape would require an adhesive as with many other components of the racquet, but it was not possible to identify these adhesives precisely. Because the adhesive mass is exceptionally small it was expected to contribute very little to the overall environmental impact of the racquet. The two materials that contribute a lot to the overall difference in racquet weight are the carbon and glass fibre sheet moulding compound (SMC) used in the frame. The final material that exhibits differences depending on design is the method B finish. This is a polyamide used only in method B to provide a better, more resilient finish to the frame décor. In the following section the environmental impact of materials used and the waste bi-products of the processes are calculated.

Table 1. Inventory of four tennis racquets.

Part Name	Material	Weight (g)			
		RADLAQ 1A	RADTT 1B	10 LAQ 2A	10TT 2B
bumper	PA 66	19	19	21	21
nylon parts	PA 6	23	23	23	23
lead tape	lead	37	37	8	8
frame rubber	nitrile rubber	1.8	1.8	1.8	1.8
neck foam	PUR hard	6	6	6	6
GF lams	SMC GF50	64.6	64.6	7.8	7.8
CF lams	SMCCF50	148.3	148.3	178.5	178.5
grips	PUR flexible	89	89	89	89
adhesive tape	PET	0.5	0.5	0.5	0.5
attach tape	PVC soft	1	1	1	1
string	polyester	19.8	19.8	19.8	19.8
string	PA 6	39.4	39.4	39.4	39.4
back paper	paper	5.6	5.6	5.6	5.6
TT finish	PA 6	N/A	16	N/A	16
spray paint	N/A	N/A	N/A	N/A	N/A
cardboard	cardboard	87.2	87.2	87.2	87.2
pack plastic	PET bottle grade	21.8	21.8	21.8	21.8
racquet cover	LLDPE	21.8	21.8	21.8	21.8

2.3. Impact Assessment

The impact assessment stage involves technical, quantitative and qualitative efforts to determine the effects or impacts of the inventory, i.e. what sort damage the use of resources or output of emission has on the environment. The main purpose of this stage is to allow comparison of processes, emissions and other waste on the same scale, thus determining an overall effect of the product. The principle aim is thus to provide a quantitative indicator of environmental impact. However, inconsistencies and uncertainties combined with the subjective nature of some impact assessment methods means that it is not widely accepted as an absolute indicator of environmental impact, but in fact only as a relative value that can be used to compare like products or concepts.

The IdeMat database provides a number of environmental impact indicators. The EI-99 method used in this work is shown in Figure 2 running right to left.

The first step in the EI-99 impact assessment method is a process known as fate analysis, which involves converting the mass outputs of inventory into concentrations, thus gaining a spatial and temporal component. For example, X kg CO_2 from a process becomes

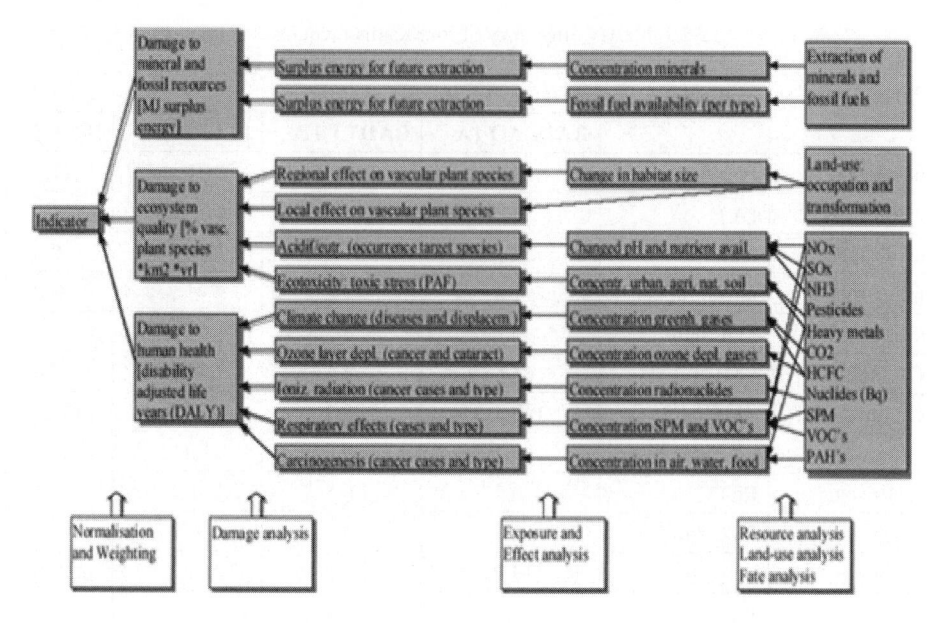

Figure 2. Graphical representation of the EI-99 method.

a concentration in a known location which disperses at a specific rate over a known time. The results of the fate analysis are then analysed for exposure and effect. This means that the consequence of the inventory concentrations are linked to "real" effects, such as the number and types of cancers in humans, or the increased energy required to extract further minerals. The results of exposure and effects analysis (damage indicators) are then treated under the heading "damage analysis". This enables the damage indicators, which are grouped according to their damage category, to be combined resulting in three figures. Finally the figures for the damage categories are combined and weighted in the EI-99 case to provide a single value. The database thus provides a single figure for each material and process that can be used to assess the environmental damage of the entire product when summed. Different indicator methods and ways of using the midpoint data are available.

The EI-99 scale is chosen so that the value of 1 pt is representative of one thousandth of the yearly load of a European citizen (Goedkoop & Spriensma, 2001). The panel of environmental experts that developed EI-99 found damage to human health and ecosystem quality to be approximately twice as important as damage to resources, which means the scores are weighted accordingly. For further information on the definitions of terminology, modelling techniques, uncertainties and assumptions employed in the development of EI-99 refer to the developers' reports (Goedkoop & Spriensma, 2001).

Before the environmental impact of the composite racquets is examined a discussion of how the uncertainty is dealt with is given. Firstly there are the indicators of uncertainty provided within IdeMat for the EI-99 data. For all data a whole number between 1 (lowest reliability and completeness) and 5 (highest reliability and completeness) is given for its reliability and completeness. These are the quantified views of the developers with regards to reliability of the sources they used and the entirety of the damage considered.

Table 2. Impact assessment results and uncertainty inputs.

Component	1A (EI-99 mPt)	1B (EI-99 mPt)	2A (EI-99 mPt)	2B (EI-99 mPt)	−Std Dev (%)	+Std Dev (%)
bumper	12.77	12.77	15.9	15.9	5	5
frame	181.96	196.36	156.14	171.07	20	10
endcap	7.51	7.51	7.51	7.51	6	6
grip	37.19	37.19	37.19	37.19	8	8
handle	10.52	10.52	10.52	10.52	10	10
strings	67.58	67.58	67.58	67.58	6	6
packaging	63.3	63.3	63.3	63.3	10	10
	380.83	395.23	358.14	373.07		

Secondly there is uncertainty with regards to the relevance and accuracy of the data used compared to the actual materials and processes that form the product life cycle; this will be known as "analyst's uncertainty". For example, the EI-99 is specific for Europe and thus applying it to manufacturing in Asia (where the racquets were constructed) can not be considered ideal. Unfortunately there are no impact assessment methods available for Asia at present, thus there is a degree of geographical or spatial uncertainty within the data. Furthermore, there is the uncertainty that the analogous materials or processes used are accurate for representing those that are actually present in the racquets. In the same manner as the database indicator, the analyst's indicator was given for each process and material between 1% for minimum uncertainty in the data and 100% for maximum uncertainty. Monte Carlo analysis was used in this work to treat the uncertainty associated with this LCA analysis. The LCA software GaBi 4 was used for this purpose. Mote Carlo analysis basically involves taking a set of random numbers from within a defined range and using them to create a probability distribution. The software requires each parameter included in the Monte Carlo analysis to be input along with a standard deviation range. All of the data was given a normal distribution or a continuously uniform distribution. In order to manage the numerous parameters (materials and processes) from the LCA in a Monte Carlo analysis, a reduced set was used. Although the reduced set includes fewer parameters all the data identified earlier has been included. Table 2 shows the impact assessment results for different tennis racquets and components with their respective uncertainty inputs.

3 Discussion of Results

Figure 3 shows a comparison of the environmental impact determined for the four tennis racquets considered. Racquet 1B is clearly the worst in terms of its environmental impact, followed by the other model 1 constructed by method A, and then the two model 2 s in the same order.

Figure 4 shows how the various components contribute to the overall environmental impact. The frame has the largest contribution to each racquet's overall environmental

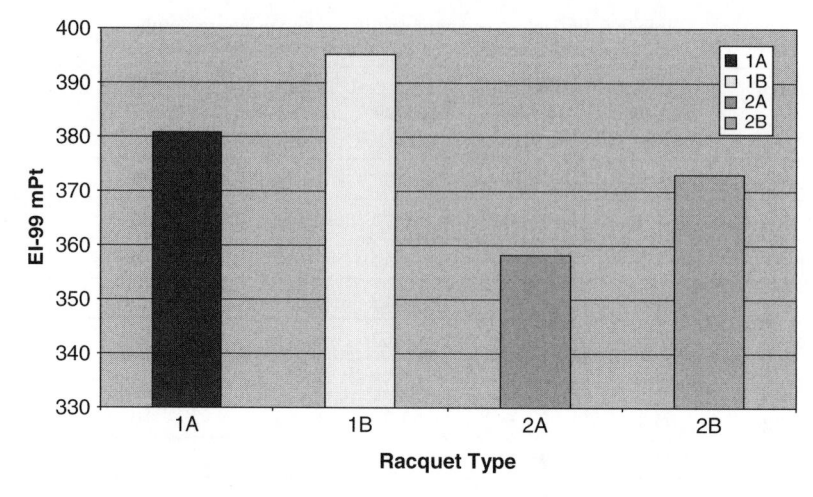

Figure 3. Comparison of environmental impacts of four composite tennis racquets.

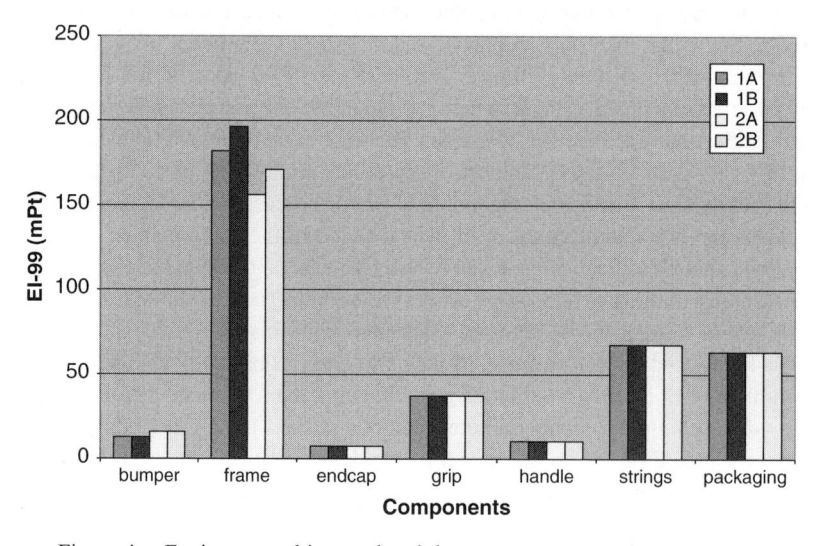

Figure 4. Environmental impact breakdown across racquets' components.

impact. The remaining components (endcap, bumpers, handle) of the actual racquets contribute little, with the separate components such as grip, strings, and packaging contributing slightly more. To put the impact in some perspective the work of Goedkoop and Spriensma (2001) should be considered. They state that 1 EI-99 pt is approximately equal to 1/1000th of the average European citizens' yearly environmental load, which equates to 114.2 mPts per hour. Hence, the racquets analysed can be equated with the average environmental load of a European citizen in 3 hours.

The two B models have extra material as part of the construction process to provide a better finish, which contributes to the higher environmental impact of the frames than their

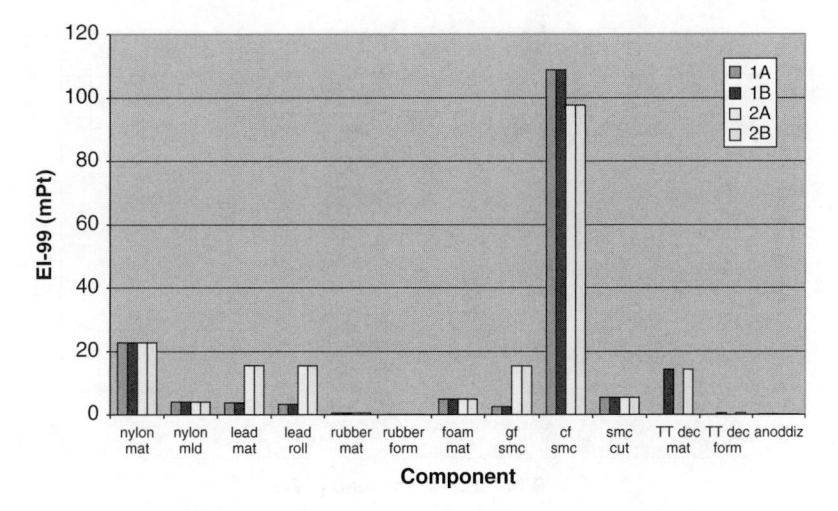

Figure 5. Environmental impact breakdown for racquet frames.

model A counterparts, but only by 20 mPts approximately. This is clearly not the largest contribution to the frames' environmental impact as seen in Figure 5 which shows the environmental impact for the parts that make up the frame, with materials and secondary processes included.

The study found that the most significant material for all racquets is the CF SMC, although this figure has high uncertainty as discussed earlier. The second highest contributors to the frames' environmental impact are lead alloy used for additional weight in model 1 racquets and nylon used in the blow moulding process of all racquets. The nylon material plus the moulding process equals 26 mPt, and in model 1, where 37 g of lead alloy is present, the lead material plus the rolling process equals 31 mPt. Both of these materials with their respective secondary processes seem to be important. However, it was noted that the human toxicity effects were not included in the EI-99 data for lead, which was represented in the reliability and completeness of data. Accordingly, this was included in a positive skew of the uncertainty factor for the frame, although overall it was still modelled with a negative skew because of the uncertainty of the CF.

The uncertainty assessment conducted using the Monte Carlo method indicates that the calculated values are mainly lower than the actual mean, which is understandable considering the standard deviation range of the frames. Figure 6 shows comparative results for the four tennis racquets considered using the Monte Carlo analysis.

4 Conclusion

The outcomes of this research identified specific design features of tennis racquets considered that should be looked at in the future from the perspective of design for environment. The study found that the racquet frame represents the most important component with respect to the associated environmental impact. The strings, packaging and the grips are also important. Although these components contribute much less to the overall environmental impact during the racquet life cycle, they have the important property of easy disposal because they are easily detached from the racquet by the user. This means that compared

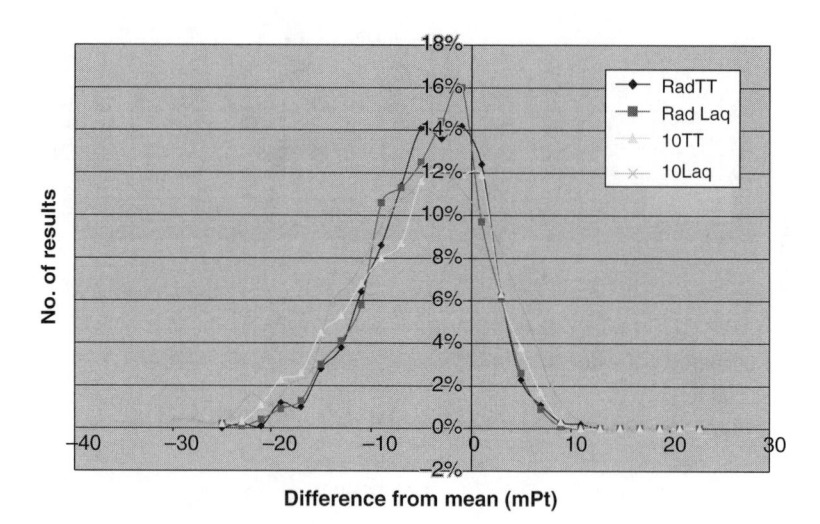

Figure 6. Comparison of streamlined LCA results for 4 racquets by Monte Carlo analysis.

with the racquet frame these components are easily separated and potentially recycled, which could improve the environmental design of the racquets. Design for end of life incorporates ideals such as these. There are already environmentally friendly initiatives in the design of strings and particularly packaging plastics. Gosen's Bio-gut Multi-oil 16 biodegradable tennis strings (http://www.gosen.jp/), made partly from corn starch are an environmentally friendly option that would reduce the environmental impact considerably. Chlorine free polythene packaging is suitable for recycling as is cardboard and paper, both of which would improve the environmental design of the product system.

The frame is the most complicated of the components and not easily recycled. This means that the most common form of racquet disposal is most probably through municipal waste at present. More work is required to evaluate the various end-of-life strategies for a tennis racquet which should enable the feasibility of frame recycling to be considered. Specifically, the recycling chain would have to be analysed and specialist composite recycling plants would have to be involved, which is difficult because there are only few in existence. Ideally this may also lead to improvements in the recyclability of the frame. Recycling of fibre reinforced composites uses solvents to dissolve the resin matrix. Anastas & Lankey (2000) note that it may be possible to develop and use less chemically hazardous solvents in conjunction with developments of polymer matrices.

The easiest and probably the most environmentally friendly end-of-life strategy is reuse. In this sense racquets are well designed, because the components that wear out can be easily replaced by the user. However, most users will prefer to buy a new design after some time, which is probably the main reason for the disposal of most racquets that is driven by current consumer practices. One aspect of reuse that could be explored is the use of unwanted equipment by disadvantaged communities once life with the principal owner is completed. In this way the equipment manufacturer can contribute to sustainable development in ways that are incredibly important although difficult to measure.

In evaluating LCA as a tool for use in sports and leisure goods industry several points have arisen. Many of these can relate to the method itself and the three spheres of knowledge identified by Hofstetter (1998). These are the technosphere, ecosphere and valuesphere. They represent the varying subjectivity present in LCA:

- Technosphere encompasses the description of the product life cycle; the emissions from the process and the allocation of procedures in terms of causal relations.
- Ecosphere encompasses the modelling of changes that affect and damage the environment e.g. processes such as global warming.
- Valuesphere involves the modelling of the seriousness of environmental changes such as global warming, as well as the modelling choices made in Technosphere and Ecosphere.

The valuesphere is an area that has attracted much interest recently, with discussion over eco-design in a global content being prominent. Many LCA professionals recognise that adhering to a standard methodology is important for transparent and comparable investigations. However, in the increasingly global world market place, it is apparent that different impact assessment methods may be required for the different phases of the lifecycle. For instance, a product is manufactured in Asia, but used in Europe and disposed of in America. The development of different weighting for different geographical areas is a very important issue, but is in direct conflict with some of the aims that exist at the moment, namely making inventory databases broadly applicable and widely available along with impact assessment methods so that product comparisons and assessments are more easily made. Also fundamental to this is the issue of weather it is possible for sensitive information on materials and properties to become widely available. If environmental legislation is going to become widely used and appropriate then these are some of the issues that must be considered.

References

Abele E., Anderl R. and Birkhoefer H. (2005) *Environmentally friendly product development. Methods and tools*. Springer, New York.

Anastas P.T. and Lankey L. (2000) Life cycle assessment and green chemistry: the yin and yang of industrial ecology. *Green Chemistry*, 2, 289–295.

Goedkoop M. and Spriensma R. (2000) The Eco-Indicator 99. A damage orientated method for life cycle impact assessment. Methodology report 3rd Edition (online). Available: http://www.pre.nl/eco-indicator99/ei99-reports.htm (accessed: June 2007).

Hofstetter P. (1998) *Perspectives in LCA: a structured approach to combine models of the technosphere, ecosphere and valuesphere*. Kluwer Academic Publishers, Boston.

International Olympic Committee (2005) Olympic movement's Agenda 21. Sport for sustainable development. IOC Sport and Environment Commission (online). Available: http://multimedia.olympic.org/pdf/en_report_300.pdf (accessed: June 2007).

International Standards Organisation (1997) 14040 series: Life Cycle Assessment. ISO.

Racquet Tech (2003): Factory tour: the making of a racquet. March, p10–13 (online). Available: www.racquettech.com (accessed: June 2007).

Subic A. (2005) A lifecycle approach to sustainable design of sports equipment. In Subic A. and Ujihashi S. (Eds.) *The impact of technology on sport*. ASTA, Melbourne, pp. 12–19.

Subic A. and Paterson N. (2006) Lifecycle assessment and evaluation of environmental impact of sports equipment. In: Moritz E.F. and Haake S. (Eds.) *The Engineering of Sport 6, Vol. 3 – Developments for Innovation*. Springer, Munich, pp. 41–46.

GOING GREEN: THE APPLICATION OF LIFE CYCLE ASSESSMENT TOOLS TO THE INDOOR SPORTS FLOORING INDUSTRY

A.W. WALKER

The Louis Laybourne Smith School of Architecture and Design, The University of South Australia, Adelaide, Australia

In recent decades, an awareness of environmental issues has increased within society encouraging people to become more aware of the impact that human activity is having upon both the natural and built environments. There is also increasing awareness of the external and internal pressures being placed upon organizations to acknowledge, analyze and report upon environmental issues. The resulting general trend is for organizations to take more responsibility for the services, products and processes that they are responsible for. Specifying and selecting sports surfaces, involves the simultaneous evaluation of a multitude of performance factors. Increasingly, environmental impact is becoming an important component in such surface evaluations. Architects, specifiers and facility operators require objective, consistent and comprehensive data detailing the environmental impacts of materials and their associated manufacturing processes. The broad aim of this paper is to analyze the environmental impacts of various constructional styles of indoor playing surface on global warming, through the application of Life Cycle Assessment. To calculate the ecological footprint of an organisation's resource use activities, it is necessary to track how consumption patterns ultimately lead to emission impacts. The approach used has involved a process based analysis, which looks at each individual process in the playing surface supply chain, required to manufacture, install, maintain and ultimately dispose of various constructional styles of indoor sports surface.

1 Introduction

The consequences of human impact on the environment are becoming more prominent. In just the last year, a number of influential reports have been published which have heightened government and public awareness of environmental issues impacting upon the environment and in particular their potential to influence climate change. In Australia these impacts are now being felt in the form of changing rainfall patterns and increasing average temperatures.

1.1. *Climate Change*

The Stern Review claims that carbon emissions have already increased global temperatures by half a degree Celsius and if no action is taken on emissions, there is more than a 75% chance that global temperatures will rise between two and three degrees Celsius within the next 50 years (Stern, 2006). To stabilize temperatures at manageable levels, emissions would need to be stabilised over the next 20 years and fall between 1% and 3% after that. The options for change could be reduced if consumer demand for heavily polluting products and services is also reduced (Stern, 2006).

We have been exceeding the Earth's carrying capacity for the past 20 years (Living Planet Report, 2006). Consumption must be balanced with the Earth's capacity to regenerate and absorb waste. Climate-changing green house gas emissions now account for close to 50% of our global footprint and the choices that we are making in our day-to-day lives now, will shape our opportunities far into the future. The cities and power-plants that we build today, will either lock humanity into dangerous over consumption or begin to propel our generation and future generations towards a truly sustainable future. The buildings we create can be future-friendly, or not. Urban infrastructures become traps if they can only operate on large footprints. In contrast, future-friendly infrastructure – cities designed as resource efficient, with carbon-neutral structures can support a high quality of life with a small footprint. The longer infrastructure is designed to last, the more critical it is to ensure that we are not building a destructive legacy that will undermine our social and physical wellbeing (WWF International, 2006).

Considerable attention is being placed on the materials we use to build our cities in the form of green building materials and products but this raises many difficult questions. What makes a given material or product green and how do we equate the relative greenness of different products? In an ideal world, the life-cycle impacts of all materials would be fully defined, so that the facility specifier can clearly and unambiguously see which material is better from an environmental point of view.

1.2. Indoor Playing Surfaces

The quantity and quality of sports facilities encourage community exchanges, sport and recreation and ultimately the health of the community. Timber indoor sports playing surfaces are the norm throughout most of the world and especially at the highest level of indoor sport. Timber is also a renewable resource with negative net carbon emissions (Petersen & Solberg 2004) but there is also the possibility that the environmental burden produced while maintaining and disposing of a wood playing surface may outweigh the environmental load of its synthetic alternatives, as installing and maintaining a sports surface brings with it heavy environmental loads in the form of periodic resurfacing, sealing and disposal (Nebel *et al.*, 2006). The environmental and financial cost of installing and maintaining timber playing surfaces is resulting in the increasing popularity of indoor playing surface alternatives.

Synthetic surfaces are continuing to gradually replace timber in many parts of the world as synthetic surfaces avoid many of timber's drawbacks, but could a synthetic surface potentially be more environmentally benign than timber itself? This paper aims to begin to answer this question by looking at both timber and synthetic surfaces used in sports applications, from the perspective of Life Cycle Assessment.

2   Life Cycle Assessment Methodical Framework

The Methodology used in this study is consistent with the methodology for Life Cycle Assessment as described in the ISO 14040 series of Standard Documents. However, the study departs from being a strict LCA by including a calculation of the global warming potential (GWP) effects for the various surfaces (Petersen & Solberg, 2004). Estimates of the GHGs emitted during the life cycle of 4 individual synthetic flooring systems were compared in relation to a natural wooden surface. A life cycle approach (from raw material acquisition through manufacturing, transportation, use and maintenance, and end-of-life disposal) was followed to determine the boundaries and elements attributable to each systems (Jonsson *et al.*, 1997).

A Life Cycle Assessment consists of four independent elements:

1. Definition of goal and scope (ISO 14040)
2. Life cycle inventory analysis (ISO 14041)

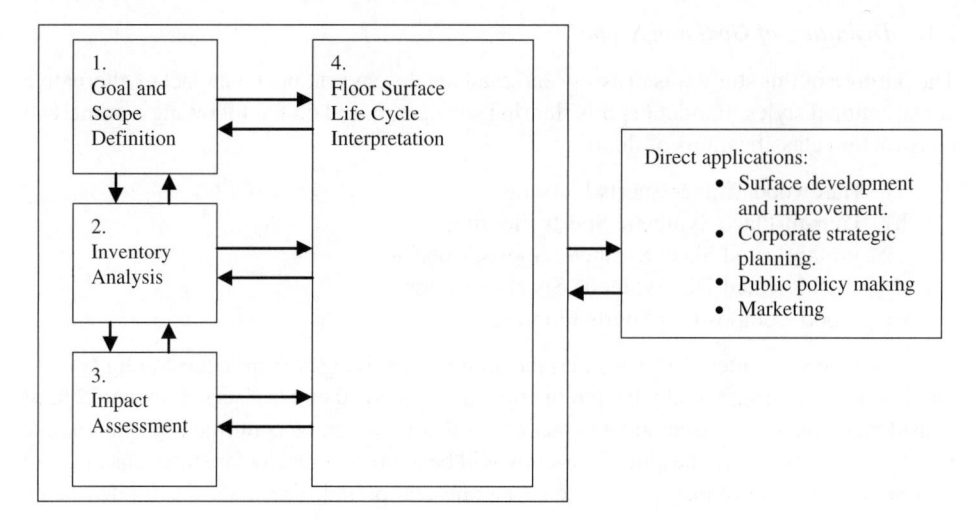

Figure 1. The four elements of an LCA and its direct applications.

3. Life cycle impact assessment (ISO 14042)
4. Life cycle interpretation assessment (ISO 14043) (LeVan, 1995)

The definition of the study's goals and scope (1) includes decisions about the functional units which will form the basis of the comparison, the flooring system to be studied, system boundaries, allocation procedures, assumptions made and limitations.

The life cycle inventory analysis (2) involves product data collection and calculation procedures to quantify the total system inputs and outputs that are relevant from an environmental point of view. Given the serious consequences resulting from climate change, the environmental focus will be climate change potential. The system boundaries are defined and relevant process flow charts are drawn.

The life cycle impact assessment (3) is a quantitative and/or qualitative process which characterizes and evaluates the significance of the potential environmental impacts using the results of the life cycle inventory analysis. One important goal of the life cycle impact assessment is to combine outputs with comparable effects such as all greenhouse gases.

The life cycle interpretation (4) involves the development of conclusions based on the life cycle inventory analysis and the life cycle impact assessment. As an outcome of the interpretation stage, recommendations can be formulated which can be directed towards manufacturers or policy makers in which options for reducing the environmental burdens of the flooring system under study, can be identified and evaluated (Nebel et al., 2006).

The life cycle of a flooring surface can be divided into the following three stages:

- Floor Production- resource extraction, processing of raw materials, surface manufacture and installation
- Floor Use- maintenance of the surface
- Floor Disposal- processing of the disposed surface.

Each stage may consist of a number of processes which each uses one or more inputs from previous processes and gives outputs to one or more next processes. Each input can be followed upstream to its origin and each output downstream to its final end. The total of connected processes is called the product system, process tree or life cycle.

2.1. Definition of Goal and Scope

The purpose of this study was to assess and evaluate the environmental impact of alternative constructional styles of indoor sports flooring surfaces, based on the following commercial construction classifications of flooring:

a) Hardwood Timber Sports Flooring
b) Pour-in-Place Synthetic Sports Flooring
c) Prefabricated Sheet Synthetic Sports Flooring
d) Prefabricated Tile Synthetic Sports Flooring
e) Wood Composition Sports Flooring

It is widely assumed that as wood is the product of a living carbon sequestrating system, that it will be environmentally benign in comparison to synthetic surfaces. It is therefore of considerable interest to ascertain whether or not this view can be confirmed by quantitative Life Cycle Assessment. The aim of the study will be to present results for emissions of CO_2e for one square meter of playing surface over a one year period. The study was limited in the following respects:

- It is assumed that there is no recycling or recovery of the floors as this is currently the case in most countries.
- The environmental impact of daily/weekly cleaning was omitted, as this would be similar for all of the surfaces regardless of the constructional style.
- In each case, specific manufacturers or manufacturers associations were chosen as the main sources of data.

The study was limited in the following respects:

- Only indoor playing surfaces were considered in the study.
- Since the focus of this study is on global warming, the scope has been limited to CO_2 equivalent data.
- It was assumed for all calculations that recycling or recovery of the playing surfaces would be limited to incineration, land-fill or recycling.
- Given that all systems require a similar concrete substructure preparation, this component was excluded from the scope of the study.
- Comparisons between the five project systems detailed above are based on a 1 m² area over one year, with the expected life of the product factored into the use/maintenance phase.

2.2. Life Cycle Inventory Analysis

A flow chart was developed, defining each of the flooring surface's system boundaries and then data on the environmental load was gathered for each life cycle. In this study, the necessary information was gathered from company literature and other LCA studies. The results of the inventory analysis were calculated for each surface's life cycle per square meter.

2.3. Life Cycle Impact Assessment

GHGs are quantified at the component level (Franklin Associates, 2006). This approach requires information regarding the quantity of raw material used in the manufacture of each component and the GHG emission factor (CO_2e/tonne) associated with the component (IPCC, 2006). In this project the focus of the study has been Global Warming Potential and therefore the account has been limited to CO_2, CH_4 and N_2O which are the three main GHGs resulting from combustion (Schipper, 2006). These three gasses have different lifetimes and greenhouse strengths (IPCC 1995). Once emissions and sequestrations for each component have been calculated, they can be summed for an overall measure of GHG emissions for each of the flooring systems over the one year period in question. Where relevant and possible, Australian GHG emission factors were applied for each component of the flooring systems under review.

3 Life Cycle Interpretation and Conclusions

Total GHG emission factors of the five competing systems are still is the process of being assessed. LCA results alone however will not lead specifiers to make environmental choices as a surfaces performance, cost and aesthetic expectations must also be taken into account. In addition, one should always be careful in drawing conclusions from a single case study such as this, as the data used will vary greatly from one geographic and climatic location to the next and different assessment methods may give different and contradictory answers (Jonsson, 2000).

References

Franklin Associates (2006) Life Cycle Inventory of Five Products Produced from Polylactide (PLA) and Petroleum-Based Resins. Athena Institute International.
IPCC (1995) Radiative Forcing of Climate Change and an Evaluation of the IPCC IS92 Emission Scenarios. Cambridge University Press.
IPCC (2006) 2006 IPCC Guidelines for National Greenhouse Gas Inventories. IPCC National Greenhouse Gas Inventories Programme. In: Eggleston S., Buendia L., Miwa K., Ngara T. and Tanabe K. (Eds.), Intergovernmental Panel on Climate Change. The Institute for Global Environmental Strategies (IGES) on behalf of the IPCC. Hayama, Japan.
Jonsson A. (2000) Tools and methods for environmental assessment of building products – methodological analysis of six selected approaches. Building and Environment, 35/3, 223–238.
Jonsson A., Tillman A. and Svensson T. (1997) Life Cycle Assessment of Flooring Materials: Case Study. *Building and Environment*, 32/3, 245–255.
LeVan S. (1995) Life Cycle Assessment: Measuring Environmental Impact. 49th Annual Meeting of the Forest Products Society, Portland, Oregon.
Nebel B., Zimmer B. and Wegener G. (2006) Life cycle assessment of wood floor coverings – A representative study for the German flooring industry. *International Journal of Life Cycle Assessment*, 11/3, 172–182.
Petersen A.K. and Solberg B. (2004) Greenhouse gas emissions and costs over the life cycle of wood and alternative flooring materials. *Climatic Change*, 64/1–2, 143–167.
Schipper M. (2006) Energy-Related Carbon Dioxide Emissions in U.S. Manufacturing, Energy Information Administration.
Stern N. (2006) Stern Review: The Economics of Climate Change. London, The Treasury.
WWF International (2006) Living Planet Report 2006. C. Hails (Ed.), WWF International.

ns and
3. Sports Medicine, Exercising and Clinical Biomechanics

SPORTS TECHNOLOGY IN THE FIELD OF SPORTS MEDICINE

B. TAN

Changi Sports Medicine Centre, Changi General Hospital, Singapore

The scope of Sports Medicine has broadened, and so has its offerings. This could not have happened without advances in sports technology and its applications. Sports or medical technology contributes to Sports Medicine from diagnostics, to management, prevention, and performance enhancement. This paper highlights some of the many contributions that sports technology has made to enhance the practice of Sports Medicine. It is hoped that with such examples, more will follow to facilitate healthcare professionals in better managing competitive and recreational athletes.

1 Sports Medicine

Sports Medicine (or Sports Medicine and Sports Science) is as an umbrella term comprising Sports Medicine proper as well as Sports Orthopaedic Surgery, Sports Physiotherapy, and the Sports Sciences. The latter can be subdivided into five disciplines (Figure 1). Sports Physicians, unlike Sports Orthopaedic Surgeons, deal mostly with the non-surgical aspects such as the management of injuries that are treated non-surgically, exercise testing, exercise prescription, anti-doping, and the management of medical conditions related to sports and exercise.

2 Sports Technology Applications in Sports Medicine

To see the role of technology in medical practice, perhaps it is best to broadly divide the practice of Sports Medicine into (1) injury surveillance (monitoring) and prevention, (2) diagnostics, (3) injury management, and (4) performance enhancement.

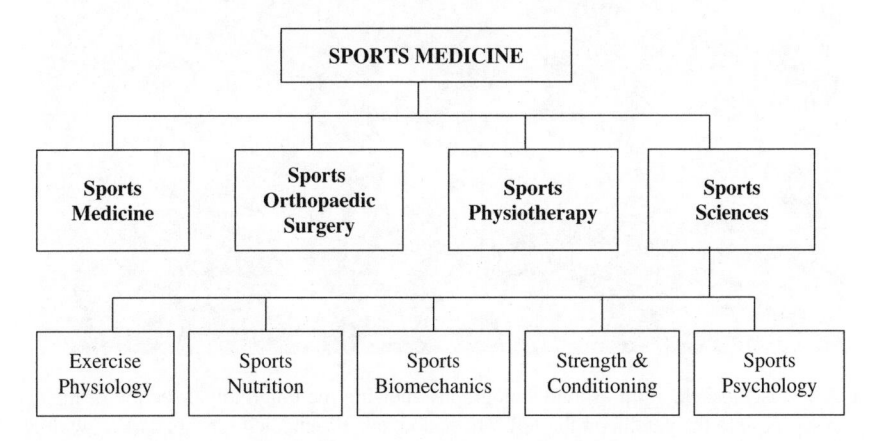

Figure 1. The various disciplines that fall under the broad umbrella of "Sports Medicine."

2.1. Injury Surveillance and Prevention

One of the strategies to prevent injuries is to identify predisposing factors or subclinical conditions that will later lead to symptoms. Examples include hamstring tightness leading to lower back injuries, pes cavus (high medial arch) leading to tibial stress fractures, and lesions in the patellar tendon leading to symptomatic patellar tendinosis.

Sports technology helps in identifying these predisposing factors. The last example illustrates this. The use of bedside musculoskeletal ultrasound sonography has become increasingly popular among sports physicians. It is especially effective for visualizing lesions and pathologies in tendons and ligaments, and these form the bulk of sports injuries. In musculoskeletal applications, a linear array transducer is used, and edematous, injured soft tissues appear as hypoechoic areas. Chronic, degenerative lesions may contain calcified deposits, and these have a hyperechoic appearance (Figure 2). It has been found that in asymptomatic soccer players who have patellar tendinosis picked up by pre-season ultrasound sonography had a 17% chance of developing symptomatic jumpers' knee over the next 12 months (Freberg & Bolviq, 2002). This presents an opportunity for early intervention such as training modifications, quadriceps stretching and strengthening (to improve shock absorption and hence reduce loading of the patellar tendon), and perhaps "pre-emptive" extra corporeal shock wave therapy (ESWT).

2.2. Diagnostics

How sports technology plays a role in diagnosing sports injuries is obvious, especially in the area of diagnostic imaging. While X-rays help diagnose bone injuries, many injuries in athletes involve soft tissues such as muscle and tendons. Such injuries are better visualized through magnetic resonance imaging (MRI) or musculoskeletal ultrasound sonography (US).

MRIs have been a godsend to Sports Medicine, as they are useful for imaging both bony and soft tissue lesions. Bony injuries ranging from marrow edema (Figure 3) to stress fractures, to frank fractures are easily picked up by MRIs. X-rays and CT scans do not detect

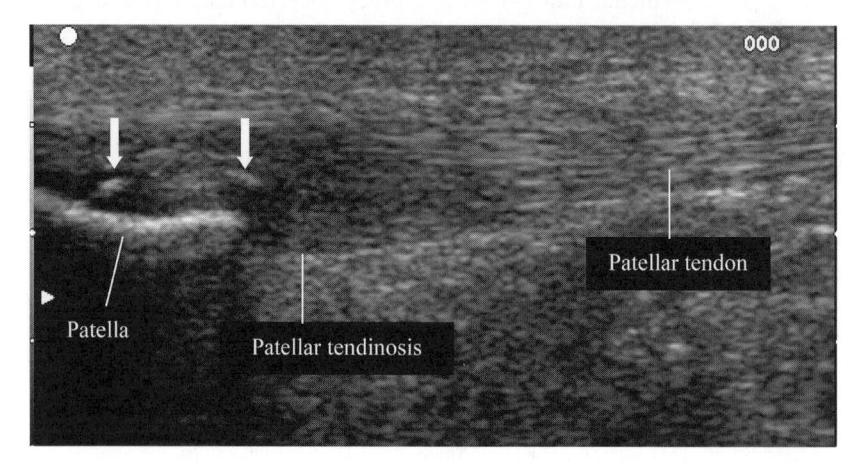

Figure 2. Musculoskeletal ultrasound sonography showing the longitudinal section of the patellar tendon inserting into the patella on the left. The lesion is hypoechoeic (appears darker) with calcifications (arrows).

bone marrow edema. Other than tendon and bone lesions, MRIs can also detect cartilage lesions such as a torn meniscus (Figure 4) and labrum, as well as hyaline cartilage thinning and ulcers (as in osteoarthritis).

Apart from imaging the fetus in a mother's womb, ultrasound sonography is also highly effective in musculoskeletal imaging. Ligaments and tendons, in particular are well visualized through ultrasound sonography. So too are cysts, haematomas (Figure 5), blood vessels, and nerves. As ultrasound waves are not able to penetrate bone, they are limited to imaging only certain soft tissue injuries. Nevertheless, the range of injuries it is able to detect is quite broad, from rotator cuff tendinopathy, to muscle tears, tendon tears, plantar fasciitis, patellar tendinosis, retrocalcaneal bursitis, and much more. A great advantage of US is that it can be done as a bedside procedure, thus offering immediate diagnosis and correlation of ultrasound findings with physical signs by the examining physician. The convenience also extends to ultrasound-guided procedures, making it more accurate and safer to aspirate

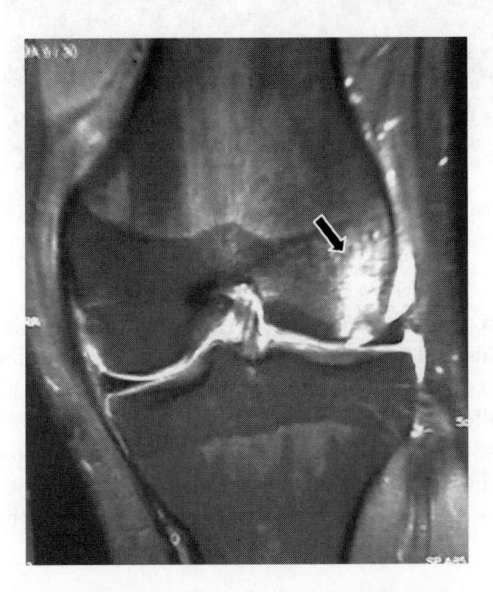

Figure 3. Magnetic resonance imaging of the knee, with a distinctive white lesion in the bone (arrow), highlighting an area of bone marrow edema which would not have been visible on X-rays.

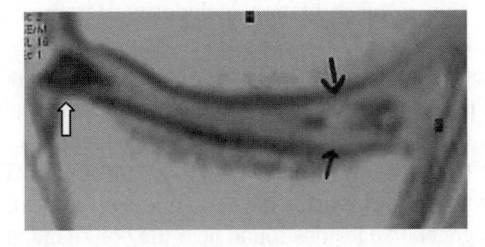

Figure 4. Magnetic resonance imaging showing a complex tear of the posterior horn of the medial meniscus of the knee (black arrows). The normal and intact anterior horn (white arrow) appears as a black triangle.

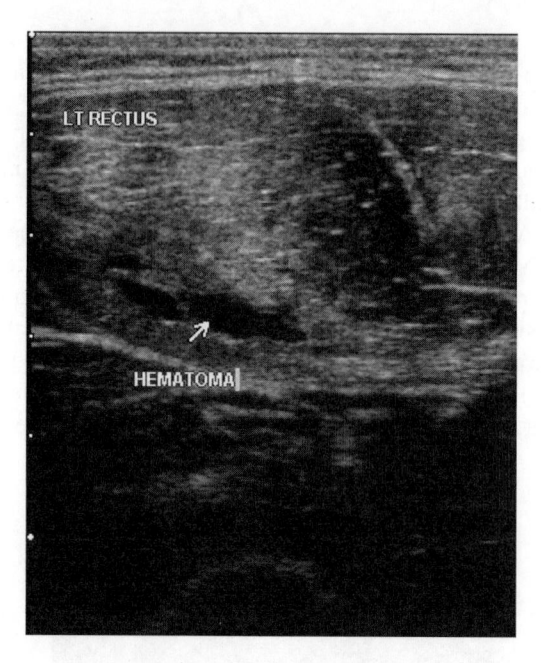

Figure 5. Ultrasound image showing the cross section of the rectus femoris muscle in a soccer player, with a haematoma. The tear would not have been detectable by physical examination.

or infiltrate lesions that are deep or in close proximity to critical structures (e.g. arteries and nerves). US imaging also allows quantitative measurements of lesions (e.g. diameter of a haematoma or thickness of a swollen plantar fascia), thus allowing accurate monitoring of such lesions during follow up visits.

Motion analysis software (e.g. siliconCOACH) helps to diagnose faulty technique or gait that may lead to injury. An example is a patient who has recurrent tibialis posterior tendinopathy secondary to overpronation that occurs during running but not during standing and walking.

2.3. *Injury Management*

Here, there are many examples of how technology has helped. Physiotherapy modalities that have been a great help in rehabilitation include interferential treatment, therapeutic ultrasound, iontophoresis, etc.

Extra corporeal shock wave therapy (ESWT), which employs shock waves to treat bone-tendon junction lesions provide an increasing popular treatment modality for injuries such as plantar fasciitis (Figure 6), Achilles enthesiopathy, patellar tendinosis, tennis elbow, golfer's elbow, and rotator cuff tendinopathy (Chung & Wiley, 2002). ESWT, which evolved from extra corporeal shock wave lithotripsy (the blasting of kidney stones), employs shock waves that are focused onto degenerative lesions found near the bone-tendon junctions. The shockwaves break up calcifications and promote healing of the degenerative lesions.

As mentioned earlier, ultrasound-guided injections and aspirations offer accuracy and safety. Yet another example is the isokinetic dynamometer, which measures muscle strength during concentric or eccentric activity. It detects and quantifies imbalances either between

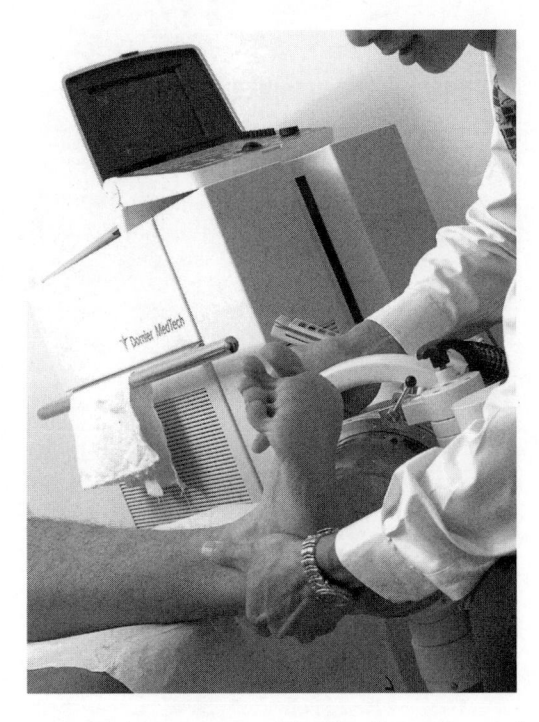

Figure 6. Focal extra corporeal shock waves are "fired" at the swollen portion of the plantar fascia under ultrasound guidance.

left and right sides of the body or between agonist and antagonistically acting muscles, thus allowing target-setting and accurate monitoring during rehabilitation.

2.4. *Performance Enhancement*

The list of examples of technology in performance enhancement is endless. It ranges from the simple nasal strip (to open the nasal passages) to million-dollar motion analysis hardware and software. Metalyzers measure resting metabolic rate, maximal oxygen consumption, and anaerobic threshold, and are thus useful for endurance athletes.

3 Conclusion

Sports technology is now an integral part of all aspects of Sports Medicine. Being a fast-evolving field, the role of sports technology becomes even more important in improving clinical care.

References

Chung B. and Wiley J.P. (2002) Extracorporeal Shockwave Therapy – A Review, *Sports Medicine*, 32/13, 851–65.
Freberg U. and Bolviq L. (2002) Significance of ultrasonographically detected asymptomatic tendinosis in the patellar and achilles tendons of elite soccer players: a longitudinal study, *American Journal of Sports Medicine*, 30/4, 488–91.

BIOMECHANICAL FACTORS IN MILD TRAUMATIC BRAIN INJURIES BASED ON AMERICAN FOOTBALL AND SOCCER PLAYERS

M. ZIEJEWSKI,[1] Z. KOU[2] & C. DOETKOTT[3]

[1]*Mechanical Engineering and Applied Mechanics Department, North Dakota State University, Fargo, USA*
[2]*Department of Radiology, Wayne State University School of Medicine, Detroit, USA*
[3]*Research Analyst, Information Technology, North Dakota State University, Fargo, USA*

The complete biomechanical assessment of head impact includes not only investigations on players' head kinematics and brain tolerances, but also improvements in the design of protective headgear for head impact sports. Although these issues have been studied, there are related areas that have not been addressed in the literature. The purpose of this paper is to address some of the related issues and to demonstrate their relevance. The objectives of this study were 1) to quantify the brain injury likelihood for the struck players in an American football game and 2) to determine the effect of shock attenuating material, with different shear moduli, on the brain tissue deformation in head impact sports. An analysis was performed based on a subset of original National Football League (NFL) data for struck players only. The univariate analysis for each biomechanical parameter and the injury risk curves for selected biomechanical parameters were determined. Characteristics of the helmet material were assessed using a finite element (FE) elliptical head model. Cumulative strain damage measure (CSDM) data sets, corresponding to the strain levels of 0.01, 0.02, and 0.03, were utilized. Injury risk curves for HIC, SI, peak translational acceleration and peak rotational acceleration indicated a significant difference between the struck NFL players and analysis based on of all of the players. For all studied biomechanical parameters, injury risk was higher for struck NFL players. The percentage of reduction in brain volume, experiencing the strain level of 0.03, was reduced with the decrease of shear moduli. The highest reduction (24%) can be seen with the lowest shear moduli of 2.0 MPa. Injury risk is higher for struck NFL players in comparison to the striking players. For helmeted struck players, the threshold of brain injury risk is suggested as HIC-110, SI-130, peak translational acc.-55 g; and peak rotational acc.-3300 rad/s^2. The headgear made of shock attenuating material, with low shear moduli, can reduce the strain on brain tissue for players of head impact sports.

1 Introduction

There are several biomechanical risk factors associated with direct head impact to a helmeted NFL football player. They include factors such as the level of preparedness of the players, location of the head strike, etc. The combination of different factors defines a unique scenario for each impact condition. It is necessary to understand individual effects of each of the recognized biomechanical risk factors.

In 1994, the Commissioner of the NFL formed a committee on mild traumatic brain injuries (MTBIs) in response to safety concerns regarding brain injuries occurring in professional football (Pellman, 2003; Pellman *et al.*, 2003). Pellman *et al.* (2003) suggested that brain injury was primarily associated with translational, or linear, acceleration resulting from direct impacts on the facemask, or side; or from falls where the impact occurred on the back of the helmet. They suggested the use of severity index (SI), or head injury criteria (HIC), as the primary biomechanical predictors of MTBI (Pellman *et al.*, 2003). In addressing their original research questions, Pellman and co-workers performed some analysis with differentiation between the struck and the striking players. The NFL data is uniquely suitable to formulate a subset that allowed for looking at the effect of an impact on a struck player with impact to the front, side and back of the head, as opposed to a striking player with impact to the crown of the helmet.

The complete biomechanical assessment of head impact includes not only investigation on players' head kinematics and brain tolerances, but also improvements in the design of protective headgear for head impact sports. Although these issues have been studied, there are related areas that have not been addressed in the literature. The purpose of this paper is to address those related issues and to demonstrate their relevance.

2 Methods

Data from a study conducted in 1997–2002, and available from the NFL, have been utilized. The NFL database consists of 182 cases with 31 of the cases having been fully reconstructed (Newman *et al.*, 2003). The analysis of the NFL data included a univariate analysis of each biomechanical parameter and the injury risk curves for selective biomechanical parameters. The purpose of the univariate analysis of each biomechanical parameter was to delineate which parameters were statistically significant in the prediction of brain injury. Due to the continuous manner of independent variables (biomechanical parameters) and the binary manner of dependent variables (brain injury, or not), logistic regression analysis was employed using SAS software version 9.0 (SAS Institute, Cary NC). The effect of the mechanical characteristics of the helmet material has been assessed in conjunction with the head impact in a soccer study (Danescu *et al.*, 2003, Ziejewski *et al.*, 2001). For the purpose of this analysis, the FE elliptical head model was employed using the National Highway Traffic Safety Administration (NHTSA) brain model forehead impact for calibration purposes (Bandak, 1996). The details of the FE brain model was previously described (Ziejewski & Song, 1998a; Ziejewski & Song, 1998b).

To assess the effect of shock-attenuating material, with shear moduli from 2.0 MPa to 5.5 MPa on the rate of brain tissue deformation, complete CSDM data sets, corresponding to the strain level of 0.01, 0.02, and 0.03, were considered. The CSDM, which measures the cumulative effect of linear and angular acceleration of the head, is based on the maximum principal strain, and is calculated from an objective strain tensor to be obtained by integrating the rate of the deformation tensor that accounts for large rotations.

3 Results

The univariate analysis of each biomechanical parameter, for the struck players, is presented in Table 1. By setting p threshold as 0.05 based on Wald χ^2 test, we found that only the angular velocity change was not significant.

Although several parameters seem significant in the prediction of brain injury, not all parameters are independent of each other. Some of them are significant just because they are highly correlated with a prime parameter. From a biomechanical perspective, we may choose to look at a variety of different parameters to evaluate the potential of the brain injury.

Table 1. Univariate logistic regression.

Parameter	Coefficient (β)	Standard error	Wald χ^2	P value	−2LLR χ^2	P value	Odds ratio	95% CI
Intercept	−2.4970	1.8777	–	–	–	–	–	–
Velocity	0.4749	0.2330	4.1537	**0.0415**	4.9745	**0.0257**	1.608	1.018–2.538
Intercept	−1.4095	1.1697	–	–	–	–	–	–
SI	0.0109	0.00559	3.7903	**0.0516**	10.6595	**0.0011**	1.011	1.000–1.022
Intercept	−1.5863	1.2400	–	–	–	–	–	–
HIC	0.0146	0.00759	3.7169	**0.0539**	11.3059	**0.0008**	1.015	1.000–1.030
Intercept	−2.9414	1.8138	–	–	–	–	–	–
Peak linear acc	0.0562	0.0250	5.0326	**0.0249**	8.6374	**0.0033**	1.058	1.007–1.111
Intercept	−4.8347	2.7978	–	–	–	–	–	–
Linear velocity change	1.0569	0.5073	4.3411	**0.0372**	8.7803	**0.0030**	2.878	1.065–7.777
Intercept	−2.3449	1.6914	–	–	–	–	–	–
Peak angular acc	0.000705	0.000327	4.6347	**0.0313**	6.6627	**0.0098**	1.001	1.000–1.001
Intercept	−0.0868	1.2039	–	–	–	–	–	–
Angular velocity change	0.0501	0.0400	1.5740	**0.2096**	1.8827	**0.1700**	1.051	0.972–1.137

From the univariate logistic regression, four biomechanical parameters, namely, HIC, SI, peak translational acceleration and peak rotational acceleration, were selected for further analysis. For each parameter, an S-shape injury risk curve was determined as shown in Figure 1 through 4. For comparison, the injury risk curves predicted by each parameter for struck players, and for all players, are plotted.

Based on each injury risk curves, a set of quantified injury threshold is presented in Table 2.

For all studied biomechanical parameters, injury risk is highest for the struck players. Without a doubt, a study based on players underestimates the injury likelihood for the struck players.

4 The Influence of Headgear Shock Attenuating Material

The objective of this study was to evaluate the effects of minor impacts as opposed to catastrophic impacts. Particular attention was placed, therefore, on a brain strain level of less than 0.04. The volume fractions of critical elements for 0.03 strain levels, resulting from an

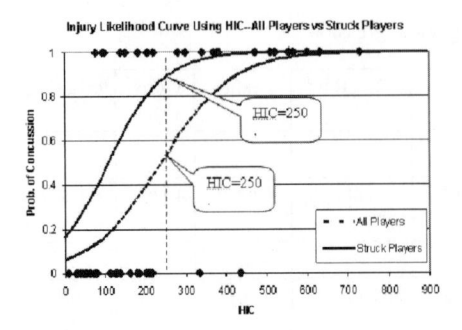

Figure 1. Injury risk curve predicted by HIC.

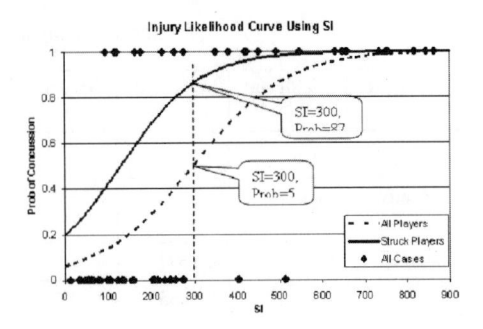

Figure 2. Injury risk curve predicted by SI.

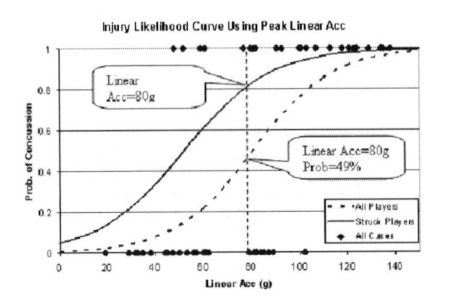

Figure 3. Injury risk curve predicted by peak transitional acceleration.

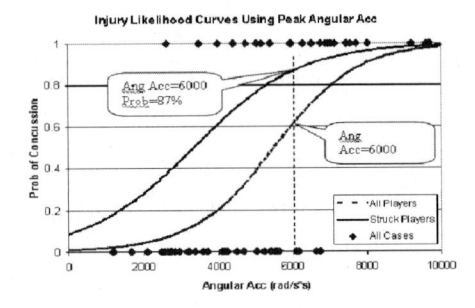

Figure 4. Injury risk curve predicted by peak rotational acceleration.

Table 2. Injury Threshold Table.

Biomechanical paramater	Probability of brain injury risk							
	50%		80%		90%		95%	
	Struck players	All players	Struck players	All players	Struck players	All players	Struck players	All players
HIC	110	240	200	360	260	440	310	500
SI	130	300	260	450	330	540	400	620
Peak translational acc. (g)	55	85	76	105	93	115	105	130
Peak rotational acc. (rad/s*s)	3300	5500	5300	7100	6600	8000	7500	8800

impact at 9.0 *m/sec,* using 2.0 *MPa* to 5.5 *MPa* shear modulus for the test material, is shown in Figure 5.

Table 3 summarizes the CSDM for selected material properties and strain levels. The CSDM represents the overall volume of the brain experiencing a particular level of strain.

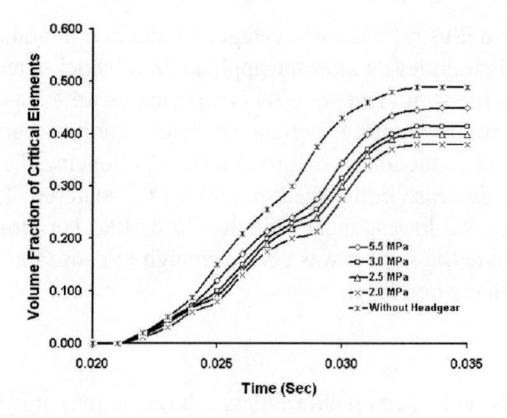

Figure 5. The brain volume fraction subjected to 0.03 strain levels after the head impact by a soccer ball without headgear and with headgears of shear moduli of 2.0, 2.5, 3.0 and 5.5 Mpa.

Table 3. The CSDM data as a function of the shear moduli of headgear material at different strain rates.

Shear modulus (Mpa)	Brain tissue strain level	Cumulative strain damage measure (%) No Shock attenuating material	Cumulative strain damage measure (%) With shock attenuating material	Reduction (%)
2.0	0.01	100	100	0
	0.02	97.5	87.5	10
	0.03	57.5	43.5	24
	0.04	27.5	23	16
2.5	0.01	100	100	0
	0.02	97.5	92	6
	0.03	57.5	45	22
	0.04	27.5	24	13
3.0	0.01	100	100	0
	0.02	97.5	93.5	4
	0.03	57.5	46.5	19
	0.04	27.5	24.8	10
5.5	0.01	100	100	0
	0.02	97.5	97.3	0
	0.03	57.5	52	11
	0.04	27.5	26	6

The data in Table 3 also illustrates the percentage of reduction in brain volume while experiencing the indicated strain levels after the application of shock-attenuating material. For comparison purposes, the strain level of 0.04 is reported as well. Based on the results, the highest reduction in brain volume, experiencing deformation, occurs at a strain level of 0.03. The percentage of reduction in brain volume, experiencing the strain level of 0.03, was reduced with the decrease in the shear moduli of the material. The highest reduction (24%) can be seen for the lowest shear moduli of 2.0 MPa. For strain levels at 0.01, no reduction occurred since the impact was severe enough to cause the entire elliptical head model to experience that strain level.

5 Discussion

Since the NFL data was collected on elite athletes, during games, it is reasonable to assume that the athletes had higher than average musculature of the neck, as well as a high level of tension. This assumption is further supported by the results of the analysis that linear and angular acceleration were highly correlated. One cannot overlook, however, the impact scenario for the weaker, or fatigued, athlete. This factor can substantially change the head kinematics with a significant increase in the angular component of the head motion and that would influence the mechanisms of the brain injury. The role of the rest of the body on head acceleration is significant, and thus, needs to be included in assessing the risk of brain injury from a given event. In a braced condition, with the entire body being correctly positioned and the muscles being tense, the player's body mass can counteract the impact of the force. By resisting the force, the head and the brain accelerate less dramatically. This explains in part, for instance, why a properly executed header, in a soccer game, can help reduce the potential for brain injury (Tysvaer *et al.*, 1991). In fact, a study on American college soccer players (Boden *et al.*, 1998) found that, although head-to-ball contact was the second most common mechanism of concussion, deliberate heading resulted in no observed conclusions.

6 Conclusions

1. Injury risk is higher for struck NFL football players in comparison to striking players.
2. For helmeted struck NFL football players, the brain injury threshold is suggested as HIC-110; SI-130; peak translational acc.-55 g; and peak rotational acc.-3300 rad/s^2.
3. The headgear made of shock attenuating material, with low shear moduli, can reduce the strain on brain tissue for players of head impact sports.

References

Boden B.P., Kirkendall D.T. and Garett W.E. (1998) Concussion Incidence in Elite College Soccer Players. *American Journal of Sports Medicine,* 26: 238–241.

Danescu R.M., Ziejewski M. and Stewart M. (2003) Practical Parameter for Characterizing the Head-to Ball Impact and Measuring the Effectiveness of Protective Headgear in Soccer. In: Hubbard M., Mehta R.D. and Pallis J.M. (Eds.) *The Engineering of Sport 5.* International Sports Engineering Association, Sheffield, UK.

Hosmer D.W. and Lemeshow S. (2000) *Applied Logistic Regression*, 2nd edition. John Wiley & Sons Inc., New York.

Ommaya A.K. (1985) Biomechanics of Head Injury: Experimental Aspects. In: Nahum A.M. and Melvin J (Eds.), *The Biomechanics of Trauma*, Appleton-Century-Crofts, Norwalk, Connecticut, pp. 245–270.

Pellman E.J. (2003) Background on the NFL's research on brain injury in professional football. *Neurosurgery*, 53: 797–798.

Pellman E.J., Viano D.C., Tucker A.M., Casson I.R. and Waeckerle J.F. (2003) Brain injury in professional football: reconstruction of game impacts and injuries. *Neurosurgery*, 53: 79–14.

Tysvaer A.T. and Lochen E.A. (1991) Soccer Injuries to the Brain: A Neuropsychologic Study of Former Soccer Players. *American Journal of Sports Medicine,* 19: 56–60.

Ziejewski M., Swenson R., Schanfield P. and Gormley M. (2001) A Biomechanical Examination of the Efficacy of Soccer Protective Headgear to Reducing Trauma to the Head from Low Impacts. *The Brain Injury Association's 20th Annual Symposium*, Atlanta, GA.

Ziejewski M. and Song J. (1998) Generation of Integrated Skull-Brain Helmet. *5th International LS-DYNA Users Conference*, Detriot MI.

Ziejewski M. and Song J. (1998) Assessment of Brain Injury Potential in Design Process of Children's Helmet Using Rigid Body Dynamics and Finite Element Analysis. *ATB User's Group Conference Sponsored by US Air Force Armstrong Laboratory*, Dayton OH.

EVALUATION OF NON-PHARMACOLOGIC INTERVENTION FOR PARKINSON PATIENT USING GAIT ANALYSIS: A CASE STUDY

R. GANASON,[1] S. JOSEPH,[1] B.D. WILSON[1] & V. SELVARAJAH[2]

[1]*Human Performance Laboratory, Sports Biomechanics Centre, National Sports Institute of Malaysia*
[2]*Baton Adventures, Malaysia*

A healthy active male (57 years) was diagnosed with symptoms of Parkinson disease stage II (Classification by Hoehn and Yahr) 4 years back. The subject walked with a high degree of shuffling of the feet when referred to the Human Performance Laboratory for gait assessment prior to a twenty (20) week's physical training intervention. A follow-up assessment was conducted five months after the intervention. The Helen Hayes Full Body protocol was administered for marker placement. Familiarization with walking on the gait walk way was provided before data were collected. The main objective during pre test was to localize dysfunction of the patients gait to aid the design of intervention; whereas the post test was conducted to document the effect of intervention on the patient's gait. Post test results show large increases in the patient's stride length and stride velocity, and a more upward shoulder orientation.

1 Introduction

Patients with Parkinson's syndromes (PD), notwithstanding the etiology of the disease, show combinations of clinical symptoms of rigidity, bradykinesia, akinesia as well as tremor. They also may experience disorder of posture, balance, and gait which often results in falls (Aita, 1982). The Parkinsonian gait is characterized by hypokinesia, associated with a reduction of stride, cadence and increase of stride, stance and double support duration (Knutsson, 1972). Most studies dealing with gait analyses on Parkinsonian patients are undertaken to quantify spatial, temporal and kinematics characteristic of their Parkinsonian gait (e.g. Morris *et al.*, 1998; Zijstra, 1998).

In almost all studies clinical scores (Hoehn & Yahr, 1967) are used to quantify the severity of the disease. It seems that PD patients are able to generate rhythmic gait under special conditions (Morris *et al.*, 1998). However, PD patients experienced marked deterioration in their gait pattern when they were required to perform either a motor or cognitive secondary task at the same time as walking (O'Shea *et al.*, 2002). Physical therapy plays a major role with PD patients to help them walk with normal step size and speed (O'Shea *et al.*, 2002).

This study has two purposes, first to identify and describe parameters associated with dysfunction during gait to help design an intervention in addressing symptoms; and secondly to measure and document the effects of the intervention on these identified parameters.

2 Materials and Methods

2.1. *Case History*

A healthy active male (57 years) who participates in regular weight training and swimming (30 minutes of gym training and 1 km swim per day) was diagnosed with symptoms of

Parkinson disease stage II (Classification by Hoehn & Yahr, 1967) 4 years back. The patient was on non-vegetarian diet and never had any significant changes in his movements in varied weather conditions. The present treatment therapy of the patient included one and half dopamine tablets (75 mg) per day. The patient was referred for gait analysis before and after a strength training intervention. At the time of the pre-test the subject had a high degree of foot shuffling in gait and no history of injury. Tremors in the right upper extremities and at the right thigh were observed when the subject rested on a chair. Tremors seemed to be more one sided on the right extremities rather than on the left extremities. Severity and frequency of the tremors were not severe in nature.

2.2. *Assessment and Protocol*

A pre and post gait test were administered with data collected using a Motion Analysis Inc system with six fixed infra-red Eagle cameras and two AMTI force platforms in a gait way. Clinically validated Orthotrak software was used to analyse the data. The patient preparation was done using the Helen Hayes Full Body marker protocol with familiarization trials given to the subject prior to data collection to minimise learning effects and ensure that the patient walks at his normal walking speed.

2.3. *Intervention*

A twenty (20) weeks physical training intervention in two phases was administered after the preliminary gait test. The first phase of the training programme was twelve (12) weeks of physical conditioning with a training frequency of two (2) times per week. This was to enable the patient to be familiarized with the variety of exercise techniques and movements for flexibility and strength development. The recommended flexibility exercises included static and facilitated stretching for tight muscles causing stiffness. These exercises were concentrated on the flexor muscles (i.e pectoralis, abdominalis, iliopsoas, hamstrings) as the patient often seems to be in a hip flexed position. Strengthening exercises were focused on the extensor muscles (i.e trapezius, latisimus dorsi, and gluteus) as they showed extreme weakness. The training programme in this phase also includes swimming, multiple exercises in water, and slope walking. All these exercises were to ensure a base level of overall fitness while focusing on correcting the forward flexion in gait problem.

After the completion of the 12 weeks of conditioning, the patient progressed to an eight (8) week resistance training program. The resistance training program consisted of ten different exercises at intensities of 60% of 1 RM with each prescribed exercise performed for fifteen repetitions. The exercise routine was one set of the 10 exercises for the flexor muscles (tight muscles) and two sets for the extensor muscles (weak muscles). Training frequency was two (2) times per week and all these exercises conducted began with the load of the patient's own body weight.

3 Results of the Study

Subject anthropometric measurements of the lower limbs are presented in Table 1.
There were no marked differences between the left and right side measurements
There were large increases in step length (more marked for the left side than the right), stride length, forward velocity and cadence post test compared to pre test values indicating the subject was adopting a "more normal" gait.

Table 1. Anthropometrics measurements of lower extremities (cms) in relaxed standing posture.

Area	Right (cm)	Left (cm)
Thigh circumference	53.5	52.7
Calf circumference	34.9	35.7
Knee diameter	12.4	12.5
Ankle diameter	7.2	7.4
Feet length	27.2	27.0
Feet width	9.8	9.9

Table 2. Kinematic and temporal data (mean ± standard deviation of 6 strides) and percent changes between Pre and Post-Therapy Intervention.

Parameters	Right side		Left side	
	Pre therapy	Post therapy	Pre therapy	Post therapy
Step length (cm)	65 ± 1	71 ± 1 (9.5%)	58 ± 3	74 ± 4 (27.5%)
Stride length (cm)	124 ± 4	145 ± 5 (16.6%)	123 ± 4	143 ± 3 (15.9%)
Forward velocity (cm/s)	103 ± 4	142 ± 5 (37.6%)	104 ± 2	142 ± 3 (36%)
Cadence (steps/min)	99 ± 2	116 ± 1 (16.6%)	101 ± 3	116 ± 2 (14.9%)

Measures of knee and arm joints motion data are represented in Figure 1 and 2 respectively.

The subject exhibited a higher degree of knee flexion in the post test than the pre test during pre swing and weight bearing.

Graph clearly reveals improvement in shoulder flexion; and a much balanced and improved elbow extension in post test.

4 Discussion

The intervention of a two-phased, twenty (20) week conditioning and physical training have contributed markedly to changes in the patient's gait. Post data revealed an increase in stride velocity by minimum 36% and stride length by 16% in both strides. These findings are further supported with increases in knee flexion and shoulder flexion, elbow extension during both strides that suggest a more confident/aggressive gait of the patient. The marked improvement was in contrast to pre test gait where the patient exhibited minimal flexion in the knee during pre swing and load bearing phases and exhibited bradykinesia (general slowness in movement) during weight bearing and push off phases.

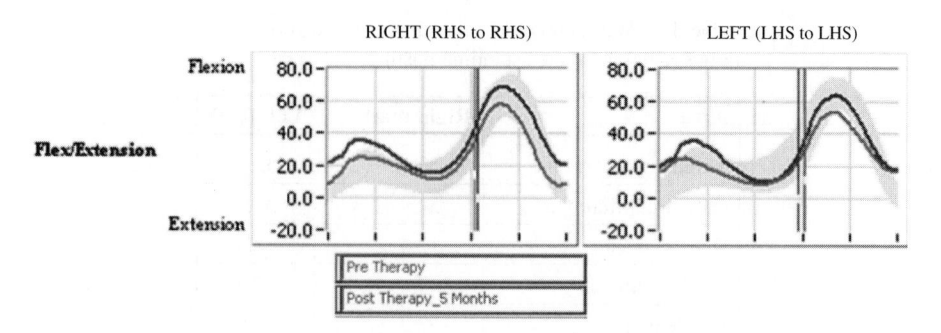

Figure 1. Knee joint angles (deg).

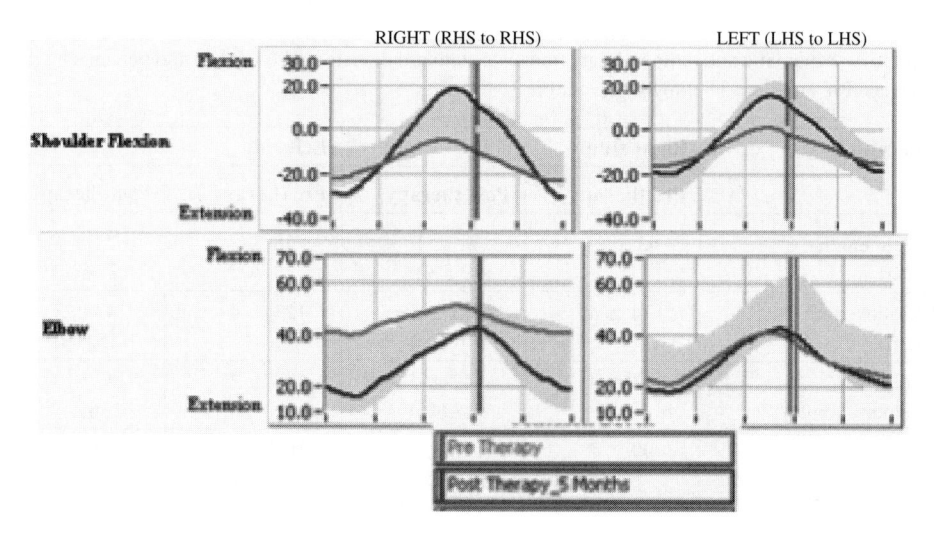

Figure 2. Arm joint angles (deg).

5 Conclusion

The primary goal in the treatment of PD is to reduce the intensity of the symptoms, as there is no cure for Parkinson's disease. The goals of treatment of Parkinson's disease may be considered as two-fold: (1) short-term treatment to alleviate symptoms and reverse functional disability; and (2) long-term treatment to maintain effectiveness and limit the complications of therapy. This study has used gait analysis to show the effectiveness of a 20 week physical training therapy on alleviating symptoms and restoring some normality to the patient's gait.

References

Aita J.F. (1982) Why patients with Parkinsonian's disease fall. *JAMA*, 247, 515–516.
Hoehn M. and Yahr M. (1967) Parkinsonism: onset, progression and mortality. *Neurology*, 17/5, 427–442.

Knutsson E. (1972) An analysis of Parkinsonian gait. *Brain*, 95, 475–486.
Mitoma H. (1998) Kinematics and EMG pattern of Parkinsonian gait. *Nippon Rinsho*, 55/1, 163–167.
O'Shea S., Morris M.E. and Iansek R. (2002) Dual task interference during gait in people with Parkinson disease: effects of motor versus cognitive secondary tasks. *Physical Therapy*, 82, 888–897.
Zijlstra W. (1998) Voluntary and involuntary adaptation of gait in Parkinson's disease. *Gait and Posture*, 7, 53–63.

THE TIME-COURSE OF ACUTE CHANGES IN ACHILLES TENDON MORPHOLOGY FOLLOWING EXERCISE

S.C. WEARING,[1,2] J.E. SMEATHERS,[2] S.R. URRY[2] & S.L. HOOPER[1]

[1]*Centre of Excellence for Applied Sports Science Research, Queensland Academy of Sport, Brisbane, Australia*
[2]*Institute of Health and Biomedical Innovation, Queensland University of Technology, Brisbane, Australia*

In comparison to bone and muscle, little is known about the acute effect of strain stimulus on the morphology of human tendon, *in vivo*. The purpose of the current study was to employ sonographic imaging to investigate the acute change in the diameter of the Achilles tendon in response to a bout of heavy resistive exercise. In contrast to previous research, tendon diameter was markedly reduced immediately after exercise. Although tendon dimensions returned to pre-exercise levels within 24 hours, the primary recovery time appeared to occur between 6 and 12 hours post-exercise. The findings have potential applications for the design of physical activity and rehabilitation programs, however further research is required before exercise targets can be optimized for individual athletes.

1 Background

Mechanical strain is central to both Wolff's law and Frost's "Mechanostat" model of bone. However, few empirical studies have considered the adaptation of soft connective tissue structures, such as tendon, to load. Although there is evidence that markers of tendon inflammation and collagen synthesis may peak as late as 24 hours post-exercise (Langberg *et al.*, 2002; Miller *et al.*, 2005), there has been minimal systematic research conducted on the acute effect of strain stimulus on the *in vivo* properties or morphology of human tendon. In one of the few studies performed thus far, Kubo *et al.* (2002) noted a 27% decrease in the Achilles tendon stiffness immediately following five minutes of isometric plantarflexion exercise and hypothesized that mechanical loading invoked a biphasic response in tendon, in which post-exercise remodeling resulted in a transient period of mechanical weakness prior to full recovery. While the study did not include measures of tendon morphology to support their claim, a recent investigation employing magnetic resonance imaging (MRI) has shown a transient increase in Achilles tendon volume (12%–17%) and intratendinous signal (23%–31%) 30 minutes after exercise (Shalabi *et al.*, 2004a; Shalabi *et al.*, 2004b). The authors proposed that the greater dimensions and increased water content served a protective mechanism, which effectively lowered the stress within the tendon during exercise. However, this is not a biologically plausible mechanism, as an increase in the water content of tendon would not influence the tensile stress borne by the collagen fibres, but would influence tendon diameter and apparent stiffness (modulus). Moreover, given that both experiments employed cross-sectional study designs, the time-course of the adaptive change in tendon morphology to exercise remains unknown. The aim of the current study, therefore, was to describe the time-course of the acute change in the sagittal thickness of the Achilles tendon in response to a bout of resistive exercise.

2 Methods

2.1. Subjects

A convenience sample of 8 healthy adult males (mean age 42 ± 10 years, height 1.79 ± 0.06 m and weight 78.1 ± 11.1 kg) from university staff and students participated in the study. Volunteers were excluded from entering the study if they reported a history of Achilles tendon pain within the previous 24 months, or presented with a medical history of inflammatory joint disease or surgery involving the lower limb. Participants were requested to refrain from vigorous physical activity for 24 hours prior to and throughout the testing period. The study received clearance from the University Human Research Ethics Committee and adhered to the National Statement on Ethical Conduct in Research Involving Humans.

2.2. Methods

Study participants underwent sonographic examination of the Achilles tendon by a single operator experienced in musculoskeletal imaging. Achilles tendon diameter was determined from sagittal sonograms of the tendon acquired with a 10–5 MHz linear array transducer (LOGIQ Book XP, GE Healthcare). Images were obtained using a stand-off pad while the participant lay prone with the ankle dorsiflexed at 90° to the leg. B mode ultrasound scans of the Achilles tendon were acquired and cine-images were stored to PC using Digital Imaging and Communications in Medicine (DICOM) format for post-processing. The sagittal thickness (diameter) of the Achilles tendon was measured at a standard reference point, 1 cm from the proximal calcaneal insertion. The coefficient of variation for repeated measures of tendon dimensions using this technique is 1.1% (Ohberg et al., 2004). Following baseline measures, subjects completed a series of bilateral standing ankle plantarflexion exercises (five sets at 15–20 repetitions/set) at 250% bodyweight. The exercise was performed from full ankle plantarflexion to full dorsiflexion with the knee in an extended position (Yanagisawa et al., 2003) and was estimated to induce tensile loads similar to that calculated for the Achilles tendon during running (Scott & Winter, 1990). Tendon measures were acquired immediately on completion of the exercise and repeated at 3, 12 and 24 hours post-exercise.

3 Results

Figure 1 demonstrates the change in Achilles tendon diameter over 24 hours following exercise. While tendon diameter decreased (~15%) immediately after resistive plantarflexion exercise, it returned to pre-exercise levels within 24 hours. The primary recovery time, the time to return to 63% of pre-exercise dimensions, occurred between 6 and 12 hours post-exercise.

4 Discussion

In contrast to the findings of previous studies employing magnetic resonance imaging (Shalabi et al., 2004a; Shalabi et al., 2004b), the current study observed a marked decrease in tendon thickness immediately following a bout of heavy resistive plantarflexion exercise. Based on the work of Kastelic and Baer (1979), collagen realignment and a reduction in the crimp angle could account for only up to 5% of the observed reduction in dimensions. We propose, therefore, that the acute changes in tendon diameter post-exercise likely reflect fluctuations in the fluid content of the tendon matrix. *In vitro* studies have shown that water

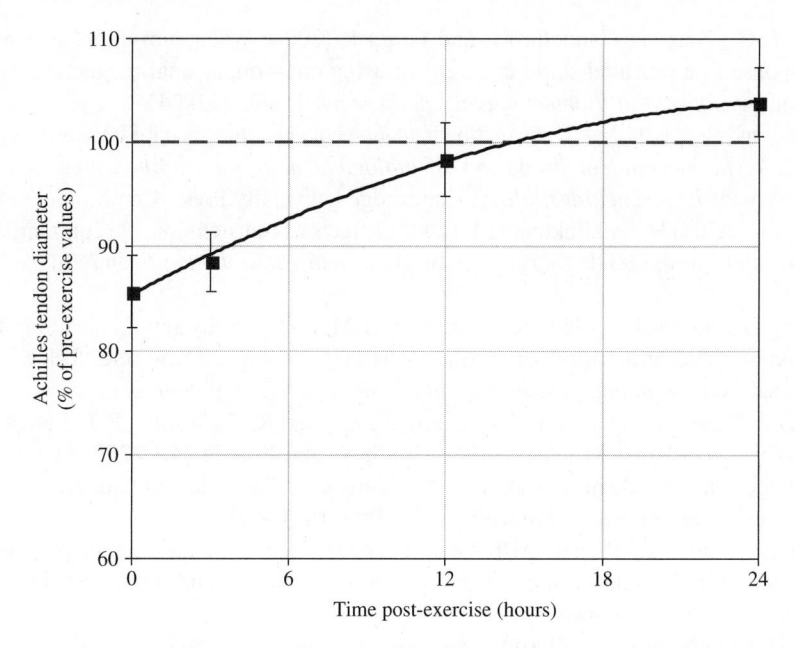

Figure 1. Change in tendon diameter post-exercise, expressed as a percentage of pre-exercise values.

is lost from animal tendon when exposed to either static or cyclic loading (Hannafin & Arnoczky, 1994). Static tensile loading is thought to promote the rapid unbinding of water from tendon matrix and the radial extrusion of water from the tendon core (Han *et al.*, 2000; Wellen *et al.*, 2005). Unloading, in contrast, is thought to permit the return of water to the tendon core via an osmotic driven diffusion process (Helmer *et al.*, 2006). Such changes in the fluid content following exercise would alter the cross sectional area and apparent elastic modulus of tendon but would not affect the tensile stress borne by collagen fibres. Further research evaluating fluid diffusion and the time-course of the acute adaptive response of tendon to a varied strain stimulus is required before exercise and rehabilitation programs can be tailored to individual athletes.

Acknowledgements

This research was funded by a grant from the Queensland Academy of Sport. The authors would like to thank Prof. JH Evans for his technical advice.

References

Han S., Gemmell S.J., Helmer K.G., Grigg P., Wellen J.W., Hoffman A.H. and Sotak C.H. (2000) Changes in ADC caused by tensile loading of rabbit achilles tendon: evidence for water transport. *Journal of Magnetic Resonance,* 144/2, 217–227.

Hannafin J.A. and Arnoczky S.P. (1994) Effect of cyclic and static tensile loading on water content and solute diffusion in canine flexor tendons: an in vitro study. *Journal of Orthopaedic Research,* 12/3, 350–356.

Helmer K.G., Nair G., Cannella M. and Grigg P. (2006) Water movement in tendon in response to a repeated static tensile load using one-dimensional magnetic resonance imaging. *Journal of Biomechanical Engineering,* 128/5, 733–741.

Kastelic J. and Baer E. (1979) Deformation in tendon collagen. In Vincent J.F.V. and Currey J.D. (Eds.) *The mechanical properties of biological materials: 34th Symposium of the Society for Experimental Biology.* Cambridge University Press, Cambridge.

Kubo K., Kanehisa H. and Fukunaga T. (2002) Effects of transient muscle contractions and stretching on the tendon structures in vivo. *Acta Physiologica Scandinavica,* 175/2, 157–164.

Langberg H., Olesen J.L., Gemmer C. and Kjaer M. (2002) Substantial elevation of interleukin-6 concentration in peritendinous tissue, in contrast to muscle, following prolonged exercise in humans. *Journal of Physiology,* 542/3, 985–990.

Miller B.F., Olesen J.L., Hansen M., Dossing S., Crameri R.M., Welling R.J., Langberg H., Flyvbjerg A., Kjaer M., Babraj J.A., Smith K. and Rennie M.J. (2005) Coordinated collagen and muscle protein synthesis in human patella tendon and quadriceps muscle after exercise. *Journal of Physiology,* 567/Pt 3, 1021–1033.

Ohberg L., Lorentzon R. and Alfredson H. (2004) Eccentric training in patients with chronic Achilles tendinosis: normalised tendon structure and decreased thickness at follow up. *British Journal of Sports Medicine,* 38/1, 8–11.

Scott S.H. and Winter D.A. (1990) Internal forces of chronic running injury sites. *Medicine and Science in Sports and Exercise,* 22/3, 357–369.

Shalabi A., Kristoffersen-Wiberg M., Aspelin P. and Movin T. (2004a) Immediate achilles tendon response after strength training evaluated by MRI. *Medicine and Science in Sports and Exercise,* 36/11, 1841–1846.

Shalabi A., Kristoffersen-Wiberg M., Svensson L., Aspelin P. and Movin T. (2004b) Eccentric training of the gastrocnemius-soleus complex in chronic achilles tendinopathy results in decreased tendon volume and intratendinous signal as evaluated by MRI. *The American Journal of Sports Medicine,* 32/5, 1286–1296.

Wellen J., Helmer K.G., Grigg P. and Sotak C.H. (2005) Spatial characterization of T1 and T2 relaxation times and the water apparent diffusion coefficient in rabbit Achilles tendon subjected to tensile loading. *Magnetic Resonance in Medicine,* 53/3, 535–544.

Yanagisawa O., Niitsu M., Yoshioka H., Goto K. and Itai Y. (2003) MRI determination of muscle recruitment variations in dynamic ankle plantar flexion exercise. *American Journal of Physical Medicine and Rehabilitation,* 82/10, 760–765.

DEVELOPMENT OF PORTABLE EQUIPMENT FOR CARDIORESPIRATORY FITNESS MEASUREMENT THROUGH A SUB-MAX EXERCISE

K.R. CHUNG, S.Y. KIM, G.S. HONG, J.H. HYEONG & C.H. CHOI
Fusion Technology Center, Korea Institute of Industrial Technology, Cheonan, Chungnam, South Korea

The purpose of this study was to develop an accurate and easily administrative submaximal test for the prediction of maximal oxygen uptake by using measured pulse rates during the submaximal exercise. Twenty-two healthy subjects (22 men, mean ± SD age: 20.2 ± 3.3 years, height: 173.5 ± 7.2 cm, weight: 67.8 ± 7.7 kg) participated in the study. Each subject completed a 30-squat submaximal exercise test and a treadmill VO_2max test (graded exercise test; GXT) on the same day, with a 20-minute rest period between each test. Multiple regression analyses were used to develop an equation by using the measured heart rates during the submaximal exercise test. Only variables related to factors with heart rate were included to the independent variable as predictive factors in the analysis. The regression equation developed from the submaximal test was as follows: VO_2max (ml/kg/min) = 83.893 + 6.102 × (HR1) − 10.328 × (HR2) + 3.921 × (HR3) + 6.214 × (HR12) + 3.947 × (HR23) + 0.065 × (HR13) with R^2 = 0.535 and SEE = 3.0 ml/kg/min. Predicted and measured mean of VO_2max was 51.3 ± 3.6 ml/kg/min and 51.3 ± 4.4 ml/kg/min. The correlation coefficient between predicted peak oxygen consumption (VO_2max) and measured one was r = 0.82, and there were no significant differences ($p < 0.01$). The predicted VO_2max from submaximal exercise test (30-squat) showed good reliability and concurrent validity with the measured VO_2max from GXT. Results indicate that the developed portable fitness test equipment can be used for measuring cardio-respiratory fitness by estimating VO_2max with simply designed sub-maximal exercise test protocol.

1 Introduction

Cardiorespiratory endurance, one of the health-related components of physical fitness, is the ability of circulatory and respiratory systems to supply energy during physical activity. It involves the ability of the heart and lungs to supply oxygen to the working muscle for an extended period of time, which is a good indicator of aerobic capacity and health-risk status (Blair *et al.*, 1989). It is also related to the body's ability to use the large muscles for prolonged periods of dynamic, moderate-to-high intensity exercise.

In order to assess aerobic power, maximal oxygen uptake capacity (VO_2max) is widely used as an indicator of the power, and many methods, both directly and indirectly, to measure the VO_2max were developed (Keren *et al.*, 1980). Direct measurement of VO_2max has been considered to be the most accurate means of determining aerobic fitness level using graded exercise test (GXT); however, it contains many restrictions such as long measurement time, high measurement cost, dangerous factors to perform labs, etc (Matthew *et al.*, 2004). Due to these limitations of direct measurement method predicting VO_2max, many methods for indirect measurement, submaximal exercise test protocols, have been developed, and numerous studies have shown the correlation between predicted and measured VO_2max utilizing some of indirect assessment methods: 1-mile walking test, 12 minute Run test, Harvard step test, Astrand Rhyming cycle ergometer test, YMCA cycle ergometer test etc (Astrand & Rhyming, 1954; Kline et al., 1987; Swank et al., 2001). However, most of standardized maximal and submaximal exercise tests and related endurance tests are needed test equipments such as treadmill, cycle ergometer, test step, elliptical trainer, gas analyzer etc.

The purpose of this study was to develop an accurate and easily administrative submaximal test for the prediction of maximal oxygen uptake using measured pulse rates during the submaximal exercise (30-squat).

2 Methods

2.1. Participants

Healthy community-dwelling people (22 men) aged from 18 to 29 years were recruited on a volunteer basis. The physical characteristics of the subjects are presented in Table 1. Subjects signed an informed consent and completed a health-history questionnaire, who confirmed the screening criteria (e.g., diagnosis of cardiovascular or musculoskeletal problems) before participating in the study.

2.2. Exercise Testing

The subjects were instructed to refrain from severe activity 12 hours before testing and to avoid eating food and smoking 4 hours before testing. Each subject completed the submaximal exercise test (30-squat) and maximal exercise (VO_2max) test on the same day.

Submaximal exercise test. Before the commencement of 30-squat submaximal test, the subjects were instructed how to execute the test and allowed to warm-up for 3~5 minutes. After the warming-up, subjects took a rest until recovering the steady state condition, which was checked by monitoring heart rate. With beeper sound of a metronome subjects started 30-squat exercise and each squat (1.5 sec) was controlled by the metronome to ensure a consistent squat among subjects. During the submaximal exercise test heart rate was measured at 3 times, before and after the 30-squat execution. First measurement (HR1) was completed in the resting state, before the exercise. Second measurement (HR2) was done after the 30th squat, and third measurement (HR3) was finished after a 50-second rest period followed by the completion of second measurement.

Table 1. Physical characteristics of subjects ($N = 22$).

Variable	Mean	SD	Minimum	Maximum
Age (years)	20.2	3.3	18	29
Weight (kg)	67.8	7.7	58	90
Height (cm)	173.5	7.2	157	184
HRmax (bpm)	194.1	6.1	181	204
VO_2max(ml/kg/min)	51.3	4.4	43.3	59.7

Maximal exercise test. Peak oxygen consumption (VO$_2$max) was measured by using modified Bruce protocol on a motorized treadmill. Before the commencement of the maximal exercise test, subjects practiced walking at 1.7 mph at 0% grade for 3 minutes, after which the protocol began. The protocol began with walking at 1.7 mph at 10% grade for 3 minutes (stage 1), and the stage increased with 0.8~0.9 mph in speed, 2% in grade until subject's exhaustion. Test was considered maximal when at least two of the following criteria were achieved; 1) failure to maintain the work rate, 2) maximal respiratory exchange ratio (RER) ⩾1.1, 3) no further increase (a plateau) in VO$_2$ with increases in workload.

2.3. Statistical Analyses

In order to develop a regression model of cardiorespiratory fitness estimation, maximal oxygen consumption (VO$_2$max) was adopted as a dependent variable, and height, weight, body mass index, heart rate were adopted as an independent variables. Among the independent variables heart rate was divided into 6 different independent variables as the followings to analyze subject's physiological responses more precisely to the suggested certain conditions during the submaximal test. HR1, HR2, and HR3 are from the three measured heart rates during the 30-squat test, and HR12 (HR2-HR1), HR23 (HR2-HR3), and HR13 (HR3-HR1) are from calculated heart rates among the three measured heart rates.

Multiple regression analysis was performed to develop an equation for estimating VO$_2$max. Before the regression analysis Pearson *r* was calculated to analyze the correlation between independent variable such as height, weight, body mass index, heart rates of abovementioned each and dependent variable such as VO$_2$max. All analyses were performed with Statistical Package for the Social Sciences, Version 12.0 (SPSS Inc., Chicago, IL).

3 Results

The descriptive data used as independent variables from the result of submaximal exercise test to develop an equation in multiple linear regression analysis for estimating maximal

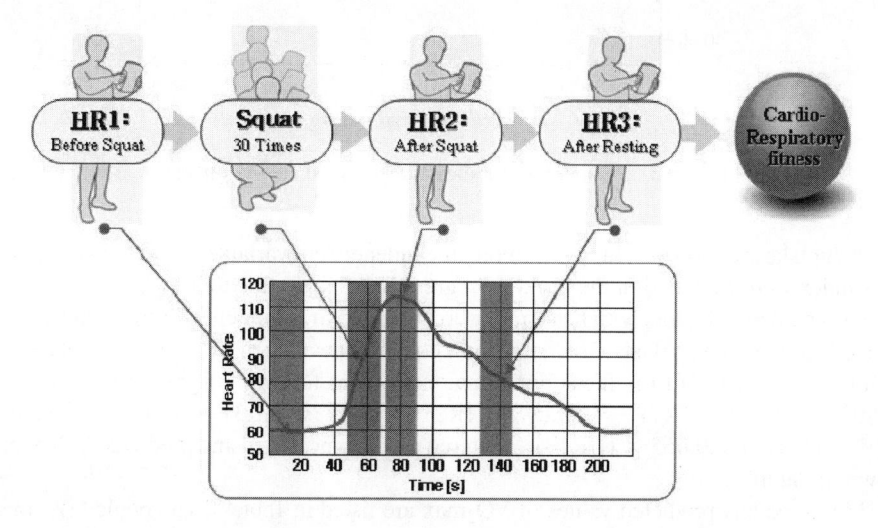

Figure 1. Procedures of 30-squat exercise and heart rate measurement.

Table 2. Independent variables included in the multiple regression analysis.

Variable	B	Sig T	R^2	Sig F	Std. error
Constant	83.893	0.000			
HR1	6.102	0.007			
HR2	−10.328	0.003			
HR3	3.921	0.110	0.535	0.005	3.00
HR12	6.214	0.007			
HR23	3.947	0.109			
HR13	0.065	0.455			

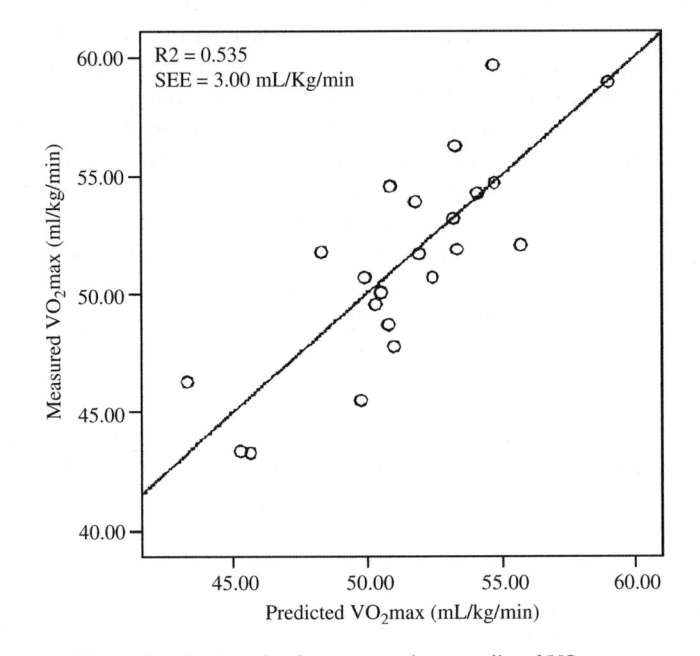

Figure 2. Scatter plot for measured vs. predicted VO$_2$max.

oxygen uptake are shown in Table 2. Among the independent variables height, weight, body mass index were excluded in the multiple regression analysis due to low correlation. Only variables related to factors with heart rate were included to the independent variable as predictive factors in the analysis. The regression equation developed ($R^2 = 0.535$ and SEE = 3.0 ml · kg^{-1} · min^{-1}) from the submaximal test was as follows: VO$_2$max (ml/kg/min) = 83.893 + 6.102 × (HR1) − 10.328 × (HR2) + 3.921 × (HR3) + 6.214 × (HR12) + 3.947 × (HR23) + 0.065 × (HR13). Scattered plot for measured and predicted VO$_2$max is shown in Figure 2.

Measured and predicted values of VO$_2$max are listed in Table 3. The pooled (22 men) mean predicted VO$_2$max using the 30-squat test was 51.3 ± 3.6 ml · kg^{-1} · min^{-1}, which was

Table 3. Measured and predicated values of VO_2max ($N = 22$).

Test	M	SD	Range
Measured	51.3	4.4	43.3~59.7
Predicted	51.3	3.6	43.3~58.9

similar to the measured VO_2max (51.3 ± 4.4 ml $\cdot$ kg$^{-1}\cdot$min^{-1}). Pearson's correlation coefficient between measured and predicted VO_2max values was $r = 0.817$ at the 0.01 level.

4 Conclusion

In this study we developed a portable equipment which can measure the heart rate changes during a designed exercise test procedure, "30-squat exercise test", and performed a preliminary clinical study to develop a simplified method to estimate the maximal oxygen uptake capacity. By multiple regression analysis with heart rates measured at three stage (resting, right after the given exercise, and after 50s additional resting) we derived a regression equation to estimate VO_2max. Results show that the mean of estimated values of the test group is closely related to the mean of measured VO_2max of the group, with no significant differences. To verify the validity of usefulness of developed fitness test, we tried to classify the cardio-respiratory fitness level of each subject with percentile values (10 grades) according to the ACSM references. 36.4% (8 subjects) of subjects showed the same 'estimated' fitness level compared to the "measured" one. 59.1% (13 subjects) of subjects registered only one grade difference between the two assessments, and only 4.9% (1 subject) of subjects registered two grades difference. This indicates that the developed simplified exercise test can be used for determination of cardio-respiratory fitness level indirectly.

Acknowledgements

This study was supported by an internal grant funding from the Korea Institute of Industrial Technology. The authors thank our colleagues in the Department of Health and Exercise Science of Namseoul University for their efforts in collecting data.

References

Astrand P.O. and Rhyming I. (1954) A nomogram for calculation of aerobic capacity (physical fitness) from pulse rate during submaximal work. *Journal of Applied Physiology*, 7, 218–221.

Blair S.N., Kohl H.W., Paffenbarger R.S., Cark D.G., Cooper K.H. and Gibbons L.W. (1989) Physical fitness and all-cause mortality: a prospective study of healthy men and women. *Journal of the American Medical Association*, 22, 2395–2401.

Dalleck L.C., Kravitz L. and Roberg R.A. (2006) Development of a submaximal test to predict elliptical cross-trainer VO_2max. *Journal of Strength and Conditioning Research*, 20/2, 278–283.

Keren G., Magaznik A. and Epstein Y. (1980) A comparison of various methods for the determination of VO_2max. *European Journal of Applied Physiology*, 45, 117–124.

Kline G.M., Porcarik J.P., Hintermeister R., Freedson P.S., Ward A., McCarron R.F., Ross J. and Rippe J.M. (1987) Estimation of VO_2max from a one-mile track walk, gender, age, and body weight. *Med. Science of Sports Exercise*, 19, 253–259.

Matthew D.B., William F.B., Diego V.H., Linda G., Galila W. and Michael L.P. (2004) Cross-validation of the YMCA submaximal cycle ergometer test to predict VO_2max. *Research Quarterly Exercise and Sport*, 75/3, 337–342.

Swank A.M., Serapiglia L., Funk D., Adams K.J., Durham M. and Berning J.M. (2001) Development of a branching submaximal treadmill test for predicting VO_2max. *Journal of Strength and Conditioning Research*, 15, 302–308.

GENETIC PROGRAMMING FOR KNOWLEDGE EXTRACTION FROM STAR EXCURSION BALANCE TEST

S.N. OMKAR,[1] K.M. MANOJ,[1] M. DIPTI[1] & K.J. VINAY[2]
[1]*Indian Institute of Science, Bangalore, India*
[2]*Indian Institute of Technology, Chennai, India*

Athletes are subjected to extreme physical conditions on various parts of the body depending on the nature of the sport. Timely checkup and suitable actions are essential to maintain the required level of fitness. Star Excursion Balance Test (SEBT) is a functional test to assess the dynamic balance and lower body stability of athletes. The knee condition plays a significant role on stability, and hence on the outcome of the SEBT results. The SEBT results being high dimensional, gathering information about the knee condition from it becomes very difficult for physiotherapists. Knowledge extracted from the existing data will assist the doctors to analyze the SEBT results and diagnose the athletes better. This paper presents an evolutionary approach to evolve an expression that can deduce useful information from the SEBT results. Genetic Programming (GP) is used to evolve expressions in terms of the parameters of the SEBT test. These results reduce the complexity by recognizing the critical variables and aiding the physiotherapists. Based on the expressions obtained, cluster plot analysis is also carried out and the results are presented.

1 Introduction

Humans use three basic mechanisms to obtain a sense of balance in daily life. The three mechanisms (visual, vestibular, and proprioceptive) interact to maintain posture and impart a conscious sense of orientation. A defect in one of these systems decreases the subject's overall ability to adjust to incongruous stimuli. Proprioceptive function can be tested by a number of balance tests. Star Excursion Balance Test (SEBT) is one such functional test that is used to assess dynamic balance and lower body stability. It integrates a single-leg stance of one leg with maximum reach of the other leg. Efficacy and reliability of this test has been established previously (Hertel *et al.*, 2006; Olmsted *et al.*, 2002). The SEBT involves a participant to maintain a base of support with one leg, while maximally reaching in different directions with the opposite leg, without compromising the base of support of the stance leg (Hertel *et al.*, 2006; Gribble & Hertel, 2003). As SEBT tests involves only standing and stretching, they offer a simple, reliable, low-cost alternative to more sophisticated instrumented methods that are currently available. It has been shown that the result pattern for a healthy ankle is different than an injured/ recovered ankle (Olmsted *et al.*, 2002; Gribble & Hertel, 2003; Hertel *et al.*, 2006) and that the SEBT results are significantly influenced by the health of the ankle (Hertel *et al.*, 2006). Since SEBT is a functional test or an index of lower body stability, it is evident that the health of the knee also plays an important role in influencing the results obtained. In this paper, the implication of SEBT results concerning the knee's health is being considered. Since the data is high dimensional (eight), it is too intricate for an individual to analyze the data efficiently without any aid. Knowledge Extraction can give fairly good simulations and predictions for such domains. It implies that the system is able to deduce some characteristics and relations which are not visible at first glance. It is this information or the hidden pattern in the presented data that we want to know and express in words. This is the main intention behind extracting knowledge from raw data, which is by nature, very hard to contemplate.

Knowledge Extraction is an interdisciplinary procedure focusing upon methodologies for discovering and extracting implicit, previously unknown and potentially useful knowledge (rules, patterns, regularities as well as constraints) from data. Ideally this process results in a symbolic description, which closely mimics the behavior of a system in a concise and comprehensible form. In modern world, data is abundantly available in various formats (text, image, audio and video), and has been gathered and stored in massive databases or data warehouses. The challenges lie in extracting knowledge from this data and use it for predicting trends and improving decisions. This simply reflects the importance of Knowledge Extraction methodologies and their applications in the present scenario (Facca & Lanzi, 2005; Remma & Alexandrea, 2002).

Knowledge can be extracted by implementing various methods like fuzzy based models (Castellano *et al.*, 2005), artificial neural networks (Duch *et al.*, 2000), Ant Colony Optimization (Parpinelli *et al.*, 2002), Genetic algorithm (Chen & Hsu, 2006) and Genetic programming.

Genetic Programming (GP) introduced by Koza and his group in 1992 (Koza, 1992) is very popular for its ability to learn relationships hidden in data and express them automatically in the form of mathematical expressions. GP has already spawned numerous interesting applications such as (Poli, 1996; Esparcia *et al.*, 1997; Kishore *et al.*, 2000).

Knowledge extraction from the SEBT results using GP could offer an elucidation by providing a symbolic link between inputs i.e. SEBT results and outputs/ classes. The extraction of easily interpretable knowledge in the form of an expression from the large amount of data measured in SEBT is well desirable.

The original idea and goal of this research paper is to use the expression evolved by GP in order to serve as an aid to physiotherapists and doctors during their normal everyday practice to analyze their subjects better. When the number of parameters being considered is high, recognizing of significance of critical variables is quite difficult. Further, determining the individual effects of these parameters on the outcome becomes too complex. The expression evolved by GP zeros in on critical variables and conveys a lot of information and will prove to be very useful to the professionals in their daily practice.

2 Material

2.1. *Subjects*

A star board was constructed on level ground, in a yoga centre facility as explained by Hertel *et al.* (2000).

Fifty subjects from various walks of life enrolled in a yoga centre volunteered to undergo SEBT. Two groups- healthy and unhealthy are formed based on their knee condition. Volunteers were grouped into unhealthy, if they have had (1) at least one episode of an acute knee injury but none within the past 6 weeks or, (2) multiple episodes of the knee giving way within the past 12 months. Healthy group consisted of volunteers who had no history of injury to either knee.

2.2. Knowledge Acquisition

A verbal and visual demonstration of the SEBT test procedure is given to each subject by the examiner. Each subject is allowed to perform 6 practice trials in each of the 8 directions for each leg to become familiar with the task, as recommended by Hertel *et al.* (2000).

It is to be noted that the trunk is kept in upright position i.e., the pelvic and pectoral girdle balance is maintained. The trunk should not tilt either forward or sideward during the test.

Normalization is performed by dividing each excursion distance by the participant's leg length, and then by multiplied by 100 (Gribble & Hertel, 2003). Normalized values can thus be viewed as a percentage of excursions distance in relation to a participant's leg length. The normalized excursion distances in each of the eight directions are tabulated with their appropriate group. An extract from the knowledge repository is shown in Table 1.

2.3. Genetic Programming for Two Class Problem

Genetic Programming (GP; Koza, 1992) is an evolutionary approach which applies the Darwin's principle of survival of the fittest to a population of parametric solution to a given problem. GP evolves a population of computer programs, which are possible solutions to a given problem. GP uses biologically inspired operations like reproduction, crossover and mutation to generate the next generation. The solution is evolved through the generations.

For a two class problem, a possible classifier or an individual is generally represented by a single tree (T) (Koza, 1992). The single tree representation of the classifier is sufficient for a two-class problem. For a pattern

$$\text{if } T(x) \geq 0, x \in \text{class 1}$$

$$\text{else, } x \in \text{class 2}$$

Table 1. Extract from knowledge repository.

A	AM	M	PM	P	PL	L	AL	CLASS
71	77	71	60	66	60	49	60	1
66	68	77	87	77	77	41	56	1
65	65	59	65	65	65	27	54	1
71	82	87	87	82	77	51	56	1
78	73	58	73	73	68	39	49	1
69	0	64	0	78	78	0	0	2
67	72	67	56	67	50	33	50	2
64	53	48	53	53	32	0	0	2
65	65	60	50	60	55	35	55	2
53	35	0	0	0	0	0	47	2

Classes: 1 – Healthy, 2 – Unhealthy

The function set and terminal set used here are as follows:

$F = \{+, -, *, /\}$ and $T = \{\text{SEBT parameters}, R\}$, where R contains randomly generated constants.

3 Simulation Results and Discussions

The data repository is populated with the normalized SEBT results and the spurious data is filtered and discarded. The data repository is then divided into two groups: training data and testing data. Almost 60% of the data is selected randomly for training and the rest is used for testing.

GP evolves an expression using the training dataset which is composed of one or more SEBT parameters, operators from the function set F and some constants. Each run of the GP generates different expressions which efficiently classifies the data. Out of these, we present some of the expressions which are composed of ⩽2 terminals so that the expression is concise and easily comprehensible. A few of these expressions are listed in Table 2. Each expression can be evaluated based on how well it can classify the testing samples.

The expression is applied to the data and classification matrix is generated to evaluate the performance.

The training comes to an end only when the overall efficiency of classification of the training data is 100%. Next, the classification matrix for the testing data is created to assess impartiality, i.e. how the expression has performed on data other than the set used to evolve it. The classification matrices for classification of the testing data using the first two expressions are listed in Tables 3 and 4. Similar efficiencies were observed with the other two expressions also.

Table 2. Simple expressions evolved from GP.

Exp No.	Expressions evolved by GP
Exp-1	AM + PL − 128
Exp-2	(PL − 56) * AL − 74
Exp-3	(PM − PL + 57) * (PL − 58)
Exp-4	PL * L * (PL − 56) − 75

Table 3. Classification matrix for Exp-1.

Exp-1: AM + PL − 128			
	Class 1	Class 2	Individual efficiency (%)
Class 1	11	1	91.66
Class 2	1	12	92.30
Overall efficiency = 92			

Table 4. Classification matrix for Exp-2.

Exp-2: (PL − 56) * AL − 74			
	Class 1	Class 2	Individual efficiency (%)
Class 1	12	0	100.0
Class 2	2	11	84.6
Overall efficiency = 92			

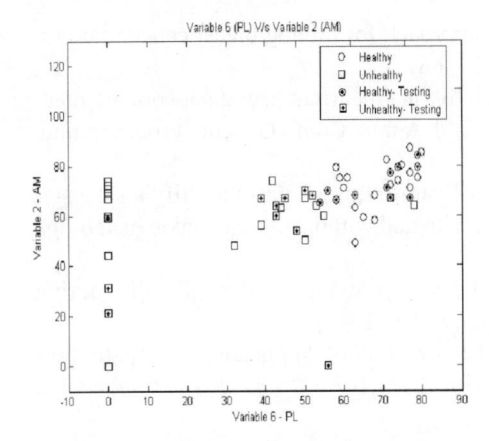

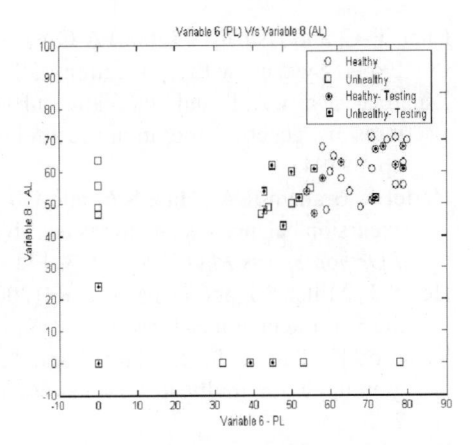

Figure 1. Cluster plot of Postero-Lateral V/s Antero-Medial leading to two separate clusters.

Figure 2. Cluster plot of Postero-Lateral V/s Antero-Lateral leading to two separate clusters.

It is to be noted that the simplest of the expressions consist of only two terminals. This means at a time, only two terms are sufficient to classify the given case as healthy or injured. The cluster plot is generated with these two variables that form the expression. This information can graphically be depicted on the cluster plot. These plots for each expression listed in Table 3 are shown in Figures 1 and 2.

From the graph, two separate clusters can be observed as expected, with a few overlaps. Thus, given a particular case, its class, healthy or not, can be determined from this graph depending on which cluster it belongs to. Hence, this plot along with the expressions can be used as an aid to determine the health of the knee.

It is to be noted that term Postero-Lateral appears in all the expressions. Along with Poetero-Lateral, any of the above mentioned parameter can be used to classify the knee as healthy or not. As the critical parameters have been recognized, SEBT can be carried out for those parameters alone. This will considerably reduce the time involved in SEBT test and prove advantageous for both subjects and the practitioners.

4 Conclusion

In this paper, Genetic Programming has been implemented successfully for classifying the SEBT results. We have evolved concise, simple, comprehendible expressions from GP. The results show that GP is able to classify SEBT results quite efficiently. The simplicity of the expressions generated and the subsequent ease of classification of testing data based on that are notable.

The extraction of simple expression was intended to allow a person with typical end-user skill level to utilize this with minimal assistance to draw considerable information about the health of the knee. The outcome from the simulation shows that the proposed approach of knowledge extraction is effective.

References

Castellano G., Castiello C., Fanelli A.M., and Mencar C. (2005) Knowledge discovery by a neuro-fuzzy modeling framework. *Fuzzy Sets and Systems,* 149/1, 187–207.

Chen T.-C. and Hsu T.-C. (2006) A GA based approach for mining breast cancer pattern. Expert Systems with Applications, 30/4, 674–681.

Esparcia-Alcazar A. I. and Sharman K. (1997) Evolving recurrent neural network architectures by genetic programming, in Proc. 2nd Annu. Conf. Genetic Programming, pp. 89–94.

Hertel J., Braham R.A., Hale S.A. and Olmsted-Kramer L.C. (2006) Simplifying the star excursion balance test: analyses of subjects with and without chronic ankle instability. *J Orthop Sports Phys Ther.*, 36/3, 131–7.

Hertel J., Miller S.J. and Denegar C.R. (2000) Intratester and intertester reliability during the Star Excursion Balance Test. *J. Sport Rehabil,* 9, 104–116.

Kishore J.K., Patnaik L.M., Mani V. and Agrawal V.K. (2000) Application of genetic programming for multicategory pattern classification. *IEEE Trans. Evol. Comput.,* 4, 242–258.

Koza J.R. (1992) *Genetic Programming: On the Programming of Computers by Means of Natural Selection.* MIT Press, Cambridge, MA.

Olmsted L.C., Carcia C.R., Hertel J. and Shultz S.J. (2002) Efficacy of the star excursion balance tests in detecting reach deficits in subjects with chronic ankle instability. *J Athl Train.,* 37/4, 501–506.

Parpinelli, R.S., Lopes, H.S. and Freitas A.A. (2002). Data mining with an ant colony optimization algorithm. *IEEE Trans. Evol. Comput.,* 6, 321–332.

Gribble P.A. and Hertel J., (2003) Considerations for normalizing Measures of the Star Excursion Balance Test. *Measurement in Physical Education and Exercise Science,* 7/2, 89–100.

Poli R. (1996) Genetic Programming for image analysis, in Proc. 1st Int. Conf. Genetic Programming, Stanford, CA, pp. 363–368.

Duch W., Adamczak R., Grabczewski K. and Jankowski N. (2000) Neural methods of knowledge extraction. *Control and Cybernetics*, 29/4, 997–1018.

AN ANALYSIS OF SUN SALUTATION

S.N. OMKAR

Department of Aerospace Engineering, Indian Institute of Science, Bangalore, India

Sun salutation is a part of yoga. It consists of a sequence of postures done with synchronized breathing. The practice of few cycles of sun salutation is known to help in maintaining good health and vigor. The practice of sun salutation does not need any extra gadgets. Also it is very much aerobic and invigorates the body and the mind.

1 Introduction

Many metals when combined appropriately and in right proportion yield an alloy, which has better properties. Best of orchestra is a harmonious combination of various instruments. Likewise, a sequence of postures when performed in synchronization with the breathing becomes sun salutation. This sequence consists of 10 postures performed in a single, conscious, graceful flow. As the rays of the sun reaches every part of the globe, these sequence of postures ensures that the internal energy reaches every part of the body. Hence the name sun salutation. Sun salutation does not need any gadgets and can be done in a limited frame of time and space. Sun salutation is an excellent yogic method to loose obesity. It has a deep impact on metabolism and helps one to keep fit in a safe way.

Research that has evaluated the energy expenditure of yoga indicates that yoga is essentially equivalent to moderate forms of exercise. (DiCarlo *et al.*, 1995). The available evidence suggests that the practice of yoga may be associated with an improvement in cardio- respiratory fitness (Tran *et al.*, 2001), and both muscular strength and endurance (Birch, 1995). While promising, these studies have involved only a few postures or exercises. A single study that evaluated heart rate for standing yoga postures found lower heart rate and higher rate of perceived exertion for the yoga posture sequence compared to treadmill walking (DiCarlo *et al.*, 1995). All of this suggests the need for research on fitness-related outcomes associated with yoga practice (Virginia *et al.*, 2007).

In this paper we look at one of the most popular set of yoga sequence known as sun salutation. The kinesiology of sun salutation and its impact on weight management makes it very attractive for all practitioners. The postures and brief procedure of sun salutation is given in Table 1.

2 Method

2.1. *Participants*

A total of 20 participants (17 male and 3 female) took part in this program. The average age of the participants is 42.1818 (SD = 8.5). Their mean body mass index is 23.7227 (SD = 2.617). All the participants are doing yoga practice, in particular the sun salutation, for at least past one year. All these participants are regular students practicing yoga

Table 1. Sun salutation-postures and brief procedure.

No.	Photo	Method	Movements
I		Stand with feet together, pull both the knee caps and chest up. Press both the palms and keep the forearms parallel to the floor. Stay with relatively deep inhalation and exhalation.	Medial rotation of shoulder, elbow flexion, medial rotation of radio-ulnar joint, extension of the wrist, knee extension.
II		With deep inhalation stretch both hands upward, exhale and bend back. Keep the legs and the arms straight.	Shoulder flexion, back extension, knee extension, Shoulder girdle elevation and adduction.
III		Inhale come up and bend forward with exhalation. Keeping the legs straight, rest both palms by the side of the foot with middle fingers and big toes in the same line. Avoid lumbar flexion.	Hip flexion, wrist extension, knee extension, shoulder girdle abduction.
IV		While exhaling, stretch right leg back. The gap between left calf and thigh should be as minimum as possible.	Right hip extension, right foot extension (plantar flexion). Left knee flexion, left hip flexion, left foot flexion (dorsiflexion), wrists in extension,

(*Continued*)

Table 1. (Continued).

			shoulder and radioulnar joint internal rotation.
V		As you exhale take the left leg also back. Spine should be parallel to the floor and entire body is supported by palms and base of the metatarsals.	Wrist extension, foot extension, dorsiflexion of the toes, knee extension, shoulder and radioulnar joint internal rotation.
VI		Flex the elbow and slowly descend down with exhalation. Keep the body nearly parallel to the floor.	Elbow flexion, wrist extension. Dorsiflexion of the toes, knee extension, radioulnar joint internal rotation.
VII		Rest the knee on the floor, exhale and bend back. Thrust the chest forward and keep the elbows closer. Stretch the foot backwards.	Back extension, wrist flexion, elbow flexion, foot extension, shoulder and radioulnar joint internal rotation, shoulder slight extension.
VIII		Fold the toes in, inhale and straightening the arms rise the hips up. Keeping the leg straight with exhalation stretch the spine and rest the head on the floor. Stretch of the Achilles' tendon and hamstring muscles may be noted.	Wrist extension, shoulder and radioulnar joint internal rotation, Flexion of shoulders, flexion of hips, dorsiflexion of foot, knee extension.
IX	Similar to posture IV	As you exhale, take the right leg forward to assume the posture similar to posture IV.	Similar to posture IV
X	Same as posture III	Same as posture III	Same as posture III
	Same as posture I	As you inhale, come to posture I	Same as posture I

Steps I to X in table 1 constitutes one cycle of sun salutation.

Table 2. Three variations considered for the study.

Case	Number of cycles	Average duration (in minutes)
1	120	35
2	144	43
3	144	52

with the author. Their consent was obtained on being informed about the purpose of the study.

2.2. *Instrumentation*

The primary data of interest was heart rate and the calories burnt. Wireless heart rate monitors that use chest electrodes are considered reliable and valid measures of heart rate (Achten & Jeukendrup, 2003). All participants wore Polar S610 heart rate monitors for each yoga session. Each participant wore the chest electrode transmitter and a wrist receiver which recorded average heart rate, maximum heart rate and calories burnt for the entire session

2.3. *Protocol*

The session begins everyday at 4.50 AM. Few breathing and simple stretching exercises are done for the first ten minutes. Later the cycle of sun salutation begins. At the beginning of the first cycle the participant is asked to start the recording. The specified number of sun salutation is done continuously at the set pace. At the end of the sun salutation the participant switches off the recording and relaxes for 10 minutes. The recordings are repeated for each participant over a number of days. The number of sun salutations and the duration taken are varied. The three variations are considered as shown in Table 2.

3 Results and Discussion

The sun salutation is performed every day. Each day one of the participants uses the chest electrode and the wrist watch to measure the calories burnt. Also each participant is evaluated over a number of days for consistency in the readings. The participants are categorized into three weight classes namely 55 kg, 65 kg and 77 kg. In the case of one of the participants the readings varied from the rest in the same age group and it has not been considered. Three variations of sun salutation are done as per the said protocol. The amount of calories burnt during each session classified by the weight group is given in Table 3.

It can be seen that the amount of energy expended increases by about 15% from case 1 to case 2. By reducing the pace of sun salutation, the calories burnt further increases as shown in the Table.

In one cycle of sun salutation there are 10 postures and associated with each posture is one cycle of breathing. Therefore breathing rate becomes a vital parameter in all the three cases considered. The breathing rate associated with each case is given in the table 4 below.

This kind of hyperventilated breathing is also known to be effective in mitigating stress related sympathoexcitation. Hyperventilation induced chemoreflex not only induces powerful

Table 3. The energy expenditure for each weight group.

Mass (Kgs)	Energy expenditure (Kcal)		
	Case 1	Case 2	Case 3
55	185	220	259
65	275	320	345
77	380	435	475

Table 4. The breathing rate for each given case.

Approximate breathing rate (Cycles/min)		
Case 1	Case 2	Case 3
34	34	28

generalized autonomic changes but also induces local circulatory changes (Steurer et al., 1995). Raghuraj et al. (1998) demonstrated that a type of rapid breathing (Kapalbhati) enhances the sympathetic drive to myocardium, enhanced the sympathovagal balance towards sympathetic side. This was also accompanied by decreases in vagal tone.

Finally, we look at the type of movements that the osteoligamentous and muscular system experiences during sun salutation. The main movements occurring during the sun salutation is given in Table 1. This shows that most muscles get a good work out.

4 Conclusion

Sun salutation is a holistic work out. It beneficially influences the physiological, psychological and neurological systems. It is an easy and practical tool for anyone who is health conscious.

References

Achten J. and Jeukendrup A.E. (2003) Heart rate monitoring: applications and limitations. *Sports Medicine*, 33/7, 517–538.

Birch B.B. (1995) *Power Yoga*. Fireside, New York.

DiCarlo L.J., Sparling P.B., Hinson B.T. Snow T.K. and Rosskopf L.B. (1995) Cardiovascular, metabolic, and perceptual responses to hatha yoga standing poses. *Medicine, Exercise, Nutrition and Health*, 4, 107–112.

Raghuraj P., Ramakrishna A.G. and Nagendra H.R. (1998) Effect of two related yogic breathing techniques on heart rate variability. *Indian Journal Physiol Pharmacology*, 42/4, 467–472.

Steurer J., Kaplan V., Vetter W., Bollinger A. and Hoffmann U. (1995) Local blood flux in skin and muscle during voluntary hyperventilation in healthy controls and patients with hyperventilation syndrome. *Int. J. Microcirc. Clin. Exp.*, 15/6, 277–82.

Tran M.D., Holly R.G., Lashbrook J. and Amsterdam E.A. (2001) Effects of hatha yoga practice on the health-related aspects of physical fitness. *Preventive Cardiology*, 4 /4, 165–170.

Virginia S. Cowen T. and Adams B. (2007) Heart rate in yoga asana practice: A comparison of styles. *Journal of Bodywork and Movement Therapies*, 11, 91–95.

DEVELOPMENT OF A FORCE FEEDBACK SYSTEM FOR EXERCISING

J.Y. SIM,[1,2] A.C. RITCHIE[1,2] & F.K. FUSS[1]

[1]*Sports Engineering Research Team, Division of Bioengineering, School of Chemical and Biomedical Engineering, Nanyang Technological University, Singapore*
[2]*School of Mechanical and Aerospace Engineering, Nanyang Technological University, Singapore*

Conventional exercise machines are mostly bulky and do not incorporate feedback control. A prototype of the proposed system consisting of a controller circuit, a DC motor and a power source was built and tested. Two different modes of exercise were developed in this prototype, namely the variable resistance mode and constant force mode. Both modes of exercise were tested and found to be functioning well. The successful demonstration of the basic functions of the system proved that the proposed system is a viable option and ought to be further explored.

1 Introduction

Exercise machines are a billion-dollar industry worldwide, and are particularly popular in wealthy countries such as the United States (Li & Horowitz, 1997a,b). Most of the exercise machines currently available allow users to exercise in a single degree of freedom motion and exercises are repetitive in nature. Exercise is achieved by providing resistance to the user and there are three main categories of resistance: the weight-based, friction based, and damper-based machines (Li & Horowitz, 1997a,b).

There are two main types of control that can be applied to exercise machines: active and passive. In active control systems, the net flow of energy is from the system to the user which means that the system will automatically provide the resistive force to the user. However, this requires prior knowledge of the user's biomechanical behaviour such as strength or maximal power output which differs between individuals. These machines are typically found in specialised rehabilitation clinics, as a trained technician is required to operate the system safely. Without such expertise, an incorrect amount of resistance may be provided, which may cause injury to the user.

In passive control systems, the net flow of energy is from the user to the system. These systems are considered to be safer due to the flexibility of control by user on the amount of resistance that is desired. Most "smart" exercise machines that have been developed use this control system. Examples include multigyms, exercise bicycles, rowing ergometers, and stair-climbers.

One major weakness of conventional exercise machines is that they only offer resistance to the user. There is no control on the amount of resistance that is optimal for the user, and usually no way to set the resistance dynamically during exercise. For example a captive-weights multigym provides a large range of resistance to the user but there is no control on the optimal resistance tailored to the user's requirement. In other words, users have to use their own judgement and this is not ideal for users with insufficient knowledge of sports science and physiology. Improper amounts of exercise may result in over-use injuries to the muscles, and will not achieve the optimal training goals.

The aim of this study was to develop a technology demonstration of an affordable exercise machine with two operational modes: fixed manually adjustable resistance, and variable resistance depending on the angular velocity of movement.

2 Method

The source of resistance in the exercise machine prototype developed in this study is a DC motor which is controlled by a microcontroller. The concept uses signals from the microcontroller to vary the speed or torque of the motor, providing the appropriate amount of resistance to the user who is to act against the torque of the motor.

Since the physical profile of each individual is unique, there is a need to identify the physiological profile of each user in order to know the optimal level of exercise for each individual. Therefore, it was proposed to incorporate a profiling or calibration unit in the control system which will collect data from the user through a series of exercise at the beginning of each work out.

The collection of data can be done by reading of the back-electromotive force (back EMF) generated by the user in exercising against the torque of the motor. This can be done through the ADC ports of the microcontroller. The data collected can then be used to calculate the level of resistance to be provided to allow user to exercise at a maximum level of power output.

To control the torque of the motor, a signal from the microcontroller to the motor is required. The signal is generated by using the pulse-width modulation (PWM) function. The controller moderates the pulse width of the signal by means of a "chopper" to adjust the actual amount of voltage that is "seen" by the motor, thus producing variable torque and speed.

The effective voltage is the average or mean voltage, V_{mean} which can be calculated as follows:

$$V_{mean} = \frac{1}{t} \int_0^t f(t)\, dt \tag{1}$$

$$= \frac{1}{t} \left\{ \int_0^{DC} V_{max}\, dt + \int_{DC}^1 V_{min}\, dt \right\} \tag{2}$$

$$= DC\,(V_{max}) + (1 - DC)(V_{min}) \tag{3}$$

$$\text{If } V_{max} = +5V \text{ and } V_{min} = 0V,\, V_{mean} = DC\,(V_{max}) \tag{4}$$

Therefore, as the pulse width of the signal is increased or decreased, a higher or lower motor speed or torque can be achieved, therefore providing resistance that the user requires.

3 Conceptual Design

The outline of the system is shown in Figure 1. There are four separate units or modules in the system with different functions. The user input/output unit for example serves as a

Development of a Force Feedback System for Exercising 89

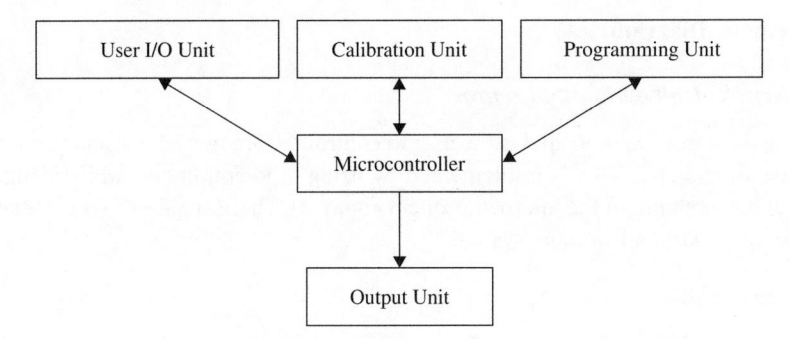

Figure 1. Outline of the system.

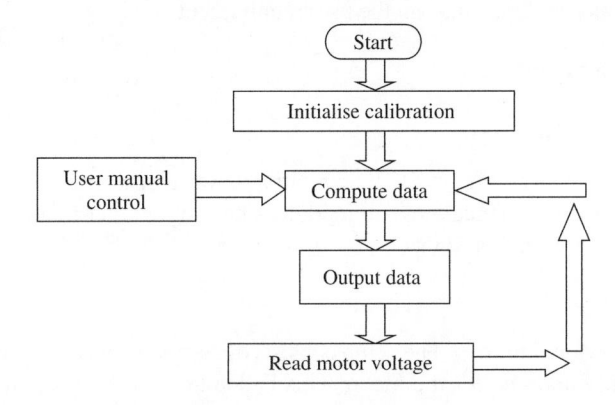

Figure 2. System operation.

communication unit between user and system. Functions such as display of instruction or status of exercise to the user would be implemented through this unit. The calibration unit would be used to map the user's power-velocity profile to determine the maximal amount of resistance required for optimal exercise or maximal power output. The programming unit would be used to load different programs in order to test different functions and would only be used in the development stage of the system. Lastly, the output unit is used to produce the control signal which regulates the amount of resistance that is provided to the user.

4 System Operation

The outline of the system operation is shown in Figure 2. After the machine is turned on, the calibration steps are initialised. User will go through a series of calibration steps in order to map their profile. Once the data has been collected and processed, the system will then compute the required resistance to be provided to the user and generate the output signal. As the exercise routine is being carried out, the performance of the user will be fed back to the system in order to make the necessary changes.

5 Schematic Diagram

5.1. *User I/O Unit – Manual Control*

The manual control function allows the user to control the amount of resistance throughout the course of exercise. This is implemented by using a potentiometer which controls the voltage at the input pin of the microcontroller (Figure 3). There are also 2 separate switches which serve to start and stop the system.

5.2. *Output Unit*

The output unit (Figure 4) consists of an operational amplifier coupled with a mechanical relay that switches the power to the motor on and off at a high frequency. The input of this unit is the PWM signal generated by the microcontroller. The output unit controls the speed or torque of the motor, depending on the user requirement.

6 Fabrication and Testing

6.1. *Basic System Circuitry*

System circuitry shown in Figure 5. The circuitry shown consists of only the control system, output module and manual control module. The programmer module is excluded as would be done for the final product.

6.2. *System Setup*

System setup shown in Figure 6. The setup consists of 2 separate power sources, one for the controller circuitry and one for the motor. This is due to the difference in power requirements and input voltage for the controller and motor.

6.3. *Testing Results*

The system was found to perform satisfactorily for speed control when unloaded. Initial difficulties when generating resistance to movement in active control were found to be due to a limitation on the power delivered by the power supply and were solved by using a separate power supply unit to power the motor, as shown in figure 6.

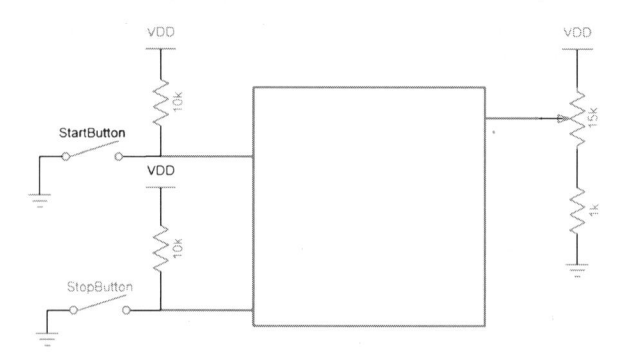

Figure 3. Block diagram of the manual control.

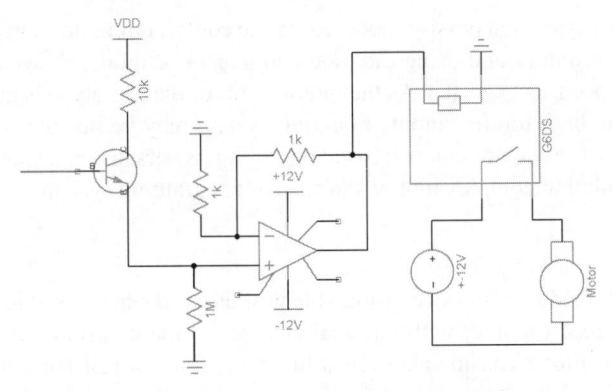

Figure 4. Block diagram of the output unit.

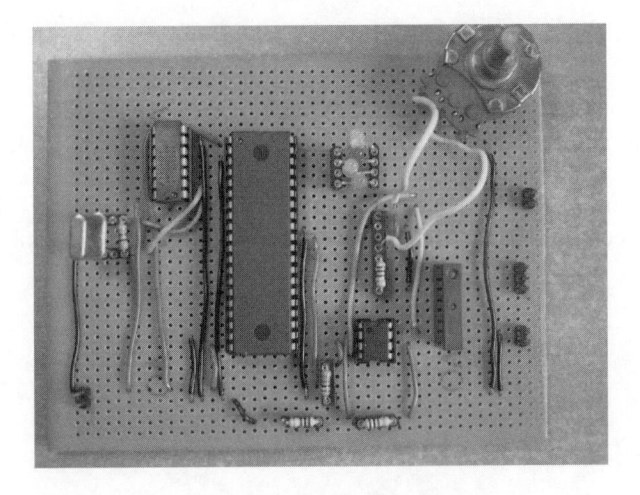

Figure 5. System circuitry.

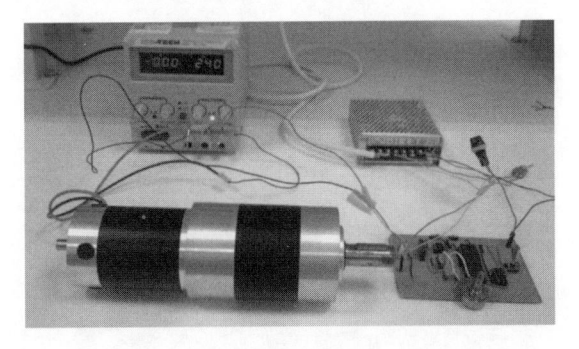

Figure 6. System setup.

Testing of the system in passive mode, using the control circuit to switch between open and closed circuit output and using the motor in generator mode, showed an increase in resistance with speed, as expected. As the motor used for the prototype is heavily geared to give a low-speed, high torque output, a gearing system may be necessary for future systems. This mode of resistance control, based on timing, is versatile and shows considerable promise for simulating complex motions in specialized training systems.

7 Discussion

The system developed here shows considerable promise as it can be used for strength training or for endurance training, with minimal changes to the control circuitry. The use of electronic control allows considerable versatility in the force or resistance regime, which is particularly useful in exercises such as the leg-press where the available force generated by the user varies considerably with position. The system also offers promise as a simulator for assessing the condition of athletes under controlled positions, as part of a process of selection trials.

References

Li P.Y. and Horowitz R. (1997a) Control of smart exercise machines – Part 1, Problem Formulation and non-adaptive control. *IEEE/ASME Transactions on Mechatronics*, 2, 237–247.

Li P.Y. and Horowitz R. (1997b) Control of smart exercise machines – Part 2, Self – optimizing control. *IEEE/ASME Transactions on Mechatronics*, 2, 248–257.

FORMALISED REQUIREMENTS DOCUMENTATION OF A NOVEL EXERCISE SYSTEM USING ENTERPRISE MODELLING CONCEPTS

A.A. WEST, J.D. SMITH, R.P. MONFARED & R. HARRISON

Wolfson School of Mechanical and Manufacturing Engineering, Loughborough University, Loughborough, Leicestershire, England

The purpose of this research was to evaluate and select a suitable technique for identifying, capturing and presenting information relating to computer integrated exercise systems. Existing systems modelling techniques were found to be complex, unstructured and semantically imprecise. A review of general systems modelling approaches resulted in the detailed analysis of enterprise modelling techniques. The Computer Integrated Manufacturing Open Systems Architecture (CIMOSA) architectural framework was selected as the candidate enterprise modelling technique and the process modelling diagrams developed at Loughborough University used to provide the graphical notation for system representation. These models were successfully used in the development of novel hardware and software systems for a state of the art exercise machine.

1 Introduction

The purpose of this study was to identify, analyse and implement a system modelling technique for illustrating the operational aspects of a novel computer integrated exercise system. The design requirements of state of the art exercise systems are becoming increasingly complex and therefore it is important that a detailed plan is established before development begins in order to produce an effective and efficient solution.

2 Current Approach to Exercise System Modelling

Basic flow charts have long been the preferred modelling method for the documentation and visualisation of systems in general and exercise related systems in particular (Cohen *et al.*, 2004; Hickman, 2004; Watterson *et al.*, 2007). Flow chart diagrams are easy to understand as the process, decision and data constructs are widely recognisable and the process route is clearly mapped. In addition, flow chart modelling is an easy method to learn and allows information to be captured quickly and relatively efficiently as the exact modelling methodology is not rigidly defined. However, the generality of the constructs used in flow chart models and the lack of a formal approach to model generation is a major weakness of this method. Differentiation between processes/activities and their respective operational levels is not possible and the inability to define task distribution makes it difficult to give a structured hierarchical overview. Thus, it is clear that in order to improve the modelling of exercise system information, a more structured framework is required.

3 Selection of a Candidate Modelling Framework

An enterprise modelling approach was identified as the most suitable solution for modelling an exercise system following a critical review of current modelling techniques.

Within enterprise modeling, the operational actions performed by the system are formalised using activity models while the available resources and their configuration are presented in the resource model. Organization models can be used to formalise the structure of the exercise system and decision making models are constructed to provide additional behavioural details (Vernadat, 1996). Using enterprise modelling techniques in such a manner should ensure that a standardised set of models is produced which communicate system knowledge in a clear, well structured manner that supports decision making during the development process. The flexibility of enterprise models allows changes to be made efficiently and thus alternative solutions can be evaluated to produce a robust system.

4 Evaluation of Enterprise Modelling Frameworks

Enterprise modelling techniques have been classified (see Figure 1) to ensure that a reasoned choice can be made when selecting a particular framework. A simple selection framework proposed by Augilar-Savén (2004) introduced a two axis classification procedure. This scheme orders each approach according to its change capabilities, "model change permissiveness", and groups the model using its perceived purpose. The main enterprise modelling frameworks identified by Augilar-Savén (2004) have been classified using the selection framework in Figure 1 together with the additional approaches identified during the literature review.

The shaded area in Figure 1 represents the required modelling focus for a computer integrated exercise system. From the diagram it is clear that the CIMOSA architectural framework is the only enterprise modelling technique which fulfils this particular specification for

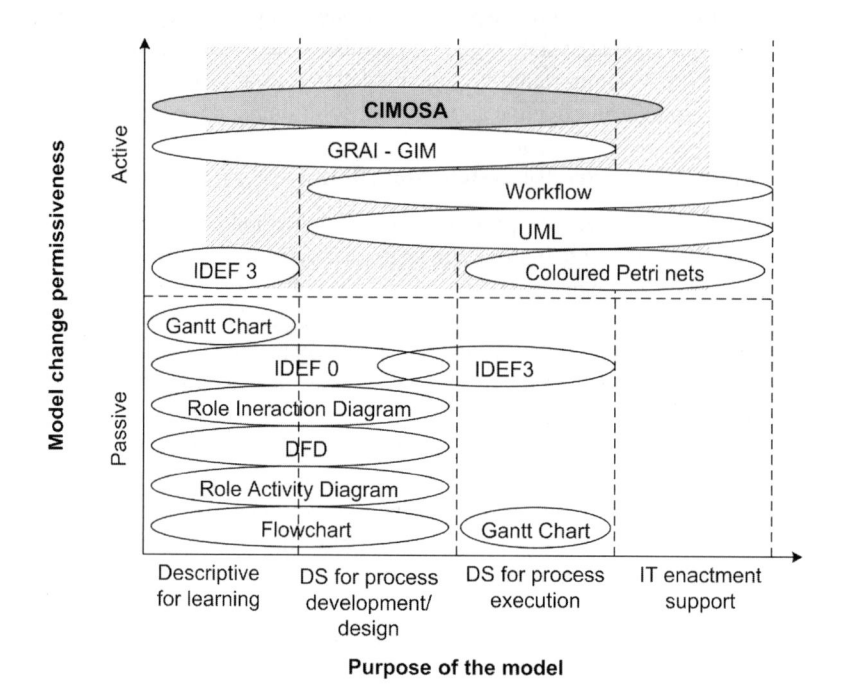

Figure 1.   Enterprise model selection framework.

an exercise system model in a single integrated method. A method with high change permissiveness such as this is highly desirable when developing an exercise system as the range of operational features can change as advance training techniques, user interfaces and system integration facilities are constantly advanced. The broad modeling purpose offered by CIMOSA allows information regarding exercise and training techniques to be presented in a formalized manor such that mechanical, electrical and software design, analysis and implementation can be completed with a single consistent modeling approach. The selection framework developed by Augilar-Savén (2004) is an assessment template using the two basic criteria identified. Increased suitability is likely to be achieved if the modelling technique is also assessed against additional criteria such as those discussed by Fox (1994) and Vernadat (1996). These additional characteristics; generality, scope, granularity, precision, efficiency, perspicuity, transformability, consistency, completeness and scalability where used to the improve the appropriateness of the enterprise modelling technique selected. From this appraisal, the CIMOSA architecture was selected as the approach most suitable for modelling a computer integrated exercise system. CIMOSA concepts allow all of the necessary model views to be created in a structured manner using a well defined methodology.

5 CIMOSA Modelling Formalisms

The CIMOSA framework has been utilised by European and International organisations as a basis for determining and formalising a standard approach to enterprise modelling. This approach, whilst providing a detailed decomposition of the business at a high level does not specify the basic modelling constructs which represent the fundamental process elements. Research conducted at the MSI Research Institute (Loughborough University) has lead to the development of a hierarchical diagram structure (Aguiar, 1995), which utilises a formal set of modelling constructs, see Figure 2, for representing a wide range of different systems (Monfared, 2002; Monfared, 2006).

Four types of diagram; the context diagram, interaction diagram, structure diagram and activity diagram have been created which illustrate the details and interaction between domain processes, businesses processes and enterprise activities. Together with the basic

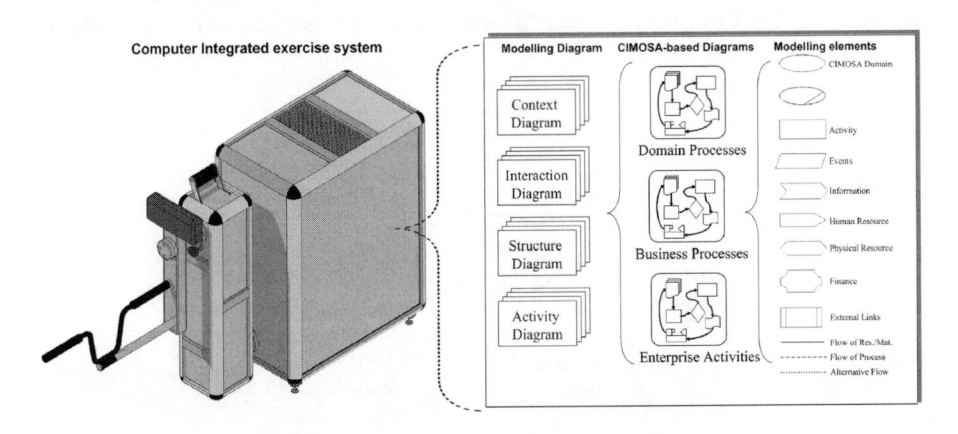

Figure 2. CIMOSA modelling constructs.

modeling elements, these diagrams represent alternative, but complimentary views of the enterprise and demonstrates how CIMOSA concepts can be implemented graphically.

6 Discussion of the Resultant Exercise System Models

Using the proposed process modelling approach and the CIMOSA architectural framework, a set of diagrams were developed which illustrate the operation of a computer integrated exercise system. The context diagram is used to define the domains to be modelled using CIMOSA formalisms. The overall context diagram for the computer integrated exercise system is illustrated in Figure 3.

The system is built around the fundamental processes involved in preparing, participating and completing any type of exercise routine. Integration of computer control, monitoring and visualisation increases the domain complexity which has been decomposed into seven key domains; Setting Up, Exercise Configuration, Monitoring Exercise, Exercise Session, Interactive Help, Safety and Emergencies and Administration that interact to create the overall system dynamics. Example structure and activity diagrams for the exercise process domain are illustrated in Figure 4.

Structure diagrams provide a graphical method for identifying and structuring enterprise activities and business processes within a domain process. The information presented is not time-based, and therefore the flow of the operations must be defined in a specific sequencing diagram. This type of sequencing data is represented in the activity diagram. Figure 5 shows how this approach has been implemented in the development of a computer integrated exercise system HMI software application. The screen shots provided, represent the information and resource interfaces developed for the business processes and enterprise activities identified in the Administration Domain Process (DP1). Software design and development has been based on the structure, interaction and activity diagrams and thus the relationships in Figure 5 indicate how system construction is directly related to the models created in Figure 4.

An initial welcome screen prompts the user to insert their storage device or enter a password in order to access the central database (EA111 and EA112). Once the system has uploaded the necessary data, key information is presented in the navigation area of the screen to notify the user that the system has accepted the stored details (EA113). A user profile manager has been developed to guide users who are new to the system through the registration procedure (EA121). Once the basic data has been entered the user can access more advanced

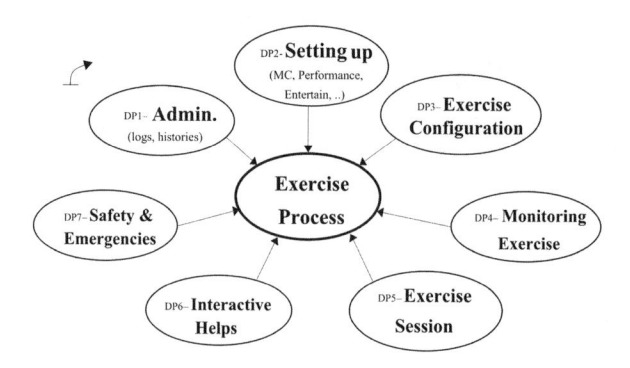

Figure 3. Exercise system CIMOSA context diagram.

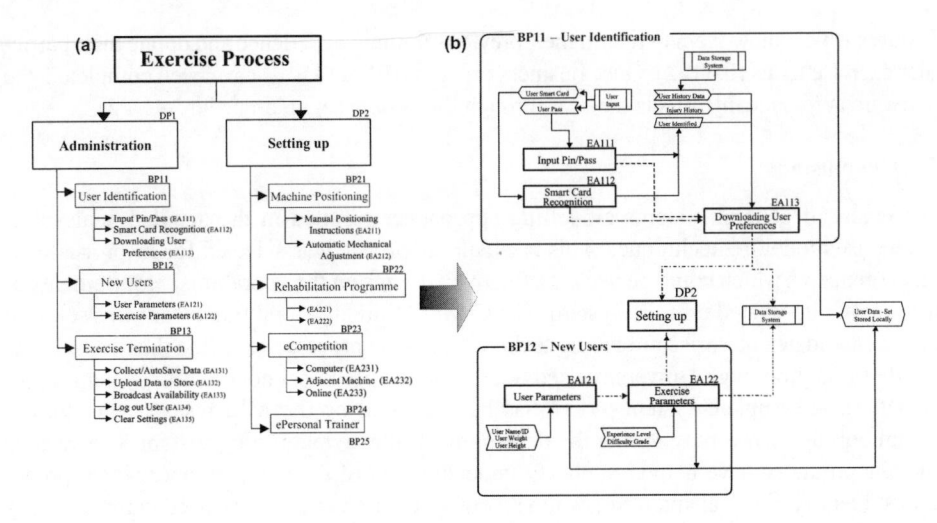

Formalised Requirements Documentation of a Novel Exercise System 97

Figure 4. CIMOSA structure (a) and activity (b) diagrams of a computer integrated exercise system.

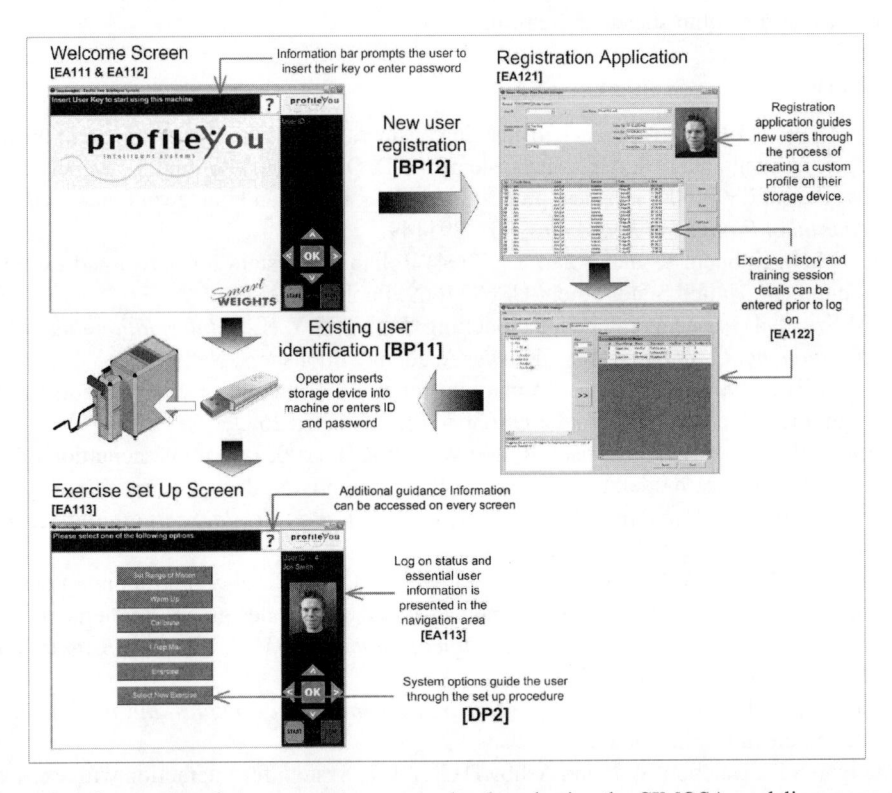

Figure 5. Screen shots from an exercise system developed using the CIMOSA modeling concepts.

features which allow them to record their previous training experience and define their particular exercise focus (EA122). Once business process BP11 or BP12 have been completed, the software automatically directs the user through the system set up procedures (DP2).

7 Conclusions

A review of general systems modelling approaches resulted in the detailed analysis of enterprise modelling techniques. This particular approach was selected for its process oriented concepts which could be used to effectively represent the operational requirements of a computer integrated exercise system. The CIMOSA architectural framework was selected as the candidate enterprise modelling technique and the process modelling diagrams developed at Loughborough University used to provide the graphical notation for system representation. A complete system model has been created by the MSI research institute at Loughborough University during the development of a novel training system. The various models produced have helped to clarify understanding of the complex operational procedures, identify physical and information resource requirements and assisted in the development of the system behaviour. The CIMOSA architectural framework has given a formal structure to the model and the object oriented constructs have resulted in the identification and re-use of common components in different domain processes. The practical development of both hardware and software systems have been based on the detailed system models created at the initial stages of the project.

References

Aguiar M.W.C. (1995) An approach to enacting business process models in support of the life cycle of integrated manufacturing systems. Ph.D. Diss., Loughborough University, UK.

Aguilar-Savén R.S. (2004) Business process modelling: review and framework. *International Journal of Production Economics*, 90, 129–149.

Cohen M.A., Kobuchi K. and Folan K. (2004) Reliability system for networked exercise equipment. United States Patent US6827669, Published 7 Dec.

Fox M.S. (1994) Issues in enterprise modelling. In: Nof S.Y. (Ed.) *Information and collaboration models of integration*. Kluwer Academic Publishers, Netherlands.

Hickman P.L. (2004) Method and apparatus for remote interactive exercise and health equipment. United States Patent US6808472, Published 26 Oct.

Monfared R.P., West A.A., Harrison R. and Weston R.H. (2002) An implementation of the business process modeling approach in the automotive industry, *Proceedings of the Institution of Mechanical Engineers (IMECHE), Part B, Journal of Engineering Manufacture*, 216, 1413–1427.

Monfared R.P., West A.A., Harrison R. and Lee S.M. (2007) Improving train maintenance through process modeling and component-based system design and implementation, *Proceedings of the Institution of Mechanical Engineers (IMECHE), Part B, Journal of Engineering Manufacture*, 221, –.

Vernadat F.B. (1996) *Enterprise modelling and integration, principles and applications.* Chapman and Hall, London.

Watterson S.R., Dalebout W.T. and Ashby D.C. (2007) System for interaction with exercise device. United States Patent US7166062, Published 23 Jan.

4. General Motion Analysis

A LOW COST SELF CONTAINED PLATFORM FOR HUMAN MOTION ANALYSIS

N. DAVEY,[1,3] A. WIXTED,[1] Y. OHGI[2] & D.A. JAMES[1,3]

[1] *Centre for Wireless Technology Griffith University, Brisbane, Queensland, Australia*
[2] *Graduate School of Media and Governance, Keio University, Kanagawa, Japan*
[3] *Centre of Excellence for Applied Sport Science Research, Queensland Academy of Sport, Brisbane, Queensland, Australia*

In this paper we describe a low cost sensor platform for use in human motion investigations. The platform is based on a readily available low cost microprocessor and features on-board memory, inertial sensors, USB communications and provision for additional sensors. Using a custom designed operating system it is easily customised by someone who has only limited programming experience. In the second part of our paper we present some simple applications for investigating human motion using the on-board sensors. These applications show the user how to collect and download data and then introduces basic data processing techniques to process and analyze the data for key signatures and markers using readily available tools such as Matlab and Mathematica or their freeware equivalents.

1 Introduction

Athletic and clinical testing for performance analysis and enhancement has traditionally been performed in the laboratory where the required instrumentation is available and environmental conditions can be easily controlled. Today however we see the emergence of small, portable technologies being applied to the sporting environment. One such technology that has seen rapid development in recent years is in the area of inertial sensors. These sensors respond to minute changes in inertia in the linear and radial directions. Such sensors have been utilised in a variety of sports related monitoring projects such as monitoring elite swimmers (Ohgi, 2002; Davey *et al.*, 2004), monitoring rowing (Lai *et al.*, 2005), athlete gait analysis (Billing *et al.*, 2003, Channells *et al.*, 2006) and a variety of other uses including estimating athlete energy expenditure (Wixted *et al.* 2006), assessing limb segment acceleration (Kavanagh *et al.*, 2006), studying swordsmanship and multi-limb motion monitoring (James *et al.*, 2005), combative sports (Partridge *et al.*, 2005), and others. Both the architecture and implementation of an operating system suitable for use in such sensor systems has been utilised as a teaching tool at University and by other organisations.

To a certain extent however the use of these sensors has been constrained to research teams with access to engineering and sport science expertise. Concurrent with the emergence of Sports Engineering as a discipline in its own right has seen the availability of such instrumentation become more common amongst research teams and in the commercial world. Commercial devices are now becoming available albeit at significant capital cost. These devices often combine inertial sensors with foot pressure sensors, GPS for positioning and other specialist sensors such as strain gauges for rowing.

This paper presents a "no frills" sensor system at low component cost of approximately $AUD100 for researchers to experiment without committing huge resources to the endeavor. It has been designed to make speculative data collection activities as simple as possible with little or no programming knowledge required. Example data analysis routines are presented to show the reader what is possible, but without being prescriptive in nature.

2 Sensor Platform

A data acquisition platform, designed for data acquisition in sporting applications must of necessity, be of a small size and lightweight, yet capable of extended operational periods. Development of such a data acquisition platform should utilise existing technologies to minimise implementation costs, provide easy upgrade paths and ensure parts are readily available. Where possible, tools developed for such an acquisition platform should be portable, low cost and be widely adopted so that specialist expertise is not required. Data formats should be non-proprietary and easily read into popular data analysis platforms such as spreadsheets, graphing packages and mathematical tools such as Matlab, Mathematica or freeware equivalents like Scilab.

2.1. Hardware

At the core of the sensor platform is an RISC based Atmel ATmega324P microprocessor, a very popular processor used for embedded applications all over the world. The Atmel product series provides a comprehensive range of microprocessors with varying features and processing power, many of the product family is pin compatible allowing for easy migration should specific applications require it. The microprocessor features 8 channels for analogue input (for sensors) and 32 Kbytes Flash memory for programming together with 1 KByte EEPROM and many digital I/O lines (32) and SPI bus for communicating with peripheral devices. Additionally the processor can operate at variable clock rates from KHz up to 20 MHz allowing for extremely low power operation or high performance operation (up to 10 MIPS processing throughput).

The platform contains 2 Mbytes of on board memory for data storage, this equates to over an hour of triaxial accelerometer data at 100 samples per second in storage mode or more if compression techniques are applied. The platform includes a Freescale tri-axial accelerometer, which features programmable gain with up to $\pm 6\,g$ range so that an appropriate range for specific activities of interest can be investigated with good resolution. An on-board USB chip and connection socket allow for easy communications with a PC, RF communications can also be added. External analogue and digital connection sockets for additional sensors and peripherals help allow the platform to be more easily customizable for specific applications. Currently the platform is powered by 2 AA battery cells for convenience though higher density solutions such as Lithium-ion replacements could also be used where size restraints exist. Figure 1 shows the board layout for the platform.

Implementation of the data acquisition, storage and communications must be carefully implemented to ensure sample time is accurate, we chose to implement a basic operating system from which appropriate system calls and user procedures could be used.

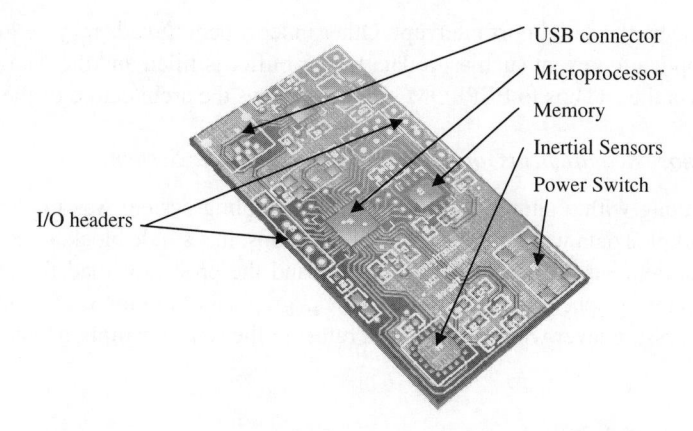

Figure 1. Sensor platform board layout showing key components.

2.2. Operating System and Configuration

The design of a data acquisition platform requires a system capable of performing sensing, data acquisition, signal processing, communications and wireless network management on a limited power budget. This degree of system complexity favors implementation of a loosely-coupled, low-power, embedded, Real Time Operating System (RTOS) on the data acquisition platform (Wixted *et al.*, 2002). The RTOS is loosely coupled to the hardware by providing appropriate hardware abstraction thereby minimising dependencies on specific hardware and allowing for periodic technology updates. The RTOS minimises its own power consumption and at the same time assists application developers in minimising application power use. The operating system is by nature embedded, since the required small size precludes anything more sophisticated. Finally, the RTOS must have real-time components and processes since the data acquisition platform is monitoring real-time events and may be required to operate within a data network.

The operating system (OS) presented here is designed to exploit the power efficiency of the processor. The OS is a simple implementation using two schedulers and some independent threads managing specific tasks. The schedulers operate background, low priority tasks and foreground real time tasks. Tasks in the Real-Time Schedule (RTS) are typically repetitive tasks managing any synchronized function such as data acquisition. RTS tasks are typically atomic or near atomic style tasks. For non atomic real time tasks the RTS can start an independent thread of control. Tasks in the RTS can add tasks to the Background Schedule (BGS) such as data processing, packing and storage.

2.2.1. Independent Threads of Control

Depending on implementation, background tasks run either on a round-robin cooperative scheduler or on a scheduler designed to run a task to completion and then remove it from the list. The RTS runs independent of the BGS. Long running tasks are run via the BGS or run on an independent thread of control. Typical independent threads include serial communications where a communications thread is initiated by a scheduled task and is then allocated time by the serial port interrupt. Other independent threads may be initiated by a particular application event such as: a data logger buffer is filled and the data needs to be transferred via the I^{2}C bus to EEPROM. Figure 2 shows the architecture of the RTOS.

2.2.2. *Memory Requirements and Processing Load – Basic System*

The RTOS along with a simple 3-channel 50 Hz sampling system was implemented and analysed. Sampled data was output serially at 38.4 kbps, in a single block, once per second. Memory requirements are identified in Table 1 and the processor load for two different processor operating speeds and RTOS clock-ticks graphed in Figure 3. The load is broken down by processing layer where application refers to the sensor sampling task.

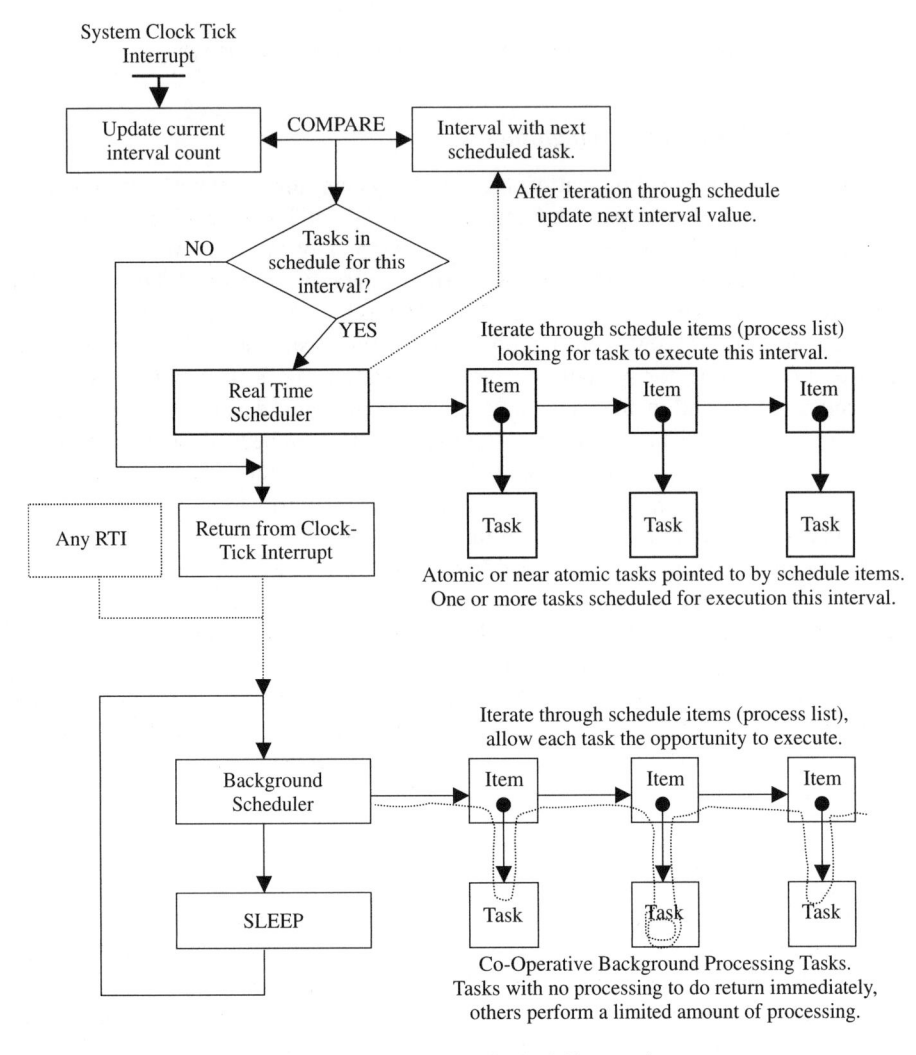

Figure 2. Architecture of scheduling services.

Table 1. Basic system memory requirements.

Component	Memory requirement
Program	1748 bytes (ROM)
Global variables	99 bytes (RAM)
Constants	40 bytes (ROM)
Stack depth	128 bytes (max)
Stack size	256 bytes
Heap size	512 bytes
Total ROM	1788 bytes
Total RAM	995 bytes

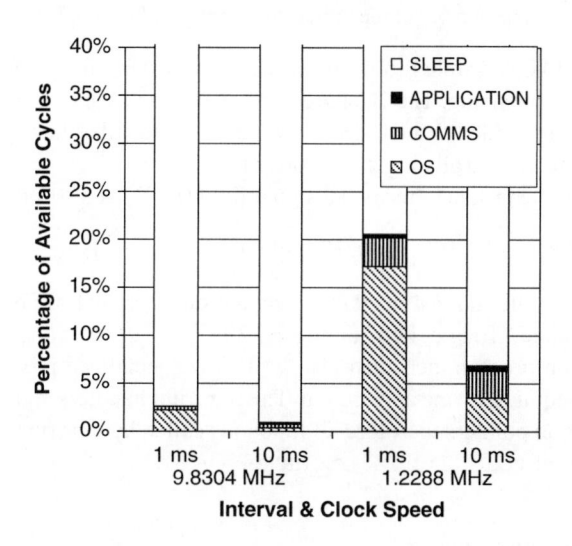

Figure 3. Processor load in a 50 Hz 3 channel implementation of the RTOS, for two crystal frequencies and two primary interval timings.

2.2.3. Using the RTOS

The primary interval timer is the heart beat of the OS. It determines when the timing interrupt is triggered to execute the task schedulers. The frequency of the interrupt is set by two values, a clock divider and a compare register and is set in a configuration file for the operating system.

The system maintains two lists of tasks. The RTS is executed based on primary interval while the BGS is executed in the system idle time. The RTS is for tasks that are either small simple tasks, such as toggling an I/O line, or setup and initialisation routines that configure a separate interrupt routine. A task should only be in the RTS if it needs to occur at a specific time and/or rate. The BGS should be used for long running tasks (such data packing and storage) or tasks that are not time critical.

Configuration of the platform is by adding tasks to the schedulers. The system operates two task schedules, a Real Time Scheduler and a Background Scheduler. Tasks are added to either the RTS or BGS using the addScheduleItem function. The addScheduleItem function has the following syntax:

addScheduleItem (<sch>,<nextival>,<ivalsize>,<func>,<priority>, <misc>,<rcount>);

where:

<sch>	Schedule to add task to, either RTSHED or BGSHED
<nextival>	The interval on which the task is executed
<ivalsize>	The interval to wait before repeating execution (0 to execute once)
<func>	Function to be executed
<priority>	Execution priority of function, 0 being highest and 255 lowest.
<misc>	Provision for used defined data.
<rcount>	The number of repetitions to execute the function for. (-1 repeat forever)

Where items could be scheduled to occur at the same instant, the priority flag is used to ensure the correct sequencing. The Background Scheduler is a simpler version of the Real Time Scheduler. Items added to the BGS are executed in the order they appear in the schedule list, the ordering is controlled by the priority flag.

A simple example task is to turn on a signal light (LED) at a specific time.

addScheduleItem (RTSHED,1000,0,Toggle_Led, 30, 1,-1);

This command will add task to RTS, executed on interval 1000 once only, calling a built in function called "Toggle_Led" at priority 30.

The application programmer must also write any specialized processing modules or hardware drivers required by the application. The interrupt handlers and schedulers call the functions by function pointers therefore all functions called by interrupt handlers or schedulers are of the form: `void function_name (void)`.

3 Data Collection and Analysis

Sensor data once collected by the platform is stored on flash memory and downloaded for subsequent analysis. There are many commercial and freeware software packages available for analysis of this kind of time series data sets. In general, collected data sets are large by conventional spreadsheet standards, typically generating many thousands of lines of data per minute of recording. Thus spreadsheet applications are less suitable, instead more powerful algebraic, numeric, statistical analysis software suites are preferred. Table 2 shows some typical commercial and license free software commonly available that the authors feel is suitable for such motion analysis research.

In the accompanying workshop, we will use Scilab a freeware application similar in functionality to the well known Matlab programming and analysis environment. Inertial sensor (tri-axial accelerometer) data is first obtained from the sensor platform as comma separated Variable (CSV) file and then imported into Scilab. Typical data analysis for inertial sensor data can be classified broadly into two categories, time domain analysis and frequency domain analysis. Normally an investigation begins with the data visualisation or "eyeballing" of the acquired acceleration in the time domain visual validation. Figure 4 illustrates typical vertical

Table 2. Available software for time series data analysis.

Software	Latest version	Licence	URL
Matlab	6.3	Commercial	http://www.mathworks.com/
Scilab	4.1	Freeware	http://www.scilab.org/
FreeMat	3.0	Freeware	http://freemat.sourceforge.net/
Octave	2.9.10	Freeware	http://www.gnu.org/software/octave/
Mathematica	5.2	Commercial	http://www.wri.com/
Maxima	5.11.0	Freeware	http://maxima.sourceforge.net/
S-PLUS	7	Commercial	http://www.insightful.com/
R	2.5.0	Freeware	http://www.r-project.org/
Igor Pro	6	Commercial	http://www.wavemetrics.com/

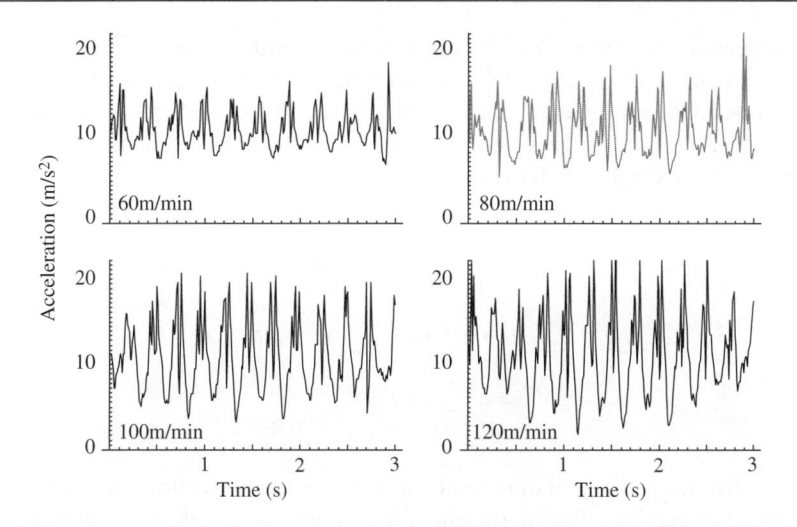

Figure 4. Vertical accelerations collected from a subject walking at four different speeds.

(z-axis) acceleration during walking at four different speeds. This data was collected using accelerometers attached at the subject's lower back (L5) with the z-axis aligned with the spine. It is important to note that in general the z-axis acceleration will not equal the vertical acceleration with respect to the global co-ordinate system due to variations in subjects gait.

3.1. Filtering Techniques

Acquired acceleration combines many signals other than those of principal interest which are often attributed to artifacts related to movement and a variety of other noise sources. By removing the high frequency component many such noise sources are eliminated, a digital low pass filter is a simple way to do this. A moving average is a simple digital filter for such smoothing. First and second ordered moving average methods have proven to be popular and are low numerical "cost" to implement. For example in the Eq. (1), a second order five point moving average filter requires only two pre and post datapoints (x_j) to produce smoothed acceleration (y_i), where (m) and (n) are the order of the filter and number of data length.

$$y_i = \frac{1}{n} \sum_{j=-m}^{m} x(i+j) \quad (1)$$

$$i = m+1, m+2, \ldots, n-m$$

A simple moving average is one the simplest class of filters referred known as Finite Impulse Response (FIR) filters and works well for the statistical analysis. However its frequency characteristics need to be considered so that important information is not lost. For instance the optimum cut off frequency is critical for preserving and enhancing the quality of our experimental data. The relationship between sampling rate (f_s), cut off frequency (f_c) and order of the moving average (m) is shown the equation as follows.

$$m = 0.443 \frac{fs}{fc} \quad (2)$$

Manipulating these variables can help optimize for order, sampling frequency and cut-off frequency for your experimental requirements. Simple moving average methods have identically weighted coefficients for each filter element. A weighed moving average method is an improved version of the simple moving average method. Definition of the general weighed moving average is as follows.

$$y_i = \frac{1}{W} \sum_{j=-m}^{m} w(j)x(i+j)$$

$$i = m+1, m+2, \ldots, n-m \quad (3)$$

$$W = \sum_{j=-m}^{m} w(j)$$

W is called the coefficient of normalization and contains the time-weighted values for each element of the filter. The coefficient of weight vectors and the normalisation are presented in Table 3. The table shows from second to sixth ordered weighed average smoothing method based on a polynomial approximation. This approximation is named Savitzky & Golay (1964) method though there are other weighting methods as well.

Table 3. Coefficients (W) of Savitzky-Golay moving average smoothing for different filter orders (m).

m\W	−6	−5	−4	−3	−2	−1	0	1	2	3	4	5	6
2					−3	12	17	12	−3				
3				−2	3	6	7	6	3	−2			
4			−21	14	39	54	59	54	39	14	−21		
5		−36	9	44	69	84	89	84	69	44	9	−36	
6	−11	0	9	16	21	24	25	24	21	16	9	0	−11

For more higher ordered smoothing the following equations allows us to obtain these coefficients (W) and their normalised weight value (k):

$$W(m) = \frac{(4m^2 - 1)(2m + 3)}{3}$$
$$k(m, j) = \frac{3m(m + 1) - 1 - 5j^2}{W(m)} \quad (4)$$

The Infinite Impulse Response (IIR) digital filter is also commonly used, it has some advantages in gain and phase performance when examining certain signals. In the field of biomechanics the Butterworth IIR digital filter has been a popular choice (Winter, 1990).

3.2. Feature Analysis

Impact motion during the walking, running or jump-like movements causes large amplitude shifts in acceleration that can be used for detecting and synchronizing activity signatures. Thus we can extract single gait cycle data and identify individual subject's steps using basic peak detection techniques. These include the thresholding methods, local max-min identification methods and differential smoothing methods. The differential smoothing method has some unique characteristics with its output data. That is the smoothed data is also differentiated by the filter. Unlike threshold methods, the differential smoothing method requires no maximum and minimum thresholds for the single cycle extraction, which is advantageous for widely varying signals. This is the principle advantage of the differential smoothing filter as it is able to identify the transitional change of the acceleration at the impact moment by using its differentiated value. Equation(5) allows us to determine the coefficients of an arbitrary ordered differential smoothing function.

$$k(m, j) = \frac{3j}{(2m + 1)(m + 1)m} \quad (5)$$

For example, a five point ($j = 1..5$) coefficient vector for a second ordered ($m = 2$) differential smoothing function would be:

$$k = \{-1/5, -1/10, 0, +1/10, +1/5\} \quad (6)$$

3.3. Frequency Domain Analysis

In addition to the data smoothing and feature analysis methods, there are many other considerations for inertial sensor analysis in the time domain. For example, auto-correlation (self-correlation) and cross correlation functions are typical methods in comparing or overlaying different time series data inter and intra subject to compare differences in gait, anthropomorphic fatigue injury and even technique. All of the time domain techniques have their equivalent methods that can be employed in the frequency domain by Fast Fourier Transforms (FFT). Many analysis techniques can perform faster with simpler implementation in the frequency domain. Together time domain and frequency domain techniques can be used to extract performance characteristics and when combined can produce powerful tools for analysis.

4 Conclusions

This paper presents a low cost platform for human motion analysis. Previously demonstrated in a number of athletic disciplines it is presented together with a simple programming environment for rapid customization. A number of simple numerical technical techniques for analysis by experienced and new investigators have been presented along with typical data that might be yielded.

In an accompanying hands-on workshop at APCST2007, the authors will make available devices for the use of participations and cover basic operation, configuration and data collection. Using readily available tools participants will be introduced to a number of data processing techniques. As an example of the platform, techniques and skills developed cyclic human motion such as walking monitoring and also, the monitoring of the vertical jump will be conducted. As an example application of ballistic non-cyclic human movement, the use of inertial sensors for jumping will be investigated to estimate human power output.

Acknowledgements

The authors wish to acknowledge funding support from the Queensland Academy of Sport Centre of Excellence for Applied Sport Research and the financial assistance of Keio University and Griffith University in providing support for academic exchange between the two institutions.

References

Billing D.C., Filipou V., Hayes J.P., Nagarajah C.R. and James D.A. (2003) Development and Application of a Wireless Insole Device for the Measurement of Human Gait Kinematics, *Proceedings World Congress on Medical Physics and Biomedical Engineering*, Melbourne, Australia, 24–29 Aug.

Channells J., Purcell B., Barrett R. and James D.A. (2006) Determination of rotational kinematics of the lower leg during sprint running using accelerometers, *Proc. SPIE*, 6036, 6–9, 1–13, In: Nicolau D.V. (Ed.); *BioMEMS and Nanotechnology II*; SPIE-International Society for Optical Engineering, Bellingham, WA.

Davey N., Anderson M. and James D.A. (2004) An accelerometer based system for elite athlete swimming performance analysis. *Proceedings of SPIE*, 5649, 409–415, Sydney, Australia.

James D.A., Jaffari I., Kavanagh J.J. and Barrett R. (2005) Instrumentation for Multi-Limb Motion Monitoring Using Tri-Axial Accelerometer Nodes on a Wireless Network. In: Subic A. and Ujihashi S. (Eds.) *The Impact of Technology on Sport*. ASTA (Australasian Sports Technology Alliance), Melbourne, Australia, pp. 361–6.

James D.A., Uroda W. and Gibson T. (2005) Dynamics of Swing: A Study of Classical Japanese Swordsmanship using Accelerometers. In: Subic A. and Ujihashi S. (Eds.) *The Impact of Technology on Sport*. ASTA (Australasian Sports Technology Alliance), Melbourne, Australia, pp. 355–60

Kavanagh J.J., Morrison S., James D.A. and Barrett R. (2006), Reliability of segmental accelerations measured using a new wireless gait analysis system. *Journal of Biomechanics*, 39, 2863–72.

Lai A., James D.A., Hayes J., Rice A. and Harvey E.C. (2005) Validation of a theoretical rowing model using experimental data. *Proceedings of the 20th Congress of the International Society of Biomechanics*, Cleveland, USA, 31 July–5 Aug.

Ohgi, Y. (2002) Microcomputer-based Acceleration Sensor Device for Sports Biomechanics – Stroke Evaluation by using Swimmer's Wrist Acceleration, *Proceedings of IEEE Sensors* 2002, pp. 699–704.

Partridge K., Hayes J.P., James D.A., and A. Hahn, "A wireless-sensor scoring and training system for combative sports", *Proc. SPIE*, Vol. 5649, pp. 402–408, *Smart Structures, Devices, and Systems II*; Said F. Al-Sarawi; Ed. Feb. 2005.

Savitzky A. and Golay M.J.E. (1964) Smoothing and Differentiation of Data by simplified Least Squares Procedures. *Analytical Chemistry*, 36, 1627–1639.

Winter, D.A., 1990, *Biomechanics and motor control of human movement*, 2nd edition, Wiley-Interscience.

Wixted A.J., James D.A., Thiel D.V., 2002, "Low Power Operating System and Wireless Networking for a Real Time Sensor Network", *Proceedings of IEEE International Conference on IT and Applications*, Bathurst, 25–29. Nov. paper 112.

Wixted A.J., Thiel D.V., Hahn A., Gore C., Pyne D. and D.A. James, "Measurement of Energy Expenditure in Elite Athletes using MEMS based inertial sensors", *IEEE Sensors Journal*, Vol 7, No. 4, pp. 481–8, April 2007.

AN EFFICIENT AND ACCURATE METHOD FOR CONSTRUCTING 3D HUMAN MODELS FROM MULTIPLE CAMERAS

C.K. QUAH,[1] A. GAGALOWICZ[2] & H.S. SEAH[1]

[1]*Nanyang Technological University, School of Computer Engineering, Singapore*
[2]*INRIA, Domaine de Voluceau, France*

To achieve accurate and reliable tracking of human motion for bio-kinematics analysis, a 3D human model very similar to the subject is essential. In this paper, we present a novel approach to (1) precisely construct the surface shape of the whole human body, and (2) estimate the underlying skeleton. We make use of multiple 2D images of the subject in collaboration with a generic anthropometrical 3D model made up of regular surfaces and skeletons to adapt to the specific subject. We developed a three-stage technique that uses the human feature points and limb outlines that work together with the generic 3D model to yield our final customized 3D model. The first stage is an iterative camera pose calibration and 3D characteristic point reconstruction-deformation algorithm that gives us an intermediate 3D model. The second stage refines the intermediate 3D model by deformation via the silhouette limbs information, thus obtaining the surface skin model. In the final stage, we make use of the results of skin deformation to estimate the underlying skeleton. The color texture for visualization is obtained by computing the projected texture coordinates from the 3D model. From our final results, we demonstrate that our system is able to construct quality human model, where the skeleton is constructed and positioned automatically.

1 Introduction

Accurate 3D modeling of the human athlete is a very essential component for sports motion analysis. Existing commercial motion capture systems require the athletes to wear some form of sensors or markers on their body. These objects severely hinder the athletes' motion and are not suitable to be used during competitions. When only video cameras are used, it is crucial to use a 3D human model that is very similar to the individual human subject. Otherwise, accurate and reliable 3D acquisition is not possible.

The objective of this research is to (1) construct accurately the surface mesh of the whole human body, (2) estimate the underlying skeleton, and (3) map the color texture for 3D visualization from arbitrary virtual views. Although many techniques for building 3D objects had been proposed, however, they are either too expensive, difficult to setup or unable to fit into the operating environmental requirements. The method that we present makes use of images of the subject acquired from wide baseline views, and in collaboration with a generic anthropometrical 3D model built from a regular surface mesh and skeleton to adapt to the specific subject. Since our setup requires no prior camera calibration and minimal human interaction, operation is simple, cheap and efficient.

In section 2, we review the existing methods. In section 3, we explain the steps of our algorithm, and then our results and conclusions in section 4.

2 Background and Existing Methods

The existing vision-based reconstruction systems that mainly deal with constructing the surface skin model fall into the 2 categories: (1) 3D laser-scanner systems, and (2) passive multi-camera systems. The commercial 3D laser-scanner systems capture the entire surface of the human body in about 15 to 20 seconds with resolution of 1 to 2 mm. However the drawbacks of such a device are (1) highly priced at about few hundred thousands of dollars, and (2) require the subject to stay still and rigid for the whole duration of scanning (about 15 seconds for full body coverage) which is quite constrictive in practice. On the other hand the passive multi-camera systems are much cheaper and video cameras are more easily available. Most of the existing methods e.g. (Weik, 2000), (Mikic *et al.*, 2003) make use of visual-hull related approaches requiring (1) the subject to be segmented from the image background, and (2) the cameras to be calibrated beforehand using calibration tools. Shape-from-silhouette approaches also give rise to "blocky" results if there are insufficient views, shown in the theoretical proof by (Laurentini, 1995). More recent approaches e.g. (Remondino, 2004) propose 3D reconstruction from un-calibrated views use feature correspondents, require the subject to remain still for about 40 seconds during the capturing the whole body. Moreover, the reconstructed model could contain non-manifold problems e.g. holes and open edges.

There are researches that attempt to estimate more precisely the joint locations. They are usually done using commercial motion capture systems e.g. (Silaghi *et al.*, 1998), (O'Brien *et al.*, 2000). However, all these methods require tedious post-processing to clean up the motion capture data. More recent approaches (Theobalt *et al.*, 2004) attempted to estimate the skeleton from sequence of volume data of rigid bodies. However, the resultant skeleton is an estimated stick-figure-like structure. These structures do not contain sufficient anatomical details for realistic character animation and skinning. Another alternative is to acquire the human skeleton via X-ray. However X-ray devices are not easily available. Moreover, tedious post-processing may be required to integrate the data from both the cameras and X-rays.

3 Human Model Reconstruction

3.1. *Framework*

Our proposed human model construction starts from a generic human model in a stanza position (Figure 1). The surface of our model is made up of about 17000 vertices and 34000

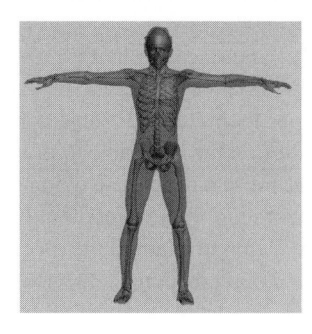

Figure 1. Generic model.

Figure 2. Features points on 3D generic model corresponding on the 2D image.

An Efficient and Accurate Method for Constructing 3D Human Models 115

triangular faces. Inside this surface is the underlying generic skeleton. The anatomic features of the subject are used for deforming this generic model to produce a customized model. The strategy of our framework is motivated by the method in (Roussel & Gagalowicz, 2003), which was used for face reconstruction.

The block diagram of our model construction approach is shown in Figure 3. In this setup, the subject's body is used as the calibration tool. The 3D generic model guides the camera calibration, which, in turn, allows 3D point reconstruction to yield the camera poses and produces the customized 3D model. Our image acquisition for all the views is instantaneous. The testing data are the images acquired from wide baseline camera views provided by MIRAGES, INRIA. Figure 2 shows the example of the selected feature points on the 2D images corresponding to the 3D points. These correspondences can be established via an interactive point-matching tool to ensure that they are all correct, so that the calibration is always stable. In our set-up, we utilized a set of 32 surface characteristic points that provide an over-determined information and view coverage for camera calibration and reconstruction of points.

3.2. Camera Calibration & 3D Feature Reconstruction

Camera calibration is done by using the POSIT (pose iteration) method (Dementhon & Davis, 1995) to obtain the positions and orientations. Using the calibrated camera parameters and the 3D/2D correspondences, we perform 3D point reconstruction to deform the 3D

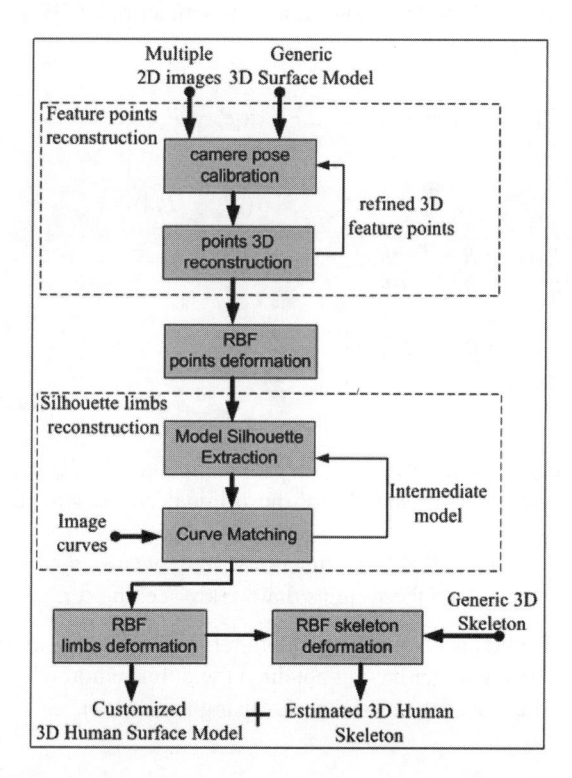

Figure 3. Block diagram for 3D human model construction.

characteristic points toward the new positions. The 3D point reconstruction is achieved by triangulating the projected rays from the characteristic image points. This algorithm takes into account that the rays will not intersect when the calibration is not perfect by minimizing the sum of square of distances to the projected rays from all the possible views. We only reconstruct the points seen in more than one image.

The camera calibration and 3D feature reconstruction are both carried out at the same time iteratively. Using the subject's 2D characteristic points from the images in collaboration with their respective correspondents on the 3D generic model (Figure 3), we iterate the process comprising the camera calibration and 3D generic model point deformation (3D reconstruction) until convergence is attained. At the start the 3D characteristic points of the generic model do not project correctly during the early iterations. As the process iterates, these 3D characteristic points will converge together with the camera poses on their (interactively) assigned 2D projections in the images. We will obtain the calibrated camera poses and a set of sparse reconstructed 3D model points R_i. We also have the original set of 3D points from the initial generic model P_i. Using P_i and R_i we can form a set of deformation vectors $\overrightarrow{P_i R_i}$.

3.3. Interpolating the Deformation & Intermediate Results

The reconstructed characteristic 3D model points are very sparsely distributed, therefore are not sufficient to represent the complete 3D model. So, we make use of the sparse points with the aid of the generic 3D model to complete the 3D model deformation via interpolation. The interpolation is done by using radial basis functions (RBF). We can write the equation of a linear system as:

$$\begin{bmatrix} \sigma(|P_1 - P_1|) & \sigma(|P_1 - P_2|) & \cdots & \sigma(|P_1 - P_N|) \\ \sigma(|P_2 - P_1|) & \ddots & & \vdots \\ \vdots & & \ddots & \\ \sigma(|P_N - P_1|) & \cdots & \cdots & \sigma(|P_N - P_N|) \end{bmatrix} \begin{bmatrix} A_{x1} & A_{y1} & A_{z1} \\ A_{x2} & A_{y2} & A_{z2} \\ \vdots & & \\ A_{xN} & A_{yN} & A_{zN} \end{bmatrix}$$
$$= \begin{bmatrix} \overrightarrow{PR}_{x1} & \overrightarrow{PR}_{y1} & \overrightarrow{PR}_{z1} \\ \overrightarrow{PR}_{x2} & \overrightarrow{PR}_{y2} & \overrightarrow{PR}_{z2} \\ \vdots & & \\ \overrightarrow{PR}_{xN} & \overrightarrow{PR}_{yN} & \overrightarrow{PR}_{zN} \end{bmatrix} \qquad (1)$$

Where:

1. $\overrightarrow{PR}$ are the set of deformation vectors computed via 3D reconstruction of the characteristic points. P and R are the original and reconstructed characteristic points.
2. $\sigma(|P_i - P_j|)$ are the radial basis functions. Here we use $\sigma(|P_i - P_j|) = |P_i - P_j|$.
3. A (i.e. A_{xi}, A_{yi}, A_{zi}) are the weights that we are seeking for.

The weights A can be obtained by solving equation (1) using simple linear algebra method like the LU decomposition. After having obtained the deformation weights A, we can then use them to deform the rest of the model points using the Eq. (2):

$$F_{x,y,z}(P) = \sum_{i=1}^{N} A_{xi, yi, zi} \bullet \sigma(|P - P_i|) \qquad (2)$$

An Efficient and Accurate Method for Constructing 3D Human Models 117

Figure 4. Intermediate model – not totally fitted.

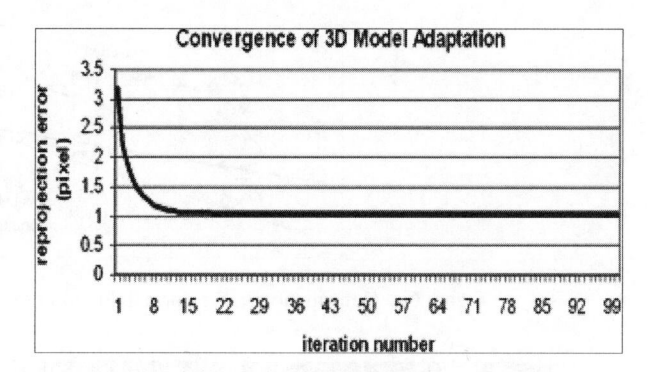

Figure 5. Plotting reprojection error vs no. of iterations.

where P is the set of 3D points from the generic model that we need to deform.

Up to this point, we have an intermediate surface model (Figure 4). We can notice from the results that projected local model silhouette limbs of the reconstructed model do not overlay exactly onto the images e.g. on the inner legs of the subject. Figure 5 shows the reprojection error of the feature points in pixels plotted against the number of iterations, which indicates convergence after about 30 iterations. The reprojected mean-square error at convergence is about 1.1 pixels with a standard deviation of 0.9 pixels.

3.4. Surface Model Refinement

This module uses the silhouette curve from the subject in the real images to improve the result of the intermediate model obtained above. We will automatically extract the silhouette edges of the intermediate model from calibrated views, and then deform the respective silhouette curves of subject in the real images via contour matching algorithm.

The silhouette of the intermediate model is extracted by using the method suggested in (Roussel and Gagalowicz, 2003). The silhouette in the real images can be extracted via image edge detection (Canny, 1986) when the operating scene is a well-controlled environment. Whereas in highly cluttered environments, contours in the real images may be extracted by using curve digitizing tool available in common commercial software like Photoshop. Matching of silhouette contours that preserve the correct topological sequence, is done by recursively sub-dividing the model curve at half curve length and seek for the closest point on the image curve.

3.5. Final Model

The zero-crossings from the matching correspondents of the contour points between the intermediate model and the real images are used to form the deformation vectors (as in section 3.2). Then the RBF interpolation (same as section 3.3) will use these deformation vectors (Figure 6) and those obtained from feature reconstruction in section 3.2 to complete the surface skin of the final model. This RBF is also used to estimate the internal skeleton of the subject by morphing from the generic skeleton. The color information on the 3D model can be visualized via the texture coordinates computed by back-projecting the 3D vertices of the model onto the real images using the calibrated camera poses.

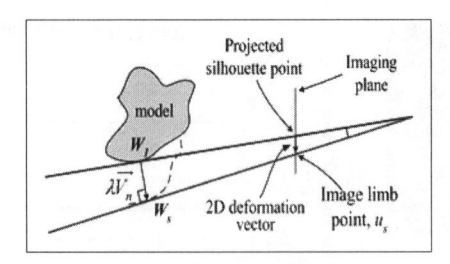

Figure 6. Deformation in the normal direction for model refinement.

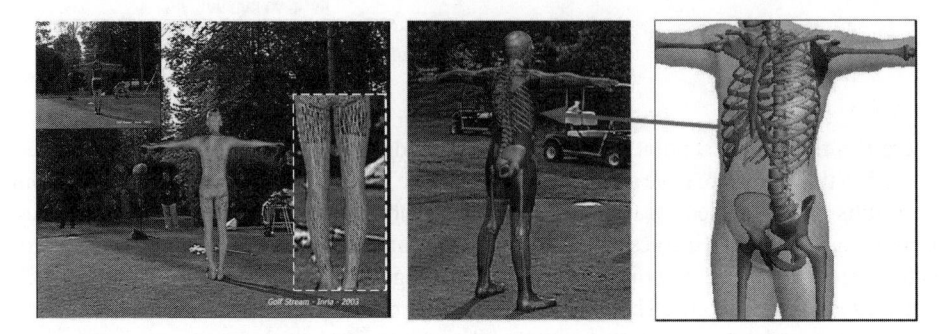

Figure 7. Final results of 3D human reconstruction. Pictures of Results showing different views, with subjects' skin and skeleton.

4 Results and Conclusions

In our setup, we used at least 4 images for reconstruction. Our algorithm was implemented using C++ (without optimization) running on a Pentium 4. The whole reconstruction process takes about 5 minutes. Figure 7 shows the results of the final model superimposed onto the real images. As we can see from the results, refining the intermediate model by using silhouette curve improved the results tremendously. The average error of the final model reprojected onto the testing images is about 0.5 pixel.

This algorithm had been tested on subjects of different shape and size. From our final results, we demonstrate that our system is able to build quality human model, where the skeleton is constructed and positioned automatically. This method has also been adapted into the French Golf-Stream project that leads towards the motion analysis of the golfers.

References

Canny J. (1986) A Computational Approach to edge detection. *IEEE Trans. PAMI*, 8/6, 679–698.
Dementhon D.F. and Davis L.S. (1995) Model-based object pose in 25 lines of code. *International Journal of Computer Vision*, 15, 123–141.
Mikic I., Trivedi M., Hunter E. and Cosman P. (2003) Human body model acquisition and tracking using voxel data. *International Journal of Computer Vision*, 53, 199–223.
Laurentini A. (1995) How far 3D shapes can be understood from 2D silhouettes. *IEEE Trans. PAMI*, 17, 188–195.

O'Brien J.F., Bodenheimer Jr R.E., Brostow G.J. and Hodgins J.K. (2000) Joint parameter estimation from magnetic motion capture data. *Proc. of Graphics Interface 2000*, Montreal, Canada, 53–60.

Remondino F. (2004) 3-D reconstruction of static human body shape from an image sequences. *Computer Vision and Image Understanding*, 93, 65–85.

Roussel R. and Gagalowicz A. (2003) A Morphological adaptation of a 3D model of face from images. *MIRAGE 2003 Conf.*, INRIA Rocquencort.

Silaghi M-C., Plankers R., Boulic R., Fua P. and Thalmann D. (1998) Local and global skeleton fitting techniques for optical motion capture. *Modeling and Motion Capture Techniques for Virtual Environments, Lecture notes in artificial intelligence*, 26–40.

Theobalt C., Aguiar E., Magnor M., Theisel H. and Seidel H-P. (2004) Marker-free kinematic skeleton estimation from sequence of volume data. *Proc. ACM Virtual Reality Software and Technology*, Hong Kong, 57–64.

Weik S. (2000) A passive full body scan using shape from silhouette, *Proc. ICPR 2000*, Barcelona, Spain, 99–105.

MARKER-LESS 3D VIDEO MOTION CAPTURE IN CLUTTERED ENVIRONMENTS

C.K. QUAH,[1] A. GAGALOWICZ[2] & H.S. SEAH[1]

[1]Nanyang Technological University, School of Computer Engineering, Singapore
[2]INRIA, Domaine de Voluceau, France

Measuring the joint kinematics of an athlete is a key component for sports biomechanical studies. In our work, we used only video cameras and do not need the subject to wear any markers or sensors on their body. The motivation of using a marker-free motion capture set-up is driven by the need to acquire video images non-intrusively in cluttered and outdoor environments, e.g. during tournaments. We proposed a new analysis-by-synthesize method that built on the concept of collaboration between computer vision and computer graphics to capture human movements. Unlike common vision-based tracking, we do not use the noise-sensitive image segmentation and skeletonization. Our algorithm automatically realizes the colour or texture onto an animatable 3D human model of our subject. Our computation will synthesize the 3D puppet postures such that it minimizes the differences between the synthesized movements and real athlete's motion. This is achieved by using the simulated-annealing algorithm to compute the degree-of-freedoms of the joint kinematics. The results show that our method is able to track the motion of the arms that appear highly articulated and qualitatively small in the images. It can also operate in cluttered environments.

1 Introduction

Capturing 3D human movements that can be used for bio-mechanical analysis requires the sensor systems to be able to follow the motion trajectory of the performer. The operating environment may be highly cluttered. The non-intrusive factor is also very important so that it can be applied in sports competitions. This implies that existing commercial motion capture (mocap) methods that require the subject to wear sensors are not suitable, hence constraining us to just video cameras. We may also have to face the situations when minimal cameras can be used due to constriction and limited resources. This prompted us to draw expertise from both computer vision and computer graphics techniques to accomplish our task. The computer graphics area will provide the knowledge for 3D modeling, animation and synthesis, while the computer vision area provides the learning and analysis capabilities.

2 Background

Large numbers of works on tracking and analysis of human motion using computer vision techniques had been proposed over the years. A comprehensive review of vision-based motion capture can be found in (Moeslund et al., 2006) where they extended the earlier survey of (Moeslund & Granum, 2001). The algorithms, technologies and equipment setups that had been used for motion capture are driven by the main application areas: surveillance, interaction/control, or analysis. The type of motion capture applications determines the main performance requirements, in terms of robustness, speed and accuracy. For example, in human motion analysis applications, accuracy is very important and a certain amount of robustness for tracking is required so that it can cope with the cluttered environment for reduction in post-processing time.

In systems that require accuracy and robustness, some kinds of human model are usually used to facilitate the tracking. The work of (Kakadiaris & Metaxas, 1998) and (Gerard & Gagalowicz, 2003) had indicated that the process of tracking is very sensitive to the shape parameters used. Therefore, it is inappropriate to use, for example, a generic "averaging human" model for accurate and precise tracking of human with different shape and size. We also have to take note that mathematical skeletons extracted from an object using image skeletonization (Wingbermuhle *et al.*, 2001) routines are not the same as the anatomical skeleton. Moreover, 2D and 3D image skeletonization algorithms are very sensitive to noise.

Nearly every vision-based mocap follow the steps: (1) segmentation of subject from the rest of the image, (2) these segmented images are transformed into some kinds of higher level representation to suit a particular tracking algorithm, and (3) how the subject should be tracked from one frame to the next. In these kinds of framework, many proposed algorithms rely heavily on the image segmentation, which is a very crucial part of the system, and to their modeling. In addition, some assumptions regarding the background scene had been assumed e.g. constant and low cluttered background.

3 Human Motion Tracking

3.1. *Framework*

We propose a model-based analysis-by-synthesis framework for human motion capturing (see block diagram on Figure 1). The concept of this work is built on the work by (Gerard & Gagalowicz, 2000) for tracking rigid objects. Prior to this framework, a textured 3D puppet model that closely resembles the subject was built (Quah *et al.*, 2007) to facilitate the motion tracking. Our method does not require image background segmentation or *skeletonization*, which are very sensitive to noise.

Our algorithm starts with automatic realization of the colour/texture onto the puppet model from its initial pre-positioned posture. Then, our computation will synthesize the 3D puppet movements such that they minimize the differences between the synthesized movements and real athlete's motion. This is achieved by using the simulated-annealing algorithm to search through the various probable degree-of-freedoms of the joint kinematics. The joint kinematics then drives the skin of the model puppet. The solution to the joint kinematics is taken when the skin puppet posture yield the smallest error when superimposed onto the

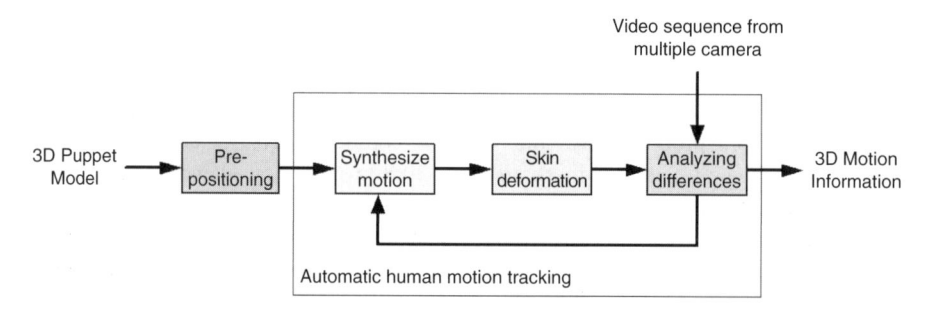

Figure 1. Model-based framework for human motion capture.

video containing the subject. The colour texturing onto the puppet is analyzed and update once every few frames so that the synthesis is not influence by the changing articulated posture and illumination variations. In this project, we implemented the rigid skin deformation. The rendering for motion synthesis is the most computational intensive module and it is sped up by using the graphics processor unit (GPU).

3.2. Human Kinematics Chain

The posture of the subject is represented by a kinematics chain linking various body parts. Figure 2 shows the direction of the kinematics chain of the human arm, where the shoulder joint is the parent to the elbow joint and so on. Each articulated node on the human body part can be modeled by the Euler rotational angles; a ball joint will have 3 degrees-of-freedom (DOF) while a hinge joint will have 1 DOF. The forearm may also be approximated by an additional twisting function.

3.3. Minimization for Instantaneous Human Kinematics

A numerical optimization algorithm is used for realizing the automatic human motion tracking loop in Figure 1. We use the simplex simulation annealing (Press *et al.*, 1992) to search for the best-fit posture by minimizing the errors between the synthesized image texture and the real images. Simulated-annealing minimization is able to bail out of local minimum, unlike some gradient-based approaches, which always converges to local minimum. Please refer to (Press *et al.*, 1992) for more details on simplex simulation annealing. Our minimization takes the forward kinematics chain to optimize for the posture of the subject hierarchically starting from the parent joint and replicates it to their respective children joints.

Let us consider the estimation of the human posture from one frame to the next. The simplex algorithm may be expressed as:

$$P_i = P_0 + \lambda_i * E_i \qquad (1)$$

where P_i is the respective vertices in the multi-dimensional space representing each synthesized postured, P_0 is the initial pose, E_i are orthonormal vectors and the λ_i are initialization values chosen to be large enough to enable the optimizer to skip over the local minima. The simplex procedure will deform with respect to the error function as it traps the potential solution in its convex hull before eventually converge to it. This deterministic simplex computation is coupled with a standard Metropolis approach so that after a convex hull deformation, a stochastic search is performed to overcome the possible local minima.

The error function that we like to minimize is written as:

$$\text{Err}(P) = \frac{1}{N} \sum_{r \in R} |I_{\text{synt}}(r) - I_{\text{real}}(r)| \qquad (2)$$

$I_{\text{synt}}(r)$: R, G, B values of the synthesized image which is a function with respect to the synthesized 3D model posture represented by P and its texture.
$I_{\text{real}}(r)$: R, G, B values of the real image.
R is the region of image that is encompassed by the synthesized object.
N is the total number of pixel in the region R.

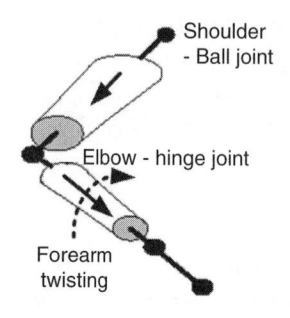

Figure 2. Kinematics chain of arm.

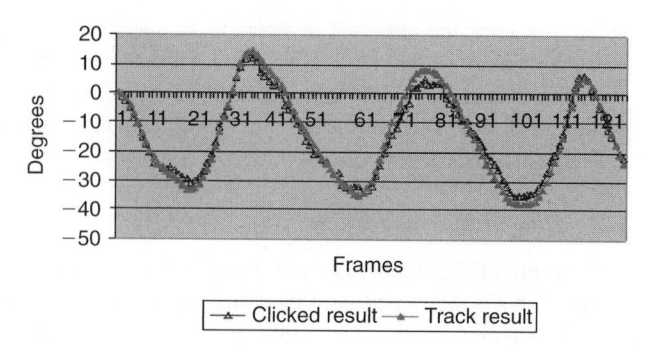

Figure 3. Plot of elbow bending angles using our tracking method versus the manually positioned angles.

Figure 4. Tracking in cluttered background.

4 Implementation, Results and Conclusions

This motion capture system is developed using C++. The texture synthesis rendering is done by using the Wildmagic 3D game engine (Eberly, 2005) that utilizes the GPU hardware acceleration driven by OpenGL.

For testing, we concentrate on tracking the motion of the arms since they usually appear highly articulated and qualitatively small in the images. All the equipment that we used comprised of cheap off-the-shelf devices.

Figure 3 shows a typical measurement over a segment of image sequence tracking the elbow bending angles using our method versus the manually positioned angles (done by using 3DS Max for manual adjustment frame-by-frame). We have tested our algorithm under different scenarios e.g. cluttered environment, changing background and self-occlusion. In addition, we have tested our algorithm using only a single camera view most of the time (at times, we used at the most 2 cameras) and show that good tracking results can be achieved. In Figures 4, 5 and 6, the meshes of the upper-arm and forearm are superimposed onto sequence of images to visualize the tracking results. The top images show the results for the camera view that we used for tracking. The bottom images show the results on another camera from the same videoing scene that we used for verification and visualization.

Figure 5. Tracking with self-occlusion.

Figure 6. Tracking in cluttered outdoor environment.

We had also observed that good motion tracking requires at most 1600 iterations of motion synthesis for each frame, real-time mark-less motion capture can be achieved by executing on a few computers and GPUs.

Acknowledgements

The first author would like to thank Mr. Ta Huynh Duy Nguyen for assisting in the image acquisition.

References

Eberly D. (2005) 3D Games Engine Architecture. *Elsevier, Morgan Kaufmann Publisher,* 2005.
Gerard P. and Gegalowics A. (2000) Three dimensional model-based tracking using texture learning and matching. *Pattern Recognition Letters*, 1095–1103.
Gérard P. and Gagalowicz A. (2003) Human Body Tracking using a 3D Generic Model applied to Golf Swing Analysis, MIRAGE 2003 Conference, INRIA Rocquencourt, France.
Kakadiaris I. and Metaxas D. (1998) 3D human body acquisition from multiple views. *Internation Journal of Computer Vision*, 30(3), 191–218.
Moeslund T.B. and Granum E. (2001) A survey of computer vision-based human motion capture, *Computer Vision and Image Understanding*, 81, 231–268, 2001.

Moeslund T.B., Hilton A. and Krüger V. (2006) A survey of advances in vision-based human motion capture and analysis, *Computer Vision and Image Understanding*, 104, 90–126.

Press W.H., Teukolsky S.A., Vetterling W.T. and Flannery B.P. (1992) Numerical Recipe in C, the art of scientific Computing. 2nd Edition, *Cambridge University Press*.

Quah C.K., Gagalowicz A. and Seah H.S. (2007) An Efficient and Accurate Method for Construction 3D Human Models from Multiple Cameras. In: Fuss F.K., Subic A. and Ujihashi S. (Eds.) *The Impact of Technology on Sport II*. Taylor & Francis Group, London.

Wingbermuhle J., Liedtke C.-E. and Solodenko J. (2001) Automated Acquisition of Lifelike 3D Human Models from Multiple Posture Data. Intl. Conf. on Computer Analysis of Image and Patterns, CAIP2001, 400–409.

ND # 5. Apparel and Sport Surfaces

IMPROVING THE UNDERSTANDING OF GRIP

S.E. TOMLINSON, R. LEWIS & M.J. CARRÉ

Department of Mechanical Engineering, University of Sheffield, Sheffield, UK

This paper introduces the initial findings from frictional tests of the human finger contact on steel, glass and rubber, as used in the manufacture of rugby balls. The materials were tested using a bespoke finger friction rig, with which the normal and frictional forces are measured when a finger is dynamically moving along a material. The results showed the coefficient of friction of the rugby ball material is much greater than that of glass and steel. The results also show and highlight the differences in the relationship between normal force and coefficient of friction for the viscoelastic and non viscoelastic surface materials, neither of which show linear relationships due to the viscoelastic nature of the finger in contact with them.

1 Introduction

The ability of a rugby player to handle a ball well can be the difference between a good or bad pass. The handling performance of a rugby ball can be quantified, in one aspect, by the coefficient of friction of the fingers with the ball material. The coefficient of friction is determined by the surface texture and also the surrounding conditions, such as temperature and the presence of moisture.

Rugby balls are made from rubber and therefore have viscoelastic properties. They also have a surface of pimples; these can be round, square, large, small, densely or sparsely populated. The fact that the materials are viscoelastic adds complexity to the situation since the majority of previous work on human skin friction has been carried out on non viscoelastic materials.

There are 3 mechanisms of friction; adhesion, deformation and hysteresis. Adhesion is where the asperities of the two surfaces form local welds when brought together and the frictional force is the force required to shear these junctions. Deformation is the energy dissipation due to the deformation of the material, in viscoelastic materials, this results in hysteresis; energy absorption resulting in a delayed response to an applied force.

Previous research has shown the coefficient of friction, between the finger and non viscoelastic material, to vary with normal force, hydration and age (to an extent). The effect of age is only seen above the age of 50 years for glass and 70 years for sandpaper (Asserin *et al.*, 2000), which is a relatively insignificant factor when considering professional sportsmen. Gender and race showed not to have an effect (Sivimani *et al.*, 2003; Spurr, 1976). The effect of normal force for the skin on non viscoelastic materials has been shown to be linear at forces between that of 0–30 g (Asserin *et al.*, 2000). However, other work found there to be a nonlinear relationship, due to testing at higher normal forces. This non-linear relationship can be approximated by equation (1) (Comaish & Bottoms, 1971; Zatsiorsky, 2002). The nonlinearity is due to the finger pad becoming stiffer when a larger normal force is applied.

$$F = \mu N^n \tag{1}$$

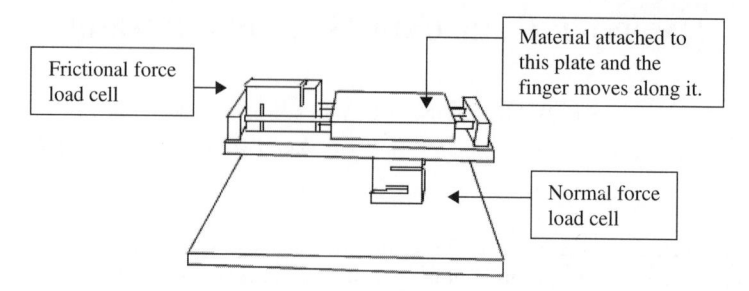

Figure 1. Illustration of the finger friction rig.

where F = frictional force, N = normal force and n = a constant less than 1. In most tests where n is quantified it is suggested to have a value of approximately -0.3.

The contact area changes depending on the applied force, due to the finger not being rigid. This variation in area and also the variation from person to person affect the coefficient of friction, the extent of which is not yet known.

2 Method

A friction rig has been designed to measure the normal force and the frictional force when a finger is run along a surface. The surface can also be changed. An illustration of the setup is shown in Figure 1. The rig consists of two load cells attached to a flat plate; one of these measures the frictional force and the other the normal force.

A series of evaluations have been done to ensure that the tests carried out on the rig represent as closely as possible the contact experienced in a game of rugby and produce consistent results. This included validating which finger to use on the rig, the correct speed to use and the area of the finger to use.

The area of the finger to be used in tests was determined by painting the hands of a rugby player blue and delivering him a slow pass. The area of the finger used in the catch was then imprinted on the ball. The player was then asked to print varying areas of the finger in a pressing action onto some paper. This enabled the area of the fingers printed on the ball to be correlated to a pressing action, making this the standard area to be used in tests. In this particular set of tests the 32 volunteers, (6 female and 26 males aged 20–49 years), ran there fingers at a slow constant speed, along 3 different materials; glass, steel and smooth rugby ball rubber. Smooth rubber was used in this instance for a direct comparison between rubber and the smooth surfaces of the glass and steel. The volunteers were asked to vary the force applied to the material; they applied the greatest force possible, the lightest force possible (with the specified area of contact) and then three intermediate levels of force.

3 Results

3.1. *Comparison to Previously Measured Coefficients of Friction*

Initial analysis carried out compared the measured results with those of previous work; the plot of this comparison is shown in Figure 2. This comparison shows that the coefficient of friction on glass measured by other researchers is lower than that measured in these tests. The main reason for this is the other glass tests (Koudine *et al.*, 2000; Prall, 1973; Johnson

Improving the Understanding of Grip 131

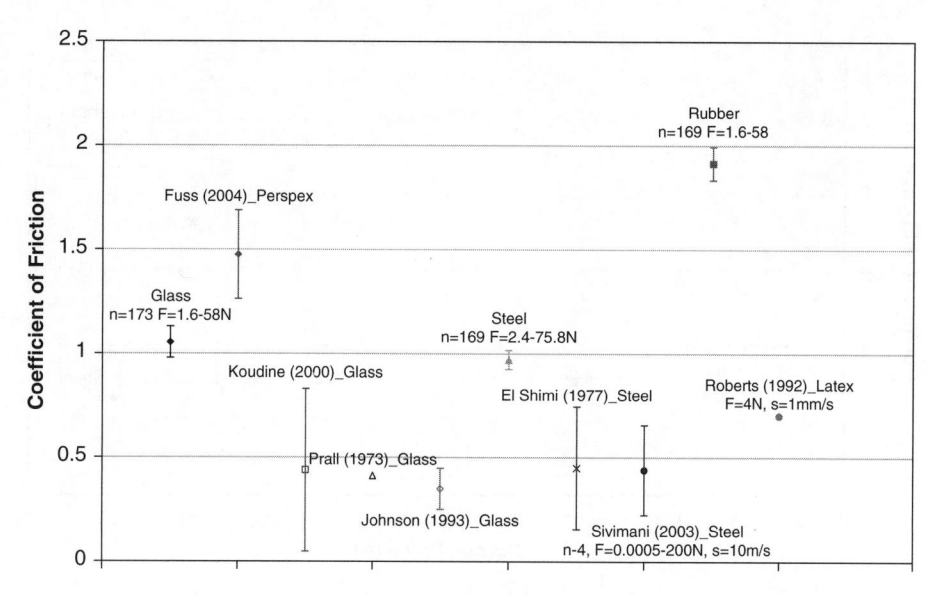

Figure 2. Measured coefficients of friction over full spectrum of forces and the coefficients of friction found in the literature.

et al., 1993) were all done on different areas on the body, mostly the forearm, not the finger pad. The skin on the finger pad has much more pronounced ridges so the surface characteristics are very different, changing the coefficient of friction. The sample size of people tested in these experiments is greater than the majority of the tests in the literature. The perspex tested by Fuss *et al.* (2004) against the finger, shows a higher coefficient of friction than that measured with the glass. Perspex is used as a substitute for glass, however the molecular structure and physical properties are different so differences in the coefficients of friction would be expected.

The previous steel coefficients of friction are also lower than the results gained in this test. Again the main reason of this is because the site of testing for the literature values is not the finger pad. Roberts measured the coefficient of friction of the finger against latex gloves. The coefficient of friction of this rubber is a lot lower than the rubber tested in these experiments; however the properties of the two rubbers are very different so there is no surprise to see a difference.

3.2. *Comparison of Rubber to Standard Materials*

The coefficient of friction between the finger and rubber is greater than that between a finger and steel or glass, as illustrated in Figure 2. This can be seen to be significantly greater, almost twice as high. The coefficient of friction was found to be slightly higher for glass than for steel, however this is not a statistical difference when compared over the full range of forces.

3.3. *Coefficient of Friction with Variable Normal Force*

The coefficient of friction was found to decrease with increasing normal force for the finger pad contact when tested on polycarbonate (Bobjer *et al.*, 1993) and acrylic (Han *et al.*,

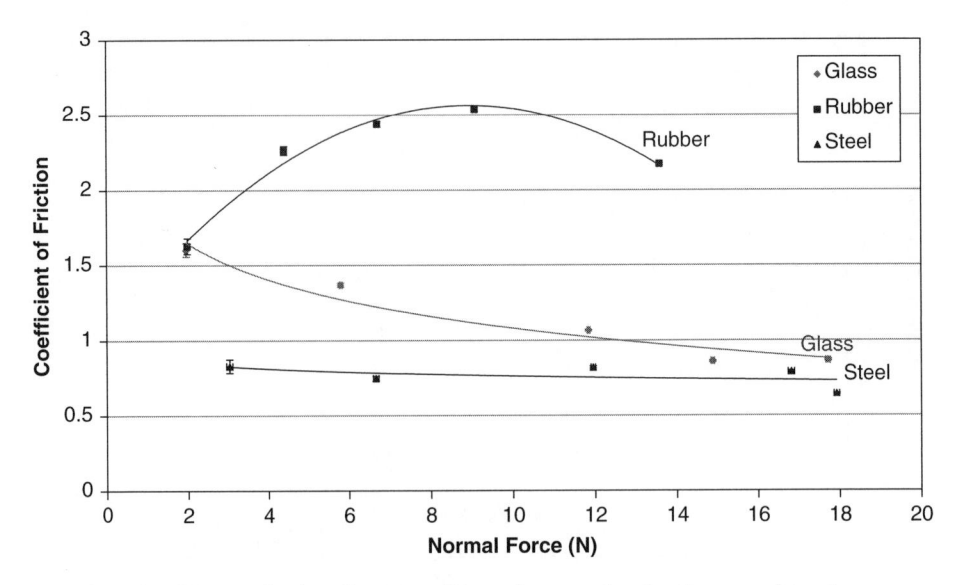

Figure 3. Measured values from one of the volunteers showing the general trend seen.

1996). The coefficient of friction for glass decreases exponentially with normal force, as shown in Figure 3, agreeing with this previously found trend. However, the coefficient of friction for the steel-finger contact seems to be fairly constant across all forces, this could be because the extent of the exponential decrease is less due to the lower coefficient of friction. The coefficient of friction varies from person to person as does the extent of the decreasing trend in coefficient of friction, so more work needs to be done on the effect of other factors such as area, to gain a fully integrated solution.

The main interesting factor that can be seen in these results is that the rubber follows a different trend to that of the standard engineering materials. Here it can be seen that the coefficient of friction increases with increasing normal force, to a point, at which it starts to decrease again, following a more parabolic trend. The main reason for this difference is that the rubber is a viscoelastic material, so therefore has very different material properties to that of either glass or steel. The rubber will also have variable stiffness, like the finger, so hysteresis is a friction mechanism associated with both rubber and fingers.

4 Discussion

Skin has varying properties depending on where it is on the body. This difference is clearly seen when comparing the skin of the palm to the forearm. Not only does the skin on the palms have more pronounced ridges, it also has a much thicker epidermis. These differing characteristics change the values of the measured coefficient of friction. This is illustrated when the results of these tests are compared to those of previous studies. This difference is also due to the test equipment; much of the work on the forearm was carried out using a probe. The hysteresis friction mechanism has a large influence on the coefficient of friction measured, so changing the profile of deformation by using a probe instead of a flat surface will change the coefficient of friction. There is also a large difference in coefficient of friction from person to person. This is due to several points. Firstly the characteristics of the skin vary from person to person; they may have dryer hands and everyone has a different pattern of undulations. The area used will also have a large effect. Skin is a viscoelastic material; the frictional properties of contacts with viscoelastic materials are dependant on area. The effect of area may be both the size of the finger and also the difference in the way it deforms with an applied load. This area effect is related to the force applied, so further analysis will be carried out on these results to find the effects of area from person to person and also the change in area with applied load for a single person.

The results showed that for glass, the coefficient of friction decreases with increasing normal force. This is the trend shown in the literature (Bobjer et al., 1993; Han et al., 1996). The coefficient of friction decreases because the skin becomes stiffer. This reduces the hysteresis effect so the coefficient of friction decreases. Increasing area will increase the adhesion mechanism, however this is to a much lesser extent than the decrease in hysteresis. The rubber showed a different trend, here the coefficient of friction increased with increasing normal force and then began to decrease after a certain load. The exact relationship differs from person to person, but they all follow this same trend. Further analysis will be carried out to try and quantify this relationship. The section of increasing coefficient of friction can be accounted to the larger contact area increasing the adhesion and therefore frictional force. That is to say that in this section the adhesion mechanism is more dominant than the hysteresis mechanism. The section of decreasing coefficient of friction can be explained in the same way to that on glass, however it is to a much greater extent because both materials are viscoelastic so the effect of reducing this mechanism is greater.

5 Conclusion

These experiments clearly show that the frictional properties and rules for rubber are very different to that of standard engineering materials. The analysis so far, goes some way to explain the trend of frictional force with normal force, however further analysis will be carried out to quantify this relationship. Once the relationship of the normal force on the rubber in the test situations has been completed this can be applied to an actual game of rugby by recording the forces involved in different passes and then using the force relationship to calculate the friction involved in that instant for the pass or throw concerned.

References

Asserin J., Zahouni H., Humbert P., Couturaud D. and Maougin D. (2000) Measurement of the friction coefficient of the human skin in vivo. Quantification of the cutaneous smoothness. *Colloids and Surfaces B: Biointerfaces*, 19, 1–12.

Bobjer O., Johansson S.E. and Piguet S. (1993) Friction between hand and handle. Effects of oil and lard on textured and non textured surfaces: perception of discomfort. *Applied Ergonomics*, 24/3, 190–202.

Comaish S. and Bottoms E. (1971) The Skin and Friction: Deviations from Amonton's laws, and the effects of hydration and lubrication. *British Journal of Dermatology*, 84/1, 37–43.

El-Shimi A.F. (1977) *In vivo* skin friction measurements. *Journal of the Society of Cosmetic Chemists*, 28, 37–51.

Fuss F.K., Niegl G. and Tan A.M. (2004) Friction between hand and different surfaces under different conditions and its implication for sport climbing. In: Hubbard M., Mehta R.D.

and Pallis J.M. (Eds.), *The Engineering of Sport 5*, Vol. 2, International Sports Engineering Assocation, Sheffield, UK, pp. 269–275.

Han H.Y., Shimada A. and Kawamura S. (1996) Analysis of friction on human fingers and design of artificial fingers. *International conference on robotics and automation*. Minnesota, April.

Johnson S.A., Gorman D.M., Adams M.J. and Briscoe B.J. (1993) The friction and lubrication of human stratum corneum. *19th Leeds-Lyon Symposium on Tribology*. Elsevier, pp. 663–672.

Koudine A.A., Barquins M., Anthoine P.H., Auberst L. and Leveque J.L. (2000) Frictional properties of the skin: proposal of a new approach. *International Journal of Cosmetic Science*, 84, 37–43.

Prall J.K. (1973) Instrumental evaluation of the effects of cosmetic products on the skin surfaces with particular reference to smoothness. *Journal of the Society of Cosmetic Chemistry*, 24, 693–707.

Roberts A.D. and Brackley C.A. (1992) Friction of Surgeons' gloves. *Journal of Physics: Applied Physics*, 25, A28–A32.

Sivamani R.K., Wu G.W., Gitis N.V. and Maibach H.I. (2003) Tribological testing of skin products: gender, age and ethnicity on the volar arm. *Skin research and technology*, 9, 299–305.

Sivamani R.K., Goodman J., Gitis N.V. and Maibach H.I. (2003) Coefficient of Friction: Tribological Studies in Man – An Overview. *Skin Research and Technology*, 9, 227–234.

Spurr R.T. (1976) Fingertip Friction. *Wear*, 39, 167–171.

Zatsiorsky V.M. (2002) *Kinetics of Human Motion*. Leeds, Champaign.

IONISED SPORTS UNDERGARMENTS: A PHYSIOLOGICAL EVALUATION

A.R. GRAY, J. SANTRY, T.M. WALLER & M.P. CAINE

Progressive Sports Technologies Ltd., Innovation Centre, Loughborough University, Loughborough, Leicestershire, UK

Sports apparel is becoming increasingly technical from both a feature and fabric perspective. A propriety technology has now made it possible to ionically treat garments. Existing literature suggests exposure to negatively charged particles may be beneficial to sports performance. Six recreational games players and six university 1st team players were recruited to quantify the influence of ionic garments (I) and non ionic garments (N) on discrete physiological parameters associated with rugby. Mean power was found to be higher (2.7%, $P < 0.05$) when wearing garment I than garment N in a second Wingate test. Additionally all rugby players tested achieved higher mean power outputs in this second test. During rest minimum heart rate was found to be lower (53 bpm, $P < 0.05$) wearing ionised garments than control (56 bpm). Other parameters of performance and recovery which included the measurement of haemodynamics, pain perception during sub-maximal arm crank ergometry and maximal power output during incremental cycle ergometry were not found to be significantly different. This is the first study of ionically treated garments and so further research is required to substantiate current findings and more clearly elucidate the effects of ionised garments in a sport specific setting.

1 Introduction

Sports apparel is becoming increasingly technical from both a feature and fabric perspective. Advancements in apparel technology have focused mainly on the enhancement of thermoregulation and muscular compression. A propriety technology has now made it possible to ionically treat garments, though benefits afforded to a wearer remain unknown. Existing literature has documented benefits of negatively charged particles via inhalation, in parameters including increased work capacity (Minkh, 1961), favourable haemodynamic (Ryushi et al., 1998), heart rate (Yates et al., 1986) and lactate responses, as well as improvements in various psychological state indices (Fornof & Gilbert, 1988). However conflicting research exists and any mechanisms in action remain unclear.

The present study aimed to investigate the influence of negative ions, via ionically treated garments, on a wide range of physiological performance and recovery parameters.

2 Methods

Following informed consent, twelve male participants (22 ± 2.8 years, 85.7 ± 14.7 kg, $10.0 \pm 2.8\%$ body fat; mean $\pm$ SD) consisting of recreational games players (n = 6) and university first team rugby players (n = 6), took part in a double blind randomised cross-over study.

Before testing participants were familiarised to equipment and requested to refrain from alcohol and caffeine for at least 24 hours. Dietary intake was recorded and matched for both trials 24 hours prior testing, while water was provided ad-libitum during the trials. A minimum of 72 hours was required between each trial.

On arrival, height, mass, and percentage body fat were measured using a stadiometer, weighing scales and skin fold callipers, respectively. Garments were then worn, where trial order of use was determined independently to ensure counter balance and double blind design remained.

Participants then assumed a semi-recumbent position for 15 minutes on a physiotherapy bed. Systolic and diastolic blood pressure, along with heart rate were monitored at 3 minute intervals (Vital Signs Monitor, Hunleigh Healthcare). A 3 minute low intensity warm-up was then performed on the cycle ergometer (894E, Monark), followed by three vertical jump repetitions with hands on hips at all times. An approximate 90° knee joint angle was maintained for 1 second before jumps.

A 15 second maximal effort Wingate test utilising a resistance equal to $0.075\,\mathrm{kg \cdot kg}$ body mass^{-1}, was performed on a cycle ergometer (894E, Monark). After a 2 minute recovery a second identical 15 second Wingate test was performed. Mean power, peak power, time to peak power and fatigue were measured. Heart rate was recorded throughout (Team System, Polar). Participants performed three further repetitions of vertical jumps using the same technique as described earlier. A 10 minute semi-recumbent recovery period was then followed.

Participants performed a five minute arm-crank ergometer test (modified cycle ergometer RB1, Reebok) where load was $0.07\,\mathrm{kg \cdot kg}$ body mass^{-1}. Localised muscular pain perception was measured using a 0–10 point pain scale self assessment method (Cook et al., 1997), along with heart rate at 1 minute intervals.

Finally, a five minute recovery was allowed before participants performed an incremental ($25\,\mathrm{W \cdot min^{-1}}$) ramped cycle ergometer (839E, Monark) test, where participants were asked to cycle starting at 100 W until exhaustion. Maximal power and heart rate were measured.

Participants then immediately assumed a semi-recumbent position during recovery, systolic and diastolic blood pressure and heart rate were measured at 3 minute intervals as before. Post testing self assessment of localised muscle soreness was made by a 0–10 scale chart (Thompson et al., 1999) for 72 hours post testing at 24 hour intervals.

3 Results

3.1. *Mean Power*

It was found that average power achieved between the second of the two Wingate tests was significantly higher (730 ± 114 W) when garment I was worn in comparison to garment N (711 ± 126 W). Across the entire cohort this equated to an increased performance of 2.7 ± 3.3%. Of the six rugby players tested, all showed improvements in mean power in the second Wingate test when wearing garment I, equating to an increase of 3.2 ± 2.5%. Only four of the six recreational players were observed to have higher mean power when wearing garment I, equating to a mean improvement of 2.1 ± 4.1% which was not significantly different to garment N.

3.2. *Minimum Resting Heart Rate*

It was also found during the 15 minute semi-recumbent resting period before testing that minimum heart rate was significantly higher while wearing garment I (56 ± 8 bpm) than garment N (53 ± 7 bpm).

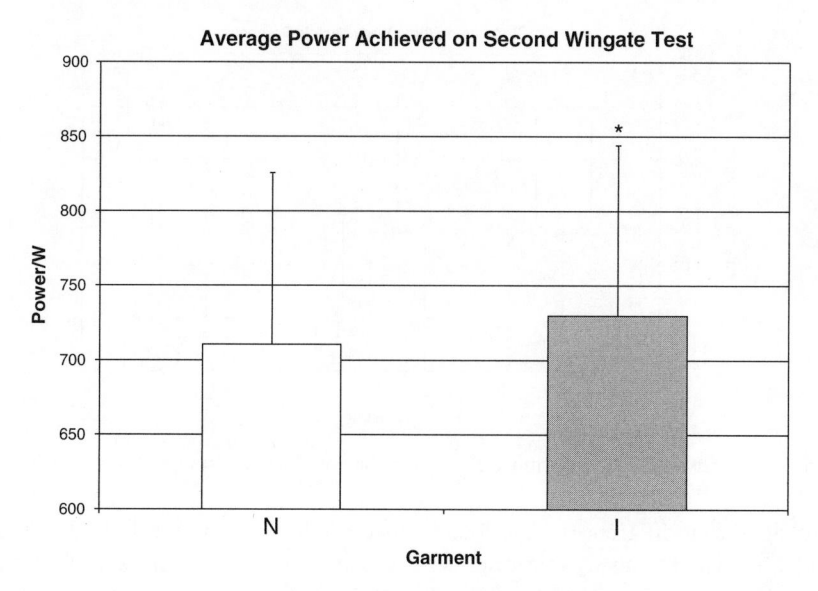

Figure 1. Mean power over 15 seconds on second Wingate test (n = 12, *P < 0.05).

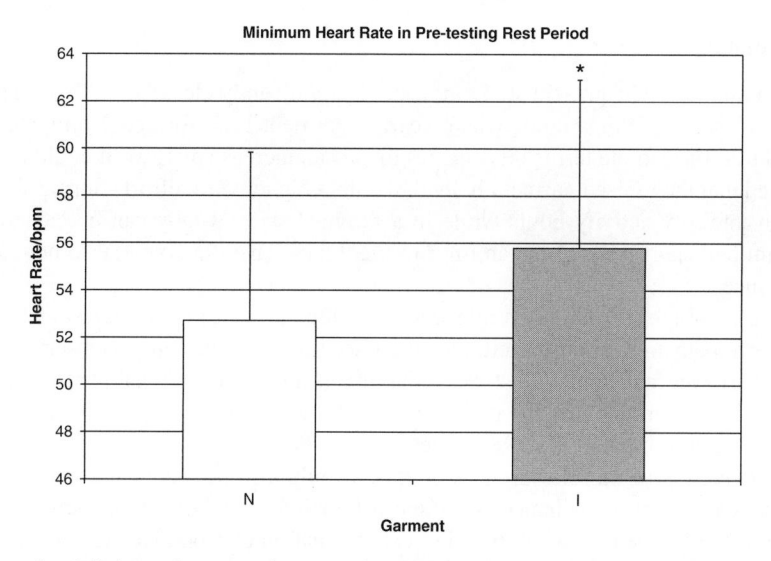

Figure 2. Minimum heart rate in pre-testing 15 minute rest period (n = 12, *P < 0.05).

3.3. *Muscle Soreness and Muscle Pain*

Whole body muscle soreness parameters measured in the 72 hour period post testing were not found to be significantly different. However on day 2 it is of interest to note the graphical representation of reduced whole body muscle soreness.

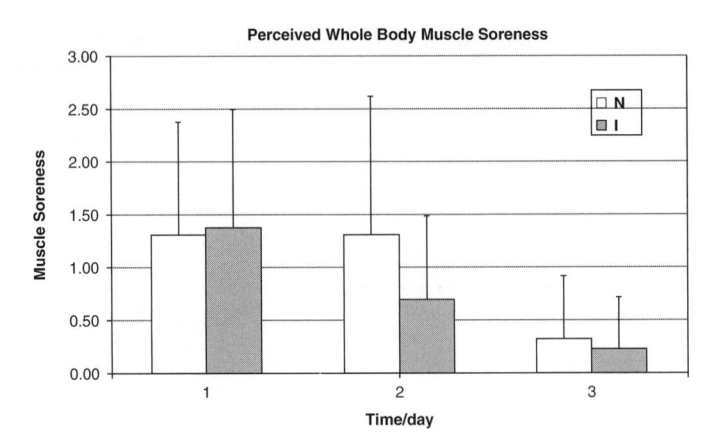

Figure 3. Perceived whole body muscle soreness for over 3 days post test (n = 12, *P < 0.05).

No significant differences were found between other parameters. Indeed it was is of particular interest that no significant effect was shown in peak power, rate of fatigue and time to peak power achieved on the Wingate test. No difference was found in pain perception on the arm-crank ergometer, or maximal power and heart rate achieved on the incremental cycling test. Finally no differences were found for recovery parameters of blood pressure and heart rate post testing.

4 Discussion

The main finding of the present study indicates a significantly elevated (2.7%) mean power during a 15 second Wingate test when wearing garment I, performed 2 minutes after an identical test. In absolute terms this equates to a mean increase of 19 W. It could be broadly implicated that the ionised garment helped maintain higher work efforts during short duration high intensity activity bouts while in a fatigued state. Rugby can be categorised as an intermittent sport where such an improvement may have the potential to benefit player performance.

The elevation of resting heart rate when wearing garment I is of interest and it may be that this sympathetic response facilitates the raised mean power output observed wearing garment I. However it should be stressed that this is an unconventional parameter. Indeed, this finding is in conflict with the main literature base that suggests negative ion treatment reduces heart rate (Inbar *et al.*, 1982; Yates *et al.*, 1986).

The reduction of muscle soreness on day 2 could also be associated with the resting sympathetic response, by enhanced clearance of noxious metabolites. In a study by Iwama *et al.*, (2002) negative ions were linked to the attenuation of blood lactate concentration in a primary healthcare setting. Future studies which initially induce raised levels of muscle soreness than those seen in the present study are required along with the measurement of blood parameters to more clearly elucidate the effects garment I may have on recovery.

In conclusion, the present study is the first in the literature where negatively charged ions were delivered to the body through an undergarment. Results are of interest, demonstrating a need for future studies to confirm current findings and whether they translate to a game setting.

Acknowledgements

The authors wish to acknowledge the financial support from Canterbury of New Zealand Ltd, the manufacturers of the garments used in the study.

References

Fornof K.T. and Gilbert G.O. (1988) Stress and physiological, behavioural and performance patterns of children under varied air ion levels. *International Journal of Biometeorology*, 32, 260–270.

Inbar O., Rotstein A., Dlin R., Dotan R. and Sulman F.G. (1982) The effects of negative air ions on various physiological functions during work in a hot environment. *International Journal of Biometerology*, 2, 153–163.

Iwama H., Ohmizo H., Furuta S., Ohmori S., Watanabe K., Kaneko T. and Tsusumi K. (2002) Inspired superoxide anions attenuate blood lactate concentrations in postoperative patients. *Critical Care Medicine*, 30/6, 1246–1249.

Minkh A.A. (1961) The effect of Ionised Air on Work Capacity and Vitamin Metabolism. Proceedings of the International Conference of Ionisation of the Air.

Ryushi T., Kita I., Sakurai T., Yasumatsu M., Isokawa M., Aihara Y. and Hama K (1998) The effect of exposure to negative air ions on the recovery of physiological responses after moderate endurance exercise. *International Journal of Biometeorology*, 41, 132–136.

Thompson D., Nicholas C.W. and Williams C. (1999) Muscular soreness following prolonged intermittent high-intensity shuttle running. *Journal of Sport Sciences*, 17/5, 387–395.

Yates A., Gray F.B., Misiaszek J.I. and Wolman W. (1986) Air ions: past problems and future directions. *Environment International*, 12, 99–108.

THE INFLUENCE OF HUES ON THE CORTICAL ACTIVITY – A RECIPE FOR SELECTING SPORTSWEAR COLOURS

J. TRIPATHY,[1] F.K. FUSS,[1] V.V. KULISH[2] & S. YANG[1]

[1]*Sports Engineering Research Team, Division of Bioengineering, School of Chemical and Biomedical Engineering, Nanyang Technological University, Singapore*
[2]*School of Mechanical and Aerospace Engineering, Nanyang Technological University, Singapore*

Few studies exist on the influence of colours on sports performance. So far, red, associated with dominance, seems to provide a higher chance for winning in male contestants. In this preliminary study, we investigated the fractal dimensions of EEG waves resulting from exposure to different colours. Green, cyan and yellow produce higher fractal dimensions than red, magenta and blue. The difference between magenta/blue and green/yellow/cyan as well as between green and red is significant at $p < 0.05$, and between green and blue even significant at $p < 0.01$. We suggest applying these colours for sportswear, which provoke higher fractal dimensions of brain waves and thus positively influence the cortical activity.

1 Introduction

Colouration of sportswear has repeatedly been reported to influence the outcome of sportive contests. Hill & Barton (2005a) investigated the outcome of four combat sports in the Athens 2004 Olympics, and concluded that wearing red is consistently associated with a higher probability of winning. In these combat sports, namely boxing, tae kwon do, Greco–Roman wrestling and freestyle wrestling, the contestants were randomly assigned red or blue outfits or body protection. In all 4 sports, the chance to win when wearing a red outfit was higher compared to a blue one. The ratio of red to blue wins in all 4 sports combined amounted to 55:45.

Hill & Barton (2005a) attributed this result to the correlation of red coloration with male dominance and testosterone levels. For example, anger is associated with a reddening of the skin due to increased blood flow. Hence, increased redness during aggressive interactions may reflect relative dominance.

Rowe *et al.* (2005) analysed sportswear colours other than red, specifically the outcome of judo matches in the Athens 2004 Olympics. In judo, one player wears blue and the other wears white. The result clearly showed that wearers of blue outperformed wearers of white.

Rowe *et al.* (2005) suggested that the outfit colour affects opponent visibility, which is crucial for avoidance and interception, and for anticipating behaviour. Thus, visual abilities could influence sporting performance, including being able to follow rapidly moving objects and perform fast visual searches. The hue, saturation, brightness and contrast of an object (or opponent) could enable it to be picked out against its background. According to Rowe *et al.* (2005) these factors are critical for combat sports and for detecting teammates on the field of play. In judo, the white judogi is likely to be perceived as brighter than the blue and may have higher contrast against the background. Rowe *et al.* (2005) hypothesised that men wearing blue may therefore have a visual advantage in being able to anticipate their (white) opponents' moves.

Hill and Barton (2005b) objected that this visibility explanation is unlikely in a situation where contestants fight at close quarters in brightly lit arenas, as in these combat sports.

Nevertheless, the influence of visibility cannot be denied, as generally known from the "grey strip" affecting the performance of Manchester United (Stephenson, 1999). Stephenson (1999) investigated the visibility of colours and sportswear, and found that bright red could generally be detected against a crowd background over a 120 degree radius, whereas pale grey could only be detected over a 40 degree radius. Patterned (particularly striped) shirts could be detected over a wider radius than plain strips (Stephenson, 1999). In other words, the Manchester United players would find it harder to spot fellow team members wearing the grey strip than players wearing plain or patterned strips in other colours – because the plain grey strip merged with the crowd in the background (Stephenson, 1999).

Hill & Barton (2005a) also report a preliminary analysis of the results of the Euro 2004 international soccer tournament, in which teams wore shirts of different colours in different matches. They compared the performance of five teams that wore a predominantly red shirt against their performance when wearing a different shirt colour (four played their other matches in white, one in blue). Hill & Barton (2005) found that all five had better results when playing in red, thereby again challenging the visibility hypothesis.

Summing up, two hypotheses exist as to the influence of colours on the outcome of contests: 1) red associated to dominance, and 2) the contrast to the background. It is well known, that performances are better in extreme situations, e.g., in danger, when the brain is in an alert state. Thus, the activity level of the central nervous system must have an influence on raising the performance. The logical hypothesis is that the brain has a specific response to particular colours. Fractal dimensions are a highly selective technique to distinguish signals. Such dimensions also reflect on the "randomness" of a signal over a certain time period. The aim of this study was to investigate whether different colours affect the fractal dimensions of EEG waves, and thus have an influence on brain activity.

2 Method

EEGs are potential fluctuations recorded from the scalp due to the electrical activity of the brain. Hence, EEGs can be viewed as temporal sequences (time series). One common practice to distinguish among possible classes of time series is to determine their so-called correlation dimension. The correlation dimension, however, belongs to an infinite family of fractal dimensions (Hentschel & Procaccia, 1983). The concept of generalized entropy of a probability distribution was introduced by Rényi (1955). Based on the moments of order q of the probability p_i, Rényi obtained the following expression for entropy

$$S_q = \frac{1}{1-q} \log \sum_{i=1}^{N} p_i^q \qquad (1)$$

where q is not necessarily an integer and log denotes $\log_2$. Note that for $q \to 1$, Eq. (1) yields the well-known entropy of a discrete probability distribution (Shannon, 1998).

$$S_1 = -\sum_{i=1}^{N} p_i \log p_i \qquad (2)$$

The Influence of Hues on the Cortical Activity

The probability distribution of a given time series can be recovered by the following procedure. The total range of the signal is divided into N bins such that

$$N = \frac{V_{max} - V_{min}}{\delta V} \qquad (3)$$

where V_{max} and V_{min} are the maximum and the minimum values of the signal achieved in the course of measurements, respectively; δV represents the sensitivity (uncertainty) of the measuring device. The probability that the signal falls into the ith bin of size δV is computed as

$$p_i = \lim_{N \to \infty} \frac{N_i}{N} \qquad (4)$$

where N_i equals the number of times the signal falls into the ith bin. On the other hand, in the case of a time series, the same probability can be found from the ergodic theorem, that is

$$p_i = \lim_{T \to \infty} \frac{t_i}{T} \qquad (5)$$

where t_i is the time spent by the signal in the ith bin during the total time span of measurements T. Further, the generalized fractal dimensions of a given time series with the known probability distribution are defined as

$$D_q = \lim_{\delta V \to 0} \frac{1}{q-1} \frac{\log \sum_{i=1}^{N} p_i^q}{\log \delta V} \qquad (6)$$

where the parameter q ranges from $-\infty$ to $+\infty$. Note that for a self-similar fractal time series with equal probabilities $p_i = 1/N$, the definition Eq. (6) yields $D_q = D_0$ for all values of q (Schroeder, 1991). Note also that for a constant signal, all probabilities except one become equal to zero, whereas the remaining probability value equals unity. As a result, for a constant signal, $D_q = D_0 = 0$. The fractal dimension

$$D_0 = -\frac{\log N}{\log \delta V} \qquad (7)$$

is the Hausdorff-Besicovitch dimension (Mandelbrot, 1983).
Note also that

$$D_\infty = \lim_{\delta V \to 0} \frac{\log p_{max}}{\log \delta V} \quad \text{and} \quad D_{-\infty} = \lim_{\delta V \to 0} \frac{\log p_{min}}{\log \delta V} \qquad (8a,b)$$

such that $D_{-\infty} \geq D_{\infty}$. In general, if $a < b$, $D_a \geq D_b$, such that D_q is a monotone non-increasing function of q (Schroeder, 1991). For a given time series ("signal"), the function D_q, corresponding to the probability distribution of this signal, is called the *fractal spectrum*. Such a name is well-justified, because the fractal spectrum provides information about both frequencies and amplitudes of the signal. Indeed, for two probability distributions, a larger value of a fractal dimension of a given order corresponds to the presence of more pronounced spikes (sharper spikes, less expected values of the signal) than in the signal for which the value of the fractal dimension of the same order is less. Furthermore, signals with a wider range of fractal dimensions, $D_{-\infty} - D_{\infty}$, can be termed more fractal than signals whose range of fractal dimensions is narrower, so that signals with the zero range are self-similar fractals. In other words, the range of a fractal spectrum is a value associated with the range of frequencies in the signal. Now, if the unexpectedness of an event is defined as the inverse of the probability of this event, then steeper spectra correspond to the signals in which unexpected values are more dominant, whereas flatter spectra represent those signals in which less unexpectedness occurs.

3 Experimental

The signal for the brain's response to various colours was done through standard EEG equipment (MindSet 24 Topographic Neuro-mapping Instrument by Nolan Computer Systems LLC). The brain responses to 8 colours were stored for each subject. The colours chosen, along with their RGB values are given in Table 1. The specialty of these colours lies in their always being rendered correctly on a monitor, regardless of the colour resolution of the display card. Furthermore, the gray-values of red, green, blue, yellow, cyan, and magenta are equal. During the experiment, Adobe Photoshop was used to let the subjects see the colours one after another. The monitor was calibrated in advance, such that the colours are displayed correctly. A blackened cardboard box was used to let the subject's eyes experience light only from the computer screen, by restricting ambience light out of the box. Twenty test persons (Asians, Indians and Caucasians) participated in this study in a very calm, noise free location. Additionally, the test subjects wore ear muffs to shield them from any disturbing sound. The machine was established as far away as possible from any electronic noise source like a

Table 1. RBG values of the colours investigated.

R-value	G-value	B-value	Colour	Symbol
0	0	0	Black	K
255	255	255	White	W
255	0	0	Red	R
0	255	0	Green	G
0	0	255	Blue	B
255	255	0	Yellow	Y
0	255	255	Cyan	C
255	0	255	Magenta	M

computer CPU/monitor. The subjects were asked to relax and any movement was strongly discouraged. The subjects were given 2 minutes in between each colour and encouraged to be as "thought-less" as possible during the experiment. The subject's eyes were ensured to be as close to the holes on the cardboard as possible to eliminate external visual interference. The MindMeld 24 was run for 1 minute for each colour. The colours were displayed for 1 minute on the monitor, in random order. All experiments were carried out on the same time of the day.

The fractal spectrum of the colours was calculated for the first and last 40 seconds, based on the method and procedure developed by Kulish *et al.* (2006, described in detail above), to study the effect of the colours on the human brain. The fractal dimension used in this study is the Hausdorff-Besicovitch dimension. The mean Haussdorf dimension of each test person was set to 1, and the relative Hausdorff dimension was calculated, for better comparison.

4 Results

The results are displayed in Figures 1–3. Minimum fractal dimension was found for magenta and maximum for green. The difference between the Hausdorff dimension of green and blue was significant at $p < 0.01$. No statistical difference was found between black and white ($p = .236$). The colours RGBCYM show a cyclic behaviour when plotted with respect to the colour circle (Figures 2 and 3).

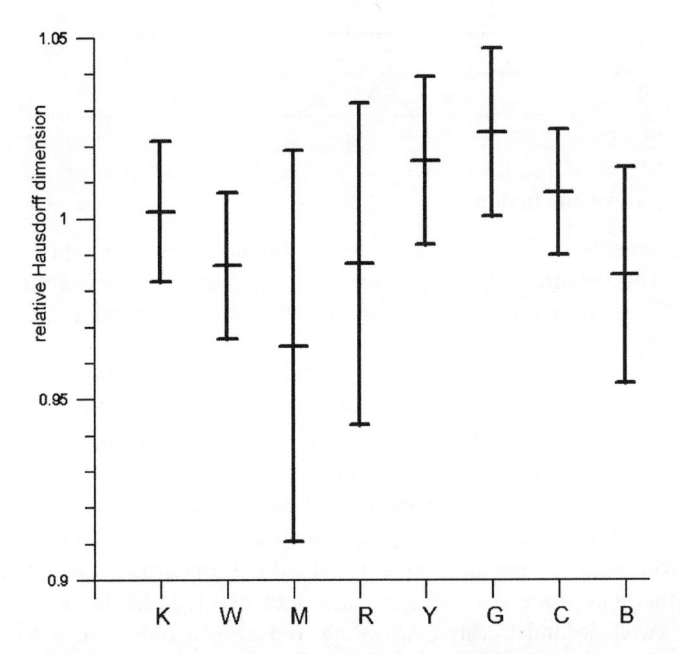

Figure 1. Colours (c.f. Table 1) and their relative Hausdorff dimensions (mean ± standard deviation).

Figure 2. Polar plot of mean Hausdorff dimension.

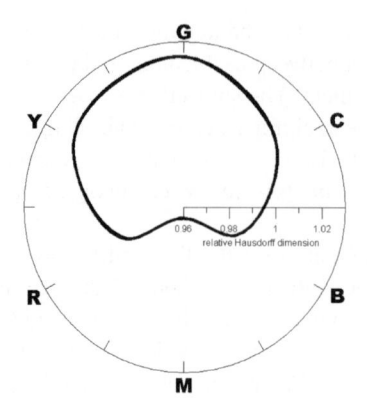

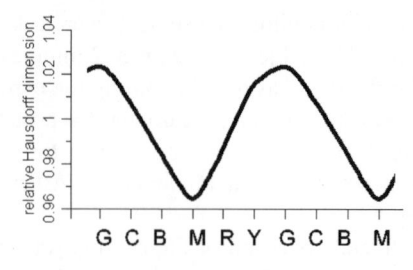

Figure 3. Linear plot of mean Hausdorff dimension.

Table 2. p-values (grey cell: $p < 0.05$, black cell : $p < 0.01$).

p	M	R	Y	G	C	B
M	–	*.02*	*.041*	*.013*	0.38	0.146
R		–	.108	*.037*	.177	.728
Y			–	.338	.167	*.021*
G				–	*.018*	**.004**
C					–	*.011*
B						–

5 Discussion and Conclusions

The fractal geometry of EEG waves is clearly influenced by different hues. The trend becomes clear from Figure 2: the most chaotic behaviour of EEG waves results from green. The Hausdorff dimension drops from green over yellow and red to magenta, as well as from green over cyan and blue to magenta.

In contrast to Hill and Barton (2005a) and Rowe et al. (2005) the fractal dimension of EEG waves showed no difference between red and blue or white and blue. It has to be considered in this context, that the fractal dimension reflects how chaotic and self-similar the EEG waves are, and how active the cortex is.

Hill & Barton (2005a,b) hypothesise that red is related to dominance and reject the visibility hypothesis of Rowe et al. (2005). The influence of visibility is ruled out in wrestling and boxing, where sportswear does not cover shoulders and arms. Hill & Barton (2005b) go a step further and prove that the dominance factor of red affects males only, whereas females (tae kwon do and freestyle wrestling, red against blue, and judo, blue against white) do not show any significant differences. A serious flaw, however, in all these studies, is that the investigators did not consider the current (at the time of the competition) individual performance (e.g., world ranking) of the contestants. Taking this data into account might change the significance of the results.

Independent of any hypothesis, it cannot be denied that colours have a psychological influence on sports. Our preliminary study proves that the level of cortical activity is clearly influences by colours. Thus, we suggest taking colours provoking cortical activity with a higher Hausdorff dimension for sportswear, especially yellow, lime, green, turquoise and cyan. This is especially suitable for team sports, provided that the competitor is not influenced positively as well. Further research, however, is necessary to understand the influence of colour on performance, e.g. whether the "yellow shirt" positively influences the leader of the Tour de France.

References

Hill R.A. and Barton R.A. (2005) Red enhances human performance in contests. *Nature*, 435, 293.

Rowe C., Harris J.M. and Roberts S.C. (2005) Sporting contests: Seeing red? Putting sportswear in context. *Nature*, 437, E10.

Stephenson G. (1999) How science, not superstition, persuaded Manchester United to change its strip. *Research Intelligence*, Issue 1, June 1999, The University of Liverpool (online). Available: http://www.liv.ac.uk/researchintelligence/issue1/manunit.html (Accessed: May 2007).

Hill R.A. and Barton R.A. (2005) Sporting contests: Seeing red? Putting sportswear in context (reply). *Nature*, 437, E10–E11.

Hentschel H.G.E. and Procaccia I. (1983) The infinite number of generalized dimensions of fractals and strange attractors. *Physica*, 8D, 435–444.

Rényi A. (1955) On a new axiomatic theory of probability. *Acta Mathematica Hungarica*, 6, 285–335.

Shannon C.E. (1998) *The Mathematical Theory of Communication*. University of Illinois Press, Champaign, IL.

Schroeder M.R. (1991) *Fractals, Chaos, Power Laws*. Freeman, New York.

Mandelbrot B.B. (1983) *The Fractal Geometry of Nature*. Freeman, New York.

Kulish V., Sourin A. and Sourina O. (2006) Human electroencephalograms seen as fractal time series: Mathematical analysis and visualization. *Computers in Biology and Medicine*, 36, 291–302.

THE DEVELOPMENT AND ENVIRONMENTAL APPLICATIONS OF THE GOINGSTICK®

M.J.D. DUFOUR[1] & C. MUMFORD[2]

[1]*Cranfield University, Cranfield, Bedford, UK*
[2]*New Zealand Sports Turf Institute, Palmerston North, New Zealand*

This paper describes the development by Cranfield University of the GoingStick® and its introduction to the British horse racing industry through the TurfTrax Group. Analysis of the data produced from three important performance characterization trials show that the GoingStick® can be used to measure objectively the going ground condition on British racecourses.

1 Introduction

Injury to horse and jockey can be linked to the state and consistency of the ground around a racecourse (Henley *et al.*, 2006; McCrory *et al.*, 2006). This directly affects the decision of the trainer/owner to run a horse in a race particularly where the conditions are either very soft or very hard. It is in the racecourse's best interest to maintain larger race fields since there is a direct relationship between the size of the race field and the revenue generated (Winter, 1998). The value of the British horseracing industry represents two-thirds of the £9 billion turnover to the betting industry and over £100 million in prize money available to owners of racehorses (Mintel, 2005). The most commonly used method to report on the condition of the ground has been based on "Clerks of the Course" walking the racecourse, pushing a walking stick into the ground to provide an opinion as to the state of the "going" – a description of the ground condition with five intermediate stages between the extremes of "hard" and "soft". The need for an objective measurement has been driven by long term frustrations regarding the accuracy and consistency of reports published by racecourses.

2 Design

To provide an objectively-based assessment of the ground condition the GoingStick® design had to reflect the existing test carried out by the "Clerk of the Course" which in turn emulated the interaction of the hoof of the horse with the ground. It also had to present the end-user with information that was comparable to the existing categories of going. As a result the GoingStick® was developed by Cranfield University to measure both the penetration resistance (the amount of force required to push the tip into the ground) and the shear resistance (the energy needed to pull back to an angle of 45° from the ground). These two measures taken in combination produce a "GoingIndex" which represents the going – the firmness of the ground and level of traction experienced by a horse during a race (Table 1).

3 Operation

To generate a penetration reading, the GoingStick® is inserted into the ground at its tip using the tee-bar to apply vertical pressure (Figure 1, left). Once inserted, the device is released to stand freely before the device indicates that the shear reading can be taken. This is made in one smooth movement, sweeping the handle back to an angle of 45° away from the vertical plane (Figure 1, right). The device is removed from the ground and the tip cleared of any adhering soil before being re-inserted for the second and third insertions required to generate an integrated "GoingIndex" (Table 1).

4 Developmental Testing

To function on a racecourse the GoingStick® has to be able to withstand heavy use on a regular basis, produce consistency of readings across models leaving the TurfTrax production

Table 1. The TurfTrax "GoingIndex" expressed by the GoingStick®.

Going	"GoingIndex"
Hard	13.0–15.0
Firm	11.0–12.9
Good to firm	9.0–10.9
Good	7.0–8.9
Good to soft	4.0–6.9
Soft	3.0–4.0
Heavy	1.0–2.9

Figure 1. Operating the GoingStick® at Huntingdon racecourse. On the left a penetration reading is being taken; on the right a shear reading is being taken.

line and be operationally effective over a range of temperatures. Three trials were undertaken to assess the performance characteristics of the sticks:

1. Penetration and shear repeatability
2. Comparison of readings between sticks
3. Temperature induced variation.

4.1. Penetration and Shear Repeatability Trials

Loads between 5 kg and 25 kg were randomly selected and applied to the tip of a GoingStick® blade to simulate penetrative forces in three replicates. A 5 kg load was applied to the side of the blade to simulate translational shear forces with the trial repeated five times. Values of penetration and shear were recorded from the digital display of the stick and after each replicate the stick was reset.

4.2. Comparison Trials

Four GoingSticks® were randomly selected from the TurfTrax production line. The trials were conducted in a controlled soil environment (18°C and 35% RH) at the soil dynamic facilities at Cranfield University. The controlled soil environment contained a sandy loam soil which was rolled to achieve three uniform densities between a softer state of 1.31 Mg m^{-3} and a harder state of 1.52 Mg m^{-3} (Table 2). A regular grid pattern of test points was established on the central 70% of the soil surface of the controlled soil environment to minimize the effect of the retaining walls on the penetration and shear tests. At each test point dry bulk density was determined in accordance with Smith & Thomasson (1982) using an undisturbed soil sample collected with a bulk density ring (20 × 54 mm). Penetration resistance measurements were also taken at each test point in 10 mm increments to a depth of 100 mm using a cone penetrometer (12 mm). Three measures of penetration and shear resistance were taken using the sticks in accordance with the manufacturer's instruction manual (TurfTrax, 2004).

4.3. Temperature Trials

Each GoingStick® was placed in a controlled temperature environment chamber for 12 hours prior to each trial to attain three constant temperatures representing a cold day at +1°C, a normal day (controlled soil environment facility) at +18°C and a hot day at +30°C that are fairly typical of British climatic conditions. Mechanical loads (5 kg for penetration, 2 kg for translational shear) were applied to the sticks whilst still inside the chamber to test the consistency of penetration and shear. The same mechanical loads where also applied to the sticks after they were transferred from an ambient temperature of +18°C into the controlled environment chamber at +1°C to establish the effect of a thermal gradient on the sticks performance.

5 Results

5.1. GoingStick® Penetration and Shear

Linear regression was used to analyze the results. A significant relationship exists between the loads applied and the penetration reading produced by the GoingStick® (Figure 2a).

The stick also consistently achieved shear readings of 1.56 MPa each time the 5 kg load was applied.

5.2. *GoingStick® Comparison*

Values of penetration resistance using the GoingStick® were converted to MPa using a calibration curve. The stick consistently recorded higher values than the penetrometer (Table 2) and there was variation between the sticks but the values did follow the trend of the penetrometer results (Figure 2b) showing that the sticks can differentiate between the different soil states tested.

5.3. *Temperature*

The constant temperature trials saw no significant variation in "GoingIndex" readings over the three temperatures tested. Repeat tests at one-minute intervals for the thermal gradient trial showed no significant variation and the total range was less than 0.5 "GoingIndex" units.

6 Environmental Applications

To give a better assessment of the going around a racecourse a zoned going grid can be created based upon an electromagnetic induction (EMI) scan to characterize the underlying

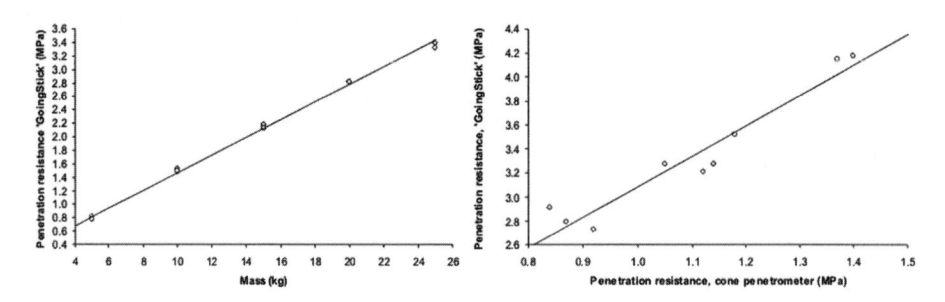

Figure 2. Linear regression analysis of the penetration trials. (a) Penetration repeatability trial ($R^2 = 0.9974$; $p = <0.001$; $y = 0.1295x + 0.1861$). (b) Comparison trial ($R^2 = 0.9276$; $p = <0.001$; $y = 2.5445x + 0.5485$).

Table 2. Summary of GoingStick® comparison results.

Surface category	Harder	Intermediate	Softer
Dry bulk density (Mg m^{-3})	1.52	1.44	1.31
Cone penetrometer: mean penetration (MPa)	1.37	1.18	0.87
GoingStick: GoingIndex	12.5	10.5	8.8
Mean penetration (MPa)	4.15	3.52	2.79
Mean shear (MPa)	0.95	0.78	0.77

soil structure. A going map can be constructed by taking the average GoingStick® reading for each zone and ascribing a predefined colour for the "GoingIndex" (Figure 3).

The segmentation of a racecourse into such zones can be used as a tool to guide management regimes – in particular irrigation, so that the optimum level of going for racing (good and good-to-firm) can be achieved. With the GoingStick® being used as the mandatory method for declaring the going at all 58 British turf racecourses from March 2007 (Associated Newspapers, 2006) there is now a means of creating more uniform ground conditions by matching the moisture content of the drier and wetter parts of the racecourse (Chivers, 1999).

Since a large number of British turf racecourses abstract water for irrigation purposes, in order to retain existing abstraction licenses or make successful applications for abstraction licenses they need to demonstrate that they are managing water-use efficiently in accordance with the Water Resources Act 2003 (OPSI, 2003). Work by Mumford (2007) has shown a change in going can be achieved using the GoingStick® in conjunction with the zoned going grid as a guide to help calculate and manage the irrigation required (Figure 4).

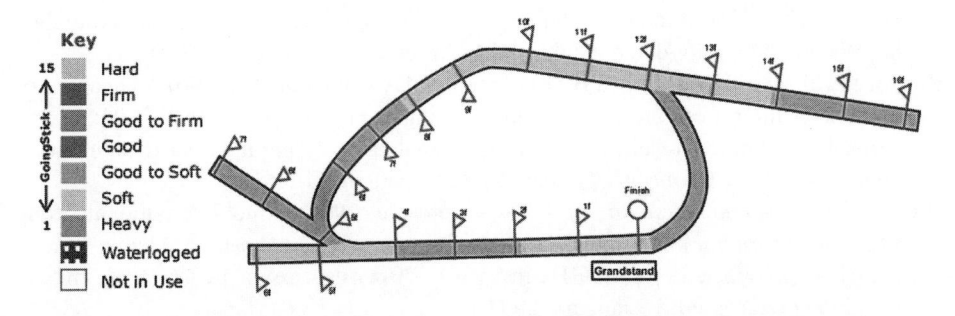

Figure 3. Going map for York racecourse, Friday 7th October 2006 © TurfTrax.

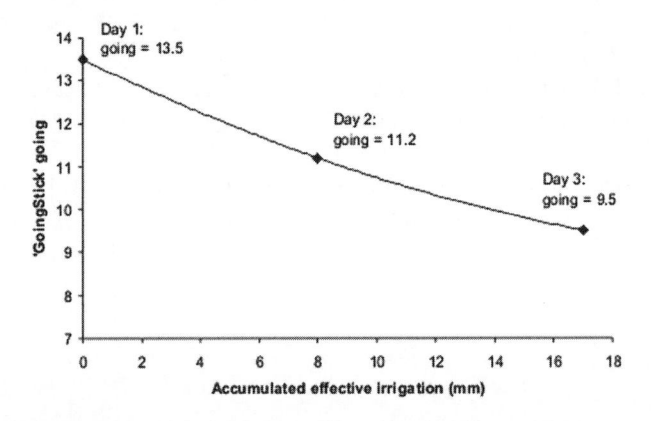

Figure 4. Change in going in response to accumulated effective irrigation over a three day period at Leicester racecourse (Mumford, 2007).

7 Conclusions

The controlled environment tests discussed in this paper show that the GoingStick® can objectively assess the going ground condition on British racecourses. Two years and more than 60,000 readings recorded at a variety of racecourses (HRA, 2003) has resulted in the GoingStick® being adopted for all British racecourses. The use of the GoingStick® to manage irrigation strategies has a clear environmental application.

References

Associated Newspapers (2006) *GoingStick introduced as standard*. Associated Newspapers Ltd, London (online). Available: http://www.mailonsunday.co.uk /pages/live/articles/sport/horseracing.html?in_article_id=423564&in_page_id=1967. (Accessed: 12 January, 2007).

Chivers I.H. (1999) Prescription surface development: racetrack management. In: Aldous D.E. (Ed.) *International Turf Mangement Handbook*. CRC Press, London.

Henley W. E., Rogers K., Harkins L. and Wood J.L.N. (2006) A comparison of survival models for assessing risk of racehorse fatality. *Preventive Veterinary Medicine*, 74, 3–20.

HRA (2003) *Jockey Club Announce Introduction of New Device For Measurement of Going*. The Horseracing Regulatory Authority, London (online). Available: http://www.thehra.org/doc.php?id=28624 (Accessed: 7 October, 2007).

McCrory P., Turner M., LeMasson B., Bodere C. and Allemandou A. (2006) An analysis of injuries resulting from professional horse racing in France during 1991–2001: a comparison with injuries resulting from professional horse racing in Great Britain during 1992–2001. *British Journal of Sports Medicine*, 40/70, 614–618.

Mintel (2005) *Dog and Horse Racing – UK – September 2005*. Mintel International Group Ltd, London (online). Available: http://academic.mintel.com/sinatra/academic/search_results/show&&type=RCItem&page=0&noaccess_page=0/display/id=125557 (Accessed: 15 January, 2007).

Mumford C. (2007) The optimization of going management on UK racecourse using controlled water applications. Eng.D. Diss., Cranfield University, Cranfield, UK.

OPSI (2003) *Water Act 2003: Chapter 37*. Office of Public Sector Information (online). Available: http://www.opsi.gov.uk/ACTS/acts2003/20030037.htm (Accessed: 17 February, 2007).

Smith P. D. and Thomasson A. F. (1982) Density and Water-Release Characteristics. In: Avery B.W. and Bascomb C.L. (Eds), *Soil survey laboratory methods*. Soil Survey Technical Monograph, No. 6, Silsoe.

Turftrax (2004) *The TurfTrax GoingStick: Operations manual*. Oakley, Bedfordshire.

Winter P. (1998) Racing into the future to keep the Going good. *Turfgrass Bulletin*, 200, 13–16.

ANALYSIS OF THE INFLUENCE OF SHOCKPAD PROPERTIES ON THE ENERGY ABSORPTION OF ARTIFICIAL TURF SURFACES

T. ALLGEUER,[1] S. BENSASON,[1] A. CHANG,[2] J. MARTIN[2] & E. TORRES[1]

[1]*Dow Europe GmbH, Horgen, Switzerland*
[2]*The Dow Chemical Company, Freeport, Texas, USA*

Sports surfaces as artificial turf or indoor sports floors see a rapid growth over the last years due to the improved player safety and game consistency provided by such surfaces. A big portion of such performance criteria are provided by the shock absorbency components integrated into such floorings in the form of granular infill or shock pads / underlays. The selection of such components obviously is a key factor for success, this work documents the efforts to develop performance models for different foam systems in order to provide appropriate selection tools. This report documents a scientific study of the influence of the foamed shockpad design parameters on turf system energy absorption performance. Different foam compositions were screened for their performance in stress response, elastic recovery and creep resistance as shockpads using lab tests in compression along with FQC (FIFA Quality Concept) tests both on the foams and on an artificial turf system. These properties were generally found to depend mainly on foam density and less on polymer density or degree of crosslinking. FQC test results demonstrate how a well designed shockpad can help in maintaining a uniform performance across the artificial turf surface overruling the effect of infill variations.

1 Introduction

Artificial turf has come a long way in 42 years. From its beginnings in 1965 with the installation of the first artificial turf playing surface in the Houston (USA) Astrodome stadium, technology has revolutionized the quality of artificial playing surfaces making possible the exceptional performance of today's "third generation" artificial pitches. Most "third generation" pitches consist of a carpet with tufted polyethylene (PE) yarns, with an infill of ground styrene-butadiene rubber (SBR) in combination with silica sand, and are installed on prepared fields. There is currently a trend in using more environmentally friendly non-SBR infills for replacing the current SBR based solution. The performance of many of these non-SBR based infill solutions makes it necessary for employing an energy absorption layer, also called shockpad, underneath the carpet. Selection of appropriate foamed shockpad material, density and thickness, is guided in part by performance requirements on shock absorption as outlined in FIFA Quality Concept (FQC) for Artificial Turf (FIFA, 2006). Moreover, design of new systems also requires consideration of long term durability and performance of the shockpads. Summarized herein are the initial findings of a scientific study of the influence of the foamed shockpad design parameters on turf system energy absorption performance.

2 Experimental

Various foam candidates were subjected to a series of quasi-static lab tests in compression, along with standard FQC tests on energy absorption.

Table 1. Foam samples.

Uncrosslinked	ρ (kg/m³)	Crosslinked	ρ (kg/m³)	Elastomeric	ρ (kg/m³)
PE(p) 33-10	33	PE(x,p) 33-10	33	PE(x,e) 124-10	124
PE(p) 45-10	45	PE(x,p) 45-10	45	SBR 513-8	513
PE(p) 64-10	64	PE(x,p) 64-10	64	PU 297-10	297
PE(p) 144-51	144	PE(x,p) 144-15	144		

2.1. Materials

A variety of closed-celled PE foams were selected to examine the effect of foam density and crosslinking. The former influences deformation properties significantly, while the latter could influence elevated temperature response. In addition, an ethylene-based crosslinked elastomer foam, along with a PU and an SBR shockpad were included. Samples are described in Table 1. Specimens were labeled based on polymer type, foam density provided by the supplier (kg/m³) and thickness (mm). In brackets, the foam type is specified as "*x*" for crosslinked, "*p*" for plastomeric and "*e*" for elastomeric.

2.2. Compression Test Methods

Stress-strain curves in compression were used to examine deformation behavior of the foams. Specimens of 5 × 5 cm were loaded into an Instron at a strain rate of about 1 min^{-1}. For experiments at 65°C an environmental chamber was used. Using the same set-up, hysteresis tests were performed with 10 cycles to examine the energy absorption and return characteristics of the foams. In this case, specimens were loaded to 0.39 MPa, the stress estimated for foot impact of athletes based on work by Hennig & Milani (1995). Compressive creep and recovery tests were performed also with the same geometry for 12 hours at a stress of 0.16 MPa, which simulates the pressure exerted by a light truck positioned on a field during maintenance or installation of an artificial turf surface.

2.3. FIFA Quality Concept Test Methods

Measurements of force reduction (FR), energy restitution (ER) and vertical deformation (VD) of both shockpads and infill/carpet/shockpad systems were done as described by the "FIFA Quality Concept Handbook of Test Methods for Football Turf" (FIFA, 2006). Specifically, the new testing apparatus described in Annex A4 of the handbook was used.

3 Results and Discussion

3.1. Compression Test Results

Compressive stress-strain behavior was measured in order to highlight the effects of foam density, crosslinking, material type, with select results illustrated in Figure 1. The curves feature typical regimes of linear elasticity, an intermediate region indicative of elastic collapse, followed by a steep densification zone, as reviewed by Gibson & Ashby (1997). Clearly,

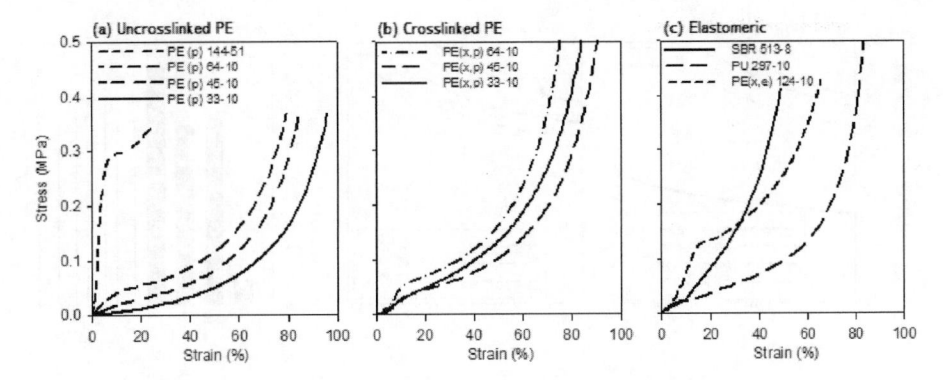

Figure 1. Compressive stress-strain behavior of selected foams at 20°C.

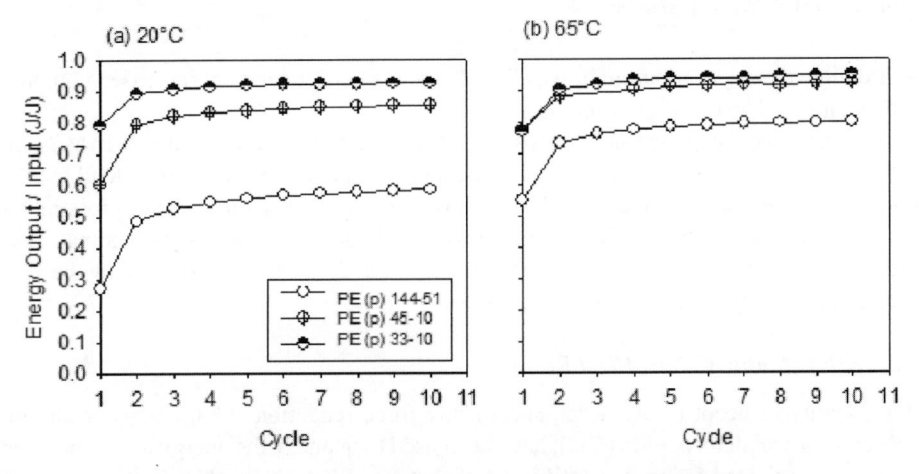

Figure 2. Energy output/input for selected foams at 20 and 65°C.

with decreasing foam density, the foams enter densification region at lower stresses. In light of stress levels expected in use, the density of the foam thus has to be picked such that the foam does not deform much beyond the plateau region. Also evident from Figure 1 is that crosslinking does not influence load-deflection curves significantly, as would be expected considering the insensitivity of the modulus to crosslinking at ambient conditions.

Cyclic tests were performed in stress-controlled mode to simulate repeated deformations applied by players, estimated at 0.39 MPa. Such tests provide insights into recovery characteristics, as well as hysteretic energy loss in each deformation cycle. The area under the loading curve, energy input, is indicative of energy absorption and the area under the unloading curve is the energy output. The ratio of output to input energy should be indicative of energy restitution in dynamic tests defined in the FQC. As shown in Figure 2, the ratio increased after the first cycle and in subsequent cycles it tended to plateau. Thus, the response exhibits an initial "breaking-in" period followed by stable performance. Higher energy output/input ratio was generally found for lower density foams which became more compliant and the effect of crosslinking was minor. When temperature was increased to 65°C, a shift

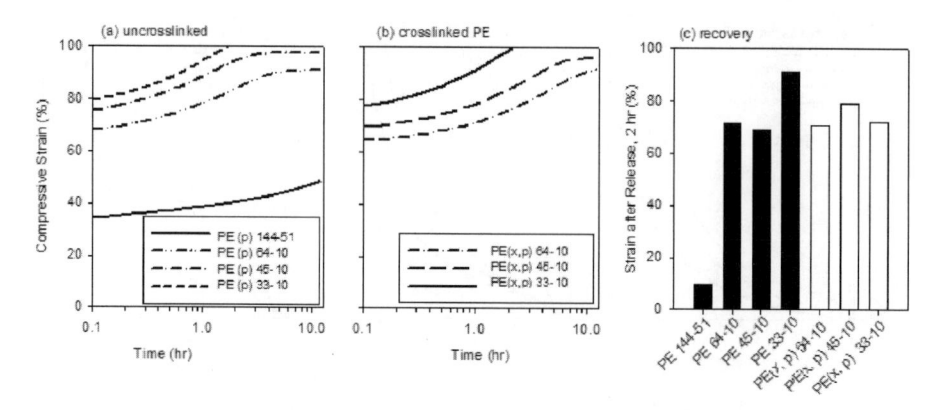

Figure 3. Compressive creep behavior at 0.16 MPa and 65°C for (a) uncosslinked and (b) crosslinked foams, and (c) recovery thereafter.

toward higher energy output/input was found, likely a consequence of increased compression leading to densification of foams.

Compressive creep measurements shown in Figure 3 were performed at 0.16 MPa and 65°C, as an estimate for highest service temperature. Samples were kept under load for 12 h followed by a period of 2 h recovery. For the lower density foams, the effect of crosslinking was not significant. However, PE(p) 144–51 foam did not reach a plateau indicating a relatively higher creep resistance, thanks to the lower initial deformation produced by the load. The strain recovery led to similar conclusions.

3.2. FIFA Quality Concept Test Results

FIFA Quality Concept (FQC) tests, specifically force reduction (FR), energy restitution (ER) and vertical deformation (VD), have been used to quantify the energy absorption performance of selected foams and artificial turf systems. FIFA 1-star classification specifies for the system an FR range of 55–70% and 4–9 mm for VD. While currently there is no specification on ER, low values from 30 to 45% are desired. Figures 4 to 6 show the average of the 4th and 5th impact results. The T-lines found above the FR and VD bars show the starting point (result from 1st impact) as an indication of how the properties of the shockpad degraded after repeated impacts.

Figure 4 shows the FQC results measured directly on shockpads. Samples are grouped as "uncrosslinked", "crosslinked" and "plastomeric vs elastomeric". Also to show the effect of "pre-compression" or severe usage, tests were performed with selected shockpads pre-compressed between two heated rolls, denoted with "**" in Figure 4. Note that elastomeric foams did not deform permanently with the same method. The difference between uncrosslinked and crosslinked samples at same foam density, although present, is not significant. FR increases and ER decreases with increasing foam density while VD does not change significantly. The key finding is the parity of performance of plastomeric and elastomeric foams despite thinner gauge for the elastomeric sample.

Figure 5 shows the FR, ER and VD results for the set of foams above, now measured as part of an artificial turf system. In this case, the system was made with 20 mm of coated sand infill evenly scattered in a 35 mm pile height PE-yarn carpet and was same in all tests.

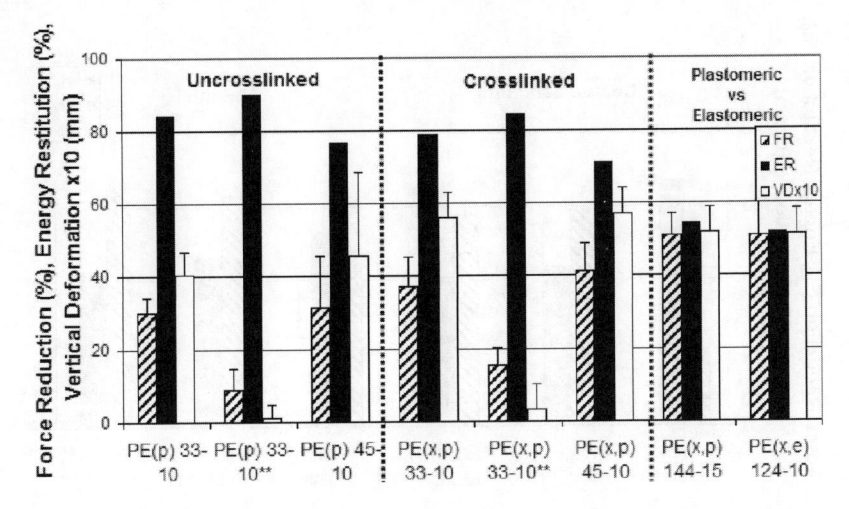

Figure 4. FR, ER and VD measured on different shockpads according to FQC Tests.

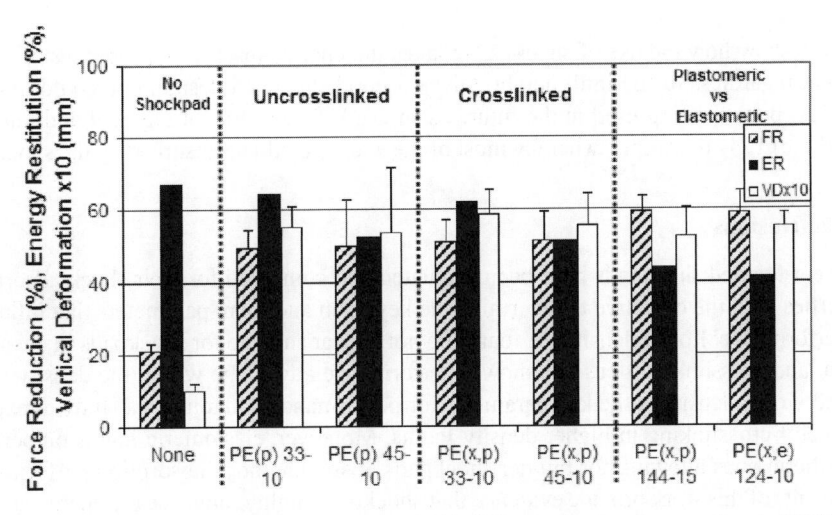

Figure 5. FR, ER and VD measured on systems with different shockpads according to FQC Tests.

Coated sand was used because it has low energy absorption capability when compared to SBR, thus the role of a shockpad is accentuated. Tested as part of the system, the difference between uncrosslinked and crosslinked shockpads is attenuated. Furthermore, for these low density foams, FR falls outside the FIFA 1-star requirement and ER is high. The higher density plastomeric and elastomeric foams do fall within the FIFA 1-star requirement.

To determine the importance of the infill versus the shockpad, the "best" shockpad from Figure 4, namely PE(x,e) 124-10, was used in systems with different infill heights. The same coated sand infill used previously was compared against SBR infill, which is best-in-class for energy absorption. In all cases, the free pile height was 15 mm. The results plotted in

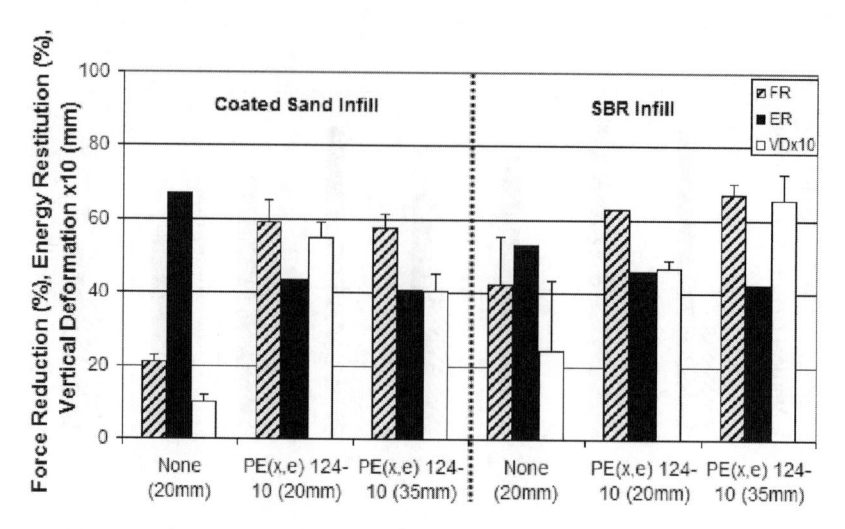

Figure 6. FR, ER and VD measured on systems with different infill types/heights according to FQC Tests.

Figure 6 show how the use of shockpads makes the energy absorption properties of a field uniform regardless of the infill height. Given a good shockpad design, one could think of infill-less pitches being used in the future as shown by the results in Figure 4 and Figure 6 for PE(x,e) 124–10 sample, whereby most of the energy could be absorbed by the shockpad.

4 Conclusions

Different foamed shockpads have been evaluated and compared for their shock absorption properties with the objective to determine the key resin and foam parameters that influence the performance. Lower density PE foams do not appear suitable for shockpads in the application, and crosslinking does not show a performance advantage within the density range studied. Foam density is the key parameter for performance. Ongoing is a study to explore the role of crosslinking in higher density foams. Moreover, elastomeric foams outperform plastomeric ones allowing for thinner shockpads at similar shock absorption performance. The results of this work provide evidence that shockpads highly influence the energy absorption property of an artificial surface, sports surfaces with no or non-shock absorbing infill can be designed and based on raw material design and foam characteristics, desired performance characteristics can be predicted. It also became evident that only a limited correlation exists between the quais-static lab tests and the performance relevant system tests.

References

Hennig E.M. and Milani T.L. (1995) In-shoe pressure distribution for running in various types of footwear. *J. Applied Biomechanics,* 11/3, 299–310.

FIFA (2006) *FIFA Quality Concept, Handbook of Test Methods for Football Turf.*, p. 49 (online). Available: http://www.fifa.com/documents/fifa/FQCturf/FQC_Requirements_manual_March_2006.pdf (accessed May 2007).

Gibson L.J. and Ashby M.F. (1997) *Cellular Solids: Structure and Properties.* Cambridge Universtiy Press, Cambridge.

6. Gait, Running and Shoes

DEVELOPMENT OF A NOVEL NORDIC-WALKING EQUIPMENT DUE TO A NEW SPORTING TECHNIQUE

A. SABO, M. ECKELT & M. REICHEL

University of Applied Science, Technikum Wien, Sports-Equipment Technology, Vienna, Austria

As a consequence of our intensive Nordic walking research from the last three years, it was found out that the use of special shoes is essential. The pole length, pole construction and the construction of the handle have a biomechanical influence. Furthermore there should be a difference in Nordic walking techniques. There is one fundamental technique from Finland, which is more or less a sporting one. But in Austria Nordic Walking is a health-movement and so, with the knowledge of the physio-therapeutics, a new, matching technique, which also requires new equipment (handle, strap...) have been developed. Our last research covered the influence of Nordic walking on the prevention of muscular tensions in the upper range of the back (M. Trapezius...). The subjects had an office job and had to work almost all day long in a sitting position in front of a computer. The subject's muscular activity was measured with EMG before and after a training cycle, in which they executed the finish technique of Nordic walking. Results show that the subjects trained muscles which are parts of the flexorloop. The muscular tensions became lower but they did not disappear. So a new Nordic walking technique was developed. This new technique trains the muscles of the extensorloop, which is not possible with conventional Nordic walking equipment, in particular with the conventional handle and strap. Therefore a new "strap system" was developed. This "strap system" does not only differ from the conventional handles and straps in form and look, there is also a difference in the transmission of force and the performance of the technique. With that new "strap system" and the new technique it is possible to train the muscles of the extensor loop effectively, which results in a prevention of muscular tensions in the upper range of the back and a prevention of muscular dysbalances.

1 Introduction

In the last few years Nordic Walking has become more and more popular and many people are trying out this new sport. But like many other sports it is very important to differentiate between correct shoes, poles and techniques. When we talk about Nordic Walking it has to be clear that this sport can not be compared with other common sports like biking, running, basketball and so on. The primary target group for Nordic Walking are people, who would not do any other sports or can not do any other sports because of the strains. But as a consequence of our intensive research we discovered that for these people and people with muscular dysbalances there should be a difference in techniques. One technique should be a sportive one, which trains the cardiovascular system, and the other should be a kind of physio-therapeutic one, which trains the back muscles effectively. The main differences in these two techniques are the working muscles, which we distinguished in the flexorloop for the sportive technique and the extensorloop for the therapeutics one. This paper describes the research about the difference in these techniques and the new equipment, which is needed for the therapeutic one (Schmölzer, 2003; de Marées & Mester, 1991; Dobner & Perry, 2000; Martin *et al.*, 1993).

2 Methods

In the fundamental research from the year 2004–2005 in which we did a study on the possible prevention of muscular tensions in the upper range of the back by exercising Nordic Walking (Wirhed, 1994; Benninghoff, 1994). The subjects in this research had an office job and had to work almost all day long in a sitting position in front of a computer. The subject's muscular activity was measured by electromyography (EMG) at the following positions:

- the upper range of the M. Trapezius
- the lower range of the M. Trapezius
- M. Deltoideus
- M. Pectoralis, as an antagonist

The results from this study showed (Figure 1), that after the training there was higher muscular activity in the lower ranger of the M. Trapezius, which was desired. But we found out, that there was also higher muscular activity in the M. Pectoralis, which was undesirable. So we analyzed the entire movement and so we found out that there should be a new Nordic Walking technique. The fundamental technique from Finland particularly trains the flexor loop (M. Biceps – M. Pectoralis). Our aim was to reduce muscular tensions in the upper range of the back and that Nordic Walking is a prevention for muscular dysbalances. So we talked to physiotherapists and with our combined knowledge we developed a new Nordic Walking technique, which trains the extensor loop, and new Nordic Walking equipment, which is needed to perform this technique.

Our new equipment differs from the conventional equipment in the construction of the handle and the strap.

The new handle:

- The shape supports the stabilisation of the wrist (20–25° dorsal flexion of the hand; 15–20° radial duktion of the hand). So the exposure at the moment when the pole contacts the ground should permit an insurance of the wrist.

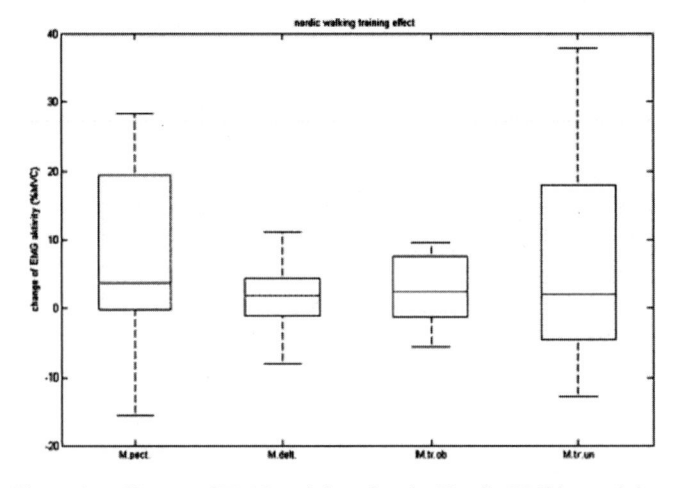

Figure 1. Change of EMG activity after the Nordic Walking training.

- The point for the transmission of the force in combination with the shape of the strap supports the stabilisation and is the crucial factor for the activation of the extensorloop.

The new strap:

- The new strap has additional padding on the edge of the hand.
- The shape of the new strap supports the correct performance of the new technique and the activation of the muscles of the extensorloop.

The subjects in the study with our new Nordic Walking equipment were Nordic Walking instructors (NWO) and people who got an insertion in Nordic Walking. The subject's muscular activity was measured with EMG at the following positions:

- the upper range of the M. Trapezius
- the lower range of the M. Trapezius
- M. Deltoideus
- M. Pectoralis
- M. Triceps
- M. Biceps
- M. extensor carpi
- M. flexor carpi

In additional to the measurements during Nordic Walking we measured the maximum voluntary contraction (MVC) for every muscle at the beginning, the end and when the subject changed the equipment.

The relevant range of the EMG data for the study is the phase of pushing with the pole from the beginning until the pushing is completed. These two points are obvious in the videos and in the data of the acceleration sensor (Figure 2), which was placed on the Caput ulnae.

The processing of the EMG raw data effected with a Matlab application (Matlab 7.0.4, The MathWorks Inc., Natick, Massachusetts) (Figure 3) in three steps (Stallkamp, 1998). The first step consists of signal processing by a Butterworth filter. This step is very important because otherwise the motional artefacts would be parts of the further steps and would have an effect on the final results. These motional artefacts were filtered out with a high pass filter and the cut-off-frequency for the filter was found out with the analysis of the data from MVC measurement. The second step was the demodulation of the filtered signal. The third step consists of placing an envelope on the signal. The face under the envelope was

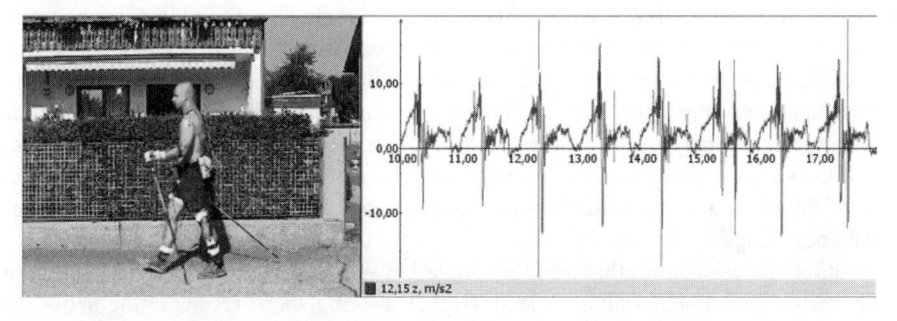

Figure 2. Video recording synchronously to the data of the acceleration sensor.

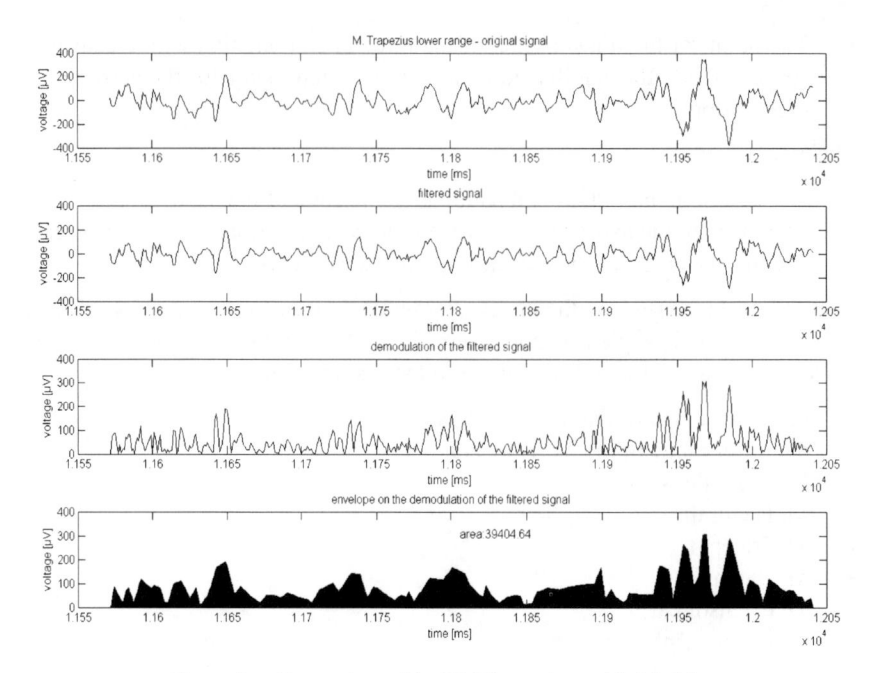

Figure 3. Processing of the EMG raw data with Matlab.

calculated and represents the muscle work, which is fundamental for the comparison of the signals from the other subjects with different Nordic Walking equipment.

3 Results

In 2004 and 2005 a letter of recommendation for the sale of Nordic Walking equipment was written (VSSÖ, 2005). It is based on results of our Nordic Walking studies from the last three years and cooperation with many companies, which produce Nordic Walking equipment (e.g. Leki, Exel, Fischer, Lowa, Salomon). This research concerned the Finnish technique. Based on the results inter alia we described how long the poles should be, which in the following text is called the "ideal" pole length.

Resultant of Nordic Walking with the new system and the new technique there was an increase of the muscular activity in the, based on the muscle loops and indicating a possible prevention of muscular dysbalances, important muscles (the lower and upper range of the M. Trapezius). 50% of the subjects show an increase of the muscle activity in the lower range of the M. Trapezius during Nordic Walking with the "ideal" pole length and the new system. Increase of muscular activity in the upper range of the back was observed in 25% of the subjects during Nordic Walking with the "ideal" pole length and the new system. 25% of the subjects showed an increase of muscular activity in the lower and in the upper range of the M. Trapezius while Nordic Walking with the new system and longer poles ("ideal" pole length +5 cm).

Motion analysis shows that a bigger angle between the legs, therefore a greater step length, results in a higher walking speed. Higher walking speed results in the majority of cases (laterality) in a higher heel pressure.

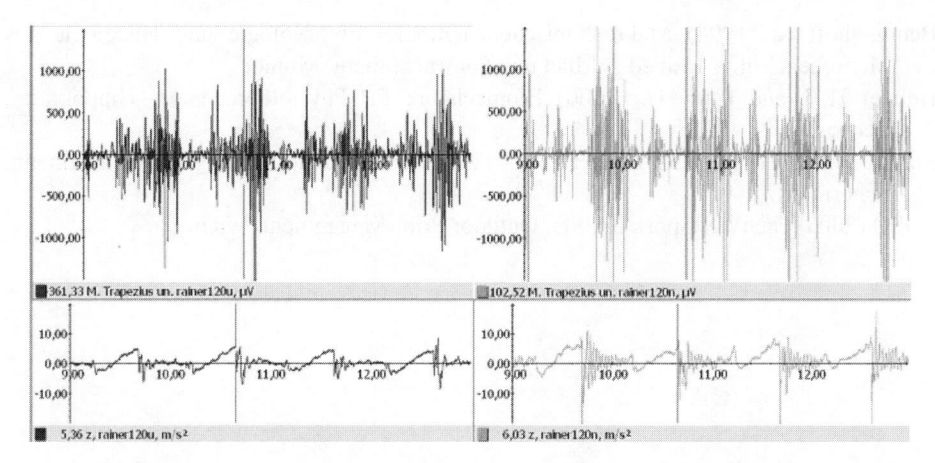

Figure 4. Comparison of the EMG raw signal of the new Nordic Walking equipment (left side) and common Nordic Walking equipment (right side) in the beginning of the pushing phase.

4 Discussion

Collected data show that Nordic Walking with the new system tends to result in a more effective training of the muscle groups, which are part of the extensor loop and longer poles are not essential for the new technique.

Figure 4 shows the difference between the EMG signal of one subject's muscular activity during Nordic Walking with the Finnish technique and the new technique respectively. EMG signal during Nordic Walking with the new system shows an activity of the muscle prior the pole-ground-contact. This can be interpreted as an initial tension, which can also result in injury- and impairment prevention.

5 Conclusion

Nordic Walking should be subdivided into two techniques. One should be in a sportive way and the other in a therapeutic way. The therapeutic one should result in an effective training of the muscles, which are part of the extensor loop. This training should result in prevention of muscular tensions and prevention of muscular dysbalances in the upper range of the back. Thus Nordic Walking with the new technique performed correctly is able to contribute to public-health in Austria.

References

Schmölzer B. (2003) Nordic Walking – Überprüfung der Funktion von Stöcken und der Effektivität des Bewegungsablaufes. Diploma Diss., University of Salzburg, Austria.
Stallkamp F. (1998) Dreidimensionale Bewegungsanalyse und elektromyographische Untersuchung beim Inline-Skating. Diploma Diss., Wilhelms-Universität Münster, Germany.
Marées H. de and Mester J. (1991) 2.Auflage. Sportphysiologie I. Diesterweg Sauerländer, Frankfurt am Main.
Wirhed R. (1994) Sport- Anatomie und Bewegungslehre. 2nd ed., Schattauer, Stuttgart.

Benninghoff A. (1994) Makroskopische Anatomie, Embryologie und Histologie des Menschen. Vol 1, 15th ed., Urban und Schwarzenberg, Munich.

Dobner H.-J. and Perry G. (2000) Biomechanik für Physiotherapeuten. Hippokrates, Leipzig.

Martin D., Carl K. and Lehnertz K. (1993) Handbuch Trainingslehre. Hofmann, Schorndorf.

VSSÖ (2005) FachWeltSport. 3/2005. Outdoor Print-Managment, Wien.

PATTERN RECOGNITION IN GAIT ANALYSIS

A.A. BAKAR,[1] S.M.N.A. SENANAYAKE,[1] R. GANESON[2] & B.D. WILSON[2]

[1]*Monash University Sunway Campus, Jalan Universiti, Petaling Jaya, Malaysia*
[2]*National Sports Institute, Center of Biomechanics, Sri Petaling,*
Kuala Lumpur, Malaysia

This paper discusses the application of pattern recognition in gait analysis. Movement patterns of sports performers for gait are acquired from optical motion capture system. The gait data appear as joint angles. Artificial Neural Networks (ANN) is trained with the existing database which contains gait data of 10 female athletes and 10 male athletes for three sports. Only useful parameters are selected for training the ANN. The parameter selection is followed by automatic determination of the frame window and data extraction.

1 Introduction

Gait data normally appear as a time series pattern. Analysis of this time series pattern may reveal any anomalies that appear in the data and helps in injury prevention and rehabilitation processes. In general, ANN is widely used in biomechanics (Barakna, 2006). There exist base line data of gait patterns based on general western population. However, gait pattern for athletes, especially in the Asia region, might be different from those existing databases. Therefore, analysis of movement patterns of athletes in Asia should not be based on the general western population database.

There is a need to create a gait database for athlete in Asia and for different sports. In order to handle this large database, an analysis system must possess the ability to recognize gait patterns for different age, gender and sports (Canosa, 2002).

2 Software and Equipment Used

2.1. *Motion Capture/EVaRT*

The walking gait cycle of each athlete was captured using 6 infra-red optical cameras. Each camera has a selectable frame rate from 1 to 2000 Hz and has a high quality 35 mm lens for low optical distortion. It has 237 LEDs for brighter and better light uniformity. The cameras are connected with a multi-port Ethernet switch and provide power for the cameras. The system is connected to motion capture software called EVaRT, a product of Motion Analysis Corporation. The images captured by the cameras are converted onto binary format by the software and reproduced in EVaRT. After post-processing the captured motion, a simulation of the walking gait can be seen in EVaRT in a 3D environment. The system also calibrates the cameras according the volume of space needed for the capture of the motion. A four point L-shaped calibration device was used to define the XYZ axes and a 500 mm wand is used for camera linearization.

2.2. OrthoTrak

The walking gaits of the athletes were saved in binary format; these are the positions of each marker (placed on the body) relative to the calibrated volume for each frame of capture for the entire trial. The binary file is uploaded into a clinical gait measurement, evaluation and database management system called OrthoTrak. OrthoTrak was developed by Motion Analysis Corporation as an instrument for clinical observation and studies, primarily in the walking gait cycle of a patient. The software can reproduce the gait simulated in EVaRT and from that the characteristics of the gait cycle such as heel strike and toe off is shown. For the purpose of this research, OrthoTrak is able to extract the joint angle values of the body and these values will be used.

2.3. JavaNNS

Neural network architecture with suitable topology was created in simulation software called JavaNNS (Neural Network Simulator). JavaNNS is a downloadable freeware that was developed by the University of Tübingen. It is the successor of the SNNS (Stuttgart Neural Network Simulator), a universal simulator of neural networks for UNIX workstations and Unix PCs, developed by the University of Stuttgart. JavaNNS's computing kernel is based on SNNS, but now with a new developed, comfortable GUI written in Java. The compatibility with SNNS is therefore achieved while the platform-independence is increased.

2.4. MATLAB

MATLAB is technical computing software that deals with algorithms, analysis and software simulation, dealing mostly with complex mathematical equations and functions. It has a variety of toolboxes or add-ons, to its main software. MATLAB was used to arrange the data pattern for the neural network in JavaNNS (Fischer et al., 2001).

3 Gait Data Collection

The athletes were required to wear a minimum amount of clothing i.e. shorts or tights for the lower body and bare-bodied for the men. Reflective markers were placed at strategic points of the body for walking gait analysis. The positions of the markers were based on a marker placement system called the Helen Hayes system. After the motion capture system has been calibrated, the athletes were to stand in the middle of the volume space with their arms out-stretched by their side and their legs apart at shoulder width. This position is known as the "T-Pose". 26 markers were placed on the body for identification of each limb and body positions (excluding the head). The purpose of this exercise is for the OrthoTrak system to recognize and calculate the exact joint position of the limbs as well as the other parts of the body in relation to each other. 4 of the markers will be removed (the left and right medial knees and ankles) during the capture of the walking gait.

The protocol for the capture of the walking gait was simply letting the athlete walk continuously through the calibrated volume. The athlete will walk towards the end of the volume, turn around, walk to the other end and continue the process. They were not told when the actual capture was taken so as to capture their natural state of walking without them thinking about it. A minimum of 3 trials (or captured movements) were taken and saved in the computer.

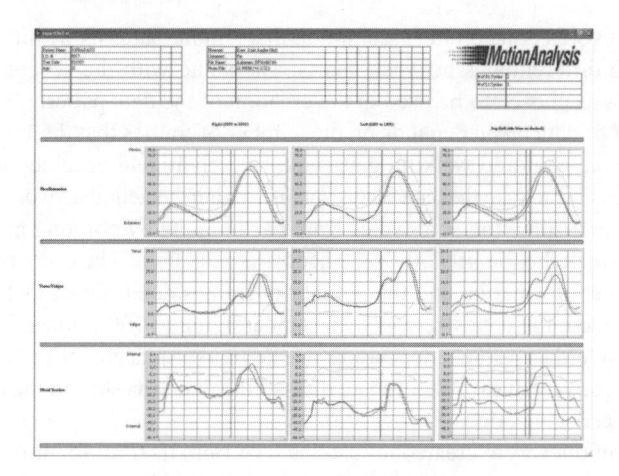

Figure 1. Knee flexion, abduction and rotation chart from OrthoTrak.

After the athletes' walking gaits have been saved, the post-process is then done. There are times when the intended markers do not appear in the captured frames. The exercise of post-processing is done to "clean" the captured gait video without altering the actual gait itself. Links were created between markers to represent the limbs and body areas. The clean-up video is then saved in binary format.

The "cleaned" binary format motion capture video is "open" in OrthoTrak. In order for OrthoTrak to read the data, it requires 2 different files. The first file is the "T-Pose" file mentioned earlier. The software will take this file as the "calibration" of the athlete, using the data as the left and right static trial. The next file loaded is the desired captured walking gait that is to be analyzed. The software has a function "Event Editor" where the data from the MAC system is reproduced and simulated. Joint angle graphs can also be plotted and displayed with the norms that has been compiled by Motion Analysis in OrthoTrak, e. g. see figure 1.

A total of 30 different joint angle values were extracted from walking gait :

Left and Right $\begin{cases} \text{Shoulder} & \text{Adduction, Flexion} \\ \text{Hip, Knee, Ankle} & \text{Rotation, Abduction, Flexion} \\ \text{Elbows} & \\ \text{Trunk, Pelvis} & \text{Lateral Tilt, Forward Tilt, Rotation} \end{cases}$

4 Gait Data Analysis

The total number of athletes' data used is 20; 10 males and 10 females. They were divided accordingly to their gender and further divided into the sports they were from. A total of 3 sports were identified at the time of this study – Athletics, Karate, and Racquet (racquet consists of 2 sports – squash and badminton).

Each of the joint angle value had to be studied and analyzed carefully for any subtle changes in pattern from the graphs. These changes or variations provide the unique quality that is inherent to the sport in which the athlete is training. It basically means that each athlete walking gait is directly connected to the sport he or she was trained in. To simplify the process, similar patterns from each athlete in each sport are identified and noted as shown in the table 1. This was done on the assumption that similar patterns within a sport are also the factor which differentiates between sports. The value with the acceptable number of similar patterns was chosen to be used in the creation of the data pattern.

The amount of data taken depends on the number of frames that the walking gait cycle occupied. Since the captured walking gait cycle of each athlete occupied a different time period, it is not possible to take the data as it is and use it train the neural network. The OrthoTrak software is able to display the gait cycles from each athlete by pointing at the frame time (or number) in which the toe lifts off or the heel strike the floor. The cycle time then can be obtained from this information and the number of frames determined. Data from two gait cycles were obtained; the data size was then standardizing to 150 frames.

The joint angle data values, depending on the market placement or type of measurement, contain negative values. To eliminate this, all the data was shift to the positive area by simply adding a constant value.

MATLAB routines were created to read the data from its CSV (comma delimited) format and re-arranged it to a format where the JavaNNS is able to understand. The routines also normalized the data. 3 types of data were created : Single Values (Male and Female), Left & Right Values (Male only) and Re-arranged Values (Male only).

The Single Values were created originally from the data analysis mentioned earlier. The Left & Right Values were created as a set of left and right values from the Single Values data. The trunk and pelvic values were not used in this case. The Re-Arranged Values were created as a testing data pattern. It is basically the Single Values data re-arranged. The routines also perform a sample size pick from the original data. The Males data have a sample size pick of every 3 frames while the Females were every 2 frames. The data is then reduced to 50 and 75 values respectively, and this corresponds to the number of inputs in the neural network architecture.

Table 1. Mala Data Grouping.

Left		Shoulder		Hip			Knee			Ankle		
		Add	Flex	Rot	Abd	Flex	Rot	Abd	Flex	Rot	Abd	Flex
Athletics	Hazwan	√		√		√	√					√
	Kalvindran	√				√		√	√			√
	Robani			√			√	√				
Karate	Puvan	√	√			√	√	√	√			√
	Mahen		√	√		√	√	√	√	√		√
	Lim	√	√	√						√		
Raqcuet	Bhee	√					√					√
	Kimlee		√				√	√	√			√
	Kuanas	√		√	√	√		√	√			
	Cwei		√		√							

5 Gait Pattern Recognition

The Elman NN created for male data patterns is a 50 input-3 output, 2 layer back-propagation network with 8 hidden neurons and same number of context neurons. The Jordan NN is similar except the number of context neurons is equivalent to the output because it receives feedback from the output layer. Data from the Male athletes were used to train these networks. The functions of the neurons have been pre-determined and the weights initialized for the network by JavaNNS.

The Elman NN created for female data patterns is a 75 input-2 output, 2 layer back-propagation network with 16 hidden neurons and same number of context neurons. The Jordan NN is similar except the number of context neurons is equivalent to the output. Data from the Female athletes were used to train these networks. The functions of the neurons have been pre-determined and the weights initialized for the network by JavaNNS.

Even though the error graph for the networks does not reach zero, the total percentage of gait pattern recognition is between 85% and 91%. A series of five sets of training and testing was set up to check the validity of the results. The tests done were: Single Values – train and test, L & R Values – train and test, L & R Values train, Single Values test (pelvis and trunk values taken out), Single Values train, Single Values re-arranged data test and L & R Values train, L & R Values re-arranged data test.

6 Conclusions

Table 2, 3 and 4 demonstrate that pattern recognition using Elman NN architecture was successful. The total gait pattern recognition for each sport is between 85% and 91%. The results

Table 2. Elman NN, Male, *Single Values*, 132 pattern data (Training & Testing) 50 i/p, 3 o/p, 8 hidden layer, JE Backpropagation (η: 0.2; d_{max}: 0.1; forceT: 0.5) Cycles = 10,000.

Error Graph/Analyzer	Athletics	Karate	Racquet	Total %	$\Sigma e^2 = 0$?
T1 / R1	41	44	44	97.7	Yes
T2 / R2	0	20	44	48.5	Yes
T3 / R3	28	44	44	87.9	Yes
T4 / R4	41	44	44	97.7	Yes
T5 / R5	19	44	44	81.2	Yes

Table 3. Elman NN, Male, *L&R Values Train* 168 pattern, *Single Values Test* 108 pattern 50 i/p, 3 o/p, 8 hidden layer, JE Backpropagation (η: 0.2; d_{max}: 0.1; forceT: 0.5) Cycles = 10,000.

Analyzer	Athletics	Karate	Racquet	Total %	$\Sigma e^2 = 0$?
R1	21	36	36	86.1	Yes
R2	21	36	36	86.1	Yes
R3	0	0	36	33.3	Yes
R4	21	36	36	86.1	Yes
R5	22	36	36	87.0	Yes

Table 4. Elman NN, Male, *L&R Values Train, L&R Values Re-Arranged Test*, 168 patterns 50 i/p, 3 o/p, 8 hidden layer, JE Backpropagation (η: 0.2; d_{max}: 0.1; forceT: 0.5) Cycles = 10,000.

Analyzer	Racquet	Athletics	Karate	Total %	$\Sigma e^2 = 0$?
R1	38	46	55	82.7	Yes
R2	49	41	54	85.7	Yes
R3	49	54	54	93.5	Yes
R4	51	24	56	78.0	Yes
R5	49	33	56	82.1	Yes

obtained are better than conventional software used. Therefore, recurrent neural network architectures such as Elman NN are good candidates for sports gait performance analysis.

References

Barakna A. (2006) Real Time Soccer Gait Recognition Using Wireless Sensors and Vision System, Final Year Project Thesis, Monash University Malaysia.

Canosa R.L. (2002) *Simulating Biological Motion Perception Using a Recurrent Neural Network*, Rochester Institute of Technology, Rochester (online). Available: http://www.cs.rit.edu/~rlc/SimBioMotion.pdf (Accessed: May 2007).

Fischer I., Hennecke F., Bannes C. and Zell A. (2001) *JavaNNS Manual User Manual V1.1* (online). Available: http://www-ra.informatik.uni-tuebingen.de/software/JavaNNS/welcome_e.html (Accessed: May 2007) http://www-ra.informatik.uni-tuebingen.de/software/JavaNNS/manual/JavaNNS-manual.pdf.

ANALYSIS OF THREE-DIMENSIONAL PLANTAR PRESSURE DISTRIBUTION USING STANDING BALANCE MEASUREMENT SYSTEM

Y. HAYASHI, N. TSUJIUCHI, T. KOIZUMI & A. NISHI

Department of Mechanical Engineering, Doshisha University, Kyotanabe, Kyoto, Japan

This paper describes the analysis of three-dimensional plantar pressure distribution using a standing balance measurement system consisting of instrument that distributed 6×4 three-axis force sensors and software that displays and preserves the output of the sensor elements. Experimental results prove the effectiveness of the system. Moreover, by analyzing the measurement data, it was shown that those data measured by the system become an effective index in rehabilitation, medical treatment, etc.

1 Introduction

Recently, we are becoming more and more interested in medical treatment and rehabilitation in an aging society. An evaluation function that uses the latest movement analysis system and applications are noticed in the fields of medical treatment and rehabilitation, especially the development of clinical walking analysis research for persons who have troubles walking in daily life.

The bottom of a person's foot grips the floor for balance, and the action force and action moment work at the foot bottom when he maintains posture and when he moves. They are important indices in the evaluation and the medical attentions of standing pose balance and gait disturbances. Therefore, it is necessary to measure force not only in the vertical direction but also in the shear direction and also to measure the distribution information on force in the three directions that act on the foot bottom when the force that acts on the foot bottom is measured by pattern analysis, etc. A lot of equipment and techniques to measure the floor reaction force have been researched and reported (Yamato *et al.*, 2001, Ito *et al.*, 2003). Some have already been commercialized and used in the fields of rehabilitation and sports engineering, etc. However, no floor reaction force meter that can measure distribution information on force in three directions exists.

In our research, we propose an instrument that can measure the distribution information force in three directions that work at the foot bottom of the bearing area by concentrating on the interaction of the foot bottom and the floor when a person stands and exercises. That is a measurement system that can measure the plantar pressure distributed by 6×4 three-axis force sensors and software that displays and preserves the output of the sensor element. A time change of the force at the foot bottom is needed as a vector by outputting each sensor element. Moreover, the action vector is three-dimensionally displayed whose data can be intuitively recognized three dimensionally a using the DLL function that is built in OpenGL. By the person's getting on the measurement instrument in various posture, and analyzing the measurement value in three-dimensions, it is shown that the relation between each subject and plantar pressure, between each postures and plantar pressure, and so on. And, it is shown that those data become an effective index in rehabilitation, medical treatment, etc.

In experiment, it will be shown that the measurement system that could measure the action force of the foot bottom as distribution information on force in three directions. Moreover, the effectiveness of the system will be proven by indicating values of force to work at foot bottom in each posture. And, by analyzing the measurement data, it will be shown that those data measured by the measurement system become an effective index in rehabilitation, medical treatment, etc.

2 Standing Balance Measurement System

The standing balance measurement system is composed of an instrument part that can measure the plantar pressure and a PC part. The instrument is composed of sensor unit where 24 triaxial force sensors are arranged. A PC has software installed that displays and preserves the output from the three-axis force sensors.

2.1. Standing Balance Measurement Instrument

Figure 1 shows a general view chart of the instrument. It is 300 [mm] × 560 [mm] × 150 [mm], and 24 (6 × 4) three-axis force sensors are arranged in the sensor unit, as shown in Figure 2. Each sensor element is sequentially numbered from the upper left to the lower right. The width of the sensor unit is large enough to be covered by the toes. It is possible to measure the action force on the foot bottom as distribution information on force in three directions when a person gets on the sensor unit. The amplifier board and an analog to digital conversion device are built into the measurement instrument, and a LAN sends output to a personal computer. The unit includes a fan, it is preventing heat from influencing system.

2.2. Triaxial Force Sensor

The shape of the triaxial force sensor that composes the measurement instrument is shown in Figure 3, and its size is 20 [mm] × 20 [mm] × 5 [mm]. The sensor detects action force as electric resistance changes of the distortion gauge attached to the surface. The bridge circuit is united to the distortion gauge in each axis, and the action force is determined by measuring the differential motion voltage. Moreover, a pressure plate is fixed to the upper part of the sensor unit with screws, and rubber is glued to it. The Z-axis force is made to be a uniformly distributed load on the upper board. The load rating of the triaxial force sensor is the X and Y-axis are 250 [N], and the Z-axis is 500 [N].

Next, the interference correction of the triaxial force sensor is shown. The output of each sensor element causes an almost linear mutual interference in each axis for the action force. Then, the interference correction matrix shown in Eq. (1) is suggested.

$$\boldsymbol{f} = \begin{pmatrix} f_x \\ f_y \\ f_z \end{pmatrix} = \begin{pmatrix} a_{11} & -a_{12} & -a_{13} \\ -a_{21} & a_{22} & -a_{23} \\ -a_{31} & -a_{32} & a_{33} \end{pmatrix} \begin{pmatrix} \varepsilon_x \\ \varepsilon_y \\ \varepsilon_z \end{pmatrix} = \boldsymbol{A}\boldsymbol{\varepsilon} \quad (1)$$

where ε is a measurement value before correcting mutual interference, f is a measurement value after correcting mutual interference, and A is transformation matrix. When E is a measurement value matrix before correcting mutual interference, and F is a measurement value matrix after correcting mutual interference, the generalized inverse E^+ of E is used, Eq. (2) is obtained from Eq. (1).

$$A = FE^+ \qquad (2)$$

The direction of forces X, Y, and Z is a load and is removed from each sensor element, and transformation matrix A is obtained from Eq. (2), and mutual interference correction is done from the output at that time in Eq. (2). As the result, the accuracy of the sensor element was about 1% RO in the X and Y-axis and about 3% RO in the Z-axis.

2.3. Data Processing Software

Data from the measurement instrument is acquired by using Microsoft Visual Basic, which also does the interference correction. It displays the data under the sampling time, updating the action force in the direction of three axes in 12 sensor elements. The data of the action force can be preserved as csv files in arbitrary time. We made a Microsoft Visual C++ system including a DLL function, which is included in the Open GL (Sakai, 2000). By using the picture box from the Visual Basic side, we can get display data as the vector of the action force in each sensor element and the vector of the action resultant force in a center of pressure of the sensor unit. Figure 4 shows the one example. However, the blue lines are force vectors to work in each sensor element, and green line is the resultant force vector to work in the center of pressure of the sensor unit. By clicking on the box at the picture's bottom (Dolly, Tumble, Pan, Zoom, Tilt, High), we can enlarge or reduce, move left or right, or rotate data. We can see action force everywhere. Moreover, the tool used for the program developed in this research is Visual Studio 6.0, and the OS is Windows 2000.

3 Experiment

Now we show, through experiments, how the measurement system can measure the action force of the foot bottom as distribution information on force in three directions.

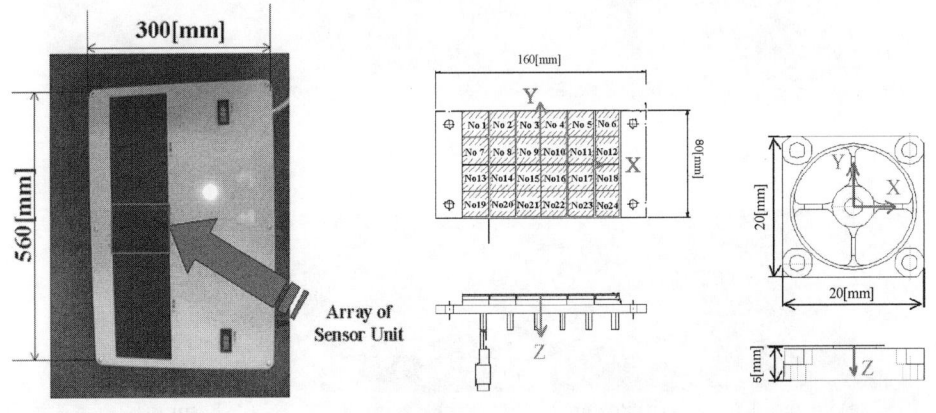

Figure 1. Measuring instrument.

Figure 2. Array of sensor unit.

Figure 3. Structure of triaxial force sensor.

3.1. Experimental Condition

Subjects: 12 persons (males of standard proportions in their twenties), sampling frequency: 50 [Hz], low-pass filter: 5 [Hz].

3.2. Experimental Method

Data for each of four postures are acquired as follows. We set the standard position, namely that which crosses the base of the big toe of the right leg, with sensor elements at position nos.8, 9, 14, and 15. The left leg is put on the stand so that the width of the foot becomes 15 [cm].

Posture 1: The subject stands with both feet in a natural posture.
Posture 2: With the heel not lifted and not twisting the waist, the head is inclined forward to the maximum after posture 1.
Posture 3: All weight is slowly moved to the right leg after posture 1. The left leg is only placed on the stand to maintain balance and does not exert any force.
Posture 4: The subject stands on tiptoe only with the right leg after posture 3. As in posture 3, the left leg is only used to maintain balance.

Figure 5 shows the appearance of each posture.

4 Experimental Results and Considerations

Figures 6 each show a 3D image of postures 2 and 4 for one of the experimental subjects. The length of the vectors in postures 4 have obviously expanded more than in posture 1, and it can be seen that force was greatly applied at the part of the sensor No.13–24 in posture 4. Moreover, because the vectors have inclined in each posture, it is understood that not only is the system measuring force in the normal direction, but also force in the direction of the shearing. In addition, the position of the center of pressure was changed according to each posture. These vector values are different for each of the 12 subjects, and characteristic values were seen in each.

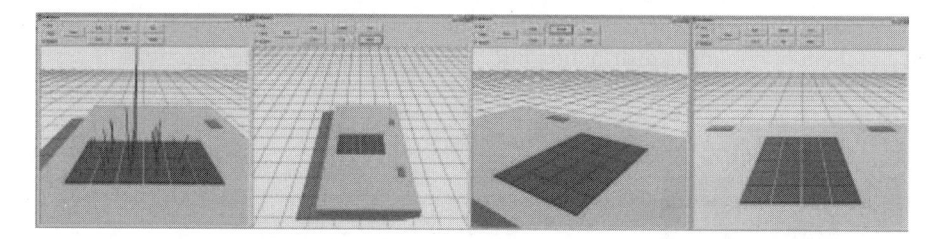

Figure 4. 3D screen.

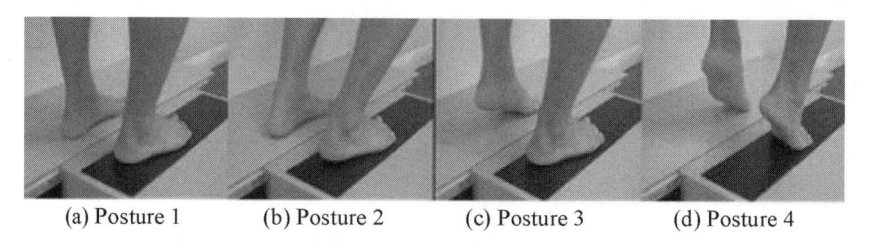

(a) Posture 1   (b) Posture 2   (c) Posture 3   (d) Posture 4

Figure 5. Appearance of each posture.

Analysis of Three-Dimensional Plantar Pressure Distribution 179

Next, we classified the Areas of the sensor elements into four groups: Area 1 (No.1, 2, 7, 8), Area 2 (No.3–6, No.9–12), Area 3 (No.13, 14, 19, 20), and Area 4 (No.15–18, No.21–24), as shown in Figure 7. We calculated the load fraction in each block in each posture and then plotted the results in Figure 8 about the X-axis and in Figure 9 about the Y-axis. It should be noted that both Figures 8 and 9 show the averaged results for the 12 subjects. Figure 8 shows that Area 2 and Area 3 are minus direction under the all postures, and Area 4 are plus direction under the all postures. This shows that force works inwards from right

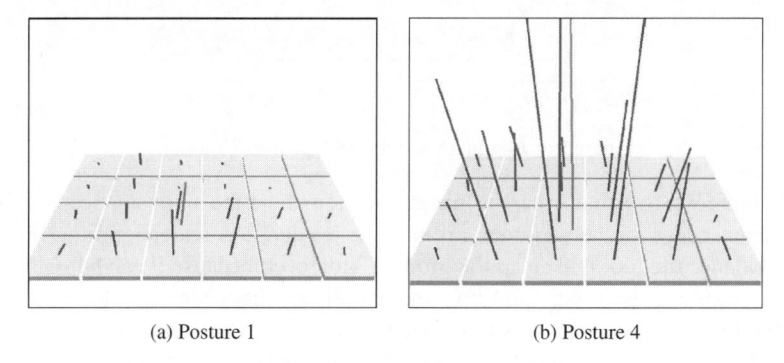

(a) Posture 1   (b) Posture 4

Figure 6.   3D image.

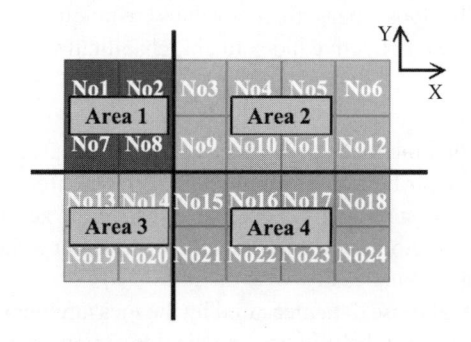

Figure 7.   Block separation for sensor element.

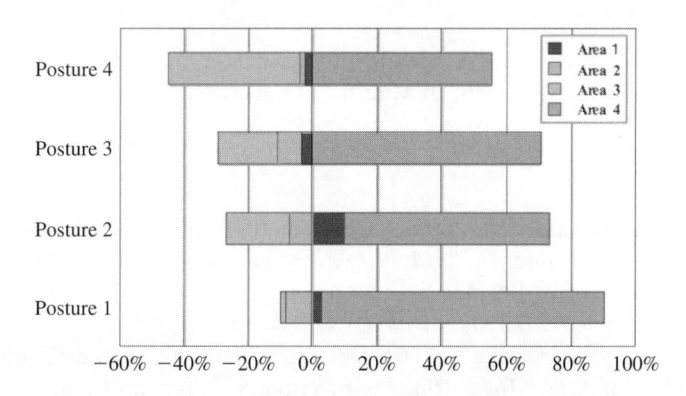

Figure 8.   Loading ratio for each posture (X-axis).

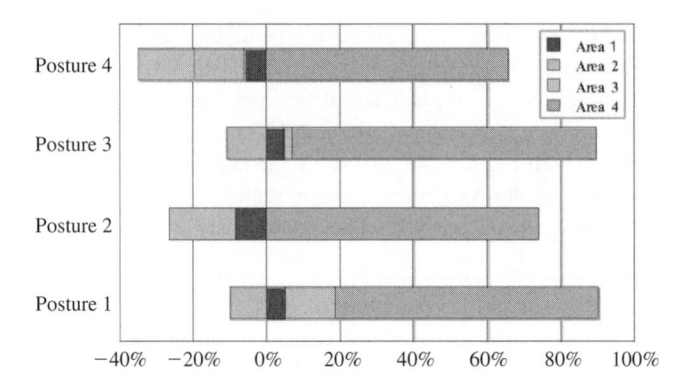

Figure 9. Loading ratio for each posture (Y-axis).

and left. Figure 9 shows that Area 2 is minus direction under the all postures, and Area 4 are plus direction under the all postures. From these results, it is understood that in order to maintain balance the foot must grip the ground. Moreover, because it can be seen that both along the X-axis and the Y-axis for Area 3 there is a large change in the value for the change in posture, it is understood that this is a place where it is easy to control posture well.

Therefore, it was shown that the measurement system could measure, for the first time, the action force of the foot bottom as distribution information on force in three directions. In addition, it was shown that those measurement values are unique to each person and posture and will therefore become an effective index to aid rehabilitation, medical treatment, etc.

5 Conclusion

(1) It was shown that the measurement system could measure the action force of the foot bottom as distribution information on force in three directions.
(2) Our proposed measurement system can visualize the action force working at the foot bottom as vectors by using Open GL and using a three-dimensional graph that we can understand easily.
(3) It was shown that those data measured by the measurement system would become an effective index in rehabilitation, medical treatment, etc.

Acknowledgements

This study was supported by the Academic Frontier Research Project on "New Frontier of Biomedical Engineering Research" of Ministry of Education, Culture, Sports, Science and Technology.

References

Ito S., Saka Y. and Kawasaki H. (2003) A Consideration on Control of Center of Pressure in Biped Upright Posture: *The Institute of Electronics, Information and Communication Engineers D-II*, J86/3, 429–436.
Sakai K. (2000) *OpenGL 3D Programming*, CQ Ltd.
Yamato J., Shimada S. and Otsuka S. (2001) Development of Gait Analyzer Using Large Area Pressure Sensor Array: *The Institute of Electronics, Information and Communication Engineers D-II*, J84/2, 380–389.

THE USE OF MICRO-ELECTRO-MECHANICAL-SYSTEMS TECHNOLOGY TO ASSESS GAIT CHARACTERISTICS

J.B. LEE,[1] B. BURKETT,[1] R.B. MELLIFONT[1] & D.A. JAMES[2]

[1]*Centre for Healthy Activities Sport and Exercise (CHASE), University of the Sunshine Coast Queensland Australia*
[2]*Centre for Wireless Monitoring and Applications, Griffith University, Brisbane Queensland Australia*

There are many forms of human movement and motion monitoring. Micro-Electro-Mechanical-Systems is one such method, these include Accelerometers and Gyroscopes. Ten healthy adult's gait patterns were monitored at seven different speeds. These were compared to walking patterns of a race walker. Light gates were used to control and measure walking velocities. Data was collected from triaxial accelerometers and rate gyroscopes positioned on the foot and sacrum. Preliminary results indicate that wireless inertial sensor data correlate well with conventional kinematic measures without the usual restrictions imposed on subjects during testing. These results also indicate that the use of microtechnology inertial sensors as a valuable tool in the analysis of human movement.

1 Introduction

Assessment of human movement and motion has been carried out in a multitude of ways. It includes functional and kinematic monitoring (Mayagoitia, et al. 2002; Uswatte, et al. 2006) as well as assessment of physical activities (Bassett, 2000; Freedson & Miller, 2000; Plasqui & Westerterp, 2005; Pfeiffer et al., 2006; Plasqui & Westerterp, 2006). The array of systems can be simple equipment such as pedometers, through to sophisticated computer aided systems. Some popular methods for biomechanical analysis include optical motion analysis tools (Mayagoitia et al., 2002). Many of these techniques and equipment are restricted to laboratory settings or limited use for field application (Luinge & Veltink, 2005). Accelerometers offer opportunities for longitudinal studies by monitoring human movement (Luinge & Veltink, 2005). The use of accelerometry is considered to be a method that can be used for human movement measurement that is unobtrusive to subjects being tested (Uswatte et al., 2006). Accelerometers and rate gyroscopes (gyros) belong to a technological group known as Micro-Electro-Mechanical-Systems (MEMS) in particular, inertial sensors.

Accelerometers have been found to be accurate at detecting steps taken at low velocities when compared to pedometers (Le-Masurier and Tudor-Locke, 2003). In a comparative study, accelerometers and gyros were analysed against kinematic data of walking gait measured by a Vicon® 3D camera system (Mayagoitia et al., 2002). The authors reported accelerometer and gyro results that closely matched the camera system especially at lower velocities. At higher velocities an increase in errors (<7% of total range) were found which the authors hypothesised was due to high impact from heel strike (Mayagoitia et al., 2002). The study fixed the accelerometers to a metal strip which was placed on the shank. Direct attachment to the segment may have given more reliable results. The study also neglected to assess flexion at the ankle and regarded the shank and foot as a single segment. This may be an oversight for analysis of gait as other studies have identified ankle flexion as a common variable with people who have stability and gait problems (Okada et al., 2001; McGibbon & Krebs, 2004; Amiridis et al., 2005). It has also been shown that chronic joint conditions have a major detrimental affect on gait (McGibbon & Krebs, 2004). Analysis of ankle movement may be an area of closer research focus. Inertial sensors placed on the feet have been used to determine gait parameters during treadmill walking (Sabatini et al., 2005).

The purpose of this study was to assess comfortable gait speed of a small population from a university community to compare and contrast to gait patterns of a race walker.

2 Methods

Ten healthy adults from a university community freely consented to participate in the general population (GP) gait analysis. One race walker (RW) freely consented to participate in the comparative study. All participants provided written consent prior to testing. Anthropometric details of lower limb dimensions were recorded.

The GP group were asked to walk a distance of 10 metres (m) at seven various velocities in increments of 1 km/h. These ranged from 1 km/h to 7 km/h and were each repeated three times. This range was applied to ensure comfortable and maximal walking velocities were captured. All participants stated that 5 km/h was the most comfortable speed, therefore this was chosen for comparison against race walking.

The RW performed 3 repeats of 10 m walking trials at comfortable, maximal normal walking and race walking pace. Appropriate warm-up was allowed prior to each new condition.

Data collection was via two inertial sensors each containing a tri-axial accelerometer, KXM52 – 1050 (Kionix, NY, USA) and a single ordinal rate gyro, KGF01 – 300 (Kionix, NY, USA). Accelerometers were calibrated by orientation of each relative accelerometer against gravity to produce $+1\,g$ then rotated 180° to produce $-1\,g$ (Lai, James et al. 2004). Gyro calibration was carried out by rotating the instrument from a horizontal position to 90° around the axis of the gyro and zeroed to ensure the known rotation. Inertial sensors were positioned on the sacrum (S1) and on the centre of the dorsum of the foot. Placement of the sensors was directly onto the skin with double sided tape, with a secondary tape over the instrument to ensure secure positioning. For capture of foot swing, orientation of the gyro was in the sagittal plane i.e. in the direction of forward movement.

For the GP group electronic Smartspeed™ lightgates (Fusion Sport, Bris, Aust) were positioned to a 10 m configuration with a series of five gates at 2.5 m intervals. The gates were programmed to pace the subjects at each velocity. Subjects were asked to keep pace with the gate signals in an attempt to keep a constant velocity. The lightgates were altered to record time for the RW.

Processing analysis was carried out using MATLAB™ (The MathWorks U.S.A.).

3 Results

Figures 1 & 2 are typical traces from foot and sacral sensors collected at 5 km/h. Figure 3 shows typical gait cycles from a race walker. These are displayed as examples of identifiable events during the gait analysis.

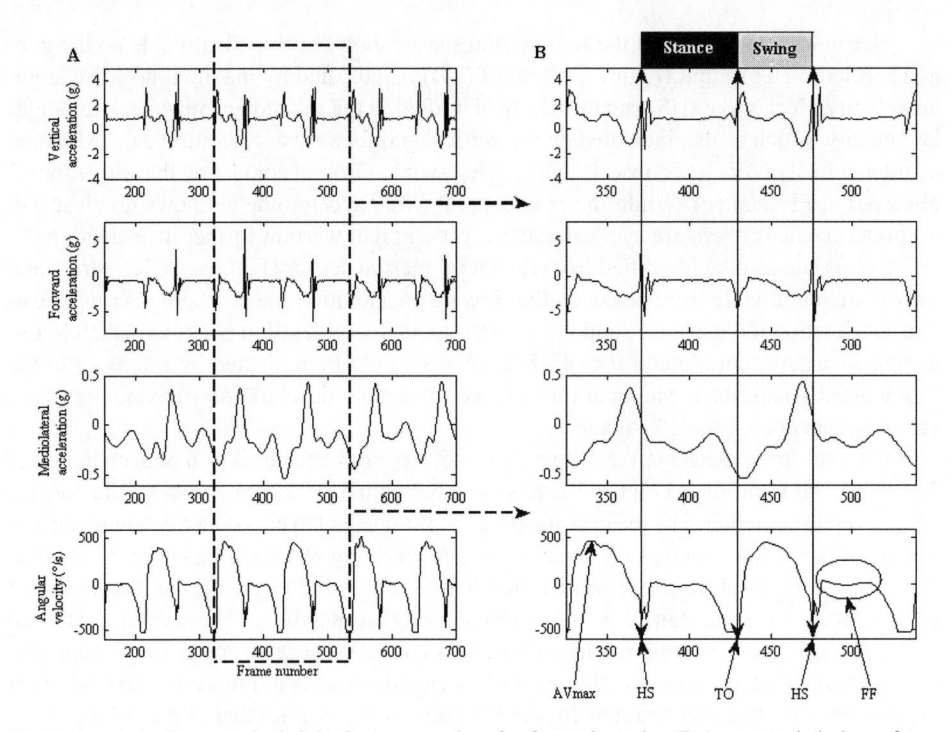

The Use of Micro-Electro-Mechanical-Systems Technology to Assess Gait Characteristics   183

Figure 1. **A** depicts a typical right foot sensor data for five gait cycles. **B** An expanded view of two strides highlighting events and phases. AV_{max} = maximal angular velocity, HS = heel strike, TO = toe off, FF = flat foot.

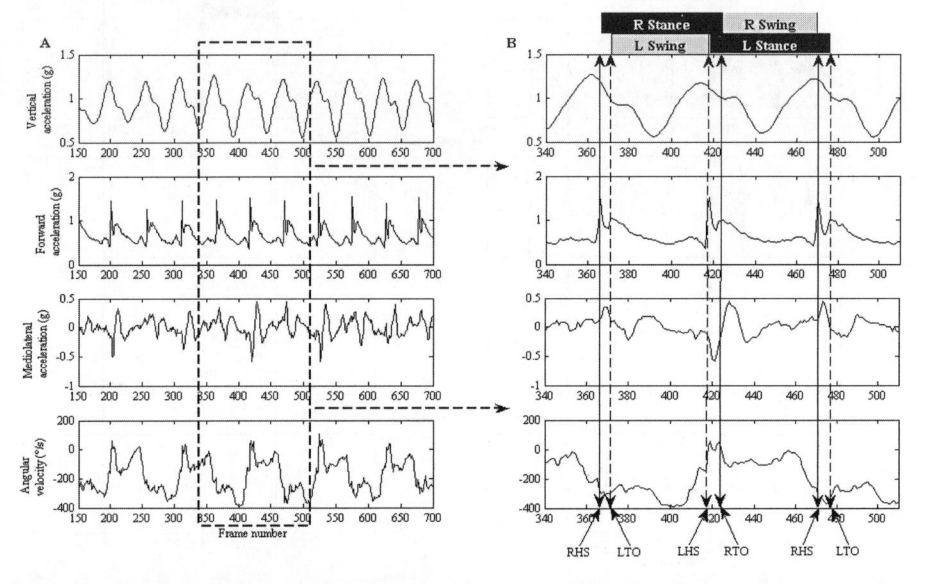

Figure 2. **A** Typical sacral sensor data for five gait cycles. **B** An expanded view of two strides highlighting events and phases. LTO = left toe off, LHS = left heel strike, RHS = right heel strike, LTO = left toe off, RTO = right toe off, RHS = right heel strike.

The combined outputs of the foot inertial sensor shows events identifiable on the gyro and accelerometer output (Figure 1). Toe off (TO) is indicated by maximal negative angular velocity. Heel strike (HS) can be clearly identified as the second negative peak for angular velocity, which is also indicated on the vertical and forward acceleration data. Maximal angular velocity (AV_{max}) is observed during leg swing. Ground contact by the whole foot is observed at FF (stance). While the mediolateral (ML) accelerometer shows no clear gait event occurrences, there are cyclical patterns present that warrant further investigation.

Events can also be identified from the sacral inertial sensor (Figure 2). Heel strike and toe off are identifiable from peaks on the forward acceleration trace, where maximal positive acceleration indicates HS, and the smaller positive acceleration peak immediately following indicates contralateral toe off. Side of HS would be indicated by reference to the foot inertial sensor data. Maximal rate change of vertical deceleration (maximal negative value) occurs at approx 45% of stance.

The heel strike to heel strike of two right foot steps of a race walk can be seen in Figure 3. A similar gyro profile to Figure 1 is observed but with a 19% greater range. The positive AV_{max} was 16% greater. The mean stride time for a single gait cycle (stance & swing phases) was 0.6 s for the RW, whereas GP mean stride time walking at 5 km/h was almost double at 1.1 s. Time for the stance phase was shorter than for the swing phase for the race walker at 0.28 seconds (s) (46% stance) & 0.33 s (54% swing) respectively. The GP times showed swing (0.5s, 45%) shorter than stance (0.6s, 55%), & RW's accelerometer traces appear to be in opposite direction compared to the GP acceleration traces in Figure 1. Maximal mean positive and negative accelerations for the RW were over 3.5 times that of the GP group.

Data from the RW's sacral accelerometer is currently under investigation.

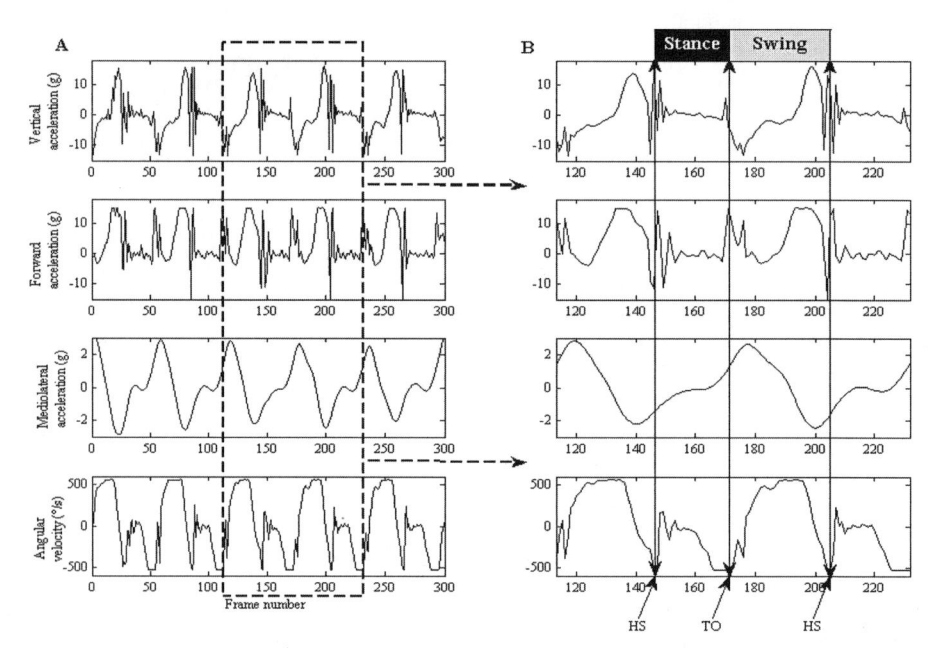

Figure 3. **A** depicts a typical foot sensor data for five gait cycles of a race walker. **B** An expanded view of two strides highlighting events and phases. HS = heel strike, TO = toe off.

4 Discussion

Similarity in gait profiles from gyro data were found when comparing normal walking to race walking. This walk profile is similar to at least one other study (Sabatini *et al.*, 2005). Less time between heel strike and toe off was observed for the RW compared to the GP group. Figure 1 shows a steady plateau during foot contact with the ground for normal walking, which lessens and tends not to have a definite plateau during race walking (Figure 3). It would be assumed race walking would require less absolute ground contact time and is evident by a reduced percentage of stance (less than half the time taken by the GP group). It is evident from the inertial sensor data that greater forces are involved during race walking compared to comfortable walking. Therefore movement analysis may be possible with further development of data interpretation.

This study has shown events of a gait cycle can be identified from data collected from inertial sensors placed on the back (sacrum) and feet. Angular velocity can be measured for foot swing as well as the period of ground contact time of the foot. Benefits are apparent for race walking, where athletes can have a sensor placed on their back. Figure 2 indicates heel strike occurs prior to toe off of the contralateral foot therefore indicating double limb support. Analysis of data would (most likely) show flight time (both feet off the ground at the same time) occurring if heel strike occurred after contralateral toe off. A key objective for RWs is to minimise time spent in double limb support (Simoneau, 2002), which could accurately be measured with inertial sensors.

The ML accelerometer data from the foot sensor (Figure 1) which coincides with ground contact time may be indicating movement from supination to pronation during the step. The maximal positive ML acceleration indicates lateral movement of the foot towards the end of the swing phase and may indicate swing foot (due to hip circumduction). The ability to assess ML movement would be beneficial in gait analysis studies as this can be a characteristic that can adversely affect walking (Simoneau, 2002), however further video analysis is required for confirmation. The sacral trace for ML acceleration is variable (Figure 2). Even so, the maximal negative acceleration (deceleration) or maximal positive acceleration immediately following HS precedes contralateral TO and is possibly indicative of propping of the hip between these two events. This may be an area for further investigation to determine whether differences between male and female ML sway can be seen.

In conclusion, inertial sensors can determine event differences between gait cycles for walking. The potential information that is captured by this technology needs to be further investigated and understood. This may lead to the use of this technology in all facets of gait analysis. Inertial sensors offer opportunities to study human movement away from the laboratory environment for real life non-simulated analysis. The ability to assess populations including athletes, general populations, elderly, occupational, and during injury/disease rehabilitation will become easier and more efficient with the use of inertial sensors.

References

Amiridis I.G., Arabatzi F., Violaris P., Stavropoulos E. and Hatzitaki V. (2005) Static balance improvement in elderly after dorsiflexors electrostimulation training. *European Journal of Applied Physiology*, 94/, 424–433.

Bassett D.R. (2000) Validity and reliability issues in objective monitoring of physical activity. *Research Quarterly for Exercise and Sport*, 71/2 suppl, s30–36.

Freedson P. and Miller K. (2000) Objective monitoring of physical activity using motion sensors and heart rate. *Research Quarterly for Exercise and Sport*, 71/2 suppl, s21–29.

Lai A., James D.A., Hayes J.P. and Harvey E.C. (2004) Semi-automatic calibration technique using six inertial frames of refernce. *SPIE* 5274/, 531–542.

Le-Masurier G.C. and Tudor-Locke C. (2003) Comparison of pedometer and accelerometer accuracy under controlled conditions. *Medicine and Science in Sports and Exercise*. 35/5, 867–71.

Luinge H. and Veltink P.H. (2005) Measuring orientation of human body segments using minature gyroscopes and accelerometers. *Medical & Biological Engineering & Computing*, 43/, 273–282.

Mayagoitia R.E., Nene A.V. and Veltink P.H. (2002) Accelerometer and rate gyroscope measurement of kinematics: An inexpensive alternative to optical motion analysis systems. *Journal of Biomechanics*, 35/4, 537–542.

McGibbon C.A. and Krebs D.E. (2004) Discriminating age and disability effeccts in locomotion: Neuromuscular adaptations in musculoskeletal pathology. *Journal of Applied Physiology*, 96/, 149–160.

Okada S., Hirakawa K., Takada Y. and Kinoshita H. (2001) Age-related differences in postural control in humans in response to a sudden deceleration generated by postural disturbance. *European Journal of Applied Physiology*, 85/, 10–18.

Pfeiffer K.A., McIver K.L., Dowda M., Almeida M.J. and Pate R.R. (2006) Validation and calibration of the actical accelerometer in preschool children. *Medicine and Science in Sports and Exercise*, 38/1, 152–157.

Plasqui G. and Westerterp K. (2005) Accelerometry and heart rate as a measure of physical fitness: Proof of concept. *Medicine and Science in Sports and Exercise*, 37/5, 872–876.

Plasqui G. and Westerterp K. (2006) Accelerometry and heart rate as a measure of physical fitness: Cross-validation. *Medicine and Science in Sports and Exercise*, 38/8, 1510–1514.

Sabatini A.M., Martelloni C., Scapellato S. and Cavallo F. (2005) Assessment of walking features from foot inertial sensing. *IEEE Transactions on Biomedical Engineering*, 52/3, 486–494.

Simoneau G.G. (2002). Kinesiology of walking. In: Neumann. D.A (Ed.). *Kinesiology of the musculoskeletal system*. St Louis, Mosby.

Uswatte G., Giuliani C., Winstein C., Zeringue A., Hobbs L. and Wolf S.L. (2006) Validity of accelerometry for monitoring real-world arm activity in patients with subacute stroke: Evidence from the extremity constraint-induced therapy evaluation trial. *Archives of Physical Medicine and Rehabilitation*, 87/10, 1340–1345.

SMART FLOOR DESIGN FOR HUMAN GAIT ANALYSIS

S.M.N.A. SENANAYAKE, D. LOOI, M. LIEW, K. WONG & J.W. HEE

Monash University Sunway campus, Jalan Universiti, Petaling Jaya, Malaysia

The smart floor design in this research is divided into three main sections; *the platform, the data acquisition circuit and the software architecture*. The platform is the place where all the activities are done, such as running, jumping and walking. The function of the data acquisition circuit is the reading and switching of sensors value and finally the software architecture is responsible for the conversion of readings for required values and also for the data logging. The prototype built is capable to quantify the forces exerted on each essential part of the foot such as the heel, middle foot and the toes. In addition, it provides qualitative information of the human motion. This includes information such as whether the person is exerting the correct amount of force on each part of the foot and also be able to determine whether the person has got a good gait posture by analyzing the distribution of force on the foot.

1 Introduction

By analyzing the force exerted and the pattern of movement of athletic on the floor, it's possible to determine what type of floor is appropriate for sports, or even what type of shoes and support needed to ensure athletic legs are protected from strain and force acting on them. All the data from analysis can help to prevent crucial injuries and help to improve the quality of a particular sport.

The design of smart floor in this research is based on the results achieved by previous researchers as discussed here. *Future Computing Environment Smart Floor* developed at Georgia Institute of Technology (Orr, 2000) is equipped with load cells and tiles are laid on it. A Hidden Markov Model (HMM) is trained using a few users footfall forced signatures; the force measured by a person walking on the tile is then compared to the trained signatures. *Design of a Pressure Sensitive Floor for Multimodal Sensing* developed at Arizona State University, USA (Srinivasan *et al.*, 2005) is made out of lightweight floor mat with a matrix of sensor with a spacing of 10 mm apart, which is about one sensor per centimeter square. The mats are made identical so that further expansion is possible. The sensors used are made using a pressure sensitive polymer between conductive traces on sheets of Mylar. *MIT Magic Carpet* developed at Massachusetts Institute of Technology; Australia (Paradiso *et al.*, 1997) is basically to analyze movement of a performer, e.g. dancer. This magic carpet is fitted with 16 × 32 grids of piezoelectric wires which are used to sense the foot pressure. The signal from each wire is buffed by a high-impendence operational amplifier. A 68HC11 microprocessor is used to scan the carpet using a multiplexer which able to scan up to 64 wires 60 times per second.

The aim of this project is to design and build a smart floor system which gives quantitative and qualitative analysis on forces exerted by a human on the floor during various movements in different sports. The smart floor takes forces exerted from foot and sends to the computer for analysis. This is done using data acquisition devices and sensors fitting on the floor. Therefore, smart floor built in this research is divided into three main sections; *the platform, the data acquisition circuit and software architecture.*

2 Designing Smart Floor for Human Gait

Designing of smart floor in this research is to improve and to provide enhanced and advanced capabilities in order to overcome drawbacks of previous researchers' achievements as discussed in the introduction. Prototype of smart floor built follows layered architecture. Further, it uses Force Resistive Sensors (FRS) as main focus is the analysis of forces exerted on the foot in various sports. FRSs are compact sensors and are designed using three layers; flexi substrate, spacer adhesive and flexi substrate with printed interdigitizing electrodes. The sensor is made of force sensitive material and whenever any force is applied to the sensor, the resistance of the sensor changes. With this, the force applied on each sensor can be easily calculated using a derived formula.

2.1. *Layered Architecture of the Platform*

Layered architecture of the platform consists of four layers, each with different purposes. This leads the selection of the most appropriate materials that suit the functionality of each layer.

The first layer is the surface where a person performs actions such as jumping, walking and running on it. Hence, the first criterion in choosing material for the first layer is a smooth hard surface. The material must be able to absorb certain amount of forces, so that lesser forces are transmitted to the lowest layer, where the sensors are placed. The group has a list of materials in consideration for first layer, which includes wooden board, floor mat for house use and yoga mat. Wooden board consists of a hard surface, but it is not compressible and does absorb large amount of forces. Most of the floor mats used in houses do not have a smooth surface, in order to prevent it from slipping easily; however the absorption of forces is appropriate. Yoga mat appears to be the best choice for the first layer, because it suits both the criterions mentioned above, e.g. see Figure 1a and 1b.

The second layer is an important element for absorption of forces. Too much force from jumping and running may exceed the maximum range of values of the sensors and that might cause damages to the same. Thus, the material selected for second layer must be soft and compressible, with the ability to take in forces. Choices for materials are sponges, shoe mat and yoga mat. The sponges include those for car wash, which is harder and less

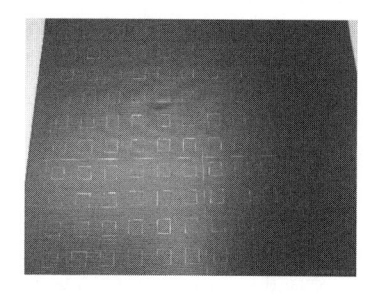

Figure 1a. Base platform.

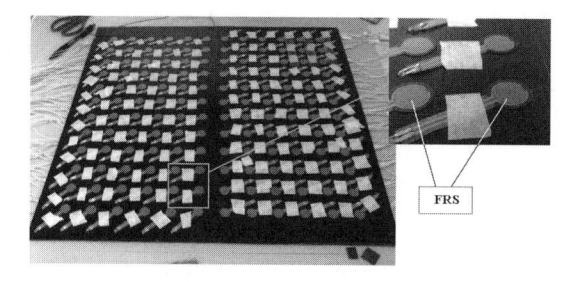

Figure 1b. Positioning of FRSs.

compressible; and sponges for dish wash which is much softer. Car wash sponge is thick hence absorb too much force, while dish wash sponges are highly compressible, which may cause instability of the Smart Floor. Shoe mat does not meet the criterion to take in forces, as it is thin. Yoga mat again stands up to be the best choice with suitable thickness and force absorption capability, e.g. see Figure 2a and 2b.

Third layer of the design has the function to hold the sensors which are allocated on the lower layer in place. Holes will be cut on the material according to the size of the sensing part of the sensor to secure them. This layer should be as thin as possible, because the sensor is only 2 mm thick. The material selected for this layer is mouse pad. Mouse pad has the thickness of about 2 mm to 3 mm, and can be easily cut. Firmness of mouse pad also holds the sensors firmly in the desired location, e.g. see Figure 3.

Finally, the forth layer is served as the base of the Smart Floor, which shall be hard and strong. Choices for the base layer includes wooden board and bristle board. Both of them meet the criterion, but wooden board is much heavier than bristle board which is a plastic material. Bristle board is chosen as it is easy to resize and does not require additional time for fabrication. Since there is an intention to replicate and to concatenate multiple force platforms, bristle board is the ideal choice, e.g. see Figure 4a and 4b.

2.2. Building Data Acquisition Circuits

The whole smart floor sensing system is divided into 4 quadrants where each quadrant holds 36 sensors. These 36 sensors are arranged in a 6 × 6 matrices arrangement. There are

Figure 2a. Use of mouse pads in layer 2.

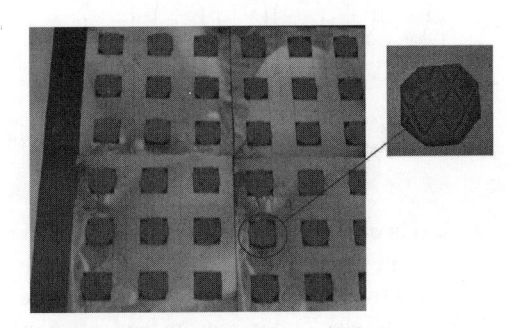

Figure 2b. Hexagon yoga mat pieces.

Figure 3. Mouse pads organized as quadrants in the layer 3.

Figure 4a. Bristle board used for layer 4.

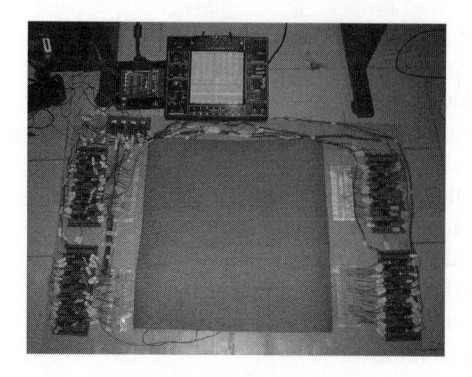

Figure 4b. Prototype of smart floor.

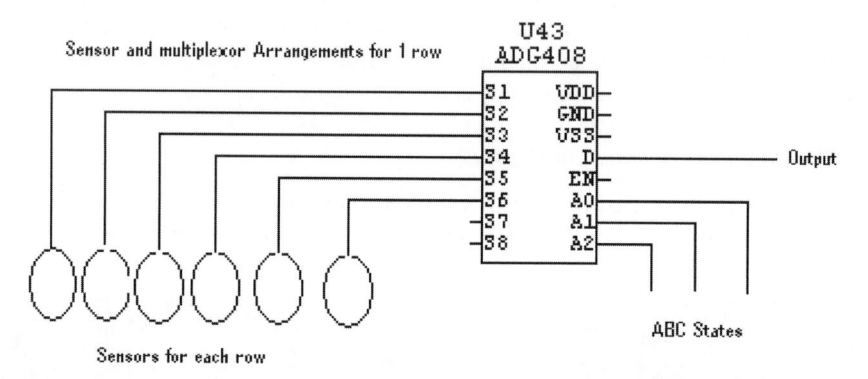

Figure 5. Column selector.

two levels of multiplexers where the 1st level acts as the row selector and the 2nd level as the column selector.

Each row contains 6 sensors and they are controlled by one multiplexer which acts as the column selector. Six sensors feed into 6 inputs of the multiplexer and there are 3 inputs for the ABC states and a common output, e.g. see Figure 5.

Each quadrant contains 6 rows of sensors similar to column selector discussed above which mean there are 6 multiplexers for each quadrant. These 6 multiplexers are then controlled by another multiplexer who acts as the row selector. In other words, there is only one output for each quadrant. The ACB states inputs for all multiplexers are shared. Therefore, only three digital outputs are required and in turn six multiplexers read data from sensors simultaneously.

The data acquisition circuit discussed above is responsible of collecting readings from the sensors. The circuit also acts as switching device from sensors to sensors to ensure all 144 readings from sensors are obtained. Main function of Data Acquisition card (DAQ) connected to the computer, which is the NI PCI 6070 DAQ card, is to collect data from the above explained quadrants and to feed to the computer for processing. Therefore, overall preprocessing data acquisition circuits discussed above is ready to read data from sensors.

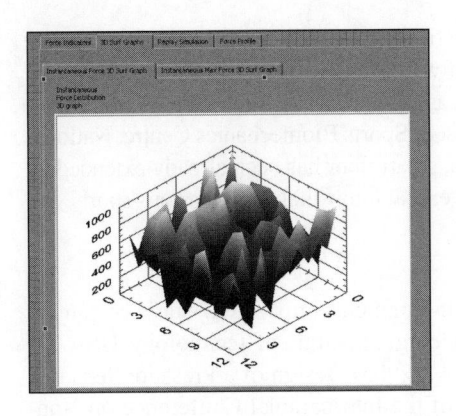

Figure 6a. Instantaneous force on jumping. 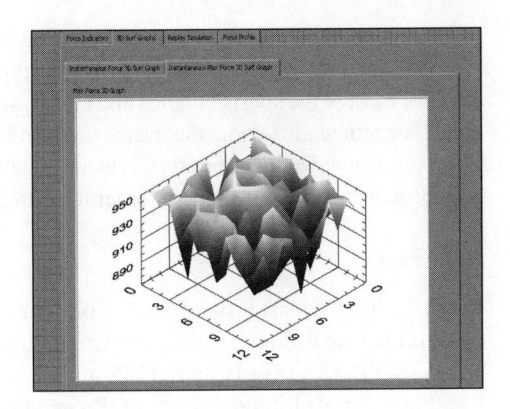   Figure 6b. Maximum instantaneous force.

2.3. Software Architecture

LabView version 7.1 is used to design the overall architecture of the software in order to acquire data, to process data and to visualize outputs obtained graphically. The software architecture designed is divided into five main software interfaces which consist of user friendly Graphical User Interfaces (GUI): *Simulation, User input, Scanning, Activities computation, Reading and simulation replay.*

Stack sequence structure is used to implement all GUIs mentioned above in order to facilitate troubleshooting and debugging efficient and effectively. The software/hardware codesign of smart floor is responsible to run all GUIs designed such that reading and processing of sensor data provide quantitative and qualitative analysis of human gait. The output of the prototype built is based on the activities realized on the smart floor. Force distribution based on the activity, Vertical Ground Reaction Force distribution based on the quadrant, characteristics of human hanging during jumping in various sports, etc are visualized as outputs and based on a specific sport extension of the software interface is viable upon addin modules to the existing software interface while remain the existing software. Moreover, layered architecture allows the extension of hardware easily too.

3 Activities on the Smart Floor

Main activities tested using the platform are static (initial position of the player), walking, running and jumping. Software interface allows the calculation of all derived parameters and visualizes all relevant data graphically. As an example, instantaneous force during jumping and as well as instantaneous maximum force exerted during jumping activity both shall be represented based on the sport carried out at a given time, e.g. see Figure 6a and 6b. Further, software interface provides the option to replay the activity done before with the facility of control knob such that sports trainer can judge the behaviour of the human gait in a specific duration. Attractive feature given by the prototype is the details related to hanging period of human gait during jumping.

4 Conclusions

The smart floor built is successfully tested in various activities. Since software/hardware co-design allows to store the activity pattern and to replay based on the stored information, prototype shall be used in most of human gait activities. Based on the analysis of output obtained and easy extendibility of hardware and software, demonstrates that smart floor is feasible for most of the sports in fields and courts, such as athletics, badminton, tennis, basket ball, etc. It is worth while to note that based on the request of Sports Biomechanics Centre, National Sports Institute, National Sports Council, Malaysia, smart floor has been already extended its area of sensing to wider scope in order to address vertical jump analysis in various sports.

References

Orr R.J. (2000) Smart Floor: Future Computing Environments http://www-static.cc.gatech.edu/fce/smartfloor/. College of Computing, Georgia Institute of Technology, USA.

Srinivasan P., Qian G., Birchfield D. and Kidané A. (2005) Design of a Pressure Sensitive Floor for Multimodal Sensing. Proceedings of the International Conference on Non-visual & Multimodal Visualization, London, UK, July 4.

Paradiso J., Abler C., Hsiao KY. and Reynolds M. (1997) The Magic Carpet: Physical Sensing for Immersive Environments. In: Proc. of the CHI '97 Conference on Human Factors in Computing Systems, ACM Press, NY, pp. 277–278.

EFFECT OF ROCKER HEEL ANGLE OF WALKING SHOE ON GAIT MECHANICS AND MUSCLE ACTIVITY

S.Y. AN & K.K. LEE

Department of Sports Science Kookmin University, Seoul, Korea

This study was to investigate the effects of rocker heel angle during walking on gait mechanics and muscle activity of lower extremity. While fifteen healthy men walked with two pairs of different rocker heel shoes (15° and 20°) and a pair of normal running shoes at 1.33 m/s on the treadmill, the joint kinematics and EMG signal were simultaneously recorded for 1 minute. Temporal variables of gait pattern and ankle, knee, hip and trunk angle were analyzed using 3D motion analysis at 100 Hz. Muscle activity of rectus femoris, tibialis anterior, biceps femoris and medial gastrocnemius were analyzed at 1000 Hz. To compare statistical difference of each shoe, the one-way ANOVA with repeated measures was conducted. Knee joint angle was increased in 15° rocker shoes than 20° rocker shoes at right toe off. Also knee joint in 15° rocker heel shoes was more flexed than in other shoes. But ankle dorsiflexion angle was significantly decreased in 15° rocker shoes compare to normal shoes at left heel strike while 20° rocker shoes reduced dorsiflexion angle compared with normal shoes at right toe off. Muscle activity of lower extremity statistically changed in both rocker shoes during gait cycle on treadmill.

1 Introduction

Stability of human is capability to maintain the orientation and position in the specific area. So walking is essential to maintain of stability and balance in order to offer a basic movement and body locomotion. Several strategies can improve stability during locomotion including shoe constructions that provide artificial surfaces into uneven surfaces. Recent results indicated that lower extremity muscles were in more activity to keep the stability in a given unbalance situation than normal situation (Nigg *et al.*, 2006). Some researches reported the functional walking shoes could help the reduction of sports injury and contribute to the rehabilitation of knee or ankle joint (Nigg *et al.*, 2006; Romkes *et al.*, 2006). The walking shoe was found lower ground reaction force and loading rate than a normal shoe because of a curved motion of a gentle slope with rounding outsole (Choi, 2003). A research showed that ankle joint angle was significant difference between pre and post training for 12 weeks from Park *et al.* (2006). According Vernon *et al.* (2004), ankle plantarflexion was decreased in initial stance phase and joint load was decreased by reducing of lower limb moment. A shoe can be different to the role and function depending on structure of outsole. In fact, scientific determination and evaluation is placed under the lack of circumstances in the rocker heel angle of most walking shoes developed in Korea. Thus the angle of optical rocker shoe is necessary for walking shoe design to examine. Also we need to investigate effects on a trunk and lower limb among the difference rocker heel shoes. This study was aim to investigate the effects of rocker heel angle during walking on gait mechanics and muscle activity of lower extremity.

Figure 1. Experimental shoes (left: Control shoe, center: 15° rocker shoe, right: 20° rocker shoe).

2 Methods

2.1. Subject

Fifteen healthy men (24.3 ± 1.3 years, 174.6 ± 3.6 cm, and 67.7 ± 6.2 kg) volunteered to participate in this study. Subjects with history or physical finding of any lower extremity orthopedic abnormalities, including foot abnormalities, were excluded. All subjects had never put on the rocker heel shoes before and gave informed written consent.

2.2. Experiments

Kinematics data were recorded and analyzed by using the 3D motion capture system (LUKOtronic, Austria) to measure the trunk tilt and joint angle of the right lower limb. Three-dimensional coordinate system was defined as the Y axis pointed in the vertically upward ground, the Z axis pointed in the direction of forward progression, and the X axis pointed in the outside of vector. And the lower extremity muscle activities were simultaneously recorded and assessed by using Electromyography (WEMG-8 System, Laxtha, Korea).

Three identical pairs of shoes were prepared for test and selected randomly for each subject. Two of these pairs was 15° and 20° rocker heel shoe (reots black, Stafild) and the other shoes used a normal running shoe (RBK-59725, Reebok) shown in Figure 1. Before data collection, reflective markers and EMG electrode were placed on the lower extremity. Markers were attached on right side of the body over the humeral head, the major trochanter, the lateral femoral condyle, the lateral malleolus, the rear foot and the fifth toe. To achieve an optimal EMG signal and low impedance, hair was removed and skin cleaned before the electrode placed in the skin. Four EMG electrodes were placed on the skin overlying the muscle belly of the rectus femoris (RF), tibialis anterior (TA), biceps femoris (BF) and medial gastrocnemius (GM) of the left lower limb.

Each subject performed totally three trails with two pairs of different heel rocker shoes and a pair of normal running shoes. The test was measured at speed of 1.33 m/s for 1 minute during walking on a treadmill. The data was recorded when the participants completed one good trail for each condition.

2.3. Data Analysis

Kinematics data were collected at sampling rate of 100 Hz. The raw data were filtered using zero-lag quadratic low-pass Butterworth filter with a cut-off frequency of 6 Hz. Kinematics variable included the degree of angular displacement at the trunk, hip, knee and ankle joint in the sagittal plane at stance phase.

EMG data from the left leg were sampled using bipolar silver-chloride electrodes at 1000 Hz. EMG raw signals were filtered using low pass Fast Fourier Transform(FFT)-filtering with a cut-off frequency of 50 Hz and full-wave rectified. The data obtained during gait was smoothing and the integral EMG value was calculated on each gait cycle. Ten gait cycle IEMG value were averaged and all data were normalized to 100% time of one gait cycle. In order to compare the IEMG value across shoe condition and each muscle, each IEMG was normalized to the average from percentage reference voluntary contraction based on total IEMG value of the normal shoe condition.

Statistical analysis was performed using SPSS version 12.0. To compare the effects of each shoe condition we used the one-way ANOVA with the repeated measures. Bonferroni's paired test was applied as post-hoc test in order to find out significant difference between 15° rocker shoe and 20° rocker shoe. The level of statistical significance was 0.05.

3 Results

3.1. *Temporal Parameters*

The swing phase time was significantly increased in the 15° rocker heel shoes about 0.02 second compared to the 20° rocker heel shoes ($p < .05$). But no significant changes were observed in stride and stance phase time.

3.2. *Kinematics*

The trunk tilt did not show any significant difference. Hip, knee and ankle joint showed the significant differences from midstance to toe-off (Figure 2). The hip joint was significantly more extended in midstance in 15° rocker heel shoes compared to the two other pair of shoes. The knee extension was increased significantly in the 15° rocker heel shoes compared to the 20° rocker heel shoes at right toe-off. Significant differences were found in knee maximum flexion and ROM ($p < .05$). Ankle dorsiflexion clearly decreased in the 15° rocker heel shoes at left heel strike and in the 20° rocker heel shoes at right toe-off ($p < .05$). Ankle maximum angle and range of motion (ROM) also significantly decreased in the 20° rocker heel shoes compared to the normal control shoes ($p < .05$).

3.3. *EMG*

Muscle activity of the tibialis anterior increased significantly in the both rocker shoes compared to the normal shoes during the gait cycle ($p < .05$) (Table 1, Figure 3).

The medial gastrocnemius showed increased muscle activity in the 20° rocker heel shoes compared to the normal shoes ($p < .05$). But no significant differences in IEMG were observed among the shoes during the stance phase. In swing phase, muscle activity of rectus femoris increased in the 15° rocker heel shoes by about 9% and biceps femoris increased in the 20° rocker heel shoes by about 17% ($p < .05$). The maximum peak time of tibialis anterior showed a delay of approximately 23.8% in the 15° rocker heel shoes ($p < .05$). The first peak time did not show any statistically significant difference.

3.4. *Coordination*

The gait pattern and the movement coordination between trunk and lower limb joint was different in two kinds of rocker shoes on the treadmill (Figure 4). Both rocker heel shoes decreased the trunk movement compared to the normal shoes and we observed that anterior

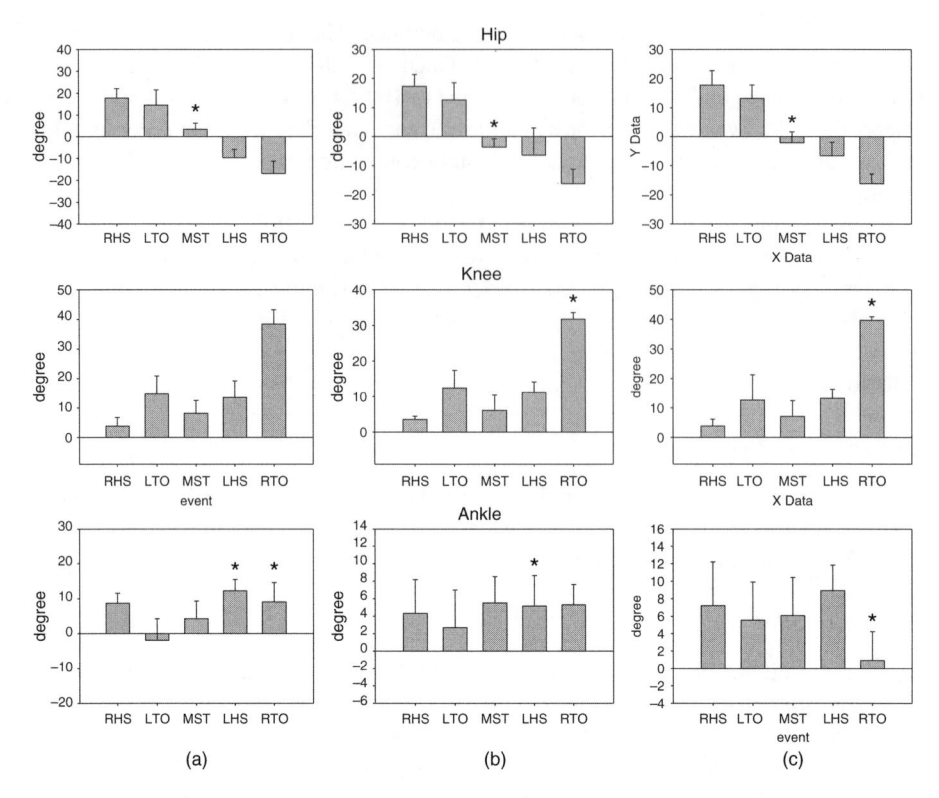

Figure 2. Joint angle, (a) control shoe, (b) 15° rocker heel shoe, (c) 20° rocker heel shoe; bars from left to right: right heel strike, left toe off, midstance, left heel strike, right toe off.

trunk tilt suddenly happened due to extension of hip joint motion. In the case of the 20° rocker heel shoes during terminal stance, the trunk rapidly performed a posterior tilt. The 15° rocker heel shoes reduced hip and knee joint range of motion and increased ankle range of motion. We found a pattern of knee flexion maintaining ankle dorsiflexion. The knee joint suddenly changed between flexion and extension in the 20° rocker heel shoes during initial stance. The overall variability of the 15° rocker heel shoes increased in the first half of the stance phase and the variability of the 20° rocker heel shoes increased in the terminal stance phase. The coordination decreased in the joints of the lower limb.

4 Discussion and Conclusion

This study was designed to compare normal running shoe with two rocker shoes of different angle. Even though a number of statistically significant changes was observed in gait kinematics and muscle activities, it is difficult to conclude that a rocker heel shoe would be better. This is due to the fact that the kinematic variables were investigated in the sagittal plane only and this study did not consider forces and moments. Nevertheless, this study showed a change in gait pattern when comparing the rocker shoes with the normal shoe. It has been shown that rocking movement at knee and ankle joint results into increased muscle activity of the gastrocnemius and tibialis muscles and the co-contraction of these muscles could

Effect of Rocker Heel Angle of Walking Shoe on Gait Mechanics and Muscle Activity 197

Table 1. Muscle activity and peak time.

Muscle	Shoes	Stride %RVC	F	p	Stance %RVC	F	p	Swing %RVC	F	p	Max. peak time %Time	F	p
RF	A	11.72±4.28			13.15±3.97			11.26±2.95			15.63±17.26		
	B	23.14±13.31	4.55	.05	20.58±10.50	2.97	.12	20.29±10.53	4.95	.02	*33.82±18.90	3.47	.06
	C	20.04±12.54			18.92±11.87			19.72±14.52			18.22±17.34		
TA	A	29.82±11.20			28.41±7.13			36.16±13.54			19.30±23.42		
	B	44.17±11.00	10.13	.01*	43.96±26.11	2.26	.17	52.25±22.19	4.07	.07	32.49±24.87	3.73	.05*
	C	42.57±13.13			32.63±10.06			53.86±24.02			43.15±19.27		
BF	A	29.70±18.84			19.97±8.20			34.64±12.48			17.89±17.91		
	B	40.17±23.48	4.23	.06	23.99±11.97	3.08	.11	54.60±28.47	6.94	.01*	27.73±21.57	0.91	.42
	C	44.39±29.27			32.64±23.56			51.65±24.97			18.33±16.08		
GM	A	28.75±9.47			38.46±8.55			17.94±5.60			34.87±4.98		
	B	46.25±18.93	7.69	.01*	51.64±18.85	3.79	.08	25.59±16.59	1.37	.28	35.01±11.78	1.23	.32
	C	46.73±18.26			55.98±27.68			27.51±21.56			39.52±8.07		

* Significance at p < .05

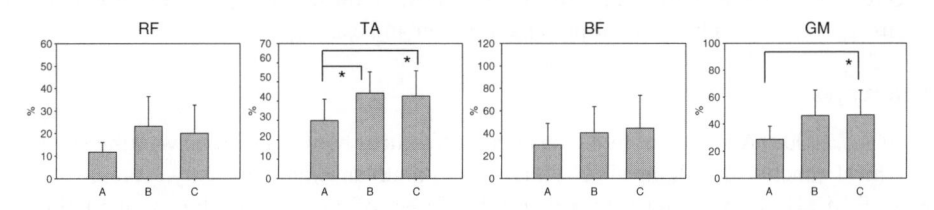

Figure 3. Muscle activity during one stride; bars from left to right: A: normal shoe, B: 15° rocker heel shoe, B: 20° rocker heel shoe; RF: rectus femoris, TA: tibialis anterior, BF: biceps femoris, GM: medial gastrocnemius.

provide stability. Nigg et al. (2006) claimed that these shoes produced changes in kinematics, kinetic and EMG characteristics that seem to be advantageous for the locomotory system. Romkes et al. (2006) claimed that these shoes should be used cautiously in patients with knee problems.

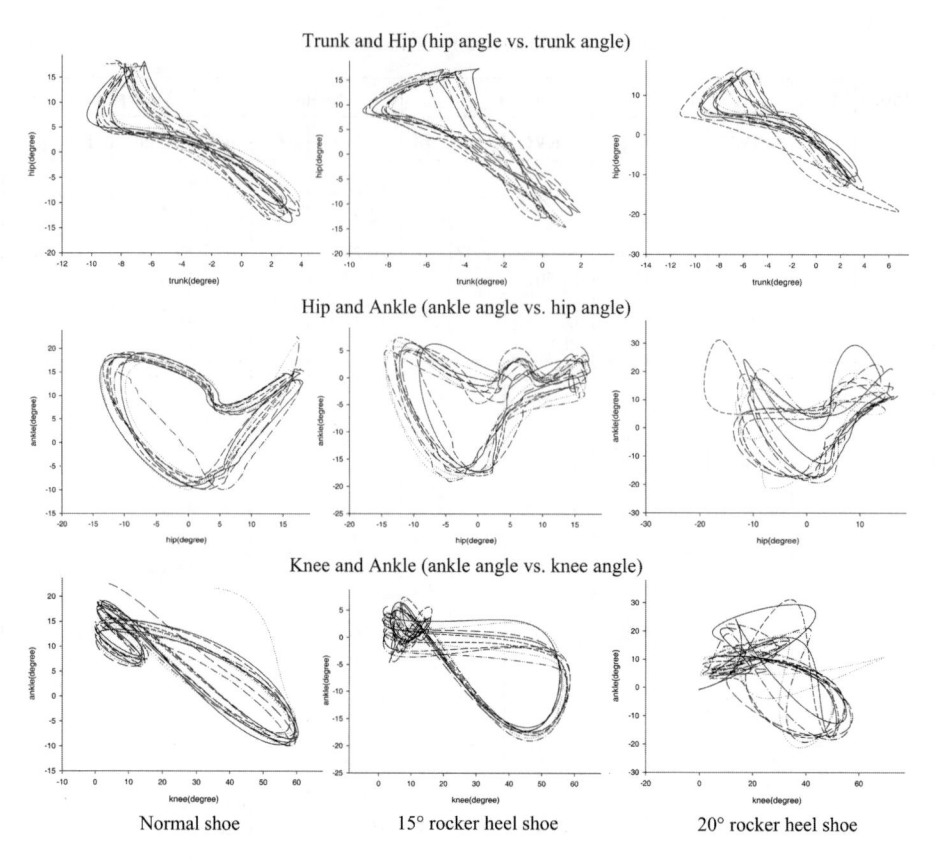

Figure 4. The coordination of trunk and lower extremity.

In conclusion, this study has found that there are different joint angles, muscle activity, and gait patterns of the lower limb in each kind of shoes. These rocker heel conditions affected the lower extremity and the whole body. A further study has to be conducted regarding long-term changes during rehabilitation and exercise.

References

Choi K.J. (2003) Analyses of biomechanical differences between general walking shoe and mBT functional walking shoe. Ph.D Diss. Sungkyunkwan University, Korea.
Nigg B.M., Hintzen S. and Ferber R. (2006) Effect of an unstable shoe construction on lower extremity gait characteristics, *Clinical Biomechanics*, 21/1, 82–88.
Park K.R., An S.Y. and Lee K.K. (2006). Effects of 12-week wearing of the unstable shoes on the standing posture and gait mechanics. *Korean Journal of Sport Biomechanics*, 16/3, 165–172.
Romkes J., Rudmann C. and Brunner R. (2006) Changes in gait and EMG when walking with the Masai Barefoot Technique. *Clinical Biomechanics*, 21, 75–81.
Vernon T., Wheat J., Naik R. and Pettit G. (2004) Change in Gait characteristics of a normal, Healthy population due to an unstable shoe construction. Sheffileld Hallam University, UK (unpublished technical report).

KINEMATICS OF THE FOOT SEGMENTS DURING THE ACCELERATION PHASE OF SPRINTING: A COMPARISON OF BAREFOOT AND SPRINT SPIKE CONDITIONS

B.J. WILLIAMS, D.T. TOON, M.P. CAINE & N. HOPKINSON

Wolfson School of Mechanical and Manufacturing Engineering, Loughborough University, Leicestershire, UK

The winning margin in an elite sprint race is often milliseconds, meaning that cutting edge technology is becoming increasingly important in footwear design. However, there is dearth of data comparing results of barefoot and shod sprinting on the foot segment kinematics. The current study explored the effect of sprint spikes on the kinematics of the foot segments during the stance phase of the sprint running gait, focusing upon data obtained during the acceleration phase of a maximal sprint. It was hypothesised that during ground contact sprint spikes would reduce the rate and range of metatarsophalangeal joint (MPJ) extension, and increase the posterior sole angle at both touchdown and take off, relative to barefoot conditions. High-speed digital video was used to capture foot-ground contact from medial and lateral aspects for 4 nationally competitive sprinters. The data presented shows that sprint spikes reduce the rate and range of MPJ flexion and extension. The angular range observed during the first 50% of ground contact at 10 m was 7.5° ± 4.0 and 3.0° ± 1.8 for the barefoot and shod conditions respectively. The angular range observed during the second 50% of ground contact at 10 m was −12.7° ± 6.3 and −9.0° ± 4.5 for the barefoot and shod conditions respectively. The magnitude of these effects was specific to the individual athlete. Shoe conditions did not affect posterior sole touchdown angle, but posterior sole take off angle was consistently reduced.

1 Introduction

There are few published studies on sprint shoe mechanical properties and their influence on athletic performance. Stefanyshyn & Fusco (2004) demonstrated that sprint performance can be improved if the stiffness of standard running spikes is increased. The authors concluded that in order to maximize performance, individual tuning of shoe stiffness to the athlete's particular characteristics is required. A recent study (Krell & Stefanyshyn, 2006) using footage obtained from the 2000 Olympic Games in Sydney revealed that the kinematics of the foot segments and the metatarsophalangeal joint (MPJ) are related to sprinting performance. The authors also suggest that shoe design may influence the kinematics of the foot segments, and thus by deduction, sprint performance. Therefore, the current study was designed to investigate the effect of sprint spikes on the kinematics of the foot segments during the stance phase of the sprint running gait cycle.

It was hypothesised that, relative to barefoot conditions, wearing of sprint spikes would have an effect on the kinematics of the foot segments during the acceleration phase of maximal velocity sprinting. Specifically, it was proposed that during ground contact sprint spikes would reduce the rate and range of MPJ extension, and increase the posterior touchdown and take off angles, in relation to barefoot conditions.

Table 1. Mean (±SD) test subject data (n = 4).

Male		Female	
Age (yrs)	21.5 ± 1.5	Age (yrs)	21.0 ± 0
Mass (kg)	79.5 ± 4.9	Mass (kg)	52.9 ± 4.4
Height (cm)	182.9 ± 10.8	Height (cm)	158.8 ± 1.8
100 m PB (s)	10.72 ± 0.12	100 m PB (s)	11.69 ± 0.10

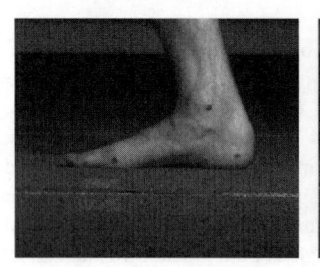

Figure 1. Marker placement – barefoot and sprint spikes.

2 Method

Athlete testing was carried out at the Loughborough University indoor High Performance Athletic Centre, with data from 4 nationally competitive sprinters (2 male, 2 female) analysed and presented. Before participating, athletes provided informed consent. Participant characteristics are shown in Table 1. The testing protocol was designed in accordance with procedures outlined by Krell & Stefanyshyn (2006). In the shod condition, adidas Demolisher (2005) sprint spikes were employed. This shoe was selected because it was found to have the highest bending stiffness coefficient of a selection of eleven currently available sprint spikes (Toon et al., 2006).

The foot segments were recorded throughout a single ground contact from both medial and lateral aspects during the acceleration phase, 10 m into a maximal 100 m sprint from starting blocks. This footage was obtained using a high-speed digital camera (Photron Fastcam – Ultima APX 120 K) recording at 1000 frames per second with a shutter speed of 1/2000s. The camera was positioned 0.07 m off the ground, focused at a point 2.90 m away in the centre of the test lane such that the field of view was 0.71 m wide, following a consistent calibration technique. The participants were asked to perform 2 successful captures whilst recording from each aspect (medial and lateral) for both barefoot and shod conditions. The resulting kinematic sagittal plane data was acquired through a process of manual digitisation using Image Pro Plus 5.0.2 software, with each individual frame being analysed in sequence. This method was facilitated through the placement of markers at predefined biomechanical positions to enable measurement of angular movement about the MPJ, shown in Figure.1.

Data were exported to Microsoft Excel for analysis and conversion into graphical format. The MPJ was defined as the angle between the superior surface of the rearfoot and the superior surface of the forefoot in the saggital plane. The angle change at the MPJ of the foot was measured throughout stance phase. Angular motion at the MPJ was categorised into three phases; initial plantar flexion, dorsi flexion and final plantar flexion. Initial plantar flexion is the first period of angular motion occurring immediately after ground contact until maximum extension. The next phase is dorsiflexion, occurring from maximum extension of the MPJ through to maximum flexion. Following this is the final plantar flexion phase, occurring from maximum flexion throughout toe-off until the foot leaves the ground. Posterior and anterior sole angles at both initial ground contact and take-off were recorded. The medial and lateral aspects of the MPJ were combined and modelled as a single ideal hinge joint (Krell & Stefanyshyn, 2006).

Data consistency was assessed through re-digitising 10% of high-speed video files captured and comparing the resulting data with the corresponding original digitised data sets. Root mean square values of the MPJ angle were calculated to be 0.92° for the initial and repeated digitisation. Precision of the manual digitisation process was calculated by determining the smallest possible change in MPJ angle. Consecutive frames with one pixel of heel movement were selected from each camera set-up and precision was defined as the mean smallest measurable change in MPJ angle between frames, and was found to be 0.43°.

3 Results and Discussion

Wearing of sprint spikes reduced the rate and range of MPJ flexion and extension. The angular range observed during the first 50% of ground contact at 10 m was 7.5° ± 4.0 and 3.0° ± 1.8 (mean ± s.d.) for the barefoot and shod conditions respectively. The angular range observed during the second 50% of ground contact at 10 m was −12.7° ± 6.3 and −9.0° ± 4.5 (mean ± s.d.) for the barefoot and shod conditions respectively. The magnitude of these effects was specific to the individual athlete.

The results corresponding to angular range of the MPJ are shown in graphical form in Figure 2 to a 95% confidence level for the true difference of the means. The dotted line and the solid line represent the barefoot and the sprint spike conditions respectively.

It was hypothesised that MPJ extension and flexion would be lower in the sprint spike condition compared to the barefoot condition. This was based on the fact that the sprint spikes used in this study have a stiff sole unit and would therefore have a limiting effect on the angular range about the MPJ due to the high forces required to bend the sprint spikes through the same angular range as that occurring in the barefoot condition. The plotted data shows that the angular range is larger in the barefoot condition compared to the sprint spike condition throughout the initial period of plantar flexion, the period of dorsiflexion and also in the final phase of plantar flexion, which concurs with the hypothesis stated. For the initial period of plantar flexion, the largest difference in angular range between the sprint spike and barefoot conditions during acceleration was 14° ± 1. For the period of dorsiflexion, the largest difference in angular range between the sprint spike and barefoot conditions during acceleration was −23° ± 2. For the final period of plantar flexion, the largest difference in angular range between the sprint spike and barefoot conditions during acceleration was 9° ± 1.

The results for the mean posterior sole angles at both touchdown and take-off for each subject at 10 m during both footwear conditions are listed in Table 2. It was hypothesised that the sprint spike condition would cause the athlete to touchdown and take-off with larger posterior sole angles. This hypothesis was based on the fact the sprint spikes used in

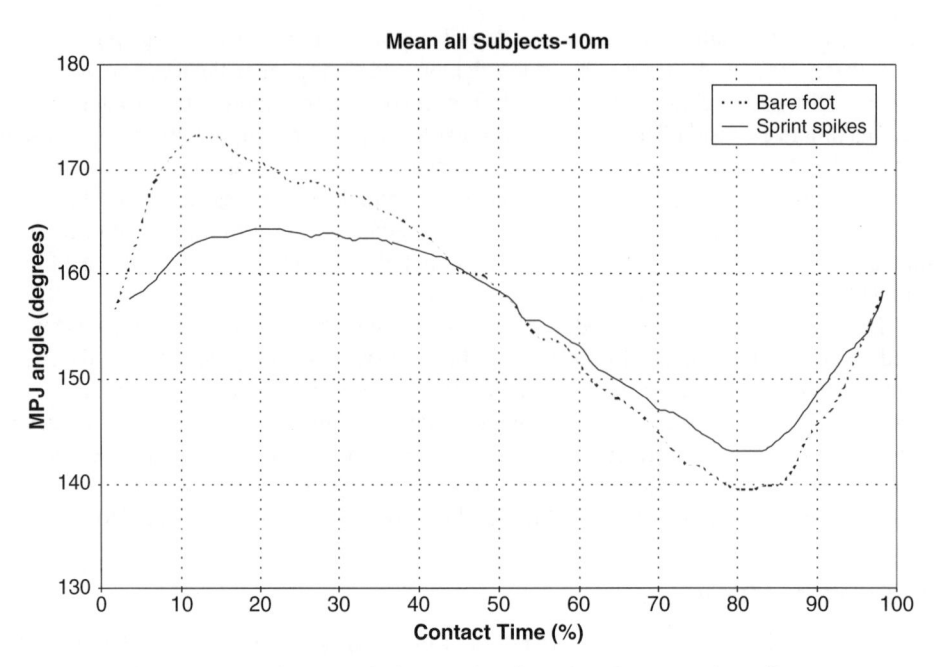

Figure 2. Angular range during acceleration – barefoot vs. sprint spikes.

this study have high bending stiffness through the sole unit, and in-built heel rise and toe spring, which have the effect of forcing athletes up onto their forefoot during sprinting.

The posterior sole angle data at touchdown was significantly different between sprint spike conditions and barefoot conditions, although a consistent trend is not present across all participants. Conversely, posterior take off angle was consistently reduced when wearing sprint spikes, reaching significance levels for participants 1 and 4. The mean posterior sole touchdown angle for males was 16.5° ± 3.2 and 19.4° ± 1.9 in the barefoot and sprint spike conditions respectively. For females the mean posterior sole touchdown angle was 6.9° ± 2.7 and 10.1° ± 8.2 in the barefoot and sprint spike conditions respectively. The mean posterior sole take off angle for males was 85.8° ± 5.0 and 79.1° ± 2.9 in the barefoot and sprint spike conditions respectively. The mean posterior sole take-off angle for females was 83.9° ± 3.6 and 79.3° ± 3.0 in the barefoot and sprint spike conditions respectively. The results for the posterior sole angle at touchdown follow a trend which suggests that sprint spikes increase this angle at touchdown during acceleration from starting blocks, reaching significance for two of the three participants for whom this is evident. However, past investigation into barefoot and shod running along with the evolution of efficient sprinting technique has shown that foot placement is actively controlled during free flight to allow the preloading of the gastrocnemius (Mero et al., 1992). Concurring with this, the results of the current study show that prior to ground contact, foot placement and posterior sole angle at touchdown are governed by an athlete's individual kinematic preparation during free flight and not by the footwear condition.

Krell & Stefanyshyn (2006) speculated that shoe design may play a role in influencing the posterior touchdown angle, further stating that a stiff contoured midsole would cause

Table 2. Posterior sole angles at 10 m (*significant p < 0.05).

Participant	Condition	Initial contact angle (°)	s.d.	Take off angle (°)	s.d.
1	barefoot	17.3*	3.2	87.1*	4.5
	sprint spikes	20.2*	1.6	80.6*	2.5
2	barefoot	15.7	3.8	84.5	6.4
	sprint spikes	18.6	2.2	77.5	3.0
3	barefoot	8.9*	2.8	82.4	4.8
	sprint spikes	18.3*	1.0	79.9	2.4
4	barefoot	4.8	0.8	85.4*	2.4
	sprint spikes	1.9	1.0	78.7*	4.2

the athlete to touch down with a large posterior sole angle. The results from the current study show that stiff sprint spikes do not necessarily cause an athlete to contact the ground at touchdown with larger posterior sole angles. In conflict with the original hypothesis the posterior sole angle data shows that sprint spikes reduce this angle at take-off in relation to the barefoot condition for both male and female subjects, reaching significance on three of eight occasions. These findings demonstrate that the influence of the sprint spike condition on foot kinematics are only realised throughout ground contact where the foot and shoe work as a system.

4 Conclusion

The current study provides data for the comparison of barefoot and shod sprinting with respect to the kinematic behaviour of the foot segments. The results presented show that sprint spikes have an inherent controlling effect over the foot throughout the stance phase of the sprint running gait cycle, for maximal acceleration from starting blocks. The effect is visible in that the barefoot condition exhibits a larger rate and range of MPJ flexion and extension in comparison to the shod condition. These findings concur with the hypothesis. The posterior sole angle at take-off was consistently reduced in the shod condition, which similarly, can be attributed to the additional stiffness of the sprint spike controlling the movement of the foot segments about the MPJ during ground contact.

The potential influence on sprinting performance of the reduced angular rate and range at the MPJ throughout extension and flexion, and the decreased posterior sole take-off angle, in sprint spikes is still not fully understood. A better appreciation of these parameters would ensure that future sprint spikes are optimised for performance. It is also evident from this study that individual differences exist between the kinematic variables of each athlete when introduced to the same footwear condition, suggesting that shoe selection is specific to the functional requirements of an individual athlete. The current study is part of a larger investigation into the start, acceleration and constant maximal velocity phases of sprinting, examining a more comprehensive range of parameters including angular velocity and individual subject analysis.

Acknowledgements

The authors would like to thank the Loughborough University High Performance Athletics Centre for the use of its facilities, and the athletes who agreed to participate in this study.

References

Krell J.B. and Stefanyshyn D.J. (2006) The relationship between extension at the metatarsophalangeal joint and sprint time for the 100 m Olympic athletes. *Journal of Sports Sciences*, 24, 175–180.

Mero A., Komi P.V. and Gregor R.J. (1992) Biomechanics of sprint running. A review. *Sports Medicine*, 13/6, 376–392.

Stefanyshyn D.J. and Fusco C. (2004) Increased bending stiffness increases sprint performance. *Sports Biomechanics*, 3, 55–66.

Toon D., Kamperman N., Ajoku U., Hopkinson N. and Caine M. (2006) Benchmarking stiffness of current sprint spikes and concept selective laser sintered nylon outsoles. In: Moritz E.F. and Haake S. (Eds.) *The Engineering of Sport 6, Vol. 3 – Developments for Innovation*. Springer, Munich, pp. 415–420.

7. Ball Sport – Golf

NON-LINEAR VISCOELASTIC PROPERTIES OF GOLF BALLS

F.K. FUSS

Sports Engineering Research Team, Division of Bioengineering, School of Chemical and Biomedical Engineering, Nanyang Technological University, Singapore

Based on a stress relaxation following the logarithmic law, a wide variety of golf balls was analyses as to viscosity constant (the multiplier of LNt), the stiffness parameters at different cross head speeds, conventional Atti and Riehle numbers, and first peak frequency of impact sound. The Atti number correlated highly with the peak frequency. Solid core golf balls are more viscous than wound balls. The force-deflection curve of solid core balls follows a parabolic function, the one of wound balls an exponential function. Based on the logarithmic law, the stiffness can be extrapolated linearly with respect to the logarithm of deflection rate, and peak forces and contact time were calculated at an impact speed of 50 m/s.

1 Introduction

The golf ball is considered the best-selling sporting good of the world. "Few sport items of such apparent simplicity have undergone more study and analysis than the golf ball" (Penner, 2003). Apart from aerodynamical properties, like drag and lift, related to roughness and dimple pattern, structural mechanical properties are confined to a single value, the PGA (Professional Golfers' Association) or Atti compression number, which is still determined by a device patented in 1942 (Atti, 1942). Golf ball compression is a measurement that defines how much a golf ball deforms under load. The amount of compression gives a player the desired feel. If a player wants a softer feel, he goes for a softer compression ball. Typical compression ratings are between 60 and 100+, with most players using a 90-compression ball as a compromise.

Additionally to the Atti compression tester, other methods are in use (Dalton, 2002), like the Riehle compression test (compression at 200 pounds), and other load/displacement tests. All these tests are carried out at different loads and displacement rates. Equations for conversion of other test methods into the usual Atti compression number are available (Dalton, 2002; Sullivan *et al.*, 2003; Nesbitt, 2006). In the Atti testing device, a plunger compresses a ball against a spring. The displacement of the plunger is fixed, and the deformation of the spring is measured. Consequently, the Atti numbers result neither from a fixed load (F) nor from a fixed deflection of the ball. According to Atti's original patent (Atti, 1942), the spring is preloaded by 200 pounds. Furthermore, a 50-compression ball is measured at 350 pounds, a 100-compression ball at 400 pounds. As ball and spring are in series, they share the same load F. Thus, the parametric relationship between deflection of the ball x, force applied F, and Atti A (Atti, 1942, and Nesbitt, 2006) and Riehle R number (Sullivan *et al.*, 2003) is:

$$\text{Atti: } F = A + 300; \quad x = (200 - A)/1000; \quad F = 500 - 1000x \tag{1}$$

$$\text{Riehle: } F = 200; \quad x = R/1000 \tag{2}$$

where the units of force and deflection are pound and inch respectively.

According to Sullivan et al. (2003), the relationship between Atti A and Riehle R is:

$$A = 160 - R \qquad (3)$$

From Eqs. (1) and (2) it becomes clear that Eq. (3) can never be a linear, even if the force-deflection function is linear ($F = kx$), which is not applicable to spherical objects.

The latter eqn results into

$$R = 200\,(200 - A)/(A + 300) \qquad (4)$$

The sum of R and A is between 135.7 and 150 for Atti numbers between 50 and 100. If the force-deflection function is a simple parabolic one ($F = kx^2$), then R is:

$$R = (200 - A)\,(200/(A + 300))^{0.5} \qquad (5)$$

The sum of R and A is 163.4–170.7 for Atti numbers between 50 and 100.

In a previous study (Fuss et al., 2005) it was determined that the compression number indicated on the ball carton does not always reflect the real compression number, mainly because different companies use different testing methods, as well as Eq. (3), which, according to Eqs. (4) and (5) is incorrect. It was suggested, that the stiffness as a function of deflection is a better parameter to classify golf balls. The present study will take the previous suggestion a step further, by considering non-linear visco-elasticity of golf balls.

As mentioned above, the amount of compression gives a player the desired feel. Yet, it seems, that the impact sound is more related to the feel than to the effect of the mechanical impact on the musculo-skeletal system. It becomes clear from Snell's study, quoted by Morrice (2005) that expert golfers distinguished the ball type by whichever sound they heard, and not by any other sensory impressions.

Impact sounds in golf putting were measured by Barrass et al. (2006). Although the authors did not specifically calculate the correlation between the peak frequency of the sound spectrum and the compression of the 5 balls investigated, their results roughly suggest that the harder the ball is, the higher is the peak frequency of the impact sound. Shannon and Axe (2002) investigated the frequency modes of the impact sound, but did not correlate them to the stiffness of the balls.

The impact of a golf ball is usually modelled by applying linear models, consisting of 2- or 3-elements (springs and dashpot), e.g. Kelvin-Voight, Zener, Kelvin-3-element model (Ujihashi, 1994; Lieberman & Johnson, 1994; Cochran, 2002, Tanaka et al., 2006). Even when attributing non-linear properties to the springs and dashpot, the models nevertheless still remain linear, or pseudo/quasi-linear, as the overall stiffness of the different models shows either a minimal value of >0 (Kelvin-Voight), or both minimum (>0) and maximum values ($<\infty$; Zener and Kelvin-3-element models), which hardly represents reality.

The aim of this study was to address the problems mentioned before, by determining the following parameters in different of golf ball models (Table 1), be it range or tournament balls: force-deflection, stiffness-deflection and stiffness-force relationship at different deflection rates, stress relaxation, stiffness parameters (initial stiffness and nature of the stiffness-deflection function), peak frequency of the impact sound, as well as conventional Atti and Riehle compression numbers for reference. Finally, an impact model will be developed based on non-linear visco-elastic properties.

Table 1. Golf Balls analysed; balls conforming (C) or not conforming (NC) with the prescribed specifications were classified according to the USGA list (USGA, 2007). The quality grades distinguish between 0 = new, + = used once or twice, + + = used 2 months at a driving range. The symbols for determinig the construction type, were adopted from the USGA list (USGA, 2007), the information given on the ball carton, (P = piece, c = cover, SC = solid core, W/LC = wound + liquid core, W/SC = wound + solid core); n = number of balls analysed; abbr. = abbreviated model name.

Model/ Brand/ working name	Abbr.	Company/ Manufacturer	Origin	Construction	C/NC	Grade	n
Awesome Distance	AD	Top-Flite/ Spalding	US	2P-SC-1c	C	0	12
Buttery Feel	BF	Top-Flite/ Spalding	US	2P-SC-1c	C	0	12
HVC SoftFeel	SF	Titleist/ Acushnet	US	2P-SC-1c	C	0	12
TourDistance	TD	Titleist/ Acushnet	US	3P-W/LC-1c	NC	0	12
Chinese Range Balls	RB	—	CN	2P-SC-1c	NC	0 (49), + + (12)	61
White Korean Range Balls	RW	—	KR	2P-SC-1c	NC	0	4
Yellow Korean Range Balls	RY	—	KR	2P-SC-1c	NC	0	4
Dura Range	RR	Seoul Nassau Co.	KR	2P-SC-1c	NC	0	4
Golden Ram	GR	RAM Golf Corp.	US	2P-SC-1c	NC	+	1
Pinnacle Gold	PG	Pinnacle/ Acushnet	US	2P-SC-1c	C	+	1
Yakui	YA	—	??	2P-SC-1c	NC	+	1
Pro Special	PS	Srixon	US	2P-SC-1c	NC	+	3
Optima HP	OH	PGF International	AU	2P-SC-1c	NC	+	1
Precept Lady	PL	Precept/Bridge stone Golf	US	2P-SC-1c	C	+	1
Prostaff Platinum Pure Distance	PP	Wilson	US	2P-SC-1c	NC	0	1
Nike One Platinum	NP	Nike	US	4P-SC-3c	C	0	12
Nike One Black	NB	Nike	US	3P-SC-2c	C	0	12

(continued)

Table 1. (continued)

Model/ Brand/ working name	Abbr.	Company/ Manufacturer	Origin	Construction	C/NC	Grade	n
Pro V1x	PV	Titleist/ Acushnet	US	3P-SC-1c	C	0	12
Penfold Tradition	PE	Penfold	UK	3P-W/LC-1c	NC	+	1
T301 Black	MzK	Mizuno	JP	3P-SC-1c	C	0	4
D301 Red	MzR	Mizuno	JP	3P-SC-1c	C	0	4
C301 Blue	MzB	Mizuno	JP	3P-SC-1c	C	0	4
S301 Green	MzG	Mizuno	JP	3P-SC-1c	C	0	4
Professional 90	PRN	Titleist/ Acushnet	US	3P-W/LC-1c	NC	+	9
DT Wound 90	DTN	Titleist/ Acushnet	US	3P-W/SC-1c	NC	+	9
DT Wound 100	DTH	Titleist/ Acushnet	US	3P-W/SC-1c	NC	+	1
Hi Brid	HB	Srixon	US	3P-W/SC-1c	NC	+	1
Hi Brid Tour	HBT	Srixon	US	3P-W/SC-1c	C	+	3

2 Experimental

2.1. *Golf Balls Analysed*

The golf balls used in this study are listed in Table 1. The balls were selected to compare range and tournament balls, softer and harder balls, as well as solid core and wound (solid core or liquid core) balls, including the Srixon Hi-Brid balls, which show features of both solid core and wound balls ("hybrid" construction), namely a thin layer of wound rubber thread and a large solid core.

2.2. *Stress Relaxation Tests*

For both stress relaxation and compression tests, an Instron material testing machine (model no.: 3366) was used. For the stress relaxation tests, the balls were pre-loaded (F_0) between 1.2 and 8.8 kN at a deflection rate of 500 mm/min, and the decreasing load F was measured for $t = 3600$ seconds. As the stress relaxation generally and consistently showed a linear behaviour when plotting F against the natural logarithm of the relaxation time $LN(t)$, a logarithmic law was selected to model the non-linear visco-elastic properties:

$$F = A + B\,LN(t) \tag{6}$$

The parameters, A and B, however, depend on the initial load F_0.

$$F/F_0 = A_F + B_F \operatorname{LN}(t) \tag{7}$$

The following models were tested: AD, BF, SF, TD, NB, NP, PV, RB (hard, medium, and soft balls), DTN, PRN, HBT, MZk, MZr, MZb, and MZg. In Eq. (7), the coefficients A_F and B_F, are functions of F_0, which in turn, is a function of the deflection x of the ball; subsequently, for impact modelling, A_x and B_x, will be defined as functions of x, as the specific constitutive equation of $F = f(x)$ is decisive to derive the constitutive equation of the ball impact.

2.3. Compression Tests

For the compression tests, the balls were loaded up to 9 kN with crosshead speeds of 500, 160, 50, 16, and 5 mm/min. From the load–deflection curve, the stiffness was calculated, which is the deflection derivative of the load. Stiffness-deflection, and stiffness-load curves were used to determine the nature of the constitutive equation for each ball model (e.g., polynomial, exponential, etc.).

2.4. Peak Frequency of the Impact Sound

The golf balls were dropped from a height of 2 m on a ceramic tile rigidly connected to a massive concrete surface. The impact sound was recorded by a Zen Neeon MP3 player in mp3-format. While playing a file with Windows Media Player, it was simultaneously re-recorded in Matlab by converting it into a wav-file at 11025 Hz. After applying FFT to the first 8 ms of the impact sound, the power spectrum was plotted and the first prominent frequency peak (lowest frequency mode) determined.

2.5. Atti and Riehle Compression and Statistics

Atti and Riehle compression numbers were determined independently by intersecting the load-deflection curves (deflection rate 500 mm/min) with Eqs. (1) and (2). Statistical calculations comprised mean, standard deviation and coefficient of determination of stiffness parameters (initial stiffness and nature of the stiffness-deflection function), peak frequency of the impact sound, and Atti and Riehle compression numbers.

3 Results

3.1. Stress Relaxation

The coefficients A_F and B_F, as functions of F_0, are shown in Figures 1 and 2. A_F and B_F change linearly with F_0. The higher the parameter B_F, the more viscous is the ball.

Solid core balls (including the "hybrid" Srixon Hi-Brid ball) are clearly more viscous than wound balls (DTN, TD, and PRN). Among the wound balls, the wound solid core (DTN) balls are more viscous than the liquid filled balls (TD and PRN). In solid core balls, B_F increases and A_F decreases with F_0. Wound balls show the opposite behaviour: B_F decreases and A_F increases with F_0. Wound balls, excluding Hi-Brid, have higher A_F-values at higher F_0. The Srixon Hi-Brid ball shows the same behaviour as solid core balls and thus has to be regarded a solid core ball.

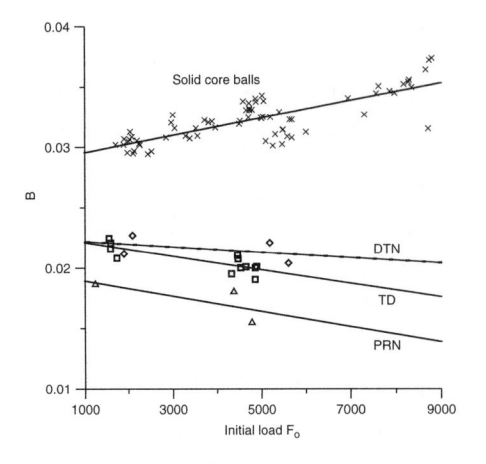

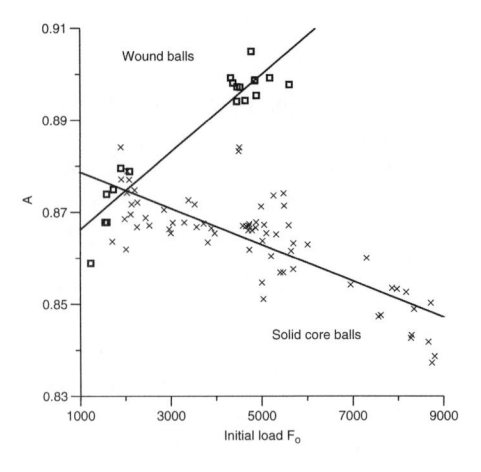

Figure 1. B_F vs. F_0.

Figure 2. A_F vs. F_0.

3.2. Compression Tests

Atti and Riehle compression numbers are shown in Figures 3 and 4. Eq. (3) is incorrect, as the sum of Atti A and Riehle R is not 160. Correlating Atti and Riehle numbers in the balls tested, the regression is linear with a coefficient of determination of $r^2 = 0.986$.

$$A = 164.4184168 - 1.150265948\,R \qquad (8)$$

The force F – deflection x and stiffness k – deflection x behaviour is different in solid-core balls (including Srixon Hi-Brid) and in wound balls. Solid-core balls show a linear stiffness with *deflection*, comparable to a hard spring. Thus, the force-deflection curve can be treated as a 2nd order polynomial or parabolic curve (Figure 5):

$$F = A_1\,x^2 + A_2\,x \qquad (9)$$

$$k = 2A_1\,x + A_2 \qquad (10)$$

Wound balls, on the other hand, show a linear increase of stiffness with *force*. As the stiffness is the deflection derivative of the force (Figure 6), a linear relationship between a function and its derivative is established by an exponential law. Thus, the following equation can be derived:

$$F = e^{D+Cx} - e^D \qquad (11)$$

$$k = C\,e^{D+Cx} \qquad (12)$$

For comparing the different models of golf balls, the following paramerters, in addition to Atti and Riehle compression, were determined:

- initial stiffness k_0, at $x = 0$; solid core balls: A_2; wound balls: $k = C\,e^D$.
- gradient of stiffness (in solid-core balls only): $2\,A_1$ (2nd derivative of Eq. (9)).

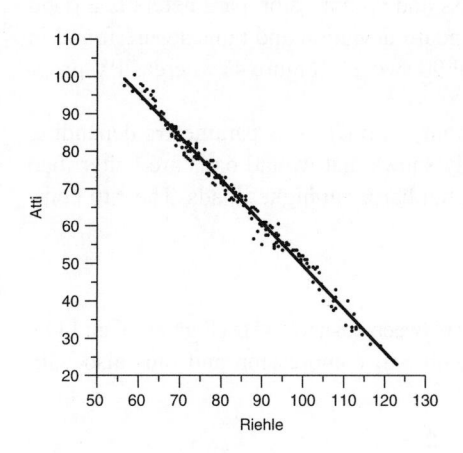

Figure 3. Atti vs. Riehle compression (compressed at 500 mm/min).

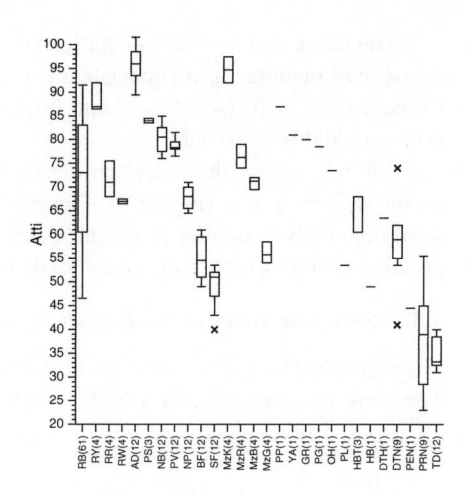

Figure 4. Range of Atti comression of different balls (number of balls in parentheses).

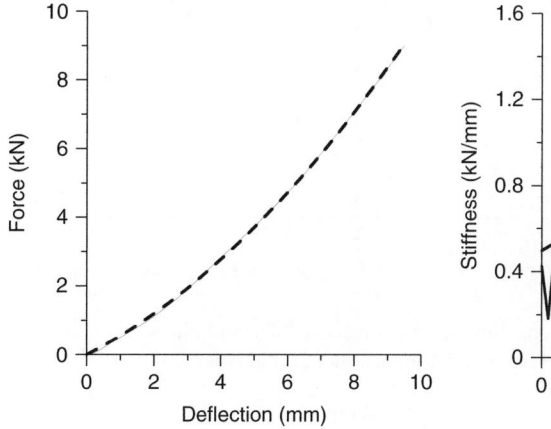

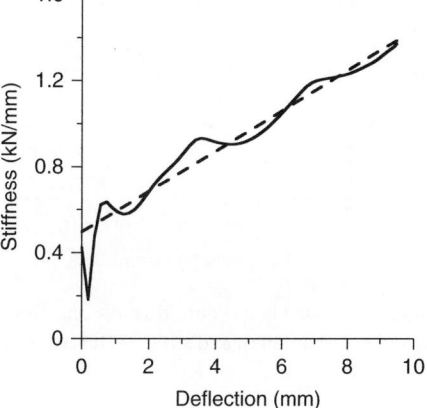

Figure 5. solid core ball (hard Chinese range ball); force-deflection curve (500 mm/min) with parabolic fit (Eq. (9), dashed line), and stiffness-deflection curve with linear fit (Eq. (10), dashed).

Comparing the Atti compression to the initial stiffness k_0 reveals that wound balls, Srixon Hi-Brid, and Chinese range balls have a smaller k_0 compared to other solid core balls (Figure 7). The 2 clusters are even completely separated (dashed line in Figure 7); only one single Chinese range ball is located on the higher k_0 side. The thin would layer in Hi-Brid balls affects the initial stiffness k_0, whereas the parabolic stiffness behaviour classifies Hi-Brid clearly as solid core ball.

Comparing the Atti compression to the stiffness gradient of solid-core balls (Figure 8) shows that Hi-Brid (HB, HBT) and Chinese range balls (RB), as well as OH, MzR, PS, PP, and AD have a higher stiffness gradient compared to other balls. Peak stiffness gradients are found in AD and in harder Chinese range balls.

The range and standard deviation of stiffness and compression parameters is a good measure of manufacturing consistency. High standard deviation and ranges were found in Chinese range balls (solid core) and Professional 90 (wound; Figure 4) whereas PV properties are highly consistent.

All balls tested showed visco-elastic behaviour, with stiffness parameters depending on the deflection rate (Figure 9). Figure 9 clearly shows that wound balls are softer than solid-core balls at smaller loads and deflections, but harder at higher loads. The Atti compression (Figure 4) represents only loads < 1 kN.

3.3. *Peak Frequency of the Impact Sound*

The peak of the lowest frequency mode was located between 2.8 and 5 kHz (Figures 10 and 11). The peak frequency shows a high correlation with Atti compression and thus also with

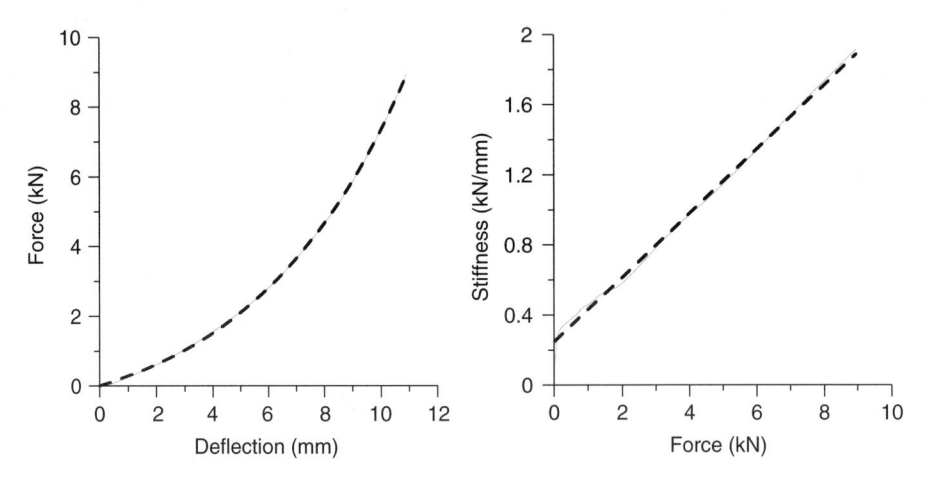

Figure 6. wound ball (Penfold Tradition); force-deflection curve (500 mm/min) with exponential fit (Eq. (11), dashed line), and stiffness-force curve with linear fit (Eq. (12), dashed line).

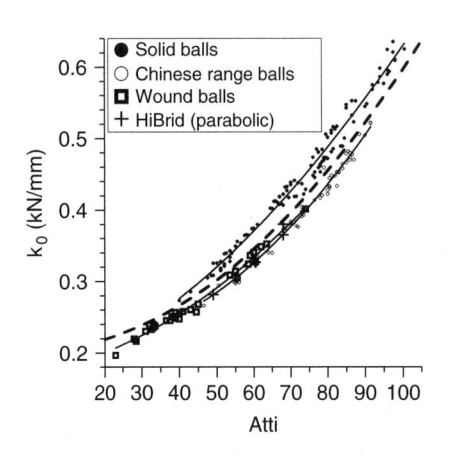

Figure 7. Initial stiffness k_0 vs Atti compression (compressed at 500 mm/min).

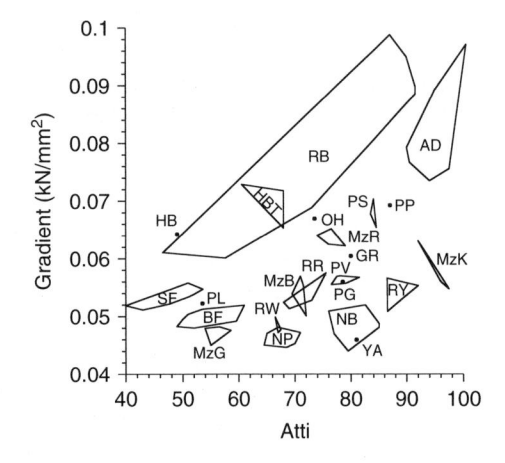

Figure 8. stiffness gradient vs Atti compression in solid-core balls (compressed at 500 mm/min).

initial stiffness k_0; only AD balls have a higher frequency peak than other balls with comparable Atti and k_0 values.

4 Constitutive Equations of the Logarithmic Law and Extrapolation to Higher Compression Speeds

Rewriting the stress relaxation equation, Eq. (6) with parameters A and B as functions of the deflection x:

$$F = x_0 A_x + x_0 B_x \, \text{LN}(t) \tag{13}$$

where x_0 is the constant deflection, applied by a Heaviside function $H(t)$:

$$x = x_0 \, H(t) \tag{14}$$

Taking Laplace transform of Eqs. (13) and (14):

$$\hat{F} = x_0 \frac{A}{s} - x_0 B \left(-\frac{\gamma}{s} - \frac{\text{LN } s}{s} \right) \tag{15}$$

$$\hat{x} = \frac{x_0}{s} \tag{16}$$

where the caret (^) denotes the transformed parameter, and γ is the Euler-Mascheroni constant (0.577215665…).

Substituting Eq. (16) into Eq. (15), we get the constitutive equation of the logarithmic law of viscoelasticity:

$$\hat{F} = \hat{x}A + \hat{x}B \left(\gamma + \text{LN } s \right) \tag{17}$$

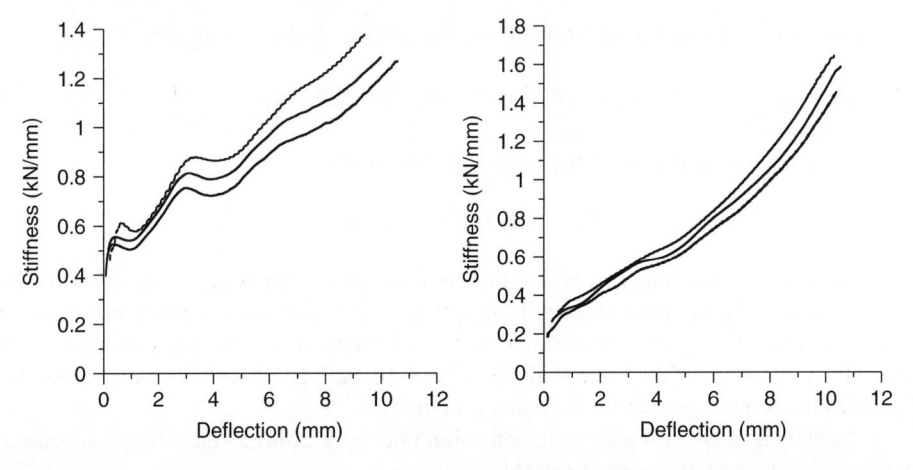

Figure 9. Mean stiffness vs. deflection at 3 different deflection rates (500, 50, and 5 mm/min); the stiffness increases with the deflection rate; left: hard Chinese range balls (solid core), right: Professional 90 (wound).

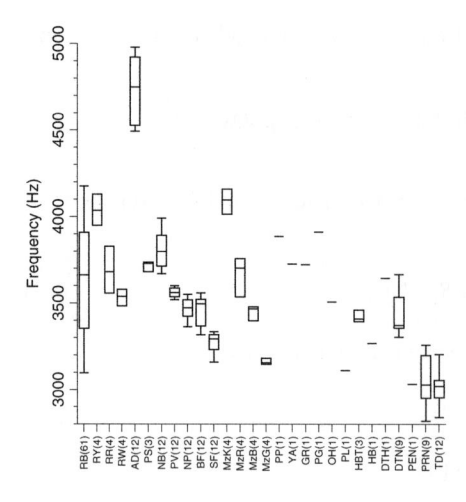

Figure 10. Range of lowest frequency mode of different balls (number of balls in brackets).

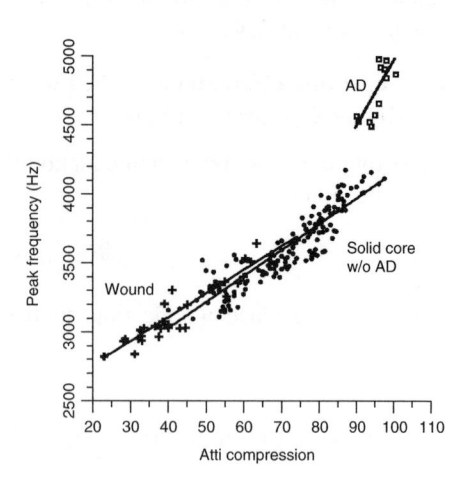

Figure 11. Atti vs. peak frequency.

In order to establish the relationship between force and deflection rate, as well as stiffness and deflection rate, we apply a ramp function:

$$x = \dot{x}_0 t \quad \text{or} \quad \hat{x} = \frac{\dot{x}_0}{s^2} \tag{18}$$

where $\dot{x}_0$ is the constant velocity or deflection rate of the cross head.

Substituting Eq. (18) into Eq. (17)

$$\hat{F} = \dot{x}_0 \frac{A}{s^2} - \dot{x}_0 \frac{B}{s}\left(-\frac{\gamma}{s} - \frac{\text{LN } s}{s}\right) \tag{19}$$

After taking inverse Laplace transform, rearranging, and replacing t by $x / \dot{x}_0$:

$$F_x = xA + xB - xB\text{LN}x + xB\text{LN}\dot{x}_0 \tag{20}$$

The deflection derivative of the force F is the stiffness k:

$$k_x = A - B\,\text{LN}_x + B\,\text{LN}\dot{x}_0 \tag{21}$$

In a purely elastic material, where the viscosity parameter is zero, Eq. (21) becomes $k = A$, which is the stiffness of a Hookean spring. The reason for this result becomes evident, when considering, that the constant x_0, initially applied for stress relaxation, does not reflect the true nature of the force-deflection function, be it polynomial or exponential. Yet, Eq. (21) allows deriving two fundamental principles:

– Replacing $\dot{x}_0$ by ν, the variable deflection rate, by keeping x, the deflection constant (x_i), i.e. exchanging constant and variable:

$$k_\nu = A - B\,\text{LN}x_i + B\,\text{LN}\nu \tag{22}$$

results into k as a function of ν, and reveals that k is a log function of ν, comparable to F, which is a log function of time in the original stress relaxation equation Eqs. (6), (7), and (13). Eq. (22) allows extrapolating the stiffness of different velocities at the same deflection. Applying Eqs. (21) and (22) to derive k as a function of both x and ν is not valid, as the initial condition is based on a constant deflection rate. Nevertheless, k as a function of both x and ν helps to approximate impact conditions (see below) if the constitutive transformed impact equation is not entirely analytical.

– The term A in Eq. (21) represents the stiffness of the system under purely elastic conditions. If the stiffness is not constant, then A can be replaced by the actual stiffness of the system, which, in the specific case of solid-core and wound golf balls is given in Eqs. (10) and (12) respectively.

Substituting these 2 equations into Eq. (21):

$$k_x = 2A_1 x + A_2 - B\,\mathrm{LN} x + B\,\mathrm{LN}\dot{x}_0 \tag{23}$$

$$k_x = Ce^{D+Cx} - B\,\mathrm{LN} x + B\,\mathrm{LN}\dot{x}_0 \tag{24}$$

The corresponding force equations thus are:

$$F_x = A_1 x^2 + A_2 x + xB - xB\,\mathrm{LN} x + xB\,\mathrm{LN}\dot{x}_0 \tag{25}$$

$$F_x = e^{D+Cx} - e^D + xB - xB\,\mathrm{LN} x + xB\,\mathrm{LN}\dot{x}_0 \tag{26}$$

The latter two equations allow establishing two further principles:
– deriving the parameters A_1, A_2, C, and D at any x and any $\dot{x}_0$, by taking the force deflection curve obtained at a specific $\dot{x}_0$, subtracting the viscous terms (containing B) and calculating a parabolic or exponential fit, by satisfying the following 2 eqns:

$$F_x - xB + xB\,\mathrm{LN} x - xB\,\mathrm{LN}\dot{x}_0 = A_1 x^2 + A_2 x \tag{27}$$

$$F_x - xB + xB\,\mathrm{LN} x - xB\,\mathrm{LN}\dot{x}_0 = e^{D+Cx} - e^D \tag{28}$$

B is known from the stress relaxation equation Eq. (13).
– Replacing $\dot{x}_0$ by x/t and taking Laplace transform of Eqs. (25) and (26) defines the constitutive equations of non-linear force-deflection curves:

$$\hat{F} = (\hat{x} * \hat{x}) A_1 + \hat{x} A_2 + \hat{x} B (\gamma + \mathrm{LN}\, s) \tag{29}$$

$$\hat{F} = e^D \left(L(e^{Cx}) - 1/s \right) + \hat{x} B (\gamma + \mathrm{LN}\, s) \tag{30}$$

where $L(e^{Cx})$ is the transformed e^{Cx}, and the asterisk (*) denotes a convolution.

5 Impact Modelling

For impact modelling, we need to consider

- a single-sided compression (in contrast to compression testing, which is double sided), and
- the force equilibrium of the reaction force at the contact area and the inertial force of the decelerated or accelerated object, which is its mass times the acceleration of its centre of mass COM.

The COM is no longer at the geometrical centre of the undeformed object, as the COM moves during impact. Thus, the stiffness determined from double-sided compression testing, cannot be simply multiplied by 2, as suggested by Cross (1999). Considering an object, consisting of infinite parallel layers of infinitesimal thickness, during impact, it becomes evident, that the layer in contact with the frame against which the object impacts, experiences the highest force on either side of this layer, whereas the layer most distant from the frame experiences zero force, if its mass is negligible. Thus, the force increases linearly from the most distant to the layer at the impact. Thus,

$$F_L = L F_R \quad (31)$$

where F_L is the force exerted to a specific layer, L is the position of the layer between L_0 (most distant layer) and L_1 (length of the object, layer at contact), and R_F is the overall reaction force during impact.

The displacement of each layer is dx:

$$dx = \frac{F_L dL}{EA} \quad (32)$$

where E and A_c are modulus and cross sectional area respectively. Substituting Eq. (31) into Eq. (32), and solving for the deflection x:

$$x = \frac{F_R}{EA} \int_{L_0}^{L_1} L\, dL \quad \text{and} \quad x = \frac{F_R}{EA}\left[\frac{L^2}{2}\right]_{L_0}^{L_1} + C \quad (33)$$

C is zero, as $x = 0$ if $F = 0$. Solving for deflections on either side of the COM, i.e. for L between 0 and 0.5 as well as between 0.5 and 1, results into ¾ compression on the contact side of the COM, and ¼ compression on the other side. Thus, the deflection between the impact zone and the COM is ¾ compared to twice ½ in compression testing. The stiffness consequently increases by 4/3 or 1.333. This applies to spherical objects as well, when considering the object to consist of infinite longitudinal columns, which become compressed at different times.

The impact is modelled by considering 3 forces, the aforementioned reaction force F_R of the impacting object, its inertial force F_I, and an applied force F_A which accelerates the object to the initial velocity at impact, v_0. F_A is applied immediately before impact by a unit impulse acceleration (Dirac delta function $\delta_{(t)}$), resulting into a unit step velocity

(Heaviside function) if the object would continue to move freely. The unit step is brought to the specific initial velocity by multiplying by v_0. Thus:

$$v = v_0 \, H(t), \; a = v_0 \, \delta(t), \; F_A = am = v_0 \, \delta(t) \, m \tag{34}$$

The inertial force F_I is

$$F_1 = am = \ddot{x} m \tag{35}$$

The overall force equilibrium is:

$$F_I + F_R = F_A \tag{36}$$

Substituting the transformed Eqs. (34) and (35), as well as Eq. (17) into Eq. (36), we get the constitutive equation of the non-linear visco-elastic logarithmic law impact:

$$s^2 \hat{x} m + \hat{x} A + \hat{x} B \left(\gamma + \mathrm{LN} \, s \right) = m v_0 \tag{37}$$

Dividing by m and solving for $\hat{x}$:

$$\hat{x} = v_0 \frac{1}{s^2 + \dfrac{A}{m} + \dfrac{B\left(\gamma + \mathrm{LN} \, s\right)}{m}} \tag{38}$$

If the object is purely elastic, then the viscosity parameter B becomes zero and the inverse Laplace transform of Eq. (38) becomes

$$x = v_0 \sqrt{\frac{m}{A}} \sin\left(\sqrt{\frac{A}{m}} t \right) \tag{39}$$

which is the constitutive equation of undamped oscillation
The transformed force equation is:

$$\hat{F} = v_0 \frac{A + B\left(\gamma + \mathrm{LN} \, s\right)}{s^2 + \dfrac{A}{m} + \dfrac{B\left(\gamma + \mathrm{LN} \, s\right)}{m}} \tag{40}$$

If $B > 0$, Eqs. (38) and (40) can be solved numerically, by applying Weeks' (1966) or Piessen & Huysmans' (1984) routine for inverse Laplace transform.

As mentioned earlier, Eqs. (23)–(26) cannot be applied for numerical solution of deflection and force during impact. Nevertheless, if the force-deflection function is non-linear and thus Eq. (38) is no longer applicable, these 4 equations are useful for approximation of deflection and force during impact. This was tested by comparing the result of Eq. (38) and of Eqs. (20)–(22) after inserting them into the impact equation. The difference between Eq. (20), $F = f(x, v)$, and Eqs. (21)–(22), $k = f(x, v)$, is that the former consists of a non-linear damper, i.e. $F = xB \, \mathrm{LN} \dot{x}_0$, which accounts for the non-conservative energy, whereas the latter is purely elastic (kinetic energy equals elastic energy, and the impulse, the area under the force-time curve, equals the linear momentum $2mv_0$). The approximation was tested in extreme viscous conditions, where $A = B$; in golf balls, B is at least 20 times smaller than A. A good approximation is obtained when modelling only the first half of the impact, by taking the mean result of Eq. (20) and Eqs. (21)–(22). For approximating the contact time during impact, the result of Eqs. (21)–(22), $k = f(x, v)$, is sufficient.

The impact at $v_0 = 50$ m/s was approximated for 2 ball models, AD, and TD (Figure 12), the latter is harder than AD at higher loads. In AD, contact time and peak force are 0.54 ms and 13.55 kN respectively, in TD 0.7 ms and 12.78 kN respectively.

5 Discussion

Even though Hi-Brid is a wound ball with a large solid core, it turned out to behave like a typical solid-core ball. This is due to the fact, that the layer of wound rubber thread is extremely thin. PRN, DTN, and HBT were sawn into 2 halves; the diameter of all 3 balls is ≈42.5 mm, the diameter of the core is ≈37.5 mm in HBT and ≈27 mm in DTN and PRN (in the latter, the wall thickness of the hollow, liquid filled core is ≈ 2 mm), the thickness of the cover material is ≈ 2 mm HBT and PRN, and ≈2.5 mm DTN, and thus the thickness of the wound layer is ≈0.5 mm in HBT and ≈5.25 mm and ≈5.75 mm in DTN and PRN respectively. In addition, wound balls have a smaller Atti number than solid core balls, which is not reflected on the ball carton. Titleist (2004) claimed that TD's Atti number is 90, whereas the mean of the measured value is 35 (Figure 4). Equally, PRN (Professional 90), DTN and DTH (DT Wound 90 and 100), do not have 90 or 100 Atti numbers, as suggested by the name, but rather 37.7, 58.2 and 63.5 respectively. Nevertheless, wound balls are far stiffer at higher loads than solid core balls, thus compensating the initial low stiffness at smaller loads by still offering a better feel. The Atti number of RB (Chinese range balls) is too inconsistent for usage at driving ranges, which confuses the beginner and prevents him form understanding the feel. The high 1st frequency peak of AD balls might be due to a missing first frequency mode is in these balls.

The results of the impact modelling (Figure 12) match closely the experimental results of Ujihashi (1994) and Cochran (2002). The disadvantage of the logarithmic law, however,

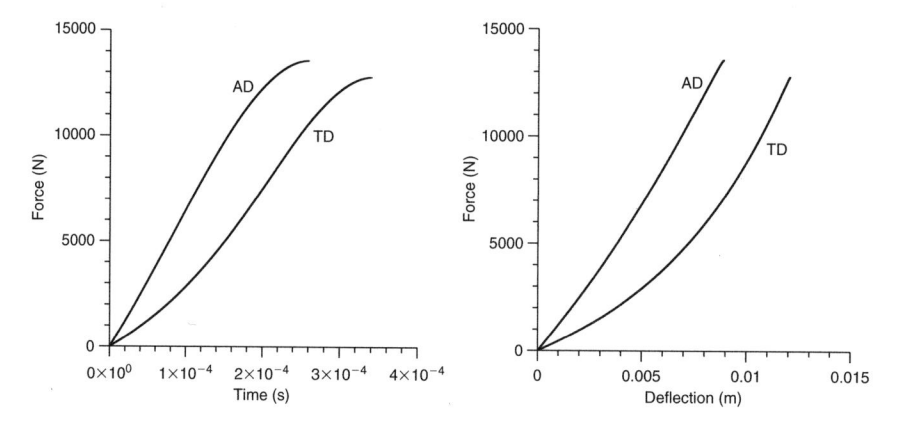

Figure 12. First half of the impact of AD and TD balls.

is that, in contrast to the results of the logarithmic model, the approximation produces an instantaneous force of $-\infty$ at $\nu = 0$. Nevertheless, it is useful to model the first half of the impact.

References

Atti R. (1942) Golf ball testing machine. United States Patent 2278416, published 7 Apr.

Barrass D.F., Roberts J.R. and Jones R. (2006) Assessment of the impact sound in golf putting. *Journal of Sports Sciences*, 24/5, 443–454.

Cochran A.J. (2002) Development and use of one-dimensional models of a golf ball. *Journal of Sports Sciences*, 20, 635–641.

Cross, R. (1999) The bounce of a ball. *Am. J. Physics*, 67, 222–227.

Dalton J. (2002) Compression by any other name. In: Thain E. (Eds.) *Science and Golf IV*. Routledge, London, p. 319–327.

Fuss F.K., Tan M.A. and Sim J.Y. (2005) Structural properties of golf balls: what is the Atti compression number really worth? In: Subic A. and Ujihashi S. (Eds.) *The Impact of Technology on Sport*. ASTA (Australasian Sports Technology Alliance), Melbourne, Australia, pp. 343–348.

Lieberman B.B. and Johnson S.H. (1994) An analytical model for ball barrier impact. Part 1: models for normal impact. In: Cochran A.J. and Farrally M.R. (Eds.) *Science and Golf II*. Spon, London, pp 309-314.

Morrice P. (2005) The search for feel: you know it when you've got it, but what is feel, really? *Golf Digest*, 2005/6 (online). Available: http://www.golfdigest.com /instruction/index.ssf?/instruction/gd200506feelsound1.html (accessed: June 2007).

Nesbitt R.D. (2006) Golf ball with sulfur cured inner component. United States Patent 7041008, published 9 May.

Penner A.R. (2003) The physics of golf. *Rep. Prog. Phys*. 66, 131–171.

Piessens R. and Huysmans R. (1984) Algorithm 619: Automatic numerical inversion of the Laplace transform. *ACM Trans. Math. Softw.*, 10/3, 348–353.

Shannon A. and Axe J.D. (2002) On the acoustic signature of golf ball impact. *Journal of Sports Sciences*, 20, 629–633.

Sullivan M.J., Nesbitt R.D. and Binette M.L. (2003) Low spin golf ball comprising silicone material. United States Patent 6634963, published 21 Oct.

Tanaka K., Sato F., Oodaira H., Teranishi Y., Sato F. and Ujihashi S. (2006) Construction of the finite-element models of golf balls and simulations of their collisions. *Proceedings of the Institution of Mechanical Engineers, Part L, Journal of Materials: Design & Applications*, 220/1, 13–22.

Titleist.com (2004) personal communication.

Ujihashi S. (1994) Measurement of dynamic characteristics of golf balls and identification of their mechanical models. In: Cochran A.J. and Farrally M.R. (Eds.) *Science and Golf II*. Spon, London, pp 302–308.

USGA (2007) Conforming Golf Balls, effective June 6, 2007 (online). Available: http://www.usga.org/equipment/conforming_golf_ball/gball_list.pdf (regularly updated on the first Wednesday each month; accessed: June 2007).

Weeks W.T. (1966) Numerical inversion of Laplace transforms using Laguerre functions. *J. ACM* 23/3, 419–429.

THE INFLUENCE OF WIND UPON 3-DIMENSIONAL TRAJECTORY OF GOLF BALL UNDER VARIOUS INITIAL CONDITIONS

T. NARUO[1] & T. MIZOTA[2]

[1]*Mizuno Corporation, Osaka, Japan*
[2]*Fukuoka Institute of Technology, Fukuoka, Japan*

Aerodynamic forces and torque acting on the ball were measured under various flight conditions in a wind tunnel flow. Using the aerodynamic force coefficients, mathematical calculation of flight trajectory equation was made by time integral calculus, and 3-dimensional flight trajectory, changes in velocity as well as rotation velocity were obtained. Furthermore the logarithmic law was applied to trajectory formation of a golf ball in order to include influence of atmospheric boundary layer. The trajectory formation considering atmospheric boundary layer was verified by two experiments. One of the experiments was conducted to measure wind velocity distribution. By this experiment, logarithmic law could be verified. Other experiment was conducted in order to verify 3-dimensional flight trajectory applied logarithmic law. Many golf balls were hit under various initial conditions of golf ball (initial velocity, launch angle, spin rate) by a professional golfer using various golf clubs. Initial conditions just after golf balls launched were measured by launch monitor and wind velocity distribution in the direction that golf ball flied was measured. In addition, golf ball trajectory under atmospheric boundary layer was calculated by using measured initial launch conditions and wind velocity distribution. As a result, the calculated result of drop positions by trajectory analysis agreed with the actual measured data. So far, we know the influence of wind upon trajectory only qualitatively. Flight trajectory considering natural wind was calculated by using 3-dimensional flight trajectory formation applied logarithmic law under various initial conditions in detail. As a result, we could get knowledge about 3-dimensional flight trajectory influenced by natural wind. Such as we found out that there are conditions that distance reduces by tailwind and showed the conditions. In addition, we could get quantitative data: the influence of spin rate and launch angle including wind upon distance. Moreover we analyzed the quantitative influence of spin rate and launch angle including wind upon curving e.g. hooks and slices.

1 Introduction

Golf is a sport played outdoors under natural conditions. Especially, the effect by the wind on the results of the game is not small. Skilled professional golfers, through experience, will have a feel for the force and direction of the wind and set up strategies in club selection and the shot to be made. However a quantitative study concerning the effect of wind on the trajectory of the ball under various conditions is not found. Mizota *et al.* (2002) and Naruo & Mizota (2005) proposed a method of analyzing the trajectory under the atmospheric boundary layer. Upon confirming the validity of the method on analyzing the trajectory (Naruo & Mizota, 2006), the method was used to analyze the effect of the wind on trajectory under various launch conditions.

2 Method of 3-D Trajectory Analysis of Golf Ball under Atmospheric Boundary layer

Aerodynamic forces were measured by the wind tunnel and ball rotating device as shown in Figure 1.

In this experiment, spin parameter is made to change over a wide range of 0.03 to 1.13 by changing flow velocity U and spin rate N. The wind velocity was changed to a maximum of 44 m/sec and rotations to a maximum of 10,000 rev/min.

The rotation axis inclination θ is defined to construct 3-dimensional trajectory analysis. By this inclination, lift is inclined. A force in the right or left direction acts on the ball. So θ is the factor to decide the side deviation distance Y.

And the logarithmic law is applied to analyze trajectory of golf ball under atmospheric boundary layer. This logarithmic law closely matches the experimental value in the wind tunnel boundary layer and it is said that it also coincides with the observation results in surface boundary layer having smooth surfaces like those of grasslands.

Figure 2 shows the wind velocity vertical distribution diagram used to embrace the effects of the atmospheric boundary layer. Where V_0 is the velocity at a height of 10 m above ground, V_Y is the average velocity at a height of Y, V^* is the friction velocity, and τ_0 is the shearing stress at ground surface.

Also, κ is Karman constant with an approximate value of 0.4, while Y' is the roughness constant, but here 0.09 is used assuming that the periphery of the course consists of thick grown grass, the components of the wind are calculated from the angle λ of the velocity components (clockwise direction from X axis considered as +). The aerodynamic force applied on the ball is obtained from the relative speed of the ball and air current, while the position of the ball is obtained from the absolute speed. This is repeated until the ball lands on the ground.

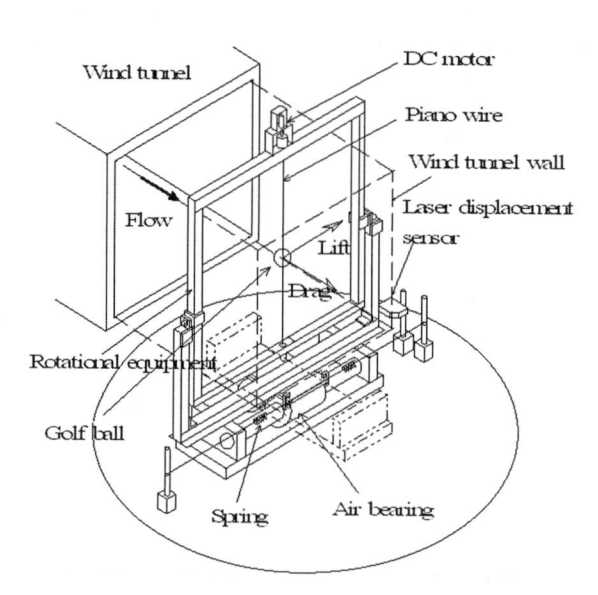

Figure 1. Wind tunnel and rotating device.

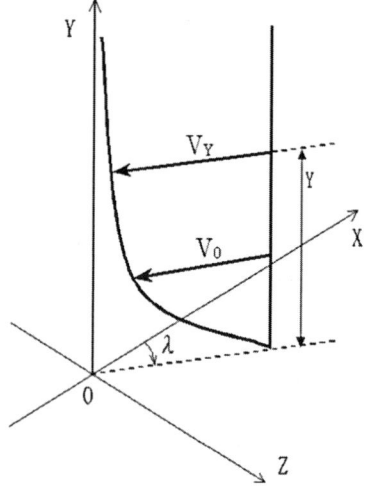

Figure 2. Wind velocity vertical distribution.

3 Comparison between Trajectory Analysis and Ejection Experiment Results

To verify the accuracy of this trajectory simulation considered the effects of the atmospheric boundary layer, the initial conditions of the golf ball hit by a professional golfer were measured and using the values, the trajectory was calculated and the final distance was compared with actually measured values. The initial conditions are initial velocity, launch angle, and spin rate.

Moreover four poles equipped with vanes anemometers at the height of 5.5 m were installed at every 50 m from the hitting position towards the flight direction and the wind direction and the velocity were measured as shown in Figure 3. Figure 4 shows a comparison of the actual measurements and the calculated values of flight distance. The measurements with considering the wind are shown by the solid square, while the results without consideration of the wind are shown by circle. When the actual measurement and the calculated value coincide, the data is appeared on y = x. When minimum square approximation is conducted on the result in which wind is not considered, y = 0.945x is obtained, but when the wind is considered, the result becomes y = 0.981x, i.e. it approaches y = x. Comparison of numerical analysis results and actual measured values and an extremely favorable matching of results was seen.

4 Analysis of the Influence of Wind upon 3-Dimensional Trajectory under Various Initial Conditions

Using the above mentioned analysis method, the effect of the wind on trajectory of the ball was analyzed under various launch conditions. The present calculations were all made with a ball velocity of 60 m/sec. 60 m/sec is equivalent to the speed of the ball when the ball is hit by the center of the driver head face by a golfer with a head speed of 40 to 42 m/sec that is an average of amateur male golfers in Japan.

Firstly, the flight distance was calculated with various launch angles and spin rates with a tail wind of 5 m/sec and a headwind of 5 m/sec. Distance was calculated for all combinations, with the launch angle of the ball in 0.5 degrees increments within a range of 5 to 30 degrees, and with the spin rate in 100 rev/min increments within a range of 700 to 6,000 rev/min.

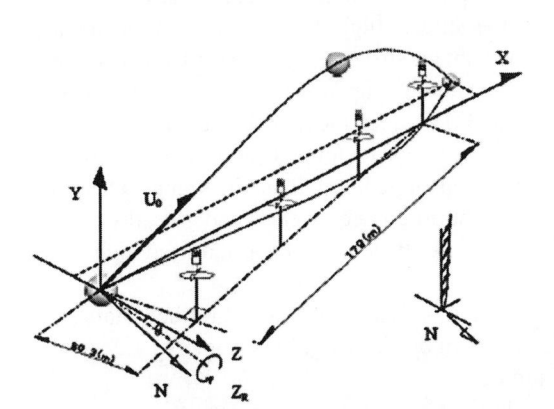

Figure 3. Ejection experiment.

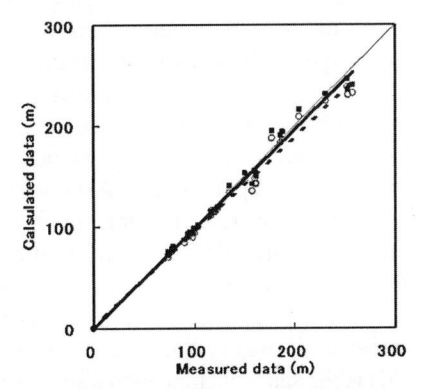

Figure 4. Comparison between calculated results and experimental results.

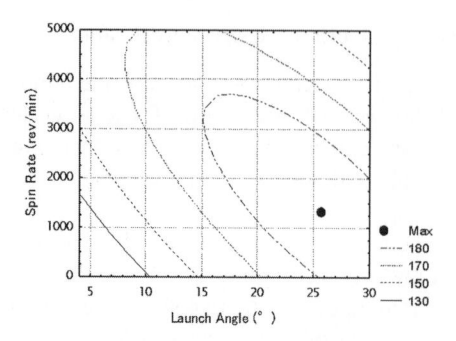

Figure 5. Carry distance with a headwind of 5 m/s.

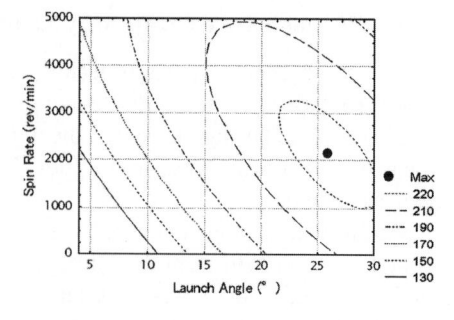

Figure 6. Carry distance with a tailwind of 5 m/s.

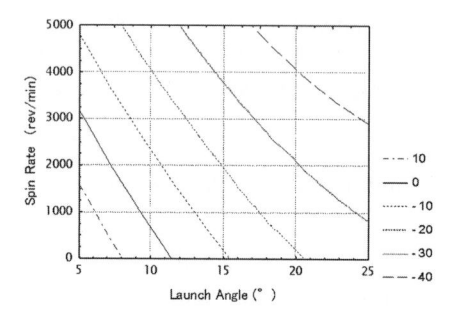

Figure 7. The difference in flight distance with a headwind and a tailwind.

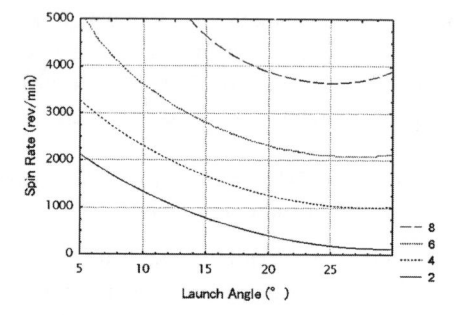

Figure 8. The side deviation distances with a headwind.

Furthermore, a secondary smoothing process was performed and a contour drawing was made. The results in case of a headwind are shown in Figure 5 while the results in case of a tailwind are shown in Figure 6. The respective longest flight distances are marked with a solid circle. Figure 7 shows the difference in flight distance with a headwind and a tailwind. As seen from Figure 5 and Figure 6, it can be seen that the launch angle hardly has any effect in regard to conditions for obtaining the longest flight distance under tailwind and headwind conditions. In contrast the effect by spin rate becomes greater under a headwind. However, as seen in Figure 7, with roughly any spin rate, a difference of 10 yard in flight distance is seen between a headwind and a tailwind condition when the launch angle is increased 5 degrees from 3 degrees. The spin rate difference of 2000 rev/min is equivalent to a 10 yard difference in flight distance. Also, when the launch angle is small and in a range where the spin rate is low, there is a zone in which the flight difference reverses under headwind and tailwind conditions. This is to say, the flight distance becomes longer in a headwind compared with the flight distance in a tailwind.

This is because in a flight distance in which sufficient height cannot be obtained, a larger lift is obtained in a headwind. This is known by skilled golfers from experience and this is the same phenomenon seen in ski jumps.

Next, with the rotating inclination angle θ set at 5 degrees, side deviation distances were calculated under various conditions with a headwind of 5 m/sec and a tailwind of

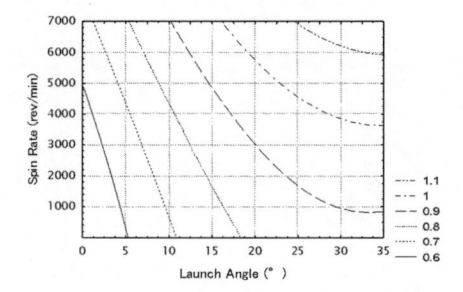

Figure 9. The ratio of side deviation distances with a tailwind to a headwind.

5 m/sec. Figure 8 shows the side deviation distances with a headwind. Roughly, in all areas, side deviation distance becomes greater as the launch angle becomes higher and as spin rate becomes greater. However with a launch angle of 24 degrees or more, the affect by the angle is hardly seen and side deviation distance is greatly dependant on spin rate.

Figure 9 is the ratio of side deviation distances with a tailwind to a headwind. In roughly all areas, side deviation distance is greater under a headwind compared with a tailwind. This is because with a headwind, relative flow velocity increases, lift becomes greater, and the driving force to curve the ball sidewise becomes greater.

Moreover as launch angle and spin rate are becoming greater, the ratio of side deviation distances with a tailwind to a headwind becomes closer to 1.

5 Conclusion

The analyzed trajectory of golf ball under atmospheric boundary layer applied the logarithmic law was verified by comparing the calculated data and the actual measured data. The effect of the wind on the trajectory was analyzed under various launch conditions. As the result, various phenomena known from experience were analyzed quantitatively.

References

Mizota T., Naruo T., Shimozono H., Zdravkovich M. and Sato F. (2002) *3-Dimensional trajectory analysis of golf balls*, Science and Golf 4, E.&F.N.Spon, London, pp. 349–358.

Naruo T. and Mizota T. (2005) *Trajectory analysis of golf ball under atmospheric boundary layer*, In: Subic A. and Ujihashi S. (Eds.) *The Impact of Technology on Sport*. ASTA (Australasian Sports Technology Alliance), Melbourne, Australia, pp. 253–260.

Naruo T. and Mizota T. (2006) *Experimental verification of trajectory analysis of golf ball under atmospheric boundary layer*. In: Moritz E.F. and Haake S. (Eds.) *The Engineering of Sport 6, Vol. 1 – Developments for Sports*. Springer, Munich, pp. 149–154.

THE INFLUENCE OF GROOVE PROFILE; BALL TYPE AND SURFACE CONDITION ON GOLF BALL BACKSPIN MAGNITUDE

J.E.M. CORNISH[1], S.R. OTTO[2] & M. STRANGWOOD[1]
[1]*University of Birmingham, Edgbaston, UK*
[2]*R & A Rules Ltd, St Andrews, Fife, UK*

The oblique impact behaviour of a range of golf balls (multi-piece solid construction balls representing a variety of core and cover Shore D hardess values) has been investigated. The balls have been fired from a gas cannon, at a speed of $30\,\text{m s}^{-1}$, at a range of plates with varying groove profiles at loft angles from 35–70°. Testing was carried out using both dry plates and plates covered with moistened newsprint to simulate wet conditions. The ball speed, launch angle and backspin magnitude post-impact has been recorded using a launch monitor. Within the range of lofts that have been tested, spin rates have varied significantly between dry and wet conditions; differing groove depths and ball construction. In wet conditions the backspin maximum at higher loft angles ($\geqslant 50°$) is significantly reduced to around 2000–3000 rpm, compared with around 4000–10000 rpm (depending on ball type) in dry conditions. This is consistent for most grooves due to the presence of the intermediary layer which decreases the friction co-efficient and increases ball sliding. It has been seen that for the three-piece balls, the spin magnitude is increased in wet conditions for the deeper U-shaped grooves compared with shallower U-grooves at lower loft angles ($\approx 35°$). At higher loft angles in wet conditions the spin values seen are similar for all ball types (≈ 1500–2500 rpm). These variations in backspin have resulted from deformation of the cover into the grooves providing forces parallel to the inclined plate face and the effects of newsprint compressed into the groove during impact.

1 Introduction

The cover hardness of a golf ball has previously been related to its backspin rate for impacts with obliquely angled plates with controlled groove patterns (Sullivan & Melvin, 1994; Monk, 2006). Similar studies revealed that, in grooved versus un-grooved conditions, a softer covered two-piece ball had a larger drop off in backspin rate in un-grooved conditions than a soft covered multi-piece ball with a hard mantle layer (Cornish *et al.*, 2006). This suggests that grooves have more influence on backspin rate for certain ball types and that ball construction will also have an influence on the backspin rate.

A report by The R & A and the United States Golf Association (R&A/USGA, 2006) investigated the effect of groove shape in dry conditions and found there to be no significant difference in spin rate between the groove types, at loft angles of 20 to 60°, for a 3-piece golf ball. In wet conditions the backspin rate was greater for the U-shaped grooves than the V-shaped grooves at loft angles of 20 to 60°; it was suggested this could be explained by a greater ratio of groove cross sectional area to spacing between grooves (R&A/USGA, 2007). It is the aim of this work to investigate the role of ball construction and groove profile on the backspin rate of a number of commercially available golf ball types in both dry and simulated wet conditions.

Table 1. Construction and Shore D hardness values for all ball types used in the study.

Ball	Ball Type	Construction	Core Hardness	Mantle Hardness	Cover Hardness
(Ball A)	Distance	2-Piece	37	N/A	63
(Ball B)	Spin	2-Piece	41	N/A	52
(Ball C)	Distance/Spin	3-Piece	54	66	57

Table 2. Groove characteristics used in study.

Groove ID	Shape	Depth (mm)	Width (mm)
U Base	U	0.51	0.9
V Base	V	0.51	0.9
U Deep	U	1.02	0.9

2 Experimental

Three ball types were tested and are detailed in Table 1. Golf balls were fired from an ADC Supercannon 2000 at an inbound speed of approximately 30 m s^{-1} at angled plates with effective loft angles of 35–70°. The groove profiles on these plates varied in shape (V–U) and in depth (Table 2), whilst the groove separation and edge radius (rounding at the top of the groove) was kept constant at 3.5 and 0.254 mm respectively. The depths of the grooves on plate U Deep exceed the limits set out in the Rules of Golf (R&A, 2004).

Inbound speeds were recorded using a pair of light screens, the outbound velocity, launch angle and backspin rate post impact was measured using a proprietary launch monitor. Golf ball hardness was measured using a Mitutoyo Shore D Durometer with a steel cone and tip radius of 0.1 mm. Due to the complexities of using grass, wet impacts were simulated by placing moistened newsprint on the grooved plate prior to impact (R&A/USGA, 2006). High-speed camera images were taken using a Phantom 7.1 high-speed camera operating at 33,500 fps.

3 Results and Discussion

Initial results investigated the effect of groove shape on backspin rate (Figure 1) at higher loft angles (50–70°). In dry conditions, all balls showed an initial increase in backspin rate with increasing loft angle, up to a maximum, followed by a decrease as loft angle increased further. Ball C peaked at a loft angle of around 60° whilst the angle at which peak backspin occured decreased to around 58° and 54°, for balls B and A respectively. It should be noted that the angle at which these peaks have been determined is purely indicative. This trend qualititatively, but not quantitatively, follows the variation in Shore D value for the core and the difference in Shore D between the cover and outer core (2-piece)/mantle (3-piece). This suggests a ball with a harder core and a smaller difference between cover and subsurface layer Shore D value reaches a peak backspin rate at a higher loft angle. Thus, hardness values are a guidance to ball behaviour, but do not fully account for backspin generation. Within experimental scatter there

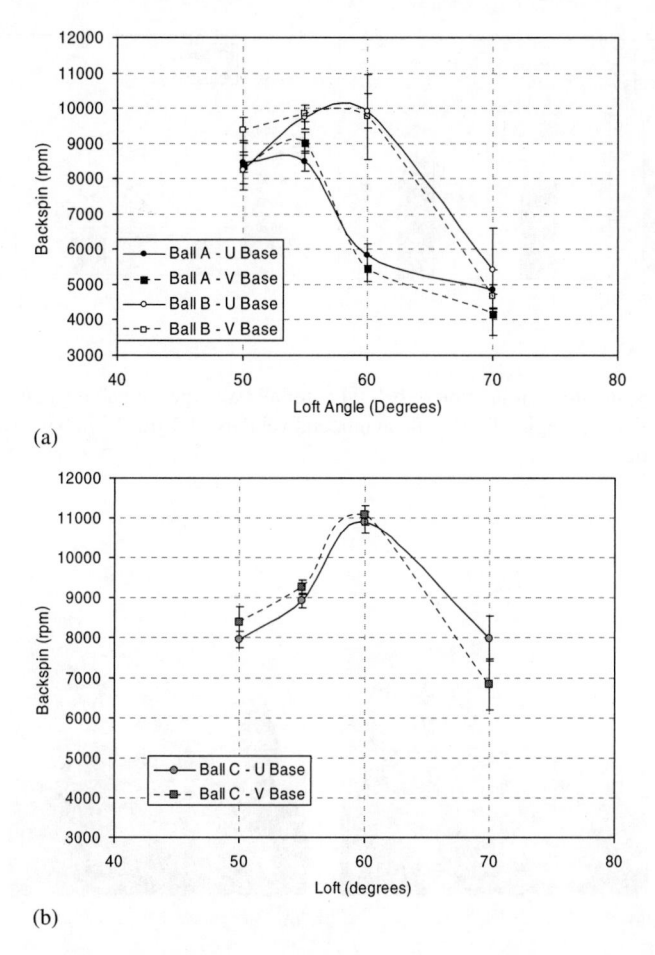

Figure 1. Backspin magnitude against loft angle for a) Ball A and Ball B and b) Ball C in dry conditions for plates U Base and V Base.

is no significant difference between backspin generated for U- and V- grooves for any ball type. In simulated wet conditions the backspin rate for all ball types decreased monotonically from around 2500–3500 rpm to around 1500–2000 rpm as loft angle increased from 50 to 70°. High loft angles give reduced normal forces on impact resulting in lower ball compression, which may reduce the differences seen between ball types. Hence, work in this paper focuses on U-shaped grooves at a loft angle of around 35°.

Figure 2a shows that, in dry conditions; the backspin rate for all balls is increased by around 1000–1500 rpm for the plate with deep U-grooves (U Deep) compared with that having shallower U-grooves (U Base). This suggests a greater interaction of the ball with the grooved plate for U Deep compared with U Base, due to a greater depth for cover penetration. This would be consistent with the cover deforming to fill the groove in the U Base condition and would correspond to all ball types approaching a limiting friction co-efficient.

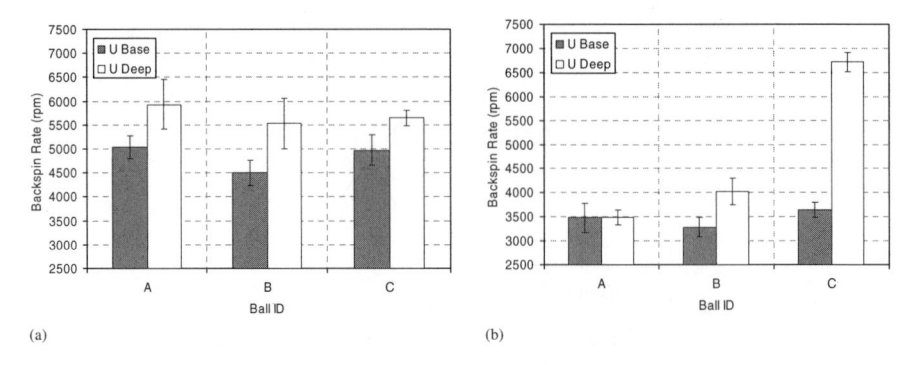

Figure 2. Backspin rate as a function of ball type, for all ball types impacting plates U Base and U Deep at an effective loft angle of 35° with an inbound velocity of 30 m s^{-1} in both a) dry and b) wet impacting conditions.

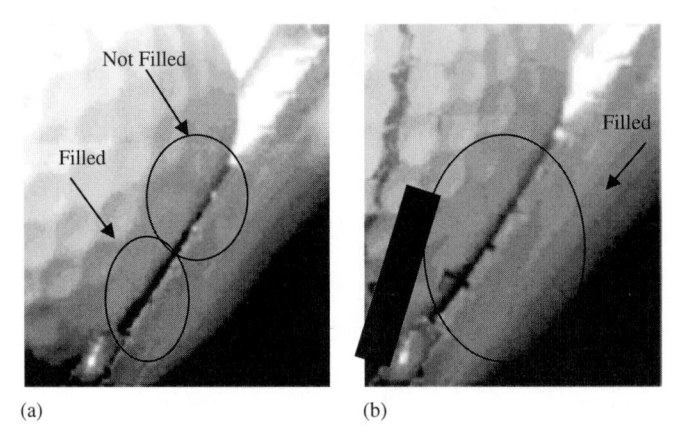

Figure 3. Shows the level of interaction for a) Ball A and b) Ball C impacting plate U Base at an inbound speed of 30 m s^{-1} with an effective loft angle of 35°.

Figure 3 shows that this hypothesis is partially correct as there is complete filling of some grooves during impact for the balls with the U Base plate. However, as seen in Figure 3a, not all grooves are filled across the ball/plate contact area for Ball A, whereas they are for Ball C, Figure 3b. This would suggest greater effective friction for Ball C than Ball A and in turn an increased backspin rate, but, as seen in Figure 2a, this is not the case. The two-dimensional imaging of Figure 3, does not indicate the width of the ball filling the groove, which will affect the torque acting on the ball and may be greater for more compressible balls, such as Ball A. In dry U base conditions it would appear that the contact area and groove filling combine to give similar torques for Balls A and C with slightly smaller values and a lower backspin rate for Ball B. Deepening the grooves (U Deep) should have little effect on contact area and so the increase in backspin rate would indicate that the covers can deform further into the groove increasing the ball/plate interaction and so effective friction co-efficient. Balls A, B and C give slightly different increases in backspin, despite the same change in groove depth.

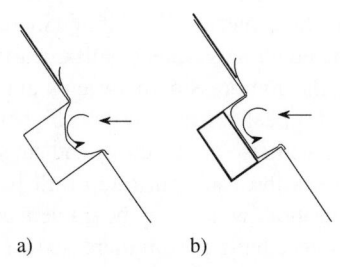

Figure 4. Schematic diagram of the penetration of a three-piece golf ball into a deep U-groove, in a) dry conditions and b) simulated wet conditions showing the difference in sidewall interaction due to the presence of a groove bottom.

Balls A and B exhibit increases in backspin rate of around 1000 rpm, this would suggest greater deformation of Balls A and B into the U Deep grooves compared with the U Base grooves. The increased penetration into the U Deep grooves for Ball C is less giving a smaller increase in backspin rate ($\approx$ 650 rpm). These trends do not correlate with any individual Shore D value, but do follow the Shore D differences between cover/sub surface layer. It is expected that cover thickness will also play a part and investigation into combined thickness/hardness parameters is ongoing. High-speed video imaging confirms that none of the balls fill any of the U Deep grooves during impact.

In simulated wet conditions, the trend with groove depth is reversed with Ball C showing significant increases in backspin rate ($\approx$ 3000 rpm) for U Deep compared with U Base. Ball A shows no change in backspin rate between the two groove types, whilst Ball B shows a smaller increase ($\approx$ 700 rpm) than in the dry. This behaviour again correlates qualititatively with the differences in Shore D between ball layers, but in the reverse order to dry conditions. The contact area is unlikely to change between dry and wet conditions, but the use of a moistened layer of newsprint means that material will be forced into the grooves along with the cover. In this case, Ball A would be expected to loosely compress the newsprint into the grooves so that the contact area and depth are the same, leading to a similar torque as seen in the dry conditions. The cover of Ball B would compress the newsprint into the groove more, so that penetration is increased for U Deep compared to U Base but not to the same extent as in dry conditions. The softer cover of Ball B compared with Ball A would be consistent with this behaviour. The increased strain acting on the softer cover for Ball C due to a hard mantle layer would suggest that the cover is compressed (with the newsprint) into the groove to a greater extent due to strain not being dissipated through the core (as with Ball B). The cover for Ball C will then be strained laterally (Figure 4), resulting in a greater interaction with the groove sidewall (due to the presence of a groove bottom) and hence torque than in dry U Deep conditions.

4 Conclusions

The backspin rate of a number of commercially available multi-piece golf balls have been determined impacting angled plates (with different groove shapes and depths). The results have indicated a number of parameters which influence the backspin rate of the golf ball:

- Surface condition: the presence of a wet interlayer decreases the backspin rate of all ball types at higher loft angles (50–70°).

- Ball construction: the three-piece ball exhibited increased ball/plate interaction in wet conditions compared with two-piece balls due to an increased lateral strain on the cover induced by the compression between mantle and newsprint. In dry conditions the loft angle for peak backspin rate is higher for balls with a smaller difference in Shore D value between the cover and subsurface layer.
- Groove depth: increases the backspin rate of all balls in dry conditions due to increased cover penetration, which may be influenced by ball compressilbility. In wet conditions three-piece balls exhibit increased backspin due to increased compression of the newsprint into the groove.

References

Cornish J., Monk S., Otto S. and Strangwood M. (2006) Factors Determining Backspin from Golf Wedges. The Engineering of Sport 6, Munich, Springer.

Monk S.A. (2006) The Role of Friction Coefficient on Launch Conditions in High-Loft Angle Golf Clubs. PhD Diss., University of Birmingham.

Sullivan M.J. and Melvin T. (1994) The Relationship Between Golf Ball Construction and Performance. In: Cochran A.J. and Farrally M.R. (Eds.), *Science and Golf II – The Proceedings of the World Scientific Congress of Golf*, E & FN Spon, St Andrews, Scotland.

R & A Rules Ltd (2004) The Rules of Golf, Williams Lea Group.

R & A/USGA (2006) Interim Report: Study of Spin Generation.

R & A/USGA (2007) Second Report on The Study of Spin Generation.

EXPERIMENTAL AND FINITE-ELEMENT ANALYSES OF A GOLF BALL COLLIDING WITH SIMPLIFIED CLUBHEADS

K. TANAKA[1], H. OODAIRA[1], Y. TERANISHI[2], F. SATO[2] & S. UJIHASHI[1]
[1]*Tokyo Institute of Technology, Tokyo, Japan*
[2]*Mizuno Corporation, Osaka, Japan*

The purpose of this study is to construct a Finite-Element (FE) model which can accurately simulate the behaviour of a golf ball upon impacting with clubheads and to investigate the influence of different impact conditions on the rebound behaviour of a golf ball and clubheads after impact, by conducting FE analyses of the ball colliding with simplified clubheads. The simplified clubheads were designed based upon the mass, volume and position of the centre of gravity of commercial clubheads; 4 titanium alloy circular hollow bodies of constant mass, with increasing loft angle, were manufactured. The FE models of the clubheads, with linear elasticity, were constructed from 8-node solid and shell elements. The FE model of the ball consisted of 8-node solid elements, and the material model was expressed as a hyper-elastic/viscoelastic model. Impact experiments were also conducted to confirm the accuracy of the FE models, by comparing the results of the experiments to those of the FE analyses. The results of the impact simulations closely matched the experimental results. The impact behaviours were analysed by varying the impact point of the ball colliding with the clubheads. The impact point where the rebound velocity of the ball became the largest was located below the sweet spot (on the sole of the clubhead), and this impact point depended on both the position of the sweet spot and the loft angle.

1 Introduction

High-performance golf equipment, especially drivers with a high coefficient of restitution (COR), have contributed to increasing the driving distance and improving the trajectory of golf balls upon impact. Concerned that this high-tech equipment is outdating golf courses and ruining the game, the Royal and Ancient Golf Club of St Andrews together with the United States Golf Association have determined the requirement for players and manufacturers to conform to new rules, setting an upper limit for the COR, thereby disallowing the use of drivers with high CORs. Therefore FE models whose CORs can be precisely estimated are in demand in order to create high-performance golf clubs which can meet these new rules.

The regulation of the COR was a turning point in the development of driver heads; the design and development of driver heads has been led previously only by the launch condition of the golf ball after impact, most notably the launch angle and spin rate of the ball, as well as the launch velocity. The initial launch condition of the ball depends greatly on the interactions between the mechanical properties of the clubhead and the ball. Therefore, it is important to grasp precisely what factors affect the behaviour of the golf ball during and after impact with the clubhead.

The purpose of this study is to construct a finite-element (FE) model which can accurately simulate the behaviour of golf impacts, and to investigate the influence of different

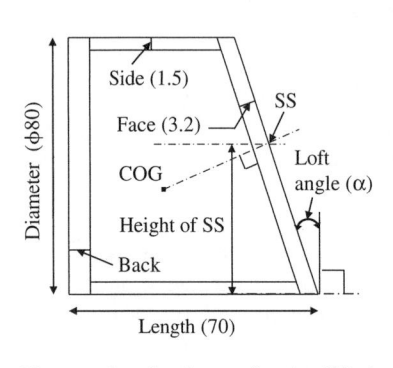

Table 1. Mechanical properties of simplified clubheads.

	Model 0	Model 10	Model 20	Model 30
Loft angle [°]	0	10	20	30
Mass [g]	214	217	214	222
Volume [cm^3]	352	316	279	236
Ratio of SS	0.50	0.55	0.58	0.58

Commercial number 1 wood (driver);
Mass: 180–210 g
Volume: 380–460 cm^3
Ratio of SS: 0.51–0.63

Figure 1. Section of simplified clubhead.

impact conditions on the rebound behaviour of a golf ball and clubheads after impact, by conducting FE analyses of the ball colliding with simplified clubheads.

2 Impact Experiment between the Golf Ball and the Simplified Clubheads

2.1. *Design and Manufacture of the Simplified Clubheads*

The clubheads, which were simplified in their shape, were designed and manufactured to investigate the interactions between the mechanical properties of the clubhead and the rebound behaviour of the ball after impact, as shown in Figure 1 and Table 1.

The simplified clubheads were designed based upon the mass, volume and position of the centre of gravity (COG) of commercial clubheads, so that a typical golf impact would be realized. 4 titanium alloy circular hollow bodies of constant mass, with increasing loft angle, were manufactured. This constant mass with changing loft angle was made possible by altering the thickness of the back face (1.5–3.0 mm). The ratio of the sweet spot (SS) was defined as the ratio of the height of the SS to the diameter of the clubhead. This ratio was used as the mechanical property representing the position of the COG of the clubheads.

2.2. *Experiments of Ball/Clubhead Impacts*

Experiments were conducted in order to obtain a typical impact between the ball and the clubhead, as shown in Figure 2. The three-piece type ball was fired from an air gun, in a strain-free and non-rotational condition, and collided with the face of a freely-supported clubhead. The impact velocity of the ball varied (30, 40, and 50 m/s). The initial impact position was the SS of the clubhead, and was then varied within the range of 20 mm from the SS at ±5, ±10 mm intervals.

The behaviour of the ball and clubheads during and after impact was recorded using a high-speed camera (20,000 fps, shutter speed 1/100,000 s). The velocities of incidence and rebound, rebound angle and spin rate of the ball, and the launch velocity and angle of the clubhead were ascertained from the photographs taken by the high-speed camera, as shown

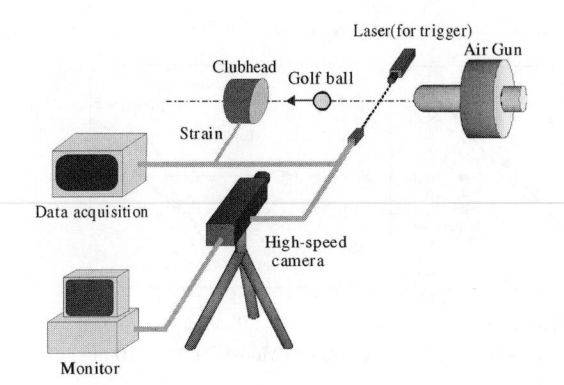

Figure 2. Experimental apparatus.

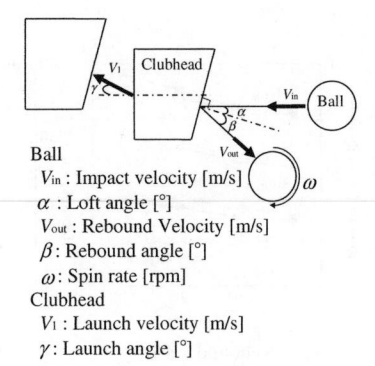

Ball
V_{in} : Impact velocity [m/s]
α : Loft angle [°]
V_{out} : Rebound Velocity [m/s]
β : Rebound angle [°]
ω : Spin rate [rpm]
Clubhead
V_1 : Launch velocity [m/s]
γ : Launch angle [°]

Figure 3. Impact behaviour of ball and clubhead as ascertained from high-speed camera.

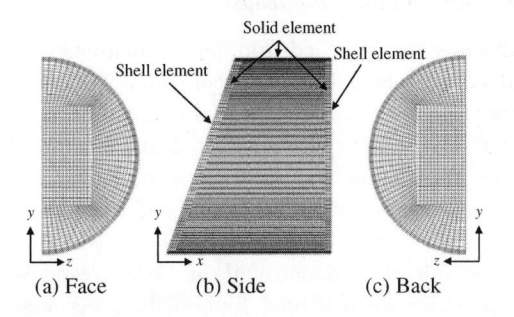

(a) Face (b) Side (c) Back

Figure 4. FE model of a simplified clubhead (Model 20).

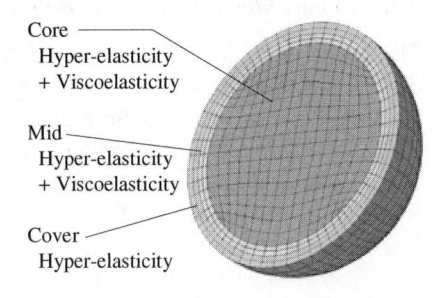

Figure 5. FE model of golf ball.

in Figure 3. In addition, the strain responses of the impact and back faces were measured by strain gauges affixed to the clubhead.

3 FE Analyses of Ball Colliding with Simplified Clubheads

3.1. *Construction of the Ball and Clubhead FE Models*

FE models of the clubheads, with linear elasticity, were constructed from 8-node solid and shell elements, as shown in Figure 4. The impact simulations were carried out using a half model, with plane symmetry applied.

The FE model of the ball, which consisted of an outer cover, mid and core (same as those used in the experiments), was constructed from 8-node solid elements, as shown in Figure 5. The material model of the cover was expressed as a hyper-elastic model, whereas the material models of the mid and core were expressed as a viscoelastic model with hyper-elasticity (Tanaka et al., 2006). In these simulations, the Mooney-Rivlin model was used for the hyper-elastic model, and the three-element model was used for the viscoelastic model (The Japan Research Institute, 2003).

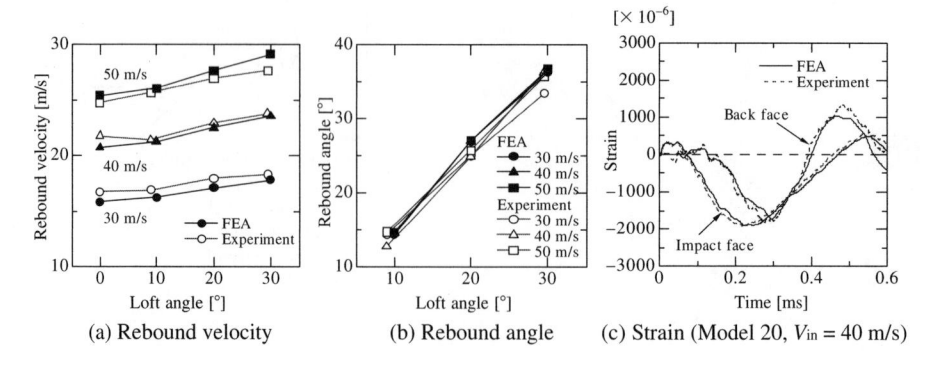

Figure 6. Results for the experiments and the FE analyses of the constructed models.

3.2. Impact Simulations of the Ball Colliding with the Clubheads

The impact behaviour of the ball and clubheads was obtained from the simulations conducted with these FE models, and the results of the FE analyses were compared with those from the experiments to confirm the accuracy of the FE models. The commercial FE code LS-DYNA was used for the computer simulations. Friction between the ball and the impact face of the clubhead was expressed by Coulomb friction, and the friction coefficient was found to be 0.3, based on the previous work conducted by Nakasuga and Hashimoto (Nakasuga & Hashimoto, 1997).

Figure 6 shows the results for the rebound velocity, rebound angle and strain response (at the impact and back faces for Model 20, at a velocity of 40 m/s), for both the experiment and the FE analyses, for an impact at the SS of the clubhead. The simulation results for the rebound velocity and angle of the ball (Figure 6a and b) closely match with the experimental results for a range of impact velocities (30–50 m/s) and loft angles (0–30 degrees). The simulation results for the strain response of the clubhead (Figure 6c) also agree with the experimental results. This indicates that the simulation models are able to accurately express the impact behaviour of the ball and clubhead in the typical range of a ball/clubhead impact.

4 Effects of the Impact Condition on the Behaviour of the Ball and Clubhead

The impact behaviours were analysed by varying the impact point of the ball (using the FE models) to investigate the effects of the impact condition on the launch condition of the ball. Figure 7 shows the influence of changing the impact point of the ball on the rebound velocity, rebound angle and spin rate, at a velocity of 50 m/s. The x-axis represents the distance between the SS and the impact point. "0" represents an impact point agreeing with the SS, while a negative x-axis value means that the impact point is situated on the sole rather than the SS. A positive x-axis value means that the impact point is situated on the crown of the clubhead.

The rebound velocity of the ball tends to increase with the loft angle, regardless of any change to the impact point (Figure 7a). The surface area of the impact face increases with the loft angle, and the impact face becomes more flexible as a result. Deformation of the

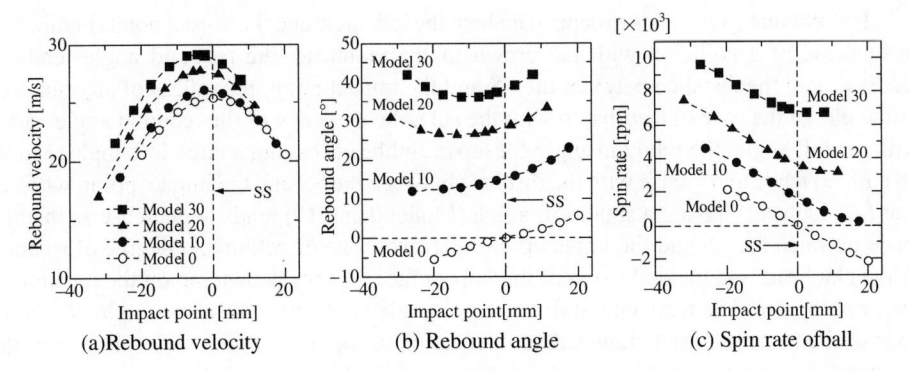

Figure 7. Results for the FE analyses of the collision between the ball and clubheads with a changing impact point ($V_{in} = 50$ m/s).

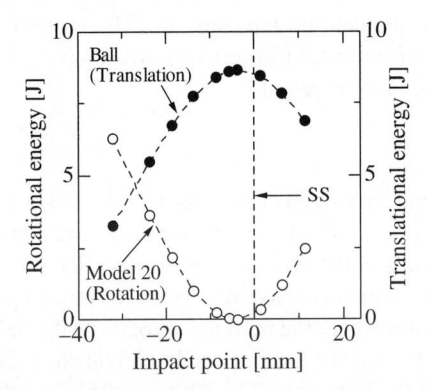

Figure 8. Relationship between the translational energy of the ball and the rotational energy of Model 20 with a changing impact point ($V_{in} = 50$ m/s).

ball during impact is suppressed as the face becomes more flexible, and the energy loss of the ball due to viscosity decreases. As a result, the total energy loss decreases because the ball material is more energy-absorbent than the metallic material of the clubhead. Therefore, the rebound velocity increases with the loft angle.

The impact point, where the rebound velocity becomes the largest, varies with the clubheads and moves towards the sole rather than the SS as the rebound angle increases. Figure 8 shows the relationship between the translational energy of the ball and the rotational energy of Model 20 after impact, with a changing impact point. This result indicates that the translational energy of the ball negatively correlates with the rotational energy of the clubhead, and that the tendency of the translational energy agrees with that of the rebound velocity (Figure 7a). Decreasing the clubhead's rotational energy leads to an increase in the rebound velocity of the ball, and it is estimated that the impact point with the least rotation of the clubhead is related to the loft angle and the position of the SS. Therefore, the impact point where the rebound velocity becomes the largest depends on the loft angle and the position of the SS.

The rebound angle tends to depend on both the loft angle and the impact point (Figure 7b). In the case of a collision with the crown of the clubhead, the rebound angle tends to increase with the distance between the SS and the impact point, regardless of any range of loft angle. In the case of a collision with the sole, the tendency of the rebound angle varies with the loft angle; the rebound angle for those clubheads having a large loft angle (Model 20 and 30) tends to increase with the distance between the SS and the impact point, whereas that for clubheads having a small loft angle (Model 0 and 10) tends to decrease as the distance between the SS and the impact point increases. The direction and degree of rotation of the clubhead, and the deflection of the impact face all vary depending on the position of the impact point. The tendency of the rebound angle is assumed to be caused by the influence of the rotation of the clubhead and deflection of the impact face. The spin rate of the ball tends to decrease as the impact point moves from the sole towards the crown (Figure 7c). This tendency seems to be caused by a gear effect after impact. It is estimated that ball sliding and ball rolling occurs on the face during impact. Deflection of the impact face, in addition to ball deformation, also occurs and this complex interaction, as well as the gear effect, is believed to be dependent on the spin rate. Therefore, it is necessary to quantify this complex behaviour during impact in order to investigate the relationship between the clubhead and the spin rate in more detail.

5 Conclusions

FE models of a golf ball and simplified clubheads, which can simulate the behaviour of typical golf impacts, were constructed. The results for the impact simulations closely matched with the experimental results in the typical range of a ball/clubhead impact.

The impact behaviours were analysed by varying the impact point of the balls using the FE models. The impact point where the rebound velocity of the ball became the largest was located below the SS (on the sole of the clubhead), and this impact point depended on both the position of the SS and the loft angle. The rebound angle and the spin rate seemed to be dependent mainly on the loft angle and the deflection of the impact face.

References

Nakasuga M. and Hashimoto R. (1997) Measurement of tangential force of golf ball impact (in Japanese), Proceedings of the Japan Society of Mechanical Engineers, 97-10-2, 42–45.

Tanaka K., Sato F., Oodaira H., Teranishi Y., Sato F. and Ujihashi S. (2006) Construction of the finite-element models of golf balls and simulations of their collisions, *Proceedings of the Institution of Mechanical Engineers, Part L: Journal of Materials: Design and Applications*, 220/1, 13–22.

The Japan Research Institute, Limited (2003) *LS-DYNA ver.970 USER'S MANUAL Volume II* (in Japanese).

THE DYNAMIC CHARACTERISTIC OF A GOLF CLUB SHAFT

M. SHINOZAKI[1], Y. TAKENOSHITA[2], A. OGAWA[2], A. MING[2], N. HIROSE[3] & S. SAITOU[3]

[1]*Setagayaizumi High School, Tokyo, Japan*
[2]*The University of Electro-Communications, Tokyo, Japan*
[3]*Mamiya-OP Co., Ltd., Saitama, Japan*

Jerk, time derivative of acceleration, is a physical value which evaluates feeling by human. Toward the quantitative evaluation of dynamic characteristic of four axes golf clubs, jerk is proposed as a new performance index in this paper. The comparison between the four axes club and conventional club on jerk value and transfer function are investigated by experiments. Experimental results show that there exists difference between the four axes club and conventional club.

1 Introduction

A CFRP shaft for the golf club is laminated structure by carbon fiber prepreg sheets, and the characteristic of it is variable by the quality, placement and a pair of orientation angles of fibers. The CFRP shaft composed of a three axes fabric prepreg sheet has higher torsion and bending rigidity than that of a conventional shaft ("Normal" shaft) (Matsumoto & Nishiyama, 2000). Now, the golf club shaft composed of four axes fabric prepreg sheets, which will be called as "4-Axes" shaft in this paper, is used for further rigidity and lightness (Gakken, 2005). Many golfers take in the difference of the dynamic characteristic between "4-Axes" and "Normal", even if the static characteristic is the same. But the quantitative evaluation is a problem in the design and the fitting of clubs.

By now, the quantitative evaluation using FEM for normal golf club shafts has been done by many researchers (e.g. Friswell *et al.*, 1998). In the case of "4-Axes", it is difficult to evaluate the dynamic characteristic by the method, because the golf club has a complicated structure with different kinds of composite materials. Therefore, the aim of this study is to clarify the difference of dynamic characteristic between "4-Axes" and "Normal" by experiments. We pay our attention to jerk, which is said to be concerned with ride comfort and sensitivity of a person, and compare four golf clubs with the evaluation of the dynamic characteristic of these shafts by experiments with the step base excitation.

2 Jerk and the Dynamic Characteristic

2.1. Jerk

Jerk is the derivative of acceleration with respect to time. In comparison with displacement, velocity and acceleration, jerk shows a big value at a state of transient motion such as at the start or stop of a motion. Jerk is one of the quantities of sensitive physics for the human body.

2.2. Related Studies about the Dynamic Characteristic by Jerk

For improvement of ride comfort of vehicles such as a train, a car and an elevator, jerk is used for an evaluation function for motion control with acceleration. The related studies are as follows.

Hitachi Ltd. suggested the effectiveness of the control method using a jerk sensor for a "Motion control system", and pointed out that jerk feedback can improve control performance (Tsuchiya et al., 1997; Yamakado & Kadomukai, 1998). Mazda motors Ltd. measured few values of the jerk that differentiate values of accelerometer on the floor of a car and developed reduction technology of a clutch engagement shock using an evaluation function including both jerk and acceleration (Hujikawa et al., 2002).

On the other hand, measurements of vibration characteristic through behavior of the shaft by strain gauges, behavior of a swinging club by DLT method (high speed video camera photography) and modal analysis have been performed (Matsumoto & Nishiyama, 2000; Friswell et al., 1998; Koizumi et al., 1998; Takemura et al., 1996), but the evaluation of dynamic characteristic by jerk has not been performed yet.

2.3. Influence of Jerk on Dynamic Characteristic of a Golf Club

Suppose that the characteristic of a shaft has an element depending on jerk by a special material or feedback control. We define the element as Eq. (1). By substituting it into the motion equation of a club shown in Eq. (2), Eq. (3) is obtained.

$$F_0 = k_j \dddot{x} \qquad (1)$$

$$m\ddot{x} + c\dot{x} + kx = F - F_0 \qquad (2)$$

$$\dot{A}/F = s^3/(k_j s^3 + ms^2 + cs + k) \qquad (3)$$

By Routh-Hurwitz criterion for stability, we can obtain the design condition of the jerk coefficient, which makes the club with stability, as Eq. (4).

$$① \; k_j, m, c, k < 0, \quad ② \; H_2 = \begin{vmatrix} m & k \\ k_j & c \end{vmatrix} = mc - k_j k > 0 \quad \therefore 0 < k_j < \frac{mc}{k} \qquad (4)$$

In the case of a real club, there exist infinite vibration modes. For simplicity, consider a model with a concentrated mass and an equivalent rigidity. The parameters of two clubs which are used for simulation and experiment are shown in Table 1.

By substituting the values of parameters of 4-Axes 1-Layer club in Table 1 into Eq. (3) and changing the jerk coefficient within the stable range shown in Eq. (4), step response can be derived. The jerk output is shown in the left side of Figure 1 and the sum of squared jerk is shown in the right side of the figure. As a result, by increasing the jerk coefficient, the sum of squared jerk is in a tendency to decrease. It shows that the jerk can be adjusted by introducing an element related to the jerk. But it should be noticed that the sum of squared acceleration, the sum of squared velocity and the sum of squared displacement will increase if the jerk coefficient is increased.

Table 1. The specification of a pair of clubs.

		4-Axes 1-Layer club	Normal 1-Layer club
Natural frequency	f_n [Hz]	4.1125	4.05
Jerk coefficient	k_j [N/(m/s^3)]	0.0010628	0.0009755
Mass	m [kg]	0.309	0.31
Damper constant	c [N/(m/s)]	0.7196	0.6537
Spring constant	k [N/m]	209	208

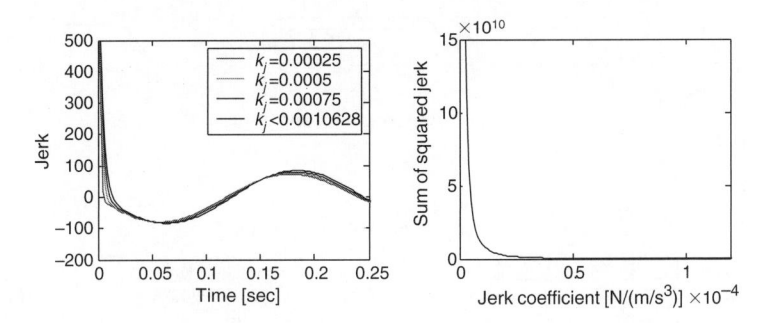

Figure 1. Simulation results of jerk and the sum of squared jerk in the step response.

3 Experiments

3.1. Experimental Setup and Evaluation Method

As shown in Figure 2, the golf club is fixed to a rotation motor by the grip part of a golf club. And the base excitation is realized by controlling the motion of the motor. A force sensor is used to measure the force and moment between the motor and the grip part. And accelerometers are set to the grip part and head to measure the input and output accelerations to the shaft.

Jerk is obtained by differentiating a value of the accelerometer with respect to time on each part.

3.2. Clubs for Measurements

In experiment, four kinds of golf clubs in total are used. And the clubs can be classified roughly to two sets. One set is the group of a club using one layer of four axes fabrics ("4-Axes 1-Layer" club) and a normal club with same static characteristics ("Normal 1-Layer" club). The other set is a group of a club using two layers of four axes fabrics ("4-Axes 2-Layers" club) and a normal club with same static characteristics ("Normal 2-Layers" club).

The grip part is given with a step excitation of 5° by the motor. The excitation direction is assigned as X and the plumb upswing direction is assigned as Z. The measurement lasts to 6.4 s with 16,384 points of data. Ten measurements are performed for each club.

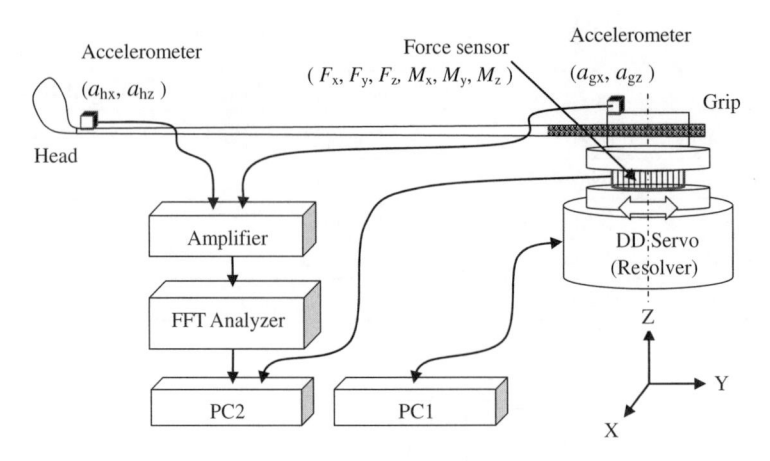

Figure 2. Experimental apparatus.

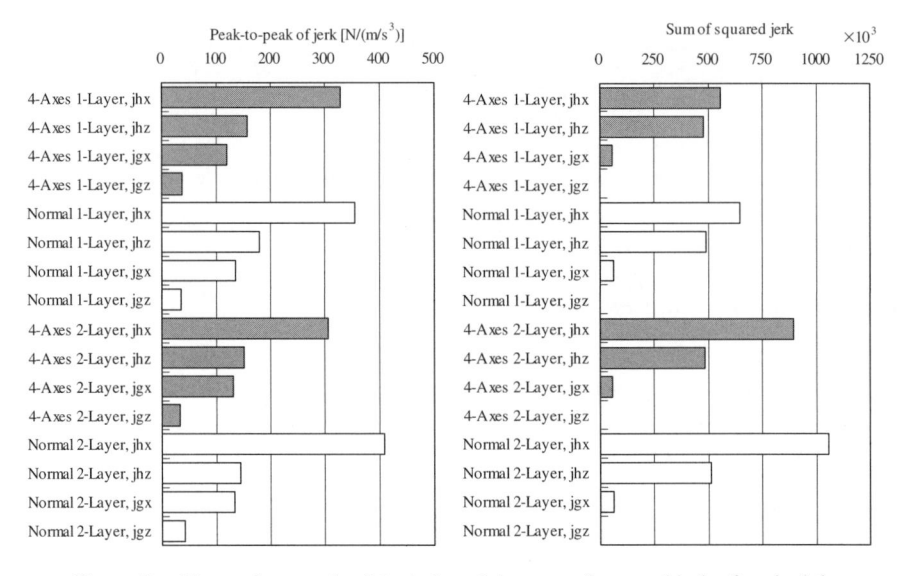

Figure 3. The peak-to-peak of the jerk and the sum of squared jerk of each club.

4 Results

4.1. *The Evaluation by Value of Jerk*

As the experimental results, the peak-to-peak value of jerk and the sum of the squared jerk are shown in Figure 3. In the figure, "g" means the grip part; "h" means the head side. In comparison of the peak-to-peak value of jerk and the sum of the squared jerk, "4-Axes" clubs are almost smaller than "Normal" clubs for each layer group.

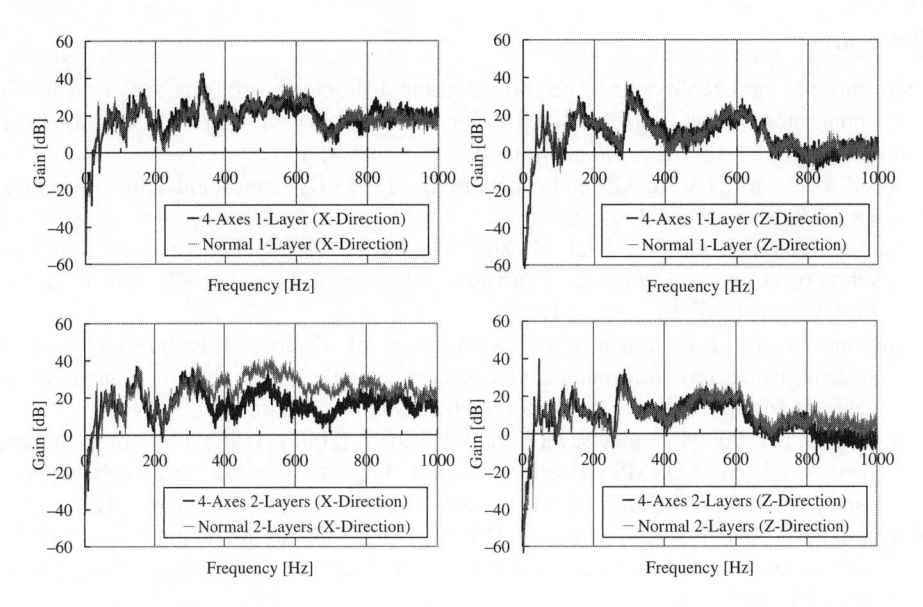

Figure 4. The transfer function j/F of four clubs (upper: one layer group, lower: two layers group, left: X-direction, right: Z-direction).

4.2. *The Transfer Function*

The transfer function j/F, j/F' and a/F are inspected for whether it is effective as an evaluation method of sensitivity. As the transfer function j/F of the head side, Figure 4 shows an experiment result of four clubs. "4-Axes 1-Layer" club and "Normal 1-Layer" club are at the upper, "4-Axes 2-Layers" club and "Normal 2-Layeres" club are at the lower.

"4-Axes 2-Layes" club shows lower gain from about 300 Hz in X-direction and from about 650 Hz in Z-direction than "Normal 2-Layeres". About j/F', the result was similar to j/F, and same as a/F.

5 Conclusion

Toward the quantitative evaluation of dynamic characteristic of four axes golf clubs, jerk is proposed as a new performance index in this paper. The comparison between the four axes club and normal club on jerk value and transfer function are investigated by experiments and following results are obtained.

1. Four axes fabric clubs indicated smaller value of jerk than normal clubs.
2. Lower gain was observed for the club using two layers of four axes fabric for force-jerk transfer in higher frequency band.

Further work about the relation between the results and human feeling will be done.

References

Matsumoto K. and Nishiyama T. (2000) Vibrational Properties of Triaxial Woven Fabric Composites for Golf Club Shafts. *JSME Symposium on Sports Engineering*, 00–38, 91–94.

Gakken Co., Ltd. (2005) *Encyclopedia Golf Tune up 2004*, p.52.

Friswell M.I., Mottershead J.E. and Smart M.G. (1998) Dynamic models of golf clubs, *Sports Engineering*, 1/1, 41–50.

Tsuchiya T., Yamakado M., Ishii M. and Sugano M. (1997) Fundamental Study on Vibration Control Using the Derivative of Acceleration, "Jerk", Sensor. *JSME Transactions C*, 63/614, 112–117.

Yamakado M. and Kadomukai Y. (1998) A Study of Motion Evaluation and Control Systems using Jerk Information (1st Report, A Jerk Sensor and its application to Vehicle Motion Control System), *JSME Transactions C*, 64/619, 135–141.

Hujikawa T., Miwade H., Yanagisawa M. and Katsui T. (2002) Analysis of clutch engagement shock by ADAMS/Driveline (online), http://www.mscsoftware.com/support/library/conf/adams/japan/2002/papers/MAZDA.pdf (Accessed: 1 March 2007).

Koizumi T., Tsujiuchi N. and Murai N. (1998) A Study of the Matching Characteristics between Dynamic Performance of Golf Club. *JSME Symposium on Sports Engineering*, 98/31, 87–90.

Takemura S., Matsumoto T., Horii H. and Mouri M. (1996) Torsional Vibration Property of PITCH-based CFRP Golf Club Shafts. *JSME Symposium on Sports Engineering*, 96/20, 198–192.

ADVANCED MATERIALS IN GOLF DRIVER HEAD DESIGN

H.G. WIDMANN[1], A. DAVIS[1], S.R. OTTO[2] & M. STRANGWOOD[1]

[1]*Department of Metallurgy and Materials, University of Birmingham, Edgbaston, UK*
[2]*R & A Rules Ltd, St Andrews, Fife, Scotland*

Bulk metallic glasses, such as Vitreloy 105 ($Zr_{52.5}Cu_{17.9}Ni_{14.6}Al_{10}Ti_5$ at.%) offer high yield strengths combined with low elastic moduli in the region of 1520 MPa and 107 GPa respectively. These are superior to those of titanium-based alloys, which are currently used in the manufacture of golf driver heads. The production of complex shapes (such as the crown) and their joining to fabricate hollow driver heads requires thermo-mechanical exposure. The crystallisation kinetics of Vit 105 were characterised by isothermally annealing fully amorphous samples in the supercooled liquid region (420–435°C). The Avrami exponents, based on Differential Scanning Calorimetry (DSC) results, were determined as 1.6 to 2.3 and 3.1 to 4.8 for the first and second events respectively. This correlates to nucleation of nanocrystallites and initial growth followed by growth or polymorphic transformation (consistent with tetragonal Zr_2Ni phase from X-ray diffraction analysis) in the second event. The Avrami exponents derived from the change in hardness agree for the first event, but not for the second stage, where morphology of the larger crystalline regions has a greater effect on mechanical properties than DSC energy release.

1 Introduction

Golf driver heads are generally made up of three components being the face, crown and sole, relating to the impact site, top and bottom respectively. The crown and sole are separated by a skirt for some constructions. Previous research (Strangwood, 2003) has shown that the performance of a driver head is mainly determined by the elastic deformation of the crown, with the properties of the face and sole playing a lesser role for current designs of clubs. The joint between the face and crown has been determined to be important for deformation transfer (Adelman et al., 2006). Non-optimal joints (welds in the previous research) were defined as those giving greater constraint, either geometrically (wide/thick) or mechanically (higher modulus). A better performing driver hence needs a suitable crown material connected to the face with a joint having similar dimensions and properties.

The main requirements of a crown material are low elastic modulus, E, combined with a high yield strength, σ_y, maximising the performance index, σ_y/E. Metallic glasses offer superior performance indices to crystalline materials currently used in commercial driver heads, such as Ti-6Al-4V (alpha + beta – Ti) [σ_y/E : 9] or metastable beta – titanium [σ_y/E : 11]. Early metallic glasses were made by rapid quenching at cooling rates of 10^5–10^{6}°C s^{-1}, limiting sample thickness and hence their application. In recent years, the development of multi-component alloys has permitted the production of amorphous alloy sections in bulk (maximum thickness up to 10 cm). Amorphous zirconium-based alloys offer a highly suitable performance index in the region of 19. These alloys have previously been used as face inserts in driver heads, which were determined to have a poor performance due to the poor adhesive bond (large constraint) between the face and the crown, which prevented deformation of the crown resulting in a poor performance.

Production of a crown from Zr-based metallic glasses (such as Vit 105) would require shaping of cast thin plates. Homogeneous deformation of metallic glasses has been predicted to take place at 0.6 of the glass transition temperature, T_g (Argon, 1979). The deformation behaviour of Vit 105 has been determined to be strain rate dependent, behaving like a Newtonian fluid at low strain rates ($<10^{-3} s^{-1}$) and becoming non-Newtonian at high rates ($>10^{-3} s^{-1}$) being associated with crystallisation (Nieh, 2002). Shaping a plate of Vit 105 into a crown requires low strain rates to minimise crystallisation. The present work describes the crystallisation kinetics for Vit 105 under thermal exposure and describes a suitable processing window for the shaping-procedure of Vitreloy 105 ($Zr_{52.5}Cu_{17.9}Ni_{14.6}Al_{10}Ti_5$ at. %).

2 Experimental

Vit 105 $Zr_{52.5}Cu_{17.9}Ni_{14.6}Al_{10}Ti_5$ ingots were prepared by arc melting of high purity elements (Zr: 99.94, Cu: 99.99, Ni: 99.98, Al: 99.98, Ti: 99.80%) in a titanium-gettered argon atmosphere on a water-cooled copper hearth. Ingots of 9 to 10 grams mass were prepared. An initial ingot was prepared by melting the appropriate amounts of zirconium and nickel to reduce the required heat of melting due to the low vapour pressure of aluminium. These ingots were re-melted with the corresponding amounts of copper, titanium and aluminium and turned over three times. For the last melting procedure, the alloy was heated for 120 seconds to ensure a homogeneous melt. The power was then halved for a further 30 seconds followed by switching off the arc allowing rapid cooling to room temperature. The cooling rate was determined using a Minolta-Land Cyclops 52 optical pyrometer (emissivity factor: 0.34), to be $27°C\ s^{-1}$ between 700–600°C. The cross sectional area of the ingot was determined to consist of a crystalline layer close to the copper hearth and a large amorphous area fraction (0.8) at the centre of the ingot. Ingots were mounted in Durofix-2 and cut into 1 mm thick slices using a SiC cutting blade on an Accutom-50 operating at 3000 rpm and a feed rate of 0.030 and $0.020\ mm\ s^{-1}$ through the Durofix-2 and ingot respectively. Crystalline areas were removed from each slice leaving fully amorphous samples. Annealing of amorphous samples was carried out in a Netzsch Pegasus 404C differential scanning calorimeter (DSC) under a flowing argon atmosphere (100 ml min^{-1}). The equipment was calibrated by melting In, Sn, Zn, Al, Ag, Au and Ni in alumina pans and recording melting temperatures and enthalpies of fusion. The temperature error within the supercooled liquid region of Vit 105 is $\pm 0.7°C$. T_g and crystallisation temperature, T_x, were determined by heating a sample to 750°C at a heating rate of $10°C\ min^{-1}$. Isothermal annealing was carried out at temperatures of 420, 425, 430 and 435°C. A relation between time and hardness was created for annealing temperatures of 420, 425, 430 and 435°C for approximately 10, 17, 28, 53, 74 and 90% crystallinity. Hardness was measured on a Mitutoyo microhardness machine at a load of 1000 g. The indent dimensions were measured on a Leica microscope using KS 300 software. The elastic modulus, E, was calculated from Equation 1:

$$E = \rho c^2 (1 + v)(1 - 2v)/(1 - v) \tag{1}$$

where ρ is the density, c is the speed of sound and v is Poisson's ratio (assumed 0.36 for BMGs (Szuecs et al., 2001)). The speed of sound was measured using a Panametrics ultrasonic gauge (model 25DL). X-ray diffraction was performed on a Philips X'Pert XRD system operating at 40 kV. 2θ values between 20 and 100° were examined for an incident radiation wavelength, λ, of 1.5406 Å (Cu K_α). Transmission electron microscopy (TEM) samples were prepared by the ion beam thinning method. TEM was performed in a Philips CM 20 operating at 200 kV.

3 Results and Discussion

The mechanical properties of Vit 105 were determined to be a Vicker's hardness of 505 and an elastic modulus of 107 GPa in the as cast amorphous state resulting in a σ_y/E performance index of 14. The fully amorphous nature of the samples was determined using X-ray diffraction and transmission electron microscopy as shown in Figure 1.

T_g and T_x were determined to be 398 and 446°C, respectively. The DSC trace for different ingots produced in the present study is shown in Figure 2, showing the consistency of the amorphous phase formation for different ingots.

The processes occurring during the early crystallisation stages of Vit 105 are still unclear. Some authors having quoted spinodal decomposition into two amorphous phases prior to nucleation and growth of crystalline phases characterised by in situ small- and wide angle x-ray scattering experiments (Wang et al, 2003), whereas others determined primary nano-crystallisation without previous phase separation by small angle x-ray scattering and transmission electron microscopy (Revesz et al., 2001). Kuendig et al. determined the nucleation of Ti-enriched nanocrystals in the first crystallisation event followed by the nucleation of tetragonal Zr_2Ni. Despite the uncertainty about the early crystallisation path, authors agree that the main equilibrium phase is tetragonal Zr_2Cu (Nieh et al., 1999).

The isothermal kinetics of Vit 105 were characterised by annealing samples at temperatures ranging from 420 to 435°C. The DSC traces are shown in Figure 3(a). All curves show two exothermic processes occurring, which are faster at higher temperatures. The crystallised area fraction, x, can be determined by integrating the area above the graph as a fraction of the total energy release. The crystallisation kinetics can be described by two constants, k and n, using the Avrami equation:

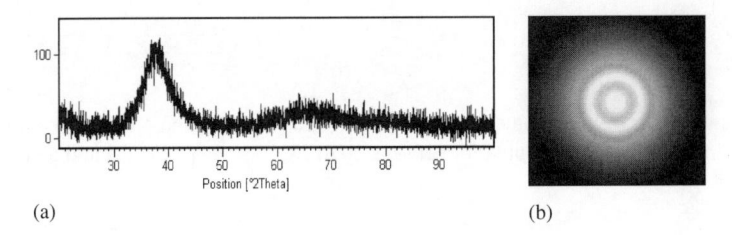

$$1 - x = \exp(-kt^n) \qquad (2)$$

Figure 1. (a) XRD trace showing the lack of long range order, (b) SADP showing amorphous rings.

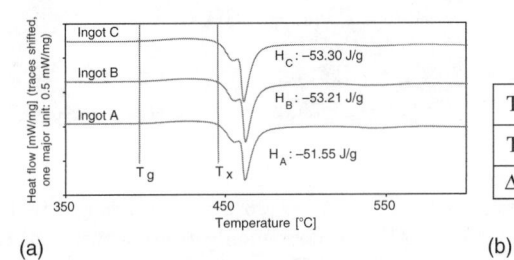

	Ingot A	Ingot B	Ingot C
T_g [°C]	399.4	396.4	398.1
T_x [°C]	446.5	446.9	445.9
ΔH [J/g]	−51.6	−53.2	−53.3

Figure 2. (a) DSC traces for the fully amorphous sections used in the present study, (b) Table showing T_g, T_x and ΔH for respective ingots.

where t is the time. The Avrami exponent, n, for the two different crystallisation events reflects separate mechanisms of crystallisation shown by the different gradients, Figure 3 (b). The first event (value of $n = 1.6$–2.3) reflects mainly nucleation followed by diffusion-controlled growth ($n = 3.1$–4.8) in the second event, which may involve polymorphic transformation. X-ray analysis of the samples revealed no appreciable peaks for the initial crystallisation event due to small crystal size, as shown in Figure 4(a) for a sample having transformed by 26% at an annealing temperature of 420°C. The TEM image in Figure 4(b) shows the small size of the nuclei. TEM – SADP showed slight contrast differences around a largely amorphous ring, Figure 4(c). Due to the small size, EDX analysis was not possible.

The activation energy of crystallisation, E_a, can be determined using the Arrhenius plot, which is based on the DSC data:

$$t(x) = Ce^{\frac{-E_a(x)}{RT}} \tag{3}$$

where C is a constant, R is the ideal gas constant ($8.314\,\text{J mol}^{-1}\,\text{K}^{-1}$) and T is the temperature. The activation energies for the first and second event are 235 to 288 and 207 to 224 kJ mol^{-1}.

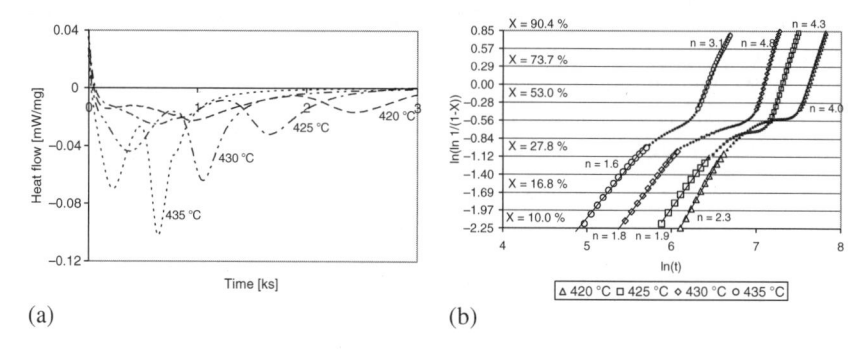

Figure 3. (a) DSC traces for isothermal annealing at 420, 425, 430 and 435°C; (b) Avrami plots for given annealing conditions computed for between 10 to 28% for the first event and 53 to 90% for the second event.

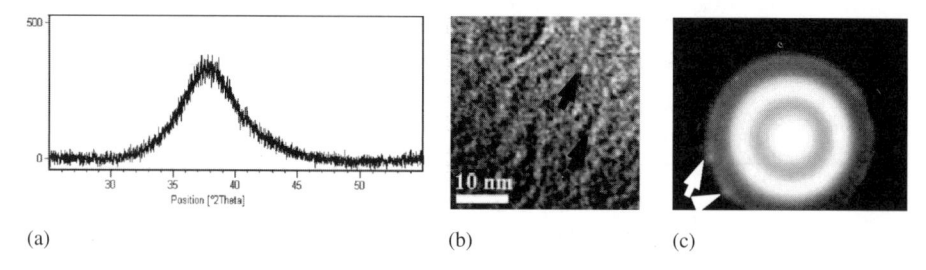

Figure 4. (a) XRD trace for a sample annealed at 420°C for 1140 s (26% transformation); (b) TEM image showing a sample annealed at 430°C for 480 s (23% transformation) (arrows showing nuclei); (c) SADP of respective TEM image (arrows showing contrast differences in the halo resulting from the presence short range order).

The activation energies for diffusion of nickel and aluminium in $Zr_{46.8}Ti_{8.2}Cu_{7.5}Ni_{10}Be_{27.5}$ were determined to be 266 kJ mol^{-1} and 396 kJ mol^{-1}, respectively (Macht et al., 2001). Due to copper and nickel being neighbouring elements on the periodic table and having similar atomic radii, they are expected to have similar activation energies. The activation energy for diffusion of titanium is likely to be similar or higher than that for aluminium due to its larger atomic radius. Nucleation in the first event is expected to be dominated by the diffusion of copper or nickel, and could be of the tetragonal Zr_2 (Ni,Cu) type, which is in contrast with Ti-rich nano-crystals detected by Kundig. The activation energy of the second stage is lower than that determined for any individual diffusion process. This would suggest that development of crystalline regions (with a higher Avrami exponent) in the second crystallisation stage is not controlled solely by the energetics of long range diffusion of a single species. X-ray diffraction analysis of the samples annealed to in second crystallisation event showed the presence of the metastable, tetragonal Zr_2Ni phase (space group I4/mcm, a = 6.486 Å, c = 5.279 Å), which has commonly been observed during the crystallisation of Zr-based metallic glasses containing low levels of oxygen impurity. Ti-, Al- and Cu-depleted crystals of the tetragonal Zr_2Ni phase have been determined for annealing a sample at 420°C for 3000 seconds (Kundig et al., 2005). The formation of the fcc phase of Zr_2Ni (space group Fd3m, a = 12.270 Å) was not detected in the current project and has been linked to the devitrification of oxygen-rich Zr-based glasses (Barrico et al., 2001) causing a reduction in the glass forming ability.

The increase in hardness for the first crystallisation event followed by a lower rate of hardening is shown in Figure 5. Avrami analysis of the hardness data, Figure 6, shows that the exponents, n, (1.3–1.8) agree with the exponents measured using the DSC data for the first event. For this stage of crystallisation the hardness increases largely in line with volume fraction crystallised (to which DSC signal is proportional). The correlation between hardness and volume fraction is lost, however, for the second crystallisation event. For volume fractions greater than 0.4 the hardening curves are shallower, Figure 5, which indicates a change in the morphology of the crystalline regions. This would be consistent with a change from nucleation to growth-dominated crystallisation, i.e. increased size for a smaller number of crystalline regions. These would give less effective strengthening, but are likely to cause greater reduction in toughness as cracked crystalline regions would be larger; this effect needs to be verified. The change in hardening response is established in the Avrami exponent values, Figure 6, which are 0.6–1.8 based on hardness c.f. 3.1–4.8 for DSC measurements. The requirement for high σ_y / E values

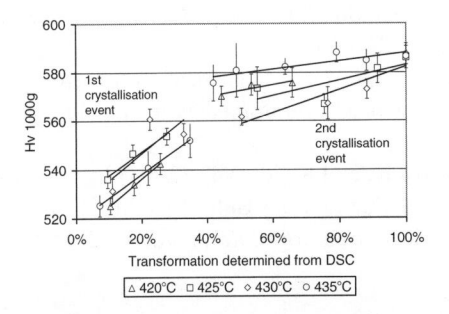

Figure 5. Hardness variation versus % transformation.

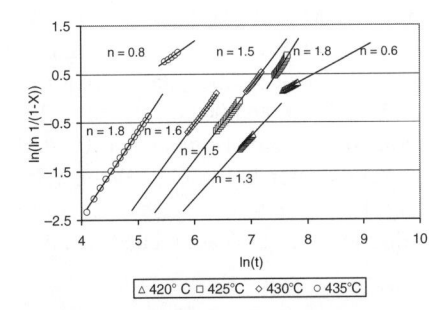

Figure 6. Avrami plot for hardness values taken for first crystallisation event.

would give a maximum processing window of 430 s at 435, which may be expanded to 1430 s at 420°C if hardening through the first crystallisation stage can be utilised.

4   Conclusions and Further Work

Vitreloy 105 ($Zr_{52.5}Cu_{17.9}Ni_{14.6}Al_{10}Ti_5$ at. %) offers highly suitable mechanical properties for application in golf driver head design with a performance index of 14. For the production of a crown for a driver head, the kinetics of devitrification have been studied in the temperature range 420–435°C. It was found that the initial crystallisation event was mainly the nucleation followed by growth or polymorphic transformation of crystals consistent with the tetragonal Zr_2Ni phase in the second event. The characterisation of the development of the fracture toughness and elastic modulus requires the production of larger sections, which is work to be performed.

References

Adelman S., Otto S. and Strangwood M. (2006) Modelling Vibration Frequency and Stiffness Variations in Welded Ti-Based Alloy Golf Driver Heads, The Engineering of Sport 6, Munich, Springer.

Argon A.S. (1979) Plastic deformation in metallic glasses. *Acta. Metall.*, 27, 47–58.

Baricco M., Spriano S., Chang I., Petrzhik M.I., and Battezzati L. (2001) "Big cube" phase formation in Zr-based metallic glasses. *Materials Science and Engineering A*, 304–306, 305–310.

Kundig A.A., Ohnuma M., Ohkubo T. and Hono K. (2005) Early crystallization stages in a Zr-Cu-Ni-Al-Ti metallic glass. *Acta Materialia*, 53, 2091–2099.

Macht M.P., Naundorf V., Fielitz P., Ruesing J., Zumkley T. and Frohberg G. (2001) Dependence of diffusion on the alloy composition in ZrTiCuNiBe bulk glasses. *Materials Science and Engineering A*, 304–306, 646–649.

Nieh T.G., Mukai T., Liu C.T. and Wadsworth J. (1999) Superplastic behavior of a Zr-10Al-5Ti–17.9Cu-14.6Ni metallic glass in the supercooled liquid region. *Scripta Materialia*, 40, 1021.

Nieh T.G., Schuh C., Wadsworth J. and Li Y. (2002) Strain rate-dependent deformation in bulk metallic glasses. *Intermetallics*, 10, 1177.

Revessz A., Donnadieu P., Simon J.P., Guyot P. and Ochin P. (2001) Nanocrystallization in a Zr57Ti5Cu20Al10Ni8 bulk metallic glass. *Philos. Mag.*, A81, 767.

Strangwood M. (2003) *Materials in Golf*. In: Jenkins M. (Ed.) *Materials in Sports Equipment*, Woodhead Publishing Ltd, Cambridge, UK.

Szuecs F., Kim C.P. and Johnson W.L. (2001) Mechanical properties of Zr56.2Ti13.8Nb5.0Cu6.9Ni5.6Be12.5 ductile phase reinforced bulk metallic glass composite, *Acta. Mater.*, 49, 1507–1513.

Wang X.L., Almer J., Liu C.T., Wang Y.D., Zhao J.K. and Stoica A.D. (2003) In situ Synchrotron Study of Phase Transformation Behaviors in Bulk Metallic Glass by Simultaneous Diffraction and Small Angle Scattering. *Phys. Rev. Lett.*, 91, 265501-1.

THE INFLUENCE OF DIFFERENT GOLF CLUB DESIGNS ON SWING PERFORMANCE IN SKILLED GOLFERS

N. BETZLER[1], G. SHAN[2] & K. WITTE[3]

[1]Napier University, School of Life Sciences, Edinburgh, UK
[2]University of Lethbridge, Dept. of Kinesiology, Lethbridge, Canada
[3]University of Magdeburg, Institute of Sport Science, Magdeburg, Germany

The objective of this study was to quantify the effects of 17 different golf club designs on kinematic and kinetic measures of swing performance. Ten subjects performed three golf swings with seven different types of drivers and ten different #6 irons. Kinematic analysis of data gathered with a motion capture system (Vicon v8i, 12 cameras, f = 250 Hz) showed a significant correlation between mechanical club parameters (club length, club weight) and swing characteristics (impact velocity, launch velocity). Four swings of one of the subjects were modeled as a full-body motion using MSC.ADAMS and its LifeMOD plug-in. This model successfully combined motion capture data with a club model that included a flexible shaft and a clubhead based on 3D scans. It was further demonstrated by this model how club characteristics can affect shoulder joint torque profiles.

1 Introduction

1.1. Background

Golf is an extremely popular sport discipline throughout the world and played by millions every year. The majority of golfers invest vast quantities of money on their equipment, choosing with great care from a wide range of products (Proctor, 2002). Based on the variability of human motion, it is reasonable to hypothesize that some golfers perform better with certain clubs, and that their swing characteristics can change depending on which club they use. One method that can provide an insight into the interaction of athletes and their equipment is biomechanical full-body modeling. This approach has been used to analyze some aspects of the golf swing (Nesbit et al., 1994; Tsunoda et al., 2004; Nesbit, 2005), however only two previous studies (McGuan, 1996; Kenny et al., 2006) have addressed how changes in mechanical club parameters affect the player.

1.2. Objectives

The primary aim of this study was to quantify the effects of different golf clubs on kinematic characteristics of the golf swing. A further aim was to create a full-body model of a golf swing in order to determine how different clubs affect kinetic swing variables.

2 Methodology

2.1. Data Collection

Full-body motion of ten skilled golfers was recorded when performing three full golf swings using seven drivers and ten #6 iron clubs (Vicon v8i, f = 250 Hz, 12 cameras, see Betzler *et al*.

2006). Two force plates recorded the ground reaction forces produced during the swing (KISTLER 9826AA, f = 1050 Hz). The subjects' anthropometric variables (mean ± standard deviation) were as follows: age 34.8 ± 13.4 years, height 1.82 ± 0.03 m, weight 91.9 ± 14.6 kg. Mean handicap was 11.7 ± 10.5, and golf experience was 16.3 ± 11 years. Club face decals (Golfworks Inc., Canada) were used to register the impact position of the ball. The swing parameters "impact velocity", "launch angle", "launch velocity" and "ball direction" were determined from the motion capture data. Validation procedures showed that launch condition parameters obtained with the motion capture system correlated well with measurements taken with alternative devices (photo cells, radar) (Betzler et al., 2006). Inter-marker distance measurements (Chiari et al., 2005) allowed estimating the static and dynamic accuracy of the motion capture system as 0.15 mm and 0.35 mm, respectively.

2.2. Development of a Full-Body Model

From one subject, four swings were selected to be used in a full-body model (MSC.ADAMS/LifeMOD) (McGuan, 1996). Four different clubs were included in the model, of which each shaft was represented by finite-element models based on results of three point bending tests (Mase, 2004). In order to determine the centre of gravity and the inertia tensor of each clubhead, the clubhead geometry was characterized using a CAD model based on 3D scans (OptoShape, Massen GmbH). Static, kinematic and kinetic outputs of one of the simulations were compared to data recorded during the respective practical tests for model validation purposes. While previous authors suggested attaching the

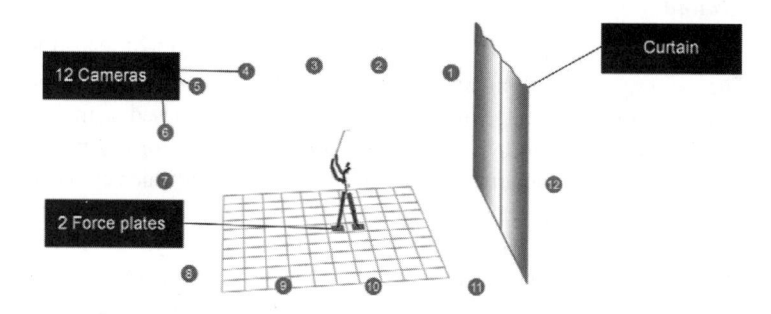

Figure 1. Data collection.

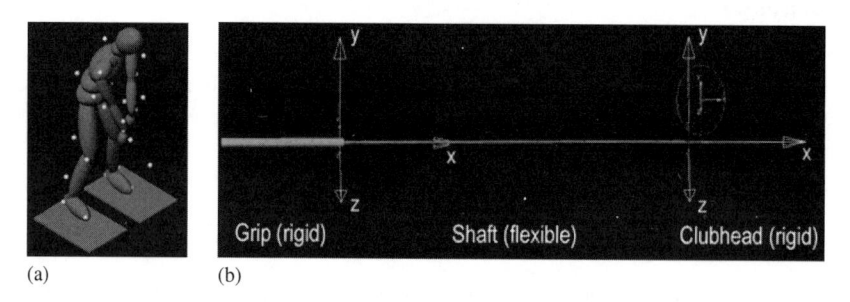

Figure 2. (a) Body model with motion capture markers and force plates and (b) club model.

model's feet rigidly to the ground (Nesbit *et al.*, 1994, BRG 2005), the present study took an alternative approach by using contact definitions that allowed free foot motion.

3 Results and Discussion

3.1. *Kinematic Analysis based on Motion Capture Data*

A typical example of a result data set is displayed in Figure 3, which shows the impact velocities achieved by each subject with each club (each line represents the results of one subject). It can be seen from Figure 3 that inter-subject differences were much greater than inter-club differences. The impact velocities of the drivers were consistent. With the iron clubs, however, small, distinct club effects on impact velocities were apparent. Impact velocity for iron I01 and I06 was reduced in comparison with every other club. It was further observed that the majority of players achieved peak impact velocities with iron I09.

Additional swing parameters, including ball launch velocity and launch angle (data not shown), showed no clear patterns depending on the driver used, whilst slower impact velocities observed for irons I01 and I06 resulted in lower ball velocities and higher launch angles.

By a pair-wise comparison of swing results obtained from clubs that were identical in all but one mechanical parameter, it was intended to isolate the effect of individual mechanical parameters, such as shaft stiffness, loft angle, shaft material or shaft length. Using a long (38") or a short (37¼") iron had a significant effect on impact velocity ($p < .01$; clubhead and shaft type of both clubs were identical). Players achieved higher impact velocities with the long club, which is in agreement with previous studies looking at drivers with different lengths (Mizoguchi *et al.*, 2002; Chen *et al.*, 2005; Kenny *et al.*, 2006). Furthermore, impact positions were significantly closer to the heel for the long iron ($p < .05$). No previous study reported variations in impact positions depending on shaft length. A possible reason for these differences is that players did not adjust their body position to consider changes in club length, therefore explaining why balls were hit towards the heel of the club using the longer club. In all other paired comparisons (regular vs. stiff shaft, 9.5° vs. 10.5° loft angle, steel vs. graphite shaft), swing parameters were not significantly different. This is probably due to the relatively small range of differences between the pairs of clubs and the presence of inconsistencies within the swing motion of the players. In case of iron shaft

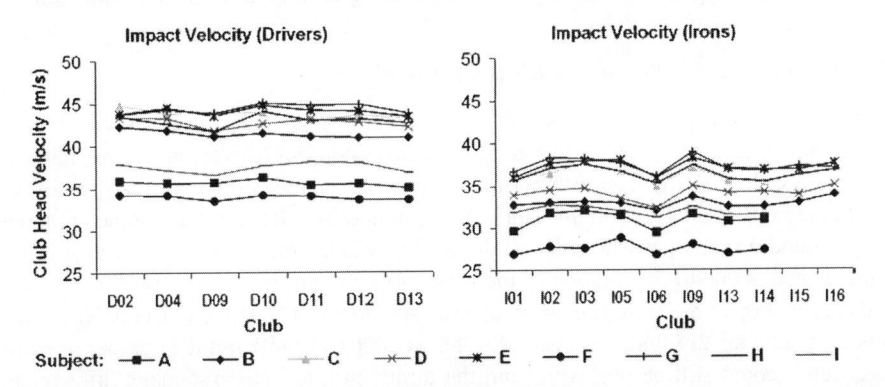

Figure 3. Clubhead velocities achieved by all subjects with the clubs tested. Each line represents one subject and each dot the individual mean result for one club.

Table 1. Pearson correlations of selected mechanical parameters with swing parameters

Drivers	Club weight		Club length		Club COG		
	p	$r_{Pearson}$	p	$r_{Pearson}$	p	$r_{Pearson}$	n
Impact velocity (normalized)	<.01	−.27	<.01	.44	<.01	.21	260
Impact position	n. s.		n. s.		n. s.		261
Ball velocity (normalized)	n. s.		n. s.		n. s.		261
Launch angle	n. s.		n. s.		n. s.		260
IRONS	Club weight		Club length		Club COG		
	p	$r_{Pearson}$	p	$r_{Pearson}$	p	$r_{Pearson}$	n
Impact velocity (normalized)	<.01	−.59	<.01	.34	<.01	.54	376
Impact position	<.01	.15	<.01	−.17	<.01	−.17	377
Ball velocity (normalized)	<.01	−.42	<.01	.40	<.01	.41	377
Launch angle	<.01	.17	<.01	−.14	<.01	−.14	377

material (steel vs. graphite), this indicates that the shaft material could potentially affect the feel of the golf shot rather than being directly related to performance variables.

To determine whether there were any correlations between mechanical parameters and swing parameters, the recorded swings were pooled, forming a driver and an iron group of 261 and 377 data sets, respectively. The results (see Table 1) indicated that there were a number of significant correlations. One possible interpretation for these results is that longer clubs allowed a higher impact velocity to be created, however only as long as the club's inertia was kept constant through reducing the club weight at the same time. Based on these findings, it could be suggested golfers trying to achieve higher impact velocities with their irons should use longer, lighter clubs.

3.2. Kinetic Analysis Based on Full-Body Model

After applying a low-pass Butterworth filter (cut-off frequency: 5 Hz) it was possible to compare the joint torque histories for four different club conditions (driver: stiff vs. regular shaft, iron: long vs. short shaft). The pattern of the curves shown (see Figure 4) was typical for all arm torque profiles analyzed in this study. It was found that the shoulder torque component acting around an axis perpendicular to the sagittal plane contributed the greatest level of torque, which is due to the rotation of the upper arm around this axis supporting the downward acceleration of the club. The observed peak shoulder (90 Nm) and wrist (30 Nm) torques supported previous 2D studies (Neal & Sprigings 2000). Despite limited number of simulations, trends could still be observed from the torque profiles. No systematic differences in joint torques between the two driver conditions existed as the two drivers differed only in their shaft stiffness. In contrast to this, the simulation results suggested that the player applied a

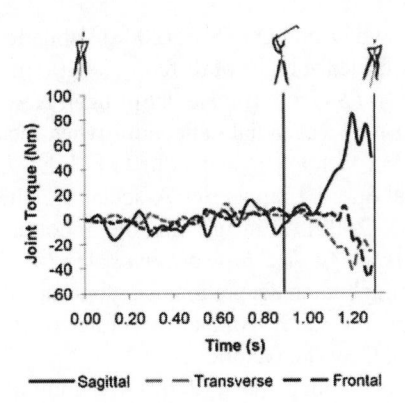

Figure 4. Examples for joint torques (left shoulder) as estimated by the full-body model after applying a Butterworth filter (5 Hz cut-off frequency).

higher peak torque to accelerate the longer iron compared to the shorter iron. Similar results have been reported previously for drivers (Kenny *et al.*, 2006).

4 Conclusion

The aim of this study was to quantify the effects of different golf clubs on golf swing performance. Kinematic analysis showed that mechanical parameters (length, weight) are significantly correlated with performance indicators (impact velocity, launch velocity). Several other club variations did not lead to significant changes in the performance parameters, which indicated that small variations in club characteristics have no significant effect on individual swing results.

In addition to this, a 3D full-body model successfully combined motion capture data, mechanical test results, FE analyses and CAD modeling results, and further included a new and more realistic approach of defining the foot-ground contact. Initial results from this model indicate that variations in shaft stiffness have no major effect on shoulder and arm joint torques. In contrast to this, an increased iron shaft length was associated with increases in joint torque maxima for two sample golf swings.

References

Betzler N., Kratzenstein S., Schweizer F., Witte K. and Shan G. (2006) 3D motion analysis of golf swings: Development and validation of a golf-specific test set-up. Poster presentation given at the 9th ISB symposium on the 3D Analysis of Human Movement, Valenciennes, France, 28–31 June.

BRG, Inc. (2005) Manual – LifeMOD Biomechanics Modeler (online). Available: http://www.lifemodeler.com/Downloads/LM_Manual.pdf. (accessed 22 Nov, 2005).

Chen C., Inoue Y. and Shibata K. (2005) Study on the interaction between human arms and golf clubs in the golf downswing. In: Subic A. and Ujihashi S. (Eds.) *The impact of technology on sport*. Australasian Technology Alliance, Melbourne.

Chiari L., Della Croce U., Leardini A. and Cappozzo A. (2005) Human movement analysis using stereophotogrammetry. *Gait and Posture*, 21, 197–211.

Kenny I., Wallace E.S., Brown D. and Otto S.R. (2006) Validation of a full-body computer simulation of the gold drive for clubs of differing length. In: Moritz E.F. and Haake S. (Eds.) *The Engineering of Sport 6. Volume 2*. Springer, New York.

Mase T. (2004) Correcting for local radial deflections when measuring golf shaft flexural stiffness. In: Hubbard M., Mehta R.D. and Pallis J.M. (Eds.) *The Engineering of Sport 5. Volume 1*. International Sports Engineering Association, Sheffield.

Neal R.J. and Sprigings E.J. (2000) An insight into the importance of wrist torque in driving the golfball. *Journal of Applied Biomechanics*, 16, 356–366.

McGuan S. (1996). Exploring human adaptation using optimized, dynamic human models. Presentation given at the 20th Annual Meeting of the American society of Biomechanics, Atlanta, Georgia, October 17–19.

Mizoguchi M., Hashiba T. and Yoneyama T. (2002) Matching of the shaft length of a golf club to an individual's golf swing motion. In: Ujihashi S. and Haake S.J. (Eds.) *The Engineering of Sport 4*. Blackwell, Oxford.

Nesbit S.M., Cole J.S., Hartzell T.A., Oglesby K.A. and Radich A.F. (1994) Dynamic model and computer simulation of a golf swing. In Chochran A.J. and Farraly M.R. (Eds.), *Science and Golf II*. E & FN Spoon, London.

Nesbit, S.M. (2005). A three dimensional kinematic and kinetic study of the golf swing. *Journal of Sports Science and Medicine*, 4, 499–519.

Proctor S. (2002) Economic contribution of golf to the UK economy. In: Thain E. (Ed.) *Science and Golf IV*. Routledge, London.

Tsunoda M., Bours R.C.H. and Hasegawa H. (2004) Three-dimensional motion analysis and inverse dynamic modeling of the human golf swing. In: Hubbard M., Mehta R.D. and Pallis J.M. (Eds.) *The Engineering of Sport 5. Volume 2*. International Sports Engineering Association, Sheffield.

SKILL ANALYSIS OF THE WRIST TURN IN A GOLF SWING TO UTILIZE SHAFT ELASTICITY

S. SUZUKI[1], Y. HOSHINO[2], Y. KOBAYASHI[2] & M. KAZAHAYA[1]
[1]*Faculty of Engineering, Kitami Institute of Technology, Kitami, Japan*
[2]*Graduate school of Engineering, Hokkaido University, Sapporo, Japan*

This study examined skill of the wrist turn in a golf swing analytically and experimentally. It is observed that the swings of professional and expert golfers include the wrist turn as a "natural" or "late" release. Since dynamic boundary condition of the wrist strongly affects the shaft deformation during the down swing, the relationship between the timing of the wrist turn and the shaft vibration was examined. It was demonstrated that "natural release" at the zero-crossing point of the bending vibration of the shaft and "late hitting" could efficiently increase the head speed at impact by employing a three-dimensional dynamic model. Furthermore, An expert's skill of the wrist turn was experimentally verified by measuring the movement of the wrist and the dynamic deformation of the shaft during the down swing.

1 Introduction

Several previous studies on golf swing analysis have focused on improving the performance of golfers and golf clubs. It is generally known that professional and expert golfers try to match shaft flexibility with their swing style, and that wrist turns in expert swings tend to appear as a "natural" or "late" release. Jorgensen (1970) obtained numerical solutions for the effect of a delayed wrist turn. Pickering & Vickers (1999) numerically show the effect of "natural release" and "late hit" using a two-dimensional rigid double pendulum model. These results suggest that delaying the wrist turn will enhance the club head speed at impact. Contrary to these reports, the results of the simulation performed by Springs & Mackenzie (2002) indicate that there is only a small advantage in employing the delayed release technique. However, this simulation employs an unrealistic resistive torque of the wrist to delay the wrist turn. The objective of this study is to demonstrate the effect of natural delayed release of the wrist in conjunction with utilizing shaft elasticity, which has not yet been studied. Thus, it is expected that the results of this study will contribute improving the performance of golfers and golf clubs by examining the relationship between the timing of the wrist turn and the shaft vibration. It is demonstrated that "natural release" with no acceleration torque at the shaft's zero-crossing point for bending vibration and the "late hitting" can efficiently increase the head speed at impact. Furthermore, the skill of the wrist turn for a long drive is experimentally verified by measuring the movement of the wrist and the dynamic deformation of the shaft for various levels of golfers.

2 Modeling

In the golf swing analysis, a three-dimensional dynamic model was simplified as shown in Figure 1. The down swing was assumed to occur in one plane so that the double pendulum

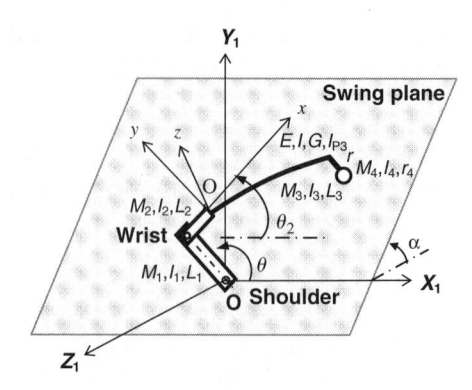

Figure 1. Simplified dynamic model for computing.

model composed of two coplanar rigid segments, which represent the left arm and the hand-grip part, could be employed. The tilt angle of the swing plane α is set at $\pi/3$ radians for the driver shot. The grip part also rotates around the longitudinal axis to account for the supination of the forearm. In the shaft vibration analysis, the shaft was assumed to be Bernoulli-Euler beam, therefore, in-plane and out of plane bending vibration and torsional vibration were examined using a uniform continuous cantilever model. A spherical club head is set at the tip of the shaft with an offset symbolized by r.

3 Skill Analysis

3.1. *Motion Setting*

θ_1 and the relative angle between the center axes of the arm and the grip (i.e., π-θ_1 + θ_2) are initially set to $\pi/2$ radians. At the start of a swing, the shoulder is accelerated by joint torque and the wrist is fixed in order to maintain the initial angle only by the geometric constraint (i.e., the wrist is the same as a hinge joint with a stopper). After starting, the wrist is naturally released at the appropriate point by the change of interference between the joints. Thus, in the latter half of the swing, both the shoulder and wrist joints become free. In the final state, the center axes of the arm and grip should become straight without an adjustment of joint torque.

3.2. *Natural Release*

The wrist is naturally released at the zero-crossing point of the bending deformation of the shaft vibration with a large change of interference between the joints. Expert golfers place a special emphasis on selecting a shaft whose flexibility compliments their swing style. Therefore, it is expected that "natural release" is closely related to the bending vibration during the swing. Figure 2(a) shows ten different release points that coincide with various phases of the shaft vibration. The time from the zero-crossing point of vibrational displacement at a tip of the shaft y_p to the minimum point were equally divided into ten release points. Efficiency index τ that can evaluate the proficiency degree of these swing motions is calculated as the

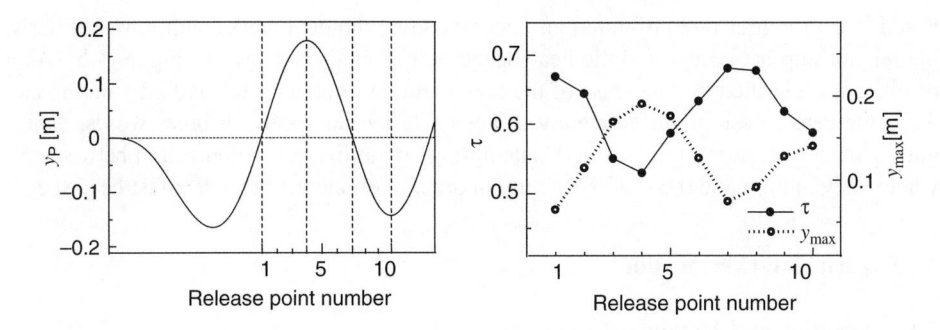

Figure 2. Effectiveness of natural release with a flexible shaft (a) Release points based on shaft deformation (b) Variation of efficiency index and maximum shaft deformation.

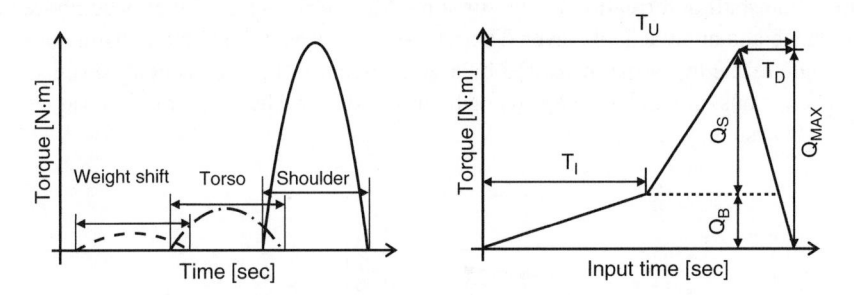

Figure 3. Acceleration torque of the shoulder joint (a) Composite acceleration torque (b) Simplified two-step modulation torque.

ratio of kinetic energy of the head at impact to the work of the shoulder. Figure 2(b) shows τ and the maximum displacement of the shaft at impact zone y_{max} for each release point under the same work of the shoulder. It was demonstrated that τ becomes large with the "natural release" at the zero-crossing point, and that the maximum displacement simultaneously decreases. This indicates that much of the shaft's elastic strain energy is transformed into the kinetic energy of the club head by the zero-crossing release.

3.3. Late Hitting

The expert is capable of maintaining the initial wrist angle longer than the beginner during the down swing. Therefore, delaying the zero-crossing point of the shaft vibration is most important for efficiently achieving a fast head speed. In order to delay the zero-crossing, acceleration of the shoulder joint has to increase during the acceleration of the arm. However, it is extremely difficult to increase the acceleration by a single movement of the shoulder. Therefore, it is expected that expert golfers delay the zero-crossing by utilizing their whole body motion. Thus, the acceleration torque is assumed to consist of the composite effects of weight shifting and torso and shoulder rotation, as shown in Figure 3(a). Then, the acceleration torque was simplified to a two step modulation torque, as shown in Figure 3(b). Validity of this torque function was examined by comparing a simple trapezoidal and triangular function. Figure 4(a) shows the relationship between the shoulder joint angle at the wrist release

θ_r and Q_{MAX} for three types of functions under the same shoulder work conditions. Similarly, the relationship between τ and the head speed at impact V_h is shown in Figure 4(b). As a result, it was clarified that increase of the acceleration torque with whole body motion can delay the zero-crossing and efficiently achieve a fast head speed. In other words, expert golfers practice "natural release" and "late hitting" by utilizing harmonization between the whole body motion and the shaft flexibility in order to efficiently achieve a fast head speed.

4 Experimental Verification

4.1. Experimental Method

In order to demonstrate the validity of the skill analysis results, it was experimentally verified whether the experts practiced "late hitting" by delaying the zero-crossing. As shown in Figure 5, the shaft deformation is measured by "Zero-cross sensor" that is composed of a strain gauge and an attachable layer. The release point was defined by a change of radial-ulnar angle of the left wrist measured by a goniometer. In the experiment, seven subjects who have various official handicap swung their own driver. The swing motion was captured by VICON system.

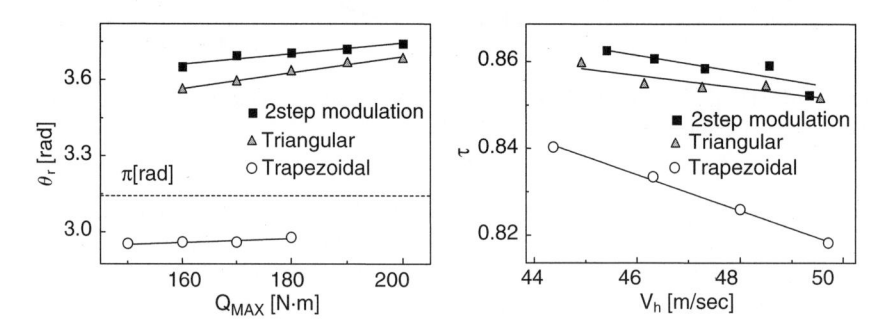

Figure 4. Effectiveness of the whole body motion (a) Comparison of release angle (b) Comparison of the efficiency index.

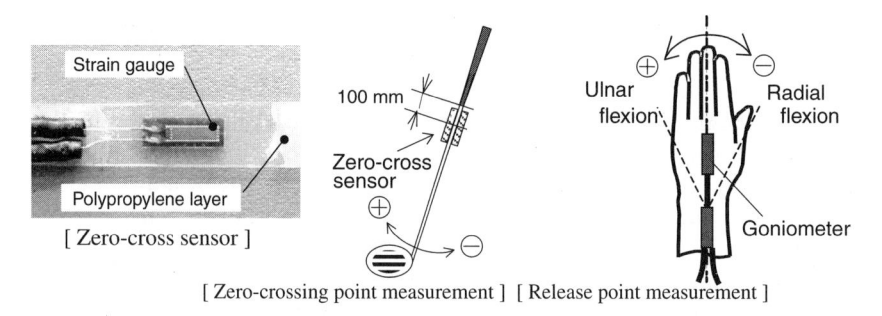

Figure 5. Experimental equipment.

4.2. Skill Evaluation

As shown in Figure 6(a), the time from the start to the release point of the wrist, to the zero-crossing point and to the impact are represented by Tr, Tz and Ti, respectively. Each subject swung each club and hit a light plastic imitation ball three times, and rate of the release time to the zero-crossing time Tr/Tz, the rate of Tz to downswing time Tz/Ti and the rate of Tr to downswing time Tr/Ti were averaged and compared in Figure 6 (b).

It was demonstrated that the zero-crossing point was later and the release point was closer to the zero-crossing in the down swing as the skill level of the golfer increased. However, every subject naturally released the wrist at the zero-crossing point. This suggests that the shaft vibration strongly affects the movement of a wrist joint. As a result, the analytical results on the skill of wrist turn were experimentally verified. Figure 7 shows the results of swing motion analysis measured by VICON for examining the skilful timing of weight shift, torso twist and shoulder rotation. The timing that was calculated by the second difference was marked with a black point in the figure. It was also demonstrated that an expert golfer can gradually move a body from the lower part to the upper part during the down swing. This suggests that the expert can increase the acceleration of the shoulder in the latter half of the downswing, and can delay the zero-crossing point.

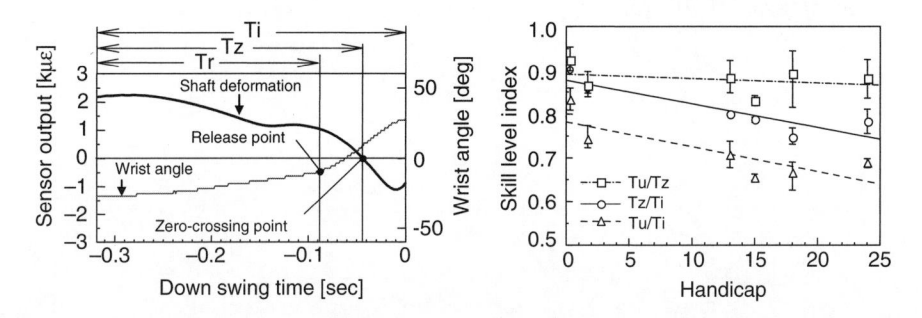

Figure 6. Experimental results of skill evaluation (a) Skill level parameters (b) Comparison of skill level index.

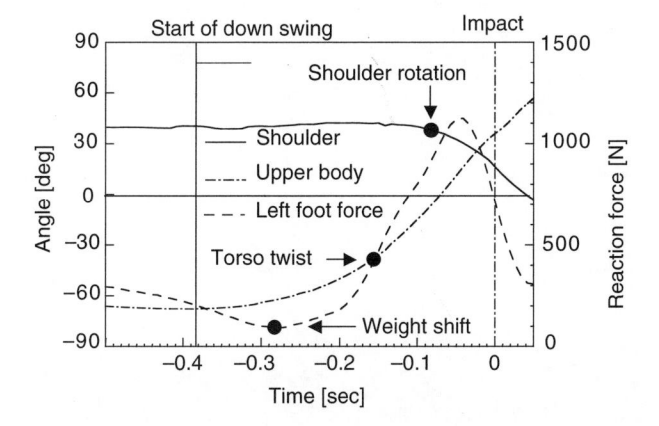

Figure 7. Skilful whole-body swing motion (Hcp = 0.3).

5 Conclusions

Techniques employed by expert golfers including "natural release" and "late hitting" were examined using a simplified dynamic model and were verified by experiments. As a result, it was analytically demonstrated that "natural release" at the zero-crossing point of the bending vibration of the shaft and "late hitting" achieved by delaying the zero-crossing point could efficiently increase the head speed at impact. Finally, an expert's skill was experimentally verified by measuring the movement of the wrist and the dynamic deformation of the shaft during the down swing.

References

Jorgensen T. (1970) On the dynamics of the swing of a golf club. *American Journal of Physics*, 38, 644–651.

Pickering W.M. and Vickers G.T. (1999) On the double pendulum model of the golf swing. *Journal of Sports Engineering*, 2, 161–172.

Springs E.J. and Mackenzie S.J. (2002) Examining the delayed release in the golf swing using computer simulation, *Journal of Sports Engineering*, 5, 23–32.

Suzuki S. (2003) Skill analysis of golf swig. *Journal of the Japan Society for Simulation Technology*, 22/1, 10–15.

MOTION ANALYSIS SUPPORTED BY DATA ACQUISITION DURING GOLF SWING

A. SABO[1], M. BALTL[2], G. SCHRAMMEL[1] & M. REICHEL[1]

[1]*University of Applied Science, Technikum Wien, Sports-Equipment Technology, Vienna, Austria*
[2]*Austrian Golf Association, Vienna, Austria*

The adjustment between momentum behavior and the correct club choice is a highly individual issue. The momentum behavior is affected by coordinative abilities, muscle control and anthropometric requirements. The most important point with the club choice is the correct flex and torque performance. To analyze and contain the influence and description values of the golf swing, a complete measurement station was built at the Technikum Wien (Department of Sports-Equipment Technology) in Vienna. In addition to the collection of biokinematic, biodynamic and anthropometric data, material technological components like flex and torque performance can also be examined. In order to evaluate the determined data synchronously, different measuring systems such as protractors, acceleration transducers, resistance strain gauges, electromyography, force surface plates, qualitative and quantitative pressure pick-offs as well as 2D/3D high frequency film/video picture measurements were linked in diverse combinations. If the shaft is too stiff or too soft, optimal transfer of energy does not takes place on the golf ball at impact. Bending the wrist leads to an additional acceleration phase. Reflex-steered players are able to bend their wrists later and need a stiffer shaft. The bending movement control made by the central nervous system in the wrist can be made later and results in the player needing a softer shaft. Those in this setup of determined data should support and promote Pros and Beginners in their personal training program. On the basis of early measurement results, the coach is able to prepare the future training program. Most importantly, it depends on whether the training form, the material or both require renewal.

1 Introduction

Because of the fast movements during a golf swing Golf coaches (independent from their educational level) are not able to acquire all the movement parameters correctly. The ability of perception is affected by the visual ratings of the coaches. This ability is limited. Through the saccadic eye movement (short, fitful eye movements) no information reception occurs for a short time. The estimation of body angles and pressure distributions, respectively, during the golf swing is not possible (Ballreich & Kuhlow-Ballreich, 1992; Klaas, 2004).

The setup, built at the Technikum Wien, Department of Sports Equipment Technology, should assist the coach in his work. The assignment of measurement and information reception of the trainer allows a impartial acquisition of many biomechanical movement parameters. The combination of determined data can fit the individual training program for every player.

The development of the setup underlies continual improvement. Aim of this poster presentation is to give a short survey of the used measurement and motion capture systems. Different systems were merged to measurement packages to acquire more parameters in one swing.

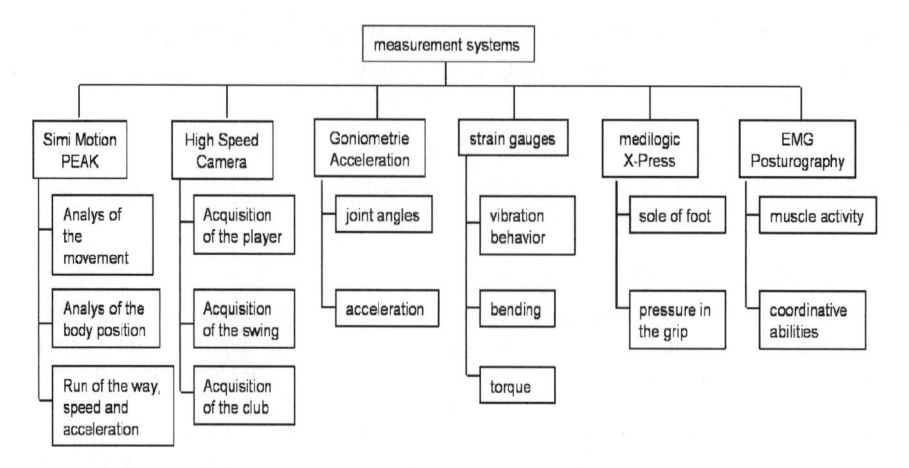

Figure 1. Overview of the used measurement systems.

2 Measurement

2.1. *Measurement & Parameters*

Figure 1 gives a short overview about the used measurement systems and measured parameters.

2.2. *Materials and Methods*

To acquire qualitative and quantitative pressure distribution on sole of feet and palms two different systems were used. Xpress measured the qualitative pressure distribution with 8 FSR-sensors in the left palm (hypothenar, first phalanx-middle finger, second phalanx-thumb, third phalanx-forefinger) and right palm (first phalanx-middlefinger, first phalanx-forefinger, hypothenar, thenar). Because data is recorded on the sound track (sample rate 25 Hz) video and determined data can be evaluated synchronously.

Medilogic provides quantitative pressure distribution on the sole of feet with up to 240 SSR-sensors. In addition to acquire the pressure of the whole sole (Medilogic default), each sensor is acted with MATLAB. So it is possible to part each sole in different sections, for example forefoot, metatarsus and heel.

To hit the ball in the right moment it is necessary to bend the wrist correctly. The bending of the wrists was acquired using Twin Axis Goniometer SG65 (BIOMETRICS) in two planes (flexion, abduction). The sensors were fixed on the first phalanx of the thumb and last part of the radius.

Acceleration at the wrist and the club head were measured with a 3D acceleration transducer with a possible load up to 50 g.

To get the behaviour of the shaft during the swing resistive strain gauges were placed at the flexpoint of three different shafts (regular flex, stiff flex and a steel shaft). With the system Telemyo IV we record the muscle activity during the swing. The electrodes were positioned on the m. pectoralis major, m. latissumus dorsi and the m. rhomboideus. Additional electrodes were placed on the paravertebral muscles and the M. obliquus externus abdominis.

Coordinative abilities of the player were acquired with the posturography. The postural loop includes the balance, the visual system and the sensorimotor functions of the body.

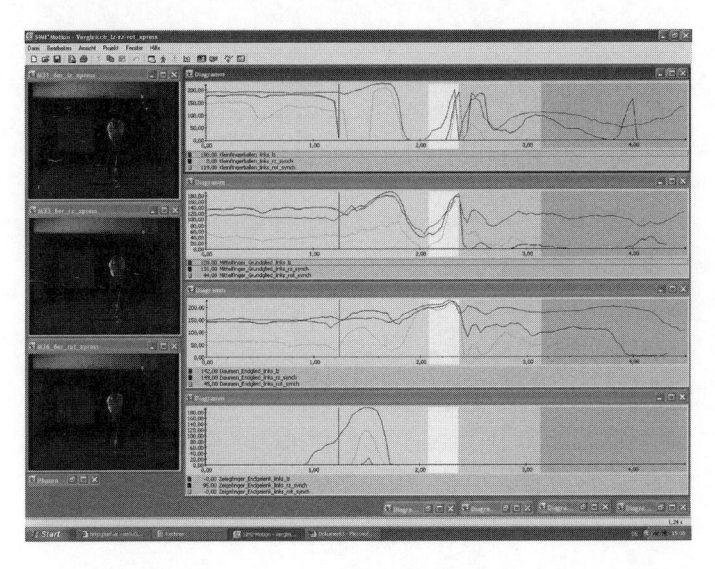

Figure 2. Pressure distribution in the palm of the left hand dependent on three different swing types (more power in the left hand, more power in the right hand and power out of the rotation of the body).

This system works with four force measurement plates, which reflect the balance movements of the player at eight different positions. The positions go from eyes open-hard underground to eyes closed-soft underground).

One force measurement plate (OR 6–7, AMTI) were used to measure the ground reaction forces in x, y und z direction.

To evaluate the fast motions during the golfswing a high speed camera (Vosskühler HCC-1000) was used. This camera provides a resolution from 1024×1024 pixel recording 462 frames per second up to 1024×256 pixel recording 1825 frames per second. For 2D views the camera was positioned in front of the player. 3D views were accomplished with 4 Sony DCR-TRV 75E Camcorders (positioned in intervals of 45° around and above the player).

All data and videos were prepared to import them into the Motion Analysis System SIMI-Motion. Synchronisation between data and video occurred with a light barrier or with different motions or actions caused by the player.

2.3. Measurement Packages

Another aim of the setup was to create measurement packages acquire more than one parameter during the swing. With the available measurement two packages were built. The first includes the high speed camera, Medilogic, resistive strain gauges and SIMI Motion 2D/3D. The second includes the high speed camera, medilogic, the goniometer and the acceleration transducers and SIMI Motion 2D/3D.

3 First Valorizations

The first valorizations are shown in Figures 2 to 5. The used systems are XPress, goniometry, Medilogic and the measurement with resistive strain gauges.

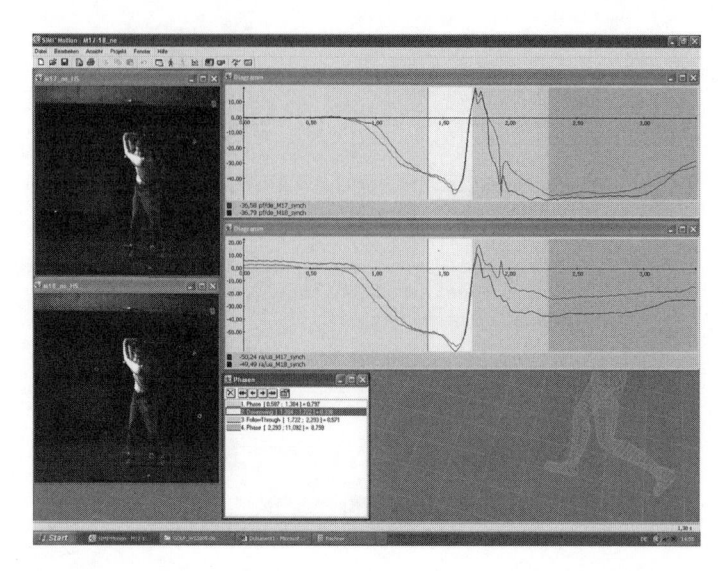

Figure 3. Measuring the bending of the wrist in two planes (upper picture-abduction, lower picture-flexion).

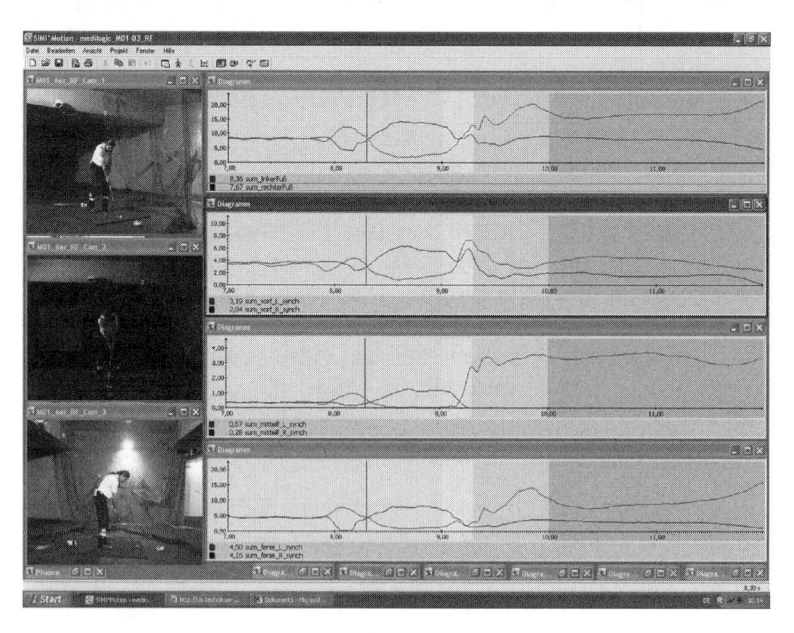

Figure 4. Qualitative pressure distribution of selected parts of the sole (pictures top down: whole sole of the feet, sum of the forefoot, middle foot and the heel).

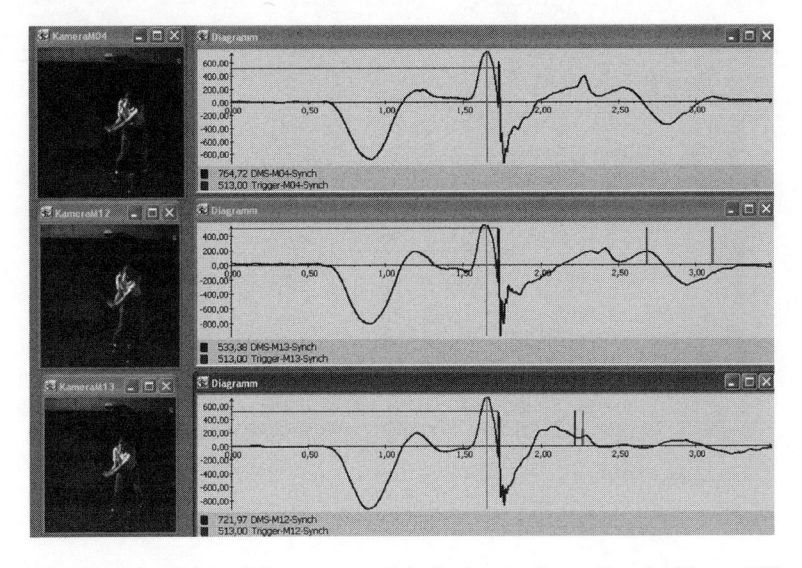

Figure 5. Comparison of three different types of shafts (upper picture-Regular Flex, middle picture-Stiff Flex and lower picture-Steel shaft).

4 Further Work

Further work is underway to validate the first results of the different systems and the measurement packages. Additionally, it is necessary to find out weak spots in matters of attaching sensors, right illumination and camera positioning. The influence of a couple of systems like resistive strain gauges, EMG or goniometry to the player should be reduced.

A long-term goal is to develop this setup for different institutions. Target groups are particularly departments (golf academies, federal training centers) and golf courses. Depending on financial potentials the setup could be used stationary or mobile.

References

Ballreich R. and Kuhlow-Ballreich A. (1992) *Biomechanik der Sportarten, Biomechanik der Sportspiele*,Vol 3, part 1. Enke, Stuttgart, 30–108

Klaas J. (2004) Die Grenzen einer Bewegungsanalyse beim vollen Golfschwung unter Berücksichtung der Ausbildungsqualifikation. Diploma Diss., Golf-University Paderborn

AUTOMATIC DIAGNOSIS SYSTEM OF GOLF SWING

M. UEDA[1], Y. SHIRAI[2], N. SHIMADA[2] & M. OONUKI[1]

[1]*SRI R&D Ltd, Chou-ku Kobe Hyogo, Japan*
[2]*Ritsumeikan University, Kusatsu, Shiga, Japan*

An easy and economic system to automatically diagnose golf swings has been developed. The system takes 180 photos and selects 12 images required for diagnosis of the golfer's swing. Swing positions are detected by extracting the body silhouette in selected images and identifying colored marks on clothing. Results of the analysis are then compared with a standard swing. The system then generates suggestions for improving the golfer's swing, as well as, practice method recommendations.

1 Introduction

An important part of playing golf is ability to hit the ball with distance and accuracy. A good score depends on the golfer's skill as well as high performance equipment. To improve skill, the golfer may take lessons from an instructor. However, there are many golfers who cannot take lessons because of lack of time or the high cost. To reduce the cost and time for improving a golfer's game, a system that automatically diagnoses the golfer's swing has been developed. Swing analysis requires identification of the desired points of reference in the golfer's swing or "form." A computer analysis of the form is then automatically generated. Image analysis systems have been used in various types of sports to analyze form. However, it is difficult to analyze the form by hand. Recently, a motion capture system (Nakagawa & Moorhouse, 2004) was developed which accurately analyzed the form automatically. The measurement accuracy of this system is high, and the system is suitable for research on golf swing kinematics. However, since two or more special cameras are needed, it is difficult to set up the system in a typical golf retail shop. Moreover, this system can only analyze the position where an image analysis mark is attached to the golfer. A motion capture system which employs one camera (Fukui *et al.*, 2002) was developed. This system requires that all the positions in the swing image to be identified. The types of swings that can be analyzed are limited since it is difficult to specify the swing positions and their colors may darken or become invisible while the swing is being made. The system proposed in this paper utilizes two cameras and specially marked clothing. The system does not require that all the positions (such as joints) in the swing image to be recorded, but selects the set of images necessary to make the swing diagnosis. It then combines and processes two or more images to estimate the positions required to make the analysis, even if the marks are hidden. The unique feature of the system is that it can generate the required points of reference in locations where there are no markers. In addition, the system makes swing and corrective practice recommendations based on the Dunlop Training Program (Sumitomo Rubber Industries, 1994).

2 Automatic Diagnosis System Components

2.1. *Hardware*

Figure 1 shows the major components of the automatic diagnosis system. A camera is set up on the front and side positions (opposite from the position from where the ball flies). A light is set up above each camera. The system employs cameras that capture 60 images a second with a shutter speed of 1/500 second and can take swing images for 3 seconds. The system measures the initial conditions of the hit ball, for example, the head speed, the ball speed, and ball spin rate, etc. The system then computes the trajectory of the ball from the measured data and aerodynamic characteristics. 150 images are taken before the impact time and 30 images after impact. The camera activation signal is triggered by a sensor that measures the golf club head speed.

2.2. *Swing Analysis Process*

1. Figure 2 shows the special jacket required to analyze the golf swing. The golfer wears the jacket with the attached colored marks to detect the required analysis points, for example, shoulders, and elbows, etc. Two colored marks, in contrasting colors, are attached to the golfer's pants to specify the position of the golfer's knees.
2. The golfer uses a golf club attached with 3 colored marks and hits a ball about 5 times. The system measures the initial conditions of hit ball and captures the swing images at the same time.
3. The system estimates swing positions by processing the captured images. The swing is analyzed and swing recommendation and practice methods are generated.
4. The golfer is given the diagnosis result sheets, or the CD-ROM saved data within about 15 minutes.

3 Golf Swing Image Processing

3.1. *Image Analysis*

Based on the Dunlop Training Program, the system selects a set of images necessary to diagnose the swing. Figure 3 shows a set of typical images.

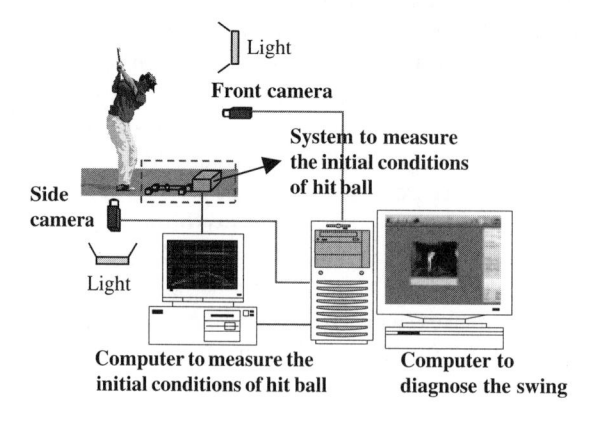

Figure 1. Major components.

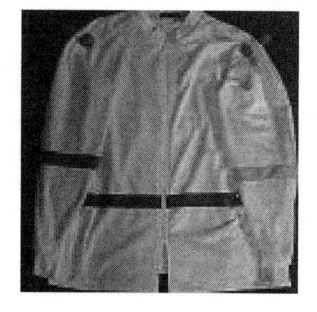

Figure 2. Special clothing.

3.2. Selection of Check Images by Detecting Colored Marks

The system detects 3 colored marks attached to a golf club by using Hue, Saturation, and brightness colorspace, estimates of the shaft orientation in each image from the detected marks, and then selects "check-images" from 180 images according to the estimated shaft orientations. All 3 marks may not be seen in all the swing images. For example:

- When all marks are hidden in golfer's body.
- When 3 marks are connected in the image, only one mark is detected.
- When the distance between the mark and the camera is too far, the marks in the image become too small to be recognized.

In these cases, the system discontinues tracking of the marks, and selects check-images by interpolating the difference between frames.

3.3. Selection of Check Images by Using Template Matching Processing

In order to select images with Take-Back and Downswing of the left arm in the horizontal direction, the system tracks the left arm region in the swing images by using SAD (Sum of

(a) Address (b)Take-Back golf shaft in the navel's direction (c)Take-Back golf shaft in the horizontal direction (d)Take-Back left arm in the horizontal direction

(e) Top (f) Downswing left arm in the horizontal direction (g) Downswing golf shaft in the horizontal direction (h) Impact

(i)Follow-Through golf shaft in the navel's direction (j) Follow-Through golf shaft in the horizontal direction (k) Follow-Through left arm in the horizontal direction (l) Finish

Figure 3. Swing image analysis.

Absolute Difference) method. For example, the process which selects the image with Take-Back left arm in the horizontal direction is as follows:

1. An intensity template is made on the left arm from the image with the Take-Back golf shaft in the vertical direction by using the positions of left shoulder and the grip (Fig.4a).
2. The template made in the previous frame is moved to the grip position presumed from 3 shaft marks.
3. The template is maintained within ±2 pixels from the grip position and is rotated clockwise by 0 to 9 degrees around the grip position. Therefore, 250 templates are generated (Fig.4b). The SAD Value is calculated by using one template in the present frame and each template made in the previous frame.
4. The template with the minimum SAD value is regarded as the template in the next frame.
5. Processing steps 2 to 4 are repeated, and when the template is parallel to ground (horizontal line), this image is regarded as the image with Take-Back left arm in the horizontal direction. (Fig.4c).

3.4. *Position Specification Required to Diagnose the Swing*

In order to specify the position necessary to diagnose each check-image, the system executes the following three steps to process the images.

1. The system first extracts the silhouette by the background difference method. The system takes 30 background images, and generates an average background image. Using this image, the background subtraction is computed for each of the check-images. The color-sensitive background difference in this system is given by:

$$D_1(c_1, c_2) = \sqrt{(r_1 - r_2)^2 + (g_1 - g_2)^2 + (b_1 - b_2)^2} \quad (1)$$

$$D_2(c_1, c_2) = \sqrt{(H_1 - H_2)^2} \quad (2)$$

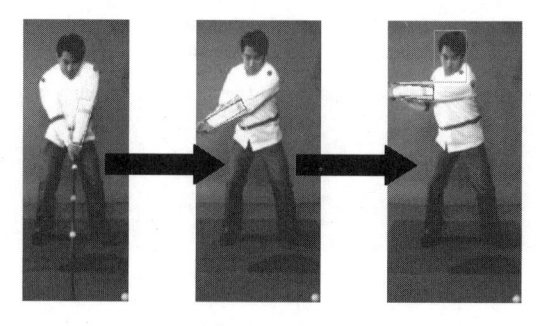

Figure 4. Selection of images with Take-Back left arm in the horizontal direction; (a) take-back golf shaft in the vertical direction, (b) the made template, take-back left arm in the horizontal direction.

where c_1, c_2 are the pixels where the colored background difference calculation is performed, and ri, gi, bi are the values of RGB elements. Hi is the value of Hue calculated from the value of RGB elements. Equation (1) is used when the system calculates the color-sensitive background difference of the areas in the golfer's vicinity of one's feet and equation (2) is used when the system calculates the colored background difference of the other areas.

2. The outline of the body is extracted as the boundary of the silhouette derived by Step 1. The curvature of the outline is given by:

$$C = \tan^{-1}\left(\frac{y_0 - y_{-k}}{x_0 - x_{-k}}\right) - \tan^{-1}\left(\frac{y_k - y_0}{x_k - x_0}\right) \qquad (3)$$

where (x_0, y_0) are coordinates of the points that calculate the curvature, (x_k, y_k), (x_{-k}, y_{-k}) are coordinates of the "k" distance from (x_0, y_0). The edge of the contour is calculated by the Sobel method. When the outlines of the curvature are within ±0.17 rad and is continuous within five pixels or more, the system regards the outlines as a straight-line.

3. Based on the outline of the body, the curvature, straight-lines and the edge of the contour, the system estimates the required positions. The system can estimate the required positions by not only by processing one image but also combining and processing two or more images, even if some marks have not been attached or if the marks are hidden.

- The positions of shoulders are specified by detecting the colored marks in the estimated range where a right and left shoulder would appear to exist by using the outline of the silhouette. The method of image processing uses the HSV element within the range. If the colored mark cannot be detected, a right and left shoulder is estimated by using the curvature. (Fig.5a)
- The width of the stance is estimated by using the outline of the silhouette and the left leg contour. (Fig.5b)
- The line of the spine is estimated by using the outline (in an oval frame). The starting point of the spine line is estimated by detecting the face. [Fig.5(c)]
- The right toe, the right ankle and the left toe in the side image are estimated by using the curvature. [Fig.5(d)]
- The position of the belt (navel) in the front image is estimated by using the side image because in the front image the colored mark is hided by the arms.
- The right kneecap is estimated by using a straight-line in the vicinity of a right knee.

Fig.6 shows the positions specified by the above-mentioned processing method. By using these positions, the system analyzes the swing form, for example, the width of the stance, and the distance between the end of grip and the body.

4 Golf Swing Diagnosis

The golf swing is analyzed using the points that characterized the swing using 34 points in the front direction and 34 points in the side direction. The system compares the result of

(a) Color (b) Edge and outline
(c) Outline (d) Curvature

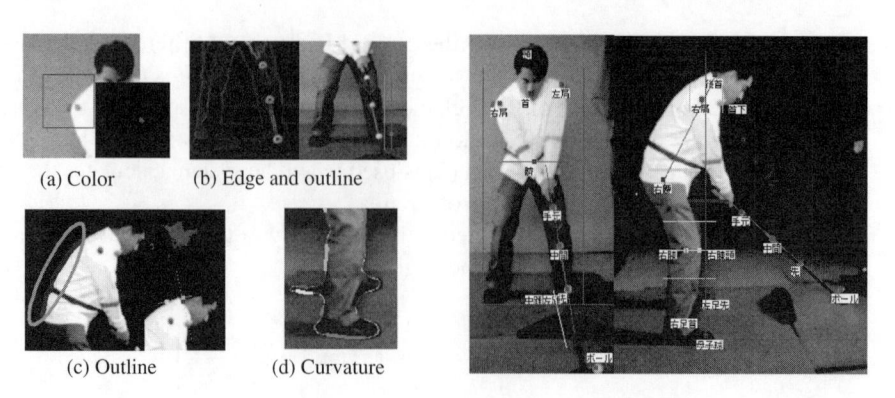

Figure 5. Extraction method.

Figure 6. Position necessary to diagnose the address image.

analysis with the standard swing. Since it is difficult for golfers to improve their swings if the system provides too many recommendations, the system prioritizes the recommendations and corresponding practice methods. In addition, the golfers can enjoy the process as though it were a game, because the system can show the score based on the result of the analyzed swing and the ball trajectory.

5 Conclusion

A system has been developed to analyze a golfer's swing. The image analysis system automatically provides feature points for diagnosis of golf swing and then generates recommendations for practice. The system will enable golfers who are not able to take lessons from an instructor to learn from the system and see improvement in their game.

References

Nakagawa K. and Moorhouse I. (2004) The application to sport of an automatic 3D movement analysis system, *Japan Journal of Biomechanics in Sports & Exercise,* 8/3, 193–200.

Fukui T., Onishi T. and Morozumi T. (2002) Motion capture method of use the human body model, 8th Symposium on Sensing via Information (SSII2002).

Sumitomo Rubber Industries (1994) *Dunlop Training Program.*

KINEMATIC ANALYSIS OF GOLF PUTTING FOR ELITE & NOVICE GOLFERS

J.S. CHOI[1], H.S. KIM[1], J.H. YI[1], Y.T. LIM[2] & G.R. TACK[1]
[1]*Biomedical Engineering, Konkuk University, Chungju, Korea*
[2]*Sports Science, Konkuk University, Chungju, Korea*

Putting is more delicate motion than any other strokes in golf. One of the keys for putting stroke is an accurate ball impact. Putter head control is an important factor at ball impact. For accurate ball impact, the upper extremities looks like a pendulum motion during entire putting phase. Therefore, the purpose of this study was to compare and analyze angular motion of shoulder, elbow and wrist which has direct effect on putter head motion. Participants consisted of two groups based on their playing ability: 10 elite golfers (handicap $\leq$ 2) and 10 novice golfers (handicap $\geq$ 25). 1 m $\times$ 10 m artificial putting surface was set up for the experiment. During actual putting experiment, 3D motion analysis system with 4 high-speed Falcon digital cameras was used. 3D data (120 Hz) was collected for each subject performing 5 trials of putts from each of these distances: 1 m, 3 m and 5 m. Entire putting stroke was divided into back swing, through swing, and follow through phase. From this study, it could be concluded as follows: First, as the putting distance increased, jerk cost function at putter head increased for two groups. Only mediolateral directional component of jerk cost function of novice group increased statistically significantly compared with elite group, which means that the mediolateral directional component of the jerk cost function at putter head might be the quantifiable index for distinguishing the playing level of golf putting. Second, there existed a clear difference in the trajectory of shoulder line between two groups. As for novice group the rotational center did not converge into one point, for elite group the rotational center converged into precise single point. Third, there is a clear difference pattern in anterior-posterior directional movement at shoulder between two groups. Additionally the mediolateral directional movement of left shoulder of novice group is much greater than that of elite group.

1 Introduction

In golf, stroke divided into driving and putting stroke. Putting stroke was considered as another golf game. Putting is important to have good scores. Gwyn & Patch (1993) reported that putting stroke is about 40% of whole golf shot. However, generally most golfer practices driving stroke which is only 19% per round (Wiren, 1992). Since it is hard to keep consistency of putting stroke performance, most golfer has difficulties on putting stroke. Several studies about putting have been carried out so far. Paradisis (2002) reported the phase time of putting stroke and the displacement of putter. Delay (1997) studied the motor control of putting stroke from the control of force point of view.

Our previous study has reported that the difference between elite and novice golfers was the motion of putter head (Tack 2005, 2006). Putter head control is an important factor at ball impact. For accurate ball impact, the upper extremities looks like a pendulum motion during entire putting phase. Therefore, the purpose of this study was to compare and analyze angular motion of shoulder, elbow and wrist which has direct effect on putter head motion. Additionally, it was assumed that there exist smoothness difference between elite and novice golfers during putting. Difference in smoothness between golfers can be quantified by jerk cost function. Since the jerk cost function can be an index for the smoothness of the given movement (Viviani, 1995), the putting stroke between two groups was evaluated by this.

2 Methods

The participants consisted of two groups based on their playing ability: 10 elite golfers (handicap ≤ 2, average age 22.3) and 10 novice golfers (handicap ≥ 25, average age 26.4). All of them were male subjects. 1 m × 10 m artificial putting surface was set up for the experiment. The participants tried to perform a putting stroke as accurately as possible in order to reach the target hole. During actual putting experiment, 3D motion analysis system (Motion Analysis Corp., USA) with 4 high-speed Falcon digital cameras was used. The 3D kinematic data (sampling frequency of 120 Hz) were collected for each subject performing 5 trials of putts from each of these distances (random order): 1 m, 3 m and 5 m. Data from 3D motion system were filtered with simple moving average method to remove noise. It was enough to use this simple filtering method since putting stroke was relatively slow and small movement. Entire putting sequence was divided into 3 phases: back swing, down swing, and follow through phase.

- Back Swing Phase (BS) : address to back swing top
- Down Swing Phase (DS) : back swing top to ball impact
- Follow Through Phase (FT) : ball impact to follow through end

In order to improve accuracy and consistency of putting performance, it is important to control of motion at putter head (Delay, 1997). Since upper limb has direct effect on the movement of the putter, linear and angular motion of shoulder, elbow and wrist joints was evaluated between two groups at anterior-posterior (AP), mediolateral (ML), vertical (V) direction and the resultant. Definition of upper limb angles is given in Figure 2.

In order to compare the smoothness between two groups during putting, the jerk cost function of all markers was evaluated. However, according to the minimum jerk theory, it is enough to calculate the jerk cost function at the end trajectory (at putter head) of the movement. Thus the jerk cost function at the putter head was compared between two groups. Jerk cost function (JC, unit: $m^2 s^{-5}$) which is used as an index for the quantitative measure of given movement is defined as follows; where r is the position vector, T is the duration of movement (Hreljac, 2000).

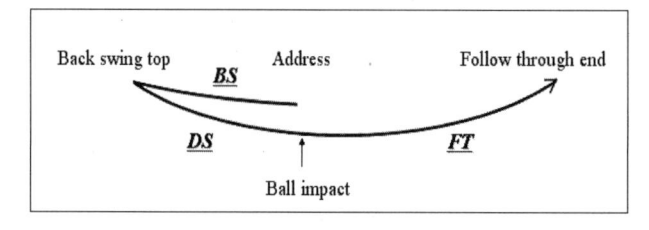

Figure 1. Putting stroke phases according to the trajectory of putter head.

$$JC = \int_0^T \left(\frac{d^3r}{dt^3}\right)^2 dt \qquad (1)$$

3 Results and Discussion

Since our previous study has reported that the difference between elite and novice golfers was the motion of putter head (Tack, 2005, 2006), in this study the results which explained this motion were reported.

Figure 3 shows the displacement (actual moving distance of the putter during putting stroke) and the jerk cost function at putter head between two groups. Left in Figure 3 is the

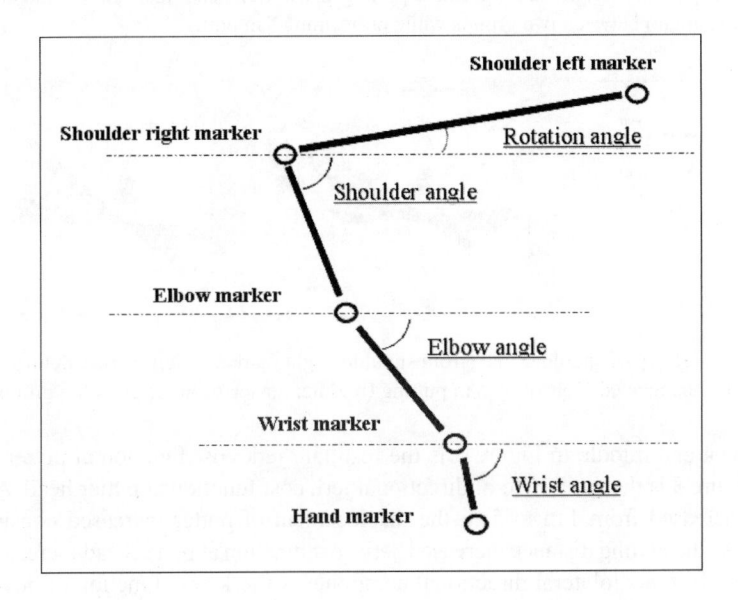

Figure 2. Definition of upper limb angles used in this study.

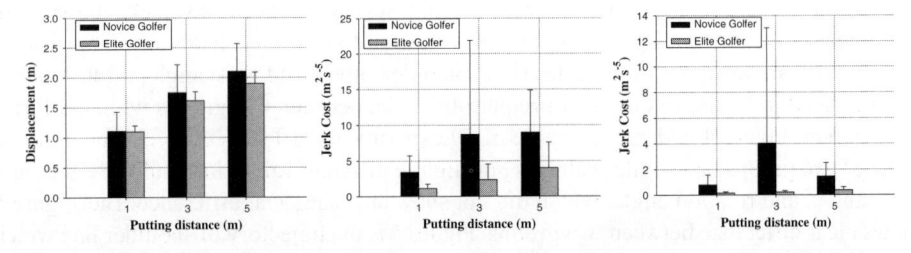

Figure 3. Relationship between displacement (left), jerk cost function (middle: resultant, right: mediolateral directional component) at the putter head and the putting distance.

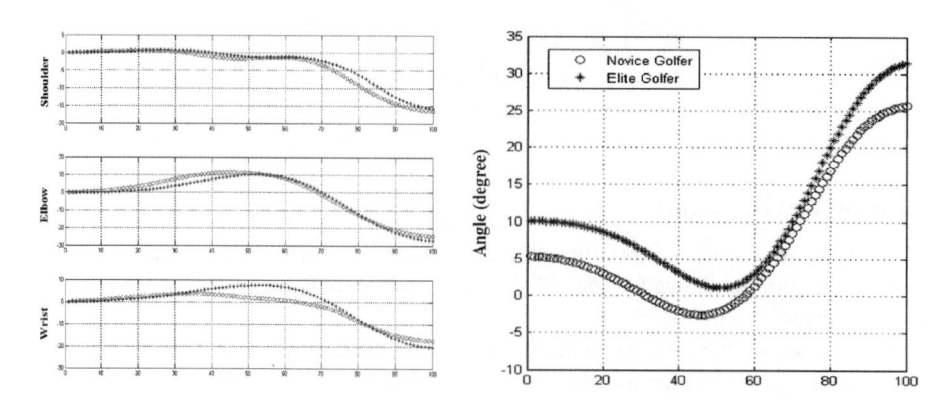

Figure 4. Upper limb angles during entire putting phase (left: shoulder, elbow and wrist angle, right: rotation angle) between two groups while performing 5 m putt.

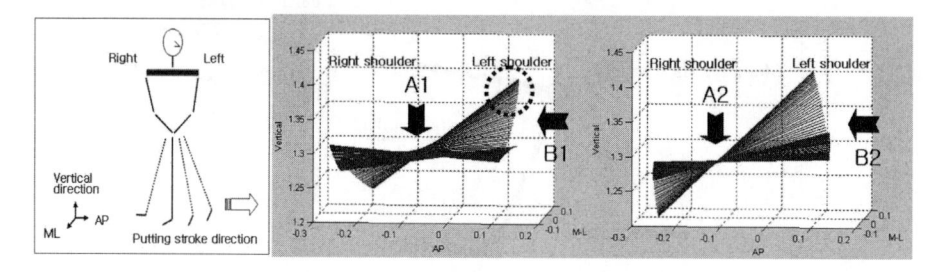

Figure 5. Trajectory of shoulder line (from shoulder right marker to left marker defined in left figure) which is the repeated plots of 50 5 m putting (middle: novice group, right: elite group).

displacement and middle in Figure 3 is the resultant jerk cost function at putter head and right in Figure 3 is the mediolateral directional jerk cost function at putter head. As putting distance increased from 1 m to 5 m, the displacement of putter increased consistently as expected. As the putting distance increased, jerk cost function at putter head increased for two groups, too. But mediolateral directional component of jerk cost function of novice group increased statistically significantly compared with elite group ($p < 0.05$), which means that the mediolateral directional component of the jerk cost function at putter head might be the quantifiable index for distinguishing the playing level of golf putting. From this, it can be said that the elite group who has consistent and accurate stroke showed more smooth movement than the novice group during putting stroke. Conversely the smooth movement of the elite group during putting turned out to be more accurate and consistent.

Figure 4 shows upper limb angles (left: shoulder, elbow and wrist angle, right: rotation angle defined in Figure 2) during entire putting stroke between two groups while performing 5 m putt. Figure 4 is the mean value of 50 experiments (10 subjects × 5 times). Mean value of two groups had similar patterns but slightly different values in shoulder, elbow and wrist angle, and rotation angle, which did not show any statistical difference. But Figure 5 shows clear difference between two groups. Figure 5 is the trajectory of shoulder line which is the repeated plots of entire 50 5 m putting experiments. Shoulder line is defined in left Figure 5, which connects reflective marker of both shoulders.

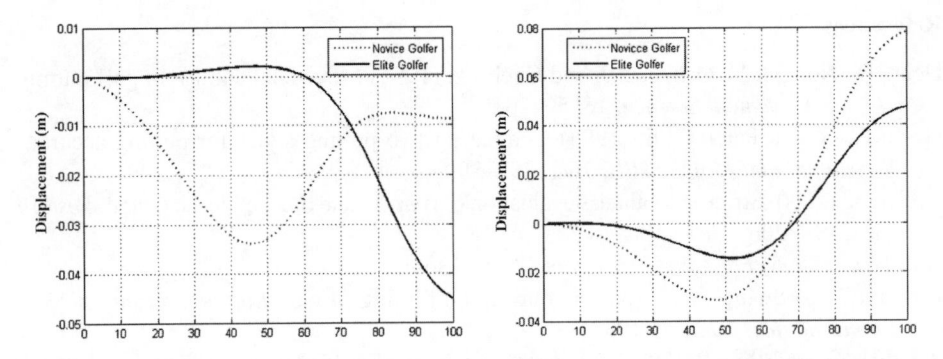

Figure 6. Movement of left shoulder marker during entire putting stroke between two groups (left: anterior-posterior direction, right: mediolateral direction), which is the mean of 50 5 m experiments.

A1 and A2 point in Figure 5 is the center of axis of shoulder rotation. Whereas for novice group the rotational center did not converge into one point, for elite group the rotational center converged into precise single point as shown in middle and right figure of Figure 5. The trajectory of left shoulder marker showed clear difference between two groups (B1, B2 and checked region).

Figure 6 shows the movement of left shoulder marker during entire putting stroke between two groups (left: anterior-posterior direction, right: mediolateral direction), which is the mean of 50 5 m experiments. There is a clear difference pattern in anterior-posterior directional movement between two groups. Additionally the mediolateral directional movement of left shoulder of novice group is much greater than that of elite group. This means that novice group could not maintain pendulum motion during entire putting phase. These phenomena magnified as the putting distance increased.

4 Conclusion

This study investigated the kinematic differences between elite and novice golfers during putting stroke. From this study, it could be concluded as follows: First, as the putting distance increased, jerk cost function at putter head increased for two groups. Only mediolateral directional component of jerk cost function of novice group increased statistically significantly compared with elite group, which means that the mediolateral directional component of the jerk cost function at putter head might be the quantifiable index for distinguishing the playing level of golf putting. Second, there existed a clear difference in the trajectory of shoulder line between two groups. As for novice group the rotational center did not converge into one point, for elite group the rotational center converged into precise single point. Third, there is a clear difference pattern in anterior-posterior directional movement at shoulder between two groups. Additionally the mediolateral directional movement of left shoulder of novice group is much greater than that of elite group. Further studies concerning EMG analysis at lower back, joint torque values through the musculoskeletal modeling etc. are needed to clarify the current understanding of the putting mechanism.

References

Delay D., Nougier V., Orliaguet J. and Coello Y. (1997) Movement control in golf putting, *Human Movement Science,* 16, 597–619.

Gwyn R. G. and Patch C. E. (1993) Comparing two putting styles for putting accuracy, *Perceptual and Motor Skills*, 76/2, 387–390.

Hreljac A. (2000) Stride smoothness evaluation of runners and other athletes, *Gait & Posture*, 11/3, 199–206.

Kim H. C. (2005) *Par saves short game.* NexusBooks.

Park J. (2000) Swing Time Analysis during the Putting Stroke, *Korean Journal of Sport Biomechanics*, 9/2, 187–193.

Paradisis G. and Rees J. (2002) Kinematic analysis of golf putting for expert and novice golfers, Available at: http://coachesinfo.com/category/golf/60/, (Accessed: 2007/2/20)

Tack G. R., Choi J. J., Yi J. H., and Lim Y. T. (2005) Identifying Critical Kinematic Parameters for Better Golf Putting, XXIII ISBS Meeting, Beijing, China.

Tack G. R., Kim H. S., Choi J. J., Yi J. H., and Lim Y. T. (2006) *A Study on the Grip Force During Putting Stroke,* XXIV ISBS Meeting, Salzburg, Austria.

Viviani P. and Flash T. (1995) Minimum-Jerk, Two-Thirds Power Law, and Isochrony: Converging Approaches to Movement Planning, *Journal of Experimental Psychology*, 21/1, 32–53.

Wiren P. (1992) *Golf, building a solid game*, Englewood Cliffs: Prentice-Hall, 189–201.

DEVELOPMENT OF WIRELESS PUTTING GRIP SENSOR SYSTEM

H. S. KIM[1], J. S. CHOI[1], J. H. YI[1], Y. T. LIM[2] & G. R. TACK[1]

[1]*Biomedical Engineering, Konkuk University, Chungju, Korea*
[2]*Sports Science, Konkuk University, Chungju, Korea*

It was reported that the putting stroke accounts for 40~50% during a golf rounding both novice and elite golfers. There are lots of variables to affect control of ball movement during golf putting. Among them, a grip force during putting stroke is one of the important variables. However, there is not much quantitative evidence from published literatures. To quantify the grip force, wireless putting grip sensor system was developed and used to putts by elite (handicap ≤ 2) and novice (handicap ≥ 25) golfers. But the system has some limitations such as constraint of hands from co-axial cable, sweat and uncomfortable feeling. To improve these, the measurement and sensor module was re-designed. To get rid of a discomfort from co-axial cable, the micro-processor was changed from ATMEGA128 (Atmel Corp., USA) to nRF24E1 (Nordic semiconductor, Norway). A subject gains more freedom from co-axial cable by changing micro-processor. Moreover, the measurement module was minimized and could be operated with less battery consumption. For better performance the measurement module was separated into analog and digital parts. To be free from sweat, a thimble and golf glove was used. FSR (Interlink Electronic, USA) sensors were used to detect grip forces. The newly developed system consists of 16 FSR sensors and software for post-processing developed by using LabVIEW 7.1 (National Instrument Inc., USA). The developed system and 3D motion analysis system (Motion Analysis Corp., USA) were synchronized by developed software. The system is expected to have light weight, long term measurement, and more comfort during actual putting experiment.

1 Introduction and Motivation

Golf, baseball, tennis, cricket and hockey are ball game with striking devices. In such a game, there are lots of variables to affect the control of ball movement such as direction, speed, rotation, distance and so on. Among these variables, a grip force is one of the significant parameters. If the grip forces are inconsistent during the whole swing or impact, the ball does not go as he/she wishes. Despite of the significance of the grip force, little study about the grip force has been carried out so far (Schmidt, 2006).

In a golf event, there are two kinds of strokes, swing and putting. In particular, the putting stroke accounts for 40 ~ 50% during a golf rounding both novice and elite golfers. There are lots of variables to affect the control of ball movement during golf putting. Likewise, the grip force is one of the important variables. However, there is not much quantitative evidence from published literatures at each putting phase (Delay, 1997; Gwyn, 1993). To quantify the grip force, a wireless putting grip sensor system was developed and used to putts by elite (handicap ≤2) and novice (handicap ≥25) golfers (Tack, 2006; Kim, 2006). The results of the data indicated that there was a significant difference in the variation in grip force between two groups. But the system has some limitations such as constraint of hands from co-axial cable, sweat and uncomfortable feeling. In this paper, we improved these limitations. The measurement and sensor board was re-designed through the change of micro-processor and wireless communication method and glove. The newly developed system has a light weight, long-term measurement, and more comfort during putting experiment.

2 Measurement System

The newly developed grip force measurement system for putting stroke has several differences compared with the previous system. A main difference is that left and right sensor systems are separated. So the limitations from co-axial cable could be improved. By changing the micro-processor, current consumption is much less than old one. Figure 1 is the overview of the newly developed grip force measurement system.

2.1. Grip Sensors

During the putting stroke, grip force should be as light as possible and keep the force at a constant throughout the putting stroke. So the sensitive pressure sensors are needed to detect the change in grip force during the putting stroke (Kim, 2006). A detectable pressure sensor is designed by using the FSR sensor (Interlink Electronics, USA) as shown in Figure 2. Sensing area of the FSR sensor has epoxy resin bump for better interface between hand and putter. The sensor is connected to the electronic circuit using a thin co-axial wire (0.4 mm). Synthetic leather protects the FSR sensor from external mechanical stress and forms a bending string. The sensors are fixed on the sensing points of the finger-tips and the palms using the strip-velcro and the glove respectively as shown in Figure 3 and 4.

2.2. Grip Board and Wireless Communication

The grip board integrates four sections: power, signal condition, micro-processor and radio. Power section consists of step-up regulator. The regulator supplies a +3.3 V power

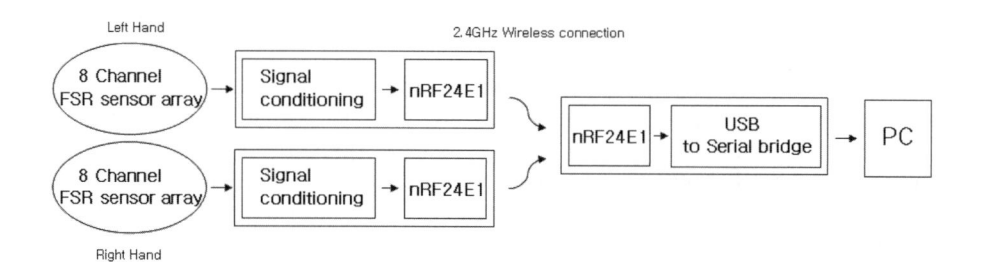

Figure 1. Overview of newly developed grip force measurement system.

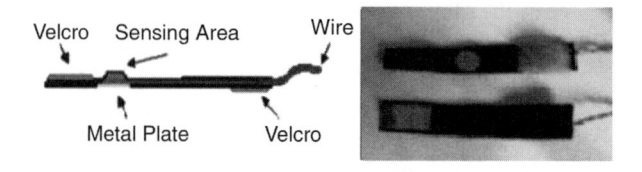

Figure 2. Detectable pressure sensor.

from two AAA NiMH re-chargeable battery to grip board (it could be operates only one battery). Signal condition section amplifies a grip force signal from grip sensor array. It was organized two low-power rail-to-rail op amps. Micro-processor nRF24E1 was used to convert the grip force analog data to digital data through analog to digital conversion and to send the data to PC. The nRF24E1 (Nordic Semiconductor datasheet, 2006) is an nRF2401 2.4 GHz radio transceiver with an embedded 8051 compatible microcontroller and a 10-bit 9 input (8 channels are analog inputs and other 1 channel is used for monitoring a supply power) 100 kSPS AD converter and supplied by one voltage with range 1.9 V to 3.6 V. The nRF24E1 supports the proprietary and innovative modes of the nRF2401 such as ShockBurst and Duo Ceiver™. At radio section, ShockBurst™ and DuoCeiver™ modes were used. The ShockBurst™ technology uses on-chip FIFO to clock in data at a low data rate and transmit at a very high rate (1 Mbps) offered by the 2.4 GHz band, thus enabling extreme power reduction. The DuoCeiver™ technology provides 2 separate dedicated data channels for RX and replaces the need for two, stand alone receiver systems. By using these two modes, the grip board reduced current consumption significantly in wireless communication (see Table 1) and two boards (left and right grip board) was successfully separated.

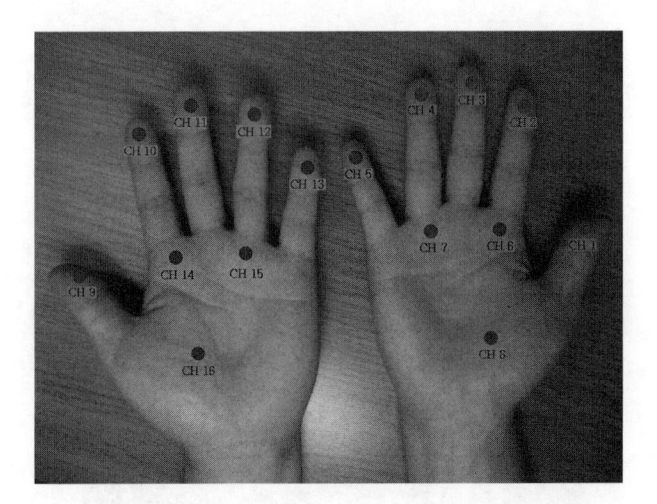

Figure 3. Sensing points and channel numbers (Sensing points are selected based on golfers' advice).

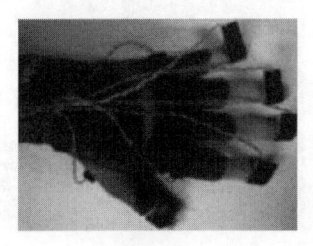

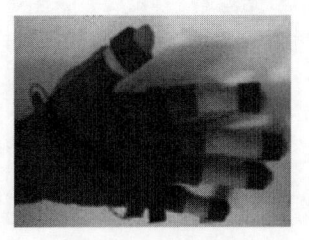

Figure 4. A hand with sensors and glove.

Table 1. Comparison between previous and newly designed system (* Weights are measured except battery pack).

	Previous system	**Newly designed system**
Micro-processor	Atmega128 (AVR)	nRF24E1(8051)
Package type	64pin TQFP(14 × 14 mm)	36pin QFN(6 × 6 mm)
Wireless link	Bluetooth(ACODE-300)	nRF2401 2.4 GHz RF transceiver
Dimension	8.5(W) × 5(L) × 1.4 cm (H)	5.8 (W) × 4.5(L) × 0.1 cm (H)
Weight	37.18 gram	9.14 gram (left board only)
DC current consumption	Radio : 100 mA (TX mode)	Radio : 16 mA (TX mode)
	Core : 20 mA	Core : 3 mA
AD characteristics	8Ch, 10bit, 15kSPS	9Ch, 10bit, 100kSPS
Battery	4 AAA NiMH Re-chargeable batteries	2 or 1 AAA NiMH re-chargeable batteries

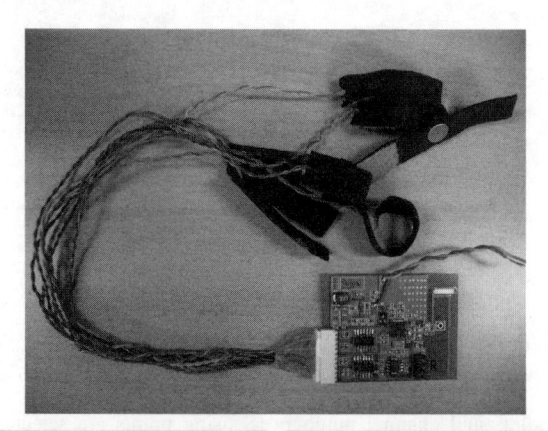

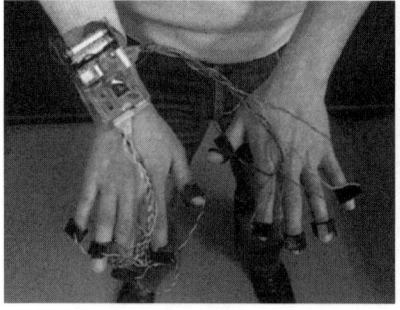

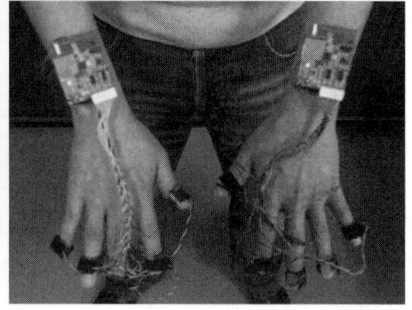

Figure 5. A Grip board with a grip sensor (left: previous, right: new).

Figure 6. User interface of Grip force recorder.

Additionally, a stand alone receiver system was developed. It consists of two parts, nRF24E1 and USB to serial bridge. Micro-processor nRF24E1 receives the grip force data by using two modes and convert it to serial data. The serial data is transferred and converted of data between USB and RS-232. Serial data speeds are catered for 57,600 baud rates.

2.3. Display and Data Record

The user interface of the data record software was improved. The time set function for record data in fixed time is added and the time could be set freely. If the record data button is pushed, grip force data is stored (default data file type is *.xls) and synchronize with 3D motion analysis system (Motion Analysis Corp., USA). Software was developed by using LabVIEW 7.1 (National Instrument Inc., USA).

3 Results and Discussion

The newly developed system was evaluated.

From the results, the newly developed system is better than previous system. Dimension is smaller, especially board height had decreased about over 10 times. Weight is decreased over 4 times. So subject can not feel the board. We think that experiment with previous system had an effect on the results of grip force data because of some limitations. Since the board is changed, the following conditions for experiment could be obtained: little effect of the sensor system on putting and subject's unawareness of the experiment due to the un-tethered light-weight sensor system.

4 Conclusions

The new grip force measurement system was developed in this study. Compare with the previous system, the system is light-weight, less-size, possible to long-term measurement due to the ShockBurst™ mode, and more comfortable due to the DuoCeiver™ mode during actual putting experiment.

Acknowledgements

This research was supported in part by the Ministry of Commerce, Industry and Energy (MOCIE) through Busan Techno Park.

References

Delay, D., Nougier, V., Orliaguet, J. P. and Coello, Y. (1997) Moment Control in Golf Putting, *Human Movement Science*, 16(5), 597–619

Gywn, R.G. and Patch, C.E. (1993) Comparing Two Putting Styles for Putting Accuracy, *Perceptual and Motor Skills*, 76(2), 387–390

Kim, H. S., Yi, J. H., Tack, G. R., Choi, J. S. and Lim, Y. T. (2006) Wireless Grip and Acceleration Measurement System for Putting Stroke Analysis, WC2006, 14, 646–648

Schmidt, E., Roberts, J., and Rothberg, S. (2006) *Time-Resolved Measurements of Grip Force During a Golf Shot.* In: Moritz, E. and Haake, S. (Eds.), *The Engineering of Sports 6*, Springer, 57–62

Tack, G. R., Choi, J. J., Yi, J. H. and Lim, Y. T. (2005) Identifying Critical Kinematic Parameters for Better Golf Putting, XXIII ISBS Meeting, Beijing, China

Tack, G. R., Kim, H. S., Choi, J. J., Yi, J. H. and Lim, Y. T. (2006) A Study on the Grip Force During Putting Stroke, XXIV ISBS Meeting, Salzburg, Austria

Nordic Semiconductor, Norway. Available: http://www.nordicsemi.no/files/Product/data_sheet/Product_Specification_nRF24E1_1_3.pdf.

Nordic Semiconductor, Norway. Available: http://www.nordicsemi.no/files/Product/white_paper/ Easy_nRF_RS232_to_USB_upgrade.pdf.

Nordic Semiconductor, Norway. Available: http://www.nordicsemi.no/files/Product/white_paper/nRF240x-ShockBurst-feb03.pdf.

Nordic Semiconductor, Norway. Available: http://www.nordicsemi.no/index.cfm?obj=product&act5display&pro=79#.

LEADERSHIP BEHAVIOR AS PERCEIVED BY COLLEGIATE GOLF COACHES AND PLAYERS IN TAIWAN AND THE RELATIONSHIP TO BASIC PERSONALITY TRAITS

B.F. CHANG[1], I.C. MA[2] & Y.J. JONG[3]

[1]*Department of Physical Education, National Taichung University, Taiwan*
[2]*Office of Physical Education, Chung Shan Medical University, Taiwan*
[3]*Department of Physical Education, National Chiayi University, Taiwan*

A tremendous amount of research has been conducted related to leadership, job satisfaction, and role expectation. A lesser amount has been conducted on leadership behavior and personality traits as perceived by athletic team coaches and members, especially in golf. The study of personality traits found in golf team members is critical for a coach's effective leadership and team success. The purpose of this study was to explore the possible relationship between leadership behaviors demonstrated by golf coaches and their basic personality traits, and the basic personality traits of athletes among collegiate golf teams in Taiwan. The sample population of this study included 29 golf coaches and 236 golf athletes in 29 institutions with collegiate golf teams within Taiwan. Data were collected through the use of a self-designed demographic information questionnaire, the revised *Leader Behavior Description Questionnaire, Form XII*, and the revised *Emotions Profile Index*. Primary findings to emerge from this study included the following: (a) golf coaches regarded the leadership behavior to be used significantly more often than did the athletes; (b) there were significant relationships between the team members' perceived leadership behavior of their coaches in Initiating Structure and the personality traits of the team members in the dimensions Dyscontrolled, Depressed, and Distrustful, respectively. In addition, there were significant relationships between the team members' perceived leadership behavior of their coaches in Consideration and the personality traits of the team members in Trustful, Timid, Distrustful, Controlled, Aggressive and Bias dimensions respectively.

1 Context of the Study

Behavior can be viewed as a function of role expectations and individual need dispositions. Both of these are further influenced by the institution and the interaction between the institutional and individual dimensions. Therefore, the study and understanding of in congruency between role expectations and individual need is important. A multitude of studies with various approaches have attempted to assess either directly or indirectly this in congruency between the individual and organization, and the possible effects upon productivity, turnover, absenteeism, group cohesion, morale, and job satisfaction (Daniel, 1975).

A successful leader should have a firm belief in her or his ability, confidence in her or his philosophy, a strong sense of purpose, and a genuine respect of self. Sound leadership can often be identified by the way the administrator behaves in person-to-person relationships. An effective administrator seeks to integrate the welfare of the organization with the needs of the people. Successful administration is marked by good teamwork, a balance of responsibility and authority, integrity, consideration for others, and satisfaction of the needs for self-realization (Frost et al., 1988).The lead research on emotions was initially addressed to the study of anxiety and psychological factors inhibiting performance (Hackfort & Spielberger, 1989; Jones & Hardy, 1990).

Anxiety has been equated with a loss of emotional control and, therefore, viewed as something that will have a deleterious impact on sport performance. However, after almost two decades of extensive research, questions remain about the usefulness of assessing anxiety in sport apart from broader considerations of other emotional states (Mark & David, 1999). Nesti & Sewell (1997) have suggested that more might be learned by focusing on the meaning that anxiety has for an individual and, following existential psychology, viewing anxiety as something that should be faced constructively rather than avoided at all costs. From this study, coaches in Taiwan will be able to see the relationship between the leadership styles and the personality traits of athletes who play golf.

A tremendous amount of research has been conducted related to leadership, job satisfaction, and role expectation. A lesser amount has been conducted on leadership behavior and personality traits as perceived by athletic team coaches and members, especially in golf. The study of personality traits found in golf team members is critical for a coach's effective leadership and team success.

The knowledge of differences between coaches' and team members' perceptions, and the relationship between coaches' leadership behavior and athletes' basic personality traits may help coaches adopt more appropriate leadership behaviors in working with their team members.

The study of the basic personality traits found in team members is critical for a coach's effective leadership. The knowledge of differences between coaches' and team members' perception, and the relationship between coaches' leadership behavior and athletes' basic personality traits may help coaches adopt more appropriate leadership behaviors in working with their team members.

This study was limited by the following factors:

1. Questionnaires were mailed and the concern exists that the answers would not be written and returned by the golf coaches and athletes.
2. Conclusions of this study are only applicable to collegiate golf teams in Taiwan.
3. Of the subjects selected to participate in this study, only those who returned the questionnaires by the deadline were included in the study.
4. The study is limited by the accuracy and completeness with which the subjects respond to the questionnaires.

2 Literature Review

Leadership has been studied in many different ways, depending on the researchers' definitions of leadership and methodological preferences. One line of research on leadership effectiveness was the focus on the personal attributes of leaders, the so-called trait approach. Another was a behavioral approach in which the behaviors of effective and ineffective leaders were compared, usually based on the reported perceptions of others. Some researchers have employed a situational approach which assumes that different behavior patterns or trait patterns were effective in different situations, and that the same pattern was not optimal in all situations (Hartman, 1999).

Personality traits found to be especially relevant for leadership effectiveness include high energy and stress tolerance, self-confidence, internal locus of control orientation, emotional maturity, personal integrity, socialized power motivation, moderately high achievement orientation, and low need for affiliation (Bass, 1990).

Leadership behavior has been studied for many decades. Much of research was conducted regarding leadership behavior in the fields including business, education, industry, service trade, or government. Extended studies which were related to emotional expression and personality trait were also mentioned in this chapter. Besides, behavioral theories of leadership, and the instruments used in this study, the *Leadership Behavior Description Questionnaire* and the *Emotion Profile Index* were introduced.

3 Methodology

The population for this study included golf team coaches and their team members in those institutions including colleges and universities that have student golf teams within Taiwan. The sample consisted of up to 161 coaches and all team members of each golf team, depending upon the result of the census survey. Exact sample size was determined following a survey of interest and total possible sample.

The data for the study were collected through the use of a self-designed demographic information questionnaire, the *Leader Behavior Description Questionnaire*, and the *Emotions Profile Index*.

Null Hypotheses

1. There is no significant difference between the self-perceived leadership behavior of the coach and the team's perceived leadership behavior of the coach.
2. There is no significant difference in athlete's self-perception or in the athletes' perception of coaches, and the coach's self-perception regarding personality traits.
3. There is no significant relationship between the leadership behavior of the coach and the personality traits of the coach.
4. There is no significant relationship between the athletes' perceived leadership behavior of the coach and the athletes' self-perceptions of their own personality traits.

Each set of questionnaires was coded for the purpose of sending follow-up requests to those not responding to the initial request and to match the coach with his or her own team members. The coach was requested to distribute the questionnaires and cover letters to his or her team members who has sent back the informed consent forms. Upon completion of the survey, each team member and coach returned their surveys directly to the researcher in separate self-addressed stamped envelope.

The research used the paired-sample *t* tests to verify the difference between the perceived leadership behavior of the coach and team's perceived leadership behavior of the coach. The paired-sample *t* tests to verify differences in personality traits of players who perceive the coach the same as the coach and those players who perceive the coach differently.

The research used the Pearson product-moment correlation coefficients to verify the relationship between the leadership behavior of the coach and the coaches' personality traits and the Pearson product-moment correlation coefficients to verify the relationship between the team's perception of the coach's leadership behavior and the team's personality traits.

A simple multiple regression analysis was performed to examine the significance of leadership behavior and selected demographic variables in predicting golf athletes' personality traits.

4 Results

The population of this study included 29 golf coaches and 236 golf athletes in 29 institutions within Taiwan with collegiate golf teams. They were all invited to participate in this investigation. Within the allowed time, 28 (96.6%) golf coaches and 186 (78.8%) golf athletes responded to the survey. Within the survey, 27 (93.1%) coaches and 168 (71.2%) athletes completed a useful survey.

Hypothesis 1 stated that there was no significant difference between the self-perceived leadership behavior of the coaches and the athletes' perceived leadership behavior of the coaches. Independent t tests were performed on the composite means for the leadership behavior regarding Initiating Structure and Consideration. According to the results of data analyses, there were significant differences between the team's perceptions and coach's' self-perceptions regarding the leadership behavior in Initiating Structure (t (193) = 5.149, $p < .001$), and in Consideration (t (193) = 5.901, $p < .001$). Golf coaches regarded the leadership behavior in Initiating Structure and Consideration to be used significantly more often than did the team. Therefore, hypothesis 1 was rejected.

Hypothesis 2 stated that there was no significant difference in athlete's self-perception or in the athletes' perception of coaches, and the coach's self-perception regarding personality traits. Independent t tests were performed on the composite means for the percentile of personality traits.

Results of analyses indicated that there were no significant differences between the self-perceptions of coaches and athletes regarding the personality traits in Trustful, Timid, Depressed, Distrustful and Aggressive dimensions, respectively. There were significant differences between the self-perceptions of coaches and athletes regarding the personality traits in Dyscontrolled, Controlled and Gregarious dimensions, respectively. Athlete's self-perceived personality traits in Dyscontrolled ($M = 35$) and Controlled ($M = 47$) dimensions were significantly stronger than those of coaches ($M = 21, 34$).

Hypothesis 3 stated that there was no significant relationship between the leadership behavior of the coach and the personality traits of the coach. Pearson product-moment correlation coefficients were performed on the composite means of the leadership behavior and the personality traits of the coach.

The results showed that there was a significant negative relationship between coaches' self-perceived leadership behavior in Initiating Structure and coaches' self-perceived personality traits in Distrustful dimension, $r = -.654, p < .001$. As well, there was a significant negative relationship between coaches' self-perceived leadership behavior in Consideration and coaches' self-perceived personality traits in Distrustful dimension, $r = -.432$, $p < .05$.

Hypothesis 4 stated that there was no significant relationship between the athletes perceived leadership behavior of the coach and the athletes' self-perceptions of their own personality traits. Pearson product-moment correlation coefficients were performed on the composite means of the team members' perceived leadership behavior of their coaches and the personality traits of the team members.

The results showed that there were significant relationships between the team members' perceived leadership behavior of their coaches in Initiating Structure and the personality traits of the team members in Dyscontrolled ($r = .172, p < .05$), Depressed ($r = -.295$, $p < .001$), and Distrustful ($r = .284, p < .001$), respectively.

5 Discussion

No matter what the athletes' perceptions of their coaches, athlete's self-perceptions or coach's self-perceptions, the personality traits of coaches and athletes were very similar; both of them had high emotions in Timid, Depressed, Distrustful and Aggressive dimensions, and low emotions in Trustful, Dyscontrolled and Gregarious dimensions. In general, the personality traits of the golf coaches and athletes tended to be cautious, careful, anxious, sad, gloomy, stubborn, resentful, rejecting of people, and aggressive. In addition, the low Bias scores showed that athletes and coaches were described in socially undesirable ways.

Basically, it was believed that these emotions are not positive and good for a person, especially, a golf coach or an athlete. These factors could become barriers for coaches to lead their golf team members and for athletes to become elite. So the coaches and athletes in this study should improve their personality traits as indicated above in order to make progress in their leadership behavior or competitive performance.

In the comparisons of athletes' perceptions and coaches' self-perceptions regarding Initiating Structure and Consideration of leadership behavior, golf coaches regarded the leadership behavior in Initiating Structure and Consideration to be used significantly more often than did the athletes. Coaches reported that they often clearly defined their own role, while letting athletes know what was expected. Athletes disagreed, however. The average score of their perceptions of Initiating Structure was between occasionally and often. Coaches also regarded that they often contributed to the comfort, well-being, status, and value of the athletes. Athletes did not totally agree. The average score of athletes' perceptions of Consideration was between occasionally and often.

The evidences above indicated that golf athletes did not seem to be satisfied with their coach's leadership behavior in Initiating Structure and Consideration, while golf coaches seemed to be confident in their performance. Coaches did not completely let their team members understand the leadership behaviors they performed and this could be a barrier in making progress in the skill and ability toward competitive performance. Coaches need to demonstrate a leadership behavior that is more clear and understandable to the athlete.

According to the analyses of Pearson product-moment correlation coefficients, there were significant negative relationships between coaches' self-perceived leadership behavior in Initiating Structure and Consideration, respectively, and coaches' self-perceived personality traits in Distrustful dimension. Coaches' leadership behavior in Initiating Structure and Consideration of their athletes was deeply and negatively affected by their own emotion in the Distrustful dimension. In this study, the percentile of coaches' Distrustful emotion was very high; therefore, coaches need to improve this aspect of emotion in order to enhance their leadership behavior as perceived by their team members, and thus make a breakthrough for the competitive performance.

The correlations of athletes' perceptions of leadership behavior and their personality traits showed that there were a significantly positive relationship between the team members' perceived leadership behavior of their coaches in Initiating Structure and the personality traits of the team members in Dyscontrolled ($r = .172, p < .05$) and Distrustful ($r = .284, p < .001$) dimensions, and a significantly negative relationship in Depressed ($r = -.295, p < .001$) dimension. The more Dyscontrolled and Distrustful emotions of coaches that athletes perceived, the more Initiating Structure athletes reported. The more depressed emotion of coaches that athletes perceived, the less Initiating Structure athletes reported.

References

Bass B.M. (1990) *Bass and Stogdill's Handbook of Leadership: Theory, Research, and 11 Managerial Applications (3rd ed)*. Free Press, New York.

Daniel J.V. (1975) Faculty job satisfaction in physical education and athletes. In: Zeigler E.F. and Spaeth M.J. (Eds) *Administrative theories and practice in physical education and athletes*, 152–155.

Frost R.B., Lockhart B.D. and Marshall, S.J. (1988) *Administration of physical education and athletics: Concepts and practices*. W.C. Brown, Dubuque, Iowa.

(1989). Sport related anxiety: Current trends in theory and research. In: Hackfort D., and Spielberger C.D. (Eds.), *Anxiety in sports: An international perspective* 261–267. Hemisphere, New York.

Hartman L. (1999) A psychological analysis of leadership effectiveness. *Strategy & Leadership*. 27/6, 30–32.

Mark N. and David S. (1999) The importance of anxiety and mood stability in sport. *Journal of Personal and Interpersonal Loss*, 4, 257–267.

8. Ball Sport – Cricket

NON-LINEAR VISCOELASTIC PROPERTIES AND CONSTRUCTION OF CRICKET BALLS

B. VIKRAM & F.K. FUSS

Sports Engineering Research Team, Division of Bioengineering, School of Chemical and Biomedical Engineering, Nanyang Technological University, Singapore

The mechanical properties and the internal constructions of different brands of cricket balls were studied and compared. Compression and stress-relaxation tests showed that most brands were inconsistently manufactured. Different specimens of the same brand in some cases had visibly different constructions, which altered their mechanical behaviour. Viscoelastic properties of cricket balls follow the power law, with a viscosity constant, the exponent in the power law, between 0.05 and 0.08. Consistently manufactured balls show a clear increase of the stiffness with increasing deflection rate. The stiffness of cricket balls varies, resulting into harder and softer balls. The lack of standardisation and testing of cricket balls is thus likely to have an impact on the game.

1 Introduction

Cricket is one of the top-tier sports and yet very little methodological scientific research has gone into the study of cricket balls. The condition in which the cricket ball is used has dramatic consequences for the outcomes of the game – be it its surface and seam for aerial movement, especially for the swing of the ball, be it its hardness, determining the ease or difficulty with which it can be hit. Cricket ball technology, in contrast to golf balls, is highly underdeveloped, as most cricket balls are still hand-made.

Carré *et al.* (2004) investigated the impact of a cricket ball, and discussed that the stiffness might increase with the deflection rate. Subic *et al.* (2005) performed compression testing in the Kookabura Tuf Pitch cricket ball at different deflection rates and found an exponential relationship between the loading velocity and the ball stiffness.

The aim of this study is to investigate the mechanical properties and the internal constructions of a few brands of cricket balls. Furthermore, the non-linear viscoelastic properties of cricket balls will be compared among different brands. The experimental data from mechanical testing and the observations made from internal constructions will be applied to propose a standard for quality control.

2 Experimental

2.1. Cricket Balls Analysed

The balls studied are listed in Table 1. All balls were new, and were tested only once, either for stress relaxation, or for compression.

2.2. Stress Relaxation Tests

For both stress relaxation and compression tests, we used an Instron material testing machine (model no.: 3366). For the stress relaxation tests, the balls were pre-loaded (F_0) between

Table 1. Cricket balls investigated and their details.

Brand	Model	Colour	Country of origin	Construction	Abbreviation
Kookaburra	Special Test	red	Australia	machine made, molded cork-rubber centre, tension machine wound*, 2 pieces	Kr
Gray-Nicolls	Super Cavalier	white	Pakistan	hand made, woolen twine and cork layers surrounding a central core of cork-rubber, 4 pieces	Gw
Regent	Match	red	India	4 pieces	Rr
Regent	Match	white	India	2 pieces	Rw
SG/ Sanspareils-Greenlands	Tournament	red	India	hand made, high quality naturally seasoned centre, encasing layers of cork, tension wound under pure wool, 4 pieces	Sr

* the tension winding was missing in the balls analysed

1.5 and 6 kN at a deflection rate of 500 mm/min, and the decreasing load F was measured for $t = 3600$ seconds. As the stress relaxation generally and consistently showed a linear behaviour when plotting natural logarithm of LN(F) against the natural logarithm of the relaxation time LN(t), a power law was selected to model the non-linear visco-elastic properties:

$$F = At^{-B} \quad \text{or} \quad \text{LN}(F) = A - B\,\text{LN}(t) \qquad (1)$$

The parameter A, however, depends on the initial load F_0. B is the load-independent viscosity constant. Thus,

$$F/F_0 = A_F t^{-B} \qquad (2)$$

The planes of loading were: perpendicular to the plane of the seam (plane A), and parallel to the plane of the seam (plane B). The following models were tested: Kr, Gw, Sr (in each model 4 balls in plane A and 1 in plane B), Rr and Rr (3 balls each in plane A).

2.3. Compression Tests

For the compression tests, the balls were loaded up to 9 kN with crosshead speeds of 500, 160, 50, 16, and 5 mm/min. From the load – deflection curve, the stiffness was calculated, which

Figure 1. B vs. F_0.

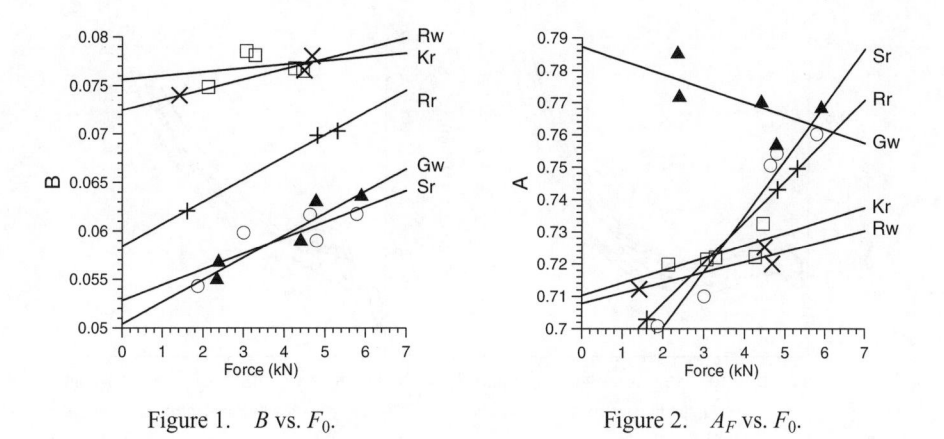

Figure 2. A_F vs. F_0.

is the deflection derivative of the load. For each model, 3 balls were tested at 500 mm/min, and at least one ball at the 4 slower deflection rates, depending on the consistency.

2.4. Ball Construction

All balls used for compression testing were also analysed by examining their internal construction. After splicing their seams open, the nature of their anatomy, the materials used and the properties of those materials were all noted. For example, whether there were layers of cork and woolen twine enclosing the internal core; the size, texture and other properties of the internal core; and the consistency of the construction across different specimens of the same brand and model. Details of ball construction were correlated to the compression data.

3 Results

3.1. Stress Relaxation

The coefficients A_F and B, as functions of F_0, are shown in Figures 1 and 2. In contrast to Eq. (2), A_F and B did not turn out to be constant and rather change linearly with F_0. The higher the parameter B, the more viscous is the ball. B generally increased with F_0; Kr and Rw are the most viscous balls, Sr and Gw the least, and Rr showed intermediate viscosity (Figure 1). A_F showed a steep increase with F_0 in Sr and Rr, a slight increase in Kr and Rw, and a decrease in Gw (Figure 2). There was no difference between stress relaxation in plane A and plane B (Kr, Gw, Sr).

3.2. Compression Tests and Ball Structure

Kr showed a higher stiffness at higher deflection rates, which was expected as $B > 0$. In all other balls, however, this result was slightly diluted by the poor consistency and non-uniform construction. The individual results for all balls tested are as follows:

Kookaburra Kr: all balls showed a uniform construction. There were no layers of tension wound wool inside the leather contrary to the claims of the manufacturer (Kookaburra, 2007). The stiffness was highly consistent and thus showed a clear dependence on the

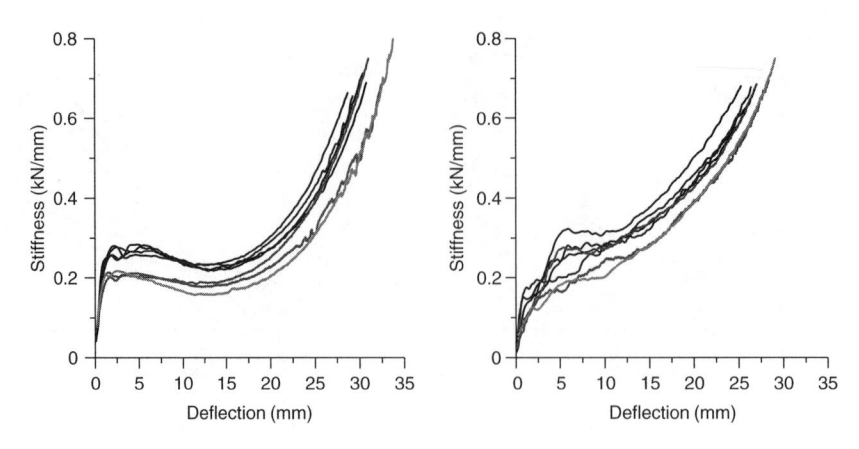

Figure 3. Kr: stiffness vs. deflection (left: plane A, right: plane B), at 5 deflection rates.

deflection rate. When compressing in plane A, the stiffness dropped after an initial peak, in contrast to compression in plane B (Figure 3).

Gray-Nicolls Gw: all balls had layers of woolen twine along with cork that padded an inner core of molded rubber-cork. Surprisingly, the cores were not of uniform size. There were two distinct size ranges – a smaller set of cores (4.4 ± 0.2 cm) and a larger set of cores (5.7 ± 0.3 cm). For those with smaller cores, the cork shavings were arranged in a more irregular and haphazard manner. There was no difference in stiffness between compression in plane A and B. However, the balls with smaller cores were significantly stiffer than those with larger cores (Figure 4), forming two separated clusters of inconsistent stiffness curves. This was verified by t-test applied to the two sets of variables – the gradient of the stiffness-deflection curve and the size of the core ($p = 0.0016$): small cores, stiffness gradient: $29.4 \pm 5.37 \, N/mm^2$, large cores, stiffness gradient: $17.7 \pm 4.11 \, N/mm^2$.

Regent white Rw: all balls had a construction similar to that of Kr – a large core of rubber-cork without woolen twine windings. The stiffness behaviour was highly irregular and inconsistent (Figure 5), without any distinct difference between plane A and B. The different hardness of the cores was also verified by auditory tests as the pitch of the impact sound was correlated with the hardness.

Regent red Rr all balls had a smaller central core covered by layers of cork twined with wool. The cores were of varying hardness which was distinctly felt when they were sawn in half. The harder cores were more difficult to cut. The cork in the cores was mixed with a white or silvery substance. The stiffness behaviour was highly inconsistent (Figure 6) without any clear difference between plane A and B.

Sanspareils-Greenlands red Sr: though all balls were of the same model (SG Tournament), we could identify two distinct types of constructions: one set of balls had neat layers of cork in baseball construction (comparable to the 2 leather pieces of a baseball) with woolen twine covering a rubber sphere; the 2nd set of balls had cork packed in a rough manner with woolen twine covering a core made of cork with varying degrees of woodiness, and ranging from spherical to completely irregular shape. The balls with rubber cores were clearly less stiff than the ones with cork cores, forming two distinctly separated clusters (Figure 7). There was no difference between A and B.

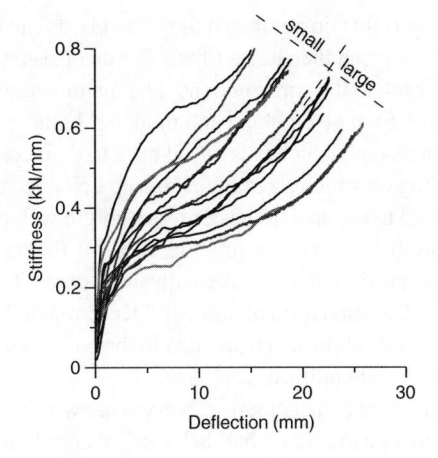

Figure 4. Gw: stiffness vs. deflection at 5 deflection rates (small, large = core diameter).

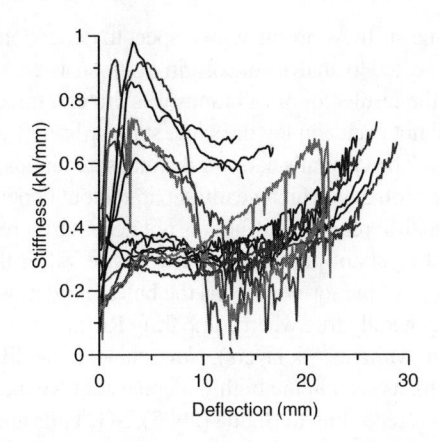

Figure 5. Rw: stiffness vs. deflection at 5 deflection rates.

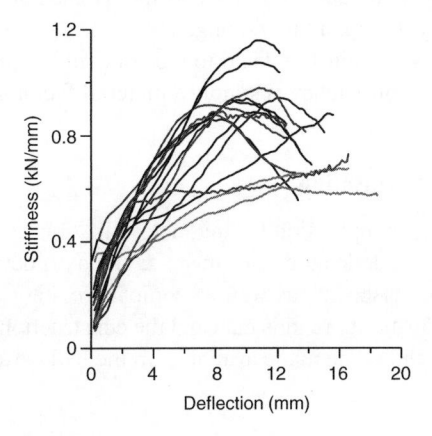

Figure 6. Rr: stiffness vs. deflection (5 defl. rates).

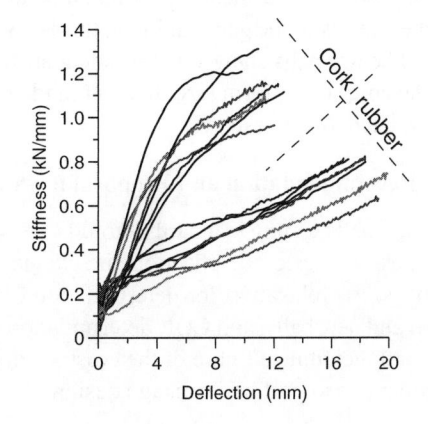

Figure 7. Sr: stiffness vs. deflection (5 defl. rates).

4 Discussion

Excepting the Kr balls, none of the other brands showed any degree of consistency in stiffness across different specimens. Even the Kr balls did not have the construction that was promised by the manufacturers. A certain degree of variation in properties is only natural considering the biological origin of the materials used. Nevertheless, the fact that the machine-made Kr balls were quite consistent shows that the manufacturing process is important for consistency. If the quality of the materials or the manufacturing process are not standardised, a high degree of inconsistency is going to be the result. Therefore, it is recommended that standardised testing of the different balls is carried out by the governing bodies of the sport. The stress-relaxation and compression tests performed in this study can be used for that purpose. Randomly picked specimens from a batch of newly manufactured balls can be tested to see if they provide the same stiffness profile with respect to deflection. Balls that do not match those standards can be dissected to analyse the origin of inconsistency which can then be rectified. The inconsistency in the results for most brands was directly traced back to the constructions. The manufacturers did not even employ the same standardised process for a given brand and model of balls.

The Gw balls had two distinct size ranges for its core which was traced back to their compression data which neatly separated out depending on which size range the core belonged to. The difference in stiffness is not necessarily related to the size of the core, as it might well be, that layers of pure cork and wool are stiffer than the cork-rubber mixture used for the core. More of the former means the ball is harder, which is the case for balls with small cores. This is generally true when comparing Rr (layers of cork around a small core) and Kr (cork-rubber core without cork layers). Nevertheless, the difference might also be related to the cork-rubber ratio, as seen in the highly inconsistent Rw balls despite uniform core size.

According to Bhatia (2005), SG "balls are checked exhaustively at every stage within the factory ... A minor element out of place could change the way a ball behaves". Accordingly, SG claim that "all SG. cricket equipment is manufactured to strict quality controls" (SG Cricket, 2007). Our experiments proved those claims untrue, as the 2 Sr subtypes contained either rubber or cork cores, resulting into different stiffness. This clearly indicates a failure of quality control and inconsistent manufacturing techniques. For a brand that is used at the highest levels of the game in India, this is a very disappointing finding.

The Rw balls showed inconsistent stiffness, obviously related to the cork-rubber ratio of the cores of uniform size. In this brand, the inconsistency is simply a material factor and not related to the design.

5 Recommendation and Proposal for Standardisation

Cricket balls testing standards should extend to compression testing, determination of the stiffness at a specific deflection and standardized deflection rate (mean ± standard deviation), stress relaxation for determination of the viscosity, as well as comparison between used and new balls, and COR determination. Furthermore, this data and the construction of the ball including a figure of the cross section should be made available on the ball carton, comparable to golf ball package design.

References

Bhatia R. (2005) Cherry Pickin'. *Wisden Asia Cricket*, 2005/6 (online). Available: http://content-sl.cricinfo.com/ci/content/story/210922.html (accessed June 2007).

Carré M.J., James D.M. and Haake S.J. (2004) Impact of a non-homogeneous sphere on a rigid surface. *Proc. Instn. Mech. Engrs. Part C: J. Mechanical Engineering Science*, 218, 273–281.

Gray-Nicolls Sports Ltd. (2007) Company Website (online). Available: www.gray-nicolls.co.uk (accessed June 2007).

Kookaburra Sports Ltd. (2007) Company Website (online). Available: http://www.kookaburra.biz (accessed June 2007).

SG Cricket Ltd. (2007) Company Website (online). Available: www.sgcricket.com , http://www.sgcricket.com/html/GUARANTY.HTM (accessed June 2007).

Subic A., Takla M. and Kovacs J. (2005) Modelling and analysis of face guard designs for cricket. *Sports Engineering*, 8, 209–222.

NON-LINEAR VISCOELASTIC IMPACT MODELLING OF CRICKET BALLS

B. VIKRAM & F.K. FUSS

Sports Engineering Research Team, Division of Bioengineering, School of Chemical and Biomedical Engineering, Nanyang Technological University, Singapore

In this study, the cricket ball is modelled based on a non-linear visco-elastic power law, derived from stress relaxation behaviour. Due to the polynomial nature of the stiffness-deflection relationship of cricket balls, peak forces and contact time can only be approximated, however, with good conformity with the numerical solution of non-linear viscous power law model with constant elastic component. The peak forces and contact time in the cricket balls investigated range between 4.2–7.4 kN, and 1.2–1.9 ms at an impact velocity of 15 m/s, and between 13.3–27 kN, and 0.9–1.7 ms at 45 m/s.

1 Introduction

Ball impact is of general interest in sports, and has been, for example, extensively studied in golf balls. In contrast to golf balls, where the only important impact occurs between ball and club, impacts of cricket balls extend to turf, bat, and protective apparel like face guards. Cricket balls are modelled either as a single spring-damper system (Carré et al., 2004) or using FEM (Subic et al., 2005). Developing a model for cricket balls from first principles is incredibly tedious considering its multi-tiered layered structure (Subic et al., 2005). Carré et al. (2004) studied the impact of a cricket ball on a rigid surface and developed an empirical model consisting of a Hertzian spring and a non-linear damper in parallel. Subic et al. (2005) performed compression testing in the Kookabura Tuf Pitch cricket ball at different deflection rates for FEA of face-guard designs, and found an exponential relationship between the loading velocity and the ball stiffness. Vikram & Fuss (2007) analysed stress relaxation as well as compression at 5 different deflection rates in five ball models, and found that the stress relaxation of cricket balls generally follows the power law.

The aim of this study is to develop a mathematical model of the non-linear viscoelastic behaviour of cricket balls. This model is further applied to model the impact behaviour of the cricket ball with a rigid surface.

2 Mathematical Modelling

2.1. *Constitutive Equations and Relationship between Stiffness and Deflection Rate*

Starting at the stress relaxation following a power law

$$F = At^{-B} \tag{1}$$

where t is the relaxation time, A is the multiplier, and B is the viscosity constant.

Considering A as a function of the initial load F_0, and the latter a function of the deflection x, $F_0 = f(x_0)$, then we get:

$$F = F_0 A_F t^{-B} \qquad (2)$$

$$F = x_0 A_x t^{-B} \qquad (3)$$

The viscosity constant B of the power law is theoretically independent of F and x. In reality, however, B is not constant and is rather a function of load (Vikram & Fuss, 2007), and consequently, of deflection.

From Eq. (3), the stiffness – deflection rate relationship is calculated, comparable to the procedure developed by Fuss (2007) for the logarithmic law. In Eq. (3), x_0 is the constant deflection, applied by a Heaviside function H(t), the basic requirement for stress relaxation:

$$x = x_0 \, \mathrm{H}(t) \qquad (4)$$

Taking Laplace transform of Eqs. (3) and (4):

$$\hat{F} = x_0 A \frac{\Gamma(-B+1)}{s^{-B+1}} \qquad (5)$$

$$\hat{x} = \frac{x_0}{s} \qquad (6)$$

where the caret (^) denotes the transformed parameter, and Γ denotes a gamma function.

Substituting Eq. (6) into Eq. (5), we get the constitutive equation of the power law of viscoelasticity:

$$\hat{F} = s^B \hat{x} \, A\Gamma(1-B) \qquad (7)$$

Taking a closer look at Eq. (7), it becomes clear that:

- $A\Gamma(1-B)$ is a constant (constant C)
- B, the viscosity constant, ranges between 0 and 1: $0 \leq \phi < 1$. If $B = 0$, then Eq. (7) becomes $F = xA$, a purely elastic object in terms of a Hookean spring. If $B \rightarrow 1$, then the object becomes maximally viscous, with $\Gamma(1-B)$ approaching $+\infty$; thus, a solution of Eq. (7) for B = 1 does not exist, as $\Gamma(0)$ is not defined (transition from $+\infty$ to $-\infty$).
- The load F is a fractional derivative, specifically the "**B**th derivative", of the deflection x (times constant C), as the ith derivative of a function $f(t)$ becomes $s^i F(s)$... after Laplace transform.

In order to establish the relationship between load and deflection rate, as well as stiffness and deflection rate, we apply a ramp function:

$$x = \dot{x}_0 t \qquad \text{or} \qquad \hat{x} = \frac{\dot{x}_0}{s^2} \qquad (8)$$

where $\dot{x}_0$ is the constant deflection rate.

Substituting Eq. (8) into Eq. (7)

$$\hat{F} = \dot{x}_0 A \frac{\Gamma(-B+1)}{s^{-B+2}} \tag{9}$$

Applying the recursion formula of the Gamma function, we obtain:

$$\hat{F} = \dot{x}_0 \frac{A}{1-B} \frac{\Gamma(-B+2)}{s^{-B+2}} \tag{10}$$

Taking inverse Laplace transform, rearranging, and replacing t by $x/\dot{x}_0$:

$$F_x = x^{1-B} \dot{x}_0^B \frac{A}{1-B} \tag{11}$$

The deflection derivative of the load F is the stiffness k:

$$k_x = A\, x^{-B}\, \dot{x}_0^B \tag{12}$$

The stiffness – deflection rate relationship results after exchanging the variable and the constant: the variable deflection becomes a specific deflection x_i, and the constant deflection rate becomes the independent variable v, the deflection velocity.

$$k_v = A\, x_i^{-B}\, v^B \tag{13}$$

Equation (13) has the same structure as Eq. (1): the independent variable to the power of B times a constant. Thus, the relationship between stiffness and deflection rate follows a power law too. Eq. (13) allows extrapolating the stiffness k for higher velocities at a specific deflection. In a purely elastic material, where the viscosity parameter B is zero, Eq. (13) becomes $k = A$, which is the stiffness of a Hookean spring. The reason for this result becomes evident, when considering, that the constant deflection x_0, initially applied for stress relaxation, does not reflect the true nature of the force-deflection function, which is a 2nd order polynomial in the Sanspareils-Greenlands "Tournament" ball with rubber core (SGr), and a higher order polynomial in the Kookaburra "Special Test" ball (Vikram & Fuss, 2007). Applying Eqs. (12) and (13) to derive k as a function of both x and v is not valid, as the initial condition is based on a constant deflection rate $\dot{x}_0$. Nevertheless, k as a function of both x and v helps to approximate impact conditions if the constitutive transformed impact equation is not entirely analytical.

The term A in Eq. (12) represents the stiffness of the system under purely elastic conditions. If the stiffness is not constant, then A can be replaced by the actual stiffness of the ball, which, in the specific case of the SGr ball with rubber core is given in Eq. (14).

$$k_x = (2A_1 x + A_2) x^{-B}\, \dot{x}_0^B \tag{14}$$

The corresponding force equation thus is:

$$F_x = x^{1-B} \dot{x}_0^B \left(\frac{2A_1 x}{2-B} + \frac{A_2}{1-B} \right) \tag{15}$$

Replacing t by $x/\dot{x}_0$ and x by $t\dot{x}_0$, taking Laplace transform, and replacing $\dot{x}_0$ by $\hat{x}\,s^2$ and $\dot{x}_0^2$ by $0.5(\hat{x}*\hat{x})s^3$, where the asterisk (*) denotes a convolution, we get the constitutive equation of the power law with a parabolic force – deflection function:

$$\hat{F} = s^B[(\hat{x}*\hat{x})A_1\Gamma(-B+2) + \hat{x}\,A_2\Gamma(-B+1)] \tag{16}$$

2.2. Impact Modelling

The stiffness of a purely elastic object during impact is different from the stiffness obtained from compression tests. The non-linear stiffness of cricket balls, obtain in the study by Vikram & Fuss (2007) will be multiplied by 1.333, according to Fuss (2007).

The force equilibrium of an impacting object is (Fuss, 2007):

$$F_I + F_R = F_A \tag{17}$$

where F_A is an applied force which accelerates the object of mass m through a unit impulse Dirac delta function $\delta(t)$ to the desired impact velocity v_0: $F_A = a\,m = v_0\,\delta(t)\,m$; F_I is the deceleration or acceleration of the object after impact: $F_I = a\,m = \ddot{x}\,m$; and F_R is the reaction force at the contact between object and frame, the transformed equation of which is Eqs. (7), and also (16).

Taking Laplace transform of Eq. (17):

$$s^2\hat{x}m + s^B\hat{x}\,A\Gamma(-B+1) = mv_0 \tag{18}$$

Dividing by m and solving for $\hat{x}$:

$$\hat{x} = v_0 \frac{1}{s^2 + s^B \dfrac{A\Gamma(1-B)}{m}} \tag{19}$$

If the object is purely elastic, then the viscosity parameter B becomes zero and the inverse Laplace transform of Eq. (19) becomes

$$x = v_0\sqrt{\frac{m}{A}}\sin\left(\sqrt{\frac{A}{m}}\,t\right) \tag{20}$$

which is the constitutive equation of undamped oscillation.

The transformed force equation, after substituting Eq. (19) into Eq. (7), is:

$$\hat{F} = v_0 A\Gamma(1-B)\frac{s^B}{s^2 + s^B \dfrac{A\Gamma(1-B)}{m}} \tag{21}$$

Again, as mentioned above, it becomes evident, that F is the Bth derivative of x, times $A\Gamma(1-B)$. If $B > 0$, the Eqs. (19) and (21) can be solved numerically, by applying Piessen & Huysmans' (1984) routine for inverse Laplace transform.

In a parabolic force-deflection curve, Eq. (19), according to Eq. (16), becomes:

$$s^2 \hat{x} m + s^B \left((\hat{x} * \hat{x}) A_1 \Gamma(-B + 2) + \hat{x}\, A_2 \Gamma(-B + 1)\right) = m v_0 \qquad (22)$$

Equation (22) cannot be applied for numerical solution of deflection and force during impact. However, if the force-deflection function is non-linear (e.g., parabolic), Eqs. (11)–(15) are useful for approximation of deflection and force during impact, when making stiffness k and load F functions of deflection x and velocity v: $F, k = f(x, v)$. Strictly speaking, this procedure is not valid, as Eqs. (11)–(15) were derived based on a constant deflection rate $\dot{x}$. Nevertheless, these equations serve for approximating force and deflection. This was tested by comparing the result of Eqs. (19) and (21) to the result of and of Eqs. (11)–(13) after inserting them into the impact equation Eq. (17). The problem is that Eqs. (11)–(13), converted into $F = f(x, v)$ and $k = f(x, v)$, are purely elastic, even if Eq. (11), converted into $F = f(x, v)$, seems to consist of a non-linear damper of the power law, $F = \dot{x}_0^B$, multiplied by a Hertzian spring, $F = A(1-B)^{-1} x^{1-B}$. It is important to notice, that the forces of spring and damper are multiplied, and not added, as in a parallel arrangement. Thus, the damper cannot model the non-conservative energy, as the force of a typical damper is negative in the 2nd half of the impact, when the direction of the velocity is reversed; multiplied by the still positive spring force would result into an overall negative reaction force.

The approximation was thus tested only for the first half of the impact, generally for $0 \leq B \leq 0.1$, and specifically for B-values of cricket balls, ranging between 0.05 and 0.08 (Vikram & Fuss, 2007). The purpose was not to model the overall deflection and force functions with time and the force-deflection function, but rather to get an approximation of the peak force as well as the contact time of the impact. For the model, Eqs. (19) and (21) were solved numerically by Piessen & Huysmans' (1984) routine for inverse Laplace transform. The results were compared to the ones of the approximation, when substituting F_R in Eq. (17) by Eq. (11). The contact time, taken from the force-time results, were the same in model and approximation, with a negligible error. The peak forces of the approximation were higher then in the model, and overestimated by 0–5%. Thus, the peak forces had to be divided by $1 + 0.5285B - 0.125B^2$.

Finally, the approximation was applied to find the peak forces and contact time of "Kookaburra Special Test" ball (impact perpendicular and parallel to the seam plane) and "SG/Sanspareils-Greenlands Tournament" ball. In the latter model, the 2 varieties, with cork and rubber core, occurring naturally in this model due to inconsistent manufacturing process (Vikram & Fuss, 2007), were analysed separately.

The equations for deriving the polynomials independent of the deflection rate, were:

$$\text{SG with rubber core:} \quad \frac{k_x}{(v^B x^{-B})} = (A_1 x + A_2) \qquad (23)$$

$$\text{SG with rubber core:} \quad \frac{k_x}{(v^B x^{-B})} = (A_1 x^3 + A_2 x^2 + A_3 x + A_4) \qquad (24)$$

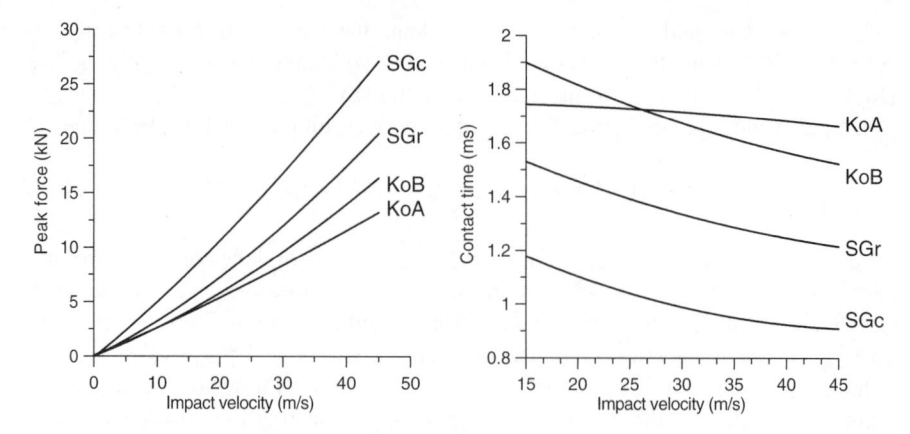

Figure 1. Peak force and contact time as functions of the impact velocity; SGc and SGr = SG balls with cork and rubber cores respectively, KoA and KoB = Kokaburra ball, impact perpendicular and parallel to the seam plane.

$$\text{Kookaburra: } \frac{k_x}{(v^B x^{-B})} = (A_1 x^5 + A_2 x^4 + A_3 x^3 + A_4 x^2 + A_5 x + A_6) \quad (25)$$

where the polynomials fit functions applied to find the coefficients A_n are on the right side of Eqs. (23)–(25). The coefficients were averaged and the overall stiffness equation was determined, after rearranging Eqs. (23)–(25). The overall stiffness, a function of x and v, was multiplied by 1.333 (see above), and the peak forces (divided by the correction factor) and the contact time were determined numerically for impact velocities ranging between 0 and 45 m/s.

3 Results

The peak forces and contact time as a function of impact velocity are shown in Figure 1. As expected from the different stiffnesses (Vikram & Fuss, 2007), the cork-core SG ball produced the highest impact forces.

4 Discussion

In contrast to Subic et al. (2005), who found an exponential relationship between the loading velocity and the ball stiffness, our model is based on a power law. This means, that the relationship between the loading velocity and the ball stiffness follows a power law as well. An exponential relationship results generally from spring-damper systems. The ball model investigated by Subic et al. (2005), Kookaburra Tuf Pitch, and the Kookaburra Special Test, analysed by us (Vikram & Fuss, 2007) are identical, according to the manufacturer (Kookaburra, 2007), only the Tuf Pitch's leather cover is selected and finished for greater resistance to abrasion.

Considering the high impact forces of cricket balls, it is not further astonishing that such impacts can even be lethal, as seen in the accident of Raman Lamba (Saeed, 1998; Cricinfo, 2007). The knowledge of peak forces during impact are decisive for designing protective equipment, e.g., by FEA, as shown in the study by Subic et al. (2005).

References

Carré M.J., James D.M. and Haake S.J. (2004) Impact of a non-homogeneous sphere on a rigid surface. *Proc. Instn. Mech. Engrs. Part C: J. Mechanical Engineering Science*, 218, 273–281.

CricinfoIndia (2007) Raman Lamba – player info (online). Available: http://content-www.cricinfo.com/india/content/player/30745.html (accessed: June 2007).

Fuss F.K. (2007) Non-linear viscoelastic properties of golf balls. In: Fuss F.K., Subic A. and Ujihashi S. (Eds.) *The Impact of Technology on Sport II*. Taylor & Francis Group, London.

Kookaburra Sports Ltd. (2007) Company Website (online). Available: http://www.kookaburra.biz (accessed June 2007).

Piessens R. and Huysmans R. (1984) Algorithm 619: Automatic numerical inversion of the Laplace transform. *ACM Trans. Math. Softw.*, 10/3, 348–353.

Saeed H. (1998) Indian batsman Lamba dies after sustaining head injury (24 Feb 1998) (online). Available: http://content-www.cricinfo.com/india/content/story/75247.html (accessed: June 2007).

Subic A., Takla M. and Kovacs J. (2005) Modelling and analysis of face guard designs for cricket. *Sports Engineering*, 8, 209–222.

Vikram B. and Fuss F.K. (2007) Non-linear viscoelastic properties and construction of cricket balls. In: Fuss F.K., Subic A. and Ujihashi S. (Eds.) *The Impact of Technology on Sport II*. Taylor & Francis Group, London.

AERODYNAMICS OF CRICKET BALL – AN UNDERSTANDING OF SWING

F. ALAM, R. LA BROOY & A. SUBIC

School of Aerospace, Mechanical and Manufacturing Engineering, RMIT University, Bundoora, Melbourne, Australia

The aerodynamic properties of a cricket ball largely depend on seams, surface roughness and the actions of the bowlers. Asymmetric airflow over the ball that is bowled fast, causes flight deviation (swing) and unpredictability in flight. Swing makes it difficult for a batsman to hit the ball and prevent getting bowled. Although some studies on swing have been conducted, the mechanism of reverse swing is not well understood. Hence the primary objective of this work was to understand the swing mechanisms affecting a cricket ball when bowled fast. The study is part of a larger research program including flow visualisation around a scaled cricket ball. A set of real cricket balls was also used to measure aerodynamic forces, using a six component force sensor in an industrial wind tunnel. These forces and moments were measured over a range of speeds and seam orientations.

1 Introduction

Cricket is one of the most popular and widely watched games in the world. Over 60 countries of the former British Empire are involved with a potential viewing audience of over 1.5 billion people. Recently China has adopted a plan to participate in the game by 2012. Then, cricket will be the 2nd most viewed game following various forms of football in the world. Cricket's popularity has already moved outside the boundary of the former British colonial countries. In any event making up the game, a bowler bowls a ball to a batsman who attempts to hit the ball and score runs whilst protecting three standing stumps sited at the extremity of a narrow strip of turf. The primary objective of a bowler is to bowl the ball such a way so as to deceive the batsman into making a false stroke resulting in the ball hitting the stumps or forcing the batsman to mis-time his shot and hit the ball on the full to a fielder of the opposing team. In order to deceive the batsman, the bowler has to manipulate the ball's trajectory and landing position on the wicket. As a cricket ball has to be projected through the air as a three dimensional body, the associated aerodynamics play a significant role in the motion of the ball.

A cricket ball is constructed of a several layers of cork tightly wound with string. The ball is covered with a leather skin comprising 4 quarters stitched together to form a major seam in an "equatorial" plane. Moreover the quarter seams on both halves of the ball are internally stitched and juxtaposed by 90 degrees. The seam comprises six rows of stitches with approximately 60 to 80 stretches in each row. A cricket ball at a mass of 156 gm is much heavier than a tennis ball. However, the prominence of the seam and mass can slightly vary from one manufacturer to another. At present, over a dozen firms manufacture cricket balls, under the auspices of the International Cricket Council, overseeing the game at the highest level (Test cricket).

The aerodynamic properties of a cricket ball are affected by the prominence of the seam, the surface roughness of the ball in play and the launch attitude of the ball by the bowler. Asymmetric airflow over the ball then causes the flight deviation (swing). The role of this research is to explain the mechanism of swing and de-mystify the unpredictability of the ball's trajectory. The sideway deviation of the ball during the flight towards the batsman is called swing. There are various types of swing: conventional swing and reverse swing. Conventional swing results in the ball experiencing a sideways force directed away from the shiny half of the ball. Such a force is achieved by maintaining laminar boundary layer of air flowing over the shinny or smooth half with a turbulent boundary layer of air flowing over the rough half. Roughness over one half of the ball is a result of its natural deterioration during play whilst a shiny side is maintained by polishing the ball when the opportunity presents itself to the fielding team.

Conventional swing can be achieved in at least two ways:

(a) by angling the seam to the batsman and with the mean direction of the flight so that one side experiences laminar (smooth) airflow and other side experiences turbulent airflow caused by the angulation of the seam itself. The points on each side of the ball half where the flow separates is asymmetrical, generating aerodynamic pressure variations with a component transverse to the ball's motion causing eventual trajectory deviations. Generally, the ball is pushed towards the half where the airflow is remains attached.
(b) by bowling a deteriorated ball possessing shiny and rough halves to a batsman. However, by aligning the ball's seam under some angles, the bowler can generate different type of swings which will be discussed later.

Generally, if the ball moves in a direction away from the bat, the deviation is called *out* swing. Conversely when the ball moves towards the batsman the resulting sideway deviation is called *inswing*. Figure 1 illustrates a typical swing of a cricket ball to a right-handed batsman.

Reverse swing is generally achieved when the airflow becomes turbulent on both sides of the ball. Here the turbulent airflow at sides separates earlier on one side than the other. The phenomenon usually occurs with a ball that is bowled fast. It is not clear where the limits of velocity exist for this type of swing. A comprehensive study is required to answer this question. In reverse swing, unlike conventional swing, the ball deviates toward the rough side of the ball. Usually, any swing makes difficult for the batsman to hit the ball with his bat and guard the stamps. Traditionally reverse swing occurs when one half of the ball is has been naturally worn significantly. In most cricket matches, the phenomenon of reverse swing occurs after 40 or more overs (one over consists of a set of six bowled balls). The mechanism for a reverse swing is complex and still not fully understood due to the degree of ball's surface roughness and required seam alignment angles with the mean direction of the flight. Although, some studies by Mehta (1985, 2000), Mehta & Wood (2000), Barrett & Wood (1996), Wilkinson (1997) and Barton (1982) were conducted to understand the aerodynamics of cricket ball, a comprehensive study to understand the gamut of complex aerodynamic behaviour resulting a wide range of swing under a wide range of wind conditions, relative roughness, seam orientations and seam prominence is yet to be conducted. Therefore, a large research project has been undertaken in the School of Aerospace, Mechanical and Manufacturing Engineering, RMIT University to understand the overall aerodynamic behaviour of the cricket ball both experimentally as well computationally. The work presented here is a part of this large research project. In order to understand the general behaviour of the airflow around a cricket ball, a large cricket ball (450 mm in diameter) was

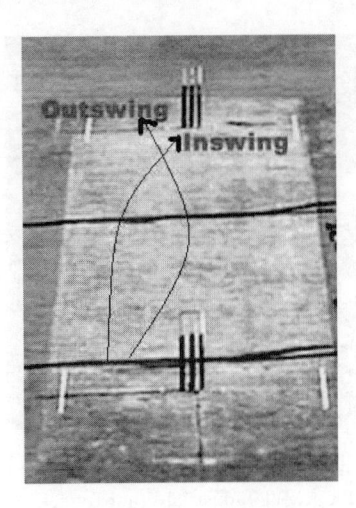

Figure 1. A schematic of outswing & inswing (Baig, 2007).

Figure 2. A typical Test cricket ball.

constructed to visualise airflow using smoke trails. Three brands of First-Class cricket balls were used (see Figure 2) including a One-Day International ball (see Figure 3) were used to measure the aerodynamic properties under range of wind and spin conditions.

2 Experimental Procedure and Results

The RMIT University Industrial Wind Tunnel was used to visualise the airflow around ball and measure the aerodynamic properties (drag, lift and side force and their corresponding moments) using a six-component force sensor under a range of wind speeds (40 km/h to 140 km/h in increments of 20 km/h). Seam orientations (steady conditions-no back spin involved) and with back spin conditions (100 rpm to 700 rpm in increments of 50 rpm). In order to visualise airflow, a large diameter (450 mm) ball was constructed and a six-rows of artificial seams was imposed. The larger diameter ball was photographed to check flow characteristics and quantify the flow effects with seam orientation angles with wind directions. Smoke was used to see the airflow trail around the ball. Artificial roughness was also created on the ball. Airspeeds ranged from 10 km/h to 40 km/h for the smoke flow visualisation. The RMIT Industrial Wind Tunnel is a closed return circuit wind tunnel with a turntable to simulate the cross wind effects. The maximum speed of the tunnel is approximately 150 km/h. The dimension of the tunnel's test section is 3 m wide, 2 m high and 9 m long and the tunnel's cross sectional area is 6 square m (more details about the tunnel can be found in Alam, 2000)).

Three different types of balls were tested. However the results of these tests were not included in this paper. Only the results from the flow visualisation are presented here. The simplified large ball was mounted on the side wall of the wind tunnel as shown in Figure 4. Six rows of seam were replicated using a Silicone-rubber compound and the surface roughness at one side of the ball was also replicated using the same material. Airflow characteristics under various seam orientation are shown in Figures 5 to 9. Figure 5 shows the flow separation occurring at the ball's apex (90 degrees) whilst the seam remains parallel to the wind direction (zero seam angle with the horizontal axis) reaffirming the classical flow

Figure 3. A typical One Day cricket ball.

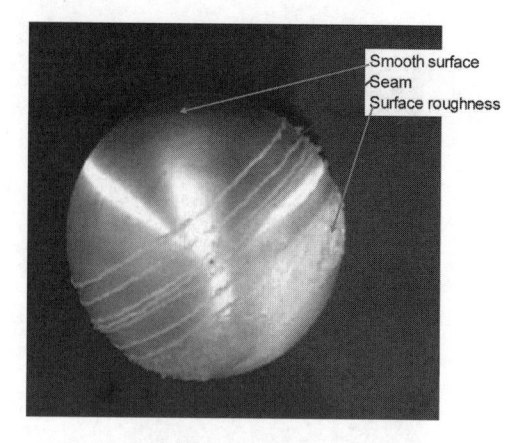

Figure 4. A simplified 450 mm diameter cricket ball in the RMIT Industrial Wind Tunnel.

Figure 5. Seam orientation parallel to flow direction (0 degrees).

Figure 6. Seam orientation approximately 30 degrees with flow direction.

separation point from a sphere. Here, the seam does not play any role in triggering flow separation at all. However, when seam is angled to the flow direction (shown here at approximately 30 degrees with surface roughness elements located upstream of the seam, the flow no longer separates at the apex (as it was in Figure 5), but accelerates and separates at around 30 degrees past the apex. The surface roughness and the seam then enable turbulence and enhance the delayed flow separation.

When seams are artificially placed at approximately 70 degrees to the flow direction (see Figure 7), the airflow separation is still delayed but not as much as was the case in Figure 6. Here the seams and surface roughness locations are close to the natural trigger of flow separation (see Figure 5). In Figure 8, the seams and surface roughness location trigger the natural flow separation as they are located at the critical zone (apex) and the airflow separates earlier than in the case of Figure 5. The airflow characteristics in Figure 9 show

Aerodynamics of Cricket Ball – An Understanding of Swing

Flow separation location

Figure 7. Seam orientation approximately 70 degrees with flow direction.

Flow separation location

Figure 8. Seam orientation approximately 90 degrees with flow direction.

Flow separation location

Figure 9. Seam orientation parallel to flow direction (30 degrees).

the similar pattern as in Figure 5. In this case, both surface roughness and seams do not play any role in the flow separation at all.

3 Conclusions and Future Work

The following conclusions were drawn from the work presented here:

- The airflow around a cricket ball is complex due to the surface roughness and seams.
- The side-ways deviation largely depends on the seam orientation to the flow direction:
- The seam location close to the apex of the ball triggers early flow separation.
- The seam location close to the mean direction of the airflow (horizontal axis) enhances the delayed flow separation.

- Further flow visualisation is required to quantify the exact location using a real cricket ball.
- The quantification of deviating side forces is required to support the aforementioned conclusions.

Acknowledgements

The research was supported by Cricket Australia.

References

Alam F. (2000) The effects of Car A-Pillar and Windshield Geometry on Local Flow and Noise. PhD Diss, Department of Mechanical and Manufacturing Engineering, RMIT University, Melbourne, Australia.

Baig M.R. (2007) The Basics of Swing Bowling (online). Available: http://www.cricket-fundas.com/cricketcoachingjan0907swing.html (accessed: June 2007).

Barrett R.S. and Wood, D.H. (1996) The theory and practice of reverse swing. *Sports Coach*, 18, 28–30.

Barton N.G. (1982) On the swing of a cricket ball in flight. *Proceedings of the Royal Society of London, Series A*, 379, 109–131.

Mehta R.D. (1985) Aerodynamics of Sports Balls. *Annual Review of Fluid Mechanics*, 17, 151–189.

Mehta R.D. (2000) Cricket Ball Aerodynamics: Myth vs Science. In: Subic A. and Haake S. (Eds.) *The Engineering of Sport 4*, pp 153–167, Blackwell Science, Oxford, 2000.

Mehta R.D. and Wood D.H. (2000) Aerodynamics of the Cricket Ball. *New Scientist*, 87, 442–447.

Wilkinson B. (1997) *Cricket: The Bowler's Art*. Kangaroo Press, Kenthurst, Australia, pp. 1–191.

ANALYSIS OF CRICKET SHOTS USING INERTIAL SENSORS

A. BUSCH & D.A. JAMES

Centre for Wireless Monitoring and Applications, Griffith University, Brisbane, Australia

Using bat-mounted, tri-axial accelerometers, data is collected from multiple positions on a cricket bat at a high frequency. Using this information, a number of useful characteristics of a batsman's swing can be extracted, including shot power, point of impact, bat angle, and angular velocity. From these characteristics it is possible to provide an objective, statistical evaluation of performance in both training and match situations, enabling fast and efficient feedback to be provided to the athlete. The sensors are lightweight and completely self-contained, allowing for minimum inconvenience to the batsman.

1 Introduction

Striking a cricket ball effectively is a complex movement involving many parameters. Measuring the quality of such shots has to date been a purely subjective measure based on visual assessment by an expert observer. Such metrics are prone to errors due to differences in observers, the inability of the human visual system to accurately perceive high-speed motion, and other human factors. It is also not unusual for different batsman to have markedly different batting techniques, which further reduces the accuracy of human assessment.

By taking advantage of the advancements in microelectronics and other micro technologies it is possible to build instrumentation that is small enough to be unobtrusive for a number of outdoor sporting applications with comparable precision to laboratory based systems (James *et al*., 2004). One such technology that has seen rapid development in recent years is in the area of inertial sensors. These sensors respond to minute changes in inertia in the linear and radial directions. These are known as accelerometers and rate gyroscopes respectively. When combined with absolute positioning technologies such as magnetometers and even GPS, laboratory equivalent performance analysis can be obtained.

Accelerometers and gyroscopes have in recent years shrunk dramatically in size as well as in cost (~$US20). This has been due chiefly to industries such as the auto-mobile industry adopting this technology in airbag systems to detect crashes. Micro electromechanical systems (MEMS) based accelerometers like the ADXLxxx series from Analog Devices (Weinberg, 1999) are today widely available at low cost. The use of accelerometers to measure activity levels for sporting (Montoye *et al*., 1983), health and gait analysis (Moe-Nilssen and Helbostad, 2004) is emerging as a popular method of biomechanical quantification of health and sporting activity. With the increasing capacities of portable computing, data storage and battery power due to the development of consumer products like cell phones and portable music players this is set to be a rapidly developing technological area in health and sport.

2 The Monitoring Platform

The cricket bat sensor is based heavily on the nCore, a scalable, generic monitoring platform developed using the work by James *et al*. (2004). This platform contains two embedded accelerometers each of which is capable of measuring acceleration forces of ±10 g in two perpendicular directions. This enables collection of data in full three dimensional space. These are true DC accelerometer devices, meaning that they will report a static 1g response due to gravity if oriented vertically. Whilst this poses a number of challenges when processing some types of data, it is necessary in order to obtain accurate reading in slowly varying systems.

The PCB is then attached directly to the top part of the back of the bat using screws, ensuring a stable connection and thus a good transfer of inertial forces. Using the external input pins on the nCore board, two more accelerometers are attached to the bottom part of the bat, providing the same three axis readings at that location. The other system components such as the battery and memory card are also firmly affixed to the bat in a convenient location.

3 Data Collection and Extraction

Data was collected during a number of experiments from subjects of varying skill levels. Six separate streams of data are collected during each session, those being the x, y, and z axes at both the top and bottom of the bat. In this context, the x axis refers to forces along the length of the bat, while the y axis represents forces operating in the direction of the edges of the bat. The z axis corresponds to forces operating perpendicular to the face of the bat, that is, in the direction of a typical swing.

Each subject, after an adequate warm-up period, was first asked to perform a number of practice shots for the off drive, straight drive and on drive, without actually hitting a ball. Following this, the batsman faced a set number of balls to which various shots were played, and relative statistics for each shot noted. Such data included the assessed power of the shot, the direction of ball travel, approximately where on the bat the ball struck, and the elevation of the shot. An independent assessment of the shot quality was also recorded. In order to create a standardized testing environment, a belt-driven bowling machine was used during testing, meaning that each batsman faced bowling of almost the same speed and direction. The data from a typical shot is shown in figure 1. Features which can be readily identified from this example are the time during which the swing occurs, evident from the large spike in both the x and z axes, the actual time of ball impact, evidenced by a small dip in these same axes during the swing, and the bat tapping on the ground, which is shown as small spikes immediately prior to the swing.

Of particular interest in these acceleration traces is the z axis. The large spike on this graph does not represent the swing itself, but rather the deceleration of the bat after it has reached maximum velocity. The smaller negative peak immediately prior to this is the actual forward acceleration from which the shot power is generated.

4 Calculating Shot Characteristics

Previous research has shown that the use of inertial sensors attached to a sporting implement can gather significant performance related data (James et al., 2005). Although the motion of a cricket bat during a shot is typically not well constrained as is the case in sports such as golf, such instrumentation can still enable the collection of a number of useful shot parameters. To date, this research has concentrated solely on the so-called "straight bat" shots, during which the bat remains in an approximately vertical position. Such shots include the on drive, straight drive, and cover drive, and are amongst the most commonly played cricket strokes in most forms of the game. By limiting analysis to these shots, it can be assumed that all bat motion

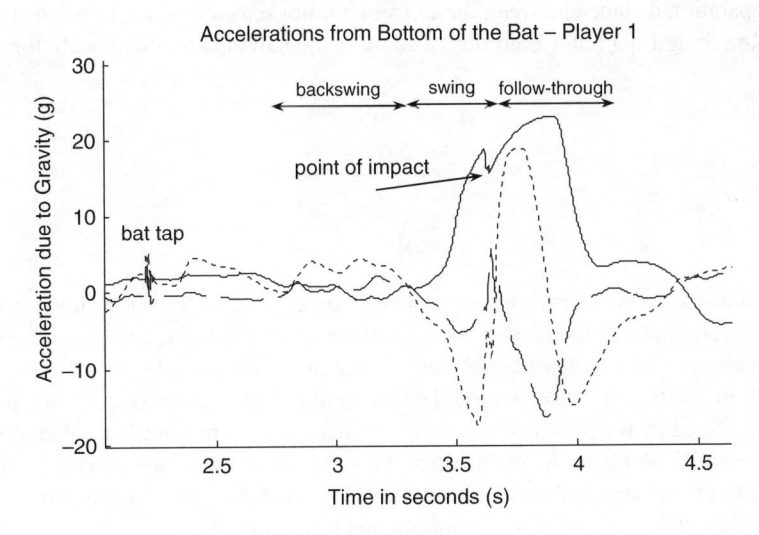

Analysis of Cricket Shots using Inertial Sensors 319

Figure 1. Data extracted from a typical shot. Note the small dip in acceleration in the *x* and *z* axes (solid and dotted lines respectively) at the tie of impact, and the large spike during the swing. Twisting of the bat at the point of impact and follow-through is noted in the *y* axis (dashed line).

during the period of interest is limited to a single plane, allowing a number of useful assumptions to be made when processing the data.

4.1. Bat Angular Position and Velocity

Calculating absolute velocity or position values using inertial sensors is a difficult task, due to the need for at least one integration and the resulting summation of errors. When the object to be measured is also subject to rotational motion, as is the case with a cricket bat, gravitational forces are also varying with respect to each axis, meaning that unless the exact angular position is known at each instant, additional errors are introduced. As a cricket swing can be characterized as a roughly circular motion, angular velocity is also a very important quantity when assessing characteristics such as shot power and elevation. The calculation of angular velocity is also considerably simpler than for absolute velocity, due to the presence of two sensors on the bat.

Assuming a roughly circular swing motion, one method of estimating the angular velocity of an object is to measure the resulting centrifugal force. This is a real force, as the accelerometers are measuring forces from a rotating reference frame, and is given by

$$a_c = \omega^2 \mathbf{r} \qquad (1)$$

where *a* is the centrifugal acceleration, ω is the angular velocity of the bat, and **r** is vector representing the distance from the centre of rotation. Due to differing methods of playing shots between batsmen, the centre of rotation, and thus **r**, is often quite different, stemming from rotational forces generated by the wrists, elbows and shoulders. Using the position of the accelerometers on the bat, it is possible to estimate these values using Eq. (1) above and the known separation distance between the top and bottom sensors, which in our test case was 0.48 m. Rewriting Eq. (1) for both sets of sensors and solving simultaneously for ω gives

$$a_{ct} = \omega^2 \mathbf{r}$$
$$a_{cb} = \omega^2 (\mathbf{r} + d)$$
$$\omega = \sqrt{\frac{a_{cb} - a_{ct}}{d}} \quad (2)$$

where a_{ct} and a_{cb} are the centrifugal acceleration forces (x axis) at the top and bottom of the bat respectively, and d is the known distance of separation. Solving these equations simultaneously gives and estimate of both ω and $\mathbf{r}$, quantities which are both useful in assessing the quality of a batsman's shot. It should be noted that the estimates of centrifugal acceleration are affected by the presence of gravity, which is not a constant force due to the rotation of the bat. Due to the identical orientation of the sensors, however, the gravitation component of the centrifugal acceleration should be equal for both top and bottom sensors, meaning that it will cancel due to the subtraction term in Eq. (2).

Another method of estimating the angular velocity of the bat is to use angular kinematics. From these equations, it can be shown that the angular acceleration α, is given by:

$$\alpha = \frac{(a_t - a_b)}{d} \quad (3)$$

where a_t and a_b are the acceleration values in the direction of bat motion (nominally, the z axis) at the top and bottom of the bat, respectively, and d is the distance separating them. Integrating the angular acceleration with respect to time will give the instantaneous angular velocity, and if required, a second integral will provide an estimate of the angular position of the bat. Due to the inevitable drift when performing integration using inertial sensors, the estimates of angular velocity achieved using this technique cannot be considered accurate in an absolute sense. Using heuristic measures to zero the angular velocity at known points in the swing such as the tapping of the bat can greatly improve these estimates, effectively resetting the drift at short intervals.

4.2. Bat Twist

During a typical cricket shot, the angle of the bat relative to its forward motion is quite important. In order to maximize the chance of successfully striking the ball, the face of the bat should be perpendicular to the direction of the ball, allowing for the maximum possible contact area. In order to direct the ball, however, the batsman must often change this direction. For example, when playing the off drive, the bat will be oriented in order to face towards the "off side", that is to the right side for a right-handed batsman. Conversely, when playing an on drive, the face of the bat will be more closed, and thus angle more to the left. This can be accurately determined using the inertial sensors by detecting the amount of acceleration in the y axis during the swing. As the bat face is opened, the component of the swing acceleration along the y axis is increased, while closing the face will have a corresponding negative effect on this value. In this way, the angle of the bat during the swing can be accurately determined

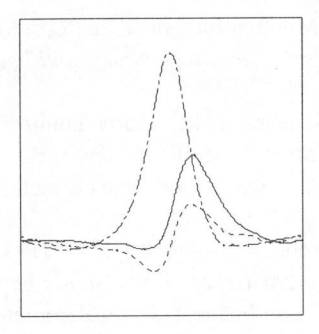

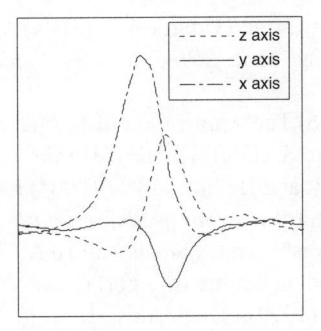

Figure 2. Accelerometer responses for a simulated on drive (left) and off drive (right). Note the difference in the y-axis response (solid line), representing the twisting of the bat during the shot in opposite directions.

by the relative strength of the y axis. This is illustrated in figure 2, which shows two example swings for a batsman playing both an off drive and on drive. This figure clearly shows the difference in bat angle between the top types of shot, and was repeatable for all batsman tested during the experiment when playing straight bat shots.

5 Conclusions and Future Research

In this paper we have presented work showing the usefulness of inertial sensors in helping to characterize cricket shots. Using two tri-axial bat-mounted accelerometers, we have shown that it is possible to estimate the angular velocity and bat twist at any instant during the swing, as well as detect the time of impact and other important events. Conducting field tests on a variety of subjects has shown that these results are highly repeatable across a wide range of batting standards and conditions. It is expected that future research will enable to extraction of such features as shot power, shot direction and angle of elevation.

In order to assess the accuracy of such measurements, future research will also utilize a high-speed motion tracking system, which is capable of recording human movements from multiple angles in a laboratory setting. Although it is not possible to utilize such equipment in a match environment, it will serve to verify and further refine the data obtained by inertial sensors, and provide a more accurate basis upon which future research can be conducted.

Acknowledgements

The authors would like to thank Queensland Cricket and the Cricket Australia Centre of Excellence for providing access to various facilities during the data collection phase of this research. Vernon D'Costa also provided invaluable assistance during the construction and testing of the electronics and during data collection. Parts of this research were funded by a Griffith University Research Grant, and this assistance is gratefully acknowledged.

References

James D., Uroda W. and Gibson T. (2005) Dynamics of Swing: A Study of Classical Japanese Swordsmanship using Accelerometers. In: Subic A., Ujihashi S. (Eds) *The Impact of Technology on Sport,* Japan: ASTA, pp 355–60.

James D., Davey N. and Gourdeas L. (2004) A Modular Integrated Platform for Microsensor Applications. *Proc. SPIE, Microelectronics: Design, Technology, and Packaging*, 5274, 371–78.

James D. (2006) The Application of Inertial Sensors in elite sports monitoring. In: Moritz E.F., Haake S. (Eds). *The Engineering of Sport 6*, Springer, New York.

Moe-Nilssen R. and Helbostad J. L. (2004) Estimation of gait cycle characteristics by trunk accelerometry, *Journal of Biomechanics*, 37/1, 121–126.

Montoye H., Washburn R., Smais S., Ertl A., Webster J.G. and Nagle F . (1983), Estimation of energy expenditure by a portable accelerometer, *Med. Sci. Sports Exerc.*, 15, 403.

Weinberg H. (1999). Dual Axis, Low g, Fully Integrated Accelerometers, *Analogue Dialogue* 33, Analogue Devices (publisher).

HIGH SPEED VIDEO EVALUATION OF A LEG SPIN CRICKET BOWLER

A. E. J. CORK, A. A. WEST & L. M. JUSTHAM

Sports Technology Research Group, Wolfson School of Mechanical and Manufacturing Engineering, Loughborough University, Loughborough, Leicestershire, England

The skilled leg spin bowler is able to bowl a number of variations of their standard leg break delivery. The bowler is able to impart different spin orientations onto the ball and alter the trajectory of the ball by manipulating the wrist, hand and forearm during the bowling action. The results of a focused study of a National standard leg spin bowler using two orthogonal high speed video cameras sampling at 2000 frames per second is described in this paper. Analysis is later carried out to quantify the initial launch characteristics of four delivery variations; leg break, wrong'un, flipper and slider.

1 Introduction

Cricket is a team sport where the focus narrows down to the duel between batsman and bowler for each delivery during a match. In order to execute a shot well a batsman must be able to predict accurately the length, direction and pace of the delivery from the movements of the bowler and his judgment of the initial launch characteristics of the ball. One aim of the bowler is to trick the batsman into playing a false stroke by releasing deliveries with different launch characteristics and similar bowling actions (Renshaw & Fairweather, 2000), thus reducing the information available to the batsman or at least leaving the crucial information as late as possible during the bowling action (Gibson & Adams, 1989).

In order to greater understand the demands placed upon batsmen when making decisions on which shot to play to a delivery it is important to record and take measurements from the various bowling actions of cricketers. More precisely, it is important to understand the physical differences that occur in bowlers' actions when bowling variations of their standard deliveries. This knowledge will support research into the visual cues used by batsmen of different capabilities (i.e. elite to recreational) and enable these cues to be represented in automatic training systems to enhance their visual skills, with an emphasis on advance cue usage as this will improve the probability of a player progressing from intermediate to expert level (Penrose & Roach, 1995). If such a training system is to reproduce human realistic deliveries it is important to measure the capabilities of human bowlers and the initial launch characteristics they impart onto the cricket ball.

The aim of this paper is to present differences in the launch characteristics of the cricket ball at release when selected variations of leg spin deliveries are bowled by elite cricketers. There is currently no literature detailing the initial launch characteristics of a cricket ball during various types of deliveries. One National standard Leg Spin bowler is focused upon with measurements of ball spin rate, resultant velocity, vertical launch angle, release position and pitching length of the ball taken from a selection of delivery types. These variations were (1) standard leg spin, (2) wrong'un (googly), (3) flipper and (4) slider as these are considered the most common variations utilized by leg spin bowlers.

(1) Leg Spin   (2) Wrong'un   (3) Flipper   (4) Slider

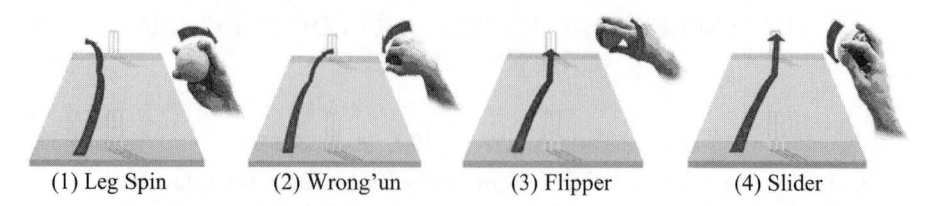

Figure 1. The four variations of delivery focused upon in this paper. (BBC Sport, 2006).

Leg spin bowlers aim to spin the ball away from a right handed batsman after the ball has bounced. The action to produce a standard Leg Spin delivery for a right handed bowler is a turning of the first three fingers over the top of the ball from left to right (Philpott, 1978) as a batsman would view, bringing the hand down the inside of the ball (Brayshaw, 1978). The first of the variations seen here is the wrong'un where the ball is released with the same action but from the back of the hand to impart spin in the opposing orientation to that seen in the leg spin delivery. For the second, the flipper delivery, the bowler holds the ball between the two first fingers and thumb and squeezes the ball out by "snapping" the fingers together to impart backspin onto the ball resulting in a low, skidding delivery. Finally, for the slider delivery, the ball is released with the same action as the standard leg break, however, the wrist angle is changed in order to impart more topspin onto the ball. This results in a more "looping" ball flight with the ball bouncing higher, faster and with less lateral deviation than a leg spin delivery. Figure illustrates these four variations with the grip from a batsman's perspective and expected ball trajectory demonstrated.

2 Method

Player testing was conducted at the National Cricket Centre, Loughborough. The player was asked to bowl each of the four previously mentioned deliveries. The action of the bowler was recorded throughout the delivery stride using two Photron high speed video cameras. The pitching position of each delivery was also recorded by a third high speed camera and the whole session was recorded by two standard 50 Hz video cameras sampling from an umpire's and a batsman's viewpoints respectively.

A specifically designed test bed was assembled to house two orthogonal high speed video cameras. Both Photron Ultima APX High Speed Video cameras were set up to film at a frame rate of 2000 frames per second which enabled data to be recorded at a resolution of 1024×1024 pixels. When using these cameras there is a trade off between sampling rate and image size. For the purposes of these tests the largest field of view possible was captured with the highest time resolution. The cameras were synchronized and recording was initiated using an SV TTL trigger when the bowler broke a laser beam positioned across the bowling crease. This position was chosen to ensure that data was recorded from a field of view encompassing the initiation of the delivery stride through to the bowler's follow-through. One camera was positioned to record the action from side on standing at head height on a standard Manfrotto tripod along the line of the popping crease (High Speed Camera 1, Figure 2), the second recorded the action from above housed on a 5 metre long beam suspended over 3 metres from the floor along the line of the popping crease (High Speed Camera 2, Figure 1). Figure 2 is a scaled CAD representation of the test setup used in the indoor facility.

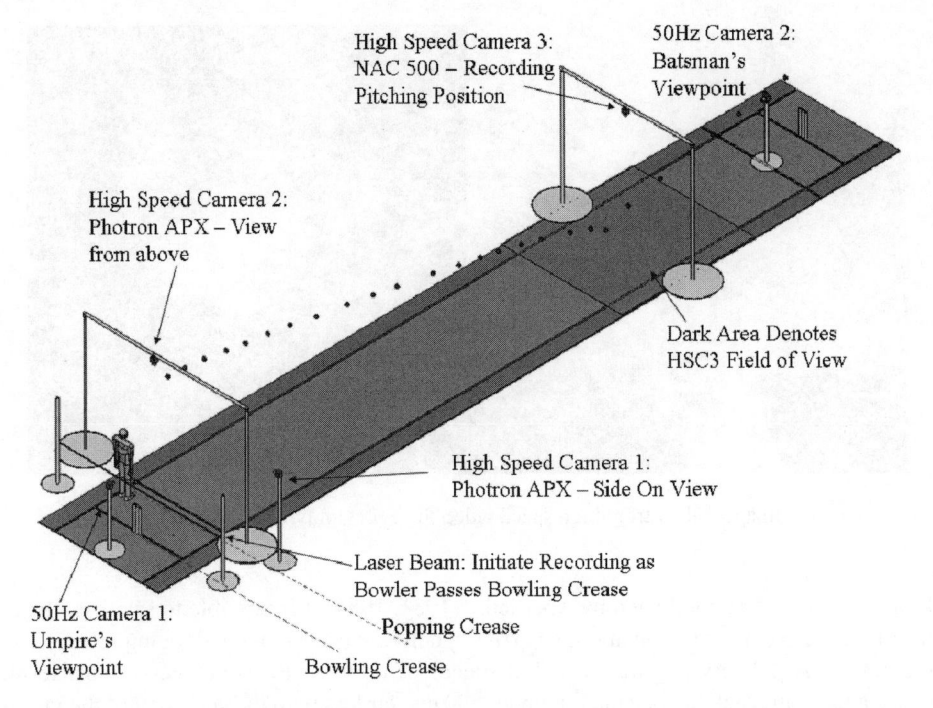

Figure 2. Setup of apparatus for indoor player testing session created using CAD software.

A target length of delivery was ascertained for a good length ball from the player's coach. A third high speed camera was positioned above the pitch focused upon the designated good length pitching area. This camera was used to record the ball as it hits the pitch for the first time post release from the bowler. The camera selected for this was the NAC 500 High Speed Video Camera. This model enabled sampling at 500 frames per second and would allow continuous recording of data onto a standard VHS tape. The tapes were later digitized in preparation for analysis. The continuous recording capability of the NAC 500 meant that it required far less maintenance during testing although this was offset by the reduction of 75% less in time resolution. This sampling rate was however adequate to enable ball/pitch impact position and deviation measurements to be determined.

Further to the high speed cameras used during testing, two additional 50 Hz cameras (both Canon XM1) were used to record the testing session. These cameras were placed at an Umpire's standing position and a Batsman's position. These cameras gave an overview of the testing session and an insight into the views of each delivery from a batsman's and an umpire's point of view.

The cricket balls used in the testing session all conformed to international standards to allow for consistency and represent similar match conditions for the subjects. However, additional markings were placed onto the balls to divide the surface into quadrants for the purpose of later analysis, specifically, the spin rate of the ball.

3 Results and Discussion

Data from the deliveries bowled will be analysed with the objective of establishing the areas of a delivery that batsmen could use to differentiate between delivery variations.

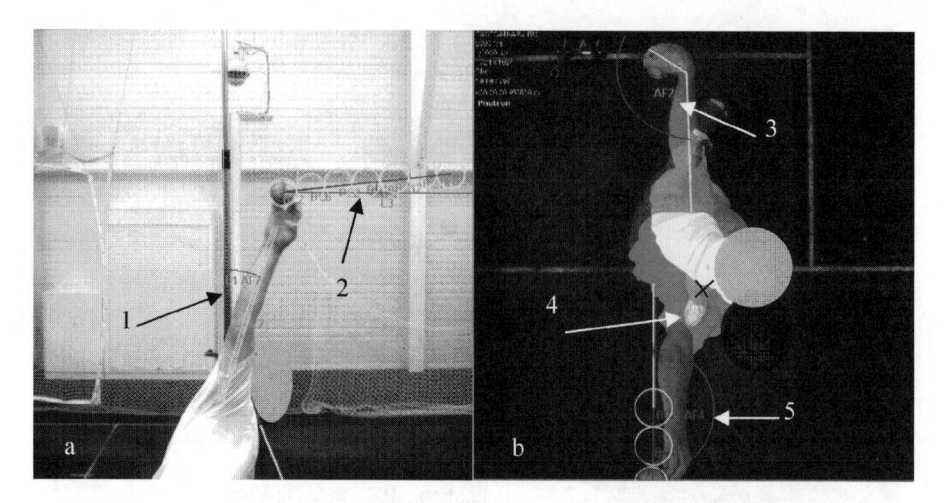

Figure 3. Images taken from high speed video footage with typical measurements taken.

From previous work conducted by Abernethy (1981, 1984) it is possible to say that a batsman must derive information about an arriving delivery prior to the ball being released by the bowler. A delivery can be bowled at speeds of up to 44.8 m/s at International level, hence a ball can reach the batsman in under 500 ms, far less than the sum total of the visual reaction time of the batsman and the movement time for the lower extremities and bat (Muller *et al.*, 2006; typically 200 ms and 700 ms respectively (Glencross & Cibich, 1977; Abernethy, 1981; McLeod, 1987).

Below in Figure 3 are two still images of the bowler at release during the leg spin delivery. The images have example measurement markings of the bowler's arm angle at release (1), vertical launch angle (2), maximum wrist angle during the bowling action (3), front foot position(heel) (4) and horizontal launch angle (4). The images are taken from the high speed footage recorded from side on by high speed camera 1 (a) and from above by high speed camera 2 (b).

For the purposes of this paper, the key differentiators between each delivery type have been analysed, these being spin rate, resultant speed and vertical launch angle, these data are summarised in Table 1. In addition, the horizontal launch angle, release height and width and the pitching length of the ball from release are quantified. The time period between the bowler's front foot impact with the floor (FFI) and the release of the ball has been calculated. Muller *et al.*, (2006) reported that highly skilled batsmen can pick up information pertaining to ball type during this period. Two final measures have been calculated that are particularly pertinent when analysing from a batsman's viewpoint, the maximum wrist angle (the angle created between the forearm and hand) as this is seen to be a key differentiator when judging ball type (Muller *et al.*, 2006) and the orientation of spin imparted onto the ball, which, provided the seam is predominantly revolving around the same axis can give a batsman information pertaining to ball type.

There appear to be few clear differentiating values for a batsman to use when judging delivery type. Table 1 shows that the bowler in question was able to impart the wrong'un delivery with significantly greater levels of spin (31.3 rps) than the flipper delivery released

Table 1. Average results of ball launch characteristics by a Leg Spin bowler.

Ball Type	Spin Rate (rps)	Resultant Speed (m/s)	Vertical Launch Angle (Degrees from horizontal)	Horizontal Launch Angle (Degrees from Normal)	Pitching Length (Metres)	Release Height (Metres from floor)	Release Width (Metres wide of stumps)	Time From FFI to Release (Seconds)	Max. Wrist Angle (Degrees)	Spin Orientation (Degrees from Normal)
	±0.31	±1.11	±1.07	±0.22	±0.48	±0.005	±0.08	±0.002	±2.56	±7.53
Leg Spin	26.2	19.1	6.4	180.4	15.2	2.08	0.82	0.115	126	144
Wrong'un	31.3	18.0	4.2	180.2	13.3	2.02	0.85	0.114	132	212
Slider	19.3	18.0	6.3	180.9	14.6	2.06	0.62	0.116	114	168
Flipper	25.0	20.8	1.8	181.5	17.1	2.03	0.82	0.111	130	-4

with 19.3 rps. This may be due to a lack of wrist involvement during the bowling action which can add a greater amount of spin imparting torque to the ball. For the batsman however, viewing from approximately 18 metres away it would difficult to differentiate between leg spin, flipper and wrong'un on spin rate alone. The 12 rps difference between wrong'un and slider may however give him enough information.

The flipper would be regarded as the easiest to differentiate from the other three deliveries, the ball was released by the bowler with a flat trajectory, only 1.8°, 4.6° lower than the most flighted delivery, the leg spin. The speed of the ball may also have given a batsman clues as to which delivery was approaching, 2.8 m/s faster than he was releasing the slider. Couple these pieces of information together and the batsman would have a good idea of the delivery type.

It should be noted that the slider delivery was released from 0.20 metres closer to the stumps than any other delivery released by the bowler. Given the width of the bowling crease from the centre of the stumps to the edge of the pitch is 1.32 metres, this is quite a large distance and would have given the batsman an early indication of the delivery type to follow. The slider delivery is often used by bowlers when looking for an LBW (leg before wicket) decision, by moving closer to the stumps it means the ball will travel on a straighter path down the centre of the pitch than if the ball was released wider. The chances of an LBW is likely to be increased as the ball is bound to pitch within the extremities of off and leg stump.

The wrist angle appears to give the greatest clue of ball type. The angle between the forearm and hand (wrist flexion) vary between the wrong'un and slider deliveries by 17.83°, this is when the maximum wrist angle during the bowling action is recorded, couple this with the change in orientation of the hand through pronation of the radioulnar joint for the wrong'un and the batsman is likely to have a good idea of which delivery is coming. However, when you consider that the time interval the batsman has to view these subtle variations is approximately 0.114 seconds (average FFI to release) then the task undertaken by the batsman appears a great deal harder once more.

References

Abernethy B. (1981) Mechanisms of Skill in Cricket Batting. *Australian Journal of Sports Medicine*, 13, 3–10.
Abernethy B. and Russell D.G. (1984) Advance Cue Utilisation by Skilled Cricket Batsmen. *The Australian Journal of Science and Medicine in Sport*, 16/2, 2–10.
BBC Sport (2006) (online). Available: www.bbc.co.uk/sport/cricket (accessed: May 2007).
Brayshaw I. (1978) *The Elements of Cricket*. Griffin Press Limited, Adelaide.
Gibson A.P. and Adams R.D. (1989) Batting Stroke Timing With a Bowler and a Bowling Machine: a Case Study. *The Australian Journal of Science and Medicine in Sport*, 21/2, 3–6.
Glencross D.J. and Cibich B.J. (1977) A Decision Analysis of Game Skills. *Australian Journal of Sports Medicine*, 9, 72–75.
McLeod P. (1987) Visual Reaction Time in High-Speed Ball Games. *Perception*, 16, 49–59.
Muller S. Abernethy B. and Farrow D. (2006) How Do World Class Cricket Batsmen Anticipate a Bowler's Intention? *The Quarterly Journal of Experimental Psychology*, 59/12, 2162–2186.

Penrose J.M.T. and Roach N.K. (1995) Decision Making and Advanced Cue Utilisation by Cricket Batsmen. *Journal of Human Movement Studies*, 29/5, 199–218.

Philpott P. (1978) *Cricket Fundamentals*. Everbest Printing Co. Ltd. Hong Kong.

Renshaw I. and Fairweather M.M. (2000) Cricket Bowling Deliveries and the Discrimination Ability of Professional and Amateur Batters. *Journal of Sports Sciences*, 18, 951–957.

AN ANALYSIS OF THE DIFFERENCES IN BOWLING TECHNIQUE FOR ELITE PLAYERS DURING INTERNATIONAL MATCHES

L.M. JUSTHAM, A.A. WEST & A.E.J. CORK
*Wolfson School of Mechanical and Manufacturing Engineering,
Loughborough University, Loughborough, Leicestershire, England*

Three types of cricket match are played during international competitions: the five day test match, the One Day International (ODI) and the Twenty:20 match. In test match cricket there is no limit to the number of deliveries which may be bowled during each innings whereas ODI matches are limited to 50 overs per innings and Twenty:20 matches are limited to 20 overs. It has been hypothesized that players will alter their bowling style depending on the type of match being played i.e. the speed and pitching position of the ball on the pitch. An investigation has been carried out using player performance analysis data colleted using the Hawk-Eye ball tracking system during the 2006 playing season. These data have shown that, contrary to the original hypothesis, bowlers do not alter their bowling style. However for fast and medium paced bowlers the number of runs that the batsman was able to score against them in the limited overs matches was higher than during the test matches.

1 Introduction

Cricket is a sport which has been played for over five centuries and is established in almost 100 countries. There are three types of match which are played at an international level: (i) the traditional five day test match where each team has two innings and no limitation in the number of overs which may be delivered, (ii) The one day international (ODI) which is a limited overs match and each team is constrained to fifty overs of deliveries and (iii) the new format "Twenty:20" match where each team is constrained to twenty overs of deliveries and the match is completed within a few hours.

The bowler delivers the ball to the striking batsman in a head to head battle between the two individual players, if the batsman strikes the ball then he is able to run while the fielding side attempt to catch, run or stump him out. It is widely agreed that there are three classes of bowling delivery (i) fast bowling, (ii) medium paced or swing bowling and (iii) spin bowling. Fast bowlers routinely deliver the ball at speeds in excess of 80 mph and rely largely on the speed of the ball to outwit the batsman and enforce errors in their judgment. Medium paced or swing bowlers deliver the ball between 65 mph and 80 mph. They carefully position the protruding seam of the cricket ball such that the air flow around it causes a sideways drifting motion during flight. Spin bowlers, who deliver the ball at speeds between 40 mph and 60 mph, impart a fast rifle style spin onto the ball which causes sideways deviation during the ball's flight and an unpredictable break from the bounce. The spin is predominantly created using the fingers for off-spin and the wrist for leg-spin (Abrahams, 2004).

The parameters which define bowling are not quantitatively understood and coaching manuals rely on qualitative descriptions to illustrate each type of delivery (Andrew, 1989;

Bradman, 1984; Khan, 1989). Ongoing research work at Loughborough University has resulted in a detailed classification of cricket bowling being developed (Justham, 2007). Within this classification the speed of the ball at release, the angle of the release trajectory, the spin rate and direction of spin have been investigated and quantified for the three main bowling types and a selection of the most common variation deliveries. The research presented in this paper is an extension to this classification and is focused on match based performance data which has been collected for three international standard players during the three different types of match play.

2 Data Collection Procedure

Hawk-Eye is a ball-tracking system which uses images from three fixed orthogonal cameras to track the ball from the moment of release by the bowler to the moment of impact with the bat. The system is used during match play as a commentary tool but also post-match by coaches as a performance analysis device. The speed of the ball at release, its pitching position and where it would have passed the stumps in the absence of a batsman is recorded and can be used to monitor a bowler over the course of the entire match. Data from a selection of matches can then be compared to determine whether the bowling style changed or varied for different match situations (Hawk-Eye, 2006).

Within this paper Hawk-Eye data, which was collected during the Pakistan tour of England during July and August 2006 (ECB, 2006), has been used to compare variations in bowling deliveries during the five-day test series, the ODI series and the Twenty:20 matches which were played. Three international players have been used for this analysis: (i) a right arm fast bowler, (ii) a right arm medium paced bowler and (iii) a right arm off-spin bowler, The main focus of the analysis has rested on the speed of the ball at release and the pitching line and length of the ball at the bounce.

3 Results

Data was collected for each of the three bowlers over a five match test-match series, a six match ODI series and a Twenty:20 match, see Table 1. The average ball release speed, measured in mph, has shown that for the fast and medium paced bowlers the fastest average ball release speed occurs during the test match series and the slowest average ball release speed occurs during the Twenty:20 match. However for the case of the spin bowler, the fastest ball release speeds occur during the Twenty:20 match and the slowest ball release speeds occur during the test series.

The average ball pitching length and line are both measured in yards from the batsman's centre stump. For the pitching line, a negative number denotes that the ball pitches to the left of the stump whereas a positive number denotes a pitching line to the right of the stump. For the fast bowler the pitching line remains similar in each of the three matches whereas the pitching length is at its shortest (furthest away from the batsman) during test matches and at its fullest (closest to the batsman) during the Twenty:20 match. This corresponds to the fastest ball deliveries being pitched further away from the batsman. For the medium paced and spin bowlers the shortest pitching length occurred during the ODI matches and the longest occurred in the Twenty:20 match. Although for the case of the spin bowler the variation in pitching length was less than 1 m throughout all three matches.

Table 1. The average ball speed (mph) and ball pitching position (yards from the batsman's centre stump) for the three international bowlers under consideration.

Player	Match	Ball release speed		Ball pitching length		Ball pitching line		Run rate
		Mean	St. Dev.	Mean	St. Dev.	Mean	St. Dev.	
Fast	Test	85.92	7.92	8.01	2.03	−0.26	0.18	0.68
	ODI	83.69	4.49	7.60	2.26	−0.25	0.17	0.74
	20:20	81.13	1.90	7.26	1.14	−0.29	0.22	1.09
Medium	Test	72.65	4.54	6.68	1.78	−0.34	0.17	0.64
	ODI	72.03	7.64	5.82	2.71	−0.33	0.23	0.60
	20:20	67.17	3.00	7.68	1.51	−0.28	0.24	1.58
Spin	Test	50.64	6.54	5.06	3.70	−0.48	0.10	0.46
	ODI	51.69	10.33	4.54	2.39	−0.38	0.24	0.88
	20:20	52.88	1.31	5.27	1.69	−0.32	0.25	0.47

The run rate is a measure of how many runs were scored by the batsmen facing the bowler with respect to the number of deliveries bowled. For the fast bowler the run rate was at its lowest during the test matches (0.68) and at its highest in the Twenty:20 game (1.09). For the medium paced bowler the run rate was similar for both the test and ODI matches (0.60 and 0.64 respectively) but higher, at 1.58, in the Twenty:20 match. The higher run rate during the Twenty:20 match is expected because the batsman must attempt to score as many runs as possible over the course of the limited number of deliveries. However for the spin bowler the run rate is higher, at 0.88, during the ODI matches and comparable with the test match run rate during the Twenty:20 match at 0.46 and 0.47 respectively.

The ball pitching length with respect to the speed of the ball at release for each of the three bowlers has been compared in Figure 1. This has shown that the fast bowler, Figure 1a, routinely delivers the ball at a speed between 80 mph and 94 mph but has a "slow ball" variation where the ball is released between 60 mph and 75 mph. Similarly for the medium paced bowler the delivery speed is between 60 mph and 80 mph with two slow ball variations occurring around 50 mph and between 60 mph and 70 mph, Figure 1b. The spin bowler does not appear to use ball speed as a delivery variation and the ball is released between 45 mph and 55 mph throughout, Figure 1c. There is a large variation in the length where the ball pitches for the fast bowler with deliveries varying from two yards to thirteen yards from the batsman's stumps. For the medium paced bowler most deliveries are bowled with a pitching length between two yards and nine yards from the batsman's stumps. The spin bowler delivers the ball most fully between two yards and seven yards from the batsman. The pitching line has been compared with the pitching length in Figure 2. For all three bowlers the delivery line has been to the left of the stumps. The fast bowler, whose deliveries are quickest, bowls closest to the centre stumps. The pitching line for the medium paced bowler is similar to the fast bowler; but it is slightly wider. The spin bowler shows the greatest variation in pitching line which is caused by the ball deviation in flight due to the rifle style spin imparted onto it.

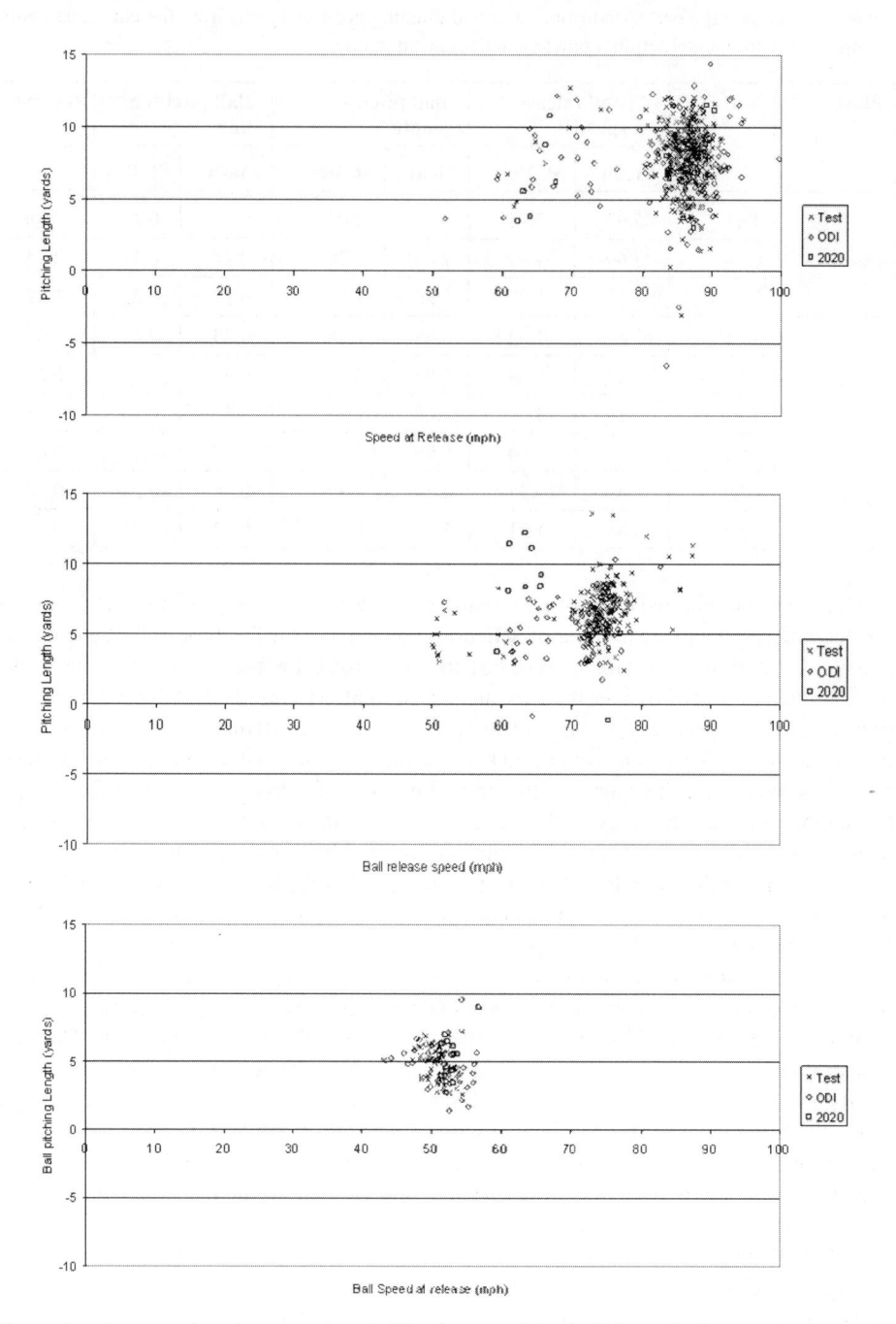

Figure 1. A comparison between the pitching length of the ball from the batsman's stumps (in yards) and the speed of the ball at release (mph) for the (a) fast, (b) medium and (c) spin bowlers.

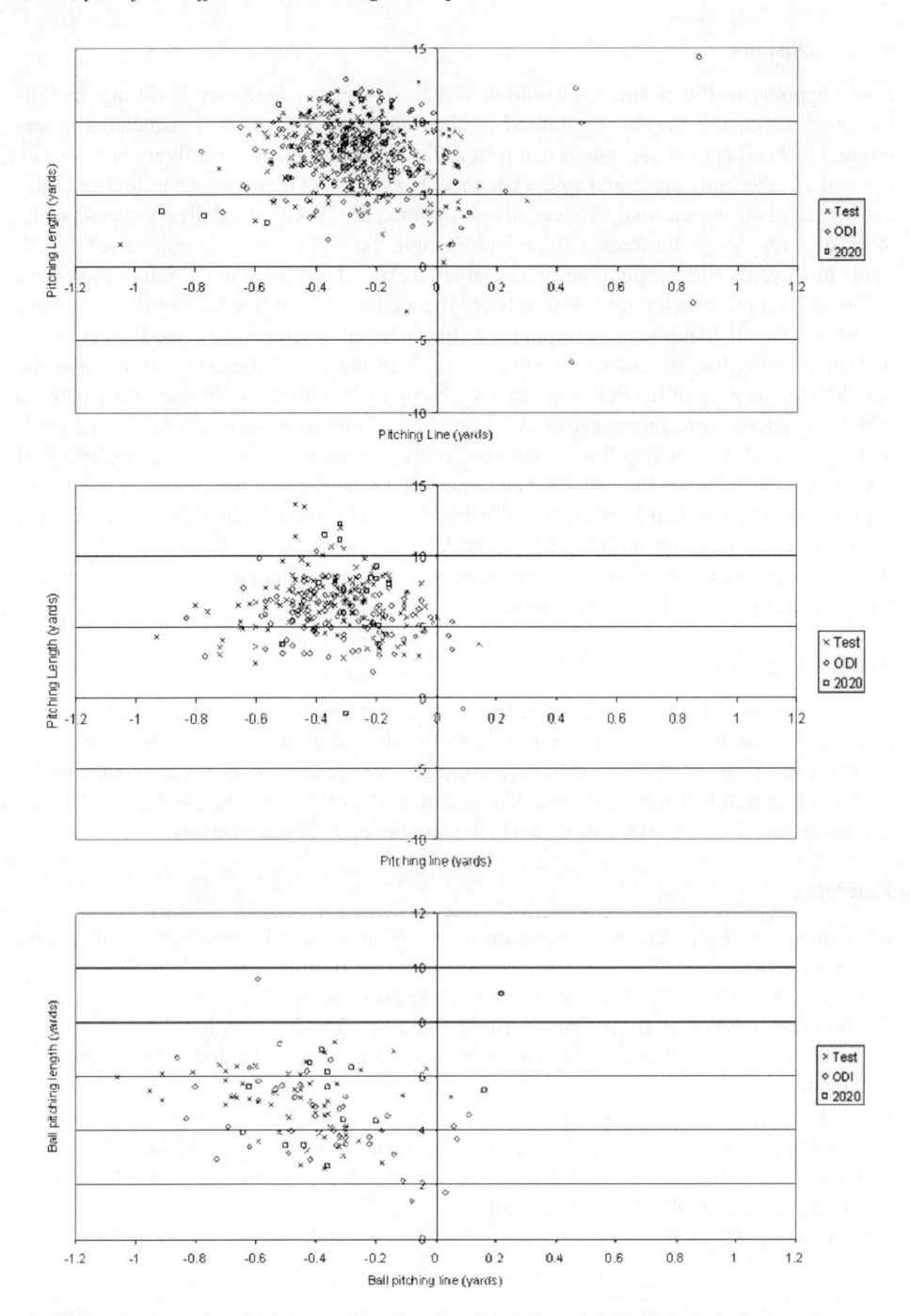

Figure 2. A comparison between the pitching length and line of the ball from the batsman's stumps (in yards) for the (a) fast, (b) medium and (c) spin bowlers.

4 Conclusions

It was hypothesized that bowlers would deliver the ball more aggressively during the limited overs matches as they have a limited number of deliveries to get the batsmen out. It was thought that this aggression would manifest as larger variations in the delivery characteristics and a higher run rate scored by the batsmen. However this hypothesis has not been fully substantiated during the analysis carried out. For the pace bowler the delivery speed for the stock delivery ranges between 80 mph and 94 mph, the ball pitching length varies from 2 yards to 13 yards and the pitching line is centered around 0.2 yards to the left of the centre stump in all three match types. Similarly for the medium paced bowler, the stock delivery speed is between 70 mph and 80 mph, the delivery length ranges from 2 yards to 10 yards and the pitching line is centered around 0.3 yards to the left of the centre stump. For the spin bowler the pace of the delivery increases by up to 2 mph for the limited overs matches which suggests a more aggressive delivery but the pitching length remains between 2 yards and 7 yards and the pitching line is centered around a point which is 0.4 yards to the left of the stumps regardless of the match being played. It has been concluded that bowlers do not significantly change their bowling style during different types of match play. However the batsman's run rate varies with the type of match. This suggests that the batsman's playing style becomes more aggressive in limited overs matches, although further investigations are required to substantiate this hypothesis.

Acknowledgements

The authors would like to acknowledge the support of the Engineering and Physical Sciences Research Council (EPSRC) of Great Britain and the Innovative Manufacture and Construction Research Centre (IMCRC) at Loughborough University for financial support during this research. They would also like to extend their thanks to the England and Wales Cricket Board for the Hawk-Eye data which was supplied for this analysis.

References

Abrahams J. (2004) Personal Communication: A generalized classification of cricket bowling.
Andrew K. (1989) *The Skills of Cricket*. Crowood Press, Ramsbury.
Bradman D. (1984) *The Art of Cricket*. Hodder and Stoughton, London.
ECB (2006) Personal Communication: Donation of Phototgraphs and data for research publication.
Hawk-Eye (2006) The manufacturers website for the Hawk-Eye Sports Tracking System (online). Available: https://www.hawkeyeinnovations.co.uk (accessed May 2007).
Justham L. (2007) The design and Development of a Novel Training System for Cricket Ph.D. Diss., Loughborough University, UK.
Khan I. (1989) *Imran Khan's cricket skills*. Hamlyn, London.

ENGINEERING A DEVICE WHICH IMPARTS SPIN ONTO A CRICKET BALL

L.M. JUSTHAM, A.A. WEST & A. E. J. CORK

Wolfson School of Mechanical and Manufacturing Engineering, Loughborough University, Loughborough, Leicestershire, England

Cricket bowling machines are employed within a training environment and used by many players. Current machines operate as mechanical ball launching devices which do not create technically correct bowling deliveries, nor do they use a proper cricket ball as standard. A research project based at Loughborough University to develop a novel training system which will revolutionize the use of bowling machines in training is currently being undertaken. The system being developed addresses the problems associated with current bowling machine technologies and is a computer controlled and programmable system which is able to accurately recreate the launch characteristics of any common bowling delivery. Player testing, using elite level cricket bowlers, has been carried out to determine the control and repeatability required by the training system. This has shown that they are able to control the speed of the ball to within ± 1.2 m/s and the spin rate imparted to within ± 5 rps. Controlled testing has been carried out using the prototype novel training system and has shown similar capabilities, with the speed of the ball controllable to ± 1.16 m/s and the spin imparted controllable to ± 4 rps.

1 Introduction

During a cricket match the bowler delivers balls to the striking batsman and tries to force them into making errors which results in them getting out. Therefore to practice different shot selections and batting techniques is a major aspect of training and involves the batsman facing numerous deliveries to become accustomed to the types of cues and information provided by the bowler during the run-up and delivery. It is important for the batsman to practice facing real bowlers but during an intensive training session the bowlers are at risk of over training and becoming fatigued, injured or developing errors within their technique. Therefore bowling machines are used extensively as they provide an endless supply of deliveries without any of the risks faced by real bowlers.

However there are major limitations to current bowling machine technologies as they do not generally use real cricket balls, nor do they have the functionality to create every common bowling delivery with the same speed and spin characteristics that a real player would impart onto the ball. These issues are particularly pertinent when considering spin bowling deliveries which require a rifle style spin to be imparted onto the ball in a direction transverse to the direction of motion. The development of a training system which is able to re-create any common bowling delivery is the focus of ongoing research at Loughborough University. The design and development of the system has been carried out using a systems engineering approach and the House of Quality, which is part of the Quality Function Deployment method, has been used to identify the requirements of the training system from the point of view of the system stakeholders (ReVelle, 1998). Forty seven requirements which are divided into four key categories have been identified within the Voice of the Customer (VOC) component of the planning matrix (Justham, 2006). The research presented in this paper is focused on the development of the component which imparts rifle style spin onto the cricket ball which allows technically correct spin bowling deliveries to be recreated.

2 Determination of the System Requirements

In cricket there is limited quantitative information regarding the launch characteristics of the ball and that which is available is focused upon pace and swing bowling deliveries (Wilkins, 1991). Therefore to ensure the correct speed and spin was imparted onto the ball by the machine, player testing was carried out to measure how the bowler releases the ball. Testing was carried out at the England and Wales Cricket Board National Cricket Centre (ECB-NCC), which is an indoor training facility based at Loughborough University, to measure the release characteristics of the ball using high speed video (HSV) filming. A Photron Ultima APX Fastcam high speed video camera was used and images were recorded at frame rates of 10,000 fps and 2000 fps. The images were digitized and analyzed to calculate the speed of the ball at release, its initial flight trajectory, the rate of spin and the orientation of the spin on the ball. The results obtained were used to classify bowling, see Table 1 for the results obtained for spin bowling, and to quantify the requirements of the new training system (Justham, 2007). The off-spin and leg-spin deliveries are the stock deliveries of the finger and wrist spin bowler respectively. The arm-ball and straight-ball are the most common variations of the off-spin bowler whereas the googly, top-spinner and flipper are the common variations of the leg-spin bowler.

For all seven delivery types the release trajectory is up to 10 degrees above the horizontal which means the ball loops upwards and creates a curved trajectory before pitching. The stock off-spin and leg-spin deliveries are bowled with rifle spin in a direction which is transverse to the direction of flight whereas the variation deliveries have a combination of spins which can result in the ball being delivered with a more scrambled seam. The training system must be able to recreate every combination of spin at a realistic speed and seam position. Therefore for the training system under development, four main requirements were identified for the creation of spin bowling deliveries:

1. The ability to deliver the ball at release speeds up to 60 mph
2. The ability to deliver the ball with rifle spin in both directions
3. The ability to deliver the ball with spin speeds in excess of 2000 rpm
4. The ability to deliver the ball not only with rifle spin but also with backspin, topspin or a combination of spin directions for the variation deliveries.

3 Design of the Spin Component

The training system under development has two distinct parts: (i) a mechanism to propel the ball and (ii) the spin component which imparts rifle spin. The ball propulsion mechanism was chosen to be a two counter rotating wheel design which is capable of releasing the ball at velocities in excess of 100 mph. The rotating wheels are controlled independently such that spin which is in line with the direction of motion may be imparted onto the ball. The spin component is solely responsible for the creation of rifle style spin which is transverse to the direction of motion of the ball.

Table 1. The results obtained from player testing to quantify spin bowling of elite cricketers.

Delivery Type	Speed at release (mph)	Spin rate (rpm)	Spin axis	Seam position	Release trajectory
Off-Spin	40–55	1200–1700	Rifle	Perpendicular to Flight	Up to 8 degrees
Arm-ball	45–60	750–1000	Backspin	Up to 45 deg	Up to 7 degrees
Straight-ball	45–60	900–1350	Backspin	Up to 45 deg	Up to 6 degrees
Leg-Spin	40–55	1200–1700	Rifle	Perpendicular to Flight	Up to 5 degrees
Googly	40–55	1500–2000	Rifle + Topspin	Perpendicular to Flight	Up to 5 degrees
Top-spinner	40–60	1000–1500	Topspin	Up to 45 deg	Up to 10 degrees
Flipper	40–50	1200–1700	Backspin	Up to 45 deg	Up to 3 degrees

The spin component was designed around rifling theory which is an established method of imparting spin in a direction which is transverse to the direction of motion. Rifling has been used in firearm technology since the 15th century and occurs as a result of helical grooves being cut into the barrel of a rifle (Barkla & Auchterlonie, 1971). This forces the bullet to twist as it is released from the barrel and results in an increased accuracy and range of flight (Daish, 1972). The raised areas between the grooves are known as lands and they make contact with the bullet as it is fired along the barrel length. In a rifle the grooves are fixed in direction and have a pre-specified twist rate so a similar spin rate and spin axis is imparted onto every bullet leaving the barrel. However the spin required for bowling deliveries is not in a fixed direction or at a fixed speed. The spin must be imparted in both clockwise and anticlockwise directions at speeds between 0 rpm and 2000 rpm to enable all types of deliveries to be created, see Table 1.

The design chosen for the spin component is based upon a smooth barrel which does not interact with the ball as it passes through and allows pace deliveries to be bowled without imparting any additional spin onto the ball. Portions of the barrel have been cut away to allow straight lands to protrude at a known separation such that they interact with the ball as it passes through. The barrel is then coupled to the rest of the machine such that it is able to rotate at speeds up to 3600 rpm which results in the lands appearing to the ball to have a variable helical configuration so variable spin rates can be imparted onto the ball. Rifle lands are traditionally set at a separation which is smaller than the diameter of the bullet and the number of lands within the barrel varies depending on the rifle. Four lands were chosen for the bowling machine, which mimics a number of currently manufactured rifles (SAKO Rifles, 2006), and their separation was set to 92% of the ball diameter.

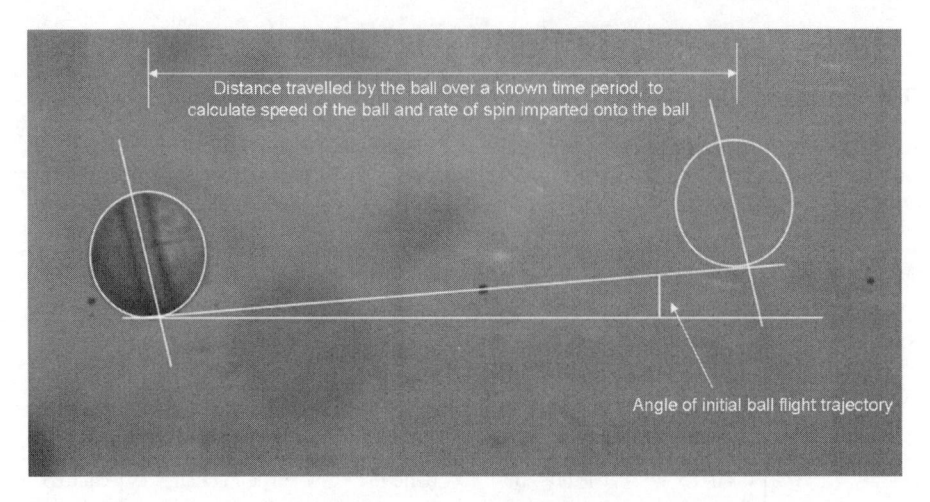

Figure 1. The image analysis process which was undertaken to determine the velocity and initial flight trajectory of the ball and the speed and direction of the spin imparted onto the ball by the spin component.

4 Testing the Spin Component

Testing was carried out in the Sports Technology Laboratory at Loughborough University. The propulsion wheels were rotated at two speeds; (i) 45 mph which is a slow spin bowling delivery and (ii) 60 mph which is a fast spin bowing delivery. The rotational speed of the spin component was varied between 500 rpm and 3600 rpm in 500 rpm increments. HSV filming was carried out using the Photron Ultima APX Fastcam camera recording at a frame rate of 10,000 fps. In the laboratory additional illumination was required and this was supplied using two Unomat LX 901GZ 1000W halogen spotlights which were set-up on tripods, one on either side of the HSV camera. The camera was positioned perpendicular to the nominal ball flight path with a 0.75 m × 0.5 m measurement region. A real cricket ball was used throughout testing and was fed into the machine at the correct orientation for spin bowling Images were colleted and analyzed using Image Pro Plus image processing software, see Figure 1. Measurements were taken with respect to the number of frames taken for the ball to travel a known distance. This enabled the velocity of the ball and its initial flight trajectory to be calculated as well as the direction and speed of the spin imparted by the spin component.

The results obtained are displayed in Figure 2 and show a linear relationship between the rotational speed of the spin component and the rate of rifle spin imparted onto the ball. At the maximum rotational speed of 3600 rpm, rifle spin was imparted onto the ball at speeds in excess of 50 rps (3000 rpm), which is 176% faster than the spin rates measured during player testing. The maximum variation in the repeat measurements taken at each rotational speed was ±4 rps which corresponds to a maximum standard deviation of 3.2, a maximum error in the mean of 1.85 and an efficiency of 70%. The speed of the propulsion wheels were varied such that the ball was released at 45 mph and 60 mph. Generally the speed of the rifle spin imparted onto the ball was quicker for the faster moving ball. However in each case the variation between the two ball release speeds was less than the ±4 rps variation observed within each measurement point.

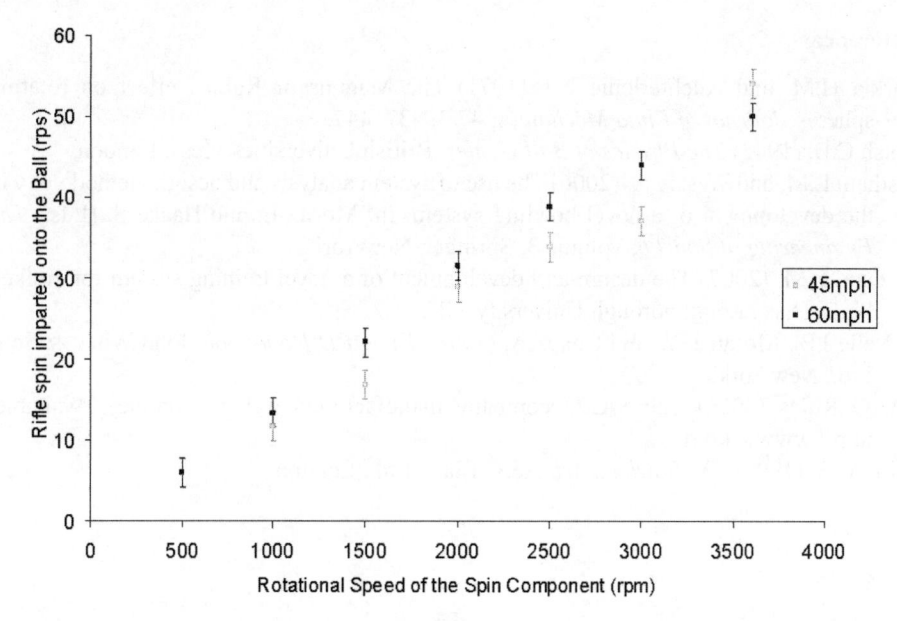

Figure 2. The results obtained from testing the spin component. A linear relationship between the rotational speed of the spin component and the rate of rifle spin imparted onto the ball can be seen.

5 Conclusions

The results obtained during the player testing have shown that elite bowlers are able to control the speed at which they release the ball to within ±1.2 m/s and the rate of spin imparted onto the ball within ±5 rps. The results obtained from the laboratory testing using the prototype training system have shown that the speed of the ball at release may be controlled to within ±1.16 m/s and the spin imparted onto the ball can be controlled to within ±4 rps using the novel training system. The direction of spin imparted onto the ball may be controlled using a combination of the counter rotating wheels and the spin component. The propulsion wheels are responsible for the impartation of backspin and topspin whereas the spin component imparts rifle spin onto the ball, these combine to form the combinations of spin required for the spin-bowling variation deliveries.

Acknowledgements

The authors would like to acknowledge the support of the Engineering and Physical Sciences Research Council of Great Britain (EPSRC) and the Innovative Manufacture and Construction Research Centre (IMCRC) at Loughborough University for funding the project. They would also like to thank Mr. Steve Carr and Mr. Andrew Hallam from the Sports Technology Research Group at Loughborough University for their help and support during the manufacture of the prototype training system. Finally they would like to acknowledge the support of the England and Wales Cricket Board who have provided elite level bowlers for the player testing and the indoor training facility as a location for testing.

References

Barkla H.M. and Auchterlonie L.J. (1971) The Magnus or Robins effect on rotating spheres. *Journal of Fluid Mechanics*, 47/3, 437–447.

Daish C.B. (1972) *The Physics of Ball Games*. British Universities Press, London.

Justham L.M. and West A.A. (2006) The use of system analysis and design methodology in the development of a novel bowling system. In: Moritz E. and Haake S. (Eds.) *The Engineering of Sport 6*, Volume 3, Springer, New York.

Justham L.M. (2007) The design and development of a novel training system for cricket. Ph.D. Diss., Loughborough University, UK.

ReVelle J.B., Moran J.W., and Cox C.A. (1998) *The QFD Handbook*. John Wiley & Sons Ltd., New York.

SAKO Rifles (2006) The SAKO company manufacturers website (online). Available: http://www.sako.fi.

Wilkins B. (1991) *The Bowler's Art*. A&C Black Ltd., London.

9. Ball Sport – Baseball

THE EFFECT OF WOOD PROPERTIES ON THE PERFORMANCE OF BASEBALL BATS

K.B. BLAIR[1], G. WILLIAMS[2] & G. VASQUEZ[3]

[1] *sports innovation group LLC, Arlington, MA, USA*
[2] *Rawlings Sporting Goods, St. Louis, MO, USA*
[3] *Massachusetts Institute of Technology, Cambridge MA, USA*

The purpose of this experiment was to determine whether the Coefficient Of Restitution (COR) between a baseball and wood plate depends on the measurable wood property, hardness. Should a performance difference exist, a more optimal type of wood could be found for baseball bats. Even small performance increases in baseball bats can have a large benefit for the batter. Seven wood plates were tested and the resulting performance was compared with the wood properties of the plates. Significant differences existed between wood plates indicating a slight trend that softer woods are higher performing.

1 Introduction

The selection of the type of wood used in baseball bats involves very little science, and few wood species have been used to make bats throughout baseball's history. Early baseball bats were made of hickory. Hickory is a very strong wood, and bats made from it are durable, however hickory is also a very dense wood. Thus the high strength and durability comes at the cost of added weight. Bat weight became an important factor for players as lighter bats can be swung faster. Ash, a lighter wood, became the wood of choice and replaced hickory. The lighter weight ash also resulted in less durable bats, but in today's game, many players are willing to make this sacrifice if it means quicker swing speed. The most recent development in wood bats has been the increasing popularity of maple bats, which are only slightly heavier than ash bats, but offer added durability and have a different feel.

For many bat manufacturers, the selection of wood for bats is much more an art than a science. The selection process consists primarily of choosing the proper weight and inspecting the quality of grain structure by eye. This paper investigates whether there is a correlation between performance and certain wood properties such as hardness. This could offer new methods of wood selection for use in bats. The coefficient of restitution (COR) of seven different wood plates was tested and compared to the hardness of the wood plates.

2 Methods

2.1. COR Testing

A test setup was constructed that was capable of measuring the COR for the collision between a baseball and a wood plate. The experimental setup was a modified version of ASTM standard F1887-02, Standard Test Method for Measuring the Coefficient of Restitution (COR) of Baseballs and Softballs (ASTM, 2002).

A two wheel Jugs pitching machine was used to fire a baseball at a wood plate from 2.2 m away at a speed of approximately 29 m/s. The ball was required to pass through a

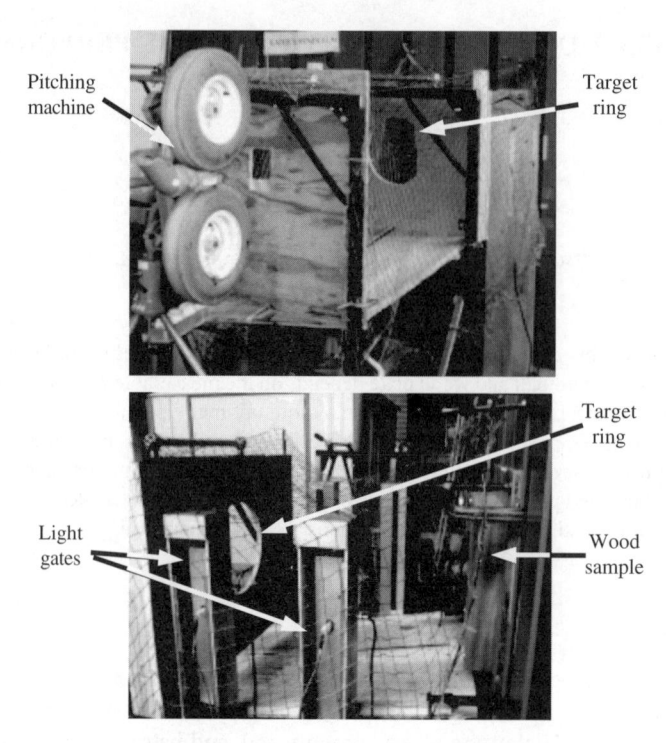

Figure 1. Photographs of the experimental setup illustrating the location of the pitching machine, target ring, light gates, and wood samples.

0.3 m target ring on the inbound path and rebound path to assure a trajectory nearly normal to the wood plate. The balls impacted the plates within a 0.15 m circle, all as specified by the ASTM standard (Figure 1).

Pre and post collision velocities were measured using infrared light gates. The light gates were spaced 0.3 m apart and were 0.3 m from the wood plate. The light gates were attached to an oscilloscope and a signal was received when the ball passed through each gate. The time difference between gates was measured on the oscilloscope and COR was computed as a ratio of the outgoing to the incoming ball velocity.

The testing took place over period of two sessions using Rawlings R100 baseballs (Rawlings, Inc., St. Louis, MO, USA). Each session consisted of each plate being impacted once with each of 20 balls. The same 20 balls were used to test each of the plates in a given session. New balls were used for each session. In the event of a misfire, the same baseball was used again, however not within two minutes. The balls were stored in the same laboratory where testing took place. Laboratory temperature and humidity varied, but generally remained close to 21 C, and between 10 and 50% humidity.

Seven plates were tested. The Ash-1, Ash-3, Mahogany, Maple, Ipe, and Poplar plates had a grain orientation parallel to the ball velocity direction. The grain orientation of the Ash-2 plate was perpendicular to ball velocity. The plates measured 40 cm tall by 40 cm wide by 7 cm thick with the exception of the maple plate, which was 6 cm thick. The plates were constructed by laminating several smaller pieces of wood together with wood glue.

Table 1. Experimental results for moisture content, hardness and COR for each sample.

Sample	Thickness cm	Width cm	Height cm	Grain Orientation	Moisture Content	Hardness kg	COR
Ash-1	7	40	40	Parallel	ambient	679	0.541
Ash-2	7	40	40	Perpendicular	ambient	831	0.536
Ash-3	7	40	40	Parallel	90 humidity	640	0.525
Mahogany	7	40	40	Parallel	ambient	633	0.532
Maple	6	40	40	Parallel	ambient	841	0.532
Ipe	7	40	40	Parallel	ambient	2060	0.528
Poplar	7	40	40	Parallel	ambient	404	0.542

Table 2. Statistical significance testing comparing the COR of the test plates.

Sample	Ash-1	Ash-2	Ash-3	Mahogany	Maple	Ipe
Ash-2	2.96					
Ash-3	−9.69	−6.14				
Mahogany	4.95	2.00	−3.78			
Maple	−5.30	−2.12	4.11	0.02		
Ipe	−7.18	−4.00	1.88	−1.86	−2.02	
Poplar	−0.94	−3.60	−9.68	−5.41	−5.73	−7.44

One plate (Ash-3) was stored in a humidity chamber that was maintained at 90–100% humidity. The other plates were kept at ambient conditions that varied, but generally remained below 50% humidity. Plates were attached to a 1.6 cm thick piece of aluminum using four bolts on 35.5 cm centers. The piece of aluminum was attached to a steel hardback with four C-clamps. The order in which plates were tested was randomized for each session.

2.2. Hardness Tests

After COR testing, the wood plates were tested in the region of impact in three locations for hardness. This test was done according to ASTM standard D143-94 (ASTM, 1994), Standard Methods of Testing Small Clear Specimens of Timber with minor changes. The force required to imbed a 1.11 cm diameter ball into each wood sample was measured.

3 Results

Table 1 shows the results of the COR and hardness tests for all sample plates. The table results show a relatively wide variation in hardness and a small variation in COR.

Statistical significance of COR was determined between each pair of plates using a 95% confidence t/z test. An absolute value greater than 1.96 means the pair of plates are

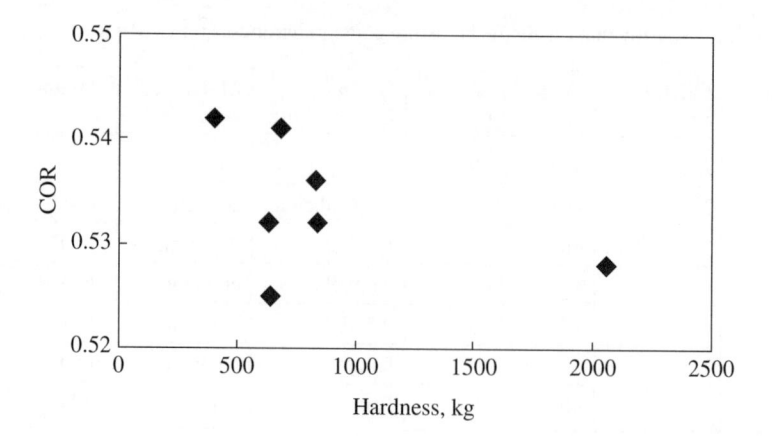

Figure 2. Graphical comparison between the hardness and the coefficient of restitution.

significantly different with at least 95% confidence. Wood pairs that were not significantly different are highlighted below.

A comparison of COR and hardness is shown in Figure 2. There does not appear to be a correlation between hardness and COR.

4 Discussion

An increase in COR means the collision is more efficient and thus the ball will be hit farther. According to Adair (2002), an increase in COR of 0.01 corresponds to an increase in hit distance of approximately 1.8 m for a collision between 38 m/s ball and 31 m/s bat. Thus, even the small changes of COR seen in this testing could have an impact on bat performance.

It is important to recognize the difference between the testing done in this experiment and the real-life hitting situation. The testing in this experiment was done with a 29 m/s ball impacting a stationary plate. This is quite different from the impacts seen in professional baseball where bat and ball velocity both consistently approach 40 m/s resulting in a relative impact velocity of 80 m/s or more. During these interactions, a cylindrical bat is being impacted with a ball. This may introduce other effects not modeled by using the flat plates. For example, the round bat and higher impact velocity may cause a larger deformation of the baseball. This becomes important because of the fact that a baseball is constructed in layers with different yarns and materials at each level.

5 Summary and Conclusion

Based on the results from this experiment, it is uncertain whether the COR of wood is directly related to hardness. Many more data points would be necessary to establish a trend, if one exists. Although this original objective was not confirmed, it is clear that the COR does in fact vary significantly among different wood species.

Simply having a high COR does not necessarily mean a type of wood is feasible to use in a baseball bat. In this experiment, poplar was both the highest performing wood

(COR = 0.542) and the softest. Unfortunately, constructing a usable bat out of this wood may not be practical because the bat would not be very durable. Even during this testing with a stationary plate and only 29 m/s pitches, there were indications of some physical deformation of the surface. Clearly, wood cannot be chosen on the basis of COR performance alone, but weight and softness considerations must be accounted for as well.

Because of wood's naturally inherent structural make-up, even within a species, variation in hardness and subsequently COR will be inevitable. However, based on the findings of this experiment, it may be worthwhile for bat manufacturers to consider not only the weight and grain structure of their bat material, but other properties that may affect COR as well.

Acknowledgements

The authors would like to acknowledge the aid given by the technical staff at MIT's Department of Aeronautics and Astronautics: Dave Robertson, John Kane, Dick Perdichizzi, and Todd Billings.

References

ASTM (2002) Standard Test Method for Measuring the Coefficient of Restitution (COR) of Baseballs and Softballs. F1887-02.
ASTM (1994) Standard Methods of Testing Small Clear Specimens of Timber. D143-94.
Adair R.K. (2002) *The Physics of Baseball*. 3rd Edition, HarperCollins Publishers Inc., New York.

DESCRIBING THE PLASTIC DEFORMATION OF ALUMINUM SOFTBALL BATS

E. BIESEN & L.V. SMITH

Washington State University, Pullman, WA, USA

Hollow aluminum bats were introduced over 30 years ago to provide improved durability over wood bats. Since their introduction, however, the interest in hollow bats has focused almost exclusively around their hitting performance. The aim of this study was to take advantage of the progress that has been made in predicting bat performance using finite elements and apply it to describe bat durability. Accordingly, the plastic deformation from a ball impact of a single-wall aluminum bat was numerically modeled. The bat deformation from the finite element analysis was then compared with experiment using a high speed bat test machine. The ball was modeled as an isotropic, homogeneous, viscoelastic sphere. The viscoelastic parameters of the ball model were found from instrumented, high-speed, rigid-wall ball impacts. The rigid-wall ball impacts were modeled numerically and showed good agreement with the experimentally obtained response. The strain response of the combined bat-ball model was verified with a strain-gaged bat at intermediate ball impact speeds in the elastic range. The strain response of the bat-ball model exhibited positive correlation with the experimental measurements. High-speed bat-ball impacts were performed experimentally and simulated numerically at increasing impact speeds which induced correspondingly increased dent sizes in the bat. The plastic deformation from the numerical model found good agreement with experiment provided the aluminum work hardening and strain rate effects were appropriately described. The inclusion of strain rate effects was shown to have a significant effect on the bat deformations produced in the finite element simulations. They also helped explain the existence of high bat stresses found in many performance models.

1 Introduction

Much of the work involving finite element modeling of a bat-ball collisions concerns bat performance. The bat is typically described elastically as either wood or aluminum, while the ball is described as a viscoelastic (Sandmeyer, 1995; Vedula, 2004) or hyper-elastic (Mustone and Sherwood, 1998, Nicholls 2003) material. Their correlations with performance are generally positive, although establishing the accuracy of the ball models has been problematic.

Only one example was found in the literature that considered bat plasticity (Mustone, 2003). Here the ball was described as a hyper-elastic material. The resulting model produced irregular ball deformations and overestimated bat performance.

Improved ball models based on the dynamic response of baseballs and softballs have produced positive performance comparisons (Duris, 2004; Shenoy et al., 2001). In this work the ball was modeled as a viscoelastic material, whose properties were tuned to the dynamic response of the ball, independent of the bat. The viscoelastic shear modulus, $G(t)$, was described using a power law according to

$$G(t) = G_\infty + (G_0 - G_\infty)e^{(-\beta t)} \qquad (1)$$

where G_∞ is the long term shear modulus, G_o is the short term shear modulus, and β is the decay constant. The bulk modulus, k, was constant.

The aim of this work was to take advantage of the progress that has been made in predicting bat performance and apply it to describing bat durability. The work considered a single wall aluminum softball bat whose plastic deformation was compared with a finite element model as a function of impact speed.

2 Ball Characterization

The response of the softball is rate dependent and requires careful characterization to accurately model its interaction with the bat. A high speed pneumatic ball cannon was used for the ball and bat tests as described in ASTM F2219. The velocity of the ball, before and after impact, was measured by the use of infrared light gates.

The hardness and elasticity of softballs is a function of the impact speed and surface geometry. A test was developed to account for these effects, where the ball was fired at a four inch long (202 mm) solid steel half cylinder 2.25 inches (57 mm) in diameter which was rigidly mounted (Duris and Smith, 2004). The impact force was measured by a group of load cells placed between the cylindrical impact surface and the rigid support.

An idealized dynamic stiffness of the ball may be obtained by equating its incoming kinetic energy with its stored potential energy at maximum displacement. If we assume the ball to act as a linear spring during deformation (a relatively good approximation) the unknown displacement can be eliminated, from which the stiffness is found from

$$k = \frac{1}{m}\left(\frac{F}{v_p}\right)^2 \qquad (2)$$

where m is the ball mass, F is the peak impact force, and v_p is the pitch speed.

The test balls used for this work were ASA certified slow pitch softballs. The dynamic stiffness (DS) of the balls was measured as a function of speed between 60 and 130 mph (27 and 58 m/s). The balls were stored and tested in a laboratory space that was held to 72° F ± 4° (22°C ± 2°C) and 40% R.H. ± 10%. The cylindrical COR (CCOR) and dynamic stiffness were found as a function of speed, as shown in Fig.1.

Since the CCOR and DS are not constant, but depend on speed, the rigid cylinder and recoiling bat test speeds must be correlated. If we assume the ball must undergo the same deformation in both cases, the recoiling, v_r, and fixed, v_f, speeds are related according to

$$v_f = v_r\left(1 + \frac{m_b r^2}{MOI}\right)^{-\frac{1}{2}}. \qquad (3)$$

where m_b is the mass of the ball, r is the distance from the pivot to the impact location on a bat, and MOI is the mass moment of inertia of the bat about the pivot.

The ball was modeled as a 12 inch (305 mm) circumference solid, isotropic sphere using 10,240 8-noded solid elements. Its time dependent properties (Eq. 1) were tailored to provide the measured DS and CCOR. The ball tuning was carried out by simulating the

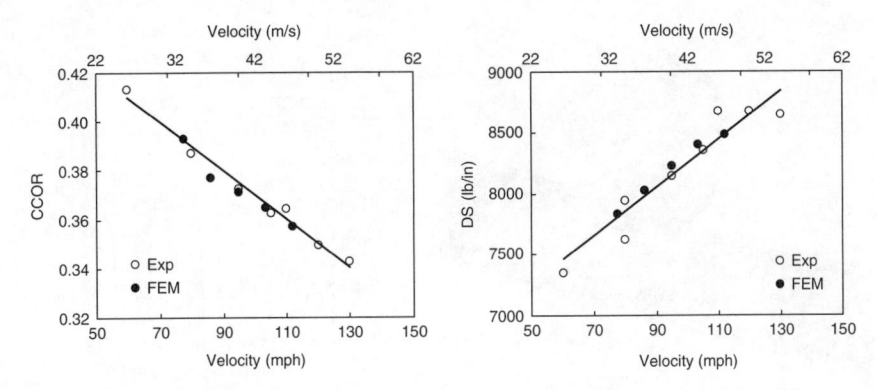

Figure 1. The cylindrical coefficient of restitution (left) and dynamic stiffness (right) as a function of impact speed (each point is an average of 6 balls).

dynamic stiffness experiment, and using Eq. 3 for the test speed. The numerical model used a ball weight of 6.745 oz (192 g).

The finite element model of the dynamic stiffness test consisted of the ball model impacting a cylinder of the same shape and size as the one used for the experiments. All nodes of the cylinder were fixed to describe the solid cylinder. Symmetry was applied to the mid-plane of the ball and cylinder to match that applied to the bat-ball case. The rebound speed of the ball was taken at its center. The impact force was taken from the sum of the nodal forces in contact during impact. The resulting *CCOR* and *DS* were compared to the experimental values.

The ball viscoelastic properties were changed in an iterative fashion until the *DS* and *CCOR* from the model matched that which was found experimentally. The finite element results are compared with the experimental values in Fig. 1.

3 Bat-ball Impact Simulations

A numeric model of a softball impact with a single-wall aluminum bat was constructed. The slow pitch bat was made by Louisville Slugger (model SB806). The bat had a nominal length of 34 inches (0.86 m), a weight of 28 ounces (795 g), and an MOI of 7029 oz in^2 (129 g m^2). The bat was made from 7046-T6 aluminum with a yield strength of 60 ksi (414 MPa), ultimate strength of 65 ksi (448 MPa), Young's modulus of 10.4 Msi (71.7 GPa), density of 0.102 lb/in^3 (2.82 g/m^3), and a Poisson's ratio of 0.30 (Suchy, 2005).

The FEA bat model is shown in Fig. 2 and used primarily thick shell elements. These eight noded elements allowed a smooth transition in wall thickness and were computationally more efficient than solid brick elements. A convergence study showed that five integration points through the thickness were sufficient. The thick shell elements had difficulty describing plasticity under flexure, however. Eight noded, constant stress, solid brick elements were, therefore, employed in the impact region.

The impact region of the bat involves significant flexure. Accordingly, multiple solid elements through the thickness were required to describe this region. A convergence study showed that four elements through the thickness were sufficient. The elastic response of the converged model was compared with experimental impact data. One comparison involved

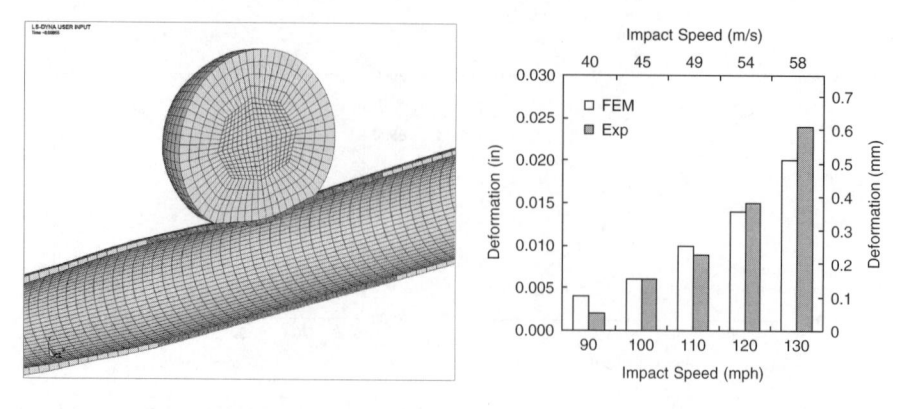

Figure 2. Mesh of bat-ball model (left) comparison of measured and predicted dent size (right).

bat performance as described in ASTM F2219. The numerical and experimental bat performance measures were within 0.9% using a batted ball speed scale.

The strain response of the model was considered by instrumenting the bat with strain gages a bat at three locations. The bat was impacted at 78.5 mph (35 m/s), 19 inches (0.48 m) from the knob, in its fully elastic range. Strains were measured in the taper and impact locations (14.2 and 19 inches (0.36 and 0.48 m) from the knob, respectively). The grips and end caps were removed to simplify the FEA of the bat as well as apply a strain gage behind the impact location.

A comparison of the experimental and FEA strains was favorable, although the responses were slightly out of phase. In the numeric model at the impact location a large strain momentarily occurred as the bat and ball came in contact. This was apparently associated with the surface contact algorithm that prevents surface penetration, and did not induce plasticity. (The momentary spike in strain was not observed outside the impact location.)

The plastic deformation was measured experimentally by comparing dent depth. This was accomplished by placing the bat in a lathe and scanning its surface circumferentially and longitudinally with a dial indicator. Once the center of the dent was located, the indicator was zeroed and measurements were taken at 1/2 inch (13 mm) increments longitudinally along the bat for 3.5 inches (89 mm) in either direction. Since the bat barrel is not a perfect cylinder, manufacturing anomalies were separated from plastic deformation. A straight line was found from the outer most measurements (3.5 inches (89 mm) from the dent) where plasticity did not occur. Dent depth was taken as the normal distance from the dent center to this line.

The deformation from the numerical simulations was determined from the outer diameter of the bat at the point of impact. These measurements were taken as the bat oscillated and were, therefore, averaged over time.

The numeric model used an isotropic power law to describe hardening. The strength coefficient and the strain hardening exponent were found from tensile stress-strain data (Suchy, 2005) as 82,657 psi (570 MPa) and 0.0684, respectively. Unfortunately, the numerical model over estimated the dent depth by a factor of 10 using this plasticity model alone.

While the plasticity of many aluminum alloys is rate sensitive, this effect has not been included for bat impacts found in the literature. These effects were incorporated into the numeric model by scaling yield strength with strain rate according to (Stranart, 2000)

$$1 + \left(\frac{\dot{\varepsilon}}{C}\right)^{\frac{1}{P}} \qquad (5)$$

where $\dot{\varepsilon}$ is the strain rate, and C and P are scaling coefficients.

While scaling coefficients for the 7046-T6 alloy could not be found, they are reported for a similar alloy, 7075-T65 as 1300 and 5 for C and P, respectively. For a 130 mph (58 m/s) impact, for example, strain rates for the numerical model were on the order of 1100 strain/s, effectively scaling the yield strength by a factor of two.

A comparison of the experimental and numeric dent sizes is presented in Fig. 2 for five impact speeds. The comparison is favorable showing that dents can be reliably predicted at a variety of speeds and develop at relatively low impact speeds.

Predictions from numerical models involving bat-ball impacts often report bat stresses far exceeding the alloy's yield strength. This apparent anomalous result contradicts the generally positive correlation obtained in performance comparisons. The contradiction appears to be addressed by consideration of strain rate effects. It also suggests that bat alloy selection should involve both static strength and strain rate dependence.

Bat durability is often described as a fatigue problem. The results clearly show here, however, that for a relative impact speed of 90 mph (40 m/s) a measurable dent would occur on the first impact, although it may not be detectable to the batter. And for a 110 mph (49 m/s) impact speed typical of a high level player, a noticeable dent is formed after only one impact. While the speeds where dents form in a bat will depend on its design, bat durability may be more suitably described using plasticity than fatigue analysis.

4 Conclusions

This work has considered experimental and finite element methods to describe the durability of a single-wall aluminum softball bat. A ball model was developed for the finite element simulations by measuring its response under controlled high speed impacts. The model was able to accurately describe ball hardness and elasticity for a number of incident speeds. The desired accuracy was only achieved, however, by tailoring the ball's response at each impact speed. The accuracy of the bat-ball model was demonstrated through strain and performance comparisons. The influence of strain hardening, mesh refinement, and strain rate were explored in the finite element bat model. Strain rate effects, which are typically not included in bat models, were shown to have a significant effect in accurately describing the plasticity observed from bat-ball impacts.

References

Duris J. (2004) Experimental and numerical characterization of softballs. MS Thesis, Washington State University.

Duris, J., Smith, L.V. (2004) Evaluating Test Methods Used to Characterize Softballs. In: Hubbard M., Mehta R.D. and Pallis J.M. (Eds.) *The Engineering of Sport 5, Vol. 2*. International Sports Engineering Association, Sheffield, UK pp. 80–86.

Mustone T.J. and Sherwood J. (1998) Using LS-DYNA to Characterize the Performance of Baseball Bats. Proceedings of the 5th International LS-DYNA Users Conference. Southfield, MI.

Mustone T.J. (2003) A Method to Evaluate and Predict the Performance of Baseball Bats Using Finite Elements. MS Thesis, University of Massachusetts Lowell,.

Nicholls R.L. (2003) Mathematical Modeling of Bat-Ball Impact in Baseball. PhD Thesis, University of Western Australia.

Sandmeyer B.J. (1995) Simulation of bat/ball impacts using finite element analysis. MS Thesis, Oregon State University,

Shenoy M.M., Smith L.V., Axtell J.T. (2001) Performance Assessment of Wood, Metal and Composite Baseball Bats, *Composite Structures*, 52:397–404.

Stranart J.C. (2000) Mechanically Induced Residual Stresses: Modeling and Characterization. Ph.D. Thesis, University of Toronto.

Suchy J. Hillerich & Bradsbury (Louisville Slugger). (2005) Private Communication.

Vedula G. (2004) Experimental and Finite Element Study of the Design Parameters of an Aluminum Baseball Bat. MS Thesis, University of Massachusetts Lowell.

A RESEARCH ABOUT PERFORMANCE OF METAL BASEBALL BATS

Y. GOTO[1], T. WATANABE[1], S. HASUIKE[1], J. HAYASAKA[1], M. IWAHARA[1],
A. NAGAMATSU[1], K. ARAI[1], A. KONDO[1], Y. TERANISHI[2] & H. NAGAO[2]

[1]*Hosei University, Department of Mechanical Engineering, Faculty of Engineering,
Tokyo, Japan*
[2]*Mizuno Co. Ltd., Osaka, Japan*

Many baseball players emphasise sweet spots at metal baseball bats to improve their flying distance. The sweet spot is also the one of important factors to specify bats. However, many players tend to recognize the point based on their subjective feeling. In this paper, we applied Experimental Modal analysis to find out the position of sweet spot with numerical data. By applying the numerical and experiment method, we investigated the position for center of percussion and of sweet spot.

1 Introduction

1.1. Background

Many types of sporting goods are developed and improved using the new materials, and new design methods by sporting goods companies. Many researchers also study many types of sporting goods (Hiroshi & Nagamatsu, 1979). In particular, metal baseball bats are used by amateur players such as baseball players at high school. In each use of bats, the companies sell many types of bats, for example, middle balanced bats for alley hitters and top balanced bats for long-ball hitters. For baseball metal bats, the batting sound, hand numbness and repulsion are related to vibration. Baseball players emphasize the batting performance. Each batting performances are determined by each sweet spots. Regarding sweet spots, with their subjective feelings, the players recognize the point based on the point for batting, the flying distance and the point which has the highest reflection coefficient in the bat. In the development and design phase of the production, designers and developers at companies have designed by improve the characteristics for current commercialized metal bats (Adair & Nakamura, 1996).

1.2. Purpose of the Study

This paper assumes that the sweet spot on bats is the most reflection point. The paper describes the characteristics for sweet spots in two bats and the difference between types of bats based on Experimental Modal Analysis (Nagamatsu, 1998). Furthermore, the paper mentions whether for specifications for commercially-supplied bats are closely related to the results by Modal Analysis.

2 Experiments

2.1. Products Under Experiments

Figure 1 shows specifications for each bat in the experiments. To compare the difference each specifications and vibration characteristics, a bat named VS for alley hitters and a bat named VK for long-ball hitters are applied to the experiments.

Figure 2 shows the hard ball used in the experiments. The mass is 149.9 g, the maximum outer diameter is 74.78 cm and minimum outer diameter is 7.27 cm. Figure 3 shows the laser speed sensor. Due to the distance between two laser sensors, each passing times by the ball are different at each sensors. An oscilloscope can acquire two types of voltage patterns from these sensors while the ball is passing between the two sensors. Therefore, the ball velocity can be calculated using the time difference.

Figure 4 shows high speed video camera. The camera takes pictures of a ball at the time of impact on the bat. This allows to measure reflection speed of the ball.

Figure 5 shows a shooting machine and a ball is set at the battery position on the machine. First, the machine accumulates carbon dioxide in the accumulator. Second, the machine makes it possible to shoot a ball at the constant speed by open an air driven valve of it.

2.2. Preliminary Experiment

Laser sensors and high speed camera measured the shooting ball speed simultaneously to evaluate these performances as sensors. Figure 6 shows the results for preliminary experiments. The horizontal axis indicates a velocity measured by the high speed camera. The vertical axis indicates a measured velocity by the laser sensors. Figure 6 shows a variance between a result by laser sensor and by a high speed camera. Figure 6 indicated that the high speed camera is suitable to measure reflection speed for bats.

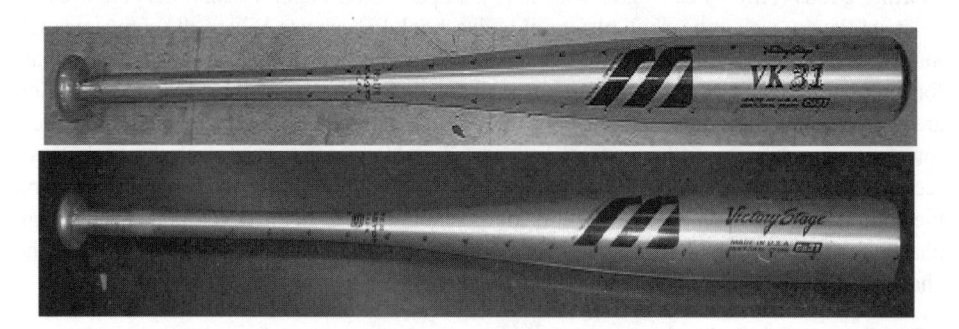

Figure 1. Metal Bats.

Table 1. Bats Specifications

Classification	Length [mm]	Mass [g]	Passage position [mm]	Blow central position [mm]	Center of gravity position [mm]	Moment of inertia [kg · m]	Maximum outer diameter [mm]	Average thickness [mm]
VK	839.0	903.0	165.0	195.3	306.3	5.34E-02	66.6	2.98
VS	832.5	897.5	161.5	194.7	297.0	4.91E-02	66.7	2.93

3 Definition

3.1. *Sweet Spots*

Some baseball players find the sweet spot on bats based on their each subjective feeling. In other words, each player tends to decide different positions. Therefore, in this paper, the sweet spot is defined as the maximum reflection point, because experimental numerical data helps us to understand the characteristics for each bat.

3.2. *Center of Percussion*

A center of percussion exists on any bat. The center of percussion is a fixed point at the knob on the bat. When the ball hits the bat, translational motion energy and rotary motion energy are generated on the bat. However, if two motions are coupled at a particular percussion point, the bat has a fixed point at the knob. This point is called center of percussion.

Figure 2. Hard ball.

Figure 3. Laser sensor.

Figure 4. High speed video camera.

Figure 5. Shooting machine.

4 Experiment

4.1. *Experiment Modal Analysis*

Adair & Nakamura (1996) mentioned that bats have no oscillating points in spite of the vibrating bats. To investigate this point, we conducted experimental modal analysis. In the experiment, eighty eight points were excited by an impulse hammer and the data at one point is measured by FFT analyzer. Table 2 shows the node positions which are identified by Experimental Modal Analysis. The below points indicate the distance from the tip of bats to the node.

4.2. *Measurement of Reflection Coefficient*

Using the ball shooting machine, we conducted experiments to measure reflection coefficients and determine the position of sweet spot on VK and VS bats. As a result, regarding VK and VS bats, the position for sweet spot was determined by the experiment using the ball shooting machine.

Collision test was conducted five times at every measurement point and the average is reflection coefficient. A pressure for shooting ball is set as 2.0MPa. Shooting ball velocity is 60 ± 5 km/h.

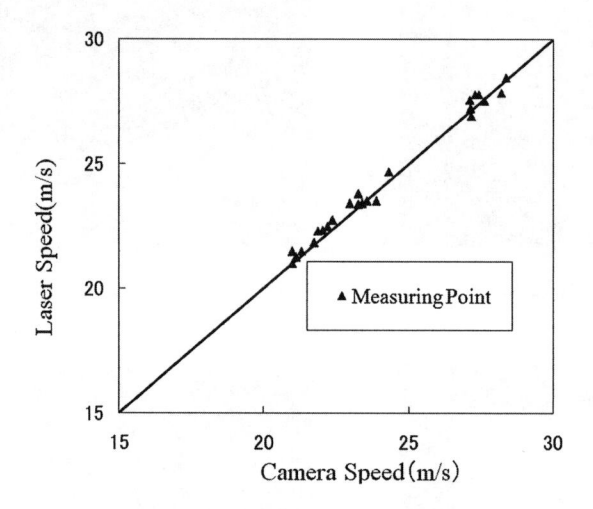

Figure 6. Comparison with camera speed and laser speed.

Table 2. Node positions

(mm)	1st	2nd
VK	165.0	119.0
VS	161.5	119.0

The measurement range is from a tip of bat 0 mm to a center of gravity on each bat. The reflection coefficients are calculated by the following equation. In Eq. (1), the first term is the velocity ratio before and after collision. The second term is the relative velocity of bats.

$$e = -\frac{Vl_o}{Vl_i} + \frac{V_t + L_f \cdot \omega_t}{Vl_i} \quad (1)$$

Vl_i : Speed before a collision of a ball
Vl_o : Speed after a collision of a ball
Vt : Translation speed after a collision of a bat
L_f : A center of gravity and distance among a collision department
ω_t : Angular velocity of a bat

5 Results

Figure 7 shows that the maximum reflection coefficient and velocity ratio of VK is larger than that for VS. Table 3 shows distances between the sweet spot and bat tip.

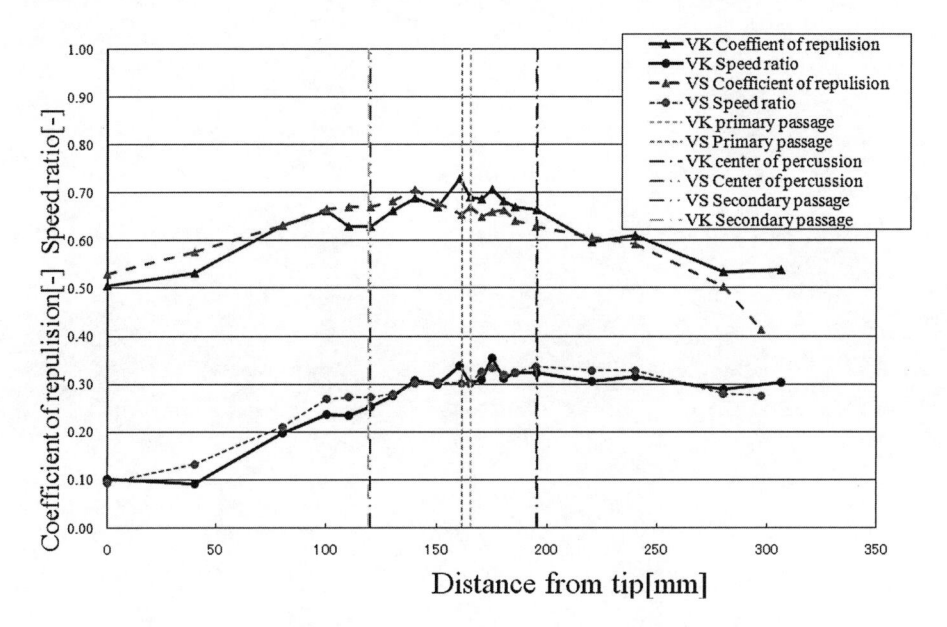

Figure 7. Coefficient of restitution and the speed ratio of VK and VS.

Table 3. Distance from the tip

(mm)	2nd	sweet spot	1st	center of percussion
VK	119.0	160.0	165.0	195.3
VS	119.0	140.0	161.5	194.7

6 Conclusion

The characteristics for sweet spot positions are different from each types of bat. All sweet spots for them exist between the second node and first node around the center of percussion. In particular, regarding VS, high reflection coefficients are measured around the range on the bat.

References

Nagamatsu A. (1998) *Introduction to modal analysis*. Corona Company, Tokyo, Japan.
Hiroshi A., Nagamatsu A. (1979) *Forming a new industry dynamics*. Youkendou, Tokyo, Japan.
Adair R.K., Nakamura K. (1996) *Physics of baseball*. Kinokuniya, Tokyo, Japan.

ANALYSIS OF BASEBALL BATS PERFORMANCE USING FIELD AND NON-DESTRUCTIVE TESTS

R.W. SMITH, T.H. LIU & T.Y. SHIANG

Institute of Sports Equipment Technology, Taipei Physical Education College, Taiwan

Currently the most common and excepted methods for testing bats are those approaches according to ASTM and NCAA. Both of these methods are done in a lab setting, in order to control the environment. The ASTM method measures Bat Performance Factors (BPF) of the bat, and the NCAA method measures Ball Exit Speed Ratios (BESR) of the bat. While these methods are proven effective, they neglect to take into account the actual environment when the bat is used in the field. The purpose of this research is to draw a correlation between BESR and BPF (bat performance factor) of three types of wood materials found in field tests, with values of a natural frequency and MOEd (dynamic modules of elasticity) found in the lab. By analyzing correlations between the field tests and non destructive testing (NDT), a new index can be drawn, which then can transform the natural frequency and MOEd results into BESR and BPF results. Research results show no significant differences between the BESR and BPF when comparing the three types of wood materials. However there was a strong, positive correlation between the transverse natural frequency and the BESR and BPF of the bats. In other words, the higher frequency of the bat resulted in higher BESR and BPF values. In conclusion, the NDT method was proven to be effective both in the laboratory and in the field analysis. Future use of the NDT could be further used to create a new index to better analyze bat performances.

1 Introduction

Currently, bat performance testing is divided into two excepted standards the ASTM and the NCAA. The ASTM method measures Bat Performance Factors (BPF) of the bat, and the NCAA method uses an air cannon to propel the baseball towards a stationary bat, and then measures the velocity before and after impact to determine the BESR (ball exit speed ratio). Both of these methods are done in a lab setting which is a stable and controlled environment, making them different from field testing. Shenoy *et al.* (2001) used a constant bat swing speed to compare the response of different bat types. Shaw (2006) used laboratory and field experimental investigations to better understand the relationship of bat properties on batted-ball speeds. Many engineering and physics researchers have developed mathematical models that have enable the calculation of ball exit speeds and bat swinging speeds (Adair, 2002; Nathan, 2000). Based on this, our research used a high speed video camera system to capture the batting parameters (BPF, BESR, Bat COR) during field testing. The non destructive testing (NDT) method was also used to measure the longitudinal and transverse natural frequencies of the bat. Natural frequencies are the key to estimating longitudinal and transverse dynamic modulus of elasticity. This research discusses the correlation between the NDT method results and batting parameters.

2 Methods

Five baseball players from Taipei Physical Education College volunteered for this study. Batters tested three types of wooden bats (ASH, BEECH, and MAPLE). A pitching machine was used to propel the baseballs toward the batters bats. Pitching velocity was 100~110 kilometers per hour. The batting parameters (ex: pitching velocity, batting velocity, ball exit velocity, and impact duration) were measured using a high-speed video camera system as depicted in Figure1.

The subjects where only allowed to hit the ball in the batting zone. The high speed video camera was calibrated with a scale that was then placed vertically next to home plate. In order to ensure validity of the balls they were first tested for Coefficient of Restitution (COR) using ASTM F1887-02 standard and then tested for compression-displacement for baseballs using ASTM F1888-02 standard. Bat performance was measured using Bat Performance Factor (BPF) ASTM F1881-05 standard and then Ball Exit Speed Ratio (BESR) was tested using NCAA standards (1999).

Ball Coefficient of Restitution (Ball COR), which is the ratio of velocity rebounding from a hard surface. Consequently, the COR is equal to the square root of the proportion of the collision energy returned to the kinetic energy of the ball's flight.

$$\text{BallCOR} = \frac{(v' - V')}{(v - V)} \qquad (1)$$

$v - V$ the relative velocities before impact, $v' - V'$ the relative velocity after impact.

Bat Performance Factor (BPF), which is the ratio of BBCOR and Ball COR, also is an important index for bat performance. The BPF formula is shown as follows:

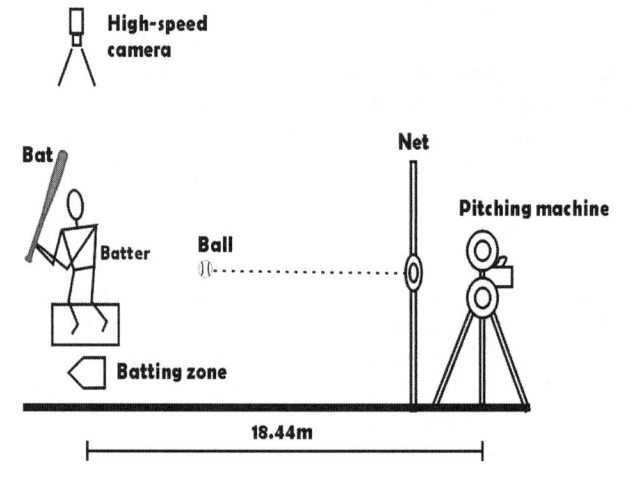

$$\text{BPF} = \frac{\text{BBCOR}}{\text{BallCOR}} \qquad (2)$$

Figure 1. Schematic of video system used to determine the batting parameters.

Figure 2. Longitudinal NDT method.

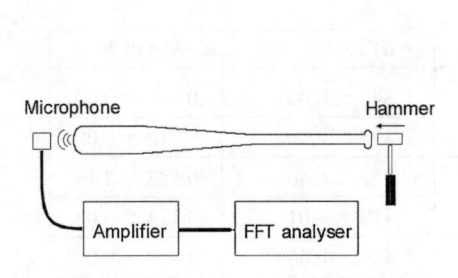

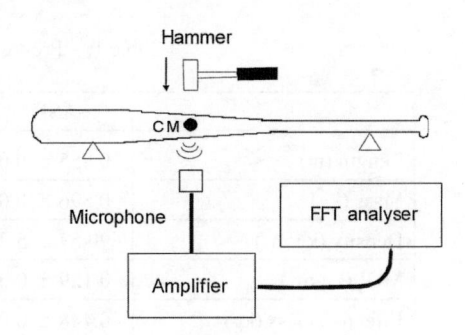

Figure 3. Transverse NDT method.

Ball Exit Speed Ratio (BESR) allows one to determine the ball exit speed v_f when the bat speed v_{bat} and the pitching speed v_{ball} are specified. The relationship between the BESR and these speeds are:

$$\text{BESR} = \frac{v_f + \frac{1}{2}(v_{ball} - v_{bat})}{v_{ball} + v_{bat}} \qquad (3)$$

The bats were also measured using the NDT stress wave test method, which allowed us to understand the natural frequencies and dynamic module elasticity of the bats. In an early study by Brody (1986) he stated that the frequency of bat oscillation was measured using a microphone, audio amplifier and an oscilloscope. A more recent study Adair (2002) pointed out that the material hickory wood has twice the elasticity modulus of ash, the stiffer hickory wood bats also vibrated with a smaller amplitude and high frequency. The NDT method applies stress waves to the wood material tested in order to analyze the natural frequency. The NDT method is shown in Figure 2 and Figure 3. This study used longitudinal and transverse first natural frequencies to estimate longitudinal and transverse dynamic modulus of elasticity. The longitudinal and transverse MOEd formula is shown as follows:

Longitudinal oscillate formula:

$$V = 2 \cdot f \cdot l \qquad (4)$$

$$E_l = V^2 \cdot \rho \qquad (5)$$

and the transverse oscillate formula:

$$E_t = \frac{4\pi^2 \cdot f \cdot l^4 \cdot A \cdot \rho}{\beta^4 \cdot I}, \qquad (6)$$

where E_l: longitudinal modulus of elasticity (GPa), E_t: transverse modulus of elasticity (GPa), V: velocity (m/s), f: frequency (Hz), l: bat length (m), ρ: bat density (kg/m^3),

Table 1. Pre-test results of bats x ± sd.

	ASH	BEECH	MAPLE
Length (m)	0.855 ± 0.001	0.845 ± 0.001	0.837 ± 0.001
Mass (kg)	0.896 ± 0.007	0.892 ± 0.009	0.918 ± 0.002
Density (kg/m^3)	690.57 ± 5.21	688.55 ± 6.46	708.23 ± 2.14
MOI (kg-m^2)	0.129 ± 0.02	0.118 ± 0.01	0.124 ± 0.02
Effective mass (kg)	0.448 ± 0.004	0.446 ± 0.004	0.459 ± 0.001
Bat recoil factor	0.325 ± 0.003	0.326 ± 0.003	0.317 ± 0.001
Ring width (lines/cm)	3.33 ± 0.416	3.86 ± 1.616	3 ± 0.529

Table 2. Parameters of field-test method (x ± sd).

	ASH (n = 30)	BEECH (n = 30)	MAPLE (n = 30)
ball velocity (m/s)	29.26 ± 1.31	29.55 ± 1.54	29.54 ± 1.841
swing velocity (m/s)	36.15 ± 2.10	35.48 ± 1.39	35.36 ± 1.42
BEV (m/s)	41.49 ± 2.03	40.63 ± 2.26	39.82 ± 1.86
Impact duration (ms)	1.5 ± 0.18	1.5 ± 0.13	1.5 ± 0.32
BESR	0.582 ± 0.029	0.579 ± 0.029	0.569 ± 0.024
BPF	0.818 ± 0.074	0.800 ± 0.072	0.793 ± 0.061
Bat COR	0.431 ± 0.039	0.422 ± 0.038	0.418 ± 0.032

A: cross-sectional area (m^2), β: boundary coefficient ; free-free fixed $\beta = 4.75$, I: inertia distance, $\dfrac{\pi \cdot D^4}{64}$ (m^4).

3 Results and Discussion

3.1. *Test Results of Balls and Bats*

The study followed ASTM standard testing methods. Thirty-six Model Ky-800 baseballs were used as the testing balls, lab test results of balls are mass (g) 145.55 ± 0.915, COR 0.527 ± 0.0057, and compression value (lbf) 239.68 ± 11.56. Three types of different wood material bats were used as the testing bats, the information of the bats are shown in table 1. Field tests and NDT tests results of bats are shown in table 2 and table 3.

The batting parameters three types of different wood material bats were analyzed using one-way ANOVA. Results indicate no significant differences in BESR and BPF among the three types of wood materials. This proves there are no batting parameter effects between the different wood materials.

3.2. The Correlation Between the Lab Test and Field Test

Smith (2001) used three bat performance methods (ASTM, NCAA and tee batting) to measure batting parameters. It was shown that BPF of the aluminum bat was 0.91, the BESR of aluminum bat was 0.73, the BPF of the wooden bat was 0.98, and the BESR of the wooden bat was 0.78.

Nathan (2003) used ASTM and NCAA bat performance methods to measure batting parameters, that showed BPF of the aluminum bat was 1.05, the BESR of the aluminum bat was 0.71, the BPF of the wooden bat was 0.96, and the BESR of the wooden bat was 0.71. Figure 4 compares each scholar's research data to our experimental result. Because the ASTM and NCAA methods use a fixed grab machine that holds the bat in place. In the BESR test results deviated about 0.25. In BPF test results deviated approximately 0.1. Based on this finding we could determine the different energy lost between the fixed grab machine and the subjects holding the bat in their hands. As we know the subjects that hold the bat in hands during the impact process result in the hands partially absorbing the energy during the hitting process. It is possible for this reason that Smith & Nathan research data recorded higher results than ours.

The BESR of NCAA has set the legal limit of 0.728 and any bat that has a ball exit speed at or below this line is legal. Likewise, any bat that produces a ball exit speed above this line is illegal. The NCAA rules state clearly that safety of the players is of the up most important and a balance between offence and defense needs to be kept (Fallon et al., 2000).

Table 3. Parameters of NDT method x ± sd.

	ASH	BEECH	MAPLE
Longitudinal frequencies (Hz)	2470.33 ± 19.00	2342.66 ± 15.30	2285.66 ± 4.51
Transverse frequencies (Hz)	163.3 ± 0.21	135.2 ± 0.72	124.23 ± 0.68
Longitudinal MOEd (Gpa)	12.32 ± 0.21	10.77 ± 0.15	10.68 ± 0.06
Transverse MOEd (Gpa)	4.49 ± 0.11	3.11 ± 0.12	2.70 ± 0.14

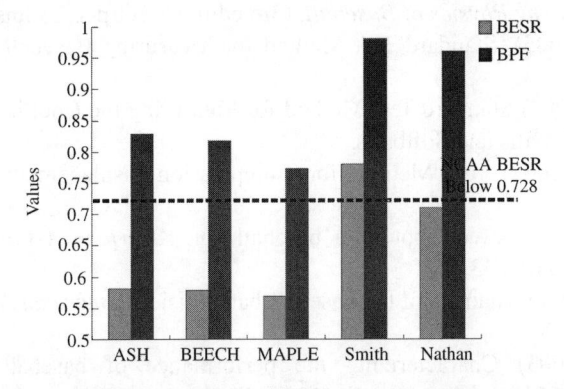

Figure 4. Comparison of BESR and BPF, the BESR value must below the limit line in order to be legal for game use.

Table 4. The correlation between the NDT method results and batting parameters.

	BESR	BPF	Bat COR
Longitudinal frequencies	.868	.852	.856
Transverse frequencies	.853	.836	.840
Longitudinal MOEd	.714	.696	.691
Transverse MOEd	.823	.808	.804

Our experiment resulted in the BESR conforming to the NCAA stipulation of being lower than 0.728.

3.3. *The Correlation Between the NDT Method Results and Batting Parameters*

There was a strong, positive correlation between the natural frequencies and MOEd, as well as the BESR and BPF of the bats. The used natural frequency to calculate MOEd, the relationship between natural frequency and MOEd were in direct proportion. The higher the MOEd value means the better elastic bending modules of the bats. Results are shown in table 4. In other words higher frequencies of the bat resulted in high values in BESR and BPF.

4 Conclusion

In conclusion there were no significant differences in batting parameters among the three types of wood materials. On the other hand there was a strong, positive correlation between the natural frequencies and MOEd, as well as the BESR and the BPF of the bats. The higher MOEd value represents a better elastic bending modules of the bats. In other words higher frequencies of the bat resulted in higher BESR and BPF values. In conclusion, the NDT method was proven to be effective and positively correlated with both laboratory and field tests. Future use of the NDT could be further used to create a new index to better analyze bat performances.

References

Adair R.K. (2002) *The Physics of Baseball*. (3rd edition). Harper Collins, New York.
ASTM F 1881 (2005). Standard Test Method for Measuring Baseball Bat Performance Factor.
ASTM F 1887 (2002) Standard Test Method for Measuring the Coefficient of Restitution (COR) of Baseballs and Softballs.
ASTM F 1888 (2002) Test Method for Compression-Displacement of Baseball and Softball.
Brody H. (1986).The sweet spot of a baseball bat. *American Association of Physics Teachers*, 54, 640–643.
Nathan A.M. (2000) Dynamics of the baseball-bat collision. *American Journal of Physics*, 68(11). 979–990.
Nathan A.M. (2003) Characterizing the performance of baseball bats. *American Association of Physics Teachers*, 71, 134–142.

National Collegiate Athletic Association Provisional Standard for Testing Baseball Bat Performance (1999). *NCAA news release* (online). Available: http://www.ncaa.org/releases/miscellaneous/1999/1999092901ms.htm

Nicholls R.L., Elliott B.C., Miller K. and Koh M. (2003) Bat Kinematics in Baseball: Implications for Ball Exit Velocity and Player Safety. *Journal of Applied Biomechanics*, 19, 283–294.

Shaw R.H. (2006) Laboratory and Field Experimental Investigations of the Relationship of Baseball Bat Properties on Batted-Ball Speed. MSc Thesis, University of Massachusetts, Lowell.

Shenoy M.M., Smith L.V. and Axtell J.T. (2001) Performance Assessment of Wood, Metal and Composite Baseball Bats. *Composite Structures*, 52, 397–404.

Smith L.V. (2001) Evaluating baseball bat performance. *Sports Engineering*, 4, 205–214.

Fallon L.P., Collier R.D, Sherwood J.A. and Mustone T.J. (2000) Determining Baseball Bat Performance Using a Conservation Equations Model with Field Test Validation. In: Subic A. and Haake S. (Eds.) *Engineering of Sport – Research Development and Innovation*, Blackwell Science, Oxford, pp. 201–212.

MODELLING BOUNCE OF SPORTS BALLS WITH FRICTION AND TANGENTIAL COMPLIANCE

K.A. ISMAIL & W.J. STRONGE

Department of Engineering, University of Cambridge, Cambridge, UK

A lumped-parameter model was used to represent compliance arising in a small deforming region around the contact point of a ball and a rigid surface. Accurate representations for nonlinear compliance of the contact region on the ball in directions perpendicular and parallel to the common tangent plane of impact were modeled from static measurements. The equations of motion at the contact point were solved using this nonlinear, hysteretic compliance model and result a more accurate response of the ball.

1 Introduction

Measurement of bounce for a batted baseball shows dependency of the rebound velocity and spin on physical properties that affect the energetic coefficient of restitution (ECOR) and compliance of the contact region. The objective of this paper is to develop a more accurate representation for compliance of the contact region on the ball that will improve calculation of velocity and spin that result from batting. These calculations can be used to determine the optimal strategy for batting long balls.

The present analysis uses lumped-parameter models to represent compliance arising in a small deforming region around the contact point of a baseball; this compliance can be represented by discrete elements oriented in directions perpendicular and parallel to the common tangent plane of impact (CTP) – components that can be identified as normal and tangential respectively. The basis of this study is introduced by Stronge (2001) through a "Linear Compliance Model". That model assumed linear deformation and hysteresis in the normal element only. Although Stronge's model shows some features characteristic of experimental results (Cross, 2002), it fails to accurately represent energy dissipation of tangential element. Experimental results by Cross & Nathan (2006) indicated that tangential compliance is significant in calculating the scattering of a baseball by a bat.

The normal compliance of a baseball was measured in a static compression test where the ball was compressed between parallel flat plates. An unloading curve has been estimated on the basis of the ECOR value and continuity at maximum loading. Results of a tangential test show dependency of the tangential compliance on normal force acting on the baseball; this results from the increasing area of contact with the normal force.

Our dynamic collision model that is based on the structure of the ball accounts for energy losses that are consequences of material hysteresis and friction. The main development of the present paper is to incorporate energy dissipation and nonlinearity in both normal and tangential elements. Significant material and structural parameters that influence response of a baseball during bounce have been identified. These have led to accurate representation of compliance in a dynamic bounce model and corresponding accurate calculations of velocity and spin of a batted baseball.

2 Measurement of Compliance

2.1. Normal Compliance

Assuming that the ball deformed symmetrically, it experienced an equal normal deflection u_3, on two sides under the action of normal force F_3. This is in contrast with impact during ball games, where deformation occurs only on one side.

Figure 1 shows force-deflection curve for a static compression test of a baseball. The maximum force is representative of a baseball struck by a bat at a relative speed of impact equal to 58 m/s (Nathan, 2000). Results show a non-linear increase of force with increasing deflection during the loading phase. Hysteresis effect of material was evident from the area enclosed by the loading and unloading curves. It represents energy losses due to internal friction of material within the baseball.

Curve fitting using power law relationship (i.e. $F_3 = k u_3^\alpha$) was used to represent experimental data of the baseball during the loading and unloading phase. A power law was suggested by the Hertz contact relation. A curve was fitted directly on the experimental data during the loading phase. The best curve fit to this data gives a force-deflection relation $F_3 = 1.02 \times 10^9 u_3^{2.49}$.

During unloading phase, curve fitting was estimated based on the normal ECOR value, e_{3E} and continuity at maximum loading (u_{3max}, F_{3max}). An unloading curve was obtained with hysteresis representing the fractional energy loss $(1 - e_{3E}^2)$. Value of e_{3E} for a baseball was obtained as a function of normal relative velocity (Sawicki et al., 2003). Method for approximating an unloading curve described above was introduced by Nathan (2000). This approximation results in an unloading curve of $F_3 = 2.92 \times 10^{23} u_3^{10.7}$ that agrees

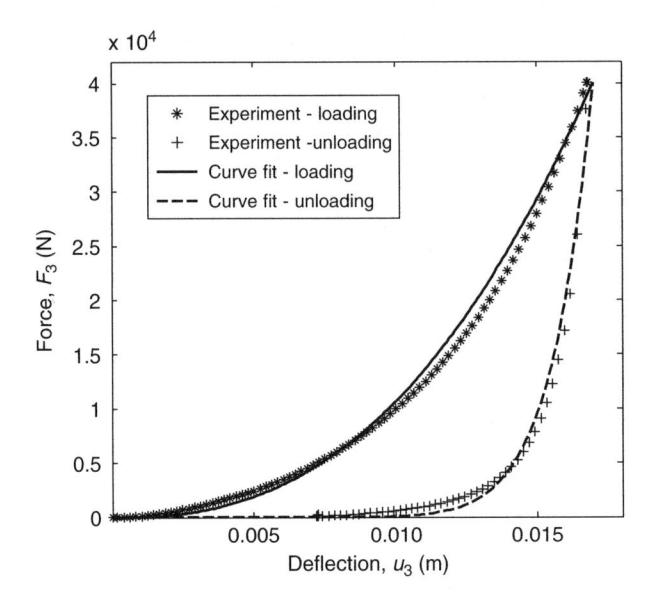

Figure 1. Normal compliance of a baseball measured in a static compression test. Also shown is best curve fit to this experimental measurement.

with the experimental unloading data. Hence, it proves that method described by Nathan (2000) is valid for estimating the unloading curve.

2.2. Tangential Compliance

Inset in Figure 2 is a sketch showing an experimental set-up for measuring tangential compliance of a sports ball. This arrangement overcomes the problem of slip at the grips. The ball was clamped between two parallel plates to avoid slipping motion at start of the test. The amount of clamping force F_3 applied on the ball can be controlled by a screw adjuster that was attached collinearly with a load cell and a moveable plate. A webbing strap was used as a mean of transferring tangential forces F_1 to the ball where it was tied as a loop around the ball. Configuration of the test set-up ensured that tangential force was acting on both sides of the ball that were in contact while the strap was being pulled by a tensile test machine. A displacement transducer was placed at the crown of the ball and used to measure deflection due to the action of tangential force. Tangential tests were conducted for the value of clamping force F_3 at 2, 4 and 6 kN.

Figure 2 shows experimental results of the tangential test in a non-dimensional form in order to compare with an analytical partial-slip solution of the contact patch (Johnson, 1985). Experimental results clearly substantiated the analytical solution. Hence, this analytical solution can be used to model tangential compliance during the loading phase.

The curve for cyclic loading-unloading was obtained by loading the baseball up to a maximum force which was less than the value that initiated continuous slip. The load was then removed to obtain unloading data. The tangential ECOR, e_{1E} can be estimated from this

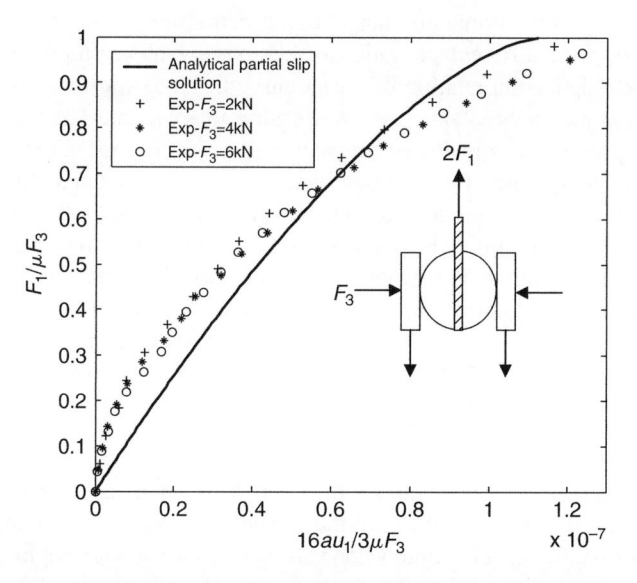

Figure 2. Experimental results of tangential test for a baseball with different clamping forces. Also shown is an analytical solution for partial-slip for contact between a spherical body and an elastic half-space (Johnson, 1985).

experimental loading-unloading data by using the inverse method as described for unloading normal compliance curve.

3 Dynamic Bounce Analysis

The dynamic bounce analysis of sports balls uses lumped-parameter representation of a small deforming region around the contact point of a ball. The equation of motion for relative acceleration at the contact point of a ball (mass M) and a rigid surface can be expressed as (Stronge, 2000);

$$\begin{Bmatrix} \dfrac{dv_1}{dt} \\ \dfrac{dv_3}{dt} \end{Bmatrix} = M^{-1} \begin{bmatrix} \beta_1 & -\beta_2 \\ -\beta_2 & \beta_3 \end{bmatrix} \begin{Bmatrix} F_1 \\ F_3 \end{Bmatrix} \qquad (1)$$

where the inertia coefficient β_i was defined as follows for a case of collinear collision between a ball of radius r, with radius of gyration $\hat{k}$, and a rigid surface;

$$\beta_1 = 1 + \frac{r^2}{\hat{k}^2}; \quad \beta_2 = 0; \quad \beta_3 = 1 \qquad (2a, b, c)$$

Friction induced during sliding between the contacting surfaces is given by Amonton's Coulomb law of friction (Johnson, 1985). During sliding motion, this law relates the tangential and normal contact forces, F_1 and F_3 respectively by a coefficient of limiting friction (COF), μ.

A compliance model as described in section 2 can be used to model the compliance of the contact region in the normal and tangential direction for estimating responses during ball games. However, different compliance relations which appropriate for low speed impact (i.e. force-deflection relation of a best curve fit up to a lower maximum F_3) were used in this dynamic analysis since we were trying to solve and compare the analytical results with the results of a low speed impact experiment (Cross, 2002). Also, effect of varying F_3 was incorporated for the compliance relation in the tangential direction. The equation of motion (Eq. 1) was solved numerically using Matlab and force-time histories results were plotted as a ratio of F_1/μ and F_3 so that both curves were overlapping during the period of sliding motion. The negative value of normal contact force $(-F_3)$ was also plotted for an envelope of normal contact force so that the value of F_1/μ will never exceed this envelope. These results were then compared to the experimental results of Cross (2002), who studied impact of sports balls on low and high friction surfaces.

4 Discussion

The results of Figure 3 show that a body will either slip or stick depending upon the ratio of F_1/μ and F_3. At an angle of 41° to the CTP, the ball will initially slide until this ratio becomes less than the normal force, at the time when subsequent stick motion begins. The tangential contact force changes direction during the stick period, which Cross (2002) described as "grip", where vibration of the ball occurs in the horizontal direction. Finally, the ball bounces with reverse slip since it is sliding in the direction opposite to initial sliding.

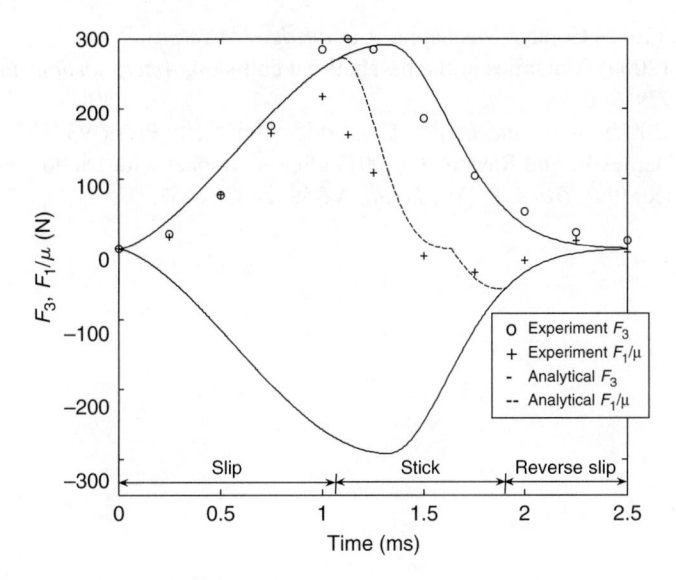

Figure 3. Analytical solution of normal (F_3) and tangential (F_1/μ) forces during low speed oblique impact of a baseball against a rigid surface ($v_1(0) = 1.698$ m/s, $v_3(0) = 1.476$ m/s $\mu = 0.46$, $\theta = 41°$, $e_{3E} = 0.6$, $e_{1E} = 0.75$). Also shown is an experimental result of Cross (2002).

The analytical results based on the measured compliance show a close approximation in terms of maximum contact forces and impact duration compared to the experimental results. Also, decay in the maximum value of F_1 at every successive cycle represents more accurate energy dissipation in the tangential element during stick motion, while friction dissipates energy only during slip motion.

5 Conclusion

Incorporation of measured compliance into the dynamic bounce analysis of sports balls gives a good approximation of the responses of the ball during short duration impact against a rough surface. Friction model provides an insight into the stick and slip motion that occurs during bounce so that velocity and spin can be calculated more accurately. However, there are still issues that need to be addressed at this point especially i) transition between slip and stick motion; ii) transition between cycles of tangential motion during stick.

References

Cross R. (2002) Grip-slip behavior of a bouncing ball, *Am. J. Phys.* 70 (11), 1093–1102.
Cross R. and Nathan A.M. (2006) Scattering of a baseball by a bat, *American Journal of Physics*, 74 (10), 896–904.
Sawicki G.S., Hubbard M. and Stronge W.J. (2003) How to hit home runs: Optimum baseball bat swing parameters for maximum range trajectories, *American Journal of Physics*, 71 (11), 1152–1162.

Johnson K.L. (1985) Contact Mechanics. *Cambridge University Press.*
Nathan A.M. (2000) Dynamics of the baseball-bat collision, *American Journal of Physics,* 68 (11), 979–990.
Stronge W.J. (2000) *Impact mechanics.* Cambridge University Press, 93–104.
Stronge W.J., James R. and Ravani B. (2001) Oblique impact with friction and tangential compliance, *Phil. Trans. R. Soc. Lond.,* A 359, 2447–2465.

THREE-DIMENSIONAL KINETIC ANALYSIS OF UPPER LIMB JOINTS DURING THE FORWARD SWING OF BASEBALL TEE BATTING USING AN INSTRUMENTED BAT

S. KOIKE[1], T. KAWAMURA[1], H. IIDA[2] & M. AE[1]

[1]*Institute of Health and Sport Science, University of Tsukuba, Ibaraki, Japan*
[2]*Polytechnic University, Sagamihara-City, Kanagawa, Japan*

The purpose of this study was to clarify the role of each upper limb during the forward swing of T-batting motion by using an instrumented bat. Seven collegiate male baseball players' motion were captured by VICON 612 system (8 cameras, 250 Hz) and kinetic data of each hand were collected by a bat instrumented with strain gauges (500 Hz). The torque about the shoulder adduction/abduction axis was the largest among all calculated torques. The sum of the positive works generated by the knob-side upper limb was considerably larger than that generated by the barrel-side upper limb, which indicated that the role of the knob-side upper limb was to accelerate the bat during the forward swing. As for the negative works, the barrel-side upper limb generated the greater sum of work, which indicated that this limb may have been used as energy absorber.

1 Introduction

Fine coordination of both the upper limbs is essential for various tasks performed in baseball batting. During this motion, the upper limbs and the bat form a closed kinematic chain loop that causes kinetic redundancy. This closed loop prevents us from calculating single hand forces and moments from only kinematic data. That is likely the reason why there are few studies on the kinetics of the upper limbs during batting motion, most of them rather being on the kinetics of the lower limbs using one (Messier *et al.*,1985) or two force platforms (Welch *et al.*,1995).

One of the functions of the upper limbs during batting is to transfer the energy generated by the lower limbs and trunk to the bat to hit the ball with large head speed and to control it to give the appropriate direction and timing. The knowledge of the patterns of the upper limb joint torques can be useful in improving batting training and coaching.

The purpose of this study was to clarify the role of each upper limb during the forward swing of T-batting motion by three-dimensionally quantifying their kinetics using an instrumented bat (Koike *et al.*, 2004).

2 Method

A bat instrumented with strain gauges (Koike *et al.*, 2004) was used to calculate the forces and moments applied by each hand during T-batting, that is, hitting a ball placed on a tee that was settled at the height of the subject's hip joint. The bat was instrumented with eleven sets of gauges that were attached to the bat over the barrel-side grip handle and to the surface of an aluminum bar inserted underneath the knob-side grip handle. A personal computer was used to store the strain gauge signals after they passed through dynamic strain amplifiers. The sampling frequency was set at 500 Hz.

Seven baseball players' motion were captured with VICON 612 system (Oxford Metrics, 250 Hz, eight cameras). Coordinate data of the body segment endpoints and the bat were obtained from markers that were placed on them. The coordinates were filtered by a Butterworth digital filter and interpolated to 500 Hz using a spline function.

Joint torques were calculated by inverse dynamics and multiplied by the joint angular velocities to obtain the joint torque powers, which in turn were integrated to obtain the mechanical joint works. The shoulder flexion/extension axis was defined as a vector crossing the center of joints of both shoulders. The shoulder internal/external rotation axis was defined as the upper arm's longitudinal axis. The cross product of these two axes defined the shoulder adduction/abduction axis.

The forward swing motion was defined from the time when the sum of the head and knob-end speeds reached 5 m/s to the impact of the bat with the ball. The downward vertical velocity peak of the bat head (V_{zmax}) divided the motion in two phases, which were called downswing and level swing phases.

3 Results

Figure 1 shows the bat head and the knob-end speeds of the two trials of a subject. The time curves of both trials showed almost identical patterns for the bat head and knob-end speeds. The bat head speed continuously increased towards the impact, whereas the knob-end speed reached its peak during the down swing phase and then decreased before the impact. That means it is a rotational motion during the level swing phase that mainly contributes to the increase of the bat head speed.

Figures 2a–c show the forces and moments exerted by the hand on the knob- and barrel-side grip handles according to the swing plane coordinate system described by Koike *et al.* (2004). The solid and dotted lines represent the values of the knob- and the barrel-side hands, respectively. The x_{sp}-axial forces of the hands showed approximately inverse patterns. The y_{sp}-axial force of the knob-side hand increased towards the impact and reached large positive values compared to other forces. The y_{sp}-axial forces of the barrel-side hand

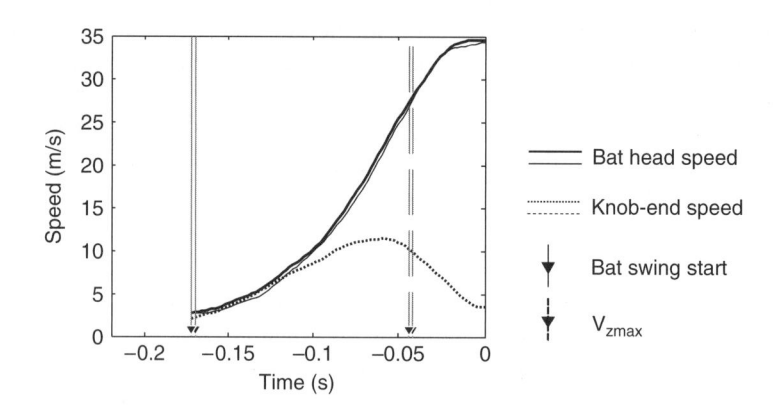

Figure 1. Time curves of the bat head and knob-end speeds of two representative trials of a subject.

showed a different pattern, reaching the peak nearby the start of the level swing phase and decreasing towards the impact. The z_{sp}-axial moment of the knob-side hand increased until about halfway the down swing phase, after which it decreased rapidly towards the impact.

The upper limb joint torques are shown in Figures 3a-g, where the joint axes are abbreviated as follows: shAA – shoulder abduction/adduction, shFE – shoulder flexion/extension,

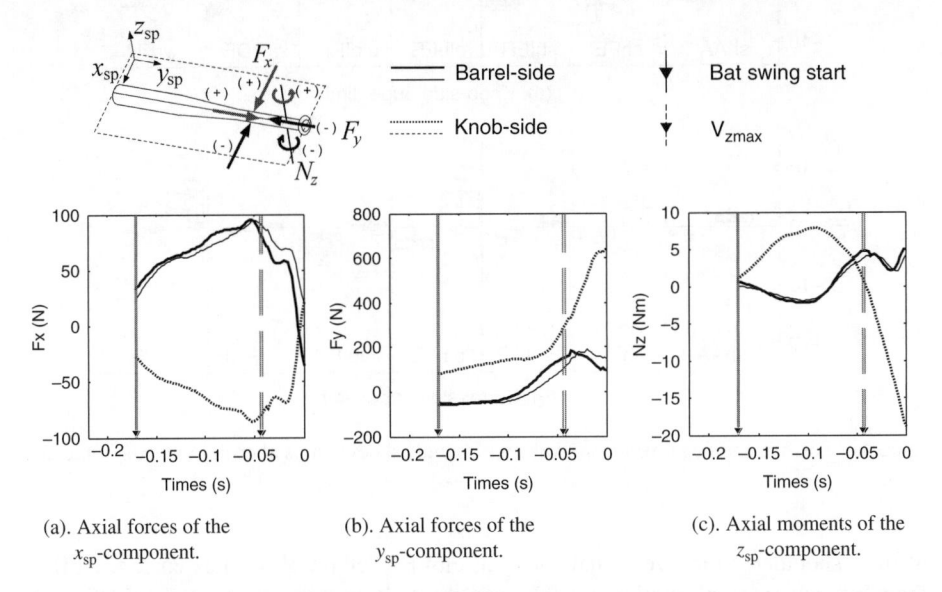

(a). Axial forces of the x_{sp}-component.

(b). Axial forces of the y_{sp}-component.

(c). Axial moments of the z_{sp}-component.

Figure 2. Forces and moments exerted by each hand in the expression of the swing plane coordinate system.

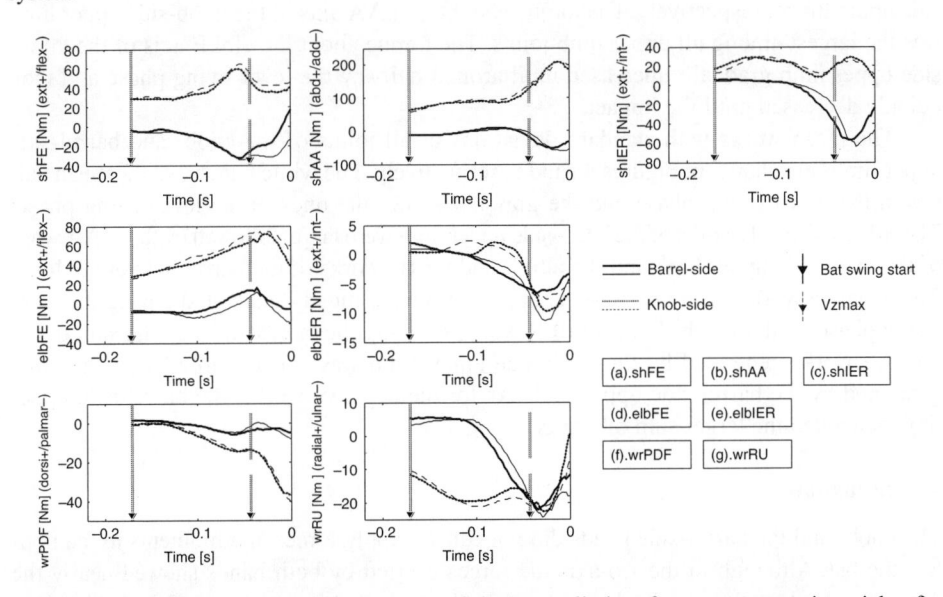

Figure 3. Time curves of the joint torques of the upper limbs of two representative trials of a subject.

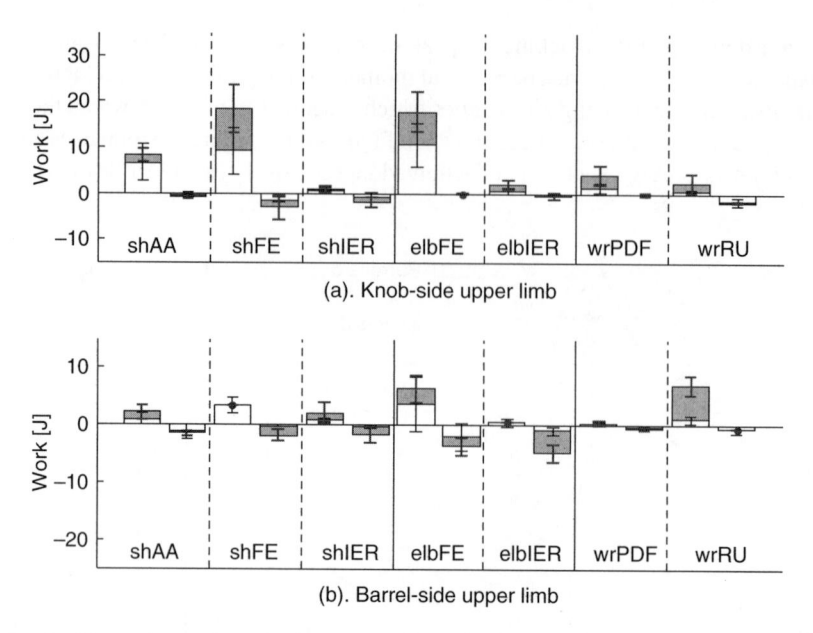

Figure 4. Positive and negative mean works with standard deviations at each joint axis of the upper limbs.

shIER – shoulder internal/external rotation, elbFE – elbow flexion/extension, elbIER – elbow internal/external rotation, wrRU – wrist radial/ulnar deviation, and wrPDF – wrist palmar/dorsal flexion. The solid and dotted lines represent the values of the knob- and barrel-side upper limbs, respectively. The torque about the shAA axis of the knob-side upper limb was the largest among all upper limb joints. The torque about the elbFE axis of the knob-side upper limb gradually increased until around halfway the level swing phase and then quickly decreased until the impact.

The mean works with standard deviations of all joints of the knob- and barrel-side upper limbs are shown in Figures 4a and b, respectively. The white bar shows the mean values in the down swing phase and the gray bar shows the ones in the level swing phase. The shFE and elbFE works of the knob-side upper limb were large, followed by the shAA work of the knob-side upper limb and the elbFE and wrRU works of the barrel-side upper limb. The negative works of the barrel-side upper limb were mostly generated during the level swing phase, with the elbFE and elbIER works showing the largest values. The sum of the positive works generated by the knob-side upper limb was considerably larger than that generated by the barrel-side upper limb. As for the negative works, the barrel-side upper limb generated the larger sum of works.

4 Discussion

The knob- and the barrel-side hands did not equally apply forces and moments to manipulate the bat. Although in the x_sp-axis the forces exerted by both hands showed nearly the same magnitude with opposite direction almost like coupled forces (Figure 2a), the forces in the y_sp-axis showed different patterns of the hands (Figure 2b). The y_{sp}-axial force of the knob-side hand was positive while that of the barrel-side hand was negative during the first half of the down swing phase. This shows that there were forces and moments that cancelled each other and did not produce movement of the bat.

It can be thought that the torque about the shoulder flexion/extension axis was small because the axis was approximately parallel to the y_{sp}-axis of the bat. On the other hand, the torque about the shoulder adduction/abduction axis of the knob-side upper limb was the largest because the y_{sp}-axis of the bat was almost perpendicular to this axis, which created a centrifugal force of the bat that the shoulder had to resist.

During the down swing phase, the positive works were mainly produced by large group muscles torques about the shAA, shFE and elbFE axes of the knob-side upper limb. Since the positive works contributed to increase the energy in the system, it can be said that the works about those axes mainly contributed to the increase of the bat head speed during the down swing phase, which was about 80% of the head speed at the impact. During the level swing phase, small group muscles torques about the elbIER, wrPDF and wrRU axes of the knob-side upper limb and about the wrRU axis of the barrel-side upper limb participated in the generation of the positive work, while those about the shFE, shIER, elbFE and elbIER axes of the barrel-side upper limb generated most of the negative work.

In other words, during the forward swing motion of baseball T-batting the upper limbs applied forces and moments on the bat with different purposes. The results indicated that torques like those about the shFE and elbFE axes of the knob-side upper limb exerted a role of main actuators in the acceleration of the bat, while others like those about the elbFE and elbIER of the barrel-side upper limb exerted a role of energy absorbers.

References

Messier S.P. and Owen M.G. (1985) The mechanics of batting: Analysis of ground reaction forces and selected lower extremity kinematics. *Research Quarterly for Exercise and Sport,* 56/2, 138–143.

Koike S., Iida H., Kawamura T., Fujii N. and Ae M. (2004) An instrumented bat for simultaneous measurement of forces and moments exerted by the hands during batting. In: Hubbard M., Mehta R.D. and Pallis J.M. (Eds.) *The Engineering of Sport 5, Vol. 2.* International Sports Engineering Association, Sheffield, UK, pp.194–199.

Welch C.M., Banks S.A., Cook F.F. and Draovitch P. (1995) Hitting a baseball: a biomechanical description. *J. Orthop. Sports Phys. Ther.*, 22/5, 193–201.

10. Ball Sport – Soccer

THE FLIGHT TRAJECTORY OF A NON-SPINNING SOCCER BALL

K. SEO[1], S. BARBER[2], T. ASAI[3], M. CARRÉ[2] & O. KOBAYASHI[4]

[1]*Yamagata Univ., Yamagata, Japan;* [2]*Univ. of Sheffield, Sheffield, UK;*
[3]*Univ. of Tsukuba, Tsukuba, Japan;* [4]*Tokai Univ., Hiratsuka, Japan*

This paper describes the aerodynamic characteristics of a non-spinning soccer ball and its flight trajectory on the basis of the aerodynamic data obtained from the non-spinning ball. It has been found that there are four velocity regions. The first region occurs at less than the critical Re, in which the aerodynamic forces are very stable. The second region is around the critical Re, in which the lift and the side forces oscillate. The third region is between just above the critical Re and $25 \text{ m} \cdot \text{s}^{-1}$, in which the aerodynamic forces are stable. The forth region occurs above $25 \text{ m} \cdot \text{s}^{-1}$, in which all aerodynamic forces oscillate markedly. The simulated flight trajectory fluctuates, but the amplitude is slightly less than observed in practice.

1 Introduction

In the case of a non-spinning soccer ball, especially in the case involving the Teamgeist (Adidas, the official World Cup ball in 2006), the flight trajectory is unpredictable (Asai *et al.*, 2007). The ball trajectory fluctuates during the flight. It is difficult for a goal keeper to save, even if the ball tends to fly to the center of the goal mouth. The reason why the non-spinning ball fluctuates during flight might be due to both the steady forces caused by the face orientation of the ball (Barber *et al.*, 2007) and the time variations of the aerodynamic forces. The former results from the time-averaged steady forces, while the later is due to the unsteady forces. In this paper, we pay most attention to the unsteady aerodynamic forces. We have carried out wind tunnel tests on a non-spinning soccer ball and simulated the flight trajectory on the basis of the unsteady aerodynamic forces.

2 Wind Tunnel Test

2.1. *Experiment*

A full-size soccer ball was employed to determine the aerodynamic forces acting on a ball in a low-speed wind tunnel with a $1.5 \text{ m} \times 1.0 \text{ m}$ rectangular cross-section. We used a commercially-available soccer ball (Teamgeist and Fevanova), together with a steel plate and a stainless steel rod. The steel plate was bent and glued to the ball. Data were acquired from a six-component strut type balance over a period of 10 seconds, using a personal computer with the aid of an A/D converter board. The sampling rate was 1024 per second. Aerodynamic force data were taken for various wind speeds, $|\vec{V}|$, between 8 and $35 \text{ m} \cdot \text{s}^{-1}$.

2.2. *Experimental Results and Discussion*

The time-averaged drag coefficients for four cases are shown in Fig. 1. The drag coefficient, C_D, is defined as the drag divided by the product of the dynamic pressure and the projectile area. The solid line denotes C_D variation on a smooth sphere (Achenbach, 1972), other

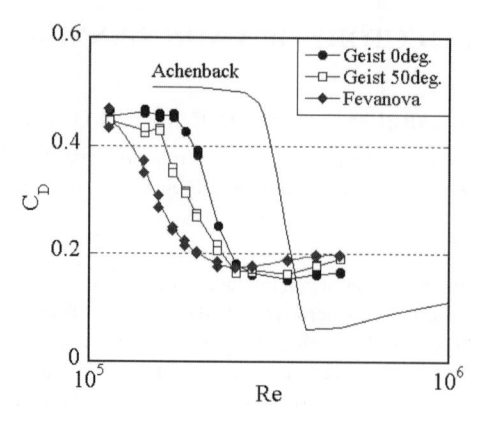

Figure 1. The drag coefficient, C_D, versus the Reynolds number, Re.

three curves denote C_D variations with soccer balls. Two of these show the case with a Teamgeist ball, while the other refers to a Fevanova ball. In the case of Teamgeist, the face orientation is also taken as a parameter. The face orientation is 0° if the valve is situated on the top against the wind, and the face orientation is 50° if the valve is situated on the right side at 50° against the wind.

The dependence of C_D on Reynolds number, Re, is qualitatively same. However, the critical Re on a soccer ball decreases when compared with a smooth sphere. The critical Re on the Fevanova ball is the lowest of the three samples, and that on the Teamgeist ball at 0° is the highest. In the super critical region, the C_D on the smooth sphere is less than 0.1. In the case of the soccer ball, C_D is also small, but has a value is slightly above that on a smooth sphere. The C_D on the Fevanova ball is the highest, while that on Teamgeist ball at 0° is the lowest. These results mean that the aerodynamic characteristics of a Teamgeist ball at 0° is the closest to the smooth sphere, while that on the Fevanova ball is the farthest away because of its geometrical features. The seam of the Fevanova ball is deeper and it is constructed from more panels than the Teamgeist ball.

The time variations of the drag, the side force and the lift are shown in Fig.2-a, b and c, respectively, for the case of the Teamgeist ball at 0°. The velocity is taken as a parameter. Since the frequency of the knuckle effect is less than 10 Hz and the natural frequency of the supporting system is about 16 Hz, the frequency components over 10 Hz were cut by the low-pass filter. It can be seen from Fig.2-a that the mean value increases with increasing the velocity, except between 10 and 20 m · s^{-1}. Since the mean value is almost constant between 10 and 20 m · s^{-1}, C_D decreases drastically in this region. Increasing the velocity, the amplitude of the drag oscillation increases as with a smooth sphere (Sawada et al., 2004). At 35 m · s^{-1}, 1.5 N are intermittently lost in a very short time in the drag. The drag oscillation makes the catch difficult for a goalkeeper because the rate of velocity decrease varies with time.

On the other hand, it can be seen from Fig.2-b and c that there are two velocity regions in which the side force and the lift apparently oscillate. One is at 13 m · s^{-1}, and the other is at 35 m · s^{-1}. The velocity of 13 m · s^{-1} coincides with the critical Re. In the critical region, the flow is changeable between laminar and turbulent flow. The flow is very sensitive. This might be the reason why the side force and the lift oscillate at around 13 m · s^{-1}.

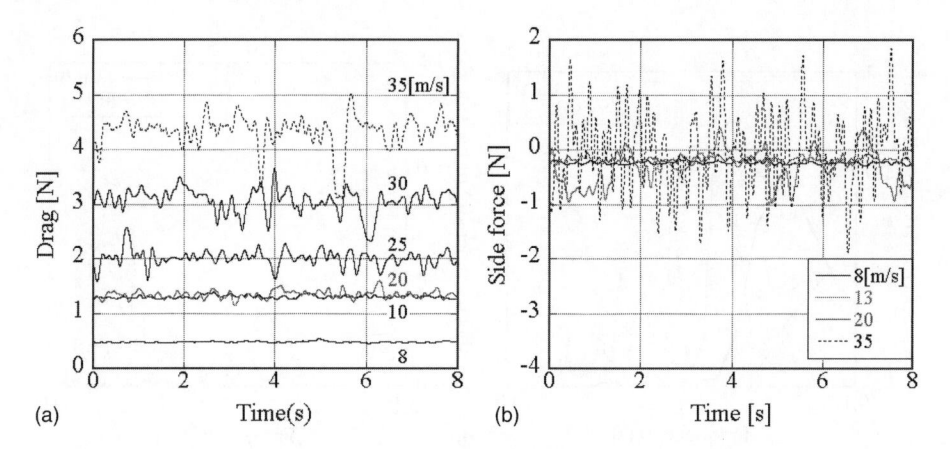

Figure 2. Time variations (a: Drag, b: Side force). The velocity is taken as a parameter.

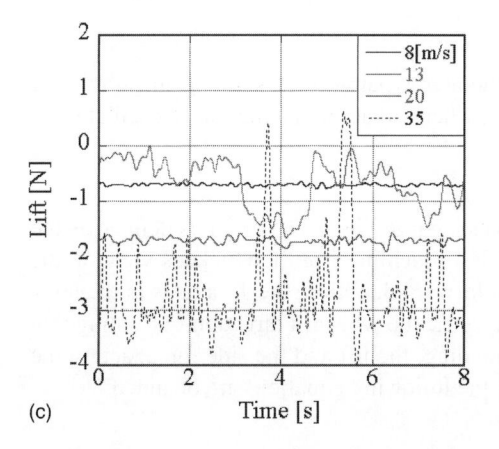

Figure 2-c. Time variations of the lift.

Figure 3. Tuft observation at 18[m/s].

The amplitude of the oscillation of both forces is larger than that of the drag, and that of the lift is the greatest within the three aerodynamic forces. The mean value of the lift is always negative. Figure 3 pictures of the tuft observation at $18 \text{ m} \cdot \text{s}^{-1}$. It can be seen that the tuft tends to be upwards behind the ball. This means that the negative lift acts on the ball. Same results for the tuft observation were obtained in all velocity ranges.

The power spectrum densities of the lift and the side forces are shown in Fig.4. Since the other velocities are negligible small, only the two velocities of 13 and $35 \text{ m} \cdot \text{s}^{-1}$ are shown. It can be seen that the power spectrum density of the lift is greater than that of the side force. The frequency of the lift is lower than the side force. These frequency components are much lower than those derived from the Strouhal number.

There are four velocity regions. The first region occurs at less than the critical Re, where the aerodynamic forces are very stable. The second region appears around the critical Re, where the lift and the side force oscillate. The third region occurs between just

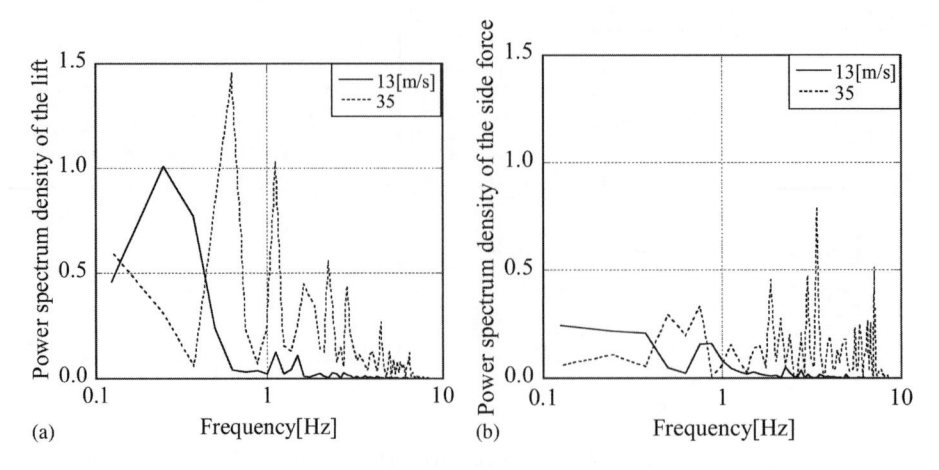

Figure 4. Power spectrum density (a: Lift, b: Side force).

above the critical *Re* and 25 m · s⁻¹, where the aerodynamic forces are again stable. The forth region appears at greater than 25 m · s⁻¹, where all aerodynamic forces oscillate.

3 Flight Trajectory

The inertial right-handed coordinate system is shown in Fig.5. The origin is defined as the point of intersection of the goal line and the left touchline from the kicker's view on the ground, where the X_E-axis is in the horizontal forward direction, the Y_E-axis is in the horizontal right direction and the Z_E-axis is in the vertical downward direction. Assuming that only three aerodynamic forces, comprising the drag, the lift and the side force, act on the ball and the ball does not rotate during flight, the following equations are obtained.

$$m\dot{U} = \frac{1}{|\vec{V}|}\left(-DU + \frac{L}{|\vec{V}|}UV - YV\right) \quad (1)$$

$$m\dot{V} = \frac{1}{|\vec{V}|}\left(-DV + \frac{L}{|\vec{V}|}VW + YU\right) \quad (2)$$

$$m\dot{V} = \frac{1}{|\vec{V}|}\left(-DW - \frac{L}{|\vec{V}|}(U^2 + V^2) + mg\right) \quad (3)$$

Here, (U, V, W) are the (X_E, Y_E, Z_E) components of the velocity vector, (D, L, Y) are the drag, the lift and the side force, m is the mass of the ball, and g is the gravitational acceleration. The velocity components (U, V, W) are obtained by integrating equations (1) through (3), to obtain the flight trajectory. It is necessary to know the aerodynamic forces $\vec{F}(t) = (D, L, Y)$ in equations (1) through (3). This was obtained using the Fourier series of the 40th harmonics

The Flight Trajectory of a Non-Spinning Soccer Ball 389

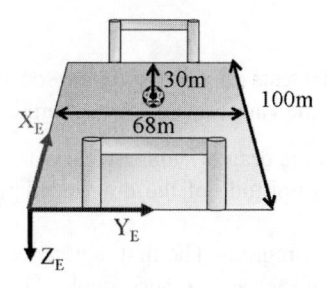

Figure 5. Coordinate system.

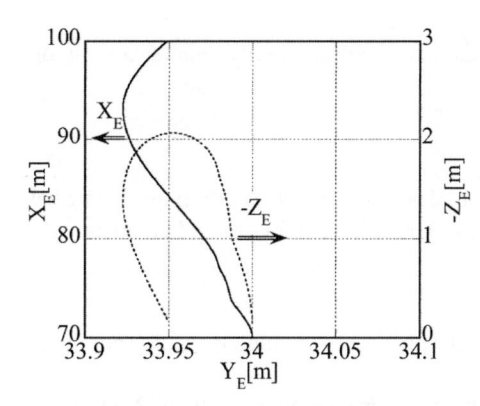

Figure 6. Flight trajectory.

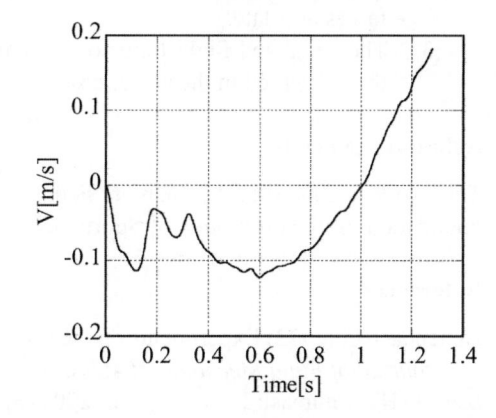

Figure 7. Velocity on the Y_E-axis.

in equation (4). Since the time variations depend on the velocity, as shown in Fig.2, Fourier coefficients, $\vec{a}_0$ through $\vec{a}_{40}$ and $\vec{b}_0$ through $\vec{b}_{40}$, are defined as a function of the velocity. Here, all Fourier coefficients were obtained for the Teamgeist ball at 0°.

$$\vec{F}(t) = \vec{a}_0 + \sum_{n=1}^{40} (\vec{a}_n \cos n\omega t + \vec{b}_n \sin n\omega t) \qquad (4)$$

The flight trajectories and the velocity on the Y_E-axis, V, are shown in Figs. 6 and 7, respectively. The initial position is assumed to be $(X_E, Y_E, Z_E) = (34, 70, -0.1098)$; i.e., 30 meters behind the goal line and at the center of both touchlines. The radius of the ball is 0.1098 m. The initial velocity components are assumed to be $(U_0, V_0, W_0) = (24.03, 0, 6.89)$; i.e., $|\vec{V}| = 25$ m · s^{-1}, while the elevation angle from the horizontal line is 16°. It can be seen from Fig.6 that the flight trajectory fluctuates at the very last moment. The ball tends towards the left in the first half, and then to the right after $X_E = 93$ m. The velocity component in the Y_E-axis (Fig.7), V, is negative in the first half, and then it becomes positive after 1 second. It appears, though, that the amplitude of the fluctuation is slightly smaller, when compared with the actual case.

4 Summary

We have carried out wind tunnel tests on a non-spinning soccer ball and simulated the flight trajectory on the basis of the time variations of aerodynamic forces. The results are:

1. The time variation of the drag is stable, when compared with the lift and the side force. However, the amplitude of the drag oscillation increases with increasing velocity.
2. There are four velocity regions. The first region occurs at less than the critical Re, where the aerodynamic forces are very stable. The second region is close to the critical Re, where the lift and the side forces oscillate. The third region occurs just above the critical Re, up to $25\,\text{m}\cdot\text{s}^{-1}$, where the aerodynamic forces are again very stable. The forth region occurs above $25\,\text{m}\cdot\text{s}^{-1}$, where all the aerodynamic forces oscillate.
3. The simulated flight trajectory fluctuates, but with a slightly smaller amplitude than observed in the actual case.

Acknowledgements

This work is supported by Inamori Foundation, The Descente and Ishimoto Memorial Foundation for the Promotion of Sports Science & Grant-in-Aid for Young Scientists (A).

References

Achenbach E. (1972) Experiments on the flow past spheres at very high Reynolds numbers, *Journal of Fluid Mechanics*, 54, 565–575.
Sawada H., Kunimasu T. and Suda S. (2004) Sphere drag measurement with the NAL 60cm MSBS, *Journal of Wind Energy*, 98, 129-136.
Asai T., Seo K., Kobayashi O. and Sakashita R. (2007) A study on wake structure of soccer ball. In: Fuss F.K., Subic A. and Ujihashi S. (Eds.) *The Impact of Technology on Sport II*. Taylor & Francis Group, London.
Barber S., Seo K., Asai T. and Carré M. (2007) Experimental investigation of the effects of surface geometry on the flight of a non-spinning soccer ball. In: Fuss F.K., Subic A. and Ujihashi S. (Eds.) *The Impact of Technology on Sport II*. Taylor & Francis Group, London.

A STUDY ON WAKE STRUCTURE OF SOCCER BALL

T. ASAI[1], K. SEO[2], O. KOBAYASHI[3] & R. SAKASHITA[4]
[1]*Tsukuba Univ., Tsukuba, Japan*
[2]*Yamagata Univ., Yamagata, Japan*
[3]*Tokai Univ., Hiratuka, Japan*
[4]*Kumamoto Univ., Kumamoto, Japan*

The purpose of this study is to discuss the aerodynamic characteristics of soccer ball using Computer Fluid Dynamics (CFD) and visualization of the vortex structure around the real flight soccer ball in high Reynolds number. An incompressible unsteady analysis was performed using the finite volume method based on fully unstructured meshes with a commercial CFD code (FLUENT6.2, Fluent Inc.). The turbulent model of this study was Large Eddy Simulation (LES) model. The drag coefficient of non-spinning soccer ball in CFD was approximately 0.19 and that of wind tunnel test was about 0.15. It was observed that the large scale fluctuation was generated in the lift coefficient. In order to visualize the flow around the soccer ball during flight, a ball was coated as uniformly as possible with titanium tetrachloride. It seemed that the Strouhal number of wake near the real fright soccer ball was about 1.0 as similar as the high-mode value of a smooth sphere. After balls undergoing a knuckle effect were airborne, large scale fluctuations of the vortex trail were observed when the St was between 0.1 and 0.01.

1 Introduction

In recent years there has been considerable focus on how a soccer ball drops and curves after a non- or low-rotating moving shot or standing free-kick. The detailed mechanism behind this so-called "knuckling effect" (Mehta & Pallis, 2001) or "knuckle effect" has hitherto remained unclear (Fig. 1); therefore, we sought to elucidate it.

Figure 1. An example image of "Knuckle Effect" in soccer using stroboscopic technique.

To date, miniature soccer balls (Matt *et al.*, 2004; Matt and Asai, 2004) and standard-size soccer balls (Asai *et al.*, 2006) have been studied in terms of their basic aerodynamic properties, and observed for surrounding flow using both wind tunnel experiments and computer-simulated Computational Fluid Dynamics (CFD) (Asai *et al.*, 2000; Barber *et al.*, 2006). However, all of these studies have focused on stationary analyses; to understand the knuckle effect, which is fundamentally a non-stationary phenomenon, it is essential to use a non-stationary analysis that incorporates a time component. In the present study, we therefore analyzed the dynamics of the wake of non-rotating soccer balls by non-stationary CFD using a combination of Large Eddy Simulation (LES) and a fluid visualization method using titanium tetrachloride. We also examined the fundamental mechanism of the knuckle effect.

2 Methods

2.1. *Computer Fluid Dynamics (CFD)*

Analyzing meshes were prepared using MSC.Patran (MSC.Software Inc.) and GAMBIT (Fluent Inc.) from Computer Aided Design Model for a soccer ball (32 ball panels type). The unstructured analyzing meshes used a triangular tetrahedral element and comprised approximately one million units (Fig. 2). The cylindrical space (2.44 m radius × 4.88 m length) was used for the analyzing meshes. The initial flow velocity was set at four conditions for this analysis: 15, 20, 25 and 30 m/s. This analysis defined the velocity inlet and the pressure outlet. An incompressible unsteady analysis was performed by digitizing a Navier-Stokes equation using the finite volume method based on fully unstructured meshes with a commercial CFD code (FLUENT6.2, Fluent Inc.). The turbulent model of this study was Large Eddy Simulation (LES) model.

2.2. *Visualization Experiment*

A visualisation experiment using titanium tetrachloride (Asai *et al.*, 2006) was also conducted in order to visualise the flow around the soccer ball during flight. A soccer ball was

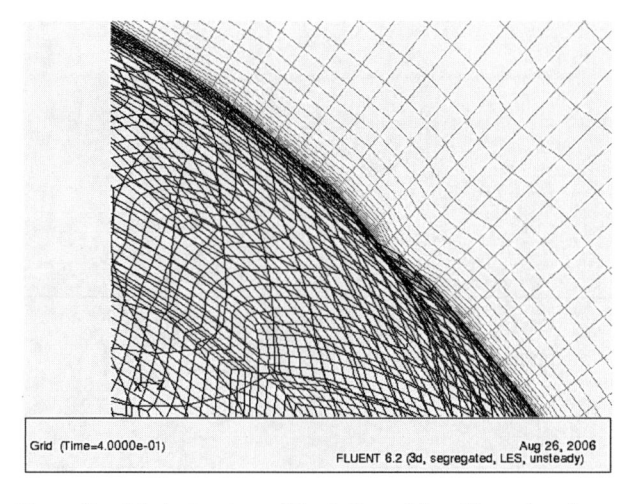

Figure 2. Mesh structure of the ball model and boundary layer.

placed directly in front of a soccer goal 15 m away and we had a subject perform a straight kick that involved virtually no rotation and a side-spinning curve kick. Both kicks were placement kicks delivered at the same velocity, as would occur in a real game. A high-speed VTR camera (Photron Ultima; Photron Limited) was set up at a midpoint between where the ball was placed and the soccer goal, and photographs were taken at 4,500 fps.

The experimental procedure was as follows. Each soccer ball was brush-painted with titanium tetrachloride, placed on a designated spot and then kicked towards a goal. As the ball flew towards the goal, the air flow around it was revealed by white smoke produced by the titanium tetrachloride. Photographs were taken using a high-speed video camera. Finally, the ball was collected and cleaned.

3 Results and Discussion

Observation of time-series data of drag coefficient and lateral force coefficient revealed an unstable early period that tended to stabilize thereafter. Although the momentum involved fine vibrations, the value was low and within a range that could be mostly ignored. In the present study, therefore, the mean drag coefficient value in the 0.2 sec period from 0.2 sec to 0.4 sec after the start of calculations was defined as the drag coefficient for that case.

The drag coefficient for CFD in the present study was about 0.19 for all cases at 15–30 m/s, and about 0.15 for 32-panel balls in the wind tunnel experiment (Fig. 3). It was therefore slightly greater in the present study than was recorded in the wind tunnel experiment. Similar to the variation seen in the wind tunnel experiment of 32-panel balls, the CFD drag coefficient varied little in response to the change in the Reynolds number. Examination of the distribution of flow velocity around the ball at a supercritical CFD range revealed that the boundary layer separation point receded to about 120 degrees from the front stagnation point (Taneda, 1978). This was a similar result to that obtained from visualization experiment images.

Examination of the CFD lateral force coefficient revealed irregular changes reaching a maximum of about 0.1, from about 0.1 s when the drag coefficient value began to stabilize.

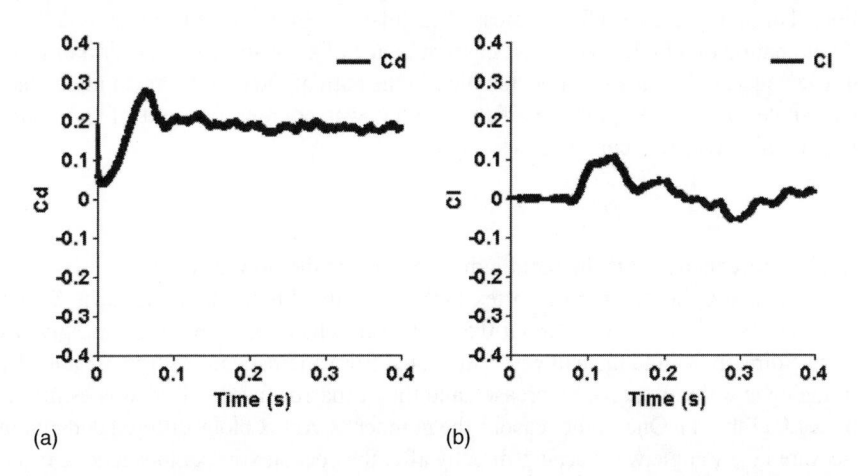

Figure 3. Drag coefficient vs time (a) and Lift coefficient vs time (b) on CFD.

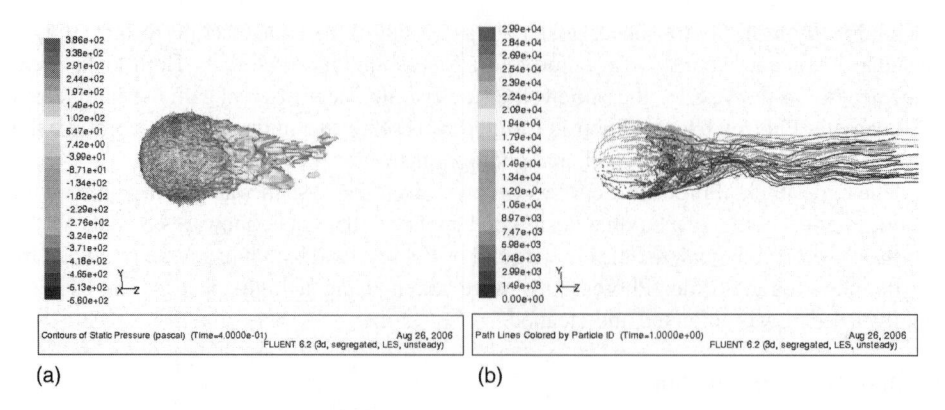

Figure 4. Contours of pressure on vorticity (a) and path lines of the ball (b) on CFD.

This trend was seen in all cases and, although the details remain unclear, it is thought that this was related to the trailing vortex structure. When the trailing vortex in CFD was displayed in terms of peripheral ball vorticity, a decrease in the wake region similar to that seen in the visualization experiment images was evident (Fig. 4). Although the overall impression was one of similarity, it cannot be determined whether or not the small-scale vortexes were similar. When the flow from the ball to a point a short distance away was examined using a pass line display, a near wake was observed, but the slightly separated far wake was reduced. Images from actual visualization experiments revealed many instances of non-symmetrical structures that incorporated a vortex loop, suggesting that for the CFD in the present study the vortex in the early stage was scattered. The precise cause of this is unclear, but it is possible that the inability of the LES model used in the present study to express retrograde transport (inverse cascade) of energy was a contributing factor. Therefore, although it is possible to use LES analysis to roughly predict the drag coefficient and separation point, predicting the trailing vortex structure, especially far wake, is more difficult. In future, attempts must be made to improve mesh quality and quantity, turbulence models, schematic calculations, and border conditions. Improving the hardware will also be important for enhancing the accuracy of calculations. For high-precision CFD, increased calculation resources will be required.

Examination of a high speed VTR camera image of a non-rotating soccer ball while in flight revealed a slightly irregular vortex blob in the path of the ball. From an image having a broad angle of view, the number of these blobs was counted per unit of time and the Strouhal number (St) was calculated (Eq. (1)).

$$St = nd / U \qquad (1)$$

Here, n is the frequency, d is the ball diameter, and U is the flow velocity.

By calculating the number of vortex blobs from this broad angle view, the St (broad angle view) was estimated to be about 0.5. However, when vortex blobs seemingly with a vortex ring directly behind the ball were calculated from an image taken from a panned, narrower angle view, the frequency increased and the St (narrow angle view) was estimated to be about 1.0 (Table 1). One of the reasons the number of vortex blobs differed depending on the measured view angle was because directly after they occur, vortex blobs tend to coalesce with time, causing the measured frequency to change (Fig. 5). Further detailed examination

Table 1. Strouhal number in narrow range images.

Trial	n (Hz)	d (m)	U (m/s)	St
A	107.1	0.22	24	0.98
B	104.7	0.22	25	0.92
C	138.5	0.22	26	1.17

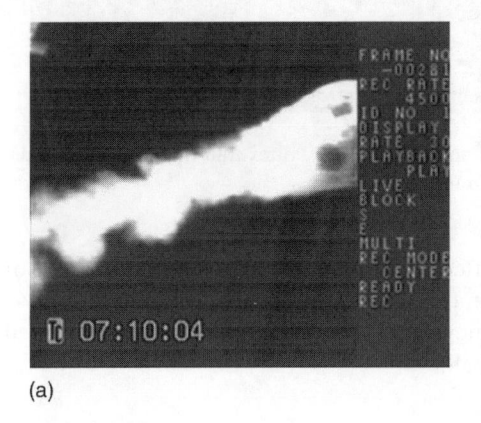

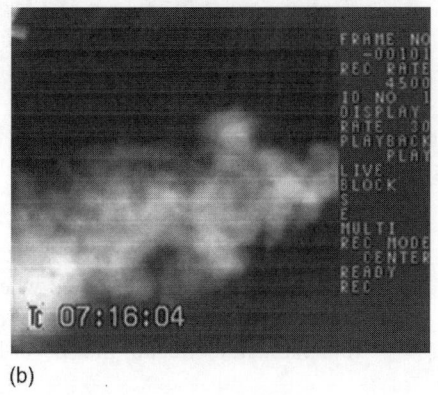

(a) (b)

Figure 5. Vortex visualization of a low-spinning Soccer ball from side view (a: Time is 0.00 s, b: Time is 0.04 s).

is required to determine which method is the most valid for identifying the St. If the frequency is calculated from a vortex blob, such as a vortex ring, directly after it occurs and the St is then estimated, the likely outcome would be a high mode value of about 1.0. This would be close to a smooth ball Reynolds number of 4×10^4 (Sakamoto and Haniu, 1995). After balls undergoing a knuckle effect were airborne, large scale fluctuations of the vortex trail were observed when the St was between 0.1 and 0.01. The precise mechanism behind the vortex fluctuation observed in the present study was unclear, but the St low-mode association will need to be investigated further. Also, it is highly possible that the amount of movement generating this fluctuation is related to the knuckle effect of the soccer ball (Fig. 6).

4 Conclusion

The purpose of this study is to discuss the aerodynamic characteristics of soccer ball using computer fluid dynamics (CFD) and visualization of the vortex structure around the real flight soccer ball in high Reynolds number. The results may be summarized as follows:

1. When the flow from the ball to a point a short distance away was examined using a pass line display, a near wake was observed, but the slightly separated far wake was reduced.
2. On the basis of calculating the frequency from a vortex blob, such as a vortex ring, directly after it occurs and the St is then estimated, the likely outcome would be a high mode value of about 1.0.

Figure 6. Flow visualization of the large scale fluctuations of the vortex trail (the trajectory of the real flight ball was not same as that of the vortex trail).

3. After balls undergoing a knuckle effect were airborne, large scale fluctuations of the vortex trail were observed when the St was between 0.1 and 0.01. Also, it is highly possible that the amount of movement generating this fluctuation is related to the knuckle effect of the soccer ball.

References

Mehta R. and Pallis J. (2001) The aerodynamics of a tennis ball, *Sports Engineering* 4(4), 177–189.
Carré M.J., Goodwill S.R., Haak, S.J., Hanna R.K. and Wilms J. (2004) Understanding the aerodynamics of a spinning soccer ball. In: Hubbard M., Mehta R.D. & Pallis J.M. (Eds.) *The Engineering of Sport 5*, Vol. 1, pp. 70–76. The International Sports Engineering Association, Sheffield.
Carré M.J. and Asai T. (2004) Biomechanics and aerodynamics in soccer. In: Hung G.K. and Pallis J.M. (Eds.) *Biomedical Engineering Principles in Sports*, pp. 333–364. Kluwer Academic Plenum Publishers, New York.
Asai T., Seo K., Kobayashi O. and Sakashita R. (2006) Flow visualization on a real flight non-spinning and spinning soccer all. In: Moritz E.F. and Haake S.J. (Eds.) *The Engineering of Sport 6*, Vol. 1, pp. 327–332. Springer, Munich.
Asai T., Masubuchi M., Nunome H., Akatsuka T. and Ohshima Y. (2000) A numerical study of magnus force on a spinning soccer ball. In: Hong Y. (Ed.), *International Research in Sports Biomechanics*. Routledge, pp. 216–223.
Barber S., Haake S.J. and Carré M.J. (2006) Using CFD to understand the effects of seam geometry on soccer ball aerodynamics. In: Moritz E.F. and Haake S.J. (Eds.) *The Engineering of Sport 6*, Vol. 2, pp. 127–132. Springer, Munich.
Taneda S. (1978) Visual observations of the flow past a sphere at Reynolds numbers between 10^4 and 10^6, *Journal of Fluid Mechanics*, 85, 187–192.
Sakamoto, H. and Haniu, H. (1995) The formation mechanism and shedding frequency of vortices from a sphere in uniform shear flow, *Journal of Fluid Mechanics*, 287, 151–171.

EXPERIMENTAL INVESTIGATION OF THE EFFECTS OF SURFACE GEOMETRY ON THE FLIGHT OF A NON-SPINNING SOCCER BALL

S. BARBER[1], K. SEO[2], T. ASAI[3] & M.J. CARRÉ[1]
[1]*University of Sheffield, Sheffield, UK*
[2]*Yamagata University, Yamagata, Japan*
[3]*University of Tsukuba, Tsukuba, Japan*

Wind tunnel tests were undertaken on several different soccer balls in order to measure the effects of surface geometry and orientation on their trajectories. The drag, lift and side forces were measured at various Reynolds numbers for six balls, which differed in manufacturing technique, panel number and shape, seam size and surface roughness. Three of the balls were oriented at five different angles about the vertical y-axis relative to the air flow. Oil flow visualisation was also carried out on various balls. The results show that the seam alignment with the air flow near the transition region is key; more alignment means more chance of the ball suddenly dropping towards the end of flight and less alignment means the ball will have a low (turbulent) drag coefficient at high Reynolds number and therefore travel further. Hence a compromise must be made in considering ball design.

1 Background

Since Newton commented on the deviation of a tennis ball (Newton, 1672), a large amount of work has been done on measuring and understanding the aerodynamic forces acting on balls and spheres. Balls have been rolled down ramps into wind tunnels (Bentley *et al.*, 1982), projected into the air and their trajectories compared to projectile theory (Carré *et al.*, 2004) and dropped vertically into wind tunnels (Davies, 1949). As technology has progressed, the most accurate method has been found to mount a ball on a force balance and place it in a wind tunnel.

Previous stationary wind tunnel experiments include tests on smooth and rough spheres (Achenbach, 1972, 1974), golf balls (Bearman & Harvey, 1976), tennis balls (Chadwick & Haake, 2000) and soccer balls (Carré *et al.*, 2005). The recent improvements in computer power have also allowed Computational Fluid Dynamics studies to be undertaken on balls such as smooth spheres (Constantinescu & Squires, 2004) and, more recently, soccer balls (Barber *et al.*, 2006).

The aerodynamics of soccer balls have not been studied to a great extent due to a number of difficulties, including the following: their relatively large size requires a large wind tunnel with a sensitive force balance; their pressurized air-filled state results in mounting difficulties, especially at a range of orientations; the small details of their surface geometry require very accurate measurements.

This work presents a new method for the detailed aerodynamics study of soccer balls and discusses the effects of surface geometry and ball orientation on the drag force coefficients (C_D) of the balls.

2 Method

2.1. *Set-up*

A closed-circuit wind tunnel based at Tokai University, Japan, with a 1.0 m × 1.5 m open working section, a 6-component force balance and a turbulence intensity <1% was used for the study. Samples were taken for 10 seconds at a rate of 1024 Hz using a personal computer with the aid of an A/D converter board. 18 different combinations of ball geometries and orientations were tested, as described in Table 1 and pictured in Figure 1b. The balls were attached to a rear-mounted L-shaped sting via a thin, curved plate that was glued securely to the rear of each ball, as shown in Figure 1a. The orientation about the vertical y-axis was altered by gluing the balls to the plate in different positions. Separate tare measurements were made and deducted from each result.

Table 1. Description of soccer balls tested.

Ball	No. panels	Panel shapes	Manufacture	Surface	Orientation (°)
1	32	Hex/pent	Stitched	Standard	0, 20, 50, 70, 90
2	32	Hex/pent	Bonded	Standard	0, 20, 50, 70, 90
3	14	Long/large	Bonded	Smooth	0, 20, 50, 70, 90
4	32	Hex/pent	Stitched	Hatched	0
5	32	Hex/pent	Stitched	Striped	0
6	20	Long	Stitched	Standard	0

(a) Wind tunnel set-up   (b) Balls 1-6 (0°)

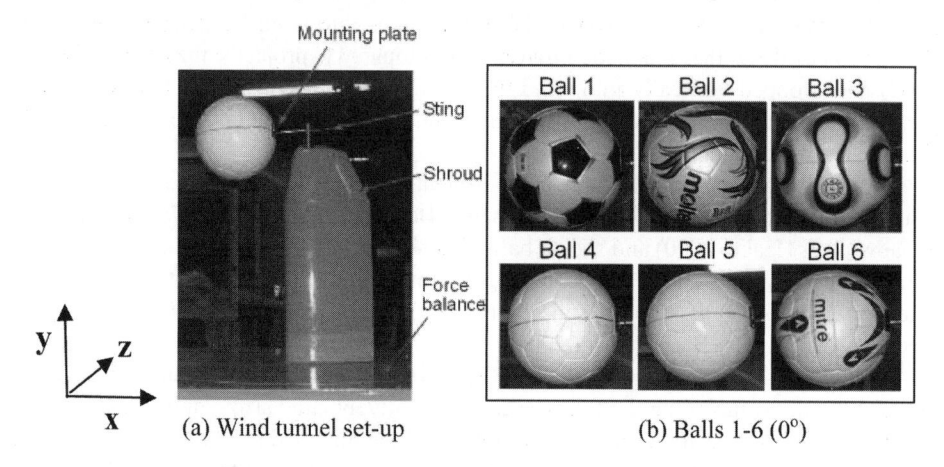

Figure 1. Wind tunnel testing details. For each ball, the drag coefficient (C_D), lift coefficient (C_L) and side force coefficient (C_S) were measured at a range of Reynolds numbers (Re). Additionally, oil flow visualisation was done on Ball 3 (at both 0° and 50°), Ball 5 and Ball 6, which were spray-painted black and coated in a white mixture of liquid petroleum and titanium tetrachloride. Photographs were taken at various wind tunnel velocities, and the results are shown in Figure 2.

2.2. Errors

The general equation for aerodynamic force, F (N), is given by $F = \frac{1}{2}\rho A C_x v^2$, where v = flow velocity (m/s), A = ball frontal area (m^2), ρ = air density (kg/m^3) and C_x = aerodynamic force coefficient. v was calculated from pressure measurements in the tunnel and the error was ± 0.1 m/s, corresponding to a percentage error range of 0.3%–2.5%. The worst-case percentage error of the force balance was estimated to be $\pm 2\%$, which included the tare error. A was found by measuring the diameter of each ball, which had a maximum error of ± 1 mm, translating to about a 0.5% percentage error. A percentage error of 0.5% was also

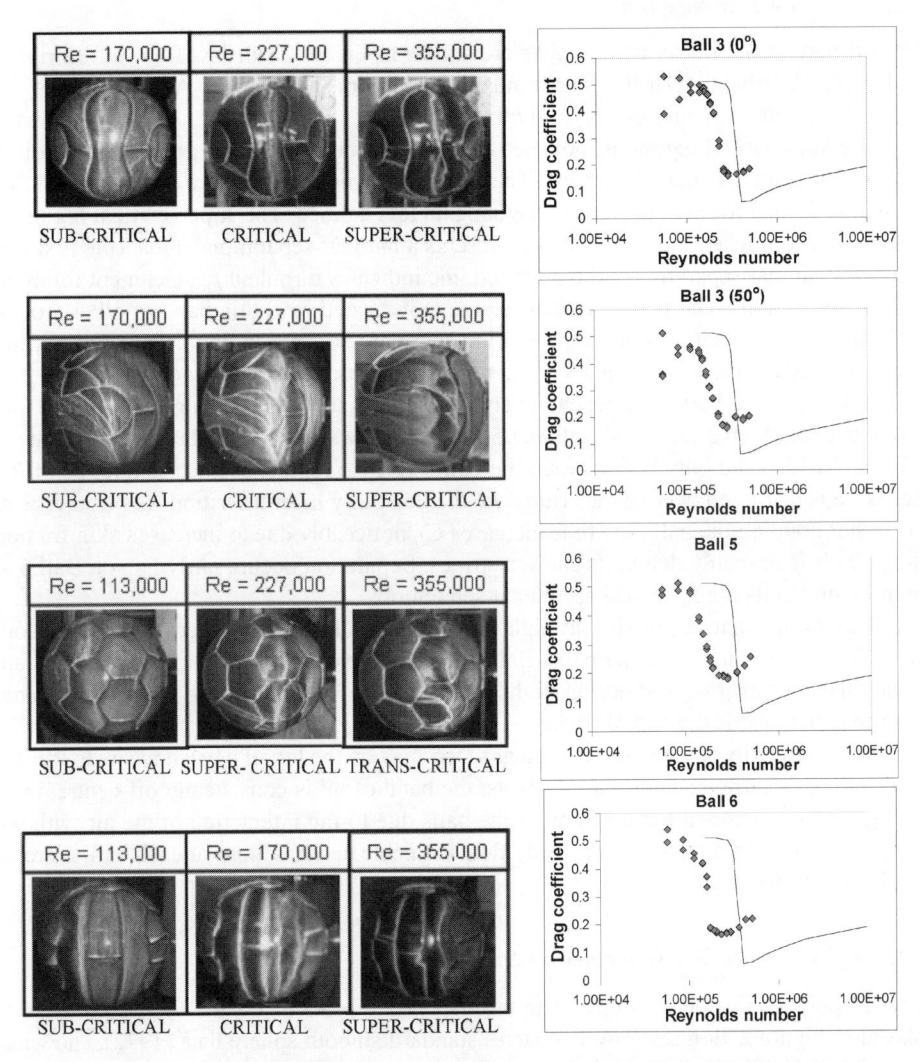

Figure 2. Oil flow visualisation and C_D curves, Ball 3 (0° and 50°), Ball 5 and Ball 6, each compared to a smooth sphere (Achenbach, 1972).

assigned to ρ. Cross-checks were done occasionally using a thermometer and obtaining the pressure reported at the local weather station, and no significant discrepancy was found.

Based on these values, the worst-case standard error in the measurement of the force coefficients was estimated as 5.4%. The repeated C_D measurements were all within 5.4% of each other. C_L and C_S were much less repeatable due to the vibrations experienced by the ball, which were caused by unsteady flow from the rear of the ball. These force coefficients are analysed in a further paper (Seo et al., 2007).

3   Results

3.1.   *Oil Flow Visualisation*

The oil flow visualisation photographs are shown in Figure 2 (airflow from left to right), and allow identification of the four standard flow regimes defined by Achenbach (1972): sub-critical, critical, super-critical and trans-critical.

The **sub-critical** regime is indicated at lower velocities by a thin band of white oil at around 80° from the stagnation point. The **critical** regime is indicated by a band of oil that is slightly further towards the rear of the ball and less defined. The **super-critical** regime is indicated by a thicker band of oil that represents a laminar separation bubble (the first line indicates laminar separation and the second line indicates turbulent reattachment followed by late separation). The **trans-critical** regime is indicated by a thin band of oil at around 110° from the stagnation point representing turbulent separation and a faint line further towards the stagnation point representing transition to turbulent flow.

Ball 3 at 0° and 50° both reach the **critical** regime at a high Re due to their small and sparsely distributed seams, and Ball 3 at 50° reaches the **super-critical** regime very late at 25 m/s. Ball 5 is the only ball to reach the **trans-critical** regime at the tested range of Re, i.e. as Re is increased beyond the critical Re the boundary layer transitions to turbulence at an earlier point on the ball (and thus increases C_D noticeably due to increased skin friction drag). This is probably due to its surface stripes. Separation occurs fairly late on Ball 6 at high Re due to its long, vertical seam near separation.

This means that in practice, at high Re, Ball 5 would have a larger C_D and therefore slow down more than the other balls. Ball 3 would be more likely to move from turbulent to laminar flow as it slowed during flight, and therefore would be more likely to suddenly change path towards the end of flight.

More generally, a seam perpendicular to the flow on the top of a ball appears to trip the flow into early turbulence. For all the tests, the band of oil is consistently off-centre, indicating that a downward force acting on the balls due to the interaction of the air with the sting. Additionally, a region of unsteady flow is visible at the bottom of each ball, increasing in size with increasing Re.

3.2.   *Effects of Surface Geometry on Drag*

The oil flow results were compared to corresponding plots of C_D vs. Re, which are displayed in Figure 2 along with Achenbach's standard smooth sphere data (1972), shown as a smooth line. All the graphs behave as expected, with a sudden drop in C_D from about 0.5 to about 0.2 at transition from laminar to turbulent flow. Transition occurs at a lower Re

than for a smooth sphere because the seams trip the flow into turbulence more readily. The flow regimes match well with the oil flow visualisation.

Similar graphs for the other balls show that balls with smaller and fewer seams exhibit transition at a higher Re and growth in C_D following transition, meaning they would travel faster at high Re and be more likely to drop suddenly towards the end of flight as they slow down and transition to laminar flow.

Smaller, bonded seams cause the airflow around a ball to transition to turbulence faster (i.e. the graph in the critical regime is steeper) and cause a more delayed rise in C_D following transition than stitched seams. This would mean that the ball would travel fast and far when kicked with a mid-range Re.

Increased surface roughness causes a sharper rise in C_D after the minimum because the flow becomes turbulent more readily and so the skin friction drag increases; a ball with increased surface roughness would slow down quickly when kicked with high Re.

3.3. *Effects of Ball Orientation on Drag*

From observing actual kicks, there is evidence to suggest a change in orientation of a ball about the vertical y-axis relative to the flow has a major effect on C_L and C_S. The main effects on C_D are due to the **alignment** of the seams relative to the flow.

Transition occurs at a smaller Re as the seam angles near transition become less aligned with the flow, meaning that a ball with lots of seams perpendicular to the flow near the transition region would travel furthest when kicked at high Re, and a ball with lots of seams in line with the flow near the transition region would be more likely to move into the laminar regime during flight. Additionally, minimum C_D and therefore skin friction drag increases as a fewer proportion of seams are aligned with the flow.

4 Conclusions

- A wind tunnel test method and an oil flow visualisation method have been established for the analysis of full-size soccer balls at different orientations to the flow;
- A smoother ball with fewer seams would travel faster at high Re and be more likely to suddenly slow and change path towards the end of its flight;
- A ball with smaller, bonded seams would travel far and fast when kicked with a fairly high Re;
- A ball with significant surface roughness would slow down more than the other balls and therefore travel less distance;
- The alignment of the seams near transition is important: more alignment means more chance of the ball suddenly dropping towards the end of flight and less alignment means the ball will have a low (turbulent) C_D at high Re and therefore travel further. Hence a compromise must be made in considering ball design;
- The proportion of aligned seams may also be important: the greater the proportion of aligned seams the faster the ball would move at high Re.

In a future paper these results will combined with the lift and side force measurements into a trajectory prediction model.

Acknowledgements

Thanks to Professor Kobayashi and Naoko Takahashi (Tokai University), Nagai-kun (Yamagata University), Molten Corporation and the University of Sheffield's Excellence Exchange Scheme.

References

Achenbach E. (1972) Experiments on the flow past spheres at very high Reynolds numbers. *Journal of Fluid Mechanics,* 54, 565–575.

Achenbach E. (1974) The effects of surface roughness and tunnel blackages on the flow past spheres. *Journal of Fluid Mechanics,* 65, 113–125.

Barber S., Haake S.J. and Carré M.J. (2006) Using CFD to understand the effect of seams on soccer ball aerodynamics. In: Moritz E. & Haake S. J. (Eds.) *The Engineering of Sport 6.* Springer, Munich, Germany.

Bearman P. W. and Harvey J.K. (1976) Golf ball aerodynamics. *Aeronautical Quarterly,* 27, 112–122.

Bentley K., Varty P., Proudlove M. and Mehta R.D. (1982) An experimantal study of cricket ball swing. *Imperial College Aero Tech. Note.*

Carré M.J., Goodwill S.R. and Haake S.J. (2005) Understanding the effect of seams on the aerodynamics of an association football. *Journal of Mechanical Engineering Science,* 219, 657–666.

Carré M.J., Goodwill S.R., Haake S.J., Hanna R.K. and Wilms J. (2004) Understanding the aerodynamics of spinning football. In: Hubbard M., Mehta R.D. and Pallis J.M. (Eds.) *The Engineering of Sport 5.* International Sports Engineering Association, Sheffield, UK.

Chadwick S.G. and Haake S.J. (2000) The drag coefficients of tennis balls. In Subic A. and Haake S.J. (Eds.) *The Engineering of Sport 3.* Blackwell, London.

Constantinescu G. and Squires K.D. (2004) Numerical investigation of flow over a sphere in the subcritical and supercritical regimes. *American Institute of Physics,* 16, 1449–1466.

Davies J.M. (1949) The aerodynamics of golf balls. *Journal of Applied Physics,* 20, 821–828.

Newton I. (1672) New theory of light and colours. *Philosophical Transactions of the Royal Society, London,* 80, 3075–3087.

Seo S., Barber S., Asai T., Carré, M.J. and Kobayashi, O. (2007) The flight trajectory of a non-spinning soccer ball. In: Fuss F.K., Subic A. and Ujihashi S. (Eds.) *The Impact of Technology on Sport II.* Taylor & Francis Group, London.

INFLUENCE OF FOOT ANGLE AND IMPACT POINT ON BALL BEHAVIOR IN SIDE-FOOT SOCCER KICKING

H. ISHII & T. MARUYAMA

Graduate School of Decision Science and Technology, Tokyo Institute of Technology, Tokyo, Japan

The first objective of this study was to investigate the influence of foot angle and impact point on ball velocity and rotation immediately after impact in the side-foot soccer kicking. The second objective was to calculate impact force during the ball impact phase. Five experienced male university soccer players performed side-foot kicks with one step approach in varying foot angle and impact point. The kicking motions were captured three-dimensionally by two high-speed video cameras at 2500 fps. Impact on the area from the center of mass of the foot to the medial malleolus produced the greatest ball velocity. Varying the foot angle with the same impact point did not affect the ball velocity. Also, Impact on the surrounding area of the metatarsal with large foot angle produced the greatest ball rotation. Even if the impact point was the same, the ball rotation increased with the foot angle. Time change of the impact force was calculated from the ball deformation, based on the Hertz contact theory. As an example, in a trial with a ball velocity of 16.3 m/s, a peak ball deformation was approximately 4 cm and a peak impact force was approximately 1200 N.

1 Introduction

The side-foot soccer kicking is the most frequently used technique during a soccer match, and in this kicking the ball is hit by the medial aspect of the kicking foot. In general, the side-foot kicking is used when precision is the main priority. Ball impact phase is important for determining ball behavior after impact. However, there are various patterns of impact in the side-foot kicking depending on the match situation and intent, therefore the mechanism of impact is complex. Although several studies have examined the side-foot kicking, most have been concerned with leg swing phase or the resultant ball velocity (Levanon & Dapena, 1998; Nunome et al., 2002). Therefore, there are few studies about the ball impact phase and the phenomena caused by impact have not yet been determined.

The first objective of this study was to investigate the influence of foot angle and impact point on ball velocity and rotation immediately after impact in the side-foot kicking. The second objective was to calculate impact force from ball deformation during the ball impact phase, based on the Hertz contact theory.

2 Methods

2.1. *Experiment*

Five experienced male university soccer players (age = 23.4 ± 0.5years, height = 171.4 ± 5.2 cm, weight = 57.6 ± 3.9 kg) participated in this study. They were instructed to perform side-foot kicking with one step approach, to a target that was positioned 4 m away. All participants performed approximately 30 kicks in varying patterns of impact. A FIFA

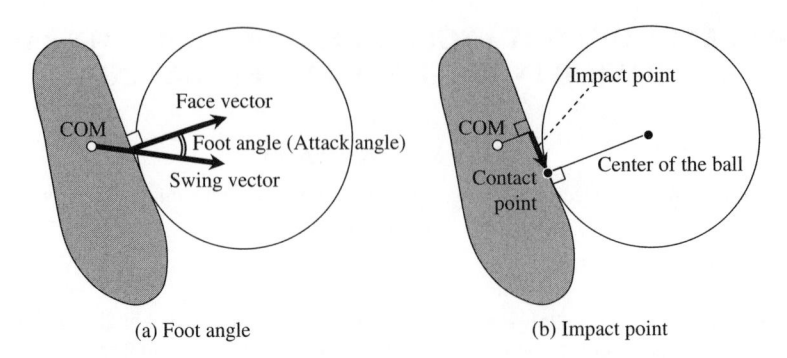

Figure 1. Definitions of the foot angle and the impact point.

approved size 5 soccer ball (mass = 432 g) was used and its inflation was controlled throughout the trials at 9.7 psi. Two electrically synchronized high-speed video cameras (MEMRECAM fx-K4, nac Inc., Japan) were used to capture the kicking motions at 2500 fps (exposure time = 1/5000 s).

2.2. Data Processing

A digitizing system (Frame-DIAS II, DKH Inc., Japan) was used to digitize markers fixed on body landmarks and the ball surface. The direct linear transformation method was used to obtain the three-dimensional coordinates of each marker. The coordinate data were digitally smoothed by a fourth–order Butterworth low-pass filter at the cut-off frequency determined by using the residual analysis.

2.3. Ball Velocity and Rotation

The position of the center of mass of the kicking foot was derived from the toe and heel markers. The position of the center of the ball was calculated as the point located at equal distances (radius of the ball = 11 cm) from the markers fixed on the ball surface. The contact point between the kicking foot and ball was calculated as the intersection point of the perpendicular line dropped from the center of the ball to the medial aspect of the kicking foot.

To express patterns of impact in this study, foot angle and impact point were defined (Figure 1). The foot angle (attack angle) was calculated as the angle between the swing vector that was the velocity vector of the center of mass of the foot and the face vector normal to the medial aspect of the foot. The impact point was calculated as the distance from the center of mass of the foot (projected onto the medial aspect) to the contact point (heel side: +, toe side: −).

The horizontal component of the ball velocity was calculated as the first derivative of linear regression lines fitted to their displacements in 10 flames after impact. The vertical component of the ball velocity was derived from the first derivative of a quadratic regression curve with its second derivative set equal to -9.81 m/s^2 fitted to its displacement in 10 flames after impact. The absolute magnitude of the ball velocity was calculated from the values of its components. The ball rotation was derived using three markers fixed on the ball surface. The influence of the foot angle and the impact point on the ball velocity and rotation immediately after impact was examined.

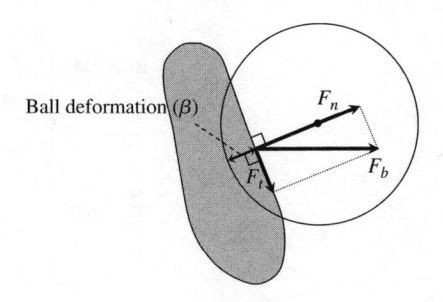

Figure 2. Definitions of the ball deformation and the impact force.

2.4. Ball Deformation and Impact Force

Definitions of the ball deformation and the impact force are shown in Figure 2. The ball deformation β was calculated by subtracting a distance between the center of the ball and the contact point from the radius of the ball (11 cm). Furthermore, the impact force was calculated from the ball deformation, based on the Hertz contact theory (Greszczuk, 1982; Queen et al., 2003; Timoshenko & Goodier, 1970). The normal impact force F_n ($= k\beta^{3/2}$) and the frictional force F_t act on the ball during the impact. Assuming that the frictional force F_t is represented by the Amontons-Coulomb law, the absolute magnitude of the impact force $|F_b|$ is expressed as:

$$|F_b| = \sqrt{F_n^2 + F_t^2} = \sqrt{F_n^2 + (\mu F_n)^2} = \sqrt{(1+\mu^2)F_n^2} = \sqrt{1+\mu^2}\, k\beta^{3/2} = N\beta^{3/2} \quad (1)$$

where k is a coefficient defined by the shapes, Poisson's ratios, and Young's moduli of the ball and foot, μ is the coefficient of friction, then N ($= \sqrt{1+\mu^2}\, k$) is the constant. N is derived using the impulse-momentum relationship of the ball. Therefore, the absolute magnitude of the impact force $|F_b|$ is given by:

$$|F_b| = \frac{m_b V_{b1}}{\int_0^{t_i} \beta^{3/2}\, dt} \beta^{3/2} \quad (2)$$

where m_b is the mass of the ball, V_b1 is the ball velocity, and t_i is the contact time between the ball and foot. Time change of the impact force $|F_b|$ for each trial was calculated using Eq. (2).

3 Results and Discussion

3.1. Ball Velocity and Rotation

In consideration of inter-trial variance in swing speed, the ratio between the ball velocity immediately after impact and the foot velocity immediately before impact (ball-foot velocity ratio) was used as an index. The influence of the foot angle and the impact point on the ball-foot velocity ratio for one subject is shown with the least squares regression surface in

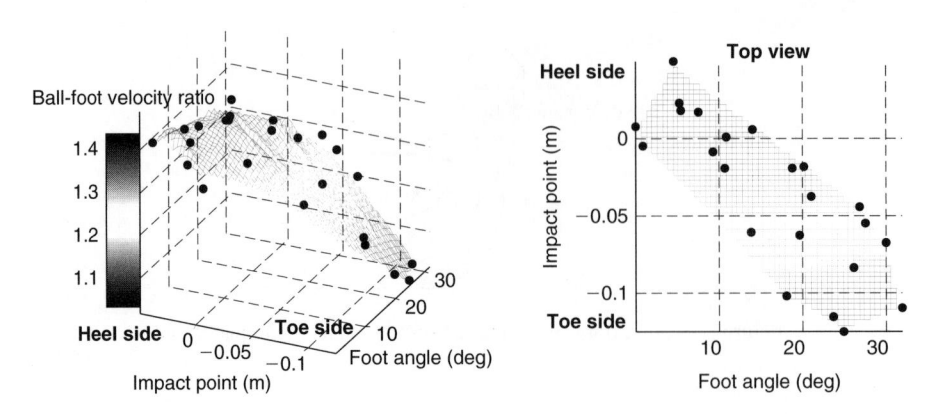

Figure 3. Influence of the foot angle and the impact point on the ball-foot velocity ratio (Sub. A).

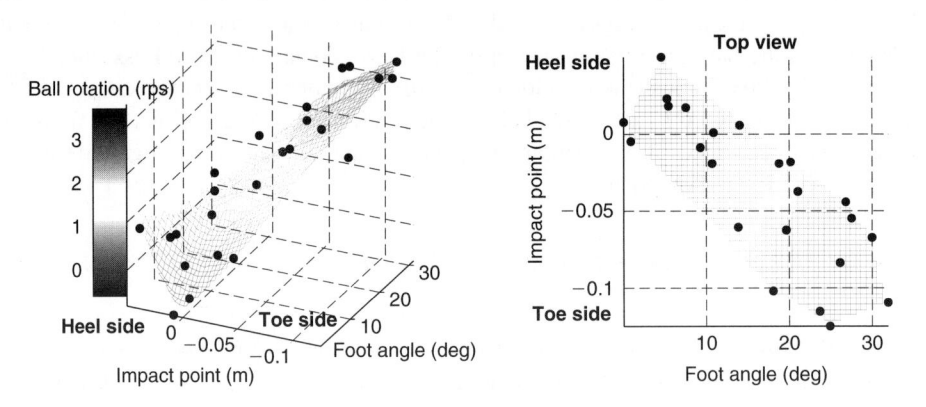

Figure 4. Influence of the foot angle and the impact point on the ball rotation (Sub. A).

Figure 3. Impact on the area from the center of mass of the foot to the medial malleolus produced the greatest ball velocity. Varying the foot angle with the same impact point did not affect the ball velocity.

The influence of the foot angle and the impact point on the ball rotation for one subject is shown with the least squares regression surface in Figure 4. Impact on the surrounding area of the metatarsal with large foot angle produced the greatest ball rotation. In contrast to the ball velocity, even if the impact point was the same, the ball rotation increased with the foot angle.

3.2. Ball Deformation and Impact Force

Time changes of the ball deformation and the impact force for arbitrary three trials are shown in Figures 5 and 6, respectively. As an example, in Trial 1 with a ball velocity of 16.3 m/s, a peak ball deformation was approximately 4 cm and a peak impact force was approximately 1200 N.

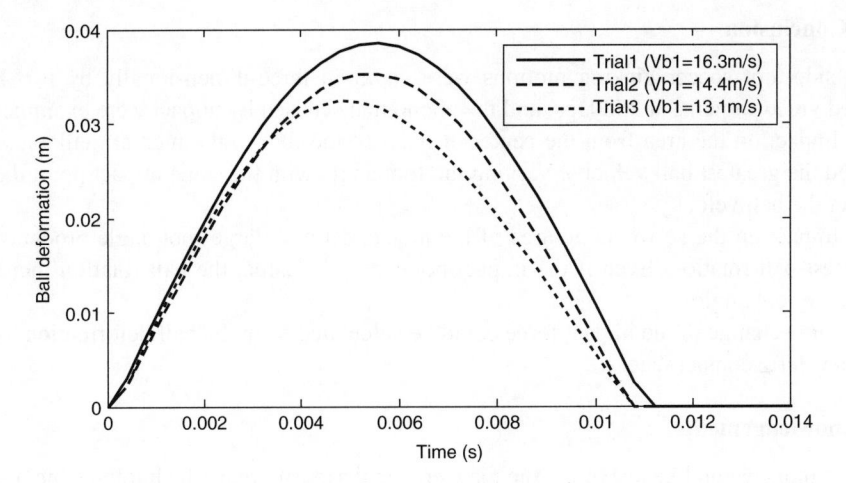

Figure 5. Examples of time change in the ball deformation during the ball impact phase.

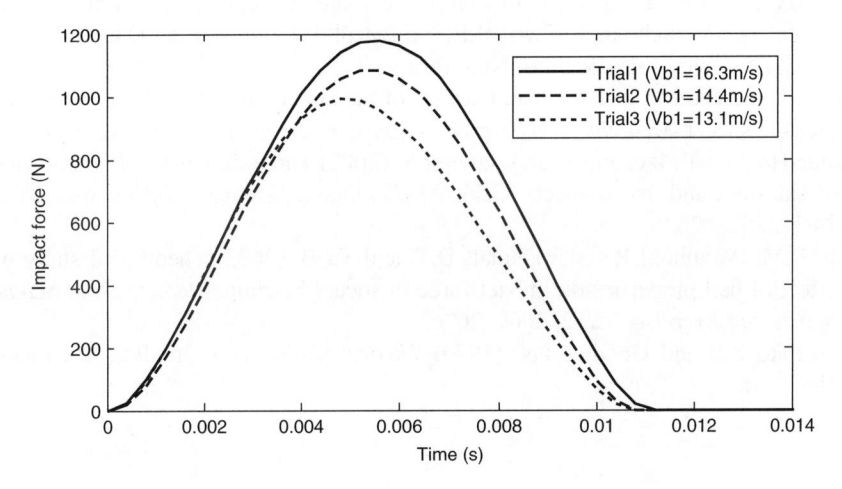

Figure 6. Examples of time change in the impact force during the ball impact phase.

The limitations of calculation approach used in this study include the following. First, based on the Hertz contact theory, both the ball and the foot were assumed to be purely elastic, hence the viscosity was neglected. Also, the frictional force was assumed to be expressed by the coulomb friction. Although these assumptions affect time change of the impact force, it is considered that they would have only a small effect. Therefore, time change of the impact force calculated in this study can be used in the kinetic analysis during the ball impact phase.

4 Conclusion

The side-foot soccer kicking motions were captured three-dimensionally by two high-speed video cameras at 2500 fps, and the phenomena caused by impact were examined.

Impact on the area from the center of mass of the foot to the medial malleolus produced the greatest ball velocity. Varying the foot angle with the same impact point did not affect the ball velocity.

Impact on the surrounding area of the metatarsal with large foot angle produced the greatest ball rotation. Even if the impact point was the same, the ball rotation increased with the foot angle.

Time change of the impact force could be calculated from the ball deformation, based on the Hertz contact theory.

Acknowledgements

The authors would like to express their sincere thanks to nac Image Technology, Inc. for use of the high-speed video cameras in this study.

References

Greszczuk L.B. (1982) Damage in composite materials due to low velocity impact. In: Zukas J.A. Nicholas T., Swift H.F., Greszczuk L.B. and Curran D.R. (Eds.) *Impact Dynamics*. John Wiley & Sons, New York.

Levanon J. and Dapena J. (1998) Comparison of the kinematics of the full-instep and pass kicks in soccer. *Medicine and Science in Sports and Exercise*, 30/6, 917–927.

Nunome H., Asai T., Ikegami Y. and Sakurai S. (2002) Three-dimensional kinetic analysis of side-foot and instep soccer kicks. *Medicine and Science in Sports and Exercise*, 34/12, 2028–2036.

Queen R.M., Weinhold P.S., Kirkendall D.T. and Yu B. (2003) Theoretical study of the effect of ball properties on impact force in soccer heading. *Medicine and Science in Sports and Exercise*, 35/12, 2069–2076.

Timoshenko S.P. and Goodier J.N. (1970) *Theory of elasticity*. 3rd Ed., McGraw-Hill, New York.

INTELLIGENT MUSCULOSOCCER SIMULATOR

S.M.N.A. SENANAYAKE & T.K. KHOO

Monash University Sunway campus, Jalan Universiti, Petaling Jaya, Malaysia

The main objective of this research is to develop an intelligent musculosoccer simulator for soccer gait performance analysis. Key phases involved in the development of the simulator are modeling musculoskeletal models of soccer players, animating musculoskeletal models based on the actions performed and applying neural networks to detect anomalies of soccer gait patterns of players. Modeling and simulation of musculoskeletal models to perform soccer actions are based on the detail analysis of human anatomy, kinematic parameters and inverse and forward dynamics of soccer player. Due to the nature of time varying patterns and track temporal patterns of soccer player, two hierarchical Elman Neural Network (ENN) architectures are proposed. First level of ENN architecture is to correctly associate between the identity of the soccer player and soccer action that is performed and it is followed by the second level of ENN, which utilizes the knowledge from the 1st level ENN to select the correct network to input the relevant data. This allows the formation of individual ENN at the second level for each test subject in order to determine anomalies of soccer players.

1 Introduction

Following the footsteps of available sports biomechanical analysis software VICON and SiliconCOACH, the main focus of this research is to apply concepts of sports biomechanics to develop tools that offer a new alternatives to soccer players in order to gauge the improvement of their performances after undergoing a stringent training regime, as well as to provide guidance in the detection or prevention of injuries, which if ignored, lead to an injury-prone and less satisfying soccer career. The difference between the tools developed in this research with other biomechanical analysis software is that intelligent features that have been incorporated into the analysis to notify soccer trainers and players if any anomalies exist based on the soccer gait patterns knowledge base built during the training and historical competitive games. Hence, this research aims to clarify concepts and methodologies used to develop an Intelligent Musculosoccer Simulator (IMS).

The soccer gait analysis is based on the concepts described in Barton & Less (1997), which utilize kinematics data to differentiate gait patterns. Joint angles obtained from the simulation stage are utilized to differentiate between different soccer gaits. However, the analytical technique is different from the reviewed article so as to introduce a new approach in gait data analysis. The inclusion of intelligent feature is based on the discussions and conclusions arrived in articles (Chua, 2001; Chua, 2001; Barton & Less, 1997).

Therefore, objectives for this research can be divided into three sections; modeling musculoskeletal models of soccer players, animating musculoskeletal models based on the actions performed and to develop an intelligent approach to detect anomalies in the gait pattern of soccer players performing specific soccer actions.

Soccer gait analysis used in this article is based on primary soccer gait actions; walking, running, passing and kicking a ball (Novacheck, 1998). These four actions are the basis to distinguish primary soccer actions as well as derived soccer actions and to detect the existence of any anomalies.

2 Modeling Musculoskeletal Models of Soccer Players

Having selected and analyzed system requirements, it was found out that in order to start execute the modeling and simulation processes in the correct manner, Motion Capture (MOCAP) data must be acquired first. MOCAP data is required during both the modeling and simulation process of the musculoskeletal models. During the modeling process, the MOCAP data are used to give the musculoskeletal model an initial pose or stance, while the simulation process requires MOCAP data to move or animate the musculoskeletal model (Chao, 2003).

Each test subject is required to perform a total of four actions, which are walking, running, passing and kicking. Three repetitions or more are required for each action as a portion of these data are used for neural network training. Hence, for each test subject, a minimum total of 12 MOCAP data are obtained. A total of 10 test subjects took part in the MOCAP data acquisition process.

The MOCAP system that was used is the Motion Analysis Corporation (MAC) MOCAP system. The test subjects were required to wear non-reflective body suits which are basically scuba diving suits while also removing any reflective material that will affect the motion capture process. After fitting into the non-reflective suit, 39 markers are attached to relevant points on the body of the test subject. The marker placement follows a standard marker protocol known as the plug-in marker gait, which is used worldwide for MOCAP applications. Besides that, in order to facilitate an easier process for placing markers on the foot of the test subject, the foot markers are pasted on shoes with different sizes as different test subjects have different foot sizes.

After having selected a modeling and simulation tool (LifeMOD) and after successfully converting the MOCAP data, the next step is to model the musculoskeletal models of each test subject. The modeling of each test subject's musculoskeletal model involves a set of processes; *body segment generation, Joint/soft tissues (muscle) creation, MOCAP data import and contact creation.*

Updating the soccer player's model posture is done based on the movement of musculoskeletal model upon 'pulling' forces provided by the spring element as it seeks to minimize the energy, e.g. see Figure 1.

The contact forces in soccer are to create contacts between the foot of the musculoskeletal model and the soccer pitch or floor. The contact surface, which is the soccer pitch, can be generated by specifying the length, width as well as its thickness. There is also an option to display the ground reaction force vector graphics during the simulation process to have a better visualization on how the ground reaction force acts on the foot of the musculoskeletal model. After the contacts have been created between the musculoskeletal model and the soccer pitch, there are arrows and wordings near the area of contacts, e.g. see Figure 2.

3 Animation of Musculoskeletal Models (Simulation)

After the modeling process is completed, the simulation process begins and the simulation procedure can be summarized into two steps, which are: *Inverse Dynamics Simulation and*

Intelligent Musculosoccer Simulator 411

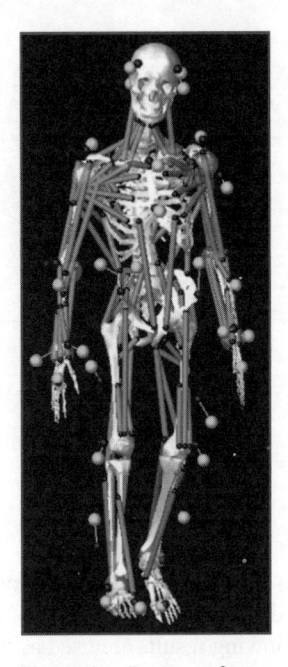

Figure 1. Posture of soccer player.

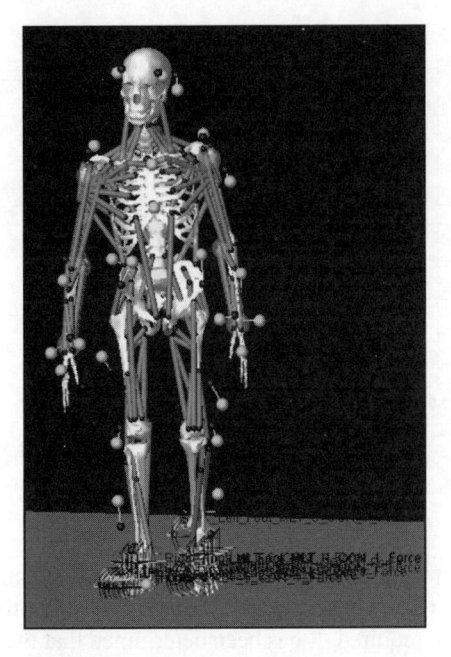

Figure 2. Musculoskeletal model with contacts.

Forward Dynamics Simulation. The forward dynamics simulation is performed only after the inverse dynamics simulation has been successfully completed. During the forward dynamics simulation, the motion agents are 'disabled' meaning that they are no longer driving the musculoskeletal model. The movement of the musculoskeletal model during this stage is generated by the joints and muscles, which have stored the joint angulations and muscle shortening and lengthening patterns from the inverse dynamics simulation. The musculoskeletal model is driven by the joint torques and muscle forces while under the influence of other external forces such as gravity and the contact between the foot and the soccer pitch.

The above simulation results range from kinematics to kinetics data. They are analyzed and then useful information is extracted as the input for the intelligent feature of this research in order to detect any anomalies in the gait patterns of soccer players. The results from the simulation are the joint angle plot in various planes (Sagittal, Frontal and Transverse plane) for the lower extremity of the body, which includes the hip, knee, and ankle joint, e.g. see Figure 3.

4 Anomalies Detection Simulator

The anomalies detection simulator utilizes the Elman recurrent neural network to differentiate between different gait patterns as well as to detect any anomalies that are present. The Elman recurrent neural network was selected because this network is meant to detect time varying patterns and learn temporal patterns found in graphs. Before the anomalies detection phase may begin, relevant features must be extracted from the results obtained from the forward dynamics simulation. Hence, processes involved in data preprocessing are discussed.

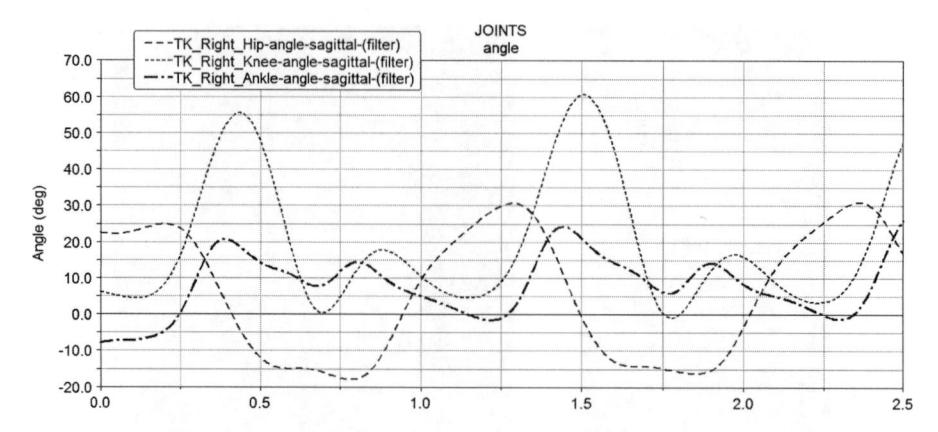

Figure 3. Hip, Knee and Ankle angle in the Sagittal Plane during Walking.

The results as obtained from the forward dynamics simulation are used as inputs to the intelligent system developed. However, features that distinguish soccer actions from the results are required. It has been determined that the following results are used as the inputs to the intelligent system: Hip Angle in the Sagittal, Frontal plane, Knee Angle in the Sagittal plane, Ankle angle in the Sagittal and Transverse plane, Upper Leg Velocity (Magnitude), Lower Leg Velocity (Magnitude) and Foot Velocity (Magnitude).

Hence, there are total of 8 features, which are extracted for each action that a test subject executes. These features undergo same feature extraction processes, which are: *specify gait cycle time, fit curve to region of interest and data normalization.*

Neural Network (NN) simulator design is made up of two phases. First, the soccer player's identity is assumed to be known. Hence, the 1st phase of the NN design is required to correctly associate between the identity of the soccer player and soccer action that is performed. After having executed 1st phase, the 2nd phase utilizes the knowledge from the 1st phase to select the correct network to input the relevant data. The 2nd phase is made up of individual networks with each test subject having 4 neural networks. The 4 personal networks are the walking, running, passing and kicking networks. The purpose of having 4 neural networks for each test subject is to facilitate the comparison of the current gait pattern with previous gait pattern. The test subject's individual neural network will be trained with past records of the same action. Hence, if the new data came in and there is a major difference between the current gait pattern and previous gait patterns, the neural network will notify the trainer that there is an anomaly detected in the soccer gait, e.g. see Figure 4.

The 1st phase of Elman neural network is tested with various combinations but only two are worth mentioning in which network 1 contains 30 neurons in the recurrent layer and network 2 with 20 neurons. The accuracy of pattern recognition from both networks is shown in Table 1.

Maximum velocity of the foot is considered for anomalies detection. The test subject's average maximum velocity is computed and if the new data has a maximum value that is 20% the average maximum velocity and the output of the NN is 1 which means the pattern of graph is normal, then it is safe to say that the test subject's performance has increased or

Intelligent Musculosoccer Simulator 413

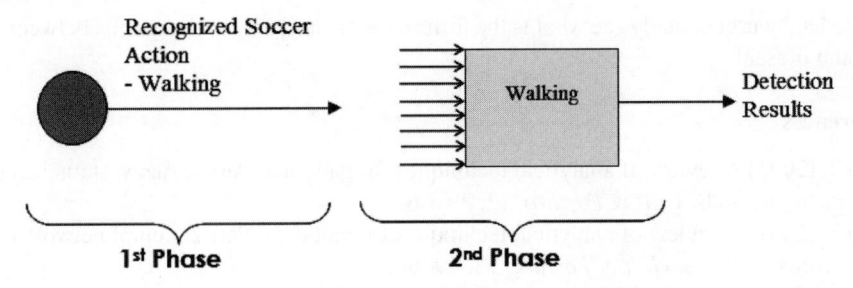

Figure 4. NN simulator design with two phases.

Table 1. Accuracy of soccer action recognition using NN.

	Network 1	Network 2
No. of Test Samples (Untrained Data)	11	11
Before considering Gait Cycle Time		
No. of correctly recognized action	7	6
Accuracy before including gait cycle time	63.64%	54.54%
After considering Gait Cycle Time		
No. of correctly recognized action	9	8
Accuracy after including gait cycle time	81.84%	72.72%

Table 2. Anomalies detection based on the maximum velocity.

NN Output	Maximum Velocity	Anomalies
1	±20% of average velocity	Performance ↑↓
0	±20% of average velocity	Abnormal Gait Pattern

decreased based on the maximum value of foot velocity. This concept is summarized into Table 1.

5 Conclusions

New concepts for anomalies detection as mentioned in section 4 can be verified by acquiring more data to test the reliability and accuracy of the proposed concept. Besides that, a Graphical User Interface (GUI) can be developed to incorporate the data preprocessing steps as well as the option to adaptively train the neural network, once a new data has been acquired. Adapting is suggested as compared to training because for adapting neural network, new input can be presented to the network at any time. However, the same is not true for the training of neural network. If a neural network is retrained, it forgets what it has previously learnt. The GUI can also provide a plotting area whereby after classifying the input data as normal or abnormal, the various angle graphs can be plotted together with previous data to let the user actually see what is the difference or similarity in the graphs between the past and present.

References

Chau T. (2001) A review of analytical techniques for gait data. Part 1: fuzzy, statistical and fractal methods. *Gait & Posture*, 13, 49–66.

Chau T. (2001) A review of analytical techniques for gait data. Part 2: neural networks and wavelet methods. *Gait & Posture*, 13, 49–66.

Barton J.G. and Lees A. (1997) An application of neural networks for distinguishing gait patterns on the basis of hip-knee joint angle diagrams. *Gait & Posture*, 5, 28–33.

Novacheck T.F. (1998) The Biomechanics of Running. *Gait and Posture*, 7, 77–95.

Chao E.Y.S. (2003) Graphic-based musculoskeletal model for biomechanical analyses and animation. *Medical Engineering & Physics*, 25, 201–212.

INTELLIGENT SYSTEM FOR SOCCER GAIT PATTERN RECOGNITION

A.A. BAKAR & S.M.N.A. SENANAYAKE

Monash University Sunway campus, Jalan Universiti, Petaling Jaya, Malaysia

The objective of this research is to use combination of sensors and vision system to determine kinematic parameters of soccer gait. In order to classify and to recognize soccer gait in real time, recurrent neural networks, especially Elman Neural Networks (ENN) are employed. ENN are widely used in recognizing time varying patterns. Motion capture and/or movements in real time are done using a vision system together with wireless accelerometers. These accelerometers are connected to a base station with 900 MHz which is capable to stream data coming from 30 feet radius. The purpose of accelerometers is to compliment the visual data that is acquired so as to give a more qualitative analysis. Nine ENN topologies are constructed for the classifications and recognition of soccer player's main gait. ENNs proposed are based on accelerometer sensors' outputs, vision system results and the combination of selective outputs of them. ENN topologies based on vision and sensors demonstrate good classification and recognition of soccer gait. Real time data streaming and capturing is done in order to test the knowledge base stored. Results show that all soccer actions are recognized with less than 18% error. Further, all derived actions of fundamental soccer actions are also included in the recognized pattern. Therefore, ENN is a good candidate to recognize soccer gait patterns.

1 Introduction

Gait is the way locomotion is achieved using human limbs. There are several types of gait such as walking, running and crawling. Gait analysis is the process of quantification and interpretation of human (including animal) locomotion.

Human motion analysis and synthesis using computer vision and graphing techniques done at Universitat de les Illes Balears uses a method or system to capture human motion by using computer vision to track and recognize each motion. The approach used in this system motion covers a general framework, which is; *Feature extraction process, Feature correspondence and High-level processing*. The system is based on a biomechanical graphical model and matching process between primitives from the model and from the sequences of images (Perales, 2001).

A new approach towards the reconstruction of the human posture from monocular video images using the Criterion Function is used to define the representation of the residuals between feature points in the monocular image and the corresponding points resulted from projecting the human model to the projection plane at Nanyang Technological University. The human body is divided into different parts and is defined as moving relative to each other by the limitations of their movements. The Criterion Function is then set by calculating this limitations and a series of equations are derived (Jianhui, Ling & Kwoh, 2005).

In this research, PEAK MOTUS vision system is used to capture motion or movements in real time together with wireless accelerometers from Microstrain. These accelerometers are connected to a base station with 900 MHz (base station is connected to the PC via USB

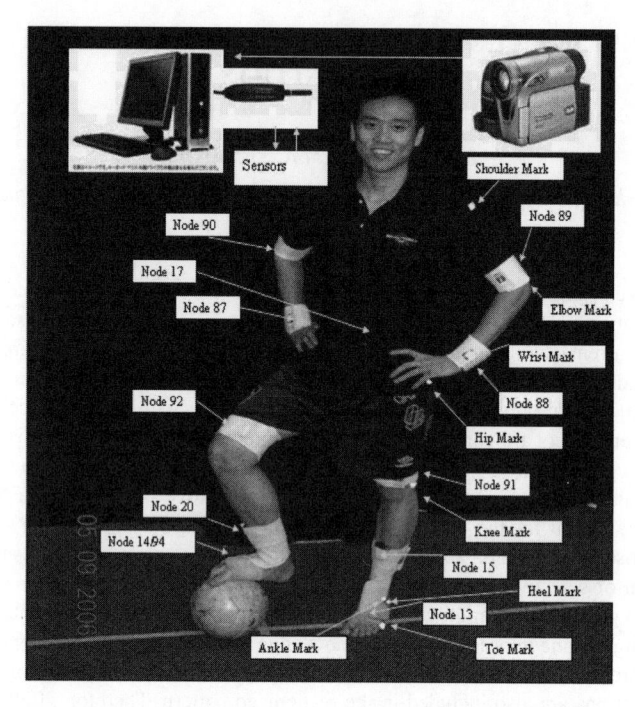

Figure 1. Body mounted sensors and markers of a human test subject.

port) which is capable to stream data coming from 30 feet radius. The purpose of accelerometers is to compliment the visual data that is acquired so as to give a more qualitative analysis. Sensors and markers are located in the test subject for data streaming using sensors and PEAK MOTUS, e.g. see Figure 1. Therefore, soccer gait patterns are prepared based on kinematic parameters obtained from sensors and vision system.

2 Gait Pattern Preparation using Accelerometers and Vision System

Soccer gait data streaming is done taking into consideration its fundamental actions; walking, jogging, passing and heading. All other actions are treated as derived actions from these fundamental actions.

2.1. *Gait Parameters based on Tri-axial Accelerometers*

As tri-axial accelerometer measurements are with respect local coordinate frames, two different measurements are taken into consideration in the pattern preparation. One is measured with respect to time and original orientation of the sensor in consideration. Secondly, the measurement is taken with respect to the sensor located at centre of mass. All these measurements are carried out setting up one of the channels of the tri-axial sensor in the direction of human motion. Therefore, gait parameters based on tri-axial accelerometers are resultant acceleration, resultant force, time and average velocity, e.g. see Figure 2 (Senanayake, 2004).

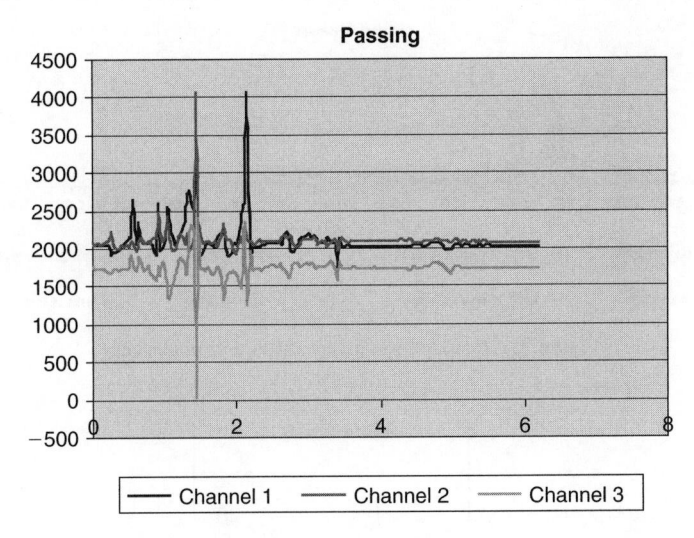

Figure 2. Acceleration in three channels of accelerometer during passing.

2.2. Gait Parameters based on Vision System

In order to generate the data, the markers on the captured video must be digitized. This is done by tracking each point for every frame. In the initial frame, the points are defined accordingly, from the hip to the toe. The system then moves to the next frame automatically and place the tracking cursor to the last position of the first point i.e. the hip. The process is then continued until all the desired frames are digitized. The points have been pre-defined in the system before the capture. This is done by defining the spatial model in the system. The spatial points are the points corresponding to the markers in the digitizing stage. From here, the user can define each point and label it accordingly. Segments can also be defined (distances between points) and label. Graphs of each test subject are plotted for comparison and analysis. The points that are digitized for analysis can be seen as measures of velocity and acceleration. The predefined arm and leg angles can also be plotted as graphs for analysis. As the graphs are analyzed, a pattern began to appear in the velocity and acceleration. Three of the points shows interesting value pattern of consistency among most of the subjects. These points are the ankle, toe and heel, e.g. see Figure 3.

2.3. Data Extraction for Gait Pattern Preparation

From the analysis, it can be concluded that the foot data is the best from the other data in classifying the gait patterns. The data is then extracted. This is done by taking samples from each test subject for every movement. In order to extract the data, a frame "window" is made according to the amount of data that can be extracted from test subjects. This window is kept to a standard amount of frames so as to keep the data from the subjects within a certain range. The frames can be read from the system itself. Since the system runs at 60 frames a second, a good set of range can be defined as most of the movements captured runs at approximately 2 to 5 seconds. When the predetermined range is determined from the graph, the frame range is recorded and uses as reference when the data spreadsheet is open.

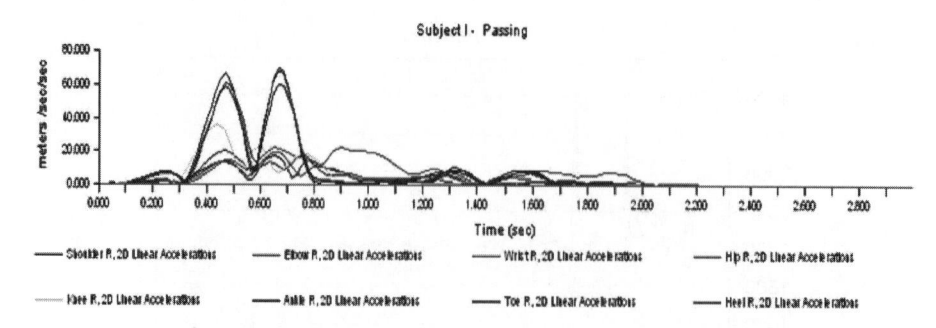

Figure 3. Acceleration using vision during passing.

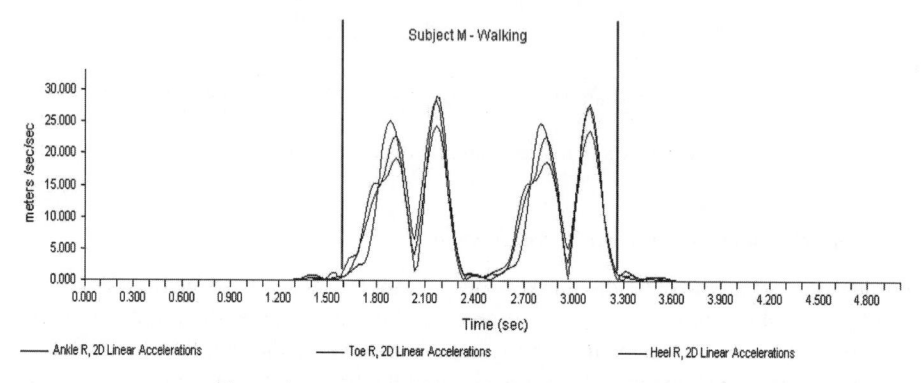

Figure 4. Data extraction using a frame "window".

So it is just a matter of extracting the ring set of data from the spreadsheet and saving it on Excel format for pattern preparation, e.g. see Figure 4.

3 Gait Pattern Classification and Recognition using ENN

The ENN is used for pattern classification and recognition (Giofsofts & Grieve, 1995). The set of patterns for training that is created before is used to train the network. In this case, 100 patterns are based on velocity and acceleration and 70 patterns based on angles. The training is done separately for each value. When the network is fully trained, randomly chosen patterns are used as input patterns to NN again to confirm that the training is successful. The trained neural network is then saved and the test pattern to test the NN is input into the network. It is proven that Velocity – 18 out of 20 patterns identified = **90%**, Acceleration – 20 out of 20 patterns identified = **100%**, Angle – 6 out of 10 patterns identified = **60%**. From this, it is concluded that the classification of the soccer gait pattern is successful.

Similarly soccer gait recognition and validation are done for chosen frames within 100 Seconds. Table 1 shows pattern recognition and validation for such frames (Canosa, 2002). Using trial and error, the best threshold value to identify faulty pattern recognition is set to 0.3. This is also justified from the fact that it has a lower percentage of faulty recognition. It is also safer to lower the threshold value so as to ensure the unclassified patterns are recognized as such and to be included in the training.

Table 1. Pattern recognition using threshold value 0.3.

Time frame	Behaviour	Presented before	Faulty patterns with THRES = 0.3
0–5s	Walk	YES	1
6–11s	Walk Opposite	NO	0
12–17s	Walk Opposite	NO	0
18–23s	Jog	YES	0
24–29s	Jog Opposite	NO	0
30–35s	STOP	NO	0
36–41s	Pass	YES	2
42–47s	Jog & Pass	NO	0
48–53s	Pass	YES	3
54–59s	Pass	YES	2
60–65s	Dribbling	NO	0
66–71s	Jog	YES	3
72–77s	Jog Opposite	NO	0
78–83s	Dribbling	NO	0
84–89s	Dribbling	NO	0
90–95s	Pass	YES	4
	Total faulty patterns		15
	Percentage (%)		18.06493506

4 Conclusions

The knowledge base formed using ENN produces positive results which signify a good stepping stone to create an accurate soccer gait pattern recognition system. The ENN architecture used for this project has proved itself to be a suitable NN as it managed to train time dependant resultant parameters such as acceleration, forces, angles, etc. Although it was time consuming to gather the data and to analyze them, once the data are prepared for the ENN simulator, the knowledge base is formed easily for soccer gait pattern recognition.

During the preliminary preparation of the knowledge base, soccer gait classification is performed on 35 human subjects. The overall system managed to classify and to recognize soccer gait functions correctly. Therefore the knowledge base of the system can be enriched by introducing more unfamiliar patterns, until it covers most of the different gait patterns of a soccer player.

The prototype developed is useful in soccer gait performance, screening for orthopaedic problems and injury prevention. The system developed so far provides a knowledge base to select players based on their abilities and performance during their training. The positions of players can be optimally assigned by recognizing their attributes through incremental and case based learning. Training and learning both are referred in ENN, training of ENN during the training of players prior to a soccer game.

References

Perales F.J. (2001) Human Motion Analysis & Synthesis using Computer Vision and Graphing Techniques. State of Art and Applications (online). Available: http://herakles.zcu.cz/workshops/presentations/Pereales02.pdf (Accessed May 5, 2007).

Zhao J.H., Li L. and Kwoh C.K. (2005) Human Posture Reconstruction from Monocular Images Based on Criterion Function. In: Hamza M. H. (Ed.) Proceedings of CGIM 2002, Computer Graphics and Imaging, Kauai, USA, 12–14 August, Acta Press, Calgary, Canada.

Senanayake A. (2004) Walking, Running and Kicking using Body-Mounted Sensors. IEEE Conference on Robotics, Automation and Mechatronics, Singapore pp. 1141–1146, December.

Giofsofts G. and Grieve D.W. (1995) The Use of Neural Networks to Recognize the Patterns of Human Movement: Gait Patterns. *Clinical Biomechanics*, 10/4, 179–183.

Canosa R.L. (2002) *Simulating Biological Motion Perception Using a Recurrent Neural Network*, Rochester Institute of Technology, Rochester (online). Available: http://www.cs.rit.edu/~rlc/SimBioMotion.pdf (Accessed: May 2007).

11. Ball Sport – Tennis and Badminton

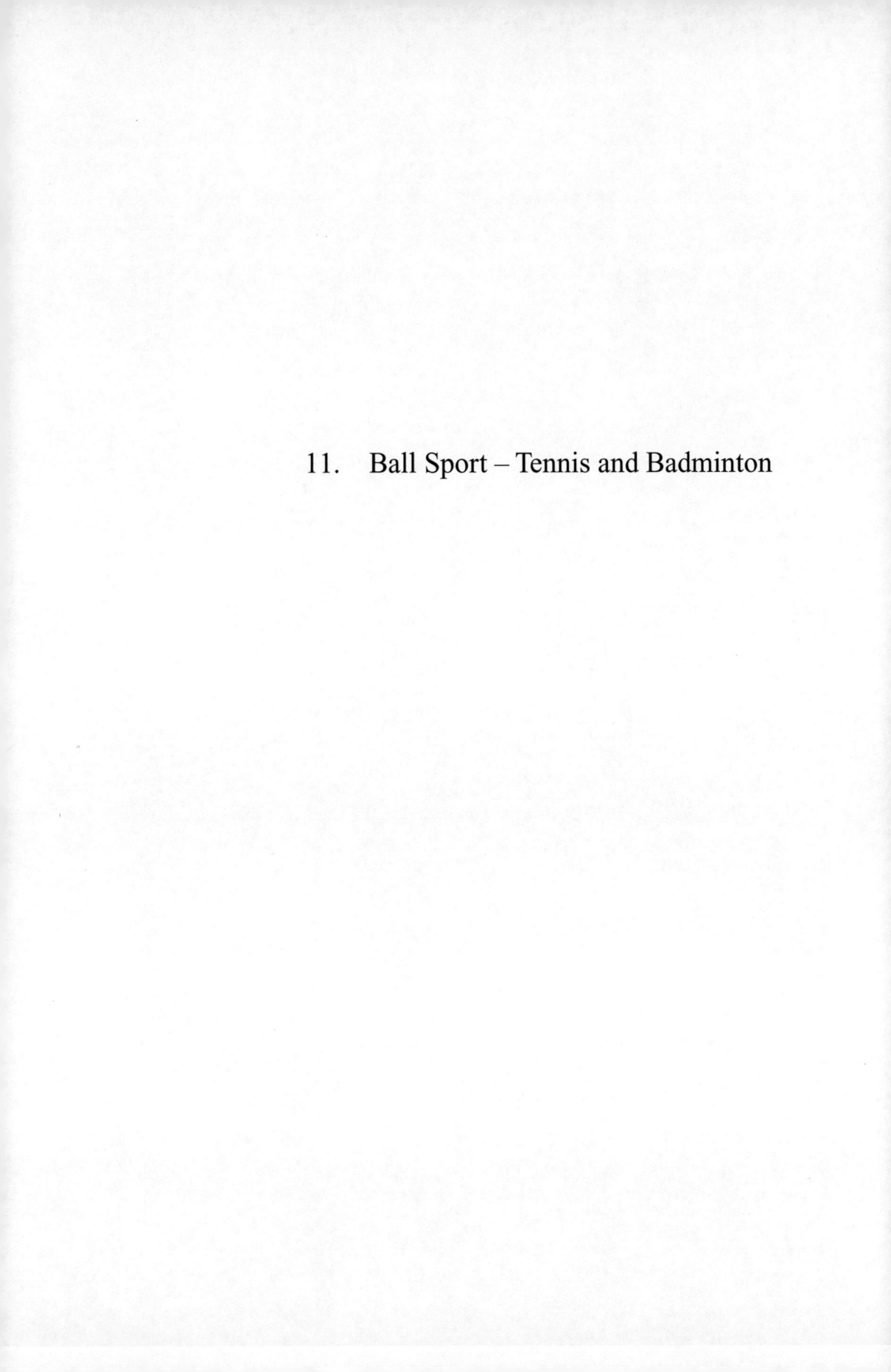

PARALLEL MEASUREMENTS OF FOREARM EMG (ELECTROMYOGRAPHY) AND RACKET VIBRATION IN TENNIS: DEVELOPMENT OF THE SYSTEM FOR MEASUREMENTS

A. SHIONOYA[1], A. INOUE[1], K. OGATA[1] & S. HORIUCHI[2]

[1]*Nagaoka University of Technology, Nagaoka, Niigata, Japan*,
[2]*Asia University, Musashi-sakai, Musashino, Tokyo, Japan*

Racket vibration and EMG (Electromyography) measurement are important factors in the evaluation of racket performance or the investigation of a sports injury such as tennis elbow. The purpose of this study was to develop a system for parallel measurements of tennis racket vibration as a physical element and an EMG as a physiological element. Until now, vibrations and iEMG could not be measured in parallel using one unit, because the natural frequency of the vibration was different from that of the EMG. In the system, both can be measured electrically over time, in parallel. This allows integration of physical and physiological data for the study of such situations as a ball hitting a racket face. Test results reveal that both integrated EMG and vibration frequency components show less reaction when the point of impact is in the sweet point or at the base of the racket, compared to the other three measurement points.

1 Introduction

The performance of a tennis racket is best evaluated when a person grips a racket, as this is closest to actual conditions of use. The vibration of a tennis racket has also been measured while suspending a racket using lines or while fixing it using a vice. However, such an approach ignores the physiological reactions of the human involved, and does not allow investigation of the mechanics of sports injuries such as tennis elbow. One attempt to remedy this involved building a non-rigid robot that could change the strength of its grip (Herbert, 1998), but more often human subjects are used. Kawazoe *et al.* (2001) have evaluated the vibration characteristics of hand-held rackets using a mood analysis. Furthermore, using this evaluation, they have optimized the racket vibration on impact by an actual hitting experiment and a simulation (Kawazoe & Tanahashi, 2002; Kawazoe *et al.* 2002). EMG (Electromyography) has frequently been used to measure muscle activity in an actual hitting experiment (e.g. Kuriyama, 1983). However, as the natural frequency of EMG is different from that of the vibration, EMG could not be measured in parallel with a physical element like vibration, using one unit.

For closer investigation of the effect of the impact of a tennis ball on a hand-held racket, a single system that can measure both vibration and muscle activity simultaneously would be extremely useful. The purpose of this study was to develop and test a system that could be used for parallel measurements of tennis racket vibration as a physical element and EMG for the physiological element.

2 Design of the System

Figure 1 shows an outline of the system for parallel measurements of tennis racket vibration (the physical element) and EMG (the physiological element). To measure the vibration

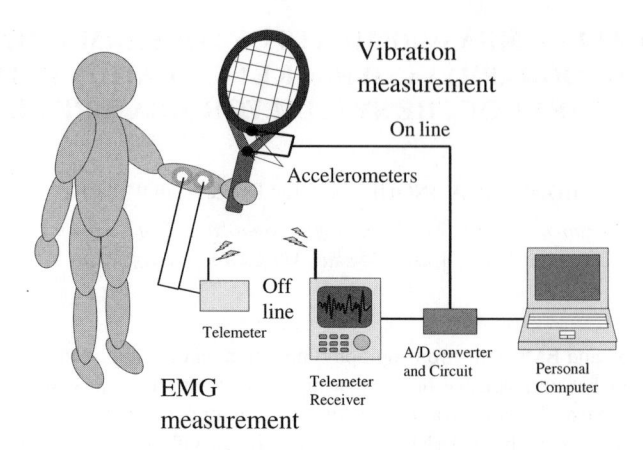

Figure 1. Outline of the system.

of a tennis racket, two accelerometers (PCB PIEZOTRONICS: type 352B10) are installed on the throat and grip of a racket. Furthermore, EMG surface electrodes are attached to the subject's forearm flexion and extension muscles. Accelerometers detect the vibrations of the throat and grip of the racket, and the EMG output signals from the flexion and extension muscles are detected by means of a telemeter system (NEC: MT11). Through a specific unit (PCB PIEZOTRONICS corporation: Signal Conditioner 480B10) and circuit (made by Elmec corporation and modified by our research group), the detected vibrations and the EMG are transformed by an A/D converter and transferred to a personal computer (IBM: Think Pad). The vibrations are transformed into power spectrums by a Fast Fourier Transform (FFT) program (Elmec Corporation: WAAP-WIN) and stored in a hard disc. The EMG are integrated to become iEMG (integrated EMG) using a code developed by our group. The vibrations and EMG are measured electrically, in parallel, and over time with a signal conditioner unit (PCB PIEZOTRONICS corporation: Signal Conditioner 480B10) and circuit (made by Elmec Corporation and modified for this study). Further information on the development of this system can be found in Shionoya *et al.* (2005).

Until now, the vibrations and the EMG could not be measured in parallel using only one unit, because the natural frequency of the vibration was different from that of the EMG. Here, we employ a damping ratio δ calculated from the vibration using the following formula:

$$\delta = \frac{1}{m} \ln \frac{a_n}{a_{n+m}}$$

where a is amplitude, subscript n refers to the wave where the calculation is started, and m is the number of waves considered after that point.

3 Measurement

To confirm the validity of the system, the vibrations of a tennis racket and the EMG were measured in parallel over time under the following conditions: (1) the subject gripped a racket with a back-hand grip at waist height with the racket face held parallel to the ground; (2) a ball

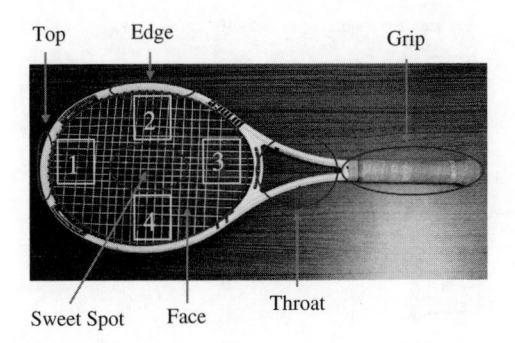

Figure 2. Ball impact points on the tennis racket face.

Table 1. Physical characteristics of rackets used for study (* Center of gravity (CG) shows the length from the top to the CG of a racket).

	Length (mm)	Weight (g)	Center of Gravity (cm)*	Natural Frequency (Hz)	Dumping Ratio
Graphite Racket	698	293	363	140	0.20
Wood Racket	683	385	360	90	0.275

was dropped onto the racket face from a height of 2 m above the racket; 3) ball impact was measured at 5 points: the sweet spot, the top and base of the racket face, and both sides of the sweet spot. Figure 2 shows the ball impact points. An accelerometer was installed on the grip and near the throat to measure racket vibration.

The vibration was transformed into the power spectrum and the dumping ratio calculated by the formula given above. Furthermore, EMG measured the forearm flexion and extension muscles of a racket gripping arm, using surface electrodes, and the iEMG was calculated. Two oversized rackets were used in this study: a graphite racket reinforced by tungsten, and a wood racket reinforced by graphite. The graphite racket is shown in Figure 2.

4 Results

Table 1 gives the physical characteristics of the rackets used in this study. The frequency components between 1.5 and 700 Hz on impact are shown in Figure 3 for each impact point of the graphite racket. For ball impact points of 1, 2 and 4 (see Figure 2), various frequency components, especially low frequency components, were detected. At point 3 and in the sweet spot, fewer frequency components were detected. The maximum amplitude frequency was 140 Hz, the natural frequency.

Figure 4 shows the vibration frequencies of the wood racket for each ball impact point. Similarly to the graphite racket, at the ball impact points of 1, 2 and 4, various frequency components were detected on impact, but at point 3 and the sweet spot, fewer were detected. Once again, the natural frequency was the maximum amplitude frequency, 90 Hz.

In Figure 5 the iEMG increase ratio (%) of the forearm flexion muscles is given for each ball impact point. The baseline is 100, which is the averaged iEMG before impact.

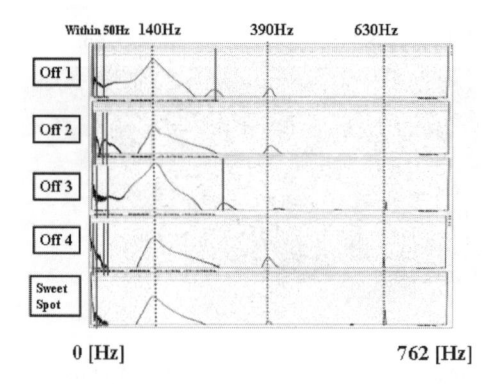

Figure 3. On-impact vibration frequency components of the graphite racket for each ball impact point.

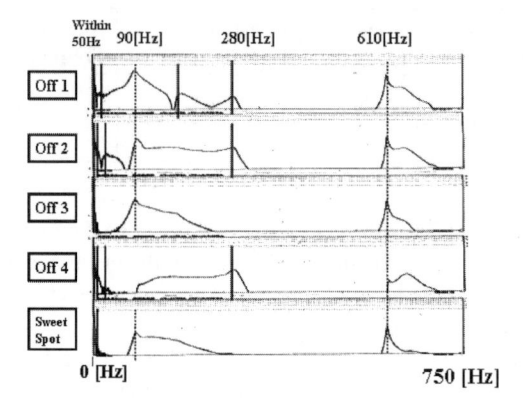

Figure 4. On-impact vibration frequency components of the wood racket for each ball impact point.

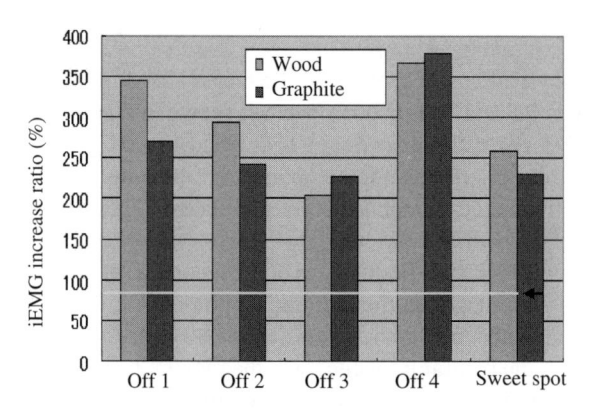

Figure 5. The iEMG increase ratio (%) of the forearm flexion muscles for each ball impact point.

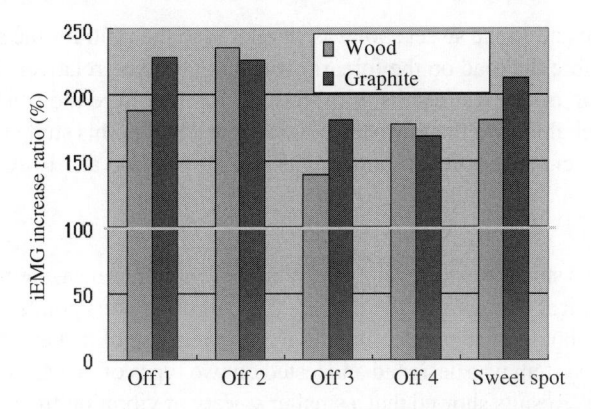

Figure 6. The iEMG increase ratio (%) of the forearm extension muscles for each ball impact point.

On impact at point 4, the iEMG increase ratios of both rackets were at the maximum value measured, while on impact at point 3, the iEMG increase ratios of both rackets were at the minimum value. Figure 6 shows the iEMG increase ratio for the forearm extension muscles: on impact at point 4, the iEMG increase ratios of both rackets were at the maximum value. On the other hand, for impact at 3 and the sweet spot, the iEMG increase ratios of both rackets were the lowest values found.

5 Discussion

The purpose of this study was to develop and test a system that combines measurement of both physical and physiological elements of the impact of a ball on various points on a tennis racket face. The system should give parallel measurements of vibration and EMG over time. The system was tested on 2 rackets, using 5 impact points. Various frequency components, and especially low-frequency components, were detected on impact at the top point (1 in Figure 2) and to the left and right side of the sweet spot (points 2 and 4). However, a smaller variety of frequency components were detected for impact at the base of the racket (point3) and in the sweet spot. This tendency was detected in both the graphite and the wood racket. Odate *et al.* (2004) report similar results in their vibration study. Kawazoe (2000) reported that the rebound power coefficients at the base of the racket face and the sweet spot were higher than those of the other points. Furthermore, the rebound power coefficient at the base was higher than that of the sweet spot. This is because a racket's center of the gravity is towards the base rather than at the sweet spot. Therefore, it is thought that impact energy is used for rolling a racket on impact of the ball at the top of the racket face and at both sides of the sweet spot, and the ball rebounded power on impact at the base or sweet spot impact.

  The iEMG increase rate on impact at the base of the racket face and in a sweet spot was smaller than that at other points. This tendency was detected in both rackets. The iEMG is almost proportional to the muscle strength. For impact at the top of the racket face and at both sides of the sweet spot, the subject gripped the racket grip more powerfully in order to decrease the rolling motion of the racket. On the other hand, the iEMG increase ratio for impact at the base and in the sweet spot was smaller than the ratio for the above-mentioned points, and therefore the load on the forearm was thought to be relatively low.

A comparison of the test results with those in the literature indicates that the results gained here are reliable, and therefore the system developed in this study is able to perform parallel measurements over time of both tennis racket vibration and EMG.

6 Conclusions

The purpose of this study was to develop a system that combines measurement of both physical and physiological elements of the impact of a ball on various points on a tennis racket face. The system should give parallel measurements over time of racket vibration and EMG of the forearm. A system was designed and tested on two types of rackets, dropping a ball on five impact points. Results showed that a smaller variety of vibration frequency components were found for impact at the base of the racket face and in the sweet point, and iEMG measurements for the same points were relatively small. These results agree with other studies, and indicate that a one-unit system has been successfully developed.

References

Herbert H. (1998) The center of percussion of tennis rackets: a concept of limited applicability, *Sport Engineering*, 1, 17–25.

Kawazoe Y. (2000) Performance prediction of large face size tennis racket with different mass and mass distribution in terms of power. *Proceedings of the Japan Society of Mechanical Engineers Symposium on Sports Engineering*, 00/38, 110–114.

Kawazoe Y., Tomosue R., Murayama T. and Yanagi H. (2001) Experimental study of the larger ball's effects on the feel at the wrist joint during the ground stroke and at the elbow joint during the service stroke in tennis, *Proceedings of the Japan Society of Mechanical Engineers Symposium on Sports Engineering*, 01/22, 69–73.

Kawazoe Y. and Tanahashi R. (2002) Prediction of various factors associated with tennis impact: Effects of large ball and strings tension. In: Ujihashi S. and Haake S. (Eds.) *The Engineering of Sport 4*. Blackwell, London, pp. 176–184.

Kawazoe Y., Tomosue R., Murayama T. and Yanagi H. (2002) Experimental study on the effects of larger tennis balls on the comfort of the wrist and the elbow. In: Ujihashi S. and Haake S. (Eds.) *The Engineering of Sport 4*. Blackwell, London, pp. 192–206.

Kuriyama S. (1983) Study on cause of tennis elbow: Stroke analysis by EMG, muscle strength evaluation by Cybex and medical care. *Japanese Journal of Sport Science*, 2/5, 356–364. In Japanese.

Maeda H. and Okauchi M. (2002) The transmission of impact and vibration from tennis racket to hand. In: Ujihashi S. and Haake S. (Eds.) *The Engineering of Sport 4*. Blackwell, London, pp. 223–230.

Odate J., Iwahara M. and Nagamatsu A. (2004) The research on vibration characteristic of tennis racket by the modal analysis. *Proceedings of the Japan Society of Mechanical Engineers Symposium on Sports Engineering*, 04/26, 56–59.

Shionoya A., Ariyoshi K., Inoue A. and Tanaka K. (2005) Experimental development of the system for measuring the vibration of sports gear with considering the man-machine system, *Proceedings of the Biomechanism Symposium*, 19, 337–346.

PARALLEL MEASUREMENTS OF FOREARM EMG (ELECTRO-MYOGRAM) AND TENNIS RACKET VIBRATION IN BACKHAND VOLLEY

A. SHIONOYA[1], A. INOUE[1], K. OGATA[1] & S. HORIUCHI[2]

[1]Nagaoka University of Technology, Nagaoka, Niigata, Japan
[2]Asia University, Musashi-sakai, Musashino, Tokyo, Japan

Racket vibration and EMG (Electromyography) measurement are important factors in the evaluation of a racket performance or a sports injury such as tennis elbow. We developed a system for parallel measurements of tennis racket vibration as a physical element and an EMG as a physiological element. In this study, the vibration frequency of tennis racket and the EMG of forearm in parallel in backhand volley on 2 rackets using 5 impact points were measured using this developed system. On every impact points, the peak frequency detected was the natural frequency of a racket in bath the throat of the racket and the wrist of the subject. Exactly, the natural vibration component of a tennis racket was mainly transmitted to wrist joint. The iEMG increase rate on impact at the base of the racket face and in a sweet spot was smaller than that at other points. This tendency was detected in both rackets. The iEMG is almost proportional to the muscle strength. For impact at the top of the racket face and at both sides of the sweet spot, the subject gripped the racket grip more powerfully in order to decrease the rolling motion of the racket. On the other hand, the iEMG increase ratio for impact at the base and in the sweet spot was smaller than the ratio for the above-mentioned points, and therefore the load on the forearm was thought to be relatively low.

1 Introduction

The performance of a tennis racket is best evaluated when a person grips a racket, as this is closest to actual conditions of use. The vibration of a tennis racket has also been measured while suspending a racket using lines or while fixing it using a vice. However, such an approach ignores the physiological reactions of the human involved, and does not allow investigation of the mechanics of sports injuries such as tennis elbow. One attempt to remedy this involved building a non-rigid robot that could change the strength of its grip (Herbert, 1998), but more often human subjects are used. Kawazoe et al. (2001) have evaluated the vibration characteristics of hand-held rackets using a mood analysis. Furthermore, using this evaluation, they have optimized the racket vibration on impact by an actual hitting experiment and a simulation (Kawazoe & Tanahashi, 2002; Kawazoe et al. 2002). EMG (Electromyography) has frequently been used to measure muscle activity in an actual hitting experiment (e.g. Kuriyama, 1983). However, as the natural frequency of EMG is different from that of the vibration, EMG could not be measured in parallel with a physical element like vibration, using one unit. Shionoya et al. (2007) developed a single system that can measure both vibration and muscle activity simultaneously.

The purpose of this study was to measure tennis racket vibration and EMG in parallel using this developed system in backhand volley as an actual hitting condition.

2 Methods

2.1. *Impulse Response of Tennis Racket*

To measure the impulse response of tennis racket, racket was suspended, and two accelerometers (PCB PIEZOTRONICS: type 352B10) were installed on the throat and grip of a racket. Accelerometers detect the vibrations of the throat and grip of the racket was detected by means of a telemeter system (NEC: MT11). Through a specific unit (PCB PIEZOTRONICS corporation: Signal Conditioner 480B10) and circuit (made by Elmec corporation and modified by our research group), the detected vibrations and the EMG were transformed by an A/D converter and transferred to a personal computer (IBM: Think Pad). The vibrations were transformed into power spectrums by a Fast Fourier Transform (FFT) program (Elmec Corporation: WAAP-WIN) and stored in a hard disc.

2.2. *Parallel Measurement of Racket Vibration and EMG in Backhand Volley*

Figure 1 shows the outline of parallel measurement of tennis racket vibration and EMG in backhand volley using a single system developed in the previous study (*Shionoya et al.*, 2007). Subject was an expert tennis player having the N prefectural ranking points in Japan. Subject gripped the racket with a backhand grip and tried backhand volley to the ball shot from the shooting machine. The interval between subject and the shooting machine was 7.0 m.

Ball impact was measured at 5 points: the sweet spot, the top and base of the racket face, and both sides of the sweet spot. Figure 2 shows the ball impact points.

One accelerometer was installed on the throat near the racket face to measure racket vibration and another one was installed on the wrist of subject to investigate, which racket vibration was transmitted to human body or not. The vibrations were transformed into the power spectrum. Furthermore, EMG measured the forearm flexion and extension muscles of a racket gripping arm, using surface electrodes, and the iEMG was calculated. Two oversized

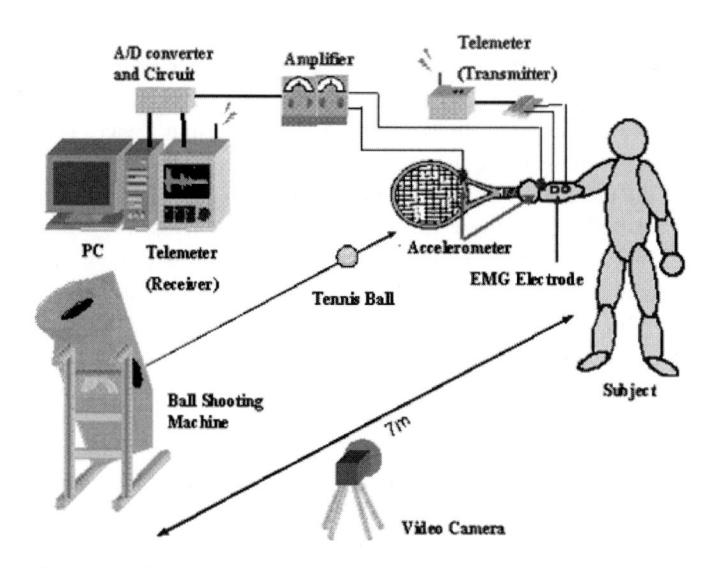

Figure 1. Outline of parallel measurement of tennis racket vibration and EMG in backhand volley.

rackets were used in this study: a wood racket reinforced by graphite, and a graphite racket reinforced by tungsten.

3 Results

Figure 3 shows the impulse response of the wood racket. A solid line is the frequency on the throat and a dot line is that on the grip of the racket. The maximum amplitude frequency of both the throat and the grip were almost 90 Hz, the natural frequency. Figure 4 shows the impulse response of the graphite racket. The maximum amplitude frequency of both the throat and the grip were almost 140 Hz, the natural frequency.

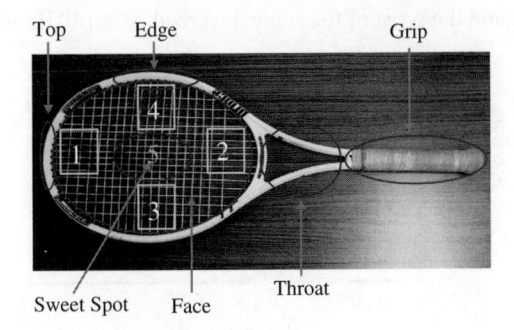

Figure 2. Ball impact points on the tennis racket face.

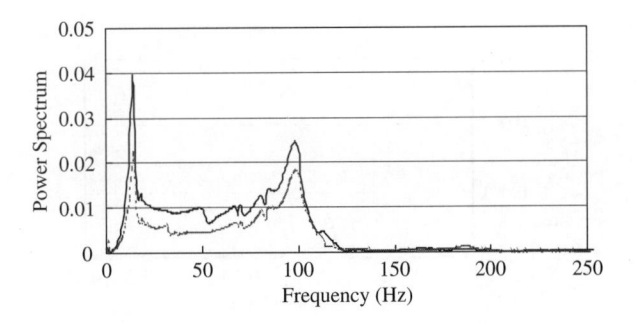

Figure 3. Impulse response of the wood racket.

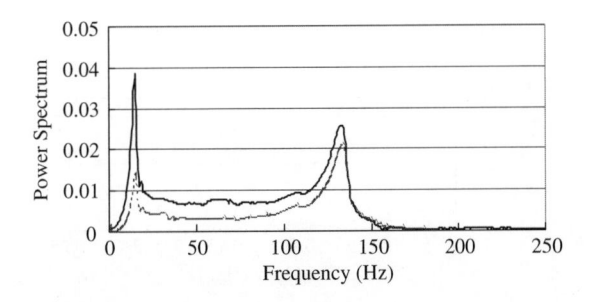

Figure 4. Impulse response of the graphite racket.

Figures 5 and 6 show examples of the wood racket vibration frequency in backhand volley. Figure 5 is in impact at the top of the racket face (point 1 in Figure 2) and Figure 6 is in impact at the base of the racket face (point 2 in figure 2). A solid line is the frequency on the throat of the racket and a dot line is that on the wrist of the subject. The peak frequency of both the racket throat and the wrist of the subject were almost 90 Hz (the natural frequency) in both impact points.

Figures 7 and 8 show examples of the graphite racket vibration frequency in backhand volley. Figure 7 is in impact at the top of the racket face (point 1 in Figure 2) and Figure 8 is in impact at the base of the racket face (point 2 in Figure 2). A solid line is the frequency on the throat of the racket and a dot line is that on the wrist of the subject. The peak frequency of both the racket throat and the wrist of the subject were almost 140 Hz in both impact points.

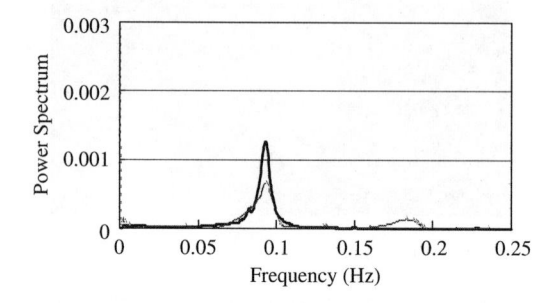

Figure 5. Wood racket vibration frequency in backhand volley in impact at the top of the racket face.

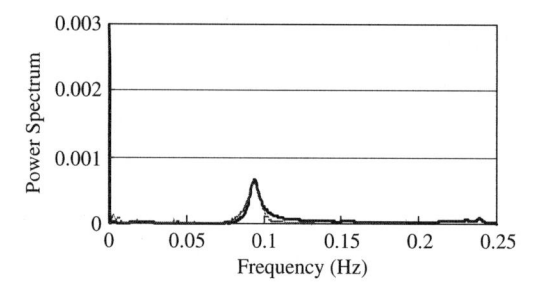

Figure 6. Wood racket vibration frequency in backhand volley in impact at the top of the racket face.

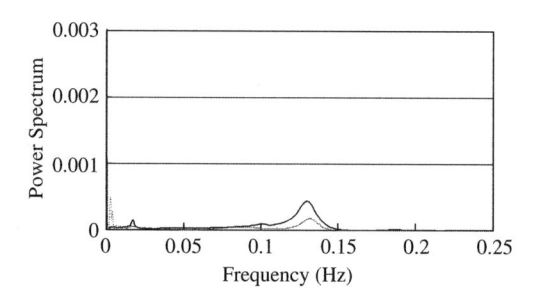

Figure 7. Graphite racket vibration frequency in backhand volley in impact at the top of the racket face.

Figure 9 shows the iEMG increase ratio for each impact point in the wood racket. In figure, the left bar in each impact point is the forearm extension muscles and the right bar is the forearm flexion muscles. The iEMG increase ratio is calculated the multiple of the averaged iEMG before impact. The iEMG increase ratio in both muscles were higher in impact at 1, 3, 4 points than that at 2 and 5 points. Figure 10 shows the iEMG increase ratio for each impact point in the graphite racket. Similarly to the wood racket, the iEMG increase ratio in both muscles were higher in impact at 1, 3, 4 points than that at 2 and 5 points.

4 Discussion

The purpose of this study was to measure tennis racket vibration and EMG in backhand volley as an actual hitting condition using a system that combines measurement of both physical and physiological elements of the impact of a ball on various points on a tennis

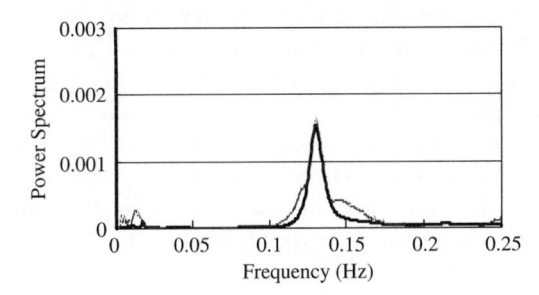

Figure 8. Graphite racket vibration frequency in backhand volley in impact at the base of the racket face.

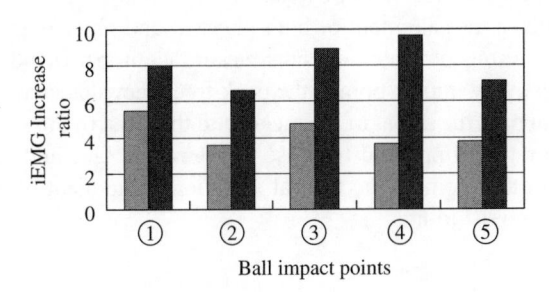

Figure 9. The iEMG increase ratio (multiple) for each impact point in the wood racket.

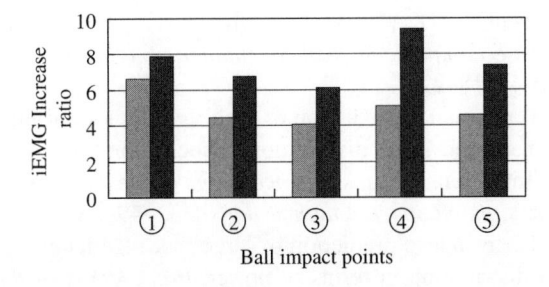

Figure 10. The iEMG increase ratio (multiple) for each impact points in the graphite racket.

racket face. The system should give parallel measurements of vibration and EMG over time. In this study, this system measured the vibration frequency of tennis racket and the EMG of forearm on 2 rackets using 5 impact points. In every impact points, the peak frequency detected was the natural frequency of a racket in bath the throat of the racket and the wrist of the subject. Furthermore, this tendency was same in both the wood and the graphite racket. Exactly, the natural vibration component of a tennis racket was mainly transmitted to wrist joint. This result was similarly to the Kawazoe reports (1998).

The iEMG increase rate in impact at the base of the racket face and in a sweet spot was smaller than that at other points. This tendency was detected in both rackets. The iEMG is almost proportional to the muscle strength. For impact at the top of the racket face and at both sides of the sweet spot, the subject gripped the racket grip more powerfully in order to decrease the rolling motion of the racket. On the other hand, the iEMG increase ratio for impact at the base and in the sweet spot was smaller than the ratio for the above-mentioned points, and therefore the load on the forearm was thought to be relatively low. This result was similarly to Inaoka report (1995). Relating to this result, Kawazoe (2000) reported that the rebound power coefficients at the base of the racket face and the sweet spot were higher than those of the other points. Furthermore, the rebound power coefficient at the base was higher than that of the sweet spot. This is because a racket's center of the gravity is towards the base rather than at the sweet spot. Therefore, it is thought that impact energy is used for rolling a racket on impact of the ball at the top of the racket face and at both sides of the sweet spot, and the ball rebounded power on impact at the base or sweet spot impact.

5 Conclusions

The purpose of this study was to measure tennis racket vibration and EMG in parallel using a system that combines measurement of both physical and physiological elements of the impact of a ball on various points on a tennis racket face in backhand volley as an actual hitting condition. In every impact points, the peak frequency detected was the natural frequency of a racket in bath the throat of the racket and the wrist of the subject. Furthermore, the iEMG increase rate in impact at the base of the racket face and in a sweet spot was smaller than that at other points. The natural vibration component of a tennis racket was mainly transmitted to wrist joint.

References

Herbert H. (1998) The center of percussion of tennis rackets: a concept of limited applicability, *Sport Engineering*, 1, 17–25.

Inaoka K. and Matsuhisa. H. (1995) Impulsive force of head and arm in Tennis, *Proceedings of the Japan Society of Mechanical Engineers Symposium on Sports Engineering,* 95/45, 18–22.

Kawazoe Y. (1998) Performance prediction of high rigidity normal face size tennis racket with different mass and mass distribution (Shock vibration of a racket grip and a player's wrist joint in tennis impact) *Proceedings of the Japan Society of Mechanical Engineers Symposium on Sports Engineering,* 98/31, 49–53.

Kawazoe Y. (2000) Performance prediction of large face size tennis racket with different mass and mass distribution in terms of power. *Proceedings of the Japan Society of Mechanical Engineers Symposium on Sports Engineering*, 00/38, 110–114.

Kawazoe Y., Tomosue R., Murayama T. and Yanagi H. (2001) Experimental study of the larger ball's effects on the feel at the wrist joint during the ground stroke and at the elbow joint during the service stroke in tennis, *Proceedings of the Japan Society of Mechanical Engineers Symposium on Sports Engineering*, 01/22, 69–73.

Kawazoe Y and Tanahashi R. (2002) Prediction of various factors associated with tennis impact: Effects of large ball and strings tension, *The Engineering of Sport 4*, 176–184.

Kawazoe Y., Tomosue R., Murayama T. and Yanagi H. (2002) Experimental study on the effects of larger tennis balls on the comfort of the wrist and the elbow In: Ujihashi S. and Haake S. (Eds.) *The Engineering of Sport 4*. Blackwell, London, pp. 192–206.

Kuriyama S. (1983) Study on cause of tennis elbow: Stroke analysis by EMG, muscle strength evaluation by Cybex and medical care. *Japanese Journal of Sport Science*, 2/5, 356–364. In Japanese.

Shionoya A., Inoue A. and Horiuchi S. (2007) Parallel measurements of forearm EMG and racket vibration in tennis – Development of the system for measurement. In: Fuss F.K., Subic A. and Ujihashi S. (Eds.) *The Impact of Technology on Sport II*. Taylor & Francis Group, London.

AN EXPERIMENTAL AND COMPUTATIONAL STUDY OF TENNIS BALL AERODYNAMICS

F. ALAM, W. TIO, A. SUBIC & S. WATKINS

School of Aerospace, Mechanical and Manufacturing Engineering, RMIT University, Bundoora, Melbourne, Australia

The aerodynamic behaviour of a tennis ball is very complex and significantly differs from other sports balls due to its surface structures (fuzz, seam orientation etc). Relatively high rotational speeds (spin) make the aerodynamic properties of tennis balls even more complex. Although several studies have been conducted on drag and lift in steady state condition (no spin involved) by the author and others, little or no studies have been conducted using computational method. Therefore, the primary objectives of this work were study the aerodynamic properties of a tennis ball using both experimental and computational methods. The Computational Fluid Dynamics (CFD) results were compared with the experimental findings. A comparison of drag coefficients was made for a range of Reynolds numbers.

1 Introduction

The popularity of ball games has been increased significantly and the trend will continue in near future. Player's individual performance is in the peak form. As viewers better performance, therefore, an alternative is to look at the way the game is played itself. The direct approach will be to build a better ball that can travel faster, further (eg. golf ball, base ball) and slower in some particular games (eg. tennis ball). The International Tennis Federation (ITF) is trying its best to slow down the speed of the ball as viewers become bored for not being able to see the flight of the ball as currently tennis ball travels very fast. This problem is especially with the top ranking male player and some women players as well. An alternative to reduce the speed of the ball is to introduce larger ball (with bigger mass), however, it may change the game itself completely. Wilson Rally 2 ball has approximately 20% larger diameter compared to Wilson DC 2 or Wilson US Open 3. A study by Alam *et al.* (2003) showed that the larger diameter Wilson Rally 2 has the similar drag coefficient as normal diameter Wilson DC 2 and Wilson US Open 3. However, the Wilson Rally 2 has the larger overall drag force due to its larger cross sectional area. In the same study, Alam *et al.* (2003) also showed that the Bartlett ball with the similar diameter of Wilson DC 2 and Wilson US Open 3 has the highest drag coefficient (over 20%) over a range of speeds. A visual inspection indicated that the Bartlet ball has a very prominent seam compared to any other ball in its category. The surface structure of a tennis ball is very complicated due to the fuzz structure (furry surface) and complex orientation of seam. The aerodynamics properties of tennis balls under steady conditions (no spin involved) has been studied by Alam *et al.* (2003, 2004, 2005), Mehta & Pallis (2001), Chadwick & Haake (2000). As the ball's flight can significantly be deviated due to spin effects (some player can introduce spin up to 6000 rpm), a comprehensive study by Alam *et al.* (2005, 2006) has been conducted. Most of these works were in experimental. The effects of seam and fuzz are believed to be dominant at a very low speed. It is generally difficult to measure these effects experimentally at these low speeds as instrumental errors are significant. Therefore the primary objective of this work was to study a tennis ball's seam effects on aerodynamic properties using CFD (computational fluid dynamics method) and compared with experimental findings. As it is very difficult to construct fuzz on tennis ball, a simplified sphere and sphere with various seam widths was studied to simplify the computational process.

2 Description of Computational Fluid Dynamics (CFD) Modelling Procedures

In computational study, commercial software FLUENT 6.0 was used. In order to understand the simplified model first, a sphere was made using Solidworks (see Figure 1). Then two simplified tennis balls without fuzz were also made which are shown in Figures 2 & 4. Two simplified tennis balls were constructed with the following physical geometry: diameter 65 mm, mass 56 grams, seam with 2 mm width, 1.5 mm depth; and 5 mm width, 1.5 mm depth respectively. All models were then imported to FLUENT 6.0 and GAMBIT was used to generate mesh and refinement. The major consideration when performing the computational analysis is to model a simulation with a reasonable amount of computing resources with some degrees of accuracy, the models for the wind tunnel was created as a

Figure 1. Full sphere 3D CAD model.

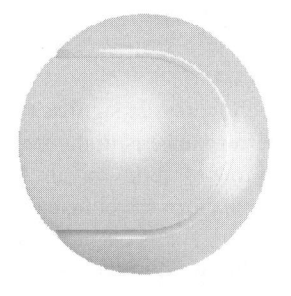

Figure 2. 3D CAD model of tennis ball with smaller seam dimension.

Figure 3. A real tennis ball.

Figure 4. 3D CAD model of tennis ball with bigger seam dimension.

Figure 5. Tennis ball with tetrahedral grid.

three dimensional sketch. A control volume was created to simulate the wind tunnel and the ball was placed in the control volume. The control volume (wind tunnel) can be scaled down to reduce the computational cost due to the fact that modelling the full scale wind tunnel with respect to the small size of the tennis ball will create the unused domain of the wind tunnel where the areas which are not important for the analysis will also be calculated if whole domain of wind tunnel is created. Therefore the reasonable size of domain will be considered to enhance the calculation speed and save the computational time and space. The sphere was used for a benchmark comparison.

A real tennis ball has a textured surface with a convoluted seam (see Figure 3). In this study only seam effects will be considered as the construction of the filament material (fuzz) of a tennis ball is difficult to construct in CAD and to mesh them. As mentioned earlier, two simplified tennis balls one with 2 mm seam width and 1.5 mm depth, and the other with seam width 5 mm and 1.5 mm depth. The RMIT Industrial Wind Tunnel (control volume) was modelled using GAMBIT thanks to its simple geometry. The dimensions of the wind tunnel are 9 m in length, 3 m wide and 2 m height. One of the reasons for modelling of RMIT Wind Tunnel was to compare experimental data that were acquired using this wind tunnel. More details about the RMIT Wind Tunnel can be found in Alam et al. (2003). Since the actual dimensions are too large compared size of a tennis ball (65 mm in diameter), the wind tunnel's dimensions were scaled down to save computational time and resources. However, special care was taken in scaling down as the fluid flow through the whole wind tunnel must be steady and not affecting the results on a tennis ball. That means the fluid flow near top and bottom at the wind tunnel must not disturb the flow on the region near the tennis ball as it can affect the accuracy of the whole calculation. The dimensions of the reduced scale wind tunnel are: 2 m long, 1 m wide and 1 m high.

The accuracy of a CFD solution is governed by the number of cells in a grid. Generally, larger the number of cells, better the results accuracy. In general the large number of cells equates to a better solution of accuracy, however it may require better computational power, time and resources. An optimal solution can be achieved by using fine mesh at a location where flow is very sensitive and relatively coarse mesh where airflow has little change. Tetrahedron mesh with mid-edges nodes was used in this study as mid-edges nodes can achieve better accuracy. Figure 5 shows a model of the tennis ball with the tetrahedron mesh. It can be seen that the tennis ball was meshed with considerably fine mesh in order to improve the accuracy of the results. Generally, the structured (rectangular) mesh is preferable to tetrahedron mesh as it gives more accurate results. However, there are some difficulties to use structured mesh in complex geometry. Therefore, in this study, all models were meshed with tetrahedron mesh.

The RMIT Industrial Wind Tunnel (control volume) was modelled using GAMBIT thanks to its simple geometry. The dimensions of the wind tunnel are 9 m in length, 3 m wide and 2 m height. One of the reasons for modelling of RMIT Wind Tunnel was to compare experimental data that were acquired using this wind tunnel. Since the actual dimensions are too large compared size of a tennis ball (65 mm in diameter), the wind tunnel's dimensions were scaled down to save computational time and resources. However, special care was taken in scaling down as the fluid flow through the whole wind tunnel must be steady and not affecting the results on a tennis ball. That means the fluid flow near top and bottom at the wind tunnel must not disturb the flow on the region near the tennis ball as it can affect the accuracy of the whole calculation. The dimensions of the reduced scale wind tunnel are: 2 m long, 1 m wide and 1 m high.

(a)   Seam Positions 1 and 2   (b)   Seam Positions 3 and 4

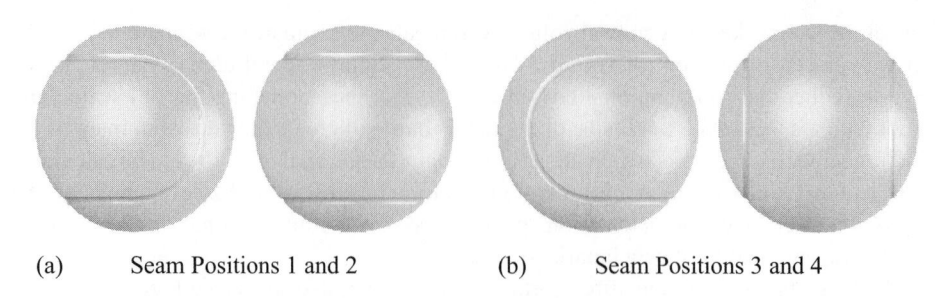

Figure 6.   Four different seam positions facing the wind.

The accuracy of a CFD solution is governed by the number of cells in a grid. Generally, larger the number of cells, better the results accuracy. In general the large number of cells equates to a better solution of accuracy, however it may require better computational power, time and resources. An optimal solution can be achieved by using fine mesh at a location where flow is very sensitive and relatively coarse mesh where airflow has little change. Tetrahedron mesh with mid-edges nodes was used in this study as mid-edges nodes can achieve better accuracy. Figure 5 shows a model of the tennis ball with the tetrahedron mesh. It can be seen that the tennis ball was meshed with considerably fine mesh in order to improve the accuracy of the results. Generally, the structured (rectangular) mesh is preferable to tetrahedron mesh as it gives more accurate results. However, there are some difficulties to use structured mesh in complex geometry. Therefore, in this study, all models were meshed with tetrahedron mesh. A total of 660,000 hybrid (fine) mesh cells were used for each model. To use fine mesh in the interested areas, sizing function in GAMBIT was used. Mesh validation was done using Examining Mesh command or "Check Volume Meshes" in GAMBIT. The four different seam positions facing the wind are shown in Figure 6. The same positions were used for experimental and computational analysis. The standard k-epsilon model with enhanced wall treatment was used CFD computational process. Other models were used to see the variation in solutions and results.

The experimental study of a series of tennis balls was conducted in RMIT Industrial Wind Tunnel (see Alam *et al.*, 2004). It is a closed test section, closed return circuit wind tunnel and is located in the School of Aerospace, Mechanical and Manufacturing Engineering. The maximum speed of the tunnel is approximately 150 km/h. The tunnel's test section is rectangular with the following dimension: 9 m long, 3 m wide and 2 m height. Tunnel's free stream turbulence intensity is approximately 1.8%.

3   Results and Discussion

Figure 7 illustrates the drag coefficient variations with speeds for various seam positions facing the wind. It is noted that there is little variation of Cd for seam positions. The Cd for a smooth sphere is also shown in Figure 7 which has similar trend as sphere with seam. However, the Cd (0.623) for sphere is slightly lower than the Cd for sphere with different seam positions (0.626, 0.630, 0.627, and 0.627 for seam position 1, 2, 3, 7 respectively). As no significant difference in Cd between simple sphere and sphere with seam was found, a tennis ball with larger seam (5 mm width) was also studied. Figure 8 illustrates the drag coefficient variations as a function of speed for two simplified tennis balls with 2 mm and

Drag Coefficient Variation with respect to speed (CFD)

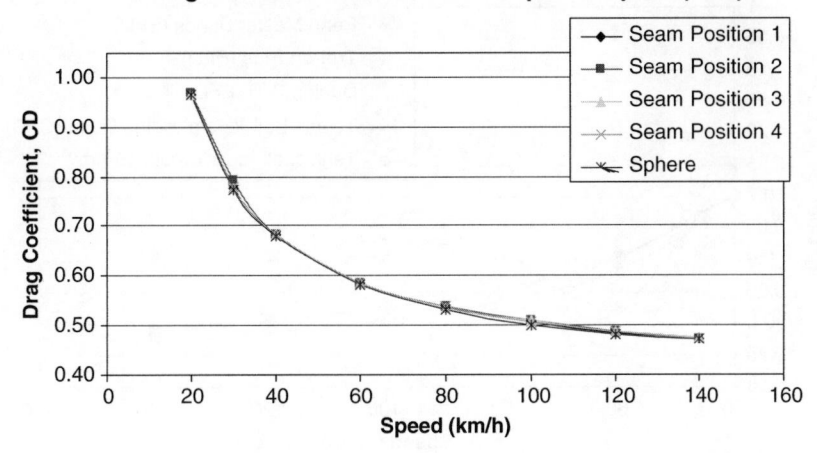

Figure 7. Drag variations with respect to the flow velocity.

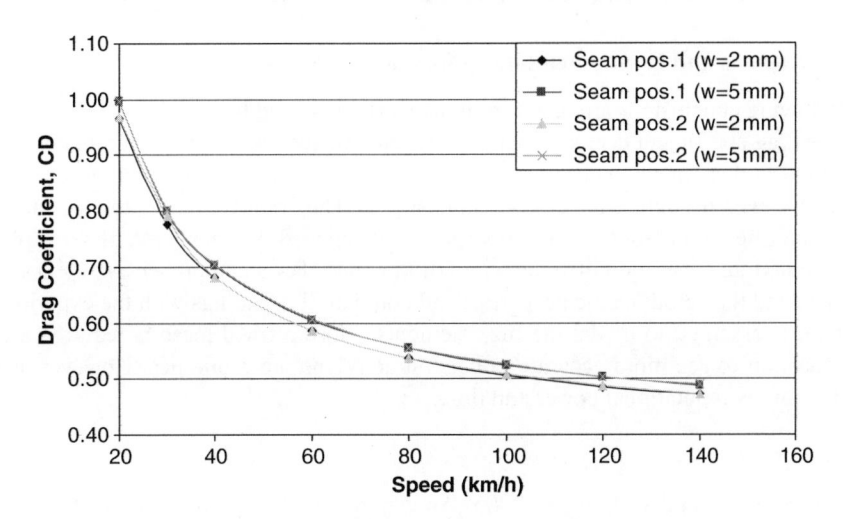

Figure 8. Drag variations with respect to different seam size and position.

5 mm seam width. The figure indicates a slight increase in drag coefficients of the tennis ball with 5 mm seam width compared to tennis ball with 2 mm seam width (0.647 and 0.649 for seam position 1 and 2 compared to 0.626 and 0.630 for the smaller seam ball). The larger seam increased approximately 3.6% drag coefficient compared to smaller seam. The CFD results compared with experimental data for a series of real tennis ball in Figure 9. As expected a significant variation in drag coefficients between CFD and EFD (experimental fluid dynamics) is noted. The tennis ball used in CFD analysis was simplified with no fuzz, whereas real balls were used in experimental (EFD) studies. However, a general trend for both CFD and EFD findings is observed. It may be mentioned that the fuzz of a tennis ball is difficult to generate in CAD. Even if the CAD model with fuzz is made successfully, meshing the model in FLUENT using GAMBIT is virtually impossible.

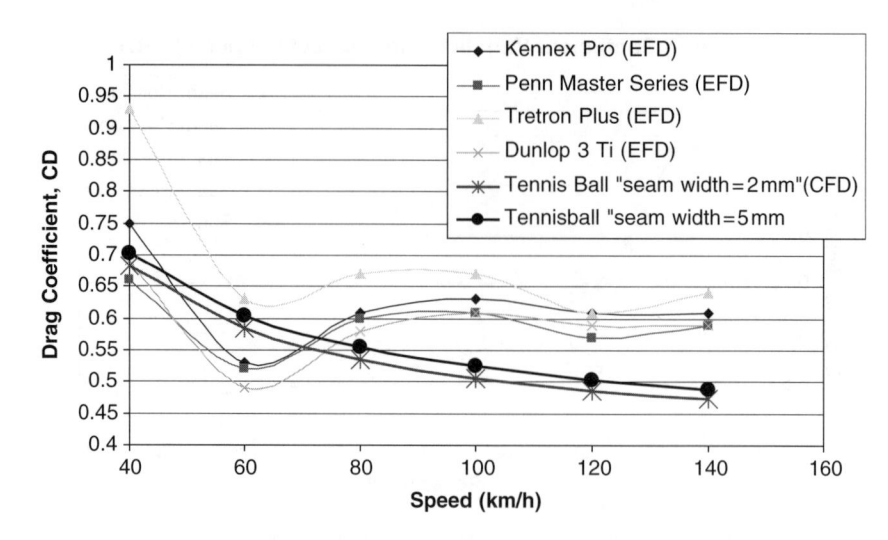

Figure 9. Drag Coefficient as a function of speed (CFD and EFD).

4 Conclusions and Recommendations for Further Work

The following conclusions are made from the work presented here:
 The drag coefficient decreases with an increase of speeds. However, it remains almost constant at higher speeds. The seam has some effects (generates additional drag) at lower speeds, however, no significant effect at high speeds. The orientation of seam does not have significant effect on aerodynamic properties. It is believed that the depth of seam plays a vital role in drag generation than the width of the seam. For a comprehensive CFD studies to understand the aerodynamic properties and compare the findings with the experimental results, it is essential to model the fuzz elements. More refined mesh is required around seam location to see more accurate effects on aerodynamic properties. However, it may require more computational power and time.

References

Alam F., Subic A. and Watkins S. (2005) An experimental study of spin effects on tennis ball aerodynamic properties. In: Subic A. and Ujihashi S. (Eds.) *The Impact of Technology on Sport*. ASTA (Australasian Sports Technology Alliance), Melbourne, Australia.

Alam F., Watkins S. and Subic A. (2004) The Aerodynamic Forces on a Series of Tennis Balls. *Proceedings of the 15th Australasian Fluid Mechanics Conference*, University of Sydney, 13–17 December, Sydney, Australia.

Alam F., Subic A. and Watkins S. (2004) Effects of Spin on Aerodynamic Properties of Tennis Balls. In: Hubbard M., Mehta R.D. and Pallis J.M. (Eds.) *The Engineering of Sport 5, Vol. 1* International Sports Engineering Association, Sheffield, UK, p 83–89.

Mehta R.D. and Pallis J.M. (2001) The aerodynamics of a tennis ball. *Sports Engineering*, 4/4, 1–13.

Chadwick S.G. and Haake S.J. (2000) The drag coefficient of tennis balls. In: Subic A. and Haake S. (Eds.) *The Engineering of Sport: Research, Development and Innovation*, pp 169–176, Blackwell Science, Oxford.

TESTING OF BADMINTON SHUTTLES WITH A PROTOTYPE LAUNCHER

J.C.C. TAN[1], S.K. FOONG[1], S. VELURI[2] & S. SACHDEVA[3]

[1]*National Institute of Education, Nanyang Technological University, Singapore*
[2]*Anglo-Chinese Junior College, Singapore*
[3]*Victoria Junior College, Singapore*

The aim of this study is to consider an alternative for testing of shuttles. This study seeks to provide a more objective test to the current subjective practice in (even the most prestigious) badminton tournaments. In this project, a prototype launching device was considered and built for testing of shuttle cocks in badminton. Using kinematic approach in motion analysis, the flight characteristics of the shuttles were quantified to test the projection capability of the launching device. The ability to discriminate the quality of shuttles was also compared with that of the traditional approach. The results showed the prototype launching device was capable of discriminating quality shuttles as categorized by the traditional approach. The launcher was also found to be consistent in projecting shuttles with similar flight characteristics.

1 Introduction

In most prestigious badminton tournaments, shuttles are tested before they can be deemed "fit" for use. According to the International Badminton Federation Laws (2005), "a shuttle is tested with a full underhand stroke, hit with an upward angle". This is also known as a serve stroke in badminton. The shuttle is projected "from one end of the back boundary line and the shuttle should land not less than 530 mm and not more than 990 mm short of the other back boundary line" (refer to Figure 1). This implies that an "accepted" shuttle should traverse a

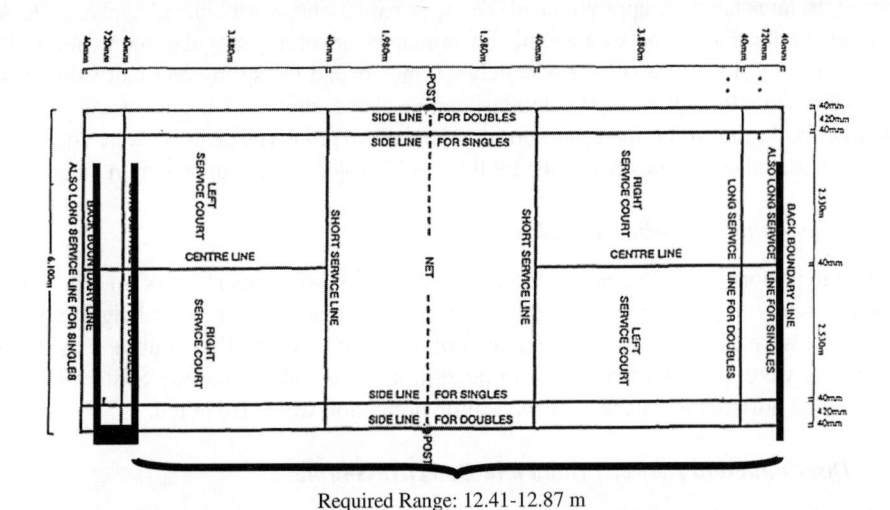

Figure 1. Requirement for shuttle testing (adapted from Badminto-Information.com, 2006).

horizontal distance between 12.41 and 12.87 m to be deemed fit for use. This approach is very subjective as it assumed that badminton players can project or strike shuttle with flawless consistency. The aim of this study is to examine the kinematics of the shuttle flight of the serve test and also attempt to suggest an objective alternative for testing of shuttles.

In order to proceed with the study, a device for launching shuttle must be conceived and developed. Lee & Srinivasan (2004) had developed a badminton shuttle launcher using the lever mechanism for examining the flight of plastic and feather shuttles. However their maximum projection distance was about 3 m. In conceiving the development of a shuttle testing launcher, the kinematics of badminton serve that is used in the current conventional testing had to be quantified so as to provide the performance target of the launching device or the launcher. The developed launcher needs to be tested for consistency in projections. In this study, two types of shuttles made of different materials (a conventional feather shuttle and a synthetic plastic shuttle) were used for this test. The developed launcher was finally assessed to ascertain if it could differentiate shuttles that were categorized according to the conventional testing.

2 Procedures & Methods

2.1. *Shuttle Kinematics Involved in the "Serve" Shuttle Testing*

Qualysis Proflex (a 3D opto-electronc kinematics analysis system) was used to determine the release velocity, height and angle of the shuttle in a serve test. The horizontal distance traveled was also measured. The results obtained (given in Section 3.1) from this experiment were then used to design a shuttle launcher.

2.2. *Launcher Design*

The "stretched spring release" mechanism was used in the launcher for projecting the shuttles. Based on the results of the kinematic study on the shuttle during the serve, the dimensions of the launcher were approximated. The approximations were obtained applying Hook's Laws in elastic energy of spring and the principles involving angular momentum. The required length of plank was approximately 2.0 meters and the spring constant value was a large value, close to 1500 N/m. The dimensions calculated suggested that the model would be extremely bulky and voluminous therefore a scaled down prototype launcher were considered instead. A scaled down launcher was built that could project shuttle up to 6 m in length.

2.3. *Consistency Test with Launcher*

Tests were performed on the launcher for consistency in projecting shuttles. In order to test for consistency in projection of shuttles, the release parameters (release height, velocity and angle) were examined with projections of a plastic and a feather shuttle. The release parameters were obtained from each projection using a Video Analysis System (Silicon Coach). The horizontal distances of the shuttle projection were also noted.

2.4. *Discrimination Between Good and Defective Shuttles*

The launcher was subsequently tested for its ability in differentiating shuttles that were deemed fit for competition. An ex-international badminton player tested 24 shuttles according to the

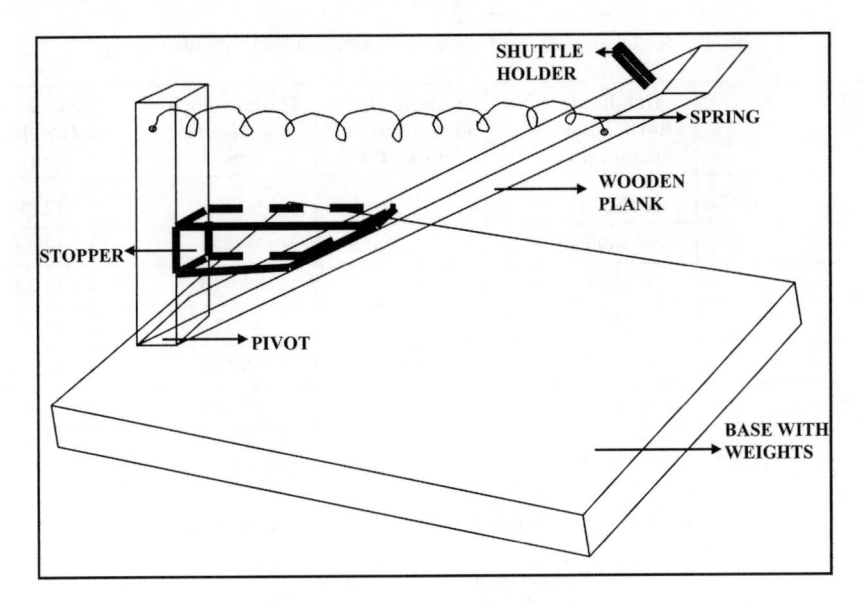

Figure 2. The proposed model of Launcher.

current International Rules of Badminton, and determined that 13 were fit (or "accepted") and 11 were defective (or "rejected"). These 24 shuttles were projected with the developed launcher. The projected distances were noted for each of the shuttle.

3 Results

3.1. Shuttle Kinematics Involved in the "Serve" Shuttle Testing

The release kinematics and the horizontal distance traversed by shuttle that is projected from a serve stroke are presented in Table 1. A total of five "serves" or trials were analysed using Qualysis Proflex. (Qualysis Proflex is a 3D opto-electronic kinematics analysis system).

3.2. Consistency Test with Launcher

A feather shuttle was launched six times. All the six launches were video recorded and analysed with Silicon Coach Video Analysis software programme. Using this programme, the shuttle's angle of release, initial velocity of the shuttle as well as the angular velocity of the plank of the launcher just after release were determined. The distance traveled by the shuttle for each trial was measured manually. This was repeated with a plastic shuttle. The results are tabulated in Table 2.

The small variations in the kinematic variables in Table 2 indicated that the launcher could launch shuttles at a certain degree of consistency. Table 3 compares the kinematics differences between a launched feather and plastic shuttle. The release parameters and the projected distances of feather and plastic shuttle are distinctively different. T-test was performed and found the flight distance between the feather and plastic were significantly

Table 1. Kinematics of shuttle release in serves used in shuttle testing.

Trial	Height of shuttle above ground (m)	Angle of projection of shuttle (°)	Initial velocity of shuttle (ms^{-1})	Distance traveled (m)
1	0.4226	31.30	50.395	12.30
2	0.5696	31.81	38.624	12.56
3	0.3863	38.29	48.920	12.42
4	0.4177	37.87	49.315	12.47
5	0.4172	38.15	36.524	12.37
Mean	0.4427	35.48	44.756	12.42

Table 2. Kinematics of shuttle released with launcher.

| Trial | Angle of projection | | Distance traveled (m) | | Release velocity (m/s) | | Angular Velocity (rad/s) | |
	Feather/Plastics		Feather/Plastic		Feather/Plastic		Feather/Plastic	
1	30°	28°	5.30	4.49	10.26	10.07	8.73	10.5
2	32°	33°	4.90	4.44	10.20	10.17	8.29	7.42
3	30°	30°	5.15	4.54	10.67	10.65	8.73	8.73
4	29°	37°	5.16	4.09	10.53	9.77	9.60	6.11
5	32°	27°	5.10	4.60	10.58	10.55	8.29	8.73
6	33°	32°	5.15	4.69	10.56	10.44	7.85	8.29
Mean	31°	30°	5.13	4.55	10.47	10.38	8.58	8.73

Table 3. Comparison of feather and plastic shuttles release.

Parameter	Feather	Plastic
Mean and standard error of shuttle released velocity shuttle (m/s)	10.47 ± 0.07	10.38 ± 0.10
Mean and standard error of distance traveled (m)	5.13 ± 0.05	4.55 ± 0.04
Mean and standard error of angular velocity of plank (rad/s)	8.59 ± 0.26	8.73 ± 0.45
Mean and standard error angle of release (°)	31.0 ± 0.6	30.0 ± 1.1

different (p < 0.01). This implies that the launcher is capable of differentiating the feather from the plastic shuttle.

3.3. Discrimination Between Good and Defective Shuttles

An ex-national badminton player tested 24 shuttlecocks using the conventional method in accordance to the rules of International Badminton Federation. Of the 24 shuttles, 13 were deemed fit or "accepted" for tournament use and 11 were "rejected". These shuttles were then projected with the launcher to ascertain if the launcher could distinguish "accepted" from the "rejected" shuttles.

4 Discussion

It was found that the manual task of testing shuttle by the current standard is to project the shuttle at velocity of about 44.7 m/s, at the angle of about 35.5° from the horizontal and at the height of 0.45 m above the ground. The ability of a badminton player to project shuttle at such release kinematics consistently is indeed questionable. A launcher was constructed to provide a more objective alternative to testing of shuttles. The constructed launcher was first tested for consistency in projecting shuttles. The results from the experiment of projecting a feather and a plastic shuttle indicated that the launcher was reasonably consistent. The standard errors for the measured kinematics were relatively small when compared to the mean value. The mean projected distance for feather shuttle was 5.13 m while the standard error of about 5 cm only is within 1% of the mean value. The plastic shuttle was projected at a mean distance of 4.55 m with a standard error of 4 cm. The differences in projected distances between the plastic and the feather shuttles were found to be significantly different (p < 0.01). In short, the developed shuttle launcher can differentiate plastic shuttles from the feather ones by the projection distances.

A subsequent test also indicated that the launcher could also distinguish the "accepted" shuttles from the "rejected" ones (categorized by the conventional serve test). The mean projected distance of the "accepted" shuttles was 4.93 m. It has a standard error of 2.3 cm. The mean projected distance of "rejected" shuttles was 4.79 m and it has a standard error of 3.2 cm. The projected distances of the two groups of shuttles were significantly different (p < 0.01). This implies that the launcher may be creditable device in testing shuttles for badminton tournament.

However, the results must be considered with caution as it must be noted that the test trials are small in number. In addition, the crude launcher could not project shuttles in excess of 12 m as the convention serve shuttle test requires. Even with these flaws, the study has indicated that it is possible to develop an objective alternative to testing of shuttles in badminton tournament. It would be ideal to have a launcher that could project shuttles in excess of 12 m. This would make the simulation of the serve in badminton more realistic. Clearly, the "stretched and release" spring mechanism in this prototype launcher may not be adequate in projecting shuttles to such distances. A different mechanism needs to be explored.

5 Conclusion

A simple "stretched and release spring" shuttle launcher was developed and its reliability in projection were tested. This launcher can also project feather shuttle at a mean horizontal distance of about 5.1 m with a standard error of 5 cm. From the projection distance, this launching device can differentiate feather from plastic shuttles and can distinguish conventionally tested "accepted" shuttles from "rejected" ones. The results implied that this prototype shuttle launcher can be a creditable alternative to testing shuttles for badminton competitions. However, the test trials were small in numbers and therefore the result must be treated with caution. A more robust and portable device could be developed so that it could ideally project shuttles beyond 12 m in order to pose a more serious alternative to the current subjective shuttle testing in badminton tournament.

References

Badminton-Information.com (2006) *Badminton Court.* (online) Available: http://www.badminton-information.com/badminton-court.html (Accessed: May 2007).

International Badminton Federation (2005) *The laws of badminton* (online). Available: http://www.badmintonengland.co.uk (Accessed: May 2007).

Lee T.C. and Srinivasan S. (2004) Trajectory of shuttle in badminton II. In: *Proceedings of Technology and Engineering Research Programme 2004*, Nanyang Technological University, Singapore, pp. 188–196.

TRAJECTORIES OF PLASTIC AND FEATHER SHUTTLECOCKS

S.K. FOONG[1] & J.C.C. TAN[2]

[1]*Natural Sciences and Science Education, National Institute of Education,
Nanyang Technological University, Singapore*
[2]*Physical Education and Sport Sciences, National Institute of Education,
Nanyang Technological University, Singapore*

In this study, the flight patterns of feather and plastic shuttles were compared. All the experiments were conducted in an enclosed hall where there was minimal air movement, comparable to that of an indoor badminton court. The shuttles were projected and their flight trajectories were captured on video using 200 Hz cameras. The captured motions of the shuttles were digitized and analyzed using the Peak Motus motion analysis system. The flight trajectories as well as the orientation of the shuttles were examined. The results indicate that there are differences in the orientation as well as flight trajectory of the two types of shuttles.

1 Introduction

The aim of this study is to compare the flight trajectories of feather and plastic shuttles. Initially, both plastic and feather shuttles were projected with a launcher and the horizontal flight distance of the shuttles were noted. Then the trajectories of the shuttles were obtained from video analysis and compared with the theoretical projectile motion of a point mass (not subjected to air-resistance). It was also compared with a simulated shuttle flight with air-resistance. Finally, the orientation angles of the shuttles throughout the flight were examined to provide insight into the trajectories of the shuttles. All the experiments were conducted in an enclosed hall where there was minimal air movement, comparable to that of an indoor badminton court. It is hoped that this study will add to the little that has been published on the technical information of shuttlecocks as observed by Cooke (2002).

2 Investigating Flight Distance of the Shuttles

A plastic shuttle and a feather shuttle were launched with almost the same initial velocity. This is achieved by the use of a launcher. The results as shown in Table 1 (Tan *et al.*, 2007) clearly show that the feather shuttle travels further than the plastic shuttle. In Section 3, we examine their trajectories after the turn over. Shuttles are noted to turn over about an axis normal to the flight after they have been struck so that the cock end of the shuttles will now lead the flight. In Section 4, the effects of the turn over on the shuttle flight were also examined.

3 Investigating the Trajectories of the Shuttles

3.1. *Methods*

In order to examine the trajectories of badminton shuttles, shuttles were projected across a distance of 4 m. The shuttles were projected and their flight trajectories were captured on

Table 1. Comparison of feather and plastic shuttles.

Parameter	Feather	Plastic
Mean and standard error of shuttle released velocity shuttle (m/s)	10.47 ± 0.07	10.38 ± 0.10
Mean and standard error of distance traveled (m)	5.13 ± 0.05	4.55 ± 0.04
Mean and standard error angle of release (°)	31.0 ± 0.6	30.0 ± 1.1

video using 200 Hz cameras. The captured motions of the shuttles were digitized and analyzed using the Peak Motus motion analysis system (1997). The initial velocities, angles and heights of projection of the shuttles were noted. Theoretical flight paths of point mass were also simulated and compared to provide an insight to how feather and plastic shuttles traverse through the medium of air in a normal indoor badminton courts environment. Four different shuttlecocks, 2 feather and 2 plastic, were studied.

3.2. Theoretical Flight Paths

a) Without considering the air-resistance, the trajectory of the projectile is a parabola given by, with u_x and u_y denoting the x- and y-components of the initial velocity,

$$y = \left(\frac{u_y}{u_x}\right) x - \frac{1}{2}\left(\frac{g}{u_x^2}\right) x^2. \tag{1}$$

b) The air-resistance R is assumed to be proportional to the square of its velocity, namely, $R = cv^2$ where c is assumed constant in this paper, even though there are evidence that it is velocity dependent (Foong et al., 2006) In this case the spatial and time coordinates of the trajectory are given by (Tan, 1987)

$$x(\theta) = x_0 - \frac{1}{g}\int_{\theta_0}^{\theta} v^2 d\theta, \quad y(\theta) = y_0 - \frac{1}{g}\int_{\theta_0}^{\theta} v^2 \tan\theta d\theta,$$

$$t(\theta) = -\frac{1}{g}\int_{\theta_0}^{\theta} v \sec\theta d\theta, \tag{2}$$

where θ is the angle between the direction of the velocity with respect to the horizontal and the velocity is given by

$$v(\theta) = v_T \sec\theta \left[\frac{1}{\alpha_0^2 \cos^2\theta_0} - \ln\left(\frac{\sec\theta + \tan\theta}{\sec\theta_0 + \tan\theta_0}\right) + \frac{\sin\theta}{\cos^2\theta} - \frac{\sin\theta_0}{\cos^2\theta_0}\right]^{-\frac{1}{2}}, \tag{3}$$

with $\alpha_0 = v_0/v_T$ denotes the ratio of the initial velocity to the terminal velocity $v_T = \sqrt{mg/c}$, and m = mass of shuttlecock and the subscript "0" denotes the initial value of the respective quantities which are required as input in order to plot the flight path.

Table 2. Terminal velocity and constant c of each shuttlecock.

Shuttlecock	Mass/g	Average terminal velocity (m/s)	$c=mg/v_T^2$ (g/m)
S1 (Feather)	5.379	6.60	1.21
S2 (Feather)	5.167	6.15	1.34
S3 (Plastic)	5.088	5.85	1.46
S4 (Plastic)	4.796	5.45	1.58

All the values are easily measured, except for terminal velocity. It appears that there is no known formula that relates the x and y coordinates of the shuttlecock directly as in case a).

3.3. Determination of the Terminal Velocity

The terminal velocity of a shuttlecock was determined by fitting its displacement-time formula to the data captured near the last 4 m of a vertical fall of 7 m (assuming an air drag of $R = cv^2$). The general displacement-time formula is given by

$$y = \frac{v_T^2}{g} \ln\left(\frac{1 + \beta_0 e^{2gt/v_T}}{1 + \beta_0}\right) - v_T t, \tag{4}$$

where v_T is the terminal velocity, v_0 is the initial velocity, and

$$\beta_0 = \frac{1 + \alpha_0}{1 - \alpha_0} \quad \text{and} \quad \alpha_0 = \frac{v_0}{v_T}. \tag{5}$$

Four trials for each shuttlecock were conducted and we obtained a consistent value for the terminal velocity of each shuttlecock. The average terminal velocities and the constant c are given in Table 2.

3.4. Observed Flight Paths

The flight paths of the 4 shuttlecocks, 2 feather and 2 plastic, were tracked by using the Peak Motus System. The observed results for the feather shuttlecock are plotted in Figure 1, while those of plastic shuttlecock are plotted in Figure 2, with the trajectories of (1) and (2) superimposed for comparison.

From the figures, it is evident that the observed trajectories of feather shuttlecock lie closer to the projectile motion compared to those of the plastic shuttlecocks. It is also clear that trajectories of plastic shuttlecock are in good agreement, especially for Shuttle 3, with the trajectories with drag as given by eq. (2).

The results so far are consistent with the finding in Section 2 that a feather shuttlecock flies further given the same initial velocity. We study the orientation of the shuttlecock in the next Section in our attempt to explain the difference in the trajectories.

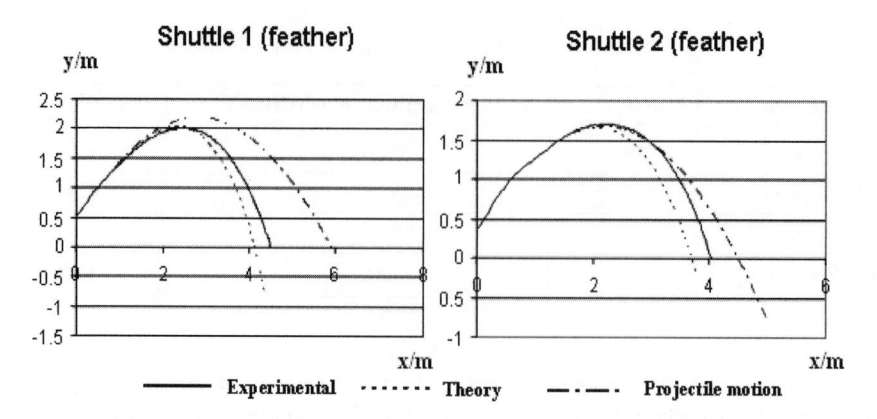

Figure 1. Comparing the observed (experimental) flight paths of feather shuttlecock with theory (eq.(2)) and projectile motion.

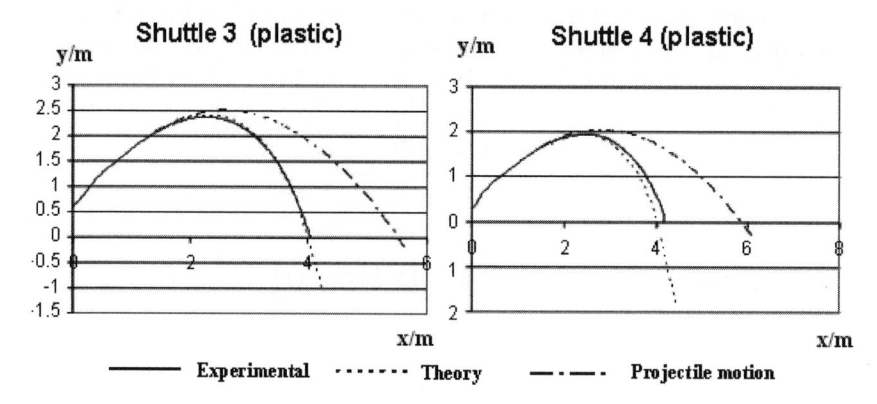

Figure 2. Comparing the observed (experimental) flight paths of plastic shuttlecock with theory (eq.(2)) and projectile motion.

4 Orientation of the Shuttlecocks

4.1. *Methods*

Three points on the images of the shuttles were digitized-the cork and the two skirt ends of the shuttle (Figure 3). A line was drawn through the midpoint of the two ends of the skirt and the centre of the cork. The line is the line of symmetry of the shuttle. The orientation angle is the angle between the line of symmetry and the horizontal, as shown in Figure 3, and is calculated via the gradient of the line.

Initially, the shuttles were projected from a hand's throw to eliminate the effect of the turnover. The trajectories were recorded on video. However, it was difficult to maintain the initial velocity and angle of release for different throws. This problem was reduced by selecting those trajectories with similar initial velocity and angle of release for further analysis. After some initial study, the shuttlecocks were launched by the use of a launcher (which is an earlier design of the one used in Section 2) in order to achieve very much the same initial

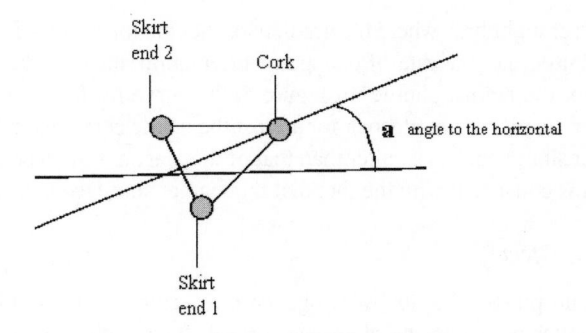

Figure 3. Definition of the angle to the horizontal

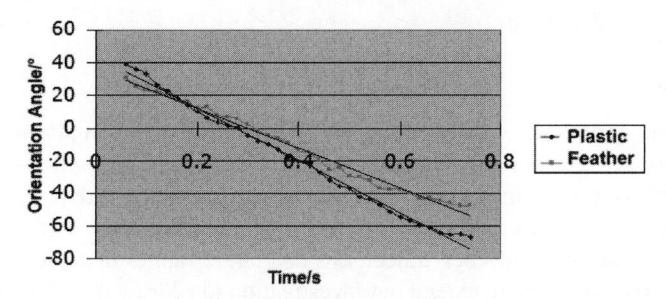

Figure 4. Orientation angle of feather and plastic shuttles plotted against time

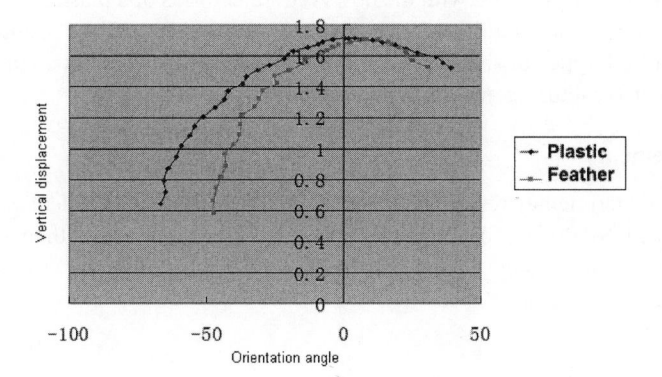

Figure 5. Feather shuttle has a smaller orientation angle than plastic shuttle. The orientation angles of the shuttles evolved from positive to negative during the flight.

velocity. However, shuttles projected this way still suffered from some degree of variation, thus only the flights with similar characteristics were selected for further analyses.

4.2. Results

4.2.1. Orientation Angle Throughout Flight

The data for the orientation angles of the feather and plastic shuttles throughout flight were obtained and plotted in Figures 4 and 5. In Figure 4, the data points for each shuttle lie approximately on a straight line, where the gradient of the line for the plastic shuttle is steeper than the feather shuttle, i.e. the rate of change of orientation angle of the plastic shuttle is greater than that for the feather shuttle. In Figure 5, the curve for the feather shuttlecock is below the curve for the plastic shuttlecock for almost the whole curve, namely the orientation angle of the feather shuttlecock is smaller than that of the plastic shuttlecock almost throughout the flight. This is consistent with the fact that the feather shuttlecock travels further.

4.2.2. Turn Over Effect

Both the feather and plastic shuttles were observed to turnover in a similar manner, with both suffered a similar reduction in velocity as a result. The feather shuttle had an average reduction of 32.6% (5.85 m/s to 3.94 m/s) while the plastic had an average reduction of 33.9% (4.91 m/s to 3.25 m/s).

5 Conclusions

Our preliminary results show that feather shuttle travels further than plastic shuttles over the range of projected distance of 4 to just over 5 m. This is consistent with another finding: a feather shuttle is less inclined to the horizontal when the shuttles are at equal heights. In Cooke (Cooke A., 2002), where the range of the projected distances is about 8 m, the opposite is reported: Plastic shuttlecock trajectories had larger ranges than the feather shuttlecock of up to 10%. We hope to extend our investigation to address these differences.

Our preliminary results also show that the trajectories with a velocity square dependence drag (Tan A., 1987) agrees with the observed trajectories of a plastic shuttlecock well, but not those of a feather shuttlecock. The pattern of turn over and the resulting reduction in velocity for both types of shuttles are similar. These results were obtained with the help of advanced video capturing and digitizing equipment.

Acknowledgements

Many of the experimental results reported here come from the student projects in the TERP-NTU and NRP-NTU programmes. We wish to acknowledge their contributions: Samuel Chong Meng Siong, Huang Jinhui, Lee Tat Chong, Shyam Srinivasan, Shimul Sachdeva and Shouri Veluri.

References

Cooke A. (2002) Computer simulation of shuttlecock trajectories. *Sports Engineering*, 5, 93–105.
Foong S.K., Tan J., Low H.S. and Ng W.L. (2006) unpublished 4th year student project on shuttle flight NIE, NTU.
Peak Performance Technologies Inc. (1997). Peak Motus Motion Measurement System – User Manual, Englewood, CO (USA)
Tan J., Foong S.K., Veluri S. and Sachdeva S. (2007) *Testing of Badminton Shuttles with a Prototype Launcher*. In: Fuss F.K., Subic A. and Ujihashi S. (Eds.) *The Impact of Technology on Sport II*. Taylor & Francis Group, London.
Tan A. (1987) Shuttlecock Trajectories in Badminton. *Mathematical Spectrum*, 19, 33–36.

SPORTS TECHNIQUE IMPARTATION METHOD USING EXOSKELETON

H.B. LIM, T. IVAN, K.H. HOON & K.H. LOW
School of Mechanical and Aerospace Engineering, Nanyang Technological University, Singapore

Every sport has its unique movement, for example, shooting in basketball, smashing in tennis and badminton, etc. The techniques used in executing these motions are different for various kinds of sports although they might share the same namesake. In order to master a sport, one must know and be able to execute most, if not all the techniques for that sport. In this paper, we focus on the smashing technique in badminton. To execute a proper smashing, proper timing and the right technique in smashing are crucial. Conventional method of learning a sport technique is coaching by an experienced player or coach. However, a beginner might find it hard to understand and learn all the necessary rotation, twisting and translation for upper limb for making a proper smashing in badminton. Typical training sessions for a beginner would involve watching numerous demonstrations, constant one-on-one coach-to-trainee guidance and corrections, and persistent practices. Imparting of the skills would therefore involve laborious personal coaching and guidance and endless practices under supervision. It is therefore almost impossible to impart to a beginner by demonstrations and guidance. This paper presents the novel method for coaching a beginner to execute a proper smashing action in badminton. Smashing actions of experienced players is recorded by a motion capture system. Analysis is carried out to translate the captured motion as recognizable input to the exoskeleton. An innovative idea for a device to train and guide the movement of amateur player is discussed in this paper. It aims to reduce the mistake of amateur player in smashing by restraining and/or guiding the motion of the player. The possibility of mapping the motion of experienced player into robot executable movement is discussed in this paper. A novel method to impart sports technique to beginner is explored. The method will greatly reduce the time required to train a player and increase the efficiency of training.

1 Introduction

Every sport has its own basic technique movement. In training of certain type of sport, the coach will be given a set of training in repeating the basic movement of that sport. For example: in badminton training, the coach will ask the player to do shadowing training in repeating all the technique movement of badminton such as backhand, smashing, etc., in basketball, the player will undergo the basic training in passing, shooting and dribbling. This entire training is in order to perfect the technique of the player and make the player to get used with the movement. In this type of training, it is hard for the beginner player to catch up as they are still new with the movement. Hence in learning this new skill, the coach will slowly guide the player by giving them the model of the movement and ask them to follow the movement accordingly. As each type of person has different type of perception for the movement, there is a need for the coach to correct the movement of each player (feedback). This process will take a lot of time and effort, as the coach has to concentrate to each of the player.

An exoskeleton is a hard outer structure, such as the shell of an insect or crustacean, which provides protection or support for an organism. In robotics, an exoskeleton is a device that is follows the structure of certain part of the body and its function to assist the movement of that part and also to protect it (Low, 2006). Several countries are currently developing the exoskeleton to serve the purpose in help the elderly and disabled people in their activity (Tsagarakis & Caldwell, 2007). Based on the characteristic of exoskeleton, we try to implement it on the sport training. The exoskeleton will assist the player in moving their body according to the correct movement.

2 Method

In this project, we will focus on the badminton and its smashing technique. The motion capture of the player is done by putting several markers on the body of the player following the location shown in Figure 1. The reference names for those positions are shown in Table 1. For capturing the motion, we are using 6 infra red cameras to cover the workspace (The area where the movement will be capture) and sampling rate of 100 Hz (each 0.01 second the data is taken). After the motions have been captured by the camera, the input will be processed in the computer. The output data consist of the XYZ-position of the marker in term of time.

3 Analysis

In this data analysis, we will concentrate in the right arm of the object, starting from the upper-arm, forehand and palm. The XYZ position of the each point of the object (Clavicle, Epi lateral, wrist and hand) will be used to obtain the angle rotation of each joint. First we will model the arm with each of the joint, where shoulder joint has 3 DOF, elbow joint has 2 DOF and wrist joint has 1 DOF. The model is shown in Figure 2.

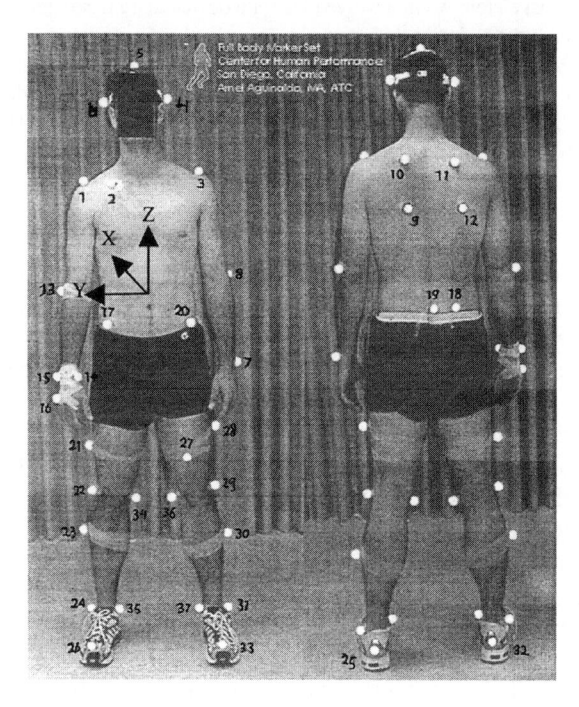

Figure 1. Full body markers set.

Using the DH parameter, we establish each coordinate system to the arm and tabulate the relationship as shown in Table 2.

where
θ_i = angle of rotation for respective system i
L_1 = the length of upper hand

Table 1. Reference name for the markers set.

1	Right Acronim	14	Right Wrist Rad	27	Left Quad
2	Right Clavicle	15	Right Wrist Uln	28	Left Thigh
3	Left Acronim	16	Right Hand	29	Left Knee
4	Left Head	17	Right ASIS	30	Left Shank
5	Top Head	18	Right PSIS	31	Left Ankle
6	Right Head	19	V. Sacral	32	Left Heel
7	Left Wrist Rad	20	Left ASIS	33	Left Toe
8	Left Epi Lat	21	Right Thigh	34	Right Knee Med
9	Left Scap Inf	22	Right Knee	35	Right Ankle Med
10	Left Scap Med	23	Right Shank	36	Left Knee Med
11	Right Scap Med	24	Right Ankle	37	Left Ankle Med
12	Right Scap Inf	25	Right Heel		
13	Right Epi Lat	26	Right Toe		

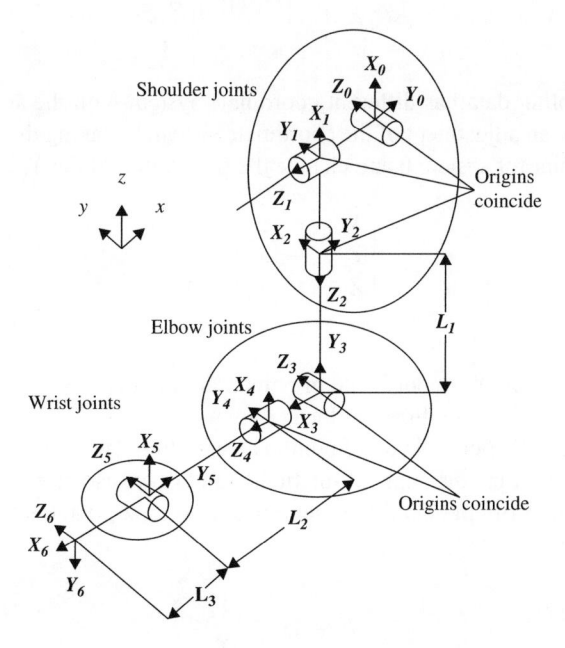

Figure 2. Analysis joints model of right arm.

Table 2. D-H Parameter link of the arm-model.

Joint i	Joint variables Θ_i (deg.)	α_i (deg.)	a_i (mm)	d_i (mm)
1	θ_1	90	0	0
2	θ_2	−90	0	0
3	θ_3	−90	0	L_1
4	θ_4	90	0	0
5	θ_5	−90	0	L_2
6	θ_6	0	L_3	0

L_2 = the length of forehand
L_3 = the length of palm
we can compute that the rotational matrix is:

$$\left[H_{(i-1,i)}\right] = \begin{bmatrix} \cos\theta_i & -\cos\alpha_i \sin\theta_i & \sin\alpha_i \sin\theta_i & a_i \cos\theta_i \\ \sin\theta_i & \cos\alpha_i \cos\theta_i & -\sin\alpha_i \cos\theta_i & a_i \sin\theta_i \\ 0 & \sin\alpha_i & \cos\alpha_i & d_i \\ 0 & 0 & 0 & 1 \end{bmatrix} \quad (1)$$

that satisfies the equation of:

$$\begin{bmatrix} X_{a,i-1} \\ Y_{a,i-1} \\ Z_{a,i-1} \\ 1 \end{bmatrix} = \left[H_{(i-1),i}\right] \begin{bmatrix} X_{a,i} \\ Y_{a,i} \\ Z_{a,i} \\ 1 \end{bmatrix} \quad (2)$$

As our sampling data has different coordinate system with the system 0 at our hand model, we will do an adjustment of the coordinate system by taking the acronym (x_0, y_0, z_0) as the $(0, 0, 0)$ point for system 0 and change the orientation of the X, Y and Z respectively.

$$\begin{bmatrix} X_{a,0} \\ Y_{a,0} \\ Z_{a,0} \end{bmatrix} = \begin{bmatrix} z_a \\ x_a \\ y_a \end{bmatrix} - \begin{bmatrix} z_0 \\ x_0 \\ y_0 \end{bmatrix} \quad (3)$$

Where:
$X_{a,i}, Y_{a,i}, Z_{a,i}$ = X, Y, Z coordinate of point a at system i
x_0, y_0, z_0 = X, Y, Z coordinate of the acronym
x_a, y_a, z_a = X, Y, Z coordinate of point a at sampling data
in order to determine the angle of rotation θ_1 and θ_2 we use inverse kinematics by substituting the position of epi lateral at system 0 $(X_{3,0}, Y_{3,0}, Z_{3,0})$ and system 3 $(0, 0, 0)$ into Eq. (2), we obtain:

$$\theta_1 = \tan^{-1} \frac{Y_{3,0}}{X_{3,0}} \quad (4)$$

$$\theta_2 = \cos^{-1}\frac{Z_{3,0}}{L_1} \qquad (5)$$

For determining the angle of rotation θ_3 and θ_4 we use inverse kinematics by substituting the position of the wrist at the system 2 ($X_{5,2}$, $Y_{5,2}$, $Z_{5,2}$) and system 5 (0, 0, 0) into Eq. (2), we obtain:

$$\theta_3 = \tan^{-1}\frac{Y_{5,2}}{X_{5,2}} \qquad (6)$$

$$\theta_4 = \cos^{-1}\frac{Z_{5,2} - L_1}{L_2} \qquad (7)$$

and the last for the angle of rotation θ_5 and θ_6 we use inverse kinematics and substituting the position of the palm at the system 4 ($X_{6,4}$, $Y_{6,4}$, $Z_{6,4}$) and system 6 (0, 0, 0) into Eq. (2), we obtain:

$$\theta_5 = \tan^{-1}\frac{Y_{6,4}}{X_{6,4}} \qquad (8)$$

$$\theta_6 = \sin^{-1}\frac{L_2 - Z_{6,4}}{L_3} \qquad (9)$$

The angle rotation of each joint can be calculated using Eq. (1)–(9). Sample plot for shoulder joint is shown in Figure 3.

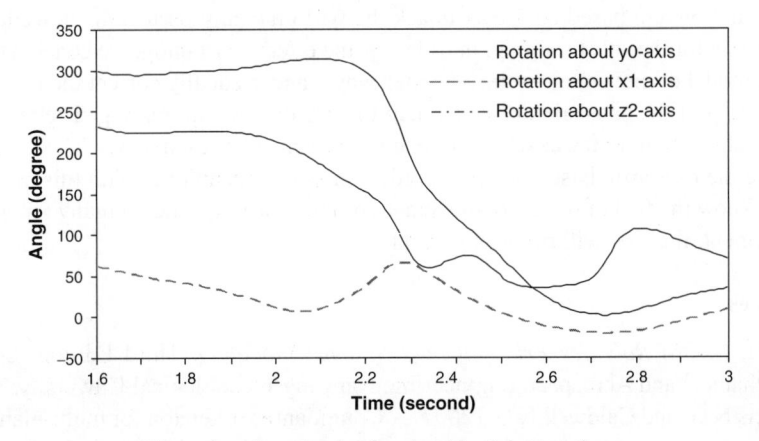

Figure 3. Sample plot for shoulder joint rotation.

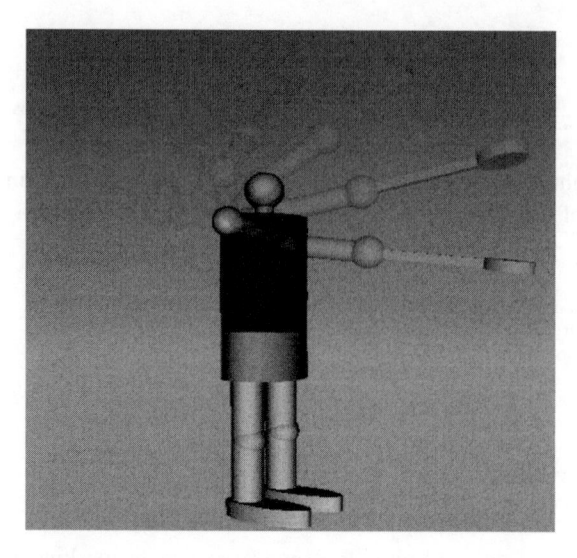

Figure 4. Simulation based on the converted data.

4 Simulation and Results

To simulate the movement of the player, we use the Simulink program. First we will construct the model of the hand with 7 –DOF. The model is constructed based on the arm model (Figure 2). By using the angles that have being calculated for each joint, the angles will be converted into Matlab file as the input for the actuator that model after an exoskeleton. The simulation is successful and it follows the motion that captured from the badminton player (see Figure 4).

5 Discussion

In learning a new skill, a person has to perform a closed loop task, where they will execute a certain motion and based on the feedback, he will gradually perfect his movement. It is also the same for the sport training, in order for the effective training, the coach has to give a sample model of movement to the beginner player and gradually correct their movement. With the help of the exoskeleton, it can help to save the time in training. The exoskeleton will be a substitution for the coach in guiding the movement of each player. This exoskeleton will move the user arm, based on the recorded smashing technique. With this method, the user will know the feel of the exact movement of the smashing, and by many repetitions of the movement; the user will master the technique.

References

Low K.H. (2006) *Robotics Principle and Systems Modeling*. Third Edition. School of Mechanical and Aerospace Engineering, Nanyang Technological University.
Tsagarakis N.G. and Caldwell N.G. (2007) A Compliant exoskeleton for multi-planar upper limb physiotherapy and training. Advanced Robotics. Invited Paper, (submitted).

BADMINTON SINGLES SIMULATION FROM THE DATA OBTAINED IN PHYSICAL EDUCATION CLASS

K. SUDA, T. NAGAYAMA, T. ARAI & M. NOZAWA

Graduate School of Decision Science and Technology, Tokyo Insutitute of Technology, Meguro, Tokyo, Japan

Game simulation is useful to make a training program. Coaches imagine matches before a tournament in their brain. This process is necessary when they plan for practice. In this study, we tried badminton singles simulation in physical education class. Students played singles matches and all the flight data were obtained. The court was divided into ten areas. Probabilties of players' selection of their flight from all the area were calculated. Badminton game simulation was done among the students. The players shots were determined with randomly picked number and the flight selection probability from the all the court areas. Twenty eight matches were played on computer. The results were compared with real matches. The results of twenty six matches were the same as the real matches. This suggests that the computer simulation of the badminton game in physical education class is successful.

1 Introduction

Game simulation is useful to make a training program. Coaches imagine matches before a tournament in their brain. This process is necessary when they make plans for practice. They may weigh on physical fitness or skill training depending on their strategy. This way is subjective and depending on the experience of the coaches. Game simulation is objective and does not depend on the experience but data. If this is possible, coaches' skill may increase.

In a big tournament, it is difficult to videotape several matches of a player because one of the players of a match who lose the match usually cannot play any more in the tournament. On the other hand, in physical education classes, it is possible to record many matches repeatedly from many students. This is good for badminton game simulation. Badminton doubles is more complicated than singles and the performance of a player depends on the combination of players, it is difficult to obtain data of personal characteristic of play from the players. So we tried badminton singles simulation in physical education class.

The objectives of this study are to simulate badminton singles matches using shuttle flight selection data of students attending physical education class.

Studies of ball game matches are mainly statistical ones such as study by Sampaio *et al.* (2006). And simulations of sports are used mainly for movement analysis (Vaughan, 1984). This study is concerning different aspect of ball game analysis.

2 Methods

Students played singles matches and the matches were videotaped with digital video camera. The data of the shuttle flight were obtained from these singles matches.

Each side of badminton court was divided into ten areas as shown in Figure 1. All flights data were recorded which included the information from and to which area shuttle

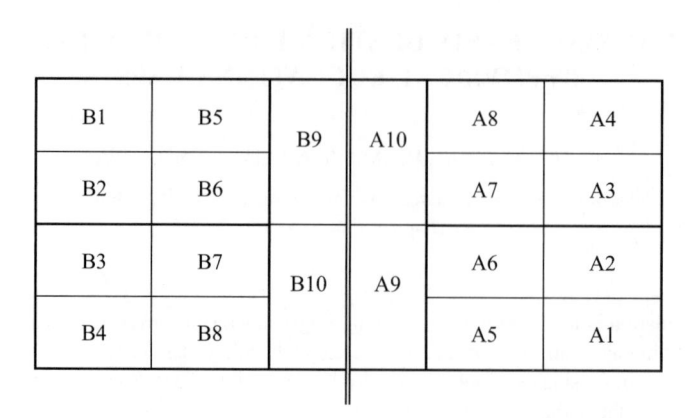

Figure 1. Division of badminton singles court.

Table 1. Example of flight selection probabilities. Probabilities were calculated for each area and each players.

From A1												
Destination of the flight (area)	B1	B2	B3	B4	B5	B6	B7	B8	B9	B10	Errors	Aces
Probability	0	0.05	0.05	0.2	0	0	0.1	0.2	0.1	0.2	0.05	0.05

flew and the errors. The positions of the players were determined with image processor (Video Tracker G260, DKH, Tokyo Japan). Errors of stroke and the forced error of the opponent were considered one of the flights. The probabilities of flight selection were calculated in each areas for all the players (Table 1). Using these data, badminton game simulation was done between the players. Players shot were determined with randomly picked number and the flight selection probability from all the court areas.

Recently badminton rules have changed to so-called rally point rule in which winner gets point whether or not the winner was the server. The old rule was used for this simulation in which the winner gets point if the player was the server of the rally. Players who got eleven points earlier than the opponent won the game because eleven point match was played in the class. Twenty-eight match simulations were played on computer. This game simulation was repeated ten thousand times for each match. We considered the winner whose winning probability was larger than 50%.

3 Results and Discussion

The winning probabilities of the players by the simulation and the results of real matches are shown in Table 2. The results of twenty-six simulation matches were the same as the results of real matches.

Although data used for the simulation is limited only flight selection, prediction performance was good. Table 3 shows the flight selection probabilities of all players from

Table 2. The results of simulation match and the real match. Numbers above the diagonal line is the winning percentage of the player in the left column by the simulation. The results of the real matches of the player in the left column are shown below the diagonal line. The open circles denote winning matches and the solid circles denote lost matches. The players in the top row are the opponents.

	A	B	C	D	E	F	G	H	I	J	K	L	M
A				88	55			77			65		
B					0			18	0				99
C						46	2	0					
D	●				9			1					
E	○*	○		○		55			5	100			
F			○○		●●				5				
G			○					41	25	100		84	
H	○*	○	○○	○			○				99		100
I		○			○	○	○			100			
J					●		●●		●				
K	●						●						
L						●	●						
M								●					

Table 3. Flight selection probabilities of all players from area A2.

	B1	B2	B3	B4	B5	B6	B7	B8	B9	B10	Error	Ace
A	0.07	0.23	0.21	0.02	0.00	0.07	0.09	0.02	0.02	0.05	0.12	0.09
B	0.00	0.00	0.06	0.06	0.11	0.17	0.22	0.00	0.00	0.11	0.22	0.06
C	0.00	0.00	0.00	0.00	0.00	0.40	0.00	0.00	0.20	0.00	0.40	0.00
D	0.17	0.17	0.00	0.00	0.17	0.00	0.33	0.00	0.00	0.00	0.17	0.00
E	0.02	0.17	0.07	0.00	0.05	0.17	0.14	0.07	0.02	0.02	0.15	0.14
F	0.13	0.11	0.03	0.02	0.00	0.08	0.20	0.06	0.00	0.03	0.22	0.13
G	0.00	0.33	0.33	0.00	0.00	0.00	0.00	0.00	0.00	0.33	0.00	0.00
H	0.04	0.12	0.08	0.12	0.00	0.17	0.13	0.04	0.04	0.04	0.17	0.06
I	0.00	0.00	0.33	0.00	0.00	0.17	0.00	0.00	0.17	0.17	0.00	0.17
J	0.00	0.00	0.00	0.00	0.00	0.20	0.07	0.00	0.20	0.07	0.40	0.07
K	0.02	0.02	0.05	0.05	0.02	0.15	0.15	0.00	0.07	0.15	0.17	0.15
L	0.00	0.04	0.04	0.00	0.00	0.37	0.11	0.04	0.11	0.04	0.22	0.04
M	0.00	0.00	0.19	0.00	0.00	0.44	0.13	0.00	0.00	0.00	0.25	0.00

area A2. In this study, players' skill levels were diverse because the players were students who attended the physical education class and some had played badminton for several years and others had played only several times. So the difference of fight selection probabilities of two players was large as shown in Table 3. Length of flight that the players could make was different from player to player. For example the some players could deliver shuttles to all the backward area (B1, B2, B3, B4) from A2. But other players could not deliver shuttle to these areas (Table 3). That seems the main reason why the prediction performance was good, even though only flight selection data were used. If the players' skill levels are higher, much more factors should be taken into consideration to do good simulation because ability to make flight from a certain area must be much more similar compared to the present study.

We divided the court into only ten areas for the convenience of identification of the players' location. Although the smaller division of the court may be useful for detailed description of flight selection, the numbers of the flight data from a certain area must decrease. Thus, more matches are necessary for obtaining data.

In this study, time of hitting was not recorded. If it is taken into consideration, players' movement speed, shuttle speed can be calculated and more precise simulation may be possible.

Players who could deliver shuttle to various areas were stronger. This means that ability to deliver shuttlecock to various areas from all the court area is very important for good badminton performance in physical education class.

References

Sampaio J., Ibanez S., Lorenzo A. and Gomez M. (2006) Discriminative game-related statistics between basketball starters and nonstarters when related to team quality and game outcome, *Percept. Mot. Skill*, 103/2, 486–94.

Vaughan C.L. (1984) Computer simulation of human motion in sports biomechanics. *Exerc Sport Sci Rev.* 12, 272–416.

12. Ball sport – Basketball, Bowling, and Hockey

A COMPARISON BETWEEN BANK AND DIRECT BASKETBALL FIELD SHOTS USING A DYNAMIC MODEL

H. OKUBO[1] & M. HUBBARD[2]

[1]*Chiba Institute of Technology, Narashino, Japan*
[2]*University of California, Davis, CA, USA*

A dynamic model is used to analyze the release conditions of direct and bank shots in basketball. The dynamic model contains six sub-models: ball-rim, ball-bridge, ball-board, ball-bridge-board, and ball-rim-board, and gravitational flight with air drag, and calculates ball deflection, contact time and strange motions such as long rolling on the rim. Each ball-contact sub-model has possible slipping and non-slipping interactions at the contact point. Margins in release angle and velocity are used to characterize robustness of capture to perturbations in these variables. Based on the margins we compare direct and bank shots as a function of the release position on the court.

1 Introduction

Basketball field shots are of two types. A direct shot is usually attempted in free throws, three point shots, and jump shots near the end line. However, players use bank shots near the hoop and around 45 degrees from the backboard surface.

Some previous studies have analyzed release conditions for bank and direct shots using simulation models. Pure flight models (Brancazio, 1981; Hay, 1993) limit themselves to swish shots that touch neither the rim nor backboard. Algebraic impact models (Shibukawa, 1975; Hamilton & Reinschmidt, 1997; Huston & Grau, 2003) are based on impulse-momentum principles. More complicated dynamic models (Okubo & Hubbard, 2002; 2003; 2004a; 2004b; 2005; 2006a; 2006b) can follow the ball through an arbitrary number of bounces anywhere on the rim or backboard and during longer rim contact periods. They are able to calculate ball deflections, contact times and strange motions such as long rolling on the rim.

In general direct and bank shots, the basketball may contact the rim, board, bridge, bridge and board, and rim and board simultaneously with slipping and non-slipping interactions. Non-linear ordinary differential equations with six degrees of freedom describe the ball angular velocity and ball center position. We calculate successful release conditions and discuss the relative advantages of bank and direct shots from a variety of release positions.

2 Dynamic Model

2.1. *Overall Model*

Our overall model for basketball shots has six distinct sub-models; gravitational flight with air drag (coefficient of air drag $C_D = 0.5$), and ball-rim (Okubo & Hubbard, 2002; 2004a), ball-board (Okubo & Hubbard, 2003), ball-bridge (Okubo & Hubbard, 2004b; 2005; 2006b), ball-bridge-board (Okubo & Hubbard, 2004b; 2005; 2006b) and ball-rim-board

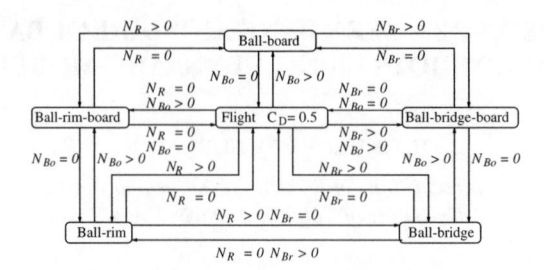

Figure 1. Overall model.

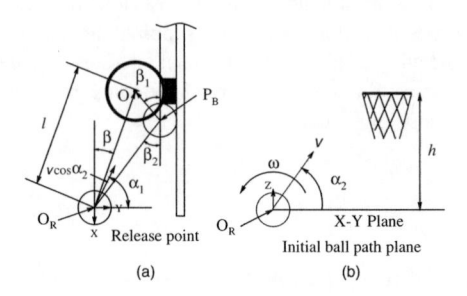

Figure 2. Geometry of general field shots. (a) top and (b) side views.

(Okubo & Hubbard, 2006a) contacts as shown in Fig. 1. We switch between the sub-models depending on the reaction forces at the contact point. Each contact sub-model has possible slipping and non-slipping motions. Switching between slipping and non-slipping is based on the contact velocity and friction forces.

2.2. Geometry of Field Shots

A right-handed coordinate system (Fig. 2) with the origin O at the center of the hoop has its X axis parallel to the backboard, Y axis perpendicular to the board and Z axis vertical. The release position O_R is ($l\cos \beta$, $-l \sin \beta$, $-h$) where β is the floor angle around the hoop center between the XY projection of line OO_R and the XZ plane, and l and h are horizontal and vertical distances from the hoop center, respectively. The release velocity is characterized by four input parameters; velocity v, angular velocity ω, lateral angle from the YZ plane α_1, and release angle α_2. In general, shots are attempted with $\alpha_1 = 90 - \beta$ deg for direct shots, and with $\alpha_1 = 90 - \beta_2$ deg and $\beta_1 \approx \beta_2$ for bank shots, where β_2 is the projected angle of incidence, β_1 is the angle between the board surface and the XY projection of line OP_B, and P_B is the location of the ball center when the ball first contacts the board. If the ball were modeled as a frictionless particle, $\beta_1 = \beta_2$ would be the projected angle of reflection. We assume, however, that the ball has non-zero rotational inertia, radial compliance, and friction, and possibly slips at the contact point; in general, therefore, the angle of reflection $\beta_1 \neq \beta_2$.

3 Numerical Simulation

3.1. Simulation Parameters

In the simulations described below we choose the parameters shown in Table 1: m, I, and R_b are the basketball mass, central mass moment of inertia, and undeflected ball radius, respectively; k and c are equivalent stiffness and damping coefficients; R_h and R_r are major and minor radii of the toroidal rim; and μ and μ_{Bo} are the coefficients of friction between the ball and rim/bridge, and board, respectively.

3.2. Capture Combinations for Angled Shots

Capture combinations in the release velocity and angle space for direct and bank shots are plotted as dark regions in Figs. 3(a) and (b), for which the initial pure backspin angular velocity is 4π rad/s, and the release position is $(3\cos 33.5, -3\sin 33.5, -0.2)$ m. The release angular velocity vector is assumed to be perpendicular to the vertical plane of the initial ball path. The lateral angles α_1 for the direct and bank shots are near 56.5 and 49.2 deg, respectively.

Shots first hit the front of the front rim in region I in Fig. 3(a). Region II is the largest for direct shots, where shots first touch inside the rim or swish. Shots first contact the back rim in region III. In the bank shots, the ball first contacts the board with $\beta_1 \approx \beta_2$. Typical shots in region IV are successful bank shots. Both direct and bank shots have one large capture region as shown in Figs. 3(a) and (b). The possibility of lucky shots with multiple bounces is very small. To be captured the bank shots need almost the same release angle as, and a slightly larger release velocity than, successful direct shots.

We define four margin parameters M_{vd}, M_{ad}, M_{vb}, and M_{ab} (Figs. 3(a) and (b)) that characterize the width and depth of the largest capture regions for direct and bank shots,

Table 1. Numerical values of simulation parameters.

m kg	R_b m	I kgm^2	k KN/m	C Ns/m	R_h m	R_r m	μ	μ_{Bo}
0.6	0.12	0.0057	47	19	0.234	0.009	0.5	0.7

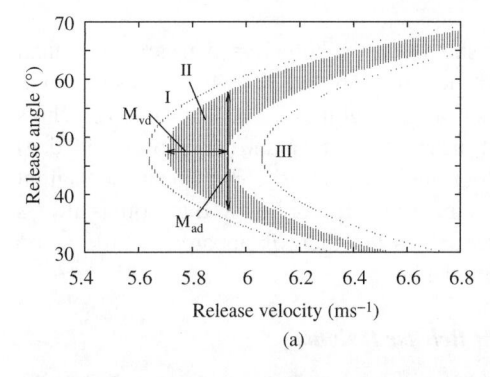

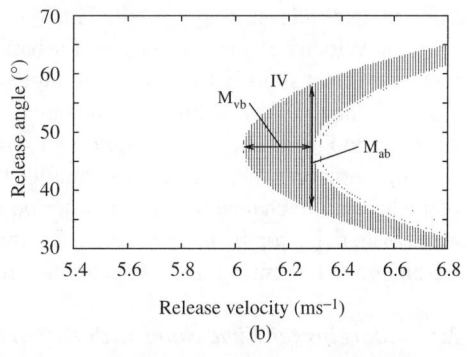

Figure 3. Capture combinations with $\beta = 33.5$ deg, $l = 3$ m, $h = 0.2$ m and $\omega = 4\pi$ rad/s for (a) direct shots with $\alpha_1 = 6.5$ deg, (b) bank shots with $\alpha_1 = 49.2$ deg.

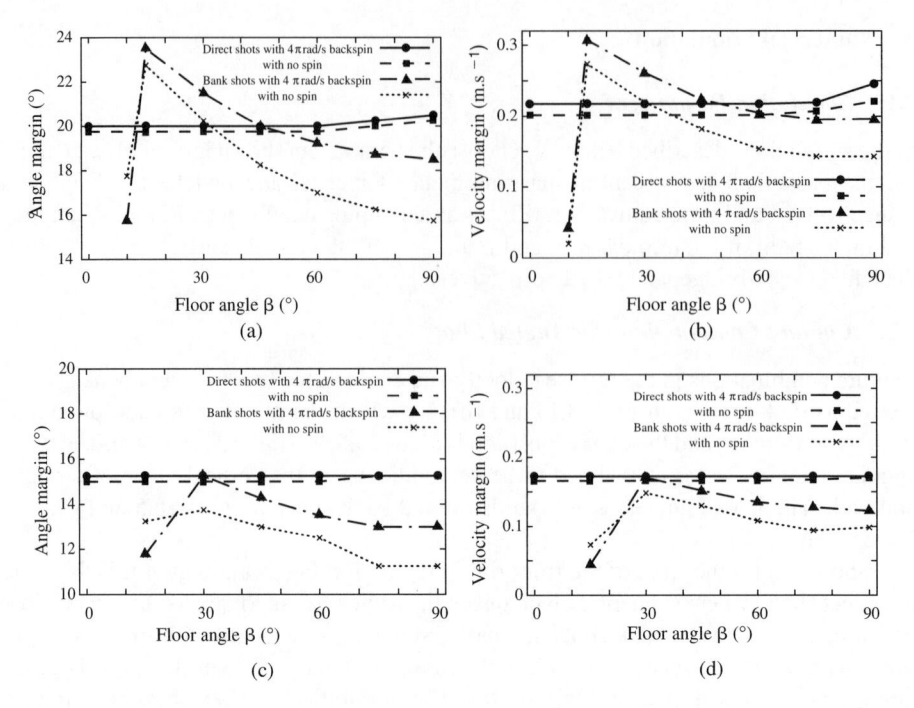

Figure 4. Angle and velocity margins as function of floor angle β for direct and bank shots with $h = 0.2$ m and (a) (b) $l = 3$ m, (c) (d) $l = 6$ m.

respectively. M_{vd} and M_{ad} are the direct shot velocity and angle margins at the minimum velocity on the right boundary of region II. M_{vb} and M_{ab} are similar margins for bank shots in region IV.

3.2. Margins for Field Shots from Different Court Positions

The margins of release angle and velocity are calculated for direct and bank shots with no backspin and a typical backspin, from release positions with different floor angles β but the same vertical release height $h = 0.2$ m.

The velocity and angle margins for bank shots near the hoop ($l = 3$ m) are larger than those for direct shots with small floor angles (β less than around 30 to 40 deg) as shown in Fig. 4(a) and (b). Direct shots have advantages near the sagittal YZ plane or for longer shots as shown in Fig. 4. The margins for direct shots near the sagittal plane (roughly $\beta = 90$ deg) are slightly larger than those with smaller floor angle because the shots near the sagittal plane have more chance for capture after bouncing off the rim or board. Backspin is always more beneficial for bank shots than for direct shots but there is apparently little or no advantage of backspin for direct shots far from the hoop.

3.3. Margins for Field Shots with Different Release Heights

Figures 5(a)–(d) show release angle and velocity margins as functions of release height for direct and bank shots with the floor angle $\beta = 30$ deg and for no backspin and a typical

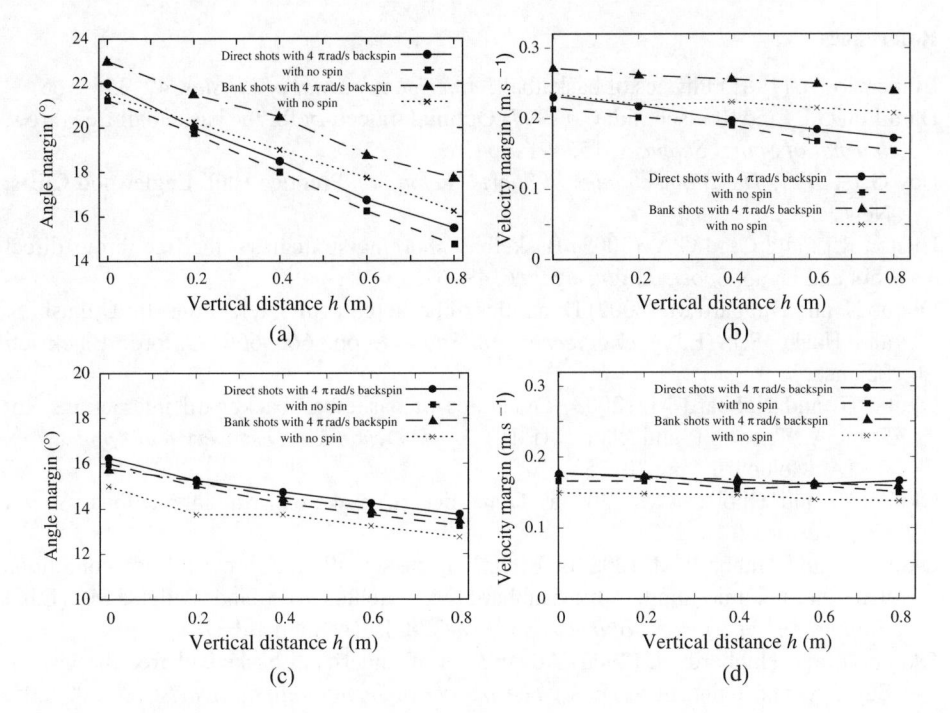

Figure 5. Angle and velocity margins as functions of vertical distance h between release height and hoop center for direct and bank shots with floor angle $\beta = 30$ deg and (a) (b) $l = 3$ m, (c) (d) $l = 6$ m.

backspin. Both types of shots near the hoop are significantly more successful with a higher release position (Figs. 5a and b). Especially of note is that a higher release point is more effective for bank shots. However shots farther from the hoop with a higher release point apparently have little or no advantage (Figs. 5c and d). Thus taller players should try bank shots near the hoop. The effect of backspin is almost the same in longer bank and direct shots with different release heights (Fig. 5d).

4 Conclusions

We have used our dynamic model to compare the advantages of direct and bank field shots. Calculations using the model show that direct shots are advantageous when the initial ball position is in or near the sagittal plane or far from the hoop. On the other hand, bank shots become useful near the hoop from directions around 45 deg from the board surface. Shots with a higher release point are advantageous near the hoop. However, for longer shots far from the hoop, higher shots have little or no advantage over shots with a lower release point. Backspin generally helps capture for both direct and bank shots, and especially for bank shots near the hoop. Although here we have used individual velocity and release angle margins as measures of shot effectiveness, the product of the two margins would probably be a better measure of capture probability.

References

Brancazio P.J. (1981) Physics of basketball, *American Journal of Physics*, 49, 356–365.

Hamilton G.R. and Reinschmidt C. (1997) Optimal trajectory for the basketball free throw, *Journal of Sports Sciences*, 15, 491–504.

Hay G.J. (1993) *The Biomechanics of Sports Technique*. Prentice Hall. Englewood Cliffs, New Jersey.

Huston R.L. and Grau C.A. (2003) Basketball shooting strategies – the free throw, direct shot and layup, *Sports Engineering*, 6, 49–63.

Okubo H. and Hubbard M. (2002) Dynamics of basketball-rim interactions. In: Ujihashi S. and Haake S.J. (Eds.) *Engineering of Sport 4*, pp. 660–666. Oxford: Blackwell Science.

Okubo H. and Hubbard M. (2003) Dynamics of basketball-backboard interactions. In: Subic A., Trivailo P. and Alam F. (Eds.) *Sports Dynamics: Discovery and Application*. ASTA, Melbourne, pp. 30–35.

Okubo H. and Hubbard M. (2004a) Dynamics of basketball-rim interactions, *Sports Engineering*, 7, 15–29.

Okubo H. and Hubbard M. (2004b) Effects of basketball free throw release conditions using a dynamic model. In: Hubbard M., Mehta R.D., and Pallis J.M. (Eds.) *Engineering of Sport 5, Volume 1*. pp. 372–378. ISEA Sheffield, UK.

Okubo H. and Hubbard M. (2005) Advantages of underhand basketball free throws. In: Subic A. and Ujihashi S. (Eds.) *The Impact of Technology on Sports*: pp. 200–205. ASTA, Melbourne.

Okubo H. and Hubbard M. (2006a) Strategies for bank shots and direct shots in basketball. In: Moritz E.F. and Haake, S. (Eds.) *Engineering of Sport 6, Volume 3*. pp. 243–248. Springer, New York.

Okubo H. and Hubbard M. (2006b) Dynamics of the basketball shot with application to the free throw, *Journal of Sports Sciences*, 24, 1304–1314.

Shibukawa K. (1975) Velocity conditions of basketball shooting, *Bulletin of the Institute of Sport Science, The Faculty of Physical Education*, 13, 59–64. (In Japanese)

Silverberg L. Tran C. and Adcock K. (2003) Numerical analysis of the basketball shot, *ASME Journal of Dynamic Systems, Measurement, and Control*, 125, 531–540.

PERFORMANCE ANALYSIS WITH AN INSTRUMENTED BOWLING BALL

L.S.A. KHANG & F K. FUSS

Sports Engineering Research Team, Division of Bioengineering, School of Chemical and Biomedical Engineering, Nanyang Technological University, Singapore

Bowling, a typical skill sport, requires exact motion timing and control of finger forces. The latter, the reaction forces between the three fingers and the ball, were individually measured with an instrumented bowling ball in bowlers with different performance levels and in different shots. The results showed, that, in better bowlers, the duration of the forward swing is relatively higher, the peak forces are relatively larger, and the impulse is relatively higher during the forward swing, with a good correlation of $r > 0.7$. The larger forces are attributed to the relatively stronger grip during the forward swing in better bowlers.

1 Introduction

Bowling is a typical skill sport. During the approach, the movement timing as well as the control of finger and thumb forces is decisive. Both movement and finger forces affect the kinematics of the ball, which is classified by three different types of shots:

1. the "straight ball" is used mostly by beginners and is done simply by throwing the ball in a straight line down the lane.
2. the "hook ball" is throwing the ball in a manner in which it moves in a smooth arcing motion down the lane. This allows the ball to enter the pocket at an angle, which helps to create more "pin action."
3. the "spin ball" is such that the ball is released to create spin on its vertical axis. The bowler attempts to utilise the deflection of the bowling ball off the head pin, subsequently, running the ball down the side of the deck so that the spin mixes up the pins to carry the strike.

For the measurement of finger reaction forces during bowling, Fuss *et al.* (2006) developed an instrumented bowling ball and tested it in one subject. The instrumented bowling ball consists of three 6DOF transducers (3 forces, 3 moments), connecting the ball to the finger and thumb holes, which were replaced by aluminium tubes (Figure 1).

In the first prototype of the instrumented bowling ball, the transducers were connected to cables, however, during release of the ball, the finger reaction forces drop to zero and thus there is no need for the ball to roll properly.

The aim of this study was to test the suitability of the instrumented bowling ball for performance analysis and to measure the finger reaction forces in bowlers of different performance levels during different shots.

2 Experimental

All experiments were carried out with the instrumented bowling ball developed by Fuss *et al.* (2006). The transducers, inserted into a commercially available bowling ball (Columbia

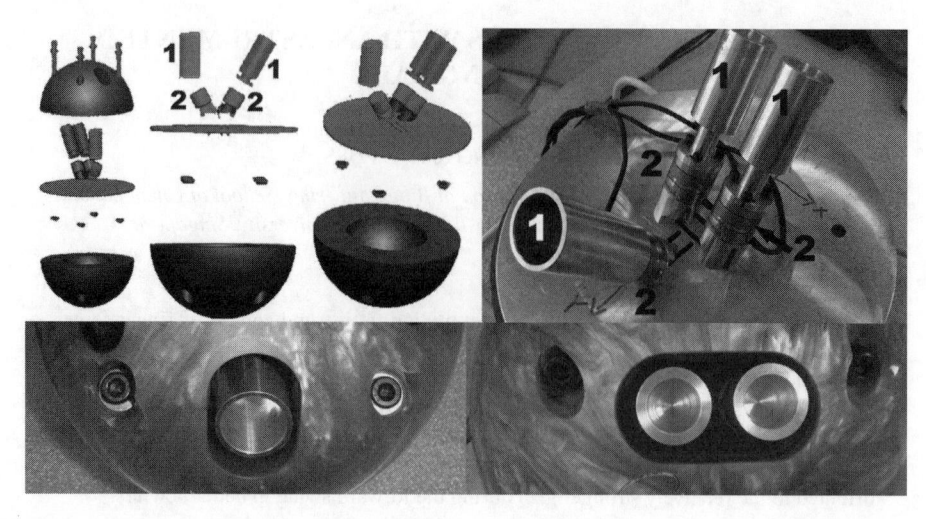

Figure 1. Prototype of the instrumented bowling ball, design, construction, and assembled ball; 1: finger and thumb holes, 2: transducers

300 Blue Dot), are 6-DOF silicon strain-gauge sensors (Nano25, ATI Industrial Automation, Apex, NC, USA). The force and moment data was recorded at 1 kHz and collected with LabView (National Instruments, USA). Nine bowlers of different performance participated in the experiments (Table 1). Each bowler performed the shots 4 times to assess repeatability and consistency.

The forces were analysed as to 3 components (x, y, z) with respect to time, resultant finger force with respect to time, and overall force applied to the ball. From the moment equilibrium, we calculated the origin (centre of pressure, COP) of the resultant finger force at the inner surface of the tube. Additionally, we determined the overall ball moment produced by the fingers about an axis perpendicular to the frame plate of the transducers. This served to quantify the twisting moment for spin shots. Finally, the force vector diagram was generated and visualised in AutoCAD 2000 (Autodesk, USA) by combining COP and resultant force, whereby force vectors applied by the ball to fingers and thumb, were displayed on the ball (comparable to a Pedotti diagram on a force plate).

The transition from backswing to forward swing was defined by the intersection points the gradients of the force-time graphs before and after dead centre. For each phase (back and forward swing) the duration was determined, and the peak forces and impulses of all 3 fingers summed up. Based on this data, the percentage of the following parameters was calculated for the forward swing (total sum of the parameters of back and forward swing = 100%): sum of peak forces of all 3 fingers, sum of impulses of all 3 fingers, time (duration) of forward swing. This data was correlated with the average score (listed in Table 1).

3 Results

The force-time graphs of all 3 fingers are shown in Figure 1. Generally, the better the bowler is, the longer is the relative duration of the forward swing (figure 3), the higher is the impulse (Figure 4) and the larger are the finger forces of the forward swing (Figure 5).

Table 1. Bowlers and their experience (IVP: Institute-Varsity-Polytechnic, JC: Junior College).

No.	Name of Bowler	Age	Types of shot specialised	Experience	Average score	High score
	Straight shot					
1	Expert 1	23	Hook shot	National bowler	198	299
2	Expert 2	23	Hook shot	Combined School bowler	180	287
3	Beginner 1	25	Straight shot	bowls seldom	90	140
4	Beginner 2	25	Straight shot	bowls occasionally	120	175
	Spin shot					
5	Expert 1	23	Spin shot	IVP bowler	193	290
6	Expert 2	24	Spin shot	IVP bowler	180	265
7	Average bowler 1	21	Spin shot	JC school bowler	160	255
8	Beginner 2	25	Straight shot	bowls occasionally	120	170
	Hook Shot					
1	Expert 1	23	Hook shot	National Bowler	198	299
2	Expert 2	23	Hook shot	Combined School bowler	180	287
8	Average bowler 1	21	Spin shot	JC school bowler	160	255
9	Average bowler 2	21	Hook shot	JC school bowler	150	203

The coefficients of correlation are listed in Table 2. Of the individual shots, the straight shot shows a high correlation ($r > 0.75$) with all 3 parameters, the hook shot with force, and the spin shot with time and impulse.

Figure 6 shows the force vector diagrams of different shots and performance levels. A clear difference between beginners and experts can be seen in the final force spike, which produces the moment about the sagittal axis, causing the ball to hook once the coefficient of friction increases on the lane. This force spike, produced by middle and ring fingers (arrow in Figure 2) is reflected in the vector diagram as well, causing the vectors to fan out (arrow in Figure 6).

4 Discussion

This study represents the first approach towards defining dynamical performance parameters of bowling, by means of instrumented sport equipment. All three parameters investigated, relative duration of forward swing, relationship between the peak forces of fingers and thumb and between the impulses of forward- and backswing, showed good correlation ($r > 0.7$) with the performance (average score) of the bowlers. The relative duration of the forward swing is longer in better bowlers. This, in turn, means that the differences in angular velocity between backswing (slower, $>50\%$) and forward swing (faster, $<50\%$) are smaller in more experienced bowlers. Thus, beginners tend to execute the forward swing far

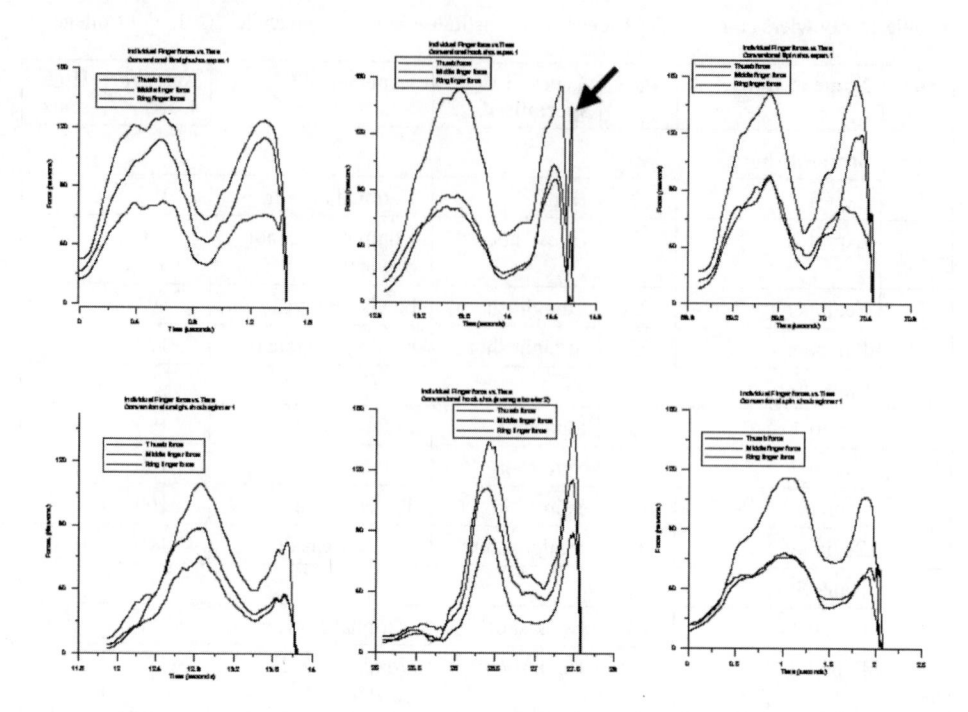

Figure 2: Force-time graphs of thumb (maximal force), middle finger, and ring finger (minimal force); top row: experts, bottom row: average bowlers and beginners; left: straight shot, middle: hook shot, right: spin shot.

Table 2. Coefficients of correlation of shots and relative parameters of forward swing

Shot	Time	Force	Impulse
straight	0.91	0.87	0.96
hook	0.14	0.77	0.46
spin	0.92	0.21	0.82
all shots	0.71	0.75	0.81

ster than the backswing. Based on this, we would expect a result, which is completely opposite to the one shown in Figure 4. The forces acting on the bowling ball during the swing are:

1. gravitational force, $F_G = m\,g$,
2. inertial force, $F_I = m\,a = m\,\alpha\,r$,
3. centrifugal force, $F_C = m\,\omega^2\,r$,
4. applied force F_A, which is grip force, or forces applied by pinching the ball.

Performance Analysis with an Instrumented Bowling Ball 477

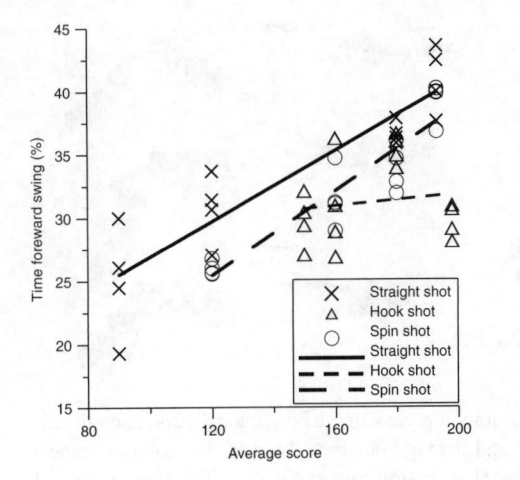

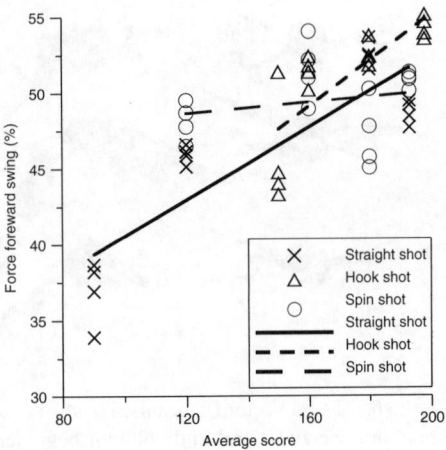

Figure 3. % time of forward swing vs. average score

Figure 4. % force of forward swing vs. score

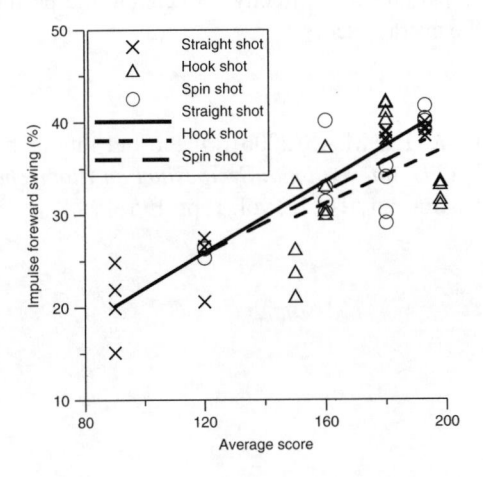

Figure 5. % impulse of forward swing vs. score

The latter forces cancel each other. The sum of all 4 forces is measured with the instrumented bowling ball, individually for each finger. The faster the motion, the higher are the angular velocity ω and acceleration α, and the higher are the forces acting on the ball. Furthermore, the higher the centrifugal force is, tending to pull the ball off the fingers, the larger is the grip force required. Figure 4 reveals the contrary. The relatively slower forward swing in better bowlers (Figure 3) shows relatively higher finger forces (Figure 5, forces >50%) during the forward swing, compared to the back swing. As F_I and F_C are definitely smaller in slower motion and as F_G is constant, the only reason for the unexpected higher force during the forward swing of better bowlers is the far larger grip force. Thus, the grip force seems to be an important performance criterion. At that stage of the research, it cannot

Figure 6: Force Vector Diagrams; 1st and 3rd column: top view of the bowling ball, 2nd and 4th column: side view; 1st and 2nd column: beginner and average bowlers, 3rd and 4th column: expert bowlers; top row: straight shot, middle row: hook shot, bottom row: spin shot; T = thumb, M and R = middle and ring finger.

be decided whether this parameter is directly influencing the performance, or only indirectly associated with the average score.

References

Fuss F.K., Kong E.C.H. & Tan M.A. (2006) Finger and thumb forces during bowling. *Proceedings of the XXIV International Symposium on Biomechanics in Sports* (ISBS 2006), Salzburg, Austria, July 14–18, vol. 1, pp 195–198.

ACCELEROMETRY IN MOTION ANALYSIS

S.M.N.A. SENANAYAKE[1], A.W.W. YUEN[1] & B.D. WILSON[2]

[1]*Monash University Sunway Campus, Jalan Universiti, Petaling Jaya, Malaysia*
[2]*Center of Biomechanics, National Sports Institute, Kuala Lumpur, Malaysia*

This paper explores the opportunity of using tri-axial accelerometers for short term supervised monitoring of sports movements. A motion analysis system of upper extremities for lawn bowling is developed in this project. Six accelerometers were placed on strategic parts of human body. One sensor was placed in front of the chest to represent the shoulder movements. Two sensors were placed on the back to capture the trunk motion. Three sensors were used to obtain the right arm movements, namely one at the back of the hand, one right above the wrist and one right above the elbow. These sensors placement were carefully designed in order to avoid restricting bowler's movements.

1 Introduction

Up to now, motion analysis methods on activities condicted outdoors are mostly depended on digitizing the cine or video record. This method is not practical for the sports training situation, because it is a time consuming process to obtain the results for sports performance. On the other hand, the acceleration signal from player's body motion is suitable for real time analysis and immediate feedback to the coach and athlete.

Motion analysis is very important for sports performance enhancement or injury prevention. Accelerometers have been proven to be a very useful for short term supervised monitoring and long term unsupervised monitoring (Mathie *et al.*, 2004). Accelerometers have the advantage of being able to be used in almost any environment and, therefore, can be used to perform motion analysis on different types of sports. Triaxial accelerometers provide information on the acceleration in three planes, namely the vertical plane (Z-axis), anteroposterior (Y-axis) and lateral direction (X-axis).

2 Equipment and Software Used

2.1. *Accelerometers*

The accelerometers used are the Microstrain Wireless Tri-axial Accelerometers. Some of the features of these accelerometers are: the size (25 mm $\times$ 25 mm $\times$ 5 mm) which makes it possible for placement at critical points of interest on the body, Compatible with most personal computers, 2 MB of memory for storage (can store 1 million data points) and Large range roughly 30 meters from base station is at line of sight, 9 Volt batteries for each accelerometer (Figure 1), The MicroStrain Base-Station (Figure 2) is used to communicate between the accelerometers and the computer used during the experiment.

The base station will be connected to the computer via USB. Elastic straps were used for attachment of sensors.

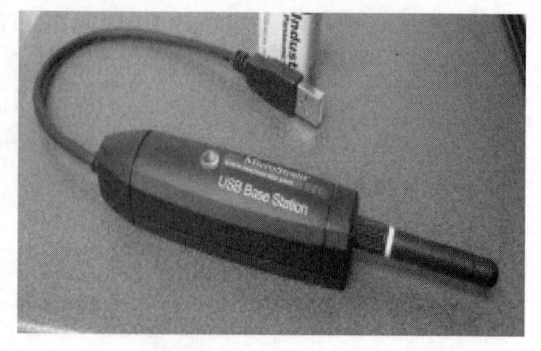

Figure 1. Accelerometer.

Figure 2. Base-Station.

2.2. *Software*

A number of different software were used:

MicroStrain Agile Link 1.4.0 software controls the accelerometers in matters such as power management, channel configuration, transfer of readings to Microsoft Excel and sensor configurations to fit user needs

The mathematical functions of Microsoft EXCEL were used for calculations of accelerations, velocities and positions in relation with time.

The functions available in National Instruments LabVIEW 8.0 enable users to design Graphical User Interface (GUI) for different purposes. In this project, this software was used to create a GUI in order to display the results obtained after processing the data.

3 Body mounted accelerometers and experiment setup

Accelerometers were placed on the human body as shown in Figure 3. The selection criteria for sensors placements were primarily based to detect motion on selected critical segments of the human body. Motion analysis done in this research was mainly of the lawn bowling action. Therefore, human motion analysis was critical for only certain body parts and they are addressed clearly in this article. Six accelerometers were used in the analysis of lawn bowling.

The Agile Link software was used to trace and to download sensor data. The accelerometers were calibrated before being used. To help ensure the experiment was conducted smoothly, a range test was conducted on each sensor triggered to store data during the experiment.

The network trigger function in the Agile Link program was used to trigger all accelerometers simultaneously.

All collected data were then transferred to a set of Microsoft Excel sheets specially written to calculate the accelerations, velocities and displacements for each accelerometer.

Once all data were complied and processed, they were stored in another Excel file directly interfaced with the designed Graphical User Interface (GUI) in LabVIEW for presentation. The GUI design is as shown in Figure 4.

The GUI allows the user to view the reconstructed human body in a stickfigure representation; to demonstrate the acceleration, velocity and displacement of each accelerometer against time. The GUI also provides details of the human subject such as name, weight, height, etc and to search the specific acceleration, velocity and distance at a specific time.

Accelerometry in Motion Analysis 481

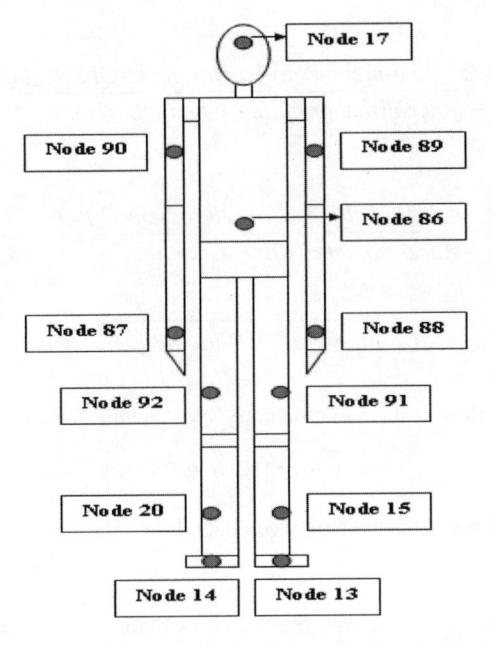

Figure 3. Positions of accelerometers for motion analysis.

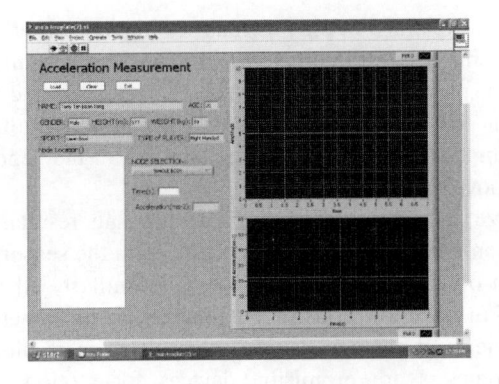

Figure 4. GUI designed using LabVIEW.

As accelerometers store data in bits, the following set of formulae was used to convert stored bit values into acceleration values.

$$Channel\_1 = \frac{(AccelerometerBits\_Channel\_1 - AccelerometerOffset\_Channel\_1)}{AccelerometerGain\_Channel\_1} \times 9.81$$

$$Channel\_2 = \frac{(AccelerometerBits\_Channel\_2 - AccelerometerOffset\_Channel\_2)}{AccelerometerGain\_Channel\_2} \times 9.81$$

$$Channel\_3 = \frac{(AccelerometerBits\_Channel\_3 - AccelerometerOffset\_Channel\_3)}{AccelerometerGain\_Channel\_3} \times 9.81$$

$$resultant = \sqrt{Channel\_1^2 + Channel\_2^2 + Channel\_3^2}$$

The trapezoidal rule was used to calculate velocity and displacement as shown below;

$$I = (b-a)*\{[f(b) + f(a)]/2\}$$

where (b −a) ~ time difference between each interval,
f(b) ~ value at b,
f(a) ~ value at a,
The first round of iterations was required to calculate velocity, whereas the second round of iterations helps to calculate the displacement of the test subject.

4 Motion Analysis using Accelerometers

Based on the formulae explained in the previous section and based on the critical motion analysis of lawn bowling, six accelerometers were chosen to be strategic (Kibele, 1998). They were; one sensor placed in front of the chest to represent the shoulder movements, two sensors placed on the back to capture the trunk motion and three sensors placed to obtain the right arm movements, namely one at the back of the hand, one right above the wrist and one right above the elbow.

In order to analyze the motion analysis of lawn bowling, resultant acceleration graphs were obtained based on the read and calculated data from the sensors. Figure 5 shows the resultant acceleration of the wrist (sensor node 88). Similarly all resultant acceleration graphs were obtained in order to reconstruct the human are movement of lawn bowling.

From the multiple tests conducted, similar acceleration results were produced that demonstrate accelerometers are promising devices for motion analysis. The graphs obtained also show logical acceleration patterns while the lawn bowling action was being performed.

There are two methods of triggering the accelerometers, the Network Trigger method (NT) and Event-driven Trigger method (EDT). These two methods were evaluated by plotting the motion on a 3D graph using a stick figure representation. Results show that EDT method is preferable because it is more reliable in triggering all sensors at the same time.

Reconstruction of human motion in lawn bowling was based on all derived parameters displacements, velocities and angles between joints. These parameters were obtained taking into consideration specific motion cycles from each of the resultant acceleration graphs already given in Figure 5 and the rest of motions from all other sensors. LabVIEW was used to reconstruct the motion of lawn bowling.

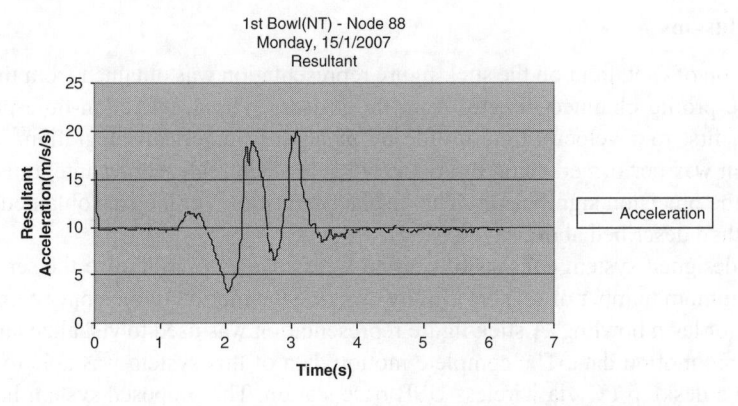

Figure 5. Resultant acceleration of wrist in lawn bowling.

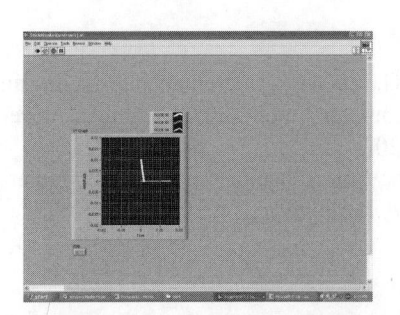

Figure 6. Lawn bowling action started.

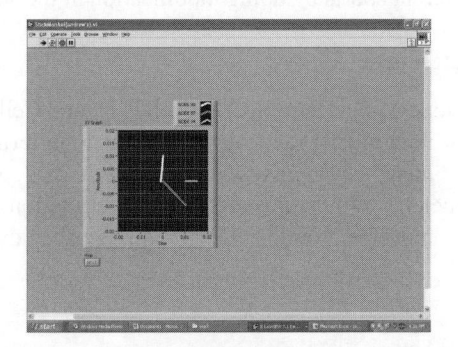

Figure 7. Motion of the lower arm backward.

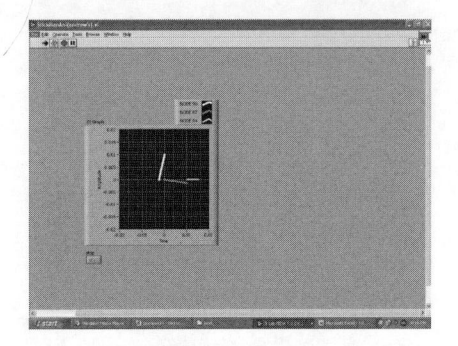

Figure 8. Motion of the elbow and arm.

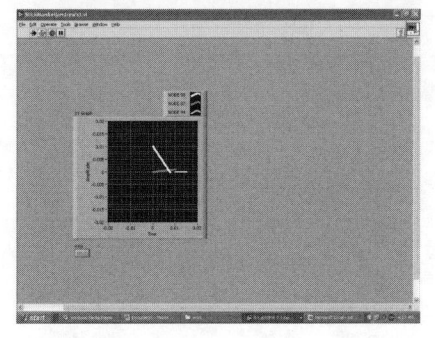

Figure 9. Motion of the arm forward.

In order to visualize graphically, discrete time intervals were defined such a way that lawn bowling actions are correctly displayed during the reconstruction of the stick figures as shown in Figures 6, 7, 8 and 9.

Intermediate frames of arm movements are not shown. Complete stick figure movement is required to understand the gait cycle of lawn bowling in a smooth continuous way.

5 Conclusions

The position of each joint on the stick figure representation was obtained from the acceleration-time profile obtained directly from the sensors. The acceleration-time profile was converted first to a velocity-time profile by means of numerical integration. Numerical integration was performed using the trapezoid rule since higher order integration did not improve the precision significantly. The displacement-time profile was obtained using the same method described above.

The designed system consists of easy-to-wear sensors to minimize the setup time. It uses a minimum number of sensors to fully describe the motion of the upper extremities of the body for lawn bowling. A stick figure representation was used to visualize and validate the acquired motion data. The complete motion data of this system was able to be downloaded to a desktop PC via wireless USB base station. The proposed system has demonstrated the use of accelerometers for motion analysis. The system can be adapted for different sports by doing modification at the software level alone.

References

Mathie M.J., Coster A.C., Lovell N.H. and Celler B.G. (2004) Accelerometry: providing an integrated, practical method for long-term, ambulatory monitoring of human movement. *Physiological Measurement*, 25/2, R1–R20.

Kibele A. (1998) Possibilities and limitations in the biomechanical analysis of countermovement jumps: A methodological study. *J. Appl. Biomech.*, 14, 105–117.

HUMAN PERCEPTION OF DIFFERENT ASPECTS OF FIELD HOCKEY STICK PERFORMANCE

M.J. CARRÉ & M.A. McHUTCHON
Department of Mechancial Engineering, University of Sheffield, Sheffield, UK

This study sought to apply the semantic differential approach of human perception analysis to measure hockey players' emotional responses to different hockey sticks in different play situations. Three phases of testing were carried out: The first phase involved players being given a set of sticks with varying physical properties. Players were asked to carry out a set of repeated ball-strikes before rating the sticks using a semantic differential questionnaire. The second phase of testing involved repeating the ball-strike study, but with the players' vision of the ball's post-hit behaviour being impaired. This was carried out to assess the need for visual cues in assessing stick performance (e.g. power and accuracy). The final phase of testing involved player perception of dribbling performance and used a different set of disguised sticks, including some with mass-distributions that had been altered. The key findings were that players associate a heavy stick with a mass-distribution concentrated towards the head-end as being well suited to hitting. However, for dribbling, players prefer a stick with a centre of mass higher up the handle. Interestingly, visual cues of post-hit ball behaviour made little difference to a player's ability to judge power and accuracy. Results from these studies have been incorporated into an all-encompassing design methodology for sports equipment.

1 Introduction

A traditional approach to improving the performance of sports equipment usually involves studying the equipment in use, simulating the important interactions and developing mathematical models which predict the effect of changing key variables (e.g. stiffness, mass-distribution etc.). These models are based on an understanding of the underlying physics of the problem. This approach is well-proven and necessary, but it often ignores how players relate to their equipment or what the effect of changing these key variables would have on their psychological relationship with their equipment. This study sought to investigate the perception of a range of hockey sticks under different aspects of use, and compare this to physical properties.

Previous authors have studied the "feel" of sports equipment, such as tennis rackets (Brody *et al.*, 2002) and golf clubs (Hocknell *et al.*, 1996), in different ways. Techniques generally include different ways of assessing and interpreting responses through questionnaires. One technique is known as "Kansei engineering" (Nagamachi, 1995) and aims to identify correlations between quantified feelings and product properties and/or design elements. Semantic pairs (e.g. good/bad) are used to by subjects to describe their emotions (Osgood *et al.*, 1967). Such an approach was used in this study.

2 Testing Carried Out

The study was made up of three phases of tests, covering different aspects of hockey stick performance. Each phase will be described and the main results presented in turn.

2.1. Phase 1

Phase 1 involved players being given a set of five disguised sticks taken from a commercially available range of designs, with varying physical properties. This part of the study has been described in detail in a previous paper (Carré and McHutchon, 2006). The sticks (1 to 5 in Table 1) varied due to their mass; moment of inertia in the swinging direction (*MOI*); centre of mass measured from the handle end (*COM*); modulus of rigidity based on a standard, three-point bend test (*EI*) and moment about the handle end, when held horizontally (M_{HE}). All the sticks were similar in length. Two were manufactured completely from composite materials (including carbon, aramid and Kevlar fibres) and the other three were wooden, reinforced with thin composite weave wrapping.

Twenty-six university-level players were asked to carry out a set of repeated ball-strikes at a target 3 m away, using a double-handed shot (both hands close to the handle end), before rating the sticks using a semantic differential questionnaire with a seven point Likert scale for each factor. The questions related to comfort, power, weight, stiffness, control and overall performance. The players could accurately rank the sticks in terms of power, which could be predicted using classical mechanics (McHutchon *et al.*, 2004). Good correlation was found between perceived weight and the moment about the handle end (giving a squared Pearson correlation coefficient, $R^2 = 0.84$). Players were unable to perceive stiffness directly, despite there being an excellent correlation between perceived power and measured modulus of rigidity, *EI* ($R^2 = 0.94$). Players seemed to align their perceptions of comfort and control, with that of overall performance and the authors suspect that the players may have been influenced by the fact that the heads of the composite and wooden sticks were not disguised adequately.

2.2. Phase 2

In an attempt to overcome these problems a second phase of testing was carried out with a slightly reduced number of sticks (1, 4, 5 and 6 from Table 1). This time, as well as being covered in tape, the stick heads were completely painted black so that based on visual inspection, the sticks could not be differentiated. The question relating to "control" was broken down into three questions relating to "manoeuvrability", "balance" and "accuracy". Players were asked to carry out the same double-handed shot as in Phase 1, but the accuracy of each shot was recorded, based on the ability to hit a target (see Figure 1).

Table 1. Properties of the sticks used in Phases 1 and 2.

Stick	Design	Mass, kg	Length, m	*MOI*, kgm²	*COM*, m	*EI*, Nm²	M_{HE}, Nm
1	Composite	0.618	0.928	0.0516	0.556	708	0.344
2	Composite	0.576	0.926	0.0510	0.578	500	0.333
3	Wood	0.611	0.930	0.0510	0.551	310	0.337
4	Wood	0.635	0.929	0.0514	0.555	311	0.353
5	Wood	0.591	0.927	0.0497	0.542	217	0.320
6	Composite	0.638	0.929	0.0524	0.572	389	0.365

This phase of the study was divided into two parts as follows:

- Part 1 used a conventional hitting exercise as per Phase 1, with 30 players (15 male, 15 female) using the new survey questions and range of sticks.
- Part 2 used "blind" testing, as illustrated in Figure 1, with another 30 players (15 male, 15 female) to examine whether post-strike observations of the ball are used when perceiving such factors as "power" and "accuracy". Each player took position behind a screen and hit the ball underneath it, thus severely reducing the distance over which they could visually follow the ball, post- impact.

After examining all the data from both parts of Phase 2, it was clear that perceived power was aligned with mass-related stick properties that affect actual power (according to a mechanical analysis of a ball-strike). A *COM* location nearer the head end, indicated a stick of greater perceived power, as shown in Figure 2. This was statistically significant at level $p < 0.05$ for both sets of tests ($R^2 = 0.94$ for part 1 and $R^2 = 0.88$ for part 2, the "blind" tests). Both sets of tests also showed that a high power rating was associated with a high measured moment of inertia.

Figure 1. "Blind" tests carried out in the second part of Phase 2.

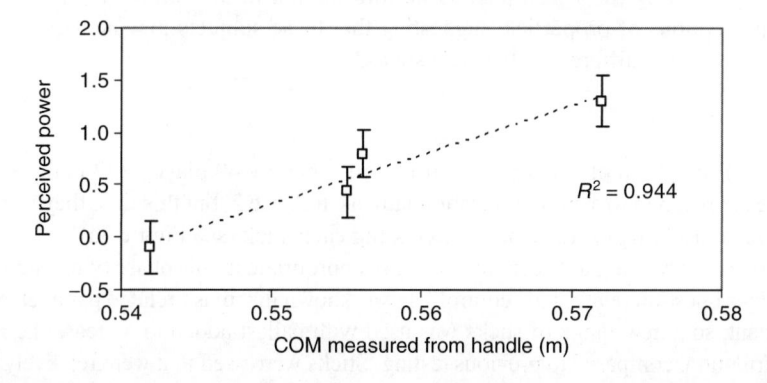

Figure 2. Correlation between perceived power and COM location for the first part of Phase 2 (error bars related to ± one standard error).

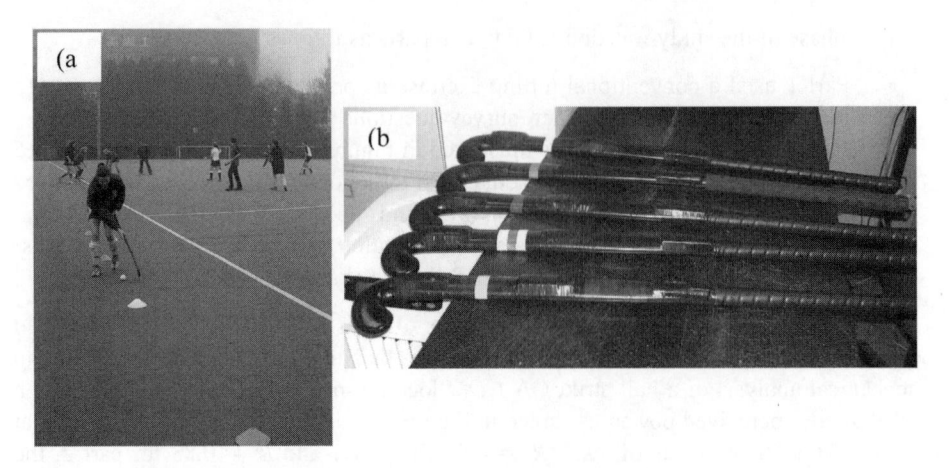

Figure 3. (a) Dribbling exercise and (b) range of sticks used in Phase 3.

Interestingly, compared to Phase 1, there was no observed correlation between modulus of rigidity and perceived power, indicating that the results from Phase 1 may have been affected by players being able to tell between the composite and wooden sticks. It is possible that their preconceptions of these sticks, skewed the results. Statistical analysis showed the subjects in Phase 2 were able to tell the difference between physical properties such as mass and mass distribution, but not stiffness. Therefore, it is likely that they used the mass-related properties to gauge their perception of power.

The subjects were very good at assessing "accuracy" and generated excellent correlations ($p < 0.05$) between perceived accuracy and the measured ability to hit the target, even when they could not observe the path of the ball, post-strike ($R^2 = 0.94$ for part 1 and $R^2 = 0.98$ for part 2). However, there were less meaningful data related to "manoeuvrability" and "balance" with no consistent correlations found across both sets of tests. It is likely that hitting was not the best activity to make these sort of judgments.

Various subject data was studied in detail and it was found that gender, playing position and the type of stick currently used by the subjects had no obvious effect on perceptions. Interestingly, limiting the visual post-strike information in the "blind" tests, led to more extreme perceptions of properties, suggesting that these subjects may actually have been more sensitive to the differences between sticks.

2.3. *Phase 3*

A third and final phase of testing was introduced, whereby 40 players (32 male, 8 female) were asked to respond to a similar questionnaire as in Phase 2, but this time they were asked to carry out a dribbling exercise before assessing each stick (see Figure 3a).

The question relating to "accuracy" was not appropriate in this phase of testing so it was replaced by a question relating to "control". It was known that mass-related parameters would be important, so a new range of sticks was used with ballast added to increase the range of mass distribution, compared to previous testing. Sticks were used that were relatively light so that when extra mass was added they would not seem overly different to that which players were used to (see Table 2). Three ballast locations were chosen; near the head-end, close to the

Table 2. Properties of the sticks used in Phase 3.

Stick	Design	Ballast	Mass, kg	Length, m	COM, m	EI, Nm²	M_{HE}, Nm
7	Wood	None	0.582	0.928	0.542	253	0.316
8	Composite	None	0.591	0.929	0.588	442	0.347
9	Composite	COM	0.656	0.929	0.591	415	0.388
10	Composite	Handle	0.681	0.929	0.567	440	0.386
11	Composite	Head	0.656	0.929	0.618	391	0.406

Table 3. Pearson correlation coefficients, R, for perceived factors compared with moment about the handle end, M_{HE}.

Perceived factor	Correlation coefficient, R
Good manoeuvrability	−0.935
Good balance	−0.987
Good control	−0.941
Good comfort	−0.909
Good overall performance	−0.855
Light weight	−0.999
High stiffness	−0.978

COM location and just below the handle grip. By adding either heavy, lead weights or "dummy" cardboard weights at these locations, a range of mass distributions could be achieved without the sticks appearing any different under visual observation (see Figure 3b).

As with Phase 1, subjects aligned their perception of stick weight most strongly with the moment they experienced when holding the stick horizontally. In other words, a stick that was "head-heavy" would be perceived as simply "heavy". This led to a very strong correlation between perceived weight and moment about the handle end, M_{HE} ($R^2 = 0.9998$). In fact, for this phase of testing it was the mass-related property, M_{HE}, that dominated the perceptions of all the stick properties and performance factors. A stick that had more mass near the handle end and so had a *low* value of M_{HE}, was also perceived as being *good* in terms of manoeuvrability, balance, control, comfort and overall performance as well as being light and, interestingly, of high stiffness (see Table 3). The correlation between perceived stiffness and moment about the handle end is thought to be skewed by the fact that the comparatively flexible, wooden stick used in Phase 3 also had a relatively low value for M_{HE}. There were no other statistically significant ($p < 0.05$) relationships between any of the perceived factors and the measured properties.

3 Conclusions

From the three phases of human perception described above it can be concluded that:

- Post-strike visual cues make little difference to player perception when carrying out hits.

- Players associate a heavy stick, with a centre of mass located near the head end and a high moment of inertia, as being of high hitting power.
- For dribbling, players prefer a stick with a low moment about the handle end, linked to a light stick with a centre of mass located nearer to the handle.

These last two points suggest a conflict in design properties for different aspects of hockey stick use (e.g. hitting and dribbling). It is possible that sticks could be designed to better suit certain types of player, as discussed in a previous paper by McHutchon *et al.* (2006).

Acknowledgements

The authors would like to thank Grays International for providing samples for this study.

References

Brody H., Cross R. and Lindsey, C. (2002) *The Physics and Technology of Tennis*. Racquet Tech Publishing, Solana Beach, CA.

Carré M.J. and McHutchon M.A (2006) Understanding human perception of field hockey stick performance. In: Moritz E.F. and Haake S. (Eds.) *The Engineering of Sport 6, Volume 3 Developments for Innovation*: Springer, New York. pp. 237–242.

Hocknell A., Jones R. and Rothberg S. (1996) Engineering "feel" in the design of golf clubs. In: Haake S.J (Ed.), *The Engineering of Sport*. Balkema, Rotterdam. pp. 333–338.

McHutchon M.A., Curtis D. and Carré M.J. (2004) Parametric design of field hockey sticks. In: Hubbard M., Mehta R. and Pallis J. (Eds.) *The Engineering of Sport 5, Volume* ISEA, Sheffield UK. pp. 284–290.

McHutchon M.A., Manson G.A. and Carré M.J. (2006) A fresh approach to sports equipment design: Evolving hockey sticks using genetic algorithms. In: Moritz E.F. and Haake S. (Eds.) *The Engineering of Sport 6, Volume 3 Developments for Innovation*. Springer, New York. pp. 81–86.

Nagamachi M. (1995) Kansei Engineering: a new ergonomic customer-oriented technology for product development. *International Journal of Industrial Ergonomics*, 15, 3–11.

Osgood C.E., Suci G.J. and Tannenbaum P.H. (1967) *The measurement of meaning*. University of Illinois Press, Illinois.

A BIOMECHANICAL ANALYSIS OF THE STRAIGHT HIT OF ELITE WOMEN HOCKEY PLAYERS

S. JOSEPH[1], R. GANASON[1], B.D. WILSON[1], T.H. TEONG[2] & C.R. KUMAR[3]

[1]*Human Performance Laboratory, Sports Biomechanics Centre,
National Sports Institute of Malaysia*
[2]*Sports Centre, University of Malaya*
[3]*National Women Hockey Coach, Malaysia*

The aim of the study was to examine the players' straight hit technique, in order to improve performance and reduce injury risk to the players. Fifteen National women hockey players performing straight hits at the goal while in training were analyzed with the help of video. The back swing and swing time, knee angles during the swing, contact and follow-through actions, and the pattern of movement of the rear foot in the follow through were measured. Most of the players had a cross-over rear-foot action which reduced the rotation speed of the hockey stick, and a follow through where the rear foot did not follow the overall movement, thus increasing the execution time of the hit and also reducing the ball speed. The cross-over movement also leads to front knee stress and could be a major cause for knee, ligament and muscular injuries. Correcting the course of the follow through would reduce the injury risk and also indirectly improve the speed and effectiveness of the stroke.

1 Introduction

Performance analysis has assumed an important role in the evolution of competitive sports techniques in the last few decades. It is difficult to analyze the sports performance in acyclic field sports activities compared to that of the cyclic type of activities. However, through observation methods using video and game analysis and skill analysis with various kinds of software, an overall view of the execution of various skills can be obtained. Movement descriptions and performance diagnosis on tactical, technical and physical aspects of the team and individuals' performance can then be conveyed to the coaches, trainers and athletes to improve their understanding of the genesis of performance. On this basis, sports performance, team strategy etc., can be determined and the sports training can be formulated.

Epidemiological studies have consistently shown that injuries in hockey are numerous and can be serious (Alcock *et al.*, 1997, Fox 1994, Fuller 1990, Jaimson & Lee 1989). Most serious injuries result from being struck by the stick or the ball. Overuse injuries are also frequently reported on mostly foot, ankle and knee (Hering & Nilson 1987, Freke & Dalgleish 1994). To reduce overuse injuries, while training to achieve elite performance, the coaches and trainers should look into the technical execution of skills, and amend techniques as required, so that the likelihood of injuries could be reduced and the athletes would then be in a position to perform the skills efficiently.

The straight hit is one of the skills used by the players to attempt to score a goal from open play, or from a penalty corner and is also a means to pass the ball to your team mate as quickly as possible. Video analysis of repeated hits in training will be used to describe the straight hit technique in terms of movements of the lower limbs and the possible stress on the knee joint. This analysis should help the coaches and players concerned to identify the strong and weak points of the skill and thereafter to improve their performance and minimize the risk of injuries. A complete review of the skill is beneficial for making changes within the framework of training and can lead to improvement in competition performance.

2 Materials and Methods

Fifteen players of the Malaysian national women's hockey team (15 ± 1.4 years) who were undergoing the national hockey camp in preparation for various international hockey tournaments participated in the study. The players were undergoing a normal training session in which they were asked to go for straight hit at the goal from the top of the shooting circle. There was a maximum distance of 5 m from the top of the circle where they were permitted to dribble, trap and release prior to hitting the moving ball at the top of the circle. The hitting motions were videotaped with a Panasonic model NVGS150 video camera which was set approximately perpendicular to the direction of execution of the skill. The camera was set at a nominal frame rate of 50 Hz and a shutter speed of 1/250 sec. Each player was asked to hit the ball into the goal and there was no enforcement on the style and technique of hitting except on the condition of hitting the ball into the goal from the top of the shooting circle.

The video footage of five repetitions for each individual was analyzed for various temporal and kinematic parameters using siliconCOACH Pro software. The movement pattern of the lower extremity in the follow-through process was described. The following measurements were recorded as per the descriptions below:

Right elbow angle: Angle at the time of the ball contact.
Left knee angle: Angle at the time of back swing.
Left knee angle: Angle at the time when the stick contacts ball.
Left knee angle: Angle at follow through.
Total angle: Left dorsi flexion from when leg is planted till heel raise.
Back swing time: The time taken to take a back swing either from close to the ball or up to hip height to the maximum height of the back swing.
Swing down time: The time taken when the stick starts descending after the back swing to the contact of the ball with the stick.
Total swing and strike time: The summation of back swing and swing down time.
Ball velocity: The velocity in which the ball travels towards the goal after the hit.
Angle of flight of the ball: The angle in which the ball travels to the goal immediately after the hit relative to ground level.
Ball contact in relation with the left leg toe: The distance at which the stick contacts the ball either before or ahead of the toe line.

3 Results and Discussion

The straight hit was categorized into three style variations as executed by the players. The categorization was based on distinct differences in the movement of the trailing (right) leg in the follow through for the three styles:

Technique A (Figure 1),where the right leg moves forward in the follow through:

Figure 1a. Ball contact.    Figure 1b. Stick follow through.    Figure 1c. Right leg movement.    Figure 1d. Right leg landing.

Technique B (Figure 2),where the right leg is pulled back in the follow through:

Figure 2a. Ball contact.    Figure 2b. Stick follow through.    Figure 2c. Right leg movement.    Figure 2d. Right leg landing.

Technique C (Figure 3), where the right leg makes a cross-over with the left leg at the time of contact and the cross-over remains in the follow through.

Figure 3a. Ball contact.    Figure 3b. Stick follow through.    Figure 3c. Right leg movement.    Figure 3d. Right leg landing.

In Technique A, a movement pattern similar to a running gait is observed. When the left foot lands and is planted on the ground for the execution of the stroke, the quadriceps muscle provides an eccentric force at the time of landing to stabilize the knee joint. Maintaining a knee angle facilitates the swing of the stick close to the foot, depending on length of the hockey stick used by the player. In the follow through, the right leg swings forward in a transfer of body weight, which reduces the stress on the left knee. A concentric action of the quadriceps propels the body forward in a gait-like motion.

In addition, the right leg's forward step is in the intended direction of the hit in the last phase of the swing resulting in an increased stick velocity as there is a continuous force that is brought on to the stick and the stick follows the ball for a longer (fraction) period of time. The swing of the right leg in forward direction also facilitates the trunk action contributing to greater force application.

In Technique B, while setting up for the hit, the body is brought abruptly to a halt with the planting of the left leg to the ground with a rapid eccentric firing from the quadriceps to stabilize the knee angle. The left abductors play the role of stabilizers in enabling the knee joint to be stable at the time when the stick contacts the ball. Force from the lower extremity is transmitted medially to contribute to the movement of the ball at strike. The upper trunk contributes only minutely as it does not lean forward instead permit early opening of the left shoulder.

The quadriceps acts in a series of contractions to halt the momentum of the body at the time of ball contact which stresses the lateral, medial, anterior collateral knee ligaments and the patellar tendon. The series of contractions exhibited are; an eccentric action while planting the foot, concentric action while facilitating medial rotation of the trunk during the hit, and lastly an eccentric action to halt the body momentum. During these actions, the body is pivoted on the left toe to and the trailing leg (right) is brought back in a reverse action to stabilize the body, thus indirectly reducing the velocity created by the swing of the stick with the ground reaction forces further reducing the effectiveness of the swing. The ball tends to be raised with a reduced velocity, and changing the course of hit in the last phase becomes difficult due to body drifting away from the ball. Repetitive efforts of this technique without recovery periods can lead to severe front knee and ankle joint stress.

In Technique C, the body is brought abruptly to a halt with the planting of the left leg to the ground, which leads to counter-rotatory movements (the swing of the hockey stick and the crossover movement of the right leg) occur across the left hip which acts as an axis of rotation. The abductors play a dominant role in stabilizing the left lateral knee. There is rapid eccentric action of the left quadriceps groups of muscle and the action remains till the follow through of the stick is completed. Ground reaction forces also act and push the body back.

The straight knee position of the left leg does not permit the upper body to rotate along with swing of the stick. The shoulders are kept open which increase the chance of striking the ball at its base causing it to fly rather than travel on the carpet. The counter movements further reduce the movement speed of the stick and also make the knee and ankle joint vulnerable to injuries, and also stress the quadriceps under eccentric action.

While executing straight hit by the different techniques groups, the total swing time was similar for Technique A and B (0.5 and 0 .51 s respectively and less for Technique C (0.47 s) (See Table 1). Different tendencies were noted in execution of back swing phase which originated either from over the ball or from knee height or from hip line. The end phases of back swing were either close to shoulder height or far beyond the ear lobe. Though there were variations in back swing, the down-swing mean time did not exhibit much difference among the groups. However, the Technique A players did deliver a slightly higher ball velocity (23 m/s) compared to B (22 m/s) and C (21 m/s) players due to the follow-through mode of the right leg which assisted in propelling the ball.

The left knee angle at the approach of the straight hit was similar in all the groups. But at the time of contact of the ball, Technique C and A players had a lesser average left knee angle compared to that of Technique B players which illustrates that the body weight was transferred from the right to the left leg for final conversion of the hit. Whereas, the higher left knee angle for technique B players exhibited that greater ground forces were acting

Table 1. Kinematic, temporal and angle measures (Mean ± Standard deviation).

Variables	Technique A ($n = 4$)	Technique B ($n = 6$)	Technique C ($n = 5$)
Back swing time (sec)	0.28 ± 0.04	0.31 ± 0.04	0.27 ± 0.03
Swing down time (sec)	0.22 ± 0.04	0.20 ± 0.02	0.21 ± 0.04
Total swing and strike time (sec)	0.50 ± 0.05	0.51 ± 0.04	0.47 ± 0.04
Ball velocity (m/sec)	22.9 ± 3.7	21.9 ± 1.8	21.1 ± 4.3
Right elbow angle at the time of strike (deg)	151 ± 9	153 ± 13	157 ± 11
Angle of flight of the ball after strike (deg)	6 ± 4	9 ± 4	11 ± 5
Left knee angle at top of backswing (deg)	173 ± 6	170 ± 8	173 ± 6
Left knee angle at ball contact (deg)	148 ± 6	154 ± 8	140 ± 7
Left knee angle at follow through (deg)	135 ± 14	148 ± 9	143 ± 8
Total angle left knee flexion from back swing to ball contact (deg)	28 ± 10	17 ± 11	32 ± 9
Total angle of left knee flexion/ extension from contact of the ball to follow through (deg)	13 ± 8 (KF)	6 ± 12 (KF)	2 ± 3 (KE)
Total angle of left ankle dorsi flexion after leg is planted (deg)	35 ± 13	22 ± 9	30 ± 7
Ball contact in relation with the left leg toe (cm)	4 ± 7	−3 ± 11	8 ± 13

* Spreedsheet applications structures diagrams provide a graphical method for identifying and structuring enterprise

against the body and thus nullifying the effect of transfer of body weight to the left leg to move forward.

The left knee angle at the follow through in case of Technique A exhibited a 13 degrees reduction as the right leg shifted forward, thereby reducing the stress on the left knee and ankle joints. In the case of Technique B, the left knee flexion angle reduced by 6 degrees only, thus encountering greater ground force; stressing the left lateral knee and ankle joints; and also limiting the forward body momentum. Whereas for Technique C, the left knee almost remained in the same position and the knee angle extended only by 2 degrees leading to heavy eccentric action forces on knee joint.

The total left ankle dorsi flexion from planting of left foot at the time of ball strike through to the follow through where the heel is lifted from the ground, is also indicative of the amount of stress the ankle is forced to undertake. The greater the angle of dorsi flexion means a lesser stress on the ankle joint as the joint absorbs the load over a greater time and range of motion. The Technique A players demonstrated higher angles of dorsi flexion which indicates that the ankle joint did not have much strain in the process of execution whereas, in the case of Technique B and C players there was a lesser dorsi flexion angle resulting in greater ground reaction forces.

Technique A players struck the ball much closer and almost in line with the left leg toe, whereas Technique B players struck before the ball came in line with the toe and the Technique C players much ahead of toe line. When the ball is in line with toe and the upper body is just over the ball, the hit would enable the ball to travel closer to the carpet. When the body is brought back with the pull back of right leg as in the case of Technique B, the ball is contacted much before it reaches the toe line, there is tendency for the ball to be struck low leading to a raised hit. Hitting the ball ahead of the leading leg toe also leads to stress on the lateral knee and ankle joints and the possibility of a raised hit.

4 Conclusion and Recommendation

Among the three styles of straight hit, Technique B and C players are at risk of knee and ankle injury due to sudden stop, encountering eccentric contractions and counter rotatory movements involved. Further, these techniques lead to a greater possibility of a raised hit. One of the reasons commonly noticed for players acquiring the wrong hitting technique was use of over-length hockey sticks at a young age as used by adult hockey players. Hence to compensate for the over-length stick in the hit, the players kept the ball away from the left toe and also opened up the left shoulder to enable the strike. The rear-foot crossover and the upper body pull back in the follow through were adopted to maintain equilibrium and facilitating stick movement. We suggest that the selection in choice of hockey sticks should be carefully done according to the height of the individual, their arm length and upper trunk mobility. Further, we recommend that with increase in height of a player, the length of the hockey sticks may be derived accordingly.

To reduce injury risk, specific strength training and pre-habilitation training of the knee and ankle joint should be carried out since the majority of the present players perform the straight hit with a technique which the players have acquired on their own.

From the tactical point of view the Technique B and C players are at a disadvantage as the player takes a longer time to react to events such as a rebound from the goalkeeper or other players or from the goal post, whereas for the Technique A group, their forward momentum would enable them to react to the rebounds at a faster rate.

Finally, the coach and the players should aim to learn all the straight hit techniques whereby injury risk could be reduced and the correct straight hit for different game situations can be achieved. There should be a realization that injury risk to players can be reduced by appropriate training and choice of hockey stick particularly at a young age.

References

Alcock J., Baker W., Donaldson A. and Gill L. (1997) *New South Wales youth sports injury report*. Northern Sydney Area Health Services.

Fox N. (1981) *Risks in field hockey*. In: Reilly T. (Ed.), *Sports fitness and sports injuries*. Faber and Faber, London, 112–117.

Freke M. and Dalgleish M. (1994a) Injuries in women's hockey: part one. *Sport Health* (Canberra, Aust.), 12/1, 41–42.

Freke M. and Dalgleish M. (1994b) Injuries in women's field hockey: part two on-tour. *Sport Health* (Canberra, Aust.), 12/3, 44–46.

Fuller M.I. (1990) A study of injuries in women's field hockey as played on synthetic turf pitches. *Physiotherapy in Sport* (London), 12/5, 3–6.

Herring S. and Nilson K. (1987) Introduction to overuse injuries. *Clinical Sports Medicine*, 6, 225–239.

Jamison S. and Lee C. (1989) The incidence of female injuries on grass and synthetic playing surfaces. *Australian Journal of Science and Medicine in Sport*, 21/2, 15–17.

13. Aquatics – Boating and Fishing

DIFFERENCE IN FORCE APPLICATION BETWEEN ROWING ON THE WATER AND ROWING ERGOMETERS

A.C. RITCHIE

School of Mechanical and Aerospace Engineering, Nanyang Technological University, Singapore

Indoor rowing has become a sport in its own right, and indoor rowers are popular conditioning machines in gyms and fitness clubs worldwide. This paper analyses the differences in biomechanics and system dynamics of "indoor rowing" and rowing in a boat. Due to differences in the disciplines of sculling and sweep rowing, sculling is examined here and compared with indoor rowing.

1 Introduction

"Indoor rowing", which began as a training tool for competitive oarsmen and women at times when training on the water is not available – during Atlantic crossings and when the rivers are iced over – has become a sport in its own right, with its own National championships and world records. The most successful and widely accepted indoor rowing machine is the Concept 2 indoor rower, shown in Figure 1. The rower has a sliding seat and footplate, and rows by pulling on the handle attached to a chain. The chain drives a flywheel during the power stroke, via a ratchet, and recoils under tension from a bungee cord during the recovery. The resistance and damping can be adjusted by changing the aperture area for flow entering the device, which functions as a rather inefficient centrifugal pump, and a performance monitor is provided to allow the rower to keep track of speed, distance travelled, power output, and time elapsed. Viscous drag due to the air being pumped through the flywheel enclosure provides a resistance which can be taken to be proportional to the rotational speed of the fan.

Figure 1. The concept 2 model C indoor rower used in this research.

In a single scull, the sculler sits on a sliding seat and is fixed to the boat by the shoes on the footrest. The oars (sculls) act as second class levers, with the pivot point at or near the junction between the shaft of the scull and the blade, with the effort of the sculler applied at the handle and transmitted through the oarlock (gate) to the rigger. This force serves to drive the boat during the power phase of the rowing stroke. Drag on the boat comes in the form of viscous or skin friction drag (proportional to the boat velocity), form drag due to the formation of a wake (dependent on hull form and proportional to boat velocity), and wave drag (proportional to boat length and the square root of the velocity) due to the formation of a bow and stern wave.

2 Force Application

For the rowing machine, the force is always tangential to the flywheel, hence it can be assumed that 100% of the effort of the rower is transmitted to it (Figure 2). The tension in the chain, F(t) is equal to the effort exerted by the rower.

On the water, the sculler is pulling on the oar, which in turn exerts force on the boat and the water. Since the force from the sculler's arm is not always perpendicular to the oar shaft and the force exerted on the water is not always in the direction of travel (Figure 3), there will be losses in the system and a variation in the efficiency of the exertion of force.

Figure 4 shows the various phases of the sculling stroke on the water. As can be seen, the sculler is at full compression at the catch, when the oars enter the water and the force is applied. At this point, the legs are biomechanically weakest as they are bent and the muscles used in the drive are close to their maximum length (Hall, 2007). A similar situation occurs on the rowing machine, which can be taken to be the same for leg and torso movement. However, the movements of the arms on the rowing machine are markedly different, as there is no rotation necessary in the shoulders and the handle follows a straight line, unlike the arc described by the handles of the sculls.

Additionally, the depth of the oar in the water, and the angle of the blade face at the catch, which have an enormous impact on the speed of the boat, are not simulated by the

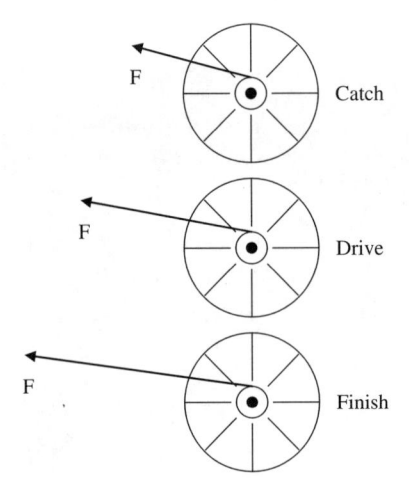

Figure 2.   Schematic showing system of force transmission from rower to ergometer flywheel.

rowing machine, hence bad technique is not punished by the machine. Rowers will say that Ergometers don't float – there is only so much that an ergometer can measure, and the ergometer is primarily used as a tool to gauge the rower's physical condition, rather than as an absolute predictor of performance.

3 System Dynamics

3.1. *Ergometer Dynamics*

The ergometer can be modelled as a damped flywheel, with the input force taken as an applied torque proportional to the motive force applied by the rower. For a flywheel with moment of inertia I, angular velocity ω, and damping coefficient c, analysis of the angular momentum gives the following relationship:

$$I\frac{d\omega}{dt} = T - c\omega \qquad (1)$$

3.2. *Dynamics of the Sculling Boat*

The single scull is a more complex system, as the hull is subject to skin friction drag, form drag, and wave drag (Rawson & Tupper, 1994). The sculler is assumed to be rowing on flat water with no wind, hence external factors such as waves, headwinds and so forth will not be considered. As the sculler sits on a sliding seat, which moves as he rows, the centre of mass of the system is constantly changing. For the purposes of the model, skin friction and form drag are lumped together. At racing speeds, single sculls travel at an average speed of just over 16 km/h or 4.44 m/s, which translates to a hull Froude number of 0.52. The critical Froude number of 0.54, which corresponds to a maximum in the wave drag resistance

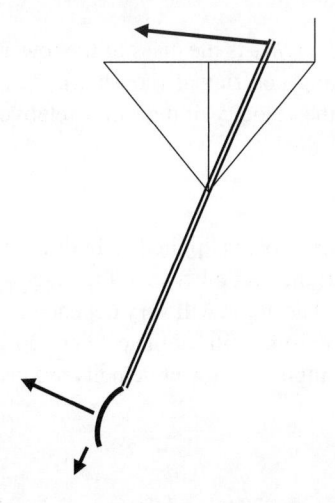

Figure 3. Transmission of force from rower to water through the oar.

Figure 4. Phases of the rowing stroke in a single scull.

curve, corresponds to a velocity of 16.54 km/h or 16.6 m/s. Conservation of momentum for the system yields:

$$\frac{d}{dt}(M_B V_B + M_R(V_B + V_R)) = F - k_{SF}V_B - k_{WD}V_B \qquad (2)$$

where M_B is the mass of the boat, M_R is the mass of the rower, V_B is the boat velocity, k_{SF} is the skin friction and form drag coefficient (constant), k_{WD} is the wave drag coefficient (dependent on V_B) and V_R is the velocity of the rower relative to the boat. V_R is assumed to be zero once the legs are straight.

3.3. Force Application

The same waveform for rower effort is applied to both the rowing machine and the boat. However, for the rowing boat, the angle between the blade and the direction of motion of the boat is taken into account. The angle will vary dependent on the gearing of the blade but for this model is taken to vary from −40° at the catch to 20° at the finish.

Inputs for applied force, angle, and rower velocity are given in Figure 5, along with the results of the model.

4 Model

The mass of the rower is taken to be 70 kg and the rower's height to be 1.8 m. The mass of the boat is 14 kg, as set by FISA (Fédération Internationale des Sociétés d'Aviron) standards

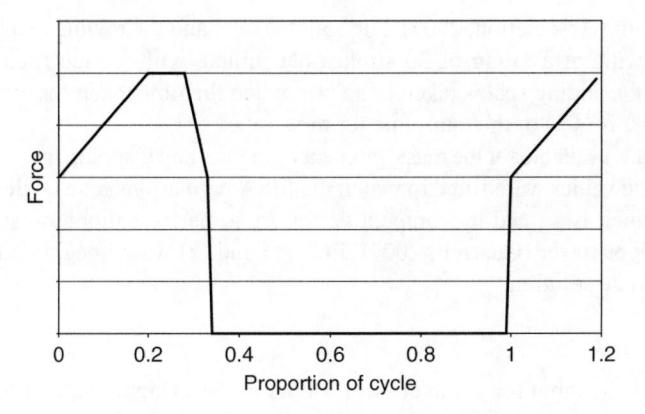

Figure 5a. Waveform of applied force in model.

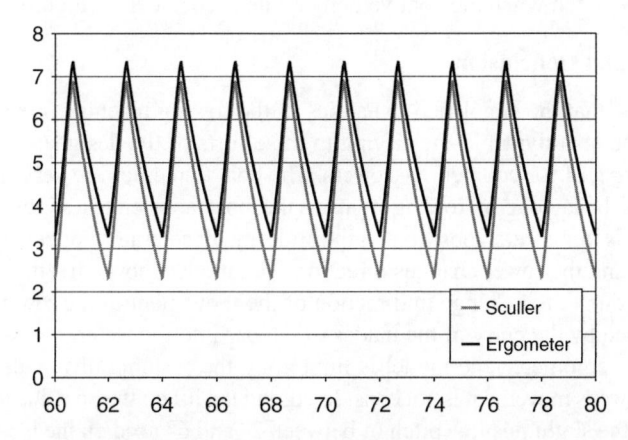

Figure 5b. Velocity of sculler (centre of gravity of sculler and boat) and velocity of ergometer in m/s at steady state.

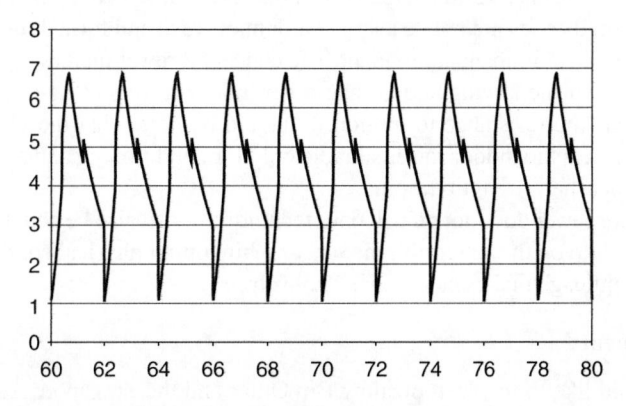

Figure 5c. Boat velocity at steady state. Note the sudden "stop" at the start of force application (catch) due to the rower's change of direction.

(Amateur Rowing Association, 2007). For both the boat and the rowing machine, the rower is taken as rowing at a rating of 30 strokes per minute with a slide:stroke ratio of 2:1. Movement of the sliding seat is taken to be 60 cm and the time taken for the legs to extend during the drive is 60% of the total time for the stroke.

As no data is available for the mass, moment of inertia, and damping factor of the ergometer flywheel, the values were fitted to match the FISA regulations for a coxless four with an 80 kg crew, which is agreed by competitive rowers to be the calibration standard for the Concept 2 indoor rower (Quarrell, 2007). Eqs. (1) and (2) were integrated numerically to give a steady state solution.

5 Results

Figure 5 shows the input force waveform for both models (Figure 5a), with the modelling results (Figures 5b and 5c). As can be seen, there is a close fit between the boat velocity and the ergometer when the centre of mass of the system (boat and sculler) is considered, but there is less of a match when the boat velocity alone is considered (Figure 5c).

6 Discussion and Conclusion

The results show that the simplified dynamics of the rowing machine provide an accurate simulation of the boat dynamics, testifying to the efforts of the designers in producing an effective training tool for competitive rowers. The boat dynamics are very interesting, and match observed behaviour of rowing boats – the boat accelerates slightly as the rower moves backwards to the catch position on the slide and decelerates noticeably as the blades enter the water and the rower changes direction. The model shows that this "check" at the catch is largely due to the change in direction of the movement of the rower's body, rather than being caused by the entry of the blades of the oar into the water.

As with the ergometer, the model is limited by the assumptions made. The pitching caused by the rower's movement is not considered, and the lift on the boat due to the buoyancy of the oars and the slight positive pitch of between 2° and 6° used on the blades (Redgrave, 1995), which reduces the drag on the boat during the drive phase, is similarly omitted. These factors will be considered as part of further research.

The model shows that the use of viscous damping on the vanes of the flywheel of the concept 2 indoor rower is a valid analogue for boat resistance, as the long and narrow hull form of rowing shells does not create a large amount of wave and form drag, with skin friction drag forming the dominant component. The concept 2 rower therefore provides a more accurate analogue of the rowing action than other indoor rowers using a friction brake or springs to provide the resistance to motion. It should be noted that the viscous nature of resistance also makes the indoor rower suitable only for conditioning and anaerobic threshold training, rather than strength training.

As the indoor rower does not punish poor technique, it can also be concluded that it can only be an adjunct to on the water training, or for maintaining physical condition. The rowing adage that "mileage makes champions" holds true.

Acknowledgements

The author would like to thank Rooizaimy Bin Omar and the Singapore Amateur Rowing Association for their assistance in this research.

References

Hall S.B. (2007) *Basic Biomechanics*, 5th Edition, McGraw-Hill, New York.

Rawson K.J. and Tupper, E.C. (1994) *Basic Ship Theory, Vol. 2*, 4th Edition, Longman, Harlow.

Amateur Rowing Association (UK) (2007) *British International Rowing Office – Rowing Guide* (online). Available: http://www.bosonmedia.co.uk/ara/gbrowing/rowing/. (Accessed: 12 June 2007).

Quarrell R. (2007) *The Rowing Service* (online) Available: http://www.rowingservice.com/. (Accessed: 13 June 2007).

Redgrave S. (1995) *Steven Redgrave's Complete Book of Rowing*, 2nd Edition, Partridge Press, London.

EFFECT OF OAR DESIGN ON THE EFFICIENCY OF THE ROWING STROKE

A.C. RITCHIE

School of Mechanical and Aerospace Engineering, Nanyang Technological University, Singapore

Oar design has changed enormously since the introduction of the "cleaver" or "hatchet" big blade oar in 1990. In this paper we examine the effect of blade design and shaft material and stiffness on the performance of oars, particularly with regard to the amount of energy absorbed by the oar as "deformation energy", both in the shaft and the blade or spoon.

1 Introduction

1.1. Blade Design

The rowing oar remained basically unchanged in form from the Seventeenth century to the mid twentieth century, with a broader, flattened blade at the end of a narrow shaft, as shown in Figure 1(a). The introduction of the tulip shaped macon oar in the 1950s was the first universally accepted change to the design, and this type of oar was prevalent until the introduction of the "big blade", first introduced by the Dreissigacker brothers (Concept 2, Vermont) in late 1991. (Miller, 2000).

1.2. Shaft Design

Up until the 1980s, most shafts and blades were made of wood, typically ash shafts with laminated plywood blades, with cross sections as shown in Figure 2(a). The shaft tapered from the thickest point, close to the handle, down to the junction between shaft and blade. Concept 2

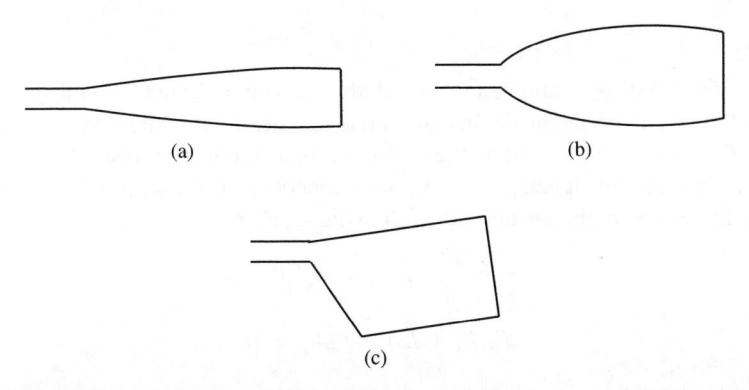

Figure 1. Blade Geometries: (a): Pencil type oar (1930s) (b): Modern Macon Oar (1980s) (c): Cleaver (Big Blade) type oar (1991-present).

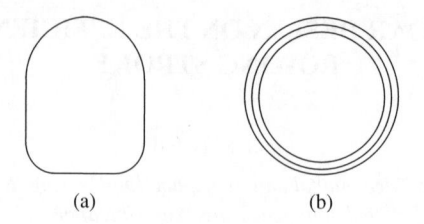

Figure 2. Cross sections of (a) wooden shaft and (b) composite shaft.

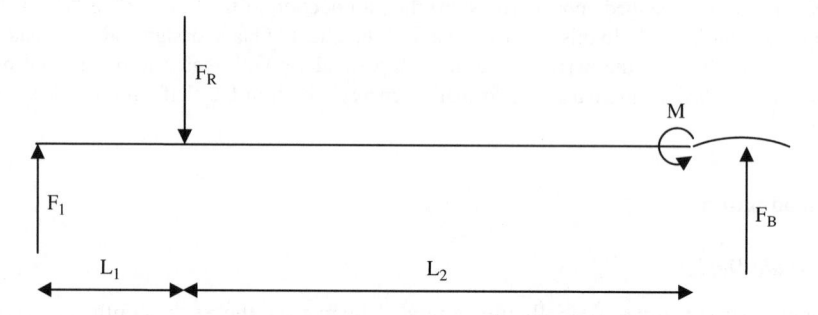

Figure 3. Model of forces acting on the blade during the drive phase of the stroke.

were the first to market with composite shafts and blades, (initially a mixture of glass and carbon fibre, made by spiral winding and later made from 100% carbon fibre, as Ultralight oars). The introduction of carbon fibre oars offered three principle advantages: lighter weight, greater consistency, and higher stiffness. Additionally, the use of mandrels and moulds for shafts and blades eliminated the need for skilled craftsmen in their manufacture. Although wooden blades remained easier to handle in very windy or rough weather (Topolski, 1989), most competitive crews rapidly switched to composite oars, due to these advantages.

The use of a wooden shaft or loom brings with it several disadvantages, as the available section is not optimal. Equivalent stiffness to even a thin section of carbon fibre would be inordinately heavy.

2 Model

A model of the force application on the oar shaft is shown in Figure 3 (Dudhia, 2007). The pressure of the water on the blade face gives rise to a bending moment M and force F_B on the blade. F_1 is the effort exerted by the rower and F_R is the force transmitted from the oar to the boat through the rigger. L_1 and L_2 are dependent on the gearing set by the crew. Resolving forces and moments gives the following relations:

$$F_1 + F_B = F_R \qquad (1)$$

$$F_1(L_1 + L_2) = F_R L_2 + M \qquad (2)$$

Analysis of the shape of puddles generated by rowers moving at steady state shows that the pivot point about which the blade rotates is located at or near the junction of blade or

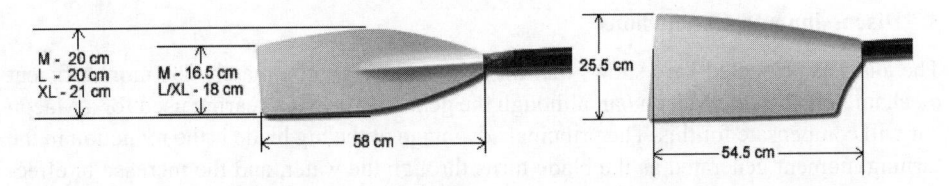

Effect of Oar Design on the Efficiency of the Rowing Stroke

Figure 4. Geometry of Macon oar (left) and Big blade smoothie (right).

Table 1. Force and area for different blade geometries.

	Area (cm²)	Force	Moment
Macon oar	1020	3.20×10^5 P	1.242×10^6 P
Big Blade oar	1132	3.33×10^5 P	1.237×10^6 P

shaft. This assumption is backed up by proprietary CFD of the pressure distribution on the face of the blade (Shaikh, 2007). In order to estimate the dependence of the transmitted moment, M, for a given value of F_B, pressure is assumed to vary linearly with distance from the pivot point and edge effects are neglected. Blade geometry was as given by the oar manufacturer, Concept 2. (Concept 2, 2007) as shown in Figure 4.

The product of pressure and elemental area was integrated analytically to give the pressure and force:

$$F = \int_0^B P(x)y(x)dx \qquad (3)$$

$$M = \int_0^B P(x)y(x)xdx \qquad (4)$$

where B is the overall length of the blade, P is the magnitude of the maximum pressure, and $y(x)$ is the height of the blade, as a function of x. For the blade dimensions given in Figure 4, the area, force, and moment are given in Table 1. Force and moment are given as a function of the pressure magnitude P, such that the pressure is given by:

$$P(x) = Px \qquad (5)$$

Taking a nominal applied force of 100 N gives a moment for the Macon oar of 38.7 Nm and for the Big Blade of 37.1 Nm. As the effort to turn the blade against this moment will not contribute to the overall movement of the boat, it can be considered as a loss. Resolving the forces in Eqs. (1) and (2) using the values for moment and force given here gives a rower effort of 268 N for the Macon blade and 258 N for the Big blade, for a motive force of 100 N applied to the water.

3 Discussion and Conclusions

The analysis presented here shows that the Big blade is approximately 4% more efficient mechanically than the Macon oar, although the generally heavier gearing used for a Macon oar will compensate for this. The principal advantage of the big blade is the reduction in the turning moment generated as the blade turns through the water, and the increase in effective area. The exact fluid dynamics of the blade are not considered here, although there has been much development effort devoted to this by manufacturers.

It is wrong to think of the blade as an aerofoil or hydrofoil, with the use of terms such as *lift* (desired) and *drag* (minimized), as can be found even in the manufacturer's publicity (Dreher, 2000). The principle by which any paddle operates is essentially induced form drag – equivalent to a parachute or sea anchor, where the movement of the blade through the water generates a wake, with an area of high pressure in front of and an area of low pressure behind the blade.

The analysis cannot take into account the "feel" of the blade design, although the lower values of pressure on the blade face translate into a slower velocity through the water and hence firmer lock at the catch.

References

Miller W. (2000) The Development of Rowing Equipment (online). Available: http://www.rowinghistory.net/Equipment.htm (Accessed: 13 June 2007).

Topolski D. (1989) *True Blue*. Doubleday Canada, Toronto.

Dudhia A. (2004) Basic Physics of Rowing (online). Available: http://www.atm.ox.ac.uk/rowing/physics/ (Accessed: 14 June 2007).

Shaikh J. (2007) Catching a Better Oar Design. *Ansys Advantage*, 1/1, 16 (online). Available: http://www.ansys.com/magazine/issues/1-1-2007-sports-and-leisure/sports-spotlight.pdf (Accessed: 14 June 2007).

Concept 2, manufacturer's data, from publicity materials.

Dreher J. (2000) Latest Development in Dreher Oars and Sculls, the Apex Blade Design (online). Available: http://www.durhamboat.com/blade_4.php (Accessed: 14 June 2007).

THE EFFECT OF ROWING TECHNIQUE ON BOAT VELOCITY: A COMPARISON OF HW AND LW PAIRS OF EQUIVALENT VELOCITY

M.M. DOYLE[1], A.D. LYTTLE[1] & B.C. ELLIOTT[2]

[1]*Western Australian Institute of Sport, Mt Claremont, Australia*
[2]*The University of Western Australia, School of Human Movement and Exercise Science, Crawley, Australia*

A new on-water rowing system was developed, in collaboration with institutions Australia wide, allowing accurate measurement of important biomechanical variables to be conducted whilst remaining highly portable. Real time data from a lightweight (LW) and heavyweight (HW) coxless pair of similar speed were examined to determine if differences in technique strategies could indicate methods to improve rowing performance. The HW crew was found to exhibit significantly higher indices of force and work, yet an average velocity equivalent to that of the LW crew. Examination of the velocity profile of the two crews revealed a noticeable difference in velocity patterns through the stroke. Subsequent analysis of the acceleration profile and the average segmental body velocities in each boat revealed different technique strategies which may indicate some underlying reasons the lightweight crew more efficiently produced equivalent boat speed.

1 Introduction

The movement of the seat in relation to boat velocity has previously been examined (Smith & Loschner, 2002), however there has been little attempt to quantify the relative contributions of the three major body segments in an on-water situation (Kleshnev, 2000). As the majority of coaching is done with no access to biomechanical measures, understanding how different methods of segment interaction can have a bearing on boat velocity may be an important aspect in the improvement of performance. Such knowledge may be readily applied as the approximation of segment velocities and contributions may be done through the use of easily accessible video or even by a trained eye.

The aim of this study was to examine several biomechanical variables in relation to boat velocity and acceleration, in conjunction with body segment movement strategies, of a LW and HW crew exhibiting dissimilar force production, yet similar average boat velocity in a race situation. It was envisioned that analysis of these variables would enable the identification of distinctive aspects of crew technique that may be significant in rowing performance. A better understanding of such strategies utilised by more efficient crews producing less force may be used in development of improved technique in all crews.

2 Methods

2.1. Data Collection

On-water biomechanical data were obtained from seven LW and seven HW men's coxless pairs. Individual LW and HW crews were then selected for comparison based on similar average velocity and temporal aspects of the stroke. These two crews were of similar experience, all four athletes competing at the 2006 Rowing World Championships.

Oarlock force data were collected using two-dimensional load transducers located within specially designed oarlocks, allowing force applied both normally and axially to the flat face of the oarlock to be measured. Horizontal oar angles were determined via a lightweight carbon arm/potentiometer arrangement which attached to the oar shaft, allowing horizontal angular displacement of the oar to be recorded, while ensuring normal range of motion of the oar about all three axes.

Seat position was determined using a drum and reel transducer similar to that previously described (Kleshnev, 2000), allowing the displacement of the rolling seat to be determined throughout the stroke. A similar arrangement was utilised to approximate the position of the 7th cervical vertebra (C7) for each athlete. This was then used to determine the position of the top of each athlete's trunk in reference to the seat movement, giving an indication of trunk flexion and extension throughout the stroke (Kleshnev, 2000).

All data were sampled at 100 Hz and transmitted via radio telemetry. Boat velocity was determined by a Rover unit (James et al., 2004), which enabled a 100 Hz accelerometer coupled with a 1 Hz GPS receiver to determine instantaneous boat velocity.

2.2. *Data Treatment*

Post collection, 15 consecutive stokes were selected for analysis from each pair and an ensemble average stroke was generated. Signal conditioning was applied only to variables that were used to derive velocity variables using a recursive Butterworth low-pass filter.

The measurement of seat position, C7 position and oar angle throughout the stroke allowed an approximation of the segmental velocity contribution to overall handle velocity to be calculated. This was achieved by first calculating the velocity of the seat, C7 and the linear velocity of the handle. The relative velocities of the trunk and arms were then calculated as follows:

$$\text{Vel}_{trunk} = \text{Vel}_{C7} - \text{Vel}_{seat} \tag{1}$$

$$\text{Vel}_{arms} = \text{Vel}_{handle} - \text{Vel}_{C7} \tag{2}$$

A segmental velocity graph was then derived. While relative contributions may differ with technique between crews, the fundamental shape and timing of the individual segments is determined by the basic biomechanical actions of the stroke.

3 Results and Discussion

The average boat velocity for all 14 crews at the self-selected race rate was $4.99 \pm 0.13 \text{ m} \cdot \text{s}^{-1}$, while that of the HW pair was $5.01 \text{ m} \cdot \text{s}^{-1}$ compared to $5.04 \text{ m} \cdot \text{s}^{-1}$ for the LW pair. Figure 1 demonstrates the velocity and acceleration profiles of the two comparison crews together with the pooled data from all crews. It can be seen that all three velocity graphs exhibit the typical form which has been previously reported in the literature (Martin and Bernfield, 1980).

To facilitate explanation of various features of Figure 1, the stroke was divided into six discrete phases, chosen to encompass specific occurrences in the velocity and acceleration curves, and are as follows;

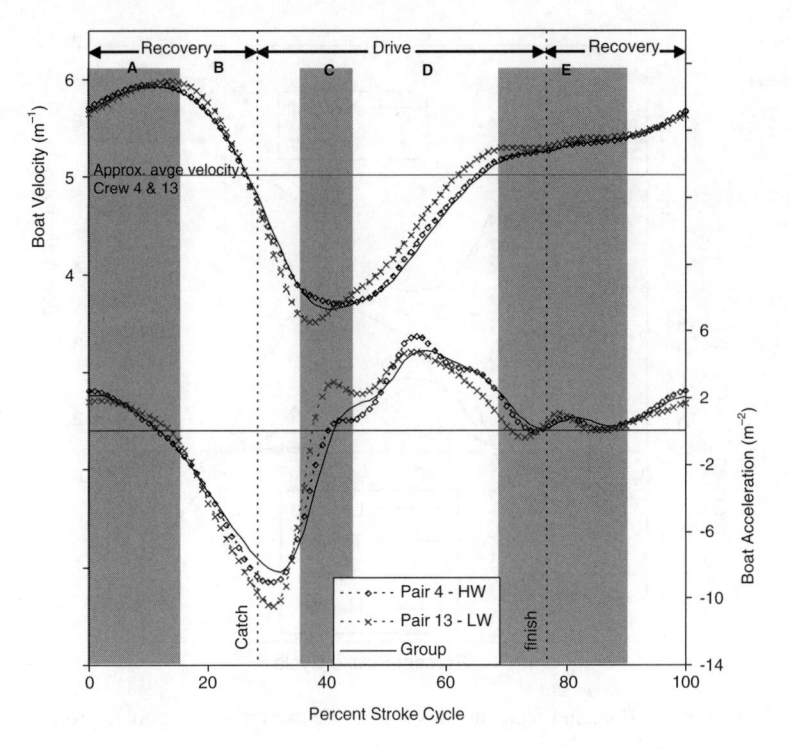

The Effect of Rowing Technique on Boat Velocity

Figure 1. Boat velocities and accelerations during a normalised stroke cycle (15 strokes).

A: Maximal Velocity Phase.   B: Boat Deceleration Phase.
C: Minimal Velocity Phase.   D: Boat Acceleration Phase.
E: Zero Acceleration Phase. F: Initial Recovery Phase.

3.1. Maximal Velocity Phase

The maximum velocity of the pooled group (5.94 m · s^{-1} at 11% of the stroke) and the HW pair (5.95 m · s^{-1} at 11%) was very similar, occurring at the same position in the stroke (11% of the stroke cycle). The maximum velocity of the LW pair, while slightly higher at 5.98 m · s^{-1}, occurred noticeably later, at 13% of the stroke cycle (1.20 S.D.'s later than the group mean).

Examination of the segment velocities (Figure 2) reveals that in this phase, the boat velocity appears to approximately mirror the pattern of seat velocity for both crews. The LW crew exhibited a seat velocity lower than that of the HW crew for the majority of the phase, only exceeding the HW pair late in the recovery. Leaving the deceleration of the seat into the catch till later in the recovery enabled them to achieve a slightly later maximum recovery seat speed of 1.40 m · s^{-1}, at 15% of the stroke cycle as compared with 1.36 m · s^{-1} at 14% for the HW pair.

3.2. Boat Deceleration Phase

From maximum velocity the speed of the boat decreased rapidly as the rowers continued to move to the stern of the boat, applying force to the feet in order to reduce their forward

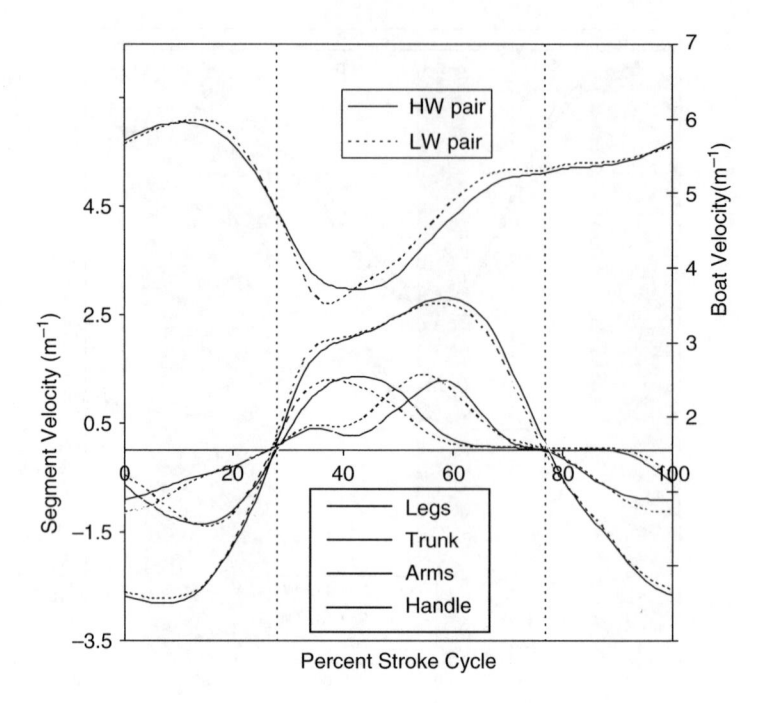

Figure 2. Boat and segment velocities of the two crews and pooled group.

body momentum in preparation for the catch. This is reflected in the seat velocity exhibiting a marked deceleration from maximal recovery velocity to a reversal of direction at the catch and rapid acceleration at the beginning of the drive phase.

The two crews showed similar velocities at the catch of $4.77 \text{ m} \cdot \text{s}^{-1}$ for the HW pair and $4.74 \text{ m} \cdot \text{s}^{-1}$ for the LW pair (95.2% and 94.0% of their average speeds respectively). This is reflected in the boat acceleration during this period, as the LW pair demonstrated a greater deceleration, reducing the boat speed more rapidly from maximum velocity.

3.3. Minimal Velocity Phase

Following the rapid deceleration of the boat, minimum boat velocity then occurred sometime after the catch. This has been attributed to the propulsive force applied early in the stroke beginning to overcome the hydrodynamic drag on the boat (Martin & Bernfield, 1980).

The minimum velocity of the LW pair occurred earlier, (37% of the stroke cycle, and was lower, at $3.51 \text{ m} \cdot \text{s}^{-1}$) than that of the HW pair, ($3.71 \text{ m} \cdot \text{s}^{-1}$ at 43%). The difference in magnitude and location of minimum velocity between the crews can be better understood by examination of the acceleration curve and segment velocities during this phase. From a greater minimum acceleration, the LW pair demonstrated a more rapid return to positive acceleration, reducing the time the boat spent at minimum velocity and ensuring their boat velocity increased noticeably before the HW pair. This is reflected by the seat velocities immediately after the catch. The LW pair exhibited a more rapid leg drive earlier in the drive phase, reaching a maximum at 38% of the stroke cycle. This occurred 5% earlier in the stroke, (approximately 10% earlier in the drive phase) than the HW pair. This earlier leg drive contributes to an earlier rise in handle speed at the beginning of the stroke for the LW combination.

3.4. Boat Acceleration Phase

While initial peak positive acceleration has been attributed to early force application, the size and magnitude of the "gap" between this and the second positive acceleration depends on the co-ordination of the movement pattern of the legs and trunk (Kleshnev, 2003). From an initial period of low positive acceleration, the HW crew exhibited a sharp rise in acceleration later in the drive phase, becoming higher than the LW crew at 52% of the stroke cycle. It can be seen that due to the earlier leg drive, the LW pair had an initial higher handle speed at the beginning of the drive phase, which dropped below that of the HW pair at approximately the same point the LW acceleration also became less. The segmental velocity graphs reveal that after the initial increase in boat speed, further increases occur at similar rates for both crews. Because the LW crew had a minimum velocity occurring earlier in the stroke, combined with a sharper minima, they reached their average velocity at 62% of the stroke, noticeably before the HW pair (at 65%).

3.5. Zero Acceleration Phase

Towards the end of the drive phase all crews show a decrease in the boat acceleration and an associated plateau in boat velocity as they approach the finish. Immediately after a brief period of boat acceleration is seen, due to negative acceleration of the mass of the rowers as they prepare for the recovery. This is followed by minimal acceleration as the initial movement of the recovery is the movement of the handle away from the release by use of the arms and the trunk, exerting little force on the boat in the propulsive direction. There was little difference between the two crews during this period, with the LW pair maintaining a slightly higher velocity due to the initial higher velocity entering the phase.

3.6. Initial Recovery Phase

The boat velocity began to increase at approximately 90% of the stroke cycle for both crews, which is reflected in the increase in boat acceleration as the seat starts to move from the bow of the boat and the mass of the rowers begins to accelerate towards the stern. The acceleration of the boat continued through the end of the stroke cycle, remaining positive until the rowers begin to decelerate towards the catch.

It was seen that the timing and magnitude of boat acceleration was reflected in the seat movement strategies of the two crews. While both crews initiated seat movement in the recovery at around 90% of the cycle, it was seen that the seat velocity of the HW crew in the initial stages of the recovery was higher than the LW crew, resulting in a higher initial acceleration of the boat in the last 10% of the stroke cycle.

4 Conclusions

It has been demonstrated in this study that the movement patterns which athletes in a coxless pair employ in their technique can have an identifiable outcome on the instantaneous velocity of the boat. This finding may have implications on the selection of heavyweight athletes. Although it has previously been demonstrated that successful HW athletes tend to be taller and larger than the average HW competitor, it should not preclude the consideration of marginally smaller athletes. These athletes, although unable to produce the equivalent amount of force, may employ a more effective technique strategy in terms of boat performance.

References

James D.A., Davey N. and Rice T. (2004) An accelerometer based sensor platform for insitu elite athlete performance analysis. In: Rocha D., Sarro P.M. and Vellekoop M.J. (Eds.), Proceedings of IEEE Sensors 2004, Third IEEE International Conference on Sensors. (pp 1373–1376). Vienna, Austria.

Kleshnev V. (2000) Power in rowing. In, Hong Y. and Johns D.P. (Eds.), Proceedings of the XVIIIth International Symposium on Biomechanics in Sport. (pp 662–666). Hong Kong: The Chinese University of Hong Kong.

Kleshnev V. (2003) *Rowing Biomechanics Newsletter*, 3(11) [online]. Available: http://www.biorow.com/RBN_en_2003.htm (Accessed: 2/4/06)

Martin T.P. and Bernfield J.S. (1980) Effect of stroke rate on velocity of a rowing shell. *Medicine and Science in Sports and Exercise*, 12/4, 250–256.

Smith R.M. and Loschner C. (2002). Biomechanics feedback for rowing. *Journal of Sports Sciences*, 20, 783–791.

OAR FORCES FROM UNOBTRUSIVE OPTICAL FIBRE SENSORS

M. DAVIS[1] & R. LUESCHER[2]

[1]*Australian Institute of Sport, Biomechanics Department, Leverrier Crescent Bruce, Australia*
[2]*Australian Institute of Sport, Physiology Department, Leverrier Crescent Bruce, Australia*

This project investigated the viability of mounting optical fibres (approximately three times the thickness of a human hair) on sculling oars to measure strain during rowing. The authors conducted successful pilot studies to measure oar force in the laboratory using optical fibres, however the results were obtained using traditional instruments mounted in large laboratory racks. Based upon our encouraging preliminary laboratory trials, the aim of this project was to miniaturise the electronics of a fibre sensor to fit within the shaft of an oar. The same technology could be applied to measure the flexure of any tube, such as in kayaking (paddle), sailing (mast and spars), pole vault and gymnastics applications. The principle of optical fibres as strain sensors is well established (e.g. Lo, 1998). Light passes down the core of a fibre via total internal reflection until it reaches a series of small etchings (called a Bragg grating); the periodicity through the grating changes as the fibre is stretched or compressed. This project customised the design of the Bragg sensor detector and laser electronics for the inside diameter of a carbon sculling oar. Custom signal processing circuitry and software was developed to allow two oars to be monitored simultaneously.

1 Introduction

The aim of this project was to miniaturise the electronics of an optical fibre sensor to fit within the shaft of an oar. The subsidiary aims were to calibrate the sensor with known masses and subsequently collect data on-water from two oars connected to data loggers. A data logger was required with a sampling frequency off at least 300 KHz and with working analog inputs. It became clear early on that the development of a high speed miniature data logger with internal signal processing and telemetry was beyond the scope of this research project.

2 Methods and Results

Within this project there were two parallel sub projects which were developed over the last eighteen months.

1. The embedding of an optical fibre sensor array within the sculling oar. The end result would need to be sturdy, durable and impervious to water and invisible to the athlete.
2. The development of a miniature Laser Controller and Scanning Detector Electronics. This unit would need to be very small, light, able to dissipate heat from the laser driver, fit inside the oar and be centred around the oarlock. The electronics' development was a considerably larger portion of the total project.

2.1. *Mounting Fibre Bragg Sensors on the Carbon Composite Rowing Oar*

During the course of this research project, optical fibres with Bragg Grating sensors were placed on a sculling oar. Three Bragg Grating sensors were embedded on the oar, with the sensors spaced at 500 mm intervals. Two sensors were placed on the tension face on the 0 deg axis to measure the bending strain of the shaft, with the third being placed at a 45 deg axis near the blade to measure torsional strain of the shaft. The surface of the oar was sanded and prepared for bonding and a small hole was drilled under the sleeve on the oar to bring the fibre from the external surface to the inside of the shaft. The fibre was protected with microbore tubing along its entire length inside the shaft and strain relief was provided for the connector by securing Aramid cord to the shaft. The sensors were positioned and tacked in place using Loctite 401Cyanoacrilate instant adhesive, ready for the glass epoxy protective layer to be applied. The protective glass layer consisted of a strip of woven Glass fibre fabric (100 g sqm plain weave) 15 mm wide cut to match the length of the external fibre. Release film and breather fabric was also cut at this time to be used during the vacuum bag curing process. Epoxy laminating resin (SP Systems Ampreg 22) was mixed as per manufacturer's instructions and the area was "wet out" around the fibre. The glass strip was applied and the entire assembly vacuum bagged as per the resin cure schedule. This process fully sandwiched a fibre between the carbon oar shaft and the protective glass layer. The shaft was then sprayed with clear polyurethane lacquer which made the fibre almost impossible to see on the shaft.

2.2. First Laboratory Trials

In the laboratory, initial trials of the sensors were performed using an Optical Spectrum Analyzer. The laser light source was connected to the fibre using a splitter to check that the three sensors were working. The oar was loaded incrementally with a series of masses, and measurements were taken of the wavelength shift of the sensor. The oar was secured in a jig that held the sleeve and the handle, with a sling tied around the blade to which the masses were attached in increments of 1.14 kg. The sensors were all reading well until after approximately 17 kg when there was a sudden decrease in signal amplitude. All connections were checked and appeared sound, it was determined that the likely cause was a micro-fracture in the fibre, resulting in attenuation of the signal. The testing had otherwise gone well with the resultant curve linear up to the point of failure.

The next key task was to determine where, how and why the failure occurred. A visible laser was used to check for discontinuities in the fibre, a power meter was also attached so we could test the fibre attenuation pre and post attachment. This evaluation led us to the belief that the flaw existed prior to mounting. Problems were also found in the way the fibre was protected in the oar cavity and the associated strain relief and termination of the connector.

2.3. Mounting of the FBG Sensor on the Second Oar

It was decided to change to a tougher Polyimide fibre coating to provide more durability for future trials. The way the fibre was housed inside the shaft was also changed. A short piece of composite tube was cut to fit into the tapered oar shaft. A bulkhead connector was then attached to this tube and a commercial fibre optic patch cord was used to provide strain relief. In this way there is less chance of damage to the bare fibre.

The visible laser was connected to the fibre to check for damage or stress on the new fibre, and the laser was left attached for the entire mounting process to highlight any problems. The layout of the sensors was planned and the shaft prepared as previously described. The fibre was placed on the surface in position and excess fibre was wound and secured within the bulkhead tube for protection. This time all three sensors were placed on the tension face on the 0 deg axis at 500 mm centre spacing between the sensors. The fibre was tacked in place with "Magic tape" to position the fibre until the epoxy was ready and removed during the epoxy adhesive application. The fibre was then sandwiched with epoxy and glass fibre cloth and vacuumed down as before. All this was performed with the visible laser attached and no problems were encountered.

2.4. Second Laboratory Trial

In the lab the fibre was connected to the Optical Spectrum Analyser and there was a clear strong signal. The oar was first flexed manually to observe the wavelength shift and then the same testing protocol as before was adopted, adding masses in 1.14 kg increments up to approximately 20 kg at the blade. This was done multiple times without any failure of the fibre, and the testing of the sensor response to load was a success. No hysteresis was observed when the test was repeated. Refer to Figure.1 for these results.

2.5. Development of Miniature Electronics Circuitry for Sensor Interrogation

After assessing a number of different system setups, the development of an electronics system which has three major components was agreed upon:

1. Laser Controller – This circuitry provides current control to the laser diode, it has closed loop control of the Peltier Cooling Pump via a thermistor inside the laser casing, and control is effected by PID circuitry. The laser controller also has a number of failsafe mechanisms. Current and optical power feedback, over and under voltage inhibits and Peltier Cooler failure inhibit. These inhibits will immediately shutdown the laser if any problems are encountered.
2. Scanning Detector circuitry – The scanning detector (Mosquito TOD050) is an optical detector with a very narrow wavelength bandwidth. By applying a logarithmic

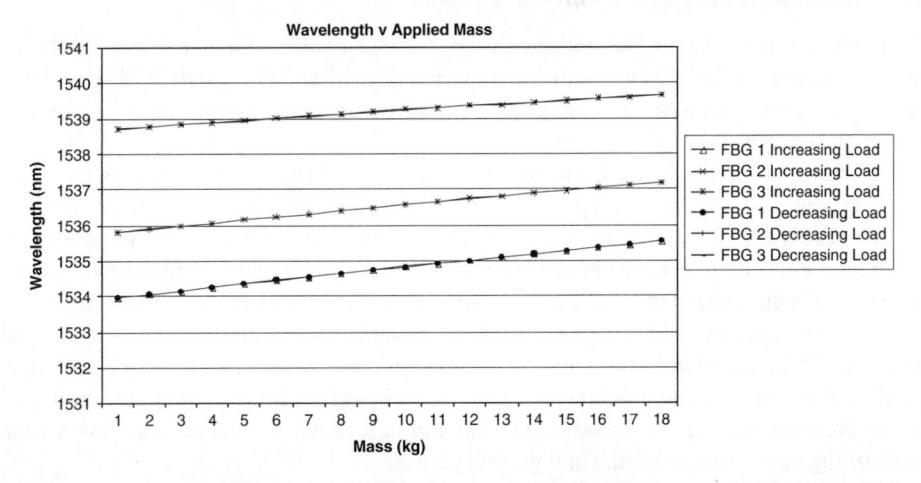

Figure 1. Force Vs Wavelength Calibration chart for Polyimide Bragg sensors. Note: Minimal hysteresis from several test cycles.

drive current to this device a narrow bandpass region can be ramped through certain optical wavelengths. In our case between 1540 nm and 1565 nm.
3. Data Logger – Due to limited resources a miniature data logger was not built. For proof of concept we used a laptop computer and a National Instruments PCMCIA data acquisition card (DAQCard-6024E). Vbragg Pty Ltd developed a program using LABVIEW software for the signal processing. Using this software the power spectrum wass subjected to a peak fitting algorithm to determine the Bragg wavelength of each reflected peak. The Bragg wavelength was then tracked and monitored during scans.

2.6. Development of Scanning Detector Circuitry

The original prototype of the Scanning Detector Interrogator was built quite large for ease of construction and to prove the concept. It was housed in a diecast box so that it could be cycled in a heat chamber and a temperature characterisation realised. The circuitry generates a 25 Hz logarithmic pulse of 6 milliSeconds duration. The amplitude of this pulse is adjustable but ideally is set to 190 milliAmps. The scanning detector (Aegis Semiconductor TOD050) uses this current to heat the substrate and ramp through the intended wavelength range. There is also circuitry to provide feedback on detector substrate temperature, a photo detector transimpedance amplifier and V-Filter output.

The final version of the Scanning Detector Interrogator circuitry was merged with the Laser Controller circuitry on to a single printed circuit board. From lab tests it seems the scanning detector device develops some self heating problems if operated above 30 Hz.

The control pulse generated by this circuitry is not quite ideal but is the best that could be achieved with analog circuitry. Preferably an on board microcontroller would be used to generate these pulses. With the processor generating the curve as recommended by the manufacturer (refer to online datasheet), it is the authors belief that the pulse frequency could be increased, possibly up to 50 Hz.

2.7. Development of Laser Controller Circuitry

Development of the Laser Controller circuitry took place in two major stages. Initially a large prototype was built incorporating a Superlum Inc. SLD-761-LP-DIL-SM super luminous diode. This prototype was large, used conventional components and it was used to prove the concept.

The original larger Laser Controller worked well but there were concerns that there might be some limitations with optical output power (i.e. Max. Optical power for this laser is 170 micro watts). It was decided that we would accommodate a higher power (10 milliWatt) super luminous diode in the later miniature Laser Controller design. The higher power laser diode used is the Superlum Inc. SLD-761-HP1-DBUT-SM (refer to online datasheet).

The new design would also power the lower powered laser if retro fitting was required. The smaller board used surface mount technology integrated circuits which are much smaller than conventional components. The circuit board itself has four circuit layers. All of the components were mounted on the top and bottom layers, the internal layers route most of the power supply wiring and ground planes.

Two high density Lithium Ion Polymer battery packs were used to supply the $+/-14.8$ volt rails. On board power supplies brought this down to regulated $+/-12$ volts. There was also a 10 volt reference for the Peltier Cooler PID controller and a -2 volt reference for biasing of circuits. The circuitry had provision for system timeout, once the oar was turned on via a magnet and reed switch it could only run for a predetermined time (i.e. 2 hours) before it automatically shuts down.

3 Conclusions

The authors believe the first failure was caused by fibre damage prior to mounting and was independent of the embedding procedure. It is also possible to place the fibre on the compression face of the oar to reduce the likelihood of micro-fractures. At this point, the authors believe that the fibre will be mounted as a secondary step to oar manufacture due to the requirement of the fibre to exit into the internal cavity of the shaft. Other composite manufacturing methods may allow the fibre to be placed within the composite material during manufacture avoiding this secondary step.

Although this force sensing system will provide reliable hysteresis-free data, there are still a number of tasks to be completed before this is a reliable on-water training aid. Once the oar is characterized against temperature recalibration will not be necessary. The laser controller still needs some refining; there are some restrictions in the laser current control on the more powerful laser device. Machining of some special heat sinks to fit inside the oar will be necessary.

The authors have begun experimenting with heat pipes; a refrigerant pipe which can remove the heat down to the oar blade where the water can be used as a heat sink.

At present the authors are using an optical coupler to split the optical signal; an optical coupler reduces the optical power level by at least half. Optical circulators have a very low optical loss; if an optical circulator were to be used there will be no need to run the laser at a higher optical power level. If the optical power is reduced this will in turn reduce the current being drawn by the laser diode, and will also reduce the amount of current used to cool the laser device. With less current being drawn, it may be possible to reduce battery size and most importantly weight, and there will also be less heat dissipated from the heat sinks.

Acknowledgements

The authors would like to thank Vbragg Pty Ltd for their support during this project.

References

Aegis Semiconductor Pty. (2006) Mosquito TOD-050 Preliminary Datasheet (online). Available: http://www.oida.org/PTAP/prototypes/us_83_1.pdf (Accessed: 2006).

Lopez-Higuera J.M. (2002) *Handbook of Optical Fibre Sensing Technology*. John Wiley & Sons, West Sussex.

Lo Y.L. (1998) Using in-fiber Bragg grating sensors for measuring axial strain and temperature simultaneously on surfaces and structures. *Optical Engineering*, 37/8, 2272-2276.

Othonos A. and Kalli K. (1999) *Fibre Bragg gratings: Fundamentals and Applications in Telecommunications and Sensing*. Artech House, Boston.

Superlum Diodes Pty. (2006) SLD-761-HP1-D134t-SM Preliminary Datasheet (online). Available: http://www.superlumdiodes.com/pdf/76hp.pdf (Accessed: 2006).

MATERIALS MODELLING FOR IMPROVING KAYAK PADDLE-SHAFT SIMULATION PERFORMANCE

P. EWART & J. VERBEEK

The Department of Engineering, Materials Division, The University of Waikato, Hamilton, New Zealand

In this work computer aided design tools are used to build iconic, semi-iconic and analogue CAD models to represent the microstructure of fibre reinforced polymer composite materials suitable for finite element simulation. The analogue method was identified as the most efficient approach due to less complex geometries with reduced pre-processing data, and reduced simulation time. The analogue method was then used to create virtual prototypes of composite kayak paddle-shafts that were 'tested' using a design analysis tool. Where elastic modulus values are required for design analysis previously developed micromechanical models are used. Simulation results were compared to real results obtained from physical testing and it was concluded that the analogue method is a favourable modelling approach in regard to composite kayak paddle-shafts. In general, these simulation tools are first pass design aids and are considered only for comparative analysis of known systems, being less than ideal for simulation of unknown systems. The use of the predictive models shown here have made more realistic simulation data possible.

1 Introduction

The use of predictive models to aid product design and development of non-standard materials needs to become an integral part of product development. Historic methods of design-build-test followed by redesign-build-test to create a product suitable for manufacture is now considered outdated. Innovative companies of today are only interested in the most direct and shortest time to market.

Virtual engineering, the workshop of the future, creates virtual prototypes in a 3D environment on a desktop computer. The virtual prototypes are tested using design analysis tools to ensure that only the best possible designs are sent for manufacture with a higher expected success rate and a lower rework rate.

In this study the objective was to determine the most efficient approach in regard to time and accuracy to model and simulate loading response of 'as manufactured' kayak paddle-shafts using COSMOSWorks (CW) FEA design analysis, and to compare the simulation data to actual experimental data.

2 CAD modelling

To use FEA to simulate equipment 'in use' a geometrical representation is required in a form suitable for the FEA solver to recognise. A CAD model made from a description of the products geometry will fulfil this requirement. On a macro-scale the paddle-shaft geometry is a simple thin walled cylinder. On a micro-scale the geometry is very complex. Figure 1a) shows a transverse section of the fibre composite microstructure with Figure 1b) showing a longitudinal section view.

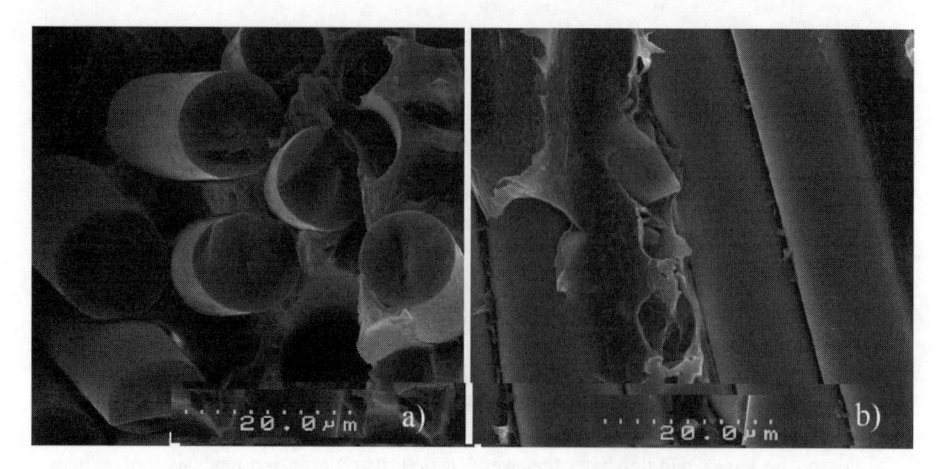

Figure 1. Fibre composite micrograph showing, a) fibre structure, b) discrete layers.

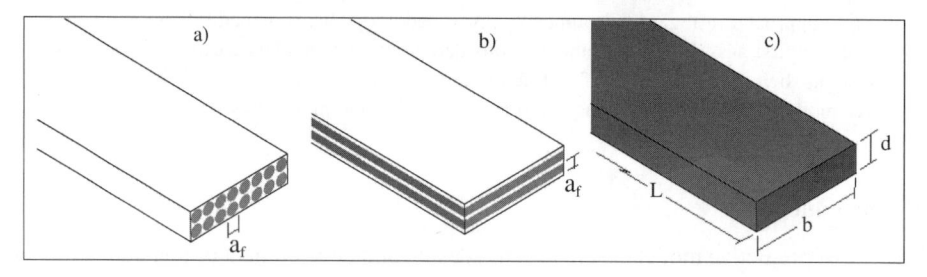

Figure 2. CAD material models a) iconic, b) semi-iconic, c) analogue.

3 CAD, materials models

To determine which approach is best for modelling composite materials the author uses three approaches determined from the micrographs of Figure 1.

These models are:

(a) Iconic, the model bares the physical attributes of the composite reinforcement and matrix. Mechanical properties are specified for each composite component, Figure 2a).
(b) Semi-iconic, the model bares physical attributes such as size but has an internal geometry represented as discrete layers of reinforcement and matrix. Mechanical properties are specified for each component, Figure 2b).
(c) Analogue, simplified geometry without the micro-physical characteristics. This approach is easier to create with less time involved in the modelling. Mechanical properties specified for the composite as a monolithic material, Figure 2c).

The CAD models are used to represent the composite material in the form of a physical test sample.

Table 1. Pre-processing data for the material model.

Model	Iconic	Semi-iconic	Analogue
a$_f$ mm		0.0250	
d mm	0.060	0.092	
L mm	0.960	1.467	
b mm	0.240	0.367	
CAD file size	211 KB	123 KB	47 KB
Pre-pro file size	15.1 MB	10.5 MB	0.8 MB
Element size mm	0.013	0.019	0.046
Element count	70870	49236	2877

4 Mechanical properties

The data required to run a simulation and analyse loading conditions using CW, is the elastic modulus, Poisson ratio, and material density. For many materials this can be obtained from the manufacturer or generic values can be used for commodity materials.

The iconic and semi-iconic approaches require mechanical properties for each individual component of the composite material. For the analogue approach mechanical properties are not readily available when the material is proprietary, or, when relevant data is not supplied. One approach to get accurate data involves physical testing. This can be time consuming and expensive especially where testing instruments or the relevant analytical skills are not readily available.

In this study two models, Eq. 1 and Eq. 2 (EV1 & EV2), are used to predict the flexural modulus of the composite used in the analogue approach (Ewart and Verbeek 2005; Ewart and Verbeek 2006). These models have been shown to be reliable for determining the flexural modulus values for some thermoset and thermoplastic reinforced composites (Ewart and Verbeek 2005; Ewart and Verbeek 2006).

$$E_C = \frac{12}{n^3} \cdot \begin{bmatrix} (\Gamma_1 - \Gamma_2 \cdot V_f^3 - \Gamma_3 \cdot V_f^2 - \Gamma_4 \cdot V_f) \cdot E_m \\ + (\Gamma_2 \cdot V_f^3 + \Gamma_3 \cdot V_f^2 + \Gamma_4 \cdot V_f) \cdot E_f \end{bmatrix} \quad (1)$$

$$E_C = E_m(\Lambda_{m1} + \Gamma_{m1}) + E_f(\Lambda_{f1} + \Gamma_{f1}) \quad (2)$$

E_C flexural modulus, $n = 5$, V_f volume fraction of fibre, E_m elastic modulus of matrix, E_f elastic modulus of fibre, The definition of $\Gamma_1, \Gamma_2, \Gamma_3, \Gamma_4, \Gamma_{m1}, \Gamma_{f1}, \Lambda_{m1}$ and Λ_f1 are given in the appendix.

5 Pre-processing

The CAD model dimensioned as shown in
Table 1, and with materials data from Eq (1) and (2) were used in CW to obtain the pre-processing data of Table 1. End restraints configured to represent a simple three point

Table 2. Composite kayak paddle-shaft material data.

Material	Shaft diam.	Shaft thickness	Fibre diam.	Density	Elastic mod.	Poisson
	mm			g/cm^3	Gpa	[-]
SP Epoxy	N/A			1.22	3.17	0.35
ADR 246 Epoxy				1.14	3.38	0.35
Hemp fibre	28.0	3.33	0.050	1.40	25.00	0.30
Glass fibre	27.9	1.25	0.015	2.60	73.00	0.20
Carbon fibre	27.6	1.10	0.008	1.61	235.00	0.10

Table 3. Post-processing data for paddle-shaft simulation and experimental.

Shaft type	V_f	Load	Modulus			Deflection			
			Exp.	EV1	EV2	Act.	Simulation		
		N	Gpa				Exp.	EV1	EV2
							mm		
Hemp/Epoxy	0.25	50	2.30	5.74	4.97	2.20	2.059	0.825	1.055
Glass/Epoxy	0.55	100	32.50	26.89	18.48	0.86	0.636	0.766	1.114
Carbon/Epoxy	0.6		67.42	94.05	40.48	0.50	0.355	0.254	0.590

bend were used with centrally applied loads. The direct sparse solver was chosen for the analysis.

6 Experimental

Kayak paddle-shafts manufactured from epoxy resin reinforced with continuous fibres of carbon, glass or hemp were tested (Table 2). The shafts were cut to 600 mm lengths, placed in a specially designed jig and loaded in a Lloyd universal tester to give central deflection values for each shaft (Table 3). Elastic modulus values were also determined from the actual test data.

7 Equipment simulation

Building the composite CAD models has shown that complexity affects pre-processing. Shown in Table 1, the element count is highest for the iconic model with the analogue model count some 25 times lower.

As the objective of the study was to determine the most efficient approach in regard to time and accuracy to model and simulate the load response of kayak paddle-shafts the analogue approach was taken. Using the paddle-shaft dimensions of Table 2 analogue models were produced with the elastic modulus values obtained from, testing the actual shaft, Eq(1) or Eq(2). The virtual paddle-shafts were then transferred to CW where loading was simulated using a simple three-point bend configuration.

8 Results and Discussion

The data in Table 3 shows the applied load values used for both the virtual shaft simulations and the actual shaft tests. The results from the simulations vary for the different shaft types. The hemp/epoxy shaft results show that both EV models produce considerably lower deflection values than using the experimentally derived data while even the actual value from testing was about 7% higher than the closest simulation result. The glass/epoxy shaft results show EV models to produce higher deflection values than the experimentally derived data, the closest simulation result to the actual test value being the EV1 model at 10% lower. The carbon/epoxy shaft shows simulation results for the EV1 model lower and the EV2 model higher than the experimental data with the EV2 model closest at 18% higher than the actual value.

Eq (1) of the EV1 models was derived for both short and continuous fibres; Eq (2) of the EV2 models was derived specific to short fibre composites. While this may put question to the use of the EV2 models it is not considered to invalidate the results at all. The variability in the results can be considered due to a number of differences in the shaft materials. The ratio between matrix and fibre modulus values increases from about 7 for the hemp to 23 for the glass and 74 for the carbon.

The hemp fibres, unlike the glass or carbon, are not manufactured specifically for structural reinforcement and so would not be optimized for this application. They merely show how sustainable engineering practices can be easily integrated into future design considerations for sports equipment.

9 Conclusion

The analogue model is shown to be a less complex and easier method of modelling composite materials. The pre-processing data in Table 1 also supports use of the analogue model with efficiencies that aid simulation. Solids modelling packages such as COSMOSWorks used for conceptual and mechanical design contain the functionality to undertake optimization and design analysis. In general, simulation tools integrated into CAD programs are first pass design aids and are considered only for comparative analysis of known systems and could be less than ideal for simulation of unknown systems. The use of the predictive models EV1 and EV2 has made more realistic simulation possible, although no clear preference is evident for either.

Acknowledgements

Kilwell fibretube, Rotorua, NZ. Sunspot Kayaks, Rotorua, NZ.

Appendix

a_{nc} non-contact region, d section depth, V_m volume fraction of matrix, V_f volume fraction of reinforcement.

$$\Gamma_1 = \frac{16}{3}\left(\frac{1}{4}n - \frac{5}{4}\right)^3 + 20\left(\frac{1}{4}n - \frac{5}{4}\right)^2 + \frac{25}{4}n - \frac{125}{6}$$

$$\Gamma_2 = \frac{443}{90720}\left(\frac{1}{4}n - \frac{5}{4}\right)^3 + \frac{169}{30240}\left(\frac{1}{4}n - \frac{5}{4}\right)^2 + \frac{117587}{181440}n + \frac{8413}{36288}$$

$$\Gamma_3 = \frac{571}{90720}\left(\frac{1}{4}n - \frac{5}{4}\right)^3 + \frac{31981}{6048}\left(\frac{1}{4}n - \frac{5}{4}\right)^2 + \frac{429043}{181440}n - \frac{303043}{36288}$$

$$\Gamma_4 = \frac{15116}{2835}\left(\frac{1}{4}n - \frac{5}{4}\right)^3 + \frac{2774}{189}\left(\frac{1}{4}n - \frac{5}{4}\right)^2 + \frac{18383}{5670}n - \frac{28891}{2268}$$

$$\Lambda_{m1} = \frac{V_m^3}{9} - \frac{2V_m^2 a_{nc}}{d} + \frac{12V_m a_{nc}^2}{d^2} - \frac{24a_{nc}^3}{d^3}$$

$$\Gamma_{m1} = \frac{8V_m^3}{9} - \frac{64V_m^2 a_{nc}}{3d} - \frac{48V_m^2 a_{nc}}{9} + \frac{8V_m^2 V_f}{3} - \frac{40V_m V_f a_{nc}}{d}$$
$$+ \frac{112V_m a_{nc}^2}{d^2} + 2V_m V_f^2 + \frac{144V_f a_{nc}^2}{d^2} + \frac{96a_{nc}^3}{d^3} - \frac{12V_f^2 a_{nc}}{d}$$

$$\Lambda_{f1} = \frac{V_f^3}{4}$$

$$\Gamma_{f1} = V_f\left(\frac{V_m^2}{3} - \frac{8V_m a_{nc}}{d} + V_m V_f - \frac{9V_f a_{nc}}{d} - \frac{6a_{nc}^2}{d^2} + \frac{3V_f^2}{4}\right)$$

References

Ewart P.D. and Verbeek C.J.R. (2005) Prediction of the Flexural Modulus of Composite Materials for Sporting Equipment. In: Subic A. and Ujihashi S. (Eds.) *The Impact of Technology on Sport*. ASTA (Australasian Sports Technology Alliance), Melbourne, Australia, pp. 446–451.

Ewart P.D. and Verbeek C.J.R. (2006) Prediction of the Flexural Modulus of Fibre Reinforced Thermoplastics for use as Kayak Paddle Blades. In: Moritz E.F. and Haake S. (Eds.) *The Engineering of Sport 6, Vol. 3 – Developments for Innovation*. Springer, Munich, pp. 107–112.

PROXIMITY SAFETY DEVICE FOR SPORTSCRAFT

S.G. O'KEEFE, B.A. MULLER, M.K. MAGGS & D.A. JAMES

Centre for Wireless Monitoring and Applications,
Griffith University, Nathan, Brisbane, Australia

An electronic device is presented that can detect the absence of a rider on a jetski or other personal watercraft. Generation of a low frequency magnetic field around the craft enables a tag worn by the rider to detect its location and transmit a RF signal back to the craft to keep the engine enabled. The functionality of the device is presented along with measurements to show the field zone created around a jetski. The aim of the device is to prevent accidents caused by runaway jetskiis and accidental lanyard activation, especially in sporting competition.

1 Introduction

Many motorized craft and vehicles used by sportspersons may become dangerous projectiles if parted from the person in control. Classic examples of this are jetskiis, personal watercraft (PWC), skidoos, motorcycles, quads etc. Currently the problem is partly ignored, or devices such as cutout lanyards are employed. These lanyards are attached to the riders body, often the wrist, and if the rider falls from the craft the lanyard pulls a plug from the craft which open circuits the ignition system, thus stopping the craft. Lanyards like this often impinge upon the freedom of the rider, especially if involved in stunts etc. Faults or damage to the craft may leave throttles open which makes the situation far more serious.

On the other hand if a lanyard were to be unexpectedly pulled during stunts or synchronized maneuvers, and the vehicle or craft suddenly losses power, then a potentially dangerous situation may be created. This situation will exist even though the rider is still on the vehicle or craft. Serious collisions may occur in this situation.

Jetskiis or personal water craft (PWC) are a classic example of water craft that use safety cutout lanyards to try to prevent runaway accidents. They typically use a wrist lanyard attached to the handlebars of the jetski. Jetskiis are used in many sporting events including speed racing, where speeds across the water in the order of 130 km/h are possible. At the other extreme freestyle jetski competitions are conducted in both surf and still water environments. Here jetskiis are jumped, spun and dived underwater. Extreme ranges of movement are required by the competitors so the wearing of lanyards is not possible.

A lanyard on a jetski can also produce a dangerous situation if accidentally pulled. Jetskiis suffer from what is called off-throttle steering loss. The jetski relies upon the steerable jet thrust to steer the craft. If the engine is accidentally stopped the jetski will lose steerage ability and may become an out of control projectile, despite the rider still being on the craft. Accidental lanyard activation may occur while performing tricks or in rough riding conditions. Runaway jetskiis can injure other competitors or spectators on the water and cases of jetskiis leaving the water and causing injury have been reported. US Coast Guard statistics report that in 2005 there were 1007 injuries and 65 fatalities reported involving PWC (US Coast Guard 2006).

This paper describes a device that has been constructed by the authors that can reliably turn off, or disable, a sportscraft should the rider be parted by more than a specified distance from the craft. The system is currently installed on stunt jetskiis performing in a theme park. In its current form the device will turn off the ignition system of the jetski should the rider be parted from the craft by more than approximately 1.5 metres. The device utilises radio signals and magnetic fields to achieve this and addresses the riderless jetski issue as well as giving riders complete freedom of movement and removing the possibility of an accidental lanyard activation.

2 Principle of operation

The device developed uses a low frequency (LF) magnetic field to define a zone, around the craft, that the rider must be inside. Previous attempts to creating a proximity detection device have often used higher frequency radio signals to produce a field, around the device, whose amplitude decays with distance. Equations 1 to 3 show the electric (E) and magnetic (H) field components produced by a loop antenna for the spherical coordinate system (ϕ,θ,R) (Cheng 1989). The imaginary operator is represented by j, β is the propagation constant, ω the frequency, μ_o the magnetic permeability and m the magnetic moment.

$$E\phi = \frac{j\omega\mu_o m}{4\pi} \beta^2 \sin\theta \left| \frac{1}{j\beta R} + \frac{1}{(j\beta R)^2} \right| e^{-j\beta R} \tag{1}$$

$$H_R = -\frac{j\omega\mu_o m}{4\pi\eta_o} \beta^2 2\cos\theta \left| \frac{1}{(j\beta R)^2} + \frac{1}{(j\beta R)^3} \right| e^{-j\beta R} \tag{2}$$

$$H_\theta = -\frac{j\omega\mu_o m}{4\pi} \beta^2 \sin\theta \left| \frac{1}{j\beta R} + \frac{1}{(j\beta R)^2} + \frac{1}{(j\beta R)^3} \right| e^{-j\beta R} \tag{3}$$

It can be seen that at distances where $\beta R = 2\pi R/\lambda \gg 1$ the dominant terms in the above equations are the *1/R* terms. This means that the decay of field strength with distance is gradual. It is therefore difficult to define a zone around the craft based on field strength when normal communication frequencies are used, due to their short wavelengths. If low frequencies LF (thus long wavelengths, λ) are employed the zone boundaries are more clearly defined by the rapid decay of the fields, ie., proportional to *1/R³*. In this case only the magnetic fields are of importance, ie., equations 2 and 3. Manipulating frequency and power of the signal transmitted from the craft can therefore control the size of the zone. For this application a radius of approximately 1.0–1.5 metres is required and a frequency of 125 kHz has been chosen. Figure 1 shows how the transmitter coils are placed on a jetski to give could coverage around the craft. Two coils have been used to create an elongated zone plus give polarization diversity due to their being placed aligned to different axis. This gives greater a greater level of safety in the prevention of false triggers of the system. The LF field is also modulated with a code to ensure other stray fields are not mistaken.

Proximity Safety Device For Sportscraft 533

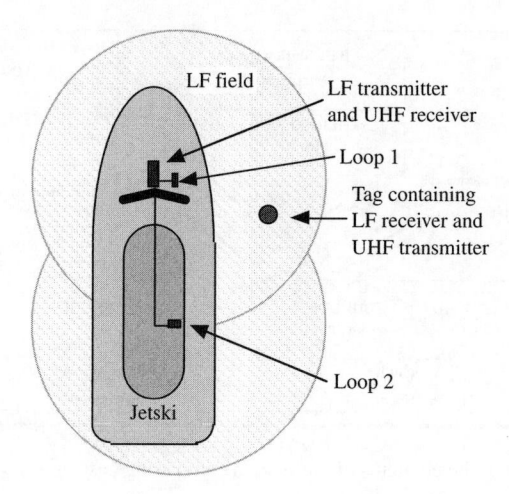

Figure 1. Placement of the base unit on a jetski showing the LF field surrounding the craft.

The rider now wears a small electronic tag which detects the encoded LF field around the jetski. Figure 2 shows an overview of the system and figure 3 shows the hardware implementation. The threshold at which the tag recognizes the LF field also determines the size of the zone around the craft. When the tag detects the LF field it responds by transmitting an ultra-high frequency (UHF) signal (915 MHz). A UHF receiver co-located which the LF generator on the craft detects the UHF transmission from the tag and will disable the craft should these transmissions cease for longer that a predetermined period, in our example 2 seconds. The UHF transmissions will of course cease if the tag is taken out of the LF field, which will occur if the rider is parted from the craft. The LF receiver antenna in the tag has a three axis loop to allow reception of any LF field polarisation. Codes transmitted by both LF and UHF transmitter prevent interference between other craft using the system. This implementation also employs 10 different UHF frequencies to allow vehicle-rider discrimination when riding multiple jetskiis close together.

A further benefit can also be implemented with this system. A UHF transmission, from a central location, containing an emergency stop code, can cause all units within range to disable their jetski. A shore-based unit (ShoreStop) performs this function and is activated by pressing an emergency stop button. This gives the marshals of an event the opportunity to prevent potentially dangerous situations by stopping all craft instantly.

The tag is battery powered and has a lifetime of at least 500 riding hours before replacement is necessary. The shorestop unit is mains powered with rechargeable battery backup has a range of up to 600 metres. The base unit can be easily retro-fitted to any jetski and is power from the jetski battery.

3 Proximity zone

The key feature of the system is the ability to create a clearly defined zone around the craft that detects proximity. Use of dual coils placed in different locations and at different orientations provides spatial and polarization diversity. Given that the tag has tri-axial reception

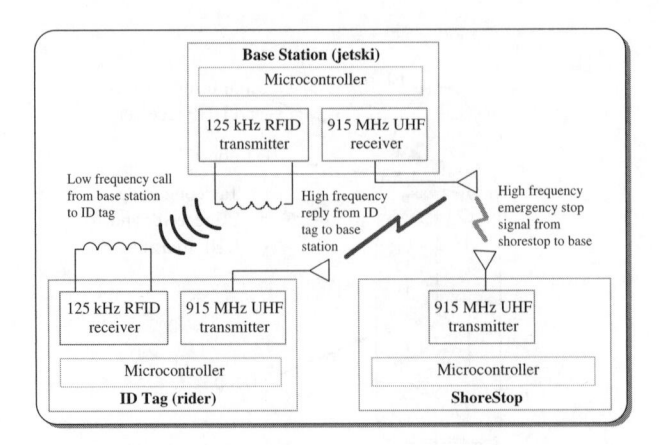

Figure 2. An overview of the elements of the system comprising base unit, ID tag and shorestop unit.

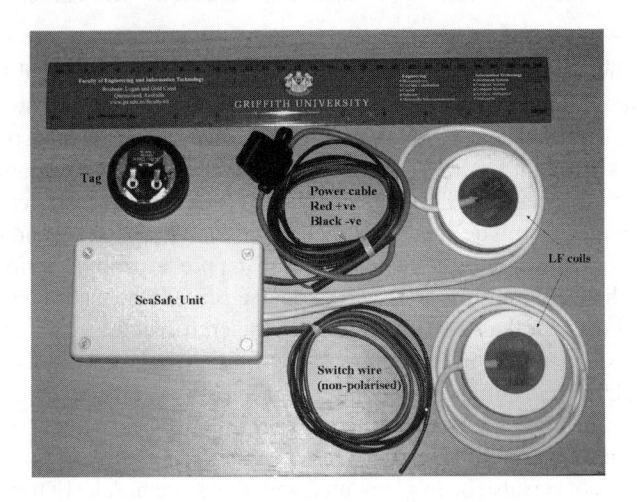

Figure 3. Photo showing the Base unit with LF coils and the tag worn by riders.

ability, this leads to improved performance and reliability for riders reaching extreme positions on the craft. In freestyle events this is a desired objective for the riders.

Measurements have been performed to ascertain the actual LF field shape around a jetski. Because the jetski is partly conductive, it will influence the zone shape. Figure 4 shows the zone created around a large 3 seat recreational jetski. This ski was fitted with two LF coils to provide a better shaped zone over such a large craft. Perturbations in the zone boundary can be observed due to metal structures in the jetski. This however does not significantly affect performance. Measurements were taken on this large jetski as it represents a worse case scenario with large metallic engine components and length of 3.3 metres. An actual tag was used to ascertain the activation boundary around the jetski. For jetskiis involved in stunts or freestyle events the zone clearance around the bow and stern of the jetski would be increased to allow for extreme rider movement.

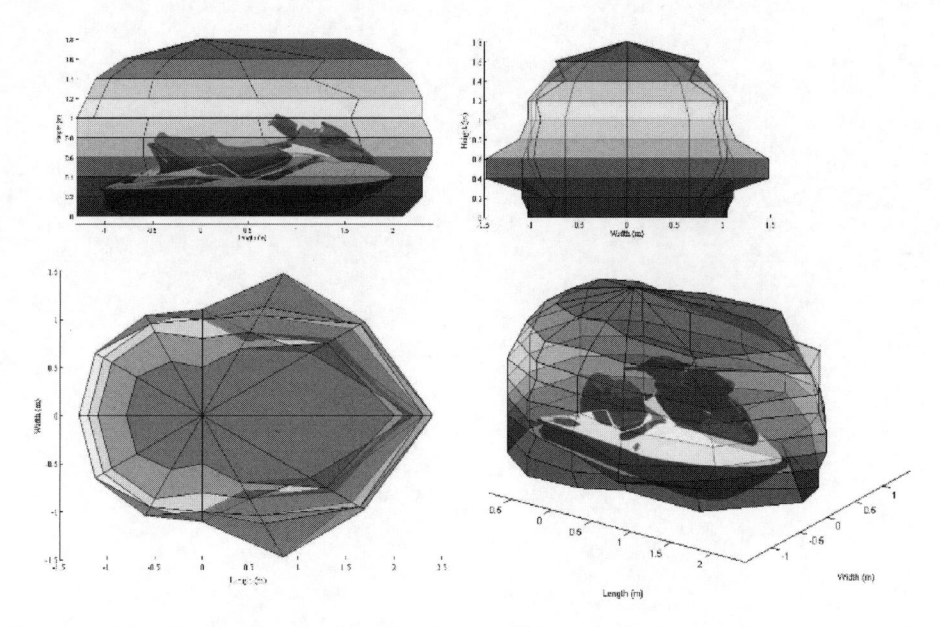

Figure 4. Plots showing the zone in which the tag will keep the jetski active. Measurements taken on a large recreational jetski fitted with two LF coils.

4 Conclusions

A system has been developed that can provide a method of disabling jetskis should the rider be thrown from the craft. The rider does not require the safety lanyard which cannot be worn in many competitive events. Riders have full freedom of movement and safety is improved for competitors and spectators alike. False activation of lanyards is also removed which again improves safety for riders. The system can easily be adapted for use in other powered and unpowered sporting craft and vehicles such as quads, motorcycles, skidoos, speedboats, sailing craft.

References

US Coast Guard (2006) Boating Statistics 2005. COMDTPUB P16754.19, Washington DC.
Cheng D.K. (1989) *Field and Waves Electromagnetics*. Addison-Wesley, Massachusetts.

ANALYSIS AND ANIMATION OF BAR FLOAT RESPONSE TO FISH BITE FOR SEA FISHING

S. YAMABE, H. KUMAMOTO & O. NISHIHARA
Kyoto University, Sakyo-ku, Kyoto, Japan

The fishing line was modeled by a multiple rigid-body link model. This model can reproduce line behaviors below the water surface. Three layouts of sub-sinkers are first examined. The free layout without the sub-sinkers yields as good responses as the other two layouts with the sub-sinkers. The float vertical response is not affected by float mass nor sinker masses, although oscillation periods become longer for a larger mass. The visibility is improved significantly by a rotational response, provided that the float is tilted by the drag force from the angler.

1 Introduction

There have been many studies investigating dynamic behavior of fly lines in the air (Gatti-Bono & Perking, 2004), whereas few studies have been made for bar float response to fish bite for sea fishing. The multi-link model in the paper reproduces the line behavior in water and the float response to fish bite is presented.

2 Bar Float Fishing Tackle Modeling

2.1. *Bar Float Fishing Tackle*

This paper considers a fishing tackle consisting of a bar float, a main line, a main sinker at the end of the main line, a hook line, sub-sinkers on the hook line, and a hook at the hook

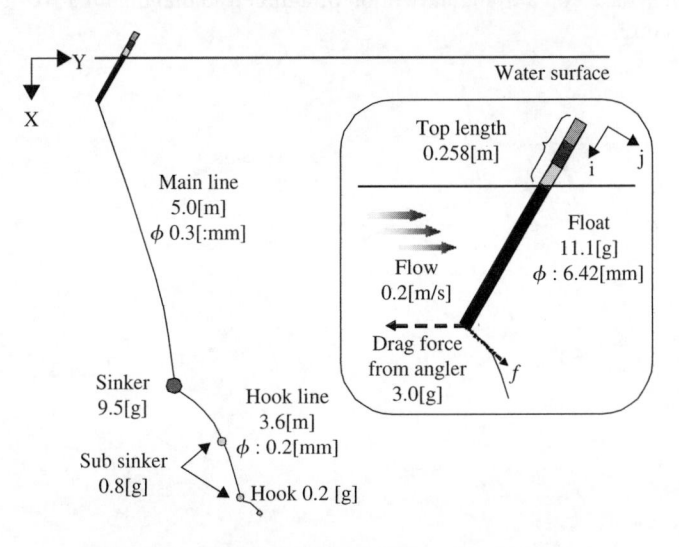

Figure 1. Bar float fishing tackle and parameters.

line end. Typical dimensions of a bar float are diameter of 6 mm, length of 1m, and mass of 10 g. The main line has length of 5.0 m and diameter of 0.3 mm. The hook line is 3.6 m in length and 0.2 mm in diameter. Both lines are made of fluorocarbon. The main sinker mass is 9.5 g and a sub-sinker mass is 0.8 g, respectively.

2.2. Modeling by Multi-Link Model

The line portion with sinker and sub-sinkers are modeled as a multi-link system that is connected to the bar float, as shown in Figure 2. A new algorithm is proposed to calculate acceleration terms satisfying Newtonian equations.

From Figure 2, the Newtonian equations for mass m_h, $h = 1\ldots n$ are given by Eq. (1) and Eq. (2) in terms of link h coordinates(i_h, j_h). The sine and cosine are denoted by $S_h = \sin\theta_h$, $C_h = \cos\theta_h \cdot \ddot{p}_{h,x}^{(h)}$ and $\ddot{p}_{h,y}^{(h)}$ are acceleration. $f_{h,x}^{(h)}$ is internal force. ω_h is angular velocity, $\dot{\omega}_h$ is angular acceleration. $F_{h,x}^{(h)}$ and $F_{h,y}^{(h)}$ are external forces, m_h is link mass, and l_h is length of one link.

$$m_h(\ddot{p}_{h,x}^{(h)} - l_h\omega_h^2) = C_{h+1}f_{h+1,x}^{(h+1)} - F_{h,x}^{(h)} + F_{h,x}^{(h)} \tag{1}$$

$$m_h(\ddot{p}_{h,x}^{(h)} - l_h\dot{\omega}_\eta) = S_{h+1}f_{h+1,x}^{(h+1)} + F_{h,y}^{(h)} \tag{2}$$

The end of the link h is a start of next link $h + 1$. Therefore, the above equations can be rewritten in terms of link $h + 1$ coordinate system(i_{h+1}, j_{h+1}).

$$m_h\ddot{p}_{h+1,x}^{(h+1)} = S_{h+1}^2 f_{h+1}^{(h+1)} + C_{h+1}m_h\ddot{p}_{h,x}^{(h)} - C_{h+1}m_hl_h\omega_h^2 + S_{h+1}F_{h,y}^{(h)} \tag{3}$$

$$m_h\ddot{p}_{h+1,y}^{(h+1)} = S_{h+1}f_{h,x}^{(h)} - S_{h+1}F_{h,x}^{(h)} + C_{h+1}F_{h,y}^{(h)} \tag{4}$$

The computational requirement varies only linearly with the number of links because the algorithm is based on a triangularization of a threefold diagonal matrix of Table 1 with a simple structure.

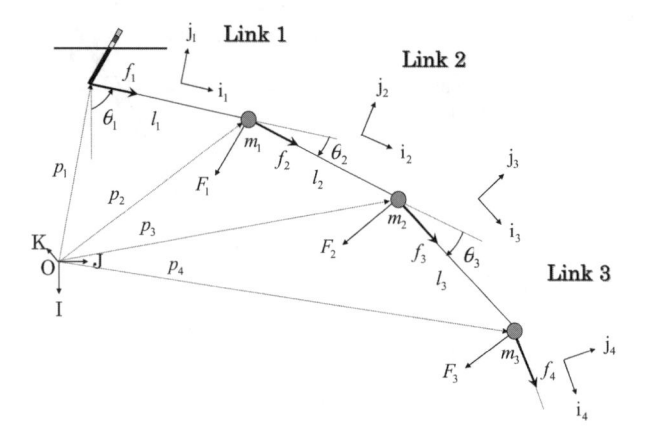

Figure 2. Modeling by multi-link model.

Eq. (1) corresponds to the 2, 4, 2n lines in Table 1. Eq. (3) corresponds to the 3, $2n-1$ lines. When $f_{h,x}^{(h)}$ become known, $\ddot{p}_{h,y}^{(h)}$ is obtained from Eq. (4). When $f_{h,x}^{(h)}$ and $\ddot{p}_{h,y}^{(h)}$ become known, $\dot{\omega}_h$ is obtained from Eq. (2).

3 Bar Float Response to Three Arrangements of the Sub-Sinkers

Three arrangements of the sub-sinkers attached to the hook line are considered to examine float responses to fish bite: 1) "free layout" without sub-sinkers, 2) "even layout" with a pair of identical sub-sinkers and 3) "uneven layout" with an upper light sinker followed by a lower, heavier sinker.

Table 1. Threefold diagonal coefficient matrix for determination of $f_{n,x}^{(n)}$, $\ddot{p}_{h,x}^{(n)}$.

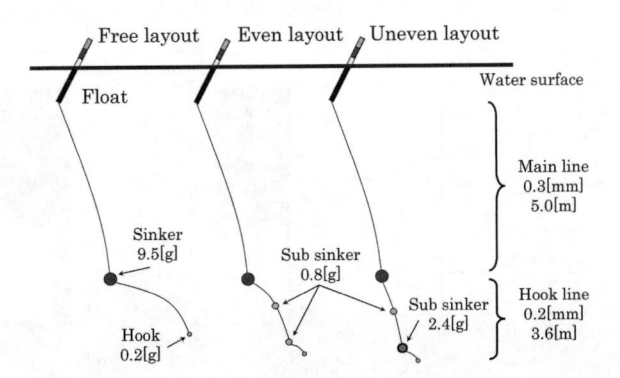

	1	2	3	4	...	...	$2n-1$	$2n$	Known values
	$f_{1,x}^{(1)}$	$\ddot{p}_{1,x}^{(1)}$	$f_{2,x}^{(2)}$	$\ddot{p}_{2,x}^{(2)}$	...	...	$f_{n,x}^{(n)}$	$\ddot{p}_{n,x}^{(n)}$	
1	$a_{1,x}$	1							$b_{1,x}$
2	1	m_1	$-C_2$						$m_1 l_1 \omega^2_1 + F_{1,x}^{(1)}$
3		$-C_2 m_1$	$-S_2^2$	m_1					$-C_2 m_1 l_1 \omega^2_1 + S_2 F_{1,y}^{(1)}$
4			1	m_2	1				$m_2 l_2 \omega^2_2 + F_{2,x}^{(2)}$
⋮				⋱	⋱	⋱			⋮
⋮					⋱	⋱	⋱		⋮
$2n-1$					$-C_n m_{n-1}$	$-S_n^2$	m_n-1		$-C_n m_{n-1} l_{n-1} \omega^2_{n-1} + S_n F_{n-1,y}^{(n-1)}$
$2n$							1	m_n	$m_n l_n \omega^2_n + F_{n,x}^{(n)}$

Figure 3. Three arrangements of the sub-sinkers attached.

A fish bite starts from 42 s, and continued 5 s. Assume no flow. Also assume no drag from the angler via the main line to the float. The float does not tilt. The line shape is downward straight. Figure 4 shows float responses for an upward bite. The free layout yields little float response to the bite, while the uneven layout shows an upward response.

Figure 5 assumes a flow and a drag from the angler. The float tilts. The bite is upward. The hook line shape is convex before the bite. The line becomes a U-shape during the bite. These shape changes are represented successfully by the link model, where the total number of links is 86. The free layout yields the largest downward response while the uneven layout maintains the upward one. The even layout appears below without being concerned with up bite. Free layout yields the smallest response.

Other hit directions are also considered. It turns out that the free layout, as opposed expectation, can produce float responses as good as the other two layouts.

4 Fishing Tackle Parameters and Float Response

This section analyzes vertical and rotational responses of the bar float to a fish bite. The fishing tackle parameters such as float mass, sinker mass are varied.

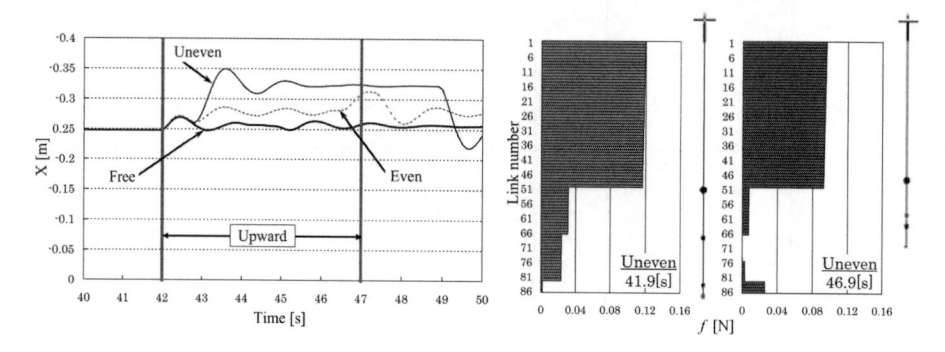

Figure 4. Float top position response to upward bite without flow.

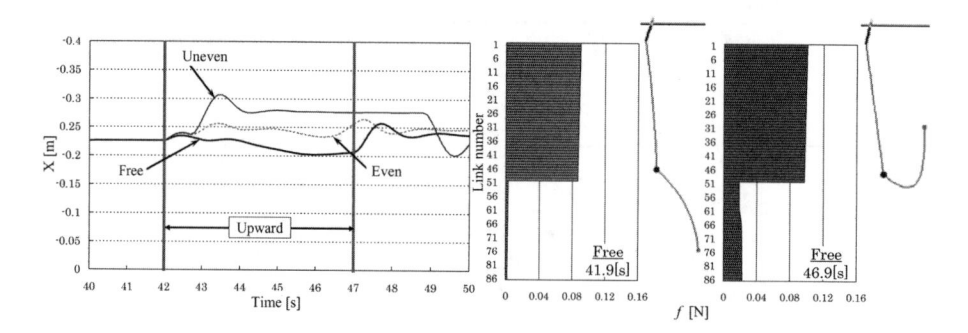

Figure 5. Float top position response to upward bite with flow.

4.1. Float Response for Different Float Masses

Table 2 summarizes float masses and lengths, given the top length of 0.258 m above the water surface.

The right-downward fish bite starts at 42 s in Figure 6. The flow is rightward. The top positions differ because the floats tilt differently. Downward float responses to the bite are similar. Oscillation periods become longer as the float mass gets larger. The damping coefficient ratios remain a similar value for the floats. This point can be reconfirmed by a theoretical analysis of a second order differential equation. Rotational float responses, improve visibility to detect a fish bite when the float is tilted by the drag force from the angler. This is especially true for a fish bit when the float is subject to the wave, as shown in Figure 7. The float is not tilted in case (a), while it is tilted in case (b). The vertical responses remain similar. Figure 8 shows response animations every 0.1 s. The visibility is improved by the visibility is improved by the rotational response of the tilted float.

Table 2. Parameters of float masses.

Parameter	Unit	I	II	III	IV
Float mass	[g]	7.3	11.1	16.8	20.6
Sinker mass	[g]	9.5			
Float length	[m]	0.82	0.93	1.11	1.23

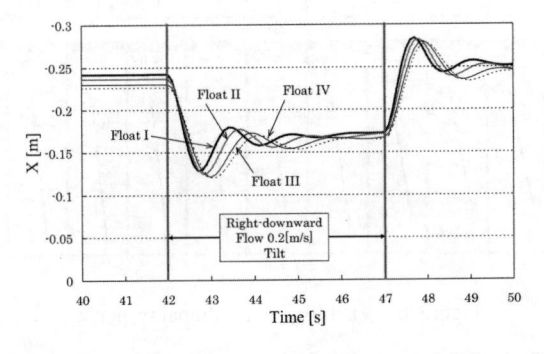

Figure 6. Float top position response to right-downward different float masses.

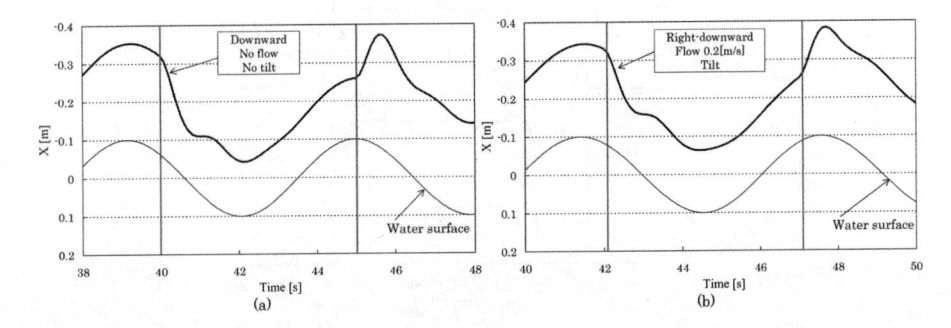

Figure 7. Float top position response during downward wave period.

4.2. *Float Response for Different Sinker Masses*

Table 3 summarizes sinker masses and float lengths, given the top length of 0.258 m. The vertical movements are not affected significantly by sinker masses, as shown in Figure 9. However, the light sinker improves visibility because the float is more easily tilted, yielding more rotations by the fish bite.

Table 3. Parameters of sinker masses.

Parameter	Unit	I	II	III	IV
Float mass	[g]	11.1			
Sinker mass	[g]	5.7	9.5	15.2	19.0
Float length	[m]	0.83	0.93	1.1	1.2

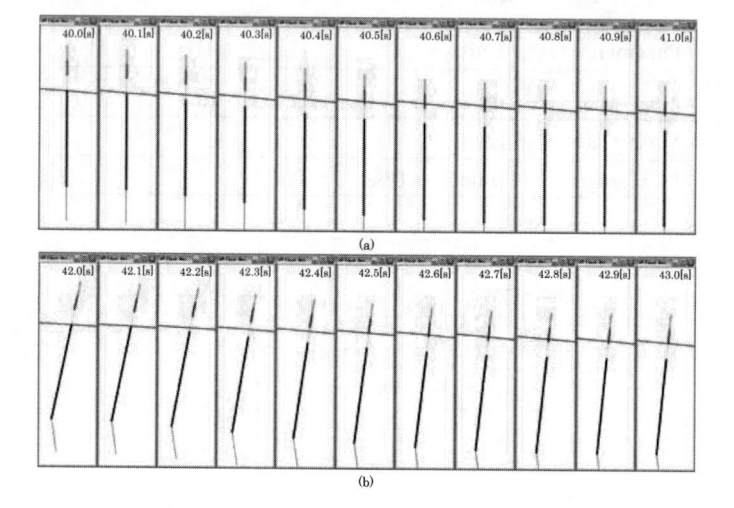

Figure 8. Visibility by rotational response.

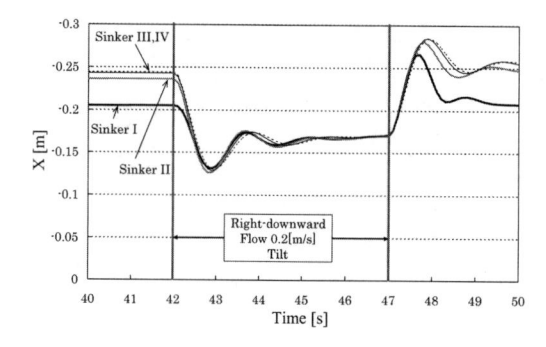

Figure 9. Float top position response to Right-downward different sinker masses.

5 Conclusions

The fishing line was modeled by a multiple rigid-body link model. The link model can reproduce line behaviors below the water surface. Three layouts of sub-sinkers are examined. The free layout without the sub-sinkers yields as good response as the other two layouts with the sub-sinkers. The float vertical response is not affected by float mass nor sinker masses. Oscillation periods become longer for a larger mass. However, the visibility is improved by the rotational response when the float is tilted by the drag force from the angler.

References

Gatti-Bono C. and Perking N.C. (2004) Numerical model for the dynamics of coupled fly line/fly rod system and experimental validation, *Journal of Sound and Vibration*, 272, 773–791.

14. Aquatics – Swimming

BIOMECHANICS OF FRONT-CRAWL SWIMMING: BUOYANCY AS A MEASURE OF ANTHROPOMETRIC QUANTITY OR A MOTION-DEPENDENT QUANTITY?

T. YANAI

*Chukyo University, School of Life System Science and Technology,
Kaizu-cho, Toyota, Japan*

The swimmer's body is acted upon by time- and position-dependent fluid forces which control linear and angular motions of the swimmer's body. In this presentation, researches on the buoyancy of human body and its influence on the front-crawl swimming are reviewed, and the current approaches for quantifying the buoyancy of a swimmer are re-examined. An early study on buoyancy of human body was found in 1757. In this study, the primary question was whether or not motionless human bodies can float in water. Similar studies were conducted in early 1900's, in which the buoyancy of human body was measured as anthropometric quantities such as the buoyant force acting on the human body submerged in water, the specific gravity of the body and the apparent body weight under water. These studies revealed that (a) not all human can float motionless in fresh water, (b) in general, women are more buoyant than men, (c) the volume of air in body cavities affects one's buoyancy, (d) buoyancy characteristics changes as children get older, and (e) the legs tend to sink from a horizontal motionless floating position. Similar approach was adopted to study if the buoyancy of a swimmer influences swimming performance. On the basis of the findings from these studies, the buoyancy was postulated to influence the swimming performances with the following mechanism: The greater the buoyancy and the lesser the leg sinking effect of buoyancy help the body to float horizontally and to reduce the immersed body cross-sectional area, both of which reduce hydrodynamic resistance, and thereby, the swimming economy would increase for better performances. In a recent study, the leg-sinking effect of buoyancy was measured as motion-dependent quantities and demonstrated that the buoyant force generated the moment that directed primarily to raise the legs and lower the head during the front-crawl swimming – the buoyancy made the leg-raising effect, rather than leg-sinking effect, during the front-crawl swimming. This counter-intuitive finding on the rotational effect of buoyancy raised a question regarding the aforementioned postulated mechanism; how the leg-sinking effect of buoyancy in a horizontal motionless floating could help the swimmer's body align horizontally during front-crawl swimming, given that the buoyancy results in the leg-raising effect on the human body during the performance of front-crawl swimming? To seek an answer for this question, the mechanism of horizontal floating of the swimmers during the front-crawl swimming needs to be re-examined in the future studies. In such studies, the buoyancy of a swimmer should not only be measured as an anthropometric quantity, but it should also be measured as a motion-dependent quantity.

1 Introduction

Front-crawl swimming is probably the fastest and the most efficient style of human aquatic locomotion. In front-crawl, the swimmer swings the arms back and forth alternately and executes near-vertical kicks at either the same or a different frequency to the stroke frequency.

These limb movements generate time- and position-dependent fluid forces which, in turn, control linear and angular motions of the swimmer's body. The time- and position-dependent fluid forces acting on a swimmer's body could be categorized into two components; hydrodynamic forces and hydrostatic, or buoyant, force. Since Houssay studied the propulsive component of the hydrodynamic forces in 1912, many researchers have devoted their energies to understand the propulsive characteristics of swimming. On the other hand, the influence of buoyancy on swimming performance has not received comparable attention and, as a result, little knowledge has been gained on it. In this presentation, the measurements of the buoyancy reported in literature and their research outcomes are reviewed to summarize the current knowledge on the influence of buoyancy on swimming performances.

2 What quantity of timber would be sufficient for a man to afloat in water? – Buoyancy measured as anthropometric quantities

An early scientific reference on the buoyancy of human body was found in 1750's (Robertson, 1757). This study was driven by the author's motivation to answer the question "what quantity of fir or oak timber would be sufficient to help a man afloat in river or seawater (Robertson, 1757, p.33)." The statement clearly indicates that the author believed a motionless human body was not floatable in water. Robertson constructed a cistern and asked ten "labouring men" who were "fortified with a large dram of brandy" to sink in the

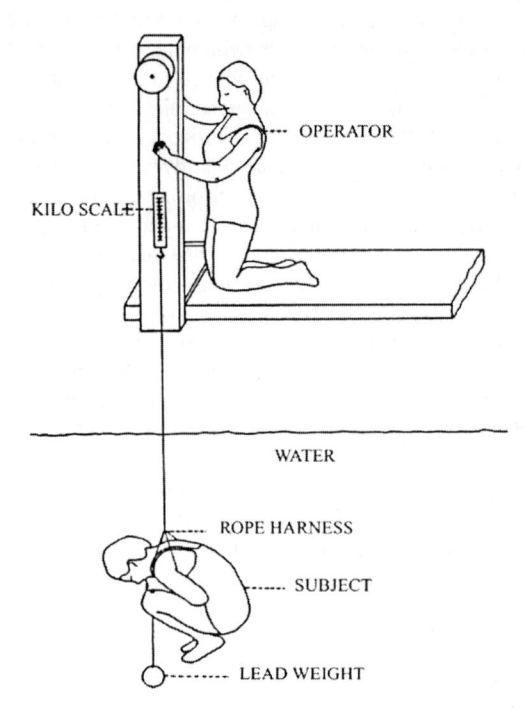

Figure 1. A method for measuring the specific gravity of human body. Reprinted with permission from *Research Quarterly for Exercise and Sport*. Vol. 8: 19–27, Copyright (1937) by the American Alliance for Health, Physical Education, Recreation and Dance, 1900 Association.

water-filled cistern. He then measured the height of the water surface to compute the volume of the subject, so that the buoyant force acting on the subject in water was determined. In the study, the buoyant force (mean: 729 N) was found to be greater than the weight (mean: 649 N) for nine out of 10 subjects, and the author concluded that (a) excepting some, "every man was lighter than his equal bulk of fresh water" and consequently, "many might be preserved from drowning" and (b) "a piece of wood, not larger than an oar, would buoy a man partly above the water." This finding was supported by a study conducted by Pettigrew (1874). He investigated the floating positions of motionless humans and claimed that everyone could float in water if breathing was held naturally and the body was relaxed.

In early 1900's, the primary question for the research continued to be whether or not a motionless human could float in water. The research in this period was conducted by using more scientifically sound approaches so as to overcome some of the obvious limitations associated with the earlier investigations. The buoyancy of both men and women of various age groups with different breathing status was studied by measuring the specific gravity of the body (Packard, 1900; Rork & Hellebrandt, 1937; Sandon, 1924), the apparent body weight under water (Cotton & Newman, 1978; Packard, 1900; Sandon, 1924), and underwater movements of the body in tuck position (Carter, 1955; Mitchem & Lane, 1968; Lane & Mitchem, 1964). In addition, the body's ability to maintain the horizontal motionless floating position was measured as the distance between the center of mass and the center of buoyancy (Rork & Hellebrandt, 1937) and also as the angle of floatation [that is the angle between the body's long-axis and the horizontal plane] (Carmody, 1965; Carter, 1955; Rork & Hellebrandt, 1937). The major outcomes of these studies were that (a) not all human could float in fresh water, (b) women were generally more buoyant than men, (c) the body became less buoyant with aging, (d) the air in lungs affected the body's buoyancy, (e) the legs tended to sink from a horizontal motionless floating position, and (f) the degree to which the leg sank from the horizontal motionless floating position depended upon the distance between the center of mass and the center of buoyancy.

Amongst the buoyancy measurements taken in the aforementioned studies, the specific gravity of the human body and the center of mass location are categorized as anthropometric

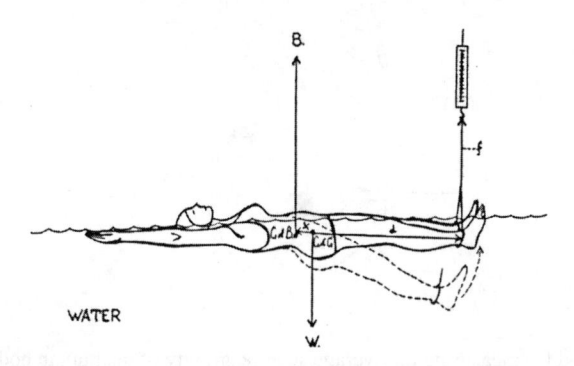

Figure 2. A method for measuring the angle of floatation and the location of center of buoyancy. Reprinted with permission from *Research Quarterly for Exercise and Sport*, Vol. 8: 19–27, Copyright (1937) by the American Alliance for Health, Physical Education, Recreation and Dance, 1900 Association Drive, Reston, VA 20191.

quantities. The buoyant force acting on a motionless body completely submerged in water and the point at which the buoyant force acts (that is, the center of buoyancy) are unique quantities that characterize the physique of the given individual (the volume of the body and its centroid) and thus, these quantities could also be categorized as anthropometric quantities. It could, therefore, be said that the analysis of the anthropometric quantities of human body have led the answer for the question – *NOT every one can float in water*.

3 Does buoyancy influence swimming performances? – Buoyancy measured as anthropometric quantities

In late 1900's, similar approaches were adopted to quantify the buoyancy of a swimmer so as to study if the buoyancy of a swimmer influences swimming performance. The buoyancy of a given swimmer was measured either as the surplus of the buoyancy over the weight of the swimmer (Chatard *et al*, 1990a-d, Capelli *et al.*, 1995) or as the leg-sinking tendency of the swimmer's body in a horizontal motionless floating (Capelli *et al.*, 1995; Gagnon & Montpetit, 1981; Pendergast *et al.*, 1977; McLean & Hinrichs, 1998; Rennie *et al.*, 1975; Zampero *et al.*, 1996). These quantities of buoyancy were investigated to examine the relations that they might have with the energy cost of swimming and the drag encountered by the swimmers. The results demonstrated that (a) the buoyant force acted more cranial to the

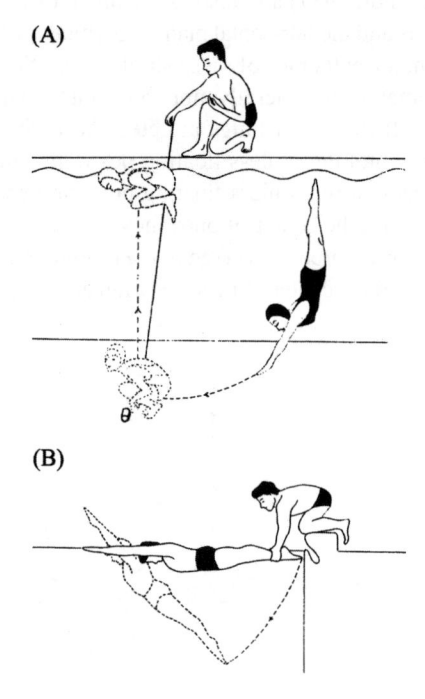

Figure 3. A method for measuring the average specific gravity of the human body (A) and the angle of floatation (B). The average specific gravity of the human body was estimated from the time taken for the body to raise to the surface in seven feet of water. Reproduced with permission from "Buoyancy and floatation," in the New Zealand Journal of Physical Education, Copyright (1955) by Physical Education New Zealand (PENZ).

body weight for a horizontal motionless floating human body in both anatomical and streamlined positions (Table 1: Gagnon & Montpetit, 1981; McLean & Hinrichs, 1998), generating a leg-sinking effect, (b) the energy cost of swimming was lower for the swimmers whose surplus buoyancy was large than those whose surplus buoyancy was small

Table 1. Positions of CB and CM (% stature) in horizontal aligned streamlined), floating position

	No-inhalation		Full-Inhalation		
	CB	CB-CM	CB	CB-CM	CM
Gagnon & Montpetit	60.4	0.3	—	—	60.1
	60.6*	0.3*	61.4*	1.1*	60.3*
McLean & Hinrichs	—	—	61.68	0.44	61.24
	—	—	—	1.0*	—

Notes:
- The measurements were taken from the feet.
- The CB stands for the center of buoyancy and the CM stands for the center of mass.
- The data with * were obtained when the subjects completely immersed in water.

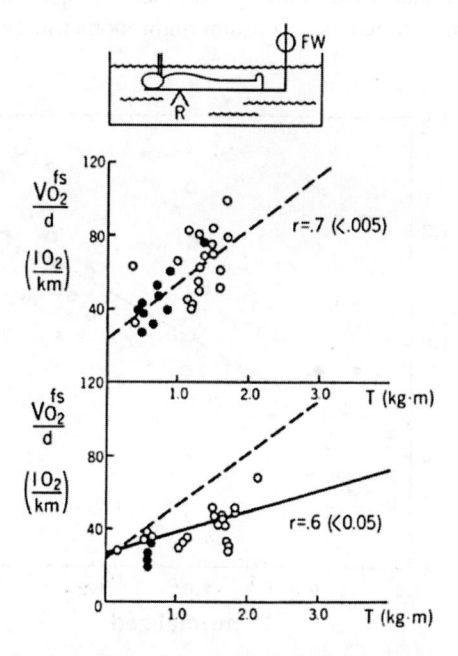

Figure 4. Energy cost of swimming (VO_2^{fs}/d), plotted as a function of the extent of the leg-sinking effect that the buoyant force makes on the swimmer in a horizontal motionless floating for males (○) and females (●). Reprinted with permission from "Quantitative analysis of the front crawl in men and women," in *Journal of Applied Physiology* (1977).

(Chatard et al., 1990), (c) the energy cost of swimming was lower for the swimmers whose leg sinking tendency was small than those whose leg sinking tendency was large (Figure 4: Pendergast et al., 1977; Rennie et al., 1975), (d) for a given swimmer, the energy cost of swimming was lower when the leg-sinking tendency was reduced by attaching an air-filled tube around the waist than when it was increased by attaching a lead-filled tube around the same position (Figure 5: Capelli et al., 1995; Zamparo et al., 1996) and (e) for a given swimmer, the resistive force acting on a swimmer during the performance was lower when the leg-sinking tendency was reduced by attaching an air-filled tube than when it was increased by attaching a lead-filled tube (Zamparo et al., 1996). These findings were generally interpreted to postulate a mechanism that relates the buoyancy and the swimming performances. The mechanism might be stated as follows: With the greater buoyancy and the lesser leg-sinking tendency, the swimmer's body can easily be aligned horizontally, which reduces the cross-sectional area of the immersed body and the hydrodynamic resistance, and thereby, the swimming economy would increase for better performances.

This postulated mechanism implies that the swimmer's buoyancy and its leg-sinking tendency of the swimmer's body in a horizontal motionless floating are the same as those during the swimming performances. By definition, the principle of buoyancy considers a submerged body in fluid *at rest*. Should we examine the derivation of the principle, however, we would notice that the fluid force of the same attributes also acts on a submerged body in moving fluid and on a submerged body in motion. It seems reasonable to apply the principle of buoyancy for analyzing the buoyant force acting on a moving body. When the principle is applied for the analysis of a moving swimmer's body, the buoyancy of the swimmer could no longer be characterized as a constant anthropometric quantity. It should rather

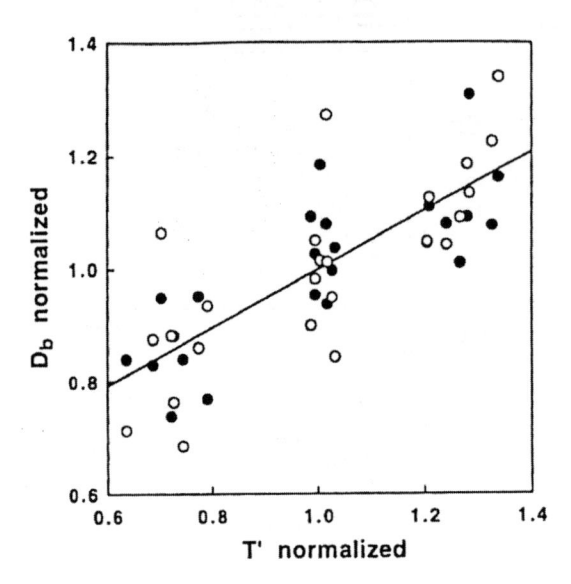

Figure 5. Normalized values of resistive force acting on the swimmer as a function of the normalized values of the leg-sinking effect of buoyancy. Reproduced with kind permission of Springer Science and Business Media from "Effect of the underwater torque on the energy cost, drag and efficiency of front crawl swimming," in *European Journal of Applied Physiology* (1996).

be characterized as a motion-dependent quantity that changes continuously throughout the stroke cycle because the magnitude and the shape of the immersed volume of the swimmer's body change continuously, changing the magnitude of the buoyant force and its point of application during the stroke cycle. In this viewpoint, the buoyancy of a swimmer during the performance could not be considered as an anthropometric quantity, but it should rather be a dynamic quantity that characterizes the swimmer's physique as well as their body movements.

4 How does buoyancy influence swimming performances? – Buoyancy measured as motion-dependent quantities

The buoyancy of a swimmer during the front-crawl swimming was determined in a study (Yanai, 2001). In this study, the swimming performances captured with a videography technique and the body dimensions modeled from the anthropometric parameters were used to compute the volume of the submerged body parts and its centroid for each field of captured images, so that the buoyant force and its point of application were determined throughout the stroke cycle. The results demonstrated that the buoyant force acted caudal to the center of mass for most phases of the front-crawl stroke cycle (Figure 6, above), causing the overall turning effect of the buoyancy to raise the legs (mean value of 22 Nm), rather than to sink the legs (Figure 6: below).

The center of mass fluctuated in the mean range of 23 mm, shifting toward the head during the recovery phase, and the center of buoyancy fluctuated in the mean range of 105 mm, shifting toward the legs at or around the initiation of the recovery phase. This shifting of the center of buoyancy occurred because the volumes of the head and the recovery arm and shoulder, all of which were located cranial to the center of mass, exited out of the water during the recovery phase and were no longer subject to the buoyant force. Due to the different shifting pattern of the two centers, the center of mass was located cranial to the center of buoyancy for most of the stroke time. This result indicates clearly that the rotational effect of buoyancy that a swimmer is subject to during the performance of swimming cannot be represented by an anthropometric quantity that quantifies the same swimmer's leg-sinking tendency in a horizontal motionless floating. With a similar approach, Yanai (2004) determined the contribution of buoyancy to the oscillatory rolling motion of the front-crawl swimmer's body about its long-axis (this rolling motion is often called body-roll). A major part of the total external torque required to generate a sinusoidal pattern of body's rolling action was found to be generated by the buoyant force. This finding was supported by a simulation study (Nakajima, 2005). These findings also suggest that the rotational effect of buoyancy that directly influences the swimming performance can neither be measured as, nor represented by, an anthropometric quantity of the swimmer.

The finding of the Yanai's study (2001) raised a question to the postulated mechanism; that is, with the greater buoyancy and the lesser leg-sinking tendency, the swimmer's body can easily be aligned horizontally, which reduces the cross-sectional area of the immersed body and the hydrodynamic resistance, and thereby, the swimming economy would increase for better performances. If the buoyant force does not generate leg-sinking effect during the performance of front-crawl swimming, how was the leg-sinking tendency of the swimmer's body in a horizontal motionless floating found to be related to the resistive force and the energy cost of swimming? Three scenarios are listed and discussed below.

The first scenario is that the swimmers whose leg-sinking tendency in a horizontal motionless floating is small may have attained a greater leg-raising effect of buoyancy during the performance of swimming than those whose leg-sinking tendency in a horizontal motionless floating is large. As mentioned earlier, the center of buoyancy shifts caudally during the recovery phase because the buoyant force acts only on the swimmer's entire body except the head and the recovery arm. This caudal shifting of the center of buoyancy (the mean range of 105 mm, Yanai, 2001) is greater than the reported distance between the

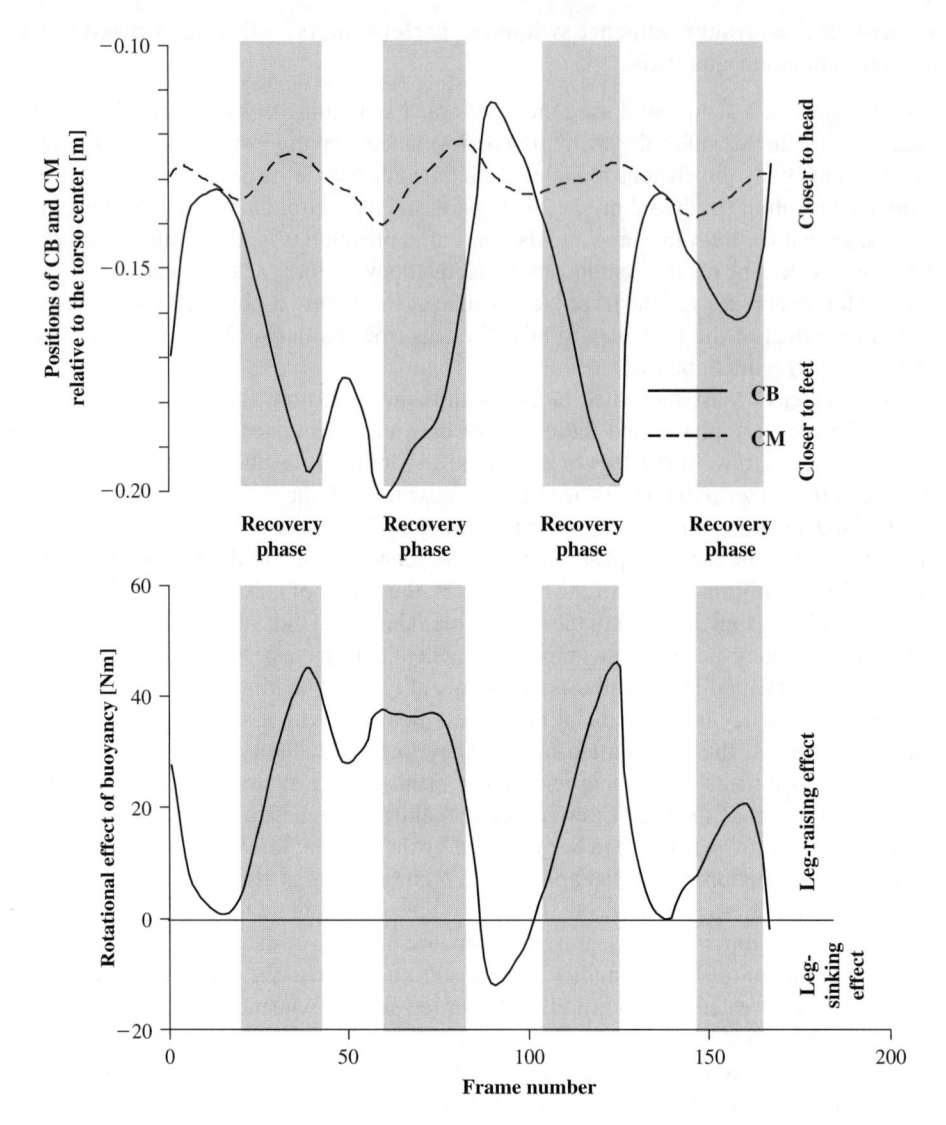

Figure 6. The positions of the center of mass (CM) and center of buoyancy (CB) observed during two stroke cycles of front crawl swimming (above) and the moment of buoyant force about the CM (below). The graphs are developed on the basis of the data presented in literature (Yanai, 2001).

center of mass and the center of buoyancy for the swimmer's body in a horizontal motionless floating (approximately 1 % of stature ≈ 18 mm, Gagnon & Montpetit, 1981; McLean & Hinrichs, 1998). For a given range of caudal shifting of the center of buoyancy during the recovery phase, those swimmers whose center of buoyancy is located close to the center of mass in a horizontal motionless floating (which results in a small leg-sinking tendency) should attain a greater moment arm of the buoyant force around the center of mass during the recovery phase of swimming, resulting in these swimmers to attain a greater leg-raising effect of buoyancy. As a result, the swimmers whose leg-sinking tendency in a horizontal motionless floating is small may be advantaged in aligning the body horizontally. This explanation may sound adequate, provided that the extent of the caudal shifting of the center of buoyancy during the recovery phase is similar among all swimmers. The additional analysis of the data presented by Yanai (2001), however, could not support this scenario (Figure 7). Further investigations are indicated to fully examine the feasibility of this scenario.

Second scenario is postulated on the basis of the data presented in literature (Schleihauf et al., 1983). The hydrodynamic forces acting on the swimmer's hands are expected to act eccentric to the center of mass of the swimmer, generating the leg-sinking moment around the center of mass (Figure 8). Based on the Schleihauf's data, the leg-sinking moment is estimated to be approximately 25–35 Nm. The leg-raising effects of the buoyant force (mean = 22 Nm, Yanai, 2001), and of the hydrodynamic forces generated by kicks (the magnitude is unknown) expectedly, counteract the leg-sinking effect of the hydrodynamic forces generated by the hands, so that the horizontal alignment of the body is maintained during the performance of front-crawl (Figure 9). The data suggest that the balance between the leg-sinking effect of the hydrodynamic forces acting on the hands and

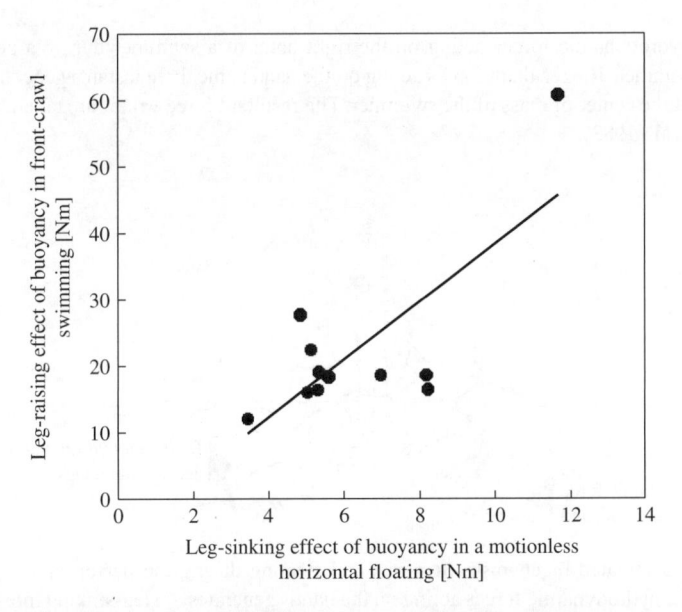

Figure 7. The relation between the leg-sinking effect of buoyancy in motionless horizontal floating and the leg-raising effect of buoyancy during the performance of front-crawl swimming, illustrated on the basis of the data presented by Yanai (2001).

the leg-raising effect of buoyancy is the primarily factor for determining the swimmer's ability to maintain horizontally aligned position during the performance of front-crawl swimming. This line of thoughts leads to the second scenario; that is, the swimmers who has less leg-sinking tendency in a horizontal motionless floating might have generated a minimum amount of leg-sinking effect of hydrodynamic forces acting on the hands, either by reducing magnitude of the hydrodynamic forces acting on the hands or by reducing the moment arm of this force around the center of mass of the swimmer's body. Adopting either one of the approaches, these swimmers are, with less effort of kicks, aligned in the horizontal position during the stroke, which may enable them to reduce the drag and the energy cost of swimming. Feasibility of this mechanism needs to be evaluated by examining the relations between the leg-sinking effect of buoyancy in a horizontal motionless floating and the leg-sinking effect of hydrodynamic forces acting on the hands.

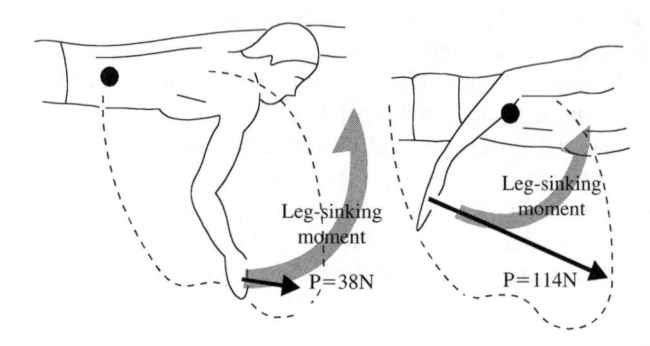

Figure 8. Hydrodynamic forces acting on the right hand of a swimmer during a performance of front crawl swimmer. The resultant forces acting on the hand at the three instants generates leg-sinking moment about the center of mass of the swimmer. The resultant force values are taken from literature (Schleihauf et al., 1983).

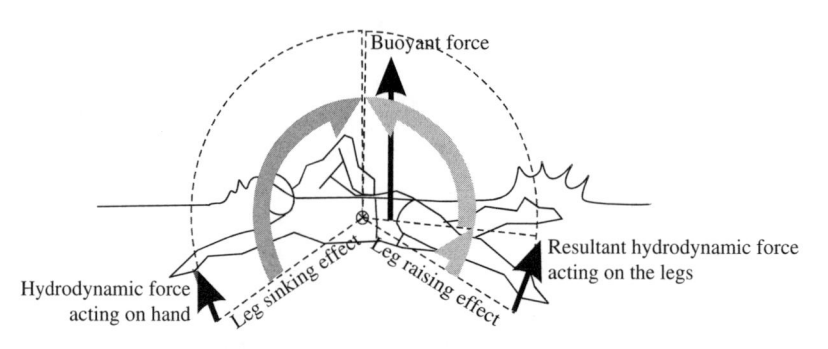

Figure 9. A postulated mechanism of horizontal floating during the performance of front crawl swimming. The hydrodynamic forces acting on the hands generates the leg-sinking moment about the center of mass of the swimmer and the buoyant force, and the hydrodynamic forces acting on the legs perhaps, generates the leg-raising moment, the balance of these moments enables the swimmer's horizontal floating during the performance of front crawl swimming.

The last scenario could be that the swimmers who had less leg-sinking tendency in a horizontal motionless floating might have been the swimmers of a small body-size, and thus the variance observed in the energy cost of swimming (Figure 4) was influenced not only by the extent of the leg-sinking tendency in a horizontal motionless floating, but also by the body size. In that study, the energy cost of swimming was expressed as the amount of oxygen consumed for the swimmer to swim 1 km of distance and the leg-sinking effect of buoyancy measured as the net value of the leg-sinking torque in a horizontal motionless floating (Figure 4). For a given distribution of body mass and volume, the leg-sinking tendency in a horizontal motionless floating is greater for a taller and heavier individual than a shorter and lighter individual, because the magnitude of buoyant force and its moment arm about the center of mass are greater for the taller and heavier individual. The energy required to swim a given distance is generally greater for a taller and heavier individual than a shorter and lighter individual if the skill levels of the two are equivalent. It is possible, therefore, that the volume of oxygen consumed by a swimmer to swim 1 km of distance might be affected primarily by the body size, rather than by the extent of the leg-sinking tendency in a horizontal motionless floating. By the same token, it is also possible that the relationship between the drag acting on the swimmer and the leg-sinking tendency in a horizontal motionless floating observed in Figure 5 might also be an effect of body size. This may be the case because the drag presented in the figure was determined on the basis of the oxygen consumptions measured at multiple speeds of swimming (Zamparo *et al.*, 1996). Further study is indicated to determine if the interrelations among the leg-sinking effect of buoyancy in a horizontal motionless floating, the drag and the energy cost of swimming observed in the literature are free from the influence of the body size.

5 Future studies on buoyancy in swimming performances

The influences of buoyancy on swimming performance reviewed in this presentation revealed that the needs for measuring the buoyancy of a swimmer have been widened from the anthropometric quantities to the motion-dependent quantities. The use of a motion-dependent quantity for studying buoyancy brought us a counter-intuitive finding on the leg-sinking/raising effect of buoyancy, which raised a question regarding the current understanding on the mechanism of horizontal alignment of front-crawl swimming. The mechanism of horizontal floating of the swimmers during the front-crawl swimming needs to be re-examined in future studies. In such studies, the buoyancy of a swimmer should not only be measured as an anthropometric quantity, but it should also be measured as a motion-dependent quantity.

References

Capelli C., Zamparo P., Cigalotto A., Francescato M.P., Soule R.G., Termin B., Pendergast D.R. and di Prampero, P.E. (1995) Bioenergetics and biomechanics of front crawl swimming. *Journal of Applied Physiology*, 78/2, 674–679.

Carmody J.F. (1965) Factors influencing the horizontal motionless floating position of the human body. *New Zealand Journal of Physical Education*, 37, 15–18.

Carter J.E.L. (1955) Buoyancy and floatation. *New Zealand Journal of Physical Education*, 6, 14–23.

Chatard J.C., Bourgaoi B. and Lacour J.R. (1990a) Passive drag is still a good evaluator of swimming aptitude. *European Journal of Applied Physiology*, 59, 399–404.

Chatard J.C., Colloimp C., Maglischo E. & Maglischo C. (1990b) Swimming skill and stroking characteristics of front crawl swimmers. *International Journal of Sports Medicine*, 11, 156–161.

Chatard J.C., Lavoie J.M., Bourgoin B., and Lacour J.R. (1990c) The contribution of passive drag as a determinant of swimming performance. *International Journal of Sports Medicine*, 11, 367–372.

Chatard J.C., Lavoie J.M. and Lacour J.R. (1990d) Analysis of determinants of swimming economy in front crawl. *European Journal of Applied Physiology*, 61, 88–92.

Cotton C.E. and Newman J.A. (1978) Buoyancy characteristics of children, *Journal of Human Movement Studies*, 4, 129–143.

Gagnon M. and Montpetit R. (1981) Technological development for the measurement of the center of volume in the human body, *Journal of Biomechanics*, 14, 235–241.

Houssay R. (1912). *Forme, Puissance et Stabilite des Poissons*. Hermann et Fils, Paris.

Lane E.C. and Mitchem J.C. (1964) Buoyancy as predicted by certain anthropometric measurements, *Research Quarterly*, 35/1, 21–28.

McLean S.P. and Hinrichs R. (1998) Sex difference in the center of buoyancy location of competitive swimmers. *Journal of Sports Science*, 16, 373–383.

Mitchem J.C. and Lane E.C. (1968) Buoyancy of college women as predicted by certain anthropometric measures. *Research Quarterly*, 39/4, 1032–1036.

Nakashima M. (2005) Mechanical study of standard six beat front crawl swimming by using swimming human simulation model [in Japanese]. *Transactions of the Japan Society of Mechanical Engineers*, 71/705B, 1370–1376.

Packard J.C. (1900) Specific gravity of the human body, *Scientific American Supplement*, 1271, 20382.

Pendergast D.R., di Prampero P.E., Craig Jr. A.B., Wilson D.R. and Rennie D.W. (1977) Quantitative analysis of the front crawl in men and women. *Journal of Applied Physiology*, 43/3, 475–479.

Pettigrew J.B. (1874) *Animal Locomotion*. Appleton and Co., New York.

Robertson J. (1757) An essay towards ascertaining the specific gravity of living men. *Philosophical Transactions*, 50, 30–35.

Rennie D.W., Pendergast D.R. and di Prampero P.E. (1975) Energetics of swimming in man. In: Clarys J.P. and Lewillie L (Eds.) *Swimming II*, pp.97–104, University Park Press, Baltimore.

Rork R. and Hellebrandt F.A. (1937) The floating ability of women. *Research Quarterly*, 8, 19–27.

Sandon F. (1924) A preliminary inquiry into the density of the living male human body. *Biometrika*, 16, 404–411.

Schleihauf R.E., Gray L. and DeRose J. (1983) Three-dimensional analysis of hand propulsion in the sprint front crawl stroke. In: Hollander P., Huijing P. and de Groot G. (Eds.) *Biomechanics and Medicine in Swimming*, pp.173–183. Human Kinetics, Champaign, IL.

Yanai T. (2001) The effect of buoyancy in front crawl: Does it really cause the legs to sink? *Journal of Biomechanics*, 34, 235–243.

Yanai T. (2004) Buoyancy is the primal source of generating bodyroll in front-crawl swimming. *Journal of Biomechanics*, 37, 605–612.

Zamparo P., Capelli C., Termin B., Pendergast D.R. and di Prampero P.E. (1996) Effect of the underwater torque on the energy cost, drag and efficiency of front crawl swimming. *European Journal of Applied Physiology*, 73, 195–201.

EFFECT OF BUOYANT MATERIAL ATTACHED TO SWIMSUIT ON SWIMMING

M. NAKASHIMA[1], Y. MOTEGI[1], S. ITO[2] & Y. OHGI[3]

[1]Tokyo Institute of Technology, Tokyo, Japan
[2]National Defense Academy of Japan, Kanagawa, Japan
[3]Keio University, Kanagawa, Japan

There seems to be a lot of people who are reluctant to swim since they cannot swim well. The objectives of this study were to propose a swimsuit for those people with buoyant material to assist the body floating, and to examine the effect of the buoyant material on swimming theoretically and experimentally. Simulation analyses and an experiment using a developed swimsuit were conducted. It was found to be effective for beginner swimmer to attach the buoyant material as much as possible. It was also found that attaching the buoyant material almost does not affect the calorie consumption.

1 Introduction

Although swimming is known to be an ideal sport for physical fitness, people have to "master" it at first. There seems to be a lot of people who are reluctant to swim since they cannot swim well. The objectives of this study were to propose a swimsuit for those people with buoyant material to assist the body floating, and to examine the effect of the buoyant material on swimming theoretically and experimentally. The swimsuit has the plate-type buoyant material with several millimeter thickness inside the cloth. Although similar swimsuits have already come onto the market, its theoretical and experimental investigations have not been reported. For the theoretical investigation in this study, our developed swimming human simulation model SWUM (SWimming hUman Model) (Nakashima et al., 2007) was employed. SWUM was designed to solve the absolute movement of the whole human body by its equations of motion, giving the inputs of the human body geometry (shape and density of 21 body segments) and the joint motion.

In this study, a preliminary experiment was firstly conducted in order to acquire the swimming motion to be input into the simulation, and to investigate the difference in swimming form between the beginner and expert swimmers. Next, attaching the buoyant materials to the swimmers' thighs in the simulation model, the effect of the buoyant material was examined by the simulation. Next, an experiment using a specially developed swimsuit, which has many pockets to put the buoyant plates in, was conducted in order to examine the validity of the simulation results. Finally, the physiological calorie consumption during swimming, which is important from the viewpoint of the sport for fitness, was estimated by the simulation, and the effect of the buoyant material on the calorie consumption was examined.

2 Preliminary Experiment

A preliminary experiment, in which the swimming motions of two males (a beginner and an expert swimmers) were shot by a waterproof camera, was conducted. The expert swimmer has an experience of the competitive swimming during his junior and senior high school days. For the beginner swimmer, it was difficult to swim 25 m to the end. The waterproof camera can be moved manually together with the swimmer. Inputting the obtained swimming motions into the simulation model SWUM, a comparison between the beginner and the expert were made by the simulation. With respect to the input method of the swimming motion, one stroke cycle with less bubble was firstly clipped from the whole video. Next, the clip video was divided into 18 frames. Then, for each frame, the joint angles were roughly determined by the operator, and input into the simulation. In addition, they were carefully adjusted again comparing the video and the simulation results. Figures 1 shows the swimming movements of the experiment and simulation. From the videos, the following findings were observed: The expert seemed to swim efficiently in the unhurried form without unnecessary strength. The beginner seemed to move only the shank and foot in the flutter kick, resulting in the inefficient kick. In order to recover this inefficiency, the beginner seemed to increase the number of times for the kicks and hand strokes. Since all his motion seemed inefficient and his body did not propel much, he rather seemed to move his limbs desperately in order not to sink. In addition, the position of his waist was evidently lower (deeper) than that of the expert.

Table 1 shows the simulation results. The stroke length is defined as the propelling distance in one stroke cycle, and is normalized by the swimmer's stature. The propulsive efficiency represents the ratio of the power contributing to the propulsive direction to all the consumed power, and defined as: $\eta = UT/P$, where η, U, T, and P respectively represent the propulsive efficiency, the swimming speed averaged in one cycle, the drag acting on the swimmer's body towed at the speed U with the gliding position, and the averaged mechanical power consumed by the swimmer. The COG position represents the center of gravity position of the whole swimmer's body in the depth direction. It is normalized by the swimmer's stature. Large absolute value with negative sign of the COG position indicates that the swimmer's body tends to sink. From Table 1, it is found that the stroke length and the propulsive efficiency of the expert become larger than those of the beginner. Therefore, it

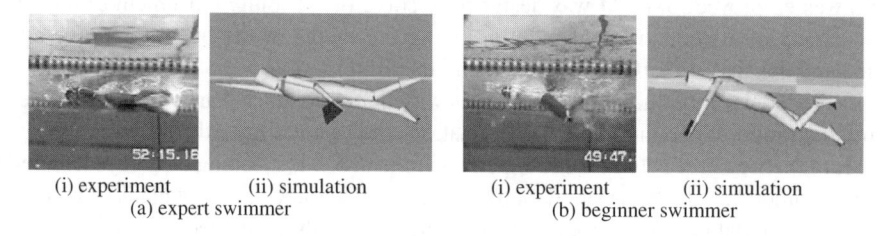

(i) experiment　　(ii) simulation
(a) expert swimmer

(i) experiment　　(ii) simulation
(b) beginner swimmer

Figure 1.　Swimming movements of the experiment and simulation.

Table 1.　Simulation results of the preliminary experiment.

	Stroke length	Propulsive efficiency	COG position
Expert	1.3281	0.1779	−0.0832
Beginner	1.0019	0.0687	−0.1027

is confirmed by simulation that the expert swam efficiently without unnecessary strength. From the simulation results of COG position, it is found that the beginner's body sinks more than the expert's one. This result also agrees with the impression of the video.

3 Examination of Effect of Buoyant Material by Simulation

The effect of the buoyant material was examined by adding buoyancy due to the buoyant material in the swimming simulation. In the simulation, buoyancy due to the swimmer's body itself acts on a tiny quadrangle as shown in Figure 2. In Figure 2, the truncated elliptic cone represents one body segment in SWUM. This cone is firstly divided for the longitudinal direction into the elliptic plates. The surface of each plate is again divided for circumferential direction into the tiny quadrangles. Buoyancy is calculated by integrating the pressure force F_b due to the gravitational force acting on the tiny quadrangle. Since the buoyant material in this study is assumed to be sufficiently thin, buoyancy due to the buoyant material is directly added in the vertical direction to the pressure force on the tiny quadrangle where the actual material is attached.

Table 2 shows the simulation results when buoyancy due to the buoyant material was 1.65 N. Buoyancy was uniformly distributed to 20 quadrangles on the front of each thigh (most front 2 quadrangle lines in the circumferential 36 division $\times$ 10 for the longitudinal division). Comparing the results in Table 2 with those in Table 1, it is found that the stroke length, propulsive efficiency, and COG position become better. Table 3 shows the simulation results of the beginner when buoyancy was increased as 5 and 8 times of 1.65 N. From the results, it is found that larger buoyancy is given, better all the indices become. Note that the similar tendency was seen in the case of the expert. Therefore, it is effective to attach the buoyant material as much as possible. In addition, the analyses for the various attaching position (thigh, shank, and waist, front, back, inner, outer) were also conducted in order to examine its effect. However, significant difference in the evaluation indices could not be seen.

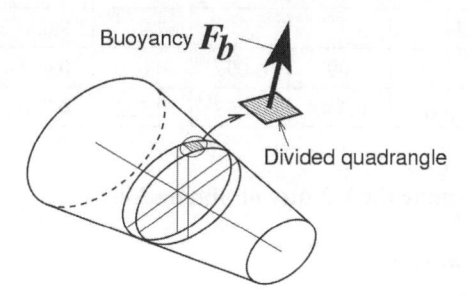

Figure 2. Calculation method of buoyancy in SWUM.

Table 2. Simulation results (buoyant force added).

	Stroke length	Propulsive efficiency	COG position
Expert	1.3356	0.1864	−0.0793
Beginner	1.0302	0.0753	−0.0925

Table 3. Simulation results (change in buoyant force for the beginner).

Buoyancy (times)	1	5	8
Stroke length	1.0302	1.0585	1.0620
Propulsive efficiency	0.0753	0.0823	0.0873
COG position	−0.0925	−0.0819	−0.0761

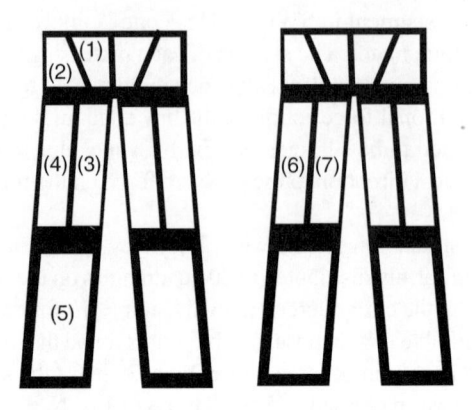

Figure 3. Pocket positions of swimsuit (left: front, right: back).

Table 4. Specification of five buoyant plates.

Plate ID	A	B	C	D	E
Volume [cm^3]	38.35	39.84	82.8	82.49	65.52
Weight [g]	3.7	3.9	7.9	8.0	6.4
Density	0.097	0.097	0.095	0.097	0.098
Buoyancy [N]	0.340	0.353	0.733	0.730	0.580

4 Experiment to Examine the Validity of Simulation

4.1. *Experimental Procedure*

An experiment was conducted in order to examine the validity of the simulation results. For this experiment, a special swimsuit which has many pockets to put the buoyant plates in was newly developed. Figure 3 shows the position of the pockets. The buoyant plate was 3 mm thickness soft polystyrene foam, and five shapes were prepared to be fit to the pockets. Table 4 shows the specifications of the five plates. Various trials were conducted changing the volume and position of the material. Table 5 shows the contents of trials. For example, the trial 1 was the case without the material. The trial 2 was the case with material of A and B in pockets (1) and (2), respectively. Trial 3 was with plates of both C and D in pocket (5). The beginner attached three same plates for each side of right and left (total

Table 5. Contents of trials.

Trial No.	1	2	3	4	5	6	7	8	9
Subject	\multicolumn{6}{} Beginner						Expert		
Pocket place	N/A	(1)(2)	(5)	(3)(4)	(3)(7)	(4)(6)	(3)(4)	(5)	(1)(2)
Buoyant plate	N/A	A B	C D	C D	C E	C E	C D	C D	A B

Table 6. Total buoyancy of experiment.

Trial No.	2	3	4	5	6	7	8	9
Buoyancy [N]	4.15	8.78	8.78	7.88	7.88	5.85	5.85	2.77

Table 7. Results of sensory evaluation (Beginner).

Trial No.	1	2	3	4	5	6
Floating feeling	50	75	85	85	75	75
Resistance	50	50	50	50	50	50
Kick	50	50	50	50	50	50
Breath	50	62.5	80	90	70	70
Total evaluation	50	75	85	85	75	75

Table 8. Results of sensory evaluation (Expert).

Trial No.	7	8	9
Floating feeling	75	75	70
Resistance	25	15	50
Kick	25	25	50
Breath	50	50	60
Total evaluation	50	75	75

of right and left becomes six), and the expert two for each (total four). Table 6 shows the total buoyancy for each trial. The swimming motion was shot by the waterproof camera, and the sensory evaluation was conducted. In the sensory evaluation, the swimming without the buoyant material was set as the reference (grade 50). The subjects were requested to answer the grade in the range of 0~100 for each item in each trial.

4.2. Experimental Results

Tables 7 and 8 show the results of the sensory evaluation. With respect to the floating feeling, both the beginner and expert felt that the body was floated more by greater volume of the

Table 9. Estimated calorie consumption.

	Beginner	Expert	Model swimming
Calorie consumption [kcal/h]	921.0	545.8	649.0

Table 10. Estimated calorie consumption of the experiment.

Trial No.	1	2	3	4	5	6
Calorie consumption	921.0	917.1	914.7	914.7	915.1	915.1

material. With respect to the breath, the beginner felt more easiness in the case of greater floating feeling. The reason for this is that he could keep his composure by being floated. The expert felt the easiness when the material was attached to the waist. With respect to the total evaluation, the beginner's results directly correspond to the floating feeling, the same as the breath. That is, floating is the most important factor for the beginner swimmer. Therefore, it is best for beginner swimmers to attach the buoyant material as much as possible. The expert, on the other hand, felt the difficulty in the resistance and kick when the material was attached to anywhere, especially to the body part far from the center of the whole body.

5 Estimation of Calorie Consumption during Swimming by Simulation

For the estimation, the metabolic equivalent (METS) was employed. One MET is the calorie consumption at rest, and is 1 kcal/kg/h. The value of METS for the crawl stroke is generally known to be 8~10. In this study, the simulation result of the model swimming by an elite swimmer (Nakashima et al., 2007) was assumed to be 10 METS. Since the weight of the body model of this simulation is 64.9 kg, the calorie consumption in one second becomes 10 kcal/kg/h × 64.9 kg × (1/3600) h = 0.180 kcal/s. On the other hand, the mechanical power consumption computed in the simulation was 202.7 W = 0.048 kcal/s. The ratio of the mechanical power consumption to the total power consumption is called "gross efficiency", and is known to be about 25% in general. The gross efficiency of the crawl stroke is calculated as: 0.048 / 0.180 × 100 = 26.8%. Using this efficiency, the calorie consumptions in one hour for the beginner and expert in our experiment were estimated by the simulation. Table 9 shows the results without the buoyant material. It is found that the calorie consumption of the beginner is 1.7 times greater than that of the expert. Table 10 shows the results of the beginner swimmer for the trials in the experiment of the chapter 4. Although the calorie consumption becomes slightly smaller due to the buoyant material, the difference can be regard as negligible. Therefore, it is effective for the beginner swimmer to attach the buoyant material to the swimsuit since it helps to swim for a long time and does not decrease the calorie consumption besides.

6 Conclusions

The main findings obtained in this study are as follows:

- From the simulation, the stroke length, propulsive efficiency, and the center of gravity position was found to become better when the buoyant material is attached

to the lower half of the body. In addition, this tendency was found to depend not on the attaching position, but on the amount of the buoyancy.
- From the sensory evaluation in the experiment, it was found to be important for beginner swimmers to attach the buoyant material as much as possible. For the expert, on the other hand, the buoyant material may rather disturb the swimming.
- From the estimation of the calorie consumption, that of the beginner was found to be 1.7 times greater than that of the expert. It was also found that attaching the buoyant material almost does not affect the calorie consumption.

Acknowledgements

This work was funded by Mizuno Sports Foundations. The authors thank Ken Matsuzaki and Naokazu Yuge from Mizuno Corporation for their devoted assistance in developing the swimsuit and conducting the experiment.

References

Nakashima M., Satou K. and Miura Y. (2007) Development of swimming human simulation model considering rigid body dynamics and unsteady fluid force for whole body. *Journal of Fluid Science and Technology*, 2/1, 56–67.

ANALYSIS OF THE OPTIMAL ARM STOKE IN THE BACKSTROKE

S. ITO

*Department of mechanical engineering, National Defense Academy,
Yokosuka, Japan*

The most common arm stroke of the backstroke in a present competitive swimming is the so-called S-shaped stroke on a side of the body. The author calculated the optimal stroke in the backstroke with a simplified physical model by fluid dynamic characteristic of a palm obtained by a wind tunnel experiment. As a result, the optimal stroke path of the maximum efficiency and the maximum thrust were obtained as a driving angle, a tilt angle, and an angle of attack of a hand. The maximum efficiency was obtained in the S-shaped pull stroke on a side of the body, while the maximum thrust was obtained in the I-shaped pull stroke parallel to the body axis.

1 Introduction

Dynamics of swimming is fluid mechanics that acts on a swimmer and the water of the circumference of the body. Counsilman (1968) proposed S-shaped pull swimming style is suitable for the front crawl in 1968. This S-shaped pull uses the resultant force of the lift and drag as an impellent. Lately, Sanders (1997) stated that competitive swimmers seem to use a drag type stroke rather than S-shaped pull stroke in his experimental research. Moreover, the author (Ito & Okuno, 2003) introduced I-shaped pull stroke obtained larger thrust force than S-shaped.

As for the backstroke, it is often considered just as a turn over style of the front crawl stroke. However, it is not correct. Because of the different movable regions of articulation humeri, the backstroke takes a stroke on a side of the body shown in Figure 1a and Figure 6a. In this study, I-shaped pull stroke was introduced in the backstroke besides S-shaped stroke, a conventional style for the backstroke. A palm plays a predominant role for an impellent in swimming according to previous studies by Hollander (1987) and Berger *et al.* (1995). The posture angles and the driving angles of a palm which produced the maximum efficiency and the maximum thrust were calculated with the experimental data.

2 Method

A replica of a hand of a good swimmer was made with plaster. The fluid dynamic characteristics of the replica was examined in a wind tunnel in the same degree of the Reynolds number, $Re = 3.4 \times 10^5$, of the inflow speed to the hand of an actual swimmer. It is possible to divide a stroke of the backstroke into three phases as shown in Fig. 1b, such as Catch (Downsweep), Pull (Upsweep), and Finish. Each phase is classified by sweepback angle ψ defined by Schleihauf. The sweepback angle ψ is an angle of inflow into a hand palm as shown in Fig. 2. It is possible to apply each phase into three kinds of degrees, 180°, 0°, and 45° respectively. Lift and drag forces were measured by three component load cells at every one degree of angle of attack from 0° to 90°. The obtained lift and drag forces were

(a) Hand path of S-shaped back crawl stroke

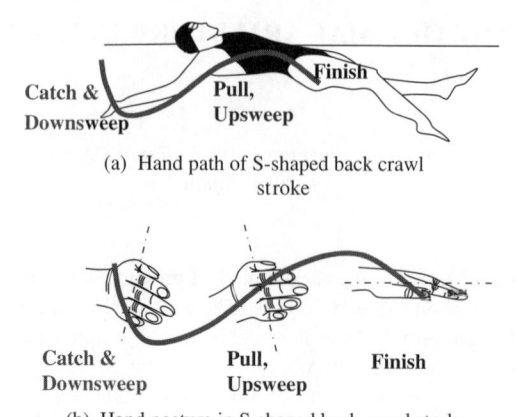

(b) Hand posture in S-shaped back crawl stroke

Figure 1. Side view of Conventional S-shaped backstroke.

calculated into lift coefficient C_L and drag coefficient C_D according to a projected sectional area of a hand palm S. Polar curves were obtained as characteristics of the hand.

An inclined hand by tilt angle θ is moved diagonally with driving velocity U and driving angle δ while a body moves with advancing velocity V shown in Fig. 3 in the backstroke. As a result, relative velocity W flows into the hand with angle of attack α. Drag force D acts opposite to a direction of W and lift force L acts perpendicular to W. Thrust force T is generated as a component of a resultant force R consisting of L and D to the advancing direction. With the above variables, geometrical relations are established with a velocity triangle composed of U, V, W and slip angle β. Furthermore, thrust force T must be equal to parasite drag of the body, D_B in a trimmed state. The following equations were derived according to the above relations:

Thrust T, power P and efficiency η can be defined with them and be normalized:

$$T = \frac{1}{2}\rho W^2 S C_R \cos(\gamma + \alpha - \theta) = \frac{1}{2}\rho W^2 S C_T \qquad (1)$$

$$P = \frac{1}{2}\rho W^2 U S C_R \cos(\gamma - \beta) = \frac{1}{2}\rho W^3 S C_P \qquad (2)$$

$$\eta = TV/P \\ = \cos(\gamma + \alpha - \theta)/\{(U/V)\cos(\gamma - \beta)\} \qquad (3)$$

where

$$L = \frac{1}{2}\rho W^2 S C_L, \quad D = \frac{1}{2}\rho W^2 S C_D \qquad (4a, b)$$

$$C_R = \sqrt{C_L^2 + C_D^2} \tag{4c}$$

$$\gamma = \tan^{-1}(L/D) \tag{5}$$

In the above variables, the following geometrical relations are established with the velocity triangle composed of U, V and W:

$$W/V = (U/V)\cos\beta - \sqrt{1-(U/V)^2 \sin^2\beta} \tag{6}$$

$$\tan(\theta - \alpha) = (U/V)\sin\beta/\{(U/V)\cos\beta - (W/V)\} \tag{7}$$

$$\delta = \theta - (\alpha + \beta) \tag{8}$$

Eqs. (6) and (7) are reduced to the following equation:

$$\beta = \sin^{-1}\{\sin(\theta - \alpha)/(U/V)\} \tag{9}$$

In a trimmed state, thrust force T must be equal to parasite drag of the body, D_B

$$T = \tfrac{1}{2}\rho W^2 S C_T = D_B = \tfrac{1}{2}\rho V^2 S_B C_{DP} \tag{10}$$

where S_B is a drag area of the driven body.

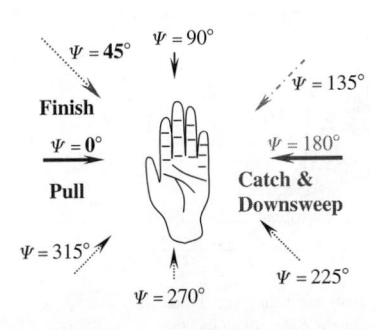

Figure 2. Sweepback angle ψ defined by Schleihauf (1979).

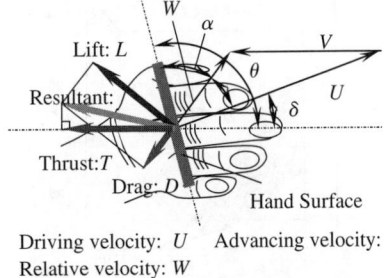

Driving velocity: U Advancing velocity: V
Relative velocity: W
Angle of attack: α Slip angle: β
Drivinig angle: δ Tilt angle of hand: θ

Figure 3. Forces acting on a hand paddle.

3 Results and Discussion

Figures 4a and b are polar curves based on two different sweepback angles, $\psi = 180°$ and $0°$ respectively, showing the relationship between the lift coefficient C_L and the drag coefficient C_D. The polar curves also correspond to a change of angle of attack α from $0°$ to $90°$ every $1°$. Markers are shown every 5 degrees in angle of attack α in the respective curves. An arc of C_{R0} is drawn from the point where the maximal thrust force occurs at angle of attack $\alpha = 90°$. The calculated thrust coefficient C_T at $\psi = 0°$ obtained from the polar curves is shown in Figure 5a. The maximum thrust point exists for the thrust coefficient, and the point is corresponding to the farthest point of drag from the origin of the polar curve, that is, the maximum resultant force point in Figure 4. During the phase from catch to upsweep, the state angle at the maximum thrust point was turned out that tilt angle θ, angle of attack α and

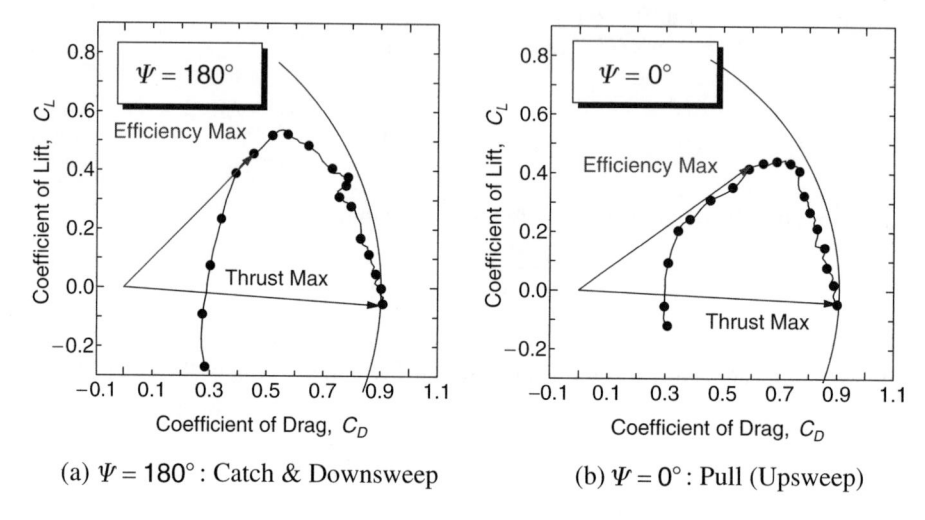

(a) $\Psi = 180°$: Catch & Downsweep

(b) $\Psi = 0°$: Pull (Upsweep)

Figure 4.   Polar curves at different sweepback angle, $\psi = 180°, 0°$
Each graph shows the maximal efficiency point and the maximal thrust point.

(a) Thrust coefficient C_T, based on driving angle δ and tilt angle θ

(b) Thrust efficiency η, based on driving angle δ and tilt angle θ

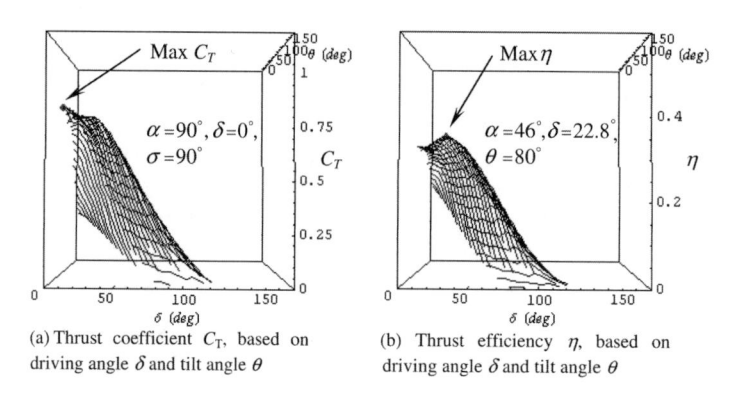

Figure 5.   Calculated results of a hand palm specification at $\psi = 0°$. Pull (upsweep) phase.

driving angle δ of a hand palm were 90°, 90° and 0° respectively. In this state, the style of swimming produced the maximum thrust in a drag type. Thrust efficiency was similarly obtained by the calculations based on the experiment. The calculation results are shown in Figure 5b. The maximum point of the thrust efficiency exists, and the point agrees with the tangential point from the origin to the polar curve in Figure 4. Figure 6 summarizes the parameter of each posture angle of the hand palm in each mode. For the maximum thrust mode shown by the dotted line, the tilt angle shows 90° in Catch and Pull phase except Finish phase. The driving angle also shows 0° in Catch and Pull phase except Finish phase. That is, it is meant that the hand palm located squarely to the driving direction should drive straight parallel to the traveling direction of the body. Figure 7 indicate the stroke path of the hand palm in the maximum thrust mode from the absolute coordinate system. The author named it as I-shaped pull stroke. For the maximum efficiency mode shown in the solid line, the tilt angles indicate that hand palm inclines less than vertical. The driving angle shows that hand palm drives diagonally from 30 to 45 degrees to the traveling direction or to the surface of the water. Considering the sweepback angle and the driving angle, the movement of the hand palm becomes S-shaped pull swimming shown in Fig. 1.

Figure 8a indicate the value of thrust coefficient in the maximum thrust mode and in the maximum efficiency mode. In the pull phase $\psi = 0°$, which relates to the propulsive force the most, the value, $C_T = 0.90$, at the maximum thrust mode is about 9% larger than the value of $C_T = 0.83$, in the maximum efficiency mode. As for the efficiency shown in Figure 8b, the thrust efficiency at each case was 34% at $C_T = 0.90$ and 37% $C_T = 0.83$, respectively. The difference of the efficiency between at the maximum efficiency and at the maximum thrust is only 3%. In the competitive swimming which requires larger thrust

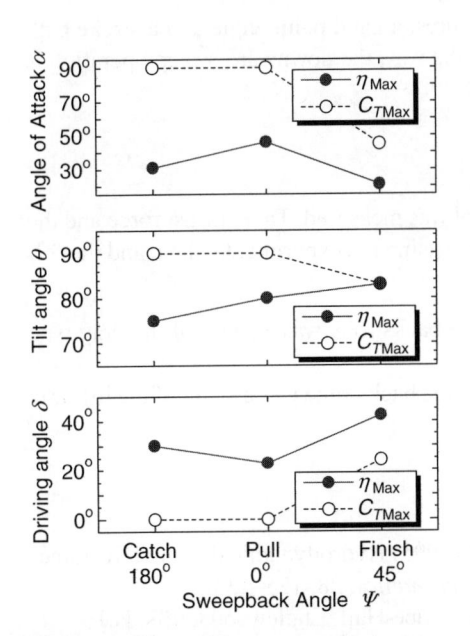

Figure 6 Transition of posture angle of the hand in each phase in the maximum thrust mode and the maximum.

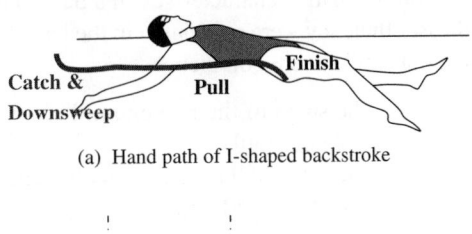

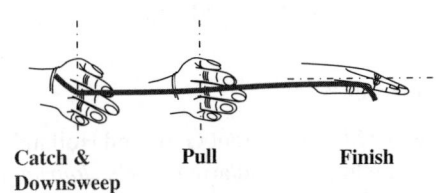

Figure 7 Side view of proposed I-shaped back stroke.

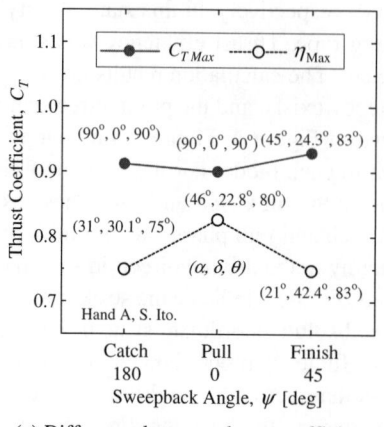

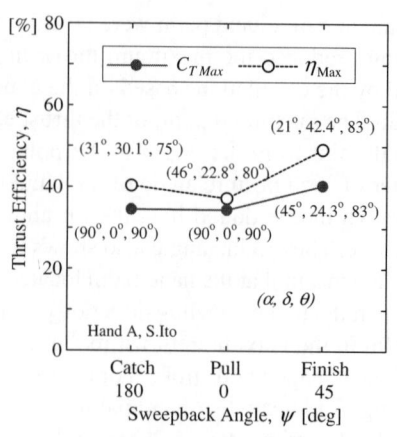

Figure 8 Difference of thrust coefficient, C_T and the thrust efficiency, η among the different sweep back angles ψ.

(a) Difference between thrust coefficient, C_T of the maximal thrust and that of the maximal efficiency among sweep back angle ψ

(b) Difference between thrust efficiency, η of the maximal thrust and that of the maximal thrust coefficient, C_T among sweep back angle ψ.

force, a swimmer should swim with an advantageous stroke in the thrust force like a drag type; I-shaped pull even if a little efficiency falls down.

When the state was concretely shown as figures, a hand palm angle and a stroke path becomes the one shown as Figure 8a and b. By the way, the downward sweep operation is inevitable because of rolling in the finish phase.

4 Conclusion

The fluid dynamic characteristic of a palm model was measured. Then, thrust force and the thrust efficiency were calculated in the backstroke using the experimental data, and the following results were obtained.

1) The swim of the maximum thrust in the backstroke was a swim of the drag type (I-shaped pull).
2) The swim of the maximum efficiency in the backstroke was a swim of the lift-drag type (S-shaped pull).

References

Berger M.A.M., Groot G. de and Hollander A.P. (1995) Hydrodynamic drag and lift forces on human hand/arm models. *Journal of Biomechanics*, 28, 125–133.
Counsilman J.E. (1968) *Science of Swimming*. Prentice-Hall, Englewood Cliffs, N.J.
Hollander, A.P. de and Ingen Schenau, G.J. van (1987) Contribution of the legs to propulsion in front crawl swimming. *Swimming Science V*, pp.17–29. Human Kinetics Publishers, Champaign, IL.

Ito S. and Okuno K. (2003) A fluid dynamical consideration for armstroke in swimming. *Biomechanics and Medicine in Swimming IX*, pp.39–44, Pub. de l'univ. de Saint-Etienne.

Sanders R.H. (1997a) Extending the "Schleihauf" model for estimating forces produced by a swimmers hand. In: Eriksson B.O. and Gullstrand L. *Proceedings of the XII FINA World Congress on Sports Medicine.* Goteborg, Sweden 12–15 April 1997, pp.421–428.

Schleihauf R.E. (1979) A hydrodynamic analysis of swimming propulsion. *Swimming III*, pp.70–109. University Park Press, Baltimore.

SWIMMING STROKE ANALYSIS USING MULTIPLE ACCELEROMETER DEVICES AND TETHERED SYSTEMS

N.P. DAVEY[1,2] & D.A. JAMES[1]

[1]*Centre for Wireless Monitoring and Applications, Griffith School of Engineering*
Griffith University, Nathan Campus, Nathan, Australia
[2]*Centre of Excellence for Applied Sport Science Research, Queensland Academy of Sport, Level Queensland Sport and Athletics Centre (QSAC), Nathan, Australia*

In swimming detailed performance analysis is performed occasionally because it typically requires specialist instrumentation setup in the pool and large amounts of time for testing and post-processing is required. Using accelerometer devices attached to the swimmers body, movement patterns can be quantified. These acceleration patterns can be used in the analysis of stroke performance. In this paper, tri-axial accelerometer devices where attached to multiple locations on a swimmer. A tethered device is used to also record the swimmers velocity. A comparison of the accelerometer data and tethered data is undertaken to examine the relationship between swimmer velocity, and acceleration. It is shown using a single axis of acceleration that there is a correlation in the timing events between acceleration data and velocity data. This correlation allows for corrections to be applied to known error in acceleration data.

1 Introduction

The measurement and analysis of sport performance is an integral aspect of training and preparation for elite athletes of all disciplines. Traditionally the majority of performance analysis occurs in a laboratory setting or during designated testing sessions, possibly in unfamiliar environments. This type of testing environment can have an affect on the athletes performance, often in a psychological (Kerr & Kuk, 2001; Kerr *et al.*, 2006) or physiological (Bonen *et al.*, 1980) sense. Therefore it is preferential to measure an athlete's performance in-situ, in their natural training or competition environment. In-situ measurement may provide additional information that can not be obtained from laboratory based testing.

Prior research using in-situ tri-axial accelerometer devices has been conducted using a single point located at the lower back centre of mass of the athlete (Davey *et al.*, 2005). This research has shown that lap time, stroke type and stroke rate information can be identified using data from this single point of reference. In this paper measurement of multiple limb segments is investigated using multiple time-synchronised in-situ tri-axial accelerometer devices. Validation and comparison with additional data collected via a tethered device (Applied Motion Research) was also undertaken.

2 Technology

The accelerometer device used in this research is the Mini-Traqua, developed by the authors in conjunction with the Australian Institute of Sport (AIS) and the Cooperative Research Centre for microTechnology. The device is a self contained water proof sensing

Figure 1. Mini-Traqua accelerometer device.

Table 1. Technical specification for Mini-Traqua.

Processor	Atmel ATMEGA128
Sensors	Kionix KXM52-1050 3axis 2 G accelerometer ADXRS150 Gyroscope
Radio	Nordic NRF2401 2.4 GHz radio with internal patch antenna
Memory	128 Mbyte Flash Memory
Inputs/Outputs	96×64 pixel monochrome LCD screen USB data transfer and charging 5 way push button

device measuring 52 mm length × 34 mm wide × 12 mm height, and weighing 22 grams. A picture of the device is shown in Figure 1. The device contains a microcontroller, tri-axial accelerometer and flash memory as detailed in Table 1. The device provides simultaneous recording of tri-axial acceleration and single axis gyroscope data to onboard memory. The device is also equipped with RF (Radio Frequency) capability to enable remote control of the device, an LCD screen to provide information, and buttons to interact with the device. The buttons on the device permit turning the device on and off, as well as controlling data recording sessions.

In this paper we demonstrate the use of multiple devices on a single athlete. To permit multiple devices to be time synchronised, a networked synchronisation protocol was also developed in the firmware. The synchronization protocol enables the simultaneous starting of data acquisition on multiple devices. The network is structured using a master/slave arrangement where the master is responsible for, and controls all device communications on the network. All devices transmit and receive on the same RF channel. Slave devices do not transmit unless instructed to by the master. The master is attached to and controlled from a PC software application.

The network is a master/slave configuration, with each device previously assigned a unique network identification (ID) number to be used on the network. The network uses a packet format, shown in Table 2. This format enables a range of different information to be exchanged between the master and the slaves.

Table 2. Network packet structure.

Packet Field Name	Size	Description
Destination ID	1 Byte	ID of receiving device
Source ID	1 Byte	ID if sending device
Command	1 Byte	Command instruction
Sequence	1 Byte	Packet sequence number
Payload	16 Bytes	Packet payload data

Figure 2. Speed Probe 5000 (picture from company website).

The ID number that a device is given has a valid range from 2 to 254, ID 1 is reserved for the master device. ID 255 is a special case, this is used as a broadcast address, that is, that if a device receives either its own ID, or ID 255 in the Destination ID field, it will receive and process the network packet. Using this feature, multiple devices can be instructed to initiate data logging, stop data logging or to place markers in the logged data.

The tethered device used in this research is the Applied Motion Research Speed Probe 5000 (Applied Motion Research), as shown in Figure 2. Tethered systems are used to measure velocity profiles for straight line motion events such as running and swimming. The device claims a resolution better than 0.1 m/s. The device uses a nylon line attached at a single point to athlete. Velocity is determined by measuring the time it takes for 1 cm of line to moved pass an optical sensor. To measure the swimmers velocity, the line from the tethered device was attached to a strap around the swimmers waist. A computer is attached to the tethered device for data storage and control. The computer software is also able to synchronise recording of video and velocity data.

3 Validation Study

The aim of this research is to investigate the relationship between acceleration signatures from different body segments and to examine the relationship between swimmer velocity, and acceleration data collected from multiple limb segments. The waist strap for the tethered device also holds an accelerometer device in position at the lower back. An accelerometer

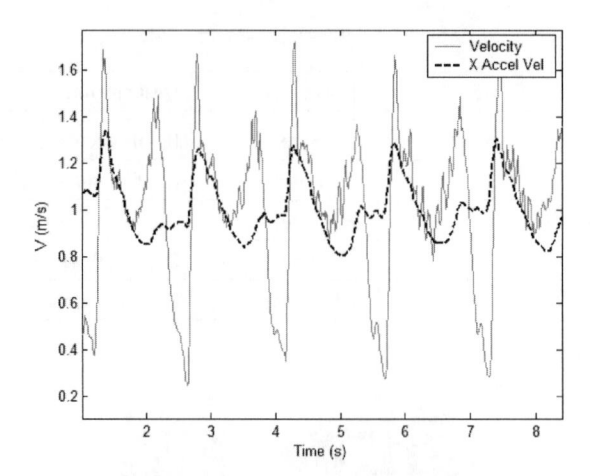

Figure 3. Breaststroke lower back forward derived velocity versus tethered device velocity.

device was also attached to the left wrist of the swimmer. This accelerometer device was attached to the outside of the wrist using a pocket and velcro straps.

One male and one female subject with no history of injury were recruited from the sub elite athlete community. Each subject provided written consent prior to participation in the study which has been approved by the Griffith University Human Research Ethics Committee. The subjects were asked to swim a series of 25 m laps at training pace. Freestyle, butterfly and breaststroke strokes were recorded.

4 Analysis and Results

In establishing a numerical relationship between accelerometer data and velocity, each stroke has different aspects to be considered. In swimming the primary direction of motion is forward, and each stroke has different characteristics in this regard. Initial investigation of the velocity showed that butterfly and breaststroke have a greater range of velocity change on a stroke to stroke basis. This is not the case for freestyle and backstroke that show a relatively constant velocity. Based on this initial assessment of velocity characteristics, breaststroke and butterfly will be further investigated.

For breaststroke and butterfly, the acceleration axis of interest is the one which is parallel with the direction of travel. As forward velocity changes, so will the acceleration pattern seen on this axis, as acceleration is a derivative of velocity from Newton's Law (see Eq. (1)).

$$Vx(t) = ([Vx(t-1) + Ax(t)] - \overline{Vx}) \times 15 + 1 \tag{1}$$

Figure 3 is a plot of the tethered device velocity data overlaid with the breaststroke forward direction accelerometer derived velocity. Eq. (1) was used to compare the data sets as shown in Figure 3. It can be seen that there is a basic correlation in the timing of events in both traces. Whilst the velocity patterns do not match, it does show that the accelerometer is seeing some velocity change. Using just a single axis of acceleration from the lower back to obtain velocity will be erroneous, as during the stroke cycle the position of the accelerometer

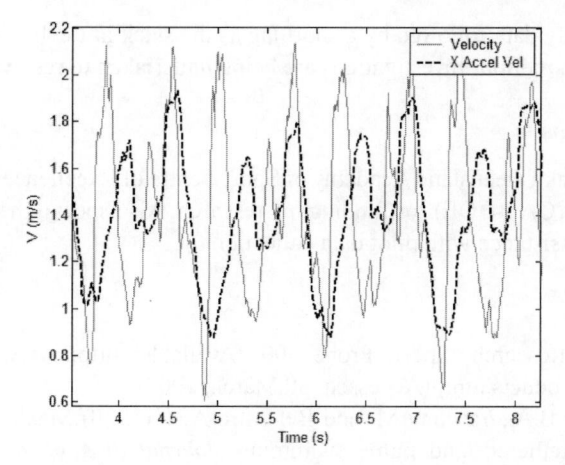

Figure 4. Butterfly lower back forward derived velocity versus tethered device velocity.

device with respect to the gravity vector changes. This positional change will combine with the dynamic acceleration and alter the derived velocity.

Acceleration data for each axis was integrated separately to get a velocity pattern per axis. When integrating accelerometer data to find velocity and displacement, drift and noise can be significant. There are numerical methods available to minimise these effects (Thong et al., 2002; Thong et al., 2004), but given the short time windows that are of interest, we consider drift not to be significant. No corrections have been applied to account for the fact the both static and dynamic accelerations will be present in the data. The derived velocity curves were then scaled to enable better comparison against with velocity data.

$$Vx(t) = ([Vx(t-1) + Ax(t)] - \overline{Vx}) \times 10 + 1 \tag{2}$$

The relationship shown in figure 4 is butterfly forward direction derived velocity from Eq. (2) and tethered device velocity. It can be seen that significant portions of the two traces do align, but the remaining portions almost appear inverted. As the influence of static acceleration in not accounted for in this data, the orientation of the accelerometer on the swimmer will be change with respect to gravity during the stroke cycle. This change in orientation will alter the derived velocity.

5 Conclusions

This research has shown that there is clear relationship between lower back acceleration and velocity, but more investigation is still required. Whilst butterfly and breaststroke where chosen for their significant changes in forward direction acceleration, the other strokes also need to be considered. Further consideration needs to be given to using more than one accelerations axis in velocity determination, as unit orientation with respect to gravity will significantly influence the derived measure.

An issue that can arise from the current tethered device configuration is that as the swimmer gets further from the unit the line can be kicked or hit. This introduces a large spike in the velocity data followed by a short lag as the slack in the line is removed by the swimmers forward motion. Investigations are being undertaken to resolve this issue.

Acknowledgements

I would like to thank Queensland Academy of Sport Centre of Excellence for Applied Sport Science Research (QAS-COE) for funding my research. Also special thanks to QAS sport science staff for assistance with pilot data collection.

References

Applied Motion Research, Speed Probe 5000 Available: http://www.appliedmotionresearch.com/products.html (Accessed: 30 March, 2007)

Bonen A., Wilson B.A., Yarkony M. and Belcastro A.N. (1980) Maximal oxygen uptake during free, tethered, and flume swimming. *Journal of Applied Physiology*, 48/2, 232–235.

Davey N.P., Anderson M.E., James D.A. (2005) An accelerometer-based system for elite athlete swimming performance analysis In: Al-Sarawi S.F. (Ed.) *Smart Structures, Devices, and Systems II*. Proc. SPIE, Vol. 5649, p. 409–415.

Kerr J.H. and Kuk G. (2001) The effects of low and high intensity exercise on emotions, stress and effort. *Psychology of Sport and Exercise,* 2/3, 173–186.

Kerr J.H., Fujiyama H., Sugano A., Okamura T., Chang M. and Onouha F. (2006) Psychological responses to exercise in laboratory and natural environments. *Psychology of Sport and Exercise,* 7, 345–359.

Thong Y.K., Woolfson M.S., Crowe J.A., Hayes-Gill B.R. and Challis R.E. (2002) Dependence of inertial measurements of distance on accelerometer noise. *Measurement Science and Technology*, 13, 1163–1172.

Thong Y.K., Woolfson M.S., Crowe J.A., Hayes-Gill B.R. and Jones D.A. (2004) Numerical double integration of acceleration measurements in noise. *Measurement*, 36, 73–92.

A SEMI-AUTOMATIC COMPETITION ANALYSIS TOOL FOR SWIMMING

X. BALIUS[1], V. FERRER[1], A. ROIG[1], R. ARELLANO[2], B. DE LA FUENTE[2],
E. MORALES[2], X. DE AYMERICH[3] & J.A. SÁNCHEZ[4]

[1]*Olympic Training Centre (CAR) of Catalunya, Catalunya, Spain*
[2]*Facultad de Ciencias de la Actividad Física y del Deporte, Granada, Andalucía, Spain*
[3]*Instituto Vasco de Educación Física (SHEE/IVEF), Navarra, Spain*
[4]*Instituto Nacional de Educación Física de Galicia, Galicia, Spain*

By appointment of the Organising Committee of the Xth World Swimming Championships the Biomechanics Department of the Olympic Training Center led the race's Competition Analysis of Semi-finals and Finals. A new tool was specially developed to ease the data video capture synchronised with the official races' timming, the data process through sophisticated image treatment algorithms, and the data hand out with the use of general and personalised spread-sheet informs. Data was published on a web site hours after the end of the competition allowing the access to this information to swimmers, coaches, media and general public.

1 Introduction

On the occasion of the Xth FINA World Championships held in Barcelona in 2003, the Biomechanics Department of the Olympic Training Center (CAR) developed a new semi-automatic tool to analyse the mean velocities, mean frequencies and mean stroke lengths of the swimmers.

All semi-finals and finals of the swimming competition were recorded and analysed. Calculated results were published the same afternoon of the competition on a specially designed web site for consultation. Swimmers, coaches, press and general public were the target users of the information shared by the research team.

The technology used was a step beyond the classic instrumentation implemented for the analysis of competition based on video recording and post-processing computer analysis. Furthermore, the structure of the final report was improved and easier to interpret by personalising the results for every swimmer competing in the races.

Four video cameras recorded the swimming pool sending their signal to four analogue-to-digital video cards. The signal was post-processed with specially designed software that automatically located the pass of the swimmers head through specific spaces of the recorded image. Small corrections had to be made in order to exactly and accurately locate the instant of the head pass. Also, software assistance was designed to ease the gathering of the stroke frequencies needed for the analysis.

The project was sponsored by CAR Catalunya, Consejo Superior de Deportes (CSD) and Barcelona'03 Organising Committee.

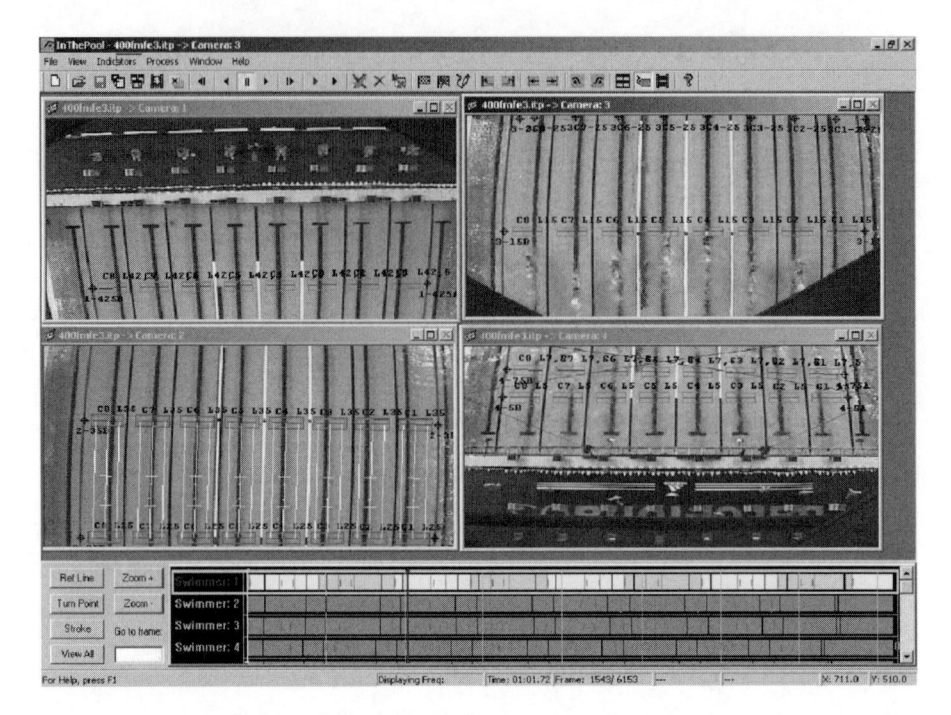

Figure 1. Computer image showing the recording frames of the 4 cameras used for analysis. The yellow zone undernieath shows the time events for every swimmer.

2 Methodology

The video recording (25 Hz) of the swimming pool was done with 4 analogue cameras (JAI Corporation CV-S3300) with Spacecom CTV – S4.8 mm 1:1.8 2/3″ lenses. The cameras were placed vertically to the pool recording 4 different transversal spaces as shown on Figure 1.

BNC cables sent the video signal to 4 IDS PIRANHA video compression boards. The SwissTiming competition race start signal was automatically registered to synchronise the video. Every wall contact was also recorded. 4 PCs Pentium 4, 2 GHz, 256 Mb RAM, 40 Gb HD were used to process data and to format the results with a specially designed software.

The objectives of the project were:

- To develop a tool to speed up the data gathering of the competition analysis
- To design a form to personalise and ease the understanding of results for its use during competition
- To improve the process required before publishing the results on the net for swimmers, coaches, press and general public

3 Results and Discussion

Results published on a specially made web site (www.car.edu/finabcn03) consisted of a 9 pages report. A first general page and eight personalised for every swimmer.

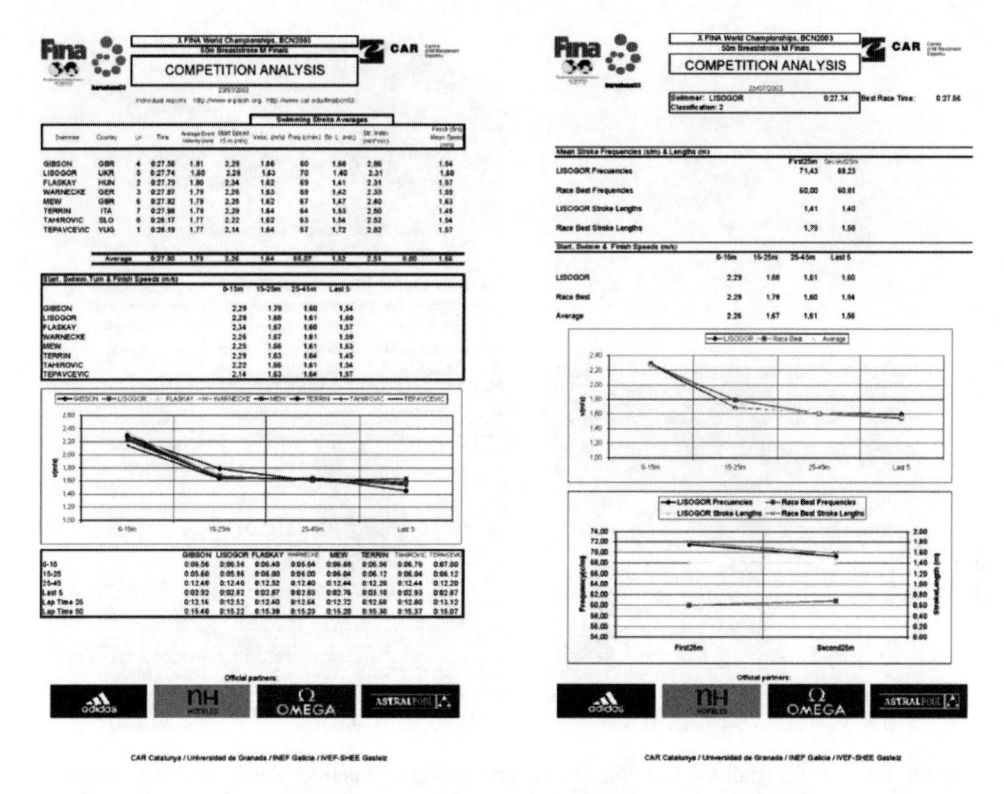

Figure 2 (Left). General information of the race for all swimmers. Figure 3 (right). Personalised information with mean velocities, stroke lengths and frequencies of the swimmer, the best swimmer and the mean values.

Figures 2 and 3 show an example of the report.

Specially made informs for all races were developed to ease the understanding of the given data. All objectives were accomplished taking into account that the tool seemed to work properly and on a good time basis after the end of the last race of every afternoon's competition. The accomplishment to personalise all informs, respecting a specially designed base, seemed to ease the understanding by coaches, swimmers and media. With respect to media Figures 4 and 5 show two examples of how media used our data to improve the quality of their information.

Acknowledgements

We would like to thank the Organising Committee of the Xth World Swimming Championships for their support and facilities to organise the Competition Analysis.

We also want to thank the sponsors of the project, the Consejo Superior de Deportes (CSD) of the Spanish Government and the Departament d'Universitats, Recerca i Societat de la Informació (DURSI) of the Generalitat de Catalunya.

Figures 4 (left) and 5 (right). Example of local press using Competition Analysis data to inform in a more qualitatively way.

References

Sánchez J.A. (2001) El análisis de la competición en natación: estudio de la situación actual, variable y metodología. Ministerio de Educación, Cultura y Deportes (Ed.) *Análisis biomecánico de la técnica de natación: programa de control del deportista de alta competición.* Madrid.

Arellano R., Ferro A., Balius X., García F., Roig A., de la Fuente B., Rivera A., Ferreruela M. and Floría P. (2001) Estudio de los resultados del análisis de la competición en las pruebas Estilo Libre en los Campeonatos de España Absolutos 1999 y 2000. Ministerio de Educación, Cultura y Deportes (Ed.) *Análisis biomecánico de la técnica de natación: programa de control del deportista de alta competición.* Madrid.

Haljand R. (2007) *LEN swimming competition analysis.* Talliinn (online). Available: http://swim.ee/competition/index.html (Accessed: 2007).

COMPUTATIONAL FLUID DYNAMICS – A TOOL FOR FUTURE SWIMMING TECHNIQUE PRESCRIPTION

M. KEYS[1,2] & A. LYTTLE[1]

[1] Western Australian Institute of Sport, Challenge Stadium, Mt Claremont, Australia
[2] Department of Human Movement and Exercise Science/Department of Civil Engineering, The University of Western Australia, Crawley, Australia

During the underwater phases of the starts and turns, elite swimmers use a variety of underwater kicking patterns with little scientific information used in their selection. This is due partly to difficulties in providing accurate and objective results as to the most efficient technique. Computational Fluid Dynamics (CFD) was developed to provide answers into problems which have been unobtainable using physical testing techniques. The current study sought to discriminate between the active drag and propulsive forces generated in underwater dolphin and freestyle kicking using the CFD technology. A 3D image of an elite swimmer was animated using results from a kinematic analysis of the swimmer performing two different patterns of underwater dolphin kick (large/slow kicks versus small/fast kicks) and the underwater freestyle kick. The CFD model was developed around this input data. The results demonstrated an advantage in using the underwater freestyle kick over either the large/slow kick or small/fast kick at $2.18 \, m \cdot s^{-1}$. A breakdown of the results of the kick cycles also highlight the potential benefits of using CFD models in technique prescription.

1 Introduction

The underwater phases of swimming form a large and important component of the total event time in modern swimming. Currently in elite competition, there exists a range of underwater technique strategies utilized by the swimmers with very little scientific rationale applied in their selection.

Previous empirical towing testing (Lyttle *et al.*, 2000) has examined the net force produced during underwater kicking due to the complexities in separating the propulsive force and active drag forces. Results were compared to prone streamlined gliding in order to prescribe an approximate velocity at which to initiate underwater kicking. The study assumed steady state (constant velocity) conditions which limited the applicability to real swimming where the body is continually accelerated and decelerated.

It has long been accepted that understanding fluid flow patterns in swimming should lead to performance enhancements. CFD was developed by engineers to numerically solve complex problems of fluid flow using an iterative optimization approach. The net effect is to allow the user to computationally model any flow field provided the geometry of the object is known and some initial flow conditions are prescribed. Using known physics, CFD allows complex fluid flow regimes and geometry to be simulated, providing visualization of the resulting variables across the entire solution domain. This can provide answers and insights into problems which have been unobtainable using physical testing techniques. As such, CFD could be seen as bridging the gap between theoretical and experimental fluid dynamics.

Table 1. Kick cycle time and centre of gravity (CG) velocity for each kick type.

	Large/Slow Dolphin kick	Small/Fast Dolphin Kick	Freestyle Kick
Kick Cycle Time (sec)	0.44	0.38	0.39
Kick Frequency (Hz)	2.27	2.63	2.56
Average CG Velocity (m·s^{-1})	2.16	2.13	1.90

The current study sought to discriminate between the active drag and propulsion (net thrust) generated during the underwater kicking phase with the ultimate goal of optimizing the underwater kicking component in swim starts and turns. The objective information gained from this type of CFD analysis can equip sports scientists with tools to more accurately provide advice on technique modifications in order to gain the extra edge at the elite level.

2 Methods

2.1. Kinematic Measurements

An elite national level swimmer was filmed underwater from a sagittal view, while performing underwater dolphin and freestyle kicks at maximal effort. The swimmer performed both high amplitude/low frequency dolphin kicks and low amplitude/high frequency dolphin kicks, as well as an underwater freestyle kick. The kinematics of these underwater kicks was of similar magnitudes to that found in current elite competition. A full 2D analysis was performed for the three selected underwater kicking scenarios and summary information is listed in Table 1.

2.2. 3D Laser Scanning of the Swimmer

A 3D mapped image of the swimmer is required for use as an input into the CFD model to describe the swimmer's geometry. The laser scanning of the selected swimmer was performed using a Cyberware WBX whole body laser scanner with a density of one point every 4 mm. Higher resolution scans were also conducted of the hands, head and feet (density of one point every 0.67 mm). This was performed given the importance of these areas in setting the initial flow conditions (in the case of the hands and head) and in developing thrust (in the case of the feet). The higher resolution scans were then aligned and merged seamlessly into the full body scan to provide more accuracy at these locations. All scans were performed with the swimmer assuming a streamlined glide position with hands overlapping and feet plantar-flexed (see Figure 1).

2.3. CFD Model Methodology

The computer simulation was performed using the CFD software package, FLUENT (versions 6.1.22 & 6.2.16). In brief, the CFD finite volume technique involved creating a domain inside which the flow simulation occurred, bounding the domain with appropriate external conditions and breaking the domain up into a finite number of volumes or cells. The governing equations of fluid flow were then integrated over the control volumes of the

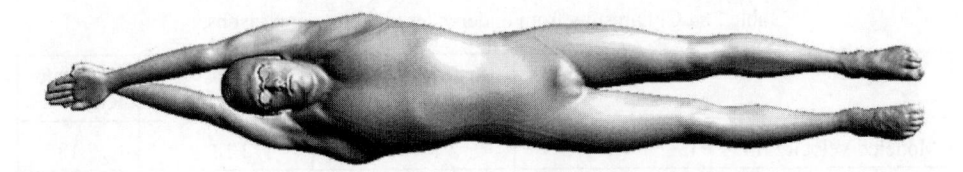

Figure 1. 3D laser scanned image of the subject.

solution domain. Finite difference approximations were substituted for the terms in the integrated equations representing the flow processes. This converted the integral equations into a system of algebraic equations that were solved using iterative methods.

This CFD study was broken into two stages; a simulation of the swimmer without limb movement (streamlined glide) and one including limb movement. The purpose of the first stage was to allow benchmarking of the swimmer's drag to experimental passive drag results. The same model was then used in the second stage with the addition of user defined functions and re-meshing to provide limb movement. This analysis was completed by breaking the limb movements down into discrete time steps and having the package solve the flow field for that position before moving on to the next position. The volume mesh was also updated at each time step with the previous flow field being used as the starting point for the next time step.

2.4. Validating the CFD Model

Although the basis of this case study was to compare three different dynamic kicking techniques, the model needs to be calibrated to show the compatibility with actual test results. Due to the unavailability of empirical testing to accurately measure active drag throughout an underwater kick cycle, the model was calibrated using steady-state tests. Repeated towing trials in a prone streamlined position indicated that the subject's passive drag were similar to the CFD results, demonstrating that the CFD predicted results were of sufficient accuracy.

2.5. CFD User Defined Functions

Using the solid-body kinematics function, user defined functions (UDF) and dynamic meshing, the body was broken into four rigid (body, thighs, shanks, feet) and three flexible sections (hips, knees, ankles). Based on the measured kinematic data of the swimmer, a mathematical curve was fitted to the rotational movements of the three main joints with global horizontal and vertical movements also modeled. Due to the accuracy of both the FLUENT software and the kinematic data, the position of the swimmer at any point in time was estimated to be within 5 mm of the actual position.

3 Results and Discussion

An output of combined pressure and viscous drag was calculated at each time step through the analysis runs. The best measurement of effectiveness of a technique is the momentum created or removed from the swimmer per cycle. This momentum can then be converted to a per-second measurement to compare different techniques. Table 2 details the momentum removed from the swimmer with higher positive values reflective of increased deceleration

Table 2. CFD results from underwater kicking comparisons.

	Large/Slow Dolphin kick	Small/Fast Dolphin Kick	Freestyle Kick
Modeled Velocity (m·s^{-1})	2.18	2.18	2.18
Total Momentum Loss per sec (Ns)	81.65	84.98	56.67
Forecast distance traveled in next sec (m)	1.73	1.71	1.87

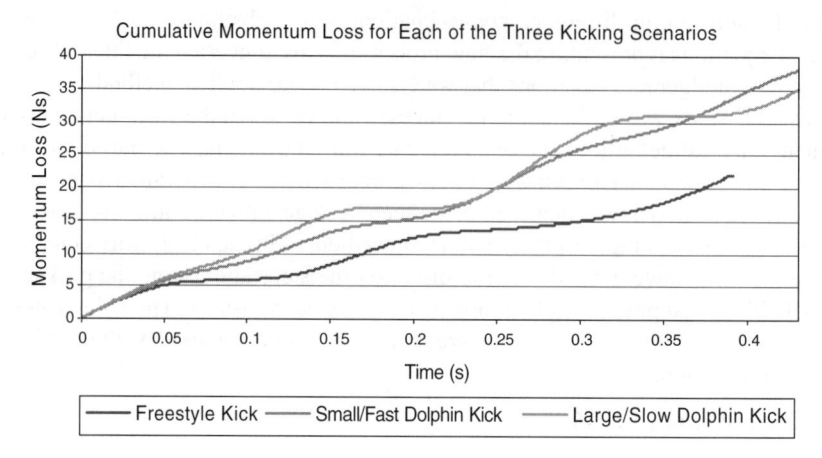

Figure 2. Graph of the cumulative momentum loss for each kicking scenario.

forces (which is also shown graphically over a relative stroke cycle in Figure 2). These momentum losses can then be extrapolated to a distance traveled in the next second of kicking based on these results to provide a more practical comparison.

At the modeled velocity of $2.18\,\text{m}\cdot\text{s}^{-1}$, the underwater freestyle kick provided the least amount of momentum loss and a greater predicted distance traveled over the subsequent second of kicking (based on a 90 kg swimmer) than either of the two underwater dolphin kicks. Further model trials at both faster and slower velocities are planned to determine whether the improved efficiency of the underwater freestyle kick is maintained over the velocity range that underwater kicking is used for. Complementary testing of energy cost for each kicking technique would also be required before recommendations of appropriate kicking styles are applied.

A major advantage of the CFD technique is the ability to differentiate what parts of the swimmer's body is creating the active drag and propulsion throughout the cycle. This allows a more effective mechanism for identifying areas of inefficiencies that can be targeted when prescribing technique modifications. Figure 3 displays a sample pressure plot that can derived from the CFD model that can graphically depict where thrust is generated throughout the kick cycle.

For the underwater dolphin kicks, the main benefit of the large kick was the acceleration that was created on both the upswing and the down-sweep. The larger kick created up to 50 N more propulsion in these acceleration phases, whilst only creating 25 N more drag in the non-acceleration phase. This extra propulsion was not coming from the feet, where

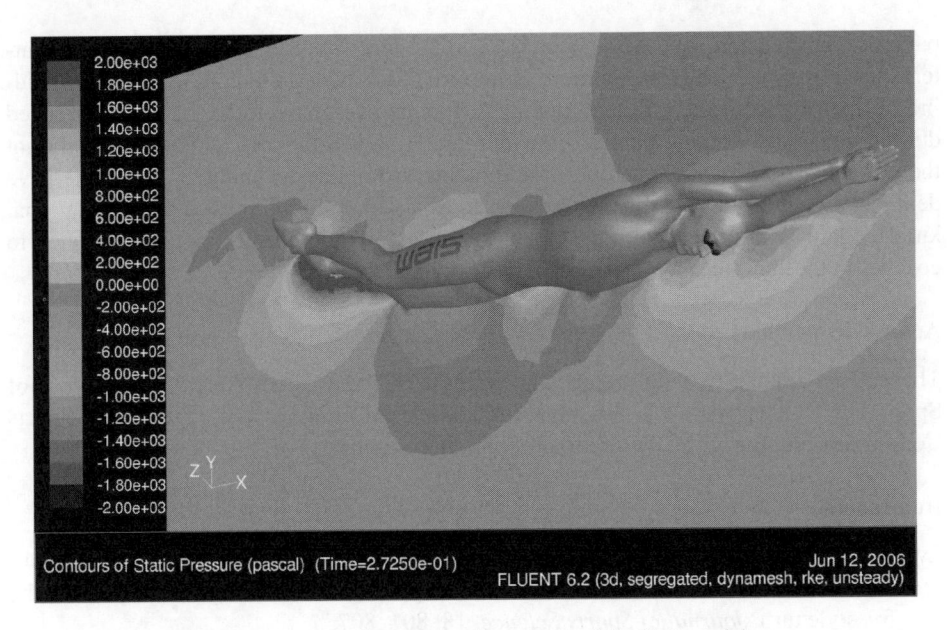

Figure 3. Sample pressure plot output of the CFD model.

the propulsive forces were only marginally greater for the large kick, but rather from the thighs and calves where much greater propulsion was generated in the large kick compared to the small kick. A major point of drag on the large kick was when the knees drop prior to the main down-sweep due to the increased frontal surface area and flow changes and created substantially more drag for the large kick model. Movement of the upper body on the large kick also generated significantly more drag in phases of the kick cycle than that of the small kick. However, in the upsweep, the body maintained sufficient momentum to offset some of the loss imposed by the high amplitude kick.

The underwater freestyle kick data showed a number of differences between the left and right leg movements within the freestyle kick. The flexibility in the right ankle was less than that for the left ankle with the range of movement for the right ankle being 57° as apposed to 69° for the left ankle. This appeared to be counteracted by the swimmer by increasing the knee bend in the right leg. The right leg knee range of movement was 45° compared to only 40° in the left leg. The results of the CFD analysis indicate that the right leg created more peak propulsion during the start of its downsweep but also creates a significantly greater drag during the upsweep of the kick. This additional drag had a greater impact on the effectiveness of the right leg, with the left leg creating almost 6.5N greater propulsion for each second, resulting mainly from the differences in net force between the left and right legs at the feet and knees. This could have resulted from the reduced flexibility of the right ankle and the impact that it appeared to have on the amplitude of the movement of the entire right leg.

4 Conclusions

Although it shows the underwater freestyle kick has produced the better results than either of the two styles of underwater dolphin kick at this velocity, this is based solely on the kicking patterns analyzed and cannot be generalized to the large number of possible kicking patterns used by swimmers. Further model trials are required to gain more of an understanding of the benefits of the different kicking technique over the velocity range experienced during underwater kicking phases of swimming. However, this case study does highlight the powerful tool that CFD can be in optimizing swimming technique. The results have demonstrated the CFD can effectively be used as a tool, both to improve the foundational knowledge of swimming hydrodynamics as well as provide useful practical feedback to coaches in the short term on technique prescription.

Acknowledgements

The authors gratefully acknowledge the support of the Western Australian Institute of Sport, The University of Western Australia and partial funding by the Australian Sports Commission via the NESC Discretionary Funding Scheme.

References

Lyttle A.D., Blanksby B.A.B., Elliott B.C. and Lloyd D.G. (2000) Net forces during tethered simulation of underwater streamlined gliding and kicking techniques of the freestyle turn. *Journal of Sports Science,* 18, 801–807.

SWIM POWER – AN APPROACH USING OPTICAL MOTION ANALYSIS

A. ONG & M. KOH
Republic Polytechnic, Singapore

This article describes a method to compute a swimmer's power, for swimmers performing drills on a swim bench, using data from an optical motion analysis system. Fifteen national age group (13–15years) swimmers participated in this study. The method uses the instantaneous force and velocity data of the center of mass (CoM) of the swimmers. We evaluated the method for the counter movement jump performed off a force plate measurement system. The results of the power values derived using data from the optical motion capture method correlated well ($r > .9$) with those from the forceplate measurement system. Between the methods, no phasal lag was observed; and, differences in derived values were not found to be significant between the two quantities computed. When applied to compute swimming power for drills performed on a swim bench, the normalized computed power values matched closely with that reported in other studies. As the set-up used in the optical motion capture system was simple and the results derived were valid, we recommend this method for monitoring power-training, even though the example shown is based on swimming. We conclude that such a method, using an optical motion system, can be an efficient way to compute athletic power and provide a means of feedback to facilitate coaches in assessing the athletes' long term development, in the absence of force plate systems.

1 Introduction

Power is a highly valued factor in training (Balčiūnas *et al.*, 2006). Hence, the accurate measurement of power will facilitate the coach and athlete in the athlete's long term developmental program (Bompa, 1999). In fact, performance increments occur during power enhancements (Trinity *et al.*, 2006). There are various protocols for measuring power (Australian Sports Commission, 2000). One of the most challenging concerns is to relate power directly with the measured performance and the actual activity, say, in a competitive environment (Cronin & Sleivert, 2005). Indirect measurements like maximum vertical and horizontal height jumped may be used to indicate power (Duncan *et al.*, 2006).

In swimming, this challenge of power measurement in the pool is exacerbated by the measurement of drag that still requires more extensive research (Toussaint *et al.*, 2004). Two methods proposed to estimate drag, namely, the MAD-system (Hollander *et al.*, 1986) and velocity perturbation method (VPM) by Kolmogorov and Duplisheva (1992) also derive swim power in the process. However, the MAD and VPM methods used average speeds in the computation. Instantaneous power, involving the corresponding force and velocity values, offers much more accuracy but may be challenging to compute. Consequently, attempts to estimate swimmers' power on land have included using instantaneous power measurements obtained from isokinetic machines (Mameletzi *et al.*, 2003) at a variety of movement speeds. However, this may not truly reflect actual performance. Biokinetic machines, on the other hand, may be used to calculate instantaneous power stroke by stroke (Potts *et al.*,

2002). Here, the swimmers lay in a prone position on a bench and pull a paddle attached to a nylon string connected to a pulley system. The velocity of the nylon is used to compute power. However, one may argue that the velocity of the nylon attached to the paddles may not be representative of the actual movement of a swimmer's centre of mass (CoM). due to the redistribution of the centre of mass towards the foot direction as the arm "pulls" towards the hips. Thus, for a power computation of a swimmer, this compensatory velocity must be considered. Otherwise, an error in the estimation of power will occur.

In the pool, the estimation of drag and the measurement of instantaneous power present equipment setup and computational challenges. On land, the challenge is in replicating in-the-water swimming conditions. Swim benches have been used to simulate in-the-pool conditions (Aujouannet et al., 2006). However, inherent errors arising from CoM movements arise as discussed above. This may be overcome with an optical motion capture system that considers the segmental movements of the body by tracking markers placed on the body. We opted for a non video-based capture system because of the significantly tedious manual digitization required from video based systems, as well as high memory storage. Also, we know of no study reportedly using the optical motion capture method to compute instantaneous power. Thus, in this paper, we will present an optical motion analysis system approach to derive the CoM kinematics and thence compute the power of swimmers on a swim bench using synchronized values of the instantaneous force on the CoM and its velocity.

2 Methods

15 developmental national age group swimmers, aged 13–15 years, participated in this study. Their average height, mass and age were 1.64 m, 48.8 kg, 13.8 years. They trained regularly and were injury-free at the time of the study. Voluntary informed consent was obtained from the parents and the swimmers assented to the tests.

Normal swimwear was worn during the trials so as not to inhibit the swimming action. Before each of the trials, the participants undertook a standardized 10 min warm-up that included light exercise of the upper body and a full body stretching routine. During testing, each swimmer each performed 4 sets of 5 counter movement jumps (CMJs) at maximal effort from a forceplate. They were allowed to swing their arms during the jumps. The data was captured and sampled at 4 different frequencies, 50, 100, 150, and 200 Hz, using six Qualisys Proflex 1000 cameras. The cameras were synchronized to a BERTEC forceplate (Type FP4060-10-1000) used to simultaneously capture the jumps kinetic parameters at 1000 Hz. A fifteen segment model (head, upper arms, lower forearms, hands, thorax, pelvis, thigh, shank and feet) was derived based on Dempster (1955) data using markers setup simplifier to other studies (Chow et al., 2006; Sujae, 2005).

The optically derived CoM velocity (VEL_c) and acceleration (ACC_c) were computed based on the first and second derivatives of the CoM displacements as captured by the Qualisys camera system using the VISUAL 3D v3.33 (C-motion Inc) software. The CoM acceleration, ACC_f, as derived from the forceplate system was also computed. The data was normalized to body weight (BW) where appropriate.

All statistical analysis were performed using SPSS V10.0. A cross correlation analysis (Mullineaux et al., 2001) was performed on the forceplate and optical motion capture systems to investigate phasal variability. RMSD and %RMSD as described by Mullineaux et al. (2001) were computed to illustrate the parameters variability. Also, the differences between two paired parameters (eg ACC_f and ACC_c) were computed. Paired t-tests were performed where appropriate.

For computation of the swim power, the 12 swimmers in prone position performed a series of ten breast-stroke arm pulls (with legs straightened and no movement) on an inclined VASA trainer (Vasa, Inc.) bench, using the fixed paddles (see Figure 1). Only data during the pull phase was recorded and the averaged values were reported.

3 Results

The typical force vs time relationship of the jump is shown in Fig. 2 where it depicts the calculated F_c and F_f values as derived from the optical motion capture and forceplate systems respectively and superimposed onto each other. The F_c and F_f values closely resemble each other for most part of the contact time except just before the toes-off slightly just after the F_{max} point.

The frame–series cross-correlation coefficients for the various captured conditions revealed peak correlations occurring invariably at frame zero. No significant difference

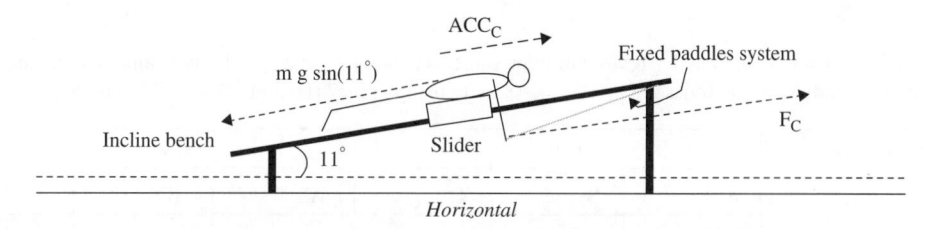

Figure 1. The body glides along the inclined bench on the smooth slider at 11° to the horizontal. The actual forces encountered along the inclined path are mainly F_c exerted by the paddles on the body and the gravitational force, m g sin(11°). F_c is the actual force effected by the body pulling on the fixed paddles. By knowing the resultant acceleration along the incline bench, ACC_c, of the body CoM we obtain $F_c = m \times ACC_C + m \times g \times \sin 11°$ where m = mass of body and g = gravitational acceleration.

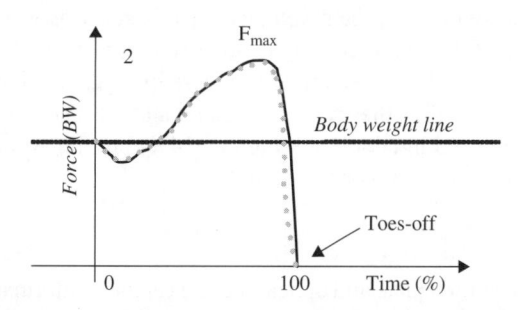

Figure 2. A typical force-time graph captured at 200 Hz normalized to BW and total contact time. F_c (dotted line) and F_f (solid line) are derived from the optical motion capture and forceplate system respectively and superimposed onto each other. Time is normalized to the start time of execution of the CMJ (0%) till toe-off (100%).

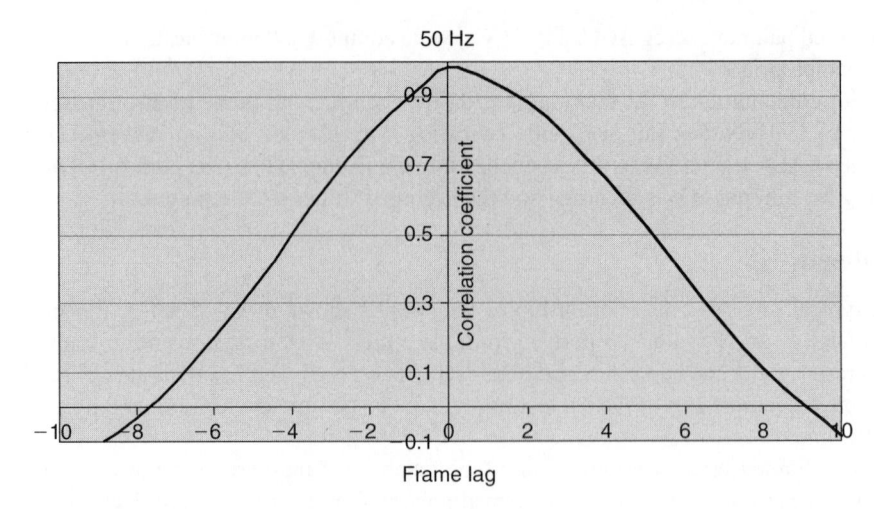

Figure 3. An example of cross-correlation coefficients of the computed F_c and F_f captured Frames lags are used instead of time lags to illustrate the actual resolution variability.

Table 1. Mean, RMSD and %RMSD of inter-subjects power (W) and speed (m/s) values calculated from this study (A) and Toussaint et al. (2004) study using "Free" (B) and "Towing" (C) methods

Mean	Speed			Power		
	A	B	C	A	B	C
	2.12	1.64	1.43	129.1	110.5	97.3
RMSD	0.11	0.1	0.09	19.1	23.6	22.4
%RMSD	5.2	6.1	6.3	14.8	21.4	23

($p > 0.01$) was found between the optical motion capture and forceplate systems. An example is illustrated in Fig. 3.

Table 1 shows the swim power and velocity for this study as well as those using the MAD and VPM methods. The velocities in the other studies were 1.43–1.64 m/s whereas it was 2.12 m/s for this study. The power values are therefore higher using the optical motion capture (A) as compared to the other two methods (B and C). The inter subjects %RMSD values for power as reflected in table 1 are high ($>14\%$) but the average %RMSD of the intra-subject power values is 4.3% for this study.

4 Discussion

The relationship between forceplate and optical motion captured information was very strong for force, peak force and impulse values ($r > .9$). Also, the paired differences is relatively low ($<6\%$) where no significant difference was observed ($p > .05$). From Figure 2, there seems to be some variability just before the toes-off (last 5%). However, no significant difference ($p > .05$; r squared $= 0.68$) was observed, indicating that the values are acceptable.

The cross-correlation results indicate that the optical motion capture and forceplate methods relate very well with each other having r values greater than .95 where the peak values occur invariably at frame zero indicating no phasal lag. This will facilitate relevant and subsequent magnitude computations requiring no phasal lag adjustments.

In this study, a full body consideration was used in 3D analysis and the results are reasonable in computing the CoM's kinematic parameters. More markers may render better estimation of segmental parameters (Palatucci, 1999). Errors do occur from individual markers identifications but the overall effect is negligible when the whole body's CoM's parameter (even the second differentials) is computed as was observed in this study.

In the swim power computation (Table 1), the swimmers in this study, averaged 129 W (SD = 19.3) with an average speed of 2.1 m/s. In Toussaint et al.'s (2004) study, the power computed was averaged at 110.5 W (SD = 24.5; 6 swimmers with about 3 trials each) where they swam at an average speed of 1.64 m/s. In this study, about 24% more speed was experienced and thus when normalized with the speed of the Toussaint et al.'s data, 104 W was obtained. Thus, power measured is within acceptable range after normalizing the speed to obtain the average value. Arm power improvement from training was found to be more than 20% (Trinity et al., 2006; Kraemer et al., 2001) and the intra-subject power %RMSD in this study is around 5%. The exact mechanism of how power relates to actual performance is not fully understood (Cronin & Sleivert, 2005). However, the optical motion capture method in this study provides a feasible means to track developmental progress in terms of monitoring improvement in power output.

The present study has not been validated outside the 50–200 Hz range and it was limited to arm movements only. Also, the effects of reducing or increasing the number of segments for power computation may need to be investigated amongst others. Further studies may be done to investigate the segmental contributions to power during swim power computation. In this study, the second derivative of the CoM's displacement, ACC, can be accurately computed. The use of these values in power computations will facilitate swim power measurements on the inclined bench. The focus of this study was on the computation of swim power using optical motion capture methods and the values obtained are valid. This enables the provision of potentially useful information to the swim coach to facilitate the swimmers' long term athlete development.

References

Ashby B.M. and Heegaard J.H. (2002) Role of arm motion in the standing long jump. *Journal of Biomechanics,* 35/12, 1631–1637.

Aujouannet Y.A., Bonifazi M., Hintzy F., Vuillerme N. and Rouard A.H. (2006) Effects of a high-intensity swim test on kinematic parameters in high-level athletes. *Applied Physiology, Nutrition and Metabolism,* 31, 150–158.

Australian Sports Commission (2000) *Physiological Tests for Elite Athletes.* Human Kinetics, Melbourne.

Balčiūnas M., Stonkus S., Abrantes C. and Sampaio J. (2006) Long Term Effects of Different Training Modalities on Power, Speed, Skill and Anaerobic Capacity in Young Male Basketball Players. *Journal of Sports Science and Medicine,* 5, 163–170.

Bompa T. (1999) *Periodization: Theory and Methodology of Training* (4th ed.). Human Kinetics, Champaign, IL.

Chow J.Y., Davids K., Button C. and Koh M. (2006) Organization of motor system degrees of freedom during the Soccer Chip: An analysis of skilled performance. *International Journal of Sport Psychology*, 37, 207–229.

Cronin J. and Sleivert G. (2005) Challenges in understanding the influence of maximal power training on improving athletic performance. *Sports Medicine*, 35/3, 213–234.

Dempster W. (1955) Space requirements for the seated operator. *WADC Technical Report*, 64–102. Wright Patterson Air Force Base, Dayton, OH.

Duncan M.J., Woodfield L. and al-Nakeeb Y. (2006) Anthropometric and physiological characteristics of junior elite volleyball players. *Journal of Sports Medicine*, 40, 649–651.

Haguenauer M., Legreneur P. and Monteil K.M. (2005) Vertical jumping reorganization with aging: a kinematic comparison between young and elderly men. *Journal of Applied Biomechanics*, 21/3, 236–246.

Hollander A.P., de Groot G., van Ingen Schenau G.J., Toussaint H.M., de Best H., Peeters W., Meulemans A. and Schreurs A.W. (1986) Measurement of active drag forces during swimming. *Journal of Sports Sciences*, 4/1, 21–30.

Kraemer W.J., Mazzetti S.A.,. Nindl B.C., Gotshalk L.A., Volek J.S., Bush J.A.,. Marx J.O., Dohi K., Gomez A.L., Miles M., Fleck S.J., Newton R.U. and Hakkinen K. (2001) Effect of resistance training on women's strength/power and occupational performances. *Med. Sci. Sports Exerc.*, 33/6, 1011–1025.

Kolmogorov S.V. and Duplisheva A. (1992) Active drag, useful mechanical power output and hydrodynamic force coefficient in different swimming strokes at maximal velocity. *Journal of Biomechanics*, 25/3, 311–318.

Mameletzi D. Siatras T., Tsalis G. and Kellis S. (2003) The relationship between lean body mass and isokinetic peak torque of knee extensors and flexors in young male and female swimmers. *Isokinetics and Exercise Science*, 11/3, 159–163.

Mullineaux D.R., Bartlett R.M. and Bennett S. (2001) Research design and statistics in biomechanics and motor control. *Journal of Sports Sciences*, 19/10, 739–760.

Palatucci M. (1999) Evaluation of an Optical Human Motion Tracking System. *Journal of Young Investigators*, 2/1. Available: http://www.jyi.org/. (Accessed: 20 Dec 2006).

Potts A.D., Charlton J.E. and Smith H.M. (2002) Bilateral arm power imbalance in swim bench exercise to exhaustion. *Journal of Sports Sciences*, 20/12, 975–979.

Sujae I.H. (2005) Understanding the biomechanics of Kuda and Sila service technique: Implications for teaching and coaching sepak-takraw. Poster presented at *United Nations International Conference on Sports and Education.*, Bangkok, Thailand, October.

Toussaint H.M, Roos P.E. and Kolmogorov S. (2004) The determination of drag in front crawl swimming. *Journal of Biomechanics*, 37/11, 1655–1663.

Trinity J.D., Pahnke M.D., Reese E.C. and Coyle E.F. (2006) Maximal Mechanical Power during a Taper in Elite Swimmers. *Medicine & Science in Sports & Exercise*, 8/9, 1643–1649.

15. Athletics and Jumping

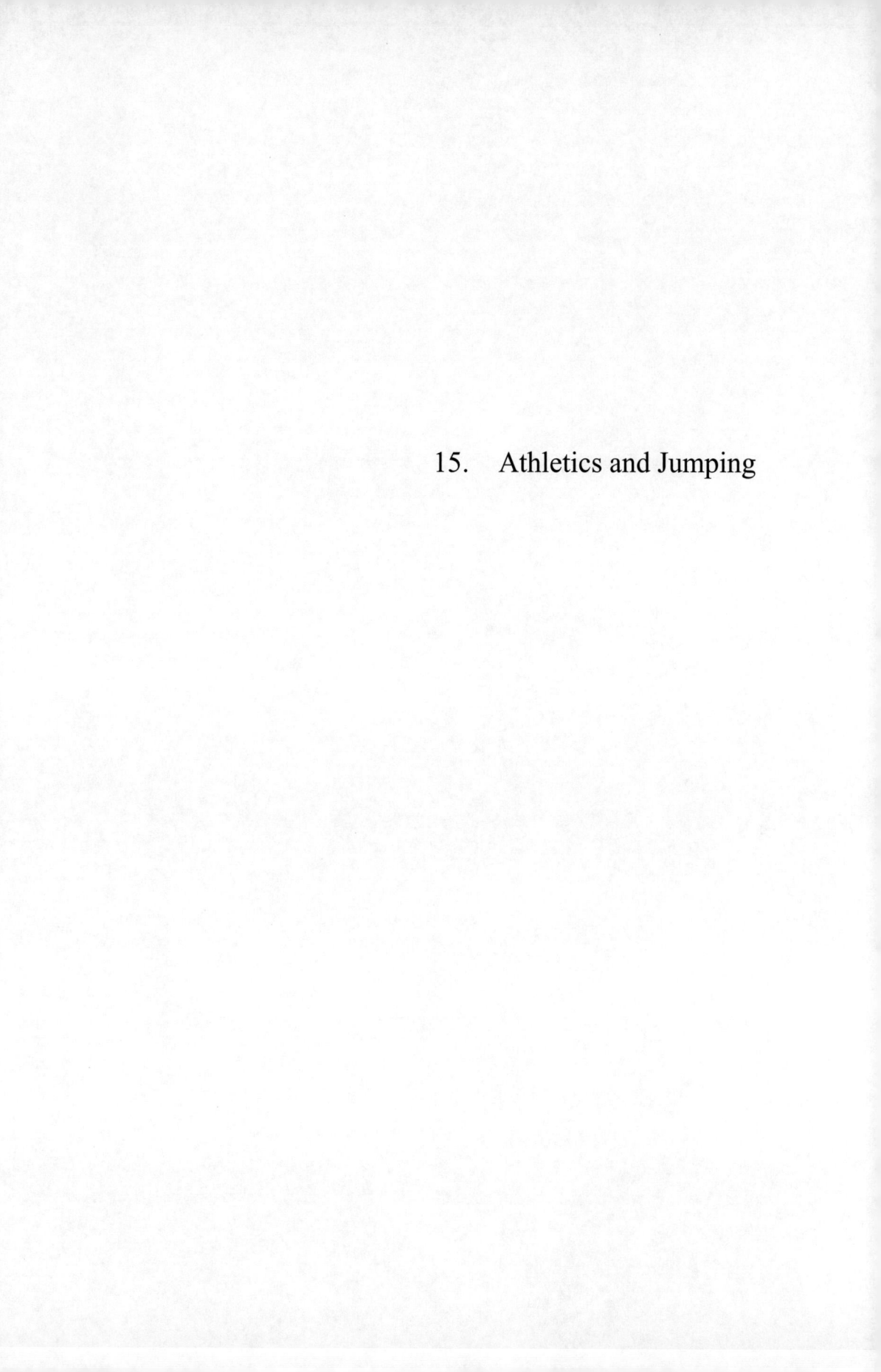

KINEMATIC ANALYSIS OF THE BEST THROWS OF THE WORLD ELITE DISCUS THROWERS AND OF THE OLYMPIC WINNER IN DECATHLON

S. VODIČKOVÁ

Technical University of Liberec, Faculty of Education, Dept. of Physical Education, Liberec, Czech Republic

This presented study focuses mainly on the analysis of discus throw at release phase at elite discus throwers and at the olympic winner in decathlon. The measurement was done on the international meeting in Czech Republic. We have performed 3D analysis when using the SIMI Motion System. The analysis of the best throw was done at all participants – 4 elite discus throwers (Olympic medalists, 1 Olympic winner in decathlon and 2 discus throwers from the Czech national team). We compared kinematic parameters at the time of release (velocity, angle of trunk, height of release and foot distance). Based on our evaluation, we assigned as a main determinant of the power the velocity at release. The optimal height of release is 90% of total body height and trunk angle is getting more reclined than at the Olympic Games in the past. We did not find the correlation between the position of the feet and thrown distance. In our next studies we would like to aim on the position of feet before the release to find out the individual differences among elite discus throwers.

1 Introduction

The performance of discus throw is determined by the distance, which the discus covers from the point determined for throwing away up to the place of landing. The distance on which the discus is moving, is dependent of physical rules, when aerodynamics plays a significant part. Biomechanical analyses of video recordings of the discus throw technique on elite sporting events prompt a lot about tendencies in throw technique. A large extent of movements from starting swings to final pull of throwing arm with a discus is typical for elite discus throwers. The differences in individual solutions are obvious not only in the throwing phase, but also during a skip and during treads down of the right and left foot (Bartonietz & Borgström, 1995).

Quantification of athletic performance during competition is rare and not well documented in the literature. Throwing has been scientifically documented for the most part under controlled conditions. Recently, however, results of analyses of throwing performances during elite competition have been documented on a large number of both male and female throwers (McCoy *et al.*, 1984; Gregor & Pink, 1985; Finch *et al.*, 1996). Some authors (Milanovic *et al.*, 1998) differentiate 5 phases of the discus throw (preparation, entry, airborne, transition and delivery. The distance of a discus throw is determined primarily by the release velocity of the discus (Dapena, 1993). According to Bartlett (1992) the release speed of a discus throwing will determine the distance of a discus when the release angle and height are held constant. According to Finch *et al.* (1996) probably other kinematic components (angle and height) had bigger influence than release velocity. A line of theoretical reasoning suggests that it would be beneficial to release a discus while the feet are still in contact with the ground, but many throwers release a discus after the feet have left

Table 1. List of Subjects (in Throwing Order).

Name	Country	Ht (cm)	Wt (kg)
Šebrle	CZE	186	87
Riedel	GER	199	115
Tammert	EST	196	115
Kövägo	HUN	196	118
Alekna	LTV	200	130

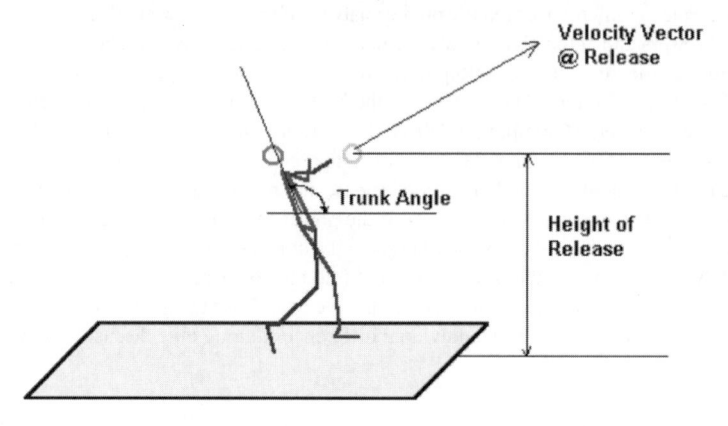

Figure 1. Biomechanical components measured at release.

the ground. The purpose of this investigation was to record the performance of all male competitors in the discus throw during the 2005 Daněk Meeting in Turnov. The two major objectives were to provide a kinematic analysis of the five male competitors (all 3 medalists from 2004 Olympic Games in discus throw and 1 Olympic winner in decathlon) and to provide information that might be used as a data base for future analyses of elite discus throwers.

2 Method

Filming procedure for later 3-D video analysis proceeded in May 2005 during Ludvík Daněk's Meeting, in which world elite discus throwers (Olympic medal holders) and athletes from the Czech national team and also the Olympic winner in decathlon take part regularly. 5 discus throwers took part in our measurement (Table 1) and each of them had 6 attempts. We have analyzed only the valid (measured) attempt. In the first phase of our study we focused on the best throws of all athletes and we compared the selected kinematic parameters.

The athletes were filmed using two digital cameras JVC 357 DVL. They were situated on a static tripod. Their optical axis formed the angle about 90 degrees. The monitored space was calibrated using calibrating block $1 \times 1 \times 2$ m. The cameras' synchronization was ensured by optical signal. The frame frequency of the cameras was 50 half frames/s. 18 points representing relative rotational centers of thrower's body segments were digitalized (Figure 1). The record from the cameras was then processed in the Simi Motion 3D software.

Table 2. Kinematic Parameters of the longest Throw.

Rank	Name	Round	Dist thrown (m)	Release Velocity (m/s)	Release Height (m)	Release Height (% of body height)	Trunk angle (deg)	Feet distance (m)
1	Alekna	4	67,91	24,37	1,75	88	113	0,85
2	Kövägo	1	65,62	24,75	1,85	94	124	1,08
3	Tammert	3	64,94	24,21	1,69	86	114	0,88
4	Riedel	4	64,15	24,37	1,82	91	115	0,8
		Mean	65,66	24,43	1,78	89,75	116,5	0,92
		SD	1,40	0,20	0,06	3,03	4,39	0,12
7	Šebrle	3	48,17	23,12	1,91	103	110	0,8

However, the analysis was limited to the best throws of all athletes. The height and velocity of the discus and the throwers' trunk angle were measured at release (Figure 1). The distance between the feet was assessed as the throwers reached the power position (i.e., position immediately after front foot contact before release). The correlation between 3D speed of discus release and final distance was calculated.

3 Results and Discussion

Release parameter data for the athletes are presented in Table 2.

It is generally recognized that during throwing events the release velocity is the most critical variable in determining of throwing distance (McCoy et al., 1984; Gregor & Pink, 1985). The velocities achieved by the world elite discus throwers (1–4) are remarkably similar (M = 24,43 m/s± 0,2). Gregor et al. (1984) report values of 24,8 m/s for men by Olympic throwers 1984. Milanovic et al. (1998) present the values of 25,4 m/s and Finch et al. (1996) report the main values of 27,3 m/s for the best throws and 27,7 m/s for the worst throws. It is higher than our results.

If we compare the whole group of the observed discus throwers – specialists (1–4) the mean velocity is a little bit lower (M = 23,96 m/s ± 0,69). In case we integrate to our evaluation also the power of the Olympic winner in decathlon (7), than the main velocity is even lower (M = 23,84 m/s ± 0,707). This verifies the relation between velocity at release and thrown distance. Significant correlation (r = 0,83) was founded between 3D projection speed and final discus distance.

A second important variable in throwing for distance is release height. The average value for the men in this study was 1,81 m, which is slightly higher than reported by Gregor et al. (1984) with 1,73 m recorded during the 1984 Olympic Games and slightly higher than the 1,80 m reported by McCoy et al. (1984) for en elite population of throwers from the United States. Milanovic et al. (1998) report the height of release between 1,65 m and 2,05 m. It was very interesting finding, that when observing the weakest power – power the decathlete Sebrle (1,91 m) – he had the highest height at release. Forasmuch as the height of release is depended on the total body height we expressed the release height as a percent of total body height. In this study release height was M = 92,71% ± 5,60 of total body height. Gregor et al. (1984) report values of 90% and McCoy et al. (1984) report values of 93% of total body height. Certainly total height of release is important, but further implications here focus on training methods and power at different arm positions for each thrower.

A third important variable is angle of release, which should not be confused with the actual pitch of the discus at release. While the position of a discus relative to the air flow around it at release is important to the aerodynamics of flight, this variable is almost impossible to measure during competition. In our study we didn't evaluate the angle of release.

A fourth parameter, trunk angle, was measured during the release. In our study trunk angle was larger than 90 degrees (M = 120 degrees ± 3,603). If we compare the previous studies (McCoy et al., 1984; Gregor et al., 1984) we can see, that trunk angle varied near vertical position (90 degrees), but we can also see more reclined position (103 degrees). Authors report, that a moderate trend appeared, indicating that as trunk angle became more reclined the release angle increased. All of the athletes in our study had trunk angle larger than 90 degrees. During the last years, since published studies, it has been proved that trunk angle is now more reclined than before.

In our study, we observed, if the feet are still in contact with the ground or not. In our study Alekna and Tammert release the discus after the feet have left the ground. Kövägo, Riedel and Šebrle release the discus when the feet are still in contact with the ground.

4 Conclusions

Based on our evaluation we conclude the following:

- release velocity is determinant of the power in discus throw;
- position of the body (trunk angle) is getting more reclined since 1984 Olympic Games;
- optimal height of release is about 90% of total body height;
- feet distance at release is not the main determinant of the power;
- contact with the ground in time of release is not necessary for the accomplishment of good power in discus throw.

A lot of coaches and athletes believe that the final part of the throw is the only important one, but we would like to find out in our next studies, whether the early part of the throw plays an important role for the thrown distance too.

References

Bartonietz K. and Borgström A. (1995) The throwing events at the World Championships in Athletics 1995 Göteborg – Technique of the world's best athletes. Part 2: Discus and javelin throw. In: *New studies in Athletica*, 1st Ed. Monaco, 4, 43–63.

Gregor R.J. and Pink M. (1985) Biomechanical analysis of a world record javelin throw : a case study. *International Journal of Sport Biomechanics*, 1, 73–77.

McCoy R.W., Gregor R.J., Whiting W.C., Rich R.G. and Ward P.E. (1984) Kinematic analysis of elite shotputters. *Track Technique*, Fall, 2868–2871.

Finch A., Haute T., Ariel G. and Penny A. (1996) Kinematic comparison of the best and worst throws of the top men's discus performers at the 1996 Atlanta Olympic Games. Available: URL: http://www.uni-konstanz.de/isbs/Abstracts/Finch_.PDF

Milanovic D., Hraski Ž., Mejovšek M. (1998) Kinematic analysis of a discus throw – a case study (online). Available: http://www.arielnet.com/topics2/Slide_Presentations/asia1/tsld069.htm

Dapena J. (1993) An analysis of angular momentum in the discus throw. In: Proceedings of the 14th Int. Congress Biomechanics, Paris France, p.105–109.

Bartlett, R. (1992) The biomechanics of discus throw. *Journal of Sport Science*, 10, 467–510.

Gregor R.J., Whiting W.C. and McCoy R.W. (1994) Kinematic Analysis of Olympic Discus Throwers. In: Martens R. (Ed.) *Biomechanics Research at the Olympic Games: 1984–1994*, 1st Ed. Human Kinetics Publishers, Inc., Champaign, p. 36–43.

ENHANCING MEASUREMENT ACUITY IN THE HORIZONTAL JUMPS: THE RIETI'99 EXPERIENCE

X. BALIUS[1], A. ROIG[1], C. TURRÓ[1], J. ESCODA[1] & J.C. ÁLVAREZ[2]

[1]*Olympic Training Centre (CAR) of Catalunya, Catalunya, Spain*
[2]*Real Federación Española de Atletismo (RFEA), Madrid, Spain*

Results in the long jump event have gradually tapered to a steady state over the last 30 years. In a bid to heighten the event's appeal, the Biomechanics Department of the (Barcelona) Olympic Training Centre (CAR) has developed a system whereby the "real" distance jumped can be automatically and instantaneously identified based on photocell detection of the precise location of the jumper's foot at take-off. This approach offers a glimpse at a future of real-time measuring systems in Track and Field.

1 Introduction

Results on the long jump event have been experiencing a progressive steady state within the last 30 years. The Biomechanics Department of the Olympic Training Center (CAR) has been developing a project with the aim to automatically and instantaneously identify the null jumps of this event and measure the real distance from the take-off. This development was intended to be an approach to the future real time measuring systems in Track and Field.

Taking into account that the technology development of this competition must be achieved without a lost of the tradition that characterises this event, we believe that a revision of the way that the measure is done nowadays can benefit the actual competition. With the new tool for the evaluation of the null jump we pretend to make the competition more dynamic and spectacular, and at the same time to reduce the unwanted discussions that arise from appreciation disagreements.

This tool, based on the implementation of infrared photocells, facilitates the athletes' take-off: instead of obliging the athletes to place their foot as close as possible to the null line, the idea is to allow the athlete to place the foot in an area where a set of photocells measure its exact position. This area (1×1 m) allows the jumper to concentrate mostly on the jump itself, and not so much on placing the tip of the foot as close as possible to the null line. The result of the jump is measured from the exact place where the tip of the foot makes contact with the floor.

With this method the winner of the long jump event will always be the athlete who jumps longer than the others. The competition itself will be more spectacular because the null jumps will be less than at the present moment, avoiding also the now typical discussions with the judges.

Finally, considering that the take-off board will not be needed with this method the risk of injury will be reduced: the take-off zone will be regular through out all its surface.

This dossier describes the experience had during the celebration of the XXIX Rieti'99 track & field meeting in Italy.

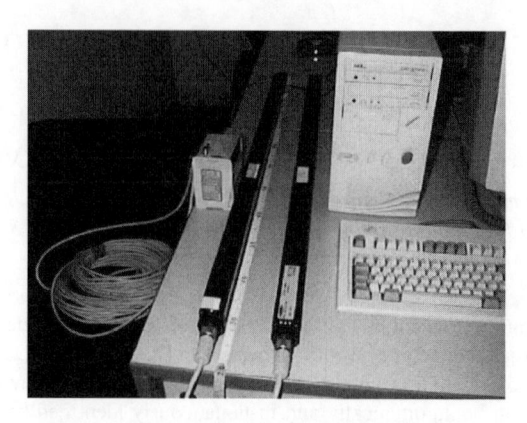

Figure 1. A-CAGETM High-Resolution MINI-ARRAYTM. Banner® plus computer.

In view of the potential improvements to the long jump event made possible by implementation of infrared photocell detection of the foot at take-off, the objectives of this project were:

1. To describe the measurement instrument.
2. To evaluate the measurement instrument in terms of:

- Functionality of the photocells. Considered as the easiness of implementation of the tool used for the measurement, and the level of inconvenience caused to the athletes and judges, if any.
- Reliability and validity of the data obtained with the tool compared to the official results. We considered the reliability to be warranted by the array manufacturer, even though we have been able to confirm its good performance in our laboratory. The validity of the tool was tested with the implementation of a video camera recording the take-off zone. The camera signal was mixed with the own made program and was used to analyse how right the cell's data was.
- Influence of the real distance results to the final classification. This influence is not relevant for the evaluation of the project yet the consequences in any event would be crucial. In this sense it has to be mentioned that with the present proposal the winner of the competition does not have to be the one who jumps further away, but the one who jumps more. From a public point of view the position reached by the winner could, in our case, be closest to the board than the position of the second classified. This topic though is not evaluated in this dossier, although the results are shown at the end.

2 Methodology

The tool used for this experience consisted on a high resolution array of photocells with the following characteristics:

A-CAGETM High-Resolution MINI-ARRAYTM. Banner® (Figure 1)

Total number of cells   384 (only 192 are operative for the project
                                      (see distribution Figure 2)

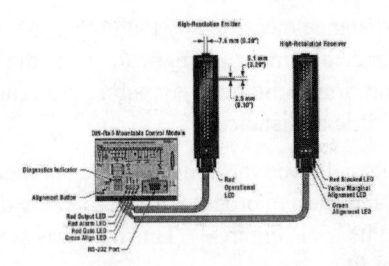

Figure 2. Distribution of the cells trough the Mini-Array. Observe the two arrays of cells (white dots) from which only one is used. This only array, the closest to the floor (8 mm), contains 192 active cells.

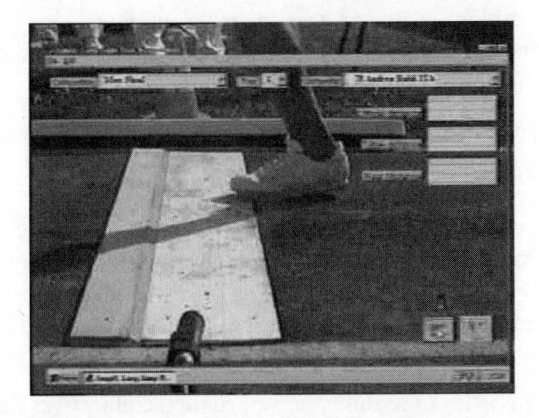

Figure 3. Custom software controlling, automatically and instantly, the real distance to the board. Notice that the figure shows the mix of the a- camera image focusing on the take-off zone, with b- instant result of the distance to the board.

Resolution (operative cells)	0.0051 m
Length of the array	0.975 m
Height of the cells from the floor	0.008 m
Wide, depth and height of the metal tubes containing the cells	1.106 × 0.0381 × 0.0381 m
Connection cables	15 m

This array consists of a pair of metal cubes (emitter/receiver) with the volume characteristics described above, connected through a 15 m cable to a Control Module (CM) that governs the set up characteristics of the cell's measurement protocol. This protocol is adaptable with a specific software provided by the array manufacturer.

All data obtained from the cells was finally controlled and stored by a custom software that managed the following information (see Figure 3):

- distance to the board: distance from the frontal tip of the shoe to the board (the distance is measured for both the valid and null jumps). This distance is obtained automatically and instantly.

- official distance: distance measured by the judges. Manually entered.
- real distance: distance obtained automatically once the official distance is entered. The result is an addition (valid jump) or subtraction (null jump) of the distance to the board and the official distance.

The custom software also allowed the control of the competition (name of the competition, name of the jumper, and trial) in order to assign to every data obtained from the cells the name of the jumper and its trial (Figure 3). This data was stored in an Access data base file easily exported to Excel files.

3 Results and Discussion

The cells were tested during the celebration of the XXIX edition of the Rieti meeting (Rieti'99). The competitions evaluated were the male and female long jump.

The results of the test are shown in Table 1 (males) and 2 (females). For the purposes of our project, 4 evaluations of the results were considered:

1. The number of correct trials recorded with our system as compared with the official results, and also reconfirmed with our video recordings.
2. The number of software errors (software bugs)
3. The number of trials mistaken due to interference. A peace of board or tartan falling in front of the take-off foot, and thus modifying the result.
4. The number of values that gave us better results than the observed with the videographical analysis. Probably due to the height of the tip of the jumper's shoes when touching the board.

Upon the characteristics of usability described on the objectives the results showed:

Functionality: the array photocells were very easy to allocate, although at this moment the involuntary impacts, due to athletes or judges crossing the run-up lane, might misplace the array null position. Securing the array to the floor would be a good solution to this problem. Telemetric communication between cells and computer facilitated the use of the tool avoiding too much people presence on the measurement zone.

Reliability: Aside from the mechanical reliability of the instrumentation (as quoted by the manufacturer), such factors as software errors (bugs) and interference of photocell function due to stray particles can also have an impact on the reliability of the measurement system as a whole. The problem concerning loose particles has been addressed by imposing a requirement upon the system: collection of data only proceeds when the number of contiguous cells obstructed is greater than 10 (a foot with a minimum length of 5-cm) in a row. If a single cell is obstructed it is not taken into account. This correction not only addresses the problem of broken pieces of board, but also the potential interceptions caused by raindrops or mosquitoes.

Validity: While the validity of the measurement tool has not been statistically confirmed, our laboratory has never observed a measurement error greater than 4 mm. During the Rieti'99 competition the official and videographical data was coherent in 87% of the registered cases. During the Rieti'00 competition the data was 100% coherent.

The impact of "real distance" results on the final classification: The comparison of official results and "real distance" results in Table 1 indicates that the measurement tool would, indeed, make a difference in the final outcome of some jumps competitions.

Table 1. Male results of the Rieti'99 competition, as measured by officials in the "standard" way (Official Results), compared with the results obtained using the photo-cell system (Real Distance Results).

Name	Official jump distance (m)	Name	Real distance jump (m)
Tarus	8.20	Tarus	8.20
Toure	7.97	Toure	7.97
Taurima	7.96	Taurima	7.96
Atanasov	7.94	Atanasov	7.95
Trentin	7.93	Trentin	7.93
Baldi	7.82	Strete-T.	7.87
Camossi	7.79	Baldi	7.84
Agresti	7.64	Camossi	7.80
Strete-T	7.63	Agresti	7.64

Table 2. Female results of the Rieti'99 competition, as measured by officials in the "standard" way (Official Results), compared with the results obtained using the photo-cell system (Real Distance Results).

Name	Official jump distance (m)	Name	Real distance jump (m)
May	6.74	May	6.87
Rahouli	6.70	Tsiamita	6.73
Vershinina	6.57	Rahouli	6.70
Tsiamita	6.55	Vershinina	6.60
Vaszi	6.49	Gotovska	6.54
Gotovska	6.43	Vaszi	6.50
Periginelli	5.33	Periginelli	5.39

The comparisons also offer insight into the jumpers' relative "efficiency". Five of the nine male competitors recorded the same distance on both columns, indicating that their foot placement on the take-off board must have been as close to the null line as allowable. Interestingly, only one of the women (RAHOULI) managed to "maximize" her official jump, indicating that the women (as a group) were not as technically accomplished as the men on this particular day.

The percentage real-distance improvement in performance obtained for the triple is 0.4%. The percentage for the men's long jump was also 0.4%, while that for the women's long jump was 1.2%.

4 Conclusions

In short, it can be said that according to the current proposal the winner of the competition does not have to be the ones who registers the farthest break in the sand. He (or she) simply has to be the one who jumps the greatest distance.

It appears that the system presented has great potential utility in the conduct of horizontal jumping events specially to contribute to a more dynamic competition by reducing the percentage of null jumps, increasing the level/result of the valid jumps and avoiding the disagreements between judges and athletes.

Acknowledgements

We would like to thank Mr. Sandro Jovanelli and his staff for the interest and help shown during the celebration of the Rieti '99 meeting, the meeting judges for their patience and co-operation, and the organisation of the European Blind Track and Field Championships held in Lisbon.

We also want to thank the sponsors of the DTL project, the International Olympic Committee (IOC), and the Organización Nacional de Ciegos de España (ONCE).

Finally we want to thank Medición y Control (Mecco) SA for their help trough out the project, to Banner Engineering (USA) for their assistance, and to Mr. Carles Turró and Mr. Ventura Ferrer (Biomecanics SCP) for the special program writing.

References

Real Federación Española de Atletismo (1987) Longitud y triple. *Cuadernos de Atletismo, 5*. ENE, Madrid.

Alvarez J.C. (1990) Tecnica y control del entrenamiento de los saltadores de longitud espanoles. In Seminario europeo de saltos. *Unisport. Seminario Europeo de Saltos*. Malaga, October 1990.

EFFECT OF JOINT STRENGTHENING ON VERTICAL JUMPING PERFORMANCE

H.-C. CHEN & K.B. CHENG

*Institute of Physical Education, Health, and Leisure Studies,
National Cheng Kung University, Tainan, Taiwan*

The objective of this study is to determine training which joint is more effective in jump height increase by computer simulation. This will be useful information for coaches and athletes in their training strategies. A planar human model with five rigid segments representing the feet, shanks, thighs, TH (trunk and head), and arms is used. Segments are connected by frictionless revolute joints representing the ball of foot, ankle, knee, hip, and shoulder. Model movement is driven by torque actuators at all joints except for the ball joint. Each joint torque is the product of maximum isometric torque and three variable functions depending on instantaneous joint angle, angular velocity, and activation level, respectively. The model can actively extend and flex these joints by changing the activation level. Jumping movements starting from a balanced initial posture and ending at takeoff are simulated. The objective is to find joint torque activation patterns during ground contact so that jump height is maximized. The simulation is repeated for varying maximum isometric torque of one joint ±20% while keeping other joint strength values unchanged. Although jump height increases/decreases with increasing/decreasing strength of each joint, the influence is different among joints. Similar to previous simulation results, strengthening knee joint to 20% is the most effective way than the same level of strengthening for other joints. In general, for the same amount of percentage increase/decrease in strength, the shoulder is the least effective joint in changing jump height.

1 Introduction

Jumping for maximum height has been studied from various approaches. Extremely simple models have been used to understand optimal running high- and long-jumping strategies (Alexander, 1990; Seyfarth *et al.*, 1999, 2000). More complex models were employed to understand strategies for maximizing jump height (Levine *et al.*, 1983a, 1983b; Pandy *et al.*, 1990; Soest *et al.*, 1993; Selbie & Caldwell, 1996). Joint kinematics, ground reaction forces (GRF), and muscle electromyogram (EMG) have been measured by numerous experimental studies (e.g. Bobbert & van Ingen Schenau, 1988).

Although various studies concerned different aspects of vertical jumping, only a few researchers investigated the effect of joint strength (resultant muscular strength) on jumping performance. A significant positive relationship between vertical jumping height and total work with hip and knee isokinetic extension moments, but low correlation coefficients between height and ankle plantarflexors moment have been found (Tsiokanos *et al.*, 2002). Training the knee extensors was shown by simulations to be the most effective in improving jumping performance among all lower extremity muscles (Nagano & Gerritsen, 2001). As previous simulated results (Bobbert & Soest, 1994), re-optimizing muscle coordination to benefit fully from training effects was shown to be necessary.

To know which joints are more crucial in jumping performance is important. However, previous researchers either focused mainly on the shoulder joint (to study the role of arms) or disregarded the effects of countermovement and arm swing in simulations (Bobbert & Soest, 1994; Nagano & Gerritsen, 2001). The purpose of this study is to investigate the influence of changing strength at each joint on vertical jumping performance. Since joint strength cannot be varied instantly and its influence cannot be isolated in real experiments, computer simulation with optimization serves as the best tool for the present study.

2 Methods

Although this study focuses on simulation, measurement of real jumping performance is necessary for model validation. Only one subject was tested because the human model is intended to be subject-specific. Three high-speed cameras (240 Hz) and a motion analysis (Motion Analysis, Eva 7.0, Santa Rosa, CA) system recorded and determined positions of six reflective markers at the fifth metatarsal, ankle, knee, hip, shoulder, and wrist.

A planar five-segment human model is used to simulate vertical jumping from initiation to takeoff. Body segments are connected by frictionless hinge joints (Fig. 1). The segments represent feet, shanks, thighs, HT (head-trunk), and the arms. Equations of motion are derived using AUTOLEV (http://www.autolev.com). Torque actuators at the ankle, knee, hip, and shoulder joints are used to drive model movement. Jumping motion starts from a balanced and nearly straight posture. The model can actively extend and flex these joints. Rather than modeling individual muscle function, these torque actuators represent total contributions of joint flexors and extensors.

Each joint torque T (effective torque on the sum of the left and right extremities) exerted is assumed to be the product of a maximum isometric torque T_{max} and three variable factors:

$$T = T_{max} f(\theta) h(\omega) A(t) \tag{1}$$

where $f(\theta)$ and $h(\omega)$ depend on joint angular position and angular velocity, respectively. $A(t)$ characterizes the coordination strategy and is termed joint activation level.

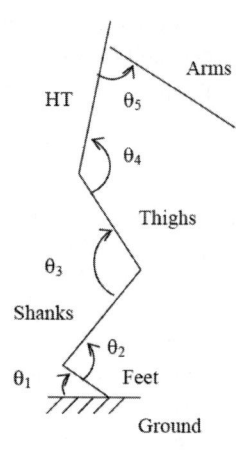

Figure 1. The five-segment (5S) human model is connected by frictionless revolute joints.

A cubic spline fit of five nodal values at equally spaced times throughout ground contact is used to represent $A(t)$. Five nodes suffice since doubling the number increases jump height by less than 1.5%. The initial nodes are fixed to correspond to the torques needed for holding the initial posture. Positive and negative $A(t)$ represent actively extending and flexing, respectively. Full-effort joint extension/flexion corresponding to $A(t) = +1/-1$. Because muscular activation cannot change instantaneously, dA/dt is constrained such that its absolute value is less than $2/0.08$ s^{-1}. This is because an activation time constant of 80 ms, near the geometric mean of muscle activation rise and decay time constants (Pandy et al., 1990), is assumed.

The joint angle factors $f(\theta)$ for the ankle, knee, and hip are taken Hoy et al. (1990). $f(\theta)$ for the shoulder is from Otis et al. (1990). The angular velocity factor $h(\omega)$ is given by (Selbie and Caldwell, 1996):

$$\begin{cases} h(\omega) = (\omega_0 - \omega)/(\omega_0 + \Gamma\omega), & \omega/\omega_0 < 1 \\ h(\omega) = 0, & \omega/\omega_0 \geq 1 \end{cases} \quad (2)$$

where $\omega_0 = \pm 20$ rad/s is maximum joint extension (positive) or flexion (negative) angular velocity, ω is instantaneous joint angular velocity (positive in extension), and $\Gamma = 2.5$ is a constant shape factor. Value of $h(\omega)$ can be increased to a saturation value 1.5 if $\omega(t)$ and $A(t)$ have different signs (eccentric muscle contraction) (Fig. 2).

Due to model simplicity and the fact that joint angle and angular velocity dependence are averaged values from other studies, exact measurement of joint T_{max} for the subject is considered unnecessary. Rather, values of T_{max} at different joints are estimated by the following optimization procedure. The first optimization finds the control variables such that the simulated joint (ankle, knee, hip, shoulder, and wrist) positions match those in the highest real jumping trial. That is, the squared differences in joint positions at all instants are minimized. In searching the optimum a difference within 1 cm is deemed acceptable and the accepted solution sets are used for the second optimization. This second optimization is performed to avoid joints being "too strong" or "too weak". Since the highest jumping trial should be nearly optimal, it is plausible to assume that by varying $A(t)$ while fixing

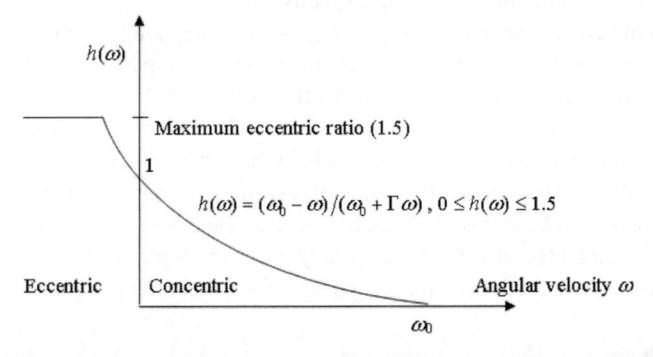

Figure 2. Angular velocity factor $h(\omega)$ for joint extension activation ($A(t) > 0$). When ω exceeds its maximum ω_0, the function is decreased to zero. When $\omega < 0$ and $A(t) > 0$ (or $\omega > 0$ and $A(t) < 0$), $h(\omega)$ increases to a saturation value 1.5, that effectively models eccentric muscle contraction.

previously determined T_{max}, the model will jump 1 cm higher. The solution sets with too strong/weak T_{max} are discarded in this way and the best estimated T_{max} are then determined after this two-step optimization procedure.

The best estimated T_{max} are used as nominal joint strength values. The strength for a single joint is then varied (while keeping other T_{max} values unchanged) and optimization for jump height is performed to examine how joint strength affects performance. Each T_{max} is varied within ±20% (Nagano & Gerritsen, 2001). Although in practice changing strength implies changing muscle mass, it is assumed that change of mass is negligible.

The control goal is to maximize jump height J_0:

$$J_0 = (y_f + v_f^2/2\,g) \qquad (3)$$

where y_f and v_f are the vertical position and velocity of jumper center of mass (c.m.) at takeoff. Since different takeoff times t_f result from different joint torque patterns (actually nodal torque activations), t_f is also a control variable (Bryson, 1999). State and control constraints are considered in maximizing J_0. In optimal torque activation calculations, nodal activation is not constrained formally, but A(t) is truncated when it lies outside [−1, 1]. Joint angle constraints to prevent joint hyperextension, and only the ankle joint angle constraint (>1 rad) was found to be active. Takeoff condition that requires zero vertical board reaction force at takeoff was also implemented.

Whenever constraints are violated, a penalty function is subtracted from the objective function J_0. Thus the problem becomes an unconstrained maximization, where possible violations of the constraints decrease the value of the objective function.

To find the global rather than a local optimum, the downhill simplex method (Nelder & Mead, 1965) with varying initial guess and re-starting the optimization from a newly found optimum was employed.

3 Results

As was mentioned previously, a two-step optimization is used for determining joint T_{max} and for model validation. Maximum torque (in N-m) determined for the ankles, knees, hips and shoulders are 500.21, 481.10, 511.02, and 179.87, respectively. The matching simulation matches the measured actual jumping motion reasonably well. Simulated and actual jump height are 1.2934 m and 1.3011 m, respectively.

The effect of percentage change in joint T_{max} on jumping height is compared (Fig. 3). In general, the shoulder is the least effective joint in changing jump height, while the ankle and knee are the most effect joints. The level of effectiveness in height increase due to joint strength increase is in the order of knee, ankle, hip, and shoulder. The level of effectiveness in height decrease due to loss of joint strength is in the order of ankle, knee, hip, and shoulder. Changing ankle and knee strength within 10% seems to have comparable effects on jump height. On the other hand, increasing strength within 2% and decreasing strength within 10% seems to have comparable effects on jump height for the hip and shoulder. However, the shoulder becomes less effective in changing height for larger strength gain/loss.

4 Discussion

Reasonably well correspondence between simulated and actual jumping movement indicates model validity. Similar to previous studies (Bobbert & Soest, 1994; Nagano & Gerritsen,

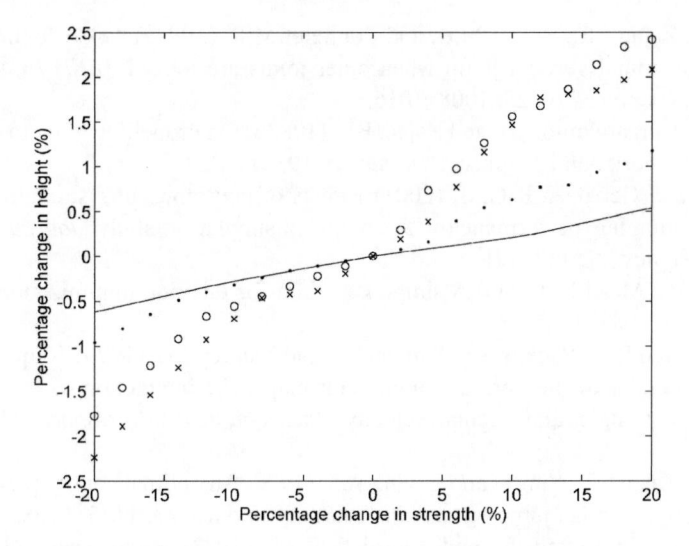

Figure 3. Dependence of percentage change in jump height on percentage change in strength for the ankle (x), knee (o), hip (•), and shoulder (—). Percentage changes are varied from the nominal values for jump height and strength.

2001), re-optimization of activation after joint strengthening is necessary. If the same activation pattern is applied to strengthened joints, jump height is decreased rather than increased. This indicates that strength training solely may not be an effective way to enhance performance unless it is combined with practicing the actual movement.

The approximately linear strength-height relation is similar to that in Nagano and Gerritsen (2001) in which the effect of all three joint extensors (ankle plantar flexor, knee and hip extensors) is considered. The 3 cm height increase for a 20% increase in knee extensors (Bobbert & Soest, 1994) is very close to that in the current study. Although jump height increase is larger due the combined effect of enhanced muscle strength, maximum contraction velocity, and activation amplitude (Nagano & Gerritsen, 2001), the influence is in the same order (knee, ankle, and hip) as the current results.

References

Alexander R.M. (1990) Optimum takeoff techniques for high and long jumps, *Philosophical Transactions of the Royal Society of London B,* 329, 3–10.
Bobbert M.F. and van Ingen Schenau G.J. (1988) Coordination in vertical jumping, *Journal of Biomechanics*, 21, 249–262.
Bobbert M.F. and van Soest A.J. (1994) Effects of muscle strengthening on vertical jump height: A simulation study, *Medicine and Science in Sports and Exercise*, 26, 1012–1020.
Bryson A.E. (1999) *Dynamic Optimization*. Menlo Park, CA: Addison-Wesley.
Hoy M.G., Zajac F.E. and Gordon M.E. (1990) A musculoskeletal model of he human lower extremity: the effect of muscle, tendon, and moment arm on the moment-angle relationship of musculotendon actuators at the hip, knee, and ankle, *Journal of Biomechanics*, 23, 157–169.

Levine W.S., Zajac F.E., Belzer M.R. and Zomlefer M.R. (1983a) Ankle controls that produce a maximal vertical jump when other joints are locked, *IEEE Transactions on Automatic Control,* AC28, 1008–1016.

Levine W.S., Christodoulou M. and Zajac, F.E. (1983b) On propelling a rod to a maximum vertical or horizontal distance, *Automatica,* 19, 321–324.

Nagano, A. and Gerritsen K.G.M. (2001) Effects of neuromuscular strength training on vertical jumping performance – A computer simulation study, *Journal of Applied Biomechanics,* 17, 113–128.

Nelder J.A. and Mead R. (1965) A simplex method for function minimization, *Computer Journal,* 7, 308–313.

Otis J.C., Warren R.F., Backus S.I., Santner T.J. and Mabrey J.D. (1990) Torque production in the shoulder of the normal young adult male. The interaction of function, dominance, joint angle, and angular velocity, *American Journal of Sports Medicine,* 18, 119–123.

Pandy M.G., Zajac F.E., Sim E. and Levine W.S. (1990) An optimal control model for maximum-height human jumping, *Journal of Biomechanics,* 23, 1185–1198.

Reklaitis G.V., Ravindran A. and Ragsdell K.M. (1983) *Engineering Optimization: Methods and Applications.* New York: Wiley.

Selbie W.S. and Caldwell G.E. (1996) A simulation study of vertical jumping from different starting postures, *Journal of Biomechanics,* 29, 1137–1146.

Seyfarth A., Friedrichs A., Wank V. and Blickhan R. (1999) Dynamics of the long jump, *Journal of Biomechanics,* 32, 1259–1267.

Seyfarth A., Blickhan R., and Van Leeuwen J.L. (2000) Optimum take-off techniques and muscle design for long jump, *Journal of Experimental Biology,* 203/4, 741–750.

Soest A.J., van Schwab A.L., Bobbert M.F. and van Ingen Schenau G.J. (1993) The influence of the biarticularity of the gastrocnemius muscle on vertical-jumping achievement, *Journal of Biomechanics,* 26, 1–8.

Tsiokanos A., Kellis E., Jamurtas A. and Kellis S. (2002) The relationship between jumping performance and isokinetic strength of hip and knee extensors and ankle plantar flexors,*Isokinetics and Exercise Science,* 10/2, 107–115.

STRENGTH OF THIGH MUSCLES AND GROUND REACTION FORCE ON LANDING FROM VERTICAL DROP JUMPS

C. KIM & J.C.C. TAN

Physical Education and Sports Science, National Institute of Education, Nanyang Technological University, Singapore

The purpose of the study was to determine a relationship between thigh muscle strength and the magnitude of vertical ground reaction force (GRF) on landing from heights. A total of 19 collegiate male students were tested on isokinetic dynamometer to measure their quadriceps and hamstring muscles strength at three angular velocities. They were also tested on drop jumps of four different heights to a force platform. The vertical GRF gradually increased ($p < .05$) and the time to peak GRF decreased ($p < .05$) as the jumping height gradually increased. Pearson's correlation analyses revealed that the combined torque of isokinetic eccentric quadriceps and hamstring muscles at 60 deg/s was the most important determinant of vertical GRF ($p < .05$). In addition, both the combined eccentric torque and vertical GRF were inversely related with the time to peak GRF at three jumping heights ($p < .05$).

1 Introduction

Landing after jumping is common in daily human activities. Jumping is an activity during which muscle contraction energy is used to elevate the body's gravitational potential energy. Strenuous take-off muscles will add high gravitational potential energy into the body, hence landing muscles must control joints to effectively dissipate the large load to ensure a safe landing. In certain sports, such as gymnastics, GRF reaches as much as 18 times body weight (BW) on single leg landing (Panzer, 1987). Nigg & Bobbert (1990) noted that it is difficult to demonstrate a causal relationship between load and injury. However, numerous studies have reported high incidence of injuries to the lower extremities in athletes who participated in sports involving frequent jumping and landing activities (Ford et al., 2003; McKay et al., 2001; Olsen et al., 2004). McKay et al. (2001), for example, reported that the rate of ankle injury was 3.85 per 1000 participations with the most common mechanism being landing (45%).

There are numerous studies attempted to identify biomechanical variables of lower extremities in landing. Those variables include mostly GRF and/or joint moments in relation to jumping heights and landing techniques (Devita & Skelly, 1992; Kovacs et al., 1999; McKinley & Pedotti, 1992), landing velocities (McNitt-Gray, 1993), surface conditions (McNitt-Gray et al., 1994), and lower extremity joints' range of motions (Zhang et al., 2000). Devita & Skelly (1992) reported that while stiff landings had less than 90 degrees of knee flexion after floor contract, soft landings had greater than 90 degrees of knee flexion. Stiff landings had lager GRFs than soft landings, and the energy was dissipated mostly by plantar flexors in stiff landing, whereas by knee and hip extensors in soft landing. McNitt-Gray (1993) tested gymnasts and recreational athletes on landings. Whereas gymnasts dissipate more energy with ankle and hip extensors, recreational athletes used greater degrees of hip flexion and a longer period of landing than the gymnasts. The possible reason suggested by the author for the different joint usage between the two groups was that recreational athletes might not be able to produce larger extensor moments at the ankle or hip during landings from great heights. This suggestion may implicate that gymnasts might have stronger ankle and hip extensors than their counterparts. However, there is a lack of study investigating the role of muscle strength in relation to biomechanical variables in landing from heights. Hence, this study is intended to identify a relationship between various thigh muscle strength components and vertical GRF (Fz) in four jumping heights.

2 Methods

2.1. Subjects

A total of 19 collegiate male students served as subjects in the study. They were all apparently healthy and did not have knee pain or past history of serious knee injuries. Their BW ranged from 56 kg to 102 kg with the average $\pm$ SD of 68.1 kg $\pm$11.1 kg. All the subjects gave informed consent prior to being tested. The test protocol was approved by the Ethics Review Board for Research Involving Human Subjects in the Physical Education and Sports Science Department in National Institute of Education, in Nanyang Technological University, Singapore.

2.2. Equipment

The Kistler Force Plate (model number: 9287BA) and BioWare Performance Biomechanical Software Analysis System (version 3.11) were used to collect data on drop jumping. A computerized strength dynamometer (model number: Cybex 6000 Extremity System) was used to measure isokinetic strengths on thigh muscles. A jumping board of different heights was built up using twenty-centimetre height steppers.

2.3. Procedures

Subjects were instructed to perform vertical drop jumps from 20 cm, 40 cm, 60 cm, and 80 cm heights. They were instructed to drop down to the force platform as safely as they could without rolling over on the force platform or on the floor. The order of the jumping heights was randomized across the subjects.

Subjects were also tested on isokinetic strengths of right quadriceps (QM) and hamstring (HM) muscles at the angular velocities of 60 deg/s, 180 deg/s, and 300 deg/s. Concentric and eccentric isokinetic strengths were measured. When subjects sat on the dynamometer, Velcro straps were tightened on right shoulder across to left flank, across lower abdomen, and left distal thigh to restrict the movements of other than right thigh muscles. Subjects completed 7 repetitions at each of the six combination conditions of three velocities and two contraction modes, each of which was preceded by 5 repetitions of familiarization trials. They were given 30 seconds of rest between a familiarization trial and an actual trial, and between velocities. They were given a minimum of 3 minutes of rest period between modes of contraction.

2.4. Data Analyses

The peak-Fz data on force plate and the peak torque data on isokinetic dynamometer were standardized to subjects' body mass for analyses. The trend analysis was employed to identify the significance in the pattern of change in the dependent variables as independent variables gradually changed. The paired t-test was used to determine the significance in the mean difference of dependent variables between two paired independent variables. The Pearson's correlation coefficients were determined to identify significant relationships. The statistical decision was made at an alpha level of .05 across all statistical analyses.

3 Results

There was no significant change in QM and HM eccentric torque as the angular velocity was increased from 60 deg/sec, to 180 deg/sec, and to 300 deg/sec ($p > .05$). However, QM concentric peak torque decreased 29.2% and 23.3% when the angular velocity increased from 60 deg/sec to 180 deg/sec, and from 180 deg/sec to 300 deg/sec, respectively. Similarly, HM concentric torque decreased 24.1% and 20% as the angular velocity increased from 60 deg/sec to 180 deg/sec, and from 180 deg/sec to 300 deg/sec, respectively. At the angular velocity of 60 deg/s, the peak torque between concentric and eccentric contractions was not significantly different in QM ($p > .05$), but was significantly greater for eccentric contraction in HM, $t(18) = 2.38$, $p < .05$. In the angular velocity of 180 deg/s and 300 deg/s, the peak torque was greater for eccentric contraction in both the QM and HM ($p < .001$) (see Table 1).

The results on jumping performances revealed that the Fz gradually decreased as the jumping height decreased from 80 cm to 20 cm by 20 cm, $F(1,74) = 22.0$, $p < .05$. Also, the trend analysis revealed that as a jumping height gradually decreased from 80 cm to 20 cm by 20 cm, the time to peak-Fz during landing on the force platform gradually increased, $F(1,74) = 18.33$, $p < .05$.

The most significant strength component in relation to the peak-Fz was the combination of eccentric QM and eccentric HM strength at 60 deg/s (CEQH60) of angular velocity. None of the strength components significantly related with the peak-Fz at 80 cm. The CEQH60 was significantly related with the peak-Fz of 60 cm, 40 cm, and 20 cm: $r = .52$, $p < .05$,

Table 1. Peak Torques of Quadriceps Muscles and Hamstring Muscles on Concentric and Eccentric Contractions at Different Angular Velocities (all the torque values were standardized to subjects' body mass).

	Angular velocity (deg/s)	QM peak torque ± SD	HM peak torque ± SD
	60	267.7 ± 36.2	146.8 ± 19.6
Concentric	180	189.4 ± 24.8	111.5 ± 22.5
	300	145.3 ± 31.1	89.2 ± 16.6
	60	260.4 ± 71.7	162.9 ± 31.5
Eccentric	180	256.1 ± 64.2	150.8 ± 27.7
	300	268.6 ± 55.1	151.2 ± 30.1

$r = .48, p < .05, r = .67, p < .05$, respectively. Time to peak force at 60, 40, and 20 cm jumping heights was inversely related with the peak force at the corresponding heights ($r = -.55, p < .05, r = -.68, p < .05, r = -.71, p < .05$, respectively). Similarly, time to peak force at 60 cm and 40 cm was also inversely related with CEQH60 ($r = -.47, p < .05$, $r = -.48, p < .05$, respectively).

4 Discussion

The results indicated that peak isokinetic eccentric torque in both the QM and HM did not change as the angular velocity increased. On the contrary, the peak concentric torque decreased as the angular velocity increased in both the QM and HM. These findings are consistent with previous reports (Baltzopoulos, 1995; Dudley et al., 1990; Kaufman et al., 1991; Kellis, 2001). Kellis (2001) reported that the decline in a maximum resultant joint moment from 30 deg/s to 150 deg/s was 27%. In the current study, the main effect of angular velocity on concentric torque was also significant in both the QM and HM. In addition, the rate of decline appeared to be similar to the value reported by Kellis (2001). The decline in the joint torque from 60 deg/s to 180 deg/s, which is only slightly different from the range used by Kellis (2001), was 29.2% in QM and 24.1% in HM.

The peak-Fz increased significantly as the jumping height increased. On the contrary, the time to peak-Fz gradually decreased as the jumping height increased. These findings are consistent with previous reports (McNitt-Gray, 1993; Song-Ning et al., 2000). For example, McNitt-Gray (1993) reported that as a jumping height increased from 0.32 m, to 0.72 m, and to 1.28 m, both gymnasts and recreational athletes showed an increased velocity at which the total body center of mass touched the ground, hence resulting in higher peak-Fz with a higher height.

Rather than single muscle strength, the combination of the eccentric strength of the QM and HM at 60 deg/s (CEQH60) was more highly related with the peak-Fz at 20 cm, 40 cm, and 60 cm jumping heights. The time to peak force at 20 cm, 40 cm, and 60 cm was inversely related with the peak-Fz at the corresponding height. In other words, subjects who produced higher peak-Fz took less time to reach the peak-Fz. In summary, subjects who were stronger at CEQH60 tended to produce higher peak-Fz and took less time to reach peak-Fz. Overall, these findings appear to be in consistent with those reported by McNitt-Gray (1993) in that stronger subjects might have produced larger extensor torque to stiff their joints earlier in the landing phase, hence producing higher peak-Fz and taking less period of time to the peak-Fz. On the contrary, weaker subjects might have used a wide range of joint motion due to the lack of capability to cope with a high extensor torque at hip and ankle joints. Further studies must be conducted to confirm these findings.

5 Conclusion

The main purpose of this study was to identify a relationship between various strength components of thigh muscles and the peak-Fz at landing from drop jumping. The results identified the CEQH60 as the most important thigh muscle strength component to determine the peak-Fz in landing from drop jump. As the CEQH60 became larger, the peak-Fz tended to became higher. At the same time, as the peak-Fz became higher, the time to peak-Fz became shorter. These findings are conjecturally consistent with previous reports, but further studies are required to confirm.

References

Baltzopoulos V. (1995) Muscular and tibiofemoral joint forces during isokinetic knee extension. *Clinical Biomechanics*, 10, 208–214.

Devita P. and Seklly W. (1992) Effect of landing stiffness on joint kinetics and energetics in the lower extremity, *Medicine & Science in Sports & Exercise*, 24/1, 108–115.

Dudley G.A., Harris R.T., Duvoisin M.R., Hather B.M. and Buchanan P. (1990) Effect of voluntary vs. artificial activation on the relationship of muscle torque to speed, *Journal of Applied Physiology*, 69, 2215–2221.

Ford K.R., Myer G.D., and Hewett T.E. (2003) Valgus knee motion during landing in high school female and male basketball players, *Medicine & Science in Sports & Exercise*, 35/10, 1745–1750.

Kaufman K.R., An K., Litchy W.J., Morrey B.F. and Chao E.Y. (1991) Dynamic joint forces during knee isokinetic exercise, *The American Journal of Sports Medicine*, 305–316.

Kellis E. (2001) Tibiofemoral joint forces during maximal isokinetic eccentric and concentric efforts of the knee flexors, *Clinical Biomechanics*, 16, 229–236.

Kovacs I., Tihanyi J., Devita P., Racz L., Barrier J. and Hortobagyi T. (1999) Foot placement modifies kinematics and kinetics during drip jumping, *Medicine & Science in Sports & Exercise*, 31/5, 708–716.

McKay G.D., Goldie P.A., Payne W.R. and Oakes B.W. (2001) Ankle injuries in basketball: Injury rate and risk factors, *The British Journal of Sports Medicine*, 35, 103–108.

McNitt-Gray J.L. (1993) Kinetics of the lower extremities during drop landings from three heights, *Journal of Biomechanics*, 26/9, 1037–1046.

McNitt-Gray J.L., Koff S.R. and Hall B.L. (1992) The influence of dance training and foot position on landing mechanics, *Medical problems of Performing Artists*, 7/3, 87–91.

McNitt-Gray J.L., Yokoi T. and Millward C. (1994) Landing strategies used by gymnasts on different surfaces, *Journal of Applied Biomechanics*, 10, 237–252.

Nigg B.M. and Bobbert M. (1990) On the potential of various approaches in load analysis to reduce the frequency of sports injuries, *Journal of Biomechanics*, 23, 2–12.

Olsen O., Myklebust G., Engebretsen L. and Bahr R. (2004) Injury mechanisms for anterior cruciate ligament injuries in team handball, *American Orthopaedic Society for Sports Medicine*, 32, 1002–1012.

Panzer V.P. (1987) Lower extremity loads in landings of elite gymnasts. *Doctoral dissertation*, University of Oregon.

Rosene J.M., Fogarty T.D., and Mahaffey B.L. (2001) Isokinetic hamstrings: Quadriceps ratios in intercollegiate athletes. *Journal of Athletic Training*, 36/4, 378–383.

Zhang S., Bates B.T. and Dufek J.S. (2000) Contributions of lower extremity joints to energy dissipation during landings, *Medicine & Science in Sports & Exercise*, 32/4, 812–819.

BIOMECHANICAL ANALYSIS OF LANDING FROM DIFFERENT HEIGHTS WITH VISION AND WITHOUT VISION

H.-S. CHUNG[1] & W.-C. CHEN[2]

[1]*Graduate institute of physical education, National College of Physical Education and Sports, Tau-Yuan, Taiwan*
[2]*Graduate institute of sports science, National College of Physical Education and Sports, Tau-Yuan, Taiwan*

Visual information has been assumed to participate in landing during our daily life. How does visual information affect our motion and cause changes in landing is important. The purpose of this study was to examine the differences of kinematic, kinetic, and EMG activity during drop landing with continuous vision and blind-condition from different heights. Twelve students served as the subjects for this study. The mean height, weight, and age were 175.1 ± 3.1 cm, 72.0 ± 6.0 kg and 23.5 ± 2.9 years. Twelve subjects performed 6 randomized sets of 3 trials after becoming familiar with the task. The 18 trials were divided in 2 visual conditions (vision and blindfold) and 3 heights of fall (20, 30 and 40 cm). AMTI force-plate (1000 Hz), Motion Analysis System (120 Hz) and EMG Bio-vision System (1000 Hz) were used to record the ground reaction force simultaneously, joint angle and EMG-signals during the landing. Maximum ground reaction force (MF; F_z), time to MF (T_{max}) and the impulse of 50 ms ($I_{50\,ms}$) normalized to weight were used as outcome measures. Joint angles were measured with 8 EVa-eagle cameras by 32 markers included hip, knee, ankle angle and the range of motion (ROM) during landing. EMG activities were recorded from rectus femoris (RF), biceps femoris (BF), tibialis anterior (TA) and soleus (SO) muscles. The selected variables were tested by two-way repeated measures ANOVA ($\alpha = .05$) and post hoc t comparisons using the Tukey-Kramer HSD method. After the treatment of the data, we have gained the results as follows: (1) F_{max} and $I_{50\,ms}$ were increased significantly with height increased. Oppositely, T_{max} were decreased significantly with height increased. (2) F_{max} and $I_{50\,ms}$ at 20 cm & 30 cm without vision were larger than those with vision. (3) There were no significant differences between vision and no vision condition in joints in each height, but the ROMs of hip and knee joint angle were increased with drop height. (4) The results of EMG signals of RF, TA, and SO after touchdown showed that the amplitude increased follow a similar linear trend whether vision was available or not, but the EMG amplitude of pre-landing didn't show the same result. Thus, it appears that vision play a dominant role during short duration falls encountered daily, but not the only one resource.

1 Introduction

All include some movements of landing type attitude in daily life and general sports events. For example: Go downstairs, get off the bus, play basketball...all can be regarded as landing movements. When high increasing the bigger strength occurs, if the insufficient muscular strength or the tactics with landing of the low limbs were not used at that time, the regular meeting caused the muscle or the joint ligament injured. For these reasons, the landing movements had been studied in humans for a long time (Greenwood & Hopkins, 1976; McNitt-Gray, 1991; Bardy & Laurent, 1998; Santello *et al.*, 2001; Liebermann & Goodman, 2006). Most interesting features of landing are the kinetic & kinematic variables

(Mc Nitt-Gray, 1991; Chang, et al., 1994; Bardy & Laurent, 1998). After feet contact ground, Electromyography (EMG) activity has also been investigated to assess the contribution of reflex mechanisms in the absorption of the impact (Duncan & McDonagh, 2000). In many research papers, athlete's low limbs were injured in landing (Gerberich et al., 1987; Ferretti & Papandrea, 1992; Bressel & Cronin; 2005).

The goal of studying how the preparatory EMG activity is modulated during the flight phase is to infer how visual information is integrated to allow a safe and effective absorption of kinetic energy after touchdown (Santello, et al., 2001; Liebermann & Goodman, 2006).In the previous study (Santello, et al., 2001) they showed that the timing and amplitude of pre-landing muscle activity was not graded in proportion to the height of drops. In those experiments subjects performed 10 drops from one height, but we seldom drop from 60 cm or 80 cm for 10 times repeated in a short time in our daily life. However, the heights from 20 cm–40 cm are the general height for us, so the aim of the study was whether without vision the skill completely degraded or whether proprioceptive and vestibular information could substitute fully for the lack of vision in adapting landing movements to different heights without practices.

2 Method

Twelve male subjects were asked to land on a force plate from drops of 20, 30, 40 cm under two experimental conditions: with continuous vision and without vision. Twelve subjects performed 6 randomized sets of 3 trials after becoming familiar with the task. The mean ± S.D. height, weight, and age were 175.1 ± 3.1 cm, 72.0 ± 6.0 kg and 23.5 ± 2.9 years. Prior to their participation in the experiments, informed written consent was obtained from the subjects.

AMTI force-plate (1000 Hz) (BP600900 2000) was used to measured ground reaction force (Fz) & I50 ms (impulse) after touch down. At the same time, eight high-speed cameras (Eagle cameras, Motion Analysis Corp., Santa Rosa, California, US) were positioned at various locations in the biomechanics laboratory to sample kinematic data at 120 Hz. Three-dimensional joint angles for the hip, knee, and ankle were determined. Before starting the experiment, 32 marks were placed on the joint of the subject. The most important five joints to calculate hip, knee, ankle joint were as follows: fifth metatarsal joint, lateral malleolus (ankle joint center), lateral epicondyle of the femur (knee joint center), greater trochanter (hip joint center) and acromion.

Surface electrodes (silver-silver chloride) with on-site preamplifiers were used to access EMG activity of the following four muscles of the dominant extremity of each subject: rectus femoris (RF), biceps femoris (BF), tibialis anterior (TA) and soleus(SO) muscles of the right leg. The electrode position for each muscle was located over the midsection of the muscle, as described by Cram et al. (1998). Before the application of an electrode, the skin was prepared by dry-shaving the area and cleansing the skin with alcohol to reduce surface impedance. A prefabricated piece of double-sided adhesive tape and conductive gel was then applied to each electrode. EMG data were sampled at 1000 Hz, amplified, and rectified with a low-pass filter at 15 Hz with a fourth order Butterworth filter. EMG data were recorded synchronously with video data and stored on the same personal computer. Data were normalized to the percentage of maximal voluntary isometric contraction (MIVC) to allow for comparison between subjects. The MIVC of EMG was recorded from each subject associated used to normalize the EMG amplitude recorded during landing. The duration of the recording was 3 seconds, with rest periods of 5 min in between sections. The EMG, force platform and Eva camera capture signal were sampled simultaneously by an analog-to-digital converter.

On arrival at the biomechanics laboratory, subjects underwent placement of the surface EMG electrodes over the four muscles as discussed earlier. Each subject was then asked to perform two maximal voluntary isometric contractions, holding each for 3 seconds. Subjects were positioned and pressure was directed as described by Kendall *et al.* (1993) for all maximal voluntary isometric contractions. After muscle contraction data were collected, the kinematic reflective markers were placed on the subject. After collection of an anatomic calibration file, which determined the location of joint centers, subjects were instructed and given an opportunity to practice the landing movement. The subjects were instructed to start and terminate the landing movement in a standing position; to take off with right foot and touch down with both feet; and finally to brake the drop smoothly as possible. During the landing, they cross their arms on their waist. The order of the drop heights and vision conditions were randomized. The blindfold condition: the subjects were asked to look at a balloon tied on a stick at 3 m height during the landing. Each subject performed three landings from each height in two conditions.

The mean of the EMG, kinematic, and kinetic values from 3 landings was calculated for each subject. The data was analyzed using two-way ANOVA with repeated measures where drop height, vision or the interaction between these two factors had a significant effect on the variables studied. When statistically significant effects were found (P < 0.05), the Tukey-Kramer HSD method was used to determine the significant differences in post hoc t comparisons.

3 Results

Table 1 shows maximum of F_z, impulse of the first 50 ms after touch down, time to MF, stiff of the lower limbs drop from three heights. This table shows that as drop height increased the MF and Impulse$_{50\,ms}$ increased significantly (P < 0.01) for both conditions. Landing without vision were characterized by significantly larger MF and Impulse$_{50\,ms}$ (P < 0.05). Time to MF decreased significantly with drop height (P < 0.01). Landing without vision were characterized by significantly larger MF and Impulse$_{50\,ms}$ except 40 cm. The stiff of lower limbs without vision was larger than with vision significantly at 30 cm.

The hip, knee, and ankle joints were calculated by digitizing markers recordings of three landings per drop flight. Using all of the data points in a two-way analysis of variance, we found the significant difference between vision and no vision in hip joint motions ($P < 0.05$; Tab. 2).

The minimum angle of hip joint during landing, subjects in vision condition demonstrated significantly more hip flexion. The range of motion (ROM) of joint angle during landing, subjects drop from the height at 40 cm were getting more hip, knee, and ankle flexion than they drop from 20 cm. However, there were no significant differences among the angles of joints at the time of touch down from different heights and vision conditions.

It was found a tendency towards increasing joint rotation with increasing drop height. (Figure 1. subject 8). In all subjects, this tendency was particularly clear in the knee and hip joints (Tab 2.). In the subject shown in Fig. 1 without vision caused a smaller hip and knee joints flexion, it shown that the trunk and lower limbs were stiffen in without vision condition.

Table 1. Max. F_z, I_{50ms}, Time to MF, stiff of the lower limbs vs. drop height.

Vision condition	Vision			No vision		
Height	20 cm[1]	30 cm[2]	40 cm[3]	20 cm[4]	30 cm[5]	40 cm[6]
MF	2.45 ± 0.32[234]	2.99 ± 0.51[13]	3.81 ± 0.43[12]	2.73 ± 0.32[156]	3.30 ± 0.42[46]	3.89 ± 0.36[45]
I_{50ms}	46.92 ± 10.51[234]	58.49 ± 10.67[135]	76.92 ± 15.20[12]	56.46 ± 13.19[156]	68.33 ± 11.16[246]	83.31 ± 13.93[45]
Time to MF	99.29 ± 44.29[34]	81.26 ± 22.80[35]	65.97 ± 10.33[12]	79.87 ± 32.06[156]	63.21 ± 16.46[24]	58.33 ± 14.64[4]
Stiffness of lower limbs	6.77 ± 2.77	6.94 ± 2.41[5]	8.00 ± 1.41	7.84 ± 2.17	8.30 ± 2.23[2]	8.49 ± 1.80

1. MF (unit) = F_z/BW Impulse$_{50ms}$ = MF*ms Time to MF = ms Stiff of lower limbs = N/mm.
2. The number on the right-upper indicated the significant difference with the other variance.

Table 2. Minimum joint angle, range of motion vs. drop height.

Vision condition	Vision			No vision		
Height	20 cm[1]	30 cm[2]	40 cm[3]	20 cm[4]	30 cm[5]	40 cm[6]
Hip joint	117.6 ± 22.3	113.9 ± 17.1	111.8 ± 18.1	126.0 ± 16.2	125.3 ± 13.8	114.7 ± 17.6
Min. angle ROM	35.3 ± 16.9	40.7 ± 12.5	43.1 ± 12.4	30.6 ± 10.6[6]	32.4 ± 10.8	41.1 ± 10.0[4]
Knee joint	97.3 ± 14.8	90.8 ± 12.1	87.7 ± 10.3	102.0 ± 13.3[6]	95.0 ± 10.1	87.4 ± 12.3[4]
Min. angle ROM	56.1 ± 14.1[3]	65.5 ± 10.4	68.9 ± 8.3[1]	54.5 ± 8.0[6]	60.9 ± 13.3	69.3 ± 8.5[4]
Ankle joint	89.5 ± 11.2	88.1 ± 8.3	86.3 ± 9.6	88.8 ± 10.4	85.7 ± 9.8	87.5 ± 10.1
Min. angle ROM	41.7 ± 11.2[3]	50.0 ± 4.5	51.4 ± 7.1[1]	43.7 ± 5.4[6]	47.3 ± 7.4	50.0 ± 7.3[4]

1. The unit of joint angle is degree (°).
2. The number on the right-upper indicated the significant difference with the other variance.

This subject fallen from 40 cm height plate got larger flexion angle than from 20 cm & 30 cm. The curves of joints angles in vision condition were more smoothly than in no vision situation. However, this feature was not typical of all of the subjects.

We found there were no significant difference among the three different heights and two visual conditions for EMG pre- landing activity amplitude ($P > 0.05$). But there had a tendency for EMG pre-landing activity amplitude to increase with increasing drop height in RF, TA, and SO. The EMG activity amplitude of RF, BF, TA, and SO was averaged over

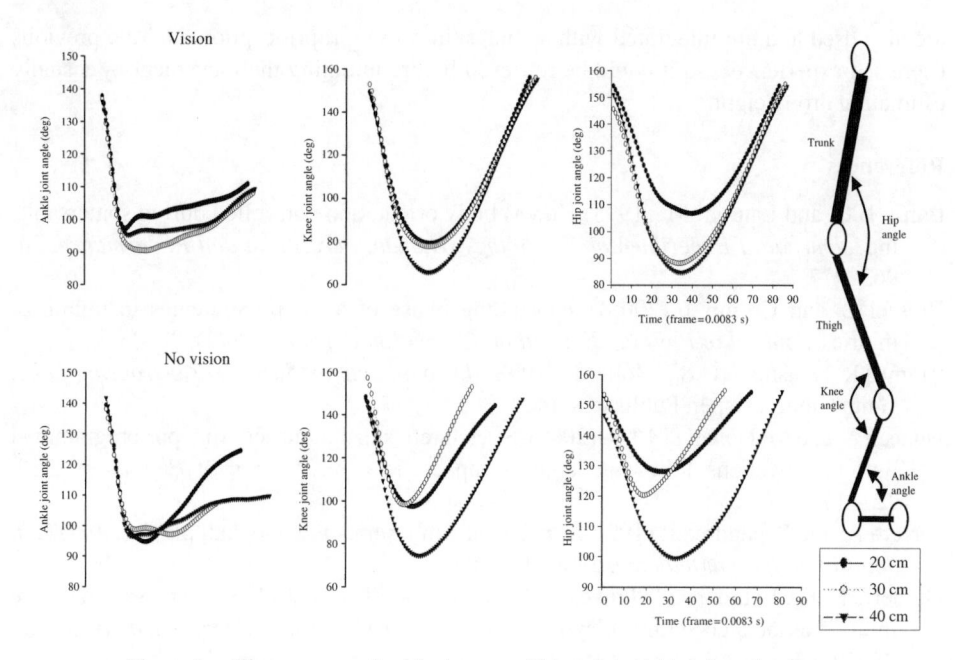

Figure 1. Time course of ankle, knee, and hip joint rotation from landing.

a period of 150 ms after touchdown. As found the pre-landing EMG amplitude, no vision did not have a significant effect on the EMG amplitude after touch down. However, in both vision conditions the EMG amplitude after feet contact tended to increase with increasing with drop height in all muscles.

4 Discussion and Conclusions

This study demonstrated height differences in the performance of drop landing, higher height showed MF and Impulse$_{50\,ms}$ increased and larger angles in hip and knee flexion during the eccentric period of landing. Landings without vision were larger values of F_z and $I_{50\,ms}$ than with vision. This implies lack of vision affected the control joints rotation after feet contact ground. These results were the same as other researchers reported. (Santello et al., 2001; Liebermann and Goodman, 2006). About the EMG activity of landing, we expected lack of vision to completely degrade the control of leg muscle activation. The EMG activity of pre-landing would be different from heights and vision conditions. However, this did not happen.

Specifically, EMG amplitude performed in the two conditions and three heights were similar scaling. The RMS values were almost the same. It may because of the subjects were physical education students, the heights were not high (20–40 cm) for them, and the gap of height were small (10 cm). So subjects adapted to different drop heights and two visual conditions.

Subjects were able to adapt the control of landing to different heights through the sensor memories associated with previously experienced impact forces, flexion angles of joints, and the speed on the flight phase. It is likely that non-visual sources of sensory information are also used and are integrated with sound, skin sense, proprioception, and the previous exercised experiences, so it could be retrieved before initiating the movement by visually estimating drop height.

References

Bardy B.G. and Laurent M. (1998) How is body orientation controlled during somersaulting? *Journal of Experimental Psychology, Human, Perception and Performance*, 24, 963–977.

Bressel E. and Cronin J. (2005) The landing phase of a jump: Strategies to minimize injuries. *Journal of Physical Education, Recreation & Dance*, 76(2), 30–35.

Cram J.R., Kasman G.S., Holtz J. (1998) *Introduction to Surface Electromyography*. Gaithersburg, Aspen Publishers, Inc.

Duncan A. and McDonagh M.J.N. (2000). Stretch reflex distinguished from pre-programmed muscle activations following landing impacts in man. *Journal of Physiology*, 350, 121–136.

Ferretti A. and Papandrea P. (1992). Knee ligament injuries in volleyball players. *American Journal of Sports Medicine*, 20/2, 203–207.

Gerberich S.G., Luhmann C., Finke C., Priest J.D. and Beard B.J. (1987). Analysis of severe injuries associated with volleyball activities. *Physician and Sports Medicine*, 15/8, 75–79.

Greenwood R. and Hopkins A. (1976). Landing from an unexpected fall and a voluntary step. *Brain* 99, 375–386.

Chang I.-J., Huang C.-F. and Jaw G.-B.(1994). Biomechanics analysis of drop landing from three different heights. *Bulletin of Physical Education National Society of Physical Education, R.O.C.*, 18, 195–206.

Kendall F.P., McCreary E.K. and Provance P.G. (1993). *Muscles, Testing, and Function*. Fourth edition. Baltimore, Williams & Wilkins.

Liebermann D.G. and Goodman D. (2006). Pre-landing muscle and post-landing effects of falling with continuous vision and in blindfold conditions. *Journal of Electromyography and Kinesiology*, [article in press, download from web].

McNitt-Gray J.L. (1991). Kinematics and impulse characteristics of drop landing from three heights. *International Journal of Sport Biomechanics*, 7, 201–224.

Santello, M., McDonagh, M.J.N. and Challis J.H. (2001). Visual and non-visual control of landing movements in humans. *Journal of Physiology*, 537, 313–327.

OPTIMAL VISCOELASTIC MODEL TO ESTIMATE VERTICAL GROUND REACTION FORCE FROM TIBIAL ACCELERATION DURING HOPPING

Y. SAKURAI & T. MARUYAMA

Department of Human System Science, Tokyo Institute of Technology, Tokyo, Japan

The purpose of this study was to seek the possibility to simulate the vertical GRF of hopping from the tibial acceleration and the viscoelastic model. First, the optimal viscoelastic model to describe landing of hopping was identified. Next, objective function using the tibial acceleration was made. The result showed that the selected viscoelastic model in this study and the tibial acceleration estimated the vertical GRF and take-off velocity of hopping. The implication of this study could be that our new method to estimate GRF may eliminate the associated disadvantages of using force plates.

1 Introduction

Loading force acting between human and the ground, called ground reaction force (GRF), can provide essential sources of information to analyze human dynamics. Force plates are believed to be the "golden standard" to measure the three orthogonal components of GRF. The use of this method, however, brings some limitations. One example is that the task may be restricted depending on the number and size of force plates which may constrain the natural movement of the subjects (Forner-Cordero et al., 2006; Oggero et al., 1999). Alternative approaches have, therefore, been considered to estimate GRF without force plates. The recent attempts include the use of positional data for vertical GRF (Bobbert et al., 1991), and pressure insoles and artificial neural networks for horizontal GRF (Savelberg & de Lange, 1999). Human body during the impact can be considered as a mechanical system with masses, springs and dampers (Liu & Nigg, 2000) and spring-damper-mass models have, therefore, been widely used to simulate human locomotion.

The acceleration of a body segment depends on GRF and the damping effects of the body. There was a correspondence in magnitude and phase between simulated acceleration of model and measured acceleration (Miyaji, 1990). The purpose of this study was to simulate the vertical GRF of hopping by using the tibial axial acceleration and viscoelastic model. Optimal viscoelastic model to best describe the landing of serial hopping was identified in the pre-processing.

2 Methodology

2.1. *Subjects and Experiment Protocols*

Eight healthy males (age: 23.4 ± 0.9 yrs, height: 1.71 ± 0.04 m, weight: 62.3 ± 8.4 kg) participated in the study. All subjects provided informed consent prior to the experiments.

After familiarization, the subjects performed barefoot hopping for eight second at a constant pace (108 jumps/min) with their hands placed on the waist.

Motion data, GRF and tibial acceleration were simultaneously recorded. The motion data were recorded at 125 Hz (1/1000 s of exposure time) with two high speed cameras (FASTCAM, Photron Inc.). GRF was measured with two force plates (Kistler type 9287BA & 9281CA, Kistler Inc.) at 1 kHz. The acceleration was recorded at 1 kHz with a miniature triaxial piezoelectric accelerometer (3053B Dytran Inc.). The accelerometer was fixed to a small plate (55 mm × 9 mm × 1 mm) with two-sided adhesive tape. The plate was firmly attached onto the left tibial tuberosity by non-woven tape (H090A & H091A, Nihon Kohden Co.) and covered by non-adhesive wrap (TJ0904, Johnson & Johnson). The accelerometer was mounted with one measuring axis set visually parallel to the tibial shaft.

2.2. Data Processing

Prior to analysis all data were filtered using weighted average. A threshold of 5 N of the left leg vertical GRF was used to identify the instance of foot-plate contacts. Motion data were converted to 1 kHz using linear interpolation.

The objective functions of several viscoelastic models were statistically compared (R). The significance level was set at 0.05.

2.3. Viscoelastic Model

To find the optimal model, a viscoelastic model of landing in running (Miyaji's model) which consisted of two masses, two linear dampers, two linear springs and a nonlinear spring (Miyaji, 1990) was first selected, and components of the model were modified. The equations for the vertical motion of the model (Figure 1) are:

$$\begin{aligned} m_1 \ddot{x}_1 &= -k_1(x_1 - x_2) - c_1(\dot{x}_1 - \dot{x}_2) - m_1 g \\ m_2 \ddot{x}_2 &= -k_2(x_2 - x_3) - c_2(\dot{x}_2 - \dot{x}_3) - m_1(\ddot{x}_1 + g) - m_2 g \\ -k_3 x_3^5 &= -k_2(x_2 - x_3) - c_2(\dot{x}_2 - \dot{x}_3) \end{aligned} \quad (1)$$

where k_3 is the nonlinear spring and g is the gravity constant ($-9.81 \text{ m} \cdot \text{s}^{-2}$).

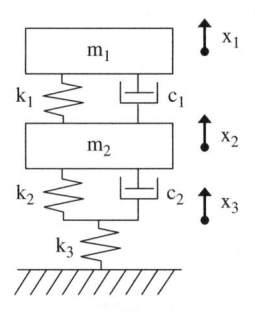

Figure 1. Viscoelastic model (Miyaji, 1990).

Six viscoelastic models (Miyaji's original model and five modified models) were used to find optimal model that can best describe the landing in hopping.

- Model 1: Miyaji's model (Figure 1)
- Model 2: upper damper (c_1) was eliminated from Model 1
- Model 3: mass of m_2 was fixed to be the weight of feet and legs of the subjects (=12.4% of body weight) in Model 1
- Model 4: upper damper (c_1) was eliminated from Model 3
- Model 5: property of nonlinear spring (k_3) was changed in relation to the displacement (x_3) cubed
- Model 6: mass of m_2 was fixed to be the weight of feet and legs (=12.4% of body weight) in Model 5.

2.4. Simulation

The equations of the model were solved using a fourth order Runge-Kutta numerical integration routine with a time step of 0.001 s. The minimum value of the objective function was found by the optimizing routine using "fminsearch" function (MATLAB).

The objective function (Objective Function 1) is expressed as:

$$\text{Objective Function 1} = \text{RSE} + \text{RERV} \tag{2}$$

where: RSE is the relative standard error of impact force and RERV is the relative error of the release velocity (Yukawa & Kobayashi, 1997). RSE was expressed as:

$$\text{RSE} = \frac{\sqrt{\frac{\sum_{i=1}^{n}(f_i - \hat{f}_i)^2}{n-1}}}{\frac{\sum_{i=1}^{n} f_i}{n}} \times 100 \tag{3}$$

where: f_i is the measured vertical GRF and $\hat{f}_i$ is the simulated vertical GRF. RERV was expressed as:

$$\text{RERV} = \left| \frac{V_{CG_{out}} - \frac{m_1 V_{1_{out}} + m_2 V_{2_{out}}}{m_1 + m_2}}{V_{CG_{out}}} \right| \times 100 \tag{4}$$

where: $V_{CG_{out}}$ is the measured release velocity of the great trochanter, $V_{1_{out}}$ is the simulated release velocity of m_1, and $V_{2_{out}}$ is the simulated release velocity of m_2.

The optimized unique set of parameters of a previous contact was used as the initial set of parameters in optimization for the subsequent contact. Initial values of acceleration, velocity and displacement were set to be $-9.81 \text{ m} \cdot \text{s}^{-2}$, the velocity of great trochanter at touchdown and zero, respectively.

3 Results and Discussion

3.1. *Optimal Viscoelastic Model*

Figure 2 shows the results of tibial acceleration time histories of a typical example during landing. Longitudinal acceleration (blue solid line) shows serial peaks with its magnitude gradually decreasing. This tendency is similar to a previous finding during running and walking (Kim *et al.*, 1994). Generally, the first peak appears following the foot contact and this value is higher than the subsequent peaks in impact acceleration. The data which did not demonstrate this typical pattern of longitudinal tibial acceleration was, therefore, considered as failure by measurement error or illegal movement, and thus be excluded.Table 1 shows the average and standard deviations of the Objective Function 1 (RSE + RERV) of all models. Model 1, Model 2, and Model 5 were significantly more accurate than the other models to simulate the landing during hopping, but these three accurate models did not differ significantly. The best optimal viscoelastic model, therefore, could not be determined from the statistical test. Model 1 was, however, selected as the optimal viscoelastic model in this study because it provided the smallest objective function value. Figure 3 shows a typical comparison between measured value in experiment and simulated value of the vertical GRF when RSE is 11.78. Simulated values did not overlap well with the measured values in the early phase after the contact. Whereas, the peak values and time to peak values resembled between the two cases. Overall, the results shown in Figure 3 and Table 1 could be interpreted as that our simulation technique with Model 1 closely estimated the measured GRF and the release velocity.

3.2. *New Objective Function*

When human body is considered as a mechanical model, m_2 in Model 1 could correspond to feet and part of legs. Acceleration of m_2, therefore, may be related to vertical tibial acceleration if subjects hopped within the same place. Another objective function (Objective

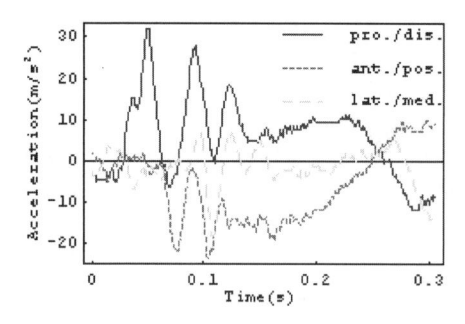

Figure 2. Typical tibial acceleration during landing.

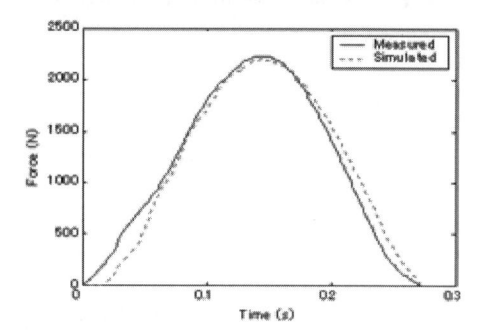

Figure 3. Measured and simulated values on vertical GRF (RSE = 11.78).

Function 2) was made by tibial longitudinal acceleration and acceleration of m₂ which was based on the simulated result using Model 1. Objective Function 2 was expressed as:

$$\text{Objective Function 2} = \left| \left(\alpha \max_{\text{acce\_m}_2} - \max_{\text{acce\_tib}} \right) - \beta \right| + \gamma \left| \hat{f}_{\text{last}} \right| + \text{rc} \times 100 \quad (5)$$

where: $\max_{\text{acce\_m}_2}$ is the peak acceleration of m_2, $\max_{\text{acce\_tib}}$ is the peak acceleration of longitudinal acceleration of tibia, $\hat{f}_{\text{last}}$ is the value of simulated vertical GRF at take-off, α, β, γ are coefficient, and rc is the number of failing to fulfill the restriction conditions. The coefficients were determined through optimization separately.

Table 1. The average and standard deviations of the RSE, RERV, and Objective Function 1 (RSE + RERV) of all models.

Model	1	2	3	4	5	6
RSE	12.2 (2.2)	11.9 (3.0)	14.8 (2.5)	147 (2.7)	11.9 (2.1)	13.4 (2.3)
RERV	3.1 (2.8)	4.7 (3.7)	4.6 (4.1)	4.4 (4.2)	4.2 (3.2)	5.0 (3.8)
Objective Function	15.3 (4.2)	16.6 (3.7)	19.4 (4.8)	19.1 (4.6)	16.1 (4.7)	18.4 (4.2)

Table 2. The average and standard deviations of RSE and the comparison of peak vertical GRF, time to peak vertical GRF, loading rate, and impulse between measured (force plate, FP) and simulated (model) conditions of all subjects. N is equivalent to the number of landing.

	RSE	Peak vertical GRF (N)		Time to peak GRF (ms)		Loading rate (BW/s)		Impulse (N · s)	
		FP	Model	FP	Model	FP	Model	FP	Model
A (n = 11)	21.5 (9.8)	2304 (151)	2271 (154)	136 (14)	118 (3)	30.2 (4.6)	33.7 (2.7)	320.6 (5.8)	294.5 (8.6)
B (n = 4)	11.3 (1.4)	2387 (130)	2367 (67)	154 (5)	152 (6)	25.2 (2.1)	25.2 (1.2)	351.8 (5.1)	343.9 (12.5)
C (n = 6)	14.6 (3.3)	2499 (160)	2729 (134)	84 (6)	85 (5)	57.3 (5.9)	62.0 (6.1)	270.3 (11.2)	270.2 (11.1)
D (n = 9)	11.5 (3.7)	1706 (126)	1805 (135)	165 (19)	159 (12)	19.8 (3.3)	21.6 (3.1)	291.7 (12.4)	289.2 (11.2)
E (n = 8)	13.2 (5.3)	2436 (56)	2301 (88)	126 (12)	129 (5)	31.4 (2.9)	28.9 (2.0)	305.9 (9.0)	310.5 (6.3)
F (n = 11)	9.5 (4.7)	2797 (247)	2859 (144)	103 (6)	105 (5)	46.4 (6.0)	46.7 (4.0)	307.4 (10.6)	309.5 (8.0)
G (n = 6)	27.1 (3.9)	2710 (201)	2619 (59.6)	164 (8)	135 (3)	21.7 (2.6)	25.4 (0.8)	424.7 (10.3)	395.9 (7.2)
H (n = 14)	14.2 (4.7)	2717 (98)	2802 (108)	112 (8)	115 (5)	35.5 (2.6)	35.6 (2.2)	364.1 (15.9)	362.1 (9.8)

The restriction conditions were as follows:

- All parameters of the Model 1 are positive.
- Mass of lower mass (m_2) falls between the weight of feet (=2.2% of body weight) and the weight of foot and legs (=12.4% of body weight).
- Acceleration of M_2 shows a typical attenuation of signal.
- Acceleration of point mass (x_3) shows a typical attenuation of signal.

In simulation using Objective Function 2, initial values of the velocity at landing were calculated using the preceding flight time given by:

$$V_{impact} = -\frac{1}{2} gt \qquad (6)$$

where: t is the flight time.

Table 2 shows RSE and peak vertical GRF, time to peak vertical GRF, loading rate, and impulse of measured and simulated (using Object Function 2) GRF of all subjects. It is important to note that the RSE appeared to increase especially when the difference in time to peak GRF between measured and simulated values was large (subject A and G). Therefore, it could be said that RSE value was increased by the unsatisfactory simulation of time to peak GRF. In fact, Objective Function 2 given in Eq. (5) considers only peak value of acceleration, and thus may have little ability to control time factor. The other six subjects, however, produced similar GRF behavior between measured and simulated cases with a concomitant small RSE value.

4 Conclusion

The main result is that the simulation using the selected viscoelastic model (Model 1) and tibial acceleration (Objective Function 2) has a possibility to estimate the vertical GRF during hopping.

References

Bobbert M.F., Schamhardt H.C. and Nigg B.M. (1991) Calculation of vertical ground reaction force estimates during running from positional data. *Journal of Biomechanics*, 24 (12), 1095–1105.

Forner-Cordero A., Koopman H.J.F.M. and Helm F.C.T. van der (2006) Inverse dynamics calculations during gait with restricted ground reaction force information from pressure insoles. *Gait & Posture*, 23, 189–199.

Liu W. and Nigg B.M. (2000) A mechanical model to determine the influence of masses and mass distribution on the impact force during running. *Journal of Biomechanics*, 33, 219–224.

Miyaji C. (1990) Landing shock during running and its simulation. *Society of Biomechanics*, 14/2, 73–77, (in Japanese).

Oggero E., Pagnacco G., Morr D.R. and Berme N. (1999) How force plate size influences the probability of valid gait data acquisition. *Biomedical sciences instrumentation*, 35, 3–8.

Savelberg H.H. and Lange A.L de (1999) Assessment of the horizontal, fore-aft component of the ground reaction force from insole pressure patterns by using artificial neural networks. *Clinical Biomechanics*, 14, 585–592.

Yukawa H. and Kobayashi K. (1997) Variability of landing shock loads and parameters during running with viscoelastic model. *Japan society of Sports Industry*, 7/1, 19–28, (in Japanese).

A PORTABLE VERTICAL JUMP ANALYSIS SYSTEM

B.H. KHOO[1], S.M.N.A. SENANAYAKE[1], D. GOUWANDA[1] & B.D. WILSON[2]

[1]*Monash University Sunway Campus, Jalan Universiti,*
Petaling Jaya, Malaysia
[2]*National Sports Institute, Center of Biomechanics,*
Sri Petaling, Kuala Lumpur, Malaysia

This paper discusses the design and development of a low cost portable vertical jump analysis system. The system consists of Data Acquisition (DAQ) device, force platforms, power supply unit and data acquisition circuit, with a Graphical User Interface (GUI). A National Instrument's™ (NI) E Series PCI DAQ card is used to acquire and to digitize the analog signals. Each force platform, with the possible extension to any number of platforms, consists of 144 Force Sensitive Resistors (FSR) with an effective sensing area of 54 cm by 54 cm. A DAQ circuit is designed to amplify signals from 36 FSRs. The amplified signals are sent to the DAQ card via four analog input ports using multiplexing technique. There are total of four DAQ circuits for each force platform. A user friendly GUI was developed using NI LabVIEW graphical programming language. The GUI has been developed to plot the force profile on 2D and 3D graphs. The GUI will compute variable such as the impulse and hang time based on the force profile.

1 Introduction

Force platforms are widely used in sports biomechanics, especially in jumping analysis, such as vertical jump, power testing etc. However, existing systems are often expensive, bulky and heavy. The bulk and weight often limits the use of a force platform. Therefore, there is a need for a portable vertical jump analysis system.

A vertical jump analysis system with two force platforms has been developed and tested. The advantages of the portable vertical jump analysis system are portability, hardware flexibility, modularity, configuration flexibility, expandability and low cost. The designed system is highly portable. It can be dissembled into 4 units for compact and easy storage. The system has high hardware flexibility. It uses the standard GPIB bus, which allows the system to be interfaced to a desktop PC, a laptop or a standalone DAQ device. This modular system is able to handle multiple force platforms by simply expanding the DAQ circuit. The DAQ circuit is specially designed to ease the expansion of the force platform in terms of size and number of platforms. The configuration of the multiple force platforms is flexible. They can be placed on the floor for jumping analysis or placed on the wall for punching and kicking analysis. Expanding the application of this system for different sports can be as simple as loading a different GUI from the main software. This system is made of low cost components and the construction of the system is carried out in the laboratory environment without the need of any special tools or equipment. The standard DAQ device and power supply can be reused for other sport analysis system.

By analyzing the force exerted and the pattern of movement of athletic on the floor, it's possible to determine what type of floor is appropriate for sports, or even what type of shoes and support needed to ensure athletic legs are protected from strain and force acting on them. All the data from analysis can help to prevent crucial injuries and help to improve the quality of a particular sport (Srinivasan *et al.*, 2005).

2   System Architecture

2.1. *Force Plates, Signal Conditioning and Switching Circuits*

The overall system of force platform mainly consists of force plates; signal conditioning and switching circuit, power supply unit and PCI 6036-E DAQ card. Each force plate consists of 144 FSRs, which are distributed over 54 × 54 cm wood board. Each force plate contains four quadrants, in each quadrant, there are 36 FSRs. Quadrant in force plate possesses similar size of 24 × 24 cm, e.g. see Figure 1. These sensors are covered by yoga mat (Paradiso *et al.*, 1997).

There are two circuit needed for force platforms; Signal conditioning circuit and main switching circuit.

In order to reduce wiring problems in signal conditioning circuit and to ease future testing and debugging processes, customized Printed Circuit Boards (PCB) were made. The design of PCB is made using Circuit Maker 2000. There are a total of 24 signal conditioning circuits and two main switching circuits.

Signal conditioning circuit mainly consists of operational amplifier and multiplexers. LM324 op-amp is used to amplify the signal from sensors. Since one force platform contains 144 sensors and it is impossible to find a DAQ card that consists of 144 inputs, CD4051 multiplexers are used to switch and read signal from one sensor to another sensor. In each signal conditioning circuit, there are two multiplexers and three operational amplifiers, which are able to handle 6 FSRs, e.g. see Figure 2.

Main switching circuit consists of four multiplexer, where each multiplexer represents one quadrant of the force plate. Main switching circuit switches and acquires signal from the quadrant and transmits them to NI-DAQ, e.g. see Figure 3.

Integration of force platforms, signal conditioning circuits and switching circuits is done, e.g see Figure 4. FSRs are connected to signal conditioning circuit. Signals received by signal conditioning circuit are amplified and sent to main switching circuit. Main switching circuit switches its states to signals received from sensors and passed them to NI-DAQ.

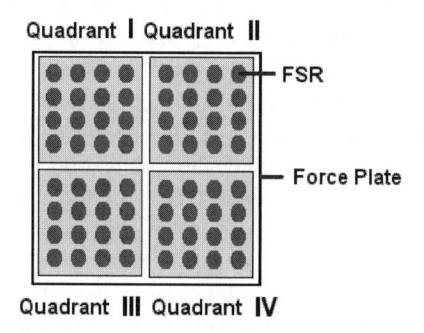

Figure 1.   Illustration of FSRs arrangements in force plate.

A Portable Vertical Jump Analysis System

In this manner, two force platforms are constructed and interfaced in order to analyze activities on multi-platforms, e.g. see Figure 5. Further, built multi-platforms prove that the idea can be extended any number of platforms with necessary voltages controlled by the software in order to power up set of platforms.

2.2. *Software Architecture of the Platform*

Software architecture is addressed taking into consideration two different software implemented using LabVIEW. The first software, called SiM-FORCE is used to collect signals

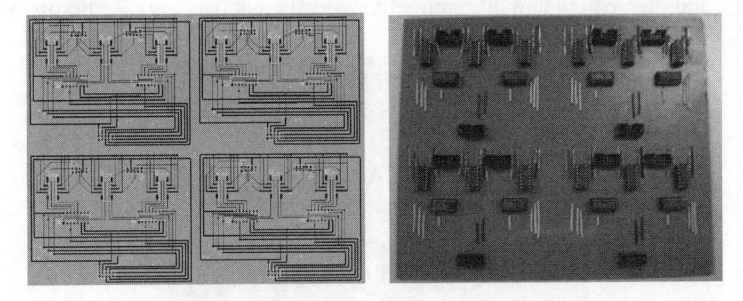

Figure 2. PCB design for signal conditioning circuit.

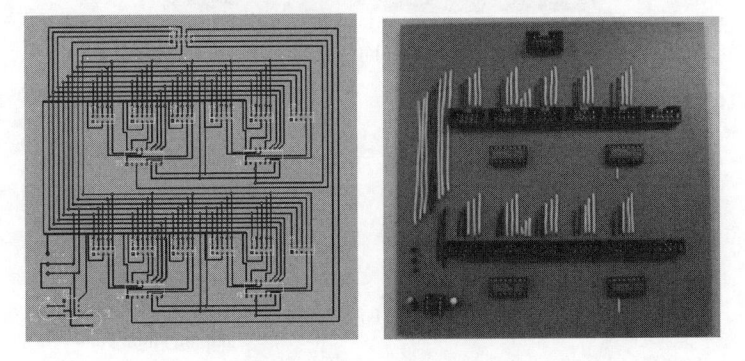

Figure 3. PCB design for main switching circuit.

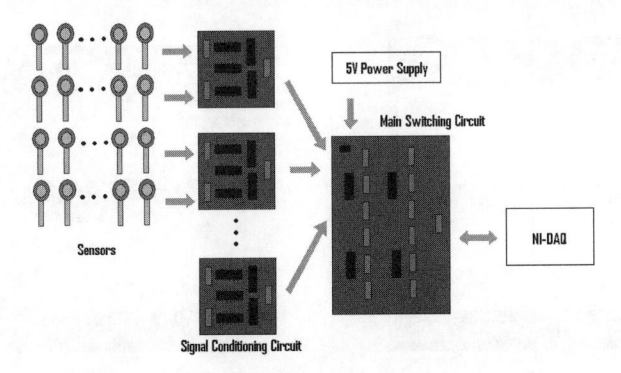

Figure 4. System integration of force platforms.

from sensors in real-time, convert them into forces and to perform force analysis. Second software, namely FORCE READER is responsible to read data already captured in SIM-FORCE and to do force analysis for sport performance.

In SIM-FORCE, user has to enter the subject's personal details i.e. name, age, height and weight. Once human subject details are provided, it will start to receive signals from DAQ card and followed by showing signals on running charts. When the data acquisition is completed, user is able to compute the jumping impulse and the jumping flight time by entering the initial time and end time, e. g. see Figure 6.

Based on SIM-FORCE software module, firstly data are acquired from DAQ card, e.g. see Figure 7 and then calculated all parameters specified above, e.g. see Figure 8.

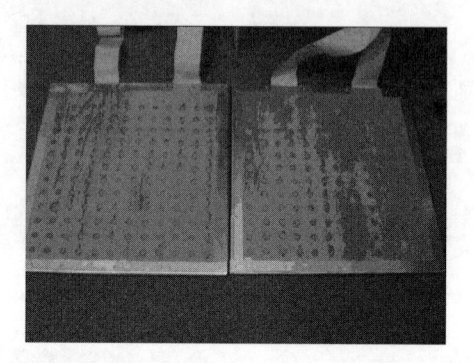

Figure 5. Multiple force platforms built and interfaced.

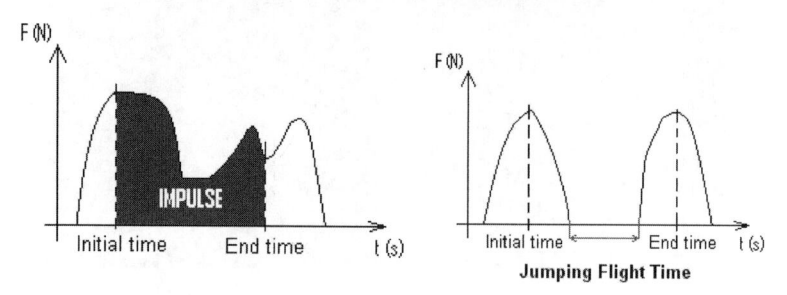

Figure 6. Initial time, end time and jumping flight time.

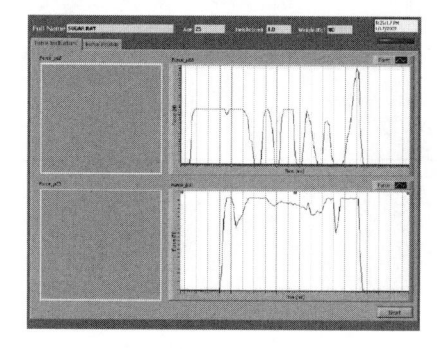

Figure 7. Data acquisition process.

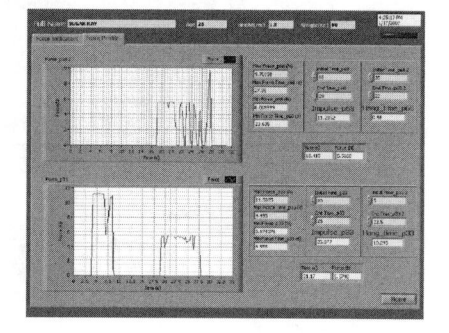

Figure 8. Force, impulse and flight time.

A Portable Vertical Jump Analysis System 641

FORCE READER possesses same functionalities as SIM-FORCE. Main difference is that the user is able to analyze sports performance of the athlete and to provide feedback on parameters measured. Therefore, FORCE-READER works offline and SIM-FORCE works in real time to acquire and to process data.

3   Activities on Force Platforms

Multi-layered force platforms developed is capable to measure; force intensity, Vertical Ground Reaction Forces (VGRF) in combined platforms and single platforms and maximum and minimum VGRFs (Orr, 2000). These measurements are taken directly from the raw data obtained from FSRs and displayed using GUI already developed with SIM-FORCE and as well as FORCE READER software.

Further, FORCE READER can calculate impulse occurred during contacts, jumping flight time, maximum height jumped and take off velocity of jumping (e.g. Figures 9 and 10).

4   Conclusions

A vertical jump analysis system with two force platforms has been developed and tested. The advantages of the portable vertical jump analysis system are portability, hardware flexibility, modularity, configuration flexibility, expandability and low cost. The designed system is highly portable. It can be dissembled into 4 units for compact and easy storage. The system has high hardware flexibility. It uses the standard GPIB bus, which allows the system to be interfaced to a desktop PC, a laptop or a standalone DAQ device. This modular system is able to handle multiple force platforms by simply expanding the DAQ circuit. The DAQ circuit is specially designed to ease the expansion of the force platform in terms of size and number of platforms. The configuration of the multiple force platforms is flexible. They can be placed on the floor for jumping analysis or placed on the wall for punching and kicking analysis. Expanding the application of this system for different sports can be as simple as loading a different GUI from the main software. This system is made of low cost components and the construction of the system is carried out in the laboratory environment without the need of any special tools or equipment. The standard DAQ device and power supply can be reused for other sport analysis system.

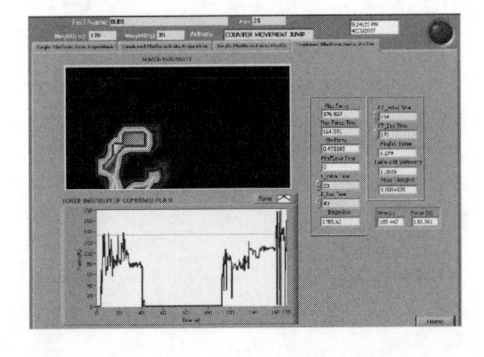

Figure 9.   VGRF analysis of platforms. 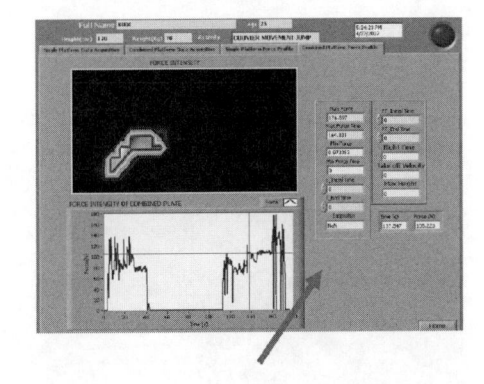   Figure 10.   Force intensity using a cursor.

References

Srinivasan P., Qian G., Birchfield D. and Kidané A. (2005) *Design of a Pressure Sensitive Floor for Multimodal Sensing.* Proceedings of the International Conference on Non-visual & Multimodal Visualization, London, UK, July 4.

Paradiso J., Abler C., Hsiao K.Y. and Reynolds M. (1997) The Magic Carpet: Physical Sensing for Immersive Environments. In: Proc. of the CHI '97 Conference on Human Factors in Computing Systems, ACM Press, NY, pp. 277–278. Available: http://acm.org/sigchi/chi97/proceedings/short-demo/jp.htm (Accessed: May 2007).

Orr R.J. (2000) Smart Floor: Future Computing Environments (online). Available: http://www-static.cc.gatech.edu/fce/smartfloor/ (Accessed: May 2007). College of Computing, Georgia Institute of Technology, USA.

THE EFFECT OF SHOE BENDING STIFFNESS ON PREDICTORS OF SPRINT PERFORMANCE: A PILOT STUDY

D.T. TOON, N. HOPKINSON & M.P. CAINE
Wolfson School of Mechanical and Manufacturing Engineering, Loughborough University, Leicestershire, England

The current pilot investigation employs two isoinertial metrics to investigate the influence of footwear on jumping and by inference sprint performance. A nationally competitive sprinter performed bounce drop jumps from a height of 0.46 m and a concentric jump with a knee angle of 120°. A control shoe of nominal stiffness and four purpose built sprint shoes were used. The sprint shoes were constructed using standard sprint spike uppers and Selective Laser Sintered (SLS) nylon-12 outsoles. The sprint shoes had bending stiffnesses that span that of current commercially available sprint spikes. Bounce drop jump performance was measured using reactive index; jump height divided by contact time. Concentric jump performance was measured using jump height. Additional parameters such as the Maximum Dynamic Strength (MDS) during takeoff and the force after 100 ms were also investigated. The results for drop jump performance show that jump height is significantly reduced in the stiffest sprint shoes compared to the barefoot equivalent (control shoe) condition. No significant differences in concentric jump performance were found between the tested footwear conditions.

1 Introduction

Stefanyshyn & Fusco (2004) investigated the affects of shoe bending stiffness on sprinting performance. The results showed that as shoe bending stiffness was increased sprint performance increased, but this relationship only held true as stiffness increased to a moderate value, after which performance decreases and this relationship no longer held. It was reasoned that that in order to maximise performance individual tuning of the athlete's shoe stiffness to the athlete's particular characteristics is required. The current pilot investigation thus sought to investigate the user/footwear interaction with respect to different shoe bending stiffness conditions. Two isoinertial metrics were used to investigate the influence of footwear on sprint performance, using sprint shoes with selective laser sintered (SLS) sole units designed to be identical with the exception of their bending stiffness. A concentric jump from 120° knee angle and a bounce drop jump were used. The two jump metrics were selected as representations of the start and maximal speed phases of sprinting respectively.

Schmidtbleicher (1992) proposed that a long stretch-shortening cycle (SSC) is characterised by large angular displacements in the joints and has duration of more than 250 ms and a short SSC has only small angular displacements and lasts 100–250 ms. The bounce drop jump can be considered as a short SSC action because of its relatively short foot support phase. Hennessey & Kilty (2001) showed high correlations of $r = -0.79$ and $r = -0.75$ between bounce drop jump performance and 30 m and 100 m sprint times respectively. The specificity of this type of jump to top speed sprinting is further justified by Mero *et al.* (1981) who reported significant correlations of $r = 0.72$ between a drop jump from 50 cm drop height and maximal speed.

Young et al. (1995) used a concentric jump from 120° knee angle to measure sprint starting performance. The authors were able to demonstrate a significant correlation of r = −0.86 between concentric jump kinetics and 2.5 m time. The authors associate the specific nature of the jump metric to sprint starting for three reasons. Firstly the knee angle of 120° is similar to the reported mean knee angle (126°) of the rear leg in the set position of the block start. Secondly, kinetics were measured under pure concentric contraction, which is the case for hip and knee extensors during the movement on the blocks. Thirdly, the total movement time of the concentric jump is consistent with the time on the blocks and during the first stride.

The bounce drop jump and concentric jump performance tests demonstrate a high specificity to the maximal speed sprinting and starting performance respectively. The jump metrics appropriately mimic the muscular contractions of the leg extensor muscles and have patterns of force production that are similar to those during the specific sprint phase. The inherent nature of the selected jump metrics, whilst being closely associated with sprint kinetics and kinematics, also lend themselves to structured lab-based investigation. This type of assessment facilitates the acquisition of detailed, controlled and repeatable measurements that are required in order to appreciate the affects of the different footwear conditions.

2 Methodology

A nationally competitive sprinter, age 19 years, height 1.83 m, mass 79 kg with shoe size UK 9 participated in the study. A health screening questionnaire was completed and informed written consent obtained. A Kistler force platform sampling at 1000 Hz was used to record the forces applied to the ground throughout the jumping actions. Bounce drop jumps were performed using a purpose built rig; drop height was fixed at 0.46 m. For the concentric jumps the jump rig was positioned such that the participant's posterior made contact with the platform ensuring the correct knee angle was achieved whilst also forcing the participant to perform a purely concentric jumping action.

A shoe with nominal bending stiffness (Vibram, Fivefingers) was used as the control shoe, replicating barefoot conditions. Four sprint shoes, size UK 9, with different bending stiffnesses spanning that of currently available sprint spikes were manufactured. Guideline mechanical properties obtained from in-house testing are listed in Table 1.

Jump tests were performed over 7 days, with at least 48 hours between tests. The participant completed their own individual warm-up common to the requirements of plyometric exercise. Jump tests were firstly performed in the control shoe, subsequent jumps were carried out in the sprint shoes, presented to the athlete in a random order.

Table 1. Sprint shoe mechanical properties.

Shoe	Mean force (N) Extension	Mean force (N) Flexion
D	9.0	7.4
C	23.7	10.8
F	24.5	14.7
A	38.0	26.1

For the bounce drop jump the drop height was fixed at 0.46 m. The participant was instructed to jump for maximum height and minimum contact time. Further instructions stressed the importance of landing from a jump in a fully extended position, completing each effort without heels forcefully striking the ground and with the hands remaining on the hips. 6 successful jumps with ground contact times of <250 ms were required in each footwear condition. Jump height was calculated from the flight time using the method of Komi & Bosco (1978). This method assumes that body positions are the same at the instants of takeoff and landing. Flight and contact times were recorded by the force platform. Reactive strength (Young et al., 1995) was calculated by dividing the jump height by the contact time. The maximum force developed during the jump take-off (maximum dynamic strength – MDS) was also recorded. The maximum dynamic strength was the maximum force achieved during the concentric phase of the jumping action.

The concentric jump test required the participant to perform 8 successful jumps from 120° knee angle. The participant was asked to stand on the force plate, sink to 120° knee angle, maintain that position for 4 seconds, then to jump as explosively as possible. Further instructions stressed the importance of landing from a jump in a fully extended position with the hands remaining on the hips. Ground reaction forces were recorded throughout the jump tests and the vertical height of the centre of mass was calculated by first determining the vertical velocity at take-off (V_{off}) by trapezoid integration according to Eq. 1 between take-off and the point 1 second prior to take-off. Where $F(t)$ is the vertical ground reaction force, g is the acceleration due to gravity and m is the body mass. Jump height is then determined from Eq. 2.

$$V_{off} = \int_1^2 (F_Z(t) - mg)dt/m \qquad (1)$$

$$h_{jump} = V_{off}^2/(2 \cdot g) \qquad (2)$$

Additional parameters measured during the concentric jump included the MDS and the force after 100 ms from the start of contraction. These parameters were selected based on their positive correlations with sprinting performance identified by Young et al. (1995). One-way analysis of variance (ANOVA) was used to compare the variability in jumping performance between the different footwear conditions for each jump metric. The 0.05 level of significance was adopted for all statistical tests.

3 Results

The results for the drop jump are shown in Table 2. The reported values are the mean of 6 successful jumps. A statistically significant difference in jump height between the control shoe and shoes F and A was observed. The participant jumped significantly lower in both shoes A and F, the two stiffest shoes tested. No other significant differences for any of the other tested variables were observed.

The results for the concentric jump are listed in Table 3. The reported values are a mean of 8 successful jumps. No significant differences in any of the measurements were observed between the footwear conditions.

Table 2. Drop jump tabulated results (*significant p < 0.05).

Condition	Jump height (m)	S.D.	Contact time (s)	S.D.	Reactivity coefficient (m/s)	S.D.	MDS (N)	S.D.	MDS/BW (N/kg)	S.D.
Control	0.339	0.027	0.182	0.016	1.873	0.232	4488.9	246.6	5.68	0.31
Shoe D	0.314	0.032	0.175	0.012	1.800	0.174	4557.7	352.2	5.77	0.45
Shoe C	0.325	0.021	0.183	0.025	1.797	0.219	4592.0	842.2	5.81	1.07
Shoe F	0.288*	0.014	0.172	0.013	1.673	0.075	4753.4	495.2	6.01	0.63
Shoe A	0.293*	0.015	0.184	0.018	1.610	0.175	4320.4	534.5	5.47	0.68

Table 3. Concentric jump tabulated results.

Condition	Jump height (m)	S.D.	MDS (N)	S.D.	MDS/BW (N/kg)	S.D.	F100	S.D.	F100/BW	S.D.
Control	0.385	0.038	2522.4	88.7	3.19	0.11	1128.1	96.9	1.43	0.12
Shoe D	0.399	0.009	2533.4	57.3	3.21	0.07	1015.7	103.0	1.29	0.13
Shoe C	0.389	0.027	2522.2	40.1	3.19	0.05	1016.3	84.1	1.29	0.11
Shoe F	0.400	0.022	2553.2	70.4	3.23	0.09	1007.2	102.4	1.27	0.13
Shoe A	0.412	0.020	2490.9	37.3	3.15	0.05	1073.6	94.4	1.36	0.12

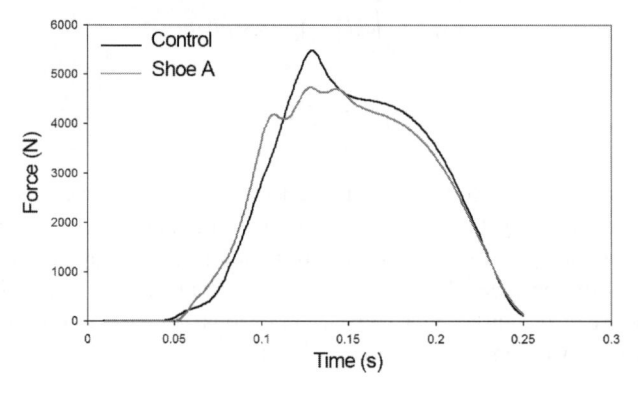

Figure 1. Drop jump force profile for the contact period.

Figure 1 shows a force trace of the contact period for the bounce drop jump (mean data of 6 successful jumps). Least stiff and most stiff footwear conditions are compared.

Figure 2 shows a force trace for the concentric jump take-off (mean data of 8 successful jumps). Least stiff and most stiff footwear conditions are compared.

4 Discussion and Conclusions

The drop jump data showed that jump height is significantly reduced in the stiff footwear conditions (shoe A and shoe F) when compared directly to the control condition (barefoot

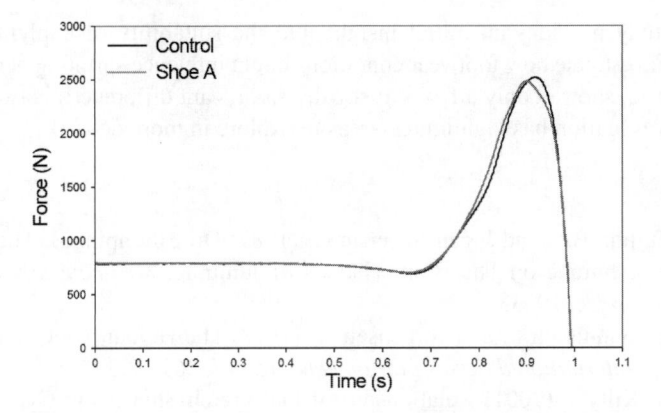

Figure 2. Concentric jump force profile for take-off.

equivalent). In agreement with this, there is also a non-significant trend in the mean data which suggests that as stiffness is increased the reactivity coefficient is decreased. Interestingly the graphical data for the drop jump, shown in Figure 1, shows that the force generated during contact in the control shoe has a distinct peak eccentric force which drops rapidly before the concentric contraction and the point of maximum dynamic strength. Bobbert et al. (1987) realised that the characteristics of the bounce drop jump mean that the time interval between the peak velocity of eccentric action and the start of concentric contraction attain a small value and therefore optimise the actions of pre-stretch and potentiation. For shoe A there is no distinct eccentric peak force, and the point of concentric contraction and hence maximum dynamic strength approximately occurs as the peak force of ground contact. The lack of distinction between eccentric and concentric contraction in the stiffer footwear condition may be detrimental to the pre-stretch and potentiation requirements of the short SSC and go some way to explaining the significantly poorer jump height performance in shoe A.

No significant differences between any of the tested variables for concentric jump performance in the different footwear conditions were found. However it is apparent from the force trace shown in Figure 2 that during take-off, in the control shoe, the concentric force generation initially rises at one constant rate and then after approximately 0.15 s into the contraction the rate of force generation increases. The distinct changes in the rate of force generation can be attributed to the separate phases of push-off. The separate phases are a consequence of the intermediate break in the foot at the metatarsophalangeal joint (MPJ). Bojsen-Møller & Lamoreux (1979) realised that the intermediate break at the MPJ is important in managing the moment force generated by the triceps surae. The MPJ reduces the demand on the triceps surae and therefore the triceps surae can generate useful forces over a longer period of time because the length of the resistance arm increases as the horizontal speed of the foot increases. The force trace for shoe A shows a constant linear increase in force production, with no distinctly separate phases. This suggests that the foot and shoe system are acting as a rigid lever generating larger moment forces about the ankle joint. Stefanyshyn & Fusco (2004) speculate that by increasing the length of the lever arm, greater moments about the ankle joint can be produced and if the triceps surae is strong enough to generate the additional force, the result would be an increase in sprint performance.

The pilot study provides an initial insight into the suitability of applying discrete jump metrics to investigate how footwear conditions might influence sprinting performance. The current data set showed only a few statistically significant differences between conditions, but the investigation has highlighted areas to explore in more detail.

References

Bobbert M.F., Huijing P.A. and Ingan Schenau G. (1987) Drop Jumping. I. The Influence of Jumping Technique on The Biomechanics of Jumping. *Medicine and Science in Sports and Exercise*, 19, 332–338.

Bojsen-Moller F., Asmussen E. and Jorgensen K. (1978) The Human Foot, A Two Speed Construction. *International Series of Biomechanics*, VI, 261–266.

Hennessy L. and Kilty J. (2001) Relationship of the Stretch-Shortening Cycle to Sprint Performance in Trained Female Athletes. *Journal of Strength and Conditioning Research*, 15, 326–331.

Komi P.V. and Bosco C. (1978) Utilization of Stored Elastic Energy in Leg Extensor Muscles by Men and Women. *Medicine and Science in Sports and Exercise*, 10, 261–265.

Mero A., Luhtanen P., Komi P.V. and Viitasalo J.T. (1981) Relationships Between the Maximal Running Velocity, Muscle Fibre Characteristics, Force Production and Force Relaxation of Sprinters. *Scandanavian Journal of Sports Science*, 3, 16–22.

Schmidtbleicher D. (1992) Training for Power Events In: Komi P.V. (Ed.) *Strength and Power in Sport*. Blackwell Scientific, London, pp. 381–384.

Stefanyshyn D.J. and Fusco C. (2004) Increased Bending Stiffness Increases Sprint Performance. *Sports Biomechanics*, 3, 55–66.

Young W., McLean B. and Ardagna J. (1995) Relationship Between Strength Qualities and Sprinting Performance. *Journal of Sports Medicine and Physical Fitness*, 35, 13–19.

THREE-DIMENSIONAL ANALYSIS OF JUMP MOTION BASED ON MULTI-BODY DYNAMICS – THE CONTRIBUTION OF JOINT TORQUES OF THE LOWER LIMBS TO THE VELOCITY OF THE WHOLE-BODY CENTER OF GRAVITY

S. KOIKE,[1] H. MORI[2] & M. AE[1]

[1]*Institute of Health and Sport Science, University of Tsukuba*
Ibaraki, Japan
[2]*Master's Program in Health and Physical Education, University of Tsukuba,*
Ibaraki, Japan

We propose a method for calculating the contributions of (1) muscular joint torques; (2) motion-dependent torques; and (3) gravitational forces to the velocity of the center of gravity of the whole body (body CG), and apply this method to a jump motion. The largest contribution to vertical body CG velocity in the support phase of the motion is found to be the ankle joint plantar torque, while the knee joint extension torque is found to contribute significantly to not only the vertical velocity but also to the horizontal velocity of body CG. While quantification of joint-torque contributions is not possible with the conventional inverse kinetic method, the proposed method clarifies the roles of joint torques in the production of body CG velocity.

1 Introduction

Since every human movement is governed by a unique equation of motion, dynamical analysis based on an equation of motion for the whole body system can be a highly effective method for understanding complex multi-joint motions such as jumping, running and throwing. Although some studies have investigated the contributions of joint torques, motion-dependent torques and gravitational forces to segment distal end-point velocities by applying equations of motion for actions such as kicking (Putnam, 1991) and throwing (Naito *et al.*, 2006), they have limited their analysis to lower or upper limbs only. For actions such as running and jumping the horizontal and/or vertical velocities of the whole body CG are of crucial relevance for performance evaluation. However, most likely due to the difficulty in deriving such equation, no study employing an equation of motion for the whole body has been seen for the analysis of sports motions.

The purpose of this study was to employ a three-dimensional equation of motion for the whole body to quantify the contribution of joint torques of the human body to the body CG vertical and horizontal velocities during a jump motion.

2 Methods

2.1 *The Equation of the 3-Dimensional Whole Body Movement*

The whole body is assumed to consist of 15 rigid segments. For each segment, the equation of motions for translational and rotational motion may be expressed as follows:

$$m_i \ddot{x}_i = f_{i+1} - f_i + m_i g \qquad (1)$$

$$\hat{I}_i \dot{\omega}_i = \overline{P_{cgA,i}} \times f_{i+1} - \overline{P_{cgB,i}} \times f_i - \omega_i \times (\hat{I}_i \omega_i) + \tau_{i+1} - \tau_i, \tag{2}$$

where m_i is the mass of the segment, x_i is the 3-dimenstional coordinate value of the CG of the segment, f_i is the joint force vector, and g is the gravitational vector of each segment., I_i is the inertia matrix of each segment, and ω_i is the angular velocity of each segment. P with subscripts cgA and cgB denotes position vectors from the CG of each segment i to the proximal and distal endpoints of the segment respectively, and τ_i is the joint torque vector.

The equations for the segment in direct contact with the ground during the support phase are as follows:

$$m_{footL} \ddot{x}_{footL} = f_{ankleL} - f_{ext} + m_{footL} g \tag{3}$$

$$\hat{I}_{footL} \dot{\omega}_{footL} = \overline{P_{cgA,footL}} \times f_{ankleL} - \overline{P_{cgCP,footL}} \\ \times f_{ext} - \omega_{footL} \times (\hat{I}_{footL} \omega_{footL}) + \tau_{ankleL} - \tau_{ext}, \tag{4}$$

where subscripts footL and ankleL represent the left foot segment and left ankle joint, and CP denotes the center of pressure at the contact point. f_{ext} and τ_{ext} are, respectively, the ground reaction force vector and external torque vector acting on the foot; they were measured with a force platform.

Combination of the above equations results in an equation for all segments:

$$M\dot{V} = PF + QT + H + G, \tag{5}$$

where M is the inertia matrix and V is a vector containing the velocity and angular velocity of each segment's CG. P is the coefficient matrix for vector F which contains all joint force vectors and external force vectors, f_{ext}. Q is the coefficient matrix for vector T which contains all joint torque vectors and external torque vectors, τ_{ext}. H is a vector containing motion-dependent torques such as Coriolis and centrifugal forces. G is the vector of the gravitational component.

In addition, assuming that every segment is connected to an adjacent segment at a joint, the geometric constraint for two segments can be expressed as follows:

$$x_i + \overline{P_{cgA,i}} - x_{i+1} - \overline{P_{cgB,i+1}} = O_{3 \times 1}, \tag{6}$$

where $O_{3 \times 1}$ is the zero vector. The constraint for the foot of the support leg under the assumption that the foot is linked with the ground at the CP is expressed as follows:

$$x_{footL} + \overline{P_{cgCP,footL}} = x_{CP} \tag{7}$$

Differentiating equations (6) and (7) yields accelerational constraint equations for all joints:

$$C\dot{V} + \dot{C}V = \ddot{\eta} \tag{8}$$

where η is the position vector of x_{cp}. Substituting eq.(8) into eq.(5), we obtain the equations of motion of the whole-body system:

$$\dot{V} = A_T T + A_V + A_G G$$
$$A_T = -M^{-1}P(CM^{-1}P)^{-1}CM^{-1}Q + M^{-1}Q$$
$$A_V = -M^{-1}P(CM^{-1}P)^{-1}(CM^{-1}H + \dot{C}V - \ddot{\eta}) + M^{-1}H$$
$$A_G = -M^{-1}P(CM^{-1}P)^{-1}CM^{-1} + M^{-1} \qquad (10)$$

where A_T is the coefficient matrix of the vector T, A_V is the vector of the motion-dependent torques, and A_G is the coefficient matrix of gravity. The first term, $A_T T$, represents the accelerations of each segment caused by joint torques, the second term, A_V, represents the accelerations caused by the motion-dependent torques, and third term, $A_G G$, represents the accelerations caused by gravitational forces.

2.2. Contributions of Joint Torques to Body CG Velocity

The three terms on the right side of eq.(10) show the relationship between the joint torques, motion-dependent torques, and gravitational forces, and the accelerations of the centers of gravity of all segments. These terms can be converted into acceleration of body CG $\ddot{x}_{cg,body}$ via the transformation matrix, S, as follows:

$$\ddot{x}_{cg,body} = S\dot{V} = SA_T T + SA_V + SA_G G$$
$$S = \frac{1}{m_{body}}[m_1 E \quad O \quad m_2 E \quad \cdots \quad O \quad m_N E \quad O] \qquad (11)$$

where m_{body} is the mass of the whole-body, E is the unit matrix, and O is the zero matrix. Integration of eq.(11) transforms body CG acceleration, $S\dot{V}$, into body CG velocity:

$$\dot{x}_{cg,body} = \int S\dot{V} dt = \int SA_T T dt + \int SA_V dt + \int SA_G G dt \qquad (12)$$

The vertical component of body CG velocity can then be obtained as follows:

$$\dot{x}_{cg,body,z} = C_{TT} + C_H + C_G$$
$$C_{TT} = [0 \quad 0 \quad 1]\int SA_T T dt$$
$$C_H = [0 \quad 0 \quad 1]\int SA_V dt$$
$$C_G = [0 \quad 0 \quad 1]\int SA_G G dt \qquad (13)$$

C_{TT} represents the vertical contribution of all joint torques to the body CG velocity, C_H is the contribution of motion-dependent torques, and C_G is the contribution of gravitational forces. Furthermore, C_{TT} can be written as the sum of individual joint torques, $C_{T,j}$.

$$C_{TT} = \sum C_{T,j}$$
$$C_{T,j} = \int a_j T_j dt, \quad a_j = [0 \quad 0 \quad 1]SA_T \qquad (14)$$

where $C_{T,j}$ is the contribution to the vertical component of body CG velocity produced by the torque at joint j. The contributions to horizontal velocity are calculated similarly.

2.3. Data Collection and Analysis

One male subject participated in the study. The subject was asked to jump from a single (left) leg after an approach run and to touch a target located at a height of 3 meters. Forty seven reflective markers were placed on the subject, and 3-dimenstional coordinate data was captured with a VICON 612 system (Oxford Metrics, 250 Hz, nine cameras) during the support phase of the jump. Ground reaction forces (GRF) were simultaneously measured with a force platform (Kistler, 1000 Hz). The subject performed 8 successful trials in which he planted his takeoff foot completely on the force platform and touched the target.

The coordinate data of the markers were filtered with a Butterworth digital filter. Then the filtered data were converted to 1000 Hz data via spline interpolation for synchronization with the GRF data. The coordinate data of the CP were approximated by fifth-order polynomial functions in order to obtain stable values when the second order differentiation of the CP data was calculated.

All joints were treated as ball joints, and torque and power of all joints were calculated by an inverse dynamics approach based on kinematic and kinetic data. The mechanical work of each joint was calculated by integration of the torque power. The contributions C_H, C_G and $C_{T,j}$ were calculated according to eqs. (13) and (14). The joint torques, mechanical works and contributions were standardized and averaged for 8 trials.

3 Results

The jump motion was analyzed from the time when the foot of the support leg contacted the ground (L-on) to the time when this foot left the ground (L-off), and this period was divided into two phases according to the time when the knee joint flexion angle of the support leg reached its largest value (MKF) as shown in Figure 1.

The mechanical work of the support leg in the direction of flexion-extension was positive at the hip joint, and negative at the knee and ankle joints throughout the 1st phase (Figure 2a). Throughout the 2nd phase the work showed an inverse pattern with respect to the 1st phase (Figure 2b). The work in all other directions (adduction-abduction and external-internal rotation) remained small. The largest values of joint mechanical work in the support phase occurred at the knee joint.

Figure 3a and b show the time-varying contributions of the support leg's joint flexion-extension torques to the horizontal and vertical velocities of body CG. The dotted line

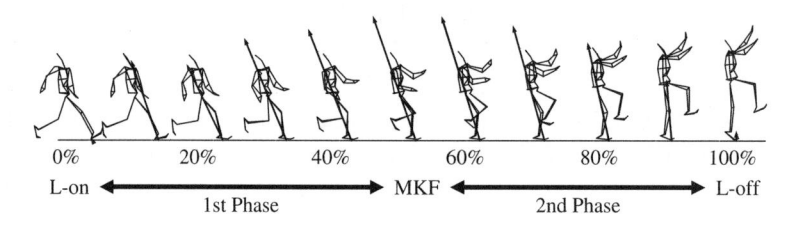

Figure 1. Stick pictures during the support phase in jump motion.

designates the total velocity of body CG. The solid lines with circles, triangles and squares designate the contributions of hip, knee and ankle joint torques, respectively, to body CG velocity.

In the first phase, the largest contribution of all joint torques to the vertical velocity of body CG was made by the ankle joint dorsi-plantar flexion torque. The hip joint torque contributed only a small amount to the vertical velocity of body CG. The contributions of joint torques about other axes (adduction-abduction and external-internal rotation) were also small.

In the second phase, the largest contribution of all joint torques to the horizontal velocity of body CG was made by the knee joint flexion-extension torque. The hip and ankle joint torques contributed only a small amount, and the contributions of joint torques about other axes remained small.

4 Discussion

The method developed in this study was designed to investigate the contributions of joint torques to the velocity of the whole-body CG in a jump motion. The main advantages of this method are: (a) The total torque term C_{TT}, the motion-dependent torque term C_H and the gravitational force term C_G of the whole body during the support phase are taken into

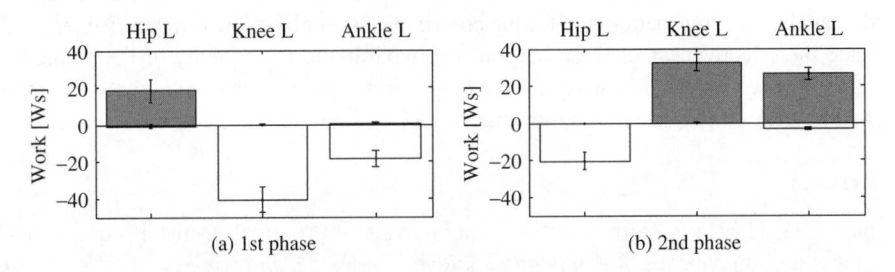

Figure 2. The mechanical work of each joint of the support leg (Flex-Ext axis).

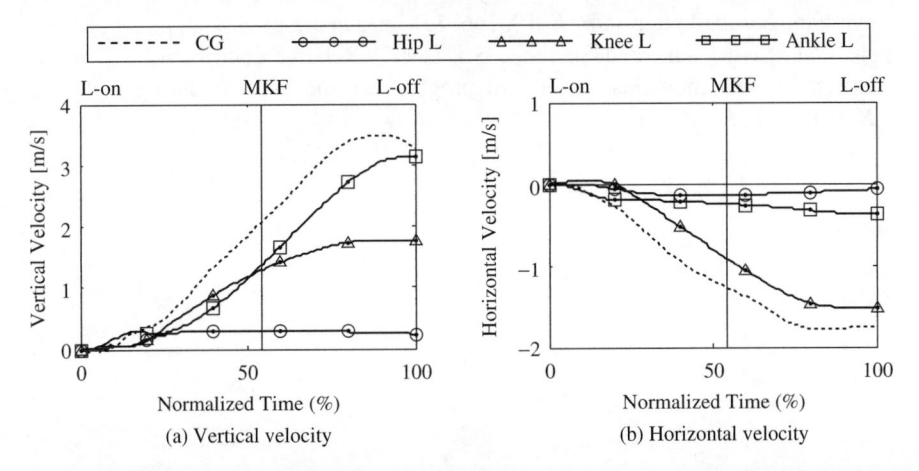

Figure 3. The contributions of the support leg's joint torques to the velocity of body CG (Flex-Ext axis).

account when calculating the contributions to the velocity of body CG; (b) the total torque term can be written as the sum of individual joint torque terms $C_{T,j}$; and (c) the effects of kinetic terms such as joint torques, motion-dependent torques and gravitational forces on accelerating body CG in any direction are quantified.

The proposed method reaches beyond the scope of the inverse kinetic method by quantifying the contributions and elucidating the effectiveness of joint torques on maximizing the velocity of the whole-body CG. The ankle joint dorsi-plantar flexion torque acted to accelerate the body CG in the vertical direction during the support phase of the jump motion. The knee joint flexion-extension torque then acted to accelerate the body CG in the vertical direction and to decelerate the body CG in the horizontal direction in the 2nd phase. This suggests that the knee joint played an important role in converting the body CG's horizontal velocity to vertical velocity.

The motion-dependent torques in the jump motion had only a small influence on the vertical velocity of body CG during the support phase, whereas substantial effects of the motion-dependent interactions between segments during a kicking motion have been reported by Putnam (1991). This can be explained by the fact that the velocity of the support leg in a jump motion is small compared to that of the swing leg in a kicking motion.

In summary, we have successfully developed and applied a method of quantifying the contributions of joint torques and other terms to the velocity of body CG. This method can be used to analyze various patterns of torque contributions employed by different subject and to evaluate the effectiveness of these patterns on maximizing the velocity of the whole-body center-of-gravity. While this study focused on a jump motion, the method can be readily applied to analysis of other sports movements such as running, throwing and striking.

References

Putnam C.A. (1991) A segment interaction analysis of proximal-to-distal sequential segment motion patterns. *Medicine and Sciences in Sports and Exercise*, 23/1, 130–144.

Naito K. and Maruyama T. (2006) A 3D dynamical model for analyzing the motion-dependent torques of upper extremity to generate throwing arm velocity during an overhand baseball pitch. *JJBSE*, 10, 146–158 (in Japanese).

Kepple T.M., Siegel K.L. and Stanhope S.J. (1997) Relative contributions of the lower extremity joint moments to forward progression and support during gait. *Gait & Posture* 6, 1–8.

THE 3-D KINEMATICS OF THE BARBELL DURING THE SNATCH MOVEMENT FOR ELITE TAIWAN WEIGHTLIFTERS

H.T. CHIU, K.B. CHENG & C.H. WANG

Institute of Physical Education, Health and Leisure Studies,
Cheng-Kung University, Tainan, Taiwan

The subjects include four elite male Taiwan weightlifters who participated in the 2006 Asian Games qualifier for the Taiwan National Team. The four subjects were the best two weightlifters in the 62 kg and 69 kg classes, respectively. The lifts chosen for analysis were the heaviest successful attempts for each subject: a 123 kg lift for subject 1 (S1, 62 kg class), a 125 kg lift for subject 2 (S2, 62 kg class), a 131 kg lift for subject 3 (S3, 69 kg class) and a 135 kg lift for subject 4 (S4, 69 kg class). Two high-speed cameras (Mega-speed MS1000, sampling rate: 120 Hz) were set on the left side of the lifters to film the trajectory of the left barbell. The angle between the optical axes of the two cameras was about 90°. The vertical barbell displacement was normalized at the height of the lowest position of the barbell during each catch phase to compare each lift. A small displacement of the barbell in the media-lateral horizontal direction showed that the barbell movement is almost in the sagittal plane for elite lifters. The horizontal travel range, maximum vertical displacement, drop displacement, and maximum vertical velocity of the barbell were 5.9 ± 2.6 cm, 101.0 ± 4.4 cm, 19.9 ± 2.3 cm and 196.7 ± 13.1 cm/s, respectively. The pathway of the barbell for S4 was significantly different from other lifters. The vertical displacement-velocity diagram showed that the four elite lifters had the same pattern for the first 60% in height from the lowest position of the barbell during each catch phase.

1 Introduction

Olympic weightlifting is a sport event that requires high-technique and stability. In the snatch lift, the barbell is lifted in one continuous motion from the platform to fully extended arms length overhead. The lifter catches the barbell overhead in a deep squat position, and then stands with the barbell in control until a "down" signal is received from the officials. In real competition, a failed attempt occurs when the barbell falls in front of or behind the weightlifter during catch phase because of improper pull force. The general motion characteristics of the snatch technique for elite weightlifters have been mentioned in previous studies (Baumann *et al.*, 1988; Gourgoulis, *et al.*, 2000). However, few studies involved three dimensional kinematics of the barbell. The purpose of this study is to describe the three dimensional kinematics of the barbell during the snatch and to try to find a standard pattern in the barbell trajectory.

2 Methods

2.1. Subjects

The subjects include four elite male Taiwan weight lifters who participated in the 2006 Asian Games qualifier for the Taiwan National Team (Table 1). The four subjects were the

Table 1. Characteristics of the weightlifters and the barbell.

Subjects	Weight category (kg)	Body mass (kg)	Height (cm)	Barbell mass (kg)
S1	62	61.72	158	123
S2	62	59.87	156	125
S3	69	68.50	162	131
S4	69	68.65	162	135

best two weightlifters in the 62 kg and 69 kg classes, respectively. The lifts chosen for analysis were the heaviest successful attempts for each subject. It is noteworthy to mention that Subject 2 has won gold medals in international weightlifting competitions.

2.2. Instrumentation and Procedure

Two high-speed cameras (Mega-speed MS1000, sampling rate: 120 Hz) were set on the left side of the lifter to film the trajectory of the left barbell. The angle between the optical axes of the two cameras was about 90°. Both video cameras used a metal framework, which formed a rectangular cube: 148 cm long, 100 cm wide, and 175 cm high, for calibration of the movement space. The three dimensional spatial coordinates of selected points was calculated using a direct linear transformation procedure using 24 control points by Kwon 3D motion analysis software. The reconstruction error was about 0.5 cm. The raw data was smoothed using 4th-order butterworth low-pass filter with a cut frequency set at 6 Hz. When the lifter was ready to start to lift the barbell off the platform, an experienced experimenter pushed a button to trigger the two synchronized cameras to film the snatch motion.

2.3. The Phases of the Snatch

The analysis focused on the snatch movement from the beginning of the barbell lift-off to the instant at which the lifter caught the barbell overhead. In previous studies, the snatch was divided into five phases: first pull, transition, second pull, turnover under the barbell, and catch phase. These phases were primarily determined by the change in the lifter's knee angle. However, since the athlete lifts the barbell as close to his body as possible to decrease the resistant loading, five new events were chosen in this study. To easily describe the barbell trajectory, the five events are defined as: the barbell lifting off the floor (LO), the barbell clearing the knee of the lifter (CK), the lifter extending his hip joints to push the bar away from his body (PB), the barbell reaching its maximum vertical height (MH), and the lifter catching the bar overhead (CB). For comparing between each lifter, the vertical position was normalized at the height of the lowest position of the barbell during each catch phase.

3 Results

3.1. Barbell Trajectory

The barbell trajectory for the heaviest successful snatch lift of each weightlifter is shown in Figure 1. A vertical reference line was drawn through the center of the barbell just prior to lift-off. The subject is standing to the right of the vertical reference line in Figure 1.

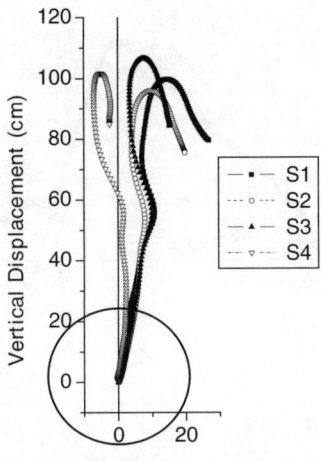

Figure 1. The barbell trajectories of the snatches for the four weightlifters.

Table 2. The kinematics of the barbell in the horizontal and vertical directions.

Subjects	Horizontal travel range (cm)	Maximum vertical displacement (cm)	Drop displacement (cm)	Maximum vertical velocity (cm/s)
S1	3.8	99.8	20.1	198.2
S2	3.9	96.0	20.5	182.5
S3	6.4	106.6	22.2	213.8
S4	9.3	101.7	16.8	192.3
Mean ± S.D.	5.9 ± 2.6	101.0 ± 4.4	19.9 ± 2.3	196.7 ± 13.1

The barbell trajectories shown in Figure 1 are all similar in one characteristic: the barbell moved toward the lifter, followed by movement away from the lifter, and finally toward the lifter again as the bar descended with the lifter moving under the bar into the catch position. The paths of the barbells for S1, S2 and S3 were similar until the barbell achieved the height of about 50 cm (prior to PB event) and all still stayed to the right of the vertical reference line projected upward from the start position of the barbell. The pathway of the barbell for S4 crossed to the left of the vertical reference line. Comparing the barbell trajectory among the athletes, the maximum height of the barbell and the catching position for S2 was lower than that of the other lifters.

3.2. Barbell Displacement and Velocity

A very small displacement of the barbell in the media-lateral horizontal direction (y-direction) showed that the barbell movement is almost on the sagittal plane for elite lifters. The horizontal travel range, maximum vertical displacement, drop displacement, and maximum vertical velocity of the barbell were 5.9 ± 2.6 cm, 101.0 ± 4.4 cm, 19.9 ± 2.3 cm and 196.7 ± 13.1 cm/s,

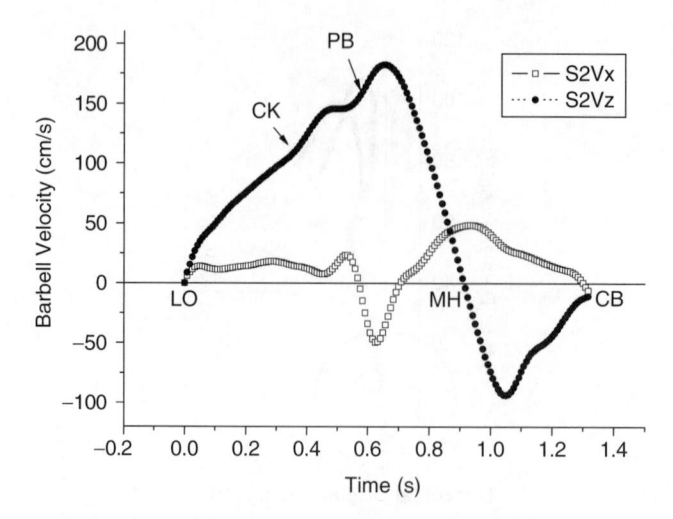

Figure 2. The vertical and horizontal velocity of barbell for subject 2 (The events were described in the text).

respectively. The vertical velocity of the barbell for S2 increased steadily to a maximum value of 182.5 cm/s (Figure 2). The others had two clear peaks of barbell velocity because of the decreasing velocity prior to the PB event.

The movement pattern of the barbell could be expressed with a phase diagram (showing the displacement-velocity relationship). Figure 3 shows the movement patterns of the barbell in the horizontal and vertical directions for the four weightlifters. To compare between the lifters, the vertical displacement was normalized at the lowest position of the barbell during the catch phase. The results of the movement patterns show that there is greater variation in the horizontal displacements of the barbell in the sagittal plane between lifters. Movement in the vertical direction shows that the four elite lifters had the same pattern for the first 60% in height from the lowest position of the barbell during the catch phase.

4 Discussion

The purpose of this study was to describe the movement of the barbell during the snatch using a three-dimensional kinematics approach. A very small displacement of the barbell in the media-lateral horizontal direction showed that two-dimensional kinematics is appropriate to describe the barbell movement of the weightlifter. Table 3 shows the comparison between the kinematics of the barbell in previous studies to the present study. Except for larger maximum vertical velocities, the results of this study are similar to those of the women weightlifters in the study of Hoover et al. (2006).

The curve of the vertical linear velocity of the barbell is important for assessing lifting technique (Baumann et al., 1988; Isaka et al., 1996). In this study, S2 pulled the bar with a more steadily increased velocity than the other lifters. It seems that skillful lifters could pull the barbell more smoothly without a marked deceleration of the barbell between CK and PB events. The movement pattern of the barbell for the four lifters' successful attempts indicated that elite weightlifters seem to have similar performances in the vertical direction prior to pushing the bar away from their bodies.

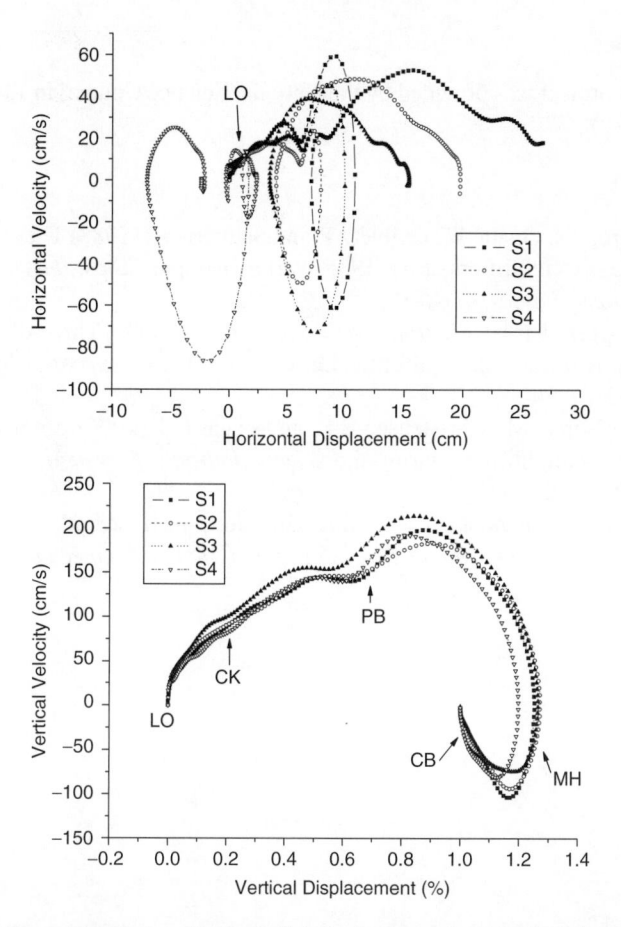

Figure 3. The movement patterns of the barbell in the horizontal and vertical directions for the four weightlifters (Note that the vertical displacement was normalized at the lowest position of the barbell during the catch phase. The events were described in the text).

Table 3. Comparison between the kinematics of the barbell in previous studies to the present study.

Investigators (approach)	Horizontal travel range (cm)	Maximum vertical displacement (cm)	Drop displacement (cm)	Maximum vertical velocity (cm/s)
Isaka et al. (1996) (2-D)	8 ~ 15	152 to 163	10.1 ~ 24.3	186*
Gourgoulis et al. (2000) (3-D)	3.17 ± 4.18	121 ± 8	13.5 ± 2.8	167 ± 10
Hoover et al. (2006) (2-D)	N/A	101.5 ± 7.4	20.7 ± 5.92	164.8 ± 19.1
Present study (3-D)	5.9 ± 2.6	101.0 ± 4.4	19.9 ± 2.3	196.7 ± 13.1

* *Note*: This was the mean maximum vertical velocity for six weightlifters. The highest value was 195 cm/s.

Acknowledgement

Funding for this project was provided by the National Science Council in Taiwan (NSC-95-2413-H-006-014-).

References

Baumann W., Gross V., Quade K., Galbierz P. and Schwirtz A. (1988) The snatch technique of world class weightlifters at the 1985 World Championships. *International Journal of Sports Biomechanics*, 4, 68–89.

Gourgoulis V., Aggelousis N., Mavromatis G. and Garas A. (2000) Three-dimensional kinematic analysis of the snatch of elite Greek weightlifters. *Journal of Sports Sciences*, 18, 643–652.

Hoover D.L., Carlson K.M., Christensen B.K. and Zebas C.J. (2006) Biomechanical analysis of women weightlifters during the snatch. *Journal of Strength and Conditioning Research*, 20/3, 627–633.

Isaka T., Okada J. and Funato K. (1996) Kinematic analysis of the barbell during the snatch movement of elite Asian weightlifters. *Journal of Applied Biomechanics*, 12, 508–512.

16. Climbing and Mountaineering

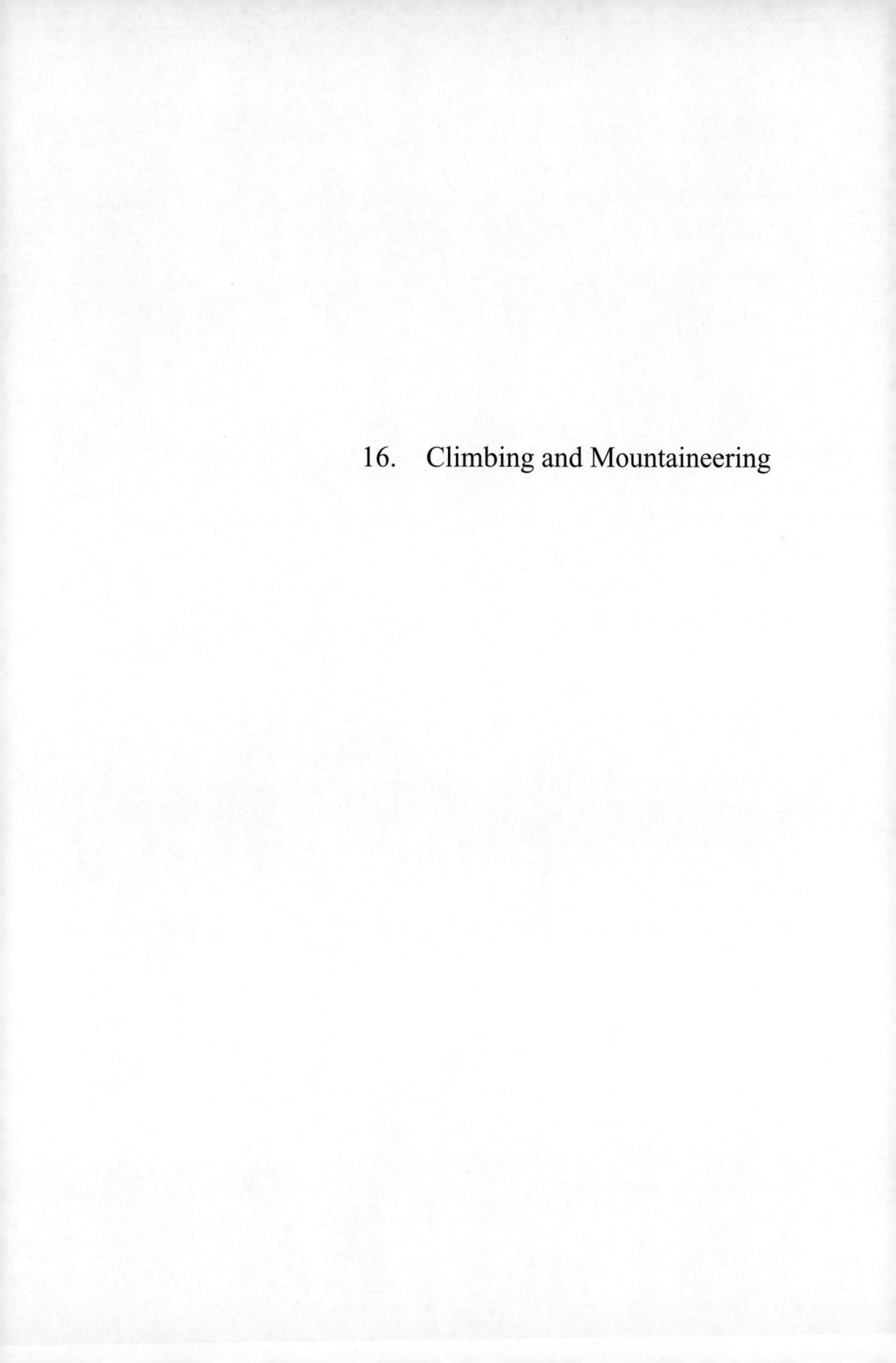

STRESS DISTRIBUTION AT THE FINGER PULLEYS DURING SPORT CLIMBING

M.A. TAN[1], F.K. FUSS[1] & G. NIEGL[2]

[1]*Sports Engineering Research Team, Division of Bioengineering, School of Chemical and Biomedical Engineering, Nanyang Technological University, Singapore*
[2]*Department of Anthropology, University of Vienna, Vienna, Austria*

The A2-pulley was modelled as a structure consisting of 19 individual elastic pulleys with elastic inter-pulley connections. The model was used to determine the force distribution along the axial pulley length with respect to different tendon angles and different phalangeal curvatures, in order to simulate different finger flexion angles in different climbing grip positions. It was shown that high finger flexion produces force spikes at the marginal fibres and should be avoided during climbing. This supports the fact that pulley injuries occur more during crimp grip than during open handgrip. Moreover, the more a phalanx is curved, the smaller is the load at the marginal pulley fibres.

1 Introduction

Pulleys are retinacular structures found on the palmar side of the finger phalanges. They maintain the flexor tendons of the hand in constant relationship to the joint axes of the fingers. This helps to preserve the efficiency of finger flexion. The A2 pulley is the longest and strongest amongst the finger pulleys (Lin et al., 1990). However, A2 pulley injuries are common with rock climbers (Klauser et al., 2002a) and the pulley injury risk is 0.75 fingers per individual per year for climbers at the beginner's stage (Niegl & Fuss, 2003).

In relation to the function of the digits, the curvature of the phalanx seems to play an important role. Arboreal primates are noticed to have phalanges which are more curved than humans. Hypothesis had been made regarding the advantages of the curved phalanx but no mechanical explanation had been given (Stern et al., 1995). As the attachment points of the pulleys are on the phalanx, the curvature is bound to affect the performance of the phalanx.

The aim of this project is to model the A2 finger pulley as a structure consisting of 19 individual elastic pulleys with elastic interpulley connections. The model will then be used to determine the force distribution along the axial pulley length with respect to different tendon angles, in order to simulate different finger flexion angles in different climbing grip positions. Furthermore, the influence of the phalangeal curvature will be investigated as well. This serves to determine, whether climbers with higher curvature have an advantage in terms of reduced stress.

2 Methodology

In order to model the A2 pulley mathematically, the followings symbols are defined:

    n   Number of pulley fibres used to model the A2 pulley
    a_1, b_1 x- and y-coordinates of the start point of the finger tendon

Set at an arbitrary point to simulate the angle of the tendon at the metacarpophalangeal joint (MCP)
a_2, b_2 x- and y-coordinates of the end point of the finger tendon
Set at an arbitrary point to simulate the angle of the tendon at the proximal interphalangeal joint (PIP)
c_i, d_i x- and y-coordinates of the attachment point of the finger pulley ith pulley fibre on the phalanx counting from the proximal end of the proximal phalanx
x_i, y_i x- and y-coordinates of the contact point between the ith pulley and the tendon
Ft   Applied force of finger tendon
Fp_i Force experienced by the ith pulley fibre
C   Stiffness of A2 pulley
k   Stiffness of individual fibre
l   Length of pulley fibre from the sagittal view
q   Interpulley stiffness. Defined as the elasticity between pulleys; (q_c) for compression and (q_t) for tension
s   Interpulley distance

The A2 and A3 pulleys are the site where the highest percentage of pulley failures occurs during climbing (Niegl & Fuss, 2003). Thus, A2 is the choice of pulley for our modelling. The A2 pulley consists of arcuate fibres and arches palmarly over the flexor digitorum profundus and the flexor digitorum superficialis tendons and attaches itself to the proximal and lateral regions of the proximal phalanx.

The data needed to simulate the A2 pulley model is as follows:

- Curvature of phalanx: included angle 18°–32° (in humans) (Susman, 1988, 2004)
- Axial width of attachment area: 15.9 mm to 20.5 mm (Doyle, 1990)
- Pulley length: 14 mm (measurement done from cadaver)
- Pulley stiffness: 117.81 N/mm to 161.87 N/mm (Lin et al., 1990).
- Tendon force: 85 N to 256 N (Marco et al., 1998)

In order to obtain a functional A2 finger pulley system, the pulley fibres were defined according to the following non-linear static conditions.

- The pulley is modelled as i equal fibres attached to the phalanx.
- The deflection point of the pulley at the free end of the ith pulley fibre is defined by the point x_i, y_i. This is where the finger tendon exerts a force on the finger pulleys.
- The attachment point of the fibre at the bone (phalanx) is defined as c_i, d_i. The tendon passes through all the pulley fibres and is linked to the adjacent deflection points of the subsequent and preceding pulley fibre by x_{i+1}, y_{i+1} and x_{i-1}, y_{i-1}.
- The pulley fibre is represented by an elastic rope of length l_i and spring constant k.
- The fibre reaction force Fp_i and the tendon force Ft on each side of the fibre are in equilibrium in x-and y-directions, thus forming two force equilibrium equations.
- The endpoint x_i, y_i of the fibre is situated at the circumference of a circle of centre c_i, d_i and radius $l_i + Fp_i/k$. This will enable us to develop a kinematic equation.
- The 3 equations form the basis of the A2 finger pulley for n number of fibres.
- There are n fibres, leading to $3n$ equations and thus we obtain $2n$ solutions. The boundary condition is that the pulley can only be stretched and Fp must be positive and only positive coordinates are attainable, resulting in only one set of solution.

Stress Distribution at the Finger Pulleys During Sport Climbing

In addition to the basic considerations above, the effects of interpulley stiffness cannot be ignored. This stiffness results in 2 forces from either side of the pulley. They can be either compressive (q_c) or tensile (q_t). The general equations to represent all the above considerations are as follows:

$$\left(\frac{(x_{i-1} - x_i)}{\sqrt{(x_{i-1} - x_i)^2 + (y_{i-1} - y_i)^2}}\right) \cdot Ft + \left(\frac{(x_{i+1} - x_i)}{\sqrt{(x_{i+1} - x_i)^2 + (y_{i+1} - y_i)^2}}\right)$$
$$Ft + \frac{(c_i - x_i)Fp_i}{l_i + \frac{Fp_i}{k}} +$$
$$\left(\frac{(x_{i-1} - x_i)}{\sqrt{(x_{i-1} - x_i)^2 + (y_{i-1} - y_i)^2}}\right) \cdot \left(\sqrt{(x_{i-1} - x_i)^2 + (y_{i-1} - y_i)^2} - s_i\right) \cdot \quad (1)$$
$$\left[\frac{q_r}{2}\left(\text{sgn}\sqrt{(x_{i-1} - x_i)^2 + (y_{i-1} - y_i)^2} - s_i + 1\right) - \frac{q_c}{2}\right] +$$
$$\left(\text{sgn}\left(\sqrt{(x_{i-1} - x_i)^2 + (y_{i-1} - y_i)^2} - s_i\right) - 1\right)\right] +$$
$$\left(\frac{(x_{i+1} - x_i)}{\sqrt{(x_{i+1} - x_i)^2 + (y_{i+1} - y_i)^2}}\right) \cdot \left(\sqrt{(x_{i+1} - x_i)^2 + (y_{i+1} - y_i)^2} - s_i\right) \cdot$$
$$\left[\frac{q_r}{2}\left(\text{sgn}\left(\sqrt{(x_{i+1} - x_i)^2 + (y_{i+1} - y_i)^2} - s_i\right) + 1\right) - \frac{q_c}{2}\right.$$
$$\left.\left(\text{sgn}\left(\sqrt{(x_{i+1} - x_i)^2 + (y_{i+1} - y_i)^2} - s_i\right) - 1\right)\right] = 0$$

$$\left(\frac{(y_{i-1} - y_i)}{\sqrt{(x_{i-1} - x_i)^2 + (y_{i-1} - y_i)^2}}\right) \cdot Ft + \left(\frac{(y_{i+1} - y_i)}{\sqrt{(x_{i+1} - x_i)^2 + (y_{i+1} - y_i)^2}}\right)$$
$$Ft + \frac{(d_i - y_i)Fp_i}{l_i + \frac{Fp_i}{k}} +$$
$$\left(\frac{(y_{i-1} - y_i)}{\sqrt{(x_{i-1} - x_i)^2 + (y_{i-1} - y_i)^2}}\right) \cdot \left(\sqrt{(x_{i-1} - x_i)^2 + (y_{i-1} - y_i)^2} - s_i\right) \cdot \quad (2)$$
$$\left[\frac{q_r}{2}\left(\text{sgn}\left(\sqrt{(x_{i-1} - x_i)^2 + (y_{i-1} - y_i)^2} - s_i\right) + 1\right)\right.$$
$$\left. - \frac{q_c}{2}\left(\text{sgn}\left(\sqrt{(x_{i-1} - x_i)^2 + (y_{i-1} - y_i)^2} - s_i\right) - 1\right)\right] +$$
$$\left(\frac{(y_{i+1} - y_i)}{\sqrt{(x_{i+1} - x_i)^2 + (y_{i+1} - y_i)^2}}\right) \cdot \left(\sqrt{(x_{i+1} - x_i)^2 + (y_{i+1} - y_i)^2} - s_i\right) \cdot$$

$$\begin{pmatrix} \dfrac{q_r}{2}\left(\text{sgn}\left(\sqrt{(x_{i+1}-x_i)^2+(y_{i+1}-y_i)^2}-s_i\right)+1\right) \\ -\dfrac{q_c}{2}\left(\text{sgn}\left(\sqrt{(x_{i+1}-x_i)^2+(y_{i+1}-y_i)^2}-s_i\right)-1\right) \end{pmatrix} = 0$$

$$(c_i - x_i)^2 + (d_i - y_i)^2 = \left(l_i + \frac{Fp_i}{k}\right)^2 \tag{3}$$

With the general equations developed and the above data collected, the A2 finger pulley can now be modelled. After applying a convergence test, 19 pulley fibres were modelled, as a compromise of sufficient accuracy (error < 8%) and acceptable processing time. In order to investigate its effects, the tendon angle and the curvature of the phalanx were varied and the corresponding pulley forces calculated and recorded.

3 Results

3.1. *Influence of Tendon Angle*

From Fig. 1 it can be seen that the curves changes from "n" shape to "u" shaped curves as the tendon angle increases from 21° to 36°. This is a gradual change and shows the modified load distribution at different amounts of flexion of the fingers. The force is initially concentrated in the central fibre for low tendon angles. This force concentration slowly moves towards the marginal fibres as the tendon angle increases. In addition, the maximum force of each curve

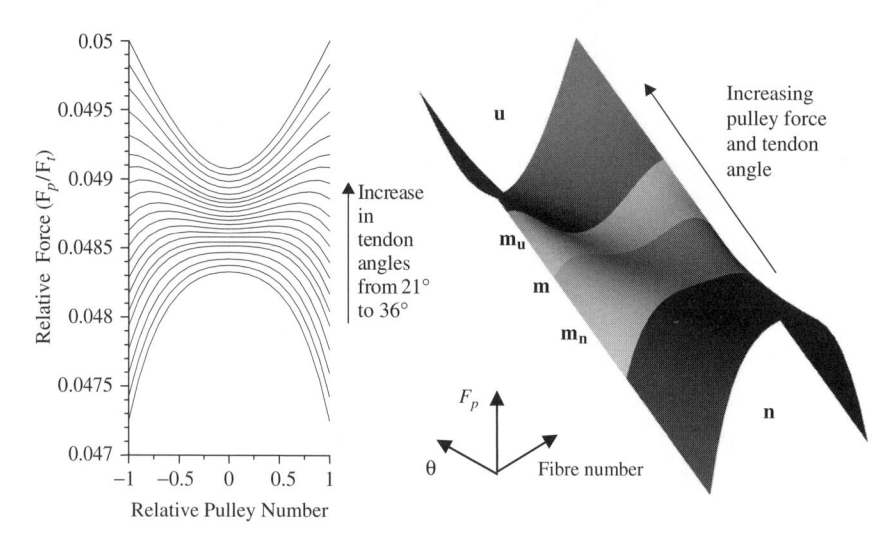

Figure 1. (a) Variation in the shapes of the force distribution curves along the phalanx as the tendon angle varies from 21° to 36°. (b) 3D plot of pulley force and tendon angle with respect to fibre number (θ = tendon angle, Fp = pulley force).

increases with an increase in tendon angle and the force distribution becomes less uniform. The gradients at the sides of the curve are steeper when the tendon angle is increased. This adds up to a combined effect of high loads at the marginal fibres at high tendon angles.

The smaller the tendon angle is (more extended fingers), the more the force distribution was "n"-shaped, meaning a lower force at the marginal fibres than at the central fibre. In higher tendon angles, the force was distributed like a "u", with peak forces at the marginal fibres. The transition from "n" to "u" produced an "m"-shaped force distribution, with 2 peaks between the marginal and central fibres. The most even force distribution proved to be in an "m" with equal forces at marginal and central fibres. The highest force of the central fibre in an "n"-distribution was still smaller than the lowest force of the central fibre in a "u"-distribution.

The marginal fibre is noticed to be undertaking a higher force in comparison to the central fibre except for tendon angles below 29°. However, at small tendon angles, the force concentration at the central fibres is low compared to forces at higher tendon angles. Thus the area of concern should be the marginal fibres where high force concentration will be experienced.

3.2. Influence of Phalangeal Curvature

Assuming a uniformly distributed force on the pulley system, the ratio between the marginal fibre and the central fibre is 1 ("m" in Figure 1b). This is to allow a basis for comparison between the different included angles (curvature of phalanx). Figure 2 shows the comparison for an included angle between 20° and 30°. In humans, the included angle is 25° ± 4° (Susman, 1988). As can be seen from Figure 2, the smaller the included angle, the higher the relative stress at the marginal fibres, compared to the central fibre. In other words, in higher included angles, the most even stress distribution ("m") occurs at higher tendon angles. However, the absolute stress increases with the included angle (Figure 3).

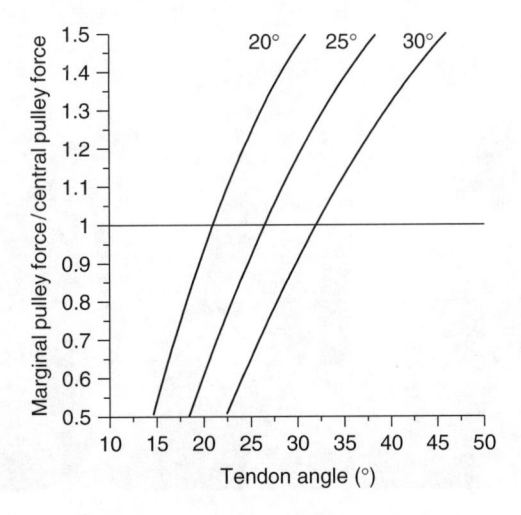

Figure 2. Force ratio of marginal and central pulley fibres vs. tendon angle.

Nevertheless, the absolute stress at marginal pulley fibres is still lower at phalangeal curvatures of an included angle of 30°. This becomes evident form Figures 2 and 3: the ratio of marginal pulley force and central pulley force is 1 at a tendon angle of 26–27° in a phalanx curvature at an included angle of 25°. The corresponding ratios of included angle of 20° and 30° are 1.32 and 0.75 respectively (Figure 2). The same ratios correspond to absolute marginal pulley forces of 1.19 and 0.82. Thus, the marginal pulley force is 1.45 times higher in the less curved phalanx (included angle of 20°) than in the one with a higher curvature (30° included angle).

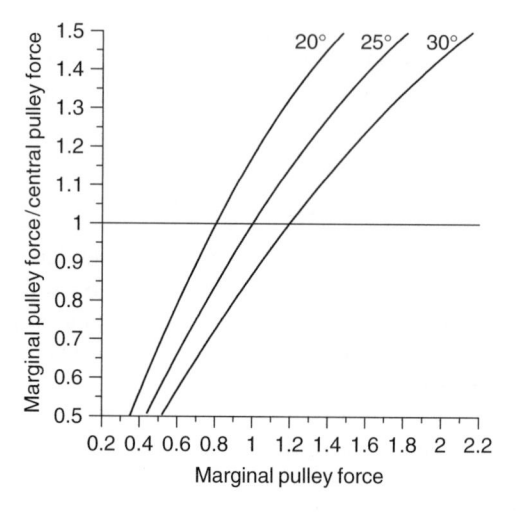

Figure 3. Force ratio of marginal and central pulley fibres vs. normalized marginal pulley force.

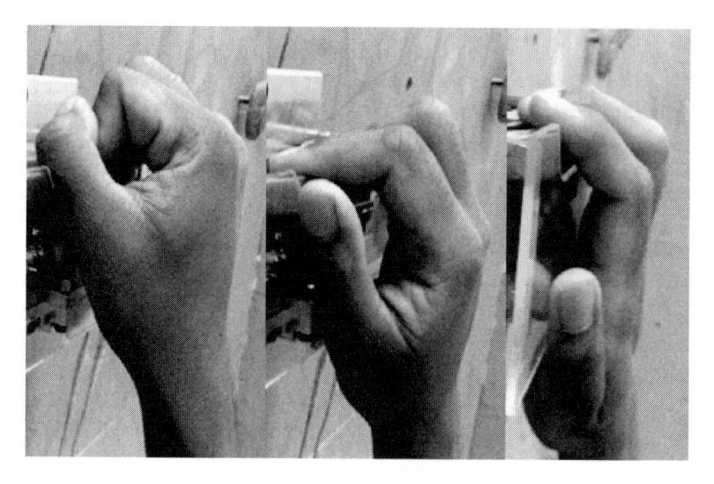

Figure 4. Closed crimp grip, open crimp, and open hand grip.

4 Discussion

The results of the study clearly show that the higher the tendon angle is, the larger is the overall pulley stress, and the more the marginal fibres are loaded. This explains the mechanism of the phenomenon, that pulleys usually fail at the marginal end, specifically at the distal one (Klauser *et al.*, 2002b), and at higher finger flexion (crimp grip). A higher phalanx curvature turned out to have specific advantage of smaller stress at the marginal fibres and more even stress distribution at higher tendon angles and finger flexions.

Based on the results of this study, climbers should refrain from adopting the crimp grip, and should train and apply the open hand grip (Figure 4). The knowledge of this fact is also important for trainers as they have to take care of teaching the beginners the problems and risks of disadvantageous grips.

References

Stern J.T., Jungers W.L. and Susman R.L. (1995) Quantifying phalangeal curvature: an empirical comparison of alternative methods. *Am. J. Phys. Anthropol.*, 97/1, 1–10.

Lin G.T., Cooney W.P., Amadio P.C. and An K.N. (1990) Mechanical properties of human pulleys. *J. Hand Surg. [Br]*, 15/4, 429–434.

Klauser A., Frauscher F., Bodner G., Halpern E.J., Schocke M.F., Springer P., Gabl M., Judmaier W. and Zur Nedden D. (2002a) Finger pulley injuries in extreme rock climbers: depiction with dynamic US. *Radiology*, 222/3, 755–761.

Niegl G. and Fuss F.K. (2003) Finger forces, -injuries, and -force distribution in sport climbers. In: Subic A., Trivailo P. and Alam F. (Eds.), *Sports Dynamics, Discovery and Application*. RMIT University Press, Melbourne, pp. 197–202.

Susman R.L. (2004) Oreopithecus bambolii: an unlikely case of hominid-like grip capability in a Miocene ape. *J. Hum. Evol.*, 46/1, 105–117.

Doyle J.R. (1990) Anatomy and function of the palmar aponeurosis pulley. *J. Hand Surg. [Am]*, 15/1, 78–82.

Marco R.A., Sharkey N.A., Smith T.S. and Zissimos A.G. (1998) Pathomechanics of closed rupture of the flexor tendon pulleys in rock climbers. *J. Bone Joint Surg. Am.*, 80/7, 1012–9.

Susman RL. (1988) Hand of Paranthropus robustus from Member 1, Swartkrans: fossil evidence for tool behavior. *Science*, 6/240(4853), 781–4.

Klauser A., Frauscher F., Hochholzer T., Helweg G., Kramer J. and Zur Nedden D. (2002b) Diagnostik von Überlastungsschäden bei Sportkletterern. *Radiologe*, 42/10, 788–98.

BIOMECHANICS OF FREE CLIMBING – A MATHEMATICAL MODEL FOR EVALUATION OF CLIMBING POSTURE

N. INOU, Y. OTAKI, K. OKUNUKI, M. KOSEKI & H. KIMURA

Tokyo Institute of Technology, Department of Mechanical and Control Engineering, Tokyo, Japan

Indoor free climbing is studied from a biomechanical engineering viewpoint. The two-dimensional human model is proposed for the first step of this study. The model is a chain of nine links that produces torques at the link joints to keep configuration of the model. The produced torques can be calculated as a static problem when the terminal force is given. We introduce an evaluation function related to muscular activity and optimize the function. The minimized value is correlated with order of sensory loads to keep postures in climbing.

1 Introduction

Free climbing is a sport that a person climbs up a rock face or an artificial wall without climbing tools except a life rope. From a biomechanical viewpoint, the free climbing is an intriguing sport that it requires skillful physical movement performing proper muscular control of body parts, which has different physical characteristics from other sports. However, there are few studies on this biomechanical viewpoint.

Our study aims to formulate a mathematical model and clarifies characteristics of free climbing. The biomechanical study is expected to contribute effective tutorial training for beginners or to suggest a new climbing strategy for experts.

2 Method

This paper focuses on examining biomechanical characteristics of climbing postures. Figure 1 shows the nine postures to examine the mechanical states with a two-dimensional model. We asked four persons to take these postures and inquired about sensory load in keeping these postures in order of easiness. This study aims to establish an evaluation method to estimate the sensory load with a mathematical model.

2.1. *Human Posture Model*

The human posture model is expressed by a series of links as shown in Fig. 2. The model consists of nine links corresponding to the following body parts: foot, lower thigh, thigh, torso including head weight, adjustable part, upper arm, lower arm, hand and finger. The adjustable part corresponds to link 5 of which length is variable because the shoulder point continually changes according to the arm direction. The both ends of the model corresponding to foot and hand freely rotate around the terminal points. Nine kinds of link models were made for each subject using the pictures as shown in Fig. 1.

To perform mechanical simulation of the model, we determine the part of mass m_i and the position of center l_{Gi} as properties of individual body segments with the following equations,

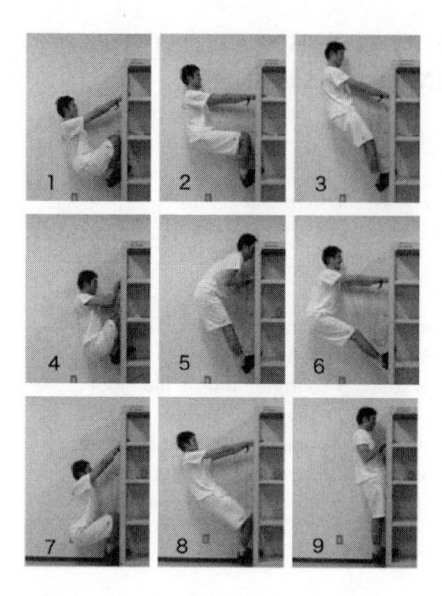

Figure 1. Nine postures for biomechanical analysis.

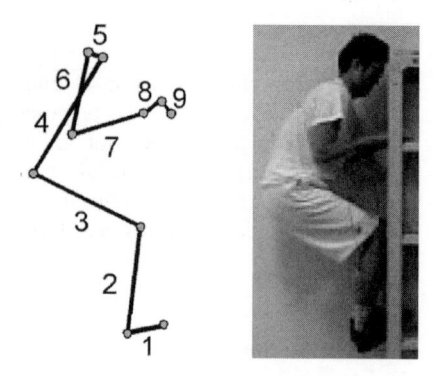

Figure 2. Link model of a human body.

Table 1   Data of human model of Subject A.

Link i	l_i [m]	l_{Gi} [m]	m_i [kg]	S_i
1	0.155	0.060	1.88	–
2	0.380	0.226	5.82	1.904
3	0.410	0.227	14.70	4.455
4	0.390	0.285	29.88	–
5	variable	–	0.00	–
6	0.218	0.085	3.17	1.807
7	0.223	0.093	1.79	1.000
8	0.120	0.080	0.84	0.250
9	0.040	–	0.00	0.250

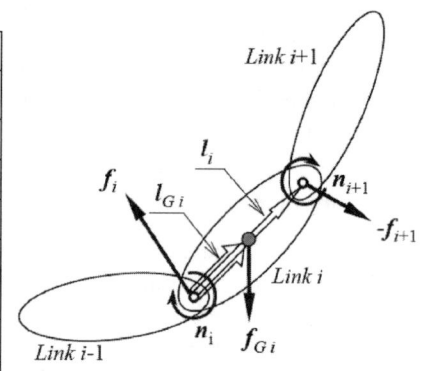

Figure 3. Mechanical balance of links.

$$m_i = A_i + B_i l_i + C_i M \qquad (1)$$
$$l_{Gi} \times D_i l_i \qquad (2)$$

where, l_i is individual body part length, M is body mass, and A_i, B_i, C_i and D_i are coefficients of standard Japanese athletes (Ae *et al.*, 1992). Table 1 shows data of the human body of subject A. S_i is mean sectional area of a body part (Link i) mentioned in the chapter 2.4. Other three subjects (subject B, C and D) are also calculated in the same way.

2.2. Calculation of Torques

Using the link model, we can obtain required torques at the joints to keep the each posture considering the static moment balance. Figure 3 shows the static balance of the model. Loading conditions at the neighboring links are expressed by the following equations,

$$f_i - f_{i+1} + f_{Gi} = 0 \tag{3}$$

$$n_i - n_i+1 - l_i \times f_i+1 + l_{Gi} \times f_{Gi} = 0 \tag{4}$$

where, f_i is force vector that the link $i - 1$ acts on the link i, f_{Gi} is gravitational force, n_i is torque vector at the joint i, l_i is vector from joint i to $i + 1$, and l_{Gi} is vector denoting center of mass of link i from joint i.

Moments at the both terminals should be also balanced. At the end of link N (in this study, $N = 9$), the moment balance should be kept as follows,

$$P_{N+1} \times f_{N+1} + \sum_i (P_{Gi} \times f_{Gi}) = 0 \tag{5}$$

where, P_{N+1} is position vector at the end of the joint N and P_{Gi} is position vector denoting center of mass of link i. As the proposed model forms a simple chain, we can calculate all torques of the links. That is, when a reaction force at the end of link 1 or 9 is given, all torques are determined. Furthermore, the equation 5 can be expressed with x and y component forces as follows,

$$(f_{N+1})_x = \frac{(P_{N+1})_x (f_{N+1})_y + \sum_i ((P_{Gi})_x (f_{Gi})_y - (P_{Gi})_y (f_{Gi})_x)}{(P_{N+1})_y} \tag{6}$$

The equation 6 means that $(f_N + 1)_x$ and $(f_N + 1)_y$ are not independent. Either force component is fixed, the other one is determined.

2.3. Measurement of iEMG

We measured integrated electromyography (iEMG) of several typical musculoskeletal muscles (e.g. biceps brachii, triceps bracii, vastus lateralis and biceps femoris). Figure 4 shows the iEMG on the four kinds of muscles for the nine postures as shown in Fig. 1. The iEMG diversely change with the postures because the iEMG correlate with muscular activities. If the musculoskeletal system performs effectively, it is expected that the iEMG correlate with torques at the corresponding joints.

2.4. Introduction of the Loading Sensory Function

To compare iEMG data with simulation results of the proposed model, we consider the relationship between a muscular force and the corresponding torque of the model. Crowninshield and Brand showed that endurance time of a muscle T could be expressed by the following equation (Crowninshield & Brand, 1981),

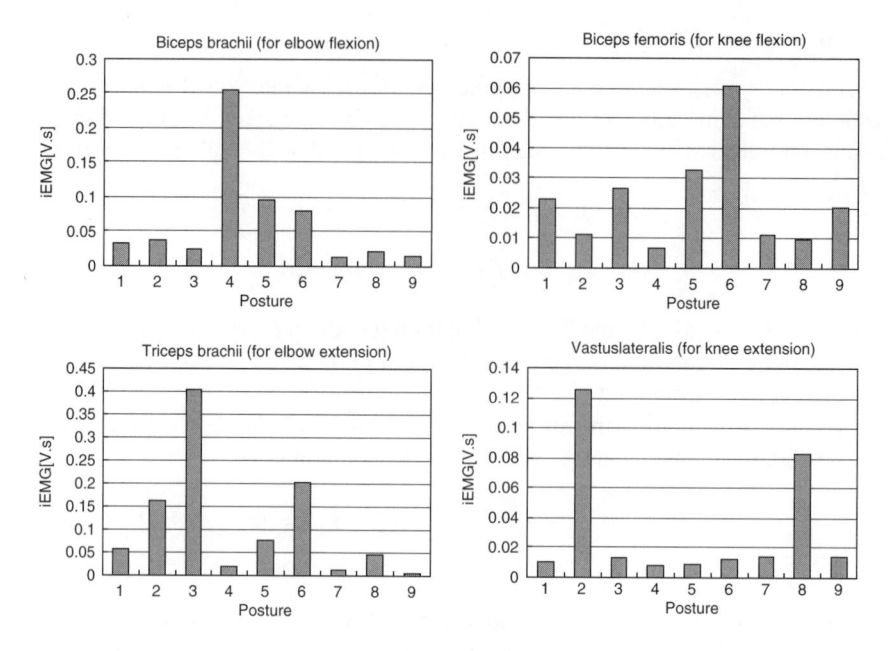

Figure 4. iEMG of four muscles.

$$T \propto (f/f_m)^{-a} \tag{7}$$

where, f is muscular force, f_m is maximum muscular force, and a is power of a number. It is reported that the power a takes between 2.54 and 3.14 (Dons et al., 1979).

Dul, Johnson and et al. examined the endurance time from a viewpoint of muscular load sharing. They introduced the minimum-fatigue criterion to maximize the endurance time (Dul et al., 1984). We propose the following evaluation function for estimating biomechanical efficiency of muscles considering that a large muscular force produces a large torque at the corresponding joint. The function is defined as

$$J = \sum_{i=0}^{N} \left(\frac{n_i}{S_i} \right)^a \tag{8}$$

where, n_i is torque produced at joint i and S_i is mean sectional area of the body part (i.e. link i) standardized by that of the lower arm (i.e. link 7) assuming that apparent density of all body parts is equal. We expect that Si is roughly proportional to the maximum torque at the joint i because S_i is proportional to the muscular area to produce the torque. For example, sectional area of an upper arm is used for producing torque at the elbow joint. The value S_i is calculated from l_i and m_i in this study. Only finger part is assumed to be 25% of hand sectional area because of the finger structure.

Here, n_i/S_i shows how much muscle i is working in keeping the posture. The total cost J is expected to be muscular efficiency as a biomechanical system. We assume the power of

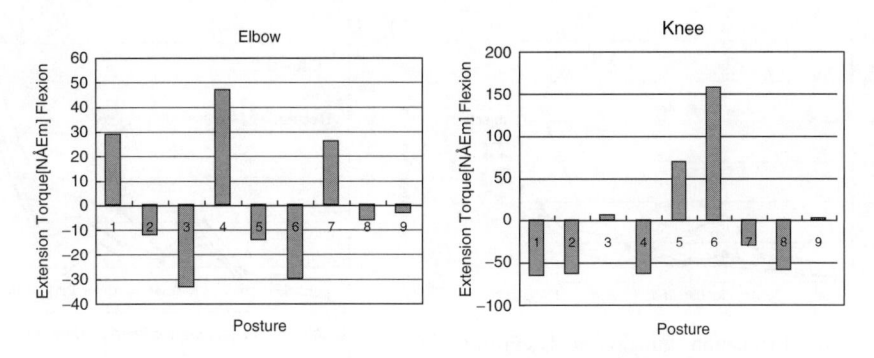

Figure 5. Calculated torques of elbow and knee joints.

n_i/S_i to be 3 (i.e. $a = 3.0$). As the result of effective muscular control, it is anticipated that our bodies minimize sensory load for the each posture. We searched torque allocation to minimize the evaluation function J while changing a reaction force at the end of the model.

3 Results and Discussion

First we compare torque allocation to minimize J with iEMG. Figure 5 shows the calculated torques at elbow and knee joints when subject A took the nine postures. Biceps bracii contributes to elbow flexion. The muscular activity produces positive torque at the elbow joint. On the other hand, triceps brachii contributes to elbow extension, and produces negative torque.

It is a strong correlation between the large iEMG and the calculated torques at the elbow joint. For example, posture 4 shows a large iEMG at biceps brachii as in Fig. 4 and a large flexion torque as in Fig. 5. It is also observed that when iEMG appeared at both muscles, the larger iEMG is dominant. We can see the difference of the iEMG is related to flexion or extension of the torque. The iEMG of biceps femoris and vastus lateralis are similarly related to the calculated torques for knee flexion and extension. The correlative relationship between iEMG and the torques to minimize the evaluation function J shows validity of the proposed method. We also confirmed that there is less correlation between iEMG and torques when the power of a number a is 1.0.

Figure 6 shows the minimized J plotted in order of sensory load of the each subject. The results show that the minimized J roughly correlates with the sensory load. But there are some disagreements with the individual. The possible reason is that we do not yet use individual muscular data because it is difficult to estimate muscular sections for the each individual. In this study, we examined affect of muscular areas in case of subject C who does not have a habit of physical exercise. We examined the minimized J changing muscular areas, and found out that lower thigh muscle (calf muscles) was sensitive to the evaluation function as shown in Fig. 7.

The evaluation function J of the posture 4 is comparatively small in this result. There are two possible reasons. The first reason is that the finger part is very sensitive to the minimized J. For more detailed discussions we need exact finger properties such as angle, length and muscular area. The second reason is affect of mechanical constraints. In case of

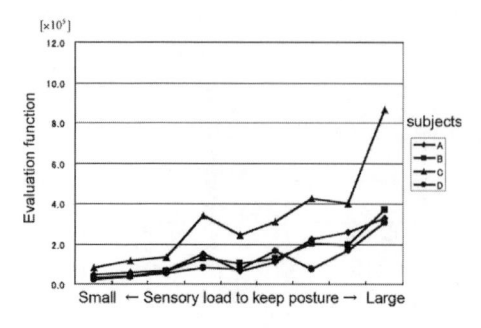

Figure 6. Evaluation function and sensory load.

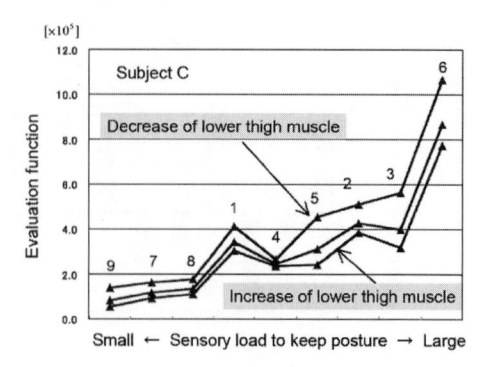

Figure 7. Effect of lower. (thigh numbers 1-9 denote the postures).

the posture 1 and 4, the actual foot (ankle) and knee joint do not need active torques to keep the angles because of limit of flexions. If we do not count the torques at foot and knee parts in the computation, the disagreement will reduce between posture 1 and 4 although the first reason is dominant to the evaluation function.

4 Conclusions

We proposed a mathematical model to estimate mechanical characteristics of free climbing. The simulation results to minimize the evaluation function well agree with actual muscular activities and sensory load to keep static postures in free climbing. For the next study, we will develop a three-dimensional link model for more feasible expression in climbing postures.

References

Ae M., Tang H. and Yokoi T. (1992) Estimation of inertia properties of the body segments in Japanese athletes. *Biomechanism*, 11, 23–33 (in Japanese).

Crowninshield R.D. and Brand R.A. (1981) A physiologically based criterion of muscle force prediction in locomotion. *Journal of Biomechanics*, 14/11, 793–801.

Dons, B., Bollerup, K., Bonde-Petersen, F. and Hancke, S. (1979) The effect of weight-lifting exercise related to muscle fiber composition and muscle cross-sectional area in humans. *European Journal of Applied Physiology*, 40/2, 95–106.

Dul J., Johnson G.E., Shiavi R. and Townsend M.A. (1984) Muscular synergism- II. A minimum-fatigue criterion for load sharing between synergistic muscles, *Journal of Biomechanics*, 17/9, 675–684.

THE FULLY INSTRUMENTED CLIMBING WALL: PERFORMANCE ANALYSIS, ROUTE GRADING AND VECTOR DIAGRAMS – A PRELIMINARY STUDY

F.K. FUSS[1] & G. NIEGL[2,3]

[1]*Sports Engineering Research Team, Division of Bioengineering, School of Chemical and Biomedical Engineering, Nanyang Technological University, Singapore*
[2]*Department of Anthropology, University of Vienna, Austria*
[3]*MedClimb, OEAV/ÖGV (Austrian Mountaineering Federation), Vienna, Austria*

An instrumented climbing wall was designed and constructed for this study, consisting of 8 holds, instrumented with 6DOF transducers. The holds are arranged such that a climber can complete a bouldering circle. In 5 climbers, the mechanical parameters as well as the fractal dimensions were determined from the force-time signals, and applied for performance analysis and route grading. Tiles difficult to hold exhibit a more chaotic force-time signal and thus have a higher normalized Hausdorff dimension. The latter correlated well with difficult holds as well as with the overall performance of the climbers investigated. Furthermore, the 3D force vector diagrams were visualized for each hold and combined into a single figure.

1 Introduction

Instrumented climbing holds have been developed, and applied, by Fuss *et al.* (2003ab, 2004) and Fuss & Niegl (2006ab) in various studies to assess the performance parameters of sport climbing. The most suitable parameters proved to be the mean and maximal reaction force, the mean and maximal substatic friction coefficient, the contact time, the impulse, and the dimensionless smoothness factor. The latter indicates how strong and irregular the force-time signal fluctuates about an ideal smooth parabolic curve of the same impulse by dividing the body weight by the mean of the absolute difference (in N) between the vertical force – time graph and the parabolic curve. The better the climber is, the smaller are the forces and impulse, the shorter is the contact time, and the larger are the friction coefficient and the smoothness factor. A single instrumented hold measures the climber's performance on this very hold only, as well as the difficulty to grip the hold. For assessing the climber's performance over a longer time, as well as the overall and changing difficulty of a route, a couple of instrumented holds in series are required (fully instrumented climbing wall). As the smoothness factor mentioned above, depends on the reference of a parabolic curve, which might not represent the ideal force application, especially, when the hold has to be held over a longer time, a new tool was sought to measure the irregular fluctuations, or chaotic behaviour, of a force-time signal. These considerations result into analysis of fractal dimensions, based on the self-similarity of the signal.

The aim of this study was to develop a fully instrumented climbing wall, design a bouldering route, and suggest a preliminary procedure for performance analysis and route grading.

2 Experimental

The fully instrumented climbing wall, developed for this study, as well as the bouldering route designed (movement sequence), is shown in Figure 1. In this wall (Figure 1), each of the 8 holds is instrumented with 6-DOF silicon strain-gauged transducers (multi-axis force/torque sensor; Theta SI 2500-400, ATI Industrial Automation, Apex, NC, USA). The data sampling frequency was set to 250 Hz. The holds were arranged such that the climbers could complete a bouldering circle. 5 climbers (4 males, 1 female) of different experience level participated in this study (red point 6a – 7a/French scale). All participants climbed the route twice for 1.5 boulder circles (1.5 circles allowed to exclude the static start position, as the test person had to climb through the starting holds again). For each hold, the following parameters were determined: mean force, maximal force and impulse (all normalised to body weight) as well as the contact time. The smoothness factor (see above) was replaced by the Hausdorff dimension, calculated according to the method described by Kulish *et al.* (2006), and applied by Kulish *et al.* (2006) and Tripathy *et al.* (2007) for EEG signal processing. As the Hausdorff dimension increases with the amplitude of the signal, it was normalised as well: 1) to the mean force of each single hold, and 2) to the average of the mean forces of

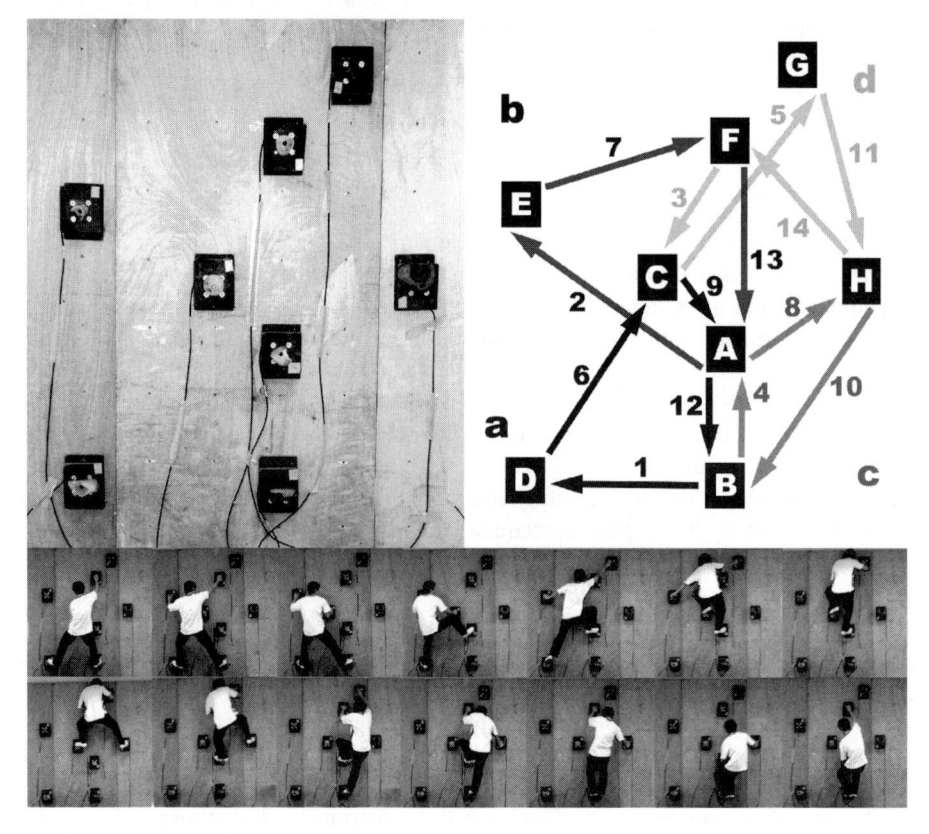

Figure 1. Wall and route design, and movement sequence (1–14); A–H: hold; a = left foot, b = left hand, c = right foot, d = right hand.

all holds. The former serves to assess the individual performance on, and difficulty of, each single hold, and the latter provides the mean performance and average difficulty. The force vector diagrams (3D force vector with time) were generated and visualised in AutoCAD 2000 (Autodesk, USA), whereby the force vectors applied by the holds to hands and feet, were displayed on the hold (comparable to a Pedotti diagram on a force plate). The vector diagrams of each hold and each limb (hands, feet) were drawn on a separate layer to avoid mixing up force vectors of 2 different limbs on the same hold at different times. The time was colour-coded in rainbow-colours from the beginning (red) to the end of a contact (magenta) between a limb and a hold.

3 Results

3.1. *Climbing Parameters*

The force, contact time, and impulse parameters did not show any significant difference between the 5 climbers. The normalized Hausdorff dimension, however, clearly shows that the 2 weaker climbers (red point 6a) have higher values. The higher the Hausdorff dimension, the more chaotic is the signal.

The left subfigure of Figure 2 displays the performance on, and difficulty of, each hold. Foot holds have small normalised Hausdorff dimensions, as they are usually loaded higher. Holds E1 (1st contact) and H2 (2nd contact) show the highest values, the former initiates the shoulder action to execute a dynamic move, the latter requires to counteract body rotation. Although the 2 weaker climbers (red point 6a) generally have higher values, one better climber (red point 6b) shows higher values in hold E1, and another one (red point 7a) in hold H2. The

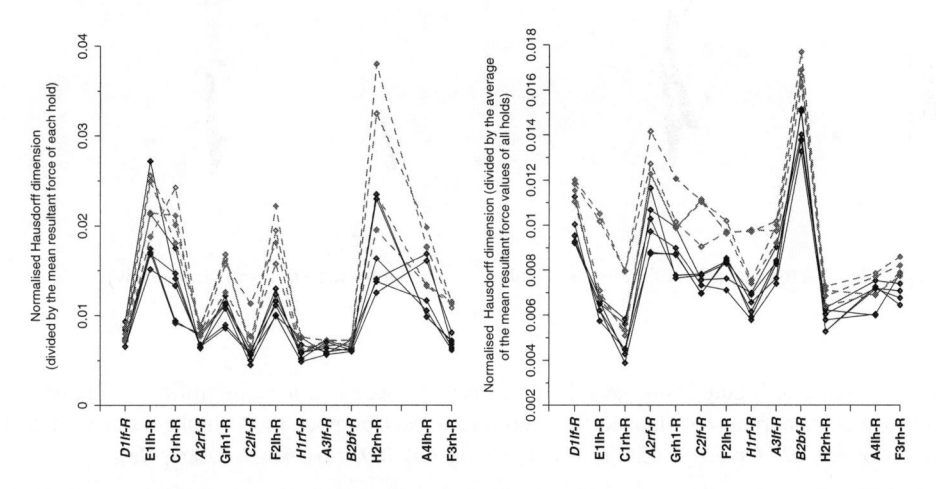

Figure 2. **a** – left side: Hausdorff dimension normalised to the mean resultant force of each single hold (performance on, and difficulty of, each hold), **b** – right side: Hausdorff dimension normalised to the average of the mean resultant forces of all holds (the mean performance and average difficulty); weaker climbers (red point 6a): dashed grey lines, better climbers (red point 6b-7a): solid black lines; code of climbing holds, e.g. D1lf-R: A–H: number of hold, 1–4: number of contact, r/l/b: right/left/both, h/f: hand/foot, R: resultant force.

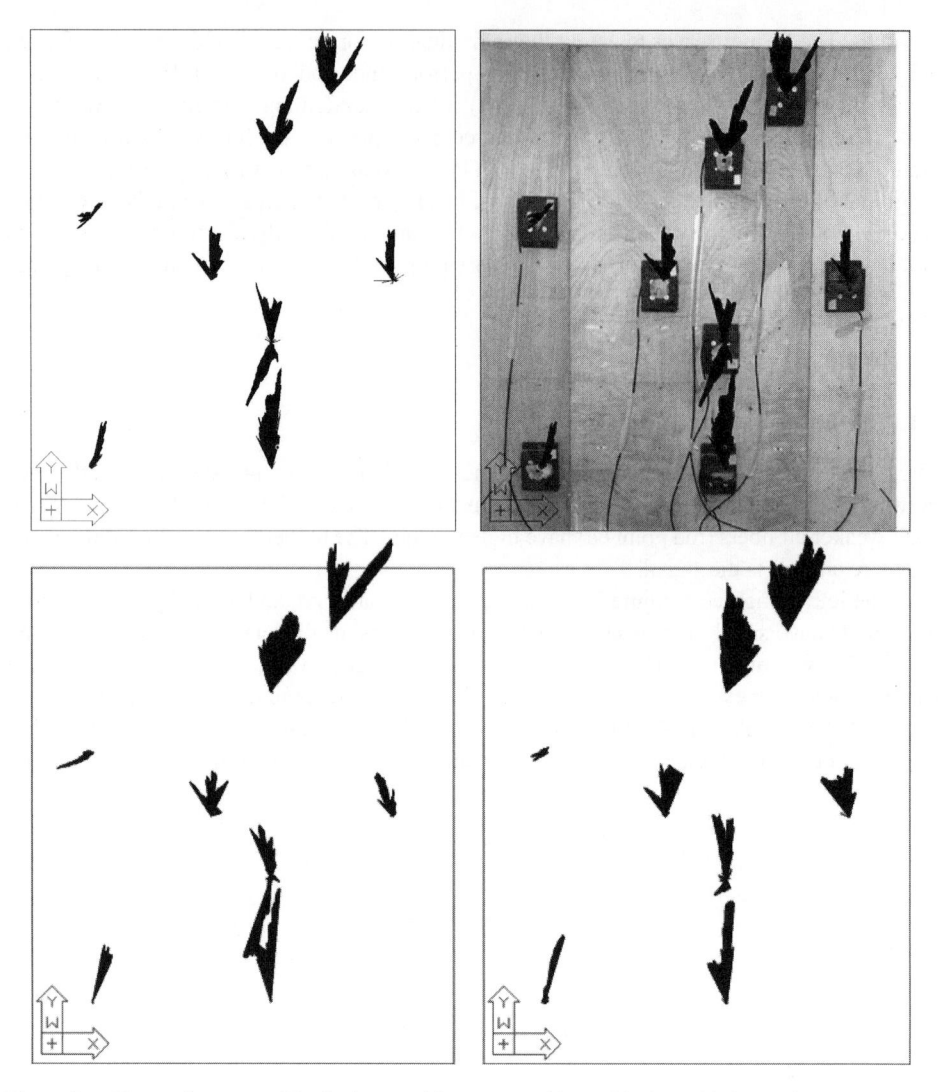

Figure 3. Vector diagrams of 3 climbers and in superposition with the wall (front view).

left subfigure of Figure 2 displays the mean performance and average difficulty. Taking the average of the normalized Hausdorff dimensions of each climber, the 5 climbers can be ranked as follows: 0.025 (rp 6b), 0.024 (rp 6b), 0.030 (rp 6a), 0.032 (rp 6a), 0.025 (rp 7a).

3.2. Vector Diagrams

The vector diagrams are shown in Figure 3. The vectors are drawn to the same scale, such that the magnitude of the forces at the different holds can be compared, in addition to the inclination of the forces.

4 Discussion

The results clearly show that the Hausdorff dimension is a suitable replacement for the smoothness factor previously applied for performance analysis (Fuss *et al.*, 2003ab, 2004; Fuss & Niegl, 2006a). It has to be tested extensively, however, as to general application for performance analysis and route grading. Nevertheless, this study presents preliminary results and suggests a specific procedure. The necessary tool for advanced performance analysis in sport climbing is the fully instrumented climbing wall. The wall developed for this project consists of eight holds, which is sufficient enough for a circular boulder and within budgetary limits. A further advantage of an instrumented wall is the visualization of forces with 3D vector diagrams, which become 4D vectors, if the time is colour-coded. These diagrams help to understand the dynamics of climbing by means of a single picture (Figure 3), whereas parameter – movement or parameter – hold diagrams (Figure 2) are difficult to read. In the latter diagrams, the Hausdorff dimension has to be normalised to the force, as the Hausdorff dimension increases with the amplitude. The normalisation to the force depends on whether a single hold or the entire route is of interest. In the former case, the Hausdorff dimension is divided by the mean force at the hold of interest (Figure 2a). As the force applied to foot holds is usually higher then the one at handholds, it is logical that footholds have a smaller normalized Hausdorff dimension. Apart from that, the performance of hands and upper limbs is more crucial in climbing. If the entire route has to be graded, or the performance throughout the route to be quantified, then the Hausdorff dimensions of the individual loads has to be divided by the average of all mean forces at the individual holds (Figure 2a). In this diagram, the Hausdorff dimension of the individual hold becomes unimportant with respect to the overall trend. Thus, a single value, namely the mean of the Hausdorff dimensions at all holds suffices to express the performance of a climber or the difficulty of a route.

These preliminary results have to be confirmed by two further experiments: 1) same route difficulty and climbers of different experience, preferably 2 groups, beginners and experts, and 2) climbers of the same experience (red point level) and the same route with two different degrees of difficulty. The latter can easily achieved by keeping the position of the holds and changing their shape, e.g. from jugs to small ledges or slopers.

References

Fuss F.K. and Niegl G. (2006a) Instrumented Climbing Holds and Dynamics of Sport Climbing. In: Moritz E.F. and Haake S. (Eds.) *The Engineering of Sport 6, Vol. 1 – Developments for Sports*. Springer, Munich, pp. 57–62.

Fuss F.K. and Niegl G. (2006b) Dynamics of Speed Climbing. In: Moritz E.F. and Haake S. (Eds.) *The Engineering of Sport 6, Vol. 1 – Developments for Sports*. Springer, Munich, pp. 51–56.

Fuss F.K., Niegl G., Boey L.W. and Liu X. (2003a) Mechanical parameters of handhold forces during sport climbing. In: Subic A., Trivailo P. and Alam F. (Eds.) *Sports Dynamics, Discovery and Application*. RMIT University Press, Melbourne, Australia, pp. 283–288.

Fuss F.K., Niegl G. and Yap Y.H. (2003b) Dynamics of sport climbing: influence of experience and training. In: Subic A., Trivailo P. and Alam F. (Eds.) *Sports Dynamics, Discovery and Application*. RMIT University Press, Melbourne, Australia, pp. 24–29.

Fuss F.K., Niegl G., Yap Y.H. and Tan M.A. (2004) Measurement of pinch grip forces during sport climbing. In: Hubbard M., Mehta R.D. and Pallis J.M. (Eds.) *The Engineering of Sport 5, Vol. 2*. International Sports Engineering Association, Sheffield, UK, pp. 262–268.

Kulish V., Sourin A. and Sourina O. (2006) Human electroencephalograms seen as fractal time series: Mathematical analysis and visualization. *Computers in Biology and Medicine*, 36, 291–302.

Tripathy J., Fuss F.K., Kulish V.V. and Yang S. (2007) The influence of hues on the cortical activity – a recipe for selecting sportswear colours. In: Fuss F.K., Subic A. and Ujihashi S. (Eds.) *The Impact of Technology on Sport II*. Taylor & Francis Group, London.

PHYSIOLOGICAL RESPONSE TO DIFFERENT PARTS OF A CLIMBING ROUTE

G. BALASEKARAN[1], F.K. FUSS[2] & G. NIEGL[3,4]

[1]*Nanyang Technological University, National Institute of Education, Physical Education and Sports Science, Singapore*
[2]*Sports Engineering Research Team, Division of Bioengineering, School of Chemical and Biomedical Engineering, Nanyang Technological University, Singapore*
[3]*Department of Anthropology, University of Vienna, Austria*
[4]*MedClimb, OEAV/ÖGV (Austrian Mountaineering Federation), Vienna, Austria*

This study investigates the fluctuations of oxygen uptake in synchrony with repetitively climbing a circular boulder on an instrumented climbing wall. Oxygen consumption (VO_2, liters · min^{-1}) was measured with a portable battery powered Cosmed K4b^2 metabolic systems via open circuit indirect calorimetry., and the movement sequence with time was identified from the force signals at the different holds. The results show that the fluctuations of VO_2(liters · min^{-1}) are synchronized with climbing and depend on the route conditions, which in turn require different climbing techniques. The cyclic behaviour of respiration provoked by a circular boulder route suggests that the oxygen uptake is influenced by the climbing route, even after the steady-state is reached.

1 Introduction

It is generally known, that cardio-respiratory parameters, oxygen uptake (VO_2, liters · min^{-1}) and heart rate (HR, beats · min^{-1}), are related to different climbing conditions. According to Mermier *et al.* (1997), HR, lactate, and VO_2 increase with the degree of difficulty of a climbing route (tested in different inclinations of the climbing wall). The relationship between HR and VO_2 is non-linear in climbing (Mermier *et al.*, 1997), compared to running and cycling. This observation was confirmed by Sheel *et al.* (2003), which he attributes to the fact that climbing requires the use of intermittent isometric contractions of the arm musculature and the reliance of both anaerobic and aerobic metabolism. According to Watts and Drobish (1998), VO_2 does not increase with the inclination of the wall (measured on a treadwall), whereas the HR increases progressively. A possible explanation of the unexpectedly missing VO_2 increase is that VO_2 peak values are lower in arm work than in leg exercise (Vokac *et al.*, 1975) whereby the finger reaction forces increase with the angle of inclination (Fuss & Niegl, 2007a). According to Booth *et al.* (1999), VO_2 and HR increase linearly with the climbing speed (measured on a treadwall). According to Watts (2004), climbers do not typically possess extremely high aerobic power, typically averaging between 52–55 ml · kg^{-1} · min^{-1} for maximum oxygen uptake (VO_2 peak). Performance time for a typical ascent ranges from 2 to 7 min and VO_2 averages around 20–25 ml · kg^{-1} · min^{-1} over this period. Peaks of over 30 ml · kg^{-1} · min^{-1} for VO_2 have been reported. VO_2 tends to plateau during sustained climbing yet remains elevated into the post-climb recovery period (Watts, 2004).

The literature does not provide any information on the changes of oxygen uptake when climbing a route with changing degree of difficulty. It was the aim of this study to analyse the respiratory response to different parts of a climbing route.

2 Experimental

In order to test repetitively and consistently changing climbing conditions, a boulder route was set by instrumenting eight holds with 6DOF transducers (multi-axis force/torque sensor; Theta SI 2500-400, ATI Industrial Automation, Apex, NC, USA). This fully instrumented climbing wall (Figure 1) was specifically designed for performance analysis in sport climbing (Fuss *et al.*, 2007b).

Three male climbers, members of the University Mountaineering Club, participated in the experiments (climbing experience 3–5 yrs, red point UIAA 7− – 7+, body mass 58–68 kg). The climbers familiarized themselves with the route first, by climbing the boulder circle once. Subsequently, the climbers were connected to a portable battery powered Cosmed K4b^2 metabolic systems (Cosmed K4b^2, Cosmed, Italy). After a 45-minute warm-up period, the Cosmed K4b$_2$ gas analysers and gas delay were calibrated before each test using room and known standard gases (O$_2$: 16% and CO$_2$: 5%). The flowmeter was calibrated using a 3.00-L syringe, correcting for temperature and barometric pressure. Before the commencement of testing, a Cosmed facemask and headset was placed over the subject's head followed by a 5-minute habituation period. Expired air was then recorded for the climbing with VO$_2$(ml · min^{-1}), VCO$_2$ (ml · min^{-1}), Ventilation (VE, ml · min^{-1}) and respiratory

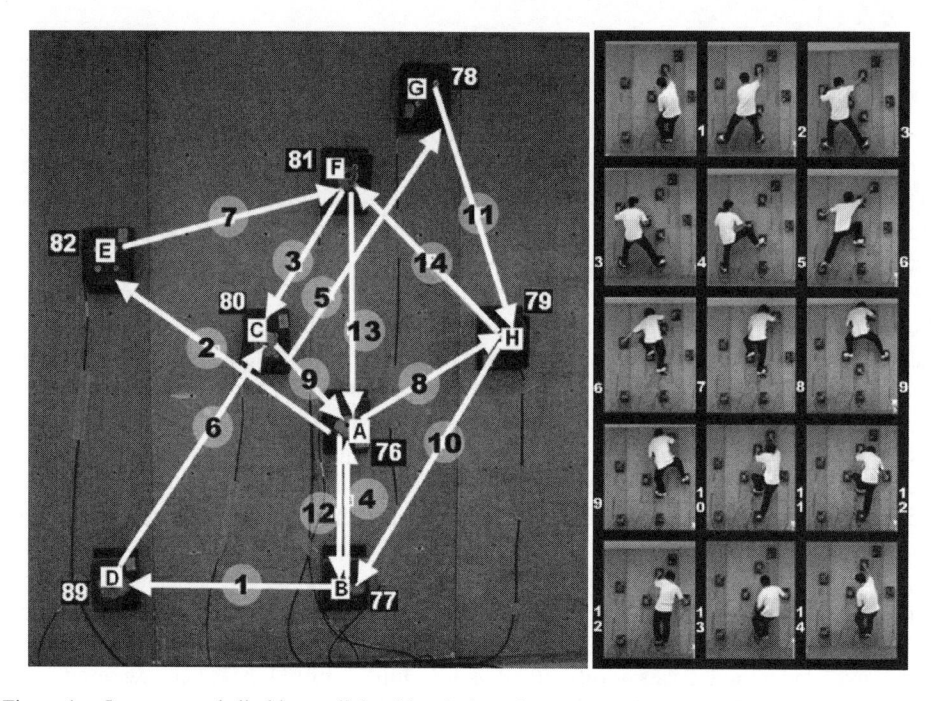

Figure 1. Instrumented climbing wall: boulder circle and movement diagram (left) and the 14 different moves (right).

exchange ratio (RER, VO_2/VCO_2) being determined via open circuit indirect calorimetry. Polar Heart Rate Monitor Transmitter (Polar Electro, Finland) system consisting of a transmitter with an elastic belt and wrist monitor were used to measure heart rate (HR)(beats · min^{-1}). The Polar Transmitter was attached to the elastic belt and secured around the chest of the subject above the pectoral muscles and the HR signals were transmitted to the Polar wrist monitor and the Cosmed K4b^2.

Oxygen uptake VO_2 (ml · min^{-1}) and HR (beats · min^{-1}) were measured continuously during climbing of 5 consecutive boulder circles. To ensure continuous climbing moves, the test persons were assisted by vocally indicating the limb to be moved and by marking the next hold by a laser pointer.

Simultaneously with recording the respiratory data, the 3D forces at the holds were measured at a data sampling frequency of 50 Hz. This served to identify the different moves and to link the time history of the respiratory data to the movement sequence. The time window of a move was defined based on the force data, between the end of contact at the preceding hold and beginning contact at the following hold. The entire boulder circle consisted of 14 moves (Figure 1). The time axis for both VO_2(ml · min^{-1}) and movement number was expressed in integer seconds. First, the VO_2(ml · min^{-1}) data was filtered with a 2nd order Savitzky-Golay filter of a window width of 7 data with respect to the time axis. Subsequently, a baseline correction was applied through a 4th order polynomial fit to account for the initial VO_2(ml · min^{-1}) rise and the steady-state plateau (Figure 2). The data after baseline correction represented the fluctuations of VO_2(ml · min^{-1}) about the initial rise as well as the steady state. The VO_2(ml · min^{-1}) data was filtered again, this time with respect to the movement axis (5 data per move, 5 × 14 moves in 5 boulder circles), with a 3rd order Savitzky-Golay filter of a window width of 51 data. The purpose of this procedure was to find a cyclic behaviour of the VO_2(ml · min^{-1}) fluctuations about the base line, and whether the cyclic behaviour is related to the sequence of the climbing moves. Based on the cyclic behaviour of VO_2(ml · min^{-1}), other respiratory parameters were tested as well. The force data was analysed as to mechanical parameters (Fuss & Niegl, 2006).

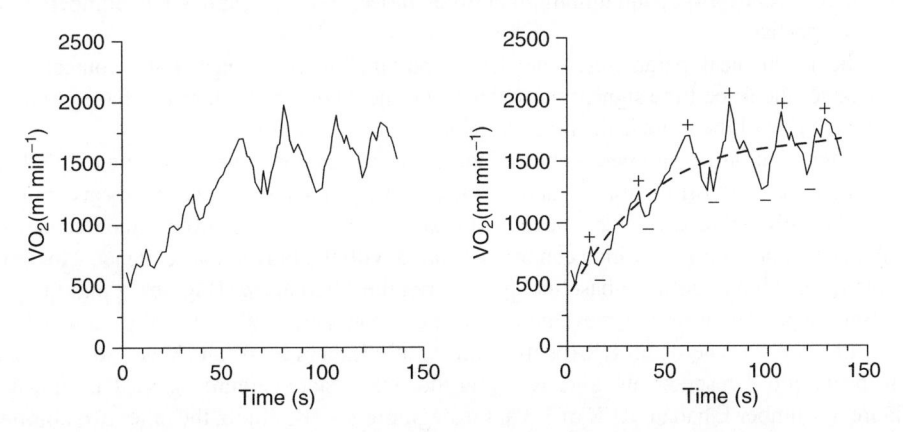

Figure 2. Climber Y, VO_2(ml · min^{-1}) vs. time (s) (unfiltered data); right subfigure: the dashed line defines the baseline, and the positive and negative VO_2 peaks are marked with "+" and "−".

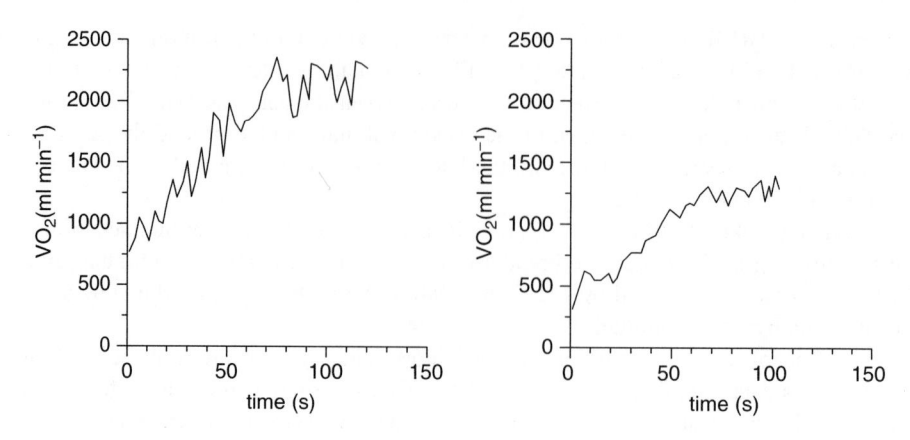

Figure 3. Climbers W and E, VO_2(ml · min^{-1}) vs. time (s)(unfiltered data).

3 Results

All climbers reached the steady-state after 60–80 seconds of climbing (Figure 2). Climber W showed the highest steady-state value of about 2100 ml/min, and climber E the smallest (1250 ml/min). Figure 2 clearly exhibits VO_2(ml · min^{-1}) fluctuations about the base line. Climber Y shows the most pronounced VO_2(ml · min^{-1}) changes, and climber E the least (Figure 4). As the climbing velocity is not necessary constant, the VO_2(ml · min^{-1}) fluctuations have to be plotted with the movement sequence to prove a dependency of the boulder route.

Figure 3 shows the mean change of VO_2(ml · min^{-1}) with respect to the movement sequence, and clearly reveals a cyclic behaviour of the VO_2(ml · min^{-1}) about the baseline. The difference between minimum and maximum is about 280 ml · min^{-1} in climber Y, 145 ml · min^{-1} in climber W, and 70 ml · min^{-1} in climber E. Interestingly, minimal and maximal values have different locations within the boulder circle: maximum between moves 13–14, 10–11, and 14–1, and minimum between moves 7–8, 3–4, and 5–6 in climbers Y, W, and E respectively.

The mechanical parameters (mean and maximal force at each hold, contact time, impulse) of the force-time signals showed any significant difference between the 3 climbers, e.g. which part of the route is the most difficult one for each climber.

Other respiratory parameters showed the same cyclic behaviour as well, namely VT (l), VE (ml · min^{-1}), VO_2(ml · min^{-1}), and VCO_2(ml · min^{-1}). However, the HR (beats · min^{-1}) was not clearly influenced by the movement sequence. Climber Y showed some fluctuations of the HR (beats · min^{-1}), which can be associated with the movement sequence. However, it was impossible to define a baseline due to irregular fluctuations (Figure 4), and thus the possible dependency on the movement sequence cannot be analysed in the same way as VO_2(ml · min^{-1}). The mean HR (beats · min^{-1}) of climbers E, W, and Y was 163, 171 and 146 beats min^{-1} respectively, after reaching the steady state (Figure 5). RER is shown in Figure 6. Climber E had an RER of 1.3 at the beginning of the climb, the other two climbers had an RER of about 0.85. Climber E was the fastest and completed the 5 boulder circles in 98 s, climber W in 112 s, and climber Y in 137 s.

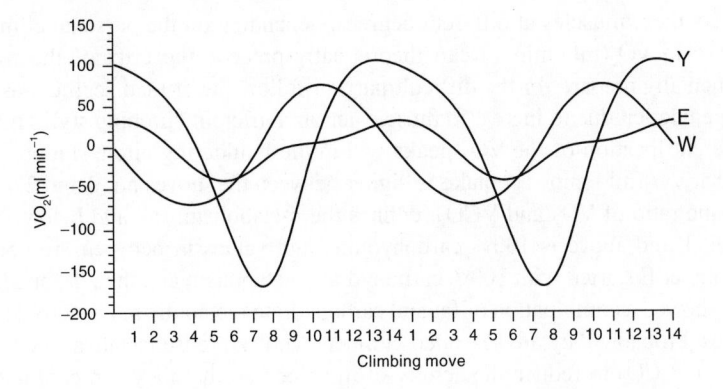

Figure 4. VO$_2$ (ml · min^{-1}) vs. climbing moves, in climber Y, W, and E.

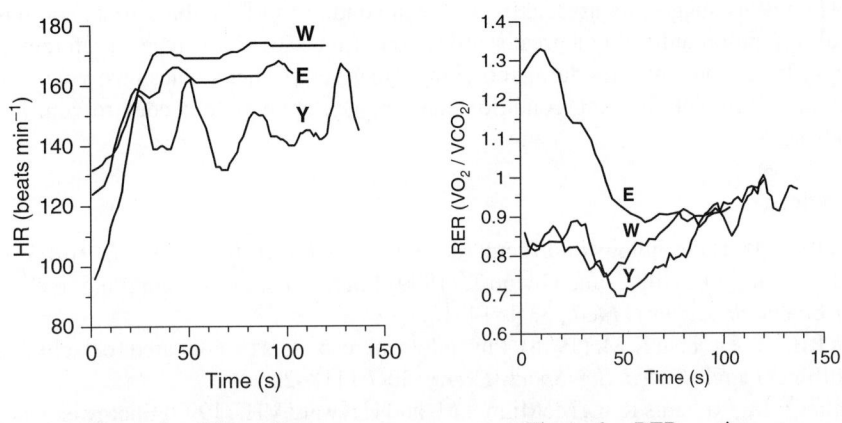

Figure 5. HR vs. time.

Figure 6. RER vs. time.

4 Discussion

The results of this study clearly prove that the physiology of respiration depends on the conditions of the climbing route. Even after reaching the steady-state, VO$_2$(ml · min^{-1}) fluctuates about the mean VO$_2$(ml · min^{-1}) in synchrony with the boulder circle. The striking differences in the mean VO$_2$(ml · min^{-1}) between the 3 climbers (Figures 2 and 3), and the missing difference in the mechanical parameters leads to the conclusion that the oxygen uptake is not directly connected to the difficulty of the route and the personal climbing performance (red-point grade) of the climber.

This conclusion is reflected in the 2 opposing results by Mermier *et al.* (1997) and Watts and Drobish (1998), stating that the oxygen uptake does increase, or does not increase with the difficulty of the route. There are obviously more factors to be considered than simply the degree of difficulty or the wall inclination. Among these factors is the action of trunk muscles as well as shoulder muscles with thoracic attachments. Contraction of these muscles influences the respiration in different ways. Climbers of the same experience level might even contract these muscles at different degrees, depending on the personal climbing style. The increase in $VO_2(ml \cdot min^{-1})$ can theoretically precede the crux of the route, if the climbers mentally prepare for the difficult part, or follow the critical segment as a reaction to the oxygen deficit due to increased muscle action. Different climbing styles might reflect in the different location of the VO_2 peaks within the bouldering circle (Figure 4). Fact is, however, that $VO_2(ml \cdot min^{-1})$ uptake is higher between the moves no. 9 and 2.

RER, the ratio of VO_2 and VCO_2, defines the metabolism: 0.7 and below is 100% fat metabolism, 1 and above is 100% carbohydrate, and values in between are combinations thereof. Climber E started with 100% carbohydrate metabolism and then dropped. The other 2 climbers showed a combination of fat and carbohydrate metabolism, i.e. aerobic and anaerobic sources. Probably they are seasoned climbers who can reuse lactate as a energy source and breathe out CO_2 to reduce the acidity of the blood so that they can continue climbing without fatigue (since lactate causes fatigue). We hypothesise that this combination of fat and carbohydrate is to forestall fatigue in seasoned climbers. However, this hypothesis needs to be confirmed by analysing lactate with a discontinuous climbing protocol.

The results suggest an association between conditions of climbing route and physiological respiration indicating a greater utilisation of muscles and maybe a preferential use of larger lower body muscles during climbing. However, further studies involving subjects undergoing different climbing conditions and oxygen uptake are needed to confirm this hypothesis.

References

Watts P.B. (2004) Physiology of difficult rock climbing. *Eur. J. Appl. Physiol.*, 91/4:361–72.

Booth J., Marino F., Hill C. and Gwinn T. (1999) Energy cost of rock climbing in elite performers. *Br. J. Sports Med.*, 33/1, 14–18.

Watts P.B. and Drobish K.M. (1998). Physiological responses to simulated rock climbing at different angles. *Med. Sci. Sports Exerc.*, 30/7:1118–22.

Mermier C.M., Robergs R.A., McMinn S.M. and Heyward V.H. (1997) Energy expenditure and physiological responses during indoor rock climbing. *Br. J. Sports Med.*, 31/3, 224–228.

Vokac Z., Bell H., Bautz-Holter E. and Rodahl K. (1975) Oxygen uptake/heart rate relationship in leg and arm exercise, sitting and standing. *J. Appl. Physiol.*, 39/1:54–9.

Fuss F.K. and Niegl G. (2006) Instrumented Climbing Holds and Dynamics of Sport Climbing. In: Moritz E.F. and Haake S. (Eds.) *The Engineering of Sport 6, Vol. 1 – Developments for Sports*. Springer, Munich, pp. 57–62.

Fuss F.K. and Niegl G. (2007a) Change of mechanical climbing parameters with inclination of the wall. Unpublished data.

Fuss F.K. and Niegl G. (2007b) The fully instrumented climbing wall: performance analysis, route grading and vector diagrams – a preliminary study. In: Fuss F.K., Subic A. and Ujihashi S. (Eds.) *The Impact of Technology on Sport II*. Taylor & Francis Group, London.

Sheel A.W., Seddon N., Knight A., McKenzie D.C. and Warburton D.E. (2003) Physiological responses to indoor rock-climbing and their relationship to maximal cycle ergometry. *Med. Sci. Sports Exerc.*, 35/7:1225–31.

EFFECT OF ACETAZOLAMIDE ON PHYSIOLOGICAL VARIABLES DURING HIGH ALTITUDE IN 15-YEAR OLDS

G. BALASEKARAN[1], S. THOMPSON[1], J. GRANTHAM[1,2] & V. GOVINDASWAMY[3]

[1]Nanyang Technological University, National Institute of Education, Physical Education and Sports Science, Singapore
[2]Sports Medicine Centre, Qatar National Olympic Committee, Doha, Qatar
[3]Computer Science and Engineering, University of Texas at Arlington, Texas, USA

Most scientific research into the physiological effects of High Altitude (HA) has focused on adults. The effect of short-term use of acetazolamide in teenagers and its effect on selected physiological parameters have not been studied. The purpose was to determine the effect of the carbonic anhydrase inhibitor, acetazolamide in 15-year olds over 17-day at HA. The physiological changes was investigated comparing a control group (n = 12) and a treatment group(n = 13). The treatment group took 250 mg of acetazolamide. Field-testing of Resting Heart Rate (RHR) and Acute Mountain Sickness (AMS) using the Lake Louise questionnaire were measured each morning and afternoon; Resting Blood Pressure (RBP) was measured each afternoon. Pre and post blood samples were taken at Sea Level (SL) one week before departure and within 40 hours on return to SL and analysed for haemoglobin(Hb) and hematocrit(Hct). Plasma Volume (PV) changes were estimated using Dill and Costill's method. RHR was significantly lower in the treatment group than the control group on days 2, 3 and 5($p < 0.05$). Systolic blood pressure was different between both groups for day 4 and 13($p < 0.05$). The groups did not differ in diastolic blood pressure on any occasion. There was no significant difference between the groups' pre and post altitude for Hb, Hct and PV. AMS scores did not differ from the two groups except on morning of day 12($p < 0.05$). This study found teenagers experienced similar degrees of AMS and physiological changes with or without acetazolamide.

1 Introduction

Acetazolamide is a sulphur-based drug that is often used to prevent or treat acute mountain sickness (AMS). Use of this drug has gained popularity especially as high altitude travelling increases and trekkers want to increase the speed of acclimatisation. (Krasney, 1993). The pharmacological effect of acetazolamide is the inhibition of carbonic anhydrase. Carbonic anhydrase is the enzyme that catalyses the conversion of carbon dioxide (CO_2) and water (H_2O) to carbonic acid (H_2CO_3). Thus, when acetazolamide is administered, there is interference with CO_2 transport. Renal excretion of bicarbonate occurs, causing intracellular acidosis, which stimulates respiration. (Hultgren, 1997; Ward et al.,1995; Pollard & Murdoch, 1997). The drug stimulates ventilation, which brings greater volumes of air into the lungs and therefore oxygen into the body and aids acclimatisation. Prevention of AMS involves sensible acclimatisation through gradual ascent. The use of the prophylaxis, acetazolamide is also effective for the first 4-6 days at high altitude. (Hultgren,1997). The effectiveness of acetazolamide in the prevention and treatment of AMS has been clearly established in adults in a number of well-controlled studies since the mid 1960's. (Greene et al, 1981; Grissom et al., 1992). No clinical trials of acetazolamide have been conducted on children or the effects of the drug on physiological parameters; however, there is no reason to suggest that its use would not be effective. (Pollard & Murdoch, 1997). Most scientific research at altitude has involved adults, mostly male and there is a dearth of research on teenagers at altitude. The purpose of the investigation is to look at effect of carbonic anhydrase inhibitor, acetazolamide on following physiological responses; resting heart rate (RHR), resting blood pressure (RBP), haemoglobin and hematocrit in 15-year-olds on a high altitude (HA) trek in Ladakh, India. The effectiveness of the prophylaxis, acetazolamide, as an aid to acclimatisation will also be investigated.

2 Methods

The subjects were 25 fifteen-year-old students, 17 females and 8 males, and all subjects are residents in Singapore. The subjects were volunteers and they were randomly assigned into a control group (n = 12) and treatment group (n = 13). Ethical approval for the study was sought from the human subjects review board at the National Institute of Education, Physical Education and Sports Science Academic Group. Informed consent was sought from the subjects and their parents or legal guardians.

2.1. *Days at High Altitude*

The treatment and control groups were monitored for changes in RHR, RBP and AMS scores during the first 13 days at altitudes of 3,500–5,100 m. The total time spent at HA was 17 days; the initial 3 days were spent at 3,500 m for purposes of acclimatisation followed by a 10-day trek.

2.2. *Acetazolamide*

The treatment group took a dosage of 125 mg of acetazolamide twice daily (6–7 am and 6–7 pm) from day one to the morning of day seven. The rationale behind this time frame being that on day 8 at HA a 4,950 m pass would be crossed followed by a descent to 3,500 m which is the same altitude that the subjects flew into on day one and both groups should have achieved similar levels of acclimatisation by this stage. In this study the dosage of 250 mg/day was recommended by Dr. J. B. West and Dr. J. Dallimore via email communication. In this study the time period of six and a half days coincided with the 3-day acclimatisation period at 3,500 m, and the first major challenge in terms of altitude gain on day 7. On day 7 the highest point was 4,950 m followed by a descent to 3,500 m.

2.3. *Resting Heart Rate, Resting Blood Pressure and AMS*

Field-testing of RHR and AMS symptoms was done twice daily, once in the morning before rising (5.30–7.00am) and once in the afternoon (5–6pm). RBP measurements were taken at the same time as afternoon RHR using an Omron automatic digital blood pressure monitor HEM-703C (Omron Asia Pacific Pte Ltd., 83 Clemenceau Ave. Singapore 239920). RHR was measured using Polar Accurex Plus heart rate monitors (Polar Electro Inc: 99, Seaview Boulevard, Port Washington, NY11050). AMS symptoms were measured in the morning and evening by the Lake Louise AMS questionnaire a self-reporting questionnaire (Sutton *et al.*, 1991) Subjects had a an hour of complete rest prior to data collection.

2.4. Blood Samples

Blood samples were taken one week prior to departure and within 40 hours on return to SL. Blood samples were analysed for Hb using a Hemoglobin Meter Hb-202 (Optima INC, Japan). A Micro-hematocrit centrifuge (Hawksley, England), measured Hct. PV changes were calculated based on Dill & Costill's plasma volume correction formula (Dill & Costill, 1974).

3 Statistical Analyses

An independent *t* test was applied to test for differences between treatment and control group for the following variables; Resting Heart Rate and resting blood pressure. AMS scores was analysed by non-parametric Mann-Whitney U test and Bonnferoni adjustment was made for multiple comparisons. A paired dependent *t* test was used on both the treatment and control groups to test for significant differences in Hb, Hct and PV both pre and post HA. Pearson product moment coefficient of correlation (r) was used to test for a correlation between AMS v HA, RHR v HA and RBP v HA. All data analysis was performed using Microsoft Excel. The level of significance for all statistical analysis was $p < 0.05$.

4 Results

RHR was always significantly different from SL ($p < 0.05$) except for the morning of day 4 (act) and the morning of day 10 (control). RHR was significantly higher in the control group on days 2, 3 and 5 during drug administration. Systolic Blood Pressure (SBP) was always significantly greater at HA for both groups ($p < 0.05$ or less) with the exception of day 10 (control and act) and day 11 (control). On day 4, SBP (control) was significantly greater than SBP (act) group; this was reversed on day 13 ($p < 0.03$). Diastolic Blood Pressure (DBP) was significantly greater than at SL on days 6,7,9,11,12,and 13 (control) and days 3 to 9 and 11 to13 (act) ($p < 0.05$). The groups did not differ significantly from each other on any occasion. Hb and Hct groups were significantly different to sea level values ($p < 0.01$). There was no significant difference between the groups' pre and post altitude for Hb and Hct. PV decreased by 18.4% ± 11.3 (control) and 14.0% ± 10.0 (act), again no significant differences were found between groups. AMS scores were recorded using the Lake Louise questionnaire throughout the stay at altitude. The groups did not differ significantly from each other except on day 11(am) when the AMS scores of the acetazolamide group were significantly greater than the control group. A trend of low scores was seen for several days in both groups after the first HA pass (4,900 m) had been crossed.

5 Discussion

5.1. *Resting Heart Rate (RHR)*

This study found RHR at HA to be consistently elevated as compared to SL values with the exception of day 4 for the acetazolamide group, which was during the drug administration period and on day 10 in the control group. Increases are more pronounced in the afternoon, which despite an hour of complete rest prior to data collection is probably related to the exertion of walking, heat exposure and dehydration earlier in the day. Malconian *et al*. (1990) studied 6 young male subjects in a hypobaric chamber over a 40-day period. Increases in

RHR were less pronounced due to the slow ascent. Reeves *et al.* studied eight subjects in a simulated climb of Mount Everest and found that resting and exercise heart rates rose with altitude. This study agrees with the finding that exposure to HA causes an elevation in RHR.

5.2. *Resting Blood Pressure (RBP)*

The results of this study found that SBP was almost always significantly greater than SL values whereas increases in DBP values were less consistent. Although not conclusive there is some suggestion that systolic pressure is affected more than diastolic pressure (Hultgren, 1997). An unpublished study by Ward & Milledge (cited in Ward *et al.*, 1995) found that on ascent to 4,300 m systolic blood pressure (SBP) rose from pre-ascent values of 103 mmHg to 121 mmHg and remained raised. They reported no change in diastolic blood pressure (DBP). Another factor in elevation of blood pressure is a decrease in atrial natriuetic peptide (ANP) secretion. ANP causes vasodilation, however, secretion of this hormone is reduced under hypoxic conditions and this may be partly responsible for increases in blood pressure (Boning, 1997b).

5.3. *Acute Mountain Sickness (AMS)*

AMS was measured using the Lake Louise self-reporting questionnaire (Sutton *et al.*,1991). No significant differences ($p < 0.05$) were found between the groups except on day 11 (am) when the control group showed significantly lower AMS scores than the acetazolamide group. At this point the drug administration period was complete. Many previous studies have found that AMS symptoms such as nausea, headache, insomnia, lassitude and dizziness are reduced in subjects taking acetazolamide. (Grissom *et al.*, 1992; Greene et al., 1981). Subjects in these studies are all adults, 12 climbers, 24 amateur climbers and 20 males, respectively.

5.4. *Blood Analyses*

Pre and post-altitude blood analyses in both groups showed a significant increase ($p < 0.05$) after 17 days at HA (3,500–5,100 m), there was no significant difference ($p < 0.05$) between the groups as expected (table 7). Hct increased by 5–7%, which is similar to a study by Stokke *et al.* (1986) where ten male subjects at a mean altitude of 4,100 m for 20 days showed an increase of 5.4% in Hct. The authors warn that at HA a decrease in PV leads to hemoconcentration, which could be misinterpreted as increased erythropoietic activity. Their data showed wide individual variation with one subject recording very high hct levels (69%), which required hemodilution. Hb concentration also increased in this study after exposure to HA by about 1g/dl or 6–7% for both groups. Other studies have shown greater increases in Hb; Boning *et al.* (1997a) reported increases of 14% 7–8 days after return to SL, however their study lasted 26–29 days at 4,700–7,600 m which is both longer and higher than this study and the subjects were 12 experienced male mountaineers.

5.5. *Plasma Volume*

Both groups showed similar decreases in PV $14.0 \pm 10.0\%$ for the acetazolamide group and $18.4 \pm 8.5\%$ for the control group as estimated by Dill & Costill's method (1974). The teenage subjects in this study had blood samples taken 40 hours after departure from HA and PV losses due to dehydration would only account for a small percentage of this. Other possible reasons for PV losses are likely to be related to HA exposure, namely plasma protein loss, increased capillary permeability and diuresis (Robach et al., 2000). The study of 10 male subjects at an average of 4,100 m for 4 weeks found an overall 4% increase in PV (Stokke et al.,1986). However, 4 out of the 10 subjects showed a decrease in PV, and it remains to be determined under what conditions (length of time and elevation) PV changes increase rather than decrease. The results of this study agree with other studies that short duration (2–3 weeks) at HA causes a significant decrease in PV.

6 Conclusions

Exposure to HA (3,500–5,100 m) in teenagers for 2–3 weeks causes significant increases ($p < 0.05$) in RHR over SL values. Taking moderate doses (250 mg/day) of acetazolamide may cause less pronounced increases in RHR. SBP shows significant increases on exposure to HA; taking acetazolamide does not significantly ($p < 0.05$) affect this increase. DBP often shows significant increases ($p < 0.05$) at HA although not as consistently as SBP. There is no evidence to suggest that taking acetazolamide affects DBP. AMS scores were not significantly different ($p < 0.05$) between the acetazolamide group and the control group during the drug administration period.AMS scores were not significantly different ($p < 0.05$) between the acetazolamide group and the control group after the drug administration period. However, both groups exhibited low scores from day 8-11 which suggests an improvement in acclimatisation; these low scores persisted until the next major gain in altitude, up to 4,800 m, took place. Hb significantly increased ($p < 0.05$) after 17 days at HA. Short-term use of acetazolamide does not affect this increase. Hct significantly increased ($p < 0.05$) after 17 days at HA. Short-term use of acetazolamide does not affect this increase. Both groups showed equivalent losses ($p < 0.05$) in PV after 17 days at HA. The opportunity to travel and trek at HA is open to all ages; however, teenagers may not react physiologically to hypoxic environments in exactly the same way as adults. A teenager's perception of AMS may be different to an adult and be affected by factors such as the ability to cope with fatigue from walking in a hot dry environment for up to 10 hours a day and dealing with extremes of temperature.

Given that the predominance of literature on HA physiology focuses on adult males it is hoped that the results from this investigation will add to the somewhat sparse literature on HA physiology in teenagers.

References

Boning D., Maassen N., Jochum F., Steinacker J., Halder A., Thomas A., Schmidt W., Noe G. and Kubanek X. (1997) After effects of a high altitude expedition on blood. *International Journal of Sports Medicine*, 18, 3, 179–185.

Dill D.B. and Costill D.L. (1974) Calculation of percentage changes in volumes of blood, plasma and red cells in dehydration. *Journal of Applied Physiology*, 37/2, 247–248.

Greene K., Kerr A.M., McIntosh I.B. and Prescott R.J. (1981) Acetazolamide in prevention of acute mountain sickness: a double-blind controlled cross-over study. *British Medical Journal*, 283, 811–813.

Grissom C.K., Roach R.C., Sarnquist F.H. and Hackett P.H. (1992) Acetazolamide in the treatment of Acute Mountain Sickness: clinical efficacy and effect on gas exchange. *Annals of Internal Medicine*, 116, 461–465.

Hackett P.H., Schoene R.B., Winslow R.M., Peters R.M., and West J.B. (1985) Acetazolamide and exercise in sojourners to 6300 m – a preliminary study. *Medicine and Science in Sports and Exercise*, 17/5, 593–597.

Hultgren H. (1997) *High altitude medicine*. Hultgren publications, Stanford, CA.

Krasney J.A. (1993) A neurogenic basis for acute altitude illness. *Medicine and Science in Sports and Exercise*, 26/2, 195–208.

Malconian M., Rock P., Hultgren H., Donner H., Cymerman A., Groves B., Reeves J., Alexander J., Sutton J., Nitta M. and Houston C. (1990) The electrocardiogram at rest and during a simulated ascent of Mount Everest (Operation Everest II). *American Journal of Cardiology*, 65, 1475–1480.

Pollard A.J. and Murdoch D.R. (1997) *The high altitude medicine handbook*. Radcliffe Medical Press, Inc.Oxford.

Robach P., Dechaux M., Jarrot S., Vaysse J., Schneider J.C., Mason N.P., Herry J.P., Gardette B. and Richalet J.P. (2000) *Journal of Applied Physiology*, 89, 29–37.

Stokke K.T., Rootwelt K., Wergeland R. and Vale J.R. (1986) Changes in plasma and red cell volumes during exposure to high altitude.*Scandinavian Journal of Clinical Laboratory Investigations*, 46, Suppl. 184; 113–117.

Sutton J.R., Coates G. and Houston C.S. (Eds.). (1991). *Hypoxia and mountain medicine*: proceedings of the 7th International Symposium. Oxford; NY: Permagon Press.

Ward M. P., Milledge J. S. and West J. B. (1995). *High altitude medicine and physiology* (2nd edition). Chapman and Hall Medical, London.

EXAMINATION OF THE TIME-DEPENDENT BEHAVIOR OF CLIMBING ROPES

I. EMRI[1], M. UDOVČ[1], B. ZUPANČIČ[1], A. NIKONOV[1], U. FLORJANČIČ[1], S. BURNIK[2] & B.S. VON BERNSTORFF[3]

[1]*Faculty of Mechanical Engineering, Center for Experimental Mechanics, University of Ljubljana, Ljubljana, Slovenija*
[2]*Faculty of Sport, University of Ljubljana, Ljubljana, Slovenija*
[3]*BASF Aktiengesellschaft, Ludwigshafen, Germany*

The experimental-analytical methodology, based on a simple non-standard falling weight experiment, which allows examination of the time-dependent behavior of ropes exposed to arbitrary falling weight loading conditions, was developed for mechanical characterization of climbing ropes. In this paper we show that the newly developed simple procedure can be successfully applied for calculation of several important characteristics, such as the impact force on the rope; maximum deformation of the rope; derivative of the (de)acceleration; dissipated energy during the loading and unloading of the rope. All those parameters can be determined just from a single dynamic response of the rope exposed to impulse loading. The ropes of three different commercial manufacturers were exposed to the same loading conditions. Calculated mechanical properties of all three ropes were then compared using the developed methodology. The obtained results indicate that ropes, which according to existing UIAA standard belong to the same quality class and are declared to have the same UIAA standard characteristics, actually exhibit significantly different behavior when exposed to the same loading conditions by using our experimental-analytical methodology.

1 Introduction

The climbing ropes are designed to secure a climber. They are designed to stretch under high load so as to absorb the shock force. This protects the climber by reducing fall forces. Ropes should have good mechanical properties, such as high breaking strength, large elongation at rupture and good elastic recovery, e.g. Jenkins (2003).

The UIAA (Union Internationale des Associations d'Alpinisme) has established standard testing procedures to measure, among other things, how a rope reacts to severe falls. The standard says little about the durability of the rope, which is more difficult to define or assess with a simplified procedures. Ropes are produced from polymeric fibers, which exhibit viscoelastic behavior. Thus, durability in this case does not mean just failure of the rope, but rather deterioration of its time-dependent response when exposed to an impact force. The experiments prescribed by the UIAA standard are not geared to analyze the time-dependent deformation process of the rope, which causes structural changes in the material and consequently affects its durability.

In this paper we present comprehensive dynamic analysis of a simple non-standard falling weight experiment, which allows examination of the time-dependent behavior of ropes exposed to arbitrary falling weight loading conditions. Developed analytical treatment is subsequently examined on commercial dynamic climbing ropes.

2 Theoretical Treatment

The time-dependent response of the rope under dynamic loading generated by a falling weight may be retrieved from the analysis of the force measured at the upper fixture of the rope, as schematically shown in Figure 1a. In such experiment a weight, m, is dropped from an arbitrary height, h. The length of a tested rope is l_0. Force measured as a function of time, $F(t)$, is schematically shown in Figure 1b.

Phase A. In this phase the weight is dropped at $t = 0$, and it falls freely until at $t = 0 = \sqrt{2h/g}$ the rope becomes straight, which is indicated in Figure 1b as point T_0. The point T_0 represents the end of the free-falling phase of the weight, and beginning of the phase B.

Phase B. At point T_0 the weight starts to deform the rope. Neglecting the air resistance the equation of motion of the weight between the points T_0 and T_5 may be written as,

$$m\ddot{x}(t) = mg - F(t). \tag{1}$$

Here m is the mass of the weight, g is the gravitational acceleration, $\ddot{x}(t)$ denotes the second derivative of the weight displacement, $x(t)$, measured from the point T_0 on. Thus, $x(t)$ represents the time-dependent deformation of the rope. Taking into account the initial conditions at point T_0: $x(t = t_0) = 0$, and $\dot{x}(t = t_0) = v_0 = \sqrt{2gh}$, solution of Eq. (1) gives the displacement and velocity of the weight as functions of time,

$$x(t) = \frac{gt^2}{2} - \frac{1}{m} \int_{t_0}^{t} \left[\int_{t_0}^{\tau} F(\vartheta) d\vartheta \right] d\tau + v_0 t, \tag{2}$$

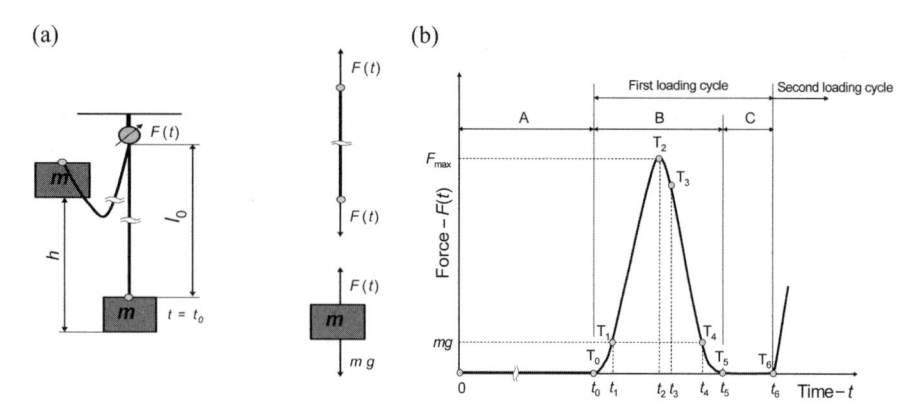

Figure 1. Schematics of the rope exposed to the falling weight (a) and force measured during the falling weight experiment (b).

$$v(t) = \dot{x}(t) = gt - \frac{1}{m} \int_{t_0}^{t} F(\tau) d\tau + v_0. \qquad (3)$$

At point T_1, where $t = t_1$, the force action on the rope becomes equal to the weight of the load, $F(t_1) = mg$. At this point the velocity of the weight reaches its maximum value, $v_{max} = v(t_1)$. The force acting on the rope reaches its maximum at point T_2, when $t = t_2$, or $F_{max} = F(t_2)$. Due to the viscoelastic nature of the rope, its maximum deformation will be delayed, and will take place at $t = t_3$, i.e., at point T_3, where the velocity of the weight is equal to zero. The maximum deformation of the rope is then,

$$s_{max} = \int_{t_0}^{t_3} v(\tau) d\tau = \int_{t_0}^{t_3} \left[g\tau - \frac{1}{m} \int_{t_0}^{\tau} F(\vartheta) d\vartheta + v_0 \right] d\tau. \qquad (4)$$

The unloading phase of the rope starts at the point T_3. The elastic component of the rope's deformation will be retrieved and will accelerate the weight in the opposite (upward) direction. At $t = t_4$, indicated as point T_4, the force acting on the weight becomes again equal to the weight of the load, $F(t_4) = mg$. At point T_5, where the force acting on the rope becomes equal to zero, $F(t_5) = 0$, the weight will start its free-fly in the upward (vertical) direction.

Phase C. Point T_5 represents the beginning of the phase C in which weight has no interaction with the rope. The weight starts to fly upwards with the initial velocity, $v(t_5)$, and returns back at point T_6. At the point T_6 starts the second loading cycle of the rope.

Energy dissipation during the rope deformation process is one of the most important rope characteristics, and should be used for comparing the quality of ropes. Force, $F(t)$, measured during the loading and unloading of the rope, may be expressed as a function of the rope deformation, $F = F(s)$. The dissipated energy within a loading and unloading cycle can be expressed as,

$$W_{diss} = \int_{t_0}^{t_3} F(\tau) \left[g\tau - \frac{1}{m} \int_{t_0}^{\tau} F(v) dv + v_0 \right] d\tau$$
$$- \int_{t_3}^{t_5} F(\tau) \left[g\tau - \frac{1}{m} \int_{t_0}^{\tau} F(v) dv + v_0 \right] d\tau. \qquad (5)$$

The experience from the car crash experiments teaches us that the time-variation of the (de)acceleration which a person has been exposed to, is much more important for its safety than the magnitude of (de)acceleration itself. Thus, the maximum of the absolute value of the derivative of the de-acceleration may be used as one of the criterions for judging the quality of climbing ropes,

$$\psi = \max \left[\frac{d^2 v(t)}{dt^2} \right] = \max \left[\frac{1}{m} \frac{dF(t)}{dt} \right] \qquad (6)$$

Thus, ropes with smaller values of ψ are better (more safe) than those with larger ψ.

3 Experiments

Experimental setup is schematically presented in Figure 2. The console is fixed at the height of 6 m above the floor. The force sensor is placed on the console. Signals from the force sensor pass through the carrier amplifier prior to being collected in digital format by the data acquisition system (DAQ). The rope is connected to the force sensor with one end and to the weight with another in such way that both ends of the rope are on the same level.

Free fall tests were conducted on specimens of three different commercial manufactures. Thickness of each rope was the same, i.e., 9.8 mm. From each of three ropes four specimens were prepared. The rope was first cut in four pieces, having the same length. Each specimen was then treated as such that nooses were sewed up on both ends of the specimen as shown in Figure 3. Each specimen had the same initial length, l_0, i.e., 3.38 ± 0.04 m. The specimens were subjected to the same room temperature conditions. Tests were conducted at temperature 26 ± 2°C and at normal outside atmospheric pressure.

Before each test, the length of the specimen was measured. At certain time the weight, which was fixed at the end of the rope, was dropped and the specimen was exposed to the impact loading. The mass of the weight was 43.85 ± 0.02 kg. For each specimen we repeated 10 falls consequently, with a waiting time between two falls of 5 minutes. The length of the specimen was measured after each fall. The measured response, i.e., force versus time, $F(t)$, was then saved for further analysis. The characteristics of the rope as described above were then calculated by using software DAR which was developed at the Center for Experimental Mechanics.

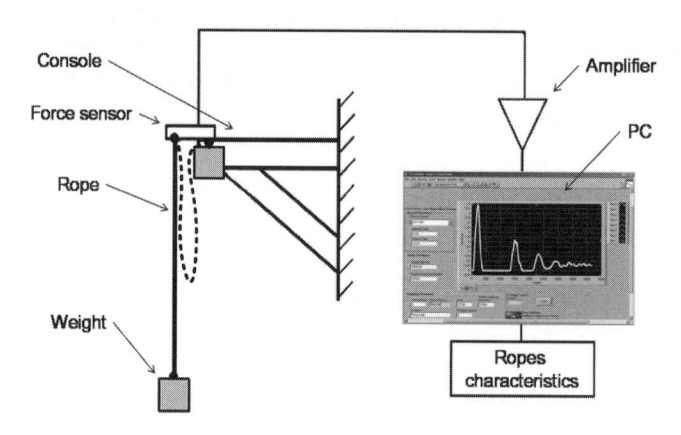

Figure 2. Schematic apparatus layout.

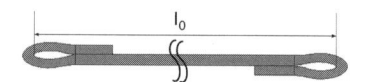

Figure 3. Specimen with nooses.

4 Results and Discussion

The newly developed experimental-analytical methodology enabled us to analyze the time-dependent behavior of ropes under impact loading. From the measured force during the fall, we have calculated the following characteristics: impact force, maximum deformation of the rope, derivative of (de)acceleration, and the dissipated energy. By using this method we compared calculated characteristics of three different ropes, which according to existing UIAA standard belong to the same quality class and are declared to have the same UIAA standard characteristics. The ropes from three different commercial manufacturers were identified as R1, R2, and R3. The average values of rope's characteristics and standard deviation were calculated from 4 measurements for each type of rope.

The comparison analysis of the time-dependent behavior of ropes when exposed to impact loading for three different commercial manufacturers is presented in Figures 4 and 5. In diagrams calculated characteristics of ropes are presented as functions of number of falls, N. Figure 4 shows the impact force, F_{max}, and the maximum deformation of the rope, s_{max}. The derivative of (de)acceleration, ψ, and dissipated energy, W_{dis}, are presented in Figure 5.

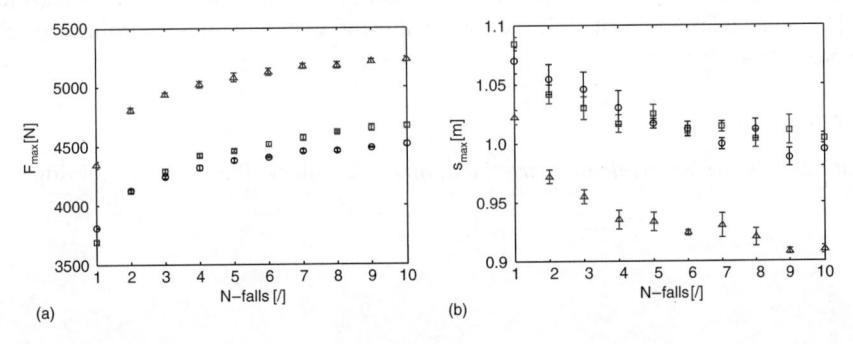

Figure 4. The impact force (a) and the maximum deformation of the rope (b) as functions of number of falls: □ – rope R1, △ – rope R2, ○ – rope R3.

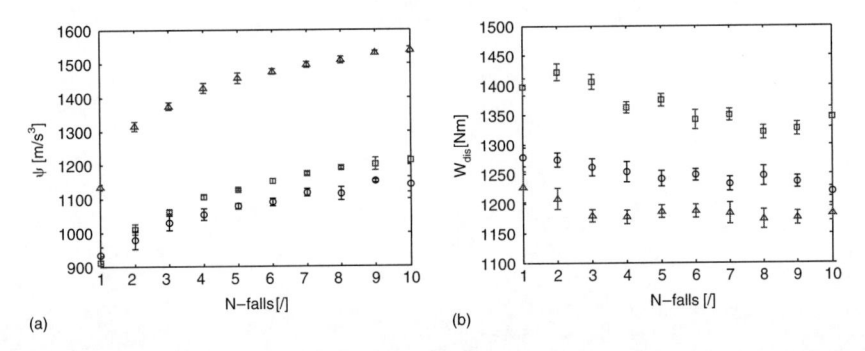

Figure 5. The derivative of (de)acceleration (a) and the dissipated energy (b) as functions of number of falls: □ – rope R1, △ – rope R2, ○ – rope R3.

From the diagrams presented above we may recognize significantly different time-dependent behavior of ropes from three different commercial manufacturers when they are exposed to the same impact loading conditions.

The rope R2 has 15% bigger impact force, 10% smaller maximum deformation and 35% bigger maximum derivative of (de)acceleration in comparison with the ropes R1 and R3 after the tenth fall. Therefore the rope R2 may be considered as more dangerous for the climbers then the two others. The dissipated energy for the rope R1 is bigger as well than for the ropes R2 and R3.

5 Conclusions

By using newly developed experimental-analytical methodology we are able to analyze the time-dependent behavior of ropes under impact loading. As the comparison analysis of three different ropes, which according to existing UIAA standard belong to the same quality class and are declared to have the same UIAA standard characteristics, shows, ropes from three different commercial manufacturers exhibit significantly different time-dependent behavior when they are exposed to the same impact loading conditions.

This work provides a foundation for the development of new methodology for testing and analyzing time-dependent mechanical properties of climbing ropes. This knowledge could be used in development of the new generation of ropes with pre-determined (mechanical) properties.

Reference

Jenkins M. (2003) *Materials in Sports Equipment*. Woodhead Publ. Ltd., Cambridge.

DEVELOPMENT OF A SHARP EDGE RESISTANCE TEST FOR MOUNTAINEERING ROPES

M. BLÜMEL[1], V. SENNER[1] & H. BAIER[2]

[1]*Technische Universität München, Institute for Sports Equipment and Materials, München, Germany*
[2]*Technische Universität München, Institute for Lightweight Structures, Garching, Germany*

Lower weight of the equipment is an advantage in competition in climbing. This is one of the reasons manufacturers of mountaineering ropes are trying to produce thinner and lighter ropes every year. The aim of this study was to develop a proposal for a new, practice relevant sharp edge test standard, which gives manufacturers an easy to use tool for the development of lighter but nevertheless safer products. On the basis of the model experiment a proposal for a new sharp edge resistance test standard could bee formulated and used for a comparison between a variety of ropes that are available on the market.

1 Background

1.1. Sharp Edge Resistance

The standard test for dynamic mountaineering ropes (EN-892/UIAA-101) puts a multiple of the load of a "normal" fall onto the rope. Never the less many experts do not see the greatest thread in these "normal" fall situations, but in downfalls in which the rope does make contact with a sharp edge.

In 2002 the UIAA resolved upon a new test standard for the sharp edge testing of dynamic mountaineering ropes (UIAA – 108), which was suspended in 2004 due to a lack of repeatability and missing relevance for the practice.

1.2. Development Trend in Mountaineering Ropes

Lower weight of the equipment is an advantage in competition in climbing. This is one of the reasons manufacturers of mountaineering ropes are trying to produce thinner and lighter ropes every year. Experts of the Mammut Sports Group AG, a leading manufacturer of climbing- and mountaineering equipment, expect that this development is critical regarding the sharp edge resistance and puts the climber at risk. To find prove for their theory, the Mammut Sports Group AG initiated this study together with the Institute for Sport equipment and Materials.

2 Methodology

In the first step of this study it was necessary to learn more about the failure mechanisms. Experts where questioned, using the Delpy Method, to identify the dangerous fall situations, in which the rope can make contact with a sharp edge. The kinematics and kinetic of these "typical" sharp edge fall situations were then analysed in a field test. The situation, in which it is most likely to damage the rope was then built up in a laboratory experiment and verified with the data acquired in the field test. The set up in the Laboratory made it possible to analyse over 200 different falls with varying fall weight and different types of ropes. The aim of the next and final step has been to find a user-friendly model experiment, which provides the same results as the test experiment but with a higher repeatability.

2.1. Field Test

Firstly the different types of stone, found in popular climbing areas in the Alps where analysed in regard of the shape and the sharpness of the edges. Calk and graphite stone showed the highest potential to produce edges sharp enough to damage the mountaineering rope. In the climbing spot at the Oberjoch in the sector of Weihar a suitable sharp edge of calk stone could be found. Using a dummy, a load cell to measure the tension of the rope and a high speed camera system the following fall situations have been analysed:

- "Pure swing fall" – The climber is at the same height as the carabineer, but not in line. During the fall and the rope is pulled over a sharp edge while the climber swings in line with the carabineer.
- "Swing fall" – same as "pure swing fall" but with the climber falls from above the carabineer. (see Figure 1)
- "Jojo-Movement" – the climber is abseiling and the robe is lying on a sharp edge. The climber will stop his downward movement from time to time causing the rope to move up and down on the edge.

The dummy used, in the experiment consisted of a truck tire field with sand. The overall mass was 80 kg. The tension in the rope during the fall situation was monitored with a load cell. A comparison between the dummy and a real person proves the dummy suitable for the test.

2.2. Laboratory Test

As the "swing fall" situation was identified in the Field Test as the one with the highest potential to damage the rope, a laboratory test stand was build up, to simulate this fall

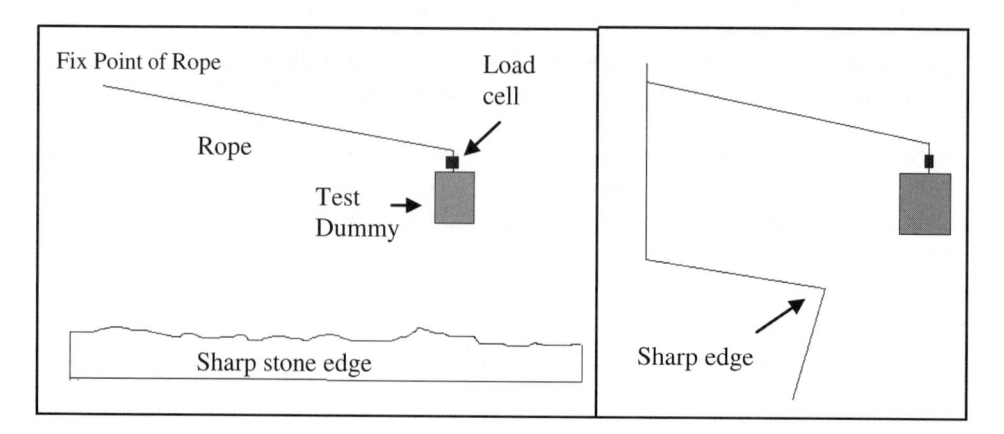

Figure 1. "Swing Fall – Situation" front and side view.

situation. In the test fall weight was increased in 5 kg steps till the rope failed. For every fall a new Rope was used. Seven different rope models, from different manufacturers where tested in this way. As in field test rope tension was recorded and the movement of the rope over the sharp edge was filmed with high speed cameras.

2.3. *Model Test*

The aim of the next and final step has been to find a user-friendly model experiment, which provides the same results as the test experiment, but with a higher repeatability.

Based on the notch bar impact test for metals the model test stand was designed (Figure 4). A specially made cutting tool made from steel cuts through the rope, which is placed in the modified notch bar impact test instead of the metal probe.

Figure 2. Sharp edge field test set up and test equipment.

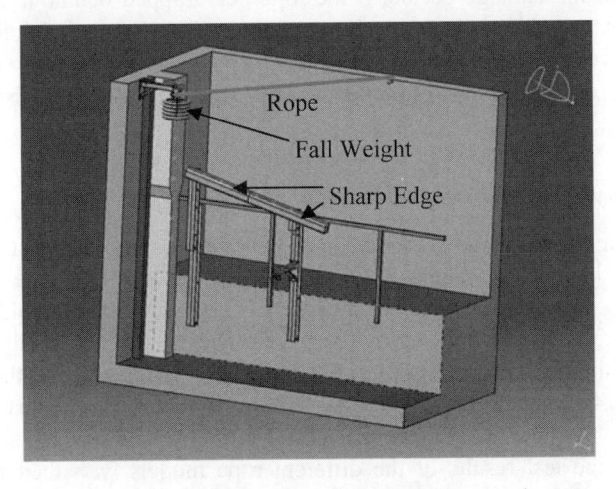

Figure 3. CAD model of laboratory test stand.

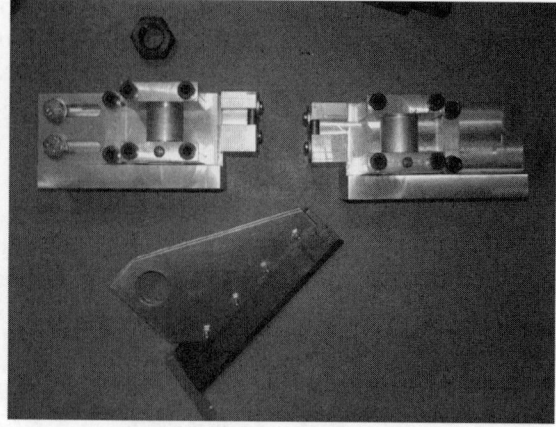

Figure 4. Model test stand with cutting tool and rope acceptance.

3 Results

3.1. Field Test

If a rope does make contact with a sharp edge during the fall, it puts the climber in danger. The "swing fall" could be identified as the situation, in which it is most likely that the rope is damaged or even cut through. The following conclusions could be drawn from the results:

- In all the analysed situations the tension in the rope was not critical for the rope (less than 3 kN).
- If the Rope does glide over the sharp edge in the direction of the rope, the damage that occurs is much lesser, than if the movement is orthogonal to the direction of the rope.
- Most of the damage occurs, if the rope gets trapped behind a jag on the sharp edge. After a certain time the climber has swung far enough that force pulling the rope over the edge is strong enough to rip the rope free. The resulting motion is very quick and mostly perpendicular to the sharp edge, causing most of the damage.

3.2. Laboratory Test

None of the tested ropes could guaranty for the climbers safety. The best half rope in the test managed to hold a fall weight of 45 kg, while the weakest half rope was already cut through by 35 kg. Single ropes performed better, but also would fail to hold a climber with 80 kg. The weakest did fail at 50 kg, while the strongest withstood a fall mass of 65 kg. Figure 5 shows the influence of different fall weights on the damage of the rope. The fall weight used was 45 kg for the top rope in the picture, 50 kg for the middle one and the rope failed at 50 kg.

Comparing the test results of the different rope models with their performance in other standard tests showed no relationship. The performance of a rope in this sharp edge

Figure 5. Influence of different fall weight on the resulting damage of the rope.

resistance test is only more or less directly proportional to the weight of the rope. A new standard for the sharp edge resistance must therefore be found. The test stand used for this laboratory experiment is not suitable for a new standard, as the test does not provide the necessary repeatability. The Results of this experiment are never the less valid, as a large number of experiments (over 200 fall tests) statistically prove the results.

The results also show that with a normal half rope used as a singe rope the risk of having an accident is much higher. If two half ropes are used together to belay the climber they provide always a higher safety standard than the single ropes tested. Tests with half ropes used as a double rope showed that the load gets equally distributed on both of the ropes, which therefore are capable of holding double of the fall weight.

The rope has to withstand a combination of different load cases. As already suspected, the cutting motion could be identified as the load case responsible for most of the damage.

3.3. Model Test

The repeatability of the results approved the test to be suitable to replace the old test standard.

Figure 6 shows how the well the different type of ropes use the material regarding sharp edge performance. The diagram shows the amount of energy needed to cut through the rope divided by its weight/meter.

4 Discussion

The model test developed in this study has the potential to become an easy to use tool for the development of new mountaineering ropes. The test proved practically relevant as its results could be verified with the results from field and laboratory test. A new sharp edge resistance standard based on this model test stand might also make climbers rethink their choice of equipment. The results of this study proved the theory of the Mammut Sports Group AG to be right. Lighter ropes can not offer the same security as heavier ropers. In fall situations in which the rope does make contact with a sharp edge, the lighter rope will fail earlier, even if its performance in the standard tests is the same. The trend to thinner and lighter ropes has therefore to be questioned.

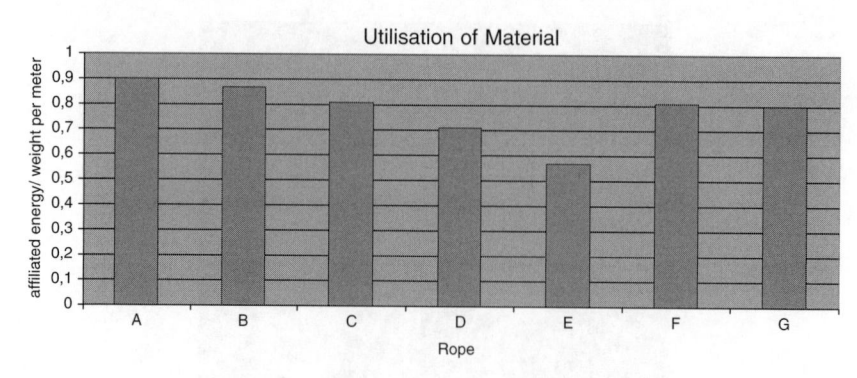

Figure 6. Utilisation of material of different rope models (A–G) in respect to their sharp edge resistance.

References

Riesch P. (2005) Ermittlung der Anforderungen und Randbedingungen für ein Prüfverfahren zur Bestimmung der Schnittfestigkeit von Bergseilen. Dipl-Ing Diss, Technical University Munich, Germany.

Schubert P. (2001) *Sicherheit und Risiko in Fels und Eis*. Vol. 1, 6th Ed., Bergverlag Rother GmbH, Munich.

Schubert P. (2003) *Sicherheit und Risiko in Fels und Eis*. Vol. 2, 1st Ed., Bergverlag Rother GmbH, Munich.

Union Internationale des Associations d'Alpinisme (UIAA, International Mountaineering and Climbing Federation; online). Available: http://www.uiaa.ch, (accessed: 8 May, 2005).

17. Martial Arts and Archery

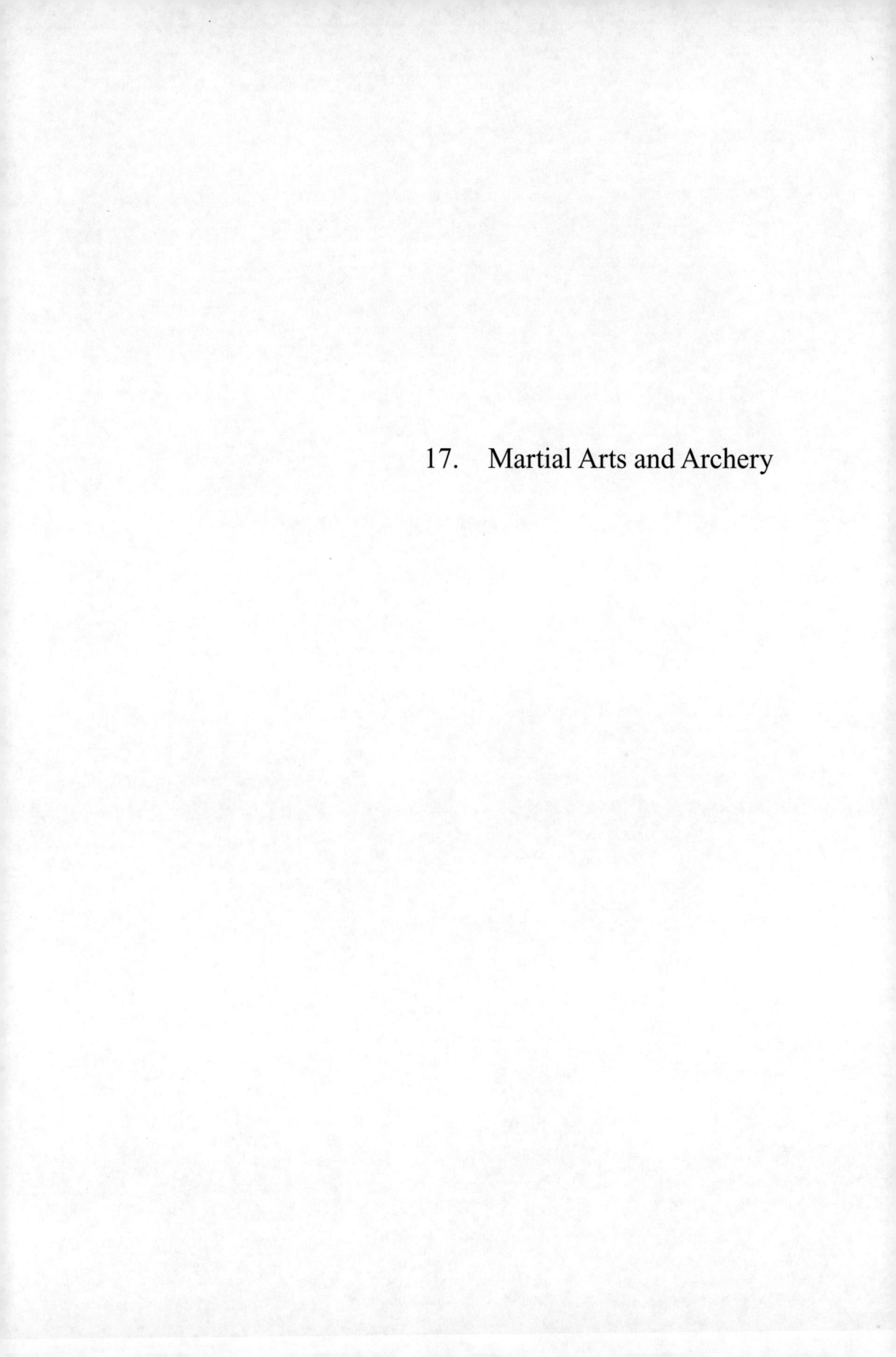

EFFECTIVE BODILY MOTION ON PUNCHING TECHNIQUE OF SHORINJI-KEMPO

T. HASHIMOTO[1], H. HASEGAWA[2], H. DOKI[2] & M. HOKARI[2]

[1]*Graduate School of Engineering and Resource Science, Akita University, Akita, Japan*
[2]*Faculty of Engineering and Resource Science, Akita University, Akita, Japan*

In this study, impact force and reaction force of pivot foot measurements and biomechanical analysis were carried out during the punching motion of SHORINJI-KEMPO to research the effective bodily motion for high impact force generation. Impact forces and kinetic parameters were measured by using strain gages and the three-dimensional rate gyro sensors. The reaction force of pivot foot was also measured by a 6-axis force sensor. The trained SHORINJI-KEMPO players showed the larger impact forces in contrast to those for the un-trained player. The impact force was affected by the difference among the motions of a waist, a shoulder and a thrusting arm. For the trained player the motions of the waist and shoulder started just before the thrusting arm began to move. Furthermore, it is important to increase the impact force that the player supports the body with the toe side and the inside of the pivot foot when the player starts the punching motion.

1 Introduction

SHORINJI-KEMPO (World Shorinji Kempo Organization) is a self-defense art using the vital spots of the body and it is possible even for a physically weaker person or female to knock down their opponent. The techniques of SHORINJI-KEMPO are divided into three categories: GOHO, JUHO, and SEIHO. For example, GOHO (hard-part) refers to attacks like kicks, punches, strikes and blows, and their defenses. SHORINJI-KEMPO is an effective application of the principles of the skeletal, biological, physical rules of the human body. For these reasons it is not just the learning of self-defense techniques, but also a very effective way of increasing health through lifelong exercise, therefore the aim of everyday practices is to achieve physical and mental improvement. Generally, the sensuous coaching based on the experience is preformed for the GOHO technique because the actual human movements in the punching motion are obviously complicated. If the quantitative description of human movements in punching motion is taken into account, it is presumable that the physical and mental improvement is easily achieved in everyday practices.

Considerable research efforts by a number of researchers have been devoted to advancing the punching technique of martial arts (Wilk *et al.*, 1983; Yoshihuku, 1984). However, the reaction force of pivot foot was not measured sufficiently for the punching motion in previous studies. In this study, the effective motions to increase the impact force and the relationship between impact forces and pivot foot reaction force were investigated for the punching motions of GOHO technique. The impact force measurements were carried out during the punching motions by using the impact force measuring device.

2 Experimental Apparatus and Method

2.1. *Experimental Apparatus*

Figure 1 shows an impact force measuring device proposed in this study. The impact force measuring device mainly consists of the iron plate with the strain gages. The size of the punching target is 100 × 100 mm. The height of the punching target can be adjusted to fit the subjects. The punching target was covered with rubber to protect the subject's fist.

It was confirmed that the impact force measurements do not affected by the rubber cover.

Figure 2 shows the cartoon depicting sensor positions (left lower thigh, left thigh, shoulder, waist, right upper arm and right forearm). The three-dimension gyro sensor was used for the motion analysis (Hokari *et al.*, 2002). Kinetic parameters at six parts of the body were measured by using the three-dimensional rate gyro sensors. This three-dimensional rate gyro sensor dimensions were 50 × 30 × 20 mm. These sensors do not disturb the punching motion, because the sensors are small. To measure the force of pivot foot, a six component contact force measurement device with dimensions of 360 × 360 mm is used. The reaction force (F_x, F_y, F_z) of pivot foot and the moment (M_x, M_y, M_z) around each axis are measured by the force measuring system as shown in Figure 2. Moreover, the horizontal distance (a_x, a_y) from the center of 6-axis force sensor to the point of reaction force application is given by

$$a_x = \frac{-F_x a_z - M_y}{F_z}, \quad a_y = \frac{-F_y a_z + M_x}{F_z} \tag{1}$$

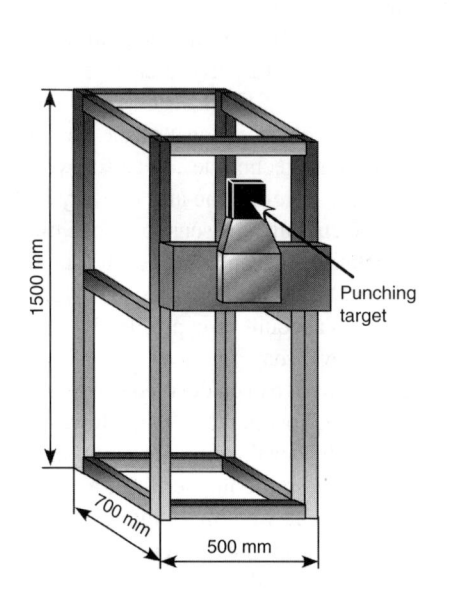

Figure 1. Impact force measuring device.

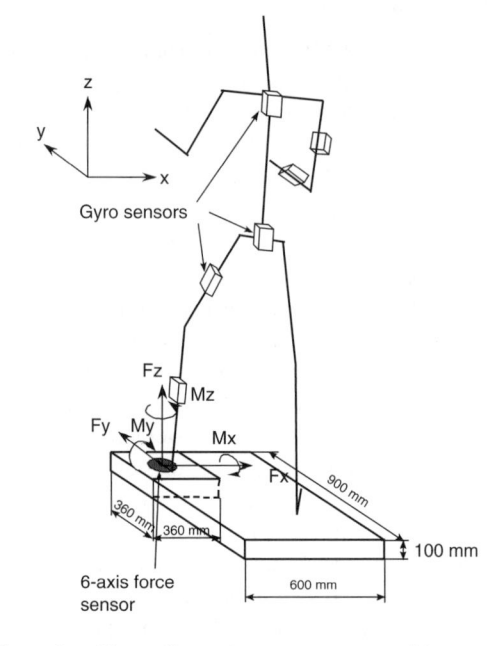

Figure 2. Three-dimension gyro sensor positions and schematic of reaction force measurements.

where a_z is the vertical distance from the center of 6-axis force sensor to the point of reaction force application. The details of the point of reaction force application are shown in Figure 3.

2.2. Experimental Method

The punching motions were adopted to analyze the SHORINJI-KEMPO motion, and two sorts of GYAKU-TSUKI (reverse punch) and UCHI (hand knife strike) motions were applied to collect date of biomechanics. GYAKU-TSUKI (reverse punch) and UCHI (hand knife strike) motions were punching motions as shown in Figure 4, respectively. In order to measure the reaction force of pivot foot, the subjects put the pivot foot on the sensor from the beginning and punched keeping their feet position. The subjects were two trained SHORINJI-KEMPO players, subjects A and B, and un-trained people, subject C.

3 Experimental Results and Discussion

To illuminate the bodily motion on punching technique, it is useful to discuss the relationship between impact forces and kinetic parameters, particularly the Tsuki motions of several subjects, because the results show the same tendency for the Tsuki and Uchi motions in this study. Table 1 shows bodily characteristics and impact force. The trained SHORINJI-KEMPO players showed the larger impact forces in contrast to those for the un-trained people. The

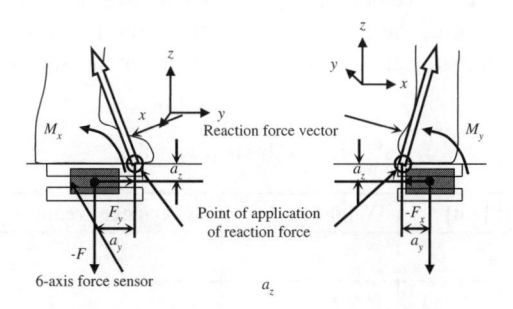

Figure 3.   Coordinate system and point of reaction force application.

(a) Tsuki motion   (b) Uchi motion

Figure 4.   Punching motions.

impact force was affected by the difference among the motions of a waist, a shoulder and a thrusting arm. For the trained players the motions of the waist and shoulder started just before the thrusting arm began to move. The two trained players, subjects A and B, indicated similar bodily attributes. However, the measured impact forces denote different value for subjects A and B. The impact forces of the subjects A and B were 1994N and 1571N, respectively (see Table 1). In order to better understand the reasons for the different impact forces, the motion of the pivot leg was measured.

Figure 5 shows the z-axis reaction force F_z of the pivot foot. In Figure 5, the solid line indicates the instant of impact. The dotted lines indicate the time when the reaction force F_z shows the maximum value before the impact. In this contact force measurement device, F_z indicates the maximum value when the subject moves the leg and steps to punch the target. Figure 6 shows the point of reaction force application of the pivot foot, which is marked by the circular symbol, when the subjects put the weight on their pivot leg. The distances a_x and a_y were calculated by using the Eg. (1). Subject A supported the body with the toe side and inside of the pivot foot. On the other hand, the subject B was supported the body with the toe side and the outside of pivot foot.

Figure 7 shows the angular velocity along the z-axis of shoulder, waist and thigh. The longitudinal solid line and the dotted lines are the same as the lines of Figure 5. For the subject A, the angular velocity of the thigh showed a minus value at point "A_1". The subject A moved the knee of the pivot leg slightly inward at the instant of impact. On the other hand, for the subject B, the angular velocity of the thigh showed a plus value at point "B_1". The subject B moved the knee of the pivot leg outward at the instant of impact. The angular velocity of waist for the subject A is lower than that for the subject B at the instant of

Table 1. Body characteristics.

	Height [cm]	Weight [kg]	Experience [year]	Impact Force [N]
Subject-A	174	67	4	1994
Subject-B	172	64	3	1571
Subject-C	172	60	0	1324

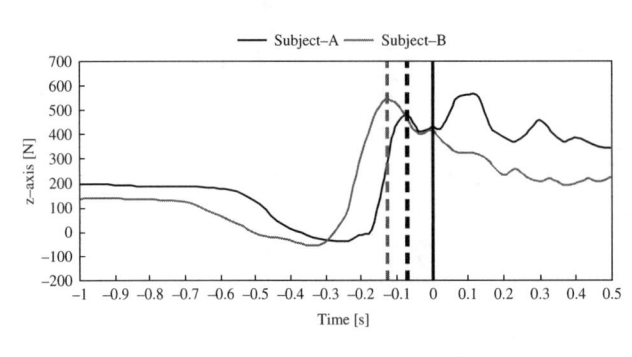

Figure 5. Reaction force F_z of the pivot leg.

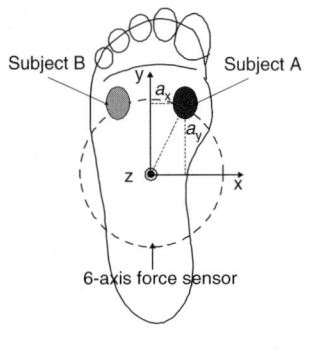

Figure 6. Point of reaction force application.

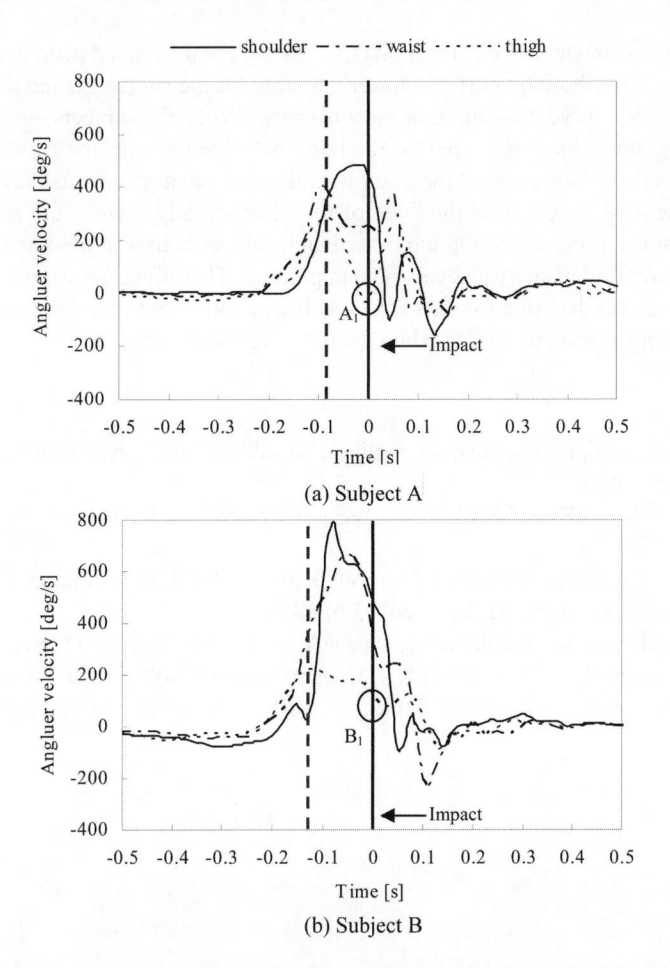

Figure 7. Angular velocity of shoulder, waist and thigh during Tsuki motion.

impact. For the subject A, the angular velocity of shoulder increases until the impact. However, for the subject B, the angular velocity of shoulder indicates the peak after the subject puts the weight on the pivot leg after which it decreases until the impact. Consequently, the subject A suppressed the rotational motion of the waist and kept the large angular rotation velocity of the shoulder because the knee of the pivot leg was moved inward at the instant of impact. On the contrary, the subject B moved the knee outward and did not suppress the rotational motion of the waist at the impact. Therefore, the impact force for the subject A is larger than that for the subject B and the impact force is affected by the pivot leg movements, which is mentioned above.

4 Conclusions

In this study, impact force and reaction force of pivot foot measurements were carried out during the punching motion of SHORINJI-KEMPO to research the effective bodily motion for high impact force generation. The trained SHORINJI-KEMPO players, subject A and B, showed the larger impact forces in contrast to those for the un-trained people. Furthermore, the impact force for the subject B was lower than that for the subject A, and the differences of motion between subjects A and B were confirmed. When the subjects put the weight on their pivot leg, the subject A supported the body with the toe side and inside of the pivot foot, and the subject B supported the body with the toe side and the outside. At the instant of impact, the subject A moved the knee of the pivot slightly inward and suppressed the rotational motion of the waist. On the other hand, subject B moved the knee outward and the rotational motion of the waist was not suppressed. Therefore, the impact force for the subject B was lower than that for the subject A. It was confirmed that the impact force during the punching motion was affected by the pivot leg movements.

References

World Shorinji Kempo Organization (online). Available: http://www.shorinjikempo.or.jp/wsko/index.html

Wilk S.R., McNair R.E. and Feld M.S. (1983) The physics of karate, American Journal of Physics, 51/9, 783–790.

Yoshihuku Y. (1984) Impact Force of Various Kinds of fighting techniques (in Japanese), Japanese Journal of Sports Science, 13/6, 485–491.

Hokari M., Watanabe K., Kurihara Y., Segawa Y. and Naruo T. (2002) Kinematic Analysis and Measurement of Sports Form – Measurement of Golf Driver Swing Form – (in Japanese), Transaction of The Society of Instrument and Control Engineers, 38/11, 922–930.

THE OPTIMUM DRIVING MODEL OF JUMP BACK KICK IN TAEKWONDO

C.-L. LEE

National Taiwan Normal University, Taipei, Taiwan

The purpose of this study is to find an optimum driving model from trunk and an attack leg in jump back kick. Six male Taewondo athletes (age: 21.2 ± 2.3 yr. old; height: 180.8 ± 4.7 cm; mass: 80.7 ± 11.6 kg) participated in this study. Two Redlake high speed cameras (60 Hz) collected the movement. Kwon 3D software was used to analyze the movement. The result shows that the trunk performs a countermovement in order to maintain the body's balance in the attack phase. When the attack leg hits a bag, the hip and knee extend. Through the analysis of trunk twist and joint extension of the attack leg, the optimum driving model of jump back kick can be analysed in this study.

1 Introduction

In Taekwondo combat, the main attack movements include round house kicks and axe kicks because these two type kicks are executed at high speed and increase chance to score. Therefore, many coaches combine these two kicking movements when training Taekwondo athletics. Using these kicks is considered the most offensive attack of athletics (Roh & Watkinson, 2002; Lee et al., 2005). However, back kick and jump back kick are the best counterattack movements against round house kick and axe kick. Therefore, back kick and jump back kick are important attack movements in Taekwondo combat.

Back kick and jump back kick are linear kicking movements in contrast to round house kick and arc kicking. This is why back kick and jump back kick are powerful attack movements. When the attack leg kicks a target, the knee joint extension will influence the attack effect in every stage of the movement. Lee & Huang (2006) compared back kick, jump back kick and 360° jump back kick. Jump back kick is better then back kick and 360° jump back kick, producing a larger attack force and angular velocity of the lower trunk. In combat situations, considering changing rules, jump back kick is a more suitable counterattack movement than back kick in the match. Because the jump back kick is not used frequently and not studied in depth, finding the correct driving model is the priority task to teach the coach and train the athletes. The purpose of this study is to find an optimum driving model from trunk twist and joint movement of an attack leg in jump back kick.

2 Methods

2.1. *Subjects*

All subjects are experienced in national or world class Taekwondo athletes, six males, age: 21.2 ± 2.3 year old; height: 180.8 ± 4.7 cm, mass: 80.7 ± 11.6 kg, learning age: 11.8 ± 4.7 years).

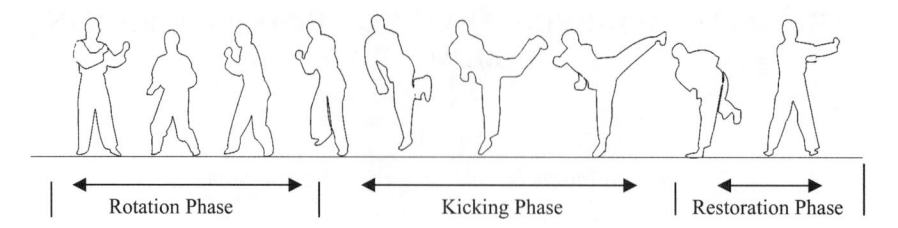

Figure 1. Illustration of jump back kick.

2.2. Research Equipment and Data Collection

The equipment used is two Redlake cameras and Kwon 3D motion system. Two Redlake (60 Hz) cameras were genlocked to collect the data of movements. The Kwon 3D motion system was used to analyze the kinematics data. The subjects performed the jump back kick (Figure 1) three times to the bag.

2.3. Definition and Data Analysis

The movement is divided into rotation, kicking and attack phases. The rotation phase is defined from the beginning of trunk rotation to take-off of the support leg. The kicking phase is defined from take-off to the initial contact at the kick bag. The attack phase starts when the leg contacts the bag. The body segment parameters were taken from Zatsiorsky & Seluynov (1983). The trunk was divided into upper and lower trunk, divided by the navel. The attack velocity is defined as the maximum velocity of the attack leg during the attack phase. The moving distance is defined as the horizontal displacement of support leg from jump to landing. The data was analyzed as to the maximal kicking velocity. The parameters analysed are: angular velocity of trunk and attacking leg as well as the attack velocity and the moving distance. The independent t-test was used to determine whether differences between mean were statistically significant. Statistical significance was set to $p < 0.05$.

3 Results

3.1. Trunk Twist

There are two patterns of trunk twist. The main difference is rotation direction of upper and lower trunk during the attack phase. Pattern I (Figure 2) shows countermovement of lower trunk in relation to the attack leg in the attack phase (one subject). Pattern II (Figure 3) shows countermovement of upper trunk in relation to the attack leg in the attack phase (five subjects).

3.2. Joint Flexion/Extension of an Attack Leg

There are two types in joint flexion/extension of an attack leg. The main difference is time sequence of when a joint is fixed in the attack phase. Type I (Figure 4) shows a fixed hip joint and then a fixed knee joint in attack phase (three subjects). Type II (Figure 5) shows a fixed knee joint and then a fixed hip joint in attack phase (three subjects).

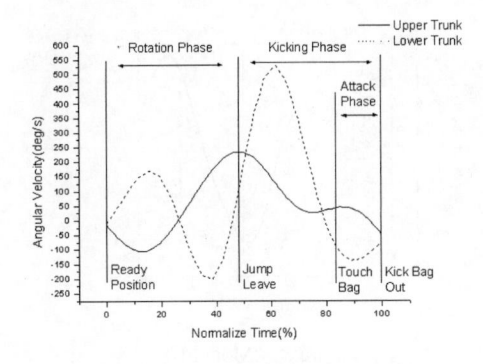

The Optimum Driving Model of Jump Back Kick in Taekwondo 717

Figure 2. Pattern I.

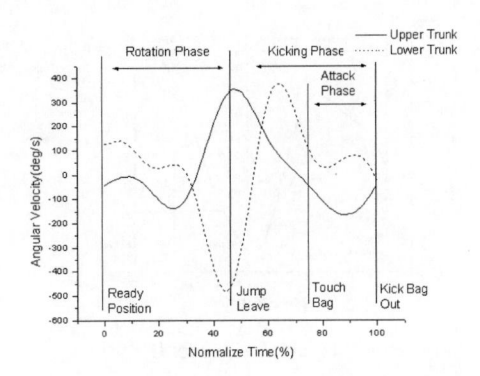

Figure 3. Pattern II.

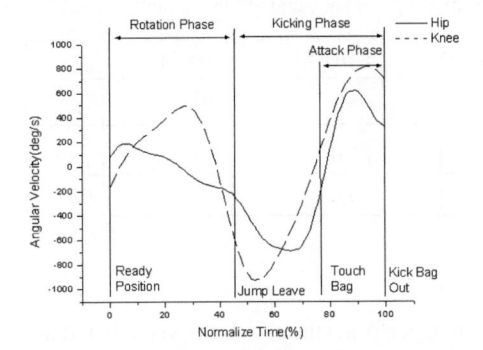

Figure 4. Type I.

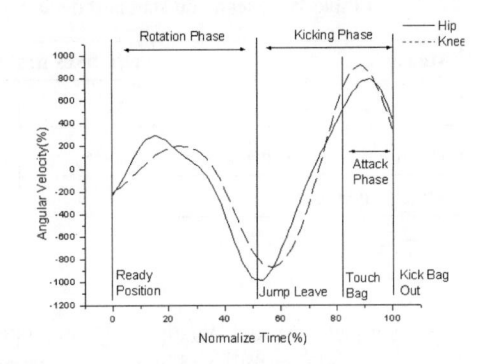

Figure 5. Type II.

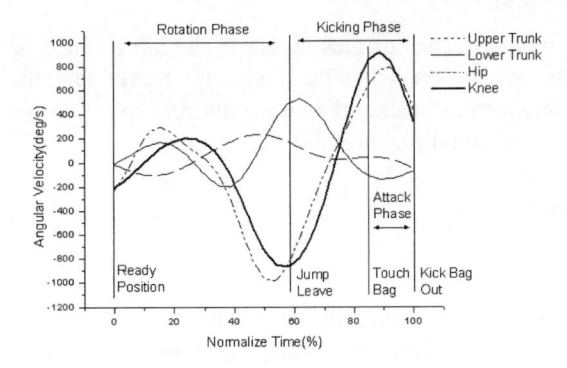

Figure 6. Model I.

3.3. Driving Model

The driving model combines the trunk twist and joint flexion/extension. Three different driving models can be distinguished. Model I (Figure 6): the lower trunk performs a countermovement and the knee is locked first followed by the hip (one subject). Model II (Figure 7): the upper trunk performs a countermovement and the hip is locked first, followed by the knee

Figure 7. Model II.

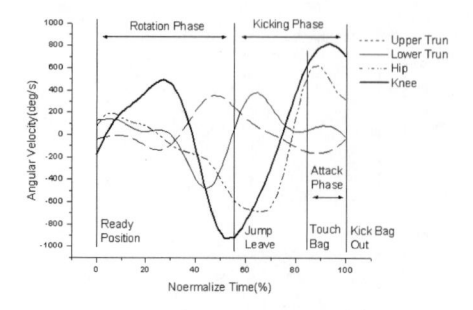

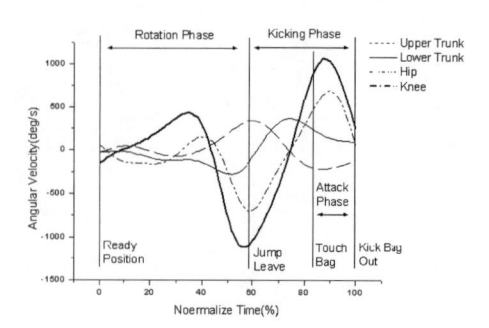

Figure 8. Model III.

Table 1. Mean and standard deviation of moving distance and attack velocity.

Measure	Hip fixes first (n=3)		Knee fixes first (n=3)		
	M	SD	M	SD	t
Moving distance (cm)	46.4	2.5	25.0	9.9	3.6*
Attack time(sec)	6.7	0.3	6.6	0.2	0.2

*p < .05

(three subjects). Model III (Figure 8): the upper trunk performs the countermovement and the knee is locked first followed by the hip (two subjects).

3.4. Statistical Test

From the time sequence of locking either the hip or knee first, the attack phase can be divided into two groups. When the hip joint is locked first, the moving distance is significantly different from locking the knee joint first ($p < .05$). However, the attack velocity shows no significant difference.

4 Discussion

4.1. Trunk Twist

When a person runs, the trunk and lower limb rotate in opposite direction about the twist axis during a single stride. This motion can be quantified as the angular momentum. In this situation, if the trunk has an angular momentum in the downward direction, and the leg has an angular momentum in the upward direction (Hrinrichs, 1987). For the same reason, the trunk and the attack leg rotate in opposite direction when attack leg hits the bag. As the trunk was divided into upper and lower trunk in this study, we can observe two patterns. In pattern I, the upper trunk's only function is to perform rotation. The lower trunk mainly controls the body rotation, and it makes a countermovement which maintains the balance in the attack phase. In pattern II, the upper trunk not only rotates but also makes a countermovement to maintain the balance in the attack phase. When the upper trunk is stable, the lower trunk starts to rotate at take-off. This is different from pattern I in which the trunk continuously rotates after leaving the ground. Pattern II has an advantage as the lower trunk can bring the attack leg to kick target. The trunk twist followed pattern II in five subjects. This study suggests that this is the optimum pattern. The trunk's main task is to keep the body in balance, and the motion of the arms is of assistive function.

4.2. Joint Flexion/Extension of an Attack Leg

From Figure 4 and 5 we can see that the main contribution is extension of the knee joint during the attack phase. Figure 4 is related to the kinetic chain which locks the hip joint and keeps the knee extended. Form Figure 5 we see that the hip joint is locked after the knee joint. Although the attack velocity shows no significant difference, the moving distance can infer that when the hip joint is locked first, the kick bag force and bag deformation is larger through knee joint extension. However, as the force measurement is difficult, we use the attack velocity as an estimate of the attack force. Further research, however, is required. When attack leg impacts the bag, the hip and knee joint will remain extended.

4.3. Driving Model

The trunk twist, combined with joint extension, leads to the three models described above. Model I is not economic as it utilizes the trunk to drive the attack leg during kicking. Model II and III are better driving models. However, considering the hip and knee locking sequence, model II seems to be better than model III. The upper trunk makes a countermovement to maintain the body balance in the attack phase. To sum up, Model II is suggested to be the optimum driving model of jump back kick.

References

Hrinrichs R.N. (1987) Upper extremity function in running. II: Angular momentum consideration. *International Journal of Sports Biomechanics*, 3, 242–263.

Lee C.L., Chin Y.F. and Liu Y. (2005) Comparing the Difference between Front-leg and Back-leg Round-house Kicks Attacking Movement Abilities in Taekwondo. Proceedings of XXIII International Symposium on Biomechanics in Sports. Vol. 2, pp 877–880, Beijing, China: The China Institute of Sport Science.

Lee, C.L. and Huang C. (2006) Biomechanical Analysis of Back Kicks Attacking Movement in Taekwondo. . Proceedings of XXIV International Symposium on Biomechanics in Sports. Vol. 2, pp 803–806, Salzburg, Austria: The University of Salzburg.

Roh, J.O. and Watkinson, E.J. (2002) Video analysis of blows to the head and face at the 1999 World Taekwondo Championships. *The Journal of Sports Medicine and Physical Fitness*, 42/3, 348–53.

.Zatsiorsky V. and Seluynov V. (1983). The mass and inertia characteristics of the main segments of the human body. In: Mastui H. and Kobayashi K. (Eds.), *Biomechanics VIII-B*. Champaign, IL: Human Kinetics.

NON-LINEAR VISCOELASTICITY OF KARATE PUNCHING SHIELDS

J.K.L. TAN & F.K. FUSS

Sports Engineering Research Team, Division of Bioengineering, School of Chemical and Biomedical Engineering, Nonyang Technological University, Singapore

Karate punching shields consist of foam material. Four shield models were investigated and tested as to stress relaxation, and stress-strain behaviour at different strain rates. Stress relaxation tests revealed that the visco-elastic behaviour of the shields and their polymer foams followed the power law. Thus, the relationship between the logarithm of the modulus and the logarithm of the strain rate is linear, and allows extrapolating the modulus to higher strain rates. Out of the four shields tested, the closed-cell foam was more viscous and also stiffer then the other three shields. The knowledge of viscosity and stiffness of foams is generally important for sports applications, as foams are widely used for shock absorption and injury prevention.

1 Introduction

Foams are cellular materials with the ability to undergo large deformation at nearly constant stress. They are used for absorbing the energy of impacts in packaging and crash protection. Such application requires knowledge of the foams properties, especially their compressive response at various impact velocities or strain rates (Ruan *et al.*, 2002).

Foam materials have a wide application in sport, e.g. in shoes, protection equipment, and artificial turfs, be it for damping, shock absorption or injury prevention. As Karate is by nature a high impact sport, foam materials are used as punching shields to cushion the impact of various movements and techniques. Karate training is usually carried out with punching shields to deliver full-power strikes while avoiding injuries of the training partner. Therefore, there is an interest to investigate the nature of the various punching shields at different speed. Several models of shields are available in the market, however, specifications as to the structural properties are neither provided by the manufacturers nor in the literature.

The aim of this study is to investigate the non-linear visco-elastic properties of four different shield models in terms of stress relaxation and modulus, and to develop a mathematical model for extrapolation of the modulus to maximal striking speed.

2 Experimental

The models investigated are shown in Figure 1; their details are listed in Table 1.

For both stress relaxation and compression tests, we used an Instron material testing machine (model no.: 3366). For the stress relaxation tests, the shields were pre-loaded (F_0) between 1 kN and 7.6 kN and the decreasing load F was measured for $t = 3600$ seconds. As the stress relaxation generally showed a linear behaviour when plotting LN(F) against LN(t), the power law was selected to model the non-linear visco-elastic properties:

$$F = A\, t^{-B} \qquad (1)$$

Figure 1. Shields investigated; from left to right: Adidas, "Blue", "Black", Victory.

Table 1. Shields investigated ("Blue" and "Black" do not have a specific brand name).

Model	Size (cm)	Thickness (cm)	Cover material	Core material
Adidas	40 × 59	122	Polymer	Open-cell polymer foam
"Blue"	24 × 47	131	Leather	Open-cell polymer foam
"Black"	350 × 470	99	Polymer	Open-cell polymer foam
Victory	380 × 570	171	Polymer	Closed-cell polymer foam

where t is the relaxation time, and A and B are constants.

Considering A as a function of F_0 or initial stress σ_0, and the latter a function of the strain ε, then we get:

$$F = F_0 A_F t^{-B} \quad (2)$$

$$\sigma = \varepsilon_0 A_\varepsilon t^{-B} \quad (3)$$

The viscosity constant B of the power law is independent of σ and ε.

For the compression tests, the shields were loaded up to 9 kN with crosshead speeds of 500, 160, 50, 16, and 5 mm/min. Each shield was tested five times at the 5 different velocities. Load and displacement (deformation) were converted into stress σ and nominal strain ε, by dividing the load by the area of the compression plate, and the displacement by the thickness of the shield. The stiffness, or tangent modulus E, of the shields was calculated from σ and ε by numerical differentiation.

From Eq. (3), the modulus – strain rate relationship is calculated, comparable to the procedure developed by Fuss (2007) for the logarithmic law and Vikram & Fuss (2007) for the power law. In Eq. (3), ε_0 is the constant strain, applied by a Heaviside function H(t):

$$\varepsilon = \varepsilon_0 \, H(t) \quad (4)$$

Taking Laplace transform of eqns 3 and 4:

$$\hat{\sigma} = \varepsilon_0 A \frac{\Gamma(-B+1)}{s^{-B+1}} \tag{5}$$

$$\hat{\varepsilon} = \frac{\varepsilon_0}{s} \tag{6}$$

where the caret (^) denotes the transformed parameter, and Γ denotes a gamma function.
Substituting Eq. (6) into Eq. (5), we get the constitutive equation of the power law of viscoelasticity:

$$\hat{\sigma} = \hat{\varepsilon} A \frac{\Gamma(-B+1)}{s^{-B}} \tag{7}$$

In order to establish the relationship between stress and strain rate, as well as modulus and strain rate, we apply a ramp function:

$$\varepsilon = \dot{\varepsilon}_0 t \quad \text{or} \quad \hat{\varepsilon} = \frac{\dot{\varepsilon}_0}{s^2} \tag{8}$$

where $\dot{\varepsilon}_0$ is the constant strain rate.
Substituting Eq. (8) into Eq. (7)

$$\hat{\sigma} = \dot{\varepsilon}_0 A \frac{\Gamma(-B+1)}{s^{-B+2}} \tag{9}$$

After applying the recursion formula of the Gamma function, we obtain:

$$\hat{\sigma} = \dot{\varepsilon}_0 \frac{A}{1-B} \frac{\Gamma(-B+2)}{s^{-B+2}} \tag{10}$$

After taking inverse Laplace transform, rearranging, and replacing t by $\varepsilon/\dot{\varepsilon}_0$:

$$\sigma_\varepsilon = \varepsilon^{1-B} \dot{\varepsilon}_0^B \frac{A}{1-B} \tag{11}$$

The strain derivative of the stress σ is the modulus E:

$$E_\varepsilon = A \varepsilon^{-B} \dot{\varepsilon}_0^B \tag{12}$$

The modulus – strain rate relationship results after exchanging the variable and the constant: the variable strain becomes a specific strain ε_i, and the constant strain rate becomes the independent variable v, the punching velocity.

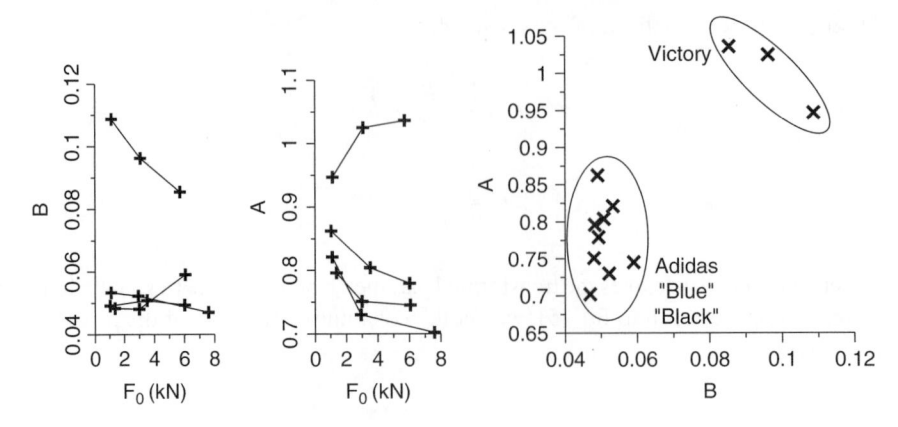

Figure 2. B and A_F vs. F_0, and A_F vs. B.

$$E_v = A\, \varepsilon_i^{-B}\, v^B \qquad (13)$$

Eq. (13) has the same structure as Eq. (1): the independent variable to the power of B times a constant. Eq. (13) allows extrapolating the stiffness or modulus E for higher velocities at a specific strain. The reason why the cross head speed was applied as a uniform absolute velocity and not as a standardized strain rate is because the maximal velocity of the fist during punching does not depend on the thickness of the shield. In experienced Karatekas (dan 1–4), the punching velocity of the fist in the Kizami-Zuki technique is 7–8 m/s (Mehanni, 2004). The stiffness of the shields investigated was extrapolated to 10 m/s, for strains of 0.05, 0.1–0.8 in 0.1 steps, and additionally for a strain of 0.9 in the Adidas shield.

3 Results

3.1. Stress Relaxation

The results of the stress relaxation tests are shown in Figure 2. The Victory shield (closed cell foam) has higher B and A_F values than the open cell foam shields.

3.2. Compression Tests

Uniaxial compression of foam shields shows the typical three regions of compression response of foam material: an initial linear elastic region, an elastic collapse plateau where the cells deform plastically, and terminal densification region (Gibson & Ashby, 1997). In Adidas and "Black" shields, the transition between the three regions is more sudden and thus clearly visible, whereas "Blue" and "Victory" shields exhibit a more gradual transition (Figure 3). This becomes evident from the modulus – strain curves as well (Figure 4).

3.3. Extrapolation to Higher Punching Velocities

The modulus of the 4 shields, extrapolated to 10 m/s is shown in Figure 5. The relationship between "Blue", "Black" and Adidas does not change distinctly, whereas the modulus of the Victory shield increased more than in the other 3 shields. This matches the result of the

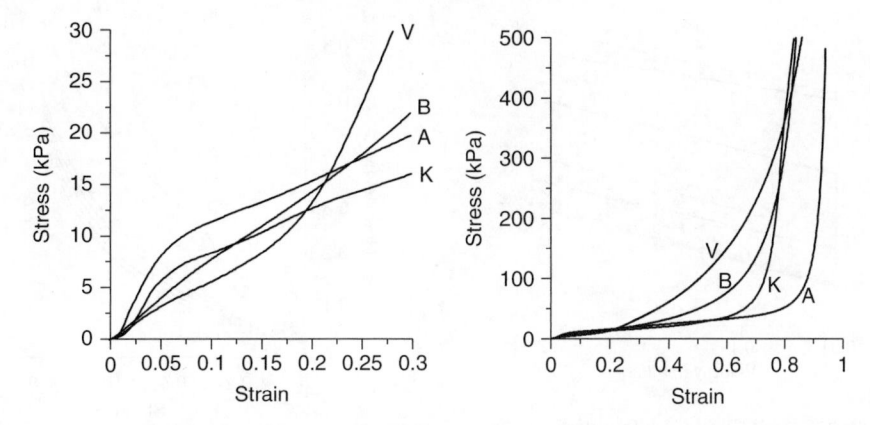

Figure 3. Stress – strain curves of the 4 shields at a deformation rate of 500 mm/min (A = Adidas, B = "Blue", K = "Black", V = "Victory").

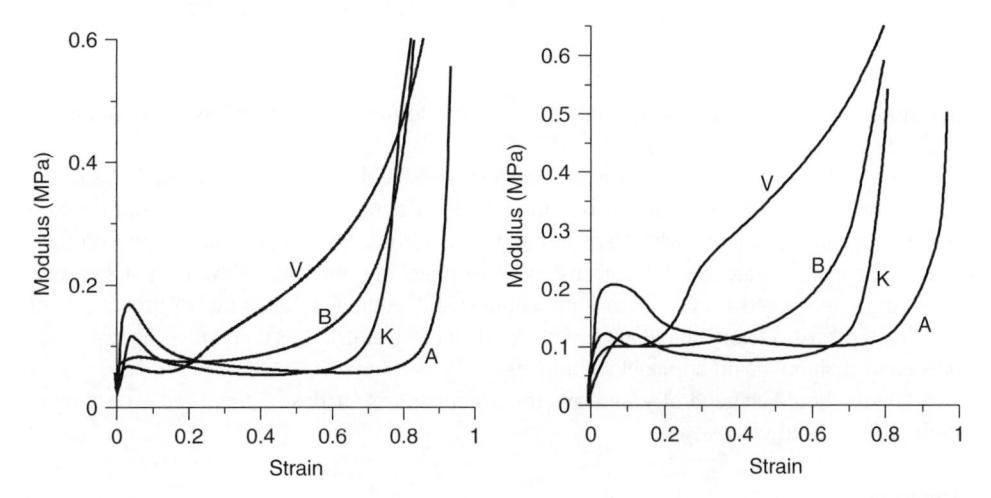

Figure 4. Modulus – strain curves (deformation rate: 500 mm/min).

Figure 5. Modulus – strain curves extrapolated to 10 m/s.

stress relaxation tests, which proved that the Victory shield is more viscous than the other 3 shields. In Figure 6, the extrapolation process is exemplified in the Adidas shield, and Figure 7 shows the final extrapolation result in the Victory shield

4 Discussion

In theory, the constants A_F and B in Eq. (2) are independent of stress and thus material constants (Findley et al., 1976). However, in the punching shields investigated, A and B are functions of the load, and thus of the initial strain.

The higher the viscosity constant B, the faster the shield relaxes, and the more the stiffness or modulus increases with the velocity or strain rate. Ruan et al. (2002) observed in

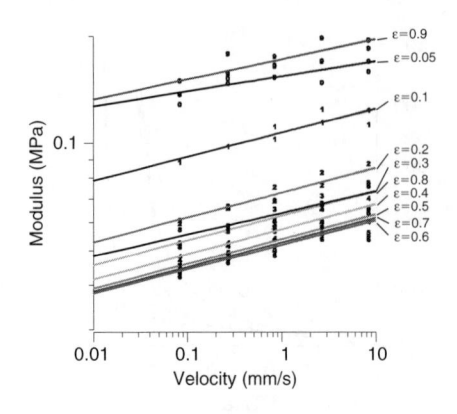

Figure 6. Power fit of the modulus at different strains (Adidas shield).

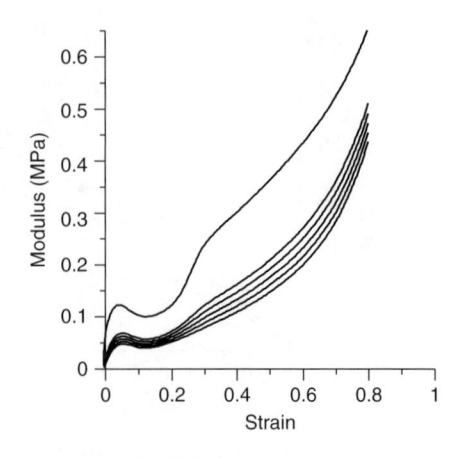

Figure 7. Modulus of the 5 cross head speeds and extrapolated to 10 m/s (Victory shield).

aluminum foams, that the strain rate dependency follows the power law, comparable to Eq. (13).

Out of the four shields investigated, the Victory shield proved to be the stiffest and the most viscous one. This knowledge is important for Karate beginners, who should develop power punching in shields, which are soft and less viscous so as to not cause injury on their wrist. As beginners are not fully accustomed to punching, they have lower wrist strength. Punching a stiffer and a more viscous punching shield would result in a higher probability of injuring the wrist. Victory shield are suited for advanced Karatekas as it provides the resistance needed to cushion the high impact techniques.

Foams should be generally, and specifically for sports purposes, classified according to their stiffness and viscosity.

References

Findley W.N., Lai J.S. and Onaran K. (1976) *Creep and Relaxation of Nonlinear Viscoelastic Materials*. Dover Publications, New York.
Fuss F.K. (2007) Non-linear viscoelastic properties of golf balls. In: Fuss F.K., Subic A. and Ujihashi S. (Eds.) *The Impact of Technology on Sport II*. Taylor & Francis Group, London.
Gibson L.J. and Ashby M.F. (1997) *Cellular Solids: Structure and Properties*. Cambridge Universtiy Press, Cambridge
Mehanni A.M.S. (2004) Kinematische und dynamische Biomechanik des Prellstosses Kizami-Zuki beim Karate. PhD Diss., University of Konstanz, Germany.
Ruan D., Lu G., Chen F.L. and Siores E. (2002) Compressive behaviour of aluminium foams at low and medium strain rates. Composite Structures, 57, 331–336.
Vikram B. and Fuss F.K. (2007) Non-linear viscoelastic impact modelling of cricket balls. In: Fuss F.K., Subic A. and Ujihashi S. (Eds.) *The Impact of Technology on Sport II*. Taylor & Francis Group, London.

BIOMECHANICAL ANALYSIS OF TAI CHI DIFFICULTY MOVEMENT "TENG KONG ZHENG TI TUI"

Y.K. YANG[1,2], W. XIE[3], D. LIM[3] & J.H. ZHOU[4]

[1]*School of Mechanical and Aerospace Engineering, Nanyang Technological University, Singapore*
[2]*Singapore National Wushu Federation, Singapore*
[3]*Biomechanics Department, Singapore Sports Council, Singapore*
[4]*Sports Biomechanics Division, Chengdu Sports University, China*

The purpose of this paper was to analyze the technical principles behind the new Tai Chi Difficulty Movement (DM) known as Flying Front Rise Kick (commonly known as "Teng Kong Zheng Ti Tui") using kinematics data. DM are technical movements that involve jumps, rotation and balance; they are harder to be executed compared to traditional TC movement. The video footages of six elite female China athletes from the "Tai Chi Quan" event were captured using three Panasonic NVGS-50 50 Hz camera during their China National Games in 2005. One trial of the "Teng Kong Zhen Ti Tui" was recorded for each athlete. The data was then processed in using Direct Linear Transformation technique on Peak Motus® motion analysis software. The DM was split into three phases. The critical factor for each phase was identified. Phase one: jump height; Phase two: preparation for landing; Phase three: landing stability. Results have shown that each critical factor was affected by different parameters such as take off angle and change in knee angle. The success of the DM may depend on more parameters than those discussed in this paper. Further investigation with kinetics data and modeling should be done to understand the importance of specific muscle groups that help in the jump.

1 Introduction

The history of Tai Chi dates back to hundreds of years ago. Recently, in 2005, new Wushu competition rules were implemented by International Wushu Federation (IWuF). The incorporation of new Difficulty Movements (DM) into traditional Tai Chi (TC) moves had brought about new challenges to athletes doing competitive TC internationally. DM are technical movements that involved jumps, rotations and balance; they are harder to execute compared to traditional TC movement. In the past, TC was about the flow of the movements and stability of the stances. Many studies have shown that TC can help improve the stability of a person (Mao *et al.*, 2006). Today the new DM requires the athletes to jump and land on single leg. Implication of Anterior Cruciate Ligament (ACL) injuries especially during landing (Boden *et al.*, 2000), has been a concern to coaches and female athletes had higher chances of ACL injuries (Nagano *et al.*, 2006). Because the new rule was implemented for only a short period, there is still no journal published regarding the DM.

Because of the sudden change in the technical requirements in competitive TC there is a need to analyze the new DM to help the athlete to master the movement more efficiently. The objective of an athlete is to master the movement in the shortest possible time and protect themselves from sustaining any injury. Hence, there is a need to identify the techniques and possible mistakes that could cause injury to the athlete. Therefore the purpose of this paper is to study the technical principles behind the TC DM Flying Front Rise Kick (FFRK) (commonly known as "Teng Kong Zheng Ti Tui") of elite female TC athletes from China. The study on female athletes was chosen because the rate of ACL injury for females in sports is higher than male (Nagano *et al.*, 2006). Understanding the principles can help new athletes know the important factors that affect the success of a FFRK. This paper will use kinematics data of the elite China athletes to identify the key parameter that affects the success of FFRK. Although other sports such as high jump (Yu & Hay, 2005) may show some similarity to the movement the author does not wish to seek reference from them to hypothesize any outcome at the beginning. This is because the author wishes to study the movement without the influence of other sports as TC DM is unique in its own and making any hypothesis before knowing the results may affect the study.

The movement of FFRK consists of a few moves. Therefore to allow the reader to understand the move better, FFRK will be broken down to 3 main phases. The first phase requires the athlete to jump using her right leg with a single step run. For the second phase, the athlete must execute an upright kick with her right leg while in the air and her feet must touch her forehead. During the process, both her legs must be straightened until she finishes the kick. The third phase is the landing. The athlete must land only with her right leg without swaying. The landing sequence must be smooth and continuous; any additional movement after a split second pause is considered swaying and points will be deducted.

The authors identify one critical point for each phase in this study. In Phase 1, a good jump height gives the athlete more time to execute the movement and it will be more impressive to the judges. In Phase 2, while in the air, the athlete must execute the kick at the right time and give enough time to prepare for landing. In phase 3, landing stability is crucial as any swaying will result in point deduction.

2 Method

Six athletes were chosen as subjects for this study. The subjects were elite athletes from China who finished among the top for Tai Chi Quan during the 10th China National Games in Nanjing. The video capture was done at the competition ground during the female Tai Chi Quan final. This video recording for each athlete was done when she was competing on the carpet and not specially executed for the study. The whole routine of all the six athletes were recorded. Their routine includes all the DM that the athlete had chosen and traditional TC movements.

The video capture was done using three Panasonic NVGS-50 cameras recording at 50 fields/second. The three cameras were placed around the competition ground as shown in the figure 1 and the synchronized signal was produced by PEAK wireless synchronization system, and transmitted to and recorded by each camera. The calibration frame was place at six different positions around the carpet from point A to F. This was done because different athletes execute the DM at difference places hence different calibration points should be used.

The frames of the FFRK for each athlete were retrieved from the three cameras and synchronized with Peak Motus® motion analysis system. Twenty one body landmarks (vertex, nose, neck, left and right shoulder, left and right elbow, left and right wrist, left and right middle finger, left and right hip, left and right knee, left and right ankle, left and right heel, left and right toe) were identified for the analysis. These points were manually digitized with

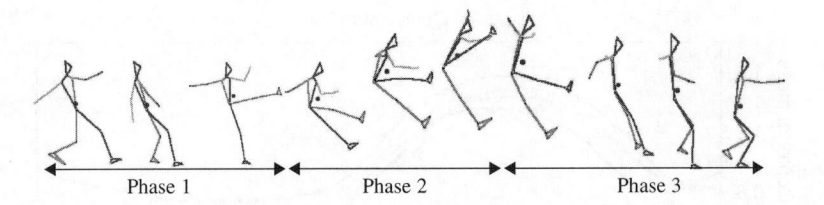

Biomechanical Analysis of Tai Chi Difficulty Movement "Teng Kong Zheng Ti Tui"

Figure 1. Phase diagram of FFRK.

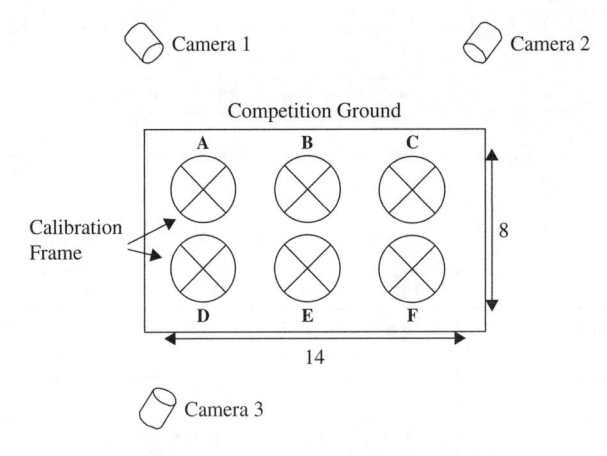

Figure 2. Recording setup at competition venue.

Peak Motus® and the DLT reconstruction method (Marzan & Karara, 1975) was used to obtain the 3-dimension coordinates the body landmarks. A Butterworth filter with cutoff frequency of 5 Hz was used to filter the unwanted noise. For phase 1, quantitative analysis was used to identify the relationship of jump height with vertical velocity and angle of take off. For Phase 2, kinematics data of the kicking leg and time difference from the maximum center of mass (COM) height to the point when the right leg touches the forehead were used for analysis. For Phase 3 the time taken for the athletes to reach their minimum COM was used to analyze stability of landing.

The small sample selected was because the author wishes to analyze only elite athletes. The main limitation was that only one trial was recorded for each athlete as it was a competition.

3 Result and Discussion

Out of the 6 athletes, only 3 of them managed to have a stable landing. The reasons for the unstable landing may be due to the jump in phase 1, the kicking in phase 2 or the landing in phase 3. After going through all the possible parameters, the author identified the main parameters in each phase that could affect the success of FFRK. A successful FFRK required a single step up take off, a front kick in the air where the toes must touch the forehead and a single leg landing without swaying.

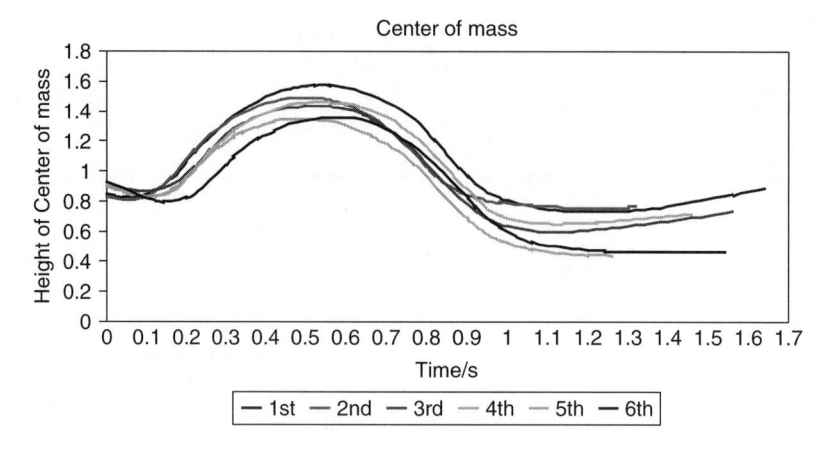

Figure 3. Change in center of mass over time.

Table 1. Kinematics of jump just before the athletes leaves the ground

	1st	2nd	3rd	4th	5th	6th
COM Velocity just before take off/ m/s	2.87	2.751	3.15	2.772	2.972	3.039
Take off angle of COM/ degree	73.1	64.4	53.17	66.2	52.8	56.4
Jump Height/m	0.394	0.294	0.246	0.28	0.231	0.262
Vertical Velocity/ m/s	2.75	2.48	2.54	2.54	2.38	2.53

3.1. Center of Mass (COM)

The graph shows how the COM of the athletes changes with time. The graph shows a consistent trend among the athletes. Their center of mass was lowered first and then increased to the maximum height when they jump and then land to a lower position than the starting position. The reason for the first change in COM was because the athletes had to prepare for the jump and had to bend their knee to preferred angle before taking off. The possible reason for them to land at a lower COM than the starting position could be to facilitate their balance.

3.2. Phase 1

The table 1 shows the kinematics data of all the athletes just before their feet leaves the ground. The author identified that jump height was a critical factor and the parameter in Phase 1 will affect the outcome of the jump height

The correlation between the jump height versus the vertical velocity ($r = 0.89$) and the angle of take off ($r = 0.92$) were positive. These show that the main parameters that affect the jump heights were the vertical velocity and angle of take off. The angle of take off determined the amount of the COM velocity converted to the vertical velocity.

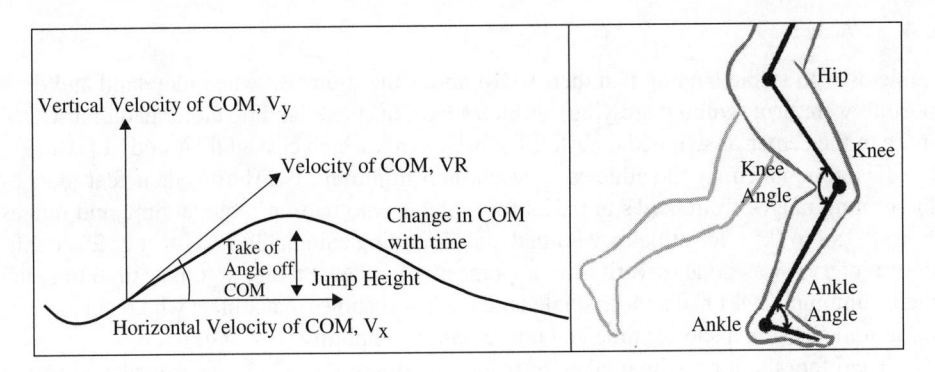

Figure 4. Diagram for definition.

Besides the factors discussed, the change of the right ankle angle may influence the take off angle. The change in ankle angles is the difference between the minimum ankle angle before take off and the ankle angle just before the athletes leave the ground.

The importance of a good jump height is to give the athlete more time to execute the movement and also impress the judges. Hence to obtain a good jump height, the athlete must attain a good vertical velocity and a good angle of take off.

3.3. Phase 2

In phase 2 when the athletes were in the air and the main movement was the kicking of right leg to the forehead without bending the knees. The maximum angular velocity of the kicking leg for the upwards movement ranges from 11.4 rad/s to 12.6 rad/s and the downward motion ranges from 12.4 rad/s to 15 rad/s. All the athletes had a higher angular velocity for the downward motion than upward motion.

The possible reasons that may cause the difference could be because the athletes were working against gravity when they kicked their leg upwards. Moreover, they must use their right leg to execute the jump first before kicking upwards. Therefore, they can only kick their right leg after they lift off from the ground. Whereas for the downward motion, gravity will facilitate the kicking motion and less energy will be needed to reach the same angular velocity as the upward motion.

Kicking the right leg fast allowed the athletes to complete the movement over the short air time. The faster they finish the movement the more time they had to prepare for landing.

The time when the right leg touches the forehead occurs when the athlete just passed their maximum COM height. The time difference, t_1, between the maximum COM height and when the right leg touches the forehead were 0.03 s and 0.045 s for the 4th and 6th athlete and the rest ranges from 0.01 s to 0.02 s. Completing the upward motion of the right leg as close to the maximum COM as possible allows more time for the athletes to prepare for landing. The unstable landing of 4th and 6th athlete could be due to the greater t_1 which resulted in less air time to prepare for landing.

3.4. *Phase 3*

Athletes with stable landing had their COM above the right feet when they land and their posture were firm (without swaying). Athletes with unstable landing either had their COM more to the center or swinged their left leg backwards which caused their body to sway.

The time taken for the athletes to reach their minimum COM after their feet touches the ground ranges from 0.14 s to 0.22 s for athletes who had unstable landing and ranges from 0.38 s to 0.5 s for athletes who had stable landing. Single Tailed T-test at 2% confidence interval was done to verify the hypothesis of athletes taking a longer time to reach their minimum COM had a more stable landing. T-test shows that athlete who reached their minimum COM in a shorter time had poorer landing stability ($p < 0.01$).

Therefore the longer time taken to reach the minimum COM was crucial to obtain a stable landing. The longer time to reach the minimum COM may give the athlete more time to adjust their COM for a stable finish.

4 Conclusion

The execution of a good FFRK requires many conditions. Each phase emphasizes on different aspects of the FFRK and every part could affect the final outcome of the jump. Overall, a well-executed FFRK requires an athlete to jump high, kick her right leg fast and have a stable single leg landing. This paper gives a general overview of the principles behind FFRK in each phase. The success of FFRK may depend on more parameters than those discussed in this paper. This paper gives a general overview of the principles behind FFRK in above phase.

References

Lees A., Vanrenterghem J. and De Clercq D. (2004) Understanding how an arm swing enhances performance in the vertical jump. *Journal of Biomechanics*, 37, 1929–1940.

Boden B.P., Dean G.S., Feagin J.A. and Garrett W.E. (2000) Mechanisms of anterior cruciate ligament injury. *Orthopedics*, 23, 573–578.

Mao D.W., Li J.X. and Hong Y. (2006) The duration and plantar pressure distribution during one-leg stance in Tai Chi exercise. *Clinical Biomechanics*, 21, 640–645.

The International Wushu Movement (2005) Rules of Taolu (online). Available: http://www.iwuf.org/indexItem.asp?column=Rules (Accessed: January 5, 2007).

Marzan G.T. and Karara, H.M. (1975) A computer program for direct linear transformation solution of the colinearity condition, and some applications of it. *Proceedings of the Symposium on Close-Range Photogrammetric Systems*, pp. 420–476. American Society of Photogrammetry, Falls Church.

Nagano Y., Ida H., Akai M. and Fukubayashi T. (2007) Gender differences in Knee Kinematics and muscle activity during single limb drop landing.*Knee*, Epub ahead of print.

Yu B. and Hay J.G. (1996) The optimum phase ratio in the triple jump. *Journal of Biomechanics* 29, 1283–1289.

ARCHERY BOW STABILISER MODELLING

I. ZANEVSKYY
Casimir Pulaski Technical University, KWFiZ, Radom, Poland

The aim of the research is to develop a method of mechanical and mathematical modelling and computer simulation of dynamic stabilisation of a bow in the vertical plane intending to get practical recommendations for the sport of archery. The behaviour of a flexible stabiliser in the main plane of the modern sport bow designed in the frame of International Archery Federation is analysed using a mechanical and mathematical model. The model is designed on the Euler-Bernoulli beam and Lagrange equations of the second kind. An engineering oriented method based on virtual modes and Rayleigh-Ritz procedure is developed to study natural frequencies of the archer-bow-stabiliser system. The results of modelling of the archer-bow-arrow system correlate with well-known results of high-speed video analysis: the process of common motion has significant non-linear character.

1 A Common Approach to the Model Design

To make a shot, a modern sport bow is situated vertically with its main plane. Despite a simple construction at first sight, bow moves surprisingly complicated. Its motion is in 3D space before, during, and after shot. Bow riser movement is significantly smaller than arrow movement, string movement, and two limbs movement in the vertical plane. All the system moves laterally too, but this movement is smaller in comparison to the movement in the main plane. Bow riser movement before the shot, i.e. before string release, is under the archer control and directly affects aiming. Movement during the shot, i.e. during string and arrow common motion is partly controlled by the archer hand holding a bow riser and partly is free. This movement affects aiming to some extent.

To avoid some part of bow riser motion, modern sport bows are supplied with a long cantilever rod (or multi-rods packet) mounted in front of the riser directly to a target. Let's consider a compound hinge and rode mechanism as a bow scheme model (Zanevskyy, 2006). A stabilizer is modelled as an elastic rod joined to the handle like a cantilever beam. According a video study, a bow stabilizer bends with the main mode of natural oscillation. Energetic methods of dynamic mechanics obtain comprehensive accuracy for mechanical engineering calculation. Therefore, we can design a mechanical and mathematical model of a bow with stabilizer using a hypothetical function of the main mode of beam (Blechman *et al.*, 1976).

At the very beginning, we assume the hypothetic function of the main mode as a function of static bend of a cantilever beam loaded by a concentrated force at the free end (Figure 1 a):

$$\eta(z) = \frac{Fz^2}{6\varepsilon}(z - 3l), \tag{1}$$

where F is a loading force; ε is distributed bend stiffness of the beam; l is length of the beam; z is longitudinal coordinate; η is transverse displacement. The error of the main natural frequency with function (1) is 0,73% (Weaver *et al.*, 1990).

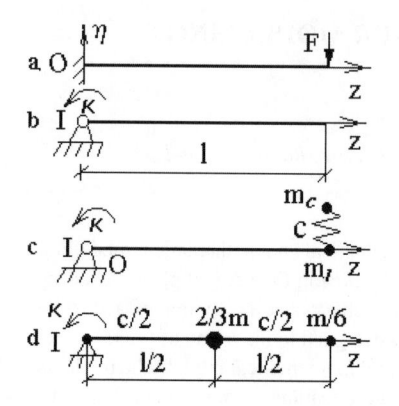

Figure 1. Scheme models of a bow stabilizer as a cantilever elastic beam (a) and as a one-end hinged elastic beam: (b) are the models 1, and 2; (c) are the models 3, 4, and 5; (d) is the model 6.

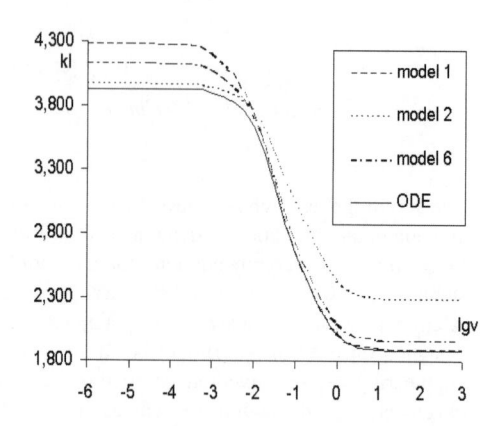

Figure. 2. Main natural frequency vs. bow and stabilizer mass-inertial parameters: ODE is considered as an exact solution (4).

As a hypothetic function of the main natural oscillation mode of one hinged beam (Figure 1 b) we assume a sum of one-half sinusoid wave and a linear function:

$$\eta = A \sin \frac{\pi z}{l} + \kappa z, \qquad (2)$$

where A is function of time; κ is an angle of sinusoid wave turn. Like (1) the function (2) satisfies only three of four boundary conditions, i.e. zero displacement and zero force moment at the pinned end of the beam and one dynamic boundary conditions, i.e. zero force moment at the free end. Another dynamic boundary condition is not satisfied, i.e. zero cross-section force at the free end. Despite this, the function (2) allows appreciated precision of the main natural frequency because the error is 1,10%.

Because there are no results on the problem of natural frequencies of one end pinned beam with a load in well-known mechanical and mathematical publications, we consider this roblem using Hamilton variation principle: $\delta \int_{t_1}^{t_2} (T - P)dt = 0$, where

$T = \frac{1}{2}\left[\int_0^l \mu \left(\frac{\partial \eta}{\partial t}\right)^2 dz + I \left(\frac{\partial^2 \eta}{\partial z \partial t}\right)^2_{z=0}\right]$ and $P = \frac{1}{2}\int_0^l \varepsilon \left(\frac{\partial^2 \eta}{\partial z^2}\right)^2 dz$ are kinetic and potential energy correspondingly; μ is distributed mass of the beam; t is time; I is moment of inertia of the load relatively the hinge axis (see Figure 2b). Placing the two last expressions of energy in the Hamilton functional, we get correspondent differential equation

$$\mu \frac{\partial^2 \eta}{\partial t^2} + \varepsilon \frac{\partial^4 \eta}{\partial z^4} = 0$$ and boundary conditions.

$$z = 0, \eta = 0, \varepsilon \frac{\partial^2 \eta}{\partial z^2} = I \frac{\partial^3 \eta}{\partial z \partial t^2}; \quad z = l, \frac{\partial^2 \eta}{\partial z^2} = 0, \frac{\partial^3 \eta}{\partial z^2} = 0. \tag{3}$$

Solutions of the problem (3) obtained using Krylov functions are roots of the determinant:

$$\begin{vmatrix} 2kl & v(kl)^4 & v(kl)^4 \\ ch(kl) + cos(kl) & sh(kl) & -sin(kl) \\ sh(kl) - sin(kl) & ch(kl) & -cos(kl) \end{vmatrix} = 0, \tag{4}$$

where $kl = \sqrt[4]{\frac{ml^3 \omega^2}{\varepsilon}}$ are dimensionless values of natural frequencies; m is mass of the beam; ω is circular natural frequencies; $v = \frac{I}{ml^2}$ is dimensionless value of moment of inertia of the load. Zero solution of the equation (4) corresponds common beam and load rotation relatively the hinge axis. When $v = 0$, we get $kl = 0; 3,927; 7,069; 10,210; 13,352; \ldots$, which are the same as known solutions for the beam with one hinged end $\frac{\pi(4i-3)}{4}$, where i is a number of natural frequency (Pisarenko et al., 1975). There is no zero solution when $v = \infty$: $kl = 1,875; 4,694; 7,855; 10,996; \ldots$, which are the same as known solutions for the cantilever beam $\frac{\pi(2i-1)}{2}$. Solution of the main frequency for different relationship of mass-inertial parameters is presented in Figure 2 (ODE line).

2 Model Versions

Using hypothetical function of the main natural mode, we can apply Lagrange equations of the second kind:

$$\frac{d}{dt}\left(\frac{\partial T}{\partial q'_i}\right) - \frac{\partial T}{\partial q_i} + \frac{\partial P}{\partial q_i} = 0, \tag{5}$$

where q_i are generalized coordinates; prefix shows a partial derivation in time, i.e. $(') \equiv \partial/\partial t$.

Model version 1 is designed as a sum of (1) and a linear function:

$$\eta = A\left(\frac{z}{l}\right)^2 \left(3 - \frac{z}{l}\right) + \kappa z \tag{6}$$

As generalized coordinates, there are A and κ. The angle of the load turning is $\left(\dfrac{\partial \eta}{\partial z}\right)_{z=0} = \kappa$. After substituting of the hypothetical function (6) in the equations of energies and then in (5), we get expressions for the main natural frequency:

$$(kl)^4 = 12 \bigg/ \left(33/35 - \dfrac{121/400}{1/3 + v}\right), \tag{7}$$

where (/) is a sign of division.

For the <u>model version 2</u>, we applied the hypothetical function (2). Like before, we get the angle of the load turn that is presented here as expression $\left(\dfrac{\partial \eta}{\partial z}\right)_{z=0} = \kappa + \dfrac{\pi A}{l}$.

Correspondent expression of the main frequency is:

$$(kl)^4 = \dfrac{\pi^4}{2} \bigg/ \left[\dfrac{1}{2} + v\pi^2 - \dfrac{(1/\pi + v\pi)^2}{1/3 + v}\right]. \tag{8}$$

In the <u>model version 3</u>, hypothetical functions are assumed with the expressions:

$$\eta_l = \kappa l; \quad \eta_c = \kappa l + A. \tag{9}$$

Virtual stiffness and mass of the beam are located at the free end (Figure 1c) using the function (1): $c = \dfrac{3\varepsilon}{l^3}$; $m_c = \dfrac{33}{140} m$. The angle of the load turning is $\left(\dfrac{\partial \eta}{\partial z}\right)_{z=0} = \kappa$.

Correspondent expression for the main frequency is:

$$(kl)^4 = \dfrac{140(1/3 + v)}{11v}. \tag{10}$$

<u>Model version 4</u> is different of the model version 3 only with the value of virtual mass $\left(m_c = \dfrac{m}{3}\right)$ that corresponds the value of moment of inertia of the beam relatively the hinge axis. Here, a formula for the main natural frequency is:

$$(kl)^4 = \dfrac{9(1/3 + v)}{v}. \tag{11}$$

<u>Model version 5</u> is a combination of two previous model versions, i.e. 3 and 4: $m_c = \dfrac{33}{140} m$; $m_l = \dfrac{41}{420} m$. General virtual mass $\left(m_c + m_l = \dfrac{m}{3}\right)$ corresponds the value of the beam moment of inertia. The main natural frequency is calculated with a formula:

$$(kl)^4 = 3 \bigg/ \left(33/140 - \dfrac{(33/140)^2}{1/3 + v}\right). \tag{12}$$

Archery Bow Stabiliser Modelling

In the model version 6 the beam mass is located in three points, i.e. at the beam-ends $\left(m_0 = m_2 = \frac{1}{6}m\right)$ and in the middle of the beam $\left(m_1 = \frac{2}{3}m\right)$. Stiffness is concentrated in two points, i.e. in the middle of the beam and at the free end: $c_1 = c_2 = \frac{6\varepsilon}{l^3}$. This is equal to the stiffness of the cantilever beam: $c = \frac{3\varepsilon}{l^3}$. Using (6), we get expression for main natural frequency:

$$(kl)^4 = (219/16)/\left(89/96 - \frac{169/576}{1/3 + v}\right). \tag{13}$$

The results on the relative accuracy for the all six model versions (7), (9–13) relatively the ODE solution (4) are grouped in the Table. From the practical point of view, appreciated results regards accuracy of the main natural frequency are obtained with the model versions 1, 2, and 6. But the best accuracy in a wide range of mass-inertial parameters of the beam and the load ($-1,5 < lgv < \infty$), we get using model version 1. Model version 2 is appreciated only for small loads ($-\infty < lgv < -3$). Using the model version 6, we get mediocre level of accuracy, and only in a narrow range of relationship ($lgv \approx -2$) accuracy is appreciated. The main natural frequency results obtained with the model versions 1, 2, and 6 and the ODE solution (4) are presented in a non-dimensional form (see Figure 2).

Taking into account a real relationship between bow and stabilizer mass-inertial parameters ($-1.5 < lgv < 2,0$) and considering the results of calculations (see Table 1 and Figure 3), we can choose the model version 1 as the best one. The smallest error of the main natural frequency result (0,16%) for this model is when relationship $v = 0,0537$ ($lgv = -1,27$).

3 Approbation of the Model

The main attempt to the dynamic problem on bow and arrow system has been founded (Zanevskyy, 2006). An example of study a modern sport bow medium parameters (WIN&WIN Recurve Bow) supplied with a light alloy rod stabiliser is calculated. The bow consists of Winact Riser (25″) and Long Limbs (70″), i.e. bow handle length is 635 mm and the whole length of the bow (measured between tips of the limbs) is 1778 mm. The standard measure of bow asymmetry in the vertical plane named "tiller" is 6 mm and bow force is 178 N. The results of solution of the problem for the parameters above are presented in the graphs (Figure 3). An arrow launches a string nock point as their longitudinal acceleration becomes zero. At the instant an arrow has the maximal longitudinal speed. The time of bow and arrow common motion is 15,8 ms. The graphs describe a process of bow stabilization in the vertical plane during bow and arrow common motion. A bow riser turn clockwise is partly compensating by stabilizer bend counter clockwise. A monotone character of these motions testifies a below resonance regime of the process.

4 Conclusions

For real relationship of bow and stabilizer mass-inertial parameters, we can get the best accuracy of the main natural frequency of the system (near 1%) using a hypothetical

Table 1. Related errors of the main frequency of a bow and stabilizer system (%).

lgv	\multicolumn{6}{c	}{Number of a model version}
	1	2	3	4	5	6
−6	9,30	1,10	1055,69	959,79	−34,62	5,26
−5	9,30	1,11	550,01	496,07	−34,61	5,27
−4	9,18	1,12	265,79	235,44	−34,58	5,20
−3	8,10	1,25	107,21	90,01	−34,25	4,61
−2	2,54	3,41	25,59	15,17	−30,66	1,82
−1	0,27	15,52	3,27	−5,30	−12,90	3,08
0	0,66	21,01	0,98	−7,40	−1,35	4,36
1	0,72	21,72	0,76	−7,60	0,51	4,52
2	0,73	21,80	0,71	−7,65	0,69	4,52
3	0,73	21,81	0,74	−7,62	0,74	4,55

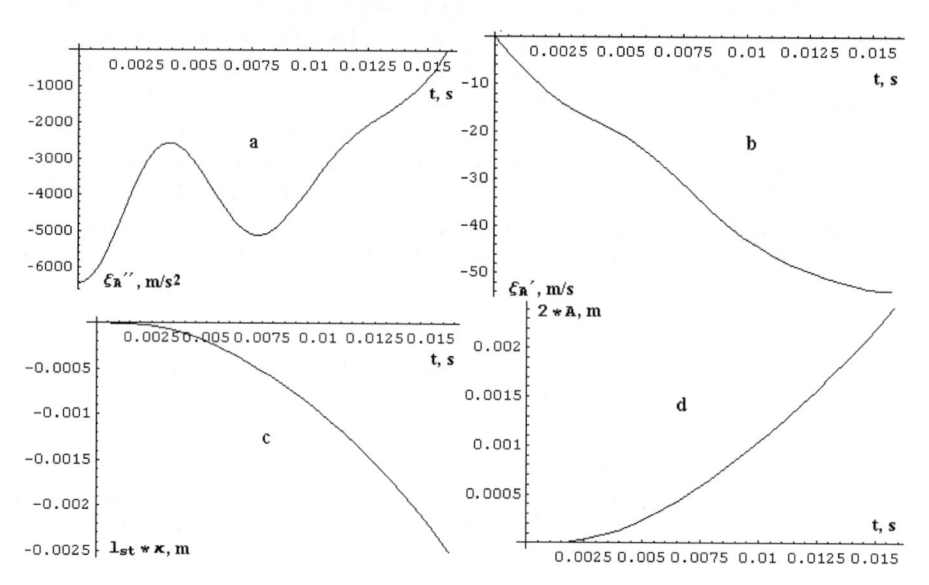

Figure 3. Kinematical parameters of the system: a – longitudinal acceleration of the arrow vs. time; b – longitudinal speed of the arrow; c – bow riser angle multiplied by stabilizer length; d – pure bend displacement of the free end of the stabilizer.

function as a combination of a linear function and the function of static bend of a cantilever beam loaded by a force at the free end. Beam stabilizer stiffness increase cause significant decrease of bow turn motion and decrease of dynamic stabilizer bend. At the same conditions, stabilizer mass increase causes decrease of bow turn motion in below the resonance zone. The attempt to the problem of sport bow stabilization in the vertical plane proposed in the paper is addressed to practical needs of applied engineering mechanics and the archery sport. The models and methods have been adapted for realization in an engineering method using well-known mathematical CAD systems.

References

Blechman I.I., Myshkis A.D. and Panovko Y.G. (1976) *Applied mathematics*. Naukova Dumka, Kiev (in Russian).
Pisarenko G.S., Yakovliev A.P. and Matvieiev V.V. (1975) *Strength of materials*. Naukova Dumka, Kiev (in Russian).
Weaver W.Jr., Timoshenko S.P. and Young D.H. (1990) *Vibration Problems in Engineering*. John Wiley & Sons, Inc., Hoboken, NJ.
Zanevskyy I. (2006) Bow tuning in the vertical plane, *Sports Engineering*, 9/2, 77–86.

MATEMATICAL MODEL OF THE AIMING TRAJECTORY

C.-K. HWANG[1], K.-B. LIN[2] & Y.-H. LIN[1]

[1]*Department of Electrical Engineering, Chung Hua University, Hsin-Chu, Taiwan*
[2]*Yuanpei University of Science and Technology, Hsin-Chu, Taiwan*

A mathematical model that can more accurately reflect the typical archery style is studied in the paper. Twelve archers from the archery team of National College of Physical Education and Sports were invited to attend this study, and each of their aiming trajectories during the last 1.5 second before releasing the arrow was recorded. Based on their time series of the aiming trajectory, we propose a linear time invariant auto-regressive (AR) process to model them. The second order of the AR model is formulated based on the recorded data of three different time periods scheduled as the last 0.5 second, one second and 1.5 second periods before releasing the arrow. The fitting errors of the AR2 model associated with three time period are also calculated to indicate the fairness of the proposed method. The aiming style may involve a constant offset term in the AR model, so the term is also added to check whether the archer has this special trend or not. The study shows that some archers have the constant offset trend, but some archers do not have it. Seven out of the twelve archers with the their best fitting error of the AR2 model are fallen into the 1.5 second period, as expected. Moreover, the archers with good performance usually fallen into the extreme cases related to the offset term added in the AR2 model.

1 Introduction

Lots of archery researches have been conducted from different approaches in order to find the key point for improving the performance of this fine and highly skilled sport. Lateral deflection of archery arrows. A whole mechanical and mathematical model of an arrow-bow motion system, which accounts for arrow deflection in the lateral plane, has been created (Zanevskyy, 2001). The model takes into consideration the mechanical properties of a string, bow limbs and a grip as an oscillator of concentrated elastic and inertial elements connected with the feathered end of the arrow. Theoretical investigation of natural modes and frequencies of bow and arrow vibration has been conducted. Data for the first four natural frequencies and modes have been obtained and practical conclusions have been drawn. As a result of modeling and computer simulations, an engineering method for matching bow and arrow parameters has been proposed. Comparative results for the wrong and right combination of these parameter values for the modern sport bow and two arrows are presented.

The aiming stability is the key factor affects the archery performance has been indicated by Shiang *et al.* (1997), and it can be determined by the size of aiming locus. They further pointed out that the aiming locus pattern is also a useful index to determine the performance. The United States Olympic Committee (1996) has defined archery fundamentals as: stringing the bow, stance, nock, set, pre-draw, draw, anchor, aim, release, and follow through. Among these fundamentals, the aim stage will also be analyzed in this paper. Furthermore, analysis of correlation between the aiming adjustment trajectory and the shot points has been studied (Lin *et al.*, 2003). In that paper, fifteen analysis units, in which each analysis unit is the average of six sampling data, are generated for analysis of vertical and horizontal deflections.

In the paper, a mathematical model is proposed to represent the aiming trajectory. That is, we propose a popular linear time invariant AR process (Ljung, 1999) that can model the recorded time series of the aiming trajectory. Individual AR model is established for the aiming trajectory along the vertical and the horizontal directions. We consider the second order of the AR model based on three different time periods. The fitting errors of the AR2 model based on three different time periods along both of the vertical and the horizontal deviations are evaluated. For most of the attendances, the longest one is the best period to obtain the better accuracy of the AR2 model as expected. The aiming style may involve a constant offset term in the AR model, so the constant term is added to check whether the archer has this special trend or not. The study shows that some archers have this kind of trend, but some archers do not have it. Similarly, the suitable time period often occurs at the 1.5 second period for most of them when the constant offset term is included.

2 Methods

Twelve archers from the man archery team of National College of Physical Education and Sports attended this experiment. In our experimental setting, a laser pen is mounted at the bow handle for capturing the aiming trajectory by using a digital video camera. Using the laser pen and the digital camera, both the vertical and horizontal aiming trajectory coordinates during a suitable period before releasing the arrow can be accurately recorded. On the other hand, the vertical and horizontal shot coordinates are captured by another camera placed in front of the target. These recorded data are then processed by APAS (Ariel Performance Analysis System) motion analysis system for studying the aiming procedure and the shot points along the vertical and the horizontal directions. The frequency of the digital video camera is set as 60 fields/second for capturing the light point trajectory, so in the fix field mode 90 fields will be recorded during the one and half second before release of the arrow.

2.1. Field Setup

The distance between start line and arrow target is 30 meters. The experiment proceeds according to the usual competition procedure. Before the test, she can shoot three arrows to adjust her bow sight. Then the test begins. She shoots thirty six arrows, that is, 3 arrows will be shot for a round and totally 12 rounds are performed. Each arrow is required to be released within 40 seconds, otherwise it is not recorded.

3 Results and Discussion

The proposed linear time-invariant AR2 model is formulated as follows.

$$x(k) = a_{x1}x(k-1) + a_{x2}x(k-2) + e(k-2)$$
$$y(k) = a_{y1}y(k-1) + a_{y2}y(k-2) + e(k-2) \qquad (1)$$

where $x(k)$ and $y(k)$ are the time series of the horizontal and vertical deviations of the aiming trajectory; a_{x1}, a_{x2}, a_{y1} and a_{y2} are the corresponding coefficients of the AR2 model; $e(k)$ are the white noise with zero mean.

In order to estimate the coefficients of a_{x1}, a_{x2}, a_{y1} and a_{y2}, we form the following vectors and matrices such as

$$\theta_x = [a_{x1}\, a_{x2}]^T,\ \theta_y = [a_{y1}\, a_{y2}]^T,\ E(k) = [e(k-2)e(k-1)\ldots e(k-2+m)]^T$$
$$A_x = [X(k-1)X(k-2)],\ A_y = [Y(k-1)Y(k-2)]$$

where $X(k) = [x(k)x(k+1)\ldots x(k+m)]^T$, $Y(k) = [y(k)y(k+1)\ldots y(k+m)]^T$ and m is the total number of applied time series. Then, Eq. (1) can be rewritten as

$$\begin{aligned}X(k) &= A_x\theta_x + E(k)\\ Y(k) &= A_y\theta_y + E(k)\end{aligned} \quad (2)$$

Moreover, the estimation of θ_x and θ_y denoted as $\hat{\theta}_x$ and $\hat{\theta}_y$ can be obtained by

$$\begin{aligned}\hat{\theta}_x &= (A_x^T A_x)^{-1} A_x^T X(k)\\ \hat{\theta}_y &= (A_y^T A_y)^{-1} A_y^T Y(k)\end{aligned} \quad (3)$$

The estimation errors $E_x = (X(k) - A_x\hat{\theta}_x)^T (X(k) - A_x\hat{\theta}_x)/(m+1)$ and $E_y = (Y(k) - A_y\hat{\theta}_y)^T \times (Y(k) - A_y\hat{\theta}_y)/(m+1)$ are evaluated and listed in Table 1. In order to check whether the offset effect exists in their aiming trajectory, we add the constant terms b_x and b_y into Eq. (1) as

$$\begin{aligned}x(k) &= a_{x1}x(k-1) + a_{x2}x(k-2) + b_x + e(k-2)\\ y(k) &= a_{y1}y(k-1) + a_{y2}y(k-2) + b_y + e(k-2)\end{aligned} \quad (4)$$

By applying the similar technique as the above, the associated estimation errors with the offset are also calculated and listed in Table 2. Additionally, the estimated constant terms b_x and b_y are also shown in Table 3.

Table 1. The fitting errors of the proposed AR2 model without the offset.

Period	Direction	Archer 1	Archer 2	Archer 3	Archer 4	Archer 5	Archer 6
1.5 second	Horizontal	0.27139	0.33798	0.28236	0.35735	0.21325	0.32211
	Vertical	0.37681	0.44707	0.29589	0.49676	0.31039	0.29083
1 second	Horizontal	0.28239	0.36617	0.28749	0.32647	0.21725	0.30561
	Vertical	0.38994	0.48683	0.27857	0.51392	0.28634	0.28264
0.5 second	Horizontal	0.2782	0.44688	0.35239	0.37896	0.2116	0.33454
	Vertical	0.44074	0.51527	0.28938	0.57836	0.26966	0.2803
Period	Axis	Archer 7	Archer 8	Archer 9	Archer 10	Archer 11	Archer 12
1.5 second	Horizontal	0.51476	0.61658	0.32616	0.31375	0.46336	1.0098
	Vertical	0.40427	0.52678	0.31849	0.31845	0.55677	0.78791
1 second	Horizontal	0.64274	0.69487	0.36856	0.30354	0.47959	1.1003
	Vertical	0.44655	0.56229	0.3416	0.30778	0.55872	0.72486
0.5 second	Horizontal	1.0553	0.78052	0.41292	0.27078	0.51932	1.1449
	Vertical	0.5876	0.65167	0.41594	0.33648	0.61209	0.74638

Table 2. The fitting errors of the proposed AR2 model with the offset.

Period	Direction	Archer 1	Archer 2	Archer 3	Archer 4	Archer 5	Archer 6
1.5 second	Horizontal	0.27138	0.33739	0.27805	0.35719	0.20572	0.3214
	Vertical	0.37356	0.44606	0.2957	0.49643	0.31024	0.29042
1 second	Horizontal	0.28238	0.36597	0.28123	0.32471	0.20626	0.30561
	Vertical	0.38743	0.48533	0.27854	0.51374	0.28221	0.28138
0.5 second	Horizontal	0.27814	0.44511	0.34122	0.37485	0.20683	0.33312
	Vertical	0.43877	0.51512	0.28822	0.57083	0.26964	0.27422
Period	Axis	Archer 7	Archer 8	Archer 9	Archer 10	Archer 11	Archer 12
1.5 second	Horizontal	0.51305	0.61635	0.32262	0.31371	0.46263	1.0097
	Vertical	0.40028	0.52651	0.31687	0.31844	0.55356	0.75776
1 second	Horizontal	0.64032	0.69485	0.36726	0.30325	0.47826	1.0998
	Vertical	0.44454	0.5612	0.3405	0.30748	0.553	0.70551
0.5 second	Horizontal	1.0359	0.78018	0.41264	0.2652	0.51721	1.1449
	Vertical	0.5791	0.65063	0.41231	0.33463	0.6077	0.72707

Table 3. The offset of the AR2 model (D = direction, H = horizontal, V = vertical).

Period	D	Archer 1	Archer 2	Archer 3	Archer 4	Archer 5	Archer 6
1.5 second	H	−0.0005805	0.01033	−0.019272	−0.003888	−0.025529	−0.023733
	V	−0.041073	−0.014966	−0.006113	−0.009744	0.003640	0.006234
1 second	H	0.0010416	0.0059878	−0.023603	−0.012531	−0.03227	−0.001829
	V	−0.037973	−0.018779	−0.002578	−0.007500	0.018714	0.010702
0.5 second	H	−0.0023238	0.018034	−0.032168	−0.019247	−0.024541	−0.030653
	V	−0.033421	−0.0060354	−0.014988	−0.048478	0.001130	0.026285
Period	D	Archer 7	Archer 8	Archer 9	Archer 10	Archer 11	Archer 12
1.5 second	H	−0.012295	0.0059099	0.017382	0.005091	0.009948	0.004111
	V	−0.078546	−0.0094663	−0.044916	−0.000966	0.054824	−0.098492
1 second	H	−0.014505	0.0019844	0.010435	0.014069	0.01313	0.013492
	V	−0.059226	−0.01917	−0.038226	−0.005206	0.067602	−0.11019
0.5 second	H	−0.040773	−0.0074621	0.004844	0.062128	0.015876	−0.001619
	V	−0.1169	−0.019238	−0.074371	−0.012973	0.068057	−0.15667

As compared the fitting errors between Tables 1 and 2, the fitting errors by taking the offset into account are better than those without considering the offset, and this result is as expected due to the additional term for data fitting. By considering the fitting error of both the vertical and horizontal deviations together, there are seven archers (archers 1, 2, 7, 8, 9, 11, and 12) whose better fitting errors are located at the 1.5 second period, but for archers

Table 4. The ranking.

Archer	1	2	3	4	5	6	7	8	9	10	11	12
Ranking	1	8	10	11	2	5	9	5	12	7	3	4

3, 4, and 6, theirs are fallen into the one second period, and the best fitting errors for archers 5 and 10 exist at the half second period.

Table 3 shows that in the one and half second period the offset b_x is insignificant for archer 1 along the horizontal deviation, and it is significant for archer 5; along the vertical deviation, the offset b_y is insignificant for archer 10 but it is significant for archer 12. Similarly, along the horizontal deviation in the one second period the offset b_x is also insignificant for archer 1, and it is significant for archer 5; the offset b_y is insignificant for archer 3 and is also significant for archer 12 along the vertical deviation. Finally, along the horizontal deviation in the half second period, the offset b_x is also insignificant for both archers 12 and 1, and it is significant for archer 10; the offset b_y is also significant for archer 12 but insignificant for archer 5 along the vertical deviation. There are some interesting findings that archers 1, 5 and 12 ranking among the top four are often fallen into these extreme cases.

4 Conclusions

In this paper, the AR2 model has been proposed to establish the time sequence of the aiming trajectory. The estimation of the associated coefficients of the model is also formulated and implemented to obtain the estimated coefficients, and the corresponding fitting errors are also calculated based on three different time periods. The additional offset term can improve the fitting errors as expected, and most of archers whose better fitting errors are located at the 1.5 second period, but there are three archers and two archers fallen into the one second period and the half second period, respectively. This phenomenon is worth for further study. Moreover, the archers with good performance usually fallen into the extreme cases related to the offset term added in the AR2 model, this situation is also worth to find the causes.

Acknowledgements

This work is supported by CHU-94-E-002, CHU-94-E-006 (Chung Hua University), NSC 95-2221-E-216-063, and NSC 93-2218-E-216-009.

References

Ertan H., Kentel B., Tumemer S.T. and Korkusuz F. (2003) Activation patterns in forearm muscles during archery shooting, *Human Movement Science*, 22(1), 34–45.
Lin K.B. and Hwang, C.-K. (2003) Analysis of correlations between aiming adjustment trajectory and target, *13th International Conference in Medicine and Biology*, Tainan, Taiwan, 132–133.
Ljung L. (1999) *System Identification*, 2nd edition. PTR: Prentice-Hall.
Shiang, T.-Y. and Tseng C.-J. (1997) A new quantitative approach for archery stability analysis, *International Society of Biomechanics Congress*, Tokyo, 142.
United States Olympic Committee (1996) *A basic guide to archery*. Glendale Calif. : Griffin Publishing.
Zanevskyy I. (2001). Lateral deflection of archery arrows, *Sports Engineering*, 4(1), 23–42.

ARCHERY PERFORMANCE ANALYSIS BASED ON THE COEFFICIENTS OF AR2 MODEL OF AIMING TRAJECTORY

K.-B. LIN[1], C.-K. HWANG[2] & Y.-H. LIN[2]

[1]*Yuanpei University of Science and Technology, Hsin-Chu, Taiwan*
[2]*Department of Electrical Engineering, Chung Hua University, Hsin-Chu, Taiwan*

In this paper, the most popular linear time invariant Auto-Regressive (AR) process is adopted to model the aiming trajectory, along the vertical and the horizontal directions. In order to analyze the archery performance, we establish the individual aiming trajectory model to reveal his particular archery style. Twelve archers from the archery team of National College of Physical Education and Sports attended this study, and each of their aiming trajectories during the last 1.5 second before releasing the arrow is recorded and modeled as the second order AR2 basis for comparison. The constant offset is also considered in the model, so there are three parameters, one constant offset term and two process coefficients of the AR2, can be utilized to study the relationship among the archery performance and them. Among these twelve archers, the larger first coefficient representing the strong relationship between the current instance and the previous instance can indicate the better archery performance. The associated poles of the AR2 model evaluated for the indication of convergent rate of the aiming trajectory are also founded to have connection with the performance. Their records show that the poles of the trajectory AR2 model for twelve archers are all real number without imaginary part, so they are expertise as expected. For the discrete time AR2 model, the farer distance from the pole to the unit circle is, the more stable of the aiming stage is. From the study, the performance is strongly related to the stability of the aiming trajectory. That is, the dominant poles (the slower model of the AR2 model) of good performance should be far from the unit circle, at least, along one of the vertical and horizontal directions. These interesting findings can be fed back as one of important guidelines to improve their performance in the future.

1 Introduction

Lots of archery researches have been conducted from different approaches in order to find the key point for improving the performance of this fine and highly skilled sport. The most important focus will be the stability of aiming style, so how to use systematic methods to evaluate it falls in the direction. A biomechanical study on the final push-pull archery has been conducted by Leroyer *et al.* (1993). The purpose of their study is to analyze archery performance among eight archers of different abilities by means of displacement pull-hand measurements during the final push-pull of the shoot. The archers showed an irregular displacement negatively related to their technical levels. Displacement signal analysis showed high power levels in both 0–5 Hz and 8–12 Hz ranges. The latter peak corresponds to electromyographic tremor observed during a prolonged push-pull effort. The results are discussed in relation to some potentially helpful training procedures such as biofeedback and strength conditioning. Landers *et al.* (1994) have examined novice archers to determine whether (a) hemispheric asymmetry and heart rate deceleration occur as a result of learning, and (b) these heart rates and electroencephalograph (EEG) patterns are related to archery performance. The electromyography (EMG) technology which measures the activation patterns in forearm muscles related to contraction and relaxation strategy during archery shooting, has been applied by Ertan *et al.*, 2003 to analyze for archers with different levels of expertise; elite, beginner, and non-archers, respectively. They found that elite archers' release started about 100 ms after the fall of the clicker, whereas for beginners and non-archers, their release started after about 200 and 300 ms, respectively.

How the novice archers apply the taught training information under different conditions and guided them to promote their motor skills required for better archery performance have investigated by Lavisse *et al.* (2000). The aiming stability is the key factor affects the archery performance has been indicated by Shiang *et al.* (1997), and it can be determined by the size of aiming locus. They further pointed out that the aiming locus pattern is also a useful index to determine the performance. Furthermore, analysis of correlation between the aiming adjustment trajectory and the shot points has been studied (Lin *et al.*, 2003). In that paper, fifteen analysis units, in which each analysis unit is the average of six sampling data, are generated for analysis of vertical and horizontal deflections. This approach can give richer information than radius due to the fact that radius is the combination of horizontal and vertical deflection.

In the paper, a mathematical model is proposed to represent the aiming trajectory. That is, we propose a popular linear time invariant AR process (Ljung, 1999) that can model the recorded time series of the aiming trajectory. The individual aiming trajectory model is established to reveal his particular archery style. The constant offset is also considered in the model, so there are three parameters, one constant offset term and two process coefficients of the AR2, can be utilized to study the relationship among the archery performance and them.

2 Methods

In our experimental setting, a laser pen is mounted at the bow handle for capturing the aiming trajectory by using a digital video camera. Using the laser pen and the digital camera, both the vertical and horizontal aiming trajectory coordinates during a suitable period before releasing the arrow can be accurately recorded. On the other hand, the vertical and horizontal shot coordinates are captured by another camera placed in front of the target. These recorded data are then processed by APAS (Ariel Performance Analysis System) motion analysis system for studying the aiming procedure and the shot points along the vertical and the horizontal directions.

A total of twelve good archers with stable archery skill attended this experiment, and the aiming trajectory of each archer during the one and half second period before releasing the arrow is recorded. The distance between start line and arrow target is 30 meters. The arrows with almost equal weights are carefully selected by an 18-meters shooting pretest to ensure uniform targeting performance. The experiment proceeds according to the usual competition procedure. Before the test, she can shoot three arrows to adjust her bow sight. Then the test begins. She shoots thirty six arrows, that is, 3 arrows will be shot for a round and totally 12 rounds are performed.

3 Results and Discussion

The proposed linear time-invariant AR2 model with the constant offset is formulated as follows.

$$x(k) = a_{x1}x(k-1) + a_{x2}x(k-2) + b_x + e(k-2)$$
$$y(k) = a_{y1}y(k-1) + a_{y2}y(k-2) + b_y + e(k-2) \quad (1)$$

where $x(k)$ and $y(k)$ are the time series of the horizontal and vertical deviations of the aiming trajectory; a_{x1}, a_{x2}, a_{y1} and a_{y2} are the corresponding coefficients of the AR2 model; b_x and b_y are the constant offset terms; $e(k)$ are the white noise with zero mean.

In order to estimate the coefficients of a_{x1}, a_{x2}, a_{y1}, a_{y2} b_x and b_y, we form the following vectors and matrices such as

$$\theta_x = [a_{x1}\ a_{x2}\ b_x]^T, \theta_y = [a_{y1}\ a_{y2}\ b_y]^T, E(k)) = [e(k-2)e(k-1)...e(k-2+m)]^T$$
$$A_x = [X(k-1)X(k-2)\ ones(m+1,1)], A_y = [Y(k-1)Y(k-2)\ ones(m+1,1)]$$

where $X(k) = [x(k)x(k+1)...x(k+m)]^T$, $Y(k) = [y(k)\ y(k+1)...y(k+m)]^T$, all elements of the vector $ones(m+1,1)$ are one, and m is the total number of applied time series using in the estimation. Then, Eq. (1) can be rewritten as

$$X(k) = A_x\theta_x + E(k)$$
$$Y(k) = A_y\theta_y + E(k) \quad (2)$$

Moreover, the estimation of θ_x and θ_y denoted as $\hat{\theta}_x$ and $\hat{\theta}_y$ can be obtained by the minimum mean square error (MMSE) criterion

$$\hat{\theta}_x = (A_x^T A_x)^{-1} A_x^T X(k)$$
$$\hat{\theta}_y = (A_y^T A_y)^{-1} A_y^T Y(k) \quad (3)$$

The fitting errors

$$E_x = (X(k) - A_x\hat{\theta}_x)^T (X(k) - A_x\hat{\theta}_x)/(m+1)$$
$$E_y = (Y(k) - A_y\hat{\theta}_y)^T (Y(k) - A_y\hat{\theta}_y)/(m+1)$$

can represent the exerting muscle strength or its stability along the horizontal and vertical direction, respectively. Therefore, both of fitting errors are the important parameters to evaluate the stability and consistency of archers during the aiming procedure, so they are calculated and listed in Table 1. Moreover, the estimated coefficients for all archers are also listed in Table 2. Additionally, the ranking order among these twelve archers is listed in Table 4 for illustrating the relationship between archery performance and the fitting error. As observed from Table 1, both of the top two ranking, archers 1 and 5, have small value of the fitting errors along both directions, so their muscle strength or stability are outstanding. However, the low ranking archer 3 also has smaller value of the fitting error than those of high ranking archers, such as archers 11 and 12. Thus, we may say that the archer with better

Table 1. The fitting errors of the proposed AR2 model with the offset.

Direction	Archer 1	Archer 2	Archer 3	Archer 4	Archer 5	Archer 6
Horizontal	0.27138	0.33739	0.27805	0.35719	0.20572	0.3214
Vertical	0.37356	0.44606	0.2957	0.49643	0.31024	0.29042
Direction	Archer 7	Archer 8	Archer 9	Archer 10	Archer 11	Archer 12
Horizontal	0.51305	0.61635	0.32262	0.31371	0.46263	1.0097
Vertical	0.40028	0.52651	0.31687	0.31844	0.55356	0.75776

Table 2. The coefficient and the offset of the AR2 model.

Coefficients	Archer 1	Archer 2	Archer 3	Archer 4	Archer 5	Archer 6
a_{x1}	0.96015	0.9823	0.61838	0.78989	0.95237	0.70255
a_{x2}	0.028338	0.015607	0.37722	0.2087	0.033628	0.2947
b_x	0.000580	0.01033	−0.019272	−0.003888	−0.025529	−0.023733
a_{y1}	0.58782	0.59964	0.32608	0.37286	0.46953	0.29698
a_{y2}	0.37976	0.39864	0.67396	0.6081	0.5187	0.63994
b_y	−0.041073	−0.014966	−0.006113	−0.009744	0.003640	0.006234
Coefficients	Archer 7	Archer 8	Archer 9	Archer 10	Archer 11	Archer 12
a_{x1}	0.63994	0.60238	0.56895	0.69537	0.59217	0.66104
a_{x2}	0.34734	0.39396	0.42751	0.30471	0.38787	0.33781
b_x	−0.012295	0.005909	0.017382	0.005091	0.009948	0.004111
a_{y1}	0.25662	0.27554	0.22217	0.29674	0.34517	0.4285
a_{y2}	0.72034	0.72035	0.76949	0.70269	0.62864	0.51785
b_y	−0.078546	−0.009466	−0.044916	−0.000966	0.054824	−0.098492

performance usually has small value of the fitting error, but small value of the fitting error may not indicate the better performance.

We now focus on the role of constant offset which represents the linear trend of the aiming trajectory along each direction as referred to Eq. (1), so it can be used to indicate or describe the particular aiming style of archers. Table 2 shows that within the one and half second period the offset b_x is insignificant for archer 1 along the horizontal deviation, but significant for archer 5. On the other hand, along the vertical deviation, the offset b_y is insignificant for archer 10 but it is significant for archer 12. There are some interesting findings that archers 1, 5 and 12 ranking among the top four are often fallen into these extreme cases. The role of constant offset is not directly related to the archery performance, and it is related to the particular linear trend of the individual aiming trajectory.

Poles of the AR2 model along the horizontal direction, λ_x, are evaluated by the following formula with its coefficients:

$$\lambda_x^2 - a_{x1}\lambda_x - a_{x2} = 0$$

Table 3. The poles of the AR2 model.

Direction	Archer 1	Archer 2	Archer 3	Archer 4	Archer 5	Archer 6
Horizontal	0.9888	0.99794	0.99681	0.99883	0.98646	0.99788
	−0.028658	−0.015639	−0.37843	−0.20894	−0.03409	−0.29533
Vertical	0.97665	0.99877	1	0.98821	0.99227	0.96212
	−0.38884	−0.39913	−0.67395	−0.61536	−0.52274	−0.66514
Direction	Archer 7	Archer 8	Archer 9	Archer 10	Archer 11	Archer 12
Horizontal	0.99058	0.99738	0.99752	1.0001	0.98568	0.99914
	−0.35064	−0.395	−0.42857	−0.30469	−0.39351	−0.3381
Vertical	0.98668	0.99761	0.9953	0.99967	0.98402	0.96509
	−0.73006	−0.72208	−0.77313	−0.70292	−0.63885	−0.53658

or equivalently

$$\lambda_x = (a_{x1} \pm \sqrt{a_{x1}^2 + 4a_{x2}})/2 \qquad (4)$$

If the poles are the complex conjugate pair $\lambda_x = \alpha \pm j\beta$, then the aiming trajectory has the oscillation with the frequency related β and the convergent rate α. On the other hand, in case of real number, it suggests that the main aiming trajectory without oscillation, and the convergent rate is dominated by the slower mode which corresponds to the value of pole closes to the unit circle $|\lambda| = 1$. Poles of AR2 model are evaluated based on Eq. (4) and listed in Table 3. The poles of the AR2 model for the twelve archers are all real number, so it indicates that their main aiming trajectories are smooth without oscillation and these archers are exporters in archery. It notes that the oscillation exists in the real individual aiming trajectory and it has been taken into account as the variation of muscle strength, that is, the fitting error in this paper. Thus, in this section we focus on the average of the aiming trajectory instead of real one.

As referred to Table 3, it can be observed that if both dominated poles ($\lambda > 0.995$) along both directions are too close to 1 (slow convergent rate), then their ranking will not be good, such as archers 2 (0.99794, 0.99877), 3 (0.99681, 1), 8 (0.99738, 0.99761), 9 (0.99752, 0.9953) and 10 (1.0001, 0.99967) whose rankings are 8, 10, 5, 12, and 7. Moreover, if the negative pole along the vertical direction is too small ($\lambda < -0.7$), then his ranking is also not so good, such as archers 7 (−0.73006), 8 (−0.72208), 9 (−0.77313), and 10 (−0.70292) whose ranking are 9, 5, 12 and 7. Archer 4 has two poles very close to one (0.99883 and 0.98821) and one very negative pole (−0.61536) along the vertical direction, so his ranking is 11. After the above ranking determination, the ranking associated with the rest archers (archers 1, 5, 6, 11, and 12) can be determined by the high value of coefficients a_{x1} and a_{y1} to indicate a high correlation between $x(k)$ and $x(k-1)$, or $y(k)$ and $y(k-1)$. Therefore, if the higher value of these coefficients is ($a_{x1} + a_{y1} > 1$), then the

Table 4. The ranking.

Archer	1	2	3	4	5	6	7	8	9	10	11	12
Ranking	1	8	10	11	2	5	9	5	12	7	3	4

better ranking is. That is, archers 1 (0.96015, 0.58782), 5 (0.95237, 0.46953), and 12 (0.66104, 0.4285) whose rankings are 1, 2 , and 4.

4 Conclusions

In this paper, the AR2 model has been proposed to establish the time sequence of the aiming trajectory. The estimation of the associated coefficients of the model is also formulated and implemented to obtain the estimated coefficients, and the corresponding fitting errors are also calculated. The fitting error is used to reveal the strength and stability of the muscle. The linear trend of the aiming trajectory is related to the additional offset in the AR2 model, and it is corresponding to the individual typical archery style. The pole of the AR2 model can be utilized to check the aiming trajectory whether the oscillation exists or not and how fast the convergent rate is. Moreover, the first coefficient of the AR2 model representing the correlation between the current time sequence and the previous one is also a useful index of the archery performance.

Acknowledgements

This work is supported by CHU-94-E-002, CHU-94-E-006 (Chung Hua University), NSC 95-2221-E-216-063, and NSC 93-2218-E-216-009.

References

Ertan H., Kentel B., Tumemer S.T. and Korkusuz F (2003) Activation patterns in forearm muscles during archery shooting, *Human Movement Science,* 22/1, 34–45.
Landers D.M., Han M., Salazar W., Petruzzello S.J., Kubitz K.A. and Gannon T.L. (1994) Effects of learning on electroencephalographic and electrocardiographic patterns in novice archers. *International Journal of Sport Psychology,* 25/3, 313–330.
Lavisse D., Devitern, D. and Perrin, P. (2000) Mental processing in motor skill acquisition by young subjects. *International Journal of Sport Psychology,* 31/2, 364–375.
Leroyer P., Van Hoecke J. and Helal J.N. (1993) Biomechanical study of the final push-pull archery. *Journal of Sports Sciences,* 11/1, 63–69.
Lin K.B. and Hwang, C.-K. (2003) Analysis of correlations between aiming adjustment trajectory and target, *13th International Conference in Medicine and Biology,* Tainan, Taiwan, 132–133.
Ljung L. (1999) *System Identification,* 2nd edition. PTR: Prentice-Hall.
Shiang, T.-Y. and Tseng C.-J. (1997) A new quantitative approach for archery stability analysis, *International Society of Biomechanics Congress,* Tokyo, 142.

18. Motor Sport and Cycling

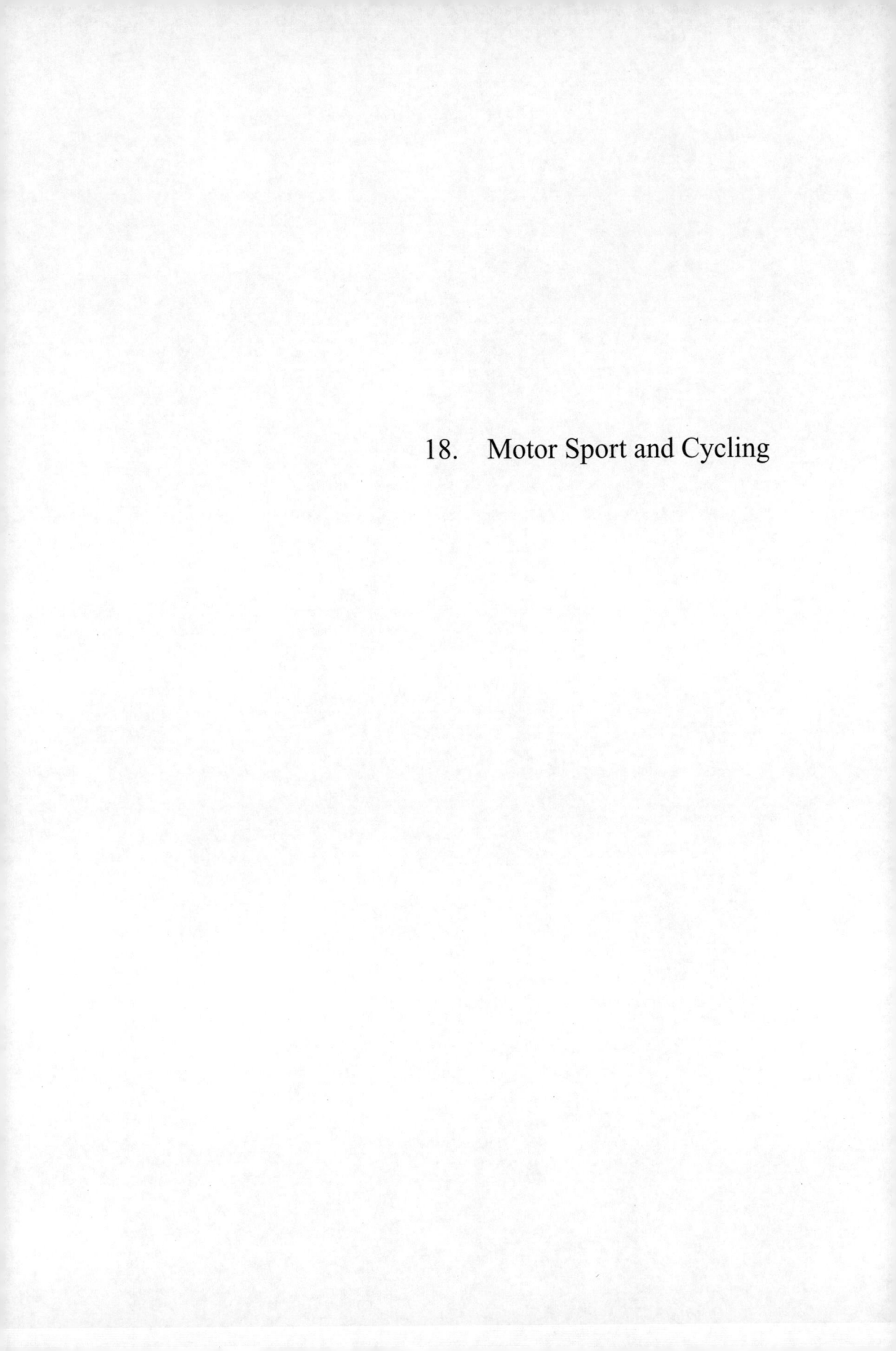

FORMULA SAE: STUDENT ENGAGEMENT TOWARDS WORLD CLASS PERFORMANCE

S. WATKINS & G. PEARSON

School of Aerospace, Mechanical and Manufacturing Engineering, RMIT University, Melbourne, Australia

Formula SAE (Society of Automotive Engineers) is the largest student-based competition in the world, involving student teams designing, building and racing a small open-wheeled racing car. Conceived in the USA in 1981, there are now over 200 university teams involved, with events in the USA, UK, Australia, Germany, Italy and Japan, and with growing interest in South East Asia. An overview of the competition and insights into how to build and manage a successful team is provided based on the RMIT team from Australia. The RMIT team started in 2000, and in a relatively few years has won the competitions in USA, UK and Australia, including concurrently setting lap records and winning the Fuel Economy Event. Strategies are outlined which foster the active student engagement required for succeeding in the intensely competitive world of FSAE.

1 Introduction – What is FSAE?

The Society of Automotive Engineers International (SAE-Int) defines FSAE as follows; "The Formula SAE® competition is for SAE student members to conceive, design, fabricate, and compete with small formula-style racing cars. The restrictions on the car frame and engine are limited so that the knowledge, creativity, and imagination of the students are challenged. The cars are built with a team effort over a period of about one year and are taken to the annual competition for judging and comparison with approximately 120 other vehicles from colleges and universities throughout the world. The end result is a great experience for young engineers in a meaningful engineering project as well as the opportunity of working in a dedicated team effort" (from http://students.sae.org/competitions/formulaseries/).

1.1. *History*

The above statement reflects the rules and numbers of teams for the major event in the USA circa 2005. However, a similar competition has been run in the USA for over two decades. The history of the competition can be traced to 1976 when SAE-Int started to run student Mini Baja competitions; off road races named from the Baja 1000 race in Mexico. The push for a road race equivalent – Mini Indy – came from the company Briggs and Stratton, who in 1979 supplied 5 hp engines and thirteen teams competed. The first FSAE competition, held in 1981, permitted any four stroke engine and limited the power via a 25.4 mm (1) restrictor – a regulation that remains to the present day. Since then there has been a single competition held every year in the USA until 2006, when two events were first run; a (new) West Coast Event near Los Angeles and an event in Pontiac, near Detroit. Over that period there were several changes to the rule – the most significant being a revision to include static events and with the overall winners being decided by the summation of points accrued with a possible total of 1000. Now events are running around the world including in the UK, Brazil, Italy, Germany, Australia and Japan, with more planned in other countries. Whilst there can be slight rule variations between countries, rules are based on the USA competition described below.

1.2. *The Competition and its Constituent Events*

The competition is made up of eight separately evaluated events – three of which are static (i.e. static studies on each car and sometime a presentation to an industry panel) and five dynamic (i.e. evaluation based on measured car performance parameters). A complete version of the rules covering the events, regulations pertaining to the cars etc. can easily be found on the web (SEA International, 2005). A brief description of each event, including the points allocated, is given below.

Presentation 75: Here one or two students from each team must present their business case to a panel for the limited volume production of their vehicle. Many aspects of the car, (e.g. design for mass manufacture) are usually presented and other aspects often include potential manufacturing plant layout and return on investment. We have found it useful to consider this as presenting a potential business to venture capitalist, emphasising confidence in achieving a good return on investment.

Design 150: "The car that illustrates the best use of engineering to meet the design goals and the best understanding of the design by the team members will win the design event". A relatively short paper document is produced by each team, backed up with design information via stand alone posters (often for each car subsystem) and/or laptop-based visuals. In some countries there is a design final after the dynamic events, where five teams selected via the initial judging round are questioned in considerable detail.

Cost Analysis 100: Each component on the car is costed (including machining, fabrication etc.) and compiled into a relatively large report. Additionally individual team members are questioned at the event for their understanding on how some bought components are manufactured. Such items could be ignition coils, rose joints, suspension springs. Teams have prior knowledge of which components will be selected.

Acceleration 75: Cars are timed over 75 metres from a standing start. Some of the faster cars can achieve times of 0 to 100 Km/h in less than four seconds.

Skid Pan 50: Here cars are assessed on cornering ability, with time being measured around a simple circular track, with minimisation of time being the objective.

Autocross 150: The cars are driven separately around one lap of a tight twisty track, with the objective again being to minimise lap time. Since the track is bend-laden and there are no long straights, designs that exhibit good handling with the ability to turn quickly do well. The nature of the tracks are such that top speeds are kept low (average speeds of about 50 Km/h), thus aerodynamic devices that are used on other types of race cars (e.g. F1 and Indy) have questionable merit. This is an area of on-going debate for some teams, with some cars opting for lightweight simple designs and no aerodynamics downforce aids, (Figure 1) and others having multi-element wings (Figure 2).

Endurance 350: A staggered start event where cars are driven for multiple laps around a tight twisty track, with the objective again being to minimise lap time.

Fuel Economy 50: Here the fuel used during the Endurance Event is measured, with the objective of minimising fuel used.

Figure 1. RMIT 2006 car.

Figure 2. Monash 2006 car.

2 RMIT Formula SAE

RMIT University has competed in Formula SAE every year since the competition was first introduced to Australia in 2000. After four years of learning the competition and establishing internal management processes and support structures, the team began travelling to international competitions in 2004. Since that time the team has been highly successful in Formula SAE competition, having won three events outright and finishing outright second on two other occasions. Our lessons learnt are presented below and whilst the intent of this paper is not to provide a detailed technical description of our cars, details are available on our website (RMIT, 2007).

2.1. *Competition Observations*

A student design team has to consider many different and often competing factors to achieve success. There are many useful resources available on the SAE-Int website (www.sae.org), including specific technical information (e.g. suspension tuning and development, Lyman,

2005) and team management information (e.g. Royce, 2005). These are generic and very useful for new teams. Aside from the various vehicle performance compromises that need to be considered (e.g. engine power versus fuel economy, chassis weight versus stiffness, suspension geometry for cornering versus straight-line acceleration, etc.), the students also need to approach practical project issues such as how their design decisions affect cost, manufacturability and marketability. It is worth noting that the first three events listed above, totalling 325 of the 1000 points, are based on reports and student knowledge, and a further 50 points are allocated to fuel economy, none of which are traditional criteria for success of a racing car. A holistic, systems approach is necessary to achieve success across the entire range of events.

Whilst the primary intent of the competition is to expose engineering students to a complete product development cycle (design, build, test), competition results tables indicate that this is not being successfully implemented in many cases. Recent competition results indicate a large proportion of teams fail to complete all the events, and in some cases fail to get a completed car to the competition at all. At the recent 2006 FSAE in Detroit, only 41 of 121 attending teams completed all events (a completion rate of 33.9%), and a further 19 teams registered for the event but failed to attend (SEA International, 2006). Analysis of further recent competition results indicates that this is typical.

In many cases, the students' perceptions are distorted somewhat by the motorsport aspect of the competition. The base task of designing and building a simple self-propelled vehicle should not be beyond the scope of a team of senior level undergraduate engineers, and the competition regulations and points structure are designed such that a team can be successful even on a modest budget. However students' prior exposure to motorsport (either directly or through the media) often leads to the impression that success is directly related to such factors as large budgets, cutting edge technology and complicated gimmickry with an emphasis on straight line speed. This philosophy is sometimes transferred to the team's FSAE project, leading to overly complex design programs and stretched time and monetary resources. This also leads to the non-performance related aspects (i.e. static events) of the competition being overlooked as the team focuses primarily on getting more speed from a technologically complex car.

The RMIT team has noted that in many cases this focus on vehicle performance and technology has had a detrimental effect, as teams design beyond their capabilities and fail to bring a tested and reliable vehicle to the event. This is evidenced in the typically poor finishing record in this competition. This observation led our team to question whether it is feasible to trade off a few potential performance points, in the quest to simplify the project and lessen the risk of non-completion in any event.

3 RMIT Strategy

3.1. *Car Design*

A Systems Engineering perspective asserts that decisions made in the conceptual phase are ultimately what drives the overall success of the project, Blanchard & Fabrycky (2006). If a project does not meet required deadlines, or if a product is not adequately tested before its release date, it can usually be tracked back to the design decisions made at the beginning of the project. It is the project team's responsibility to fully investigate and understand the resources available, (people, budget, facilities), and then make design decisions that ensure the project can be completed within these constraints. Given that typically 60% of FSAE teams fail to complete the competition, it is apparent that many teams are still not fully assessing their constraints before making their design decisions. The RMIT team develops its design concepts with strong emphasis on the early, conceptual stage of the design process. Two design dictums that the team follows are:

- The simpler, the better (or do less, and do it better)
- Every choice has a consequence

When assessing vehicle design options, emphasis is placed on simplicity so as to minimise the load on team resources and lessen risk of failure. Each component on the car comes with its own weight penalty, manufacturing issues and financial cost, and these must be considered in tandem with the component's performance capabilities. The fewer components there are on the car, the better they can be understood and the easier it is to tune and develop the car to its maximum potential. Also, it is much easier to reduce vehicle weight by disposing of components and systems altogether, than it is by redesigning components.

A wide variety of applications of high technology is possible on a FSAE car but the disadvantages can often outweigh the advantages. For instance, consider the adoption of an electronic gearshift. This may increase gear shifting speed, but consequences include additional weight and risk when compared to a simple mechanical system and will take time and money to develop reliably. Similarly aerodynamic wings offer downforce, but consequences include increased mass and MOI's, loss of fuel economy, increased vehicle sensitivity to the prevailing wind, and will again take time and money to develop and implement. In the last few years the RMIT team has downsized to a single cylinder engine that is of smaller capacity than the maximum permitted by the rules. Whilst this reduces our straight line performance (thus lowering our placing in the Acceleration Event) it has benefits our car's fuel economy and cornering performance, through weight, size and packaging advantages. Since the timelines and budgets are tight, every component design option has to be weighed against the resources required to manufacture or purchase it. It is up to the students as designers to fully understand the consequences of design decisions, and ensure that they fit in with the overall design direction of the team. One of the biggest lessons learnt is the "non-reversible nature of time and money" and the students must realise that they are not dealing with Formula One budgets and manpower!

With the above points in mind, RMIT undertook a major design re-direction in year 2003. Until that point in time the team had placed heavy emphasis on developing the engine package, to optimise straight-line performance. The team had experimented with supercharging, turbo charging, pneumatic and electric gearshifts, and twin and four cylinder engines. Each year it was observed that these development programs were costly (in terms of both time and money), and introduced great risk into the project. Two of the first three RMIT competition entries ended with engine failures at the competition before this design re-direction.

The 2003 team thus questioned whether the design emphasis on engine performance was worthwhile. Could the project be simplified, and the risk of failure be reduced, by introducing a simpler engine system? The team undertook to move away from a complex multi-cylinder design, towards a vehicle with a naturally-aspirated, small-capacity single cylinder engine. This would trade off some straight line acceleration for additional benefits in weight, size and simplicity. In particular, the single cylinder option simplified the design of the intake, exhaust and engine wiring systems, which greatly reduced manufacturing and development time.

In practice, the loss of engine performance was outweighed by gains in cornering performance, fuel economy, and reliability. This increase in reliability has particularly played a major role in the team's recent successes, as places are regularly gained through attrition of the competition.

3.2. *Team Management*

Student teams undertaking Formula SAE often see the project primarily as a technical task. However it is not so much the technical skills of the team but rather the overall team management prowess that determines the overall success. Continually high-ranking teams such as Cornell University in the USA concentrate on succession plans and the documenting of project findings to ensure ongoing success. The following outlines some of the management structures implemented at RMIT.

3.3. *Team Structure*

Our team comprises between 50–100 students, mostly from the engineering schools. Some of these students are fully committed to the project, whilst some are merely interested and just wish to learn a little about project management and automotive science. At the core of the team, it is not unknown for senior team members to spend upwards of 50 hours a week on the project. However the team also recognizes the value of those students who cannot make a major time commitment, and effort is made to accommodate these students with simple project tasks.

Our team structure is broken down as follows:

- Core Management Team – comprising Chief Engineer, Team Leader and experienced team members. This team oversees the top-level project management issues of the team, such as finance, sponsorship, time management, running team meetings, etc.
- Sub-system Design Teams – teams dedicated to the design and manufacture of particular sub-systems of the car, such as Suspension, Chassis, Drivetrain etc.
- General team membership – student members (usually junior) uncommitted to any particular role in the team, but assisting with many technical and non-technical tasks – the latter can include team clothing range, team website etc.

The Core Management Team structure is flexible to an extent, as it needs to accommodate the differing skills of team members. Generally it comprises a dedicated Chief Engineer (the design leader devoted to technical vehicle design issues, with this student often taking responsibility for chassis and using this for academic credit via final year project subject) and a Team Leader who focuses specifically on the non-technical managerial tasks. This sort of structure suits a team where there is strong technical leader who has prior automotive experience. In some years this distinction is blurred as the Chief Engineer and Team Leader work together interchangeably. Occasionally the two roles are shared across a number of students working as a core leadership team. Such a structure works where there may not be students with enough experience to take on full managerial responsibility, but are willing to work as a team and learn as they go.

The team has found that imposing a set structure on the Core Management Group can be detrimental to the team, as team members might not precisely fit the particular job roles. The primary purpose is to achieve the design tasks associated with the competition, so varying structures are tolerated as long as the core goals are being achieved.

It should be noted that whatever management structure is chosen, it is vital that the student or students taking on the Chief Engineer's role have a definite vision of the overall design direction of the car. The Chief Engineer is the technical leader of the team, and has to guide the various sub-system designers along a unified path. Disagreement on the design direction can lead to a poorly integrated vehicle and a confused project overall.

3.4. *Personnel Management*

Maintaining student interest is a key issue that needs to be addressed in managing a successful Formula SAE project team. For many students this is their first practical design project, and this can be confronting when they realise that real-world projects do not work out as perfectly as they might hope they will. This can cause great frustration to the student, and loss of interest in the project.

It is crucial to engage the students meaningfully from their first contact with the team. It has been noted that for each RMIT FSAE project, over 100 students will initially show interest in the project. This number drops rapidly to a core team of 10–20 members unless some active means is employed to retain the new students. Also, with the high turnover of students it is necessary to establish succession plans, so that knowledge is not lost with each graduating team. Some of the project management strategies that have been implemented at RMIT to address these issues include:

- Student-run technical seminars
- Training/mentoring schemes
- Organized social events
- Multi-year vehicle development plans

Student-run technical seminars: Throughout the project, senior team members and faculty advisors are encouraged to present short technical design seminars to the younger students, on a topic of their choice. Usually the topic aligns with the speaker's particular design project on the car, but can also be on other topics such as FSAE history, competitors' vehicle designs, advantages and disadvantages of the adoption of wings etc. These seminars take place during the weekly team meeting, and the team endeavours to present at least one seminar per fortnight. Note that these seminars are separate from any assessment requirements specified for a student's Final Year Design Project.

The primary benefit is increased attendance at team meetings and involvement in team activities. Regular structured seminars were first introduced in June 2006, when it was noticed that attendance at the weekly team meeting had dwindled to around 8–10 core students, which was not enough to maintain project momentum. After introduction of the seminars, regular attendance at the team meetings rose to around 30 to 40 students, across all year levels. General feed-back from the newer team members indicate that the seminars are most valuable and help them make sense of vehicle design fundamentals.

Further benefits are that team members get a more holistic perspective of the vehicle design, as they are continually exposed to design issues across the full breadth of the project, not just their own particular design sub-system. Also, the scheme enables the senior team members to practice their public speaking and presentation skills, building their confidence and aiding in their professional development.

Training/mentoring schemes: In addition to the regular short seminars above, the team has assorted mentoring and training programs that operate through the year. Senior team members hold workshops or informal sessions to train incoming members in such fields as CAD modelling, dynamometer testing and project management. Such schemes are vital to set in place succession plans, and to ensure that knowledge gets transferred from year to year. It is also gratifying for the senior team members to engage in this mentoring activity, and it builds a sense of ownership and investment in the team's long-term successes.

The team also occasionally invites speakers from academia or industry to speak on a topic of relevance or interest. Recent speakers have included an academic investigating the vehicle dynamics of "Drifting", and an industry expert on Lean Design and Manufacture. The intended goal of such activities is to make the team members feel that they are part of a dynamic and active team, and that they are benefiting from their time commitment to the FSAE team.

Social interaction: The Formula SAE project is highly demanding of a student's time, and it is important that the team members get away from the project occasionally. A number of social events are planned throughout the year, and these can range from informal barbeques at the team workshop, through to go-kart nights and organised bike rides. It is up to the team to decide what activities suit the current cohort; however such events are a vital team building activity and should not be forgotten in the rush to get a car built.

There is also regular communication and social gathering amongst the various Melbourne Formula SAE teams, and RMIT further encourages communication with other teams by hosting visiting students and international teams at the RMIT facility. For the most part, competing Formula SAE teams are most welcoming and willing to offer advice relating to their own unique experience. When beginning in FSAE it is helpful to communicate regularly with other teams, so that common pitfalls and issues can be avoided.

Multi-year Development Plans and Academic Credit: The nature of Formula SAE competition is that each year, a new group of students enters the team and a new car is designed and built. This can lead to issues in retaining acquired knowledge within the team, and also in individual teams feeling they are operating in isolation from their own university's past and future teams. The former issue can see teams make the same design mistakes year after year. The latter can manifest itself in over-zealousness as the students attempt too much and try to revolutionise the competition in just one year. Both issues hamper a team's ability to succeed.

A strategy that RMIT employs to combat this is the use of multi-year development plans. Major design changes are planned and researched in the background over a number of years, sometimes a final year project, and the team is conscious not to implement too many of these major design changes in the one year. Only changes that have proven to be entirely reliable, and that can be built to time and cost, will be incorporated on the car.

An example is the strategy undertaken over a three year cycle from 2003–2005. In 2003, the RMIT team took on a new design direction in the employment of a single cylinder engine for FSAE competition, in comparison to the more complex 4 cylinder engines used by the majority of competitors. The design philosophy was to trade off some engine

Table 1. RMIT FSAE Vehicle Performance Goals: 2003–2005 (Final achieved values in brackets).

Project year	Vehicle mass goal	Engine performance goal	Chassis construction
2003	Under 200 kg (197 kg)	50 hp (52 hp)	Full steel space-frame
2004	Under 180 kg (175 kg)	55 hp (58 hp)	CF front / steel space-frame rear
2005	Under 160 kg (154 kg)	Over 55 hp (58 hp)	Full CF composite tub

performance for the benefit of a lighter overall engine package, reduced vehicle size and simplified manufacturing. Some other teams had already experimented with this direction, although mostly these experiments were unsuccessful. The RMIT team noted that in such cases, the teams concerned tried to compensate for the lesser engine power by undergoing a massive weight saving program over the course of only one design cycle. The final result was an unreliable and unsorted car that did not operate to potential.

The 2003 RMIT team avoided this by implementing a three year development plan for the introduction of the single cylinder design concept. A new car would be built for each year's competition, but rather than attempt to build the lightest car possible in one attempt, a staged program was implemented. It was intended that this would lessen the risk of competition failure through gradually developing the vehicle and the team knowledge at the same time. The planned stages are shown in Table 1 below.

Of particular note was the staged introduction of carbon fibre (CF) composite chassis construction. It was expected that implementation of a carbon composite chassis would greatly aid the light-weight design concept, but given the changes imposed by the new engine package it was deemed wise to introduce the carbon chassis concept gradually. For the first year the team continued with a steel space-frame, a known construction process used in pre-2003 RMIT vehicles.

The strategy has been very successful, with each car running reliably and strongly in FSAE competition. Although seemingly quite heavy compared to cars of similar concept, the conservative 2003 car won the 2004 Formula Student (UK) by a 100 point margin. This was primarily due to the car running strongly and with 100% reliability, whilst many of the opposition teams suffered mechanical failures. Ongoing development saw the 2004 car come 2nd outright at the 2004 FSAE Australasian event, and the 2005 car then went on to win the 2006 FSAE Detroit event outright.

Furthermore, an ongoing cycle gives team members a sense of belonging to something greater than simply the current year's project. There is a motivation to transfer acquired knowledge to incoming team members, as the present team understands that the work they are doing currently will be contributing to future teams successes. It is worth noting that such schemes cannot be implemented if membership of the team is limited to senior or final year project students only.

Typically 40% of the students in our team will be in the final year of our degree programs (Automotive, Aerospace, Mechanical or Manufacturing) and the majority of these will be aligning their activities with the final year project for course credit. Final year project accounts for one quarter of the subjects in our final year, but FSAE nearly always takes a disproportionately large amount of time. Whilst credit can be accrued in other subjects for working on the car (and some academics recognise this, permitting aspects of student work to be used for say, a management project) it is far from universal. This causes understandable concern. We are currently considering the adoption of another "external project" course which the students can use to bring further academic credits.

4 Conclusions

In our experience FSAE has provided a motivating experience like no other, and has resulted in graduates having very strong skills in a wide range of areas, that are directly transferable to their future work opportunities. This is clearly recognised and rewarded by the automotive industries, but not always by academia. FSAE results in enhanced skills that include self and team management skills that are reinforced in a manner that does not usually occur in traditional subjects. Here the challenge to the universities is to provide adequate subject credits for the enhanced learning that occurs. Whilst the adoption of FSAE in a traditional university environment can prove problematic and at times very challenging, the outcomes are well worth it and include being the most motivating thing that one is likely to encounter in the university environment.

Acknowledgements

The authors would like to acknowledge the financial and moral support of the RMIT University and especially the RMIT team members from 2000 to 2007.

Appendix – Simon Watkins Tips for New FSAE Teams

1. Car: Keep the design relatively simple (especially for the first year) – people, especially students, are optimistic about what they can do with their time. An oft-quoted saying of mine is that what FSAE really teaches is the "non-reversible nature of time and money". Use a standard engine for first and maybe subsequent years (there are advantages in a single – one intake, one exhaust which makes tuning much quicker than a four), build a mockup chassis quickly and get students to appreciate packaging issues. Buy as many standard components as possible in early years – perhaps use a Torsen differential (centre diff from Audi Quattro) and realise that the designing and sealing of a diff housing will take time and may need technicians to assist. Use standard 13" wheels. In the first year of our competition I went to the auctions and bought a crashed Honda F2 600 cc bike. This enabled the team to utilise bits from it for the car, but also to sell bits to raise petty cash. Money will be an issue – it may cost approximately twice your initial budget! Good tyres are expensive. First year, including workshop setup, tools excluding travel etc, might cost $40,000+. This will generally come down, depending upon complexity of the car. Some Universities (including RMIT) have a TAFE (i.e. hands-on) training school and it is useful for their students to become involved with the car build.
2. Pick good student leaders for positions of Team Leader and Chief Engineer. These are key roles and need people who can not only organise and motivate themselves; they should also be able to get the best out of the team. People who promise much, but produce little, need to be rejected quickly.
3. Get the students to realise that there needs to be relatively minor development work on the engine (this is to get it to work well with a 20 mm restrictor, both on

the dyno, and also on track) but the chassis, suspension, steering, uprights, etc are their own design. This takes MUCH more input than the engine, so you need plenty of students on these areas. Usually you get too many offering to do engine work and not enough for chassis.

4. Get the students to plan and manage their work via a simple project management chart and budget. Set realistic, **immovable** deadlines and budgets. Encourage students to calibrate themselves. Get them to hold weekly management meetings with only a subsystem overview report each week and do not allow circulatory discussions regarding rules etc. Team Leader or Chief Engineer chair meetings (after education in meeting procedures from Faculty Advisor). Build a team from all years of the students – this is important to ensure continuity so there is an environment where the student novice learns from the student experienced – even with things like running meetings. All team members must know the relevant parts of the rules for their area. No questions allowed regarding rules in meetings. Focus meetings on action items and include a name and date to ensure "delivery" rather than circulatory discussion!

5. Realise that the students will be putting in enormous amounts of time and learning many things that are useful to their capabilities in industry, and try to reflect this via course credits. This is difficult but a subject devoted to this which does not need additional paperwork is useful – a major final year project is one way. Students will already be providing documents for the competition that can be used in assessment – e.g. Cost Report, design work for the Design Presentation, manufacturing work for the Presentation Event. Some will work like they have never done before. Recognize and reward, even if it is only with words.

6. Students are supposed to do all the design and tech work. This is usually not followed 100% in most universities and technicians "assist" especially in CNC machining etc, but students must have a useful workshop and thus may need some training and supervision. Problems occur when technicians finish work at 17.00, but students might start work at this time and finish at 05.00 the next day. In our early years some students used to work 80 hours with no break just to finish the car for the competition. Exams and assignments get missed, but these will be some of the most capable students you will (just!) graduate. After 80 hours without sleep students then try to learn how to drive the car at the competition; dangerous and to be avoided by planning and immovable deadlines!

7. The earlier you start the better – there is merit in getting key students in a new team to attend a competition to study good teams – not just the technical aspects but also the management ones. In your first year, obtain video footage of car designs and performances from prior competition and show to your newly-formed team. Do not underestimate the need for tight management of time and costs.

8. The teams need money and rapid access to it. This usually contradicts University bureaucracy and policy – the timelines needed to succeed in racing are different from those in large institutions. Students must become involved in raising external sponsorship and should manage this budget on their own. Do not worry – it will be very well utilised money. Most universities have rather long timescales regarding purchasing which are far too long for racing timescales. There are numerous checks in most universities that often make purchasing inefficient but

risk free. Major items with long timescales can be bought via the University, but some items (e.g. tyres, or components that fail, such as engines that blow up) often cannot be foreseen, thus you may need to have an external bank account (perhaps set up as a non-profit making organization) which the students are responsible for. In our experience the students look after monies very carefully and it is worth trusting them.

References

Blanchard B.S. and Fabrycky W.J. (2006), *Systems Engineering and Analysis*, 4th Edition, Pearson Prentice Hall, Upper Saddle River, New Jersey, USA.
Lyman S. (2005) Suspension Tuning and Development. In: the SAE-Int Collegiate Design Series, Collegiate Roadshow. http://www.sae.org/students/presentations/suspensiontd.ppt
RMIT FSAE website (2007) http://www.fsae.rmit.edu.au/.
Royce M. (2005) Team Organisation. In: the SAE-Int Collegiate Design Series, Collegiate Roadshow, http://www.sae.org/students/presentations/organization.ppt.
SEA International (2005) FSAE rules, http://students.sae.org/competitions/formulaseries/.
SEA International (2006) FSAE Results, http://www.sae.org/students/fsae2006results.xls.

SERVICE STRENGTH OF HIGH TECH BICYCLES

M. BLÜMEL[1], V. SENNER[1] & H. BAIER[2]

[1]Technische Universität München, Institute for Sports Equipment
and Materials, München, Germany
[2]Technische Universität München, Institute for Lightweight Structures,
Garching, Germany

The aim of this study is to measure the forces acting on the bicycle in a field test and to build up testing equipment for fatigue tests. The knowledge about the acting forces is also used within finite element simulations combined with optimisation tools. The possibility to perform meaningful fatigue tests on bicycle frames and components enables manufacturers to develop lighter but nevertheless safer products.

1 Background

1.1. New Challenges for the Cycling Industry

During the last five years the weight of top of the range bicycle frames and forks could be reduced without sacrificing the stiffness, which is the main quality criterion for a good handling of the bicycle. One of the main reasons for this development is the wider use of high tech materials such as carbon fibre reinforced plastics. The disadvantages of this technology are an increased development effort and higher requirements for the quality assurance for the manufacturing company.

The new materials however are not the only challenge bicycle companies are facing today. Improvements made on mountain bike suspension and brake components enable the athlete to ride their bikes more aggressively. It is expected that the typical loading pattern of modern bikes have changed. With this in mind two questions need to be answered:

1. Is the current knowledge regarding the forces acting on modern bicycles still satisfactory?
2. Can the existing standards for fatigue tests on bicycles and their components still ensure the athletes' safety?

The focus of this study is on cross country – / marathon mountain bicycles and on road racing bicycles, as these types of bicycles hold the biggest challenges for the engineer, as weight has to be kept as low as possible. Low weight directly results in an advantage in competition for the athlete.

1.2. Existing Standards for Fatigue Tests on Bicycles

The existing Standards for Bicycles in Germany are defined in DIN 79100 and ISO 4210. Last year, they were replaced by the new European Standards (EN 14764, EN 14781, EN 14766). The loads used in the different static and dynamic tests described in these standards have been determined by a comparison with existing bicycle components that proved fail save (Otto, 1995). The fatigue tests are force controlled single level tests, testing the finite fatigue life. DIN and ISO do not differentiate between the different types of bikes, while the new EN Standards do. The test for handlebars and stems has two defined load cases, but as in all the tests the forces acting in push and in pull direction are equal. The different components are tested individually as well the interconnection of the components in the complete bicycle on a roller type test stand.

1.3. *A Critical Look on the Existing Standards*

Measurements carried out by Barski *et al.* (1995) to determine the forces acting on mountain bicycles showed, that the existing standards are not as relevant for the practical use of the bicycle as necessary to ensure the user's safety. The same conclusion has been drown by Spahl (1996), Groβ (1997) and Issler (2006). Figure 1 shows Barski's results (Barski *et al.*, 1995) on the forces acting on the front fork in comparison to the test force defined in DIN 79100. Barski says that the test forces do not occur in the practical use of the bicycle and proposes that the existing standards have to be reworked.

As mentioned above the typical loading patterns of bikes have changed since then, due to improvements made in brake and suspension performance. As a result components on modern bikes have to withstand even higher forces, while the standards still have not been changed accordingly. Furthermore the existing standards do not pay enough attention to the different materials used.

2 Load Measurement

2.1. *Concept*

To ensure the safety of the rider one must be sure that the test loads are high enough for all the users of the bicycle. Figure 2 shows how different user groups put different forces on the bicycle. The bicycle components must not fail at higher loads. The aim of this study was to find the collective for the "1% Rider". The "1% Rider" is defined as the user, who puts higher forces on the bicycle than 99% of the rest.

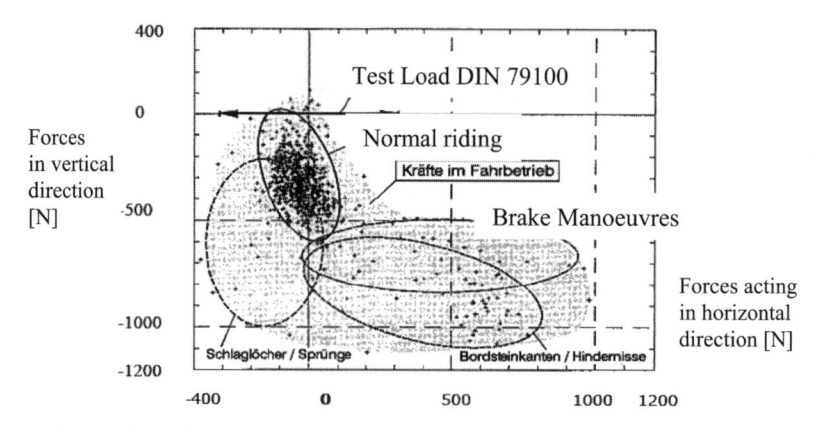

Figure 1. Comparison of the forces acting on the front fork of a mountain bicycle and the test load defined in DIN 79100 (Barski *et al.*, 1995).

Service Strength of High Tech Bicycles

The Concept to determine the load collectives is displayed in Figure 3. Amateur riders and professional sportsmen have performed extensive field tests with road race bicycles and mountain bicycles. The bikes have been equipped with special measuring equipment, as described blow, which allowed the determination of all loads between the bike, the cyclist and the ground. Most of the testing did take place on a testing ground, specially set up for this study. The test ground had the advantage, that many different riding situations (different soil conditions, uphill, downhill, drops, obstacles, ...) were located on a relatively small area, which is easy to access and most of all guaranties repeatable test conditions.

The test results were then verified with a series of tests on reference routes in the Alps, in the region of Garmisch – Partenkirchen. In the next step special occurrences (for example jumps, drops, etc.) have been analysed.

Parallel to the field tests cyclists were questioned about their riding habits, and the results again verified with video analysis on the reference routs.

The combination of the load measurements and the questionnaire resulted in new load collectives for design and testing. For easier interpretation of the measurement data, it has been synchronised with video records made during the field test.

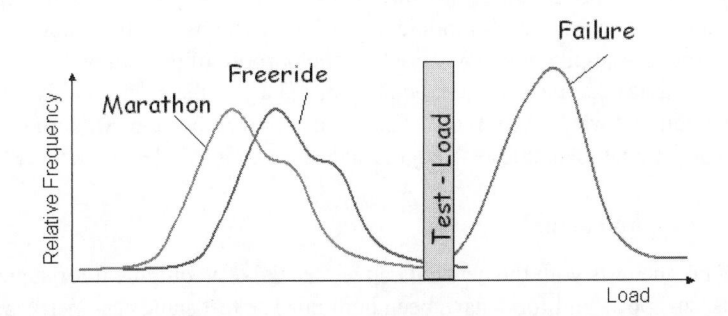

Figure 2.   Safety concept: 1% customer.

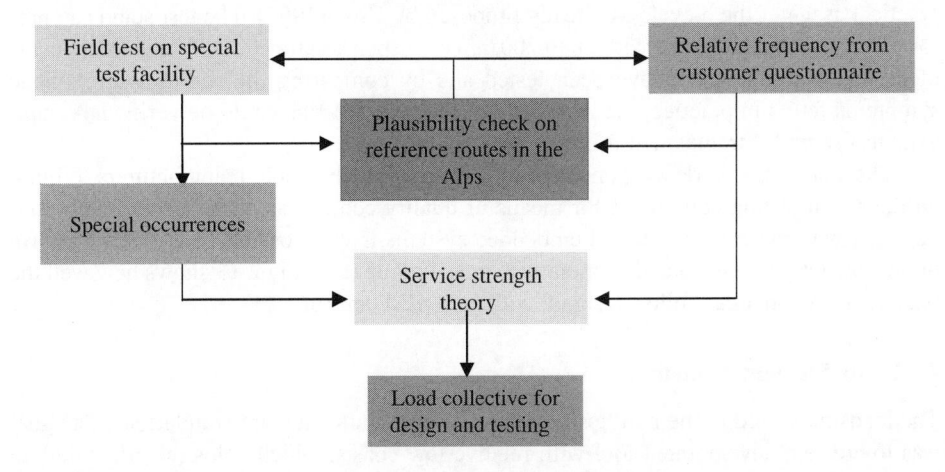

Figure 3.   Concept to determine the load collective for bicycles.

2.2. Measuring Equipment

The bicycles used for the field tests have been equipped with strain gage arrays and that measure the forces acting on the different parts. The following components have been modified for this purpose:

- Handlebar (measures forces in three directions)
- Steam (measures forces in one direction)
- Pedals (measure forces in three directions)
- Seat post (measures forces in three directions)
- Brake caliber (measures forces in one direction)
- Hubs (measures forces in three directions)

A set of measuring hubs for front- and rear wheel has been designed to act as a three component load cell. In the sum 23 different measuring points can be found on the bicycles. As all the measuring points are located on the components, the influence of different frames and forks can easily be analyzed by simply building up the frame with the measuring components.

Additionally the speed of the bicycle and the movement of the suspension elements has been monitored. From rotating parts the measurement data has been send via Bluetooth to the data logger. The Bluetooth module as well as the hubs has been developed at the Institute for Sports Equipment and Material for the purpose of this study.

In earlier studies the measurement equipment did affect the rider and his riding habits as it put an additional weight of up to 20 Kilograms onto the bicycle. Modern data logging instruments and miniature amplifiers kept the additional weight at less than three kilograms.

3 Fatigue Test Equipment

To perform fatigue tests with the new collectives special test equipment for stems, handlebars, cranks, seat posts and forks have been built up. The test stands can perform the standard tests, and realize the newly found collectives. For example, the test stand for forks can apply forces in three directions. This is necessary to perform tests that are relevant for the practical usage of the bicycle as already proposed by Groß (1997). The test stand can perform dynamic tests on forks with up to 200 mm of suspension travel. Different components of leading manufacturers have been tested and by comparing the results with damage symptoms found in practice, the behaviour of the new test rigs could be verified. A multi axial test stand for frames will follow.

The test stands work with pneumatic cylinders, which enable manufacturers to build up their own testing equipment for means of quality control, as costs are relatively low. Together with the company DHM embedded systems is was possible to not only perform multistage fatigue tests, but also operating stress fatigue tests. Figure 3 shows how well the pneumatic cylinder can follow a signal with the DHM control.

4 Finite Element Modeling

The loads measured in the field tests are also used in finite element simulations. The goal was to create a development tool with relative low coasts, which helps manufacturers to design bicycle frames more efficiently. For this purpose the finite element software of

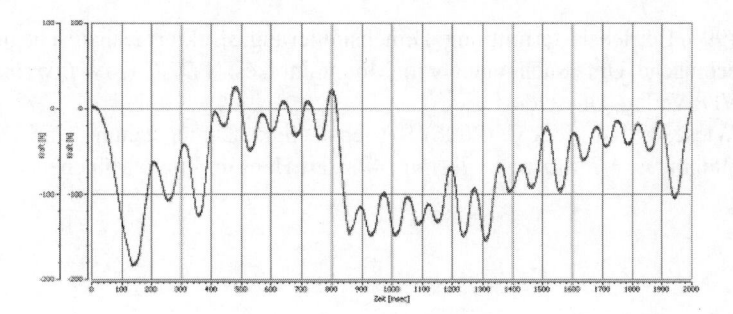

Figure 4.   Operating stress fatigue tests.

Straus 7 was used and combined with the optimization software i sight. The first simple beam models enabled us to perform a global analyses of the frame structure, which helped with the design of the multi-axial frame test stand. In the next step a more complex FE shell model has been set up, which has now to be verified with the test stand.

5   Discussion

The results of the field test lead to the same conclusion as drown from Barski *et al.* (1995), Groβ (1997) and Spahl (1996). The existing Standards have to be reworked. As expected the recent improvements made on the bicycle suspension and brake components lead to even higher loads than measured in earlier studies.

Collectives for multistage fatigue tests, which are suitable to test bicycle components of any material, have been derived from the results of the field test. To perform the tests new test rigs have been built up. The new testing methods proved practically relevant as failures found in praxis could be reproduced, which is not always the case with the old testing methods described in the existing standards. The results of this study are useful tools for the cycling industry and may lead to improvements of the existing standards.

References

DIN 79100 (2000) *Bicycles – safety requirements and test methods*; Beuth Verlag GmbH.
DIN EN 14764 (2006) *City and trekking bicycles – safety requirements and test methods*;Beuth Verlag GmbH.
DIN EN 14766 (2006) *Mountain bicycles – safety requirements and test method*, Beuth Verlag GmbH.
DIN EN 14781 (2006) *Racing bicycles – safety requirements and test methods*, Beuth Verlag GmbH.
Barski M., Groβ E. and Rieck D. (1995) Anwendung von Dehnungsmessstreifen zur Belastungsermittlung an Mountainbike.*Messtechnische Briefe* 31/2
Spahl R. (1996) *Lastkollektivbezogene Prüftechnik von Sicherheitsbauteilen an Fahrzeugen*. Shaker Verlag.
Otto M. (1995) Aktuelle Prüfungen: Reichen DIN und ISO aus? Grundlagen der Ermüdungsprüftechnik im Fahrradbau; Teil 3, *RadMarkt*, 8.
Buxbaum O. (1995) *Betriebsfestigkeit*. 2nd. ed., Verlag Stahleisen, Düsseldorf 1995.

Groβ E. (1997) Betriebslastermittlung, Dimensionierung, strukturmechanische und fahrwerkstechnische Untersuchungen von Mountainbikes; *VDI-Fortschrittsreihe* 12, Nr. 308, VDI Verlag, Düsseldorf.

Issler L., Wiese W. and Merk U. (2006) Betriebsfestigkeitsuntersuchung am Lenker eines Mountainbikes; *MP Materials Testing*, 49, Carl Hanser Verlag, München.

EFFECTS OF VENTING GEOMETRY ON THERMAL COMFORT AND AERODYNAMIC EFFICIENCY OF BICYCLE HELMETS

F. ALAM, A. SUBIC, A. AKBARZADEH & S. WATKINS

*School of Aerospace, Mehanical and Manufacturing Engineering,
RMIT University, Melbourne, Australia*

The thermal comfort and aerodynamic efficiency are becoming important design criteria for bicycle helmets. The characteristics of venting geometry, venting orientation and venting location play an important role in overall heat dissipation characteristics and aerodynamic drag. In order to design an optimal helmet for aerodynamic efficiency and thermal comfort, a comprehensive study is required. Therefore, the primary objective of this work was to study the thermal and aerodynamic efficiency as a function of venting geometry, location and orientation of a series of production helmets currently available in Australian market.

1 Introduction

Bicycle helmets are mandatory for recreational or professional bicycle riders in many countries including Australia. Over the past decade the designers and manufacturers have been engaged in helmet development that can comply with the safety standards. The regulations set the standards for impact protection to reduce head injury during serious accidents. In light of this, there is no standard concerning helmet ventilation or the effects associated with thermal stress and aerodynamic efficiency. It has only been recently that attention of such areas in helmet design has been implemented mainly pursued in the professional arena. In the pursuit for greater thermal comfort there is a trend to increase the number of ventilation openings. Unfortunately increasing the ventilation of a helmet not only degrades the structural integrity of the helmet but also increase the aerodynamic drag. Today most helmets are constructed from an external plastic liner with a foam (polystyrene) liner to protect the user from repeated impacts over the entire helmet area. Currently designed foam helmet is an excellent insulator and helps to thermal buildup within the helmet. The main proponents of ventilation from a helmet are through cooling by convection and evaporation. Airflow around the helmet will remove heat buildup and also moisture (sweat) from the surface of the rider's head. This can be very critical in humid environments. According to a survey conducted by the US Consumer Product Safety Commission in 1999, 95% of people who regularly use the helmets said that comfort or fit was an important factor. On the other hand, 18% those who do not use helmet at all said that comfort was the main reason not to wearing the helmets. Therefore, a bicycle helmet needs to be designed such a way that it can provide both thermal comfort as well as aerodynamic efficiency. Although several studies by Alam et al. (2005, 2006), Bruhwiler (2003) and Reid & Wang (2000) were conducted to measure the aerodynamic drag and temperature measurement techniques for bicycle helmets, these studies are not comprehensive and most studies except Alam et al. (2005, 2006) did not consider the aerodynamic efficiency at all. Therefore, the primary objective of this work as a part of a larger project is to study the aerodynamic and thermal efficiency (comfort) of a series of current production bicycle helmets available in Australia for recreational use under a range of wind conditions and speeds.

2 Description of Helmets, Facilities and Equipment

Five production helmets were used for this study. All helmets are different in terms of venting holes, venting location and venting geometry. All helmets were new and manufactured by Rosebank Australia. These helmets are: Blast, Mamba, Nitro, Summit and Vert (see Figure 1). The heat dissipation and aerodynamic drag were measured under a range of wind speeds (20, 30, 40, 50 and 60 km/h), yaw angles (0, ±30°, ±60° and ±90°) and pitch angles (90°, 60°, 30° and 0° from horizontal axis) using a six component force sensor and an instrumented heat pad on a dummy head respectively. Seven thermo couples were attached to a heat pad. An instrumented dummy head with the heat pad, thermo couples and helmet is shown in Figures 2 & 3. The heat pad is a thin silicon mat that has etched foil strips which are highly flexible imbedded between the silicon layers. The heat pad allows flexibility to attach to the surface of the dummy head form and generates a consistent source of heat. A mounting device for the dummy head was manufactured so that it could be mounted on the six-component force sensor to measure aerodynamic forces as well as to support the thermal measurements. Figure 3 shows the dummy head with mounting device in the test section of the wind tunnel. In order to understand the effects of no venting, the Vert helmet was modified and tested twice as standard configuration with the venting and modified configuration without the venting (venting blanked off), see Figure 1f.

The thermal and aerodynamic efficiency was measured in RMIT Industrial Wind Tunnel. It is a closed test section, closed return circuit wind tunnel and is located in the School of Aerospace, Mechanical and Manufacturing Engineering. The maximum speed of the tunnel is approximately 150 km/h. The tunnel's test section is rectangular with the

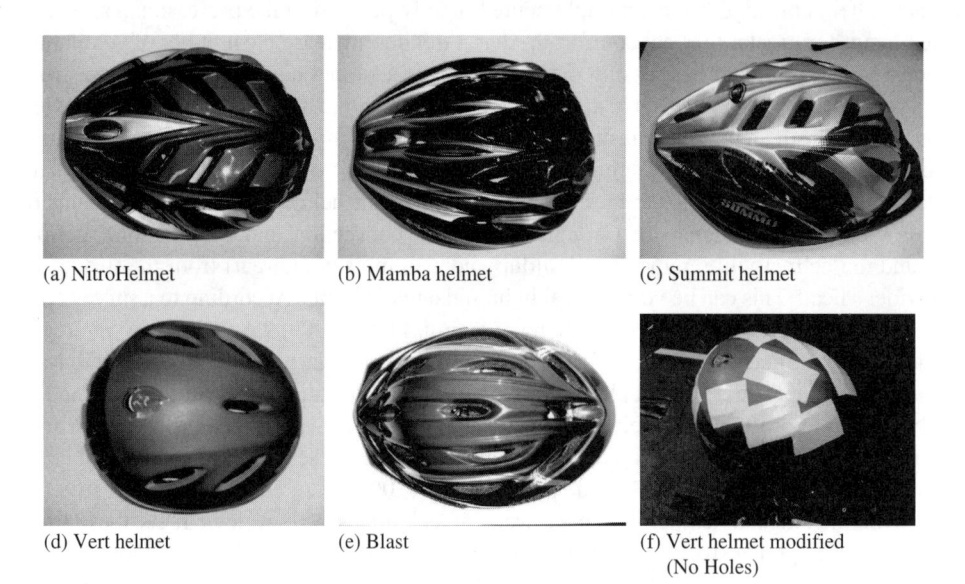

(a) NitroHelmet (b) Mamba helmet (c) Summit helmet

(d) Vert helmet (e) Blast (f) Vert helmet modified (No Holes)

Figure 1. A plan view of all helmets.

following dimension: 9 m long, 3 m wide and 2 m height. Tunnel's free stream turbulence intensity is approximately 1.8%.

Heat dissipation characteristics in terms of temperature reduction were measured by heating up the heat pad at 60°C. The selection of temperature at 60°C was arbitrarily. A thermostat was used to keep the temperature constant on the heat pad. As mentioned earlier, seven thermocouples were attached on the heat pad under the helmet to monitor the temperature drop around the head. Two thermocouples were attached at the front of the head, two on each side (left & right), two on the rear of the head and one thermocouple at the centre of the head in order to obtain a comprehensive temperature distribution (see Figure 2). The temperature drop was monitored for 10 minutes at 20 to 60 km/h with an increment of 10 km/h, yaw angles (0°, 30°, 60° and 90°) and pitch angles (90°, 60°, and 30° from vertical).

3 Description of Helmet's Venting Geometry

In order to understand the effects of geometry on aerodynamic and thermal performance, the geometry of venting (cross sectional area), venting orientation and venting heights were considered and compared results for each helmet against these three parameters.

3.1. *Effects of Venting Orientation*

The venting is generally designed to promote the transfer of heat from the head through forced convection. This forced convection is achieved by varying the inlet and outlet areas of the vents. Figure 5 shows the channeled airflow in a flute configuration to accelerate the flow over the head and achieve optimal quantities of heat removal. In this study, the mechanism that involves in increasing the airflow through the helmet is called the *Venturi Effect*. Unfortunately, the *Venturi effect* introduces airflow resistance over and through the helmet. The air velocity at the exit of the vents is defined in Eq. (1).

$$V_H = V_\infty \frac{A_1}{A_2} \qquad (1)$$

where, A_1 and A_2 are the cross sectional areas of entry and exit of venting, and V_∞ is the velocity of air at the entry of venting. The relationship describes the potential for the vents to increase the inlet velocities through the venting. It is an important parameter to

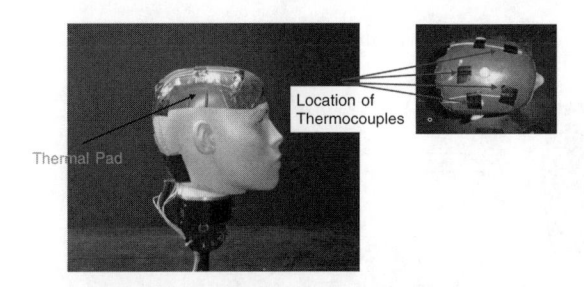

Figure 2. Experimental setup for thermal testing (dummy head, heat pad and helmet).

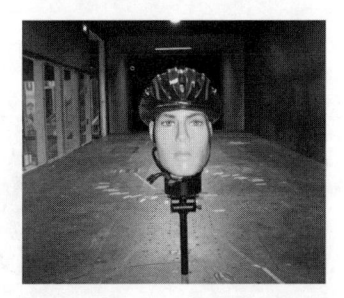

Figure 3. Experimental setup in the test section with a dummy head and helmet.

assess a helmet's potential heat removal and aerodynamic drag generated by the venting obstructions.

3.2. Effects of Venturi

The orientation of venting can have significant effects on aerodynamic and thermal performance of the helmet. The orientation of venting effectiveness can be described as a ratio of venting longitudinal length and venting lateral width times the cosine angle with the wind direction as shown in Figure 4. The venting is generally more effective when the longitudinal length (L) is parallel the incoming airflow. This is to ensure that the vents channel the flow with minimal flow disruption. Therefore, the venting system can be defined in terms of its overall effectiveness as the culmination of the venturi effect and the venturi potential to channel incoming flow which is adversely affected by the orientation to the flow. The relationship between venturi exit velocity (V_H) and venturi orientation (O_f) is shown in Eq. (2).

$$V_F = \frac{A_1}{A_2} \times O_f \qquad (2)$$

where $O_f = \dfrac{L}{W}$ for in line airflow (see Figure 4a) and $O_f = \dfrac{L}{W}\cos\beta$ for oblique airflow (see Figure 4b).

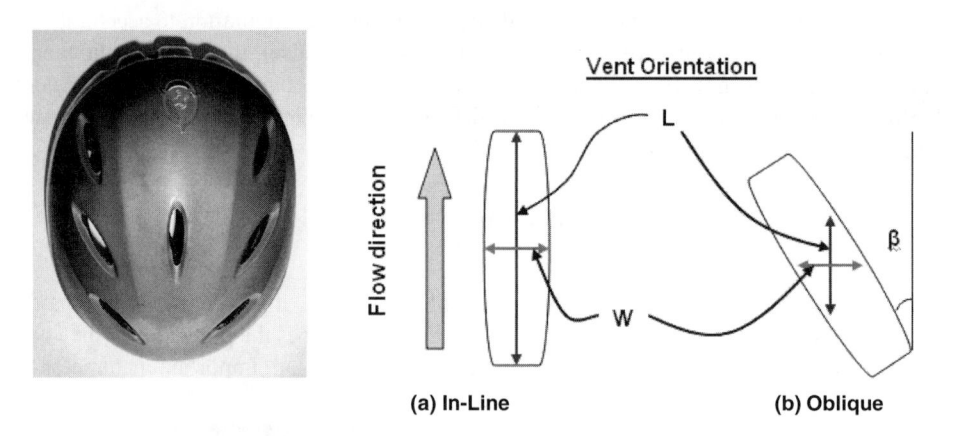

(a) In-Line (b) Oblique

Figure 4. Orientation of venting.

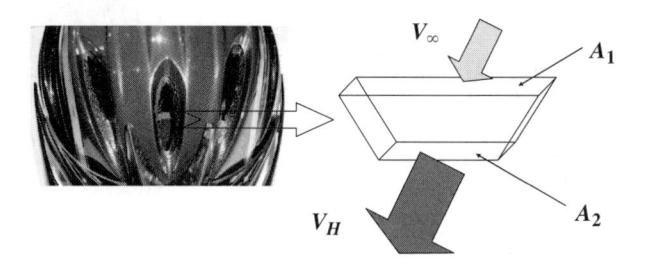

Figure 5. Venting geometry (Venturi).

3.3. Effects of Venting Height (Roughness)

The aerodynamic performance of a helmet can also be affected by the helmet surface roughness. This is generally a result of how the vents protrude from the surface of the helmet. Figure 6 shows the venting angle, which is adequate for capturing the airflow but can impede the flow over the helmet that can in turn affect the overall drag of the helmet. Figure 6 shows the definitions for the geometric relationship for the roughness factor. This concept is an important issue when designing a helmet for maximum drag reduction and can significantly impede airflow if the height deviation for each vent is large. The roughness factor is shown in Eq. (3).

$$R_f = \frac{\sum_{h_1}^{h_n} h_1 + h_2 + h_3 + \cdots + h_n}{H_T} \qquad (3)$$

The venting height deviation (roughness factor) specifies the roughness magnification of the helmet. Generally, the higher the roughness factor, the higher the aerodynamic resistance (drag). The roughness of the helmet is mostly due to the vent design and ideally needs to keep minimum in order to minimize the aerodynamic drag.

4 Results and Discussion

4.1. Effects of Venturi on Aerodynamic Drag

The non-dimensional aerodynamic drag coefficient (C_D) and the exit velocity of venturi for only 90° pitch angle is shown in Figure 7. The figure indicates a qualitative representation of the venting effectiveness and its effects on the aerodynamic performance of all helmets. With an increase of the venturi effectiveness (venturi exit velocity), the aerodynamic drag increases as well. The Mamba helmet with its highest venturi effectiveness generates largest aerodynamic drag coefficient compared to all helmets. The Vert helmet with lowest venturi effectiveness generates lowest aerodynamic drag coefficient. The results also indicate that the quantity of venting numbers along with the ratio of vent's entry and exit cross sectional areas play a dominant role in aerodynamic drag generation.

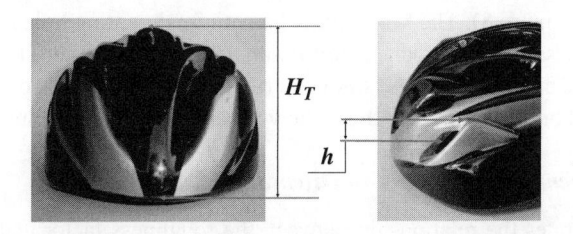

Figure 6. Venting height deviation (Roughness Factor).

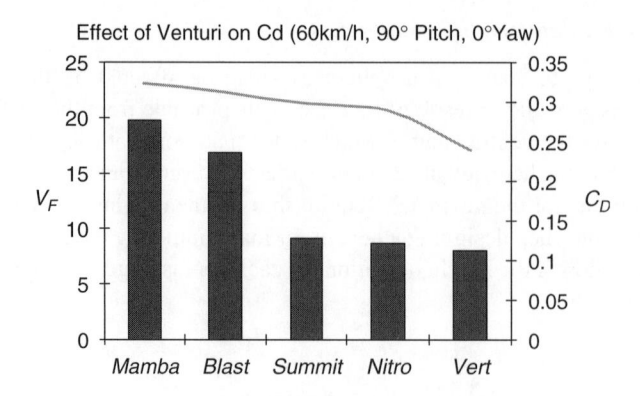

Figure 7. Effects of venturi on drag coefficient.

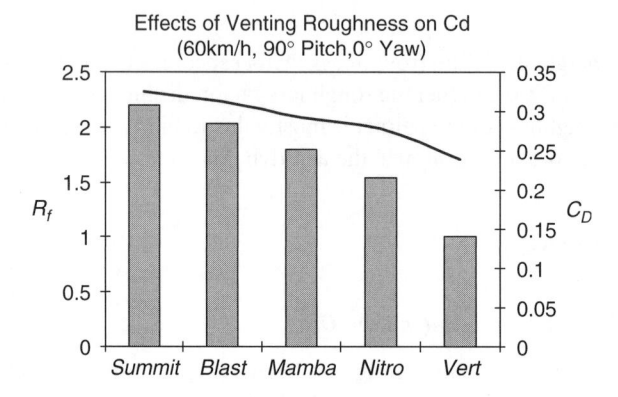

Figure 8. Effects of venting roughness on drag coefficient.

4.2. Effects of Venturi Roughness on Aerodynamic Drag

The relationship between the helmet surface roughness factor and the aerodynamic drag coefficient for all helmets is shown in Figure 8. The Summit helmet has the highest value of surface roughness factor and also it demonstrates the highest value of aerodynamic drag coefficient. Generally higher the roughness factor, greater the venting obstruction of airflow over the helmet. Figure 8 clearly indicates this trend. However, it is believed that the aerodynamic drag can be minimized with the smoother and less obtrusive vents on the helmets as there is a direct correlation between the vent roughness factor and the pressure induced drag (see Figure 8). The Vert helmet has minimal venting holes and less obtrusive (minimum roughness factor), produces minimum aerodynamic drag. The study also shows that having a large amount of vents is not so critical if the roughness of the vents is minimal as higher roughness can trigger the airflow separation over the helmet.

4.3. Effects of Venturi Roughness on Orientation Sensitivity

Figure 9 demonstrates the relationship between the roughness factor and the ratio of drag coefficients at zero and 60 degree yaw angles. The sensitivity of the helmet is based on the

**Effects of Venting Roughness on Orientation Sensitivity
(60km/h, 90° Pitch, 0° Yaw)**

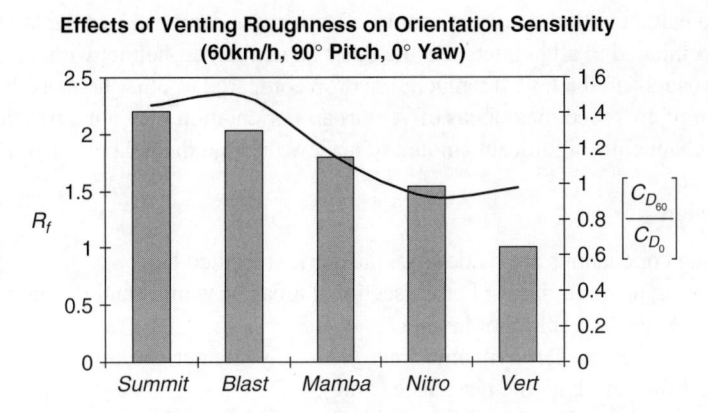

Figure 9. Effects of venting roughness on orientation sensitivity.

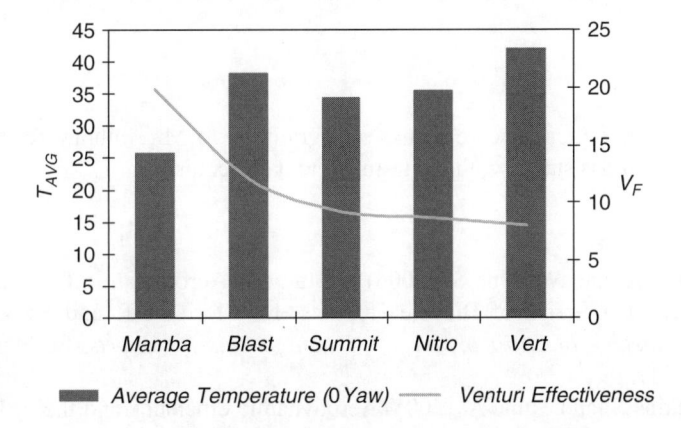

Figure 10. Temperature variation with venturi effectiveness.

ratio of drag coefficients in two different orientations. The greater the ratio; higher the possibility of venting roughness will play a major role in reducing the aerodynamic performance. The statement is evident in the Vert helmet which has no change in the orientation factor. The aerodynamic drag (due to induced pressure) will be almost independent of venting orientation if the venting roughness can be kept minimal. However, it is not the case when vent's numbers with larger dimensions are significant. This observation is evident in Summit and Blast helmets. As venting orientation is fixed, the helmet will generate induced drag under crosswind condition (under large yaw angles).

4.4. *Effects of Helmet Geometry on Heat Dissipation*

Figure 10 shows the temperature variation (heat dissipation) as a function of venturi effectiveness for each of five helmets. It is evident that there is a strong correlation with the venturi effectiveness and the amount of heat removal from each helmet. The figure demonstrates that higher the venture effectiveness, greater the temperature drop (heat dissipation).

The Mamba helmet with the highest venturi effectiveness demonstrated the largest temperature drop compared to all helmets. On the other hand, the Vert helmet with the lowest venture effectiveness had the lowest temperature drop compared to other helmets. It is believed that the ratio of cross sectional areas of venturi and orientation play a major role in removing heat by channeling significant amount of airflow through the helmet.

5 Conclusions

The following conclusions are made from the work presented here:

The venting position, ratio of cross sectional areas of venturi and geometric size play a vital role in effective cooling of helmet.

The vent increases aerodynamic drag. Generally greater the number of vents on a helmet, larger the aerodynamic resistance (drag).

The Vert helmet had performed the best in terms of the aerodynamic drag but worst in thermal performance.

The Mamba helmet is the best in thermal performance but worst in aerodynamic efficiency.

Acknowledgements

The authors express their sincere thanks and gratitude to Mr Anthony Resta and Rhys Solomons for their assistance with the testing and data acquisition.

References

Alam F., Subic A. and Watkins S. (2006) A Study of Aerodynamic Drag and Thermal Efficiency of a Series of Bicycle Helmets. In: Moritz E.F. and Haake S. (Eds.) *The Engineering of Sport 6, Vol. 1 – Developments for Sports*. Springer, Munich, pp. 127–131.

Alam F., Watkins S. and Subic A. (2005) Aerodynamic efficiency and thermal comfort of bicycle helmets, *Proc. of the 6th International Conference on Mechanical Engineering (ICME2005)*, TH-32 (1-6), ISBN 984-32-2846-4, 28-30 December, Dhaka, Bangladesh.

Bruhwiler P.A. (2003) Heated, perspiring manikin headform for the measurement of headgear ventilation characteristics. *Meas. Sci. Technol.*, 14, 217–227.

Consumer Product Safety Review (1999), Vol. 4, No. 1 by the U.S. Consumer Product Safety Commission.

Reid J. and Wang E.L. (2000) A system for quantifying the cooling effectiveness of bicycle helmets. *J. Biomech. Eng.*, 122/4, 475–460.

DESIGN OF LEISURE SPORTS EQUIPMENT AND METHODS OF SPORTS SCIENCE – AN EXAMPLE FROM CYCLING

M. MÜLLER[1], G. MECKE[1], H. BÖHM[1], M. NIESSEN[2] & V. SENNER[1]

[1]*Department Sports Equipment and Materials, Technische Universität München, Munich, Germany*
[2]*Department Performance Diagnostics, Technische Universität München, Munich, Germany*

The study was conducted to find out differences in the gross efficiency and the comfort rates between three different seat positions during cycling. The results should support the engineering design process of human powered vehicles. Two extreme ones, the road bike and the lying position, as well as the normal upright cycling position were chosen to get an idea of the optimum concerning the influencing parameters. For the gross–efficiency, oxygen consumption and power display was displayed. The comfort rates were evaluated with a questionnaire.

1 Introduction

1.1. *Sociocultural Background of the Study*

Alternative transportation systems gain importance for metropolis in the course of car free strategies for city centers and emission discussions (Jung 2006). Cycling is one of the most appropriate ways of locomotion in terms of efficiency, health and ecology (Prampero, 1986). In a feasibility study it is being tried to encounter a new concept of inner-city traffic systems, pushing the potentials of human powered vehicles . VeloVent is one possible system. The idea is to run human powered vehicles, that are bicycles and three or four wheel vehicles, in a network of tubes with transparent covers, supporting them by airflow in the direction of movement to minimize the air resistance (Müller, 2006).

1.2. *Research Question behind the Study*

Within an interdisciplinary design seminar, new human-powered vehicle concepts were designed for different transportation purposes. One of the concepts, the so called "mother-child concept" is depicted in Figure 1. But in any case, there was a lack of knowledge of how the seat position could be designed properly, that means leading to a minimum of physical stress and ergonomical discomfort.

The study therefore was started to get an idea about an optimized seat-position with respect to the physiological gross efficiency and ergonomical aspects for human powered vehicles.

2 Methods

Many studies were conducted to investigate the influence of different seat positions on the energy consumption during cycling (Gnehm *et al.*, 1997; Grappe *et al.*, 1998; Ryschon & Stray-Gundersen, 1991). In these studies the main focus was on performance sports. The positions varied only marginally for the road bikes and the aim is clearly an increase of

Figure 1. Example of a concept of human powered vehicles were the seat position plays a key role.

performance. Whereas in this study the aim was a good physiological gross efficiency in combination with a highly ergonomic seat-position. In order to get comparable results also a road-bike position investigated during the former studies was included. Furthermore two possible seat positions (lying position and upright position) for the new vehicles were analyzed. As a basis for the study design several literature source were considered regarding the biomechanics of cycling (Li & Caldwell, 1998; Weiss, 1996; Zschorlich, 2003). The three main hypotheses are:

- The seat position influences the energy consumption of the driver for a given power,
- Bad comfort rates lead to a higher energy consumption than good comfort rates do,
- A higher muscle activity leads to higher energy consumption.

Therefore the following methical approach was developed for the test design:

- Pre-test to evaluate performance of test persons (for later calculation of the power level for the main test),
- Oxygen consumption and carbon dioxide exhaust (is needed to calculate energy consumption),
- Heart-rate (to control activity),
- Electromyography (to analyse types of muscles that show activity for different seat positions),
- Lactate analysis during the tests (to evaluate intensity of tests),
- Traction power (given by dynamometer Cyclo2 and double-checked with an SRM crank).

An overall number of 17 male test subjects in the range of 20 to 36 years, one of them completed a familiarization program for the lying position to partly control disturbing variables, performed the same test for each characteristic of the variable of investigation (different seat positions). The order of testing was chosen random. The pre-test (30 seconds each power level beginning with 100 Watts and an increment of 50 Watts) was performed to failure of each test subject. The main tests were started with a warming-up of 9 minutes at no resistance and 1 minute at 30% of the maximum power (from pre-test). In addition, motion analysis and electromyography was performed during the 30% phase (Figure 2).

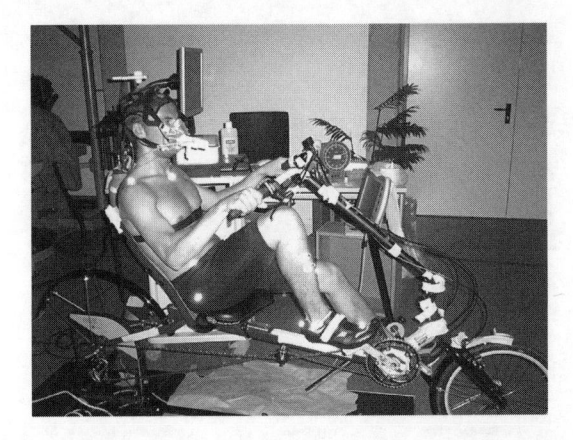

Figure 2. Test subject during the warming-up phase in lying position.

The analysis phase contained two power levels, 30% and 45% of the maximum power, each 15 minutes. The power levels should represent a typical stress of moderate activation for inner-city cycling. The cranking frequency was given with 80 rpm. Every 5 minutes lactate was taken from the test subjects, the oxygen consumption and carbon dioxide exhausts were taken permanently. The discomfort (aiming to evaluate subjective comfort rates) was estimated in terms of a questionnaire by the test subjects afterwards.

3 Results

The following way was chosen to achieve the calculation of the gross efficiency. De Marées (2003) distinguishes between gross efficiency and net efficiency. Due to the conditions in the laboratory a reliable determination of the basic metabolic rate was not possible. Hence, in accordance to some literature only the overall metabolic rate η was identified from the given power P and the overall metabolic rate B.

$$\eta = P/B \qquad (1)$$

The power P was given by the dynamometer according to 30% (P_{low}) and 45% (P_{high}) of the maximum power displayed by the test subjects in the pre-tests, respectively. The overall metabolic rate was calculated after Knechtle (2002) from the oxygen consumption and the release of carbon dioxide.

$$B\ (cal/min) = 3{,}941 * VO_2(l/min) + 1{,}106 * VCO_2(l/min) \qquad (2)$$

Figure 3 shows that the gross efficiency is slightly better for the normal bike position compared to the road bike position for both power levels. The overall efficiencies for these two positions vary little and the differences are only up to 0.81% for P_{low} and 0.41% for P_{high}, whereas the efficiency of the normal position is marginally higher. The lying position, however, shows slightly lower efficiencies of 1.9% for P_{low} and 1.6% for P_{high}, compared to the normal position.

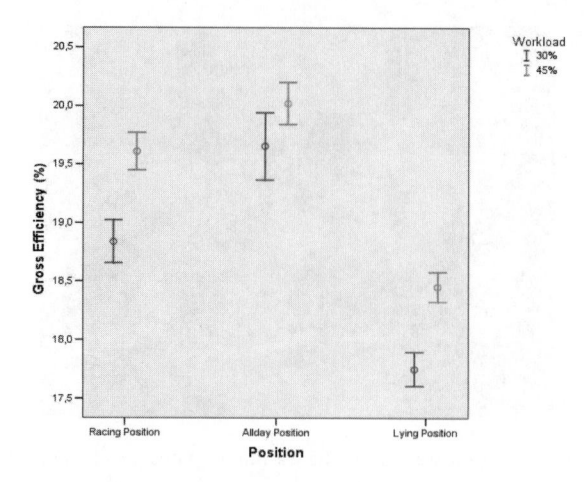

Figure 3. Gross efficiency for different seat positions with standard deviation.

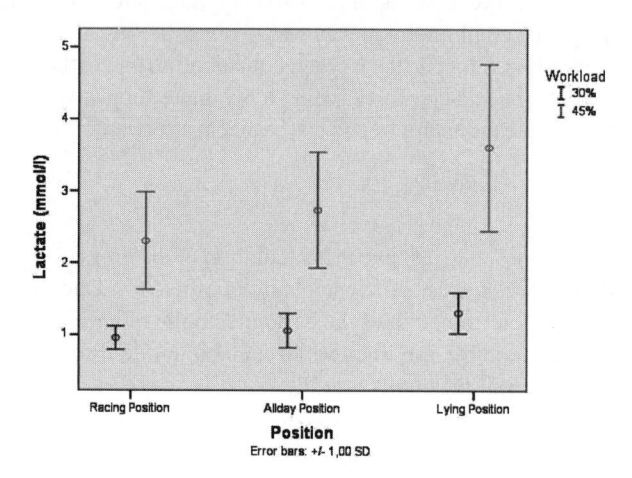

Figure 4. Mean lactate levels for the different seat positions with standard deviation.

The results in Figure 4 show the mean Lactate levels of the test subjects. The road bike position leads to a slightly lower Lactate level than the normal position (10.9% for P_{low} and 19% for P_{high}) while the Lactate for the lying position is considerably higher for both power levels compared to the Lactate display of the normal position, namely 23% for P_{low} and 32% for P_{high}.

The ergonomy questionnaire showed that the lower body part is rated being less comfortable for the lying position (Figure 5) compared to the normal position and the road bike position. For the rating, a Borg scale was used from 0 up to 50 for highest discomfort rates. The results were related to each other in a way such that the overall sum for all body parts was set to 100%. The single body parts then were related to each other according to their contribution to the overall number. Hence, the higher the percentage, the lower the comfort rates.

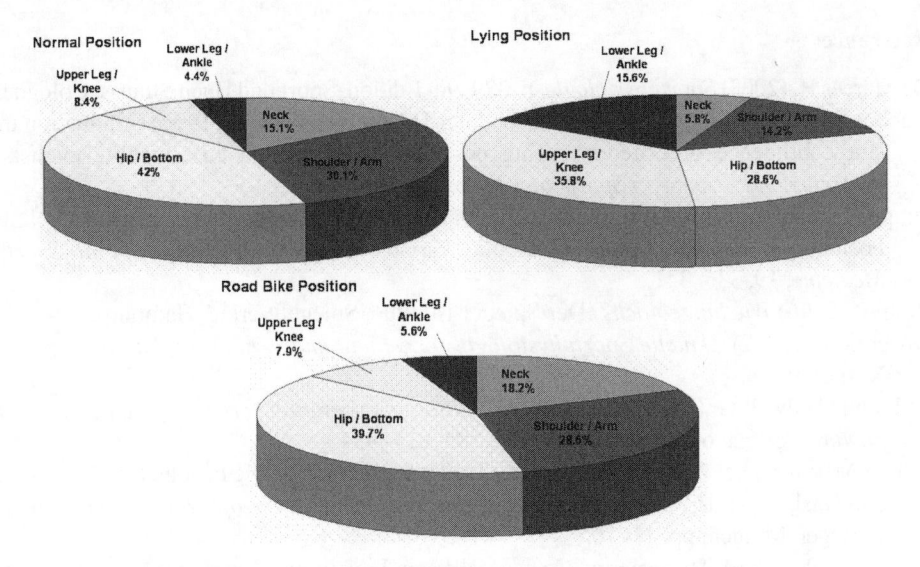

Figure 5. Comfort rates for the three seat positions and different body parts.

4 Discussion and Outlook

With respect to the hypothesis it can be concluded that the overall differences of the gross efficiency between the three seat positions are in the range of only 2%. Though, the seat position influences the energy consumption of the driver, but significance could not be proven. The results from the ergonomic rating suggest that bad comfort rates indicate a higher energy consumption or vice versa, as the lower body part being activated for the pedaling motion was rated with comfort rates considerably lower for the lying position (~28% for upper leg and knee and ~10% for lower leg and ankle) than for the other two positions. The study, however, suffers from a systematic deficiency, which is the lack of familiarization of the test subjects to the lying position. Compared to the first two positions the feet have to be kept on the pedals actively for the lying position which presumably leads to a higher activation of stabilizing muscles and therefore to a higher overall level of Lactate and a lower gross efficiency. This effect can not be quantified yet and although one test subject was familiarized to the lying position over several hours beforehand the tests to get an estimation, literature commonly recommends a period of several months of intense usage to evict influences of training effects. On the other hand the upper body of the lying position has very high comfort rates and shows that the position generally is considered as comfortable. Also the effect of the lower air resistance due to the lying position during cycling was not taken into account, but surely will lower the overall energy consumption compared to the other seat positions. It is very likely that the rating for the lying position will increase as soon as the familiarization effect can be cut off. To find out the influence of the familiarization effect, this will first be conducted as a case study to see if there might be effects worth of looking at more closely with a higher number of test subjects. Second, also the data from the EMG and the motion tracking will be analysed concerning correlations between the activation of the main muscle groups, the motion and the gross efficiency of the cyclist.

References

De Marées H. (2003) *Sportphysiologie,* p. 379, 9th Edition, Sport und Buch Strauss, Cologne.

Gnehm P., Reichenback S., Altpeter E., Widmer H. and Hoppeler H. (1995) Conduction of study. Influence of different racing positions on metabolic cost in elite cyclists. *Medicine and Science in Sports and Exercise*, 29/6, 818–823.

Grappe F., Cancau R., Busso T. and Rouillon J. (1998) Conduction of study. Effect of Cycling Position on Ventilatory and Metabolic Variables. *International Journal of Sports Medicine*, 19, 336–341.

Jung A. (2006) *Wie lange noch?,* Der Spiegel 14/2006, Spiegel Verlag, Hamburg.

Knechtle B. (2002) *Aktuelle Sportphysiologie – Leistung und Ernährung im Sport*, p. 43, Karger, Basel.

Li I. and Caldwell G. (1998) *Muscle coordination in cycling: effect of surface incline and posture*, Journal of Applied Physiology, 85, 927–934.

Müller M. (2006) VeloVent – An Inner City Traffic System for Active People, In: Moritz E.F. and Haake S. (Eds.) *The Engineering of Sport 6, Vol. 3 – Developments for Innovation.* Springer, Munich, pp. 11–16.

Prampero P. (1986) The Energy Cost of Human Locomotion on Land and in Water, *International Journal of Sports Medicine*, 7, 55–72.

Ryschon T. and Stray-Gundersen J. (1990) Conduction of study. The effect of body position on the energy cost of cycling. *Medicine and Science in Sports and Exercise*, 23/8, 949–953.

Weiss C. (1996) *Handbuch Radsport,* BLV Verlagsgesellschaft mbH, Munich.

Zschorlich V. (2003) Grundlagen der Sportbiomechanik, In: Haag H. and Strauß B.G. (Eds.), *Theoriefelder der Sportwissenschaft*, Karl Hofmann, Schorndorf.

19. Disability Sport

DISABILITY SPORTS IN SINGAPORE – PARALYMPIC MOVEMENT AND GREATER INTERNATIONAL RESPONSIBILITY

K.G. WONG

Executive Director, Singapore Disability Sports Council

With the growth of the paralympic movement around the world, Singapore has been called upon to lead and support the growth and technical development of the Asian region. Support and endorsed by the International Paralympic Committee and regions bodies, Singapore will be establishing a Disability Sports Classification and Research Centre. Besides being active internationally, Singapore also remains focused on developing disability sports locally dedicated programmes and initiatives. A key focus for Singapore is greater integration and social cohesion.

1 Singapore Disability Sports Council

1.1. Background

The Singapore Disability Sports Council (SDSC, 2007) is the only sports organisation in Singapore that reaches across all disability groups and offering sports at both elite and non-elite levels.

Although it is a registered charity, SDSC is, in fact, the national sports body for the disabled in Singapore. With a membership of 16 voluntary welfare organizations, SDSC has effectively managed to reach out to over 15,000 disabled individuals over the years.

SDSC's primary goal is to foster, through sports, the physical and mental rehabilitation of the disabled, as well as to build confidence and self-esteem and to promote team spirit and a sense of achievement. Through these aims, SDSC encourages the disabled community in Singapore to live full and independent lives.

SDSC currently performs three different roles in the local disability sports scene: that of the Singapore National Olympic Council, Singapore Sports Council and National Sports Associations. Internationally, Singapore is increasingly being viewed as the regional hub for the growth of disability sports.

1.2. History

The Singapore Sports Council for the Handicapped (now the Singapore Disability Sports Council) was jointly founded by the then Ministry of Social Affairs, Ministry of Education, Ministry of Health, and eight organizations for the disabled on 26 February 1973. Encik Othman Omar – then the Acting Permanent Secretary of the Ministry of Social Affairs – was elected as the Council's first President.

Singapore organised the first Regional Sports Meet in conjunction with the Singapore Handicapped Month in September 1973. 36 paraplegics from Australia, Indonesia, Malaysia and Singapore participated in 42 events including athletics, archery, swimming and table tennis. Farrer Park was filled with euphoria when Australian athlete, Vic Renalson, broke the world record in the discus event with a throw of 25.2 metres.

At the Commonwealth Paraplegic Games in Dunedin, New Zealand, in 1974, Singapore sent two representatives, and returned with a bronze medal for swimming. In 1975, Singapore participated in the first Far East and South Pacific (FESPIC) Games held in Japan and our athletes won three gold and 2 bronze. In a short space of time, Singapore was on the map of disability sports.

In 1985, the Council was renamed the Singapore Sports Council for the Disabled (SSCD). Notwithstanding its role as the national agency for disability sports, SSCD had no assured funding, and not even a proper office to carry out its work. Towards the end of 1994, the SSCD acquired its first office within the premises of the Handicaps Welfare Association at Whampoa Drive. In 1995, SSCD moved into its office premises, and employed two full-time staff. For the first time in 22 years, there was some semblance of a national sports body and a real sense of identity.

SSCD has come a long way since 1973. It is relentlessly pushing the frontiers of disability sports in Singapore today. In a ceremony held on 29 May 2003, the SSCD officially changed its name to the Singapore Disability Sports Council (SDSC).

1.3. *Mission and Vision*

SDSC firmly believes in the rehabilitative and therapeutic value of sports. Its programmes and activities underscore its guiding principle that **"Disability Must Never Disqualify"**.

1.4. *Programme and Function*

SDSC's programmes and activities serve to provide the disabled with opportunities to train, participate and excel in sports, opportunities to integrate into the community through recreational sports and activities and to increase public awareness of and promote support for the sporting and recreational needs of the disabled community.

SDSC's programmes include (1) **Client-Specific Programmes** that are specially tailored to meet the disabled's specific needs; (2) **Youth Programmes** that are aimed at encouraging children and students in both mainstream and special schools to take up sports; and (3) **Elite Training Programmes** (Sports Excellence) that are geared towards identifying and developing world class athletes.

SDSC has recognized the need to grow the participation base in disability sports in Singapore and encourage more disabled Singaporeans to adopt active lifestyles through sporting activities. Thus, SDSC has increased its focus on Community Sports and Participation, a component that offers opportunities for the disabled to learn a sport or enjoy recreational sporting activities.

1.5. *Activities and Achievements*

SDSC has groomed several successful icons through its programmes and selection. For example Theresa Goh who earned two world records for Singapore in a span of seven months (August 2006 to March 2007) and Muhammad Firdaus Bin Nordin, who has earned himself a Beijing Paralympic qualifying spot with a silver medal in International Paralympic Committee (IPC) Athletics World Championships Assen 2006.

While SDSC is proud of its achievements, SDSC continues to emphasize that its sports programmes are primarily to enhance the welfare and well-being of the disabled.

With the aim of integration and social cohesion in mind, SDSC relentlessly seeks new partnerships to develop disability sports. Through partnership programmes and the active involvement of both the sports community and community as a whole, SDSC hopes to bridge the gap between the disabled and able-bodied communities.

Under the aim of achieving equal opportunities for the disabled and thus social integration, SDSC is also pursuing the recognition of disabled athletes' efforts and achievements.

More recently, SDSC has established the following initiatives:

- An athlete development framework and management programme that outlines and create opportunities for individuals to take up sports and develop into elite athletes
- Established an Athlete Achievement Award (AAA) scheme – similar to the Multi-Million Dollar Award Programme (MAP) for the able-bodied, to provide equal recognition of disabled athletes' achievements
- Web-based systems to better communicate and inform members of programmes and activities
- A sports promotion and club development framework and grant to support and encourage more individuals with disability to take charge of their sports create more opportunities for sports participation
- Launched a nation wide "Learn to Play" sport promotion initiative to encourage more individuals to take up a sport

The SDSC has also been successful in the international arena, supporting regional federations in the promotion and development of disability sports in the region.
SDSC had made the following contribution

- Formed the ASEAN Disability Sailing Federation in 2006
- Donated 2 boats each to Philippines and Malaysia and helped developed their sailability programme
- Conducted a 2 week preparation programme to assist regional countries like Philippines, Cambodia, Vietnam and Malaysia to successful compete at the recent Far East Pacific (FESPIC) Games in 2006
- Conducted the Young Enabled Sailor (YES) conference and the Disability Sports Conference to promote disability sports in the region
- Successfully obtained international support to establish an international classification and research centre here in Singapore

2 Sports Programmes

2.1. *Programme*

The SDSC remains as the only association in Singapore that looks after the sports interest for all disabled Singaporeans. SDSC is also the only organisation in Singapore that is not disability specific and continues to dedicate itself towards the promotion & coordination of sports for the disabled in Singapore.

SDSC has the following focus: Disability Sports Promotion, Sports Development, National Competitions, International Competitions and Coaching.

SDSC currently supports the development and administration of 18 disability sports in Singapore. Administration includes programme development and publicity as well as the engagement of coaches and the management of local and overseas competition.

The following disability sports are currently available in Singapore: Swimming, Athletics, Sailing, Archery, Boccia, Table Tennis, Wheelchair Tennis, Wheelchair Basketball, Wheelchair Racing, Lawn Bowls, Badminton, Equestrian, Soccer, Ten Pin Bowling, Goal Ball, Shooting, Chess, and Futsal.

2.2. Coaching

As the role of *SDSC* is to promote its sports at all participation levels, from entry to elite, it is vital that appropriate coaching is available at each level, to facilitate the development of the athlete.

SDS's coaching framework serves two objectives:

1. Sports Excellence – In any model detailing an athlete's development pathway up to elite level, a "seamless" management framework is of key importance. The framework has to be able to facilitate the progress of the athlete from one level to another, so that a "quality end product" is achieved.
2. Participation – The framework has to be able to serve SDSC's objective of providing meaningful opportunities for the disabled population to participate in sports, and at the same time grow the athletes' base for sports excellence.

An established coaching framework will also provide the following:

1. Continuity within programs – Coaching continuity is important in facilitating the progression of athletes from one participation stage to another.
2. Integrated training programs – Particularly in developmental and elite athletes, technical and skills training need to be integrated into their overall training programs, which should include sports medicine and sports science elements, as well as athlete support schemes.
3. Development of expertise – The framework will facilitate the career and development of coaches with specific skills and interest in working with disability sports.

3 Paralympic Movement

The Paralympic movement has grown over the year and this is best exemplified through the phenomenal rise of the Paralympic Games. More nations competed at the Sydney 2000 Paralympics (3824 athletes, 123 nations) than in the Munich 1972 Olympic Games. In addition, through the growing attention and interest from the public, acceptance for sport for persons with a disability grew. The up coming Paralympic in Beijing bodes well for the Paralympic movement and the expansion of the Paralympics in Asia.

4 Singapore's Role in the Paralympic Movement

Singapore through it unique geographical location, multiracial and cosmopolitan society has emerged as a unique hub and platform for the further development of the Paralympics and disability sports in the region.

As a gateway for communication, expertise and funding between east and west, Singapore is well position to champion the paralympic movement in the region and Asia. Singapore has received support from the International Paralympic Committee, the Asia Paralympic Council and the Asean Para Sports Federation to establish a Classification and Research Centre and various regional sporting bodies to drive the growth of disability sports in the region.

5 Classification and Research Centre

The SDSC has, in partnership with UBS (Union Bank of Switzerland) and the Singapore Sports Council, also decided to set up a first of its kind, the **Disability Sports Classification and Research Centre** (DCRC), as part of its efforts to develop athletes and programmes and grow disability sports capabilities in Asia. The DCRC will leverage off and enhance Singapore status as a Sports Hub through undertaking research in the area of sports and equipment development. It is endorsed by the Asia Paralympic Council (APC) and the Asean Paralympic Sports Federation (APSF) and has garnered the support of the International Paralympic Committee (IPC). If successful, the IPC intends to cascade the framework for the DCRC to other regions.

5.1. *Rational and Objective of the DCRC*

There are 52 countries in Asia and Oceania. Most of these countries, including Singapore have, up till now, been more focused on economic progress. Having achieved economic stability, there are now compelling reasons for countries in the region to look into other aspects like improving the quality of life for the disabled through sports. In particular, there is a significant need for support for disability sports in several areas including: (1) Technical development; (2) Coach development; (3) Classification and (4) Training and development programmes.

The establishment of the regional DCRC would address the gaps existing in the areas above by providing the following:

a. A centralised resource and training centre;
b. Regular classification courses & workshops;
c. Research and sports development; and
d. Research and development of competition equipments.

It is responsible for establishing an athlete classification database, a framework for building up classification standards and ensuring that there are adequate qualified technical professionals for major competitions. The intention is for the DCRC to carry IPC classification standards and be run by people with disabilities from the region. It will also support the National Paralympic Councils (NPCs) in terms of providing resources, training and development programmes and workshops.

5.2. *Benefits of DCRC in Asia*

Having a DCRC located in Asia has several benefits for the region including:

a. Regular & systematic classification courses;
b. More access to classification opportunities for athletes;
c. Greater accessibility to resources and information for IPC member countries (Asia & Oceania regions);

d. A pool of IPC accredited classifiers in Singapore and the region;
 e. A resource centre for disability education & research; and
 f. Greater awareness and understanding of disability sports.

Setting up the DCRC would also generate invaluable spin offs in the form of more international competitions in the region and better public understanding of disability issues. As the DCRC will be the focus for growth of disability sports in this region, the resource centre will be the front-runner in the development of programmes and training of technical expertise.

The research into the development of sports equipment also offers the unique opportunity for commercialization that would allow Singapore to enter the lucrative sports equipment industry.

5.3. DCRC Organisation

The DCRC will be an independent entity that would be run by an international board, comprising the heads of the APC and APSF Classification and Sports Science Committees. Aside from Singapore, the other co-chair would be an IPC representative since the classification standards and codes are set by the IPC. A Secretariat will see the daily administration of the DCRC and oversee the coordination of research projects and programmes. The official launch of the DCRC is targeted for November 2007.

6 Conclusion

The DCRC will give Singapore the opportunity to be an active member in the international community and establish a foothold in the area of sports research in the region. As Singapore seeks to establish itself as a Sporting Hub, the DCRC would become one of the necessary pillars to support the overarching objective of talent and industry development in Singapore and the region.

References

SDSC (2007) Singapore Disability Sports Council (online). Available: http://www.sdsc.org.sg/ (accessed June 2007).

PERFORMANCE ANALYSIS IN WHEELCHAIR RACING

A.P. SUSANTO[1], F.K. FUSS[1], K.G. WONG[2] & M.S. JAFFA[2]

[1]*Sports Engineering Research Team, Division of Bioengineering,*
School of Chemical and Biomedical Engineering, Nanyang Technological University, Singapore
[2]*Singapore Disability Sports Council, National Stadium (West Entrance), Singapore*

Wheelchair racers have shown astonishing performance in the last few decades. Yet to date, only few studies have been carried out on wheelchair racing. Conclusive evidences from previous studies have urged more works on realistic condition. The present study aims to develop and comprehensively examine a novel method for assessing performance in wheelchair racing. In full 100-m race simulation on a track, electrogoniometers were used to detect elbow movement at a sampling frequency of 1 kHz. The new method has been verified to be reliable in assessing performance in wheelchair racing. The push frequency increases clearly during the first 40% of the race, and drops slightly during the last 10%. The maximal angular velocity of elbow extension was 26 rad/s. On average, the racer performed more than 117 pushes per minute.

1 Introduction

Starting in early 1980, the popularity of wheelchair racing (wheelchair track) has been increasing such tremendously (Cooper, 1990) that now it has become one of the most prestigious events in the Summer Paralympics. Akin to other sport branches, the tight worldwide competition demands well-built and endure athletes' performances.

Biomechanic analysis of sports to a great extent contributes in optimizing athletes' performance and preventing injuries (Zatsirosky, 2000). Yet to date, few studies have been devoted to wheelchair racing. Both coach and athlete are rarely facilitated by analysis performance study during training (Muller *et al.*, 2004).

A few laboratory studies had been conducted before 1990 analysing the performance in wheelchair racing (Coutss, 1988; Coutss & Schutz, 1988). However, conclusive evidence suggests significant differences between former and recent studies. This is possibly contributed by modern regulation amendment, mainly in wheelchair design and athlete's racing posture (Moss *et al.*, 2003; Chow & Chae, 2007).

Most of previous research was done on an ergometer or treadmill in a laboratory environment (Veeger *et al.*, 1998; Chow *et al.*, 2000; Goosey & Lenton, 2006; Cooper *et al.*, 2003; Muller *et al.*, 2004; Goosey & Campbell, 1998). However, recent findings reveal significant differences between inside laboratory and over-ground research (Cooper *et al.*, 2003; Chow & Chae, 2007). In fact, biomechanics, aiming to uncover fundamental mechanics behind human movement, should be investigated under realistic condition (Vanlandewijck *et al.*, 2001). Consequently, scientists have urged for more over-ground studies. In addition, a study shows significant differences between experienced wheelchair athletes and inexperienced control subjects (Veeger *et al.*, 1998). Therefore, it is suggested that careful consideration is crucial in selecting subjects of research.

Moss et al. (2003), in their 10-m over-ground work of sprinting start, propose that propulsion in wheelchair racing is a complex form of locomotion and cannot be exactly illustrated by propulsive and recovery phase. The latest work from Chow & Chae (2007) reveals some comparisons of speed and stroke cycle characteristics during different phases of the 100-m wheelchair racing. Both of these studies employ high speed camera and perform their experiment on the track.

Goniometers have been used extensively to study elbow movement both in normal and disability sports researches (Wright et al., 2006; Wang et al., 2005).

The aim of the present study is to develop and comprehensively examine a novel method for assessing performance in wheelchair racing. Electrogoniometers were used to detect elbow movement in a full-length (i.e. 100 m) wheelchair racing.

2 Methods

2.1. Athlete, Wheelchair and Measurement Equipment

One male subject (age = 19; body mass = 51 kg; medical condition: *spina bifida*) gave informed consent to participate in this study. The subject was an experienced and high-performance wheelchair racer (racing classification = T53, specializing in 100 m and 200 m track events, ever broke the world record and ranked world no.1 in 100 m for class T53 in 2006). The wheelchair (X-Limit, Corima, Loriol sur Drôme, France) was equipped with one front wheel (AERO wheel, Corima) and two rear wheels (4-spoke or disk type, diameter of 28″, Corima). The two rear wheels are set at a chamber angle of 12° and fitted with 0.36 m diameter push rim. It is agreed that racer adopts comfortable seating position and hand propulsion since it is his own wheelchair.

Electrogoniometers (Biometrics, Gwent, UK) were used to measure angles of elbow flexion/extension. The electro-goniometers come with a datalogger equipped with memory card which allows portable data collection.

2.2. Data Collection

Data was collected at a synthetic 400 m-athletic track usually used by the athlete for regular training. The electrogoniometers were set to frequency of 1 kHz and attached on the athlete's elbow joint by taping the two end surfaces on both the upper arm and forearm. The data logger was attached to the wheelchair's back rest.

On the start line, the transducers were normalized by instructing the athlete to fully extend his elbow and reset the datalogger (i.e. extension is 0°). This leads to convention of left elbow flexion as positive value and right elbow flexion as negative value. Starting commands used in athletics were employed by the head coach to initiate each of the race simulation. The race time were recorded with a stopwatch. The data was taken in interval of 10–20 minutes to avoid the fatigue effect. In single data collection session, only up to 7 full-length races were done.

2.3. Data Analysis

Raw data of elbow angular displacement was filtered by Savitzky-Golay filter (8th order, window width 101 data points) in Matlab. Angular displacement data was numerically differentiated with respect to time to obtain respective angular velocity. Extension movement was set to positive angular velocity. Further processing of selecting peak angular velocity (peak analysis) was done by writing Matlab programs.

The data of angular displacement and angular velocity is plotted with respect to time. The interest is narrowed only to extension angular velocity thus the negative values are eliminated. The race is divided into several phases as percentage of time records. A running average filter with window width of 35 points is applied to average the gradient of the angular velocity with respect to the angular replacement, which corresponds to the average push frequency of respective phase.

3 Results

In total of 8 sets of data was taken during the experiments, however only some of them are shown since this project only aims to introduce the method and the interpretation of the analysis. The peak of elbow flexion ranges from 85° to 95° (Figure 1). Full elbow extension is observed at the first two strokes. Along the progression of the race, elbow extension deficit reaches about 10°. The number of stroke cycles amounts to 32 peaks for 100 m. On average, the racer performs more than 117 pushes per minute.

Figure 2 shows that the peak angular velocity increases in the early stage of the race. After 5 to 6 seconds (~30–40%), the racer typically has passed 20 rad/s. The peak analysis reveals that the highest angular velocity reaches 26 rad/s (Figures 2 and 3).

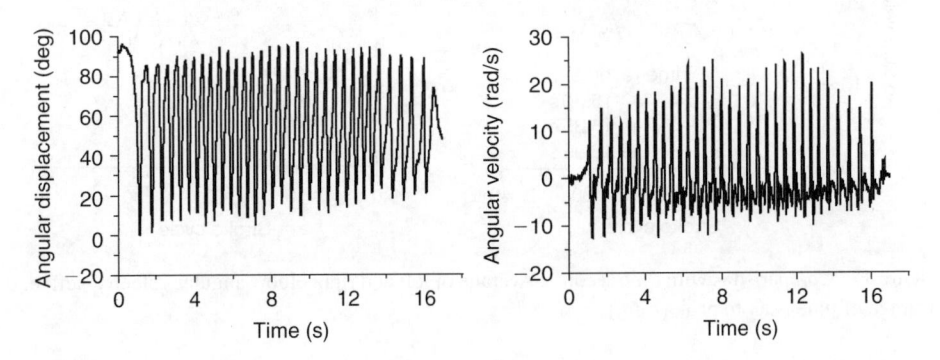

Figure 1. Angular (left) and velocity (right) displacement graph (100 m, 16.35 s, left elbow).

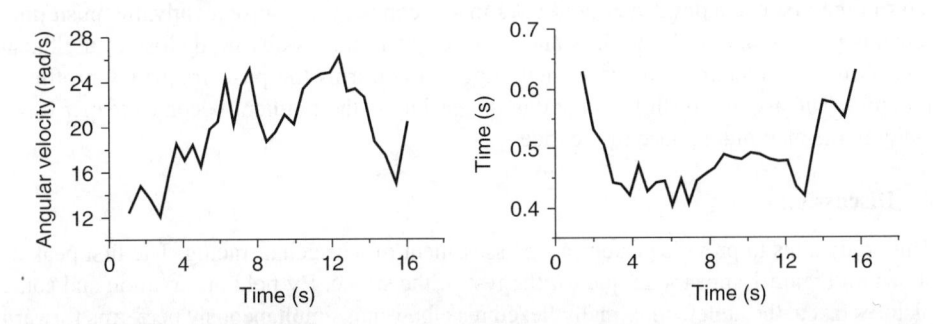

Figure 2. Results of peak analysis by Matlab programming: profile of peak angular velocity (left) and time between 2 peaks (right).

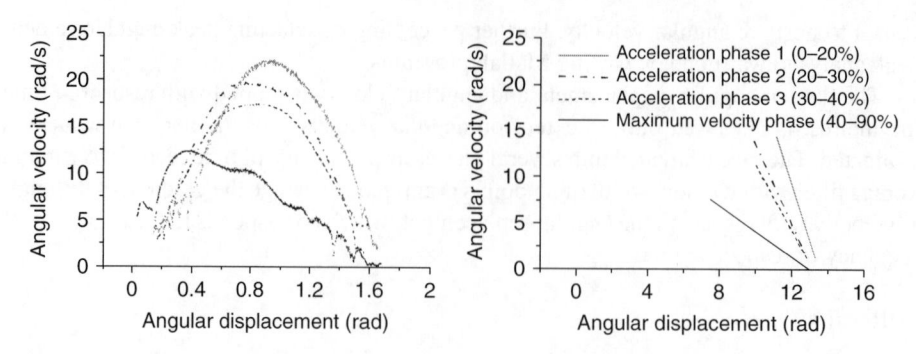

Figure 3. Graph angular velocity against angular displacement (left) and its gradient as average push frequency (right).

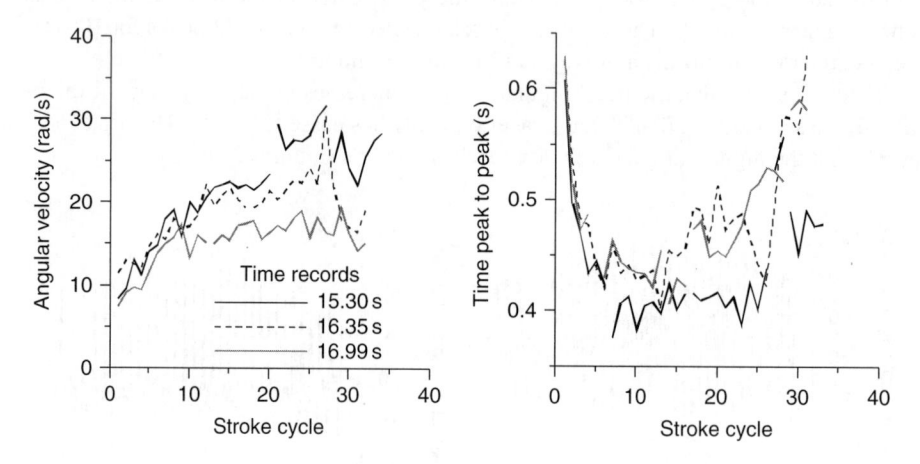

Figure 4. Comparisons with time records: Average of left and right elbow angular velocity (left) and average of time peak to peak (right).

Right after start, the peak to peak time (push time) decrease significantly from above 0.6 s in the first cycle down nearly to 0.4 s in 4 seconds time. Subsequently, the push time remains in the range of 0.4 to 0.5 s and a longer push time is observed close to the finish line. With time reciprocal to frequency, Figure 4 confirms the push frequency profile in Figure 3. The average push frequency is increasing as the gradient become steeper especially in the maximum phase of the race.

4 Discussion

This study aims to provide performance assessment of wheelchair racing. The first peak as shown in Figure 1 appears unique for the rest of the stroke. By field observation and some pictures taken, the athlete maximally flexed his elbow and simultaneously performs forward trunk movement to overcome the system inertia. This result confirms previous findings and observations by Moss *et al.* (2003).

The decreasing push time at the early stage of the race is believed to considerably contribute to the development of power and acceleration. Subsequently, this graph reveals how consistent a racer maintains the race tempo especially during the middle and last stages at a minimal stroke-time phase.

This study proposes that peak angular velocity and peak to peak time are two important factors in determining wheelchair racing performance. Figure 4 combines several data sets to show the contribution of both variables in influencing time records. Data is taken as average of left and right elbow based on symmetry of elbow kinematics during wheelchair racing demonstrated by Goosey & Campbell (1998). Through peak analysis of both left and right elbow, this present study confirms their findings.

The fastest time record (15.30 s) appears to gain apparent leading by having the highest angular velocity especially in the middle and final phases of the race. In contrast, poor performance means slow angular velocity as shown by the sample of 16.99 s race.

The present study demonstrates the presence of frequency differences along the race. This result does not confirm earlier work, suggesting that wheelchair athletes prefer to keep the same stroking rhythm when stroking with maximum effort (Chow & Chae, 2007), possibly because these authors did not analyse top athletes. Figure 4 demonstrates clearly how different push frequencies, the reciprocal push time, contribute to distinctive achievement of time records in different stages of the race. Although there is no clear separation between the 15.30 s and 16.35 s samples, the best performance nevertheless exhibits a low push time especially in the middle stage of the race.

5 Summary

The objective of this study is to develop and comprehensively examine a novel method for assessing performance in wheelchair racing. This study is performed under realistic and reliable condition. As a result, peak angular velocity and time peak to peak (push time) remain viable as two crucial factors that determine time records as an explicit manifestation of performance.

References

Chow J.W. and Chae W.-S. (2007) Kinematic Analysis of the 100-m Wheelchair Race, *Journal of Biomechanics*, in press.

Chow J.W., Milikan T.A., Carlton L.G., Chae W.-S. and Morse M.I. (2000) Effect of Resistance Load on Biomechanical Characteristics of Racing Wheelchair Propulsion over a Roller System. *Journal of Biomechanics*, 33, 601–608.

Cooper R.A. (1990) Wheelchair Racing Sports Science: A review. *Journal of Rehabilitation Research*, 27/3, 295–312.

Cooper R.A., Boninger M.L., Cooper R., Robertson R.N. and F.D. Baldini (2003) Wheelchair Racing Efficiency. *Disability and Rehabilitation*, 25 /4–5, 207–212.

Coutss K.D. (1988) Heart Rates of Participants in Wheelchair Sports. *Paraplegia*, 26/1, 43–49.

Coutss K.D. and Schutz R.W. (1988) Analysis of Wheelchair Track Performances. *Medical Science Sports Exercise*, 20/2, 166–194.

Goosey V.L. and Campbell I.G. (1998) Symmetry of the Ebow Kinematics during Racing Wheelchair Propulsion. *Ergonomics*, 41/12, 1810–1820.

Goosey-Tolfrey V.L. and Lenton J.P. (2006) A Comparison between Intermittent and Constant Wheelchair Propulsion Strategies. *Ergonomics*, 49/11, 1111–1120.

Moss A.D., Fowler N.E. and Goosey-Tolfrey V.L. (2005) The Intra-push Velocity Profile of the Over-ground Racing Wheelchair Sprint Start. *Journal of Biomechanics*, 38, 15–22.

Muller G., Odermatt P. and Perret C. (2004) A New Test to Improve the Training Quality of Wheelchair Racing Athletes. *Spinal Cord*, 42, 585–590.

Vanlandewijck Y., Theisen D. and Daly D. (2001) Wheelchair Propulsion Biomechanics: Implications for wheelchair sports. *Sports Medicine*, 31/5, 339–367.

Veeger D.J., Meershoek L.S., Van der Woude L.H. and Langenhoff J.M. (1998) Wrist Motion in Handrim Wheelchair Propulsion. *Journal of Rehabilitation Research and Development*, 35, 305–313.

Wang Y.T., Chen S., Limroongreungrat W. and Change L.S. (2005) Contributions of Selected Fundamental Factors to Wheelchair Basketball Performance. *Medicine and Science in Sports and Exercise*, 37/1, 130–137.

Zatsirosky V.M. (2003) *Biomechanics in Sport: Performance Improvement and Injury Prevention*. Blackwell Science, Oxford.

20. Winter Sports

SKIING EQUIPMENT: WHAT IS DONE TOWARDS MORE SAFETY, PERFORMANCE AND ERGONOMICS?

V. SENNER & S. LEHNER

Technical University Munich, Department of Sport Equipment and Materials, Munich, Germany

Alpine Skiing remains to be one of the most popular sports and due the economic development in the Eastern European countries and in Asia the worldwide number of active skiers still increases. Skiing can be practiced at any age, can fascinate at any experience level and its benefit for health is not questioned (although not finally quantified). However skiing is not without risk and epidemiology ranks it in the group of those leisure time sports with increased probability to suffer an injury (among other sports such as American football, soccer and horse riding). Sport engineering can contribute to more safety in skiing by improving the ski binding's release behavior, by optimizing the boot-binding interface towards better force transmission, by designing skis which are easier to handle in all situations and – last but not least – by working on improved protection gear.

This paper will give an overview of 15 years of our work in the field of skiing equipment. The overview will address the three components ski, boot and binding and cover a wide range of different aspects. It will focus on measures to reduce the risk of knee injuries, presenting some possible solutions and (the very tedious) process to evaluate their effects. Regarding the development of the conventional (force driven) ski binding the presentation will deal with the problem of inadvertent release and present an investigation that helped to better understand the state of the art regarding this aspect. The ongoing efforts to avoid lower leg injuries (mainly fractures) by proper binding setting, will be demonstrated by a recently finished study in the ski rental environment. Further one study dealing with the interaction between ski and binding will be explained. Finally the ski and the effects of its possible modifications are addressed. This includes the question, if the occurred change of ski's geometry (craving or super side cut skis) has had an impact on the injury development in the late nineties. This part will be completed by one study which examined the potential of a new edge preparation technique. The presentation will finish with a look at professional ski race and describe an equipment related intervention possibility, which is currently under investigation.

1 Introduction

1.1. *Popularity of Skiing – Skiing Epidemiology*

Alpine skiing still encounters growing popularity; the increasing welfare especially in the East European countries or in Asia contributes to this development. Approximately 8 Million Germans (almost 10% of the entire population) claim to be skiers, snowboarders or cross country skiers – it is estimated that 25 to 30% of them are practicing their winter sport activity regularly. In the United States 6.4 million people (seven years or older) have at least once participated in alpine skiing in the course of the year 2005, according to the National Sporting Goods Association). According to the World Trade Organization a total of 64 million people worldwide claim to be alpine skiers.

Although skiing is commonly regarded to be a rather dangerous sport the development of the overall injury rate tells us something different: between 1972 and 1998 it has declined by 46% as reported from the biggest epidemiologic study in the US (Johnson *et al.*, 2000). A comparable decrease can be observed in Germany, where the overall injury rate dropped by 42% between 1980 and 2005 (Gläser, 2006). Both studies and many others however state a significant increase of serious knee sprains involving the anterior cruciate ligament (ACL) within this period of time. For this reason knee injuries – accounting for 30 to 40% of all injuries in skiing – remain the major topic of concern.

1.2. *Countermeasures for Preventing Skiing Injuries*

Various countermeasures for the prevention of skiing accidents have been applied at different links in the chain of events leading to an injury. The Australian researchers Kelsall and Finch (Kelsall & Finch, 1999) have reviewed a selection of English language literature trying to quantify the effectiveness of various injury countermeasures. The most effective countermeasure they see is a properly adjusted ski binding with a potential for a 3.5-fold reduction in lower extremity injuries (Hauser, 1989).

Another ambitious effort to evaluate the effectiveness of a technical countermeasure is reported from France. Under the leadership of the "Médecins Montagne", an association of French doctors, a prospective and controlled field study is conducted since winter season 2002/03 to evaluate the effect of a binding setting that differs from the current ISO-standard (Laporte *et al.* 2003). Obviously even well established knowledge regarding personal safety equipment needs further differentiation. This is shown by Johnson and Ettlinger (2006a, 2006b) who analysed fatal head injuries reported by the in the US Consumer Product Safety Commission (CPSC) and the National Ski Areas Association (NSAA) between the seasons 91/92 and 2004/05. Their results clearly indicate that the use of helmets in alpine skiing and in snowboarding does not reduce the total number of fatal accidents (however they may reduce the severity of some type of non fatal injuries).

How tedious and time-consuming the evaluation of technological countermeasures can be is demonstrated in the study of Senner (2001). He used a computer model to quantify the effect of a release mechanism integrated into the ski boot with the objective of reducing tension force in the anterior cruciate ligament (ACL) in certain injury situations.

The current paper is giving a glance on three examples of our work in the field of technological countermeasures; it illustrates methods for the first evaluation of their effectiveness. However it should be mentioned at this point that due to ethic reasons on one side and economic reasons on the other these evaluations are not basing on controlled trials in the field. Instead computer modeling and simulation or laboratory tests with physical models in load simulators are used. Field measurements supply the necessary boundary conditions.

2 ACL Injuries – Can an Asymmetrical Twist Release Help?

2.1. *The Problem and the Idea*

Approximately 80% of all knee injuries in alpine skiing are due to mechanisms where lower leg rotation is present (Figure 1).

The orientation of the knee's anterior and the posterior cruciate ligament suggests that external and internal lower leg rotation should lead to different magnitude of the ACL tension

Figure 1. Typical injury situation in alpine skiing: knee in hyperflexion, external or internal lower leg rotation, eventually in combination with valgus or varus load and an anterior drawer. Ettlinger (1989) was the first to describe this situation calling it the "phantom foot" mechanism.

force. This is supported by Andriacchi's work (Andriacchi, 2003) who clearly demonstrated that in a deep squat position (as seen in Figure 1) the tibia internally rotates and translates anterior (relative to femur) thus tensioning the ACL. This effect increases beyond 130° flexion. Basing on these findings the idea of an asymmetrical twist release makes sense: a ski binding releasing the boot from the ski at different torque level depending on its effective direction may provide additional safety for the skier. However not only the proof of this hypothesis but also the question remains which ratio between internal and external retention force should be chosen.

2.2. The Approach

At the beginning of the investigation a load simulator (Figure 2a) was designed which allows in vitro testing of human knees under skiing typical boundary conditions. The resulting forces and moments between boot and binding are measured using a 6-component load cell. Quadriceps force is simulated by a string pulley system, additional loads (i.e. varus-valgus) can be applied. ACL tension force is measured using Markolf's bone craft technique (Markolf *et al.* 1990) as illustrated in Figure 2b. A six degree of freedom goniometer system (Kristukas, 1995) attached between femur and tibia (Figure 2c) allows a precise determination of translational and rotational displacements between these bones. Preceding the tests all cadaver knees had to go through an MRI scan which allowed the later 3D reconstruction of its individual bone architecture and its individual ligament insertion points.

As these types of experiments are rather elaborate, the next step of this research was to develop a computer model of the physical device and of the knee under load. The most challenging part, the knee model, has been developed by Lehner since the mid Nineties (Lehner, 1995). It represents a rather complex combination of several submodels, such as a multi point contact model, the specimen specific viscoelastic cartilage model, a model describing the viscolelastic properties of the ligament with its fibre bundle structure and its specimen specific cross sectional areas and insertion points. His Ph.D thesis (Lehner, 2007) deals with

Figure 2a. Load simulator device (for the tests the physical knee model seen here is replaced by a human cadaver knee).

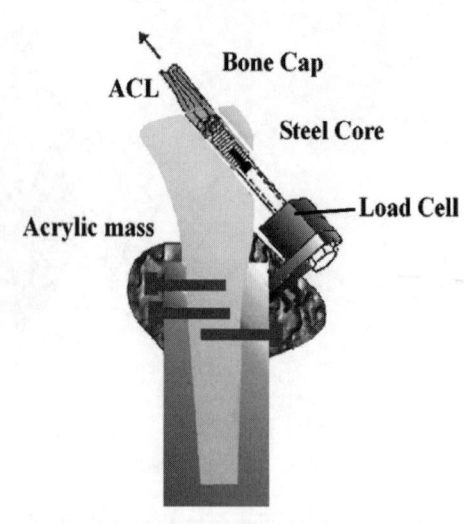

Figure 2b. Functional principle of Merkolf's ACL tension force measurement.

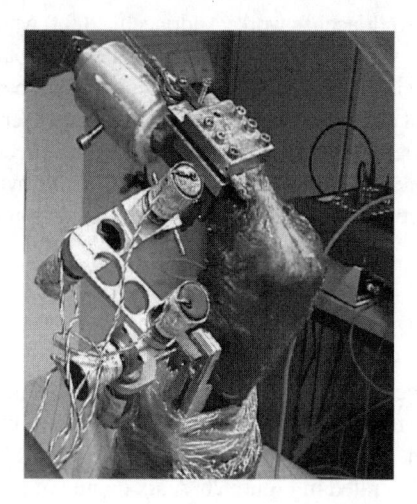

Figure 2c. Instrumented knee with patella tendon fixed to pulley string and 6 dof goniometer.

the validation of this model using the aforementioned experimental results (i.e. the measured ACL tension force) to compare with the model's output.

2.3. Major Results

The experiments and the computer simulation confirm the hypothesis attributing the ACL to different tension force depending on the direction of rotational displacement. At 120 to 130° knee flexion and at quadriceps force in (low) magnitudes around 400 N, internal rotation increases the ACL force, whereas external rotation decreases it. Higher rotational velocity seems to further increase the resulting ACL force at internal rotation with no effect at external rotation. An isolated valgus moment (in a range of 15 m) also adds to the ACL force, whereas an isolated varus moment of the same magnitude does not affect the ACL. Interesting is also the observation that the application of a constant varus moment seems to decrease the ACL force during internal rotation but increasing it at external rotation. On of the most remarkable results for future ski binding development is the fact that in several loading situations the ACL gets under tension while the binding torque still remains low. This rises the question if a moment or force driven ski binding will ever be able to significantly reduce the incidence of ACL injuries in skiing.

3 Inadvertant Release – a Problem of the Ski Binding's Heelpiece

3.1. The Problem

Surveys among expert skiers reveal that the majority of them are setting their ski bindings above the recommended values given in ISO 8061 and 11088. The reason for this is that many of them have experienced inadvertent releases which may lead to dangerous situations especially when skiing at higher velocities. ISO 9465 defines minimum requirements regarding tolerable lateral impacts. No comparable standard however exists for the binding's heel piece which does not restrict manufacturers to design ski bindings to adequate dynamic behavior. A comparison study to quantify the on market ski bindings' vertical elasticity and damping characteristics was performed.

3.2. The Approach

The study was performed with equipment of the 2001/02 winter season. Details of the study are given at Janko (2003). The heel pieces of seven different bindings mounted on the ski supplied by the manufacturer were set to an equal quasi-static release force of 1000 N (Z~8) according to the ISO standard. High speed video recordings of simulated inadvertent releases on the slope revealed that even minor vertical impacts acting in the tip area to the bottom side of the ski can open the heel piece. In the laboratory similar impacts were then applied, impact energy was stepwise increased until the binding started to release. The transition point indicates the maximum tolerable impact energy (TIE).

3.3. Major Results

The bindings in test showed significant difference regarding TIE (Figure 3). These differences can be explained with different binding design, showing clear advantages for some of the on market solutions. The results suggest that both the introduction of a new standard and more information to the distributors, retailers and the skiers themselves are needed to overcome the observed deficits regarding inadvertent release.

4 Release Binding for Skiboards?

4.1. The Problem

Skiboards are defined as skis between 70–130 cm length. These skis typically have a sidecut that makes them more like carving skis than traditional skis. A distinguishing feature for most

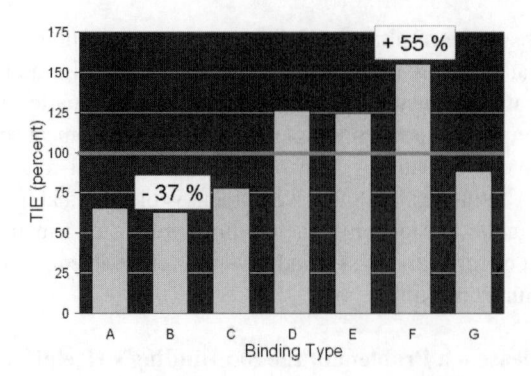

Figure 3. Tolerable Impact Energy (TIE) for impacts on ski's tip from below; related to mean of all seven bindings tested.

skiboards is that they are not typically equipped with traditional release bindings. Rather, they have fixed bindings, which do not release. Injury epidemiology for skiboards (i.e. Greenwald *et al.* 2000) reveal some varying results (which is probably due to their different populations analyzed. Agreement however exists concerning the observation that tibia fractures occur in skiboarding at a significantly higher rate than in alpine skiing. Further there is some indication for an increase of shoulder and wrist injuries compared to alpine skiing.

As with skiboards the lever arm is short, the resulting lateral or medial bending moment on the lower leg is commonly considered to be low enough not to make a release binding necessary. The above epidemiological data however seems to disagree with this idea. This raises the question, if injury risk could be lowered by equipping skiboards with release bindings.

4.2. The Approach

In winter season 1999/2000 a research project was started to deal with the above question. Details of this investigation can be taken from Senner *et al.* (2003). Skiers in alpine skiing fall very differently. Countless combinations of kinematic and dynamic parameters are possible. For a first analysis however it is enough to distinguish between the two major fall situations, forward falls (leading to a heel release of the binding) and falls with rotational components (leading to its twist release). In two steps the problem was approached. First an experimental preliminary study was performed from which some fundamental conclusions regarding the need of a twist release could be drawn. Then in a second step, a computer model was developed and used to specifically analyze the forward fall problem. For the latter a typical risk situation was selected to be simulated: an experienced skier using skiboards for the first time is approaching a mogul. Being experienced on regular skies he/she knows how to anticipate such an obstacle and his/her movements in this manoeuvre are automated. Therefore he/she would not adapt his technique (i.e. by generating additional counter movements) and he/she acts as usual. Accordingly two simulations were compared: (1) Skier with conventional skis performing a skier-typical mogul manoeuvre, and (2) Same skier with identical movements, but now using skiboards.

The computer model consisted of seven rigid segments, HAT, both thighs, lower legs, feet. The lower part of the boot is added to the foot, the boot shaft is linked to the lower leg

Figure 4a. Simulation of skier assing a mogul with traditional skis.

Figure 4b. Simulation of skier passing a mogul with skiboards.

segment. The segments are connected by frictionless hinge joints, representing the ankle, the knee and the hip. Muscle contribution is modelled by closed-loop control elements linked to each joint. These control elements use as input the joint kinematics, which has been determined in preceding field experiments. In each joint, the control loop produces just the amount of muscle torque needed to reproduce the given joint kinematics. It should be mentioned, that regardless kinematic data is used as input, the entire process remains to be a forward simulation and is not an inverse dynamics. (The equations of motion are still solved for the kinematic degrees of freedom and not for internal forces and moments. For this reason, no derivatives are needed). The model skier is "equipped" with a conventional ski or the skiboard. Both the ski (Head TR6, 203 cm) and the skiboard (Salomon Snowblade, 98 cm) are modeled as Bernoulli beam consisting of 12 sections with different geometry and material properties. The elasticity and damping parameters have been determined from two-point bending tests. The initial and boundary conditions for the forward simulation have been determined from the aforementioned studies on the slope.

4.3. Major Results

Comparing the visualizations of both simulations, differences are obvious. Figure 4 compares identical time instants of the simulation. Whereas the skier with the traditional ski is crossing the mogul as expected (Figure 4a), the skier using the skiboard is driven into a completely different kinematics (Figure 4b). This observation can be explained by the lack of the large lever arm in front of the skier to bend and absorb energy. Instead, the torso begins to rotate forward rapidly, leading to a forward fall over the tips of the skiboard. This is likely to result in an increased risk for upper extremity injuries.

The observed differences in the kinematics have noticeable effects on the resulting binding forces Looking at the time histories of the vertical force at the binding's heel piece significantly higher values can be seen for the skier with conventional skis. In this simulation the peak value comes to 1560 N, whereas for the skiboard simulation the maximum value reached 1150 N, which is more than 25% less. To further analyse the aforementioned forward fall over the tips, the angular velocity of the trunk was calculated and compared between the two models. For the skiboard the trunk's angular velocity reached more than double the value calculated for the ski-model. This means that a certain portion of the skier's linear energy has been transmitted into a rotational component, thus leading to a different trajectory of the skier's centre of gravity, more directed towards the slope's surface. To limit the trunk's angular velocity it is necessary to dramatically reduce the release value of a skiboard's heel piece. However this might increase the risk of inadvertent release in those situations where the skiboard is used in the fun park. From these observations it was concluded, that a heel piece set according to the current ISO standard will not work.

To overcome this problem a new type of binding with a horizontal release was simulated in the computer. With this innovative binding design it was possible to reduce the critical angular velocity of the trunk to 65% of the value attained with a conventional (under the present circumstances not releasing) ski binding.

References

Andriacchi T. (2003) Knee rotation in deep flexion. Keynote lecture given at the 15th International Congress on Ski Trauma and Skiing Safety, Pontresina, Switzerland, 27 April–2 May.
Ettlinger C. (1989) What can be done about knee injuries? *Skiing*, Spring 1989, 85–87.
Gläser H. (2006) Unfälle im alpinen Skisport. Zahlen und Trends der Saison 2004/2005. ARAG, Auswertestelle für Skiunfälle, Düsseldorf, Germany.
Greenwald R.M., Nesshoever M. and Boynton M.D. (2000) Ski injury epidemiology: A short-term epidemiology study of injuries with skiboards. In: Johnson R.J., Zucco P. and Shealy J.E. (Eds.) *Skiing Trauma and Safety*, Vol. 11, ASTM STP 1289, American Society for Testing and Materials, West Conshohocken, 119–126.
Hauser W. (1989) Experimental prospective skiing injury study. In: Johnson R.J., Mote, C.D. and Binet M.-H. (Eds.) *Skiing Trauma and Safety*. Vol. 7, ASTM STP 1022, American Society for Testing and Materials, West Conshohocken, 18–24.
Janko T. (2003) Dynamisches Verhalten des Fersenelements alpiner Skibindungen in Experiment und Simulation. Diploma Diss., Technische Universität München, Faculty Mechanical Engineering, Munich, Germany.
Johnson R.J., Ettlinger C.F. and Shealy, J.E. (2000) Update on Injury Trends in Alpine Skiing. In: Johnson R.J., Zucco P. and Shealy J.E. (Eds.) *Skiing Trauma and Safety*. Vol. 11, ASTM STP 1289, American Society for Testing and Materials, West Conshohocken, 37–48.
Kelsall H.L. and Finch C.F. (1999) Preventing alpine skiing injuries – how effective are countermeasures? *International Journal for Consumer & Product Safety*, 6/2, 61–78.
Kristukas S.J. (1995) A model for staying loss of knee extension. Ph.D. Diss., Faculty of the Graduate School of the University of Minnesota.
Laporte J.-D., Binet M.-H. and Bally A. (2003) Why the ski binding international standards have been modified in 2001. In: Johnson R.J., Lamont M. and Shealy J.E. (Eds.) *Skiing Trauma and Safety*. Vol. 14, ASTM STP 1440, American Society for Testing and Materials International, West Conshohocken, 64–94.
Markolf K.L., Gorek J..F, Kabo J.M. and Shapiro M.S. (1990). Direct measurement of resultant forces in the anterior cruciate ligament. An in vitro study per-formed with a new experimental technique. *J Bone Joint Surg Am.*, 72/4, 557–67.
Senner V. (2001) Biomechanische Methoden am Beispiel der Sportgeräteentwicklung. Ph.D. Diss., Technische Universität München, Faculty Mechanical Engineering,

Munich, Germany. German version available from online source: http://tumb1.biblio.tu-muenchen.de/publ/diss/allgemein.html.

Senner V., Lehner S. and Schaff P. (2003): Release Binding for Skiboards? In: Johnson R.J., Lamont M. and Shealy J.E. (Eds.) *Skiing Trauma and Safety*, Vol. 14, ASTM STP 1440, American Society for Testing and Materials International, West Conshohocken, 24–35.

Shealy J.E., Johnson, R.J. and Ettlinger C.F. (2006a) On Piste Fatalities in Recreational Snow Sports in the U. S., Journal of ASTM International (JAI), 3/5, 8 pages (pnline journal).

Shealy J.E., Johnson R.J and Ettlinger C.F. (2006b). Do helmets reduce fatalities or merely alter the patterns of death? In: Moritz E.F. and Haake S. (Eds.) *The Engineering of Sport 6, Vol. 1 – Developments for Innovation*. Springer, Munich, pp. 163–167.

AIRFOLIED DESIGN FOR ALPINE SKI BOOTS

L. OGGIANO[1], L. AGNESE[2] & L.R. SÆTRAN[3]

[1]*Norwegian University of Science and Technology, faculty of engineering science and technology, Trondheim, Norway*
[2]*Politecnico di Torino, dipartimento di ingegneria aerospaziale, Torino, Italy*
[3]*Norwegian University of Science and Technology, faculty of engineering science and technology, Trondheim, Norway*

The aerodynamic effects in sport competitions increase with the speed. In some sports like skating, skiing or cycling the athletes have an average speed from 70 km/h up to 150 km/h. The aerodynamic drag coefficient increases with the square of the speed. This means that, by reducing the aerodynamic drag coefficient, it is possible to increase the athlete's speed and improve their performances. The drag of a human body when assuming the position of an alpine skier is divided approximately in 1/3 given by the legs, 1/3 given by shoulders an arms, 1/3 given by the chest. On bluff bodies like cylinders or spheres (we can consider the human body as a bluff body) the larger part of drag is given by the difference of pressure between the front and the rear part of the body. That's due to the boundary layer separation. In this paper we focus on the reduction of drag on skiers legs by modifying the shape of the boots from a cylindrical shape to an airfoiled one. Four different solutions with different designs have been studied, each solution has been experimentally analyzed and a *CdA-Speed* curve has been obtained. By reattaching the boundary layer with a different design of the boot it has been possible to decrease from 10% up to 70% the drag coefficient. This means that a drag coefficient reduction of approximately 10% of the total drag of an alpine skier has been obtained.

1 Introduction

1.1. Background

The role of the air in sport can often be very important. In some disciplines like sailing or windsurfing, aerodynamics effects are used to generate lift and push the athletes. In some other disciplines like skiing, speed skating or cycling the air is an obstacle for the athletes and the aerodynamic effects can mostly be reassumed into drag effects.

In the second category of sports mentioned above, a large part of the external power produced by the athlete is used to overcome the drag. From the experiments carried out by Grappe *et al.* (1997) about aerodynamic resistance in cycling, 90% of the total resistance opposing motion depends on drag. From a paper written by Thompson *et al.* (2001) about drag reduction in speed skiing, the aerodynamic loads contribute more than 80% of the total drag. Reducing the drag by improving materials can then sensibly improve the athlete's performances.

The drag can be defined with the formula: $D = \rho V^2 c_d A/2$. Where A is the frontal Area, V is the wind speed and c_d is the drag coefficient.

The position assumed by the athlete during is performance has an important role in determining the aerodynamic drag. As shown by Remondet *et al.* (1997) the position can affect the drag by influencing not only the frontal area but also the drag coefficient (c_d). The c_d varies from 0.5 in downhill position to 1 in standing position.

Considering a classic downhill skiing position, the effect of each body part can be roughly divided in 1/3 given by the chest, 1/3 given by head and arms and 1/3 given by the legs. While reduce the drag on other body parts is quite hard (due to the position effect on aerodynamic resistance), reduce the drag on skier's legs is possible and 2 different solutions can be adopted:

Trip the transition to turbulent regime using roughness. Many examples about that are present in literature especially about drag reduction drag on cylinders by using rough structures (this method is practically used in the speed skater's suits).

Delay the separation keeping the flow in laminar regime. This can be done using an airfoil shape to surround the cylindrical shape of the legs. Shape the boots like an airfoil, reducing the wake dimensions and thus the drag component due to the pressure difference between front and back of the boot is what it has been done and exposed on this paper.

Some previous examples of airfoil shaped boots have been used in speed skiing, where rules are quite strict and they have been presented by Thompson et al. (2001).

1.2. Forces Acting on a Downhill Skier

There are 4 forces acting on a skier: the friction force between the skis and the snow, the gravity force gravity, the aerodynamic drag and the centripetal acceleration (this last force is 0 for on a straight path).

Considering the relations who link of all these forces, it is possible to estimate the total resistance (drag and friction) and then find a correlation between the respective influences of the two parameters mentioned above.

Considering a straight path (S=0) with 35% inclination, the influence of the aerodynamic drag is higher (55%) than the one given by the friction force (45%).

2 Experimental Setup

2.1. Wind Tunnel

The test section of the wind tunnel is 12.5 m long, 1.8 m high and 2.7 m wide. The convergent has an initial section of 20.59 m² and final section of 4.88 m² with a contraction ratio of 4:22. The wind tunnel is equipped with a 220 KW engine that drives a fan which produces a speed range between 0.5 m/s and 30 m/s.

2.2. Balance

The balance which has been used is a six components balance, produced by Carl Schenck AG, which can measure force and momentum along three directions of a preset frame of reference.

2.3. Different Design Tested

5 different versions of the boots have been tested

- normal boots (1.1)
- small spoiler down (1.2-b)

Airfolied Design for Alphine Ski Boots 815

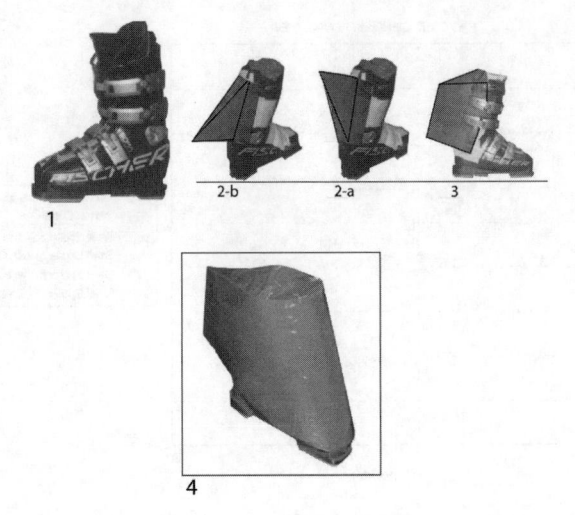

Figure 1. The 5 different versions tested.

- small spoiler up (1.2-a)
- big spoiler (1.3)
- airfoil (1.4)

The methodology used for the design followed the theoretical backgrounds about flow around bluff bodies and streamlined bodies. It has been tried to find the shape which minimize the wake (and then gives lowest drag) and at the same time the length of the spoiler.

3 Results

3.1. *Normal Boots (1)*

A preliminary test to evaluate the drag of the normal boots has been carried out. The boots have been tested with and without clamps to evaluate the effect of the clamps on the aerodynamic performances.

The effect of the clamps is quite important. Just removing the clamps it has been possible to reduce the drag of about 7% at 25 m/s.

3.2. *Short Spoiler (2-a and 2-b)*

The first attempt to improve the aerodynamic efficiency of the boots has been done adding a small spoiler behind the boots. 2 different configurations with the spoiler positioned in 2 different ways have been tested. The target was trying to "split" the wake into two smaller wakes in order to reduce the aerodynamic resistance. This test, like the previous one, has been carried out with and without clamps for both configurations.

A drag reduction of about 15% has been obtained using this small spoiler in the configuration (b) without clamps and the same result has been obtained for the configuration (a), testing both configurations without the clamps on.

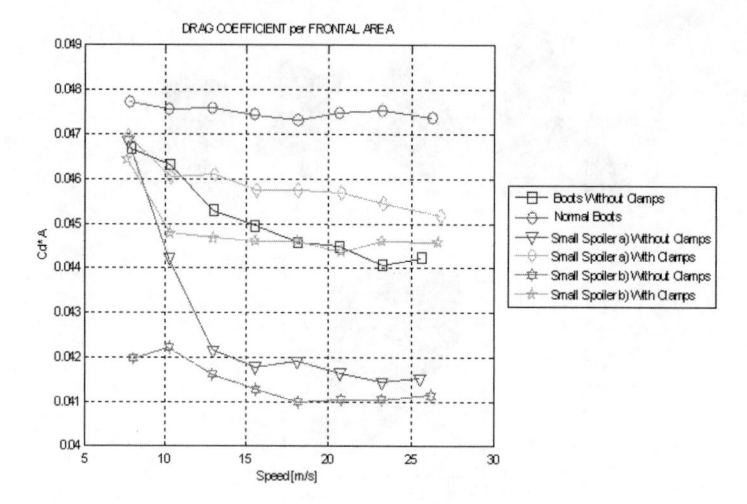

Figure 2. Short spoilers.

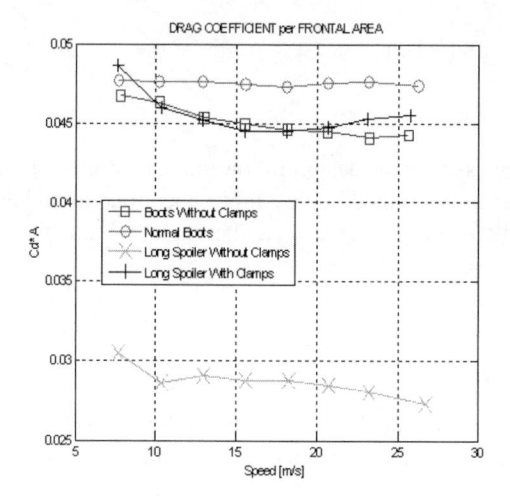

Figure 3. Long spoiler.

3.3. Long Spoiler (3)

The final target of the designing process is to get as close as possible to the airfoil shape. A bigger spoiler has been made to reduce the separation effect present in all the bluff bodies and to try to keep the flow attached to the boot, reducing the pressure drag and then reducing the total drag.

The test carried out using the balance show a drag reduction of 45% at the maximum speed for the configuration without clamps.

The aerodynamic effect of the clamps in this configuration is much higher than in the previous ones. The clamps in this case affect the flow around the boot inducing a early stage flow separation.

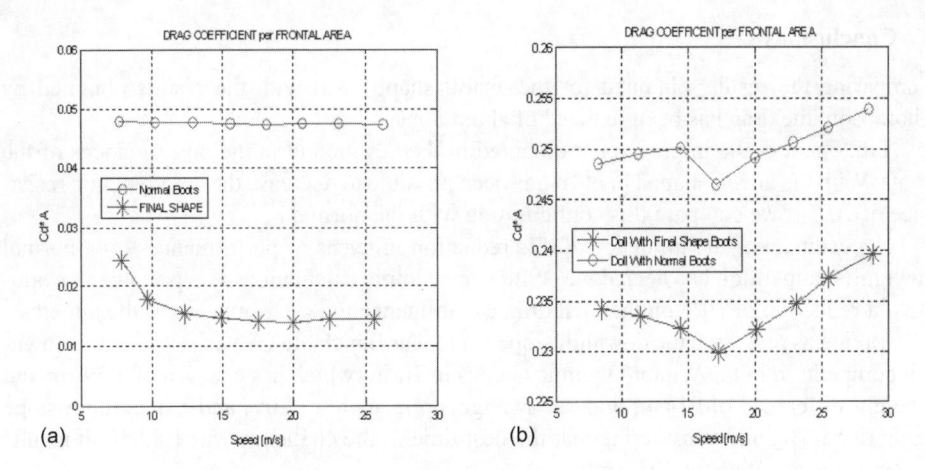

Figure 4. (a) Airfoiled shape. (b) Doll test.

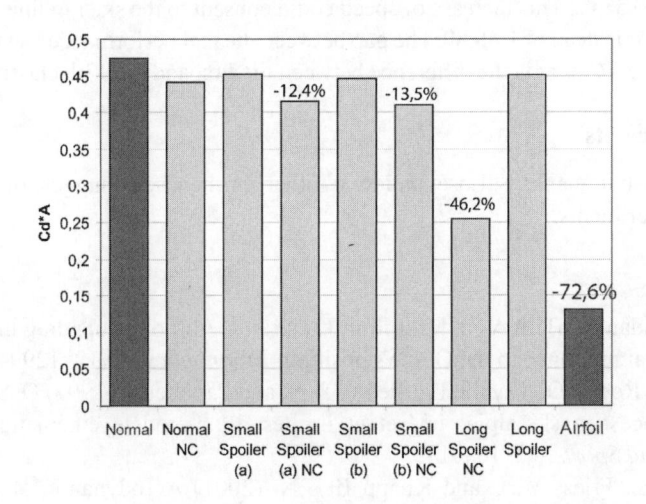

Figure 5. Drag coefficient reduction for the different boot versions at 25 m/s.

The flow separation produces a loss in term of drag of about 35% due to the increase of the wake size and consequently due to the increase of the pressure drag.

3.4. *Airfoiled Shape (4)*

To avoid the problem described above in par 3.3 (earlier separation induced by the clamps) it has been chosen to surround the boots and the clamps with a rigid cover shaped as an airfoil. The results obtained with this configuration show a drag reduction of about 70% comparing the airfoil design to the normal boots.

To evaluate the reduction on the total drag of a skier, a supplemental test mounting both the normal boots and the airfoil boots on a doll has been done.

A drag reduction of about 10% in drag has been obtained.

4 Conclusions

Comparing the results obtained for the airfoil shape (4-a) with the results obtained by Thompson, the drag has been reduced of about 50%.

Fig. 5 shows the improvement obtained in the design and in the performances of the boots. With the airfoil shaped boots it has been possible to decrease the aerodynamic resistance of 70% if we compare this configuration with the normal boots.

An estimate of the influence of this reduction in terms of performances on a normal downhill competition has been done. With some simple calculations it is possible to evaluate if a reduction of 10% on the total drag can influence the performances of the athletes.

Taking as reference the downhill slope Kandahar Banchetta G. Nasi used for the downhill competition in the Winter Olympic Games in Turin (which has a length of 3.299 m and a height difference of 914 m, with an average slope angle of 20% and a maximum slope angle of 58.5%) and considering that the best time of the challenge was 1:48:80, it results a average speed of about 30.32 m/s.

Reduce the total drag of about 10% means, in terms of speed gain, increase the speed of approximately 0.5 m/s. This increase of speed could consent to the skier to finish his run with a time of 1:47:04 instead of 1:48:80. The gap between the real performance and the calculated performance is 1'76" that is the difference between the first and the 13th classified.

Acknowledgements

The authors wish to thanks S. Løset and R. Winther for their help, suggestions and advices during the experiments.

References

Grappe F., Candau R., Belli A. and Rouillon J.D. (1997) Aerodynamic drag in field cycling with special reference to the Obree's position. *Ergonomics*, 40/12, 1299–1311.

Remondet J.P., Rebert O., Fayolle L., Stelmakh N. and Papelier Y. (1997) Optimisation des performances en ski alpin: intéret et limites d'un modèle cinématique simplifié. *Science and Sports*, 12, 163–173.

Thompson B.E., Friess W.A. and Knapp II K.N. (2001) Aerodynamic of speed skiers. *Sports Engineering*, 4, 103–112.

GROUND REACTION FORCES MEASUREMENT BASED ON STRAIN GAUGES IN ALPINE SKIING

S. VODIČKOVÁ[1] & F. VAVERKA[2]

[1]*Technical University of Liberec, Faculty of Education, Liberec, Czech Republic*
[2]*Palacky University, Faculty of Physical Culture, Olomouc, Czech Republic*

Ski experts in technique of turning on skis are dealing with problems concerning load of skis during particular kinds of joined curves. There are many theoretical works, most of them based on kinematographic or dynamographic record or on a combination of both methods. Observation of reaction forces between the ski and the plate has been performed in the past. Many studies have been performed to eliminate the number of injuries of knee-joints. However, significant changes occurred in the construction of skis and ski boots in recent years. Also a shift to the use of carving skis appeared. With such a way of skiing, bigger side-cut radius is exploited than it was with the traditional skis. This is the reason, why recently the number of studies dealing with biomechanics of alpine skiing with emphasis on measurement of trajectory and ground reaction forces. We have developed a special measuring device, which enables to measure forces in three directions and also their torsional moments around these three axis. This system works on the principle of strain gauges. During the measurement all the data are collected on Compact Flash disc with the frequency of 100 Hz. We were interested whether it is possible with help of the system, to detect interindividual deviation during a carve and thus to participate on improving of skiing technique. Our measurement refered to measurement of ground reaction forces arising during a carving turn. The investigation was carried out on a group of 6 skiers (3 racers and 3 ski teachers) with the weight of $M = 78,8 \pm 5,46$ kg. Their height was $M = 1,80 \pm 0,04$ m and age was $M = 26,5 \pm 1,61$ years. We realised that during a carving turn the strongest ground reaction forces occur during the steering phase after the fall line and during the initiation phase skis are alleviated and edging is changed. It was interesting that from the point of view of ground reaction forces there is no identical pass of right and left turn from all measured skiers. Comparison of measured left carving turn and comparison of right carving turn showed very similar cycle. Ground reaction forces measurement has already become a part of preparation of advanced national teams and it can help to improve quality of skiing not only top-class skiers, but also coaches of young gifted individuals can contribute to quick progress in technical performance of some motoric activities owing to measured data.

1 Introduction

Ski experts in technique of turning on skis are dealing with problems concerning load of skis during particular kinds of joined curves. There are many theoretical works, most of them based on kinematographic or dynamographic record or on a combination of both methods (Fukuoka, 1971; Nigg *et al.*, 1977; Müller, 1986, 1991, 1994 and others). Observation of reaction forces between the ski and the plate has been performed in the past. Many studies have been performed to eliminate the number of injuries of knee-joints (Fetz, 1977; 1991; Nachbauer, 1986; Nachbauer & Kaps, 1995; Niessen & Müller, 1999; Senner *et al.*, 2000).

However, significant changes occurred in the construction of skis and ski boots in recent years. Also a shift to the use of carving skis appeared. With such a way of skiing, bigger sidecut radius is exploited than it was with the traditional skis. This is the reason, why recently the number of studies dealing with biomechanics of alpine skiing with emphasis on measurement of trajectory and ground reaction forces. At present, there are in existence a number of objective methods enabling the determination of kinematic and dynamic parameters of a skier's drive in the turn, which have been applied in many studies (Kugovnik *et al.*, 2003; Pozzo *et al.*, 2005; Supej, Kugovnik (Nemec, 2005 and others). The basic goal of our work was construction and test of the recording device of forces between the ski and the plate during carving turn and measurement of the reaction forces.

2 Method

Based on literature published in recent years concerning ski load during course we performed an approximate calculation of dimensions, shape and stiffness of measuring elements for strain gauges application. Taking advantage of these calculations, we constructed a measuring device usable on the slope. For our study we used carving ski Blizzard SLK Kompressor of the length of 167 cm, with radius 16 m. At the original version there was a carving plate mounted under the bindings. This plate raises position of the body's center of gravity and thus facilitates bringing the ski into turning. At the ends it is supplemented with rubber blocks which limit deflection and also oscillation of the ski at its bending. Recording appliance was verified in February 2003 in Liberec on a 200 meters long sector of a slope.

The original carving plate was substituted by specially adapted element for the experiment. External shape and dimensions are similar to the original including the damping rubber elements on its ends. The new plate consists of two parts, between which the measuring elements were fixed. The bottom part creates a "tank", which is firmly fixed to the ski together with the measuring elements in the same way as on the original system but is slightly raised. A complete connection of miniature strain gauges system on the elements is made here. The strain gauges system enables measuring of all required moments of force; it is effects of force in three axes of the Cartesian system of coordinates and relevant

Figure 1. Cartesian system of coordinates.

flexural (torsional) moments around these axes (Fig.1). The upper part creating "cover" carries the bindings and is fixed to the lower part so that the system "tank" – "cover" can do mutual relative motion and thus enable measurement of force proportions.

The complete measurement system for the ride data record required to be developed in the smallest size and weight possible, shock resistant and easily placed on skier's body without limiting his/her motion. A single-chip microcontroller was chosen for this purpose and adapted specially for this case using custom-made set of amplifiers and Compact Flash Disc for data record. The system operates using battery and thanks to its compact dimensions can be placed on skier's back in a small bag.

To determine the time courses of single components of the six mentioned moments of force for each ski from the measured signals subsequent analysis and mathematical processing is necessary. The mathematical computation consists of the following steps:

1. Offset correction to get normalized scale in relation to state when the skier stands still on a flat horizontal surface
2. The strain gauges system outputs transformation to axial forces and appropriate moments including correction of cross-affection of force components in the strain gauges system and correction of vertical force with respect to weight of the skier.
3. Filtering the obtained forces data by a low-pass filter with cutoff frequency set to 10 Hz.

By linking the forces data to a picture record, information about force proportions in given moments of the skier's movement can be obtained. This picture – data synchronization is ensured by a LED diode placed on skier's forehead blinking in defined intervals. This approach gives us possibility to determine the real forces affecting the measured system, which is later subject to further analysis in connection to the picture record of skier's movement on the slope.

Forasmuch as our video recording is possible to synchronize with recording of running forces in individual turns with the help of an optical signal, we could determine the ratio of loading between an outer and inner ski in the phases of a turn.

Our measurement refered to measurement of ground reaction forces arising during a carving turn. The investigation was carried out on a group of 6 skiers (3 racers and 3 ski teachers) with the weight of M = 78,8 ± 5,46 kg. Their height was M = 1,80 ± 0,04 m and age was M = 26,5 ± 1,61 years.

3 Results and Discussion

We realised that during a carving turn the strongest ground reaction forces occur during the steering phase after the fall line and during the initiation phase skis are alleviated and edging is changed. The first annurately detected results of analysing of proportional shape of two phases of the turn indicated that the initiation phase represents approximately 40% and the steering phase 60% of the total duration of a carving turn. Interindividual variability of the ascertained data, expressed by standard deviation, is relatively very small and standard deviation fluctuates between 6–7% of duration of individual phases. In literature, we have only encountered qualitative consideration of the two phases in studies of Raschner *et al.* (2001), Müller & Schwameder (2003) and Müller *et al.* (2005), where the authors talk, on carving turn, about the initiation phase being longer than steering phase when compared

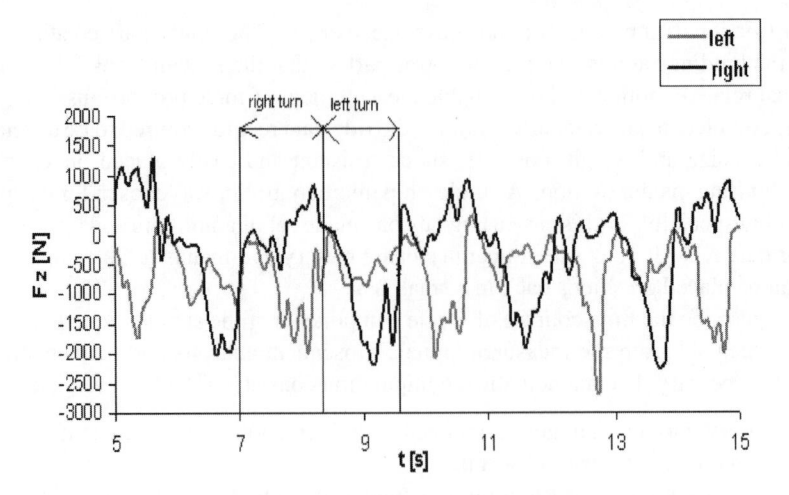

Figure 2. Ground reaction forces during left and right turns.

with a parallel turn. A rough estimate in the published graphs of Raschner et al. (2001) study is approximately 37% initiation duration phase

It was interesting that from the point of view of ground reaction forces there is no identical pass of right and left turn from all measured skiers (Figure 2).

4 Conclusions

Ground reaction forces measurement has already become a part of preparation of advanced national teams and it can help to improve quality of skiing not only top-class skiers, but also coaches of young gifted individuals can contribute to quick progress in technical performance of some motor activities owing to measured data.

References

Fetz F. (1977) *Zur Biomechanik des Schilaufs*. Inn-Verlag, Innsbruck.
Fetz F. (1991) Biomechanik alpiner Zieleinlauftechniken. In: Fetz F. and M,ller E. (Eds.) *Biomechanik des alpinen Skilaufs*, pp. 124–130. Enke, Stuttgart.
Fukuoka T. (1971) *Zur Biomechanik und Kybernetik des alpinen Schilaufs*. Limpert Verlag, Frankfurt/M.
Kugovnik O., Supej M. and Nemec B. (2003) *Biomehanika alpskega smucanja* (in Slovenian language). University of Ljubljana, Ljubljana.
Müller E. (1986) *Biomechanische Analyse alpiner Skilauftechniken – Eine biodynamische, biokinematische und elektromyographische Analyse moderner alpiner Skilauftechniken in unterschiedlichen Schnee- Gelände und Pistensituationen*. Inn-Verlag, Innsbruck.
Müller E. (1991) Biomechanische Analysen moderner alpiner Skilauftechniken in unterschiedlichen Schnee-, Gelände- und Pistensituationen. In: Fetz F. and Müller E. (Eds.) *Biomechanik der Sportarten – Biomechanik des alpinen Skilaufs*, pp. 1–49. Thieme, Stuttgart.

Müller E. (1994) Analysis of the biomechanical characteristics of different swinging techniques in alpine skiing. *Journal of Sport Sciences*, 12, 261–278.
Müller E., Schiefermüller C, Kröll J. and Schwameder H. (2005) Skiing with carving skis – what is new?. In: Bacharach D. (Seifert J. (Eds.) *Abstract Book of 3rd International Congress on Skiing and Science*, p. 1. St.Cloud State University, St. Cloud.
Müller E. and Schwameder H. (2003) Biomechanical aspects of new techniques in alpine skiing and ski-jumping. *Journal of Sport Sciences*, 21, 679–692.
Nachbauer, W. (1986) *Fahrlinie und vertikale Bodenreaktionskraft bei Riesentorlauf und Torlauf.* Eigenverlag, Innsbruck.
Nachbauer W. and Rauch A. (1991) Biomechanische Analysen der Torlauf- und Riesentorlauftechnik. In: Fetz F. and Müller E. (Eds.) *Biomechanik der Sportarten – Biomechanik des alpinen Skilaufs*, p. 50–100. Thieme, Stuttgart.
Nachbauer W. and Kaps P. (1995) Crutiate ligament forces during landing in downhill skiing. In: Häkkinen K., Keskinen K.L., Komi P.V. and Mero A. (Eds.) *Proceedings of the XVth Congress of the International Society of Biomechanics*, pp. 654–655. LIKES, Jyväskylä.
Niessen W. and Müller E. (1999) Carving – biomechanische Aspekte zur Verwendung stark tailierter Skier und erhöhter Standflächen im alpinen Skisport. *Leistungssport*, 29/1, 39–44.
Nigg B.M., Neukomm P.A. and Lüthy S. (1977) Die Belastung des menschlichen Bewegungsapparates beim Schifahren. In: F. Fetz (Ed.) *Zur Biomechanik des Schilaufs*, pp. 80–89. Inn-Verlag, Innsbruck.
Pozzo R., Canclini A, Cotelli C. and Baroni G. (2005) 3D kinematics and kineticanalysis of G-Slalom in elite skiers at Val Badia World Cup race in 2002. In: Müller E., Bachard D. and Kliggs R. (Eds.) *Skiing and Science III*, pp. 125–135. Meyer & Meyer, Oxford.
Raschner C., Schiefermüller C., Zallinger G., Hofer E., Brunner F. and Müller E. (2001) Carving turns versus traditional parallel turns a comparative biomechanical analysis. In: Müller E., Roithner R., Niessen W., Raschner C. and Schwameder H. (Eds.) *Abstract Book of 2nd International Congress on Skiing and Science*, pp. 56–57. Eigenverlag, Salzburg.
Senner V., Lehner S., Wallrapp W. and Schaff P. (2000) The boot induced ACL rupture in alpine skiing: current knowledge and feasible solutions. In: In: Müller E., Roithner R., Niessen W., Raschner C. and Schwameder H. (Eds.) *Abstract Book of 2nd International Congress on Skiing and Science*, pp. 22–23. Eigenverlag, Salzburg.
Supej M., Kugovnik O. and Nemec B. (2003) Kinematic determination of the beginning of a ski turn. *Kinesiologia Slovenica*, 9/1, 11–17.

FUSION MOTION CAPTURE: CAN TECHNOLOGY BE USED TO OPTIMISE ALPINE SKI RACING TECHNIQUE?

M. BRODIE, A. WALMSLEY & W. PAGE

Institute of Food Nutrition and Human Health, Massey University, Wellington, New Zealand

Fusion Motion Capture has been used to capture 3D kinetics and kinematics of alpine ski racing. This research has overcome the technological barriers associated with athlete performance monitoring in an alpine environment. The biomechanical analysis of a New Zealand Alpine Ski Racing team member negotiating a ten gate giant slalom course over 300 meters in length has been undertaken. Results of the analysis may provide useful design parameters to ski equipment engineers and feedback to the athletes including; limb dynamics, Centre of Mass (CoM) trajectory, CoM velocity, and external forces through augmented reality animations. In-depth analysis of the changes in net joint torques with changes in athlete posture may be useful for the coaching of athlete specific technique changes to improve performance and reduce injury potential. In addition it is possible to extract key performance indicators about the athlete's physical and physiological limits such as his mean coefficient of wind drag, and his maximum inclination angle while turning which in the future may be used to optimise an athlete's race strategy.

1 Introduction

Biomechanical analysis of alpine ski racing is difficult due to the technological barriers associated with the resolution and accuracy of 3D video analysis through large volumes. Because an improvement of as little as 100th of a second between gates is significant to race outcome, the performance enhancement of an elite athletes may involve technique adjustments that are beyond the scope and resolution of video based systems. Therefore most research to date has focused on the analysis of a short turn sequence through two or three gates representing only part of a race course. (Schiefermuller *et al.*, 2005; Supej *et al.*, 2005; Vodickova *et al.*, 2005).

The purpose of the project was to overcome the technological barriers associated with athlete performance monitoring in an alpine environment. Its success proves it is possible to capture the motion and dynamics of alpine ski racing through an entire ski run, in some cases over 1 km in length while maintaining high resolution. Previous work indicates that changes of less than 0.5° in local limb orientation can be tracked successfully (Brodie *et al.*, 2006b,c). In contrast contemporary 3D optical based systems would require many cameras to capture motion through such a large volume.

Fusion Motion Capture is a composite system utilising data from Inertial Measurement Units (IMUs), video, GPS, and an RS-Scan insole system to determine segmental and whole body kinematics and kinetics. The global motion of the subject is determined by fusing GPS data, the known location of check points (the course gates in skiing) and the double integral of Centre of Mass (CoM) acceleration. Previous research has used GPS measurements for downhill ski performance, (Ducret *et al.*, 2005) but we believe this is the first time GPS data has been fused with IMU data to obtain a continuous CoM trajectory.

Figure 1. Fusion Motion Capture Output, Giant Slalom.

The athlete's limb orientation is determined by thirteen IMU's attached to the athletes body segments. The IMU's contain 3 gyroscopes, 3 accelerometers, 3 magnetometers, and a thermometer in a 35 gram box about the size of a matchbox. The manufacturer supplied a Kalman filter algorithm which can be used to extract orientation information from the raw data; however it was found that the Kalman filter algorithm produced errors of over 20° in orientation for sustained athletic activity. Instead the author used a fusion integration algorithm suitable for measuring the athletic movements in skiing, (Brodie 2007).

In order to calculate net muscle torques around each joint centre a body model of the athlete is required. Athlete inertial parameters are obtained using the Biomechanical Man body model, (Brodie *et al.*, 2006a) constructed from 3D anthropometry using a custom built frame (see figure 2) and the scaled inertial parameters as suggested by Dumas and Reed, (Dumas *et al.*, 2006; Reed *et al.*, 1999). This system is required both to model the athlete's inertial parameters and to calibrate the attached IMUs. The local coordinate system of each IMU is mapped to the local coordinate system of the athlete's limb (to which the IMU is attached) in the calibration process.

2 Method

A member of the New Zealand national team completed three runs through a ten-gate giant slalom training course at Mt. Ruapehu Ski Area. The course was over 300 metres in length. The athlete's body segment kinematics, angular velocity and local acceleration were obtained from 13 IMUs attached to the following body segments; head, torso, pelvis, upper and lower arms, thighs, shanks, and ski boots. An RS-Scan pressure measurement system was used to determine plantar pressures. Video from a hand held digital camera was used as an external reference, and to confirm validity of the data.

The data were processed using Fusion Integration algorithms and the Biomechanical Man body model in MATLAB. The resulting data allow determination of the full kinematics and

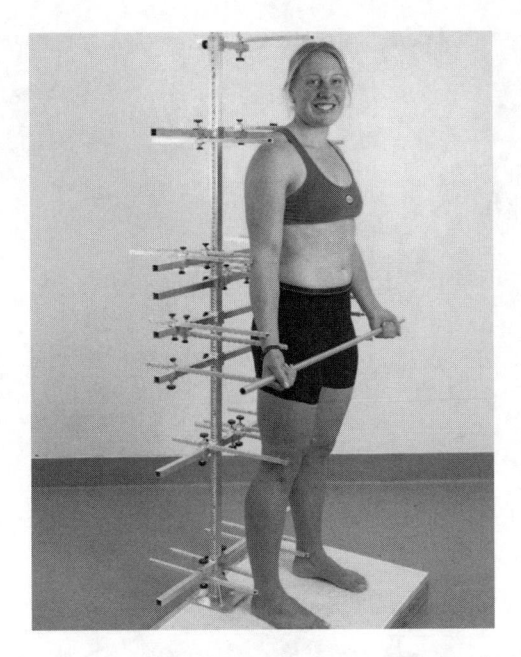

Figure 2. 3D Anthropometry and the Biomechanical Man.

kinetics of the athlete including; limb kinematics, ground reaction forces, CoM trajectory, ski orientation, ground reaction forces, net joint torques, and net joint powers.

In ski racing the dissipative forces of wind drag and ski-snow friction have a large effect on athlete performance. It was assumed that the dissipative forces could be calculated from the residual between measured ground reaction forces and the resultant external force acting on the athlete's CoM. The resultant external force was determined from the athlete's mass and CoM acceleration, which was determined from a weighted sum of the individual limb segment's acceleration measured by the IMUs. The magnitude and direction of the major component of the ground reaction force under each foot was calculated from the RS-Scan data and the measured orientation of the athlete's feet. The residual forces were assumed to be due to wind drag and ski-snow friction. Friction was modelled by equation 1, where F_{GRF} is the GRF component normal to the athlete's foot, $K_{Friction}$ is the coefficient of friction due to sliding resistance, and NV_{Foot} is the normalised velocity vector of the athlete's feet. Wind resistance was modelled by equation 2, where V is the COM velocity vector and K_{Drag} the lumped coefficient of wind drag.

$$\mathbf{F_{Friction}} = F_{GRF} K_{Friction} \mathbf{NV_{Foot}} \tag{1}$$

$$\mathbf{F_{Drag}} = -\mathbf{V}^2 K_{Drag} \tag{2}$$

3 Results

While the data from Mt. Ruapehu collected at the end of October 2006 are still being analysed, some preliminary results are available. For example, an animation of the Biomechanical

Figure 3. Diagnostic checking of the fusion motion capture system before a trial. Mt Ruapehu ski area.

Figure 4. The Near Optimum Turn. Net accelerating ground reaction forces out of the gate, (thick light vectors) followed by net retarding ground reaction forces (thick dark vectors) and both eccentric (light thin vectors) and concentric muscle torques (dark thin vectors) are visualized.

Man negotiating the giant slalom course from which a freeze-frame is shown in figure 1. The analysis will include both ground reaction forces, and net muscle torques such as the roller blading example of a near optimum slalom turn, figure 3. The animation may be used to present the complex information in an understandable way to the general audience.

A qualitative analysis has revealed that a combination of high ground reaction forces perpendicular to the CoM trajectory and high angular acceleration about the CoM trajectory results in a faster more direct route through the gates. These two parameters and the lower limb dynamics through each turn will be compared to determine the more efficient turns and discover the dynamics that produce them.

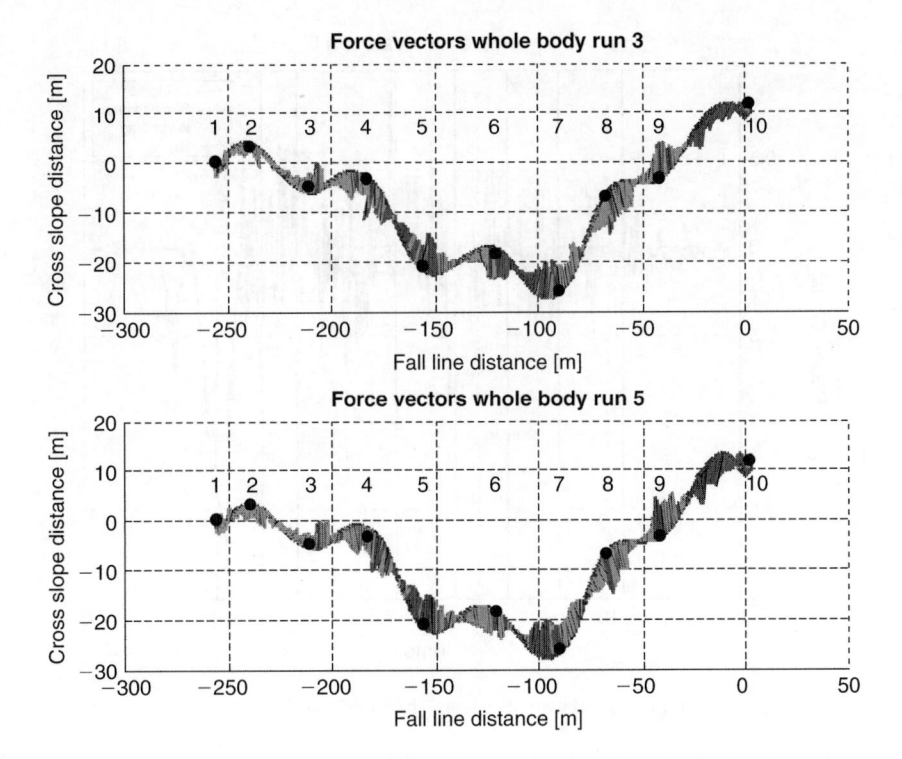

Figure 5. Force Vector Diagram, A comparison of Run 3 and Run 5.

Force vector diagrams showing the magnitude and direction of forces acting on the athletes CoM will be presented (see Figure 5). The resultant force vectors are colour coded; light for an accelerating force, and dark for a retarding force. A turn with less "dark" forces and more "light" forces is considered better. Run 5 is faster than run 3 the principal reason for this is that turn 6 of run 5 is better than turn 6 of run 3, with more accelerating forces.

Contributions of external forces to the resultant external force and athlete power will be presented, Figure 6. The power analysis shows that while gravity produces a net positive power, all other external forces including show friction, wind drag and ground reaction forces have a net negative power or slow the athlete.

4 Discussion and Conclusions

It is not yet possible to draw conclusions from the 2006 data until analysis is complete. However it is believed that the results will demonstrate the relative importance of ground reaction forces, ski-snow friction and wind drag at various stages through the course and turn. Contra intuitively it appears that while ground reaction forces are essential for changing athlete direction they do not produce a net increase in speed during downhill skiing. The progression of lower limb dynamics through each turn may also give insights into how alterations to an athlete's stance can reduce the dangerously high knee torques in ski racing, with minimal effect on performance. The measurement of joint torques and powers may be

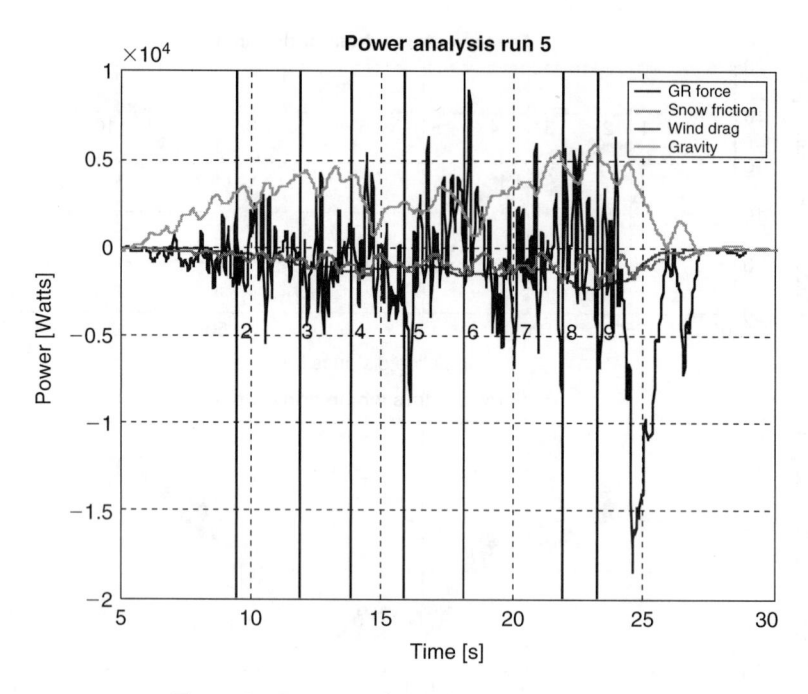

Figure 6. Power Analysis of Alpine Ski Racing.

useful for developing more specific physical training programs by indicating the dominant muscle groups and their mode of action during the turn.

Acknowledgements

The authors acknowledge the support of the New Zealand Academy of Sport Biomechanics Travel Scholarship, The Tussock Grove Hotel Ohakune, Massey University, SPARC, The Royal Society, Volkswagen NZ, and Mt Ruapehu. Thanks to Ben Griffin for his skiing expertise.

References

Brodie M., Walmsley A. and Page W. (2007) Fusion Integration: COM Trajectory from a Force Platform. *Journal of Applied Biomechanics*, accepted for publication.
Brodie M., Walmsley A., Thorpe R., Graham D., Page W. and Turner M. (2006a) 3D Anthropometry, the Biomechanical Man, and Fusion Motion Capture. *New Zealand Sports Medicine + Science Conference*, Poster Presentation.
Brodie M., Walmsley, A. and Page W. (2006b) 3D Dynamic Accuracy of Inertial Measurement Units Using Sensor Fusion. Unpublished, submitted for publication.
Brodie M., Walmsley A. and Page W. (2006c) 3D Static Accuracy of Inertial Measurement Units. Submitted for Publication.
Ducret S., Ribot P., Vargiolu R., Lawrence J. and Midol A. (2005) Analysis of downhill ski performance using GPS and ground force recordings. Science in Skiing 3. Meyer and Meyer Sport (UK) Ltd., 56–67.

Dumas R., Cheze J. and Verriest P. (2006) Adjustments to McConville *et al.* and Young *et al.* body segment inertial parameters. *Journal of Biomechanics*, in press.

Reed M.P., Manary M.A. and Schneider, L.W. (1999) Methods for measuring and representing automobile occupant posture. *SAE Technical Paper Series*, 199-01-0959.

Schiefermuller S., Lindinger S. and Muller E. (2005) The skier's centre of gravity as a reference point in movement analysis for difference designated systems. Science in Skiing 3. Meyer and Meyer Sport (UK) Ltd., 172–186.

Supej, M. Kugovnik O. and Nemec B. (2005) Advanced analysis of skiing based on 3D kinematic measurements. Science in Skiing 3. Meyer and Meyer Sport (UK) Ltd., 216–228.

Vodickova S., Lufinka A. and Zubeck T. (2005) The dynamographic and kinematographic method application for a short carving turn. Science in Skiing 3. Meyer and Meyer Sport (UK) Ltd., 247–257.

STUDY ON THE OPTIMIZATION OF A SNOWBOARD

Q. WU & S. GANGULY

Department of Mechanical Engineering & the Center for Nonlinear Dynamics and Control, Villanova University, Villanova, USA

In a recent study, Wu *et al.* (2006) have developed a new simplified mathematical model to describe the lift mechanics of downhill skiing and snowboarding, where the lift contributions due to both the transiently trapped air inside a snow layer and the solid phase (snow crystals) were determined for the first time, and a model for the stability and control of a ski/snowboard was developed. Wu *et al.*'s theory is applicable to the simple case where the planing surface is of constant width in the axial direction. However commercially available skis/snowboards have complex geometries and variable width in order to reduce the weight, improve performance while maintaining the strength and rigidity. In the current study, we shall extend Wu *et al.*'s theory to more complex planar shapes where the width of the planing surface changes with the axial location. We shall examine the performance of commercial snowboards based on our new model. The study presented herein and the previous skiing mechanics theory developed by Wu *et al.*, have laid the foundation for the optimization of a ski/snowboard from lift generation point of view.

1 Introduction

The phenomenon of downhill skiing or snowboarding, in its simplest form, refers to the motion of a human being sliding down an inclined plane on a porous medium (snow layer), as shown in Figure 1. Much of the classic literature treating the science of skiing and snowboarding focuses on the micron-thick water film that is formed on the underside of the ski or snowboard due to frictional heating (Colbeck, 1991; 1992; 1994a; 1994b; 1995). Recently, Feng and Weinbaum (2000) developed a new lubrication theory for highly compressible porous media which demonstrated that the excess pore pressure generated by a planing surface (ski/snowboard) moving on a compressible porous layer (snow) scales as $\alpha^2 = h^2/K$, where h is the layer thickness and K is the Darcy permeability; and that α is of order 10^2 or larger for humans skiing, thus, the lift forces generated can be four or more orders of magnitude greater than classical lubrication theory. The huge enhancement in the lift arises from the fact that as the porous medium (snow) compresses there is a dramatic increase in the lubrication pressure because of the marked increase in the hydraulic resistance that the fluid (air) encounters as it tries to escape from the confining boundaries through the compressed porous layer. At typical skiing velocities of 10 to 30 m/s the duration of contact of a ski or snowboard with the snow will vary from 0.05 to 0.2 s depending on the length of the planing surface and its speed. More recently, Wu *et al.* (2004a; 2004b; 2005b; 2005c) developed a novel experimental and theoretical approach to examine the excess pore pressure generation inside a snow layer on the time scale of skiing or snowboarding, which for the first time, qualitatively verified Feng and Weinbaum's theory. The fundamental insights gained from these studies on the lift generation in porous media provide a new perspective for understanding the phenomenon of skiing and have further led to

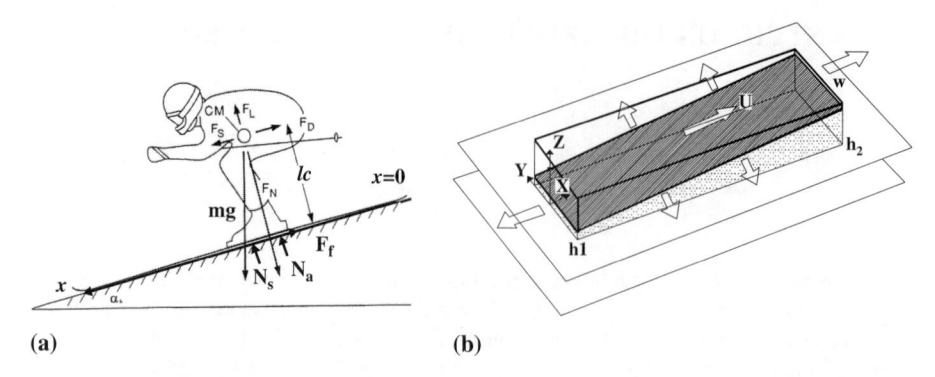

Figure 1. (a) Forces upon a skier as he/she descends down the fall line. (b) Schematic illustration of a snowboard or ski compressing a layer of snow powder.

the first realistic model for the lift mechanics of downhill skiing and snowboarding developed by Wu *et al.* (2006a; 2006b). This new skiing mechanics theory incorporates lift contribution from both the transiently trapped air and the compressed ice crystals. It captures the key physics of the stability and control during skiing or snowboarding and realistically predicts the performance of skiing or snowboarding as a function of the skier's velocity, the sliding friction between the planing surface and snow, various snow types as well as the geometry of a ski or snowboard.

Wu *et al.*'s skiing mechanics theory is applicable for the simple case where the planing surface is of constant width in the axial direction. However commercially available skis/snowboards have complex geometries and variable width in order to reduce the weight, improve performance while maintaining the strength and rigidity of the planing surface. In the current study, we shall extend Wu *et al.*'s theory to more complex shapes where the width of the planing surface changes with the axial location. A modified mathematical model will be developed where a width factor, $f(x)$, which characterizes the variation of width from the leading to the trailing edge, is introduced. We shall examine the performance of several commercial snowboards with various width factors based on our new model and optimize the shape of a given snowboard by finding the ideal width factor from lift generation point of view.

2 Formulation

For a skier/snowboarder gliding with velocity $U(U_x, U_y, U_z)$ over a snow layer as shown in Figure 1, the sudden compaction of the snow leads to the generation of pore air pressure inside the compressed layer, $\mathbf{N_a}$, as well as the solid phase lifting force from the ice crystals, $\mathbf{N_s}$. A general equation for the pore pressure distribution beneath the planing surface was developed in Wu *et al.* (2006a, 2006b),

$$\nabla^2 P = -\mu A/K - (1/K)\nabla P \cdot \nabla K, \qquad (1)$$

where μ is the viscosity of the air, A is the vertical velocity gradient, $A = -U_z(x, y, h)/h$. For the case where there is no lateral tilt, $h = h(x)$, $K = K(x)$, $A = (U/h)dh/dx$ where

Study on the Optimization of a Snowboard

$U = U_x$, and the pressure distribution in the y direction is parabolic (Wu et al., 2005a), one obtained the centerline pressure distribution beneath a ski or snowboard,

$$\frac{d^2 P_c(x)}{dx^2} + \frac{1}{K(x)}\frac{dK(x)}{dx}\frac{dP_c(x)}{dx} - \frac{8}{w(x)^2}[P_c(x) - P_0] + \frac{\mu A(x)}{K(x)} = 0, \quad (2)$$

where $w(x)$ is the local width of the planing surface, $P_c(x)$ is the centerline pressure corresponding to the cross section at the location x, P_0 is the pore pressure at the edge of the planing surface which is very close to the atmospheric pressure, and the local Darcy permeability is expressed as $K(x) = 0.077\exp[(1 - (1-1/k) \cdot (1- x/L)^{-1})\ln(K_2/0.077d^2)]d^2$, where L is the length of the ski/snowboard; d is the mean diameter of the ice crystals; $k = h_2/h_1$, h_2 and h_1 are the local thickness of the snow layer at the leading ($x = L$) and trailing ($x = 0$) edges, respectively; and K_2 is the permeability beneath the leading edge, $K_2 = 0.077\exp[(h_0/h_2)\ln(K_0/0.077\,d^2)]d^2$, where h_0 and K_0 are the undeformed thickness and permeability of the snow layer, respectively.

Wu et al. (2006a, 2006b) have examined the case when the width of the planing surface is a constant along the axial direction. However, commercially available skis/snowboards have complex geometries and variable width, as shown in Figure 2. If the nose width is W, one defines the width factor of a snowboard as $f(x) = w(x)/W$. Introducing dimensionless variables, $p' = (P - P_0)/P_0$, $p'_c = (P_c - P_0)/P_0$, $x' = x/L$, $y' = y/(W/2)$, $h' = h/h_2$ and $K' = K/K_2$, we obtain a dimensionless equation for $p'_c(x')$:

$$\frac{d^2 p'_c(x')}{dx'^2} + \frac{1}{K'(x')}\frac{dK'(x')}{dx'}\frac{dp'_c(x')}{dx'} - \frac{8}{\varepsilon f(x')^2}p'_c(x') + \frac{\theta_L}{K'(x')h'(x')} = 0, \quad (3)$$

where $\varepsilon = (W/L)^2$, $\theta_L = (\mu/P_0)(U/h_2)(dh/dx)(L^2/K_2)$. Eq. (3), subject to the boundary conditions $p'_c(0) = p'_c(1) = 0$, can be solved numerically and the 2-D pressure distribution beneath a snowboard/ski surface is determined, $p'(x', y') = [1 - (y'/f(x'))^2]p'_c(x')$. The average dimensionless pressure generated by the trapped air inside the compressed snow layer is obtained by integrating the pore pressure over the entire surface,

$$P'_{ava} = \frac{2}{3}\left[\int_0^1 p'_c(x')f(x')dx' \Big/ \int_0^1 f(x')dx'\right]. \quad (4)$$

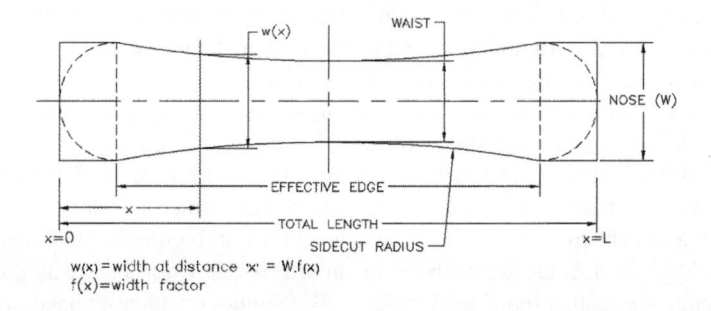

Figure 2. Geometric snowboard dimensions described using common nomenclature (top view).

The solid phase (ice crystals) lift force is obtained from the quasi-static experiments performed by Wu et al. (2005b; 2005c). This force in its dimensionless form is given by

$$p'_s(x') = P_{solid}(x')/P_0 = (P_{mg}/P_0) \cdot g(0.94[1 - \lambda/k - (\lambda - \lambda/k)x']/(\phi_0 - 0.06)), \quad (5)$$

where $P_{solid}(x')$ is the local solid phase pressure, $P_{mg} = (\mathbf{mg}\cos\alpha_h)/A_s$ where α_h is the angle of the inclined slope, A_s is the area of the snowboard, g is the empirical relation obtained in (Wu et al., 2005b; 2005c), ϕ_0 is the undeformed porosity of the snow layer, and $\lambda = h_2/h_0$ is the compression ratio at the leading edge. The average pressure generated by the solid phase is then given by:

$$P'_{avs} = \left(\int_0^L P_{solid}(x)w(x)dx / \int_0^L w(x)dx\right)/P_0 = \int_0^1 p'_s(x')f(x')dx'/\int_0^1 f(x')dx'. \quad (6)$$

Figure 1a shows the representative forces acting on a skier gliding on an inclined snow slope. The weight **mg** is resolved into two perpendicular components, $\mathbf{F_S}$ parallel to the slope, and $\mathbf{F_N}$ normal to the slope. $\mathbf{F_N}$ is to be balanced by the summation of lift forces generated by the pore air pressure as well as the solid phase force, $\mathbf{F_N} = \mathbf{mg}\cos\alpha_h = \mathbf{N_a} + \mathbf{N_s}$. If one defines $P'_{avload} = P_{mg}/P_0$, this normal force balance is equivalent to $1 = f_{air} + f_{solid}$, where $f_{air} = P'_{ava}/P'_{avload}$ and $f_{solid} = P'_{avs}/P'_{avload}$. Similarly, the sum of all torques about the center of mass (CM) must be zero too, that is, $\mathbf{N_a} \cdot (x_c - x_a) + \mathbf{F_f} \cdot l_c = \mathbf{N_s} \cdot (x_s - x_c)$, where $\mathbf{F_f}$ is the snow friction force, $\mathbf{F_f} = \eta \mathbf{N_s}$ where η is the coefficient of friction, l_c is the normal distance of the CM from the ski surface, x_c, x_a and x_s are the x coordinates of CM, center of $\mathbf{N_a}$ and center of $\mathbf{N_s}$, respectively. When a skier glides down a slope at velocity U, over an undeformed snow layer of thickness h_0 and Darcy permeability K_0, without changing the location of the skier's CM, one has to adjust the tilt angle γ (or the compression ratio from the leading to trailing edge, $k = h_2/h_1$) as well as the compression ratio at the leading edge, $\lambda = h_2/h_0$, to satisfy the force and moment balance equations.

3 Results and Discussion

For a simple rectangular ski or snowboard, the lift distribution between the trapped air and the ice crystals as a function of the planing surface dimension W/L, the velocity U, the frictional coefficient η, the location of CM (x'_c, l_c), and the properties of snow (K_0, ϕ_0, d), has been extensively studied in Wu et al. (2006a; 2006b). In the current study we shall investigate the effect of width variation in the axial direction on the performance of a snowboard. The parameters chosen are as follows: m = 80 kg, $\alpha_h = 15°$, $\eta = 0.04$, $U = 20$ m/s, $K_0 = 5.0 \times 10^{-10}$ m², $\phi_0 = 0.6$, $d = 0.42$ mm, $h_0 = 10$ cm, $x'_c = x_c/L = 0.45$, $l_c = 1.2$ m.

Table 1 lists five commercially available snowboards named A, B, C, D and E. The snowboards differ in their characteristic dimensions (refer to Figure 2 for snowboard geometry). We have numerically solved Eq. (3) and obtained the centerline pore pressure distribution beneath a snowboard surface. The results are shown in Figure 3a and the values for λ, k, f_{air}, f_{solid}, P'_{avload} and A_S are summarized in Table 1. As shown in Figure 3a pore air pressure is generated beneath a snowboard surface as one glides on a fine-grained, wind-packed snow layer at relatively high speed. And it is evident from Table 1 that approximately 50%

Study on the Optimization of a Snowboard

Table 1. Dimensions of five commercially available snowboards with corresponding performance parameters.

Snow boards	Total Length	Effective Edge Length	Nose Length	Waist Length	λ	k	f_{air}	f_{solid}	P'_{avload}	A_S
A	1.38	1.175	0.308	0.264	0.665	1.050	0.519	0.481	0.0191	0.3910
B	1.42	1.175	0.308	0.264	0.671	1.053	0.532	0.468	0.0185	0.4033
C	1.50	1.170	0.290	0.240	0.653	1.052	0.493	0.507	0.0189	0.3964
D	1.56	1.190	0.294	0.250	0.675	1.060	0.534	0.466	0.0176	0.4241
E	1.62	1.264	0.315	0.268	0.692	1.059	0.564	0.436	0.0159	0.4712

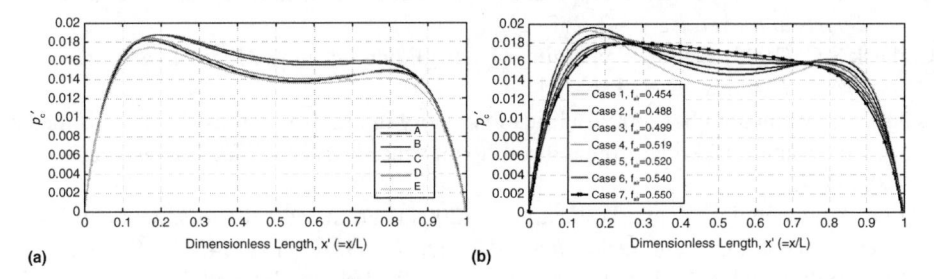

Figure 3. Centerline pore pressure distribution for (a). commercially available snowboards A, B, C, D and E. (b). Snowboard A with various waist length.

($f_{air} \approx 50\%$) of the total lift force is generated by the trapped air. There are two pressure peaks on the length of the snowboard, occurring near the nose section where the width of the snowboard is the maximum. This is due to the fact that the increased width of the snowboard at the nose prolongs the outflow of the trapped air in the lateral direction, and hence on the time scale that the snowboard is in contact with the snow, higher pore pressure is generated at these two sections. Based on similar reasoning a narrower section of the snowboard in the vicinity of the waist section that is near the center of the snowboard entails a rapid pressure relaxation and hence a drop in the pore air pressure built-up. Evidently the pore pressure reaches maximum near the trailing edge where maximum snow compression occurs. In Figure 3a, the magnitude difference in p'_c for the five listed snowboards does not indicate which one is more efficient in trapping air inside snow due to the fact that, they have different loading area A_S and thus different average loading P'_{avload}. However, the lift contribution from the air to the total lift, f_{air} which is described in Table 1, does provide us a criteria for evaluating the performance of a snowboard. For a smooth air cushioned glide one wants a high value of f_{air}. Obviously, snowboard E is better in pore pressure generation.

In order to optimize the shape of a snowboard for ideal performance, one wants to increase its value of f_{air}. In the current study, we analyze the effect of width factor on the pore pressure generation. Snowboard A is chosen and its waist dimension is varied while all other dimensions are kept as constants. Case 1 to 7 in Figure 3b correspond to the waist length of 0.22 m, 0.235 m, 0.25 m, 0.264 m, 0.275 m, 0.29 m and 0.3 m, respectively. It is evident that f_{air} increases from Case 1 through 7 due to increasing waist width and hence more trapped air inside snow.

In summary, we have developed a new realistic model for lift mechanics of downhill snowboarding, which incorporates the shape variation effect in the lift forces generation. This study is an important practical application and extension of the skiing mechanics theory developed by Wu et al. (2006a; 2006b) and would be invaluable for future snowboard design.

References

Colbeck S.C. and Warren, G.C. (1991) The thermal response of downhill skis, *Journal of Glaciology*, 37/126, 228–235.

Colbeck S.C. (1994a) A review of the friction of snow skis, *Journal of Sports Sciences*, 12, 285–295.

Colbeck S.C. (1994b) Bottom temperatures of skating ski on snow. *Medicine and Science in Sports and Exercise*, 26/2, 258–262.

Colbeck S.C. (1995) Electrical charging of skis gliding on snow, *Medicine and Science in sports and exercise*, 271, 136–141.

Feng J. and Weinbaum S. (2000) Lubrication theory in highly compressible porous media: the mechanics of skiing, from red cells to humans, *Journal of Fluid Mech*chanics, 422, 282–317.

Wu Q., Andreopoulos, Y. and Weinbaum S. (2004a) From red cells to snow-boarding: A new concept for a train track, *Physical Review Letters*, 93/19, 194501–194504.

Wu Q., Andreopoulos, Y. and Weinbaum, S. (2004b) Lessons learned from the exquisite design of the endothelial surface glycocalyx and their amazing application. In: Collins M.W. and Brebbia C.A. (Eds.) *Design and Nature II*. WIT press, UK.

Wu Q., Weinbaum S., and Andreopoulos, Y. (2005a) Stagnation point flow in a porous medium, *Chemical Engineering Sciences*, 60, 123–134.

Wu Q., Andreopoulos Y., Xanthos S. and Weinbaum S. (2005b) Dynamic compression of highly compressible porous media with application to snow compaction, *Journal of Fluid Mechanics*, 542, 281–304.

Wu Q. (2005c) Lift generation in soft porous media; from red cells to skiing to a new concept for a train track. Ph.D. Diss., City University of New York, USA.

Wu Q., Igci Y., Andreopoulos Y. and Weinbaum S. (2006a) Lift mechanics of downhill skiing and snowboarding, *Medicine and Science in Sports and Exercises* 38/6, 1132–1146.

Wu Q., Andreopoulos Y. and Weinbaum S. (2006b) Riding on Air: A new theory for lift mechanics of downhill skiing and snowboarding. In: Moritz E.F. and Haake S. (Eds.) *The Engineering of Sport 6*. Springer, New York.

COMPARATIVE EXPERIMENTS FOR SNOWBOARD VIBRATION CHARACTERISTICS

S. KAJIWARA, D. TANIGUCHI, A. NAGAMATSU, M. IWAHARA & A. KONDOU
Hosei University, Graduate School in Mechanical Engineering, Tokyo, Japan

For the snowboard made by an Austrian company and the snowboards made by a Japanese company, we propose to apply Experimental Modal Analysis to compare these numerical vibration characteristics. In the first excitation experiment, when the snowboard is supported by strapping both snowboarders' feet into the snowboard, natural frequency and dumping ratio for Japanese snowboard was increased than the Austrian snowboard. In the Second experiment, when a center part of a snowboard is supported by aluminum boards and bolts, dumping ratio for Austrian snowboard is 0.1% larger than other boards. The result shows that the Austrian snowboard absorbs vibration more than Japanese snowboards. Meanwhile, bending experiments made it possible to illustrate numerical stiffness for snowboards and their shape.

1 Introduction

owboarding is one of the most popular winter sports in the world. When the snowboarder is gliding, the board vibrates (Glenne *et al.*, 1999). For the professional snowboarders and amateur snowboarders, it is a serious problem that the vibration leads to lose player's edge control and gliding control on snow surface. In order to make the vibration characteristics clear, many studies on the problem were published by widely numerical and experimental methods. However, determining the vibration characteristics is difficult because the cause of vibration consists with many non-linear and dynamic systems such as snow surface, binding, boots and snowboarders.

By contrast, recently, many researchers and engineers started to apply modal analysis to estimate vibration behavior and make it clear. Some analyzers specialized in modal analysis are developed by instruments companies. Modal analysis has two classes, Experimental Modal Analysis and Theoretical Modal Analysis (Nagamatsu, 1993). In this paper, we propose to apply Experimental Modal Analysis to determine the vibration characteristics for snowboards.

2 Excitation Experiment

Measured object are one snowboard which is made by Austrian company, and four snowboards which are made by Japanese company. At first, when the snowboard is excited by an impact hammer at an excitation point, single-axis accelerometer measures at each Z axial measuring points. One measuring point using accelerometer was fixed on the snowboard. Sixteen excitation points are set on it. Measured signal from an accelerometer and output signal from an impact hamper are acquired by FFT analyzer. The excitation experiment by an impact hammer is conducted five times and the measuring data are averaged out.

In the first experiment, the snowboard is supported freely using rubber band. In the second experiment, to reproduce actual vibration while he is gliding on surface, the snowboard is supported by strapping both snowboarder's feet into the snowboard. The snowboard is excited by his bouncing as a natural excitation source. In the third, a tail of a snowboard is excited by an impact hammer where a center part of it is supported by aluminum boards and bolts.

Figure 1 shows sixteen excitation points and one measuring point. In Figure 1, from point No 1 to point No 16 are Z axial excitation points by an impulse hummer. A point No 17 is a measuring point which is located at a backside of No 1. Figure 2 shows a first mode shape for an Austrian snowboard by an experiment. The shape is similar to a first bending mode in one degree of freedom.

3 Result for Experimental Modal Analysis

Table 1 shows natural frequency and damping ratio by Experimental Modal Analysis. "Rubber" means a free support in which the snowboard is supported freely using rubber

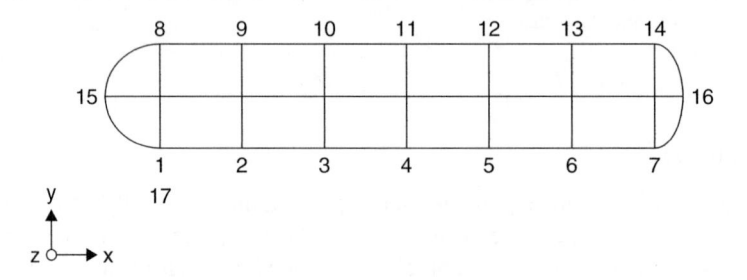

Figure 1. Measuring points and Excitation point.

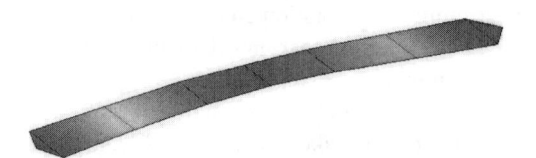

Figure 2. First Mode Shape by Experimental modal Analysis.

Table 1. Natural Frequency.

	Natural Frequency (Hz)		Damping Ratio (%)	
	Rubber	Human	Rubber	Human
A-1	15.87	18.61	0.42	1.02
O-2	16.36	18.06	0.52	3.00
O-3	16.04	16.89	0.43	3.04
O-5	15.99	18.39	0.40	1.20
O-6	16.29	18.61	0.69	1.64

band. "Human" means a ground support with the snowboarder's weight on the board. A-1 is made by an Austrian company. O-2, O-3, O-5 and O-6 are made by a Japanese company.

Regarding natural frequency, "Rubber" is not quite different from "Human". By contrast, damping ratios for "Human" varied in each case of snowboards. This experiment cannot prove that A-1 snowboard hardly vibrates in all snowboards.

Comparing "Rubber" with "Human", all "Human" of natural frequencies are more 2 Hz than "Rubber".

Dumping ratios of "Rubber" are range from 2.5 to 6 times of them of "Human". Therefore, Snow board vibration behavior is influenced by snowboarder and binding.

4 Bending experiment

Stiffness could be considered as dynamic characteristics when vibration frequency is zero. In order to compare each stiffness for the boards by the experiment, we conducted bending experiment based on JIS S7019-1987(expired in 1997). How to weight on the board is shown by Figure 3. For Load-deflection curve graph, the experimental conditions for a load are 10 kg, 15 kg, 20 kg, 25 kg and 30 kg.

5 Results for bending experiments

Table 2 shows dimensions of each snowboard. These do not have uniform width and thickness, hence Table 2 shows mean values of eleven measuring points on each snowboards. Figure 3 shows load-deflection curve based on five types of loads. Figure 4 shows the

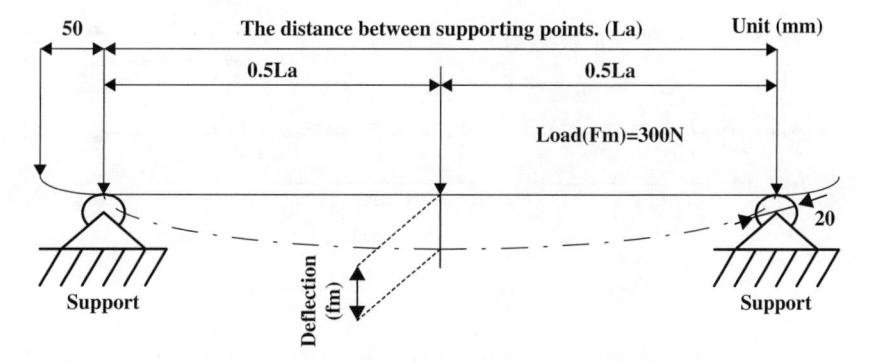

Figure 3. Ski testing method for JIS S7019-1987.

Table 2. Dimensions of snowboards.

	Width (mm)	Thickness (mm)	Weight (kg)
A-1	209.81	9.905	3.544
O-2	207.37	10.276	3.674
O-3	205.73	10.136	3.66
O-5	207.10	10.294	3.88
O-6	206.55	10.245	3.84

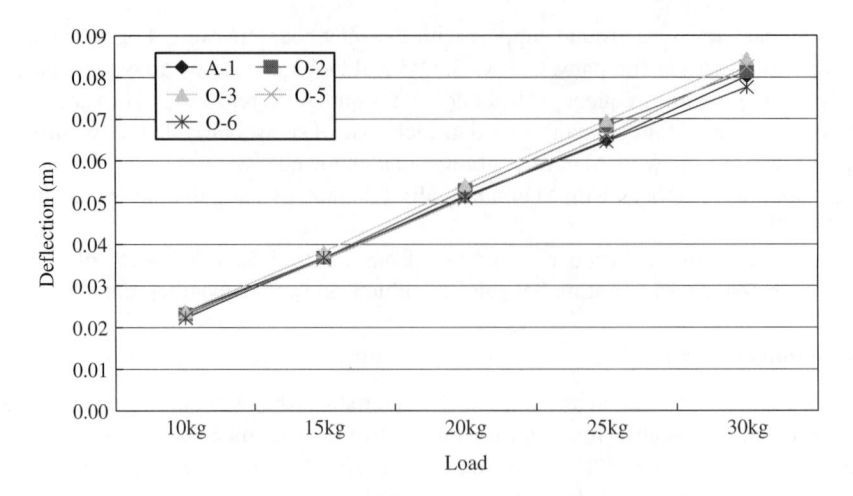

Figure 4. Load-deflection curve graph.

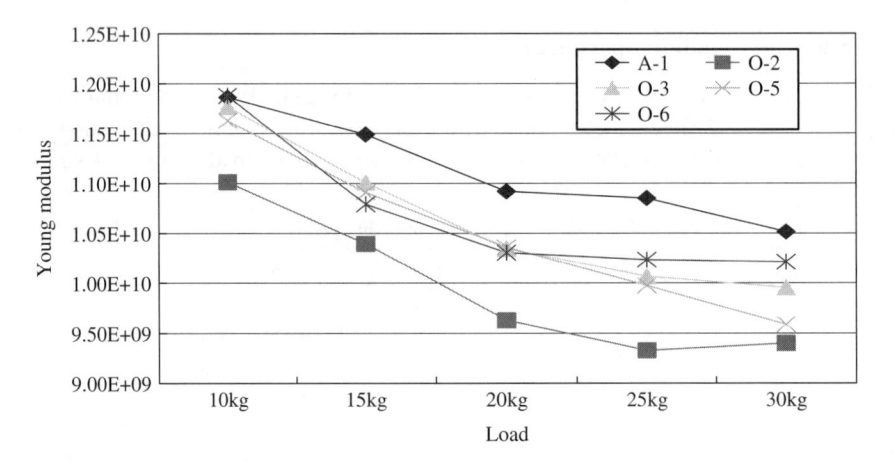

Figure 5. Deflection-Young modulus graph.

deflection – Young modulus graph which was calculated by deflections and a cross-section second coefficient.

Regarding each dimension, A-1 board is wider and thicker than other boards. For the deflection of boards, no changes exist. However, young modulus for A-I is biggest in all of boards and its stiffness per unit area of cross section is stronger than other board stiffness. The deflection of the snowboard causes the shape variation and the change of bending stiffness EI (EI = Young modulus* second moment of cross-section). Therefore, the young modulus which is the eigenvalue for the snowboard was changed.

6 Vibration characteristics for the tail of snowboard

In this study, we performed experimental modal analysis for tails or snowboards because the tail of it vibrates widely when the snowboarder is gliding. Each center of the snowboards are supported by aluminum boards and bolts. By supporting the centers, only tails can be vibrated. The experimental methodology is same as it of 2.Excitation Experiment.

7 Vibration characteristic comparison for tails of snowboards

Figure 2 shows each first bending mode. Each natural frequency is nearly 10 Hz. However, regarding dumping ratio, A-1 is 0.1% larger than other boards. Therefore, it is hardly likely that the tail of A-1 snowboard vibrates in all snowboards.

8 Conclusions

1. Experimental modal analysis made it possible to determine vibration characteristics in each snowboards.
2. The vibration is influenced great deal by nonlinear systems such as player's bodies.
3. By bending test, the shape of A-1 has different in each position on the board.
4. Regarding to result for tail of snowboards, A-1 has the biggest dumping ratio.
5. As a future issue, we focus on the consideration about vibration characteristics in each parts of the snowboard, and experimental mode analysis while a snowboarder is gliding.
   Future issues
6. Focus on the consideration in each parts of snowboards
7. Experimental Modal Analysis while a snowboarder is gliding.

References

Nagamatsu A. (1993) *Modal Analysis*. Corona Publishing Co. Ltd., Tokyo.
JIS S 7019 (1987) Testing methods for alpine skis, JIS (Japanese Standards Association).
Glenne B., Derocco A. and Foss G. (1999) Ski and Snowboard Vibration. *Sound and Vibration*, 33/1, 30–33.

AUTOMATED INERTIAL FEEDBACK FOR HALF-PIPE SNOWBOARD COMPETITION AND THE COMMUNITY PERCEPTION

J.W. HARDING[1,3,4], K. TOOHEY[2], D.T. MARTIN[1], C. MACKINTOSH[1],
A.M. LINDH[1] & D.A. JAMES[3]

[1]*Applied Sport Research Centre, Australian Institute of Sport, Australia*
[2]*Department of Tourism, Leisure, Hotel & Sport Management, Griffith University, Australia*
[3]*Centre for Wireless Monitoring and Applications, Griffith University, Australia*
[4]*Olympic Winter Institute of Australia, Australia*

No scientific research has yet targeted the athletic performance aspects or subjective judging protocols associated with elite half-pipe snowboard competition. Recently however, sport scientists from the Australian Institute of Sport (AIS) initiated a video based analysis of Key Performance Variables (KPVs) associated with elite half-pipe snowboard competition. The development of a preliminary automated feedback system based upon Micro-ElectroMechanical Systems (MEMS) sensors such as tri-axial accelerometers and tri-axial rate gyroscopes, designed to calculate objective information on these sport specific KPVs was initiated in parallel. Although preliminary, the results may provide practical benefit for elite half-pipe snowboard training and current subjective judging protocols. In light of theorised implications, this paper investigated the perception and possible social impact of these concepts on the practice community. Data was collected via semi-structured, open ended interviews with nine subjects (six athletes, two coaches, and one judge) currently involved in elite half-pipe snowboard competition. This study revealed 6 dimensions and 20 sub-dimensions relating to the practice community's perceptions of 3 major themes that emerged during interviews. The themes included: 1) State of the current subjective judging system, 2) Automated feedback and objective judging system, and 3) Future direction of the sport. There was dominant negative perception of a proposed automated judging concept based solely on objective information unless the system integrates with the current subjective judging protocol and continues to allow athletic freedom of expression and the capacity for athletes to showcase individual style and flair in elite competition. The results of this study provide the practice community an initial public forum to describe its perceptions to future automated judging concepts, nominating them to be the primary determinants of change, technological or otherwise, within their sporting discipline.

1 Introduction

Key Performance Variables (KPVs). Unlike more traditional sports such as swimming, running, and cycling, very little is known about the characteristics of half-pipe snowboarding's elite athletes or the physiological and kinematical demands of competition. Sport scientists from the Australian Institute of Sport (AIS) have recently undertaken a preliminary video based analysis of the effect sport specific key performance variables (KPVs) such as total air-time (TAT) and average degree of rotation (ADR), have on scores during elite half-pipe snowboard competition. This analysis has shown that when TAT and ADR are combined (multiple linear regression), they show a strong correlation with an athlete's subjectively judged score (r = 0.70), and subsequently account for approximately 50% of the shared variance associated with the overall score ($r^2 = 0.49$, SEE = 4.04) during Fédération Internationale de Ski (FIS) Half-Pipe Snowboard Competition Finals (n = 2) in Bardonecchia Italy, February 2005 (Harding et al., 2005; Harding et al., 2006).

Automated KPV Feedback System. The dilemma with video based analysis of sport specific KPVs is the labour intensive nature of calculation and the associated time-delay in information feedback. AIS sport scientists have therefore recently developed a method of calculating KPV information using Micro-electrochemical systems (MEMS) based tri-axial accelerometers ($ADXL_{XXX}$ series, Analogue Devices) and tri-axial rate gyroscopes (ADRXS300 gyros, Analogue Devices). It is now possible to calculate air-time during half-pipe snowboarding using 100 Hz tri-axial accelerometer data and mathematical pattern recognition algorithms. This preliminary method shows a strong correlation (r = 0.95) with a reference standard for air-time calculation (video analysis) and accounts for 90% ($r^2 = 0.90$, SEE = 0.05 s, Mean Bias = 0.01 s) of variation inherent in the air-times calculated by the reference standard. Reliability however is questionable, only correctly detecting and calculating accurate air-time for 73% (118 out of 161 aerial acrobatic manouevres) performed by elite half-pipe snowboard athletes during this study (Harding et al., 2005; Harding, 2006; Harding et al., 2006).

Proposed Automated Judging. Half-pipe snowboard competition is judged purely subjectively and recent opinion has called for a system of judging to be introduced that is based upon more accurate, reliable and stringent measures, without stifling athletic freedom of expression. Whilst AIS research presents a preliminary analysis on the effect of KPVs on athlete's scores, it also suggests that accurate and reliable quantification of these sport specific variables may prove beneficial in enhancing subjective judging protocols currently used in elite half-pipe snowboard competition. This paper proposes that a MEMS based feedback system could assist judges by providing accurate, electronic "memory boards" (a record of an athlete's run characteristics currently written by hand) by quantifying and displaying objective information on sport specific KPVs such as TAT and ADR (information currently unavailable to elite-level judges). This will enable judges to focus solely on the execution and overall composition of aerial acrobatics incorporated into a competition run.

Social Impact of Technology. The proposed introduction of a competition judging system based solely or partially upon objective information into half-pipe snowboarding (a sport habitually focused on athletic individuality and freedom of expression) will no doubt provide sport scientists (as implementers) and the practice community (those affected) with challenges. Perhaps the most significant aspect of change in sport is that any such action can dictate the future of a sport in a way that makes reversing such changes very difficult (Miah, 2000). Technology often has unintended consequences, effecting change beyond its original purpose (Tenner, 1996; Miah, 2000). As argued by Miah (2000), changes, technological or otherwise, should be preceded with some discussion about what future is sought for a specific sport and thus, where limits might be drawn on the changes. Defining who should determine the nature of a sport ought not to be a difficult issue. It seems imperative to ensure practicing communities are allowed to articulate their interests in forums that convey influence (Morgan, 1994). This paper is a result of action based research involving a progressive partnership ideology with the practice community. It has allowed the snowboarding community a forum to describe its perceptions on the impact of future automated judging concepts and possible technological change within their sporting discipline.

2 Methods

Population and Sampling. The population for this study was selected by theoretical sampling and included nine subjects (six elite athletes, two national coaches, and one elite competition judge). Subjects' level of experience in elite half-pipe snowboard competition ranged from 4–12 years.

Data Collection. The results of this study were obtained by interviewing subjects on their perceptions of how technological change would impact on the culture of their sport. The nature of the research questions directed the data collection process be in-depth and consistent with an inductive analysis (Smith & Shilbury, 2004). In-depth interviews were selected as the most suitable data collection tool. Subjects were interviewed between 16th and 20th October 2006 at the AIS. Interviews were semi-structured, involved posed, open ended questions, were recorded on video, and later transcribed word-for-word. Each subject was interviewed individually for 20–45 minutes. All subjects approached for this interview accepted.

Data Analysis. Interview transcripts were first examined to gain general familiarity. Dominant themes were noted and these formed categories. These categories became the codes by which transcripts were further interpreted. Manual coding was undertaken with a hierarchical three stage process (Strauss & Corbin, 1990) beginning with identification of "open" codes. Open codes provided a broad set of cultural categories with which to conduct subsequent reduction. "Axial" coding subsequently divided each open code into several axial codes which reflected cultural dimensions employed to measure the perception of the practice community to specific issues. Axial codes were then divided into "selective" codes. Selective codes provided researchers capacity to highlight the content of specific cultural dimensions (Smith & Shilbury, 2004). Labels were selected that best described conceptual contents of each code however; it is worth noting that labels represent researcher interpretation of content (Smith & Shilbury, 2004). The number of occasions selective codes were mentioned revealed the significance of each sub-dimension to major cultural themes. A reliability measure used by Smith & Shilbury (2004) was adopted by this study. Where Smith and Shilbury calculated an *inter-researcher* reliability score, this study used one researcher undertaking a test-retest analysis on each interview one week apart hence; an *intra-researcher* reliability score was calculated. This study generated a satisfactory *intra-researcher* reliability level greater than 90%, as recommended by Miles & Huberman (1994).

Theoretical Sampling Limitations, Informed Consent and Ethical Approval. As all subjects were of Australian nationality, this was accepted as a limitation. Future studies would benefit by increasing sample size and evaluating international practice community members. Subjects read "information to participant" and signed "informed consent" forms prior to taking part. Experimental procedures were approved by the Ethics Committee of the Australian Institute of Sport on 17th August 2006 (approval number 20060807).

3 Results

This paper investigated the snowboard practice community's perception to recent technological research. Table 1 shows the six axial codes (dimensions) and twenty selective codes (sub-dimensions) related to three major themes (open codes) that emerged during the interview process. The major themes were: 1) The State of the Current Subjective Judging System, 2) Automated Feedback and Objective Judging System, and 3) Future Direction of the Sport.

4 Discussion

Of the six axial codes the most prominent was, "weaknesses of the current subjective judging system". The most prevalent sub-dimension related to this axial code was labeled, "subjective perception of style and overall run execution". The inability of the current subjective

Table 1. Cultural dimensions (Axial and selective codes).

Strengths of the Current Subjective Judging System
1. Subjective perception of style and overall run execution 2. Allows athletic freedom of expression 3. Well trained judges with experience in the sport 4. Improvement on the gymnastics model of judging 5. Multinational judging panel
Weaknesses of the Current Subjective Judging System
1. Subjective perception of style and overall run execution 2. Expansive criteria displayed as one measurement – termed "Overall Impression" 3. Memory board short hand time constraints during judging 4. Limited access to experienced and high level judges 5. Increased population of athletes reaching "optimal performance" 6. Subjective judging error in aerial amplitude perception
Replace Subjective Judging with Automated Objective Judging
1. Removal of subjective perception of style and overall run execution 2. Removal of athletic freedom of expression 3. Generation of incorrect results 4. Separation from what is valued in the sport
Integration of Both Subjective and Objective Methods in Judging Protocols
1. Accurate KPV information perceived as beneficial 2. Objective information beneficial as judging aid (automatic memory board)
Possible Cultural Perception of Automated Judging
1. Positive if integrated with current subjective system 2. Negative if used as the only measure of performance
The Consensus on Future Direction of the Sport
1. Freedom of expression and athletic individuality culminating in well executed and stylish performances

judging system to consistently identify correct competition results was a strong perception of the practice community, mentioned by eight out of nine subjects:

...it's subjective, it's personal opinion. You might think it [a run] looks good, I might think it doesn't, and that at the moment is the weakness because that's the difference between who wins and who doesn't win sometimes. It's not the fairest way...

The axial code labeled "strengths of the current subjective judging system" was however, the second most prominent dimension. Paradoxically, the most prevalent sub-dimension associated with this axial code was also labeled, "subjective perception of style and overall run execution". All subjects mentioned this strength. The second most prevalent perception related with strengths of the current judging system was that it, "allows athletic freedom of expression":

...it let's the riders show who they are and what they can do freely. They can basically show their personal being in their style and who they are as a snowboarder. You can go at it, you can invent new tricks, and you can invent new grabs, go big, go small, and spin as fast you want. The system currently lets you do that...

That fact that paradoxically, the same sub-dimension is considered both a judging strength and a weakness offers insight into half-pipe snowboardings' underlying cultural ethos. All subjects revealed aspects of the sport they valued without posed questioning from researchers. The information emerged into a major theme labeled, 'Future direction of the sport'. Snowboard competition seems focused on "freedom of expression and athletic individuality culminating in well executed and stylish performances":

...it's a free sport driven by free people...

Snowboarding culture appears athlete focused and judging protocols have taken into account criteria that athletes value. However as performance levels increase, judging criteria may have to adapt. Miah (2000) evokes sport's overall self-annihilating teleology; that increased numbers of athletes ultimately achieve "optimal performance", thereby outgrowing the structure of the game. Sports are then forced to adopt altered performance assessment, equipment changes, or altered game rulings to again separate similarly capable athletes. Two subjects revealed this concept without posed questioning:

...the weakness now is that the sport has progressed at such a rapid rate, the majority of riders are doing the same tricks...up until 2006 it [subjective judging] worked because the level of the sport hasn't got to a level where it's an issue yet. Looking forward it is going to be an issue. It's going to be hard to judge overall impression in the next Winter Olympics, and then it is going to be even harder another 10 years down the track...

This paper has proposed that automated feedback calculating objective information on half-pipe snowboarding KPVs could benefit subjective judging protocols, and perhaps prolong the sports' possible self-annihilation (Harding, 2006; Harding *et al.*, 2006; Miah,

2000). The notion this system could judge half-pipe competition automatically, without input from subjective judges was however, vigorously opposed by all subjects:

> ...I think there's no way it could take over the judging, style and execution is such a big part of the sport...I think that would spoil snowboarding...

Judging based purely on objective information opposes what is valued, removing the two prevailing judging strengths and regimenting the competition. However, integration of objective feedback with subjective protocols was perceived positively by all subjects. The integration of technological advances without the removal of opportunities for athletic freedom of expression will maintain what the practice community values.

5 Conclusion

This is the challenge for elite half-pipe snowboarding competition in the future. While technological advancements are theorised to enhance athletic performance and judging protocols, it needs to be balanced with the culture of the sport so that athletes continue to see themselves as snowboarders with the freedom and individuality that entails. This study demonstrated there needs to be a balance between scientific advancement and practice community expectations:

> ...we do not want to be ballerinas; we want to be snowboarders...

References

Harding J.W., Mackintosh C., Martin D.T., Rosemond D., Dowlan S. and James D.A. (2005) Key Performance Variable Detection and Validation of an Accelerometer-Based Algorithm to Calculate Air-Time During Half-pipe Snowboarding. *NESC Applied Physiology Conference,* Canberra, 27–28 October.

Harding J.W. (2006) Lifestyle: AIS. *Digital Snowboard Magazine*, Issue 1.

Harding J.W., Mackintosh C., Martin D.T., Rosemond D., Dowlan S. and James D.A. (2006) Applications for inertial sensors in elite-level half-pipe snowboarding. *17th Biennial Congress of the Australian Institute of Physics*, Brisbane, 3–8 December, abstract number 484.

Miah M. (2000) New Balls Please: Tennis, technology, and the changing game. In: Haake S. and Coe O.A. (Eds.) *Tennis, Science and Technology*, Blackwell and Science, London, pp. 285–292.

Miles M. and Huberman M. (1994) *Qualitative data analysis (2nd ed)*. Thousand Oaks, Sage, California.

Morgan W.J. (1994) *Leftist Theories of Sport: A Critique and Reconstruction*. University of Illinois Press, Urbana.

Smith A.C.T. and Shilbury D. (2004) Mapping Cultural Dimensions in Australian Sporting Organisations. *Sport management Review*, 7, 133–165.

Strauss A. and Corbin J. (1990) *Basics of qualitative research: Grounded theory, procedures and techniques*. Newbury Park, Sage, California.

Tenner E. (1996) *Why Things Bite Back: Predicting the Problems of Progress*. Fourth Estate, London.

ESTIMATION OF DIRT ATTRACTION ON RUNNING SURFACES OF CROSS-COUNTRY SKIS

L. KUZMIN & M. TINNSTEN

Dept. of Engineering, Physics and Mathematics, Mid Sweden University, Östersund, Sweden

Methods for analysing impurities in snow are used in glaciology and ecological studies. However, the relationship between the dirt accumulation on the ski running surface and the concentration of pollution in the snow is not straightforward, since the interaction between the top layer of snow in the ski track and the ski running surface is responsible for the dirt accumulation on the running surface. In this paper the dirt film accumulated on the gliding surface is studied. A number of XC skis with a transparent base and a white background were examined after undergoing different treatments. Measurements of the whiteness of the running surface of the skis were carried out and glide tests were performed. The measurements and tests were repeated after skiing various distances on a ski track under varying snow conditions. The following observations were made during the study: The experimental setup could deliver a reliable value of the whiteness of the ski running surface. We achieved 0,3% standard deviation in a test on a control sample; The correlation between the ski glide and the amount of dirt is obvious and significant.

1 Introduction

The importance of keeping the ski running surface clean from any pollution in order to minimize snow-ski friction is mentioned in a number of scientific papers. Evidently, the amount of dirt that accumulates on the ski running surface is heavily dependent on the concentration of pollution in the snow.

However, our literature review discovered no studies that had investigated the contamination factor of the ski base. The lack of such an investigation may, for instance, explain a conclusion regarding the ski glide on wet snow in (Slotfeldt-Ellingsen & Torgersen, 1982). The authors believe that glide wax wears down much faster on wet snow, than on cold, dry snow.

2 Methods

2.1. General Approach

Our choice of tools, wax, skis and the procedure for ski preparation was based on direct application to cross-country (XC) skiing. Our primary experimental method was to monitor the glide variation in the case of treated skis and dry (HSS scraped) skis respectively. The absolute values of glide and surface whiteness are of secondary importance.

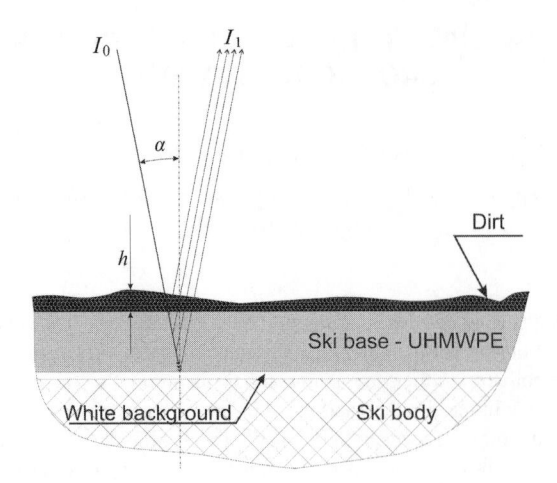

Figure 1. Beer-Lambert Law and the ski running surface.

2.2. Theory

Almost all commercially produced top level skis have a graphite base. Generally, the graphite base is a mixture of UHMWPE (ultra-high molecular weight polyethylene) and amorphous graphite (about 5%). It is very difficult (perhaps impossible) to measure the amount of dirt on the graphite base. Optically, it is not possible to see dark pollution on a black surface. Mechanically, it is hard to separate the pollution from shavings of the base material. Chemically, normal organic solvents (hydrocarbon) dissolve both the dirt and the amorphous graphite from the ski base. Therefore, we have used skis with a transparent base. Such skis were usual 15 years ago. As mentioned above, the most common type of ski base is a mixture of UHMWPE and amorphous graphite, while the transparent base in our experiments is made of pure UHMWPE. We believe, such a small amount of graphite does not significantly affect the dirt attraction pattern. Therefore, the results of our experiments on skis with a transparent base may also be applied to skis with a graphite base.

As a measurement of the rate of surface contamination build-up, we chose the whiteness rate of the ski running surface. We assumed that the transparent base and white background reflect the greater part of the incident ray, so any light loss must be the result of absorbance by the film of dirt. Our measurement method was grounded on the Beer-Lambert Law. *The Law says that the fraction of light absorbed by each layer of solution is the same.*

The absorbance A is defined as $A = \log_{10}(I_0/I_1)$, where I_0 is the intensity of the incident light, and I_1 is the intensity after passing through the material. This is shown in (Figure 1).

The equation representing the Beer-Lambert Law is very straightforward: $A = \varepsilon b c$, where ε is the molar absorptivity, b is the path length of the sample, c is the concentration of the compound in solution. In our case, ε is quite constant, c is stable too, so the path length $b = h/\cos\alpha$ is the major influencing factor, where h is the thickness of the dirt layer. The thicker the dirt layer, the greater the absorbance A, and the larger the light loss. In our real case we observed the whiteness change on a finite area, where the light absorption varied depending on both the grime thickness and the grime surface scattering.

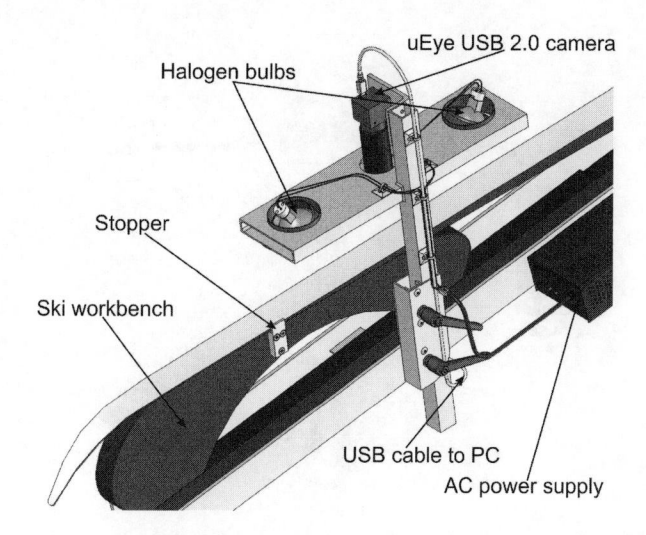

Figure 2. Dirt attraction measurement – Experimental setup.

2.3. Experimental Setup

As illustrated in (Figure 2) a uEye USB 2.0 camera acted as an image-capturing device. Two halogen bulbs provided a powerful light source. Each halogen lamp was directed to a point on the ski running surface under the camera, which gave us a very strong spotlight on the observed area. Moreover, the powerful lighting allowed us to keep the lens aperture small. Furthermore, such strong collimated light considerably improved the measurement accuracy, because the surrounding sources of light (windows, etc.) had a negligibly small influence on the total luminosity.

We used a direct current (DC) 12 V power supply with improved accuracy to eliminate the instability that may occur when using the standard alternating current (AC) 12 V from the mains power supply.

The ski was fastened to the workbench by a pivot joint in the binding, which had been mounted in advance. In addition the ski was tightly abutted on to the stopper. Such anchoring guaranteed a very accurate and repeatable positioning.

We used an "uEye Demo" as an image-capturing application The uEye Demo was configured to capture a 8 bit monochrome image with no software correction. The processing line is presented in (Figure 3). Each image was stored on the PC hard drive as a BMP 8-bit, grayscale mode file. In fact, this file is a W (whiteness) matrix of the size $m \times n$. Because the image is in a grayscale mode, each matrix element $w_{ij} \in [0.255]$.

As a whiteness value (w) we simple used the arithmetical mean of all the elements w_{ij} in the matrix W, $w = \dfrac{1}{mn} \sum_{i=1}^{m} \sum_{j=1}^{n} w_{ij}$. To realize this equation we applied a MATLAB procedure mean2 to our M-file for statistical treatment of the experimental data. We only processed the flat area of the image, which does not include the ski groove. The processed area of the ski running surface is 1013×717 pixels large (about $22{,}5 \times 17{,}5$ mm^2) (Figure 4). The area is located just in front of the pressure peak. To minimize the wearing-off effect on

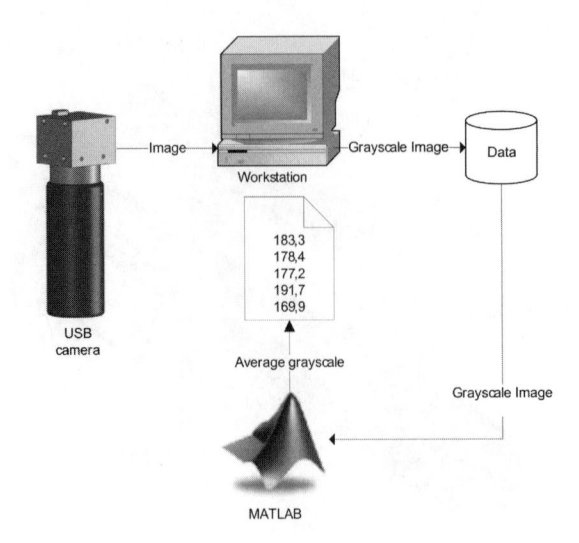

Figure 3. Analysis of observations.

Figure 4. The processed area of the ski running surface.

the inside half of ski, we marked all the skis as either left (*l*) or right (*r*) in all the pairs, and the whiteness value was calculated as ($w = 1/2(w_l + w_r)$).

3 Results

If the ski was placed in the experimental setup for a long time, the processed area became warmer and warmer, and w increased. However, when we measured the whiteness of all the skis, we measured the skis at regular intervals so that all the skis were kept at a similar temperature. In this way we obtained a quite stable measurement. In (Figure 5) the results of 40 measurements of the same sample are presented with a two-minute interval between measurements to avoid warming-up the sample. The mean whiteness was 132,8, and the standard deviation was 0,412.

The gliding abilities of used skis are very similar, but not, however, exactly equal. We therefore calculated comparative values for the waxed skis $C(s_i) = A_w(s_i)/A_r(s_i)$, where $A_w(s_i)$ is the absolute value of the parameter of a pair of treated skis, $A_r(s_i)$ of a pair of reference skis, and s_i is the distance covered.

Later on we normalized the comparative values $N(s_i) = C(s_i)/C(0)$. Therefore $N(0) = 1$, and if $N(s_i) < 1$, then the waxed skis lose some (*N*) quality faster than the reference skis after s_i km skiing, and vice versa. By linear interpolation, flat (constant) extrapolation and averaging of all the normalized comparative values we may present the principal

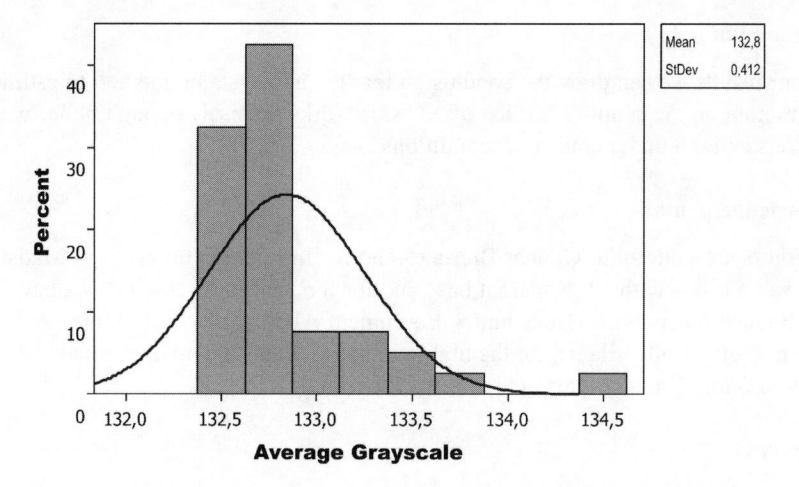

Figure 5. Distribution of the average grayscale fitted as a Gaussian distribution.

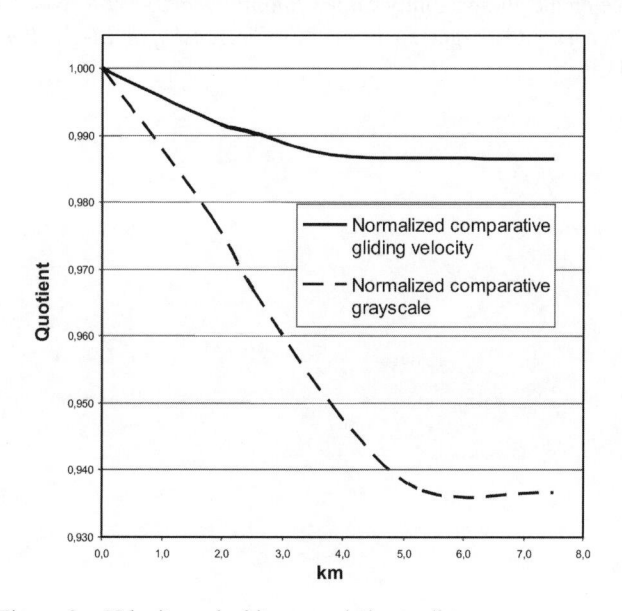

Figure 6. Velocity and whiteness relative to distance on wet snow.

trend much more visually as follows: $\bar{N}(s_i) = \dfrac{1}{m}\sum_{j=1}^{m} N_j(s_i)$, where j is a test series number, and m is the total amount of series.

The results of the comparative glide test on wet snow (for a complete description of a test procedure see (Kuzmin & Tinnsten, 2006) shows a good correlation between whiteness and velocity (Figure 6). On the other hand, on cold, dry snow grime covers the ski gliding surface utterly insignificantly, and the grayscale measurement lies inside the margin of error in the test results.

4 Discussion

From our results we can draw the conclusion that the above-stated method to estimate the dirt attraction on the running surface of XC skis works precisely enough under wet snow conditions, but not under cold snow conditions.

Acknowledgements

The authors are grateful to Gunnar Bjertnæs and Svein Inge Holtesmo from Madshus AS for the sample skis with a transparent base and for the fruitful discussions we have had, to Sven-Gunnar Johansson – Backcountry Equipment AB for the comfortable ski boots, to Torbjørn Ragg – Rottefella AS for the high-tech ski bindings and to the staff at Östersund's Ski Stadium for always preparing the ski track perfectly.

References

Kuzmin L. and Tinnsten M. (2006) Dirt absorption on the ski running surface – quantification and influence on the gliding ability. *Sports Engineering*, 9, 137–146.

Slotfeldt-Ellingsen D. and Torgersen L. (1982) *Gliegenskaper til skisåler av polyetylen*. Oslo, SINTEF.

AERODYNAMIC FORCES COMPUTATION FROM HIGH-SPEED VIDEO IMAGE OF SKI JUMPING FLIGHT

M. MURAKAMI[1], N. HIRAI[1], K. SEO[2] & Y. OHGI[3]

[1]Graduate School of Systems and Information Engineering, University of Tsukuba, Japan
[2]Faculty of Education, Art and Science, Yamagata University, Japan
[3]Graduate School of Media and Governance, Keio University, Japan

Aerodynamic forces computation from high-speed video image of ski jumping flight was attempted, of which video image was taken at Hakuba Ski Jumping Stadium in Japan. Initial 40 m part of a flight of 120 m jumping after take-off was recorded by a fixed single high-speed video camera at a frame rate of 250 frames/s. The primary purpose of a series of research of us is to extract the aerodynamic force data during real jumping flights, in particular, in the initial phase of a flight for about 2 sec. These data were difficult to measure in the wind tunnel test because of some technical difficulties.

1 Introduction

The database for the aerodynamic forces of ski jumping measured using a real sized model in a wind tunnel is now available (Seo, *et al*, 2004a), and it was used in the jumping flight optimization study (Seo, *et al.*, 2004b). However, the aerodynamic forces for the initial phase of a jumping flight immediately after a take-off could not be measured in our previous wind tunnel tests due to some experimental difficulties. In the initial phase, the ski-body angle is normally very large and rapidly changing from about 90 deg., which brings about a technical problem that the blockage rate in the test section of the wind tunnel becomes too large, as well as the difficulty in the wire support method of a jumper model. Therefore, the aerodynamic force data during initial 1 sec after take-off are the very target of this study. It is also the purpose of this field measurement to investigate whether the unsteady aerodynamic forces are of importance for ski jumping flight, because in the initial phase the ski-body angle changes rapidly and in a wide range from about 90 deg. to about 30 deg. In this study as part of a series of researches, we focused on the improvement of the computation procedure for the time series data. In fact, the original time series data include various kinds of noise, and the target aerodynamic data seem as if they are buried in the noise after the straightforward application of numerical differentiations. Therefore, introduction of numerical filtering and curve fitting based on regression analysis is inevitable for meaningful result.

2 Field Measurements

In the field measurements in the Hakuba Ski Jumping Stadium, high-speed video images of jumping flights were taken, and at the same time, an accelerometer and a gyro-sensor were installed on the back of a jumper during each jumping flight to measure the two components of the acceleration and the angular velocity of the trunk of a jumper around the horizontal axis. The flights of 120-m class jumping are selected as examination objects. The camera was set on

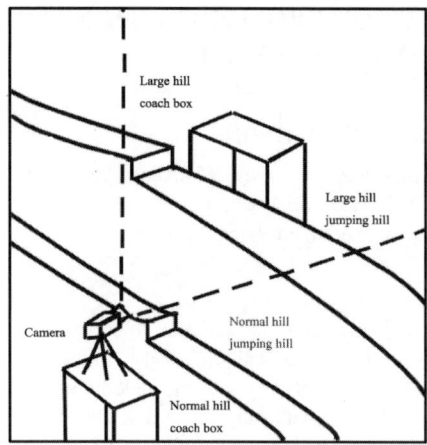

Figure 1. Spot sketch of video shooting in Hakuba.

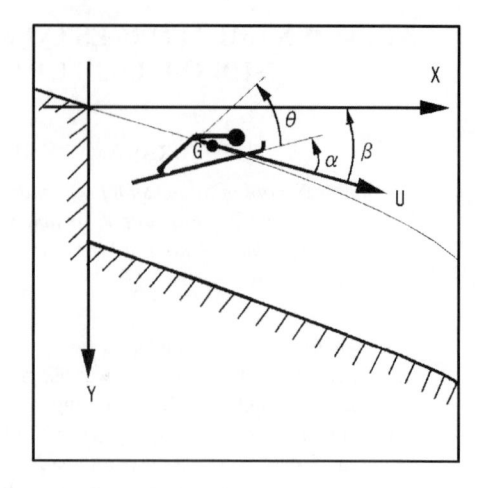

Figure 2. Notation and coordinate system.

the top of the coach tower a normal-jumping hill by the immediate side of the take-off point of Figure 1.

The location is about 15 m downward from the take-off point along the 120-m landing slope and 60 m away from the landing slope. wide-angle lens of 28 mm covers 40 m point of the landing slope from the take-off point. This roughly corresponds to a flight for initial 2 sec. The high-speed digital video camera (Photron Fastcam-X 1280PCI)was mostly operated in the mode of 1024×512 pixels at 250 frames/sec. The subjects in these measurements were members of Japanese Olympic team jumpers and a high school jumper. The effect of wind was ignored because we did not have any wind data simultaneously measured.

3 Image Data Analysis

The position of the center of gravity (x and y) of a jumper is expressed in the coordinate system indicated in Figure 2. And an example of the video image is presented in Figure 3 with having the jumper images thinned out. The origin of the coordinate system is the take-off point, and the x-axis is taken horizontally and the y-axis vertically downward. The distortion of image seen through a wide-angle lens was corrected in a digitizing stage. In order to derive the velocity and acceleration data the trajectory of the center of mass must be obtained. In this study, the center of mass was calculated in terms of the data of the waist, head and toe. Each position data was multiplied by each specific weight, and then the sum of the multiplied data was divided by the sum of the specific weights. Here, the total mass of jumper is assumed to be 70 kg. The data for specific weights were calculated by referring to Chandler (1975).

The example data in this paper are those of rather successful flight with a flight distance over 120 m. The tracking was mostly accomplished automatically with the aid of an image analysis software, partly with some manual assist for cases where the positions were hard to be distinguished because they seem as if they are blended into in the background. Since in the present measurements, a single fixed camera was used instead of adopting a three-dimensional camera system, a simple correction by a quadratic formula is applied

Figure 3. Original high-speed video image in the initial 2 sec., at Hakuba Stadium (120-m).

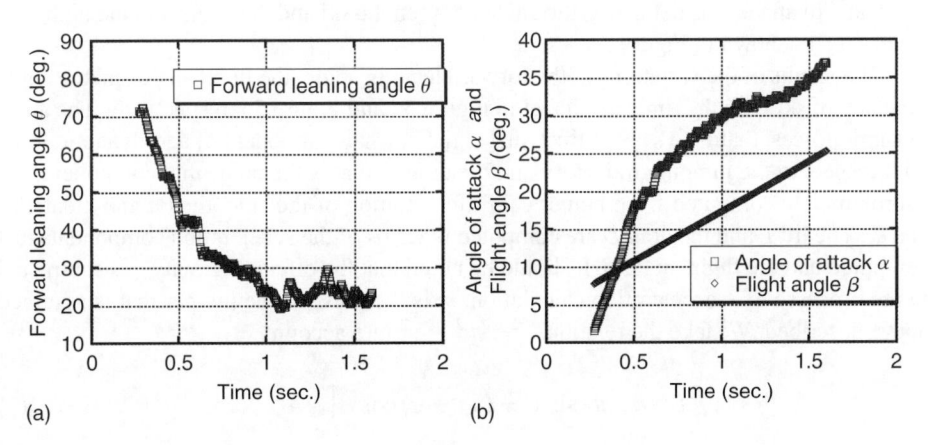

Figure 4. Time time variation result of (a) the forward leaning angle θ, (b) the angle of attack of ski α and, the flight angle β.

to the original time series data. For further data analysis, the first and the second order time derivatives are required for the velocity and acceleration for the aerodynamic forces computation. Rather higher order numerical differentiation scheme are used. For the derivation of the acceleration from each velocity data for the aerodynamic forces computation, a data smoothing procedure is required. The low pass filtering and the weighted least square regression with a higher order polynomial are used. The x- and y-velocity data, u and v, are shown in Figure 4. Gradual deceleration in u due to air drag and rather rapid acceleration in v due to gravity are clearly seen here. In this figure the low pass filtered data and the regression results are added by a solid and a broken lines respectively. In addition, the data of the forward leaning angle θ, the flight angle β of the flight trajectory U with respect to the horizontal line as defined in Figure 5 and the angle of ski with respect to the horizontal

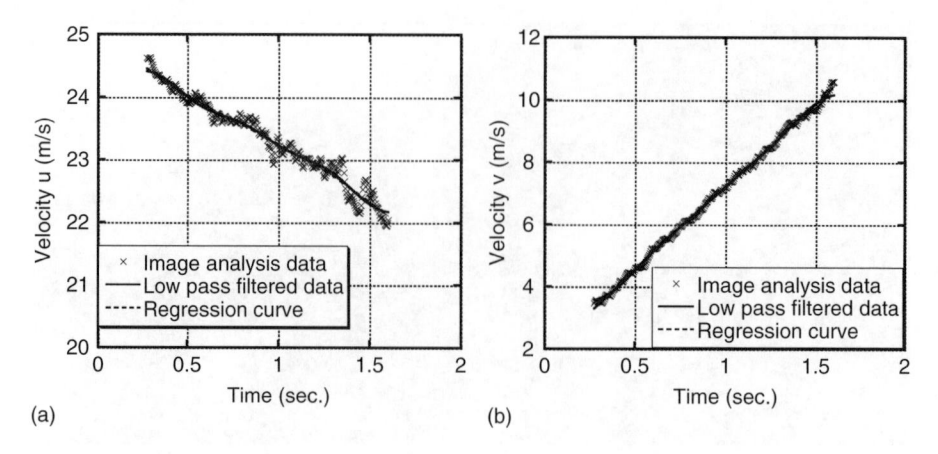

Figure 5. The x- and y-velocity data, u and v. (a) u-data, (b) v-data.

line for the later derivation of the angle of attack of ski were measured from the video image. The angle of attack α is defined as the angle between the ski and the tangent to the flight trajectory, U, as shown in Figure 2.

The results of the variations of the forward leaning angle, the flight angle and the angle of attack of ski with the time are shown in Figure 5a and Figure 5b, respectively. The aerodynamic forces, that is to say the lift L that is the normal component of aerodynamic force with respect to the jumping trajectory and the drag D that is the tangential component, are approximately computed from numerical differentiation of the smoothed u and v data, a_x and a_y. The lift L and the drag D are computed in terms of the acceleration components just computed and the flight angle β. It should be noted that the effect of gravity g is subtracted in the process of y-acceleration calculation. m is the mass of the jumper that is assumed above described. We take the relation $U^2 = u^2 + v^2$ into account

$$L = m \left| a_x \sin \beta - (a_y - g) \cos \beta \right| \tag{2}$$

$$D = -m \left| a_x \sin \beta + (a_y - g) \cos \beta \right| \tag{3}$$

For the sake of comparison of the result with the existing experimental data (Seo et al., 2004ab), the lift area S_L and the drag area S_D defined below are used instead of lift coefficient C_L and drag coefficient C_D by Tani & Miishi (1951). In these formulae,

$$S_L = 2L/\rho U^2 \tag{4}$$

$$S_D = 2D/\rho U^2 \tag{5}$$

where $U^2 = u^2 + v^2$, and ρ is the air density.

4 Discussion

The result for the time variation of S_D is shown in Figure 6, where only the reliable part of the data between 0.2 and 1.6 sec are presented. It may be concluded that a significant data

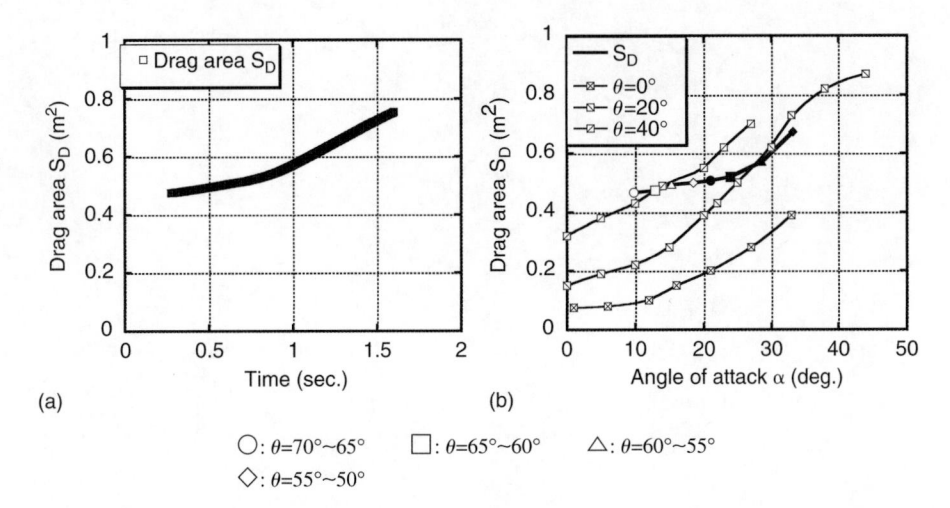

Figure 6. (a) Time variations of the drag ares S_D. (b) Time variations of the S_D plotted as a function of the angle of attack α.

of the aerodynamic forces can be extracted from the high-speed video image. It is clearly seen from this result that S_D slowly increased due to the transition to a V-style till 0.9 sec, and, however, after 1.0 sec the aerodynamic stall occurred because of large angle of attack α, when S_D rapidly increased. The data are also plotted on the S_D vs. α diagram of the wind tunnel database (Seo et al., 2004) for rather large θ in Fig. 8. It is seen the present data do not both qualitatively and quantitatively conflict with the Seo's data though they deviate slightly from the Seo's data.

References

Chandler R.F. (1975) *Investigation of inertial properties, of the human body*. Technical report AMRL-74-137, Wright Patterson Air Force Base.

Jin H., Shimizu S.,Watanuki T., Kubota H. and Kobayashi K. (1995) Desirable gliding style and techniques in ski jumping. *Journal of Applied Biomechanics*, 11, 460–174.

Seo K. Watanabe I. and Murakami M. (2004a) Aerodynamic force data for a V-style ski jumping flight. Submitted for publication in *Sports Engineering* (2004).

Seo K., Murakami M. and Yoshida K. (2004b) Optimal flight technique for V-style ski jumping. Submitted for publication in *Sports engineering*.

Tani I. and Miishi T. (1951) Aerodynamics of ski jumping. *Science* (Japanese edition), 117–52 (in Japanese).

Tani I. and Iuchi M. (1971) Flight-mechanical investigation of ski jumping. Scientific study of skiing in Japan. *Hitachi*, 35–52.

APPROACH FOR A SYSTEMATIC OPTIMIZATION OF A TWOSEATER BOBSLEIGH

M. MÜLLER & V. SENNER

Department of Sports Equipment and Materials, Technical University Munich, Munich, Germany

The present study outlines an approach for the optimization process of a twoseater bobsleigh. First, influencing parameters from the athletes' individual preferences to field conditions are discussed. The acquisition of relevant field data and synchronization issues are then presented for the steering mechanism, leading to an outlook about how the findings can be translated for the technical redesign and the validation of simulation models.

1 Introduction

Bobsleighs are an example of an almost completely competition-bound sports that nevertheless has considerable media coverage during world-cups and Olympic games. Therefore it was a matter of interest for the Bavarian Organization of Bobsleigh to initialize a research project revealing optimization potentials within the very strict international accepted regulation for bobsleighs. Of course the track times of the bobsleigh are dependent on the pilot's athleticism and ability of precise steering. But also the bobsleigh itself still holds some promising optimization potentials. An overall approach, however, can only be successful through a combination of individual parameters of the athletes and the different tracks as well as by a parametrized, technically optimized and adapted bobsleigh concept. In this paper some of the efforts that were undertaken to generate a valid data basis for further optimization process steps is presented.

2 Methods

2.1. *General Approach and Individual Influences*

One major issue for the design of sports equipment, especially when it comes to competitive sports, is the discrepancy between the optimization of measurable technical parameters and and the athlete's feeling for his piece of sports equipment. Hence, reflecting the experience gathered in this project yet, the optimization of such sports equipment is facing three main problems:

- Few athletes are capable of articulating precise technical information about the ability of their sports equipment.
- For seasonal sports it is difficult to conduct test runs with top level athletes and equipment because they obey a very packed timetable for world-cups and national competitions.
- In field tests, a high number of influencing parameters and disturbing variables have to be identified and controlled for optimization purposes.

An example should clarify that one goal of the study is to objectivize statements about sport equipment. That is, to gate out the variances in statements from test runs with athletes

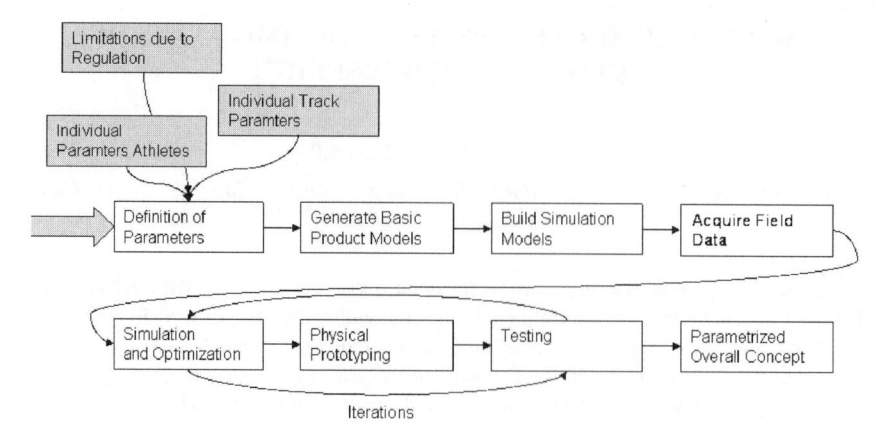

Figure 1. General approach for the optimization process of a twoseater bobsleigh.

as far as possible. Test runs were conducted by Germany's market leading company for the design of bobsleighs, implying different damping rubbers between the bobsleigh's cover and the mainframe. The athlete's were not informed about what set-up they were acutally running, whereas set-ups and drivers were mixed random. The drivers' conclusions, however, differed widely about what set-up was the best. This once again confirms that the performance of the system athlete, or even team, and equipment is not only bound to technical parameters but also to psychological factors and matters of individual feeling for the equipment. Two actions are beign tried to implement to minimize the effect of such disturbing variables for optimization purposes:

- Perform test runs with retired but very experienced athletes to minimize the variance between runs and to avoid shortage in testing-time.
- Strip, were possible, testing variables from the field to the laboratory or models and from athlete bound tests to testing machines such that conditions and influencing parameters can be controlled much easier.

The general approach is then an interative process (Figure 1), combining individual paramters, track paramters and directives from the official regulations, enabling model building, the acquistion of proper field data and simulation for the prototyping phase.

After the first goal characterized above the second goal is to verify simulation models, in this case fluid dynamics and multi body simulation, of the bobsleigh with the purpose to minimize the costs of gathering data about a certain specification. Every field test and wind tunnel study causes immense expenses that could more reasonably be fed into the actual optimization and prototyping process.

2.2. Measurement Equipment

The present study was conducted to gather basic data about the complex of loads and effects acting on a bobsleigh with the purpose to later apply the abovementioned principles. Hence, two main areas of investigation were defined:

- Aerodynamics of the cover,
- Frame, steering and suspension.

Location	Type	Sensor type	Meas. principle
Axle	Force	Strain-gauge	Resistive
Seat	Force	Strain-gauge	Resistive
Connecting-bolt front/back	Bend moment	Strain-gauge	Resistive
Steering-wire	Force	Strain-gauge	Resistive
Steering	Angle	Strain-gauge	Resistive
Axle	Angle	Strain-gauge	Resistive
Front/back part	Angle	Strain-gauge	Resistive
Back part	Acceleration	MEMS	Spring-mass
Back part	Inertial angles	MEMS	Piezo
Cover	Pressure	MEMS	Resistive

Figure 2. Parameters realized for the test runs of a twoseater bobsleigh.

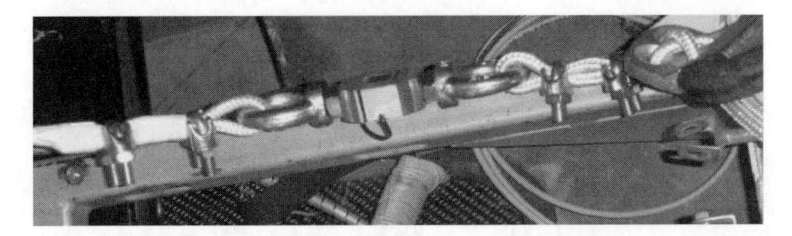

Figure 3. Load cell to detect steering forces during test runs of a bobsleigh.

The list of requirements for the measurement equipment was made from rough calculations of speed and mass as well as from the specifications found during the generation of product models (e.g. 3D scanning of the bobsleigh body) and simulation models (e.g. computational fluid dynamics model of the cover and finite elements model of the frame). Figure 2 depicts an overview about the parameters that were acquired during the test runs (including e.g. forces at the axle in 3 directions in space, pressure sensors on the cover).

For the data acquisiton state-of-the-art equipment was used to facilitate small dimensions. 32 channels can be acquired at a time, 16 with a sampling frequency of 3000 Hz using a Noraxon WLAN data logger (TeleMyo 2400 GT), 16 with a sampling frequency of 10000 Hz using a National Instruments USB data logger (USB 6211) for the channels less and more sensitive to high frequencies, respectively. Figure 3 shows one of the load cells to detect steering forces during the runs.

Furthermore an onboard camera was installed to track the test runs for later synchronization with the measurement data.

3 Results

Here, one sample of the test data is presented in the context of the optimization process. The steering mechanism beforehand was defined as a part that can be optimized regarding preciseness and smooth running. Also the lever arm ratio should be a matter of improvement. The relevant parameters are the steering force and the steering angle during the runs.

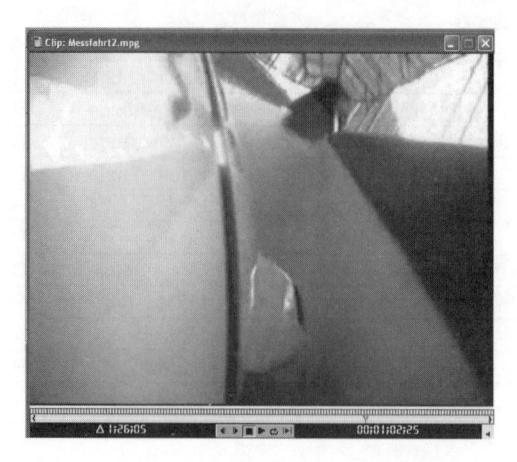

Figure 4. Freeze image from the onboard camera during a test run.

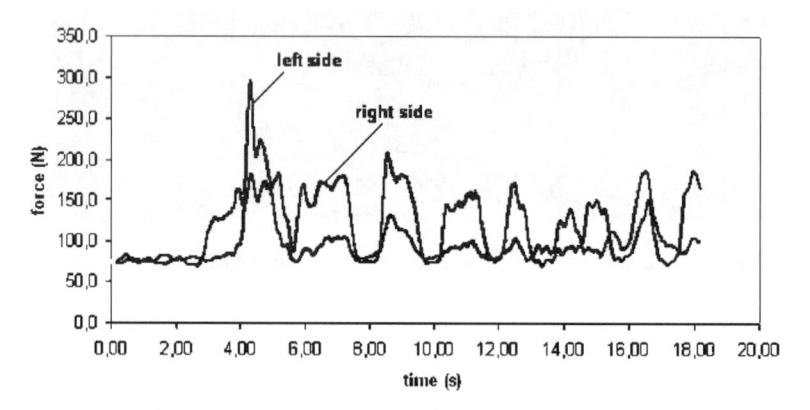

Figure 5. Steering forces of a twoseater bobsleigh.

Both were acquired and synchronized with the video data to judge about the influence of this parameters on bobsleighs running on the track as a basis for the redesign of the mechanism. The final test runs were conducted on February the 10th 2007 on the ice track in Königssee, Germany. Figure 4 shows a freeze image from the onboard camera during a test run. Figure 5 depicts steering forces for the left and right handle over a period of 18 seconds (overall track time is around 58 seconds for the test runs).

4 Discussion

Generally there are two types of data that were acquired. One is data useful for qualitative analysis of runs that directly lead to mechanical redesign (e.g. the steering conditions). Second is data that will be used for validation purposes of simulation models. The steering example shows that the athletes' individual impressions or feelings may vary from the actual physical conditions. Since there was no such investigation conducted before, the load cells were designed after the estimation of athletes about how much load they put on the steering mechanism – derived from simple experiments with weights. They expected it to be in the range of about 500 N. The actual range is up to around 300 N with a static pre-tension of

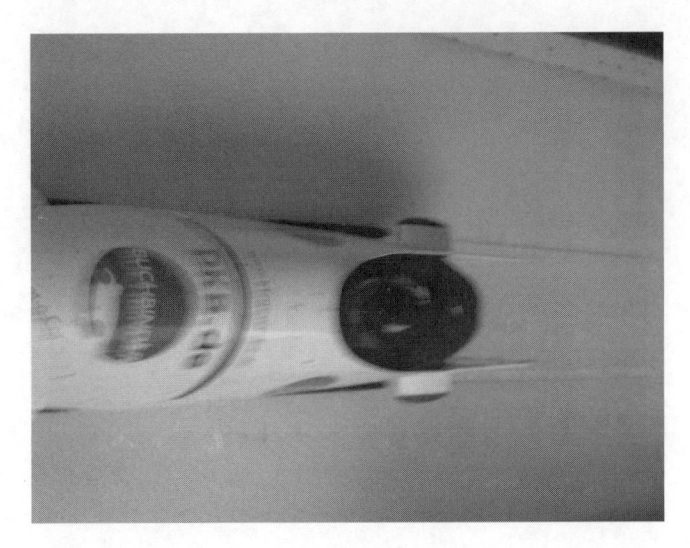

Figure 6. Bobsleigh equipped with sensors and data acquisition hardware on the track.

around 50 N, say almost 70% lower. Another issue are the obvious balancing motions in the steering mechanism to keep the bobsleigh on the track after circles or turns. They can entail contacts between the bobsleigh and the ice tunnel which leads to higher run times. The question now is how to translate these findings into a new design and if these findings could influence the tactile training habits of the athletes. A suggestion for a redesign of the steering mechanism was developed within a student project. It comprises better lever arm ratios such that less force is necessary to induce steering activity and therefore is supposed to be much more precise. It will be implemented and tested for the next season winter 2007/2008.

5 Outlook

At present the data from the test runs is still being processed. All sensors were calibrated before and data has to be processed, cut and synchronized with the video data. A major task is to find a suitable report format for force and video data. The standard graphs are not an appropriate way of presenting in the case of the pressure distribution, for instance. It is planned to set alternating bars above time, indicated in a 3D CAD model of the bobsleigh and replayed with the video at a time. The second step then is to validade or partly deny the assumptions made for the simulations models which is conducted by experts in aerodynamics and structural optimization within the team. Third and most important concerning the outcomes in competition, is the translation of the findings into a new bobsleigh concept. Germany's market leading manufacturer of bobsleighs will participate in realizing the concepts and assist with their experience and information about standardization, interfaces and regulations.

References

Hainzlmaier C. (2005) *A new tribologically optimized bobsleigh runner*, PhD Diss., Technical Universtiy Munich, Munich, Germany.

Kleemann R. (2003) *Aufbau, Validierung und Einsatz eines Messsystems für Bobkufen*, Student Project Report, Technical Universtiy Munich, Munich, Germany.

21. Youth Sports

THE DESIGN AND DEVELOPMENT OF ELECTRONIC PLAYGROUND EQUIPMENT TO INCREASE FITNESS IN CHILDREN

P.P. HODGKINS[1], S.J. ROTHBERG[1], P. MALLINSON[2] & M.P. CAINE[1]

[1]*Sports Technology Research Group, Loughborough University, Loughborough, Leicestershire, England*
[2]*PlayDale Playgrounds Ltd, Haverthwaite, Ulverston, Cumbria, England*

The concept of an electronic playground was born out of the specific need to tackle the increasing levels of childhood obesity in the UK. A worryingly high proportion of children are inactive preferring to spend their leisure time playing computer games or electronic devices; thus to engage this specific user group a whole body interactive outdoor play device was devised. The device is intended to provide "stealth fitness" with an aim to improve user's reaction time, coordination, agility and cardiovascular fitness. The current paper describes the design and development of this novel device.

1 Introduction

Over the last 20 years obesity has trebled and on current trends, it is set to become the number one cause of death in UK in the next 10 to 15 years overtaking smoking (World Health Organization, 2005). Worldwide over a billion adults and children are overweight (World Health Organization, 2004) and some experts have predicted that the current generation of children are likely to have shorter life expectancies than their parents because of obesity (House of Commons Health Committee, 2004). On present trends half of all children in England could be obese by 2020 (Lobstein *et al.*, 2005). A major concern with regard to childhood obesity is that obese children tend to become obese adults, facing increased risk of diabetes, cardiovascular disease, and many other chronic diseases (Braddon *et al.*, 1986). In fact, there is twice as much chance that an obese child will become an obese adult, than a lean child. A major cause of weight gain is the worryingly high levels of sedentary behaviour. The rise in sedentary behaviour is believed to be due to the decrease of physical activity within the transportation, domestic and recreational domains (Beaglehole *et al.*, 2002). This behaviour is exacerbated by the influence of pastimes requiring no physical exertion, such as television viewing and playing computer games. Consequently there is an increasing need to develop solutions to enhance physical play for a young "electronic savvy" user group.

Several guidelines are available that propose minimum health-enhancing physical activity recommendations for children. The UK guidelines (Biddle *et al.*, 1998) state that the primary recommendation is that all children should participate in physical activity of at least moderate intensity for 60 minutes and that children who currently do little activity should participate in physical activity of at least moderate intensity for at least 30 minutes per day. The secondary recommendation was that at least twice a weak, some of these activities should help to enhance and maintain muscular strength and flexibility, and bone health. Unfortunately the National Diet and Nutrition Survey (Gregory & Lowe, 2000) found that approximately 40 percent of boys and 60 percent of girls failed to meet the recommendation of at least 60 minutes a day of moderate intensity activity. Furthermore, in the 15 to 18 age group, 56 percent of boys and 69 percent of girls did not achieve the recommendation. Therefore the need to increase physical activity levels by enhancing the play experience is not only desirable but essential.

Play can be defined as what children do when they follow their own ideas and interests, in their own way and for their own reasons. Children, unlike adults, mainly participate in physical activities because of the intrinsic enjoyment experienced (Biddle, 1999). Playground equipment is an ideal platform and environment to enhance physical activity participation and adherence amongst children as public playgrounds are available to the vast majority of the population with most being free of charge. There is growing recognition that outdoor play experiences effectively stimulate young children's development (Henniger, 1993) and that playgrounds act as more than just areas to for physical exertion; they enhance learning and social behaviour. Gill (2004) states that disabled children should be able to use play facilities alongside others and that they should provide an inclusive environment that enables disabled people to take responsibility for themselves. Department of Health (2000) statistics show that there are 276,064 disabled children aged between 5 and 15 in the UK and in fact it is a legal requirement though the Disability Discrimination Act (DDA) 1995 to protect people with disabilities from discrimination. It is a specific aim for the device is that it must offer high levels of inclusive play.

2 Design Methodology

A systematic approach to the design process was taken ensuring a logical progression from investigation through to concept generation, initial modelling, embodiment and ultimately the final design to enable prototype manufacture. The four core design phases are outlined in the following sections.

2.1. *Concept Generation*

To innovate in the playground equipment sector, it was crucial to gain a thorough understanding of contemporary play equipment and the current trends within the market. In addition, the influences outside of the playground environment were investigated to establish an insight into popular play devices such as computer gaming which has a very high participation especially in America where 60% of the total population plays interactive games on a regular basis (Interactive Digital Software Association, 2002).

Following many brainstorms and mind mapping sessions it was decided to generate a concept based on a whole body electronic game. The chosen theme was inspired by the notion of command driven play whereby audio/light commands are given directing the user to operate a specific switch (referred to as an "activity switch"). If the activity switch is activated within a predetermined time period then a subsequent audio/light command will be given. This continues until failure. The device challenges response time and tests an individual's reflexes so that as the player improves, the game moves faster. A tally of the number of activity switches actuated is equal to the user's score. The score is then consequently an estimate of the user's ability, coordination, skill level and fitness. Along with providing individual play the concept allows assisted play, where a group of friends can help each other to achieve a high score, and competitive play, so players can play against each other promoting social interaction and cooperative behaviours. Features to encourage adherence focus on evolving and changing the game as it progresses, similar to current computer games that

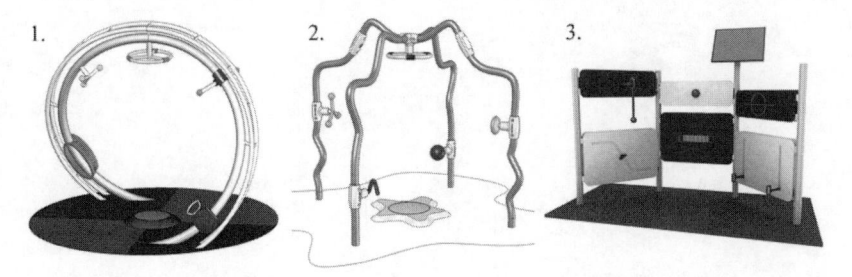

Figure 1. The three electronic outdoor playground equipment concepts.

"unlock" new levels. It is also envisaged that the device will provide "Stealth fitness" a phrase used to encompass the participation of an individual in health-enhancing physical activity without them focusing on the exertion levels required due to a distraction caused by a stimulus. More specifically for children, the dissociation for physical exercise comes from enjoyment of the activity.

2.2. Initial Modelling

Three different concepts based on the theme of whole body command driven play were developed and initial 3D CAD modelling was conducted in SolidWorks 2006 (SolidWorks Corporation, Concord, MA) to help visualise and understand how the user would interact with each design. The three concepts can be seen in Figure 1.

Concept 1 benefits from a compact space envelope and a sculptural design befitting an urban environment. Also the perimeter frame allows for climbing activities when the game is not activated. The disadvantages are the inflexible design of activity switch mounts and inclusion of a climbing frame increasing the need for floor safety provision. Concept 2 utilises activity switch mounts that allow modular framework construction and numerous possibilities for the placement and number of activity switches. In addition the activity switches are purposely varied offering an extensive array of body movements. The disadvantage is the high visual impact which could look out of place in some environments. Concept 3 utilises timber materials and incorporates panels housing the activity switches which are designed to replicate certain resistance training exercises. Although the concept looks more suited to certain environments its low visual impact and replication of exercise movements may be unappealing to some children. After evaluating the concepts by means of a matrix it was decided to embody concept 2 but alter the framework design similar to that of concept 1.

2.3. Embodiment

The selected design then underwent extensive embodiment to develop the activity switches and to define the spatial envelope requirements. The modular concept chosen utilises mount sections to position activity switches anywhere along the framework.

Each activity switch requires a specific dynamic activity to be completed designed to enhance endurance and muscular strength of the user depending on the mechanical operation of the actuator. The *rotational* activity switch (Figure 2a) is an example of a hand cycle exercise. The user is required to grasp the hand grip and complete a circular motion a set number of times for the system to register an actuation. The *linear* activity switch (Figure 2b) utilises

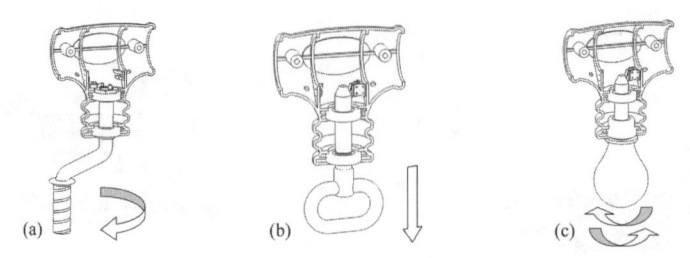

Figure 2. Internal view of rotational, linear and two degree of freedom activity switches.

the same mount but with a pull mechanism which requires the user to grasp the handle and pull away from the mount. This action can be reversed simply so that instead of requiring a pulling action, a pushing action will actuate the switch. A spring (not shown) provides the resistance which can be easily replaced depending on the resistance level required. The *two degree of freedom* mechanism (Figure 2c) works in a similar manner to that of the linear activity switch but an elastomer joint allows the hand interface to move in two rotational degrees of freedom.

The position of activity switches required careful consideration as it influences the speed of the game, the user's energy requirements, the muscle groups recruited and the accessibility for different user groups. The details presented in this section are for 12 year old children but the dimensions can be scaled to suit other ages if required. The height of the user is the most influential factor to ensure the device can be fully utilized, although to ensure inclusive play the device is programmed to eliminate the need to actuate the higher activity switches if required. It was therefore determined that the activity switches should be placed at three discrete heights. The placements are separated into three zones, low, mid and high, ensuring the user has to perform a wide range of whole body movements.

An activity switch placed at a low height needs be easily reached and activated by a 12 year old wheelchair user. Placing the activity switch as low as possible, yet at a height still accessible to a wheelchair user, also ensures that other users have to perform movements that promote high levels of flexibility. The lowest height from the ground that a 9–12 year old wheelchair user can reach is 400 mm. Therefore, this is taken as the low activity switch height. The benefit of placing the activity switch at 400 mm from ground level means it requires an able bodied user to perform a lunge or squat exercise activating a large proportion of the muscular system.

Pushing and pulling actions are generally performed most easily at between shoulder height and a little below elbow height. It has been shown that the optimum level is 70 to 80% of shoulder height, a little below elbow height (Norris & Wilson, 1995). The mid placement of the activity switch is therefore determined by the height of elbow from ground of a 50th percentile 12 year old, a distance of 935 mm (Norris & Wilson, 1995). The mid height placement of the activity switch must also be within reach of a 9–12 year old wheelchair user. The highest reach of a 12 year old wheelchair user is 1115 mm so the mid placement of 950 mm is easily within reach.

To activate a large proportion of the muscular system it is intended that the high placement of the activity switch is such that the user must jump or fully extend their reach. The position of the high activity switch is important as it is the main determinate of the typical

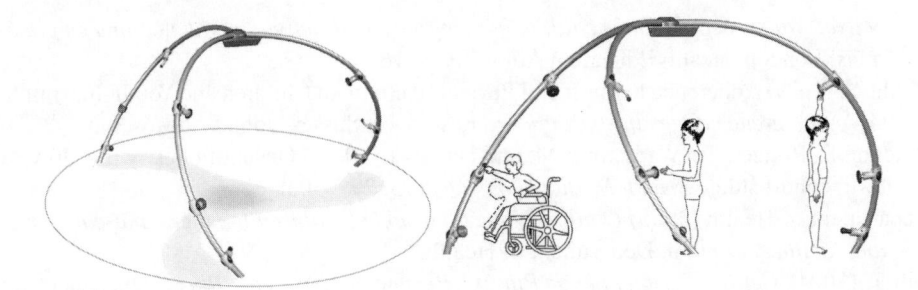

Figure 3. An isometric view and a side view of the final design incorporating 9 activity switches.

age able to play on the device. A 5th percentile 12 year old female height was used to determine the lowest acceptable height of 1.9 metres.

2.4. Final Design

The final design is presented in Figure 3. It is constructed from three curved tubular sections allowing the activity switches to be mounted off-axis at the three heights. The sculptural aesthetics provide high visual impact and look acceptable in urban environments. A working prototype has been manufactured and initial experimentation conducted to prove the principle of the device and assess the level of exertion required to advance through the game. To gather data a wireless internet connection is incorporated allowing game statistics and maintenance information to be downloaded and analysed.

3 Discussion

The design process followed has culminated in the refined final device presented. The device is envisaged to attract and engage a user group which is more accustomed to playing computer games than participating physical activity. It is thought that the device could help enhance current levels of physical activity which at present are below the requirements. The innovative step of the concept is the controllable level of exertion (similar to shuttle run training), the successive exercises, inclusion of recovery periods and the diverse range of activities required to advance through the game. Further empirical data is required to optimise enjoyment levels and total energy expenditure.

Acknowledgements

The authors would like thank Steve Hammond (Wolfson workshop technician) and Jon Trepte (System Technologies) for their electronic assistance whilst invaluable design and project support has been provided Mark Spencer (design engineer), Paul Mallinson (project manager) and John Croasdale (managing director) of Playdale Playgrounds Ltd.

References

Beaglehole R. (2002) *Oxford textbook of Public Health*, Fourth Edition: The Scope of Public Health. New York: Oxford University Press.
Biddle S, Sallis J, and Cavill N. (1998) Policy framework for young people and health-enhancing physical activity. In: Biddle S, Sallis J, and Cavill N. (Eds.). *Young and*

active? Young people and health enhancing physical activity: evidence and implications. London, Health Education Authority, 3–16.

Biddle S. (1999) Adherence to Sport and Physical Activity in Children and Youth. In: Bull,S. (ED.) *Adherence issues in sport and exercise*. West Sussex, John Wiley & Sons.

Braddon F., Rodgers B., Wadsworth M. and Davies J. (1986). Onset of obesity in a 36 year birth cohort study. *British Medical Journal*, 193: 299–303.

Department of Health (2000) *Quality protects: disabled children numbers and categories and families*. London, Department of Health.

Gill T. (2004) *Getting Serious about Play: A Review of Children's Play*. Department for Culture Media & Sport. London.

Henniger M.L. (1993) Enriching the outdoor play experience. *Childhood Education*, 70/2: 87–90.

Lobstein T., Rigby N. and Leach R. (2005) *EU Platform on Diet, Physical Activity and Health*. International Obesity Task Force EU Platform Briefing Paper. London, International Obesity Task Force.

World Health Organization (2004) Controlling the obesity epidemic. *WHO global report*. Geneva, WHO.

World Health Organization (2005) Preventing chronic diseases : a vital investment : *WHO global report*. Geneva, WHO.

THE TESTING OF ELECTRONIC PLAYGROUND EQUIPMENT TO INCREASE FITNESS IN CHILDREN

P.P. HODGKINS[1], S.J. ROTHBERG[1], P. MALLINSON[2] & M.P. CAINE[1]

[1]*Sports Technology Research Group, Loughborough University,*
Loughborough, Leicestershire, England
[2]*PlayDale Playgrounds Ltd, Haverthwaite, Ulverston, Cumbria, England*

More than half of children over the age of 6 spend less than the recommended minimum of 60 minutes a day doing moderate intensity physical activity which is the equivalent of brisk walking (Smithers *et al.*, 2000). Improving children's physical play patterns and frequency of play opportunities increases the likelihood of raising their daily energy expenditure. An interactive electronic playground device has been devised to increase fitness in children as well as agility, reaction speed and coordination. To play on the device the user is required to activate switches positioned at different heights along a tubular framework that correspond to the audio and visual commands given. The current paper describes the play testing conducted on this novel device.

1 Introduction

An interactive electronic playground device has been conceived to enhance the physical play experience of a young "electronic savy" user group. Experimental testing was therefore required to validate the concept and the spatial envelope along with providing insights into the optimal play intensity, period of play and recovery times. By altering these key parameters of the game it is thought that higher levels of enjoyment along with high levels of energy expenditure can be achieved. It is known that motivation is highest when the challenge or difficulty of the task matches personal abilities and skills (Csikszentmihalyi, 1975). This matching leads to high levels of enjoyment and engagement in the task whereas a mismatch can lead to either boredom (low challenge/high skills) or anxiety (high challenge/low skills). Effort is therefore placed at finding the optimal level of difficulty and rate of advancement through the game. The results from the experimentation are to be used to develop the final play device.

2 Proof of Principle Rig

To turn the concept into reality, an uncomplicated yet functional prototype was required. The investigation stage established that several simple electronic hand held devices are very popular amongst children aged 8 to 12 years, providing high levels of "addictive" gameplay. However, little is known about whole body electronic devices based around a comparable game format. For this reason, a test rig commanding the user to perform dynamic movements within a predetermined time was essential. To create this rig required design of a suitable framework, complete with switches positioned at specific heights controlled by software driving the game. Use of a Proof of Principle Rig (PoP-Rig) was expected to demonstrate how participants interacted with such a device and to what extent they were willing to exert themselves. To gain an understanding of these effort levels, heart rates were monitored during the test sessions and correlated with the participant's game scores.

The embodiment of the PoP-Rig consisted of five switches capable of being repositioned in different orientations and at different heights. This enabled fine-tuning of the setup ensuring an extensive array of dynamic body movements. The aim of the design was to ensure that it could be adjusted to provide various heights and widths. The PoP-Rig was designed and assembled in SolidWorks 2006 (SolidWorks Corporation; Concord MA) ensuring a satisfactory work envelope for a 5th to 95th percentile adult. It was anticipated that it would prove more beneficial to use adults rather than children for the testing as the device was at a preliminary stage, requiring ongoing design and development. The simple design allowed for quick assembly/disassembly and consisted of fourteen square tubular mild steel sections (25 mm $\times$ 25 mm $\times$ 2.5 mm) 1000 mm in length with drilled holes at 150 mm intervals for adjustment. To modify the width of the PoP-Rig the top section angles can be adjusted from 90° to 180°. It is important to note that the maximum width available is dependant upon height required.

As can be seen in Figure 1 the basic form of the PoP-Rig comprises of two archways perpendicular to each other, attached in the centre of the upper section. The placement of the five switches is critical to the physical movement patterns of the participants. To ensure participants must complete various movement patterns over various distances, each switch

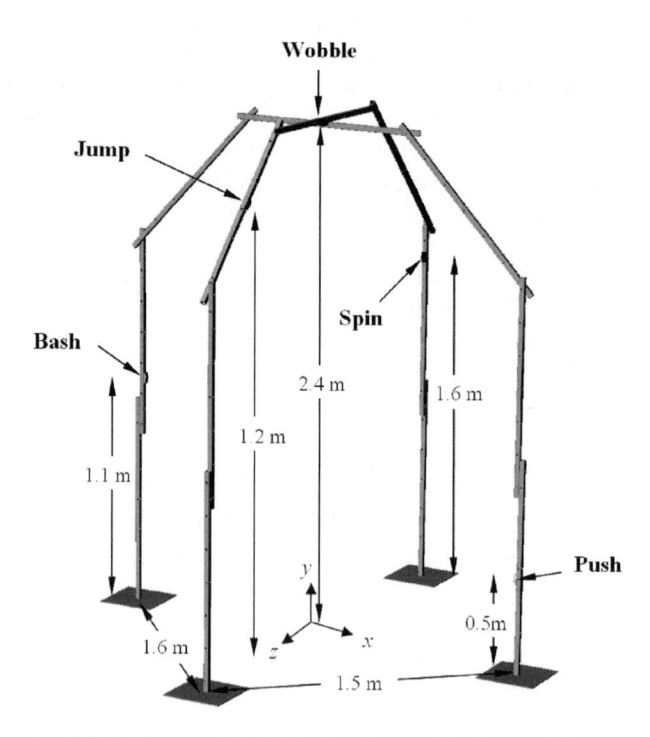

Figure 1. 3D Cad image of PoP-Rig showing positioning of all five switches.

was placed at different heights and along different sections of framework. The height of the PoP-Rig was set by the reaching distance above the head of the smallest participant who was 1.78 m tall. The five switches were given unique names so they could be easily identified and the distance between each switch combination was recorded to generate the time adjustments so that if a user has to cover a large distance an extra time period is afforded. To hit the correct switches during a game requires more than twenty five different types of movement patterns (e.g. a user will be required to move from the "push" switch to either the "spin", "wobble", "jump", "bash" or back to the "push" switch which is one of five different types of movements equating to twenty five different possible combination types).

3 Experimentation

Six, ten minute exercise sessions were conducted on the PoP-Rig. All sessions were filmed to provide an insight into how participants interacted with and around the device. Heart rates were measured using a Polar 625 × heart rate monitor (Polar Electro, Oy, Finland). Data was recorded in an Excel (Microsoft Corporation; Redmond WA) spreadsheet during the tests to document the number of games, score of each game, game period and identification of switch that the user failed on. All parameters including game scores were reset after each session to ensure consistent test conditions.

3.1. Participants

Once the PoP-rig had been manufactured, suitable participants were recruited. Testing was conducted with five healthy adult participants with a mean age of 26 ± 3.2 years. All participants had no recent history of illness or injuries, regularly partaking in exercise with good levels of cardiovascular fitness.

3.2. Experimental Protocol

The participants were asked to attach the heart rate transmitter and wrist worn receiver before the start of the experimentation. The monitor was then tested to ensure it was receiving heart beats correctly before being allowing a single practice trial to familiarise themselves with the audio commands and placement of switches. Participants were told that their scores would be recorded for the next ten minutes, in which time they should try to achieve their highest score. They were also informed that they could start as many games in that ten minutes as they like. A count down was given to the session whereby they were asked to press the start button on the watch commencing the data acquisition. Participants were told that to activate the game they should press the start button and once a fail had been committed, it could be pressed again to commence a new game returning the score to zero.

3.3. Results

An overview of the data gained from the test sessions are presented in Table 1. The sessions were video taped and reviewed after the experimentation to verify all results.

The user's score from all games were correlated to heart rate to give an indication as to how a user's heart rate increases in response to the increased intensity during a game. Table 2 gives an overview of the derived values.

Table 1. Overview of results from experimentation.

Overview of Results	Value	SD
Highest score	47	–
Game period for highest score	110 s	–
Max heart rate recorded	163 bpm	–
Mean total score	215	17.6
Mean score per game	14	1.9
Mean number of games in 600 s	16	1.5
Mean game period	39 s	3.7
Mean highest score	33	8.9
Mean highest score game period	79 s	19.8
Mean of age of participants	26	3.2
Mean heart rate	119 bpm	12.9
Mean max heart rate	140 bpm	12.9

Table 2. Overview of derived values from experimentation.

Overview of Derived Values	Value
Highest correlation coefficient	0.81
Lowest correlation coefficient	0.34
Mean correlation coefficient	0.58

3.4. *Analysis*

The mean number of games each participant accumulated in 10 minutes was 16 ± 1.5 with a mean score per game of 14 ± 1.9. This represents a high number of games with quite a low score, denoting that there was a high turnover of games during the session. As the mean game period was only 39 ± 3.7 seconds, game parameters will need to be changed in order to increase the period of each game. The highest score recorded was 47 lasting 110 seconds. The mean high score was 33 ± 8.9 with these games lasting a mean of 79 seconds. In general game scores correlate well with heart rate. The results show, as expected, when the score increases during a game, heart rate follows. The highest correlation (R value) when comparing high scores with heart rate is $r = 0.81$, the lowest being $r = 0.34$ with a mean value of $r = 0.58$. The mean heart rate for all sessions was 119 ± 12.9 bpm with the maximum heart rate averaging 140 bpm. Moderate intensity heart rate range is between 50–70% of maximum, vigorous intensity between 70–85%. Using 194 bpm as a maximum for a 26 year old estimated using 220 minus age convention (Wilmore & Costill 2004) means that moderate intensity represents between 97–136 bpm and vigorous between 136–165 bpm. The mean heart rate of 119 bpm corresponds to 61% of maximum and is classed as moderate intensity exercise while the maximum heart rate mean, achieved during the sessions, of 140 bpm is 72% of maximum and falls into the vigorous intensity category. During the sessions a rise in heart rate was universally observed suggesting that the device should produce the desired result of an aerobic based activity profile. The eventual aim of the device is, however, more subtle by encouraging adherence while maintaining high levels of enjoyment even during elevated levels of exertion. The data illustrates that the correlation between heart rate and game score shows that as the game speeds up users are willing to increase their exertion levels to a vigorous intensity. The highest correlation coefficient between heart rate and game score was during test 4 ($r = 0.81$), it was observed that this participant had a more energetic style as opposed to some of the other participants.

Users failed mainly on the "push" button, thought to be due to the low positioning. For future development a higher time adjustment period will be added to reduce premature failure. The games did not last for as long as expected, only 39 s and therefore a greater time interval will be implemented which will create a slower start but prolong the game. In addition the switches used on this PoP-Rig were too small, requiring a higher level of dexterity than if larger activity switches were used. Many users commented that the small switches actually slowing their speed of movement and if users thought they had operated the switch correctly but it did not register a hit they found it very frustrating. Effort is to be placed in the following areas to improve the current PoP-Rig:

- Increase the spatial envelope and research basis for positioning activity switches
- Use of large hand activated switches
- Data acquisition of participant gameplay performance (i.e. reaction speed)
- Audio and visual enhancements
- User feedback on perceived exertion and enjoyment levels
- Recruitment of participants aged between 10 and 12 years.

4 Future Experimentation

To measure the effect of altering specific game parameters on enjoyment levels, future experimentation will manipulate speed and length of games to generate a response in exertion levels. From these sessions enjoyment levels will be recorded to discover which type of game provides the highest level of enjoyment. To asses the level of effort a rating of perceived exertion scale specifically created for children is to be used. The children's OMNI scale developed by Robertson (2004) uses pictorial cues closely depicting the activity being performed. The user is required to select a number from the scale which ranges from 0 for "not tired at all" to 10 for "very, very tired". Studies have shown that children as young as 6 years of age can correctly use the OMNI scale and that the scale is appropriate for exercise testing, self-regulating training intensity and tracking training progress (Robertson 2004). Enjoyment is an acknowledged element of a child's motivation for partaking and continuing to partake in physical activities. To evaluate the level of enjoyment subjects experience during play with the new electronic device the scale of perceived enjoyment has been devised. This scale closely matches the format and layout of the OMNI scale of perceived exertion ensuring consistency and coherence during the user trials. The scale allows young users to quickly evaluate how they are feeling with the aid of simple diagrams and corresponding number scale.

5 Discussion

The testing performed on the PoP-Rig has provided an excellent insight into the whole body concept of command driven play. The testing allowed a substantial amount of knowledge to be acquired regarding the levels of intensity required to participate in the game, the spatial layout and the specific feedback from participants proved very constructive highlighting recommendations for the final device. The PoP-Rig demonstrates that the principle is valid although a more advanced test rig is required along with more extensive experimentation.

Acknowledgements

The authors wish to acknowledge Playdale Playgrounds Ltd, who have financially supported the project, provided design expertise and built the hardware for the rig and John Trepte of System Technologies and Steve Hammond (Wolfson electronics workshop technician who has contributed substantially to the development of the electronics and gaming software. Design and project support has been provided by Playdale Playgrounds with particular reference to Mark Spencer (design engineer), Paul Mallinson (project manager) and John Croasdale (managing director).

References

Wilmore J. and Costill D. (2004) *Physiology of Sport and Exercise*, Champaign, IL, Human Kinetics Publishers.
Csikszentmihalyi M. (1975). *Beyond boredom and anxiety*. San Francisco, Jossey-Bass.
Robertson R. (2004) *Perceived Exertion for Practitioners: Rating Effort with the Omni Picture System*, Champaign, IL, Human Kinetics Publishers.
Smithers G., Gregory J. R., Bates C. J., Prentice A., Jackson L. V., and Wenlock, R. (2000) The National Diet and Nutrition Survey: young people aged 4–18 years. *Nutrition Bulletin*, 25/2, 105–111.

VISUALIZATION OF THE HAZARDS LURKING IN PLAYGROUND EQUIPMENT BASED ON FALLING SIMULATIONS USING CHILDREN MULTI-BODY MODELS

Y. MIYAZAKI[1], S. WATANABE[2], M. MOCHIMARU[3], M. KOUCHI[3], Y. NISHIDA[3] & S. UJIHASHI[2]

[1]*Kanazawa University, Kanazawa, Ishikawa, Japan*
[2]*Tokyo Institute of Technology, Tokyo, Japan*
[3]*National Institute of Advanced Industrial Science and Technology, Tokyo, Japan*

We developed a method to construct hazard maps of playground equipment, calculated from simulations, by using computer models of children falling on a playground slide. This method makes it possible to understand the hazards of playground equipment easily. Full-body multi-body models of children, based on Japanese data, were constructed. The hazard map of a playground slide was constructed to provide an example of possible hazard maps of playground equipment. Simulations of children falling on the playground slide were carried out by using both the multi-body models for children and the playground slide models, which were constructed from CAD data of an actual slide. The calculated head injury hazard values were mapped on the playground slide model so as to easily identify hazards. As a result, the map made it possible to visually determine the hazards lurking in the playground slide.

1 Introduction

Although playground equipment makes it possible for children to play and have fun, hazards capable of causing serious injury are present. It is necessary to design safer playground equipment for children's safety.

However, there were previously some limitations when predicting the hazards lurking in playground equipment in the product design phase, because a child often behaves unpredictably in playgrounds. Therefore, it is important to clarify the hazards lurking in playground equipment, and to identify them from both accident prevention and injury mitigation points of view. It is especially effective if the hazard map can easily identify and clearly show those hazards, making it easy to understand where the risks lie.

Although research to asses the usability of a simple biomechanical model to predict the likelihood of children's injuries in playgrounds has previously been carried out (Davidson *et al.*, 2006), a method for visualizing the hazards has not been developed yet.

In order to create and present the hazard map, it is best to use real accident data containing detailed information on the injury patterns associated with both the environment and victim, for example the falling point, behavioral patterns, body shape and age. However, such precise data are not available at this time. Therefore, it is believed that an effective way to compensate for the lack of precise data is to firstly use simulation results to create the hazard map, and then to validate the map by using actual accident data.

As a large amount of simulations are required for the construction of the hazard map, a multi-body model, which features both ease of full-body modeling as well as short calculation time, is appropriate for the simulations.

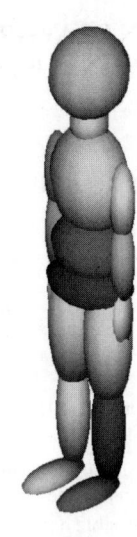

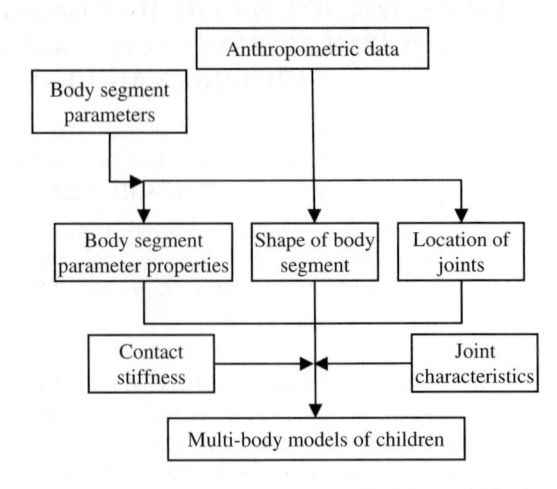

Figure 1. Multi-body model of a child.

Figure 2. Model construction method for multi-body models of children.

Therefore, the purpose of this study is to construct children's multi-body models, based on Japanese children's dimensions, and to develop a method for visualizing hazards present in playground equipment by using those simulation results.

2 Construction of Multi-Body Models

Two multi-body models for three- and seven-year old children, based on the dimensions of Japanese children, were constructed. Fig. 1 shows a model consisting of 17 ellipsoidal segments and 16 joints; the head, the neck, the thorax, the abdomen, the pelvis, the thighs, the calves, the feet, the upper arms, the lower arms, and the hands. Fig. 2 shows the process used to construct the model. The shape, location of the center of gravity of each segment, and the location of the joints were calculated from 39 different body dimensions. Although most of the body dimension data referred to the data of average 3- and 7-year old Japanese children (Consumer Product Safety Association, 1973), some data outwith the research area were obtained from the Anthrokids project (Snyder et al., 1977). The inertial properties were calculated from the body segment parameters of Japanese children (Yokoi et al., 1986).

Joint characteristics of each joint were expressed as a relationship between joint angle and passive torque as shown in Fig. 3 (Yang et al., 1997). Although the joint characteristics should be for children, the characteristics of an adult male were used due to a lack of data. The contact stiffness of each segment was expressed as a nonlinear relationship between penetration and the contact force (shown in Fig. 4). The relationships for contact stiffness were defined by using the data of a Hybrid-III dummy.

3 Construction of the Hazard Map for Playground Equipment

A visualization method capable of displaying the hazards lurking in playground equipment was developed based on the simulation results when using multi-body children's and playground equipment models. Firstly, a multi-body model of the playground equipment was

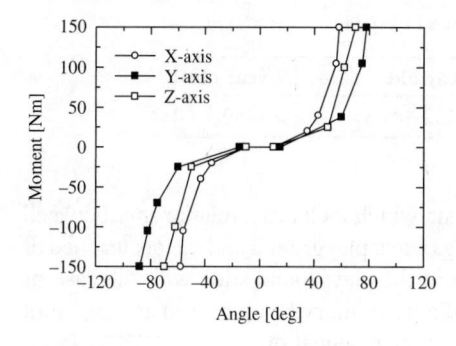

Figure 3. Moment-angle characteristics of neck joints.

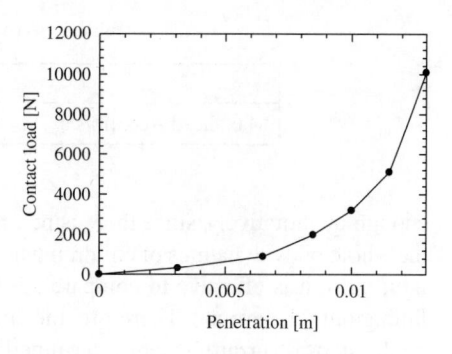

Figure 4. Contact stiffness of an arm.

Figure 5. CAD model of the slide.

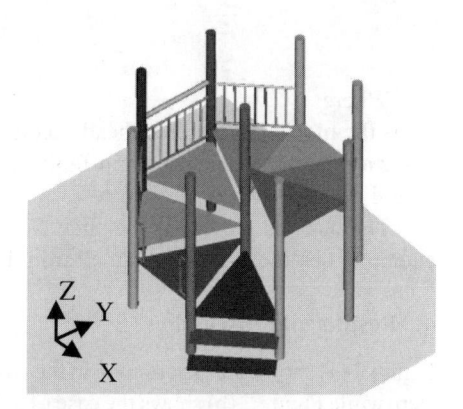

Figure 6. Simulation model of the slide.

developed from the CAD data. Several simulations were conducted to construct a virtual accident database. In this case, the multi-body model is suitable for the construction of a virtual accidents database because the execution time of each simulation is quite short.

Injury hazard values for each case were obtained from these simulation results and then visualized by mapping onto the playground model.

A playground slide was chosen to demonstrate this method. A simulation model of the slide was constructed from the actual CAD data of the slide (as shown in Figure 5). To reduce the calculation time, only the spiral staircase was modeled using finite elements and used in the simulations, as shown in Fig. 6. The steps were defined as rigid bodies since the stiffness of the steps is significantly higher than the stiffness of the head of the children's model.

Falling down the spiral staircase was simulated. Sixty-six points on each step were selected, and the multi-body models for children were made to fall from these points in four directions. The model was inclined to 20 degrees from standing to imitate a natural falling posture.

Then, the maximum resultant head acceleration response, which is used in some safety standards, was obtained from each simulation. The validity of the child model is not examined

Table 1. α values used in Eq. (1).

	3 year old	7 year old
Max. head Acc. [m/s^2]	2132.4	1989.5

enough quantitatively, since there is no precise data which includes a relationship between the whole body dynamics of children when falling onto a playground and the occurrence of injury. So, it is effective to compare the hazards in the playground slide with the case of falling onto the ground. Therefore, the chances of a head injury being caused at each point on the steps ("Hazard") were determined by the following equation,

$$H = \frac{h}{\alpha} \qquad (1)$$

Where,
– h is the maximum resultant head acceleration response observed
– α is a reference value shown in Table 1, obtained from simulations of simple falling onto a rigid ground surface.

The hazards present in the slide were visualized on the CAD data of the playground equipment by the isolines obtained from linear interpolation of the H values at each point.

4 Results and Discussion

Figure 7 (a) shows the hazard map in the case of a 3-year old child from the 3rd step to the 7th step, while Figure 7 (b) shows the case of a 7-year old. These maps demonstrate that the inside section of the spiral staircase is more dangerous than the outside. Figure 8 shows the behavior of the 3-year old model falling down from the inside section of the 6th step. In this case, the head impacts on the 3rd step without impacting on either the 5th or 4th steps, because the inclination of the staircase at the inside is much steeper than at the outside. The head also rotates when the thigh hits the 5th step, which causes a higher impact speed of the head. Furthermore, head acceleration in the case of falling in the slide was higher when compared to a normal falling situation. Additionally, the H value at both the inside of the 3rd step and the outside of the 4th step is higher because the children's model falls onto the ground surface directly due to there being no guard fences.

In the 3-year old's case, the H value was much higher at the outside of each step, although this was not observed in the case of the 7-year old, which is difficult to predict before construction of the hazard map. Fig. 9 shows a comparison of the behavior at the head contact time for the 3-year and 7-year old child models. In the case of the 3-year old, the head and pelvis collide with the step at the same time. However, in the case of the 7-year old, the head collides with the step after the pelvis contacted the step first, which makes the kinetic energy of the whole body be absorbed. As a result, the H value in the 3-year old child at this point is higher than that of the 7-year old. These results show that the danger area in the slide varies with the relative relationship of the shape and dimensions between the specific structure of the spiral staircase and children's body being different from ages, which was difficult to predict previously.

(a) 3-year old   (b) 7-year old

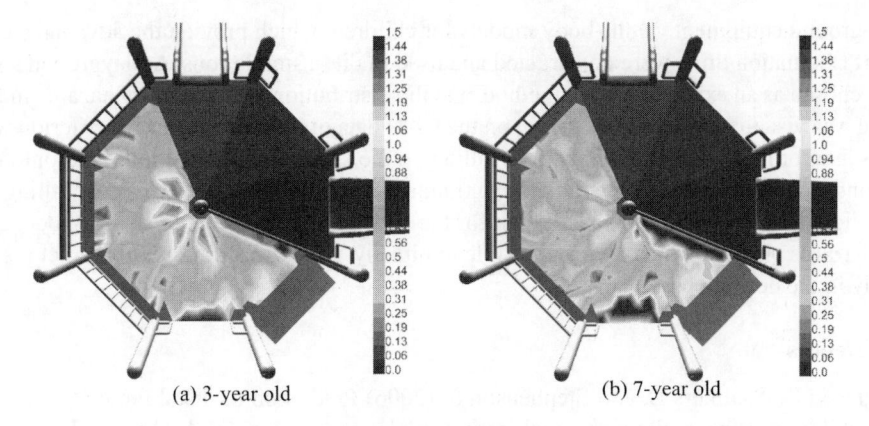

Figure 7. Hazard maps for head injury in falling.

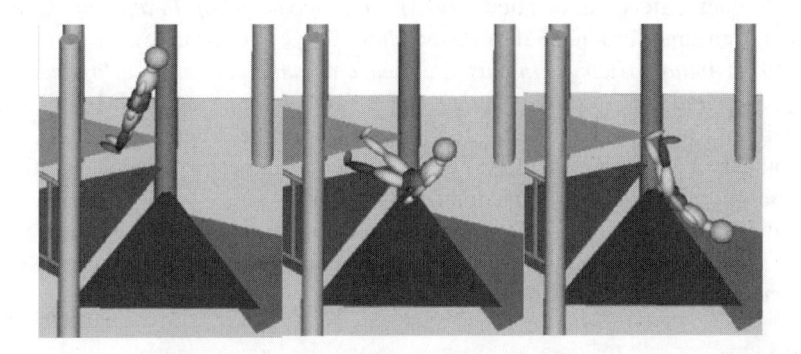

Figure 8. Behavior of the 3-year old model when falling from inside the 6th step.

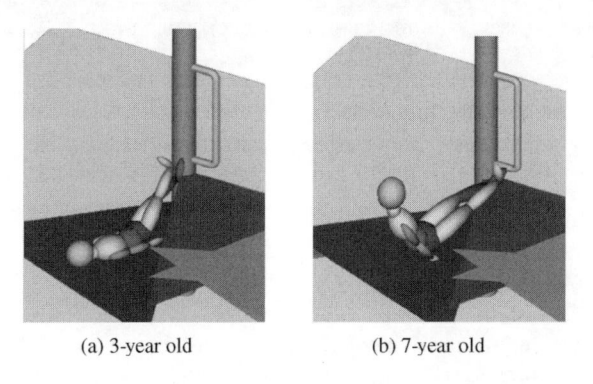

(a) 3-year old   (b) 7-year old

Figure 9. Falling behaviors from outside of the step.

5 Conclusion

A method for visualizing the hazards lurking in playground equipment was constructed based on the results of simulations conducted with computer models of both children and playground equipment. Multi-body models for children, which provide the advantage of a short calculation time, were constructed and used in these simulations. A playground slide was chosen as an example for the method, and the distribution of head injury hazards in the slide was visualized by isolines drawn on the CAD data of the slide. A hazard criterion was chosen in order to show the dangers of falling on the slide as compared to falling onto the ground, and was defined as the ratio of maximum head acceleration in the case of falling on the slide to the ground. The hazard map for the slide made it possible to easily understand the hazards present in the playground slide intuitively, which was difficult to predict previously in the design phase.

References

Davidson P., Chalmers D. and Stephenson S. (2006) Prediction of distal radius fracture in children, using a biomechanical impact model and case-control data on playground free falls. *Journal of Biomechanics*, 39, 503–509.
Consumer Product Safety Association (1973) *Anthropometry of Infants of Japanese Children*, Consumer Product Safety Association, Tokyo (in Japanese).
Snyder G. (1977) *Anthropometry of Infants, Children, and Youths to Age 18 for Product Safety Design, Prepared for Consumer Product Safety Commission*, Report UM-HSRI-77-7, University of Michigan, Highway Safety Research Institute, Michigan.
Yokoi T., Shibukawa K. and Ae M. (1986) Body segment parameters of Japanese children. *Japanese Journal of Physical Education*, 31/1, 53–66.
Yang J.K. and Lovsund P.J. (1997) Development and validation of a human-body mathematical model for simulation of car-pedestrian collisions, Proceedings of IRCOBI Conference, Hannover, 24–26 September, 133–149.

PERSONALITY TRAITS AND THEIR RELATIONSHIP TO LEISURE MOTIVATION AND LEISURE SATISFACTION IN SOUTHERN TAIWAN UNIVERSITY STUDENTS

C.H. CHEN

Department of Sports & Recreation Management, Chang Jung Christian University, Tainan, Taiwan

The primary purpose of this study was an investigation of the possible relationship of the five-factor model of personality and the leisure motivation and satisfaction. The secondary purpose was to understand the differences between demographics (gender, student classification, TV watching time, and leisure participation time) and the five dimensions of the five-factor model of personality, leisure motivation, and leisure satisfaction for university students in Taiwan. Six hundred university students were asked to participate in this study; 485 students ages 18 to 28 completed the questionnaire for a response rate of 80.8%. The results showed that (a) there were significant relationships between dimensions of the five-factor model of personality and leisure motivation, (b) there were significant relationships between dimensions of the five-factor model of personality and leisure satisfaction, (c) there was a significant relationship between leisure motivation and leisure satisfaction, (d) there were significant differences between dimensions of the five-factor model of personality and selected demographics, (e) there were significant differences between components of leisure motivation and selected demographics, and (f) there were significant differences between leisure satisfaction and selected demographics. This research contributed to the understanding of the relationships among personality, leisure motivation, and leisure satisfaction. Such an understanding is useful for guiding the design and evaluation of leisure and recreational programs.

1 Introduction

Teenagers and college-age students are experiencing rapid personality development and learning about life; they also have a great deal of curiosity and want to experience adventures. According to McCrae & Costa (1994), although the level of personality traits are very stable in adulthood (after age 30), there are significant changes between adolescence and early adulthood (with some decreases in neuroticism, extraversion and openness and increase in agreeableness and conscientiousness). Leisure activity plays a very important role in college-age students' personality. This may be the most important time of psychological and physical development in their life. Misdirected leisure time may influence personality development and learning. Participation in leisure activity that improves quality of life and leisure satisfaction becomes very important.

Because of the influence of cultural and academic pressures, the concept of education is "all work, and no play" for Taiwanese students. This leads teachers and parents to misunderstand the word *leisure*. They think leisure is a waste of time or simply meaningless activity. Therefore, they do not encourage students or children to participate in leisure activities. Tsai (2001) discovered that teenagers sometimes indulge in excessive drinking, violent behavior, or drug use, or engage in activities that have a high element of risk such as motorcycle street racing. Racing is a very serious problem among teenagers in Taiwan and is generally associated with violent behavior preceded by little or no provocation. Many youths become involved in motorcycle racing to earn attention and respect from their peers or because they are looking for excitement. Chang (1994) pointed out that one of the major contributors to this problem is lack of proper leisure experience and interests. Other possible reasons are poor academic performance or unstable home environments.

According to Maslow's (1970) study on motivation and personality, a person who reaches the self-actualization level would feel happiness, wellness, and valuation, or what Maslow called "peak experience." Teenagers participate in motorcycle racing to achieve peak experience and quickly become addicted to it (Gao, 1996b). Tragedies can occur when teenagers lack knowledge of proper leisure activities and the motivation to participate in appropriate leisure activities.

Zhong (1996) reported that college students in his study said that they lacked leisure instructors as well as leisure skills and abilities. More so, teachers do not encourage participation in leisure activities, which negatively affects participation. According to Yu (1998), overemphasis on hard work and the pressure of learning achievement can cause students to suffer many psychological disorders and develop various societal problems. Therefore, having proper leisure educational participation should be a requirement for students.

From a teacher's perspective, if there is a conflict between academics and leisure, academics should always come first. Otherwise, students who participate in leisure activities will feel guilty. Unfortunately, students usually learn how to enjoy leisure activities and experience their benefits only when starting their university education. This is why the researcher focused on the university and college education level in this research. But sadly, without a teacher's encouragement and instruction, it is difficult for students to manage their leisure activities. They do not know how to use their free time to engage in proper leisure activities and thus are limited in their options. A lack of encouragement and motivation may cause students to choose improper leisure behaviors. Therefore it is important to determine how to educate university students to participate in proper leisure values and attitudes. Teachers need to know a student's personality traits, leisure motivation, leisure satisfaction, and what kind of proper activities are available. For these reasons, this researcher hoped to be supportive of university students and teachers in finding ways to provide leisure education references and in creating better leisure opportunities for students.

1.1. *Statement of the Problem*

Research in the area of personality and leisure has been relatively scarce (Reddon *et al.*, 1996). In past leisure research, the researcher found that no attention had been given to the relationship between the five-factor model of personality and the leisure activities of university level education in Taiwan. This study investigated the five-factor model of personality and its relationship to leisure motivation and leisure satisfaction with data from 600 students enrolled in Hsing Hku University during December 2004 and January 2005.

1.2. *Purpose of the Study*

The primary purpose of this study was an investigation of the possible relationship of the five-factor model of personality to the leisure motivation and satisfaction of Taiwanese university-level students. The secondary purpose was to understand the differences between demographics and five dimensions of the five-factor model of personality, leisure motivation, and leisure satisfaction in university students in Taiwan. The findings provide theoretical and practical contributions to the academic community as well as to the leisure services profession. This research will contribute to the understanding of the relationship among personality, leisure motivation, and leisure satisfaction. Such understanding will be useful for guiding the design, implementation, and evaluation of leisure and recreational programs.

2 Methodology

2.1. Sampling

Participating in this study were 600 students enrolled during December 2004 and January 2005 at Hsing Hku University of Management in the southern region of Taiwan. The student sample was stratified by gender and class year (75 freshman males, 75 freshman females, 75 sophomore males, 75 sophomore females, 75 junior males, 75 junior females, 75 senior males, and 75 senior females). This sample was chosen using a computer-selected random sample from more than 5,000 full-time students majoring in 15 undergraduate programs.

2.2. Instrumentation

The following instruments were used for the study: the NEO Five-Factor Inventory (NEO-FFI) Form S for college students, the Leisure Motivation Scale (LMS), the Leisure Satisfaction Scale (LSS), and a demographic survey regarding gender, student classification, TV watching time, and leisure participation time.

2.3. Data Analyses

After collecting responses, the researcher scored the instruments and analyzed the data. The quantitative data were analyzed using the Statistical Package for the Social Sciences 10.0 (SPSS 10.0) computer program using descriptive and correlation analyses. Additionally, the reliability of the three instruments was examined using the Cronbach alpha coefficient test. The summary or description statistics were used to provide descriptive measures of the sample demographic variables, which included gender, student classification, TV watching time, and leisure participation time. The Pearson product-moment correlation was used to determine if there was any relationship between each dimension of the five-factor model of personality, and leisure motivation and leisure satisfaction. In addition, multiple correlation squared (R^2) and partial correlation were also tested. The one-way analysis of variance (ANOVA) was used to determine if there were differences between the selected demographics and the five dimensions of the five-factor model, leisure motivation, and leisure satisfaction of the students. The significance of the differences was tested at the .05 alpha level. A Scheffe's follow-up procedure was used to determine if there were differences in each group within each demographic variable. An alpha level of .05 was used.

3 Results and Discussion

3.1. *Responses of the Questionnaires*

Questionnaires were mailed to 600 students enrolled during December 2004 and January 2005 at Hsing Hku University of Management in the southern region of Taiwan. A total of 485 students ages 18–28 returned the completed questionnaires. Forty-eight students in the sample could not be located, and 67 students' questionnaires were not useable. The response rate of this study was 80.8%.

3.2. *Current University Students in the Five-Factor Model, Leisure Motivation, and Leisure Satisfaction in Taiwan*

In the rank of the five-factor model of personality, the highest score was the agreeableness dimension ($M = 27.64$), followed by openness to experience ($M = 27.54$), extraversion ($M = 27.39$), conscientiousness ($M = 27.22$), and neuroticism ($M = 24.14$).

For leisure motivation, the highest score was for stimulus-avoidance ($M = 30.51$), followed by intellectual ($M = 29.73$), competence-mastery ($M = 29.61$), and social components ($M = 28.99$). Therefore, the researcher assumed that students in this study had engaged in leisure activities to escape from over-stimulating life situations. This finding confirmed previous research by Hsieh (1998), who studied Taiwanese university students. The result found that Taiwanese university students had the highest score on stimulus-avoidance ($M = 31.85$), followed by competence-mastery ($M = 30.82$), intellectual ($M = 29.92$), and social subscales ($M = 29.26$).

In conclusion, the studies discussed above found that stimulus-avoidance was an important component to Taiwanese university students' motivation for participating in leisure activities. Students reported that they experienced high levels of stress arising from both their desire to meet family expectations as well as academic stresses in general. As a result, students chose leisure activities to relax and escape from this intense environment. University students in Taiwan preferred to participate in leisure activities such as singing karaoke, dancing, and watching movies…but more importantly, we must be concerned that quite often they chose inappropriate leisure activities such as motorcycle racing and drug use. As a result, access to proper leisure facilities and services must be made readily available.

On the other hand, few students reported being primarily socially motivated to participate in leisure activities. This indicated that Taiwanese university students rarely engaged in leisure activities to improve interpersonal relationships or to obtain the esteem of others. University students were more likely to be care-free and relaxed while participating in leisure activities. This lack of social motivation could be due to cultural influences. Social relationships in Taiwan continue to be very important and much care is taken to manage relationships with others. Because of this, many Taiwanese people enjoy the benefits of varied social interactions and positive interpersonal relationships. Therefore, most students were not motivated to participate in leisure activities to receive the esteem of others or to improve their social interactions.

For leisure satisfaction, the highest score was the relaxation ($M = 16.12$), followed by the psychological ($M = 14.35$), social ($M = 13.99$), educational ($M = 13.82$), physiological ($M = 12.91$), and aesthetic ($M = 12.89$) subscales. In the area leisure satisfaction, the university students who scored highest on this part sought relief from the stress and strain of life through leisure activities. This finding confirmed previous research by Hsieh (1998), who studied Taiwanese university students. His study found that Taiwanese university students had the highest score on relaxation ($M = 16.18$), followed by the psychological ($M = 14.46$), social ($M = 14.38$), educational ($M = 14.29$), physiological ($M = 13.82$), and aesthetic ($M = 13.73$) subscales. This finding also confirmed the conclusions of Huang (2003), who studied junior high school students. The fact that the results remained consistent across age groups further confirmed this finding.

In conclusion, students who scored high on the relaxation subscale indicated that Taiwanese university students sought relief from the stress and strain of life through leisure activities. This result confirmed previous findings that university students were motivated to participate in leisure activities characterized by a relaxed environment and which offered an escape from high stimulus situations. On the other hand, the aesthetic subscale had the lowest influence on leisure satisfaction. This indicated that Taiwanese university students experience less leisure satisfaction in environments that were not as aesthetically pleasing, well-planned and beautiful. Therefore, building well designed leisure facilities and providing appropriate leisure education continues to play an important role in relieving the academic and social pressures that Taiwanese university students experience.

4 Conclusions

Based on the findings, the researcher concluded:

1. There were significant relationships between dimensions of the five-factor model of personality and leisure motivation. The neuroticism dimension had a significant negative correlation with leisure motivation, while the other dimensions exhibited a significant positive relationship.
2. There was a significant relationship between dimensions of the five-factor model of personality and leisure satisfaction. The neuroticism dimension had a significant negative correlation with leisure satisfaction, while the other dimensions exhibited a significant positive relationship.
3. There was a significant relationship between leisure motivation and leisure satisfaction.
4. There was a significant difference between agreeableness and genders. Results showed that female students had higher agreeableness than male students.
5. There was a significant difference between agreeableness and student classification. Results showed that junior students had higher agreeableness than senior, freshman, and sophomore students.
6. There were significant differences within the leisure participation time with regard to the five-factor model of personality. Results showed that students who participated in leisure activities for between 4–8 hours per week had a higher neuroticism characteristic personality than the 0–3 hour(s), 9–15 hours, and more than 15 hour groups. In addition the 9–15 hours group had higher extraversion personality characteristics and a higher openness to experience than all other groups. In addition, the 9–15 hours group had a higher openness to experience personality characteristics, higher agreeableness personality characteristics than all others groups, and higher conscientiousness personality characteristics than all others groups.

7. There was a significant difference between leisure motivation and genders. Results showed that male students had higher leisure motivation than female students.
8. There was a significant difference between leisure motivation and leisure participation time. Results showed that students who participated in leisure activities between 9–15 hours per week had a higher leisure motivation than all others groups. Also a significant difference found by using Scheffe's follow up, the result showed that 9–15 hours > 0–3 hour(s) groups.
9. There was a significant difference between leisure satisfaction and genders. Results showed that male students had higher leisure satisfaction than female students.
10. There was a significant difference between leisure satisfaction and leisure participation time. Results showed that students who participated in leisure activities between 9–15 hours per week had higher leisure satisfaction than all others groups. Also a significant difference found by used Scheffe's follow-up, the result showed that 9–15 hours > 0–3 hour(s) groups.

References

Chang C.S. (1994) Educational psychology – 3 orientations (in Chinese). Taibei: East China.

Gao E.Z. (1996b) The problem of teenager motorcycle racing (in Chinese). 教育資料與研究 (*Education material and research*), 12, 78–83.

Jenkins C.D., Rosenman R.H., and Friedman M. (1967) Development of an objective psychological test for the determination of the coronary-prone behavior pattern in employed men. *Journal of Chronic Diseases*, 20, 371–379.

Maslow A.H. (1970). *Motivation and personality* (2nd ed). Harper & Row, New York.

McCrae R.R. and Costa P.T. (1994) The stability of personality: observations and evaluations. *Current Directions in Psychological Science*, 3, 173–5.

Reddon J.R., Pope G. A., Friel J.P. and Sinha B.K. (1996) Leisure motivation in relation to psychosocial adjustment and personality in offender and high school samples. *Journal of Clinical Psychology*, 52/6, 679–683.

Tsai C.S. (2001). The cause and prevention for the problem of motorcycle racing (in Chinese; online). Available: http://www.ntpu.edu.tw/law/paper/04/2000b/8871404a.PDF#search = "憩?". (Accessed: July 30, 2005).

Yu C.S. (1998) An ethnographic study on the effects of peer interaction in cooperative learning in situated learning environment (in Chinese). MSc Diss, Tamkang University, Taiwan.

Zhong Q.Z. (1996) A study of the leisure behavior for college athletes – An example of sports major student in national Taiwan college of physical education (in Chinese). MSc Diss, National Taiwan College of Physical Education.

22. Coaching Technology and Sport Education

VIDEO TECHNOLOGY AND COACHING

B.D. WILSON, A. PHARMY & M. NADZRIN

Centre for Biomechanics, National Sports Institute,
Kuala Lumpur, Malaysia

In this presentation I wish to focus on four points: First, description of the video technology that is commonly used in coaching. Second, explaination of why we use video in coaching primarily from the perspective of a sports biomechanist. Next description of how we use the technology and provide examples of the application of video technology in sports. Finally, I will reflect on some issues when considering best practice in use of video technology for feedback in coaching.

1 Introduction

A quick visit to the WWW using Google Scholar reveals in excess of one million web sources for material on "video technology and coaching". By contrast a search of the sport library databases of published literature reveals less than 1000 sources. These observations plus such phrases as "seeing is believing" and "a picture is worth a thousand words" support the contention that the use of video technology is widely promoted but not so well researched or justified. In spite of the accepted "added value" of using video when using video in a sports coaching situation the majority of studies that describe the benefits and pitfalls of video technology in coaching appear to be in the teacher training literature. The approach in this paper is a first attempt to tie the two fields together, first describing current approaches on video use in sport and then reflecting on issues that have been raised regarding video feedback.

2 What Video Technology is Used?

Taking video of a sports performance may be thought of as the process of capturing or recording multiple images, or frames, in a time sequence of the performance. The sequence of images can be replayed in a video "movie" or as single still images. By contrast multiple images not taken in a time sequence can only be viewed as still images, perhaps in a time series. The video may be recorded in analogue or digital format. A typical analogue format would be a VHS cassette or mini-cassette. Typical digital formats would be as mini-DV cassette or direct to computer via firewire, or by Cameralink or Ethernet. Digital formats have particular advantages in that the images can easily be copied and the image quality does not deteriorate. Images may also be readily manipulated in the computer for presentation of movies with headers, transitions, music etc by many available moviemaker software packages. Other examples of computer manipulation might be compression, for reduced storage space of video records, or changes in format to reduce the image recording size and increase access speeds. Compression and changes in format are likely to reduce the quality of the digital image.

Standardisation of video recording formats for has been poor so that there are multiple standards for analogue recording eg Beta, VHS, S-VHS, NTSC-VHS and PAL-VHS and for digital recording eg as AVI, WMV or DVD formats. The latter format is not compatible with many moviemaker or file manipulation software packages and hence cameras of this type are not recommended for coaching applications.

Video Camcorders: A digital video camera with an a integrated recorder and a selectable high shutter speed, commonly referred to as DV camcorder, is the "camera" of choice for sports use. Standard PAL-VHS and DV cameras operate at 25 frames per second (AKA 25 Hz). These frames can be separated into two video fields in a video player or when the video is transferred to a computer and hence the video speed is sometimes referred to as 50 Hz. Slow motion video replay is achieved very simply when a video movie is replayed as say 10 Hz rather than at normal speed of play. Slow motion replay in forward and backward directions (shuttle search), and stop-action replay are features of a video recorder/player system that are particularly useful to the coach and athlete when viewing and examining a performance. Stop action video is when the video is paused ("stopped", or "frozen") on a video field with a clean picture image shown (Clean images without blur of a fast moving object require high shutter speeds). The stop action should be field selectable to stop at video fields (50 Hz) rather than video frames (25 Hz). The greater the number of heads or driving rollers in the video recorder/player generally results in a better quality image. True high-speed cameras record images at frame rates much higher than 50 Hz. Typically a frame rate of 200 Hz might be required to record several images in the golf downswing or at ball release in 10-pin or cricket bowling. Sometimes cameras with a high shutter speed, cameras that might record a non blurred action of a fast object, are erroneously described as high speed cameras.

Video Camera: This is a stand alone image sensor requiring a separate recorder unit. An example would be the typical pencil-size security camera seen in many stations or building hallways. Typically a stand alone camera is used in sport coaching when high image quality is required, for high speed image capture to computer, or for situations requiring a very small camera. Video cameras have decreased in size as digital component circuitry has diminished, to the extent that cameras may be mounted on the performer eg head mounted or cycle mounted, or in the track or equipment used. Such small cameras can be placed in waterproof housings under-water eg "peeperscope" (Biofit, Technologies, Singapore) for synchronised swimming while the recorder is effectively the computer on the poolside.

Computer Control of Video Capture: This is a technological development which has resulted in many benefits for the coach and athlete. With a simple digital video (DV) camera connected to the computer with a firewire cable, the coach can make use of features such as multiple screen display, highlight of key frames "strobe motion" and overlaying video images. These are typical features of reasonably priced video capture and systems. Many systems such as Dartfish (www.dartfish.com) and siliconCoach (www.siliconcoach.com) offer measurement and analysis options as well as the enhanced video display features described above. Multiple camera views may be captured and displayed via firewire connection to a single laptop with commercially available software such as supplied by Simi (www.simi.com), Vicon Peak (www.inition.co.uk) or Nation Instruments Vision Master (www.ni.com/vision).

Telemetry, the Internet: Video images presented to the coach may be many kilometres away from the coach. For example video from a camera high in the stand or behind the gaol, may be transmitted by telemetry to the coach sitting on the bench at the sideline. The coach replaying and viewing the previous action may make calls for adjustments in future plays based on assessment of the telemetered video. Compressed video clips may also be readily transmitted over the internet for viewing and interpretation by the sport scientist who may not be travelling with the coach and athlete.

Table 1. Summary of coding a video record of performance for volleyball.

Quality Action	Plus +		Initiative 0		Attempt /		Minus −		Total
Reception (R)	0	0%	0	0%	1	100%	0	0%	1
Pass (P)	0	0%	2	14%	11	79%	1	7%	14
Block (B)	0	0%	4	24%	11	65%	2	12%	17
Attack (A)	4	5%	21	26%	54	67%	2	2%	81
Total	4	4%	27	24%	77	68%	5	4%	113

may then be examined to determine player strength and weaknesses. The coach my then set a priority on which players and skills should have further analysis to identify faults in the skill and hence to undertake remedial intervention. For a road cyclist, video of performance might be reviewed to examine the skills in road cycling eg to identify at which parts of the race slipstreaming was poor and to count such breakdowns in skill. At the ISN we use Sports Code (www.sportstec.com) or Utilius VS (www.ccc.software.de) to identify player skill problem areas. The Sports Code service at ISN also provides a match or game strategy analysis for the coach and athlete in field sports such hockey and court sports such as badminton and squash.

Skill Analysis: Identifying and correcting faults in skill execution. There are several well documented systematic approaches to the skill analysis process commonly referred to as *qualitative analysis* (eg Hay & Reid, 1988 or Knudson & Morrison, 2002). These approaches are advocated as the first level of application of biomechanics to sport. They focus on describing the critical features of the movement eg where should the arms be positioned at takeoff in a jump. That is, the features that are essential for correct performance of a skill. Qualitative analysis is the process of assessing whether the performer shows evidence of good use of principles of motion in execution of the skill (Knudson, 2006). Qualitative analysis uses a minimum of measurements.

Training Application: Feedback of the video to the athlete so that they can see their performance as they are performing a skill application – in weightlifting sometimes using mirrors! but also immediately after performing the skill using delayed feedback. The image can be full screen on a laptop or full size as a data projected image. Players can work on their skills with the coach providing verbal input or players can self monitor in a drills session away from the coach.

Presentation to Coaches and Athletes: Best practice is as standardised report containing video presentations. These presentation would typically contain clips of the athletes performances with comparison to their best performance or to other athletes performances. There should be centralised database of reports so that athletes performance may be tracked over time. "Expert" systems have been developed to assist with standardised reporting. The siliconCoach Wizard guides the biomechanist as to which camera views are required for analysis, which measurements to make, whether these measures fit expected ranges of motion and also provide exemplar images of best performance. Using the Wizard sport specific analyses and reports may be developed for assessment of any discrete sports skills.

Many software developers have produced sport specific software presentation and analysis packages for sports skills in response to coaches demands for the most relevant

3 Why We Use Video in Coaching

- Gives the coach an extra "tool" when working with the athlete. The lament "I have only one pair of hands" may equally be "I have only one pair of eyes"! Using video cameras the coach can be provide with several views of the athletes performance. Further, with a recording of the performance, slow motion replay and stop action, the coach may examine a performance any number of times. Unlike the performing athlete, the athlete being replayed will not tire with the repeated performance.
- Images of the performance can help the coach convince the athlete about aspects of athlete's performance. Some athletes learn best by seeing themselves perform. An athlete making small changes to technique sometimes feels that they have made large adjustments yet the coach continues to press for larger changes.
- Using the video playback options such as slow motion or freeze frame, the coach can review and analyse the performance after the event. During a performance the coach can then focus on a particular aspect of the athletes performance shutting out all other factors such as other competitors actions.
- The camera can be placed where the coach can't go. For example underwater video of synchronised swimming provides indications of how the above water positions and timing on which the performance is judged can be influenced by poor underwater stroke. In hockey a camera placed behind the goal or high in the stands provides fields of view for the coach who is restricted to sitting on the bench.
- Video can present images the coach could never see. High shutter speeds "Stop" motion, high frame rates provide more images and may help capture critical events such as moment of take-off in diving, impact in ball striking. Computer control to can present images from widely different views eg front and side view of a golf swing, overhead and side view of a discus or hammer thrower, or images of different performers on the same screen for easy comparison.

4 How We Use the Technology

Video of the performance is of use to many of the sport sciences who are part of the support team for the coach and athlete. For example, video of the warm-ups and training or competition performance can be used by the physical conditioner/trainer to check the quality of the conditioning work – does it match the movement requirements of the sport. The exercise physiologist may use the video to assess the preparation for performance – did the athlete fail in the latter stages of an event indicating aerobic system deficit. The sports psychologist might study video of the athlete for evidence of consistency in pre-competition routine or unusual stress. The nutritionist may examine the video to assess whether the athlete has good rehydration strategies during a match.

The sports biomechanics uses video technology to help the coach in several ways which can be categorised as either qualitative or quantitative approaches.

4.1. Qualitative Analysis

Match Analysis: Video is recorded of a whole match or performance. The video is then used to identify the frequency of certain events in the performance. In a hockey game the events may be the tackles made by each player, further subdivided into successful and unsuccessful tackles. Similarly for passes and other critical skills in the game. The statistics for each player

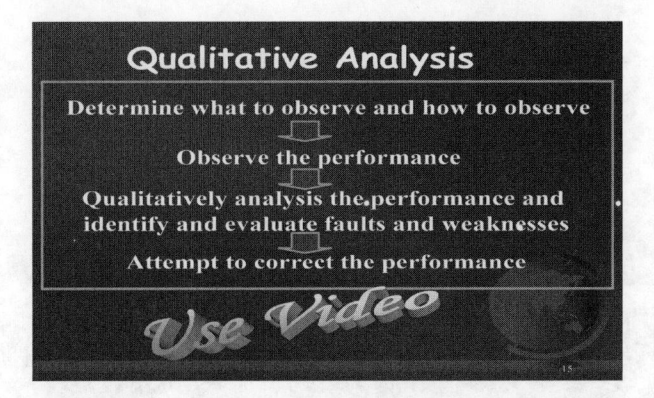

Figure 1. Steps in systematic qualitative analysis.

Figure 2. Feedback to coach via computer screen and to athlete via projected video.

and acceptable software interface. Eg. for bowling, cricket, curling, cycling, football, golf, gymnastics, and volleyball.

4.2. *Quantitative Video Analysis*

The second level of application is called *quantitative video analysis*, which involves description and explaining performance skills using measurement systems in biomechanics. A preliminary qualitative analysis is carried out to determine those variables (describing essential features), to be measured when analysing a particular skill.

Simply stated the process is to video record the performance the skill, then make measurements off the video by a process of video digitising. Measured point locations of the joints and limbs of the performer's body can be then be used to determine such things as distances moved and speeds moved by the various points. Care must be taken to obtain correctly scaled and orientated images. Mathematical analysis can then be used to determine detailed descriptions of motion and the forces and torques causing the motion. Difference between performances and performers can then be presented to the coach. For the recent World Junior Diving Competition in Kuala Lumpur, the performances of all the Malaysian

Figure 3. Multiple screen display.

divers were described and compared with their international competitors. International competitors performances in this case are the benchmark for the divers to aspire to.

Examples:

1. Performance enhancement: The analysis is of a hockey penalty corner to improve the likelihood of scoring a goal. That is to reduce the overall time of the skill and increase the speed of the drag flick. The skill is a rotational motion and requires views from multiple cameras and analysis in three dimensions (3D) to be able to measure and describe distance, angles, speeds and accelerations, the kinematics of the hockey stick and limbs of the player. Particular difficulties are video recording and calibrating in 3D. In this example three cameras were placed in the stands focusing on a 3D space frame containing a number of known points for calibration of the measurement space. An example of a simple output for the coach and player showing ball motion for various phases of the shot is as below.

More advanced approaches use automated digitising of a known size object moved around within the measurement space for calibration. Automated digitising of the performance is then usually of markers placed on the athlete. The use of markers on the performing athlete limits such automated approaches to laboratory and out-of-competition analysis provided the athlete is not hindered in their performance by wearing markers. Skills such as a golf swing, 10-pin or lawn bowling delivery performed in a limited space and with small range of motions unlikely to be limited by wearing markers, are appropriate for automated 3D motion. Such automated analyses typically include calculations of the ground reaction forces and the forces and torques at the joints driving the motion. Such kinetic analyses may result in recommendations by which the same performance outcome might be achieved but with reduced stress (forces and torques) and hence likelihood of injury at particular joints.

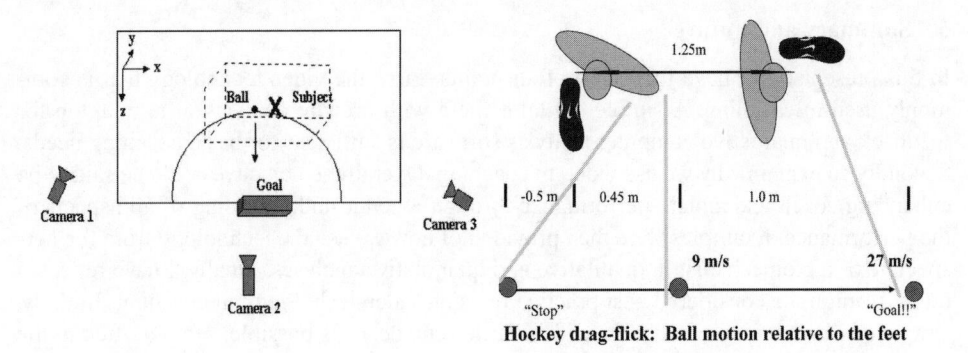

Figure 4. Cameras and calibration space for 3D and example output for coach and athlete.

2. Injury prevention: The example is of analysis of the gait of an athlete recovering from a hip injury to determine whether the athlete was sufficiently recovered to return to normal training and competition. The hurdler was video recorded while hurdling at a lower height and speed (with reduced hurdle distances) than their normal training and fastest race performances. A two-dimensional kinematic analysis was performed. Cameras were placed at the front and side of the track. The analysis consisted of determining whether the athlete performed the clearance of the hurdle with their normal range of motion and technique.

5 Quality of Feedback

Video analysis provides extrinsic feedback to the coach and athlete to add to what the athlete feels as they perform the skill. Coaches have made good use of video replay and software presentation to augment traditional coaching methods. As integration of video and computers has improved placing real-time quantitative video analysis within reach of coaches, questions remain as to what is the best form of video feedback and when is the best time to provide feedback using video technology. From our experience qualitative video feedback offered often is best accepted by the coach and athlete. Initially the presence of the camera and feedback system at weekly training sessions may be perceived as novel and may distract athletes and appear to cost the coach valuable coaching time. To win over the coach, the video feedback must address issues raised by the coach and athlete and any advice offered must be strongly grounded in biomechanical principles rather than copying an example of role model (3). The feedback format and presentation provided should match the comfort level of the coach and athlete to avoid information overload. Different feedback presentations might be prepared for coaches who are still developing their skills in use of video in coaching.

As the feedback becomes more complex, moving from qualitative to quantitative video analysis, the need for immediate feedback is reduced. Summary feedback with a video presentation and discussion in a reflective coaching session would me more consistent with what is accepted as good practice in the retention phase of skill learning (Cassidy *et al.*, 2006).

6   Summary and Future

In this presentation I have focused on four points: First, the video technology that is commonly used in coaching. A simple digital camera with firewire connection to a laptop for feedback and qualitative computer analysis software is sufficient for basic coaching needs. Secondly, to explain why we use video in coaching. Overall the objective could be said to be enhancing coach and athlete performance through a better understanding of all aspects of the performance. Examples were then provided of how we use the technology from the perspective of a biomechanist in qualitative and quantitative analyses. Finally, I have reflected on what might be considered best practice in using video technology in coaching. Initially, simple qualitative video feedback, with as little time delay as possible, offered often in the training situation appears best. This then leads to more complex delayed feedback of quantitative video analysis in relective coaching situations.

I have described technology that is currently being used and described its application in sports. However, advances in camera and computer technology are resulting in smaller and more powerful components with the potential for wider application of video in sports. To the forefront are such applications as video synchronised to transmitting GPS, video feedback through goggles in training, automated real time analysis of video of swimmers in training, and automated coding and analysis of sports court and field performances.

References

Cassidy T., Stanley S. and Bartlett R. (2006) Reflecting on video feedback as a tool for learning skilled movement. *Journal of Sports Science and Coaching*, 1/3, 277–286.
Dartfish (online), www.dartfish.com (Accesed: February, 2007).
Hay J.G. and Reid J.G. (1988) *The anatomical and mechanical basis of Human Motion*, 2nd Edition, Prentice-Hall, Englewood Cliffs, NJ, USA.
Knudson D.G. and Morrison C.S. (2002) *Qualitative analysis of Human Movement*, 2nd Edition, Human Kinetics, Champaign, Ill, USA.
Knudson, D. (2007) Qualitative biomechanical principles for application in coaching. *Sports Biomechanics*, 6/1, 109–118.
National Instruments Vision (online), www.ni.com/vision (Accessed: April, 2007).
Vicon Peak (online), www.inition.co.uk (Accessed: April, 2007).
PeeperScope. Biofit Technologies, Singapore.
SiliconCoach Ltd (online), www.siliconcoach.com.
Simi (online), www.simi.com (Accesed: April 2007).
SportsCode (online), www.sportstec.com (Accessed: April, 2007).
Utilius VS (online), www.ccc.software.de (Accessed: April, 2007).

HOW CAN VOICE-ENABLED TECHNOLOGIES HELP ATHLETES AND COACHES TO BECOME MORE EFFICIENT?

C. STRICKER[1,2] & P.H. REY[1,2]

[1]*AISTS – International Academy of Sports Science and Technology, Lausanne, Switzerland*
[2]*HES-SO Valais – University of Applied Sciences Western Switzerland, Valais, Switzerland*

The aim of this paper is to research and propose voice-enabled applications which would help athletes and coaches to become more efficient in the sport field. After an introduction to voice-enabled technologies, some examples of potential applications and a description of different ways to use speech/noise to interface with information systems will be discussed. Although the performance of voice-enabled applications today is reasonably good for home and office purposes, there are some serious challenges when it comes to sport applications. These challenges, such as noise management, the robustness of speech capture devices and user acceptance will be described in detail. This paper also presents the results of a recent experiment conducted in a sailing environment.

1 Introduction

Speech is the most natural means of human communication (Deng *et al.*, 2005). It is of higher interest to develop man-computer interface through speech. According to Cole *et al.* (1996), voice-enabled technologies that have been developed include:

- *speech recognition* : the process of converting an acoustic signal, captured by a speech device, to a set of words
- *text-to-speech synthesis* : the process of converting a sequence of text into a recognizable audio signal
- *spoken-language understanding* : the process of interpreting the meaning of key words and phrases in a speech string (Rabiner, 2003)
- *speaker identification* : the identification of a person solely by his/her voice.

Thanks to the latest research in the field and the continuous progress of computing power, performance of voice-enabled systems has improved considerably over the years. Today, speech recognition technologies are used commonly (Got, 2003). This could take the form of dictating a document instead of typing or booking a flight by calling an interactive vocal server. There are still few applications in the sports world, even if the use of voice-enabled technologies is becoming widespread in the other fields.

2 Potential Applications for Sport

During sports action, athletes frequently have their eyes and hands busy. They are therefore unable to use standard communication devices such as a keyboard, mouse or monitor. In these situations, the use of speech through a microphone to interact with information systems is ideal. Wearing a small and light speech interface device has the advantage of not inhibiting their movement.

Athletes and coaches already use microphones and two-way radios to share information with each other in many sports such as American football, rugby, soccer and cycling. In addition, there is a large amount of data already available on information system servers being used to support detailed analysis (e.g. statistical analysis of games). It would be interesting to leverage the existing use of speech devices and the existing use of information systems by combining the two. The idea would be to use a microphone to interact not only with humans but also with the computers that store the information.

One first generic use case is depicted in Figure 1: the user can ask for information via a speech interface. The information system, through the *speech recognition* engine, identifies the user's request, processes it, and sends the result back to the user. Information stored in the database is translated into speech with a *text-to-speech* engine. The information system is updated by periodically receiving information from sensors on the field. Mechanisms of *spoken-language understanding* can be integrated. Requests such as "Calories" or "Please tell me the amount of calories I've burned since the beginning of the training session" might have the same results.

A second use case consists of inputting data on an information server. Writing comments on a notepad and copying them afterwards to a computer is no longer required: coaches (ski, soccer, ice hockey, ...) can instantaneously save verbal observations on an information server. Today, the referees of the Swiss Football Federation call a vocal server after each game in order to report the game and final score that is then saved in the system.

Figure 1. Voice-enabled technologies as a way to ask information.

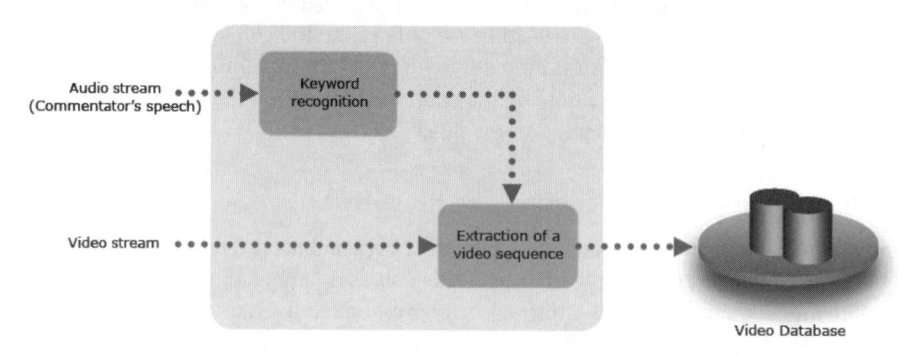

Figure 2. Voice-enabled technologies as a way to index video sequences.

There are numerous advantages: 24/7 availability, instant data availability on the Internet, time savings, and costs reduction to name a few.

The third use case consists of using voice commands to control devices and applications. Voice command is a good alternative when the user is far from another input/output interface and needs to interact quickly. By saying specific keywords, the user can modify the behavior of a device. One could imagine changing the speed of a treadmill by saying "faster" or "slower." Athletes with a disability would also benefit from such applications. Through *speaker identification*, users are given or denied access to specific commands.

One could also seek specific words in a TV/radio commentator's speech (Xiong *et al.*, 2005). This keyword identification is used to automatically index a video/audio sequence on a database (Figure 2). The *speech recognition* processor could seek the word "goal" in the commentary of a football match. After finding the keyword "goal," the system could store on the database a video/audio sequence starting 15 seconds before the seeking word and finishing 10 seconds later. This system allows users to instantly view the sequences on Internet, on their mobile phone or on TV.

3 Meeting the Challenges of Sport

The performance of voice-enabled applications is reasonably good for home and office purposes. There are still some serious challenges, however, when it comes to sport applications. The first challenge is that sport frequently takes place in noisy environments. You may have already experienced difficulties in understanding someone calling with a mobile phone from a football stadium where the crowd is yelling: the noise interference is very high. A speech recognition system faces the problems of noise interference. The surrounding noise deteriorates the quality of the speech signal. The recognition rate is significantly decreased. Because recognizers are not able to distinguish non-speech environmental sounds from acceptable speech input (Lippmann, 1997), the input signal should be as clear as possible. As a result, noise management, either with noise cancelling microphones or with Digital Signal Processors (DSP) is a major concern in developing voice-enabled technologies for sport stakeholders.

The second challenge is that speech devices are often used in very rough conditions. A microphone used by a ski coach in the mountains must work in cold and wet weather. Microphones placed in a rally car might receive shocks and vibrations and would have to resist dust and sand. Wind noise is also a factor that could dramatically decrease a microphones' accuracy, i.e. on a sailing boat. The choice of microphone needs to be made carefully depending on the conditions in which the sport takes place. Capturing speech when practicing sport requires robust, high quality, and easy-to-wear speech devices. Unfortunately, because of the extreme range of conditions, we have not yet found devices available on the market that can do the job.

The third challenge is that athletes and staff often need fast and reliable information in critical situations. As of today, even the best speech recognition system is not able to guarantee one hundred per cent recognition rate. For this reason, users seem reluctant to rely on speech interfaces for critical applications. A rally car driver would never accept, for example, to command gears, gas and brake pedals by vocal command. The risk of a crash if the system does not understand his request is too high. While humans are willing to repeat information several times to another person who does not understand them, they are less tolerant with computers (Grosso, 2003).

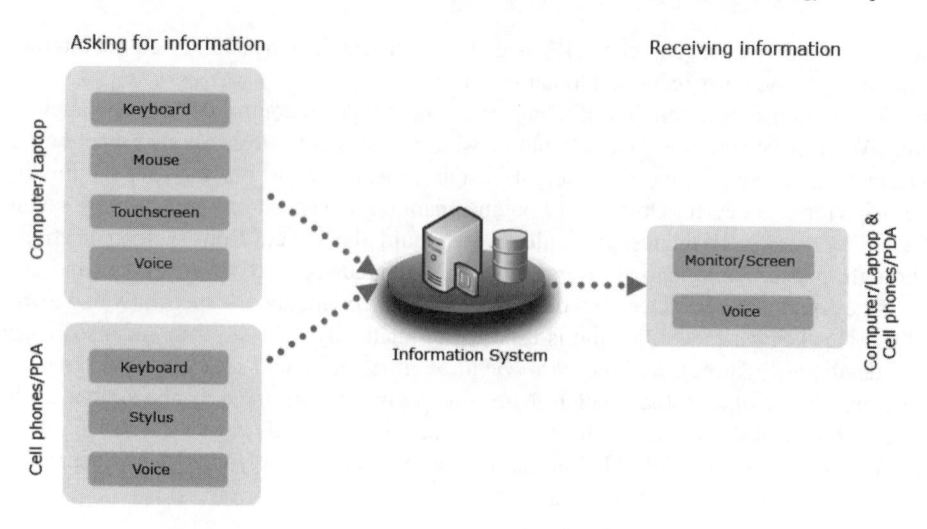

Figure 3. Multi-modal interfaces.

A fourth challenge is the importance of selecting the most appropriate user interface mode or combination of modes depending on the *device*, the *task*, the *environment*, and the *user's abilities and preferences* (Rabiner, 2002). Developers should carefully analyze the parameters (device, tasks, environment, user's abilities and preferences) in order to provide adequate voice interfaces. A well-designed voice interface for cross-country runners may be totally inadequate for swimmers. Because athletes and coaches need to access information anywhere at anytime, the system must provide multiple ways to interface (e.g. keyboards if possible, heavy duty buttons, etc.; see Figure 3).

A final challenge is that speaking consumes precious cognitive resources (Shneiderman, 2000). This often leads to a difficulty to perform other tasks at the same time. For example, studies have shown that calling with a mobile phone could affect a car driver's concentration and anticipation (Strayer & Jonhson, 2001). The drivers mind is not fully on their driving. For this reason, vocal applications should not be designed for athletes who need high concentration resources. Voice applications might distract them from what they are doing and therefore reduce performance or safety.

4 Discussion

Multi-modal interfaces used in outdoor sport environments are currently being developed in a research project applied to a sailing environment, conducted by the International Academy of Sport Sciences and Technology (AISTS), the University of Applied Sciences of Western Switzerland Valais, the College of Engineering and Architecture of Fribourg and the Dalle Molle Institute for Perceptual Artificial Intelligence (IDIAP) at the Ecole Polytechnique Fédérale de Lausanne (EPFL). The project has 2 goals:

1. The understanding of specificities and needs of sport for voice-enabled technologies.
2. The development of a speech recognition engine able to perform in a very noisy and hostile environment.

Performance of the speech recognition engine developed was proved satisfying for a small vocabulary in a quiet environment. However, speech recognition onboard with a large vocabulary is much more challenging. The noise can be such that it is not possible for humans to understand what their counterparts are saying. Do we expect to design speech recognition systems to performing better than humans? The robustness of the microphone is another important point to take into consideration. The microphone will be used in extreme conditions. The wind noise on a boat is unpredictable and could be very loud. Our equipment must be waterproof, resistant to sea-salt and resistant to shock and vibration. Microphones designed for home and call center purposes were eliminated because they are not robust enough. Moreover, it should be light and comfortable to wear. The coach has to concentrate on his job and cannot afford to be annoyed by an uncomfortable headset. For this reason, throat microphones were abandoned because of wearer discomfort. All of these constraints make it almost impossible to find the ideal microphone on the market today.

As experienced during the project, athletes and staff are ready to invest time in learning how to use new information technologies if they see the possibility to become better or to save time. Most of them are very sensitive to the latest technologies.

5 Conclusion

To reach the top, athletes progress step by step, inch by inch. The use of the latest technologies provides them with opportunities to support their progress. Based on our own experience, higher-level athletes are ready to invest time in learning how to use new information technologies if they see the possibility to become better and more efficient. A well-designed voice-enabled application allows them to save time and to access the information anywhere at anytime.

The number of potential voice-enabled applications in sport is endless. Innovations are restrained, however, because of the current limitations of voice-enabled technologies and the specificities of the sports world. If technologies continue to improve at the same rate, one can imagine that in the near future, humans will be able to converse freely with computers, even in noisy and hostile environments. Speech will become, in conjunction with other methods of interfacing, a prevalent means of communication with computers-based systems. Ultimately, users will have the last word in deciding the future success of voice-enabled technologies.

References

Cole R.A., Mariani J., Uskoreit H., Zaenen A. and Zue V. (1996) *Survey of the State of the Art in Human Language Technology*. Center of Spoken Language Understanding, Oregon Graduate Institute.

Deng L., Wang K. and Chou W. (2005) Speech Technology and Systems in Human-Machine Communication; *IEEE Signal Processing Magazine*, September 2005.

Got G. (2003) Speech Recognition's Evolution Continues, *IEEE Intelligent System*, September/October 2003, 5–7.

Grosso M.A. (2003) *The long-Term Adoption of Speech Recognition in Medical Applications*. George Washington University School of Medicine

Lippmann R.P. (1997) Speech Recognition by machines and humans. *Speech Communication*, 22/1, 1–16.

Rabiner L. (2003) The Power of Speech. *Science*, 301, 1494–1495.
Rabiner L. (2002) *Techniques for Speech and Natural Language Recognition*. Rutgers, The State University of New Jersey (online). Available: http://www.caip.rutgers.edu/~lrr/lrr talks/Challenges in Speech Recognition.ppt (Accessed: 10 February, 2007).
Shneiderman B. (2000) The Limits of Speech Recognition. *Communications of the ACM*, September 2000, Vol. 43, No. 9.
Strayer D.L. and Johnson W.A. (2001) Driven to distraction: dual-task studies of simulated driving and conversing on a cellular phone. *Psychological Science*, 12, 462–466.
Xiong Z., Radhakrishnan R., Divakaran A. and Huang T.S. (2005) *Highlights extraction from sports video based on an audio-visual marker detection framework*. ICME 2005. IEEE International Conference on Multimedia and Expo, 2005. 6–8 July 2005.

THE SINGAPORE SPORTS SCHOOL – DEVELOPING ELITE YOUTH ATHLETES

G. NAIR

Sports Science Academy, Singapore Sports School, Singapore

The development of sports in Singapore is directed by Sporting Singapore vision's three strategic thrusts – Achieving Sports Excellence, Cultivate a Sporting Culture and Create a Vibrant Sports Industry. In line with achieving Sports Excellence, is the founding of the Singapore Sports School – a specialized school catering to elite youth athletes sports needs, as well as, provision of a quality academic education. The paper will outline key developments that led to the birth of the School and an insight into the infrastructure, sports and academic programs, support services and achievements of the Singapore Sports School.

1 Brief History of Sports in Singapore

Since independence, sports have played a significant and pivotal role in the building of Singapore as a nation. In the early years, the main objective was to create an awareness of the importance of sports participation. This was followed by the setting up of the Singapore Sports Council that was responsible for the promotion of mass sports, the introduction of sports training schemes, the set-up of sports medicine services and the creation of a master plan of sports facilities. The 80's all saw a burgeoning of sports facilities with a "Sports for All" policy encouraging Singaporeans to participate in at least one sport to keep fit for life. In addition, a Sports Aid Fund was launched for assisting talented and deserving athletes.

In 1993, Sports Excellence was given a boost with the launch of a S$10 million per year program (SPEX 2000; Singapore Sports Council, 1994) with a focus on several core and merit sports. In line with this, more facilities, equipment and programs were created to provide the necessary support for elite sports group.

2 Sporting Singapore

In 2000, the Committee of Sporting Singapore (CoSS) involving members from both the private and public sectors were set up to address the following:

- Establish the vision and desired outcomes for sports in Singapore
- Identify the issues impeding the development of sports
- Formulate the development strategies for sports in Singapore; and recommend specific initiatives for the future development of sports in Singapore.

In the final CoSS Report (Ministry of Community Development and Sports, 2001), it states:

"... the academic demands in Singapore place a huge strain on students who also wish to excel in sports. Very often, students talented in sports have to give up sporting dreams to concentrate on their academic studies. In line with our Singapore 21 vision

of providing opportunities for all, there should be a system to allow for those who have the inclination for and ability in sports to pursue their athletic goals without sacrificing quality academic education. This is in line with the Ministry of Education's emphasis on ability-driven education to maximize the potential of each student by spotting and developing talent and ability for success in life. The focus of mainstream schools in general is to provide quality academic education and therefore, resources are limited for elite sports development. It is therefore proposed that a Sports School be set up to address this issue. The Sports School will have two components: an academic component similar to the other mainstream schools in Singapore; and a sports excellence training programme. The academic programme will be organized in a flexible manner to accommodate the student's training and competition schedules. The Sports School to be managed by MCDS (now known as Ministry of Community Youth and Sports, MCYS) should be modeled after the Independent Schools under the MOE system. MOE's support for the Sports School is obviously crucial."

3 Singapore Sports School

Thus, the birth of the Singapore Sports School (SSS)! Presently, there are eight core sports in the school and includes track and field, swimming, badminton, table tennis, bowling, sailing, netball and soccer. In addition, there are athletes from other sports such as golf, triathlon, gymnastics, shooting, wakeboarding and silat. There are a total of 358 athletes whose age range from 13 years to 16 years of age.

3.1. *Facilities*

The Singapore Sports School was built at a cost of S$75 million and sits on a seven hectare site located in the north of Singapore. The co-educational school was officially opened in April 2004 and provides residential boarding. The school boasts of the following facilities (Figure 1): 2 Olympic size covered swimming pools; an indoor badminton centre with 10 courts; a table tennis centre with space for 32 tables; a eight-lane 400-m synthetic running track; a synthetic soccer pitch; an indoor multi-sports auditorium; outdoor netball courts; a 12-lane bowling

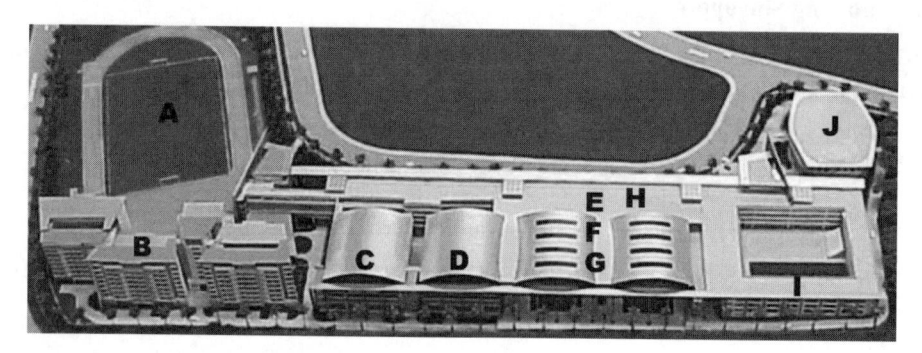

Figure 1. Plan of the Singapore Sports School; A: track & field facilities, B: school hostel, C: table tennis hall, D: badminton hall, E: sports science area, F: bowling, G: swimming pools, H: gymnasium, I: netball courts, J: auditorium.

alley with three different lane conditions; a 2-storey well-equipped weights room; a sports science centre; a dining facility; well-furnished boarding rooms and a recreation room.

3.2. Sports Coaching Support

Athletes are grouped according to their sport where a General Manager takes charge of the sport and supported by coaches and assistant coaches. The athletes are being coached by specially selected coaches who have the qualifications, as well as, experience of having coach elite youth and thus, they comprise both foreign and local coaches. In addition, consultants are brought in regularly to review existing programs and provide recommendations. Coaching of athletes under "other sports" are handled by the respective National Sports Associations. In addition, academic teachers in the school are also assigned as "teacher mentors" for a sport where they work together with the general managers and coaches to oversee the athlete progress in all areas.

3.3. Sports Science Support

The athletes also receive in-house sports science support in the areas of sports medicine, sports physiotherapy, sports physiology, sports psychology, sports biomechanics, sports nutrition and strength and conditioning. The CoSS report states that *"as technology advances, sports medicine and sports science have become an integral part of maximizing athletes" potential'*, The set-up of a sports science centre in the Sports School is surely the right direction in developing the area of sports science to reach a state of international excellence.

The sports science centre in the sports school has three main functions : provision of testing and assessment of athletes; conduct of sports science education sessions for athletes; conduct evidence-based research in the aim to enhance performance and reduce attrition due to injuries and other factors.

Prior to enrolment, all athletes undergo a complete medical screening and a sports-specific musculo-skeletal screening. Following which, a customized conditioning training program is developed for each athlete based on their strengths and weaknesses. In addition regular testing and assessment provide a yardstick of comparison and information to fine-tune training programs to enhance performance. Student-athletes are also taught topics such as injury management, nutritional strategies etc. to develop them into informed adult athletes.

3.4. Academic Education

The academic curriculum is based on the educational objectives set out by the Ministry of Education. However, the number of subjects that the students can enroll in the upper secondary level is dependent on the athlete's competition status/schedule and academic achievement. The academic curriculum in the sports school is a modular one where it enables student-athletes to meet both the demands of training and competition and academic studies. The modules are independent of one another and thus, allow students who have missed one due to overseas competitions to be able to enroll in the same module later in the year. In addition, lesson CDs and online learning resources (such as an e-learning portal) are available to allow students to continue with their studies while training or competing overseas. Each module is assessed independently and thus, there aren't the usual mid-year and end-year exams.

In terms of classroom setup, the Sports School uses a Home Room system where teachers are assigned a permanent room and where students get to experience different stimulating environments with every lesson. The concept allows teachers to plan for self-directed and student-centered work. The classroom can be set-up to cater to diverse learning needs and are kept small (15–20 students) for an optimal learning environment.

Apart from the academic subjects and to provide a holistic education, students are also requested to go through a Personal Effectiveness Program, the Arts program and a Community Involvement Program.

3.5. *Post Secondary Education*

The post-secondary education program aims to provide the students in the Sports School an academic pathway beyond their secondary education through various initiatives with tertiary institutions locally and overseas. One such initiative is the 'through-train' route. One option under this route is that the students do not need to sit for their GCE 'O' level examinations. Through a Memorandum of Understanding (MOU) signed by three institutions (Singapore Sports School, Republic Polytechnic and Nanyang Technological University), it allows a sports school student to qualify and pursue a 3-year diploma course at Republic Polytechnic and upon graduation to pursue a university degree at Nanyang Technological University via the University's Discretionary Admission Scheme. Another option is for the sports school student to pursue a 2-year Diploma in Sports Management and Exercise Science, a program developed by SSS and Auckland University of Technology (AUT). The course is conducted at SSS and allows the student to continue with the studies and train. Upon completion of the diploma, the student can pursue a degree at AUT in New Zealand. In addition, there has been collaboration with other universities in Australia, UK and USA to allow the students options to pursue their university degree.

3.6. *Achievements*

Since the inception of the school in 2004, the school has achieved both sporting and academic excellence. In sports, numerous records have been established locally, regionally and

Figure 2. 23rd South East Asian Games medalists; from left to right: seated front row: Ruth Ho (swimming, Silver); Mylene Ong (swimming, Silver); Shu Yong (swimming, Gold, Silver, Bronze); standing back row: Jazreel Tan (bowling, Silver, Bronze); Tao Li (swimming, Multiple Golds); Sean Li (sailing, Siver); Griselda Khng (sailing, Gold).

internationally. The sporting achievements include 3 World champions, 2 Asian Games gold medalists, 1 Commonwealth Games gold medalist, 2 South-East Asian Games Gold medalists (Figure 2), 1 World Youth Champion, 3 Asian Youth Champions, 2 National Champions, 11 National records, 29 National Age-group records, 12 National Schools records and 1 UK AAA U17 Indoor Record holder. In the academic field, the first cohort of student have all excelled in their "O" level exams and have placement in junior colleges and polytechnics.

4 Conclusion

Though within a short period of time the School has raised the bar for sports excellence, there is still much to be done. The School is looking into opening up to students-athletes of more sports and to have a greater synergy with national sports associations. In addition, it is exploring ways to enhance its facilities and equipment to have a technological advantage in training and education.

References

Singapore Sports Council (1994) *On Track – 21 years of the Singapore Sports Council.* Times Edition, Singapore.

Ministry of Community Development and Sports (2001) *Report of the Committee on Sporting Singapore, Towards a Sporting Singapore.* Ministry of Community Development and Sports, Singapore.

MEASURING SPORTS CLASS LEARNING CLIMATES – THE DEVELOPMENT OF THE SPORTS CLASS ENVIRONMENT SCALE

T. DOWDELL, L.M. TOMSON & M. DAVIES

School of Education and Professional Studies, Griffith University, Brisbane, Australia

This study provided the first model for the investigation of sports class learning climates, and involved a consolidation of the dimensions and items of the Perceived Motivational Climate in Sport Questionnaire-2 (Newton, Duda, & Yin, 2000) and the Classroom Environment Scale (Moos & Trickett, 1987). The development and validation of a new, unique learning climate instrument – the Sports Class Environment Scale (SCES) – was the result of this consolidation and constituted the focus of this study.

1 Introduction

The coach-athlete relationship is one of the most important influences on the athlete's motivation and performance (Mageau & Vallerand, 2003). One of the key leadership roles the coach plays in this relationship is the creation and maintenance of the sports class learning climate. Different sports class learning climates evoke different athlete perceptions about achievement, and subsequently can influence athletes' learning (Ames, 1992). The learning climate of a sports class is the relatively persistent quality of the sport training environment that is experienced by class participants and is based on the collective perception of participants' behaviour in that sports class setting. Motivational climate is a part of the learning climate and can be described as the participants' relatively persistent collective perceptions of the achievement goal structure of that setting.

Three main bodies of information formed the basis for this study's literature review: the socio-ecological approach to environmental determinants of behaviour, the historical antecedents of needs-based behaviour and environmental press, and theoretical underpinnings to class learning climate and achievement motivation climate measurement. Recent studies of motivational climate in sports have provided insight into coaching behaviour and its effect on sports class motivational climate (Ntoumanis & Biddle, 1999; Roberts, 2001). The joint influence of the environmental press (the sports class learning climate) and athletes' motivation can determine the cognitive, affective, and performance patterns regularly displayed by athletes (Ntoumanis & Biddle, 1999).

In spite of the potential value of class learning climate research to the field of sports class behavioural studies, no research has consolidated the fields of classroom learning climate research and sports class motivational climate studies. This study provides a first model for the investigation of sports class learning climates that involves a consolidation of the dimensions and items of the Perceived Motivational Climate in Sport Questionnaire-2 (Newton et al., 2000) and the Classroom Environment Scale (Moos & Trickett, 1987). The result of this consolidation is a new and unique learning climate instrument – the Sports Class Environment Scale (SCES).

2 Method

A review of historically important learning climate and motivational climate instruments, their development, validation, scoring and application in research, informed the method for developing the SCES. The process of developing this consolidated learning climate instrument began with the production of an initial scale, and was followed by a review by a panel of experts in coaching and independent university researchers in sport and physical education. The first draft of the SCES was then pilot tested with a small group (n = 41) of competitive gymnasts to prompt some changes to the scale. Initial field-testing of the second draft SCES occurred with 28 male and 180 female competitive gymnasts from 6 metropolitan and 4 regional gymnastics clubs in the state of Queensland, Australia. Exploratory factor analysis provided a revised SCES with five subscales labeled Task Involvement and Improvement; Ego Involvement and Mistakes; Coach-Athlete Communication; Effort, Order and Organization; and Affiliation. Using the revised SCES subscales as dependent variables, MANOVAs were conducted to compare club type, gender, and competitive level.

3 Results

The ten competitive gymnastics clubs in this study differed by gender and by the number of weekly training hours. At the time of this study, nine of the ten clubs participating in this study ranked in the top 10 competitive clubs in the state. Gymnasts from clubs ranked in the top four in the state normally trained 14–22 hours weekly while gymnasts from clubs ranked lower than 4th in the state normally trained between 9–18 hours weekly. The clubs in the study cohort were grouped as either "high training hours" (14–22 hrs weekly) or "low training hours" (9–18 hours weekly).

In this study, the differences between SCES responses on the subscales of Ego Orientation were related to club type. The high training hours clubs had a combination of a high task involved climate score (mean = 3.61) and a moderate to high ego involved climate score (mean = 2.21) at the same time. Two of the clubs from the lower training hours group also demonstrated this characteristic. This may be due to the fact that these clubs, like the high training hours clubs, employ professional teachers and/or tertiary educated coach practitioners. Furthermore, the study indicated that the perceptions of the learning climate in gymnastics classes as measured by Task Mastery, Ego Orientation, Effort, Order and Organisation, and Affiliation were gender-related. The means for the male gymnasts were lower than for females on Task Involvement, Effort, Order and Organisation, and Affiliation, but higher on Ego Involvement.

4 Discussion

There is some evidence that elite level athletes seem to function better when a high task mastery orientation and/or a high ego goal orientation is tempered with a high task involved class climate (Pensgaard & Roberts, 2002). It may be that highly competent athletes with either a high task mastery orientation or a high ego goal orientation are motivated in any perceived class climate, but when in a situation that threatens their perceived competence, they do better in a task involved sports class climate (Duda, 2001). All things being equal, most researchers in physical activity and sport suggest that when one is learning physical skills, being task involved (as opposed to ego involved) is motivationally conducive to learning (Roberts, 2001). This may result in greater intrinsic motivation for the athlete, and encourage adaptive behaviours by the athlete while in their sports class.

It might be expected that all gymnasts, irrespective of gender, would perceive their training class climates as more task (skill) involved and less ego involved. In this study, this was not the case. The gender differences in task involvement and ego involvement found in these gymnastics classes may reflect a gender-biased view of effort and outcome. The inverse relationship of task involved versus ego involved climate perception scores between the males and the females in these gymnastics classes tends to support the proposition that achievement goal dimensions are orthogonal (Duda & Whitehead, 1998; Roberts, 1992). Of interest is the question about achievement goal dispositions of the male and female gymnasts that predispose them to a particular achievement goal perception and behaviour. Further tests of the SCES along with measures of personal goal dispositions, such as the Task and Ego Orientation in Sport Questionnaire (TEOSQ) (Duda, 1989), used on much larger numbers of male versus female gymnasts may shed light on gender differences in perceptions of sports class learning climate.

5 Conclusion

The use of the SCES allows sports class learning climates to be displayed graphically and can give the target class and their coach timely information about the learning climate of their class as perceived by participants in that class. Learning climate intervention study is a highly pragmatic use of a valid learning climate instrument and can provide coaching professionals with useful information for monitoring and structuring the best achievement environments for their setting. Sports class climate perceptions could be most useful in intervention studies when displayed graphically, and readily indicate learning class climate at pre-test and post test.

This study breaks new ground, and may lead to novel insights into sports class learning climates. Specifically, this study has demonstrated the ability to distinctly profile the climate of sports classes and, by extension, sports club learning climates. Moreover, if there is a relationship between sports class performance/competitive outcomes and the characteristics of the attendant class learning climate, a sports class climate scale may allow a coach to easily assess the learning climate of their class and, if desired, change this learning climate to one more congruent with the desired class outcome. Because class learning climate is easier to manipulate than individual achievement goal dispositions (Whitehead, Andree, & Lee, 1997) and because perceptions of learning climate account for variance in learning outcomes beyond that attributable to student ability (Fraser, 1994, 1998, 2002), class learning climate is an important variable that should be better understood, described, developed, and manipulated. Using the SCES to measure sports class learning climates may lead to a greater understanding of effective sports classes, and of coach and athlete behaviours in those classes, and provides a first step in monitoring sports class learning climates.

References

Ames C. (1992) Classrooms: Goals, structures, and student motivation. *Journal of Educational Psychology*, 84, 261–271.

Duda J. L. (1989) The relationship between task and ego orientation and the perceived purpose of sport among male and female high school athletes. *Journal of Sport and Exercise Psychology, 11,* 318–335.

Duda J. L. (2001) Achievement goal research in sport: Pushing the boundaries and clarifying some misunderstandings. In: Roberts G.C. (Ed.), *Advances in Motivation in Sport and Exercise* (pp.129–182). Champaign, ILL: Human Kinetics.

Duda J. L. and Whitehead J. (1998) Measurement of goal perspectives in the physical domain. In: Duda J.L. (Ed.), *Advances in sport and exercise psychology measurement: fitness* (pp. 21–48). Morgantown, WV: Fitness Information Technology.

Fraser B. (1994) Research on classroom climate. In: Gabel D.L. (Ed.), *Handbook on science teaching and learning* (pp.493–541). New York: Macmillan.

Fraser B. (1998) Science learning environments: Assessment, effects and determinants. In: Fraser B.J. and Tobin K.G. (Eds.), *International handbook of science education* (pp. 527–564). Dordrecht, The Netherlands: Kluwer.

Fraser B. (2002) Learning environments research: Yesterday, today and tomorrow. In: Goh S.C. and Khine M. S. (Eds.), *Studies in educational learning environments: An international perspective* (pp. 1–25). New Jersey: World Scientific.

Mageau G.V.A.and Vallerand R.J. (2003) The coach–athlete relationship: A motivational model, *Journal of Sports Sciences, 21*, 883–904.

Moos R.H., & Trickett E.J. (1987) *Classroom environment scale manual* (2nd ed.). Palo Alto. CA: Consulting Psychologists Press.

Newton M., Duda J.L. and Yin Z. (2000) Examination of the psychometric properties of the perceived motivational climate in sport questionnaire-2 in a sample of female athletes. *Journal of Sport Sciences, 18*(4), 275–290.

Ntoumanis N. and Biddle S. (1999) A review of motivational climate in physical activity. *Journal of Sports Science, 17*, 643–665.

Pensgaard A.M. and Roberts G.C. (2000) The relationship between motivational climate, perceived ability and sources of distress among elite athletes. *Journal of Sport Sciences*, 18(3), 191–200.

Roberts G.C. (1992) Motivation in sport and exercise: Conceptual constraints and convergence. In: Roberts G.C. (Ed.), *Motivation in Sport and Exercise* (pp. 161–176) Champaign, ILL: Human Kinetics.

Roberts G.C. (2001) Understanding the dynamics of motivation in physical activity: The influence of achievement goals and motivational processes. In: Roberts G.C. (Ed.), *Advances in Motivation in Sport and Exercise* (pp.1–49). Champaign, ILL: Human Kinetics.

Whitehead J., Andree K.V., & Lee M.J. (1997) Longitudinal interactions between dispositional and situational goals, perceived ability and intrinsic motivation. In: Lidor R. and Bar-Eli M. (Eds.), *Innovations in sport psychology: linking theory and practice. Proceedings of the IX World Congress in Sport Psychology: Part II* (pp. 750–752). Netanya, Israel: Ministry of Education, Culture and Sport.

DEVELOPING MULTIMEDIA COURSEWARE IN TEACHING EXERCISE PHYSIOLOGY FOR PHYSICAL EDUCATION MAJOR

S. SETHU[1], A.S. NAGESWARAN[2], D. SHUNMUGANATHAN[3] & M. ELANGO[4]

[1]*Dr. Sivanthi Aditanar College of Physical Education, Tiruchendur, TamilNadu, South India.*
[2]*Dept. Physical Education, H. H. The Rajah's College, Pudukottai, TamilNadu, South India.*
[3]*Dept. of Physical Education, Manonmaniam Sundaranar University, Tirunelveli, TamilNadu, South India*
[4]*Dept. of Physical Education, The MDT Hindu College, Tirunelveli, TamilNadu, South India*

The purpose of the present investigation was to develop and evaluate the computer-assisted teaching in exercise physiology for physical education major. To achieve this purpose, thirty men students were selected who are studying master degree in physical education (M.P.Ed.,) randomly as subjects from Dr. Sivanthi Aditanar College of Physical Education, Tiruchendur. Their age ranged from 20 to 25 years. They were undergone computer assisted teaching for 15 working hours. All the subjects were tested on learning achievement before and after the treatment. The data pertaining to the variables in this study were examined by using one way repeated measure ANOVA. The experimental group namely the computer assisted teaching group has achieved significant improvement on learning achievement in cardiovascular system and time for learning also reduced. The subjects showed attraction towards the computer assisted teaching in classroom.

1 Introduction

The present century is rightly technological century due to the influence of advancements in the field of science and technology on the varied aspects of life, resulting in its modernization. The impact of scientific and technological advancements on education is so great that it has given rise to new discipline called Educational Technology. (Jeganath, 2003).

To-day's classroom practices are quite different from those of yesterday. Similarly, the classroom practices in the coming century may be quite different from those of today. One can easily find out the explanation for these differences in the obvious impact of technological innovation and inventions. The shape of future school, colleges and universities is bound to change radically due to technological impact in the years to come. There is a greater need to gear education and teacher education to meet the future requirements of the society utilizing the technological devices and chances. Educational technology has revolutionized the educational system. It has come in to stay for every for the enrichment of educational and instructional processes. It has greatly influenced the teaching learning process. (Halloran, 1995).

The major problem of teaching in our schools/colleges is how to accommodate instruction to individual differences of the learners. Educational technology has developed new innovate practices and strategies for this purpose. One such strategy is multimedia based modular approach. (Donnelly, 1987; Mohnson, 1995)

The present study aims at developing a multimedia courseware in teachi.ng Exercise Physiology for physical education students and finding out the effectiveness of the developed multimedia courseware. The multimedia courseware meets the requirements of individualized learning. In the present study, individualized learning is integrated together in each instructional session of three major blocks namely, cardiovascular system, Respiratory system and Exercise on Cardio respiratory system. (Powers & Howley, 1997).

2 Purpose of the Study

The purpose of the present study was to develop multimedia courseware in teaching exercise physiology for physical education major. The study consists of two parts. The first part of the study was to develop multimedia courseware in teaching selected units in exercise physiology for physical education major and the second part of the study was to find out the effectiveness of developed multimedia courseware.

3 Methodology

In Exercise Physiology, cardiovascular system, respiratory system and exercise on cardio respiratory system were selected for developing the courseware. The preparation of multimedia courseware was used only for teaching the students in physical education major. Only twenty students were randomly selected as subjects from Dr. Sivanthi Aditanar College of Physical Education, Tiruchendur, TamilNadu, studying Master's degree in Physical Education during the academic year 2004–05. The selected subjects were taught with the help of Multimedia courseware on the selected units of exercise physiology. English language was adopted to prepare the multimedia courseware.

The learning achievement was selected as dependent variable and teaching exercise physiology through multimedia courseware was considered as independent variable. Fifteen working hours were given for the subjects as learning hours. The subjects were tested thrice. Pre test was arranged before the experiment, the post test I was conducted after nine days and post test II was done after fifteen days on the learning achievement.

4 Experimental Design and Statistical Procedures

The pre and post test random group design was used as experimental design. The data pertaining to this study were examined by using one way analysis of variance (ANOVA) with repeated measures for the variables. Whenever 'F' ratio was found to be significant, the Scheffe's test was used as post-hoc test. The level of significance was fixed at 0.05 level of confidence for all the cases. (Toth-Cohen, 1995; Bukowski, 2002).

5 Analysis and Discussion of Data

In Table 1, the value required for significance at 0.05 level with df 2 and 38 is 3.24. From Table 1, the obtained F-ratio of 2004.60 for paired means is less than the table value of 3.24 with df 2 and 38 required for significance at 0.05 level of confidence. The results of the study indicate that there was significant difference among the means of three tests at different time period. To find out which of the three paired means had a significant difference, the Scheffe's post-hoc test was applied and the results are presented.

Table 2 shows that the mean differences on learning achievement scores between pre test and post test I, pre test and post test II, and post test I and post test II are 21.35, 24.35

Table 1. One Way Repeated Measure ANOVA on Learning Achievement.

Means ± Standard Deviation			Source of Variance	Sum of Squares	df	Mean Squares	'F'- Ratio
Pre test	Post Test I	Post test II					
14.45 ± 2.46	35.8 ± 3.53	39.8 ± 2.46	Between Within	5806.30 55.03	2 38	2903.15 1.45	2004.60*

(Achievement scores are in Numbers)

Table 2. Scheffe's Test for the Differences between Paired Means of Learning Achievement.

Adjusted Post Test Means			Mean Differences	Confidence Interval
Pre test	Post test I	Post test II		
14.45	35.8		21.35*	0.98
14.45		39.8	24.35*	0.98
	35.8	39.8	4.00*	0.98

*Significant at .05 level.

Table 3. Mean and Standard Deviation on Attitude of Experimental Group.

Variable	Mean	Standard Deviation	Percentage
Attitude towards Multimedia Courseware	40.43	±2.96	80.86%
Attitude towards Using Computer	42.68	±3.41	85.36%

and 4.00 respectively. The values are greater than the confidence interval value 0.98, which shows significant difference at 0.05 level of confidence.

Attitude towards multimedia courseware is 40.43 which is scored 80.86% of the total score. It is concluded that there was a positive attitude among selected subjects in using multimedia courseware as teaching and learning method and also it is noted that the multimedia courseware aided instruction proves to be effective to cater to the needs of a learner.

Attitude towards using computer is 42.68 which is scored 85.36% of the total score. It is concluded that there was a positive attitude among selected subjects in using computer as teaching and learning aids. More than 80% of the participants were satisfied with the content, the academic approach proposed and the interactive support, which is flexible and contains a self-test. The students found the CD-ROM to be a more efficient means of retaining information than classic lectures as for the evaluation of the knowledge acquired on exercise physiology.

6 Conclusions

The following conclusions were drawn from the present study.

1. It was found that learning achievement in selected units of Exercise physiology had improved significantly through multimedia courseware. Because the impact

of the multimedia courseware is very effective; the students achieved mastery as earlier than traditional method in selected units of Exercise physiology (Santar & Michaelsen, 1995; Seabra *et al.*, 2004; Buzzell *et al.*, 2002).
2. Students achieved mastery as per their own pace devoting extra interest towards the multimedia courseware which strange in a face to face class. Have an attraction towards courseware and put forth their maximum effort in learning. Students developed more interest and involve in learning activity it strengthens individualized learning irrespective of individual difference. (Halloran, 1995).

References

Bukowski E.L. (2002) Assessment Outcomes: Computerized Instruction in a Human Gross Anatomy Course. *Journal of Allied Health*, 31, 153–158.

Buzzell P.R., Chamberlain V.M. and Pintauro S.J. (2002) The Effectiveness of Web-Based, Multimedia Tutorials for Teaching Methods of Human Body Composition Analysis. *Advanced Physiology Education*, 26, 21–29.

Halloran L. (1995) A Comparison of Two Methods of Teaching. Computer Managed Instruction and Keypad Questions versus Traditional Classroom Lecture. *Computer Nursing*, 13/6, 285–288.

Jeganath M. (2003) *Modern Trends in Educational Technology*. NeelKamal Publications, Hyderabad.

Donnelly J.E. (1987) *Using Micro Computers in Physical Education and the Sports Sciences*. Human Kinetics Publishers, Champaign, Illinois.

Mohnson, B.S. (1995) *Using Technology in Physical Education*. Human Kinetics, Champaign, Illinois.

Powers, S.K. and Howely, E.T. (1997) *Exercise Physiology: Theory and Application in Fitness and Performance*. Brown & Benchmark, Madison, WI.

Santer, D.M. and Michaelsen V.E. (1995). A Comparison of Educational Interventions, Multimedia Textbook, Standard Lecture and Printed Textbook. *Arch. Pediatr. Adolesc. Med.*, 149/3, 297–302.

Seabra, D., Srougi M., Baptista R., Nesrallah L.J., Ortiz V. and Sigulem D. (2004) Computer Aided Learning versus Standard Lecture for Undergraduate Education in Urology. *Journal of Urology*, 171/3, 1220–1222.

Toth-Cohen, S. (1995) Computer-Assisted Instruction as a Learning Resource for Applied Anatomy and Kinesiology in The Occupational Therapy Curriculum. *American Journal of Occupational Therapy*, 49/8, 821–827.

Author Index

Ae, M. 377, 649
Agnese, L. 813
Akbarzadeh, A. 773
Alam, F. 311, 437, 773
Allgeuer, T. 155
Álvarez, J.C. 607
An, S.Y. 193
Arai, K. 357
Arai, T. 461
Arellano, R. 583
Asai, T. 385, 391, 397

Baier, H. 701, 767
Bakar, A.A. 169, 415
Balasekaran, G. 683, 689
Balius, X. 583, 607
Baltl, M. 265
Barber, S. 385, 397
Bensason, S. 155
Betzler, N. 253
Biesen, E. 351
Blair, K.B. 345
Blümel, M. 701, 767
Brodie, M. 825
Burkett, B. 181
Burnik, S. 695
Busch, A. 317
Böhm, H. 781

Caine, M.P. 135, 199, 643, 871, 877
Carré, M. 385
Carré, M.J. 129, 397, 485
Chang, A. 155
Chang, B.F. 289
Chen, C.H. 889
Chen, H.-C. 613
Chen, W.-C. 625
Cheng, K.B. 613, 655
Chiu, H.T. 655
Choi, C.H. 69
Choi, J.S. 277, 283

Chung, H.-S. 625
Chung, K.R. 69
Cork, A.E.J. 323, 331, 337
Cornish, J.E.M. 229

Davey, N. 101
Davey, N.P. 577
Davies, M. 917
Davis, A. 247
Davis, M. 519
de Aymerich, X. 583
de la Fuente, B. 583
Dipti, M. 75
Doetkott, C. 51
Doki, H. 709
Dowdell, T. 917
Doyle, M.M. 513
Dufour, M.J.D. 149

Eckelt, M. 163
Elliott, B.C. 513
Emri, I. 695
Escoda, J. 607
Ewart, P. 525

Ferrer, V. 583
Florjančič, U. 695
Foong, S.K. 443, 449
Fuss, F.K. 87, 141, 207, 297, 303, 473, 663, 677, 683, 721, 795

Gagalowicz, A. 113, 121
Ganason, R. 59, 169, 491
Ganguly, S. 833
Goto, Y. 357
Gouwanda, D. 637
Govindaswamy, V. 689
Grantham, J. 689
Gray, A.R. 135

Harding, J.W. 845
Harrison, R. 93

Hasegawa, H. 709
Hashimoto, T. 709
Hasuike, S. 357
Hayasaka, J. 357
Hayashi, Y. 175
Hee, J.W. 187
Hirai, N. 857
Hirose, N. 241
Hodgkins, P.P. 871, 877
Hokari, M. 709
Hong, G.S. 69
Hoon, K.H. 455
Hooper, S.L. 65
Hopkinson, N. 199, 643
Horiuchi, S. 423, 429
Hoshino, Y. 259
Hubbard, M. 467
Hwang, C.-K. 741, 747
Hyeong, J.H. 69

Iida, H. 377
Inou, N. 671
Inoue, A. 423, 429
Ishii, H. 403
Ismail, K.A. 371
Ito, S. 561, 569
Ivan, T. 455
Iwahara, M. 357, 839

Jaffa, M.S. 795
James, D.A. 101, 181, 317, 531, 577, 845
Jong, Y.J. 289
Joseph, S. 59, 491
Justham, L.M. 323, 331, 337

Kajiwara, S. 839
Kawamura, T. 377
Kazahaya, M. 259
Keys, M. 587
Khang, L.S.A. 473
Khoo, B.H. 637

Khoo, T.K. 409
Kim, C. 619
Kim, H.S. 277, 283
Kim, S.Y. 69
Kimura, H. 671
Kobayashi, O. 385, 391
Kobayashi, Y. 259
Koh, M. 593
Koike, S. 377, 649
Koizumi, T. 175
Kondo, A. 357
Kondou, A. 839
Koseki, M. 671
Kou, Z. 51
Kouchi, M. 883
Kulish, V.V. 141
Kumamoto, H. 537
Kumar, C.R. 491
Kuzmin, L. 851

La Brooy, R. 311
Lee, C.-L. 715
Lee, J.B. 181
Lee, K.K. 193
Lehner, S. 803
Lewis, R. 129
Liew, M. 187
Lim, D. 727
Lim, H.B. 455
Lim, Y.T. 277, 283
Lin, K.-B. 741, 747
Lin, Y.-H. 741, 747
Lindh, A.M. 845
Liu, T.H. 363
Looi, D. 187
Low, K.H. 455
Luescher, R. 519
Lyttle, A. 587
Lyttle, A.D. 513

Ma, I.C. 289
Mackintosh, C. 845
Maggs, M.K. 531
Mallinson, P. 871, 877
Manoj, K.M. 75
Martin, D.T. 845
Martin, J. 155
Maruyama, T. 403, 631
McHutchon, M.A. 485
Mecke, G. 781

Mellifont, R.B. 181
Ming, A. 241
Miyazaki, Y. 883
Mizota, T. 223
Mochimaru, M. 883
Monfared, R.P. 93
Morales, E. 583
Mori, H. 649
Motegi, Y. 561
Muller, B.A. 531
Mumford, C. 149
Murakami, M. 857
Müller, M. 781, 863

Nadzrin, M. 897
Nagamatsu, A. 357, 839
Nagao, H. 357
Nagayama, T. 461
Nageswaran, A.S. 921
Nair, G. 911
Nakashima, M. 561
Naruo, T. 223
Niegl, G. 663, 677, 683
Niessen, M. 781
Nikonov, A. 695
Nishi, A. 175
Nishida, Y. 883
Nishihara, O. 537
Nozawa, M. 461

Ogata, K. 423, 429
Ogawa, A. 241
Oggiano, L. 813
Ohgi, Y. 101, 561, 857
Okubo, H. 467
Okunuki, K. 671
Omkar, S.N. 75, 81
Ong, A. 593
Oodaira, H. 235
Oonuki, M. 271
Ostad-Ahmad-Ghorabi, H. 15
Otaki, Y. 671
Otto, S.R. 229, 247
O'Keefe, S.G. 531

Page, W. 825
Paterson, N. 25
Pearson, G. 755
Pharmy, A. 897

Quah, C.K. 113, 121

Reichel, M. 163, 265
Rey, P.H. 905
Ritchie, A.C. 87, 501, 509
Roig, A. 583, 607
Rothberg, S.J. 871, 877

Sabo, A. 163, 265
Sachdeva, S. 443
Saitou, S. 241
Sakashita, R. 391
Sakurai, Y. 631
Santry, J. 135
Sato, F. 235
Schrammel, G. 265
Seah, H.S. 113, 121
Selvarajah, V. 59
Senanayake, S.M.N.A. 169, 187, 409, 415, 479, 637
Senner, V. 701, 767, 781, 803, 863
Seo, K. 385, 391, 397, 857
Sethu, S. 921
Shan, G. 253
Shiang, T.Y. 363
Shimada, N. 271
Shinozaki, M. 241
Shionoya, A. 423, 429
Shirai, Y. 271
Shunmuganathan, D. 921
Sim, J.Y. 87
Smeathers, J.E. 65
Smith, J.D. 93
Smith, L.V. 351
Smith, R.W. 363
Strangwood, M. 229, 247
Stricker, C. 905
Stronge, W.J. 371
Subic, A. 25, 311, 437, 773
Suda, K. 461
Susanto, A.P. 795
Suzuki, S. 259
Sánchez, J.A. 583
Sætran, L.R. 813

Tack, G.R. 277, 283
Takenoshita, Y. 241
Tan, B. 45
Tan, J.C.C. 443, 449, 619

Author Index

Tan, J.K.L. 721
Tan, M.A. 663
Tanaka, K. 235
Taniguchi, D. 839
Teong, T.H. 491
Teranishi, Y. 235, 357
Thompson, S. 689
Tinnsten, M. 851
Tio, W. 437
Tomlinson, S.E. 129
Tomson, L.M. 917
Toohey, K. 845
Toon, D.T. 199, 643
Torres, E. 155
Tripathy, J. 141
Tsujiuchi, N. 175
Turró, C. 607

Udovč, M. 695
Ueda, M. 271
Ujihashi, S. 3, 235, 883
Urry, S.R. 65

Vasquez, G. 345
Vaverka, F. 819
Veluri, S. 443
Verbeek, J. 525
Vikram, B. 297, 303
Vinay, K.J. 75
Vodičková, S. 601, 819
von Bernstorff, B.S. 695

Walker, A.W. 37
Waller, T.M. 135
Walmsley, A. 825
Wang, C.H. 655
Watanabe, S. 883
Watanabe, T. 357
Watkins, S. 437, 755, 773
Wearing, S.C. 65
West, A.A. 93, 323, 331, 337
Widmann, H.G. 247
Williams, B.J. 199
Williams, G. 345

Wilson, B.D. 59, 169, 479, 491, 637, 897
Wimmer, W. 15
Witte, K. 253
Wixted, A. 101
Wong, K. 187
Wong, K.G. 789, 795
Wu, Q. 833

Xie, W. 727

Yamabe, S. 537
Yanai, T. 547
Yang, S. 141
Yang, Y.K. 727
Yi, J.H. 277, 283
Yuen, A.W.W. 479

Zanevskyy, I. 733
Zhou, J.H. 727
Ziejewski, M. 51
Zupančič, B. 695